Principles of Development

Principles of Development

Fourth Edition

Lewis Wolpert | Cheryll Tickle

Thomas Jessell

Peter Lawrence

Elliot Meyerowitz

Elizabeth Robertson

Jim Smith

OXFORD

UNIVERSITY PRESS

OXFORD

UNIVERSITY PRESS

Great Clarendon Street, Oxford ox2 6DP

Oxford University Press is a department of the University of Oxford.
It furthers the University's objective of excellence in research, scholarship,
and education by publishing worldwide in

Oxford New York

Auckland Cape Town Dar es Salaam Hong Kong Karachi
Kuala Lumpur Madrid Melbourne Mexico City Nairobi
New Delhi Shanghai Taipei Toronto

With offices in

Argentina Austria Brazil Chile Czech Republic France Greece
Guatemala Hungary Italy Japan Poland Portugal Singapore
South Korea Switzerland Thailand Turkey Ukraine Vietnam

Oxford is a registered trade mark of Oxford University Press
in the UK and in certain other countries

Published in the United States
by Oxford University Press Inc., New York

British Library Cataloguing in Publication Data

Data available

Library of Congress Cataloging in Publication Data

Data available

Typeset by Techset Composition Ltd
Printed in China
on acid-free paper by
C&C offset printing Co., Ltd

ISBN 978-0-19-954907-8 (pbk)
ISBN 978-0-19-955428-7 (hbk)

3 5 7 9 10 8 6 4

Preface

Developmental biology is at the core of all biology of multicellular organisms. It deals with the process by which the genes in the fertilized egg control cell behavior in the embryo and so determine the character of the animal or plant. Evolution operates by changing the development of the organism so that a better-adapted form develops. With the application of advances in cell and molecular biology and, most recently, genomics, the progress in developmental biology in recent years has been remarkable, and an enormous amount of information is now available. In this fourth edition we have included many recent advances, for example in the understanding of stem cells (Chapter 10) and expanded our consideration of development in relation to evolution (Chapter 15). We have also highlighted, wherever possible, the relevance to human embryonic development.

Principles of Development is designed for undergraduates and the emphasis is on principles and key concepts. Central to our approach is that development can be best understood by understanding how genes control cell behavior. We have assumed that students have some basic familiarity with cell biology and genetics, but all key concepts, such as the control of gene activity, are explained in the text.

Conscious of the pressures on students, we have tried to make the principles as clear as possible and to provide numerous summaries, in both words and pictures, and have tried to avoid providing too much detail. The illustrations in this book are a special feature and have been carefully designed and chosen to illuminate both experiments and mechanisms.

We have resisted the temptation to cover every aspect of development and have, instead, focused on those systems that best illuminate common principles. Indeed, a theme that runs throughout the book is that universal principles govern the process of development. At all stages, what we included has been guided by what we believe undergraduates should know about development.

We have thus concentrated our attention on vertebrates and *Drosophila*, but not to the exclusion of other organisms, such as the nematode and the sea urchin, where they best illustrate a concept. Because *Drosophila* development is so well understood and has been so influential, in in this edition, as in the previous edition, we have started the book with it rather than with vertebrates. An important feature of our book is the inclusion of plant development, which is usually neglected in general textbooks of developmental biology. There have been striking advances in the understanding of plant development in recent times, and some unique and important features have emerged.

In this edition, we have made a new separate chapter to describe the embryology and genetics of our vertebrate model organisms together with some of the important methods used to study them (Chapter 3). The mechanisms involved in early vertebrate development are then considered in the two subsequent chapters (Chapters 4 and 5). Other chapters have been extensively revised. Many new diagrams and photographs are included throughout the book.

The emphasis in the book is on early development and the laying down of body plans and organ systems, such as limbs and the nervous system, but we also include later aspects of development, including growth and regeneration. The book concludes with a consideration of evolution and development.

In providing further reading, our prime concern has been to guide the student to helpful papers rather than to give credit to all the scientists who have made major contributions: to those whom we have neglected, we apologize.

For this new edition, Cheryll Tickle has joined Lewis Wolpert as a main co-author. Each chapter has also been reviewed by a number of experts (see page xix), to whom we give thanks. We made the initial revisions, which were then deciphered, edited, and incorporated by our editor Eleanor Lawrence, whose expertise and influence pervades the book. Eleanor's input has been invaluable in ensuring that the information in the book will be readily accessible to students. The new illustrations were brilliantly drawn or adapted by Matthew McClements, who created the illustrations for the first edition.

We are indebted to Bethan Lee and Jonathan Crowe at Oxford University Press, for their help and patience throughout the preparation of this new edition.

L.W.

London
July 2010

C.T.

Bath
July 2010

 online resource centre
www.oxfordtextbooks.co.uk/orc/wolpert4e/

This book is augmented by additional online materials for both adopters of the book and their students.

For registered adopters:

Electronic artwork
Figures from the book are available to download, for use in lectures.

Journal Clubs
Discussion questions focused around primary literature articles that link to topics featured in the book, to guide the process of assimilating knowledge from the research literature.

For students:

Flashcard glossary
Flashcards, which can be downloaded to mobile devices, to help students test their recall of key terminology.

Multiple choice questions
An extensive bank of questions to help students check their understanding of concepts introduced in the book.

Web links and web activities
Links to web sites that augment the printed text, with commentary to explain the relevance of the sites to the topics in the book, and activities to encourage active engagement with the information held on these sites.

About the authors

Lewis Wolpert is Emeritus Professor of Biology as Applied to Medicine, in the Department of Anatomy and Developmental Biology, University College London, London, UK. He is the author of *The Triumph of the Embryo, A Passion for Science, The Unnatural Nature of Science,* and *Six Impossible Things Before Breakfast.*

Cheryll Tickle is Foulerton Research Professor of The Royal Society at the University of Bath, Bath, UK.

Thomas Jessell is Professor of Biochemistry and Molecular Biophysics, a member of the Center for Neurobiology and Behaviour, and a Howard Hughes Medical Institute investigator at the Department of Biochemistry and Molecular Biophysics, University of Columbia Medical Center, New York, USA. He is an author of *Principles of Neural Science* and *Essentials of Neural Science and Behaviour.*

Peter Lawrence is in the Department of Zoology, University of Cambridge and Emeritus member of the Medical Research Council Laboratory of Molecular Biology, Cambridge, UK. He is the author of *The Making of a Fly.*

Elliot Meyerowitz is the George W. Beadle Professor of Biology and Chair of the Division of Biology at the California Institute of Technology, Pasadena, CA, USA.

Elizabeth Robertson is a Wellcome Trust Principal Fellow and Professor at the Sir William Dunn School of Pathology at the University of Oxford, Oxford, UK.

Jim Smith is Director of the Medical Research Council National Institute for Medical Research, London, UK.

Eleanor Lawrence is a freelance science writer and editor.

Matthew McClements is an illustrator who specializes in design for scientific, technical, and medical communication.

Summary of contents

Contents

Reviewer acknowledgements

Many thanks to the following who kindly reviewed various parts of the book:

Michelle Arbeitman, University of Southern California

Haini Cai, University of Georgia

Scott Dougan, University of Georgia

Robert Drewell, Harvey Mudd College

Deborah Garrity, Colorado State University

Ivan Gepner, Monmouth University

Tanja Godenschwege, Florida Atlantic University

Peter Holland, University of Oxford

Douglas Houston, University of Iowa

Margaret Johnson, University of Alabama

Douglas R. Kankel, Yale University

David Leaf, Western Washington University

Barbara Lom, Davidson College

Morris F. Maduro, University of California, Riverside

Vicki J. Martin, Appalachian State University

Spencer M. Mass, SUNY New Paltz

Elliot Meyerowitz, California Institute of Technology

Andre Pires da Silva, University of Texas Arlington

Gary Radice, University of Richmond, Virginia

Ann Rougvie, University of Minnesota

Margaret Saha, College of William and Mary

Craig M. Scott, Clarion University

Diane C. Slusarski, University of Iowa

David Stein, University of Texas Austin

Leslie Stevens, University of Texas Austin

Daniel P. Szeto, Purdue University

Keiko Torri, University of Washington

Lance Urven, Marian University of Wisconsin

Matthew Wawersik, College of William and Mary

Athula H. Wikramanayake, University of Miami

Nina Zanetti, Siena College

Ted Zerucha, Appalachian State University

Figure acknowledgements

Chapter 1

Fig. 1.1 a) courtesy of Alpesh Doshi, CRGH, London; b) reproduced courtesy of the MRC/Wellcome-funded Human Developmental Biology Resource

Chapter 2

Fig. 2.1 top photograph reproduced with permission from Turner, F.R., Mahowald, A.P.: **Scanning electron microscopy of** *Drosophila* **embryogenesis. I. The structure of the egg envelopes and the formation of the cellular blastoderm.** *Dev. Biol.* 1976, **50**: 95–108. Middle photograph reproduced with permission from Turner, F.R., Mahowald, A.P.: **Scanning electron microscopy of** *Drosophila* **melanogaster embryogenesis. III. Formation of the head and caudal segments.** *Dev. Biol.* 1979, **68**: 96–109.

Fig. 2.4 left photograph reproduced with permission from Turner, F.R., Mahowald, A.P.: **Scanning electron microscopy of** *Drosophila* **melanogaster embryogenesis. II. Gastrulation and segmentation.** *Dev. Biol.* 1977, **57**: 403–416. Center photograph reproduced with permission from Alberts, B., Bray, D., Lewis, J., Raft, M., Roberts, K., Watson, J.D.: *Molecular Biology of the Cell*, 3rd edition. New York, Garland Publishing, 1994.

Fig. 2.10 reproduced with permission from Spriov, A., Fahmy, K., Schneider, M., Frei, E., Noll, Baumgartner, S. **Formation of the** *bicoid* **morphogen gradient: an mRNA gradient dictates the protein gradient.** *Dev*, 2009, **136**: 605–614.

Fig. 2.12 photograph reproduced with permission from Griffiths, A.J.H., Miller, J.H., Suzuki, D.T., Lewontin, R.C., Gelbart, W.M.: *An Introduction to Genetic Analysis*, 6th edition. New York: W.H. Freeman & Co., 1996.

Fig. 2.13 reproduced with permission from Surkova, S., Kosman, D., Kozlov, M., Myasnikova, E., Samsonova, A.A., Spriov, A., Vanario-Alonso, C.E., Samsonova, M., Reinitz, J. **Charaterization of the Drosophila segment determination morpheme.** *Dev. Biol.* 2008, **313**: 844–862.

Fig. 2.17 illustration after González-Reyes, A., Elliott, H., St Johnston, D.: **Polarization of both major body axes in** *Drosophila* **by** *gurken-torpedo* **signalling.** *Nature* 1995, **375**: 654–658.

Fig. 2.21 illustration after St Johnston, D.: **Moving messages: the intracellular localization of mRNAs.** *Nat. Rev. Mol. Cell Biol.* 2005, **6**: 363–375.

Fig. 2.22 from Martin, S.G., Leclerc, V., Smith-Litière, K., St Johnston, D.: **The identification of novel genes required for Drosophila anteroposterior axis formation in a germline clone screen using GFP-Staufen.** *Dev.* 2003, **130**, 4201–4215.

Fig. 2.26 from Yu, D., and Small, S.: **Precise registration of gene expression boundaries by a repressive morphogen in Drosophila.** *Curr. Biol*, 2008, **18**: 868–876.

Box 2D from Blagburn, J.M.: **Engrailed expression in subsets of adult Drosophila sensory neurons: an enhancer-trap study.** *Invert Neurosa.* 2008, **8**(3): 133–146.

Fig. 2.30 illustration after Lawrence, P.: *The Making of a Fly.* Oxford: Blackwell Scientific Publications, 1992.

Fig. 2.33 from Yu, D., and Small, S.: **Precise registration of gene expression boundaries by a repressive morphogen in Drosophila.** *Curr. Biol*, 2008, **18**: 868–876.

Fig. 2.38 from Goodman, R.M., Thombrel, S., Firtinal, Z., Grayl, D., Betts, D., and Roebuck, J.: **Sprinter: a novel transmembrane protein required for Wg secretion and signalling.** *Dev*, 2006, **133**: 4901–4911.

Fig. 2.41 illustration after Lawrence, P.: *The Making of a Fly.* Oxford: Blackwell Scientific Publications, 1992.

Fig. 2.47 illustration after Lawrence, P.: *The Making of a Fly.* Oxford: Blackwell Scientific Publications, 1992. Photograph reproduced with permission from Bender, W., Akam, A., Karch, F., Beachy, P.A., Peifer, M., Spierer, P., Lewis, E.B., Hogness, D.S.: **Molecular genetics of the bithorax complex in** *Drosophila melanogaster*. *Science* 1983, **221**: 23–29 (image on front cover). © 1983 American Association for the Advancement of Science.

Chapter 3

Fig. 3.3 top photograph reproduced with permission from Alberts, B., Bray, D., Lewis, J., Raff, M., Roberts, K., Watson, J.D.: *Molecular Biology of the Cell*, 3rd edition. New York: Garland Publishing, 1994.

Fig. 3.5 photographs reproduced with permission from Kessel, R.G., Shih, C.Y.: *Scanning Electron Microscopy in Biology: A Student's Atlas of Biological Organization.* London, Springer-Verlag, 1974. © 1974 Springer-Verlag GmbH & Co. KG.

Fig. 3.6 illustration after Balinsky, B.I.: *An Introduction to Embryology.* Fourth edition. Philadelphia, W.B. Saunders, 1975.

Fig. 3.14 top photograph reproduced with permission from Kispert, A., Ortner, H., Cooke, J., Herrmann, B.G.: **The chick** *Brachyury* **gene: developmental expression pattern and response to axial induction by localized activin.** *Dev. Biol.* 1995, **168**: 406–415.

Fig. 3.17 illustration after Patten, B.M.: *Early Embryology of the Chick.* New York, Mc Graw-Hill, 1971.

Fig. 3.21 top photograph reproduced with permission from Bloom, T.L.: **The effects of phorbol ester on mouse blastomeres: a role**

for protein kinase C in compaction? *Development* 1989, **106:** 159–171 Published by permission of The Company of Biologists Ltd.

Fig. 3.23 illustration after Hogan, B., Beddington, R., Costantini, F., Lacy, E.: *Manipulating the Mouse Embryo: A Laboratory Manual,* 2nd edition. New York: Cold Spring Harbor Laboratory Press, 1994.

Fig. 3.24 illustration adapted, with permission, from McMahon, A.P.: **Mouse development. Winged-helix in axial patterning.** *Curr. Biol.* 1994, **4:** 903–906.

Fig. 3.26 illustration after Kaufman, M.H.: *The Atlas of Mouse Development.* London: Academic Press, 1992.

Chapter 4

Fig. 4.14 illustration after Rodriguez, T.A., Srinivas, S., Clements, M.P., Smith, J.C., Beddington, R.S.: **Induction and migration of the anterior visceral endoderm is regulated by the extra-embryonic ectoderm.** *Development* 2005, **132:** 2513–2520.

Fig. 4.37 photograph reproduced with permission from Smith W.C., Harland, R.M.: **Expression cloning of noggin, a new dorsalizing factor localized to the Spemann organizer in Xenopus embryos.** *Cell* 1992, **70:** 829–840. © 1992 Cell Press.

Fig. 4.38 adapted from Heasman, J.: **Patterning the early zenopus embryo.** *Dev* 2006, **133:** 1205–1217.

Fig. 4.44 illustration after Beddington, S.P., Robertson, E.J.: **Axis development and early asymmetry in mammals.** *Cell* 1999, **96:** 195–209.

Box 4C adapted from Larsen, W.J.: *Human Embryology,* 2nd edn, Churchill Livingstone, 1997, p. 481.

Box 4E from Morley, R.H., Lachani, K., Keefe, D., Gilchrist, M.J., Flicek, P., Smith, J.C., Wardle, F.C.: **A gene regulatory network directed by zebrafish No tail accounts for its roles in mesoderm formation.** *Proc Natl Acad Sci* USA, 2009, **106:** 3829–3834.

Chapter 5

Fig. 5.2 photograph reproduced with permission from Hausen, P., Riebesell, M.: *The Early Development of Xenopus laevis.* Berlin: Springer-Verlag, 1991.

Fig. 5.7 from Stern, C.D.: **Neural induction: old problem, new findings, yet more questions.** *Dev,* 2005, **132,** 2007–21.

Fig. 5.9 from Kiecker, C., and Niehrs, C.: **A morphogen gradient of Wnt/beta- catenin signalling regulates anteroposterior neural patterning in Xenopus.** *Dev,* 2001, **128:** 4189–4201.

Fig. 5.12 illustration after Kintner, C.R., Dodd, J.: **Hensen's node induces neural tissue in Xenopus ectoderm. Implications for the action of the organizer in neural induction.** *Development* 1991, 113: 1495–1505.

Fig. 5.13 adapted from Delfino-Machin, M. *et. al.*: **Specification and maintenance of the spinal cord stem zone.** *Development* 2005, **132:** 4273–4283.

Fig. 5.14 illustration from Kudoh, T. *et. al.*: **Combinatorial Fgf and Bmp signalling patterns the gastrula ectoderm into prospective neural and epidermal domains**. *Development* 2004, **131:** 3581–3592.

Fig. 5.15 illustration after Mangold, O.: **Über die induktionsfahigkeit der verschiedenen bezirke der neurula von urodelen.** *Naturwissenschaften* 1933, **21:** 761–766.

Fig. 5.16 illustration after Kelly, O.G., Melton, D.A.: **Induction and patterning of the vertebrate nervous system.** *Trends Genet.* 1995, **11:** 273–278.

Fig. 5.21 illustration from Deschamps, J., van Nes, J.: **Developmental regulation of the Hox genes during axial morphogenesis in the mouse.** *Development* 2005, **132:** 2931–2942.

Fig. 5.26 illustration after Burke, A.C., Nelson, C.E., Morgan, B.A., Tabin, C.: **Hox genes and the evolution of vertebrate axial morphology.** *Development* 1995, **121:** 333–346.

Fig. 5.31 illustration after Johnson, R.L., Laufer, E., Riddle, R.D., Tabin, C.: **Ectopic expression of Sonic hedgehog alters dorsal-ventral patterning of somites.** *Cell* 1994, **79:** 1165–1173.

Fig. 5.33 illustration adapted, with permission, from Lumsden, A.: **Cell lineage restrictions in the chick embryo hindbrain.** *Phil. Trans. R. Soc. Lond. B* 1991, **331:** 281–286.

Fig. 5.34 illustration adapted, with permission, from Lumsden, A.: **Cell lineage restrictions in the chick embryo hindbrain.** *Phil. Trans. R. Soc. Lond. B* 1991, **331:** 281–286.

Fig. 5.36 photograph reproduced with permission from Lumsden, A., Krumlauf, R.: **Patterning the vertebrate neuraxis.** *Science* 1996, **274:** 1109–1115 (image on front cover). © 1996 American Association for the Advancement of Science.

Fig. 5.37 illustration after Krumlauf, R.: **Hox genes and pattern formation in the branchial region of the vertebrate head.** *Trends Genet.* 1993, **9:** 106–112.

Chapter 6

Fig. 6.5 photograph reproduced with permission from Strome, S., Wood, W.B.: **Generation of asymmetry and segregation of germ-line granules in early C. elegans embryos.** *Cell* 1983 **35:** 15–25. © 1983 Cell Press.

Fig. 6.6 illustration after Sulston, J.E., Schierenberg, E., White, J.G., Thompson, J.N.: **The embryonic cell lineage of the nematode C. elegans.** *Dev. Biol.* 1983, **100:** 69–119.

Fig. 6.7 photographs reproduced with permission from Wood W.B: **Evidence from reversal of handedness in C. elegans embryos for early cell interactions determining cell fates.** *Nature* 1991, **349:** 536–538. © 1991 Macmillan Magazines Ltd.

Fig. 6.8 illustration after Mello, C.C., Draper, B.W., Priess, J.R.: **The maternal genes apx-1 and glp-1 and establishment of dorsal-ventral polarity in the early C. elegans embryo.** *Cell* 1994, **77:** 95–106.

Fig. 6.10 from Mizumoto, K., and Sawa, H.: **Two bs or not two bs: regulation of asymmetric division by b-catenin.** *TRENDS in Cell Biology,* 2007, **Vol. 17** No. 10.

Fig. 6.12 illustration after Bürglin, T.R., Ruvkun, G.: **The Caenorhabditis elegans homeobox gene cluster.** *Curr. Opin. Gen. Dev.* 1993, **3:** 615–620.

Fig. 6.18 top and middle photographs reproduced with permission from Jim Coffman. Lower photograph reproduced with permission from Oxford Scientific Films.

Fig. 6.23 illustration from Oliveri, P., Davidson, E.H.: **Gene regulatory network controlling embryonic specification in the sea urchin.** *Curr. Opin. Genet. Dev.* 2004, **14:** 351–360.

Fig. 6.24 adapted from Oliveri, P., Tu, Q., Davidson, E.H.: **Global regulatory logic for specification of an embryonic cell lineage.** *Proc. Natl Acad. Sci. USA* 2008, **105:** 5955–5962.

Fig. 6.25 illustration from Oliveri, P., Davidson, E.H.: **Gene regulatory network controlling embryonic specification in the sea urchin.** *Curr. Opin. Genet. Dev.* 2004, **14:** 351–360.

Fig. 6.28 middle photograph reproduced with permission from Corbo, J.C., Levine, M., Zeller, R.W.: **Characterization of a notochord-specific enhancer from the Brachyury promoter region of the ascidian, *Ciona intestinalis*.** *Development* 1997, **124:** 589–602. Published by permission of The Company of Biologists Ltd. Top photograph reproduced with permission from Shigeki Fujiwara and Naoki Shimozono. Lower photograph reproduced with permission from Andrew Martinez.

Fig. 6.30 illustration from Nishida, H.: **Specification of embryonic axis and mosaic development in ascidians.** *Dev. Dyn.* 2005, **233:** 1177–1193.

Fig. 6.31 illustration after Conklin, E.G.: **The organization and cell lineage of the ascidian egg.** *J. Acad. Nat. Sci. Philadelphia* 1905, **13:** 1–119.

Fig. 6.32 Picco, V., Hudson, C., Yasuo, H.: **Ephrin-Eph signalling drives the asymmetric division of notochord/neural precursors in *Ciona* embryos.** *Development* 2007, **134:** 1491–1497.

Chapter 7

Fig. 7.4 illustration after Scheres, B., Wolkenfelt, H., Willemsen, V., Terlouw, M., Lawson, E., Dean, C., Weisbeek, P.: **Embryonic origin of the *Arabidopsis* primary root and root meristem initials.** *Development* 1994, **120:** 2475–2487.

Fig. 7.5 illustration after Friml, J., *et al.*: **Efflux-dependent auxin gradients establish the apical-basal axis of *Arabidopsis*.** *Nature* 2003, **426:** 147–153.

Fig. 7.6 adapted from Chapman, E.J. and Estelle, M.: **Cytokinin and auxin intersection in root meristems.** Genome Biol. 2009, **10:** 210. doi:10.1186/gb-2009-10-2-210

Fig. 7.7 from Friml, J., et al: **Efflux-dependent auxin gradients establish the apical-basal axis of *Arabidopsis*.** *Nature* 2003, **426:** 147–153.

Fig. 7.10 illustration after Alberts, B., Bray, D., Lewis, J., Raff, M., Roberts, K., Watson, J.D.: *Molecular Biology of the Cell,* 2nd edition. New York: Garland Publishing, 1989.

Fig. 7.14 illustration after Steeves, T.A., Sussex, I.M.: *Patterning in Plant Development.* Cambridge: Cambridge University Press, 1989.

Fig. 7.16 illustration after McDaniel, C.N., Poethig, R.S.: **Cell lineage patterns in the shoot apical meristem of the germinating maize embryo.** *Planta* 1988, **175:** 13–22.

Fig. 7.18 top panel, illustration after Poethig, R.S., Sussex, I.M.: **The cellular parameters of leaf development in tobacco: a clonal analysis.** *Planta* 1985, **165:** 170–184. Bottom panel, illustration after Sachs, T.: *Pattern Formation in Plant Tissues.* Cambridge: Cambridge University Press, 1994.

Fig. 7.20 adapted from Heisler, M.G. *et al.*: **Patterns of auxin transport and gene expression during primordium development revealed by live imaging of the *Arabadopsis* inflorescence meristem.** *Curr. Biol.* 2005, **15:** 1899–1911.

Fig. 7.22 illustration after Scheres, B., Wolkenfelt, H., Willemsen, V., Terlouw, M., Lawson, E., Dean, C., Weisbeek, P.: **Embryonic origin of the *Arabidopsis* primary root and root meristem initials.** *Development* 1994, **120:** 2475–2487.

Fig. 7.24 photograph reproduced with permission from Meyerowitz, E.M., Bowman, J.L., Brockman, L.L., Drews, G.N., Jack, T., Sieburth, L.E., Weigel, D.: **A genetic and molecular model for flower development in *Arabidopsis thaliana*.** *Development Suppl.* 1991, 157–167. Published by permission of The Company of Biologists Ltd.

Fig. 7.25 illustration after Coen, E.S., Meyerowitz, E.M.: **The war of the whorls: genetic interactions controlling flower development.** *Nature* 1991, **353:** 31–37.

Fig. 7.26 photographs reproduced with permission from Meyerowitz, E.M., Bowman, J.L., Brockman, L.L., Drews, G.N., Jack, T., Sieburth, L.E., Weigel, D.: **A genetic and molecular model for flower development in *Arabidopsis thaliana*.** *Development Suppl.* 1991, 157–167. Published by permission of The Company of Biologists Ltd. (left panel); center panel from Bowman, J.L., Smyth, D.R., Meyerowitz, E.M.: **Genes directing flower development in *Arabidopsis*.** *Plant Cell* 1989, **1:** 37–52. Published by permission of The American Society of Plant Physiologists.

Fig. 7.29 adapted from Lohmann, J.U., Weigel, D.: Building beauty: the genetic control of floral patterning. *Dev. Cell* 2002, **2:** 135–142.

Fig. 7.30 photograph reproduced with permission from Coen, E.S., Meyerowitz, E.M.: **The war of the whorls: genetic interactions controlling flower development.** *Nature* 1991, **353:** 31–37. © 1991 Macmillan Magazines Ltd.

Fig. 7.31 illustration after Drews, G.N., Goldberg, R.B.: **Genetic control of flower development.** *Trends Genet.* 1989 **5:** 256–261.

Fig. 7.35 illustration from Bl·zquez, M.A.: **The right time and place for making flowers.** *Science* 2005, **309:** 1024–1025.

Chapter 8

Fig. 8.3 from Foty, R.A., and Steinberg, M.S.: **The differential adhesion hypothesis: a direct evaluation.** *Developmental Biology* 278 (2005) 255– 263.

Fig. 8.6 illustration after Strome, S.: **Determination of cleavage planes.** *Cell* 1993, **72:** 3–6.

Fig. 8.8 from Raff, E.C., Villinski, J.T., Turner, F.R., Danoghue, P.C.J., and Raff, R.A.: **Experimental taphonomy shows the feasibility of fossil embryos.** *PNAS* 2006, **103:** 5846–5851.

Fig. 8.9 photograph reproduced with permission from Bloom T.L.: **The effects of phorbol ester on mouse blastomeres: a role for protein kinase C in compaction?** *Development* 1989, **106:** 159–171.

Fig. 8.11 from Eckert, J.J., and Fleming, T.P..: **Tight junction biogenesis during early development.** *Biochim. Biophys. Acta.* 2008, **1778:** 717–728.

Fig. 8.12 from Alberts, B. *et al.: Molecular Biology of the Cell,* Fifth Edition, Garland Science, New York, 2008.

Fig. 8.13 illustration after Coucouvanis, E., Martin, G.R.: **Signals for death and survival: a two-step mechanism for cavitation in the vertebrate embryo.** *Cell* 1995, **83**: 279–287.

Fig. 8.17 photograph reproduced with permission from Merrill, J.B., Santos, L.L.: **A scanning electron micrographical overview of cellular and extracellular patterns during blastulation and gastrulation in the sea urchin,** *Lytechinus variegatus.* In *The Cellular and Molecular Biology of Invertebrate Development.* Edited by Sawyer, R.H. and Showman, R.M. University of South Carolina Press, 1985; pp. 3–33.

Fig. 8.19 illustration after Odell, G.M., Oster, G., Alberch, P., Burnside, B.: **The mechanical basis of morphogenesis. I. Epithelial folding and invagination.** *Dev. Biol.* 1981, **85**: 446–462.

Fig. 8.21 photographs reproduced with permission from Leptin, M., Casal, J., Grunewald, B., Reuter, R.: **Mechanisms of early** *Drosophila* **mesoderm formation.** *Development Suppl.* 1992, 23–31. Published by permission of The Company of Biologists Ltd.

Fig. 8.22 illustration after Bertet, C., Sulak, L., Lecuit, T. **Myosindependent junction remodelling controls planar cell intercalation and axis elongation.** *Nature* 2004, **429**: 667–671.

Fig. 8.24 illustration after Balinsky, B.I.: *An Introduction to Embryology,* 4th edition. Philadelphia, W.B. Saunders, 1975.

Fig. 8.28 photograph reproduced with permission from Smith, J.C., Cunliffe, V., O'Reilly, M-A.J., Schulte-Merker, S., Umbhauer, M.: *Xenopus Brachyury. Semin. Dev. Biol.* 1995, **6**: 405–410. © 1995 by permission of the publisher, Academic Press Ltd., London.

Fig. 8.29 illustration after Montero, J.A., Heisenberg, C.P. **Gastrulation dynamics: cells move into focus.** *Trends Cell Biol.* 2004, **14**: 620–627.

Fig. 8.30 illustration after Wallingford, J.B., Fraser, S., Harland, R.M. **Convergent extension: the molecular control of polarized cell movement during embryonic development.** *Dev. Cell* 2002, **2**: 695–706.

Fig. 8.34 illustration after Schoenwolf, G.C., Smith, J.L.: **Mechanisms of neurulation: traditional viewpoint and recent advances.** *Development* 1990 **109**: 243–270.

Fig. 8.39 photographs reproduced with permission from Priess, J.R., Hirsh, D.I.: *Caenorhabditis elegans morphogenesis*: **the role of the cytoskeleton in elongation of the embryo.** *Dev. Biol* 1986, **117**: 156–173. © 1986 Academic Press.

Fig. 8.41 photographs reproduced with permission from Tsuge, T., Tsukaya, H., Uchimaya, H.: **Two independent and polarized processes of cell elongation regulate leaf blade expansion in** *Arabidopsis thaliana* **(L.) Heynh.** *Development* 1996, **122**: 1589–1600. Published by permission of The Company of Biologists Ltd.

Chapter 9

Fig. 9.5 adapted from Hogan, B.: **Decisions, decisions.** *Nature* 2002, **418**: 282, and Saitou, M., Barton, S. Surani, M.: **A molecular programme for the specification of germ cell fate in mice.** *Nature,* 2002, **418**: 293–300.

Fig. 9.6 illustration after Wylie, C.C., Heasman, J.: **Migration, proliferation, and potency of primordial germ cells.** *Semin. Dev. Biol.* 1993, **4**: 161–170.

Fig. 9.7 adapted from Weidinger, G., Köprunner, M., Thisse, C., Thisse, B., and Razl, E.: **Regulation of zebrafish primordial germ cell migration by attraction towards an intermediate target.** *Development* , 2002, **129**: 25–36.

Fig. 9.14 illustration after Alberts, B., Bray, D., Lewis, J., Raft, M., Roberts, K., Watson, J.D.: *Molecular Biology of the Cell,* 2nd edition. New York: Garland Publishing, 1989.

Fig. 9.19 illustration after Goodfellow, P.N., Lovell-Badge, R.: *SRY* **and sex determination in mammals.** *Ann. Rev. Genet.* 1993, **27**: 71–92.

Fig. 9.21 illustration after Higgins, S.J., Young, P., Cunha, G.R.: **Induction of functional cytodifferentiation in the epithelium of tissue recombinants II. Instructive induction of Wolffian duct epithelia by neonatal seminal vesicle mesenchyme.** *Development* 1989, **106**: 235–250.

Fig. 9.25 illustration after Cline, T.W.: **The** *Drosophila* **sex determination signal: how do flies count to two?** *Trends Genet.* 1993 **9**: 385–390.

Fig. 9.29 illustration after Clifford, R., Francis, R., Schedl, T.: **Somatic control of germ cell development.** *Semin. Dev. Biol* 1994, **5**: 21–30.

Fig. 9.31 illustration after Alberts, B., Bray, D., Lewis, J., Raft, M., Roberts, K., Watson, J.D.: *Molecular Biology of the Cell,* 2nd edition. New York: Garland Publishing, 1989.

Chapter 10

Fig. 10.3 illustration after Tijian, R.: **Molecular machines that control genes.** *Sci. Am.* 1995, **272**: 54–61.

Fig. 10.4 illustration after Alberts B., Bray, D., Lewis, J., Raff, M., Roberts, K., Watson, J.D.: *Molecular Biology of the Cell,* 2nd edition. New York: Garland Publishing, 1989.

Fig. 10.6 illustration after Alberts, B., Bray, D., Lewis, J., Raff, M., Roberts, K., Watson, J.D.: *Molecular Biology of the Cell,* 2nd edition. New York: Garland Publishing, 1989.

Fig. 10.8 illustration after Kluger, Y., Lian, Z., Zhang, X., Newburger, P.E., Weissman, S.M.: **A panorama of lineage-specific transcription in hematopoiesis.** *BioEssays* 2004, **26**: 1276–1287.

Fig. 10.10 illustration after Metcalf, D.: **Control of granulocytes and macrophages: molecular, cellular, and clinical aspects.** *Science* 1991, **254**: 529–533.

Fig. 10.13 illustration after Crossley, M., Orkin, S.H.: **Regulation of the b-globin locus.** *Curr. Opin. Genet. Dev.* 1993, **3**: 232–237.

Fig. 10.17 from Barker, N., van Esl, J.H., Kuipers, J., Kujala, P., van den Born, M., Cozijnsen, M., Haegebarth, A., Korving, J., Begthell, H., Peters, P.J., and Clevers, H: Identification of stem cells in small intestine and colon by marker gene Lgr5. *Nature*

Fig. 10.22 illustration after Taupin, P.: **Adult neurogenesis in the mammalian central nervous system: functionality and potential clinical interest.** *Med. Sci. Monit.* 2005, **11**: 247–252.

Fig. 10.27 illustration after Riedl, S.J., Shi, Y.: **Molecular mechanisms of caspase regulation during apoptosis.** *Nat. Rev. Mol. Cell Biol.* 2004, **5**: 897–907.

Chapter 11

Fig. 11.6 photograph reproduced with permission from Cohn, M.J., Izpisúa-Belmonte, J.C., Abud, H., Heath, J.K., Tickle, C.: **Fibroblast growth factors induce additional limb development from the flank of chick embryos.** *Cell* 1995, **80:** 739–746. © 1995 Cell Press.

Fig. 11.18 adapted from Wellik, D.M., Capecchi, M.R.: *Hox10* and *Hox11* **genes are required to globally pattern the mammalian skeleton.** *Science* 2003, **301:** 363–367.

Fig. 11.21 photograph reproduced with permission from Garcia-Martinez, V., Macias, D., Gañan, Y., Garcia-Lobo, J.M., Francia, M.V., Fernandez- Teran, M.A., Hurle, J.M.: **Internucleosomal DNA fragmentation and programmed cell death (apoptosis) in the interdigital tissue of embryonic chick leg bud.** *J. Cell Sci.* 1993, **106:** 201–208. Published by permission of The Company of Biologists Ltd.

Fig. 11.23 illustration after French, V., Daniels, G.: **Pattern formation: the beginning and the end of insect limbs.** *Curr. Biol.* 1994, **4:** 35–37.

Fig. 11.26 (*a*) photograph reproduced with permission from Nellen, D., Burke, R., Struhl, G., Basler, K.: **Direct and long-range action of a dpp morphogen gradient.** *Cell* 1996, **85:** 357–368. © 1996, Cell Press.

Fig. 11.26 (*b–e*) photograph reproduced with permission from Moser, M. and Campbell, G.: **Generating and interpreting the Brinker gradient in the Drosophila wing.** *Dev. Biol.* 2005, **286:** 647–658.

Fig. 11.29 photograph reproduced with permission from Zecca, M., Basler, K., Struhl, G.: **Direct and long-range action of a wingless morphogen gradient.** *Cell* 1996, **87:** 833–844. © 1996 Cell Press.

Fig. 11.31 illustration after Bryant, P.J. **The polar coordinate model goes molecular.** *Science* 1993, **259:** 471–472.

Fig. 11.40 adapted from Adler, R., and Valeria Canto-Soler, M.: **Molecular mechanisms of optic vesicle development: Complexities, ambiuities, and controversies.** *Developmental Biology,* 305: 1–13.

Fig. 11.41 photographs reproduced with permission from Gehring, W.J.: **New perspectives on eye development and the evolution of eyes and photoreceptors.** *J. Hered.* 2005, **96(3):** 171–184.

Fig. 11.49 adapted from Kelly, R.G., Buckingham, M.E.: **The anterior heart-forming field: voyage to the arterial pole of the heart.** *Trends Genet.* 2002, **18:** 210–216.

Chapter 12

Fig. 12.7 illustration after Rakic, P.: **Mode of cell migration to the superficial layers of fetal monkey neocortex.** *J. Comp. Neurol.* 1972, **145:** 61–83.

Fig. 12.8 adapted from Honda, T., Tabata, H., Nakajima, K.: **Cellular and molecular mechanisms of neuronal migration in neocortical development.** *Semin. Cell Dev. Biol.* 2003, **14:** 169–174.

Fig. 12.15 adapted from Dasen, J., Tice, B., Brenner-Morton, S., Jessell, T.: **A Hox regulatory network establishes motor neuron pool identity and target-muscle connectivity.** *Cell* 2005, **123:** 477–491.

Fig. 12.16 illustration after Alberts, B., Bray, D., Lewis, J., Raff, M., Roberts, K., Watson, J.D.: *Molecular Biology of the Cell,* 2nd edition. New York: Garland Publishing, 1989.

Fig. 12.21 photograph reproduced with permission from Serafini, T., Colamarino, S.A., Leonardo, E.D., Wang, H., Beddington, R., Skarnes, W.C., Tessier-Lavigne, M.: **Netrin-1 is required for commissural axon guidance in the developing vertebrate nervous system.** *Cell* 1996, **87:** 1001–1014. © 1996 Cell Press.

Fig. 12.28 illustration after Kandell, E.R., Schwartz, J.H., Jessell, T.M.: *Principles of Neural Science,* 3rd edition. New York: Elsevier Science Publishing Co., Inc., 1991.

Fig. 12.30 illustration after Li, Z., Sheng, M.: **Some assembly required: the development of neuronal synapses.** *Nat. Rev.* 2003, **4:** 833–841.

Fig. 12.32 illustration after Goodman, C.S., Shatz, C.J.: **Developmental mechanisms that generate precise patterns of neuronal connectivity.** *Cell Suppl.* 1993, **72:** 77–98.

Fig. 12.33 illustration after Kandell, E.R., Schwartz, J.H., Jessell, T.M.: *Essentials of Neural Science and Behavior.* Norwalk, Connecticut: Appleton & Lange, 1991.

Fig. 12.34 illustration after Goodman, C.S., Shatz, C.J.: **Developmental mechanisms that generate precise patterns of neuronal connectivity.** *Cell Suppl.* 1993, **72:** 77–98.

Chapter 13

Fig. 13.2 adapted from D.O. Morgan, *The Cell Cycle.* Oxford University Press, 2006.

Fig. 13.3 illustration after Edgar, B.A., Lehman, D.A., O'Farrell, P.H.: **Transcriptional regulation of** *string (cdc25):* **a link between developmental programming and the cell cycle.** *Development* 1994, **120:** 3131–3143.

Fig. 13.5 photograph reproduced with permission from Harrison, R.G.: *Organization and Development of the Embryo.* New Haven: Yale University Press, 1969. © 1969 Yale University Press.

Fig. 13.6 illustration after Gray, H.: *Gray's Anatomy.* Edinburgh: Churchill-Livingstone, 1995.

Fig. 13.8 from Nijhout, H.F.: **Size matters (but so does time), and it's OK to be different.** *Dev. Cell ,* 2008, **15:** 491–492.

Fig. 13.11 illustration after Walls, G.A.: **Here today, bone tomorrow.** *Curr. Biol.* 1993, **3:** 687–689.

Fig. 13.12 illustration after Kronenberg, H.M.: **Developmental regulation of the growth plate.** *Nature (Insight)* 2003, **423:** 332–336.

Fig. 13.18 illustration after Tata, J.R.: **Gene expression during metamorphosis: an ideal model for post-embryonic development.** *BioEssays* 1993, **15:** 239–248.

Fig. 13.19 illustration after Tata, J.R.: **Gene expression during metamorphosis: an ideal model for post-embryonic development.** *BioEssays* 1993, **15:** 239–248.

Chapter 14

Fig. 14.5 from Kragl, M., Knapp, D., Nacul, E., Khattak, S., Maden, M., Epperlein, H., and Tanaka, E.: **Cells keep a memory of their tissue origin during axolotl limb regeneration.** *Nature,* 2009, **460:** 60–65.

Fig. 14.11 photographs reproduced with permission from Pecorino, L.T. Entwistle A., Brockes, J.P.: **Activation of a single retinoic acid receptor isoform mediates proximodistal respecification.** *Curr. Biol.*1996, **6:** 563–569.

Fig. 14.13 illustration after French, V., Bryant, P.J., Bryant, S.V.: **Pattern regulation in epimorphic fields.** *Science* 1976, **193**: 969–981.

Fig. 14.14 illustration after French, V., Bryant, P.J., Bryant, S.V.: **Pattern regulation in epimorphic fields.** *Science* 1976, **193**: 969–981.

Fig. 14.15 photograph reproduced with permission from Müller, W.A.: **Diacylglycerol-induced multihead formation in *Hydra*.** *Development* 1989, **105**: 309–316. Published by permission of The Company of Biologists Ltd.

Chapter 15

Fig. 15.9 illustration after Akam, M.: **Hox genes and the evolution of diverse body plans.** *Phil, Trans. R. Soc. Lond. B* 1995, **349**: 313–319.

Fig. 15.10 illustration after Ferguson, E.L.: **Conservation of dorsa-ventral patterning in arthropods and chordates.** *Curr. Opin. Genet. Dev.* 1996, **6**: 424–431.

Fig. 15.11 adapted from Boisvert, C.A., Mark-Kurik, E., and Ahlberg, P.E.: **The pectoral fin of *Panderichthys* and the origin of digits.** *Nature* 2008, **456**: 636–638.

Fig. 15.12 illustration after Gerhart, J., Lowe, C., Kirschner, M.: **Hemichordates and the origins of chordates.** *Curr. Opin. Genet. Dev.* 2005, **15**: 461–467.

Fig. 15.16 illustration after Gregory, W.K.: *Evolution Emerging.* New York: Macmillan, 1957.

Fig. 15.20 illustration after Gompel, N. et al.: **Chance caught on the wing: cis-regulatory evolution and the origin of pigment patterns in *Drosophila*.** *Nature*, 2005, **433**: 481–487.

Fig. 15.22 illustration after Larsen, W.J.: *Human Embryology.* New York: Churchill Livingstone, 1993.

Fig. 15.23 illustration after Romer, A.S.: *The Vertebrate Body.* Philadelphia: W.B. Saunders, 1949.

History and basic concepts

- The origins of developmental biology
- A conceptual tool kit

The aim of this chapter is to provide a conceptual framework for the study of development. We start with a brief history of the study of embryonic development, which illustrates how some of the key questions in developmental biology were first formulated, and continue with some of the essential principles of development. The big question is how does a single cell—the fertilized egg—give rise to a multicellular organism, in which a multiplicity of different cell types are organized into tissues and organs to make up a three-dimensional body. This question can be studied from many different viewpoints, all of which have to be fitted together to obtain a complete picture of development: which genes are expressed, and when and where; how cells communicate with each other; how a cell's developmental fate is determined; how cells proliferate and differentiate into specialized cell types; and how major changes in body shape are produced. We shall see that an organism's development is ultimately driven by the regulated expression of its genes, determining which proteins are present in which cells and when. In turn, proteins largely determine how a cell behaves. The genes provide a generative program for development, not a blueprint, as their actions are translated into developmental outcomes through cellular behavior such as intercellular signaling, cell proliferation, cell differentiation, and cell movement.

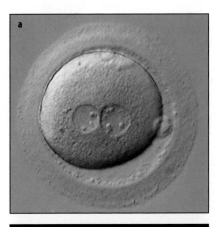

Fig. 1.1 Human fertilized egg and embryo. (a) Human fertilized egg. The sperm and egg nuclei have not yet fused. (b) Human embryo at around 51 days' gestation (Carnegie stage 20), which is equivalent to a mouse embryo at 13.5 days post-fertilization. A human embryo at this stage is around 21-23 mm long.

(a) Courtesy of Alpesh Doshi, CRGH, London. (b) Reproduced courtesy of the MRC/Wellcome-funded Human Developmental Biology Resource.

The development of a multicellular organism from a single cell—the fertilized egg—is a brilliant triumph of evolution. The fertilized egg divides to give rise to many millions of cells, which form structures as complex and varied as eyes, arms, heart, and brain. This amazing achievement raises a multitude of questions. How do the cells arising from division of the fertilized egg become different from each other? How do they become organized into structures such as limbs and brains? What controls the behavior of individual cells so that such highly organized patterns emerge? How are the organizing principles of development embedded within the egg, and in particular within the genetic material, DNA? Much of the excitement in developmental biology today comes from our growing understanding of how genes direct these developmental processes, and genetic control is one of the main themes of this book. Thousands of genes are involved in controlling development, but we will focus only on those that have key roles and illustrate general principles.

Understanding how embryos develop is a huge intellectual challenge, and one of the ultimate aims of the science of **developmental biology** is to understand how we humans develop (Fig. 1.1). We need to understand human development for several reasons. We need to properly understand why it sometimes goes wrong and why a

Box 1A Basic stages of *Xenopus laevis* development

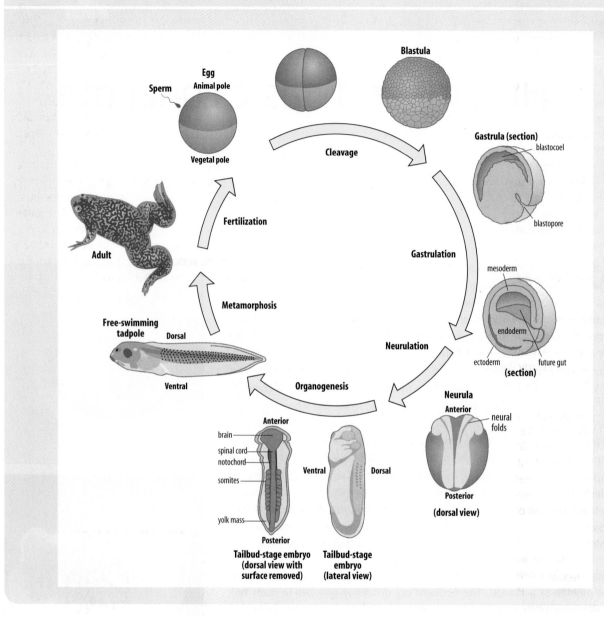

fetus may fail to be born or a baby be born with congenital abnormalities. The link here with genetic control of development is very close, as mutations in genes can lead to abnormal development; environmental factors, such as drugs and infections, can affect it too. Another area of medical research related to developmental biology is regenerative medicine—finding out how to use cells to repair damaged tissues and organs. The focus of regenerative medicine is currently on **stem cells**. These have many of the properties of embryonic cells, such as the ability to proliferate and to develop into a range of different tissues, and are discussed in Chapter 10. Cancer cells also display some properties of embryonic cells, such as the ability to divide indefinitely, and so the study of embryonic cells and their development could lead to new and better treatments for cancer, as many of the same genes are involved.

The development of an embryo from the fertilized egg is known as **embryogenesis** (Box 1A). One of the first tasks is to lay down the overall body plan of the organism, and we shall see that different organisms solve this fundamental problem in several

Although vertebrate development is very varied, there are a number of basic stages that can be illustrated by following the development of the frog *Xenopus laevis*. The unfertilized egg is a large cell. It has a pigmented upper surface (the **animal pole**) and a lower region (the **vegetal pole**) characterized by an accumulation of yolk granules.

After **fertilization** of the egg by a sperm, and the fusion of male and female nuclei, **cleavage** begins. Cleavages are mitotic divisions in which cells do not grow between each division, and so with successive cleavages the cells become smaller. After about 12 division cycles, the embryo, now known as a **blastula**, consists of many small cells surrounding a fluid-filled cavity (the **blastocoel**) above the larger yolky cells. Already, changes have occurred within the cells and they have interacted with each other so that some future tissue types have become partly specified. These are the three **germ layers**: **mesoderm**, **endoderm**, and **ectoderm**. The animal region will give rise to **ectoderm**, which forms both the epidermis of the skin and the nervous system. The future endoderm and mesoderm, which are destined to form internal organs, are still on the surface of the embryo. During the next stage—**gastrulation**—there is a dramatic rearrangement of cells; the endoderm and mesoderm move inside, and the basic body plan of the tadpole is established. Internally, the mesoderm gives rise to a rod-like structure (the notochord), which runs from the head to the tail, and lies centrally beneath the future nervous system. On either side of the notochord are segmented blocks of mesoderm called somites, which will give rise to the muscles and vertebral column, as well as the dermis of the skin (notochord and somites can be seen in the cutaway view of the later tailbud-stage embryo).

Shortly after gastrulation, the ectoderm above the notochord folds to form a tube (the **neural tube**), which gives rise to the brain and spinal cord—a process known as **neurulation**. By this time, other organs, such as limbs, eyes, and gills, are specified at their future locations, but only develop a little later, during **organogenesis**. During organogenesis, specialized cells such as muscle, cartilage, and neurons differentiate. Within 48 hours the embryo has become a feeding tadpole with typical vertebrate features.

ways. The focus of this book is mainly on animal development, in particular that of vertebrates—frogs, birds, fish, and mammals—whose early development is discussed in Chapters 3 to 5. We also look at selected invertebrates, such as the sea urchin, ascidians, and particularly the fruit fly and the nematode worm. Our understanding of the genetic control of development is most advanced in these last two animals and the main features of their early development are considered in Chapters 2 and 6, respectively. They are also used throughout the book to illustrate particular aspects of development. In Chapter 7 we look briefly at some aspects of plant development, which differs in many respects from that of animals but involves similar basic principles.

Morphogenesis, or the development of form, is discussed in Chapter 8. In Chapter 9 we look at how sex is determined and how germ cells develop. The differentiation of unspecialized cells into cells that carry out particular functions, such as muscle cells and blood cells, is considered in Chapter 10. Structures such as the vertebrate limb, and organs such as the heart, insect and vertebrate eyes, and the nervous system, illustrate the problems of multicellular organization and tissue differentiation in embryogenesis, and we consider some of these systems in detail in Chapters 11 and 12. The study of developmental biology, however, goes well beyond the development of the embryo. Post-embryonic growth and aging, and how some animals undergo metamorphosis, are covered in Chapter 13, and how animals can regenerate lost organs is discussed in Chapter 14. Taking a wider view, we shall consider in Chapter 15 how developmental mechanisms have evolved and how they constrain the very process of evolution itself.

One might ask whether it is necessary to cover so many different organisms in order to understand the basic features of development. The answer is yes. Developmental biologists do indeed believe that there are general principles of development that apply to all animals, but life is too wonderfully diverse to find all the answers in a single organism. As it is, developmental biologists have tended to focus their efforts on a relatively small number of animals, chosen because they were convenient to study and amenable to experimental manipulation or genetic analysis. This is why some creatures, such as the frog *Xenopus laevis* (Fig. 1.2), the nematode worm *Caenorhabditis elegans*, and the fruit fly *Drosophila melanogaster*, have such a dominant place in developmental biology. Similarly, work with the thale-cress, *Arabidopsis thaliana*, has uncovered many features of plant development.

Fig. 1.2 The South African claw-toed frog, *Xenopus laevis*. Scale bar = 1 cm.
Photograph courtesy of J. Smith.

One of the most exciting and satisfying aspects of developmental biology is that understanding a developmental process in one organism can help to illuminate similar processes elsewhere—for example, giving insights into how humans develop. Nothing illustrates this more dramatically than the influence that our understanding of *Drosophila* development, and especially of its genetic basis, has had throughout developmental biology. The identification of genes controlling early embryogenesis in *Drosophila* has led to the discovery of related genes being used in similar ways in the development of mammals and other vertebrates. Such discoveries encourage us to believe in the existence of general developmental principles.

Frogs have long been a favorite organism for studying early development because their eggs are large and their embryos are robust—easy to grow in a simple culture medium and relatively easy to experiment on. Embryogenesis in the South African frog *Xenopus* (Box 1A, pp. 2–3) illustrates some of the basic stages of development in all animals.

In the rest of this chapter we first look briefly at the history of **embryology**—as the study of developmental biology has been called for most of its existence. The term developmental biology itself is of much more recent origin and reflects the appreciation that development is not restricted to the embryo alone. Traditionally, embryology described experimental results in terms of morphology and cell fate, but we now understand development in terms of molecular genetics and cell biology as well. In the second part of the chapter we will introduce some key concepts that are used repeatedly in studying and understanding development.

The origins of developmental biology

Many questions in embryology were first posed hundreds, and in some cases thousands, of years ago. Appreciating the history of these ideas helps us to understand why we approach developmental problems in the way that we do today.

1.1 Aristotle first defined the problem of epigenesis and preformation

A scientific approach to explaining development started with Hippocrates in Greece in the fifth-century BC. Using the ideas current at the time, he tried to explain development in terms of the principles of heat, wetness, and solidification. About a century later the study of embryology advanced when the Greek philosopher Aristotle formulated a question that was to dominate much thinking about development until the end of the nineteenth century. Aristotle addressed the problem of how the different parts of the embryo were formed. He considered two possibilities: one was that everything in the embryo was preformed from the very beginning and simply got bigger during development; the other was that new structures arose progressively, a process he termed epigenesis (which means 'upon formation') and that he likened metaphorically to the 'knitting of a net'. Aristotle favored epigenesis and his conjecture was correct.

Aristotle's influence on European thought was enormous and his ideas remained dominant well into the seventeenth century. The contrary view to epigenesis, namely that the embryo was preformed from the beginning, was championed anew in the late seventeenth century. Many could not believe that physical or chemical forces could mold a living entity like the embryo. Along with the contemporaneous background of belief in the divine creation of the world and all living things, was the belief that all embryos had existed from the beginning of the world, and that the first embryo of a species must contain all future embryos.

Even the brilliant seventeenth-century Italian embryologist, Marcello Malpighi, could not free himself from preformationist ideas. While he provided a remarkably accurate description of the development of the chick embryo, he remained convinced, against the evidence of his own observations, that the fully formed embryo was present from the beginning (Fig. 1.3). He argued that at very early stages the parts were

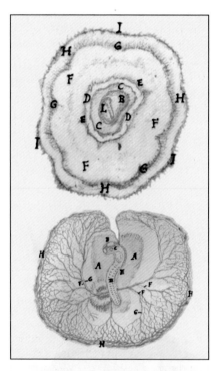

Fig. 1.3 Malpighi's description of the chick embryo. The figure shows Malpighi's drawings, made in 1673, depicting the early embryo (top), and at 2 days' incubation (bottom). His drawings accurately illustrate the shape and blood supply of the embryo.

Reprinted by permission of the President and Council of the Royal Society.

so small that they could not be seen, even with his best microscope. Other preformationists believed that the sperm contained the embryo, and some even claimed to see a tiny human—an homunculus—in the head of each human sperm (Fig. 1.4).

The preformation/epigenesis issue was vigorously debated throughout the eighteenth century. But the problem could not be resolved until one of the great advances in biology had taken place—the recognition that living things, including embryos, were composed of cells.

1.2 Cell theory changed the conception of embryonic development and heredity

The invention of the microscope around 1600 was essential for the discovery of cells, but the 'cell theory' of life was only developed between 1820 and 1880 by, among others, the German botanist, Matthias Schleiden, and the physiologist, Theodor Schwann. It recognized that all living organisms consist of cells, that these are the basic units of life, and that new cells can only be formed by the division of pre-existing cells. The cell theory was one of the most illuminating advances in biology, and had an enormous impact. Multicellular organisms, such as animals and plants, could now be viewed as communities of cells. Development could not therefore be based on preformation but must be epigenetic, because during development many new cells are generated by division from the egg, and new types of cells are formed. A crucial step forward in understanding development was the recognition, in the 1840s, that the egg itself is but a single, albeit specialized, cell.

An important advance in embryology was the proposal by the nineteenth-century German biologist, August Weismann, that the offspring does not inherit its characteristics from the body (the soma) of the parent but only from the **germ cells**—egg and sperm. Weismann drew a fundamental distinction between germ cells and the body cells or **somatic cells** (Fig. 1.5). Characteristics acquired by the body during an animal's life cannot be transmitted to the germline. As far as heredity is concerned, the body is merely a carrier of germ cells. As the English novelist and essayist Samuel Butler put it: 'A hen is only an egg's way of making another egg.'

Work on sea-urchin eggs showed that after fertilization the egg contains two nuclei, which eventually fuse; one of these nuclei belongs to the egg, while the other comes from the sperm. Fertilization therefore results in a single cell—the **zygote**—carrying a nucleus with contributions from both parents, and it was concluded that the cell nucleus must contain the physical basis of heredity. The climax of this line of research was the demonstration, towards the end of the nineteenth century, that the chromosomes within the nucleus of the zygote are derived in equal numbers from the two parental nuclei, and the recognition that this provided a physical basis for the transmission of genetic characters according to the laws developed by the Austrian botanist

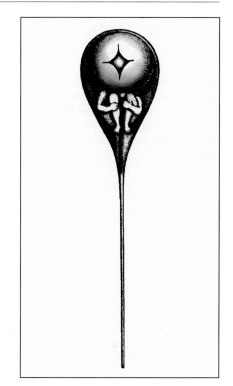

Fig. 1.4 Some preformationists believed that an homunculus was curled up in the head of each sperm.

An imaginative drawing, after Nicholas Harspeler (1694).

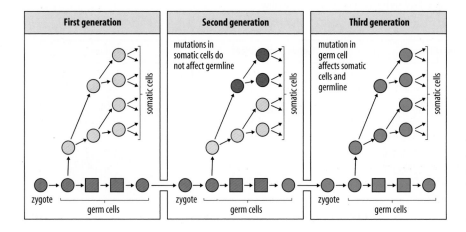

Fig. 1.5 The distinction between germ cells and somatic cells. In each generation germ cells give rise to both somatic cells and germ cells, but inheritance is through the germ cells only (first panel). Changes that occur due to a mutation (red) in a somatic cell can be passed on to its daughter cells but do not affect the germline, as shown in the second panel. In contrast, a mutation in the germline (green) will be present in every cell in the body of the new organism to which that cell contributes, and will also be passed on to future generations through the germline, as shown in the third panel.

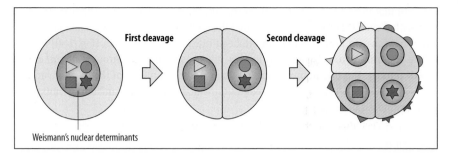

Fig. 1.6 Weismann's theory of nuclear determination. Weismann assumed that there were factors in the nucleus that were distributed asymmetrically to daughter cells during cleavage and directed their future development.

and monk, Gregor Mendel. The number of chromosomes is kept constant from generation to generation by a specialized type of cell division that produces the germ cells, called **meiosis**, which halves the chromosome number; the full complement of chromosomes is then restored at fertilization. The zygote and the somatic cells that arise from it divide by the process of **mitosis**, which maintains chromosome number (Box 1B, p. 7). Germ cells contain a single copy of each chromosome and are called **haploid**, whereas germ-cell precursor cells and the other somatic cells of the body contain two copies and are called **diploid**.

1.3 Two main types of development were originally proposed

The next big question was how cells became different from one another during embryonic development. With the increasing emphasis on the role of the nucleus, in the 1880s Weismann put forward a model of development in which the nucleus of the zygote contained a number of special factors or **determinants** (Fig. 1.6). He proposed that while the fertilized egg underwent the rapid cycles of cell division known as cleavage, these determinants would be distributed unequally to the daughter cells and so would control the cells' future development. The fate of each cell was therefore predetermined in the egg by the factors it would receive during cleavage. This type of model was termed 'mosaic,' as the egg could be considered to be a mosaic of discrete localized determinants. Central to Weismann's theory was the assumption that early cell divisions must make the daughter cells quite different from each other as a result of unequal distribution of nuclear components.

Fig. 1.7 Roux's experiment to investigate Weismann's theory of mosaic development. After the first cleavage of a frog embryo, one of the two cells is killed by pricking it with a hot needle; the other remains undamaged. At the blastula stage the undamaged cell can be seen to have divided as normal into many cells that fill half of the embryo. The development of the blastocoel, a small fluid-filled space in the center of the blastula, is also restricted to the undamaged half. In the damaged half of the embryo, no cells appear to have formed. At the neurula stage, the undamaged cell has developed into something resembling half a normal embryo.

In the late 1880s, initial support for Weismann's ideas came from experiments carried out independently by the German embryologist, Wilhelm Roux, who experimented with frog embryos. Having allowed the first cleavage of a fertilized frog egg, Roux destroyed one of the two cells with a hot needle and found that the remaining cell developed into a well-formed half-larva (Fig. 1.7). He concluded that the 'development of the frog is based on a mosaic mechanism, the cells having their character and fate determined at each cleavage.'

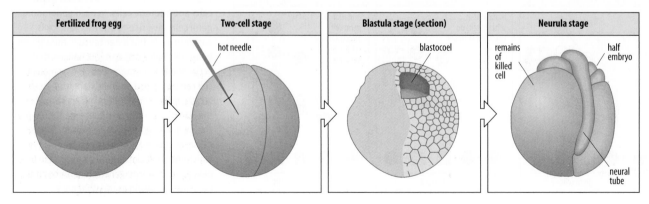

Box 1B The mitotic cell cycle

When a eukaryotic cell duplicates itself it goes through a fixed sequence of events called the **cell cycle**. The cell grows in size, the DNA is replicated, and the replicated chromosomes then undergo mitosis and become segregated into two daughter nuclei. Only then can the cell divide to form two daughter cells, which can go through the whole sequence again.

The standard eukaryotic mitotic cell cycle is divided into well-marked phases. At the M phase, mitosis and cell cleavage give rise to two new cells. The rest of the cell cycle, between one M phase and the next, is called interphase. Replication of DNA occurs during a defined period in interphase, the S phase. Preceding S phase is a period known as G_1 (the G stands for gap), and after it another interval known as G_2, after which the cells enter mitosis (see figure). G_1, S phase, and G_2 collectively make up interphase, the part of the cell cycle during which cells synthesize proteins and grow, as well as replicating their DNA. When somatic cells are not proliferating they are usually in a state known as G_0, into which they withdraw after mitosis. The decision to enter G_0 or to proceed through G_1, may be controlled by both intracellular status and extracellular signals such as growth factors. Growth factors enable the cell to proceed out of G_0 and progress through the cell cycle. Cells such as neurons and skeletal muscle cells, which do not divide after differentiation, are permanently in G_0.

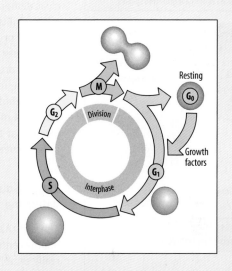

Particular phases of the cell cycle are absent in some cells: during cleavage of the fertilized *Xenopus* egg, G_1 and G_2 are virtually absent and cells get smaller at each division. In *Drosophila* salivary glands there is no M phase, as the DNA replicates repeatedly without mitosis or cell division, leading to the formation of giant polytene chromosomes.

But when Roux's fellow countryman, Hans Driesch, repeated the experiment on sea-urchin eggs, he obtained quite a different result (Fig. 1.8). He wrote later: 'But things turned out as they were bound to do and not as I expected; there was, typically, a whole gastrula on my dish the next morning, differing only by its small size from a normal one; and this small but whole gastrula developed into a whole and typical larva.'

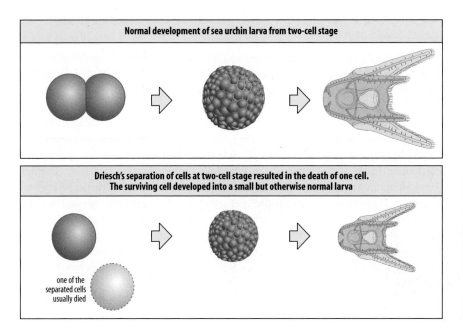

Fig. 1.8 The outcome of Driesch's experiment on sea urchin embryos, which first demonstrated the phenomenon of regulation. After separation of cells at the two-cell stage, the remaining cell developed into a small, but whole, normal larva. This contradicted Roux's earlier finding that if one of the cells of a two-cell frog embryo is damaged, the remaining cell develops into a half-embryo only (see Fig. 1.7).

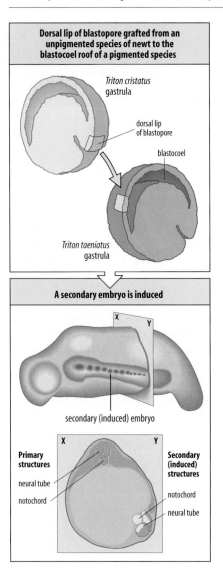

Fig. 1.9 The dramatic demonstration by Spemann and Mangold of induction of a new main body axis by the organizer region in the early amphibian gastrula. A piece of tissue (yellow) from the dorsal lip of the blastopore of a newt (*Triton cristatus*) gastrula is grafted to the opposite side of a gastrula of another, pigmented, newt species (*Triton taeniatus*, pink). The grafted tissue induces a new body axis containing neural tube and somites. The unpigmented graft tissue forms a notochord at its new site (see section in lower panel) but the neural tube and the other structures of the new axis have been induced from the pigmented host tissue. The organizer region discovered by Spemann and Mangold is known as the Spemann organizer.

Driesch had completely separated the cells at the two-cell stage and obtained a normal but small larva. That was just the opposite of Roux's result, and was the first clear demonstration of the developmental process known as **regulation**. This refers to the ability of an embryo to restore normal development even if some portions are removed or rearranged very early in development, and its basis is explained later in the chapter. We shall see many examples of regulation throughout the book (frog embryos do, in fact, regulate as well; an explanation of Roux's experiment, and why he got the result he did, is given later, in Section 4.3).

But although Weismann was wrong about nuclear determinants, there are many examples of developmentally important proteins and RNAs located in the cell cytoplasm that are distributed unequally at cell division and which make two daughter cells different from each other. These are known as **cytoplasmic determinants**.

1.4 The discovery of induction showed that one group of cells could determine the development of neighboring cells

The fact that embryos can regulate implies that cells must interact with each other, but the central importance of cell–cell interactions in embryonic development was not really established until the discovery of the phenomenon of **induction**. This is where one cell, or tissue, directs the development of another, neighboring, cell or tissue.

The importance of induction and other cell–cell interactions in development was proved dramatically in 1924 when Hans Spemann and his assistant, Hilde Mangold, carried out a famous transplantation experiment in amphibian embryos. They showed that a partial second embryo could be induced by grafting one small region of an early newt embryo onto another at the same stage (Fig. 1.9). The grafted tissue was taken from the dorsal lip of the **blastopore**—the slit-like invagination that forms where gastrulation begins on the dorsal surface of the amphibian embryo (see Box 1A, pp. 2–3). This small region they called the **organizer**, as it seemed to be ultimately responsible for controlling the organization of a complete embryonic body; it is now known as the **Spemann–Mangold organizer**, or just the **Spemann organizer**. For their discovery, Spemann received the Nobel Prize for Physiology or Medicine in 1935, the first Nobel Prize ever given for embryological research. Sadly, Hilde Mangold had died earlier, in an accident, and so could not be honored.

1.5 The study of development was stimulated by the coming together of genetics and development

When Mendel's laws were rediscovered in 1900 there was a great surge of interest in mechanisms of inheritance, particularly in relation to evolution, but less so in relation to development. Genetics was seen as the study of the transmission of hereditary elements from generation to generation, whereas embryology was the study of how an individual organism develops and, in particular, how cells in the early embryo became different from each other. Genetics seemed, in this respect, to be irrelevant to development.

The fledgling science of genetics was put on a firm conceptual and experimental footing in the first-quarter of the twentieth century by the American, Thomas Hunt Morgan. Morgan chose the fruit fly *Drosophila melanogaster* as his experimental organism. He noticed a fly with white eyes rather than the usual red eyes and by careful cross-breeding he showed that inheritance of this mutant trait was linked to the sex of the fly. He found three other sex-linked traits and worked out that they were each determined by three distinct 'genetic loci,' which occupied different positions on the same chromosome, the fly's X chromosome. The rather abstract hereditary 'factors' of Mendel had been given reality. But even though Morgan had originally trained as an embryologist, he made little headway in explaining development in terms of genetics. That had to wait until the nature of the gene was better understood.

An important concept in understanding how genes influence physical and physiological traits is the distinction between **genotype** and **phenotype**. This was first put forward by the Danish botanist, Wilhelm Johannsen, in 1909. The genetic endowment of an organism—the genetic information it acquires from its parents—is the genotype. Its visible appearance, internal structure, and biochemistry comprise the phenotype. While the genotype certainly controls development, environmental factors interacting with the genotype influence the phenotype. Despite having identical genotypes, identical twins can develop considerable differences in their phenotypes as they grow up (Fig. 1.10), and these tend to become more evident with age.

Following Morgan's discoveries in genetics, the problem of development could now be posed in terms of the relationship between genotype and phenotype: how the genetic endowment becomes 'translated' or 'expressed' during development to give rise to a functioning organism. But the coming together of genetics and embryology was slow and tortuous. The discovery in the 1940s that genes are made of DNA and encode proteins was a major turning point. It was already clear that the properties of a cell are determined by the proteins it contains, and so the fundamental role of genes in development could at last be appreciated. By controlling which proteins were made in a cell, genes could control the changes in cell properties and behavior that occurred during development. A further major advance in the 1960s was the discovery that some genes encode proteins that control the activity of other genes. A more recent discovery is that small RNAs called microRNAs are also important in the control of gene expression.

Fig. 1.10 The difference between genotype and phenotype. These identical twins have the same genotype because one fertilized egg split into two during development. Their slight difference in appearance is due to nongenetic factors, such as environmental influences.

Photograph courtesy of Josè and Jaime Pascual.

1.6 Development is studied mainly through selected model organisms

Although the development of various species has been studied at one time or another, a relatively small number of organisms provide most of our knowledge about developmental mechanisms. We can thus regard them as 'models' for understanding the processes involved, and they are often called **model organisms**. Sea urchins and amphibians were the main animals used for the first experimental investigations because their developing embryos are both easy to obtain and, in the case of amphibians, sufficiently large and robust for relatively easy experimental manipulation, even at quite late stages. Among vertebrates, the frog *Xenopus laevis*, the mouse (*Mus musculus*), the chicken (*Gallus gallus*), and the zebrafish (*Danio rerio*), are the main model organisms now studied. Among invertebrates, the fruit fly *Drosophila melanogaster* and the nematode *Caenorhabditis elegans* have been the focus of most attention, because a great deal is known about their developmental genetics and they can be easily genetically modified. Two Nobel prizes have been awarded for discoveries about development in *Drosophila* and *Caenorhabditis*, respectively. With the advent of modern methods of genetic analysis, there has also been a resurgence of interest in the sea urchin *Strongylocentrotus purpuratus* and ascidians such as *Ciona intestinalis*. For plant developmental biology, *Arabidopsis thaliana* serves as the main model organism. The life cycles and background details for these model organisms are given in the relevant chapters later in the book. The evolutionary relationships of these organisms are shown in Fig. 1.11.

The reasons for these choices are partly historical—once a certain amount of research has been done on one animal it is more efficient to continue to study it rather than start at the beginning again with another species—and partly a question of ease of study and biological interest. Each species has its advantages and disadvantages as a developmental model. The chick embryo, for example, has long been studied as an example of vertebrate development because fertile eggs are easily available, the embryo withstands experimental microsurgical manipulation very well, and it can be cultured outside the egg. A disadvantage, however, was that until very recently little was known about the chick's developmental genetics. In contrast, we know a great deal about the genetics of the mouse, although the mouse is more

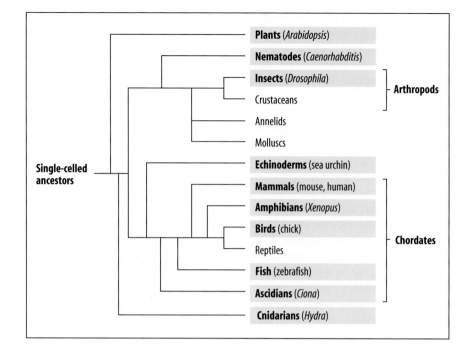

Fig. 1.11 Phylogenetic tree showing the placement of the main developmental model organisms. The organisms discussed in this book are highlighted in blue.

difficult to study in some ways, as development normally takes place entirely within the mother. It is, however, possible to culture very early mouse embryos for a little time outside the uterus and then re-implant them. Many developmental mutations have been identified in the mouse, and it is also amenable to genetic modification by **transgenic** techniques, by which genes can be introduced, deleted, or modified in living organisms. It is also the best experimental model we have for studying mammalian development, including that of humans. The zebrafish is a more recent addition to the select list of vertebrate model systems; it is easy to breed in large numbers, the embryos are transparent and so cell divisions and tissue movements can be followed visually, and it has great potential for genetic investigations.

A major goal of developmental biology is to understand how genes control embryonic development, and to do this one must first identify which genes, out of the many thousands in the organism, are critically and specifically involved in controlling development. This task can be approached in various ways, depending on the organism involved, but the general starting point is to identify mutations that alter development in some specific and informative way, as described in the following section. Techniques for identifying such genes and for detecting and manipulating their expression in the organism are described throughout the book, along with techniques for genetic manipulation.

Some of our model organisms are more amenable to conventional genetic analysis than others. Despite its importance in developmental biology, little conventional genetics has been done on *X. laevis*, which has the disadvantage of being **tetraploid** (it has four sets of chromosomes in its somatic cells, as opposed to the two sets carried by the cells of diploid organisms, such as humans and mice) and a relatively long breeding period, taking 1–2 years to reach sexual maturity. Using modern genetic and bioinformatics techniques, however, many developmental genes have been identified in *X. laevis* by direct DNA sequence comparison with known genes in *Drosophila* and mice. The closely related frog, *Xenopus tropicalis*, is a more attractive organism for genetic analysis; it is diploid and can also be genetically manipulated to produce transgenic organisms.

The complete set of genes characteristic of a particular organism is known as its **genome**. In sexually reproducing diploid species, two complete copies of the genome are carried by the chromosomes in each somatic cell—one copy inherited from the father and one from the mother. The complete DNA sequences of the genomes of many model species are now known, which helps enormously in identifying developmental genes. Genome sequences have been completed for sea urchin, *Drosophila*, *Caenorhabditis*, chicken, mouse, and human, and there is a draft genome sequence for *X. tropicalis*. The zebrafish genome is being sequenced; a complication here is that a genome-duplication event in the zebrafish's evolutionary history means that there are duplicates of at least 2900 out of the estimated 20,000 protein-coding genes in its genome.

In general, when an important developmental gene has been identified in one animal, it has proved very rewarding to see whether a corresponding gene is present and is acting in a developmental capacity in other animals. Such genes are often identified by a sufficient degree of nucleotide sequence similarity to indicate descent from a common ancestral gene. Genes that meet this criterion are known as **homologous genes**. As we shall see in Chapter 5, this approach identified a hitherto unsuspected class of vertebrate genes that control the regular segmented pattern from head to tail that is represented by the different types of vertebrae at different positions. These genes were identified by their homology with genes that determine the identities of the different body segments in *Drosophila* (described in Chapter 2).

1.7 The first developmental genes were identified as spontaneous mutations

Most of the organisms dealt with in this book are sexually reproducing diploids: their somatic cells contain two copies of each gene, with the exception of those on the sex chromosomes. In a diploid species, one copy, or **allele**, of each gene is contributed by the male parent and the other by the female. For many genes there are several different 'normal' alleles present in the population, which leads to the variation in normal phenotype one sees within any sexually reproducing species. Occasionally, however, a mutation will occur spontaneously in a gene and there will be marked change, usually deleterious, in the phenotype of the organism.

Many of the genes that affect development have been identified by spontaneous mutations that disrupt their function and produce an abnormal phenotype. Mutations are classified broadly according to whether they are dominant or recessive (Fig. 1.12). **Dominant** and **semi-dominant** mutations are those that produce a distinctive phenotype when the mutation is present in just one of the alleles of a pair; that is, it exerts an

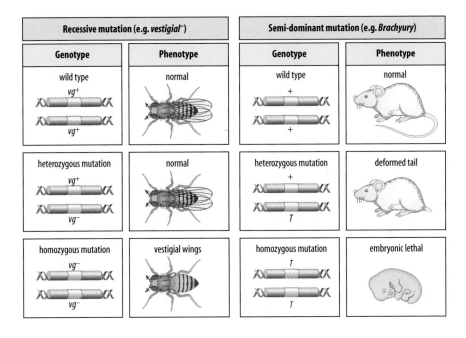

Fig. 1.12 Types of mutations. Left: a mutation is recessive when it only has an effect in the homozygous state; that is, when both copies of the gene carry the mutation. A plus sign (+) denotes the 'normal' allele (wild type), and a minus sign (–) the recessive allele. Right: by contrast, a dominant or semi-dominant mutation produces an effect on the phenotype in the heterozygous state; that is, when just one copy of the mutant gene is present. *T* denotes the mutant form of the gene *Brachyury*.

effect in the **heterozygous** state. Dominant developmental mutations can be missed as they are often lethal in the embryo. By contrast, **recessive** mutations, such as *vestigial*, which leads to the non-development of wings in *Drosophila*, alter the phenotype only when both alleles are of the mutant type; this is known as the **homozygous** state.

In general, dominant mutations are more easily recognized, particularly if they affect gross anatomy or coloration, provided that they do not cause the early death of the embryo in the heterozygous state. Truly dominant mutations are rare, however. A mutation in the mouse gene *Brachyury* is a classic example of a semi-dominant mutation and was originally identified because mice heterozygous for this mutation (symbolized by *T*) have short tails. When the mutation is homozygous it has a much greater effect, and embryos die at an early stage, indicating that the gene is required for normal embryonic development (Fig. 1.13). Once breeding studies had confirmed that the *Brachyury* trait was due to a single gene, the gene could be mapped to a location on a particular chromosome by classical genetic-mapping techniques. Identifying recessive mutations is more laborious, as the heterozygote has a phenotype identical to a normal wild-type animal, and a carefully worked-out breeding program is needed to obtain homozygotes. In mammals, identifying potentially lethal recessive developmental mutations requires careful observation and analysis, as the homozygotes may die unnoticed inside the mother.

Many mutations in invertebrates were identified as **conditional mutations**. The effects of these mutations only show up when the animal is kept in a particular condition—most commonly at a higher temperature, in which case the mutation is known as a **temperature-sensitive mutation**. At the usual ambient temperature, the animal appears normal. Temperature sensitivity is usually due to the encoded mutant protein being able to fold into a functional structure at the normal temperature, but being less stable at the higher temperature.

Very rigorous criteria must be applied to identify mutations that are affecting a genuine developmental process and not just affecting some vital but routine 'housekeeping' function without which the animal cannot survive. One simple criterion for a developmental mutation is that it leads to the death of the embryo, but mutations in genes involved in vital housekeeping functions can also have this effect. Mutations that produce observable abnormal embryonic development are more promising candidates for true developmental mutations. In later chapters we shall see how large-scale screening for mutations after mutagenesis by chemicals or X-rays has identified many more developmental genes than would ever have been picked up by rare spontaneous mutations.

Fig. 1.13 Genetics of the semi-dominant mutation *Brachyury* (*T*) in the mouse.
A male heterozygote carrying the *T* mutation merely has a short tail. When mated with a female homozygous for the normal, or wild-type, *Brachyury* gene (++), some of the offspring will also be heterozygotes and have short tails. Mating two heterozygotes together will result in some of the offspring being homozygous (*T*/*T*) for the mutation, resulting in a severe and lethal developmental abnormality in which the posterior mesoderm does not develop.

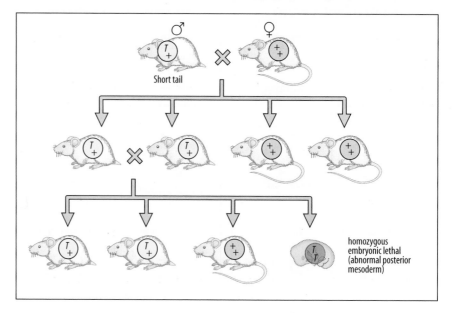

In recent years, new techniques for identifying developmental genes have appeared. These depend on knowing the existence and sequence of a gene (for example, from genomic sequences) and working backwards to determine its function in the animal, usually by removing the gene or blocking its function. Such methods are known as **reverse genetics**, as opposed to the traditional methods of **forward genetics** outlined above. Examples of reverse genetics include **gene knock-out**, in which the gene is effectively deleted from the animal's genome by transgenic techniques (discussed in Section 3.9), and **gene knockdown** or **gene silencing**, in which gene expression is prevented by techniques such as RNA interference or antisense RNA (discussed in Box 6A, p. 220).

Summary

The study of embryonic development started with the Greeks more than 2000 years ago. Aristotle put forward the idea that embryos were not contained completely pre-formed in miniature within the egg, but that form and structure emerged gradually, as the embryo developed. This idea was challenged in the seventeenth and eighteenth centuries by those who believed in preformation, the idea that all embryos that had been, or ever would be, had existed from the beginning of the world. The emergence of the cell theory in the nineteenth century finally settled the issue in favor of epigenesis, and it was realized that the sperm and egg are single, albeit highly specialized, cells. Some of the earliest experiments showed that very early sea-urchin embryos are able to regulate, that is to develop normally, even if cells are removed or killed. This established the important principle that development must depend, at least in part, on communication between the cells of the embryo. Direct evidence for the importance of cell-cell interactions came from the organizer graft experiment carried out by Spemann and Mangold in 1924, showing that the cells of the amphibian organizer region could induce a new partial embryo from host tissue when transplanted into another embryo. The role of genes in controlling development, by determining which proteins are made, has only been fully appreciated in the past 50 years and the study of the genetic basis of development has been made much easier in recent times by the techniques of molecular biology and the availability of the DNA sequences of whole genomes.

A conceptual tool kit

Development into a multicellular organism is the most complicated fate a single living cell can undergo; in this lies both the fascination and the challenge of developmental biology. Yet only a few basic principles are needed to start to make sense of developmental processes. The rest of this chapter is devoted to introducing these key concepts. These principles are encountered repeatedly throughout the book, as we look at different organisms and developmental systems, and should be regarded as a conceptual tool kit, essential for embarking on a study of development.

Genes control development by determining where and when proteins are synthesized, and many thousands of genes are involved. Gene activity sets up intracellular networks of interactions between proteins and genes, and between proteins and proteins, that give cells their particular properties. One of these properties is the ability to communicate with, and respond to, other cells. It is these **cell–cell interactions** that determine how the embryo develops; no developmental process can therefore be attributed to the function of a single gene or single protein. The amount of genetic and molecular information on developmental processes is now enormous. In this book, we will be highly selective and describe only those molecular details that give an insight into the mechanisms of development and illustrate general principles.

1.8 Development involves the emergence of pattern, change in form, cell differentiation, and growth

In most embryos, fertilization is immediately followed by cell division. This stage is known as **cleavage**, and divides the fertilized egg into a number of smaller cells (Fig. 1.14). Unlike the cell divisions that take place during growth of a tissue, there is no increase in cell size between each cleavage division; the cleavage cell cycles consist simply of phases of DNA replication, mitosis, and cell division (see Box 1B, p. 7). Early embryos are therefore no bigger than the zygote, and there is little increase in size during embryogenesis unless the embryo has access to a substantial source of material it can use to grow—for example, the large yolk available to the chick embryo and the nutrients provided to the mammalian embryo via the placenta.

Development is essentially the emergence of organized structures from an initially very simple group of cells. It is convenient to distinguish four main developmental processes, which occur in roughly sequential order in development, although in reality they overlap with, and influence, each other considerably. They are **pattern formation**, **morphogenesis**, **cell differentiation**, and **growth**. Pattern formation is the process by which cellular activity is organized in space and time so that a well-ordered structure develops within the embryo, and is of considerable importance in the early embryo. In the developing arm, for example, pattern formation enables cells to 'know' whether to make an upper arm or fingers, and where the muscles should form. There is no universal strategy of patterning; rather, it is achieved by various cellular and molecular mechanisms in different organisms and at different stages of development.

Pattern formation initially involves laying down the overall **body plan**—defining the main body axes of the embryo that run from head (anterior) to tail (posterior), and from back (dorsal) to underside (ventral). Most of the animals in this book have a head at one end and a tail at the other, with the left and right sides of the body being outwardly bilaterally symmetrical—that is, a mirror image of each other. In such animals, the main body axis is the **antero-posterior axis**, which runs from the head to the tail. Bilaterally symmetrical animals also have a **dorso-ventral axis**, running from the back to the belly. A striking feature of these axes is that they are almost always at right angles to one another and so can be thought of as making up a system of coordinates on which any position in the body could be specified (Fig. 1.15). Internally, animals have distinct differences between the left and right sides in the disposition of their internal organs—the human heart, for example, is on the left side. In plants, the main body axis runs from the growing tip (the apex) to the roots and is known as the **apical–basal axis**. Plants also have radial symmetry, with a **radial axis** running from the center of the stem outwards.

Even before the body axes become clear, eggs and embryos often show a distinct **polarity**, which in this context means that they have an intrinsic orientation, with one end different from the other. Polarity can be specified by a gradient in a protein or other molecule, as the slope of the gradient provides direction. Many cells in

Fig. 1.14 Light micrographs of cleaving *Xenopus* eggs.

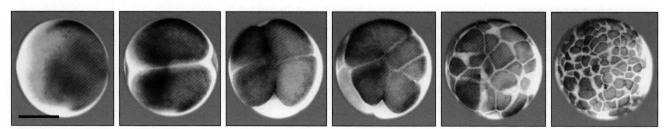

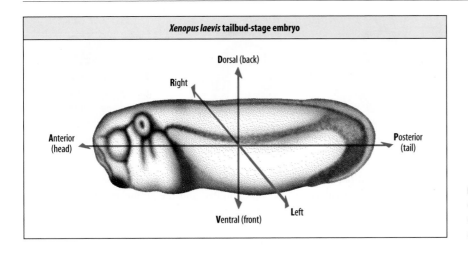

Xenopus laevis tailbud-stage embryo

Dorsal (back)

Right

Anterior (head)

Posterior (tail)

Left

Ventral (front)

Fig. 1.15 The main axes of a developing embryo. The antero-posterior axis and the dorso-ventral axis are at right angles to one another, as in a coordinate system.

developing embryos have their own individual polarity, and this can result in polarity along a sheet of cells. It is as if a direction arrow is present in all the cells, making one end slightly different from the other. This type of polarity is known as **planar cell polarity** and is discussed further in Chapters 2 and 8. Cell polarity is sometimes visible in the structures they produce—as with the epidermal hairs on the edge of the *Drosophila* wing, which all point in the same direction.

At the same time that the axes are being formed, cells are being allocated to the different **germ layers**—ectoderm, mesoderm, and endoderm (Box 1C). During further pattern formation, cells of these germ layers acquire different identities so that

---

Box 1C Germ layers

The concept of germ layers is useful to distinguish between regions of the early embryo that give rise to quite distinct types of tissues. It applies to both vertebrates and invertebrates. All the animals considered in this book, except for the cnidarian *Hydra*, are **triploblasts**, with three germ layers: the endoderm, which gives rise to the gut and its derivatives, such as the liver and lungs in vertebrates; the mesoderm, which gives rise to the skeleto-muscular system, connective tissues, and other internal organs, such as the kidney and heart; and the ectoderm, which gives rise to the epidermis of the skin and nervous system. These are specified early in development. The distinction between the different germ layers can be fuzzy and there are notable exceptions. The neural crest in vertebrates, for example, is ectodermal in origin but gives rise not only to neural tissue but also to some skeletal elements, which would normally be considered mesodermal in origin.

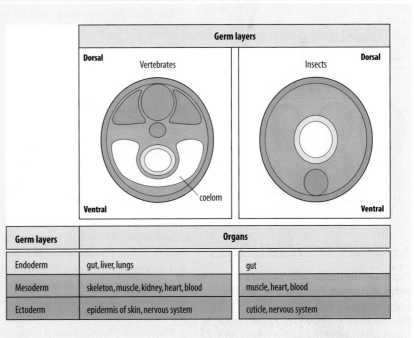

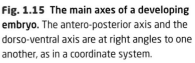

Germ layers	Organs	
Endoderm	gut, liver, lungs	gut
Mesoderm	skeleton, muscle, kidney, heart, blood	muscle, heart, blood
Ectoderm	epidermis of skin, nervous system	cuticle, nervous system

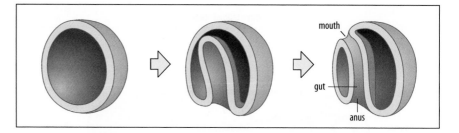

Fig. 1.16 Gastrulation in the sea urchin. Gastrulation transforms the spherical blastula into a structure with a tube—the gut—through the middle. The left-hand side of the embryo has been removed.

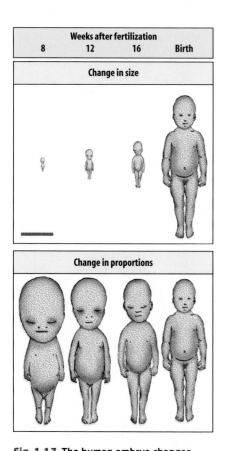

Fig. 1.17 The human embryo changes shape as it grows. From the time the body plan is well-established at 8 weeks until birth the embryo increases in length some tenfold (upper panel), while the relative proportion of the head to the rest of the body decreases (lower panel). As a result, the shape of the embryo changes. Scale bar = 10 cm.

After Moore, K.L.: 1983.

organized spatial patterns of differentiated cells emerge, such as the arrangement of skin, muscle, and cartilage in developing limbs, and the arrangement of neurons in the nervous system. In the earliest stages of pattern formation, differences between cells are not easily detected and probably consist of subtle differences caused by a change in activity of a very few genes.

The second important developmental process is change in form, or **morphogenesis** (Chapter 8). Embryos undergo remarkable changes in three-dimensional form—you need only look at your hands and feet. At certain stages in development, there are characteristic and dramatic changes in form, of which **gastrulation** is the most striking. Almost all animal embryos undergo gastrulation, during which endoderm and mesoderm move inside, the gut is formed, and the main body plan emerges. During gastrulation, cells on the outside of the embryo move inwards and, in animals such as the sea urchin, gastrulation even transforms a hollow spherical blastula into a gastrula with a tube through the middle—the gut (Fig. 1.16). Morphogenesis in animal embryos can also involve extensive cell migration. Most of the cells of the human face, for example, are derived from cells that migrated from a tissue called the neural crest, which originates on the dorsal side of the embryo. Morphogenesis can also involve developmentally programmed cell death, called **apoptosis**, which is responsible, for example, for the separation of our fingers and toes.

The third developmental process we must consider here is **cell differentiation**, in which cells become structurally and functionally different from each other, ending up as distinct cell types, such as blood, muscle, or skin cells. Differentiation is a gradual process, cells often going through several divisions between the time at which they start differentiating and the time they are fully differentiated (when some cell types stop dividing altogether), and is discussed in Chapter 10. In humans, the fertilized egg gives rise to at least 250 clearly distinguishable types of cell.

Pattern formation and cell differentiation are very closely interrelated, as we can see by considering the difference between human arms and legs. Both contain exactly the same types of cell—muscle, cartilage, bone, skin, and so on—yet the pattern in which they are arranged is clearly different. It is essentially pattern formation that makes us different from elephants and chimpanzees.

The fourth process is **growth**—the increase in size. In general there is little growth during early embryonic development and the basic pattern and form of the embryo is laid down on a small scale, always less than a millimeter in extent. Subsequent growth can be brought about in various ways: cell multiplication, increase in cell size, and deposition of extracellular materials, such as bone and shell. Growth can also be morphogenetic in that differences in growth rates between organs, or between parts of the body, can generate changes in the overall shape of the embryo (Fig. 1.17), as we shall see in more detail in Chapter 13.

These four developmental processes are neither independent of each other nor strictly sequential. In very general terms, however, one can think of pattern formation in early development specifying differences between cells that lead to changes in form, cell differentiation, and growth. But in any real developing system there will be many twists and turns in this sequence of events.

1.9 Cell behavior provides the link between gene action and developmental processes

Gene expression within cells leads to the synthesis of proteins that are responsible for particular cellular properties and behavior, which in turn determine the course of embryonic development. The past and current patterns of gene activity confer a certain state, or identity, on a cell at any given time, which is reflected in its molecular organization—in particular which proteins are present. As we shall see, embryonic cells and their progeny undergo many changes in state as development progresses. Other categories of cell behavior that will concern us are intercellular communication, also known as **cell–cell signaling**, changes in cell shape and cell movement, cell proliferation, and cell death.

Changing patterns of gene activity during early development are essential for pattern formation. They give cells identities that determine their future behavior and lead eventually to their final differentiation. And, as we saw in the example of induction by the Spemann organizer, the capacity of cells to influence each other's fate by producing and responding to signals is crucial for development. By their response to signals for cell movement or a change in shape, for example, cells generate the physical forces that bring about morphogenesis (Fig. 1.18). The curvature of a sheet of cells into a tube, as happens in *Xenopus* and other vertebrates during formation of the neural tube (see Box 1A, pp. 2–3), is the result of contractile forces generated by cells changing their shape at certain positions within the cell layer. An important feature of cell surfaces is the presence of adhesive proteins known as cell-adhesion molecules that serve a variety of functions: they hold cells together in tissues, they enable cells to sense the nature of the surrounding extracellular matrix, and they serve to guide migratory cells such as the neural crest cells of vertebrates, which leave the neural tube to form structures elsewhere in the body.

We can therefore describe and explain developmental processes in terms of how individual cells and groups of cells behave. Because the final structures generated by development are themselves composed of cells, explanations and descriptions at the cellular level can provide an account of how these adult structures are formed.

Because development can be understood at the cellular level, we can pose the question of how genes control development in a more precise form. We can now ask how genes are controlling cell behavior. The many possible ways in which a cell can behave therefore provide the link between gene activity and the morphology of the adult animal—the final outcome of development. Cell biology provides the means by which the genotype becomes translated into the phenotype.

1.10 Genes control cell behavior by specifying which proteins are made

What a cell can do is determined very largely by the proteins present within it. The hemoglobin in red blood cells enables them to transport oxygen; the cells lining the gut secrete specialized digestive enzymes; and skeletal muscle cells are able to contract because they contain contractile structures composed of the proteins myosin, actin and tropomyosin, as well as other muscle-specific proteins needed to make muscle function correctly. All these are specialized proteins that are not involved in the 'housekeeping' activities that are common to all cells and keep them alive and functioning. Housekeeping activities include the production of energy and the metabolic pathways involved in the breakdown and synthesis of molecules necessary for the life of the cell. Although there are qualitative and quantitative variations in housekeeping proteins in different cells, they are not important players in development. In development we are concerned primarily with those proteins that make cells different from one another and which are often known as **tissue-specific** proteins.

Genes control development mainly by specifying which proteins are made in which cells and when. In this sense the genes are passive participants in development, compared with the proteins they encode, which are the agents that directly determine

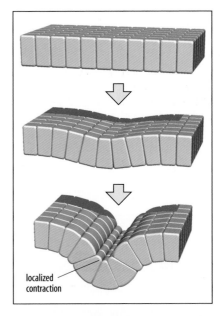

Fig. 1.18 Localized contraction of particular cells can cause a whole sheet of cells to fold. Contraction of a line of cells at their apices due to the contraction of cytoskeletal elements causes a furrow to form in a sheet of epidermis.

localized contraction

cell behavior, which includes determining which genes are expressed. To produce a particular protein, its gene must be switched on and **transcribed** into **messenger RNA (mRNA)**; the mRNA must then be **translated** into protein. The **initiation of transcription**, or switching on of a gene, involves combinations of specialized gene-regulatory proteins binding to **control regions** in the DNA.

Both transcription and translation are subject to several layers of control, and translation does not automatically follow transcription. Figure 1.19 shows the main stages in gene expression at which the production of a protein can be controlled. Even after a gene has been transcribed, for example, the mRNA may be degraded before it can be exported from the nucleus. Even if an mRNA reaches the cytoplasm, its translation may be prevented or delayed. In the eggs of many animals, preformed mRNA is prevented from being translated until after fertilization. Another process that determines which proteins are produced is **RNA processing**. In eukaryotic organisms—all organisms other than bacteria and archaea—the initial RNA transcripts of many genes can be cut and spliced in different ways to give rise to two or more different mRNAs; in this way, a number of different proteins with different properties can be produced from a single gene.

Even if a gene has been transcribed and the mRNA translated, the protein may still not be able to function immediately. Many newly synthesized proteins require further **post-translational modification** before they acquire biological activity. One very common permanent modification that occurs to proteins destined for the cell membrane or for secretion is the addition of carbohydrate side chains, or **glycosylation**. Reversible post-translational modifications, such as phosphorylation, can also significantly alter protein function. Together, alternative RNA splicing and post-translational modification mean that the number of functionally different proteins that can be produced is considerably greater than the number of protein-coding genes would indicate. For many proteins, their localization in a particular part of the cell, for example the nucleus, is essential for them to carry out their function.

Some genes, such as those for ribosomal RNAs (rRNAs) and transfer RNAs (tRNAs), do not code for proteins; in this case the RNAs themselves are the end-products. Another category of genes encodes **microRNAs (miRNAs)**, small RNA molecules that prevent the translation of specific mRNAs (see Box 6B, p. 228). Some microRNAs are known to be involved in gene regulation in development.

An intriguing question is how many genes out of the total genome are **developmental genes**—that is, genes specifically required for embryonic development. This is not easy to estimate. In the early development of *Drosophila*, at least 60 genes are directly involved in pattern formation up to the time that the embryo becomes divided into segments. In *Caenorhabditis*, at least 50 genes are needed to specify a small

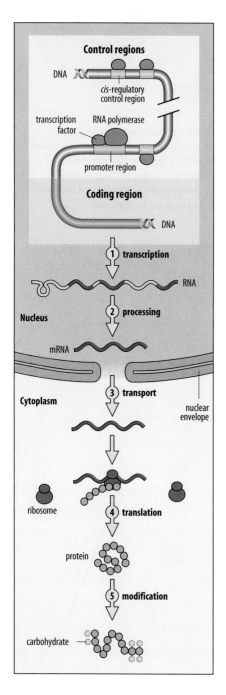

Fig. 1.19 Gene expression and protein synthesis. A protein-coding gene comprises a coding region—a stretch of DNA that contains the instructions for making the protein—and adjacent control regions. These include the promoter region, to which general transcription factors and the enzyme RNA polymerase bind to start transcription, and other *cis*-regulatory control regions at which other transcription factors bind to switch the gene on or off. These latter control regions may be thousands of base pairs away from the promoter. When the gene is switched on, the DNA sequence is transcribed to produce RNA (1). The RNA formed by transcription is spliced to remove introns and the non-coding region at the start of the gene (yellow) and processed within the nucleus (2) to produce mRNA. This is exported from the nucleus to the cytoplasm (3) for translation into protein at the ribosomes (4). Control of gene expression and protein synthesis occurs mainly at the level of transcription but can also occur at later stages. For example, mRNA may be degraded before it can be translated, or if it is not translated immediately it may be stored in inactive form in the cytoplasm for translation at a later stage. Some proteins require a further step of post-translational modification (5) to become biologically active. A very common post-translational modification is the addition of carbohydrate side-chains (glycosylation) as shown here.

reproductive structure known as the vulva. These are quite small numbers compared to the thousands of genes that are active at the same time; some of these are essential to development in that they are necessary for maintaining life, but they provide no or little information that influences the course of development. Many genes change their activity in a systematic way during development, suggesting that they could be developmental genes. The total number of genes in the nematode and *Drosophila* is around 19,000 and 15,000, respectively, and the number of genes involved in development is likely to be in the thousands. A systematic analysis of most of the *Caenorhabditis* genome suggests that approximately 9% of the genes (1722 out of 20,000) are involved in development.

Developmental genes typically code for proteins involved in the regulation of cell behavior—receptors, growth factors, intracellular signaling proteins, and gene-regulatory proteins. Many of these genes, especially those for receptors and signaling molecules, are used throughout an organism's life, but others are active only during embryonic development.

Some of the techniques used to determine where and when a gene is active are described in Box 1D (p. 20) and Box 3A (p.115). It is also possible to prevent the expression of a given gene and so prevent its protein from being made. This can be done using **antisense RNAs**, which are small RNA molecules of about 25 nucleotides complementary in sequence to part of a gene or mRNA. Such antisense RNAs can block gene function by binding to the gene or its mRNA. Artificial RNA molecules called morpholinos are used, which are more stable than normal RNA. Another technique for blocking gene expression is **RNA interference**, which operates in a similar way, but uses a different type of small RNA called small interfering RNAs (siRNAs) to exploit cells' own metabolic pathways for degrading mRNAs. These techniques are discussed in more detail in Box 6A, p. 220.

1.11 The expression of developmental genes is under tight control

All the somatic cells in an embryo are derived from the fertilized egg by successive rounds of mitotic cell division. Thus, with rare exceptions, they all contain identical genetic information, the same as that in the zygote. This is the principle of **genetic equivalence**. The differences between cells must therefore be generated by differences in gene activity that lead to the synthesis of different proteins. This is the principle of **differential gene expression**. Turning the correct genes on or off in the right cells at the right time becomes the central issue in development. As we shall see throughout the book, the genes do not provide a blueprint for development but a set of instructions. Key elements in regulating the readout of these instructions are the control regions that are located adjacent to developmental genes and genes for specialized proteins. These are acted on by **gene-regulatory proteins**, or **transcription factors**, which switch genes on or off by, respectively, activating or repressing transcription. Some gene-regulatory proteins act by binding directly to the DNA of the control regions (see Fig. 1.19) whereas others interact with transcription factors already bound to the DNA.

Developmental genes are highly regulated to ensure they are switched on only at the right time and place in development. This is a fundamental feature of development. To achieve this, developmental genes usually have extensive and complex control regions composed of one or more regulatory modules, known as *cis*-**regulatory modules** (the '*cis*' refers to the fact that the regulatory region is on the same DNA molecule as the gene it controls). Each module contains multiple binding sites for different transcription factors, and it is the precise combination of factors that binds that determines whether the gene is on or off. On average, a module will have binding sites for four to eight different transcription factors, and these factors will in turn often bind other **co-activator** or **co-repressor** proteins.

Box 1D Tracking gene expression in embryos

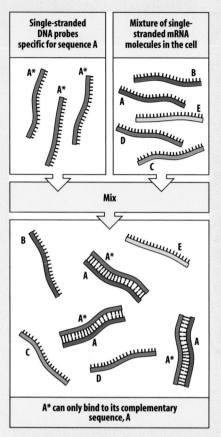

Single-stranded DNA probes specific for sequence A

Mixture of single-stranded mRNA molecules in the cell

A*   A*   A*

B   A   E   D   C

Mix

B   E   A*   A   A*   A   A*   A   C   D   A*

A* can only bind to its complementary sequence, A

In order to understand how gene expression is guiding development, it is essential to know exactly where and when particular genes are active. Genes are switched on and off during development and patterns of gene expression are continually changing. Several powerful techniques can show where a gene is being expressed in the embryo.

One set of techniques uses *in situ* hybridization to detect the mRNA that is being transcribed from a gene. If a DNA or RNA probe is complementary in sequence to an mRNA being transcribed in the cell, it will base-pair precisely (hybridize), with the mRNA (see figure, left). A DNA probe can therefore be used to locate its complementary mRNA in a tissue slice or a whole embryo. The probe may be labeled in various ways—with a radioactive isotope, a fluorescence tag (fluorescence *in situ* hybridization, FISH; see Fig. 2.10), or an enzyme—to enable it to be detected. Radioactively labeled probes are detected by autoradiography, whereas probes labeled with fluorescent dyes are observed by fluorescence microscopy. Enzyme-labeled probes are detected by the enzyme's conversion of a colorless substrate into colored product. Labeling with fluorescent dyes or enzymes has the advantage over radioactive labeling in that the expression of several different genes can be detected simultaneously, by labeling the different probes with different colored dyes, and it also means that researchers do not have to work with radioactive chemicals. To detect protein expression and localization, the detection agent is an antibody specific for the protein. Again, antibodies are tagged either with fluorescent dyes (see Fig. 2.33) or with enzymes (see Fig. 2.31). These techniques, and those for detecting mRNA, can only be applied to fixed tissue sections and embryos.

The pattern and timing of gene expression can also be followed by inserting a 'reporter' gene into an animal using transgenic techniques. The reporter gene codes for some easily detectable protein and is placed under the control of an appropriate promoter that will switch on the expression of the protein at a particular time and place. One commonly used reporter gene is the *lacZ* sequence coding for the bacterial enzyme β-glactosidase. The presence of the enzyme can be detected by treating the specimen after fixation with a modified substrate, which the enzyme reacts with to give a blue-colored product (see Fig. 5.24).

Other very commonly used reporter genes are those for small fluorescent proteins, such as green fluorescent protein (GFP), which fluoresce in various colors and were isolated from jellyfish and other marine organisms. These proteins are extremely useful as they are harmless to living cells and are readily be made visible by illuminating cultured cells, embryos or even whole animals with light of a suitable wavelength. Gene expression can therefore be tracked in living embryos and cells (see Fig. 2.22). The three scientists who discovered and developed GFP and other fluorescent proteins for use in imaging were awarded the 2008 Nobel Prize for Chemistry.

The zebrafish embryos illustrated show two different methods of detecting expression of the same gene. In the embryo in the upper photo, the expression of *Sox10* mRNA (blue) in neural crest cells has been detected by nucleic acid hybridization in a fixed embryo. The lower photo is of a live transgenic zebrafish embryo at the same stage of development, which carries a reporter transgene in which a GFP coding sequence is attached to the promoter region of *Sox10*. GFP is therefore expressed in the neural crest cells that express *Sox10*.

Photographs courtesy of Robert Kelsh.

These regulatory regions are called 'modules' as each can function somewhat independently of each other. This means that a gene with more than one regulatory module is usually able to respond to different combinations of inputs, and so can be expressed at different times and in different places within the embryo in response to developmental signals. Different genes can have the same control module, which usually means that they will be expressed together, or different genes may have modules that contain some, but not all, of the binding sites in common, introducing subtle differences in the timing or location of expression. Thus, an organism's genes are linked in complex interdependent networks of expression through their regulatory modules and the proteins that bind to them. Common examples of such regulation are **positive-feedback** and **negative-feedback** loops in which a transcription factor respectively promotes or represses the expression of a gene whose product maintains that gene expression (Fig. 1.20). Indeed, extensive networks of gene interactions for various stages of development have now been worked out for animals such as the sea urchin (see Chapter 6) and zebrafish (see Box 4E, p. 162).

As all the key steps in development reflect changes in gene activity, one might be tempted to think of development simply in terms of mechanisms for controlling gene expression. But this would be a big mistake. For gene expression is only the first step in a cascade of cellular processes that lead via protein synthesis to changes in cell behavior and so direct the course of embryonic development. Proteins are the machines that drive development. To think only in terms of genes is to ignore crucial aspects of cell biology, such as change in cell shape, which may be initiated at several steps removed from gene activity. In fact, there are very few cases where the complete sequence of events from gene expression to altered cell behavior has been worked out. The route leading from gene activity to a structure such as the five-fingered hand is tortuous.

1.12 Development is progressive and the fate of cells becomes determined at different times

As embryonic development proceeds, the organizational complexity of the embryo becomes vastly increased over that of the fertilized egg, as first recognized by Aristotle's 'epigenesis' principle outlined earlier in the chapter. Many different cell types are formed, spatial patterns emerge, and there are major changes in shape. All this occurs more or less gradually, depending on the particular organism. But, in general, the embryo is first divided up into a few broad regions, such as the future germ layers (mesoderm, ectoderm, and endoderm). Subsequently, the cells within these regions have their fates more and more finely determined. **Determination** implies a stable change in the internal state of a cell, and an alteration in the pattern of gene activity is assumed to be the initial step, leading to a change in the proteins produced in the cell. Different mesoderm cells, for example, eventually become determined as muscle, cartilage, bone, the fibroblasts of connective tissue, and the cells of the dermis of the skin.

It is important to understand clearly the distinction between the normal fate of a cell at any particular stage, and its state of determination. The **fate** of a group of cells merely describes what they will normally develop into. By marking cells of the early embryo one can find out, for example, which ectodermal cells will normally give rise to the nervous system, and of those, which, in particular, will give rise to the neural retina of the eye. However, that in no way implies that those cells can only develop into a retina, or are already committed, or determined, to do so in the early embryo.

A group of cells is called **specified** if, when isolated and cultured in the neutral environment of a simple culture medium away from the embryo, they develop more or less according to their normal fate (Fig. 1.21). For example, cells at the animal pole of the amphibian blastula (see Box 1A, pp. 2–3) are specified to form ectoderm, and

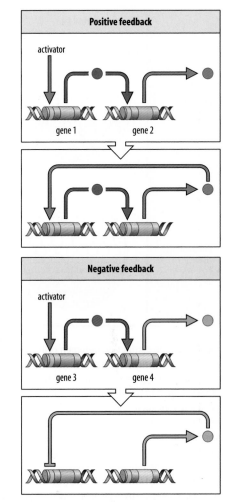

Fig. 1.20 Simple genetic feedback loops. Top: gene 1 is turned on by an activating transcription factor (green); the protein it produces (red) activates gene 2. The protein product of gene 2 (blue) not only acts on further targets, but also activates gene 1, forming a positive feedback loop that will keep genes 1 and 2 switched on even in the absence of the original activator. Bottom: when the product of the gene at the end of the pathway inhibits the first gene, a negative feedback loop is formed. Arrows indicate activation; barred line indicates inhibition.

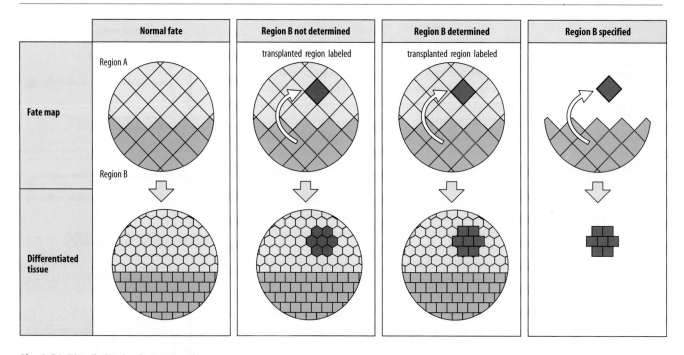

Normal fate	Region B not determined	Region B determined	Region B specified

Fig. 1.21 The distinction between cell fate, determination, and specification.
In this idealized system, regions A and B differentiate into two different sorts of cells, depicted as hexagons and squares. The fate map (first panel) shows how they would normally develop. If cells from region B are grafted into region A and now develop as A-type cells, the fate of region B has not yet been determined (second panel). By contrast, if region B cells are already determined when they are grafted to region A, they will develop as B cells (third panel). Even if B cells are not determined, they may be specified, in that they will form B cells when cultured in isolation from the rest of the embryo (fourth panel).

will form epidermis when isolated. Cells that are 'specified' in this technical sense need not yet be 'determined', for influences from other cells can change their normal fate; if tissue from the animal pole is put in contact with cells from the vegetal pole, some animal pole tissue will form mesoderm instead of epidermis. At a later stage of development, however, the cells in the animal region have become determined as ectoderm and their fate cannot then be altered. Tests for specification rely on culturing the tissue in a neutral environment lacking any inducing signals, and this is often difficult to achieve.

The state of determination of cells at any developmental stage can be demonstrated by transplantation experiments. At the blastula stage of the amphibian embryo, one can graft the ectodermal cells that give rise to the eye into the side of the body and show that the cells develop according to their new position; that is, into mesodermal cells. At this early stage, their potential for development is much greater than their normal fate. However, if the same operation is done at a later stage, then the future eye region will form structures typical of an eye (Fig. 1.22). At the earlier stage the cells were not yet determined as prospective eye cells, whereas later they had become so.

It is a general feature of development that cells in the early embryo are less narrowly determined than those at later stages; with time, cells become more and more restricted in their developmental potential. We assume that determination involves a change in the genes that are expressed by the cell and that this change fixes or restricts the cell's fate, thus reducing its developmental options.

We have already seen how, even at the two-cell stage, the cells of the sea-urchin embryo do not seem to be determined. Each has the potential to generate a whole new larva (see Section 1.3). Embryos like these, where the potential of cells is much greater than that indicated by their normal fate, are termed **regulative**. Vertebrate embryos are among those capable of considerable regulation. In contrast, those embryos where, from a very early stage, the cells can develop only according to their early fate are termed **mosaic**. The term mosaic has a long history (see Section 1.3). It describes eggs and embryos that develop as if their pattern of future development is

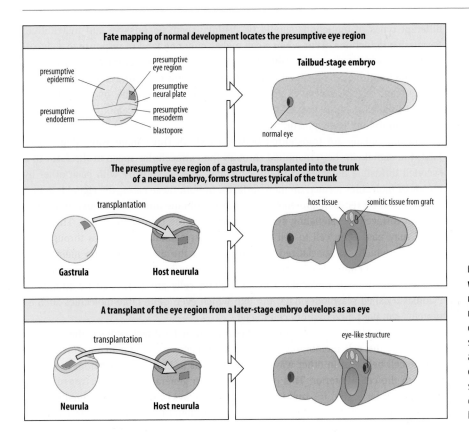

Fig. 1.22 Determination of the eye region with time in amphibian development. If the region of the gastrula that will normally give rise to an eye is grafted into the trunk region of a neurula (middle panel), the graft forms structures typical of its new location, such as notochord and somites. If, however, the eye region from a neurula is grafted into the same site (bottom panel), it develops as an eye-like structure, since at this later stage it has become determined.

laid down very early, even in the egg, as a spatial mosaic of different molecules in the egg cytoplasm. These cytoplasmic determinants get distributed in a regular fashion into different cells during cleavage, thus determining the fate of that cell lineage at an early stage. The different parts of the embryo then develop quite independently of each other. *Caenorhabditis* and ascidians are examples of embryos with significant amounts of mosaic development and they are discussed in Chapter 6. In such embryos, cell interactions may be quite limited. But the demarcation between regulative and mosaic strategies of development is not always sharp, and in part reflects the time when determination occurs; it occurs much earlier in mosaic systems.

The difference between regulative and mosaic embryos also reflects the relative importance of cell–cell interactions in each system. The occurrence of regulation absolutely requires interactions between cells, for how else could normal development take place and how could deficiencies be recognized and restored? A truly mosaic embryo, however, would in principle not require such interactions. No purely mosaic embryos are known to exist.

1.13 Inductive interactions can make cells different from each other

Making cells different from one another is central to development. There are many times in development where a signal from one group of cells influences the development of an adjacent group of cells. This is known as **induction**, and the classic example is the action of the Spemann organizer in amphibians (see Section 1.4). Inducing signals may be propagated over several or even many cells, or be highly localized. The inducing signals from the amphibian organizer affect many cells, whereas other inducing signals may be transmitted from a single cell to its immediate neighbor. Two

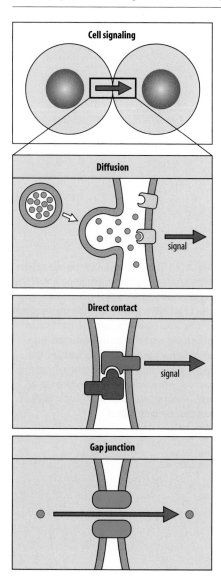

Fig. 1.23 An inducing signal can be transmitted from one cell to another in three main ways. The signal can be a secreted diffusible molecule, which interacts with a receptor on the target cell surface (second panel), or the signal can be produced by direct contact between two complementary proteins at the cell surfaces (third panel). If the signal involves a small molecule it may pass directly from cell to cell through gap junctions in the plasma membrane (fourth panel).

different types of inductions should be distinguished: **permissive** and **instructive**. Permissive inductions occur when a cell makes only one kind of response to a signal, and makes it when a given level of signal is reached. By contrast, in an instructive induction, the cells respond differently to different concentrations of the signal. 'Antagonistic' signal molecules that block induction, for example, by preventing the inducing signal reaching the cell or binding to a cell-surface receptor, are also important in controlling development.

Inducing signals are transmitted between cells in three main ways (Fig. 1.23). First, the signal can be transmitted through the extracellular space in the form of a secreted diffusible molecule. Second, cells may interact directly with each other by means of molecules located on their surfaces. In both these cases, the signal is generally received by receptor proteins in the cell membrane and is subsequently relayed through intracellular signaling systems to produce the cellular response. Third, the signal may pass from cell to cell directly. In most animal cells, this is through gap junctions, which are specialized protein pores in the apposed plasma membranes providing direct channels of communication between the cytoplasms of adjacent cells through which small molecules can pass. Plant cells are connected by strands of cytoplasm called plasmodesmata, through which even quite large molecules such as proteins can pass directly from cell to cell.

In the case of signaling by a diffusible protein or by direct contact, the signal is received at the cell membrane. If it is to alter gene expression in the nucleus, or change cell behavior in other ways, that signal has to be transmitted from the membrane to the cell's interior. This process is known generally as **signal transduction** or **intracellular signaling**, and is carried out by relays of intracellular signaling molecules that are activated when the extracellular signal molecule binds to its receptor. An example of one intracellular signaling pathway important in development is illustrated in Box 1E (p. 26)—the pathway activated by members of the Wnt family of secreted signal proteins. The basic Wnt pathway, like many others we shall encounter, is used again and again in different developmental contexts in all the animals included in this book.

Within the cell, signaling proteins and small-molecule second messengers, such as cyclic AMP, interact with one another to transmit the signal onward in the cell. Activation and deactivation of the pathway components by phosphorylation is an important feature of most signaling pathways. In the case of development, most of the signaling leads to the activation or repression of genes by switching transcription on or off. Another important target of extracellular signals in development is the **cytoskeleton**, the internal network of protein fibers whose rearrangement enables cells to change shape, to move, and to divide. Signaling pathways are also used to temporarily alter enzyme activity and metabolic activity within cells and to initiate nerve impulses in neurons. The various signals a cell may be receiving at any one time are integrated by cross-talk between the different intracellular signaling pathways to produce an appropriate response. An example of what can happen when an important developmental signaling pathway is defective is illustrated in Box 1F (p. 28).

An important feature of induction is whether or not the responding cell is able to respond to the inducing signal. This **competence** may depend, for example, on the presence of the appropriate receptor and transducing mechanism, or on the presence of the particular transcription factors needed for gene activation. A cell's competence for a particular response can change with time; for example, the Spemann organizer can induce changes in the cells it affects only during a restricted window of time.

In embryos, it seems that small is generally beautiful where signaling and pattern formation are concerned. Most patterns are specified on a scale involving just tens of cells and distances of only 100 to 500 μm. The final organism may be very big, but this is almost entirely due to growth after the basic pattern has been formed.

1.14 The response to inductive signals depends on the state of the cell

Although an inductive signal can be viewed as an instruction to cells about how to behave, it is important to realize that responses to such a signal are entirely dependent on the current state of the cells. Not only does the individual cell have to be competent to respond, but the number of possible responses is usually very limited. An inductive signal can only select one response from a small number of possible cellular responses. Thus instructive inducing signals would be more accurately called selective signals. A truly instructive signal would be one that provided the cell with entirely new information and capabilities, by providing it with, for example, new genes, which is not thought to occur during development.

The fact that, at any given time, an inductive signal selects one out of several possible responses has several important implications for biological economy. On the one hand, it means that different signals can activate a particular gene at different stages of development: the same gene is often turned on and off repeatedly during development. On the other, the same signal can be used to elicit different responses in different cells. A particular signaling molecule, for example, can act on several types of cell, evoking a characteristic and different response from each, depending on their developmental history. As we will see in future chapters, evolution has been lazy with respect to this aspect of development, and a small number of intercellular signaling molecules are used again and again for different purposes.

1.15 Patterning can involve the interpretation of positional information

One general mode of pattern formation can be illustrated by considering the patterning of a simple non-biological model—the French flag (Fig. 1.24). The French flag has a simple pattern: one-third blue, one-third white, and one-third red, along just one axis. Moreover, the flag comes in many sizes but always the same pattern, and thus can be thought of as mimicking the capacity of an embryo to regulate. Given a line of cells, any one of which can be blue, white, or red, and given also that the line of cells can be of variable length, what sort of mechanism is required for the line to develop the pattern of a French flag?

One solution is for the group of cells to acquire **positional information**. That is, each cell acquires an identity, or **positional value**, that is related to its position along the line with respect to the boundaries at either end. After they have acquired their positional values, the cells interpret this information by differentiating according to their genetic program. Those in the left-hand third of the line will become blue, those in the middle-third white, and so on. Very good evidence that cells use positional signals comes from the regeneration of amphibian and insect limbs, in which the missing regions are precisely replaced (discussed in Chapter 14).

Pattern formation using positional information implies at least two distinct stages: first the positional value has to be related to some boundary, and then it has to be interpreted. The separation of these two processes has an important implication: it means that there does not need to be a set relation between the positional values and how they are interpreted. In different circumstances, the same set of positional values could be used to generate the Italian flag or another pattern. How positional values will be interpreted will depend on the particular genetic instructions active in the group of cells and thus the cells' developmental history.

Cells could have their position specified by various mechanisms. The simplest is based on a gradient of some substance. If the concentration of some chemical decreases from one end of a line of cells to the other, then the concentration of that chemical in any cell along the line effectively specifies the position of the cell with respect to the boundary (Fig. 1.25). A chemical whose concentration varies and which is involved in pattern formation is called a **morphogen**. In the case of the French flag,

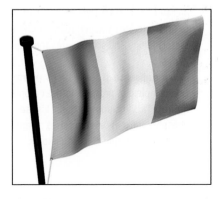

Fig. 1.24 The French flag.

Box 1E Signal transduction and intracellular signaling

Secreted signaling proteins acting on receptors in cell membranes activate numerous different intracellular signaling pathways. The example illustrated here is the important developmental signaling pathway initiated by the Wnt family of intercellular signaling proteins (pronounced 'Wints'). Named after the proteins encoded by the *Wingless* gene in *Drosophila* and the *Int* gene (now *Wnt1*) in mammals, Wnt proteins are involved in specifying cell fate in a multiplicity of developmental contexts in all multicellular animals studied so far. We shall meet many of the functions of the Wnt family throughout the book. In adult humans, Wnt is required to maintain stem cells in their self-renewing state, and incorrect control of the Wnt signaling pathway is implicated in common cancers, such as colon cancer.

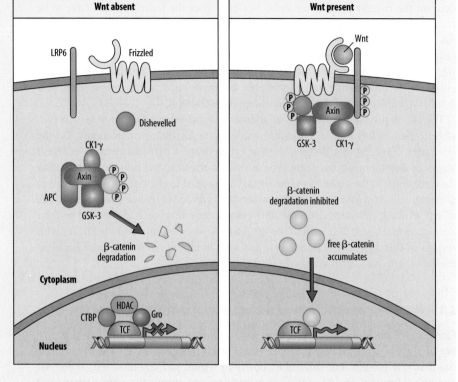

Wnt proteins can activate a number of different intracellular signaling pathways: the one we will illustrate here is involved in specifying cell fates in the early development of all animals. It is known as the **canonical Wnt/β-catenin pathway** as it was the first to be discovered and leads to the accumulation of the protein β-catenin and its action as a transcriptional co-activator. Other Wnt pathways will be described when they are encountered in the course of the book. A simplified version of the canonical Wnt/β-catenin pathway is shown in the figure. As with many developmental signaling pathways, the initial Wnt signal is converted, or 'transduced', into changes in gene expression in the nucleus. In this case, the Wnt signal prevents the degradation of the protein β-catenin in the cytoplasm, and by doing so enables β-catenin to accumulate and to enter the nucleus, where it binds to and activates transcription factors of the TCF (T-cell-specific factor) family and so activates the expression of selected target genes.

In the absence of a Wnt signal, β-catenin is bound by a 'destruction complex' of proteins in the cytoplasm (left-hand side of the figure). This includes protein kinases (CK1γ and GSK-3β) that phosphorylate β-catenin, thus targeting it for ubiquitination and degradation in the proteasome. In the absence of β-catenin

in the nucleus, transcriptional co-repressors bind to the TCF transcription factors and prevent gene expression.

Like most extracellular signals, Wnt acts on its target cells by binding to a specific transmembrane receptor protein (Frizzled) at the cell surface (right-hand side of the figure). The signal that Wnt has bound is transmitted across the membrane by Frizzled and an associated co-receptor, LRP, even though Wnt itself does not enter the cell. Activation of these receptors causes the same protein kinases that phosphorylated β-catenin in the destruction complex to now become associated with the membrane and phosphorylate the cytoplasmic tail of the activated LRP. Protein phosphorylation, which changes the activity of proteins or their interactions with other proteins, is ubiquitous in signaling pathways as a means of transmitting the signal onwards. The intracellular signaling protein Dishevelled (Dsh) and the protein Axin (which is a component of the destruction complex) are recruited to the tails of LRP and Frizzled at the membrane. This prevents the destruction complex forming and leaves β-catenin free to accumulate and to enter the nucleus. Inside the nucleus, β-catenin binding to TCF displaces the co-repressors, lifts the repression and enables the target genes to be expressed.

we assume a source of morphogen at one end and a sink at the other, and that the concentrations of morphogen at both ends are kept constant but are different from each other. Then, as the morphogen diffuses down the line, its concentration at any point effectively provides positional information. If the cells can respond to **threshold concentrations** of the morphogen—for example, above a particular concentration the cells develop as blue, while below this concentration they become white, and at yet another, lower, concentration they become red—the line of cells will develop as a French flag (see Fig. 1.25). Thresholds can represent the amount of morphogen that must bind to receptors to activate an intracellular signaling system, or concentrations of transcription factors required to activate particular genes. The use of threshold concentrations of transcription factors to specify position is most beautifully illustrated in the early development of *Drosophila*, as we shall see in Chapter 2. Other ways of specifying positional information are by direct intercellular interactions and by timing mechanisms.

The French flag model illustrates two important features of development in the real world. The first is that, even if the length of the line varies, the system regulates and the pattern will still form correctly, given that the boundaries of the system are properly defined by keeping the different concentrations of morphogen constant at either end. The second is that the system could also regenerate the complete original pattern if it were cut in half, provided that the boundary concentrations were re-established. It is therefore truly regulative. We have discussed it here as a one-dimensional patterning problem, but the model can easily be extended to provide patterning in two dimensions (Fig. 1.26).

It is still unclear how positional information is specified: morphogen diffusion is one of the mechanisms that have been proposed, but this may not be reliable enough. We do not know how fine-grained the differences in position are. For example, does every cell in the early embryo have a distinct positional value of its own and is it able to make use of this difference?

1.16 Lateral inhibition can generate spacing patterns

Many structures, such as the feathers on a bird's skin, are found to be more or less regularly spaced with respect to one another. One mechanism that gives rise to such spacing is called **lateral inhibition** (Fig. 1.27). Given a group of cells that all have the potential to differentiate in a particular way, for example as feathers, it is possible to regularly space the cells that will form feathers by a mechanism in which the first cells to begin to form feathers (which occurs essentially at random) inhibit adjacent cells from doing so. This is reminiscent of the regular spacing of trees in a forest by competition for sunlight and nutrients. In embryos, lateral inhibition is often the result of the differentiating cell producing an inhibitory molecule that acts locally on the nearest neighbors to prevent them developing in a similar way.

1.17 Localization of cytoplasmic determinants and asymmetric cell division can make daughter cells different from each other

Positional specification is just one way in which cells can be given a particular identity. A separate mechanism is based on **cytoplasmic localization** and **asymmetric cell division** (Fig. 1.28). Asymmetric divisions are so called because they result in daughter cells having properties different from each other, independently of any environmental influence. The properties of such cells therefore depend on their **lineage** or line of descent, and not on environmental cues. Although some asymmetric cell divisions are also unequal divisions in that they produce cells of different sizes, this is not usually the most important feature in animals; it is the unequal distribution of cytoplasmic factors that makes the division asymmetric. An alternative way of

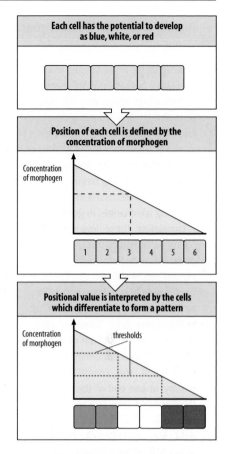

Fig. 1.25 The French flag model of pattern formation. Each cell in a line of cells has the potential to develop as blue, white, or red. The line of cells is exposed to a concentration gradient of some substance and each cell acquires a positional value defined by the concentration at that point. Each cell then interprets the positional value it has acquired and differentiates into blue, white, or red, according to a predetermined genetic program, thus forming the French flag pattern. Substances that can direct the development of cells in this way are known as morphogens. The basic requirements of such a system are that the concentration of substance at either end of the gradient must remain different from each other but constant, thus fixing boundaries to the system. Each cell must also contain the necessary information to interpret the positional values. Interpretation of the positional value is based upon different threshold responses to different concentrations of morphogen.

Box 1F When development goes awry

The signaling pathway stimulated by the small secreted protein Sonic hedgehog (Shh) (named after the Hedgehog protein in *Drosophila*, which was the first member of this family to be discovered) regulates many developmental processes in humans and other vertebrates from very early in embryogenesis onwards. In adults, Shh controls the growth of stem cells, and misregulation of this signaling pathway leads to tumor formation. Like many other signaling pathways in development (for example, the Wnt pathway discussed in Box 1E, p. 26), the end results of Shh signaling are changes in gene expression.

Unraveling these important signaling pathways can help us understand the reasons for some developmental abnormalities, and Shh is a good example. A relatively common birth defect is a condition known as holoprosencephaly, which is characterized by a spectrum of midline brain and facial abnormalities. The severity of the condition varies greatly: at one end of the spectrum the forebrain (the prosencephalon) fails to divide in the midline to form right and left brain hemispheres, there is complete absence of the midline of the face and the fetuses do not survive; at the other end, the division of the forebrain is only mildly affected and the only observable defect may be a cleft lip or the presence of one incisor tooth rather than two.

The first indications that Shh was involved in brain development came in the early 1990s, when Shh was discovered to be expressed in the ventral midline of the nervous system in mice. In 1996, clinical geneticists identified one of the genes affected in patients with inherited holoprosencephaly—it turned out to be *Shh*. At around the same time, it was found that when the *Shh* gene is completely knocked out experimentally in mice (see Section 3.9 and Box 3B, p. 118 for the technique of gene knockouts), forebrain division does not occur and the embryos have a condition known as cyclopia, with loss of midline facial structures and development of a single eye in the middle of the face—a condition representing the most severe manifestation of holoprosencephaly.

As clinical geneticists continued to identify genes involved in holoprosencephaly, more genes encoding components in the Shh pathway were implicated. In all cases, the cause of the abnormality could be ascribed to a deficiency in Shh signaling, but different genes were affected in different cases. In some groups of patients, the abnormality was caused by mutations in the *Shh* gene itself, but in others the affected genes encoded either the receptor for Shh (*Patched*) or a transcription factor at the end of the pathway (*Gli2*). Mutations in a gene encoding an enzyme required for the synthesis of cholesterol, which promotes the activity of a protein called Smoothened, which acts together with Patched to control the Shh pathway, were found in some patients. (The full Shh pathway is illustrated in Box 11C, pp. 424–425). Shh signalling takes place on cellular structures called primary cilia, which are immotile cilia present on most vertebrate cell types, and patients with genetic conditions in which formation of these structures is defective were also found to have a characteristic facial phenotype, although the alterations are very subtle and require sophisticated three-dimensional modeling to detect. Thus, the discovery of genes that produce similar symptoms in human patients can highlight genes in a common developmental pathway.

Facial defects similar to those produced by mutations in genes for Shh pathway proteins can also be caused by environmental factors. Pregnant ewes grazed in pastures in which the California False Hellebore (*Veratrum californicum*, pictured) is growing may give birth to lambs with severe facial abnormalities, such as cyclopia (the presence of a single eye in the midline of the face). The chemical in the plant that causes these defects is the alkaloid cyclopamine, and it too disrupts Shh signaling. Cyclopamine interferes with the activity of Smoothened and reduces Shh signaling. This property of cyclopamine is now being explored in a therapeutic context to develop drugs to treat cancers that display abnormal Shh signaling.

Photograph courtesy of Jerry Friedman, Copyright Creative Commons Attribution ShareAlike 3.00

making the French flag pattern from the egg would be to have chemical differences (representing blue, white, and red) distributed in the egg in the form of determinants that foreshadowed the French flag. When the egg underwent cleavage, these cytoplasmic determinants would become distributed among the cells in a particular way and a French flag would develop. This would require no interactions between the cells, which would have their fates determined from the beginning.

Although such extreme examples of mosaic development (see Section 1.12) are not known in nature, there are well-known cases where eggs or cells divide so that some cytoplasmic determinant becomes unequally distributed between the two daughter cells and they develop differently. This happens at the first cleavage of the nematode egg, for example, and defines the antero-posterior axis of the embryo. The germ cells of *Drosophila* are also specified by cytoplasmic determinants, in this case contained in the cytoplasm located at the posterior end of the egg. And in *Xenopus*, a key determinant in the first stages of embryonic development is a protein known as VegT, which is localized in the vegetal region of the fertilized egg. In general, however, as development proceeds, daughter cells become different from each other because of signals from other cells or from their extracellular environment, rather than because of the unequal distribution of cytoplasmic determinants.

A very particular type of asymmetric division is seen in stem cells. When they divide, these self-renewing cells can produce one new stem cell and another daughter cell that will differentiate into one or more cell types (Fig. 1.29). Stem cells that are capable of giving rise to all the cell types in the body are present in very early embryos, and are known as **embryonic stem cells**, whereas in adults, stem cells of apparently more limited potential are responsible for the continual renewal of tissues such as blood, epidermis, and gut epithelium, and for the replacement of tissues such as muscle when required. The difference in the behavior of the daughter cells produced by division of a stem cell can be due either to asymmetric distribution of cytoplasmic determinants or to the effects of external signals. The generation of neurons in *Drosophila*, for example, depends on the asymmetric distribution of cytoplasmic determinants in the neuronal stem cells, whereas the production of the different types of blood cells seems to be regulated to a great extent by extracellular signals.

Embryonic stem cells, which can give rise to all the types of cells in the body, are called **pluripotent**, while stem cells that give rise to a more limited number of cell types are known as **multipotent**. The hematopoietic stem cells in bone marrow, which give rise to all the different types of blood cells, are an example of multipotent stem cells. Stem cells are of great interest in relation to regenerative medicine, as they could provide a means of repairing or replacing damaged organs and are discussed in Chapter 10. For most animals, the only cell that can give rise to a complete new organism is the fertilized egg, and this is described as **totipotent**.

1.18 The embryo contains a generative rather than a descriptive program

All the information for embryonic development is contained within the fertilized egg. So how is this information interpreted to give rise to an embryo? One possibility is that the structure of the organism is somehow encoded as a descriptive program in the genome. Does the DNA contain a full description of the organism to which it will give rise: is it a blueprint for the organism? The answer is no. Instead, the genome contains a program of instructions for making the organism—a **generative program**— that determines where and when different proteins are synthesized and thus controls how cells behave.

A descriptive program, such as a blueprint or a plan, describes an object in some detail, whereas a generative program describes how to make an object. For the same object the programs are very different. Consider origami, the art of paper folding. By

Fig. 1.26 Positional information can be used to generate patterns. A good example, as shown here, is where the people seated in a stadium each have a position defined by their row and seat numbers. Each position has an instruction about which colored card to hold up, and this makes the pattern. If the instructions were changed, a different pattern would be formed, and so positional information can be used to produce an enormous variety of patterns.

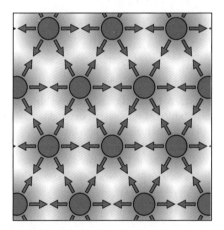

Fig. 1.27 Lateral inhibition can give a spacing pattern. Lateral inhibition occurs when developing structures produce an inhibitor that prevents the formation of any similar structures in the area adjacent to them, and the structures become evenly spaced as a result.

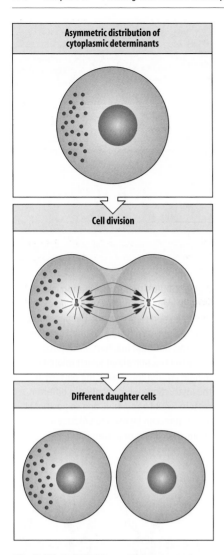

Asymmetric distribution of cytoplasmic determinants

Cell division

Different daughter cells

Fig. 1.28 Cell division with asymmetric distribution of cytoplasmic determinants. If a particular molecule is distributed unevenly in the parent cell, cell division will result in its being shared unequally between the cytoplasm of the two daughter cells. The more localized the cytoplasmic determinant is in the parental cell, the more likely it will be that one daughter cell will receive all of it and the other none, thus producing a distinct difference between them.

folding a piece of paper in various directions it is quite easy to make a paper hat or a bird from a single sheet. To describe in any detail the final form of the paper with the complex relationships between its parts is really very difficult, and not of much help in explaining how to achieve it. Much more useful and easier to formulate are instructions on how to fold the paper. The reason for this is that simple instructions about folding have complex spatial consequences. In development, gene action similarly sets in motion a sequence of events that can bring about profound changes in the embryo. One can thus think of the genetic information in the fertilized egg as equivalent to the folding instructions in origami; both contain a generative program for making a particular structure.

1.19 The reliability of development is achieved by a variety of means

Development is remarkably consistent and reliable; you need only look at the similar lengths of your legs, each of which develops quite independently of the other for some 15 years (see Chapter 13). Embryonic development thus needs to be reliable so that the adult organism can function properly. For example, both wings of a bird must be very similar in size and shape if it is to be able to fly satisfactorily. How such reliability is achieved is an important problem in development.

Development needs to be reliable in the face of fluctuations that occur either within the embryo or in the external environment. Internal fluctuations include small changes in the concentrations of molecules, and also changes due, for example, to mutations in genes not directly linked to the development of the organ in question. External factors that could perturb development include temperature and environmental chemicals.

One of the central problems of reliability relates to pattern formation. How are cells instructed to behave in a particular way in a particular position? Two ways in which reliability is assured during development are apparent redundancy of mechanism and negative feedback. **Redundancy** is when there are two or more ways of carrying out a particular process; if one fails for any reason another will still function. It is like having two batteries in your car instead of just one. True redundancy—such as having two identical genes in a haploid genome with identical functions—is rare except for cases such as rRNA, where there are often hundreds of copies of similar genes. Apparent redundancy, on the other hand, where a given process can be specified by several different mechanisms, is probably one of the ways that embryos can achieve such precise and reliable results. It is like being able to draw a straight line with either a ruler or a piece of taut string. This type of situation is not true redundancy, but may give the impression of being so if one mechanism is removed and the outcome is still apparently normal. Multiple ways of carrying out a process make it more robust and resistant to environmental or genetic perturbation.

Negative feedback also has a role in ensuring consistency; here, the end product of a process inhibits an earlier stage and thus keeps the level of product constant (see, for example, Fig. 1.20). The classic example is in metabolism, where the end product of a biochemical pathway inhibits an enzyme that acts early in the pathway. Yet another reliability mechanism relates to the complexity of the networks of gene activity that operate in development. There is evidence that these networks, which involve multiple pathways, are robust and relatively insensitive to small changes in, for example, the rates of the individual processes involved.

1.20 The complexity of embryonic development is due to the complexity of cells themselves

Cells are, in a way, more complex than the embryo itself. In other words, the network of interactions between proteins and DNA within any individual cell contains many more components and is very much more complex than the interactions between the cells of the developing embryo. However clever you think cells are, they are almost always far cleverer. Each of the basic cell activities involved in development, such as cell division, response to cell signals, and cell movements, are the result of interactions within a population of many different intracellular proteins whose composition varies over time and between different locations in the cell. Cell division, for example, is a complex cell-biological program that takes place over a period of time, in a set order of stages, and requires the construction and precise organization of specialized intracellular structures at mitosis.

Any given cell is expressing thousands of different genes at any given time. Much of this gene expression may reflect an intrinsic program of activity, independent of external signals. It is this complexity that determines how cells respond to the signals they receive; how a cell responds to a particular signal depends on its internal state. This state can reflect the cell's developmental history—cells have good memories—and so different cells can respond to the same signal in very different ways. We shall see many examples of the same signals being used over and over again by different cells at different stages of embryonic development, with different biological outcomes.

We have at present only a fragmentary picture of how all the genes and proteins in a cell, let alone a developing embryo, interact with one another. But new technologies are now making it possible to detect the simultaneous activity of hundreds of genes in a given tissue. The discipline of systems biology is also beginning to develop techniques to reconstruct the highly complex signaling networks used by cells. How to interpret this information, and make biological sense of the patterns of gene activity revealed, is a massive task for the future.

1.21 Development is intimately involved in evolution

The similarity in the developmental mechanisms and genes in such very different animals as flies and vertebrates is the result of the evolutionary process. All animals have evolved from a single multicellular ancestor and it is therefore inevitable that some developmental mechanisms will be held in common by many different animals, whereas new ones arose in different animal groups as evolution proceeded. The evolution of development is discussed in more detail in Chapter 15.

The basic Darwinian theory of evolution by natural selection is that changes in genes alter how an organism develops, and that changes in development in turn determine how the adult will interact with their environment. If a developmental change gives rise to adults better adapted to survive and reproduce in the prevailing environment, it will be selected to be retained in the population. Thus changes in development due to changes in the genes are fundamental to evolution.

Very clear examples of the workings of evolution in development can be seen in the limbs of vertebrates. In Chapter 15 we shall see how analysis of fossil evidence shows that the limbs of land vertebrates originally evolved from fins, and how the genetic and developmental basis of this evolution is being reconstructed (see Section 15.7). We shall also see how the basic pattern of the five-digit limb then evolved to give limbs as different as those of bats, horses, and humans. In bats, the digits of the forelimb grow extremely long and support a leathery wing membrane, whereas in horses, one digit of the 'hand' in the forelimb and the 'foot' in the hindlimb has become a long bone that forms the lower part of the leg and the hoof, while some other digits have been lost (see Fig. 15.14). The fossil record gives many other examples of evolutionary changes.

While some major steps in evolution can be identified, the genes involved are not easy to identify. There are also questions as to how big the genetic steps to new forms were, and there are questions about how completely novel structures evolve. But there is no doubt that it is all down to changes in embryonic development.

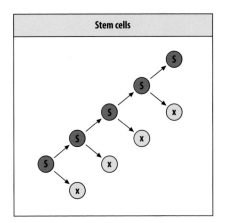

Fig. 1.29 Stem cells. Stem cells (S) are cells that both renew themselves and give rise to differentiated cell types. Thus, a daughter cell arising from stem-cell division may either develop into another stem cell or give rise to a quite different type of cell (X).

Summary

Development results from the coordinated behavior of cells. The major processes involved in development are pattern formation, morphogenesis or change in form, cell differentiation, growth, cell migration, and cell death. These involve cell activities such as intercellular signaling and turning genes on and off. Genes control cell behavior by controlling where and when proteins are synthesized, and thus cell biology provides the link between gene action and developmental processes. During development, cells undergo changes in the genes they express, in their shape, in the signals they produce and respond to, in their rate of proliferation, and in their migratory behavior. All these aspects of cell behavior are controlled largely by the presence of specific proteins; gene activity controls which proteins are made. Since the somatic cells in the embryo all generally contain the same genetic information, the changes that occur in development are controlled by the differential activity of selected sets of genes in different groups of cells. The genes' control regions are fundamental to this process. Development is progressive and the fate of cells becomes determined at different times. The potential for development of cells in the early embryo is usually much greater than their normal fate, but this potential becomes more restricted as development proceeds. Inductive interactions, involving signals from one tissue or cell to another, are one of the main ways of changing cell fate and directing development. Asymmetric cell divisions, in which cytoplasmic components are unequally distributed to daughter cells, can also make cells different. One widespread means of pattern generation is through positional information; cells first acquire a positional value with respect to boundaries and then interpret their positional values by behaving in different ways. Developmental signals are more selective than instructive, choosing one or other of the developmental pathways open to the cell at that time. The embryo contains a generative, not a descriptive, program—it is more like the instructions for making a structure by paper folding than a blueprint. Various mechanisms, including apparent redundancy and negative feedback, are involved in making development remarkably reliable. The complexity of development lies within the cells. Evolution is intimately linked to development as heritable changes in the adult organism must inevitably be the result of changes in the genetic program of embryonic development.

Summary to Chapter 1

All the information for embryonic development is contained within the fertilized egg—the diploid zygote. The genome of the zygote contains a program of instructions for making the organism. In the working out of this developmental program, the cytoplasmic constituents of the egg, and of the cells it gives rise to, are essential players along with the genes. Strictly regulated gene activity, by controlling which proteins are synthesized, and where and when, directs a sequence of cellular activities that brings about the profound changes that occur in the embryo during development. The major processes involved in development are pattern formation, morphogenesis, cell differentiation, growth, cell migration, and cell death, which are controlled by communication between the cells of the embryo and changes in gene expression. Development is progressive, with the fate of cells becoming more precisely specified as it proceeds. Cells in the early embryo usually have a much greater potential for development than is evident from their normal fate, and this enables embryos, particularly vertebrate embryos, to develop normally even if cells are removed, added, or transplanted to different positions. Cells' developmental potential becomes much more restricted as development proceeds. Many genes are involved in controlling the complex interactions that occur during development, and reliability is achieved in various ways. Thousands of genes control the development of animals and plants, and a full understanding of the processes involved is far from being achieved. While basic principles and certain developmental systems are quite well understood, there are still many gaps. Changes in embryonic development are the basis for the evolution of multicellular organisms.

▨ End of chapter questions

Long answer (concept questions)

1. The ultimate goal of the science of developmental biology is to understand human development. In your own words, discuss three areas of medicine that are impacted by the study of developmental biology.

2. The Greek philosopher Aristotle considered two opposing theories for development: preformation and epigenesis. Describe what is meant by these two ideas; refer to the concept of the 'homunculus' in describing preformation, and to the art of origami in describing epigenesis. Which better describes our current view of development?

3. Weismann's concept of 'determinants' was supported by Roux's experiments demonstrating 'mosaic development' in the frog. How would determinants lead to mosaic development? Why does Driesch's concept of 'regulation' during development better explain human twinning?

4. Pattern formation is a central process in the study of development. Describe three examples of pattern formation and explain what 'patterns' are generated.

5. Briefly define the following cell behaviors: intercellular signaling, cell proliferation, cell differentiation, cell movement, changes in cell shape, gene expression, and cell death.

6. Contrast the role of housekeeping genes and proteins during development with the role of tissue-specific genes and proteins. Which class of genes and proteins is more important during development, housekeeping or tissue-specific? Give examples.

7. All cells in an organism contain the same genes; however, they contain different types of cells. How is this paradox of 'genetic equivalence' resolved by the concept of 'differential gene expression'?

8. How do gene regulatory proteins, or transcription factors, work? What is their relationship to control regions in DNA?

9. Contrast cell fate with cell determination. What kinds of experiments allow the distinction between fate and determination to be made?

10. What are three ways in which inducing signals can be transmitted in developing embryos? In what circumstance would an inducing signal be a 'morphogen'?

11. What is the relationship between positional information and pattern formation? Give an example.

Multiple choice (factual recall questions)

NB There is only one right answer to each question.

1. Which of the following is a vertebrate 'model organism' used in the study of development?

a) *Caenorhabditis elegans*

b) *Drosophila melanogaster*

c) *Homo sapiens*

d) *Xenopus laevis*

2. Which of the following is a germ cell?

a) cells of the gonads

b) intestinal cells

c) sperm and eggs

d) zygotes

3. Which statement best describes the relationship between genotype and phenotype?

a) 'Genotype' and 'phenotype' both describe the physical properties of cells and organisms, just using different words or symbols.

b) Phenotype is entirely the consequence of a cell's genotype, which is why we now recognize the principle of 'genetic determinism.'

c) The genes found in a cell are that cell's genotype, and genes directly control a cell's phenotype.

d) The genotype determines what proteins a cell contains, and the proteins in turn determine the cell's properties, or phenotype.

4. Which would qualify as a dominant allele of a gene?

a) an allele that confers a phenotype on the organism when heterozygous; that is, when present in only one copy

b) an allele that confers a phenotype only when both copies of the gene are mutant; that is, when homozygous

c) an allele that is commonly found in wild populations, as opposed to alleles found in laboratory strains or when a pathological condition is present

d) an allele that confers a normal phenotype under some conditions, but confers a different, mutant phenotype when conditions such as elevated temperature are present

5. Which of the following basic developmental processes is most dependent on cellular movements?

a) cell differentiation

b) growth

c) morphogenesis

d) pattern formation

6. Which aspect of cellular regulation operates after a protein has been synthesized?

a) post-translational modification

b) RNA processing

c) transcriptional control

d) translational control

7. Which term could be used to refer to a segment of DNA?

a) *cis*-regulatory module

b) differential gene expression

c) gene regulatory protein

d) transcription factor

8. A morphogen is

a) a cell or group of cells that signal other cells to become determined in a specific way

b) a gene that causes a cell or group of cells to adopt a particular shape

c) a signaling molecule that causes the differentiation of neighboring cells

d) a signaling molecule that confers concentration-dependent positional information

9. What distinguishes stem cells from other cell types?

a) Although all cell types can divide, only stem cells differentiate after division.

b) Stem cells are able to divide to produce populations of cells that differentiate into a related set of cells.

c) Stem cells divide asymmetrically to give rise to one daughter cell that remains a stem cell, and a second daughter cell that differentiates into one or more cell types.

d) Stem cells have a special type of cell division in which they form small stems that then pinch off to give rise to the daughter cell.

10. How does organismal development evolve to give rise to new forms and species?

a) A structure like the fin of a fish evolves into a structure like a human hand through the needs of the organism for a hand rather than a fin.

b) Changes in genes that will be favored by natural selection will be induced, and thereby produce favorable changes in development.

c) Changes in genes will alter an organism's development, and if the new developmental patterns are favored by the environment, they will be retained through natural selection.

d) The characteristics acquired by the organism during its lifetime are passed on to the offspring; if the changes are favorable to the organism, they will eventually come to be encoded in the genes that govern development.

Multiple choice answer key

1: d, 2: c, 3: d, 4: a, 5: c, 6: a, 7: a, 8: d, 9: c, 10: c.

■ Section further reading

The origins of developmental biology

Hamburger, V.: *The Heritage of Experimental Embryology: Hans Spemann and the Organizer*. New York: Oxford University Press, 1988.

Harris, H.: *The Birth of the Cell*. New Haven: Yale University Press, 1999.

Milestones in Development [http://www.nature.com/milestones/development/index.html] (date accessed 9 May 2010).

Needham, J.: *A History of Embryology*. Cambridge: Cambridge University Press, 1959.

Sander, K.: **'Mosaic work' and 'assimilating effects' in embryogenesis: Wilhelm Roux's conclusions after disabling frog blastomeres**. *Roux's Arch. Dev. Biol.* 1991, **200**: 237–239.

Sander, K.: **Shaking a concept: Hans Driesch and the varied fates of sea urchin blastomeres**. *Roux's Arch. Dev. Biol.* 1992, **201**: 265–267.

A conceptual tool kit

Alberts, B., *et al.*: *Essential Cell Biology: An Introduction to the Molecular Biology of the Cell*. 5th edn. New York: Garland Science, 2009.

Howard, M.L., Davidson, E.H.: *cis*-**Regulatory control circuits in development**. *Dev. Biol.* 2004, **271**: 109–118.

Istrail, S., De-Leon, S.B., Davidson, E.H.: **The regulatory genome and the computer**. *Dev. Biol.* 2007, **310**: 187–195.

Jordan, J.D., Landau, E.M., Lyengar, R.: **Signaling networks: the origins of cellular multitasking**. *Cell* 2000, **103**: 193–200.

Levine, M., Tjian, R.: **Transcriptional regulation and animal diversity**. *Nature* 2003, **424**: 147–151.

Nelson, W.J.: **Adaptation of core mechanisms to generate polarity**. *Nature* 2003, **422**: 766–774.

Papin, J.A., Hunter, T., Palsson, B.O., Subramanian, S.: **Reconstruction of cellular signalling networks and analysis of their properties**. *Nat. Rev. Mol. Biol.* 2005, **6**: 99–111.

Volff, J.N. (Ed.): **Vertebrate genomes**. *Genome Dyn* 2006, **2**: special issue.

Wolpert, L.: **Do we understand development?** *Science* 1994, **266**: 571–572.

Wolpert, L.: **One hundred years of positional information**. *Trends Genet.* 1996, **12**: 359–364.

Xing, Y., Lee, C.: **Relating alternative splicing to proteome complexity and genome evolution**. *Adv. Exp. Med. Biol.* 2007, **623**: 36–49.

Box 1E Signal transduction and intracellular signaling

Logan, C.Y., Nusse. R.: **The Wnt signaling pathway in development and disease**. *Annu. Rev. Cell Dev. Biol.* 2004, **20**: 781–810.

Box 1F When development goes awry

Cohen, M.M. Jr: **Holoprosencephaly: clinical, anatomic, and molecular dimensions**. *Birth Defects Res. A Clin. Mol. Teratol.* 2006, **76**: 658–673.

Geng, X., Oliver, G.: **Pathogenesis of holoprosencephaly**. *J. Clin. Invest.* 2009, **119**: 1403–1413.

Roessler, E., Muenke, M.: **How a Hedgehog might see holoprosencephaly**. *Hum. Mol. Genet.* 2003, **12**: R15–R25.

Development of the *Drosophila* body plan

- *Drosophila* life cycle and overall development
- Setting up the body axes
- Localization of maternal determinants during oogenesis

- Patterning the early embryo
- Activation of the pair-rule genes and the establishment of parasegments

- Segmentation genes and compartments
- Specification of segment identity

The early development of the fruit fly Drosophila melanogaster *is better understood than that of any other animal of similar or greater morphological complexity. Its early development exemplifies clearly many of the principles introduced in Chapter 1. The genetic basis of development is particularly well understood in* Drosophila, *and many developmental genes in other organisms, particularly vertebrates, have been identified by their homology with* Drosophila *genes. In this chapter we look at the crucial role of maternal gene products in the earliest development of the zygote and the early syncytial blastoderm stage, and how they specify the main axes and regions of the body. We shall see how gradients of maternal morphogens specify the antero-posterior and dorso-ventral pattern of the early embryo, and how* cis-acting *regulatory elements then control the precise spatial expression of the zygote's own genes to divide the embryo into segments. Another set of genes acting along the antero-posterior axis then confers on each segment its unique character, as displayed by the appendages the segment eventually bears in the adult, such as wings, legs, or antennae.*

We are much more like flies in our development than you might think. Astonishing discoveries in developmental biology over the past 25 years have revealed that many of the genes that control the development of the fruit fly *Drosophila* are similar to those controlling development in vertebrates, and indeed in many other animals. It seems that once evolution finds a satisfactory way of patterning animal bodies, it tends to use the same mechanisms and molecules over and over again with, of course, some important modifications. While insect and vertebrate development might seem to be very different, much has been learned from *Drosophila* that has illuminated vertebrate development.

From the genome sequence there are an estimated 14,000 protein-coding genes in *Drosophila*, which is only twice the number in unicellular yeast and fewer than the 19,000 genes estimated of the morphologically simpler nematode *Caenorhabditis*. However, a recent large-scale analysis of the RNAs transcribed both during development and in adult *Drosophila* revealed a high incidence of alternative RNA splicing, which would increase the number of different proteins that could be made. In addition, around 1100 genes encoding functional RNAs, such as tRNAs and microRNAs, are present in the fly genome.

The pre-eminent place of *Drosophila* in modern developmental biology was recognized by the award of the 1995 Nobel Prize for Physiology or Medicine for work that led to a fundamental understanding of how genes control development in the fly embryo. This was only the second time that the Nobel Prize had been awarded for work in developmental biology. While insect and vertebrate development may seem to be very different, much has been learned that can be applied to vertebrate development; for example, many intercellular signaling pathways are well conserved.

The first part of this chapter describes the life cycle of *Drosophila* and its overall development. The remaining parts look at how the basic body plan of the *Drosophila* larva is established up to the stage at which the embryo becomes segmented, and how the segments are patterned and acquire their unique identities. More about *Drosophila* gastrulation, germ-cell development and sex-determination mechanisms, adult organ development, neural development, and growth and metamorphosis will be found in Chapters 8, 9, 11, 12, and 13, respectively.

Drosophila life cycle and overall development

The fruit fly *Drosophila melanogaster* is a small dipteran insect, about 3 mm long as an adult, which undergoes embryonic development inside an egg and hatches from the egg as a larva. This then goes through two more larval stages, growing bigger each time, and eventually becomes a pupa, in which metamorphosis into the adult occurs. The life cycle of *Drosophila* is shown in Fig. 2.1.

2.1 The early *Drosophila* embryo is a multinucleate syncytium

The *Drosophila* egg is rather oblong and the future anterior end is easily recognizable by the micropyle, a nipple-shaped structure in the tough external coat surrounding the egg. Sperm enter the anterior end of the egg through the micropyle. After fertilization and fusion of the sperm and egg nuclei, the zygote nucleus undergoes a series of rapid mitotic divisions, one about every 9 minutes, but, unlike in most embryos, there is initially no cleavage of the cytoplasm and no formation of cell membranes to separate the nuclei. The result after 12 nuclear divisions is a **syncytium** in which around 6000 nuclei are present in a common cytoplasm (Fig. 2.2); the embryo essentially remains a single cell during its early development. After nine divisions the nuclei move to the periphery to form the **syncytial blastoderm**. This comprises a superficial layer of nuclei and cytoplasm, and surrounds a central mass of yolky cytoplasm. The syncytial blastoderm is equivalent to the blastula stage in embryos such as *Xenopus* (see, for example, Box 1A, pp. 2–3). Shortly afterwards, membranes grow in from the surface to enclose the nuclei and form cells, and the blastoderm becomes truly cellular after 14 mitoses. Because of the formation of a syncytium, even large molecules such as proteins can diffuse between nuclei during the first 3 hours of development and, as we shall see, this is of great importance to early *Drosophila* development.

At the syncytial stage, a small number of nuclei move to the posterior end of the embryo and become surrounded by cell membranes to form the **pole cells**, which end up on the outer surface of the blastoderm (see Fig. 2.2). The pole cells eventually give rise to the germ cells from which the gametes—sperm and eggs—develop in the adult insect, while the blastoderm gives rise to the somatic cells of the embryo. This setting aside of the germline from the rest of the embryo at a very early stage is a common strategy in animal development.

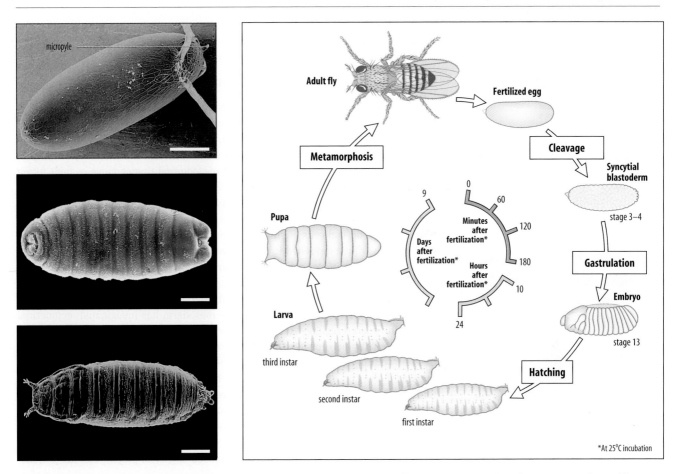

Fig 2.1 Life cycle of *Drosophila melanogaster*. After cleavage and gastrulation the embryo becomes segmented and hatches out as a feeding larva. The larva grows and goes through two molts (instars), eventually forming a pupa that will metamorphose into the adult fly. The photographs show scanning electron micrographs of: a *Drosophila* egg before fertilization (top). The sperm enters through the micropyle.

The dorsal filaments are extraembryonic structures; a *Drosophila*, second-instar larva (middle); and a *Drosophila* pupa (bottom). Scale bars = 0.1 mm.

Photographs courtesy of F. R. Turner (top, from Turner, F.R., et al.: 1976; middle, from Turner, F.R., et al.: 1979).

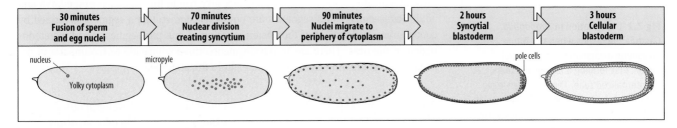

Fig 2.2 Cleavage of the *Drosophila* embryo. After fusion of the sperm and egg nuclei there is rapid nuclear division but no cell membranes surround the nucleus, the result being a syncytium of many nuclei in a common cytoplasm. After the ninth division the nuclei move to the periphery to form the syncytial blastoderm, but some

are delayed. After about 3 hours, cell membranes develop, giving rise to the cellular blastoderm. About 15 pole cells, which will give rise to germ cells, form a separate group at the posterior end of the embryo. Times given are for incubation at 25°C.

In this chapter we will be concerned with the development of the embryo to the stage at which it hatches out of the egg as a larva. Embryonic development naturally takes place inside the opaque egg case, and to study it embryos are freed from the egg case by treating with bleach and then placed in a special oil for observation.

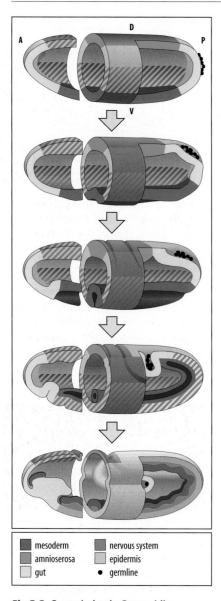

mesoderm
amnioserosa
gut
nervous system
epidermis
• **germline**

Fig 2.3 Gastrulation in *Drosophila*.
Gastrulation begins when the future
mesoderm invaginates in the ventral
region, first forming a furrow and then
an internalized tube. The cells then leave
the tube and migrate internally under the
ectoderm. The nervous system comes from
cells that leave the surface of the ventral
blastoderm and form a layer between the
ventral ectoderm and mesoderm. The gut
forms from two invaginations at the anterior
and posterior end that fuse in the middle.
The midgut region is derived from endoderm,
whereas the foregut and hindgut are of
ectodermal origin. Regions striped blue and
gray represent tissue that will give rise to
both nervous system and epidermis. The
amnioserosa, an extra-embryonic membrane,
is discussed in Section 2.5.

This can be done at any stage and they will continue to develop and even hatch. The
embryo is about 0.5 mm long.

2.2 Cellularization is followed by gastrulation and segmentation

All the future tissues, except for the germline cells, are derived from the single epithelial layer of the cellular blastoderm. For example, prospective **mesoderm** (see Box 1C,
p. 15) is located in the most ventral region, whereas the future midgut derives from
two regions of prospective **endoderm**, one at the anterior and the other at the posterior end of the embryo. Endodermal and mesodermal tissues move to their future
positions inside the embryo during **gastrulation**, leaving **ectoderm** as the outer layer
(Fig. 2.3). Gastrulation starts at about 3 hours after fertilization, when the future
mesoderm in the ventral region invaginates to form a furrow along the ventral midline. The mesodermal cells are initially internalized by formation of a tube of mesoderm, which will be described in more detail in Chapter 8. The mesoderm cells then
separate from the surface layer of the tube and migrate under the ectoderm to internal
locations, where they later give rise to muscle and other connective tissues.

In insects, as in all arthropods, the main nerve cord lies ventrally, rather than
dorsally as in vertebrates. Shortly after the mesoderm has invaginated, ectodermal
cells of the ventral region, which will give rise to the nervous system, leave the surface individually and form a layer of **neuroblasts** between the mesoderm and the
outer ectoderm. At the same time, two tube-like invaginations develop at the sites of
the future anterior and posterior midgut. These grow inward and eventually fuse to
form the endoderm of the midgut, while ectoderm is dragged inward behind them
at each end to form the foregut and the hindgut. The outer ectoderm layer develops
into the epidermis. There are no cell divisions during gastrulation, but once it is
completed, cells start to divide again. The cells of the epidermis only divide twice
more before they secrete a thin cuticle composed largely of protein and the polysac-
charide chitin.

Also during gastrulation, the ventral blastoderm or **germ band**, which comprises
the main trunk region, undergoes germ-band extension, which drives the posterior
trunk regions round the posterior end and onto what was the dorsal side (Fig. 2.4),
as described in Chapter 8. The germ band later retracts as embryonic development is
completed. At the time of germ-band extension, the first external signs of **segmentation**
can be seen. A series of evenly spaced grooves form more or less simultaneously and
these demarcate **parasegments**, which later give rise to the **segments** of the larva and
adult. Parasegments and segments are out of register, so that a segment is formed by
cooperative development of the posterior region of one parasegment and the anterior
region of the next. There are 14 parasegments: three contribute to mouthparts of the
head, three to the thoracic region, and eight to the abdomen.

2.3 After hatching, the *Drosophila* larva develops through several larval stages, pupates, and then undergoes metamorphosis to become an adult

The larva (Fig. 2.5) hatches about 24 hours after fertilization, but the different regions
of the larval body are well-defined several hours before that. The head is a complex
structure, largely hidden from view before the larva hatches. A specialized structure
associated with the most anterior region of the head is the **acron**. Another specialized
structure, called the **telson**, marks the posterior end of the larva. Between the head
and telson, three thoracic segments and eight abdominal segments can be distinguished by specializations in the cuticle. On the ventral side of each segment are belts
of small tooth-like outgrowths called **denticles**, and other cuticular structures characteristic of each segment. As the larva feeds and grows, it molts, shedding its cuticle.
This process occurs twice, each stage being called an **instar**. After the third instar,

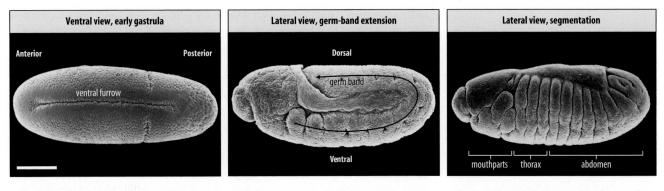

| Ventral view, early gastrula | Lateral view, germ-band extension | Lateral view, segmentation |

the larva becomes a **pupa**, inside which **metamorphosis** into the adult fly occurs and there is a major transformation of overall form.

The *Drosophila* larva has neither wings nor legs; these and other organs are formed when the larva undergoes hormone-induced metamorphosis in the pupal stage. These structures are, however, already present in the larva as **imaginal discs**, small sheets of prospective epidermal cells derived from the cellular blastoderm and usually containing about 40 cells each at the time they are formed. These discs grow by cell proliferation throughout larval life and form sacs of epithelia that fold to accommodate their increase in size. There are imaginal discs for each of the six legs, two wings, and the two halteres (balancing organs), and for the genital apparatus, eyes, antennae, and other adult head structures (Fig. 2.6). At metamorphosis, these develop into the adult organs. We will discuss the development of the imaginal discs in Chapter 11; they provide continuity between the pattern of the larval body and that of the adult, even though metamorphosis intervenes.

Fig 2.4 Gastrulation, germ-band extension, and segmentation in the *Drosophila* embryo. Gastrulation involves the future mesoderm moving inside through the ventral furrow. During gastrulation, the ventral blastoderm (the germ band) extends, driving the posterior trunk region onto the dorsal side and segmentation now takes place. Later the germ band shortens. Scale bar = 0.1 mm.

Photographs courtesy of F.R. Turner (left from Turner, F.R., et al.: 1977; middle from Alberts, B., et al.: 1994).

2.4 Many developmental genes have been identified in *Drosophila* through induced large-scale genetic screening

Valuable though spontaneous mutations have been in the study of development, suitable developmentally informative mutations are rare. Many more developmental genes have been identified by inducing random mutations in a large number of organisms by chemical treatments or irradiation with X-rays, and then screening for mutants of developmental interest. The overall aim is to treat a large enough population so that, in

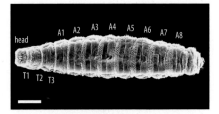

Fig 2.5 Ventral view of a *Drosophila* larva. T1 to T3 are the thoracic segments and A1 to A8, the abdominal segments. The characteristic pattern of denticles can be seen in the anterior region of each abdominal segment. Scale bar = 0.1 mm.

Photograph courtesy of F.R. Turner.

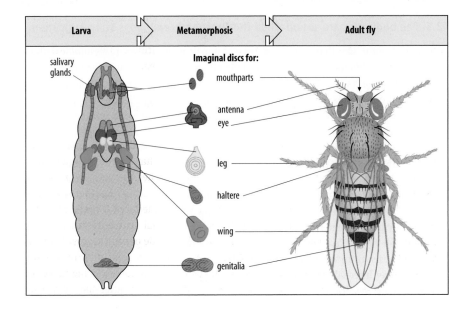

Fig 2.6 Imaginal discs give rise to adult structures at metamorphosis. The imaginal discs in the *Drosophila* larva are small sheets of epithelial cells; at metamorphosis they give rise to a variety of adult structures. The abdominal cuticle comes from groups of tissue-forming cells (histoblasts) located in each larval abdominal segment.

total, a mutation is induced in every gene in the genome. This sort of approach is best used in organisms that breed rapidly and can be obtained and treated conveniently in very large numbers.

Many of the developmental mutations that have led to our present understanding of early *Drosophila* development came from a brilliantly successful screening program that searched the *Drosophila* genome systematically for mutations affecting the patterning of the early embryo. Its success was recognized with the award of a Nobel Prize for Physiology or Medicine to Edward Lewis, Christiane Nüsslein-Volhard, and Eric Wieschaus in 1995.

In this screening program thousands of flies were treated with a chemical mutagen, and were bred and screened according to the strategy described in Box 2A (p. 42). Given the number of progeny involved, it was important to devise a strategy that would reduce the number of flies that had to be examined to find a mutation. So the search was restricted to mutations on just one chromosome at a time. As described in Box 2A, the program also incorporated a means of identifying flies homozygous for the male-derived chromosomes carrying the induced mutations and, most important, also included a way of automatically removing from the population flies that could not be carrying a mutated chromosome.

A phenotypic character of particular value in screening for patterning mutants in *Drosophila* is the stereotyped pattern of small outgrowths, or denticles, on the larval segments, as irregularities in these patterns enable mutants to be quickly spotted. In this way, the key genes involved in patterning the early *Drosophila* embryo were first identified.

One type of mutation that affects *Drosophila* embryonic development needs a somewhat different type of screen. **Maternal-effect mutations** are mutations that when expressed in the mother affect the embryo's development. They are identified as mutations that, when present in the mother, do not affect her appearance or normal physiology but have effects on the development of her progeny. As we shall see later in the chapter, some of these **maternal genes** exert their effects by being expressed in the follicle cells of the ovary, which form a bag that contains the germline-derived oocyte (the immature egg) and nurse cells. Others are expressed in the nurse cells or in the oocyte, producing developmentally important mRNAs and proteins that are deposited in the oocyte in a particular spatial pattern. This initial patterning process is crucial for correct development.

Setting up the body axes

2.5 The body axes are set up while the *Drosophila* embryo is still a syncytium

The insect body, like that of vertebrates and most of the other organisms described in this book, is bilaterally symmetrical. Like all animals with bilateral symmetry, the *Drosophila* larva has two distinct and largely independent axes: the antero-posterior and dorso-ventral axes, which are at right angles to each other. These axes are already partly set up in the *Drosophila* egg, and become fully established and patterned in the very early embryo, while it is still in the syncytial blastoderm stage. Along the antero-posterior axis the embryo becomes divided into several broad regions, which will become the **head, thorax,** and **abdomen** of the larva (Fig. 2.7). The thorax and abdomen become divided into segments as the embryo develops, while two regions of endoderm at each end of the embryo invaginate at gastrulation to form the gut (see Fig. 2.3). Each segment, and the head, has its own unique character in the larva, as revealed by both its external cuticular structures and its internal organization.

The dorso-ventral axis of the embryo becomes divided up into four regions early in embryogenesis: from ventral to dorsal these are the **mesoderm**, which will form muscles and other internal connective tissues; the **neuroectoderm**, which gives rise to the larval nervous system; the dorsal ectoderm, which gives rise to the larval epidermis;

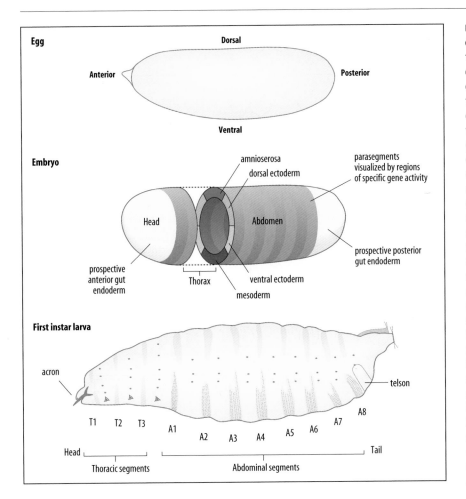

Fig 2.7 **Patterning of the** *Drosophila* **embryo.** The body plan is patterned along two distinct axes. The antero-posterior and dorso-ventral axes are at right angles to each other and are laid down in the egg. In the early embryo, the dorso-ventral axis is divided into four regions: mesoderm (red), ventral ectoderm (yellow), dorsal ectoderm (orange), and amnioserosa (an extra-embryonic membrane; green). The ventral ectoderm gives rise to both ventral epidermis and neural tissue, the dorsal ectoderm to epidermis. The antero-posterior axis becomes divided into different regions that later give rise to the head, thorax, and abdomen. After the initial division into broad body regions, segmentation begins. The future segments can be visualized as transverse stripes by staining for specific gene activity; these stripes demarcate 14 parasegments, 10 of which are marked. The embryo develops into a segmented larva. By the time the larva hatches, the 14 parasegments have been converted into thoracic (T1-T3) and abdominal (A1-A8) segments, with each segment being made up of the posterior half of one parasegment and the anterior half of the next. Different segments are distinguished by the patterns of bristles and denticles on the cuticle. Specialized structures, the acron and telson, develop at the head and tail ends, respectively.

and the **amnioserosa**, which gives rise to an extra-embryonic membrane on the dorsal side of the embryo (see Fig. 2.7). Organization along the antero-posterior and dorso-ventral axes of the early embryo develops more or less simultaneously, but is specified by independent mechanisms and by different sets of genes in each axis.

The early development of *Drosophila* is peculiar to certain types of insects, as patterning occurs within a multinucleate syncytial blastoderm (see Fig. 2.2). Only after the beginning of segmentation does the embryo become truly multicellular. At the syncytial stage, many proteins, including those that are not normally secreted from cells such as transcription factors, can diffuse throughout the blastoderm and enter other nuclei. Concentration gradients of transcription factors that provide positional information for the nuclei to interpret (see Section 1.15) can thus be set up in the syncytial blastoderm.

Early development is essentially two-dimensional as patterning occurs mainly in the blastoderm, which is essentially a single layer of nuclei, and later cells. But the larva itself is a three-dimensional object, with internal structures organized along the axis. This third dimension develops later, at gastrulation (discussed in Chapter 8), when parts of the surface layer move into the interior to form the gut, the mesodermal structures that will give rise to muscle, and the ectodermally derived nervous system.

2.6 Maternal factors set up the body axes and direct the early stage of *Drosophila* development

The earliest stage of *Drosophila* embryonic development is guided by **maternal factors**, mRNAs, and proteins that are synthesized and laid down in the egg by the mother (see Section 2.4). In all, about 50 maternal genes are involved in setting up the two axes and

Box 2A Mutagenesis and genetic screening strategy for identifying developmental mutants in *Drosophila*

The mutagen ethyl methane sulfonate (EMS) was applied to large numbers of male flies homozygous for a recessive mutation on the selected chromosome. The chromosome marked in this way is designated 'a' in the figure. A recessive mutation is chosen that gives adult flies an easily distinguishable but viable phenotype when homozygous (see Fig. 1.12). Here, we shall use the example of a mutation that gives white eyes in the homozygous state (*white⁻*), instead of the normal red eyes of the wild type.

The treated males, who now produce sperm with a variety of induced mutations on the 'a' chromosomes (a*), were crossed with untreated females that carried different mutations (*DTS* and *b*) on their two 'a' chromosomes, but were otherwise wild type. These mutations track the untreated female-derived chromosomes and automatically eliminate all embryos carrying two female-derived chromosomes in subsequent generations. *DTS* is a dominant temperature-sensitive mutation that causes death of the fly when the incubation temperature is raised to 29°C. *b* is a

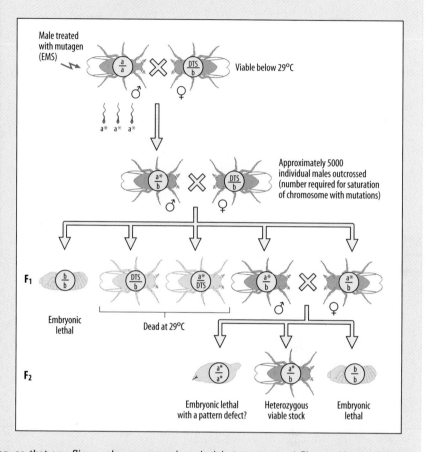

non-developmental lethal recessive mutation, so that any flies homozygous for this female-derived chromosome will die as normal-looking embryos and be automatically eliminated. The female flies also carried a balancer chromosome (not shown) that prevented recombination at meiosis; this is to prevent recombination between male-derived and female-derived chromosomes in females. There is no recombination at meiosis in male *Drosophila*.

To identify the new recessive mutations (a*) caused by the chemical treatment, a large number of the heterozygous males arising from this first cross were again outcrossed to *DTS/b* females. Of the offspring of each individual cross, only a*/b flies survive when placed at 29°C; all other combinations die. The surviving siblings were then intercrossed and the offspring of each cross were screened for patterning mutants. There are three possible outcomes: flies homozygous for the induced mutation a* (which will also be homozygous for the original mutation marking

chromosome a in males); heterozygous a* flies; and homozygous *b* flies (which die as embryos).

If a* is indeed a patterning mutation that is lethal in the larva, then the culture tube in which the cross is made will contain no adult flies with the original male phenotype—white eyes, say. Therefore tubes containing white-eyed flies can be discarded immediately, as the induced mutation a* they carry must have allowed them to develop to adulthood even when homozygous. If no white-eyed flies are present in a tube, then the homozygous a*/a* embryos may have died or been arrested in their development as a result of abnormal development, and the mutation is of potential interest. The embryos of larvae from this cross can then be examined for pattern defects, as in the example illustrated here. The phenotypically wild-type adults in this tube are heterozygous for a* and are used as breeding stock to study the mutant further. This whole program has to be repeated for each of the four chromosome pairs of *Drosophila*.

a basic framework of positional information, which is then interpreted by the embryo's own genetic program. In contrast to the maternal-effect genes, those genes expressed in the embryo's own nuclei during development are known as **zygotic genes**. All later patterning, which involves expression of the zygotic genes, is built on the framework of maternal gene products (Fig. 2.8). After development begins, the maternal RNAs

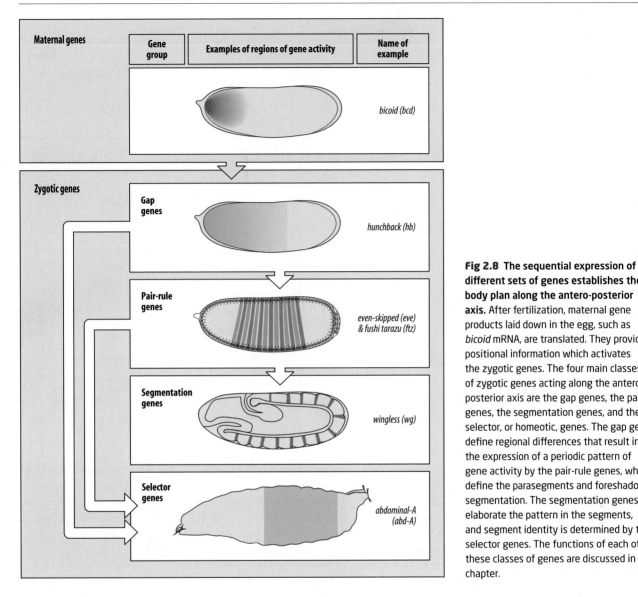

Maternal genes	Gene group	Examples of regions of gene activity	Name of example
			bicoid (bcd)

Zygotic genes			
	Gap genes		hunchback (hb)
	Pair-rule genes		even-skipped (eve) & fushi tarazu (ftz)
	Segmentation genes		wingless (wg)
	Selector genes		abdominal-A (abd-A)

Fig 2.8 The sequential expression of different sets of genes establishes the body plan along the antero-posterior axis. After fertilization, maternal gene products laid down in the egg, such as *bicoid* mRNA, are translated. They provide positional information which activates the zygotic genes. The four main classes of zygotic genes acting along the antero-posterior axis are the gap genes, the pair-rule genes, the segmentation genes, and the selector, or homeotic, genes. The gap genes define regional differences that result in the expression of a periodic pattern of gene activity by the pair-rule genes, which define the parasegments and foreshadow segmentation. The segmentation genes elaborate the pattern in the segments, and segment identity is determined by the selector genes. The functions of each of these classes of genes are discussed in this chapter.

are translated, and the resulting proteins act on the embryo's nuclei to activate zygotic genes in a specific spatial pattern along each axis, thus setting the scene for the next round of patterning.

Drosophila illustrates particularly well a general principle of development. The embryo is patterned in a series of steps. Broad regional differences are established first, and these are then refined to produce a larger number of smaller developmental domains, each characterized by a unique profile of gene activity. Developmental genes act in a strict temporal sequence. They form a hierarchy of gene activity in which the action of one set of genes is essential for another set of genes to be activated, and thus for the next stage of development to occur.

2.7 Three classes of maternal genes specify the antero-posterior axis

We will first look at how maternal gene products specify the antero-posterior axis of the embryo. The expression of maternal genes during egg formation in the mother creates differences in the egg along the antero-posterior axis even before it is fertilized. These differences already distinguish the future anterior and posterior ends of the adult. The roles of the maternal genes can be deduced from the effects of

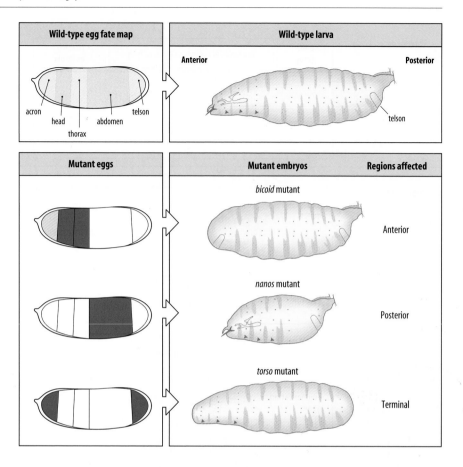

Fig 2.9 The effects of mutations in the maternal gene system. Mutations in maternal genes lead to deletions and abnormalities in anterior, posterior, or terminal structures. The wild-type fate map shows which regions of the egg give rise to particular regions and structures in the larva. Regions that are affected in mutant eggs, and which lead to lost or altered structures in the larva, are shaded in red. In *bicoid* mutants there is a partial loss of anterior structures and the appearance of a posterior structure—the telson—at the anterior end. *nanos* mutants lack a large part of the posterior region. *torso* mutants lack both acron and telson.

maternal-effect mutations (see Section 2.4) on the embryo. They fall into three classes: those that affect anterior regions; those that affect posterior regions; and those that affect both of the terminal regions (Fig. 2.9). Mutations of genes in the anterior class, such as *bicoid*, lead to a reduction or loss of head and thoracic structures, and in some cases their replacement with posterior structures. Posterior-group mutations, such as *nanos*, cause the loss of abdominal regions, leading to a smaller than normal larva, while those of the terminal class, such as *torso*, affect the acron and telson. The idiosyncratic naming of genes in *Drosophila* usually reflects the attempts by the discoverers to describe the mutant phenotype—*nanos* is Greek for dwarf, while *torso* reflects the fact that both ends of the embryo are missing. In this chapter we meet quite a number of gene names; all these are listed, together with their functions where known, in the table at the end of this chapter (p. 87).

2.8 Bicoid protein provides an antero-posterior gradient of a morphogen

Maternal *bicoid* mRNA is localized at the anterior end of the unfertilized egg during oogenesis. After fertilization it is translated, and the conventional view was that Bicoid protein diffuses from the anterior end and forms a concentration gradient along the antero-posterior axis. Very recent evidence, together with some older studies, suggests, on the contrary, that the Bicoid-protein gradient is preceded by a gradient in *bicoid* mRNA in the egg cortex, with the *bicoid* mRNA being proposed to be transported along cortical microtubules (Fig. 2.10). Translation of the mRNA gives rise to the gradient in Bicoid protein, which provides the positional information required for further patterning along this axis. Historically, the Bicoid gradient provided the first concrete evidence for the existence of the morphogen gradients that had been postulated to control pattern formation (see Section 1.15).

The role of the *bicoid* gene was first elucidated by a combination of genetic and physical experiments on the *Drosophila* embryo. Female flies that do not express *bicoid* produce embryos that have no proper head and thorax (see Fig. 2.9). In a separate line of investigation into the role of localized cytoplasmic factors in anterior development, normal eggs were pricked at the anterior ends and some cytoplasm allowed to leak out. The embryos that developed bore a striking resemblance to *bicoid* mutant embryos. This suggested that normal eggs have some factor(s) in the cytoplasm at their anterior end that is absent in *bicoid* mutant eggs. This was confirmed by showing that anterior cytoplasm from wild-type embryos could partially 'rescue' the *bicoid* mutant embryos—in the sense they developed more normally—if it was injected into their anterior regions (Fig. 2.11). Moreover, if normal anterior cytoplasm is injected into the middle of a fertilized *bicoid* mutant egg, head structures developed at the site of injection and the adjacent segments became thoracic segments, setting up a mirror-image body pattern at the site of injection. The simplest interpretation of these experiments is that the *bicoid* gene is necessary for the establishment of the anterior structures because it establishes a gradient of Bicoid protein, whose source and highest level is at the anterior end.

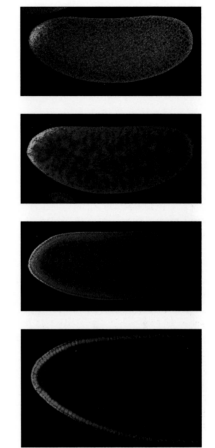

Fig 2.10 *bicoid* **RNA forms a concentration gradient in the early embryo.** Panels a to c show the formation of the gradient of *bicoid* mRNA in *Drosophila* embryos (dorsal side up and anterior to the left). *bcd* mRNA transcripts (yellow) in the cortex are detected by fluorescence *in situ* hybridization with a labeled *bcd* cDNA in (a) an unfertilized egg, and in embryos during interphase of (b) nuclear cycle 9, and (c) after onset of nuclear cycle 14. The corresponding formation of the Bicoid protein gradient can be seen in the nuclear cycle 14 embryo in (d), in which *bcd* mRNA is stained green and the Bcd protein (inside nuclei) in red.

Photographs reproduced from Spirov, A. et al.: 2009.

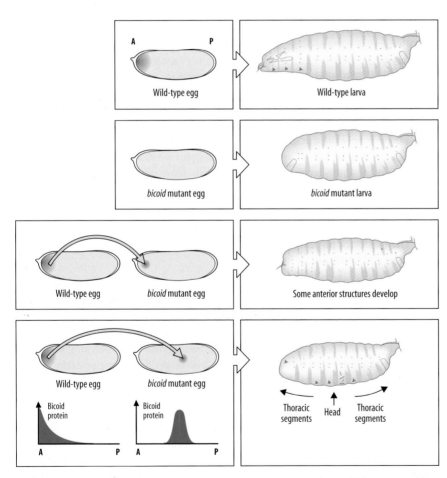

Fig 2.11 The *bicoid* gene is necessary for the development of anterior structures. Embryos whose mothers lack the *bicoid* gene lack anterior regions (second row). Transfer of anterior cytoplasm from wild-type embryos to *bicoid* mutant embryos causes some anterior structure to develop at the site of injection (third row). If wild-type anterior cytoplasm is transplanted to the middle of a *bicoid* mutant egg or early embryo, head structures develop at the site of injection, flanked on both sides by thoracic-type segments (fourth row). These results can be interpreted in terms of the anterior cytoplasm setting up a gradient of Bicoid protein with the high point at the site of injection (see graphs, bottom left panel). A, anterior; P, posterior.

bicoid mRNA was shown to be tightly localized to the anterior-most region of the unfertilized egg by the technique of *in situ* hybridization (see Box 1D, p. 20). Staining with an antibody against Bicoid protein then showed that this is absent from the unfertilized egg, but that after fertilization, *bicoid* mRNA is translated into protein. After fertilization, *bicoid* mRNA diffuses from the anterior to form an anterior to posterior gradient of RNA which is translated into an anterior to posterior gradient of Bicoid protein. Bicoid is a transcription factor and so must enter the embryo's nuclei to carry out its function of switching on zygotic genes. By the time the syncytial blastoderm is formed, there is a clear gradient of intranuclear Bicoid protein concentration along the antero-posterior axis, with its high point at the anterior end (see Fig. 2.10). In an unprecedented technical feat, the dynamics of Bicoid-gradient formation have recently been observed in living embryos by measuring the intranuclear concentrations of Bicoid protein genetically tagged with green fluorescent protein (Bicoid–GFP). The measurements indicate that the concentration of Bicoid protein inside a nucleus at a given position along the gradient is somehow maintained at a constant level, despite the increase in the number of nuclei at each mitotic cycle, and the outflow of Bicoid into the cytoplasm from nuclei when nuclear membranes break down at mitosis.

Bicoid protein acts as a morphogen, as described in more detail later in the chapter. It switches on particular zygotic genes at different **threshold** concentrations, thereby initiating a new pattern of gene expression along the axis (the concept of thresholds is discussed in Section 1.15). Thus, *bicoid* is a key maternal gene in early *Drosophila* development. The other maternal genes of the anterior group are mainly involved in the localization of *bicoid* mRNA to the anterior end of the egg during oogenesis and in the control of setting up its mRNA gradient and translation after fertilization. Given the crucial importance of *bicoid* in *Drosophila* development, it is worth noting that the *bicoid* gene is only present in a small group of the most recently evolved dipterans—such as fruit flies and blowflies. Later in this chapter we shall briefly discuss different mechanisms of early development that occur in some other groups of insects. It is not surprising that many different developmental mechanisms have evolved in such a large and diverse group of animals as the insects.

2.9 The posterior pattern is controlled by the gradients of Nanos and Caudal proteins

For proper patterning along an axis, both ends need to be specified, and Bicoid defines only the anterior end of the antero-posterior axis. The posterior end is specified by the actions of at least nine maternal genes—the posterior-group genes. Mutations in the posterior-group genes result in larvae that are shorter than normal because there is no abdomen (see Fig. 2.9). One function of the products of maternal posterior-group genes (for example, *oskar*) is to is to localize *nanos* mRNA at the extreme posterior pole of the unfertilized egg. Another action is to assemble the posterior **germplasm** in the egg. This is cytoplasm that contains the so-called germline factors; in the embryo it is incorporated into the pole cells (see Section 2.1) and will give rise to eggs or sperm.

Like *bicoid* mRNA, *nanos* mRNA is translated only after fertilization and produces a concentration gradient of Nanos protein, in this case with the highest level at the posterior end of the embryo. But Nanos does not act directly as a morphogen to specify the abdominal pattern; it has a quite different role. Its function is to suppress translation of the maternal mRNA of another gene, *hunchback*, in the posterior region of the embryo. Maternal *hunchback* mRNA is distributed throughout the embryo and starts to be translated after fertilization. However, a little later, the Bicoid protein activates expression of the embryo's own *hunchback* genes in the anterior half of the embryo. Restriction of Hunchback protein to the anterior region is crucial for correct patterning along the antero-posterior axis. To avoid interference with this patterning by maternal Hunchback protein in the posterior region, its translation is prevented

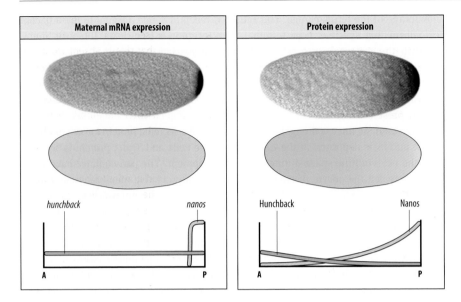

Maternal mRNA expression	Protein expression
hunchback nanos	Hunchback Nanos
A P	A P

Fig 2.12 Establishment of a maternal gradient in Hunchback protein. Left panel: in the unfertilized egg, maternal *hunchback* mRNA (turquoise) is present at a relatively low level throughout the egg, whereas *nanos* mRNA (yellow) is located posteriorly. The photograph is an *in situ* hybridization showing the location of *nanos* mRNA (black). Right panel: after fertilization, *nanos* mRNA is translated and Nanos protein blocks translation of *hunchback* mRNA in the posterior regions, giving rise to a shallow antero-posterior gradient in maternal Hunchback protein. The photograph shows the graded distribution of Nanos, detected with a labeled antibody.

Photographs courtesy of R. Lehmann, from Suzuki, D.T., et al.: 1996.

there, producing an antero-posterior gradient of maternal Hunchback protein. This is the sole task of the Nanos protein (Fig. 2.12); it prevents *hunchback* mRNA translation by binding to a complex of the mRNA and Pumilio protein, which is encoded by another of the posterior-group genes. If maternal Hunchback is completely removed from embryos, then Nanos becomes completely dispensable for antero-posterior patterning.

Evolution can only work on what is already there; it cannot 'look' at the whole system and redesign it economically. So, if a gene causes a difficulty by being expressed in the 'wrong' place, the problem may be solved not by reorganizing the gene-expression pattern, but by bringing in a new function, such as *nanos*, to remove the unwanted protein.

The fourth maternal product crucial to establishing the posterior end of the axis is *caudal* mRNA. Again, this is only translated after fertilization. *caudal* mRNA itself is uniformly distributed throughout the egg, but after fertilization, a posterior-to-anterior gradient of Caudal is established by the specific inhibition of Caudal protein synthesis by Bicoid, which binds to a site in the 3′ untranslated region of the *caudal* mRNA. Because the concentration of Bicoid is low at the posterior end of the embryo, Caudal protein concentration is highest there (Fig. 2.13). Mutations in the *caudal* gene result in the abnormal development of abdominal segments.

Soon after fertilization, therefore, several gradients of maternal proteins have been established along the antero-posterior axis. Two gradients—Bicoid and Hunchback proteins—run in an anterior to posterior direction, while Caudal protein is graded posterior to anterior. We next look at the quite different mechanism that specifies the two termini of the embryo.

2.10 The anterior and posterior extremities of the embryo are specified by cell-surface receptor activation

A third group of maternal genes specifies the structures at the extreme ends of the antero-posterior axis—the acron and the head region at the anterior end, and the telson and the most posterior abdominal segments at the posterior end. A key gene in this group is *torso*; mutations in *torso* can result in embryos developing neither acron nor telson (see Fig. 2.9). This indicates that the two terminal regions, despite their topographical separation, are not specified independently but use the same pathway.

The terminal regions are specified by an interesting mechanism that involves the localized activation of a receptor protein that is itself present throughout the

Fig 2.13 Restriction of Caudal protein to the posterior end of the embryo. Bicoid protein (green) prevents the synthesis of Caudal protein (red) at the anterior end of the embryo. The proteins were detected by fluorescent antibody staining on fixed embryos.

Photograph reproduced from Surkova, S., et al.: 2008.

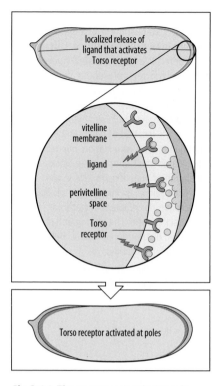

localized release of
ligand that activates
Torso receptor

vitelline
membrane

ligand

perivitelline
space

Torso
receptor

Torso receptor activated at poles

Fig 2.14 The receptor protein Torso is involved in specifying the terminal regions of the embryo. The receptor protein encoded by the gene *torso* is present throughout the egg plasma membrane. Its ligand is laid down in the vitelline membrane at each end of the egg during oogenesis. After fertilization, the ligand is released and diffuses across the perivitelline space to activate the Torso protein at the ends of the embryo only.

membrane of the fertilized egg. The activated receptor sends a signal to the adjacent cytoplasm that specifies it as terminal. The receptor is known as Torso, as mutations in the maternal *torso* gene produce embryos lacking terminal regions. After fertilization the maternal *torso* mRNA is translated and the Torso protein is uniformly distributed throughout the fertilized egg plasma membrane. Torso is, however, only activated at the ends of the fertilized egg because the protein ligand that stimulates it is only present there.

This ligand for Torso is thought to be a fragment of a secreted protein known as Trunk. *trunk* mRNA is deposited in the egg by the nurse cells and Trunk protein is secreted into the **perivitelline space** during oocyte development. The perivitelline space is the space between the oocyte plasma membrane and a protective membrane of extracellular matrix called the **vitelline membrane** or **vitelline envelope**, which surrounds the oocyte. The Trunk protein is thought to be present throughout the perivitelline space, but the Trunk fragment that acts as a ligand for Torso is generated only at the poles of the egg because the processing activity that produces it from Trunk is present only in these two regions. Crucial to this processing is the activity of a protein called Torso-like, which is produced exclusively by follicle cells adjacent to the two poles of the egg and is present in the vitelline envelope at the poles of the fertilized egg. By the time development begins after fertilization, small quantities of the Trunk ligand have been produced at the poles and are present in the perivitelline space, where they can bind Torso. Because Trunk ligand is only present in small quantities, most of it becomes bound to Torso at the poles, with little left to diffuse further away. In this way, a localized area of receptor activation is set up at each pole (Fig. 2.14).

Stimulation of Torso by its ligand produces a signal that is transmitted across the plasma membrane to the interior of the developing embryo. This signal directs the activation of zygotic genes in nuclei at both poles, thus defining the two extremities of the embryo. The Torso protein is one of a large superfamily of transmembrane receptors, known as receptor tyrosine kinases, whose cytoplasmic portions have tyrosine protein kinase activity. The kinase is activated when the extracellular part of the receptor binds its ligand, and the cytoplasmic part of the receptor transmits the signal onwards by phosphorylating cytoplasmic proteins.

The ingenious mechanism for setting up a localized area of receptor activation is not confined to determination of the terminal regions of the embryo, but is also used in setting up the dorso-ventral axis, which we consider next.

2.11 The dorso-ventral polarity of the embryo is specified by localization of maternal proteins in the egg vitelline envelope

The dorso-ventral axis is specified by a different set of maternal genes from those that specify the anterior-posterior axis. The basic mechanism is very similar to that described in the previous section. The receptor involved in dorso-ventral axis organization is a maternal protein called Toll, which is present throughout the cell membrane of the fertilized egg. The ventral end of the axis is determined by the localized production in the ventral vitelline envelope of the ligand for this receptor. The ligand is a protein fragment produced by processing of a maternal protein called Spätzle. After fertilization, Spätzle itself is uniformly distributed throughout the extra-embryonic perivitelline space. The localized processing of Spätzle is controlled by a small set of maternal genes that are expressed only in the follicle cells that surround the future ventral region—about a third of the total surface of the developing egg. The key gene here is *pipe*, which codes for an enzyme, a heparan sulfate sulfotransferase, that is secreted into the oocyte vitelline envelope by these cells. The subsequent activity of the Pipe enzyme leads, in some way not yet fully understood, to the localization of a protease activity in the vitelline envelope on the ventral side of the embryo. The processed Spätzle fragment is thus only present in the ventral perivitelline space.

Toll mRNA is laid down in the oocyte and is probably not translated until after fertilization. Although Toll is present throughout the membrane of the fertilized egg it is only activated in the future ventral region of the embryo, owing to the localization of its ligand. Toll activation is greatest where the concentration of its ligand is highest, and falls off rapidly, probably due to the limited amount of ligand being mopped up by the receptors. Activation of Toll sends a signal to the adjacent cytoplasm of the embryo. At this stage, the embryo is still a syncytial blastoderm and this signal causes a maternal gene product in the cytoplasm—the Dorsal protein—to enter nearby nuclei (Fig. 2.15). This protein is a transcription factor with a vital role in organizing the dorso-ventral axis.

2.12 Positional information along the dorso-ventral axis is provided by the Dorsal protein

The initial dorso-ventral organization of the embryo is established at right angles to the antero-posterior axis at about the same time that this axis is being divided into terminal, anterior, and posterior regions. The embryo initially becomes divided into four regions along the dorso-ventral axis and this patterning is controlled by the distribution of the maternal protein Dorsal (see Fig. 2.27).

Unlike Bicoid, Dorsal protein is uniformly distributed in the egg. Initially it is restricted to the cytoplasm, but under the influence of signals from the ventrally activated Toll proteins it enters nuclei in a graded fashion, with the highest concentration in ventral nuclei and the concentration progressively decreasing in a dorsal direction, as the Toll signal becomes weaker (see Fig. 2.15). Thus there is little or no Dorsal in nuclei in the dorsal regions of the embryo. The role of Toll was first established by the observation that mutant embryos lacking it are strongly **dorsalized**—that is, no ventral structures develop. In these embryos, Dorsal protein does not enter nuclei but remains uniformly distributed in the cytoplasm. Transfer of wild-type cytoplasm into *Toll*-mutant embryos results in specification of a new dorso-ventral axis, the ventral region always corresponding to the site of injection. This occurs because in the absence of Toll, the Spätzle fragments produced on the original ventral side diffuse throughout the perivitelline space because there is no Toll protein to bind them. When the wild-type cytoplasm is injected, the Toll proteins it contains enter the membrane

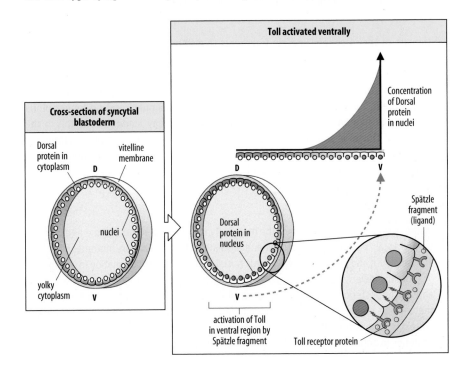

Fig 2.15 Toll protein activation results in a gradient of intranuclear Dorsal protein along the dorso-ventral axis. Before Toll protein is activated, the Dorsal protein (red) is distributed throughout the peripheral band of cytoplasm. The Toll protein is a receptor that is only activated in the ventral region, by a maternally derived ligand (the Spätzle fragment), which is processed in the perivitelline space after fertilization. The localized activation of Toll results in the entry of Dorsal protein into nearby nuclei. The intranuclear concentration of Dorsal protein is greatest in ventral nuclei, resulting in a ventral to dorsal gradient. D, dorsal; V, ventral.

Box 2B The Toll signaling pathway: a multifunctional pathway

The interaction between the Dorsal and Cactus proteins in the Toll signaling pathway of *Drosophila* is of more than local interest: Dorsal protein is a transcription factor with homology to the Rel/NFκB family of vertebrate transcription factors, which are involved in regulation of gene expression in immune responses, and the Toll signaling pathway is also used in the adult fly in defense against infection. So what might seem at first sight a rather specialized mechanism for confining transcription factors to the cytoplasm until it is time for them to enter the nucleus during embryonic development is likely to be widely used for controlling gene expression and cell differentiation.

The Toll signaling pathway is thus a good example of a conserved intracellular signaling pathway that is used by multicellular organisms in different contexts, in this case ranging from embryonic development to defense against disease. All members of the Rel/NFκB family are typically held inactive in the cytoplasm until the cell is stimulated through an appropriate receptor. This leads to degradation of the inhibitory protein, which releases the transcription factor. This then enters the nucleus and activates gene transcription (see figure). In the *Drosophila* embryonic Toll pathway, Dorsal is held inactive in the syncytium cytoplasm by the protein Cactus. When Toll is activated by binding the Spätzle fragment, its cytoplasmic domain binds an adaptor protein, dMyD88 (or Tube), which in turn interacts with and activates the protein kinase Pelle. Activation of Pelle leads, through several more intermediate steps, which have not yet been fully established, to the phosphorylation and degradation of Cactus. This releases Dorsal, which is then free to enter a nucleus.

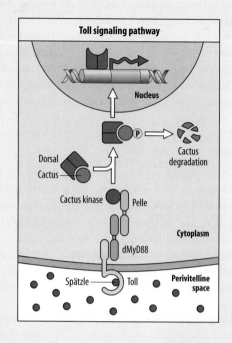

Toll signaling pathway

Nucleus

Cactus degradation

Dorsal
Cactus

Cactus kinase — Pelle

Cytoplasm

dMyD88

Spätzle — Toll   Perivitelline space

In adult *Drosophila*, the Toll receptor is stimulated by fungal and bacterial infection and the signaling pathway results in the production of antimicrobial peptides. In humans, Toll-like receptors acting by essentially the same pathway are also involved in innate immunity to microbial infection. IRAK, and IκB are the mammalian homologs of Pelle, and Cactus, respectively, and play the same roles in this pathway. In vertebrates, NFκB is also activated in response to signaling through receptors other than Toll.

at the site of the injection. Spätzle fragments then bind to these receptors, setting in motion the chain of events that defines the ventral region at the site of injection.

In the absence of a signal from Toll protein, Dorsal protein is prevented from entering nuclei by being bound in the cytoplasm to another maternal gene product, the Cactus protein. As a result of Toll activation, Cactus is degraded and no longer binds Dorsal, which is then free to enter the nuclei. The pathway leading from Toll to the activation of Dorsal is shown in Box 2B. In embryos lacking Cactus protein, almost all of the Dorsal protein is found in the nuclei; there is a very poor concentration gradient and the embryos are **ventralized**—that is, no dorsal structures develop.

Summary

Maternal genes act in the ovary of the mother fly to set up differences in the egg in the form of localized deposits of mRNAs and proteins. After fertilization, maternal mRNAs are translated and provide the embryonic nuclei with positional information in the form of protein gradients or localized protein. Along the antero-posterior axis there is an anterior to posterior gradient of maternal Bicoid protein, which controls patterning of the anterior region. For normal development it is essential that maternal Hunchback protein is absent from the posterior region and its suppression is the function of the posterior to anterior gradient of Nanos. The extremities of the embryo are specified by localized activation of the receptor protein Torso at the poles. The dorso-ventral axis is established by intranuclear localization of the Dorsal protein in a graded manner (ventral to dorsal), as a result of ventrally localized activation of the receptor protein Toll by a fragment of the protein Spätzle.

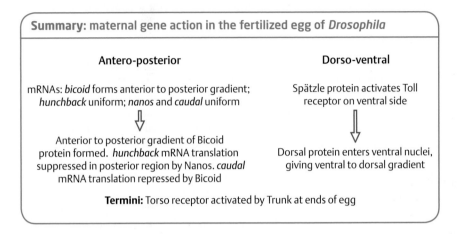

Summary: maternal gene action in the fertilized egg of *Drosophila*

Antero-posterior	Dorso-ventral
mRNAs: *bicoid* forms anterior to posterior gradient; *hunchback* uniform; *nanos* and *caudal* uniform	Spätzle protein activates Toll receptor on ventral side
⇩	⇩
Anterior to posterior gradient of Bicoid protein formed. *hunchback* mRNA translation suppressed in posterior region by Nanos. *caudal* mRNA translation repressed by Bicoid	Dorsal protein enters ventral nuclei, giving ventral to dorsal gradient

Termini: Torso receptor activated by Trunk at ends of egg

Localization of maternal determinants during oogenesis

Having considered the importance of localized maternal gene products in the egg in setting the basic framework for development, we now look at how they come to be localized so precisely. When the *Drosophila* egg is released from the ovary it already has a well-defined organization. *bicoid* mRNA is located at the anterior end and *nanos* and *oskar* mRNAs at the opposite end. Torso-like protein is present in the vitelline envelope at both poles, and other maternal proteins are localized in the ventral vitelline envelope. Numerous other maternal mRNAs, such as those for *caudal*, *hunchback*, *Toll*, *torso*, *dorsal*, and *cactus*, are distributed uniformly. How do these maternal mRNAs and proteins get laid down in the egg during **oogenesis**—its period of development in the ovary—and how are they localized in the correct places?

The development of an egg in the *Drosophila* ovary is shown in Fig. 2.16. A diploid germline stem cell in the **germarium** divides asymmetrically to produce another stem cell and a cell called the cystoblast, which undergoes a further four mitotic divisions to give 16 cells with cytoplasmic bridges between them; this group of cells is known as the **germline cyst**. One of these 16 cells will become the **oocyte**; the other 15 will develop into **nurse cells**, which produce large quantities of proteins and RNAs that are exported into the oocyte through the cytoplasmic bridges. Somatic ovarian cells make up a sheath of **follicle cells** around the nurse cells and oocyte to form the **egg chamber**. The follicle cells have a key role in patterning the egg's axes. During oogenesis, they become subdivided into functionally different populations at various locations around

Fig 2.16 Egg development in *Drosophila*. Oocyte development begins in a germarium, with stem cells at one end. One stem cell will divide four times to give 16 cells with cytoplasmic connections between each other. One of the cells that is connected to four others will become the oocyte, the others will become nurse cells. The nurse cells and oocyte become surrounded by follicle cells and the resulting structure buds off from the germarium as an egg chamber. Successively produced egg chambers are still attached to each other at the poles. The oocyte grows as the nurse cells provide material through the cytoplasmic bridges. The follicle cells have a key role in patterning the oocyte.

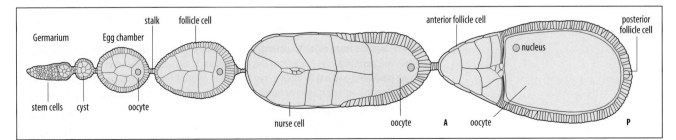

Fig 2.17 *Drosophila* oocyte development. A developing *Drosophila* oocyte (right) is shown attached to its 15 nurse cells (left) and surrounded by a monolayer of 700 follicle cells. The oocyte and follicle layer are cooperating at this time to define the future dorso-ventral axis of the egg and embryo, as indicated by the expression of a gene only in the follicle cells overlying the dorsal anterior region of the oocyte (blue staining).

Photograph courtesy of A. Spradling.

the oocyte; these subpopulations express different genes and thus have differing effects on the parts of the oocyte adjacent to them (Fig. 2.17). Follicle cells also secrete the materials of the vitelline envelope and egg case that surround the mature egg. During most of the stages of development discussed here the oocyte is arrested in the first prophase stage of meiosis (see Fig. 9.8). Meiosis is completed after fertilization.

2.13 The antero-posterior axis of the *Drosophila* egg is specified by signals from the preceding egg chamber and by interactions of the oocyte with follicle cells

The antero-posterior axis is the first axis to be established in the oocyte. The first visible sign of antero-posterior polarity is the movement of the oocyte from a central position surrounded by nurse cells to the posterior end of the developing egg chamber, in direct contact with follicle cells. This rearrangement occurs while the egg chamber is becoming separated from the germarium, and is caused by preferential adhesion between the future posterior end of the oocyte and the adjacent posterior follicle cells. Oocyte antero-posterior polarity is the result of signaling from the anterior of the older egg chamber to the posterior of the younger chamber (Fig. 2.18). Two signaling pathways are involved and act in relay. The germline cyst in the older chamber first signals to its anterior follicle cells via a widely used signaling pathway, the Delta–Notch pathway, which will be described in detail later in the book (see Box 5C, p. 190). This signaling specifies several of the follicle cells to become specialized anterior polar follicle cells. These specialized cells in turn signal to adjacent follicle cells via receptors that stimulate another intracellular signaling pathway called the JAK–STAT pathway and induce them to form a stalk between the two egg chambers. Signals, as yet unknown, from the stalk cells then cause the younger egg chamber to round up and cause the oocyte and the posterior follicle cells in this chamber to express the adhesion molecule E-cadherin, which anchors the oocyte in a posterior position. (The way in which cadherins and other adhesion molecules work is explained in Box 8B, p. 293.) Thus, an antero-posterior polarity is propagated from one egg chamber to the next. How the very first egg chamber produced gets polarized is, however, not yet understood.

Once the oocyte has become located posteriorly, the next steps in its antero-posterior polarization are mediated by a protein called Gurken, which is a member of the transforming growth factor-α (TGF-α) family (examples of the most commonly used growth factor families in *Drosophila* are given in Fig. 2.19; we shall meet

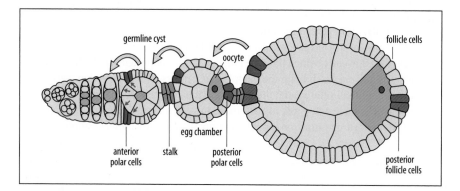

Fig 2.18 Signals from older to younger egg chambers initially polarize the *Drosophila* oocyte. As a germline cyst buds from the germarium, it signals through the Delta-Notch pathway (small red arrows) to induce the formation of anterior polar follicle cells (red). These in turn signal to the adjacent cells anterior to them and induce them to become stalk (green). The stalk induces the adjacent younger cyst to round up and the oocyte and the posterior follicle cells to produce cadherin, which positions the oocyte at the posterior of the egg chamber. The yellow arrows indicate the overall direction of signaling from older to younger egg chambers.

Common intercellular signals used in *Drosophila*		
Family and examples	**Receptors**	**Examples of roles in development**
Hedgehog family		
Hedgehog	Patched (together with Smoothened)	Patterning of insect segments Positional signaling in insect leg and wing discs (see Chapter 11)
Wingless (Wnt) family		
Wingless and six other Wnt proteins	Frizzled (together with LRP6)	Insect segment and imaginal disc specification (see Chapter 11) Other Wnts have roles in development
Delta and Serrate		
Transmembrane signaling proteins	Notch	Roles at many stages in development Specification of oocyte polarity
Transforming growth factor-a (TGF-a) family		
Gurken, Spitz, Vein	EGF receptor (receptor tyrosine kinase) One only in *Drosophila*, known as DER or Torpedo.	Polarization of the oocyte Eye development, wing vein differentiation (see Chapter 11)
Transforming growth factor-b (TGF-b) family		
Decapentaplegic	Receptors are heterodimers of type I (e.g. Thick veins) and type II (e.g. Punt) subunits Serine/threonine kinases	Patterning of the dorso-ventral axis Patterning of imaginal discs (see Chapter 11)
Fibroblast growth factor (FGF) family		
A small number of FGF homologs (e.g. Branchless)	FGF receptors (receptor tyrosine kinases) Two in *Drosophila*, e.g. Breathless	Migration of tracheal cells (see Chapter 11)

Fig 2.19 Common intercellular signals in *Drosophila*.

many of these again in vertebrates and other animals). Early in oocyte development, *gurken* mRNA is translated at the posterior end of the oocyte close to the nucleus, producing a local posterior concentration of the protein, which is secreted across the oocyte membrane. Gurken induces the terminal follicle cells to adopt a posterior fate by locally stimulating the receptor protein Torpedo, which is present on the surface of follicle cells (Fig. 2.20). Torpedo is a receptor tyrosine kinase and is the *Drosophila* equivalent of the epidermal growth factor (EGF) receptor of mammals. In response to Gurken signaling through Torpedo, the posterior follicle cells produce an as-yet-unidentified signal that induces a reorientation of the oocyte's microtubule cytoskeleton, so that about midway through oogenesis, the minus ends of most microtubules are directed toward the anterior and the plus ends toward the posterior.

2.14 Localization of maternal mRNAs to either end of the egg depends on the reorganization of the oocyte cytoskeleton

The reorientation of the oocyte microtubule cytoskeleton and the correct localization of mRNAs depends on a group of proteins known as the PAR proteins, which are themselves localized at the posterior end of the oocyte. Among other functions, these proteins are involved in determining antero-posterior polarity of cells in many different situations in animal development. They were discovered in *Caenorhabditis elegans* and are discussed in more detail in Chapter 6, in connection with their role in controlling the first asymmetric division of the fertilized nematode egg.

Microtubule reorientation is essential for the final localization of maternal mRNAs, such as *bicoid* and *oskar* at either end of the egg. *bicoid* mRNA is originally made by nurse cells located next to the anterior end of the developing oocyte, and is transferred from them to the oocyte. After the developing egg responds to the Gurken signal, *bicoid* mRNA is transported along the reoriented microtubules, probably by the motor protein dynein, to its final location at the anterior-most end of the egg

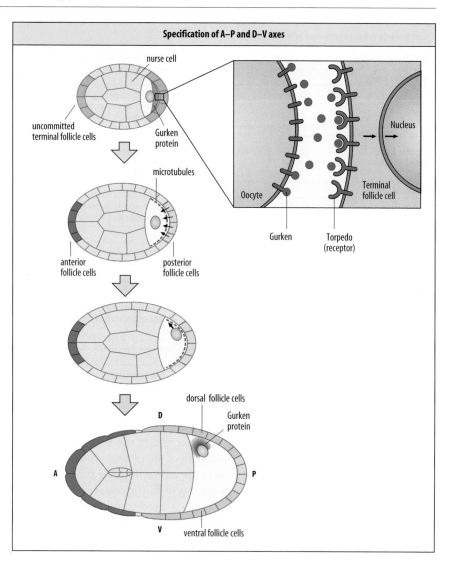

Fig 2.20 Specification of the antero-posterior and dorso-ventral axes during *Drosophila* oogenesis. The oocyte moves to the posterior end of the egg chamber and comes into contact with the polar follicle cells. It is separated from follicle cells at the anterior end (blue) by the nurse cells. *gurken* mRNA in the oocyte is localized to the posterior. It is translated and Gurken protein is secreted locally. The binding of this protein to the receptor protein Torpedo on the adjacent follicle cells initiates their specification as posterior terminal follicle cells (yellow). These send a signal back to the oocyte that reorganizes the oocyte microtubule cytoskeleton. The oocyte nucleus now moves to an anterior dorsal position and *gurken* mRNA coming into the oocyte from the nurse cells is transported to the region surrounding the nucleus. Translation of the mRNA and local release of Gurken protein specifies the adjacent follicle cells as dorsal follicle cells and that side of the oocyte as the future dorsal side.

(Fig. 2.21, first panel). Similarly, *oskar* mRNA is delivered into the oocyte by nurse cells and transported towards the posterior end of the oocyte by kinesin via another set of microtubules (see Fig. 2.20, second panel). Both these localizations require the RNA-binding protein Staufen, which can itself be observed to localize to the posterior and anterior ends of the oocyte (Fig. 2.22). Observations of *oskar* mRNA particles showed that they were moved in all directions, but with a sufficient bias towards the posterior end. One role of *oskar* mRNA and protein is to nucleate the assembly of the germplasm at the posterior end of the egg. In the embryo this cytoplasm is incorporated into the pole cells (see Section 2.1) and directs them to form primordial germ cells. The localization of *oskar* mRNA and assembly of the germplasm is also required for the subsequent localization of *nanos* mRNA to the posterior end (see Section 2.9). Unlike the transport of *bicoid* and *oskar* mRNA, *nanos* localization depends on the actin microfilaments of the cytoskeleton.

2.15 The dorso-ventral axis of the egg is specified by movement of the oocyte nucleus followed by signaling between oocyte and follicle cells

The setting-up of the egg's dorso-ventral axis involves a further set of oocyte–follicle cell interactions, which occur after the posterior end of the oocyte has been specified

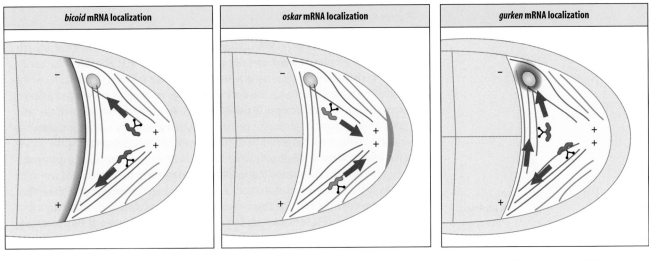

| bicoid mRNA localization | oskar mRNA localization | gurken mRNA localization |

and which depend on the previous reorganization of the microtubule array. The oocyte nucleus moves along the microtubules from the posterior of the oocyte to a site on the anterior margin (see Fig. 2.20). Gurken protein is expressed at this new site, possibly from mRNA that has been relocated to one side of the nucleus from elsewhere in the oocyte (Fig. 2.21, third panel). The locally produced Gurken acts as a signal to the adjacent follicle cells, specifying them as dorsal follicle cells; the side away from the nucleus thus becomes the ventral region by default. The ventral follicle cells produce proteins, such as Pipe (see Section 2.11), that are deposited in the ventral vitelline envelope of the oocyte and are instrumental in establishing the ventral side of the axis.

The Torso-like protein, which distinguishes the termini of the *Drosophila* embryo, is synthesized and secreted by follicle cells at both the posterior and anterior poles, but not by the other follicle cells. Torso-like protein is thus deposited exclusively in the vitelline envelope at both ends of the egg during oogenesis. After fertilization it acts together with other proteins to cause the processing of Trunk and production of the ligand for Torso (see Section 2.10).

Fig 2.21 *bicoid* and *oskar* mRNAs are localized to the anterior and posterior ends of the oocyte respectively. Localization of maternal mRNAs delivered to the oocyte by the nurse cells is by transport along microtubules. The motor protein dynein transports *bicoid* and *gurken* mRNAs towards the minus ends of the microtubules. *oskar* mRNA is transported towards the plus ends of the microtubules, possibly by the motor protein kinesin. Three different populations of microtubules are thought to be involved in the transport of *bicoid*, *oskar* and *gurken* mRNAs.

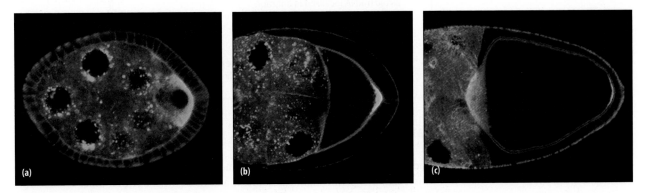

Fig 2.22 The RNA-binding protein Staufen localizes to each pole of the *Drosophila* oocyte. The location of the Staufen protein at different stages of oocyte development is made visible by linking its gene to the coding sequence of green fluorescent protein (GFP) and making flies transgenic for this construct. a, In a stage 6 egg chamber, Staufen (green) accumulates in the oocyte. The fixed egg chambers were counterstained with rhodamine-phalloidin, which labels actin filaments (red). b, In a stage 9 egg chamber, Staufen localizes to the posterior of the oocyte. c, In a stage 10b egg chamber, Staufen is localized to both the anterior and posterior poles of the oocyte.
Photographs reproduced from Martin, S.G., et al.: 2003.

Summary

Drosophila oocytes develop inside individual egg chambers that are successively produced from a germarium, which contains germline stem cells that give rise to the oocyte and nurse cells, and stem cells that give rise to the somatic follicle cells that surround the oocyte. The nurse cells provide the oocyte with large amounts of mRNAs and proteins, some of which become localized in particular sites. As a result of signals from the adjacent older egg chamber, an oocyte becomes localized posteriorly in its egg chamber as a result of differential cell adhesion to posterior follicle cells. Subsequently the oocyte sends a signal to these follicle cells, which respond with a signal that causes a reorganization of the oocyte cytoskeleton that localizes *bicoid* mRNA to the anterior end and other mRNAs to the posterior end of the oocyte, thus setting up the beginnings of the embryonic antero-posterior axis. The dorso-ventral axis of the oocyte is also initiated by a local signal from the oocyte to follicle cells on the future dorsal side of the egg, thus specifying them as dorsal follicle cells. Thus, directly or indirectly, follicle cells on the opposite side of the oocyte specify the ventral side of the oocyte by deposition of maternal proteins in the ventral vitelline envelope. Follicle cells at either end of the oocyte similarly specify the termini by localized deposition of maternal protein in the vitelline envelope.

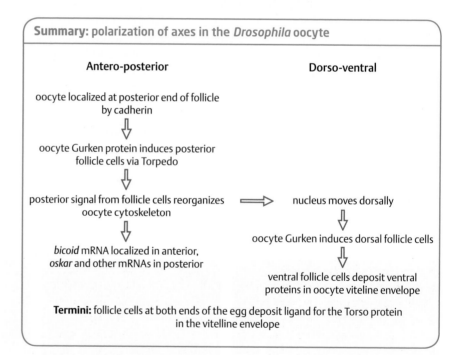

Summary: polarization of axes in the *Drosophila* oocyte

Antero-posterior	Dorso-ventral
oocyte localized at posterior end of follicle by cadherin	
⇩	
oocyte Gurken protein induces posterior follicle cells via Torpedo	
⇩	
posterior signal from follicle cells reorganizes oocyte cytoskeleton ⟹	nucleus moves dorsally
⇩	⇩
bicoid mRNA localized in anterior, *oskar* and other mRNAs in posterior	oocyte Gurken induces dorsal follicle cells
	⇩
	ventral follicle cells deposit ventral proteins in oocyte vitelline envelope

Termini: follicle cells at both ends of the egg deposit ligand for the Torso protein in the vitelline envelope

Patterning the early embryo

Understanding in such detail how the main body axes of *Drosophila* are specified was a major experimental achievement, and those who work on other animals like the frog and chick, whose developmental genetics are less well-known, are justifiably somewhat envious. We have seen how gradients of Bicoid, Hunchback, and Caudal proteins are established along the antero-posterior axis, and how intranuclear Dorsal protein is graded along the ventral to dorsal axis. This maternally derived framework of positional information is interpreted and elaborated on by zygotic genes to give each region of the embryo an identity. Most of the zygotic genes that are the first to be activated along the antero-posterior and dorso-ventral axes encode transcription factors, which are thus localized along the axes and activate yet more zygotic genes. We first consider patterning along the antero-posterior axis.

2.16 The antero-posterior axis is divided up into broad regions by gap-gene expression

Patterning along the antero-posterior axis starts when the embryo is still acellular. The **gap genes** are the first zygotic genes to be expressed along the antero-posterior axis, and all code for transcription factors. Gap genes were initially recognized by their mutant phenotypes, in which quite large sections of the body pattern along the antero-posterior axis are missing. Although the mutant phenotype of a gap gene usually shows a gap in the antero-posterior pattern in more-or-less the region in which the gene is normally expressed, there are also more wide-ranging effects. This is because gap-gene expression is also essential for later development along the axis.

Gap-gene expression is initiated by the antero-posterior gradient of Bicoid protein while the embryo is still essentially a single multinucleate cell. Bicoid primarily activates anterior expression of the gap gene *hunchback*, which in turn is instrumental in switching on the expression of the other gap genes, among which are *giant*, *Krüppel*, and *knirps*, which are expressed in this sequence along the antero-posterior axis (Fig. 2.23) (*giant* is in fact expressed in two bands, one anterior and one posterior, but its posterior expression does not concern us here).

As the blastoderm is still acellular at this stage, the gap–gene proteins can diffuse away from their site of synthesis. They are short-lived proteins with half-lives of minutes. Their distribution, therefore, extends only slightly beyond the region in which the gene is expressed, and this typically gives a bell-shaped protein concentration profile. Zygotic hunchback protein is an exception, as it is expressed fairly uniformly over a broad anterior region with a steep drop in concentration at the posterior boundary of its expression. The control of zygotic *hunchback* expression by Bicoid is best understood and will be considered first.

2.17 Bicoid protein provides a positional signal for the anterior expression of zygotic *hunchback*

The Bicoid protein induces expression of the zygotic *hunchback* genes over most of the anterior half of the embryo. This zygotic expression is superimposed on low levels of Hunchback protein expressed from the ubiquitous maternal *hunchback* mRNA, whose translation is suppressed posteriorly by Nanos (see Section 2.9).

The localized anterior expression of Hunchback is an interpretation of the positional information provided by the Bicoid protein gradient. The *hunchback* gene is switched on only when Bicoid, a transcription factor, is present above a certain threshold concentration. This level is attained only in the anterior half of the embryo, close to the site of Bicoid synthesis, which restricts *hunchback* expression to this region.

The relationship between Bicoid concentration and *hunchback* gene expression can be illustrated by looking at how *hunchback* expression changes when the Bicoid concentration gradient is changed by increasing the maternal dosage of the *bicoid* gene (Fig. 2.24). Expression of *hunchback* then extends more posteriorly because the region in which the concentration of Bicoid is above the threshold for *hunchback* activation is also extended in this direction.

A smooth gradient in Bicoid protein is translated into a sharp boundary of *hunchback* expression about halfway along the embryo. Neighboring nuclei along the antero-posterior axis experience Bicoid concentrations that are very similar, differing by only

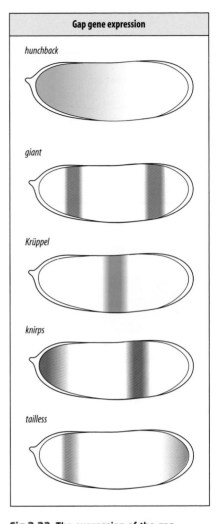

Gap gene expression

hunchback

giant

Krüppel

knirps

tailless

Fig 2.23 The expression of the gap genes *hunchback*, *Krüppel*, *giant*, *knirps*, and *tailless* in the early *Drosophila* embryo. Gap-gene expression at different points along the antero-posterior axis is controlled by the concentration of Bicoid and Hunchback proteins, together with interactions between the gap genes themselves. The expression pattern of the gap genes provides an aperiodic pattern of transcription factors along the antero-posterior axis, which delimits broad body regions.

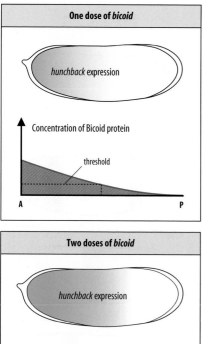

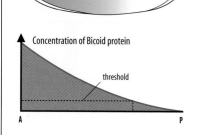

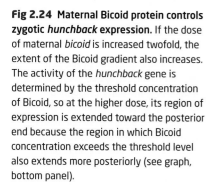

Fig 2.24 Maternal Bicoid protein controls zygotic *hunchback* expression. If the dose of maternal *bicoid* is increased twofold, the extent of the Bicoid gradient also increases. The activity of the *hunchback* gene is determined by the threshold concentration of Bicoid, so at the higher dose, its region of expression is extended toward the posterior end because the region in which Bicoid concentration exceeds the threshold level also extends more posteriorly (see graph, bottom panel).

about 10% from one cell to the next. How do nuclei distinguish such small differences and form a sharp boundary of gene expression against a background of biological 'noise', such as that resulting from small random variations in the dynamics of Bicoid behavior. This question has been addressed by using expression of a Bicoid–GFP construct (see Section 2.8) to directly observe and measure the distribution of Bicoid protein and its concentrations in nuclei at different points along the antero-posterior axis. Bicoid distribution, intranuclear concentrations at different positions, and the sharp boundary of *hunchback* expression all turn out to be highly reproducible over a large number of embryos, which suggests that background noise is in fact kept low by precise controls in the embryo. These might, for example, include communication between nuclei.

Bicoid is a member of the homeodomain family of transcriptional activators and activates the *hunchback* gene by binding to regulatory sites within the promoter region. Direct evidence of *hunchback* activation by Bicoid was originally obtained by experiments in which a fusion gene constructed from the *hunchback* promoter regions and a bacterial reporter gene, *lacZ*, was introduced into the fly genome using transgenic techniques such as P-element-mediated transformation (Box 2C, p. 60; some other related techniques for tracking gene expression and forcing new patterns of gene expression in *Drosophila* are shown in Box 2D, p. 61). The large promoter region required for completely normal gene expression can be whittled down to an essential sequence of 263 base pairs that will still give almost normal activation of *hunchback*. This sequence has several sites at which Bicoid can bind, and it seems likely that the threshold response involves **cooperativity** between the different binding sites: that is, binding of Bicoid at one site makes binding of Bicoid at a nearby site easier and so facilitates further binding.

The regulatory region of a gene such as *hunchback* is yet another example of a developmental switch that directs nuclei along a new developmental pathway. We will encounter many more examples of these transcriptional switches in *Drosophila* early development.

2.18 The gradient in Hunchback protein activates and represses other gap genes

The Hunchback protein is a transcription factor and acts as a morphogen to which the other gap genes respond. The other gap genes are expressed in transverse stripes across the antero-posterior axis (see Fig. 2.23). The stripes are delimited by mechanisms that depend on the control regions of these genes being sensitive to different concentrations of Hunchback protein, and also to other proteins, including Bicoid. Expression of the *Krüppel* gene, for example, is activated by low levels of Hunchback, but is repressed at high concentrations. Within this concentration window *Krüppel* remains activated (Fig. 2.25, top panel). But below a lower threshold concentration of Hunchback, *Krüppel* is not activated. In this way, the gradient in Hunchback protein locates a band of *Krüppel* gene activity near the center of the embryo (Fig. 2.26). Refinement of this spatial localization is brought about by repression of *Krüppel* by other gap-gene proteins.

Such relationships were worked out by altering the concentration profile of Hunchback protein systematically, while all other known influences were eliminated or held constant. Increasing the dose of Hunchback protein, for example, results in a posterior shift in its concentration profile, and this results in a posterior shift in the posterior boundary of *Krüppel* expression. In another set of experiments on embryos lacking Bicoid protein (so that only the maternal Hunchback protein gradient is present), the level of Hunchback is so low that *Krüppel* is even activated at the anterior end of the embryo (Fig. 2.25, bottom panel).

Fig 2.25 *Krüppel* **gene activity is specified by Hunchback protein.** Top panel: above a threshold concentration of Hunchback protein, the *Krüppel* gene is repressed; at a lower concentration, above another threshold value, it is activated. Bottom panel: in mutants lacking the *bicoid* gene, and thus also lacking zygotic *hunchback* gene expression, only maternal Hunchback protein is present, which is located at the anterior end of the embryo at a relatively low level. In these mutants, *Krüppel* is activated at the anterior end of the embryo, giving an abnormal pattern.

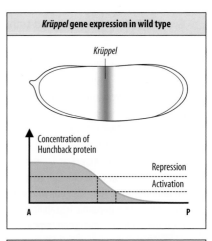

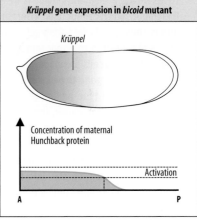

Hunchback protein is also involved in specifying the anterior borders of the bands of expression of the gap genes *knirps* and *giant*, again by a mechanism involving thresholds for repression and activation of these genes. At high concentrations of Hunchback, *knirps* is repressed, and this specifies its anterior margin of expression. The posterior margin of the *knirps* band is specified by a similar type of interaction with the product of another gap gene, *tailless*. Where the regions of expression of the gap genes overlap, there is extensive cross-inhibition between them, their proteins all being transcription factors. These interactions are essential to sharpen and stabilize the pattern of gap gene expression. For example, the anterior border of *Krüppel* expression lies four to five nuclei posterior to nuclei that express *giant*, and is set by low levels of Giant protein.

The antero-posterior axis thus becomes divided into a number of unique regions on the basis of the overlapping and graded distributions of different transcription factors. This beautifully elegant method of delimiting regions can, however, only work in an embryo such as the acellular syncytial blastoderm of *Drosophila*, where the transcription factors are able to diffuse freely throughout the embryo. This regional distribution of the gap-gene products provides the starting point for the next stage in development along the antero-posterior axis—the activation of the pair-rule genes, cellularization, and the beginning of segmentation. But before we discuss this, we return to the dorso-ventral axis at around this time point in development and look at how it is patterned further as the syncytial blastoderm becomes cellular.

2.19 The expression of zygotic genes along the dorso-ventral axis is controlled by Dorsal protein

After Dorsal protein has entered the nuclei of the syncytial blastoderm (see Section 2.12), its effects on gene expression divide the dorso-ventral axis into well defined regions. This is also the stage at which the germ layers become distinguished, as Dorsal also acts to specify the ventral-most cells as prospective mesoderm. It is estimated that 30 genes are direct targets of the Dorsal gradient. Going from ventral to dorsal, the main regions are mesoderm, ventral ectoderm, dorsal ectoderm, and prospective amnioserosa. The mesoderm gives rise to internal soft tissues such as muscle and connective tissue; the ventral ectoderm becomes the neuroectoderm, which gives rise to all the nervous tissue as well as ventral epidermis; the dorsal ectoderm gives rise only to epidermis. The third germ layer, the endoderm, which is located at either end of the embryo and which we do not consider here, gives rise to the midgut (see Fig. 2.3).

Patterning along the axes poses a problem like that of patterning the French flag (see Section 1.15). Expression of zygotic genes in localized regions along the dorso-ventral axis is initially controlled by the graded concentration of the intranuclear Dorsal protein, which falls off rapidly in the dorsal half of the embryo; little Dorsal protein is found in nuclei above the equator. In the ventral region, Dorsal protein has two main functions—it activates certain genes at specific positions in the region

Fig. 2.26 Localization of Krüppel expression by the Hunchback protein gradient. Wild-type embryos were stained with fluorescent antibodies to simultaneously detect the Hb protein gradient (red) and the expression of Krüppel mRNA (green) early in nuclear cycle 14.

Photograph reproduced from Yu, D., and Small, S.: 2008.

Box 2C P-element-mediated transformation

Transgenic fruit flies have contributed greatly to *Drosophila* developmental genetics. They are made by inserting a known sequence of DNA into the *Drosophila* chromosomal DNA, using as a carrier a **transposon** that occurs naturally in some strains of *Drosophila*. This transposon is known as a **P element**, and the technique as P-element-mediated transformation (see figure).

P elements can insert at almost any site on a chromosome, and can also hop from one site to another within the germ cells, an action that requires an enzyme called a transposase. As hopping can cause genomic instability, carrier P elements have had their own transposase gene removed. The transposase required to insert the P element initially is instead provided by a helper P element, which cannot itself insert into the host chromosomes and is thus quickly lost from cells. The carrier and helper elements are injected together into the posterior end of the egg where the germ cells are made.

As well as the gene to be inserted, an additional marker gene, such as the wild-type *white*+ gene, is added to the P element. When *white*+ is the marker, the P element is inserted into flies homozygous for the mutant *white*− gene (which have white eyes rather than the red eyes of the wild-type *Drosophila*). Red eyes are dominant over white, and so flies in which the P element has become integrated into the chromosome, and is being expressed, can be detected by their red eyes.

In the first generation, all flies have white eyes, as any P element that has integrated is still restricted to the germ cells. But in the second generation, a few flies will have wild-type red eyes, showing that they carry the inserted P element in their somatic cells.

This technique can be used to increase the number of copies of a particular gene, or to introduce a mutated gene that has its control or coding regions altered in a known way, or to introduce new genes. It is also possible to introduce genes that carry a marker coding sequence such as *lacZ* (encoding the bacterial enzyme β-galactosidase), whose expression is detectable by histochemical staining (see Box 1D, p. 20). The P element itself can also be used as a mutagen, as its insertion into a gene usually destroys that gene's function.

This approach has been adapted to large-scale screens that look systematically for genes whose overexpression or misexpression in a particular tissue causes a mutant phenotype (see Box 2D, p. 61). This is known as misexpression screening. In this case, flies expressing Gal4 in the tissue of interest are crossed with large numbers of different lines of target flies carrying random insertions of the Gal4-binding site, and the progeny screened for a mutant phenotype. This approach is a useful complement to the more conventional genetic screens described in Box 2A (p. 42), which generally detect loss-of-function mutations. If the target flies also carry a mutation in a known gene, misexpression screening can be used to identify genes whose overexpression enhances or suppresses the mutation. This approach can identify genes whose products interact directly or are part of the same pathway.

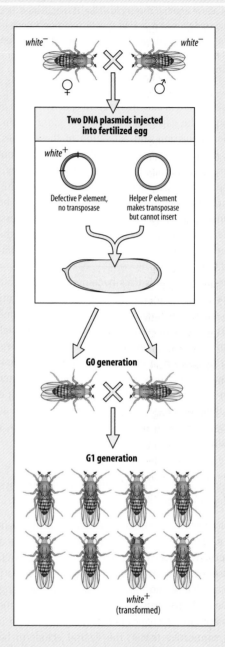

Two DNA plasmids injected into fertilized egg

white+ Defective P element, no transposase

Helper P element makes transposase but cannot insert

G0 generation

G1 generation

white+ (transformed)

Box 2D Targeted gene expression and misexpression screening

The ability to turn on the expression of a gene in a particular place and time during development is very useful for analyzing its role in development. This is called targeted gene expression and can be achieved in several ways. One approach is to give the selected gene a heat-shock promoter using standard genetic techniques such as P-element-mediated transformation (see Box 2C, p. 60). This enables the gene to be switched on by a sudden rise in the temperature at which the embryos are being kept. By adjusting the temperature, the timing of expression of genes attached to this promoter

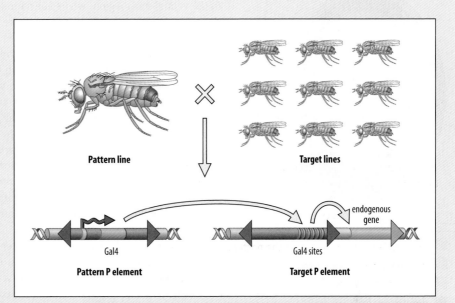

Pattern line　　　　　　**Target lines**

Gal4　　　　　　Gal4 sites

endogenous gene

Pattern P element　　　　　　**Target P element**

can be controlled; the effects of expressing a gene at different stages of development can be studied in this way.

Another approach to targeted gene expression uses the transcription factor Gal4 from yeast. This protein can activate transcription of any gene whose promoter has a Gal4-binding site. In *Drosophila*, genes with Gal4-responsive promoters can be created by inserting the Gal4-binding site. To turn on the target gene, Gal4 itself has to be produced in the embryo. In one approach, Gal4 can be produced in a designated region or at a particular time in development by introducing a P-element transgene in which the yeast Gal4-coding region is attached to a *Drosophila* regulatory region known to be activated in that situation (see top figure).

A second, more versatile, approach is based on the so-called **enhancer-trap** technique. The Gal4-coding sequence is attached to a vector that integrates randomly into the *Drosophila* genome. The Gal4 gene will come under the control of the promoter and enhancer region adjacent to its site of integration, and so Gal4 protein will be produced where or when that gene is normally expressed. A large number of *Drosophila* lines with differing patterns of Gal4 expression for different purposes have been produced by this technique. The selected target gene will be silent in the absence of Gal4. To activate it in a particular tissue, for example, flies that express Gal4 in that tissue are crossed with flies in which

the target gene has Gal4-binding sites in its regulatory region. The photo shows the use of this technique to investigate expression of *engrailed* in sensory neurons in the head of the adult fly. A *Drosophila* strain carrying an *engrailed*-GAL4 coding fusion transgene was crossed with a strain carrying a Gal4-responsive GFP transgene. In the progeny of these flies, GFP is expressed in those cells in which *engrailed* is expressed, making them glow green.

A novel pattern of gene expression can also be obtained using the Gal4 system. Using the Gal4 system, for example, the pair-rule gene *even-skipped* was expressed in even-numbered rather than odd-numbered parasegments, and this led to changes in the pattern of denticles on the cuticle.

Photograph from Blagburn, J.M.: 2008.

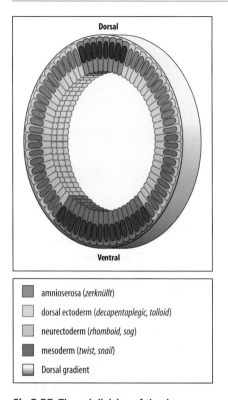

Dorsal

Ventral

- amnioserosa (*zerknüllt*)
- dorsal ectoderm (*decapentaplegic, tolloid*)
- neurectoderm (*rhomboid, sog*)
- mesoderm (*twist, snail*)
- Dorsal gradient

Fig 2.27 The subdivision of the dorso-ventral axis into different regions by the gradient in intranuclear Dorsal protein. A gradient of intranuclear Dorsal protein is formed with its high point at the ventral midline and little or no Dorsal protein in the dorsal half of the blastoderm. In the dorsal region, the genes *tolloid, zerknüllt,* and *decapentaplegic,* which are repressed elsewhere by Dorsal protein, are expressed. In the ventral half of the blastoderm, the Dorsal protein activates the genes *twist, snail, rhomboid,* and *short gastrulation* (*sog*). *twist* is autoregulatory, maintaining its own expression, and also activates *snail;* Snail protein inhibits *rhomboid* and *sog* expression, repressing these genes in the future mesoderm. *twist* and *snail* require high levels of Dorsal for their activation whereas *rhomboid* and *sog* can still be activated when lower levels of Dorsal are present.

and represses the activity of other genes, which are therefore only expressed in the dorsal region (Fig. 2.27).

In the ventral-most region, where concentrations of intranuclear Dorsal protein are highest, the zygotic genes *twist* and *snail* are activated by Dorsal protein in a strip of nuclei along the ventral middle line of the embryo; soon after this the syncytial blastoderm becomes cellular. This ventral strip of cells will form the mesoderm. The expression of *twist* and *snail* is required both for development of the cells as meso-derm and for gastrulation, during which the ventral band of prospective mesoderm cells moves into the interior of the embryo. In the future neuroectoderm, the gene *rhomboid* is activated at low levels of Dorsal protein, but is not expressed in more ventral regions because it is repressed there by the Snail protein.

The genes *decapentaplegic, tolloid,* and *zerknüllt* are repressed by Dorsal protein and so their activity is confined to the more dorsal regions of the embryo, where there is virtually no Dorsal protein in the nuclei. *zerknüllt* is expressed most dorsally and appears to specify the amnioserosa. *decapentaplegic* is a key gene in the specification of pattern in the dorsal part of the dorso-ventral axis and its role will be considered in detail in Section 2.20.

Mutations in the maternal dorso-ventral genes can cause dorsalization or ventral-ization of the embryo (see Section 2.12). In dorsalized embryos, Dorsal protein is excluded uniformly from the nuclei. This has a number of effects, one of which is that the *decapentaplegic* gene is no longer repressed and is expressed everywhere. In con-trast, *twist* and *snail* are not expressed at all in dorsalized embryos, as they need high intranuclear levels of Dorsal protein to be activated. The opposite result is obtained in mutant embryos where the Dorsal protein is present at high concentration in all the nuclei, and the embryos are ventralized; *twist* and *snail* are expressed throughout and *decapentaplegic* is not expressed at all (Fig. 2.28).

Genes whose expression is regulated by the Dorsal protein, such as *twist, snail,* and *decapentaplegic,* contain binding sites for the protein in their control regions that activate or repress gene expression at particular concentrations of the Dorsal protein. This threshold effect on gene expression is the result of the integrating func-tion of these cooperative binding sites, as discussed earlier in regard to the activation of *hunchback* by Bicoid. The ability of genes to respond in a threshold-like manner to varying concentrations of Dorsal protein is due to the presence of both high- and low-affinity binding sites for the protein in their regulatory DNA. In the most ventral part of the embryo (a strip 12–14 cells wide), where the concentration of Dorsal protein is high, the spatial extent of gene expression is delimited by the low-affinity sites, whereas high-affinity sites control expression in slightly more dorsal regions (up to 20 cells from the ventral midline) where there is less Dorsal protein. It is very likely that the threshold response involves cooperativity between the different binding sites; that is, binding at one site makes binding at a nearby site easier and so facili-tates further binding. Inhibitory interactions are also involved in delimiting areas of gene expression. For example, the regulatory regions of the *rhomboid* gene contain modules with binding sites for both Dorsal and Snail proteins; Dorsal activates while Snail represses. Snail protein thus represses the expression of *rhomboid* in the ventral region and this helps confine *rhomboid* expression to the neuroectoderm.

An approximately twofold difference in the level of Dorsal determines whether an unspecified embryonic cell forms mesoderm or neuroectoderm. Five different thresh-olds in Dorsal concentration pattern the future ventral midline and the neuroectoderm. The neuroectoderm, for example, is subsequently subdivided into three layers along the dorso-ventral axis, which will form three distinct columns in the future nerve cord (we shall return to this topic in Chapter 12). This subdivision is primarily the result of activation of genes for three different transcription factors at distinct thresholds of the Dorsal gradient, and the pattern is maintained by regulatory interactions between

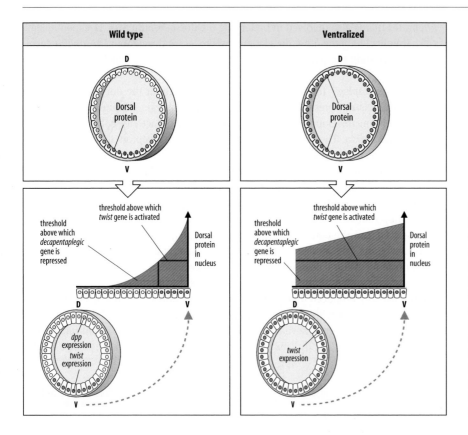

Fig 2.28 The nuclear gradient in Dorsal protein is interpreted by the activation of other genes, such as *twist* and *decapentaplegic*. Left panels: in normal embryos, the *twist* gene is activated above a certain threshold concentration (green line) of Dorsal protein, whereas above a lower threshold (yellow line), the *decapentaplegic* (*dpp*) gene is repressed. Right panels: in ventralized embryos, the Dorsal protein is present in all nuclei; *twist* is now also expressed everywhere, whereas *decapentaplegic* is not expressed at all, because Dorsal protein is above the threshold level required to repress it everywhere.

these genes and with other genes, in which genes expressed in the more ventral regions tend to repress those expressed more dorsally.

The gradient of Dorsal protein is therefore effectively acting as a morphogen gradient along the dorso-ventral axis, activating specific genes at different threshold concentrations, and so defining the dorso-ventral pattern. The regulatory sequences in these genes can be thought of as developmental switches, which when thrown by the binding of transcription factors activate genes and set cells off along new developmental pathways. The Dorsal protein gradient is one solution to the French flag problem. But it is not the whole story—yet another gradient is also involved.

2.20 The Decapentaplegic protein acts as a morphogen to pattern the dorsal region

As with the antero-posterior axis, each end of the dorso-ventral axis is specified by different proteins. The gradient of Dorsal protein, with its high point in the ventral-most region, specifies the initial pattern of zygotic gene activity, and patterns the ventral mesoderm and neuroectoderm. But the dorsal region is not similarly specified by a low-level gradient in Dorsal. Indeed, there is little or no Dorsal protein in the nuclei of the dorsal half of the embryo. The more dorsal part of the dorso-ventral pattern is determined by a gradient in the activity of another morphogen, the Decapentaplegic protein. This specifies the dorsal ectoderm and the most dorsal region, the amnioserosa.

Soon after the gradient of intranuclear Dorsal protein has become established the embryo becomes cellular, and transcription factors can no longer diffuse between nuclei. Secreted or transmembrane proteins and their corresponding receptors must now be used to transmit developmental signals between cells. Decapentaplegic is one such secreted signaling protein (see Fig. 2.19). As we shall see in Chapter 4, Decapentaplegic is a homolog of bone morphogenetic protein-4 (BMP-4), a TGF-β cytokine in vertebrates that is also involved in patterning the dorso-ventral axis. Decapentaplegic is also involved in many other developmental processes throughout *Drosophila* development, including the patterning of the wing imaginal discs, which is discussed in Chapter 11.

The *decapentaplegic* (*dpp*) gene is expressed throughout the dorsal region where Dorsal protein is not present in the nuclei. Decapentaplegic protein (Dpp) then diffuses away from this source, forming a gradient of Dpp activity with its high point in the dorsal region. The cells in the dorsal region have receptors that allow them to accurately assess how much Dpp is present and to respond by activating transcription of the appropriate genes, thus dividing up the dorso-ventral axis into different regions characterized by different patterns of gene expression.

Evidence that a gradient in Dpp specifies dorsal pattern comes from experiments in which *dpp* mRNA is introduced into an early wild-type embryo. As more mRNA is introduced and the concentration of Decapentaplegic protein increases above the normal level, the cells along the dorso-ventral axis adopt a more dorsal fate than they would normally. Ventral ectoderm becomes dorsal ectoderm and, at very high concentrations of *dpp* mRNA all the ectoderm develops as the amnioserosa.

Dpp is initially produced uniformly throughout the dorsal region just as cellularization is beginning, but within less than an hour its activity appears to be restricted to a dorsal strip of some five to seven cells and is much lower in the adjacent prospective dorsal ectoderm (Fig. 2.29). This sharp peak of Dpp concentration is not a matter

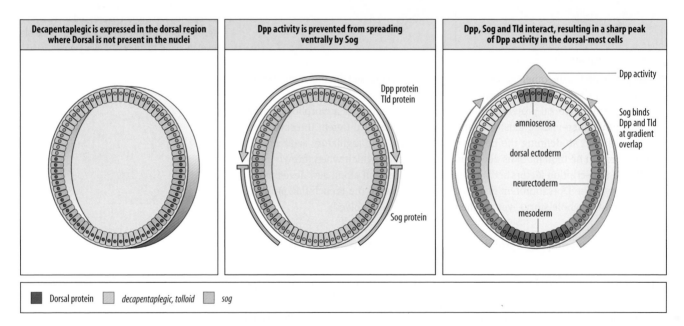

| Decapentaplegic is expressed in the dorsal region where Dorsal is not present in the nuclei | Dpp activity is prevented from spreading ventrally by Sog | Dpp, Sog and Tld interact, resulting in a sharp peak of Dpp activity in the dorsal-most cells |

Dpp protein
Tld protein

Sog protein

Dpp activity

Sog binds Dpp and Tld at gradient overlap

amnioserosa
dorsal ectoderm
neurectoderm
mesoderm

■ Dorsal protein   ▢ *decapentaplegic, tolloid*   ▢ *sog*

Fig 2.29 Decapentaplegic protein activity is restricted to the dorsal-most region of the embryo by the antagonistic activity of the Short gastrulation protein. Left panel: the gradient of intranuclear Dorsal protein leads to the expression of the gene *short gastrulation* (*sog*) in the non-neural ventral ectoderm of the early blastoderm. *decapentaplegic* and *tolloid* are expressed throughout the dorsal ectoderm. Middle panel: *sog* encodes a secreted protein, Sog, that forms a ventral to dorsal gradient, while the secreted Decapentaplegic protein (Dpp) is initially present throughout the dorsal region. The protease Tolloid (Tld) is also expressed in the same region as Dpp. Where Sog meets Dpp, it binds it and thus prevents Dpp interacting with its receptors. This prevents Dpp signaling from spreading ventrally into the neurectoderm. Right panel: Dpp bound to Sog is also thought to be carried towards the dorsal region by the developing Sog gradient, thus helping to concentrate its activity in a sharp peak in the dorsal-most region. Tld is also thought to be involved in creating the final sharp dorsal peak of Dpp activity. It binds to the Sog-Dorsal complex and cleaves Sog. This releases Dorsal, which can then bind to its receptors. In the mid-blastoderm embryo the dorsal region has been subdivided by Sog activity into a region of high Dpp signaling, which will become the amnioserosa, and a zone of lower Dpp signaling, which is the dorsal ectoderm.

After Ashe and Levine: 1999.

of simple diffusion, however; instead, it illustrates how a gradient in activity can be derived from a fairly uniform initial distribution of the protein by interactions with other proteins. In addition, Decapentaplegic activity is also thought to be modified by interactions with different forms of its receptors, further refining the activity gradient. It also interacts with collagen in the extracellular matrix, which restricts its movement and thus restricts its signaling range. Dpp is a good example of how the formation of a signal gradient is usually more complicated than simple diffusion.

Producing the sharp transition between the most dorsal ectoderm cells and those just lateral to it involves the proteins Short gastrulation (Sog), Twisted gastrulation (Tsg), and Tolloid. Sog and Tsg are BMP-related proteins that can bind Dpp and thus prevent it binding to its receptors, and in this sense are inhibitory. Tolloid is a metalloproteinase that degrades Sog when it is bound to Dpp, thus releasing Dpp.

Sog is expressed throughout the neurectoderm and binds Dpp, preventing its activity spreading into this region. Sog protein in turn is degraded by Tolloid, which is expressed throughout the dorsal region; this sets up a gradient in Sog with a high point in the neuroectoderm and a low point at the dorsal midline and ensures that Sog, along with its bound Dpp, will be shuttled towards the dorsal region. Degradation of Sog or Tsg by Tolloid releases free Dpp, which helps to generate a gradient of Dpp activity with its high point in the dorsal-most region (see Fig. 2.29). Another factor proposed to sharpen the gradient is the rapid internalization and degradation of Dpp when it binds to its receptors. Dpp activity is also affected by its synergistic action with another member of the TGF-β family, Screw, which forms heterodimers with Dpp that signal more strongly than homodimers of Dpp or Screw alone; these are likely to be responsible for most of the Dpp activity. Experimental observations and mathematical modeling suggest that heterodimers are formed preferentially in the dorsal-most region, which also helps explain the high Dpp activity here. Homodimers of Dpp and of Screw are responsible for lower levels of signaling elsewhere in the prospective dorsal ectoderm. Despite much work on this system however, the contributions of all these components to generating the gradient in Dpp activity are not completely understood.

The Dpp/Sog pairing has a counterpart in vertebrates, the BMP-4/Chordin partnership (Chordin is the vertebrate homolog of Sog), which is also involved in patterning the dorso-ventral axis. The nerve cord runs dorsally in vertebrates, however, not ventrally as in insects, and thus the dorso-ventral axis is reversed in vertebrates compared with insects; we shall see in Chapter 4 how this reversal in anatomical position is reflected in a reversal of the pattern of activity of BMP-4 and Chordin during early development.

Summary

Gradients of maternally derived transcription factors along the dorso-ventral and antero-posterior axes provide positional information that activates zygotic genes at specific locations along these axes. Along the antero-posterior axis the maternal gradient in Bicoid protein initiates the activation of zygotic gap genes to specify general body regions. Interactions between the gap genes, all of which code for transcription factors, help to define their borders of expression. The dorso-ventral axis becomes divided into four regions: ventral mesoderm, ventral ectoderm (neurectoderm), dorsal ectoderm (dorsal epidermis), and amnioserosa. A ventral to dorsal gradient of maternal Dorsal protein both specifies the ventral mesoderm and defines the dorsal region; a second gradient, of the Decapentaplegic protein, specifies the dorsal ectoderm. Patterning along the dorso-ventral and antero-posterior axes divides the embryo into a number of discrete regions, each characterized by a unique pattern of zygotic gene activity.

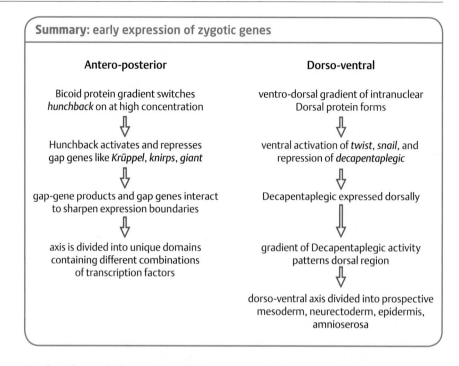

Summary: early expression of zygotic genes

Antero-posterior	Dorso-ventral
Bicoid protein gradient switches *hunchback* on at high concentration	ventro-dorsal gradient of intranuclear Dorsal protein forms
⇩	⇩
Hunchback activates and represses gap genes like *Krüppel, knirps, giant*	ventral activation of *twist, snail*, and repression of *decapentaplegic*
⇩	⇩
gap-gene products and gap genes interact to sharpen expression boundaries	Decapentaplegic expressed dorsally
⇩	⇩
axis is divided into unique domains containing different combinations of transcription factors	gradient of Decapentaplegic activity patterns dorsal region
	⇩
	dorso-ventral axis divided into prospective mesoderm, neurectoderm, epidermis, amnioserosa

Activation of the pair-rule genes and the establishment of parasegments

The most obvious feature of a *Drosophila* larva is the regular segmentation of the larval cuticle along the antero-posterior axis, each segment carrying cuticular structures that define it as, for example, thorax or abdomen. This pattern of segmentation is carried over into the adult, in which each segment has its own identity. Adult appendages such as wings, halteres, and legs are attached to particular segments (Fig. 2.30), but the discernible segments of the mature larva are not in fact the first units of segmentation along this axis. The basic developmental modules, whose definition we will follow in some detail, are the parasegments, which are specified first and from which the segments derive.

2.21 Parasegments are delimited by expression of pair-rule genes in a periodic pattern

The first visible signs of segmentation in the embryo are transient grooves that appear on the surface of the embryo after gastrulation. These grooves define the parasegments. There are 14 parasegments, which are the fundamental units in the segmentation of the *Drosophila* embryo. Once each parasegment is delimited, it behaves as an independent developmental unit, under the control of a particular set of genes, and in this sense at least the embryo can be thought of as being built up piecemeal of modules. The parasegments are initially similar but each will soon acquire its own unique identity. The parasegments are out of register with the final segments by about half a segment; each segment is made up of the posterior region of one parasegment and the anterior region of the next (see Fig. 2.30). In the anterior head region the segmental arrangement is lost when some of the anterior parasegments fuse.

The parasegments in the thorax and abdomen are delimited by the action of the **pair-rule genes**, each of which is expressed in a series of seven transverse stripes along the embryo, each stripe corresponding to every second parasegment. When pair-rule gene expression is visualized by staining for the pair-rule proteins, a striking zebra-striped embryo is revealed (Fig. 2.31).

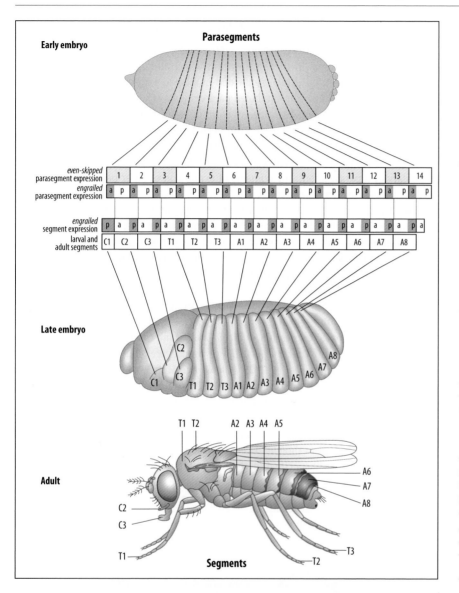

Fig 2.30 The relationship between parasegments and segments in the early embryo, late embryo, and adult fly. Initially, pair-rule genes are expressed in the embryo as stripes in every second parasegment. *evenskipped* (yellow), for example, is expressed in odd-numbered parasegments. The segmentation selector gene *engrailed* (blue) is expressed in the anterior region of every parasegment, and delimits the anterior margin of each parasegment. Each larval segment is composed of the posterior region of one parasegment and the anterior region of the next. The anterior region of a parasegment becomes the posterior portion of a segment. Segments are thus offset from the original parasegments by about half a segment, and *engrailed* is expressed in the posterior region of each segment. In this figure, a and p refer to the anterior and posterior compartments of segments or parasegments. The segment specification is carried over into the adult and results in particular appendages, such as legs and wings, developing on specific segments only. C1, C2, and C3 represent segments which become fused to form the head region. T, thoracic segments; A, abdominal segments. *After Lawrence, P.: 1992.*

Fig 2.31 The striped patterns of activity of pair-rule genes in the *Drosophila* embryo just before cellularization. Parasegments are delimited by pair-rule gene expression, each pair-rule gene being expressed in alternate parasegments. Expression of the pair-rule genes *even-skipped* (blue) and *fushi tarazu* (brown) is visualized by staining with antibody for their protein products. *even-skipped* is expressed in odd-numbered parasegments, *fushi tarazu* in even-numbered parasegments. Scale bar = 0.1 mm. *Photograph from Lawrence, P.: 1992.*

The positions of the stripes of pair-rule gene expression are determined by the pattern of gap-gene expression—a non-repeating pattern of gap-gene activity is converted into repeating stripes of pair-rule gene expression. We now consider how this is achieved.

2.22 Gap-gene activity positions stripes of pair-rule gene expression

Pair-rule genes are expressed in stripes, with a periodicity corresponding to alternate parasegments. Mutations in these genes thus affect alternate segments. Some pair-rule genes (e.g. *even-skipped, eve*) define odd-numbered parasegments, whereas others (e.g. *fushi tarazu*) define even-numbered parasegments. The striped pattern of expression of pair-rule genes is present even before cells are formed, while the embryo is still a syncytium, although cellularization occurs soon after expression begins. Each pair-rule gene is expressed in seven stripes, each of which is only a few cells wide. For some genes, such as *eve*, the anterior margin of the stripe corresponds to the anterior boundary of a parasegment; the domains of expression of other pair-rule genes, however, cross parasegment boundaries.

The striped expression pattern appears gradually; the stripes of the *eve* gene are initially fuzzy, but eventually acquire a sharp anterior margin. At first sight this type of patterning would seem to require some underlying periodic process, such as the setting up of a wave-like concentration of a morphogen, with each stripe forming at the crest of a wave. It was surprising, therefore, to discover that each stripe is specified independently.

As an example of how the pair-rule stripes are generated we will look in detail at the expression of the second *eve* stripe (Fig. 2.32). The appearance of this stripe depends on the normal expression of *bicoid* and of the three gap genes *hunchback, Krüppel,* and *giant* (only the anterior band of *giant* expression is involved in specifying the second *eve* stripe; see Fig. 2.23). Bicoid and Hunchback proteins are required to activate the *eve* gene, but they do not define the boundaries of the stripe. These are defined by Krüppel and Giant proteins, by a mechanism based on repression of *eve*. When concentrations of Krüppel and Giant are above certain threshold levels, *eve* is repressed, even if Bicoid and Hunchback are present. The anterior edge of the stripe is localized at the point of threshold concentration of Giant protein, whereas the posterior border is similarly specified by the Krüppel protein.

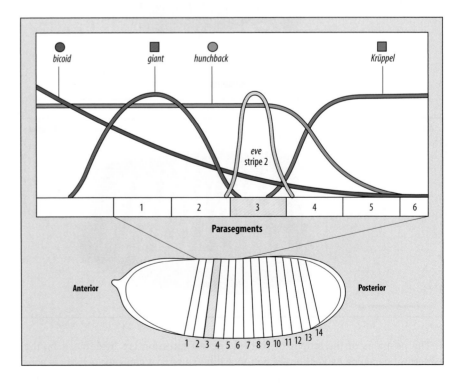

Fig 2.32 The specification of the second *even-skipped* (*eve*) stripe by gap gene proteins. The different concentrations of transcription factors encoded by the gap genes *hunchback, giant,* and *Krüppel* localize *even-skipped*—expressed in a narrow stripe at a particular point along their gradients—in parasegment 3. Bicoid and Hunchback proteins activate the gene in a broad domain, and the anterior and posterior borders are formed through repression by Giant and Krüppel proteins, respectively.

In contrast, the positioning of the third and fourth *eve* stripes depends on regulatory regions that are repressed by the high concentration of Hunchback in the anterior region. The third stripe is expressed about halfway along the embryo, where Hunchback concentration starts to drop sharply, while the fourth *eve* stripe is expressed more posteriorly, where Hunchback is at an even lower level (Fig. 2.33). The posterior boundary of the third *eve* stripe is delimited by repression by the gap protein Knirps.

The independent localization of each of the stripes by the gap-gene transcription factors requires that, in each stripe, the pair-rule gene responds to different concentrations and combinations of the gap-gene transcription factors. The pair-rule genes thus require complex *cis*-regulatory control regions with multiple binding sites for each of the different factors. Examination of the regulatory regions of the *eve* gene reveals seven separate modules, each controlling the localization of a different stripe. Regulatory regions of around 500 base pairs have been isolated that each determines the expression of a single stripe (Fig. 2.34). The *eve* gene is an excellent example of modular *cis*-regulatory regions controlling the expression of a gene at different sites.

The presence of control regions, which, when activated, can lead to gene expression in a specific position in the embryo, is a fundamental principle controlling gene action in development. Other examples of this are the localized expression of the gap genes, and of the genes along the dorso-ventral axis.

Each regulatory region of such a gene contains binding sites for different transcription factors, some of which activate the gene, while others repress it. In this way, the combinatorial activity of the gap genes regulates pair-rule gene expression in each parasegment. Some pair-rule genes, such as *fushi tarazu*, may not be regulated by the gap genes directly, but may depend on the prior expression of primary pair-rule genes, such as *even-skipped* and *hairy*. With the initiation of pair-rule gene expression, the embryo becomes segmented; it is now divided into a number of unique regions, which are characterized mainly by the combinations of transcription factors being expressed in each. These include proteins encoded by the gap genes, the pair-rule genes, and the genes expressed along the dorso-ventral axis.

The transcription factors encoded by the pair-rule genes set up the spatial framework for the next round of patterning by transcriptional activation. This involves the further patterning of the parasegments, the development of the final segmentation and the acquisition of segment identity, which is considered in the following sections.

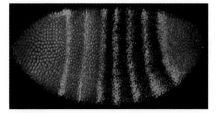

Fig 2.33 Positioning of the third and fourth *eve* stripes by the Hunchback gradient. Embryo stained with fluorescent antibodies to simultaneously detect the Hb protein (red) and the mRNA of its target gene *eve* (green). The position of the third and fourth *eve* stripes is due directly to lack of repression of *eve* by Hunchback as its gradient drops off sharply.

Photograph from Danyang, Y., and Small, S.: 2008.

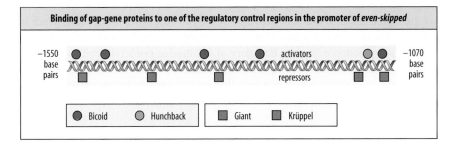

Fig 2.34 Sites of action of activating and repressing transcription factors in the control region of *even-skipped* involved in expression of the second *even-skipped* stripe. A control region of around 500 base pairs, located between 1070 and 1550 base pairs upstream of the transcription start site, directs formation of the second *even-skipped* stripe. Gene expression occurs when the Bicoid and Hunchback transcription factors are present above a threshold concentration, with the Giant and Krüppel proteins acting as repressors where they are above threshold levels. The repressors may act by preventing binding of activators.

Summary

The activation of the pair-rule genes by the gap genes results in the transformation of the embryonic pattern along the antero-posterior axis from a variable regionalization to a periodic one. The pair-rule genes define 14 parasegments. Each parasegment is defined by narrow stripes of pair-rule gene activity. These stripes are uniquely defined by the local concentration of gap-gene transcription factors acting on the regulatory regions of the pair-rule genes. Each pair-rule gene is expressed in alternate parasegments—some in odd-numbered, others in even-numbered. Most pair-rule genes code for transcription factors.

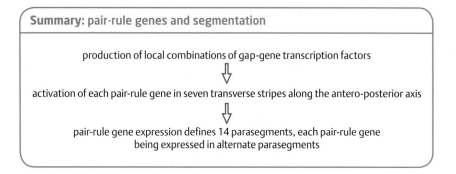

Summary: pair-rule genes and segmentation

production of local combinations of gap-gene transcription factors

⇩

activation of each pair-rule gene in seven transverse stripes along the antero-posterior axis

⇩

pair-rule gene expression defines 14 parasegments, each pair-rule gene
being expressed in alternate parasegments

Segmentation genes and compartments

The expression of the pair-rule genes defines the anterior boundaries of all 14 parasegments but, like the gap genes, their activity is only temporary. Moreover, at this time the blastoderm becomes cellularized. How, therefore, are the positions of parasegment boundaries fixed, and how do the final segment boundaries of the larval epidermis become established? This is the role of the **segmentation genes**. Unlike the gap genes and pair-rule genes, which encode transcription factors, the segmentation genes are a diverse group of genes that bear no obvious relation to each other in their protein products or mechanisms of operation.

Segmentation genes are activated in response to pair-rule gene expression. They are each expressed in 14 transverse stripes, one stripe corresponding to each parasegment. During pair-rule gene expression the blastoderm becomes cellularized, so the segmentation genes are acting in a cellular rather than a syncytial environment. One of the segmentation genes to be activated by the pair-rule genes is *engrailed*, which is expressed in the anterior region of every parasegment. *engrailed* is of particular interest as its expression delimits a boundary of cell-lineage restriction and so defines a developmental unit called a compartment, as discussed in the next section. Moreover, *engrailed* is also a **selector gene**, a gene that confers a particular identity on a region or regions by controlling the activity of other genes, and which continues to act for an extended period.

2.23 Expression of the *engrailed* gene delimits a cell-lineage boundary and defines a compartment

The *engrailed* gene has a key role in segmentation and, unlike the pair-rule and gap genes, whose activity is transitory, it is expressed throughout the life of the fly. *engrailed* activity first appears at the time of cellularization as a series of 14 transverse stripes. Figure 2.35 shows expression at the later extended germ-band stage, when part of the ventral blastoderm (the germ band) has extended over the dorsal side of the embryo. *engrailed* is initially expressed in a single line of cells at the anterior margin of each

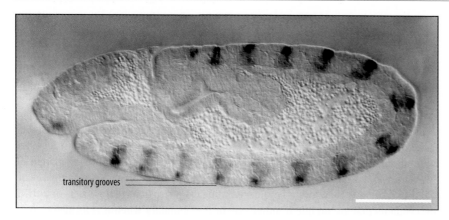

transitory grooves

Fig 2.35 **The expression of the** *engrailed* **gene in a late (stage 11)** *Drosophila* **embryo.** The gene is expressed in the anterior region of each parasegment and the transitory grooves between parasegments can be seen. At this stage of development, the germ band has temporarily extended and is curved over the back of the embryo. Scale bar = 0.1 mm.

Photograph from Lawrence, P.: 1992.

parasegment, which is itself only about four cells wide (Fig. 2.36). It is likely that this periodic pattern of *engrailed* activity is the result of the combinatorial action of transcription factors encoded by pair-rule genes, including *fushi tarazu* and *even-skipped*. Evidence that the pair-rule genes do control *engrailed* expression is provided, for example, by embryos carrying mutations in *fushi tarazu*, in which *engrailed* expression is absent only in even-numbered parasegments (in which *fushi tarazu* is normally expressed).

The anterior and posterior margins of the parasegment have a very important property—they are boundaries of **cell-lineage restriction**. Cells and their descendants from one parasegment never move into adjacent ones. Such domains of lineage restriction are known as **compartments**. A compartment can be defined as a region in

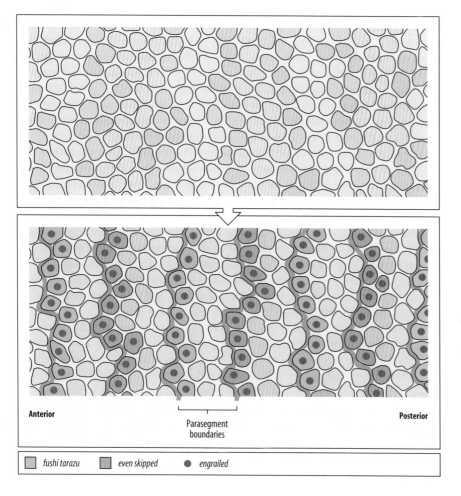

Anterior

Parasegment
boundaries

Posterior

☐ *fushi tarazu*   ☐ *even skipped*   ● *engrailed*

Fig 2.36 **The expression of the pair-rule genes** *fushi tarazu* **(blue),** *even-skipped* **(pink), and** *engrailed* **(purple dots) in parasegments.** *engrailed* is expressed at the anterior margin of each stripe and delimits the anterior border of each parasegment. The boundaries of the parasegments become sharper and straighter later on.

After Lawrence, P.: 1992.

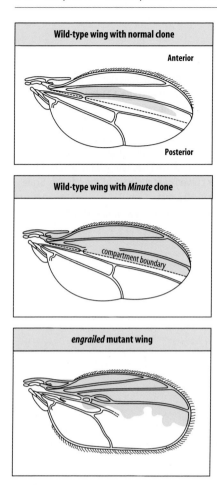

Fig 2.37 The boundary between anterior and posterior compartments in the wing can be demonstrated by marked cell clones. Top panel: in the wild-type wing, clones marked by mitotic recombination in the embryo, which gives marked cells a phenotype different from other cells of the wing, are too small to demonstrate the compartment boundary. Middle panel: the use of the *Minute* technique (see Box 2E, opposite) produces an increased rate of cell division in the marked cell and gives clones large enough to determine that cells from one compartment do not cross the boundary into an adjoining compartment. Bottom panel: in the wing of an *engrailed* mutant in which the engrailed protein is not produced, there is no posterior compartment or boundary. Clones in the anterior part of the wing cross over into the posterior region and the posterior region is transformed into a more anterior-like structure, bearing anterior-type hairs on its margin. The *engrailed* gene is required for the maintenance of the character of the posterior compartment, and for the formation of the boundary.

the embryo, larva, or adult that contains all the descendants of the cells present when the compartment is set up, and no others.

The existence of compartments can be detected by cell-lineage studies. Cell lineage can be followed by marking single cells in the early embryo so that all their descendants (clones) can be identified at later stages of development. One technique is to inject the egg with a harmless fluorescent compound that is incorporated into all the cells of the embryo. In the early embryo, a fine beam of ultraviolet light directed onto a single cell activates the fluorescent compound. All the descendants of that cell will be fluorescent and can therefore be identified. Examination of these clones, together with an analysis of *engrailed* expression, shows that cells at the anterior margin of the parasegment have no descendants that lie on the other side of that margin. The anterior margin is therefore a boundary of lineage restriction; that is, cells and their descendants that are on one or other side of the boundary when it is formed cannot cross it during subsequent development.

The lineage restriction of the anterior margin of the parasegment is carried over into the segments of the larva and the adult. But because the anterior part of a parasegment becomes the posterior part of a segment, the lineage restriction within a segment is between anterior and posterior regions. The segment is thus divided into anterior and posterior compartments, with *engrailed* expression defining the posterior compartment (see Fig. 2.30).

In a compartment, the cells may all initially be under common genetic control. Imagine a group of cells becoming divided into two separate regional populations by means of signals that turn on different genes in each region. If there is never any mixing of the descendants of the original cells across the boundary the regions are compartments.

Lineage restriction within compartments is clearly illustrated by the behavior of cells in the wing of the adult fly. It is easier to distinguish lineage restriction in adult structures than in embryonic and early larval structures, because in adults there has been considerable cell division since the initial event that determined the compartment in the embryo. Compartments are made visible by making genetically mosaic flies composed of two distinguishable kinds of cells (Box 2E, p. 73). Single cells in the embryonic blastoderm or in larval epidermis are given a distinctive phenotype by X-ray-or laser-induced mitotic recombination. The fates of all the descendants of this marked cell are then followed. Their behavior depends on the stage of development at which the founder nucleus was marked. Descendants of nuclei marked at early stages during cleavage become part of many tissues and organs, but those of nuclei marked at the blastoderm stage or later have a more restricted fate. They are found only in the anterior or the posterior part of each segment (or of an appendage like a wing), never throughout the whole segment.

Cells of imaginal discs divide only about 10 times after the cellular blastoderm stage; thus clones of experimentally marked cells are small, even in the adult, and it is not easy to detect a boundary of lineage restriction (Fig. 2.37, top panel). Clone size can be increased by the *Minute* technique, which results in the marked cell dividing many more times than the other cells (see Box 2E, p. 73). A single clone of such cells can almost fill either the anterior or posterior part of the wing, and this makes the boundary that the cells never cross more evident: this boundary separates the anterior and posterior compartments (see Fig. 2.37, middle panel). In a normal wing, each compartment is constructed by all the descendants of a set of founding cells. The compartment boundary is remarkably sharp and straight and does not correspond to any structural features in the wing. These experiments also show that the pattern of the wing is in no way dependent on cell lineage. A single marked embryonic cell can give rise to about a twentieth of the cells of the adult wing or, using the *Minute* technique to increase clone size, about half the wing. The lineage of the wing cells in each case is quite different, yet the wing's pattern is quite normal. This is very different from the nematode *C. elegans*, where the cell lineages in development are constant.

Box 2E Genetic mosaics and mitotic recombination

Genetic mosaics are embryos derived from a single genome but in which there is a mixture of cells with rearranged or inactivated genes. In flies, genetic mosaics can be generated by inducing rare mitotic recombination events in the somatic cells of the embryo or larva. The original method was to induce chromosome breaks using X-rays just after the chromosomes have replicated to form two chromatids; this results in the exchange of material between homologous chromosomes as the break is repaired. Today, mitotic recombination is induced in strains of transgenic flies that carry the gene for the yeast recombinase FLP and its target sequence, *FRT*, on their chromosomes. Activation of the recombinase will result in recombination involving the *FRT* sequence. A mitotic recombination event can generate a single cell with a unique genetic constitution that will be inherited by all the cell's descendants, which in flies usually form a coherent patch of tissue.

Easily distinguishable mutations like the recessive *multiple wing hairs* can be used to identify the marked clone; if a cell homozygous for this mutation is generated by mitotic recombination in a heterozygous larva, all of its descendant cells will have multiple hairs (see figure). Marked epidermal clones made by this method are usually small, as there is little cell proliferation after the recombination event. Larger clones can be made using the *Minute* technique. The cells in flies carrying a mutation in the *Minute* gene grow more slowly than those in the wild type. By using flies heterozygous for the *Minute* mutation, clones can be made in which the mitotic

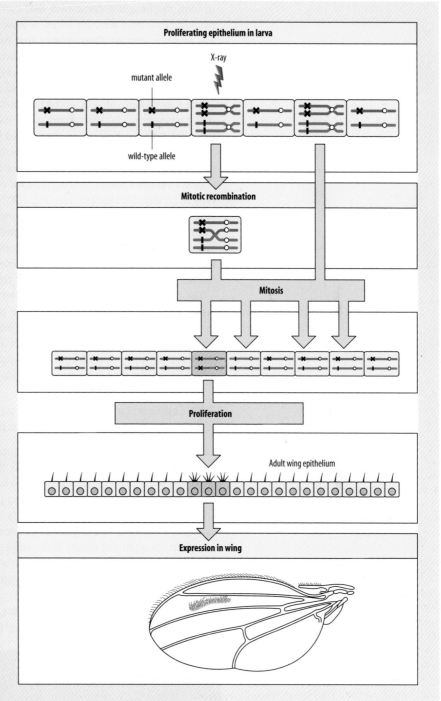

recombination event has generated a marked cell that is normal because it has lost the *Minute* mutation, and so is wild type. This normal cell proliferates faster than the slower-growing background and thus large clones of the marked cells are produced (see Fig. 2.37). The mitotic recombination technique has many applications. If clones of marked cells are generated at different stages of development, one can trace the fate of the altered cells and thus see what structures they are able to contribute to. This can provide information on their state of determination or specification at different developmental stages. The technique can also be used to study the localized effects of homozygous mutations that are lethal if present in the homozygous state throughout the whole animal.

Illustration after Lawrence, P.: 1992.

The compartment pattern in the adult *Drosophila* wing is carried over from its initial specification in the imaginal discs. When epidermal cells are set aside in the embryo to form the imaginal discs, each disc carries over the compartment pattern of the parasegments from which it arises. A wing disc, for example, develops at the boundary between the two parasegments that contribute to the second thoracic segment. Thus, the wing is divided into an anterior compartment and a posterior compartment, the compartment boundary (the old parasegment boundary) running in a straight line more or less down the middle of the wing.

The specification of cells as the posterior compartment of a segment (the anterior of a parasegment) initially occurs when the parasegments are set up, and is due to the *engrailed* gene. Expression of *engrailed* is required both to confer a 'posterior segment' identity on the cells and to change their surface properties so that they cannot mix with the cells adjacent to them, hence setting up the parasegment boundary.

Direct evidence for the role of *engrailed* comes from the behavior of clones of wing cells in an *engrailed* mutant (see Fig. 2.37, bottom panel). In the absence of normal *engrailed* expression, clones are not confined to anterior or posterior parts of the segment and there is no compartment boundary. Moreover, in *engrailed* mutants, the posterior compartment is partly transformed so that it comes to resemble the pattern of the anterior part of the wing. For example, bristles normally found only at the anterior margin of the wing are also found at the posterior margin.

The *engrailed* gene needs to be expressed continuously throughout larval and pupal stages and into the adult to maintain the character of the posterior compartment of the segment. Thus, *engrailed* is also an example of a selector gene—a gene whose activity is sufficient to cause cells to adopt a particular fate. Selector genes can control the development of a region such as a compartment and, by controlling the activity of other genes, give the region a particular identity.

2.24 Segmentation genes stabilize parasegment boundaries and set up a focus of signaling at the boundary that patterns the segment

Each larval segment has a well-defined antero-posterior pattern, which is easily seen on the ventral epidermis of the abdomen: the anterior region of each segment bears denticles, while the posterior region is naked (Fig. 2.38). The rows of denticles, of which there are six types, make a distinct pattern. Mutations in segmentation genes often alter the denticle pattern, and this is how such genes were first discovered. For example, a mutation in a segmentation gene known as *wingless* (named after its phenotype in the adult fly) results in the whole of the ventral abdomen being covered in denticles, but in the posterior half of each segment the denticle pattern is reversed. In this mutant, the anterior region of each segment has been duplicated in mirror-image polarity, and the posterior region is lost.

The larval denticle patterns depend on the correct establishment and maintenance of the embryonic parasegment boundaries. Establishment of the parasegment boundary

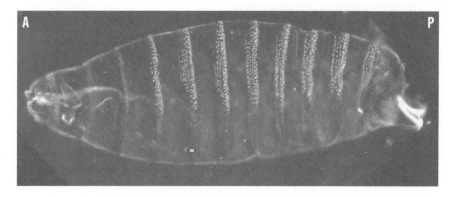

Fig 2.38 Each segment of the *Drosophila* larva has a characteristic pattern of denticles on its ventral surface. Denticles are confined to the anterior regions of segments, with each segment having its own pattern.

Photograph from Goodman, R.M., et al.: 2006.

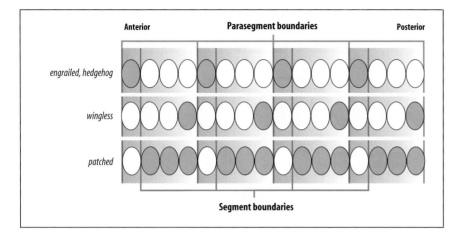

Fig 2.39 The domains of expression of the segmentation genes. At the cellular blastoderm stage, the *engrailed* gene is expressed at the anterior margin of each parasegment together with *hedgehog*, while *wingless* is expressed at the posterior margin. After gastrulation, *patched* is expressed in all of the cells that express neither *engrailed* nor *hedgehog*. About two cell divisions occur between the time that the parasegments are delimited and the time the embryo hatches.

depends on an intercellular signaling circuit being set up between adjacent cells, which delimits the boundary between them. This circuit primarily involves the segmentation genes *wingless, hedgehog,* and *engrailed,* which are expressed in restricted domains within the parasegment (Fig. 2.39). *engrailed,* as we saw earlier, encodes a transcription factor, while Wingless and Hedgehog are secreted signal proteins that act via receptor proteins on cell surfaces to alter patterns of gene expression. **Wingless** is one of the prototype members of the so-called **Wnt family** of signal proteins, which are key players in development in vertebrates, as well as invertebrates, and are involved in determining cell fate and cell differentiation in many different aspects of development. **Hedgehog** similarly has homologs in other animals, including vertebrates.

We have seen how the pair-rule genes first delimit the parasegments by inducing the expression of *engrailed* at the anterior margin of each parasegment (see Fig. 2.35). Cells expressing *engrailed* now also express the secreted signal Hedgehog, which activates and maintains expression of the signal protein Wingless in the row of cells immediately anterior to the Engrailed-expressing cells. The general signaling pathway by which Hedgehog acts is shown in Fig. 2.40. The secreted Wingless protein in turn provides a signal that feeds back over the boundary to maintain *hedgehog* and *engrailed* expression, thus stabilizing and maintaining the compartment boundary (Fig. 2.41). At parasegment boundaries, and in many of their other developmental

Fig 2.40 The Hedgehog signaling pathway. Left panel: in the absence of Hedgehog, the membrane protein Patched, which is the receptor for Hedgehog, inhibits the membrane protein Smoothened. In the absence of Smoothened activity, the transcription factor Cubitus interruptus (Ci) is held in the cytoplasm in two protein complexes—one associated with Smoothened and another with the protein Suppressor of fused (Su(fu)). In the absence of Hedgehog, Ci in the Smoothened complex is also phosphorylated by several protein kinases (glycogen synthase kinase (GSK-3), protein kinase A (PKA) and casein kinase 1 (CK1)), which results in the proteolytic cleavage of Ci and formation of the truncated protein CiRep. This enters the nucleus and acts as a repressor of Hedgehog target genes. Right panel: Hedgehog (Hh) binding to Patched lifts the inhibition on Smoothened and blocks production of CiRep. Ci is released from both its complexes in the cytoplasm, enters the cell nucleus, and acts as a gene activator. Genes activated in response to Hedgehog signaling include *wingless* (*wg*), *decapentaplegic* (*dpp*), and *engrailed* (*en*).

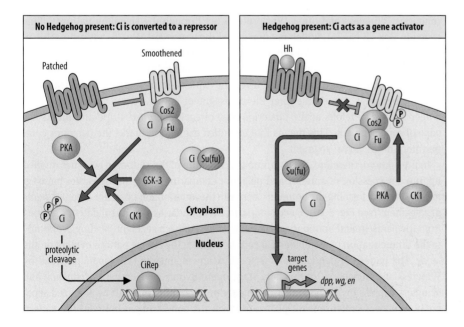

Fig 2.41 Interactions between *hedgehog*, *wingless*, and *engrailed* genes and proteins at the compartment boundary control denticle pattern. Top left panel: the *engrailed* gene, which encodes a homeodomain transcription factor, is expressed in cells along the anterior margin of the parasegment, which delimits the future boundary of the posterior compartment in the segment; these cells also express the segmentation gene *hedgehog* and secrete the Hedgehog protein. Hedgehog protein activates and maintains expression of the segmentation gene *wingless* in adjacent cells across the compartment boundary. Wingless protein feeds back on *engrailed*-expressing cells to maintain expression of the *engrailed* gene and *hedgehog*. These interactions stabilize and maintain the compartment boundary. Top right panel: in mutants where the *wingless* gene is inactivated and the Wingless protein is absent, neither *hedgehog* nor *engrailed* genes are expressed. This leads to loss of the compartment boundary and the normally well defined pattern of denticles within each abdominal segment. Bottom left panel: in the wild-type larva the denticle bands in the ventral cuticle are confined to the anterior part of the segment and are dependent on the activity of *hedgehog* and *wingless* genes. Bottom right panel: in the *wingless* mutant, denticles are present across the whole ventral surface of the segment in what looks like a mirror-image repeat of the anterior segment pattern.

Photographs from Lawrence, P.: 1992.

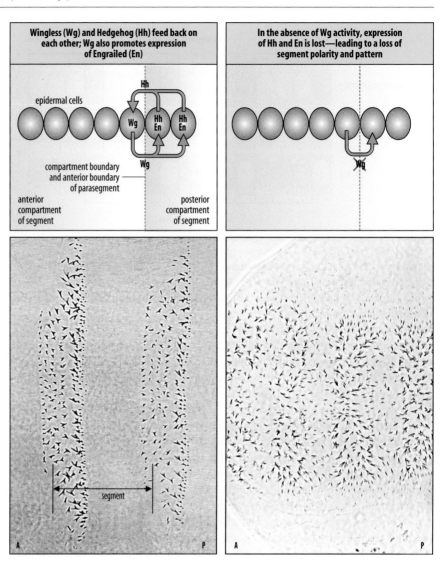

functions, Wingless and other Wnt proteins act through the signaling pathway illustrated in Box 1E (p. 26), which is often called the **canonical Wnt/β-catenin pathway** (β-catenin is called Armadillo in *Drosophila*). A computer model of the Wingless–Hedgehog–Engrailed circuitry shows that it is remarkably robust and resistant to variation, such as variations in the levels of expression of the genes governing the circuit's behavior. Some time after the parasegment boundary has been established, a deep groove develops at the posterior edge of each *engrailed* stripe and marks the edge of each segment. This means that *engrailed* expression marks the posterior compartment of the larval segment.

In the embryo, therefore, the continuous epithelium of the abdomen becomes divided up into a succession of anterior and posterior compartments, with the earlier parasegment boundary forming the boundary between the anterior and posterior compartments of a segment (see Fig. 2.30). Once compartment boundaries are consolidated, signaling from the parasegment boundary sets up the pattern of each segment, leading eventually to the differentiation of the epidermal cells to produce the cuticle pattern evident on the larva. The pattern of denticle belts is specified in response to the signals provided by Wingless and Hedgehog. At this later stage of embryonic development, Hedgehog and Wingless expression are no longer dependent on each other and can be analyzed separately. Wingless protein moves posteriorly over the compartment boundary to pattern the posterior compartment of the segment, and also moves anteriorly to pattern the anterior part of the segment immediately anterior to the cells in which it is expressed. Wingless moves a shorter distance posteriorly than anteriorly, as it is more rapidly degraded in the posterior compartment, and so its effects on patterning extend over a longer range in the anterior compartment. In the anterior compartment, Wingless represses genes required for denticle formation, and the cells that lie immediately anterior to the Wingless-expressing line of cells are specified as epidermal cells that will produce smooth cuticle, while the anterior extent of Wingless signaling delimits the posterior edge of the denticle belts in that segment. In mutants that lack *wingless* function, denticles are produced in the normally smooth regions (see Fig. 2.41).

Hedgehog signaling does not pattern the posterior segment compartment in which Hedgehog is actually expressed, as posterior compartment cells are 'blind' to its action. The Hedgehog signals move anteriorly to maintain Wingless function and also posteriorly to help pattern the anterior compartment of the next segment. The signals provided by Hedgehog and Wingless lead, by a complex set of interactions, to the expression of a number of genes in non-overlapping narrow stripes that specify the larval segmental pattern of smooth cuticle or specific rows of denticles.

Patterning within compartment boundaries is also visible in the pattern of the epidermis on the abdominal segments of the adult fly (Fig. 2.42). The epidermis of each segment is subdivided into a number of types—at least five in the anterior compartment and three in the posterior compartment—which are distinguished by the presence or absence of pigmentation and the positioning of bristles and hairs.

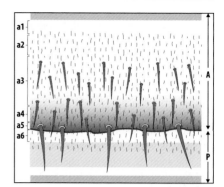

Fig 2.42 The epidermis of each segment in the abdomen of the adult *Drosophila* is divided into an anterior and a posterior compartment. A number of different regions, a1 to a6, can be distinguished in the anterior compartment, characterized by different bristles, pigmentation, and gene expression. A, anterior compartment; P, posterior compartment.

2.25 Insect epidermal cells become individually polarized in an antero-posterior direction in the plane of the epithelium

The insect epidermis in each segment not only becomes patterned into bands of epidermal cells of different types, but each cell acquires an individual antero-posterior polarity, reflected in the fact that the hairs and bristles on the adult *Drosophila* abdomen all point backwards (see Fig. 2.42). Many different types of cells have polarity, which simply means that one end of the cell is structurally or functionally different from the other. For example, polarization is clearly seen in cells undergoing chemotaxis in response to a gradient of a chemoattractant, where the advancing end of the cell is quite different from the posterior end. This type of cell polarity is called **planar cell polarity**, to distinguish it from the apical–basal polarity often present in an epithelial sheet. Planar cell polarity is also clearly visible in the *Drosophila* wing, which is composed of a thin sheet of epithelium. The wing epidermal cells all bear hairs and these hairs form at the ends of the cells away from the body—the **distal** ends of the cells—and all point away from the body (see Fig. 2.37). Another example is the insect compound eye, where the individual units (ommatidia) in each half of the eye point away from the midline (see Chapter 11), which is the result of the establishment of planar polarity in the developing photoreceptor cells in the eye imaginal disc. In vertebrates, examples of planar polarity occur in the stereocilia in the inner ear and in the scales of fish. Planar cell polarity is also important in morphogenetic cell movements such as the tissue extension that occurs at gastrulation in *Xenopus*, when the spherical frog blastula is being converted into the more elongated gastrula; this will be discussed in Chapter 8.

The mechanism by which planar cell polarity is established is not yet completely understood. Current hypotheses propose the existence of an informational gradient across the tissue that sets the overall direction of polarity, while local signaling between one end of a cell and the other, and between one cell and its neighbor are thought to set the polarity of the individual cell (see Box 2F, p. 78). In the case of the adult insect abdominal epidermis, for example, the signaling gradients set up at compartment boundaries have been proposed to have a role in establishing planar cell polarity, as well as in patterning the segment.

Box 2F Planar cell polarity in *Drosophila*

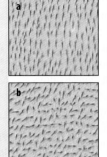

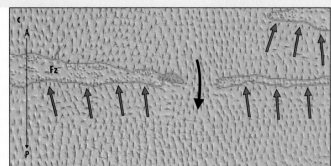

Cells in an epithelium have a distinct apical–basal polarity, but more relevant to patterning in development is polarity in the plane of the epithelium—**planar cell polarity**. This is visible in the epidermis of the adult fly, where the abdominal epidermis has hairs all pointing in a posterior direction (see figure, panel a), and the epidermis of the wing has hairs all pointing in a distal direction—away from the proximal base of the wing and towards its tip. Another example of planar cell polarity in *Drosophila* is the orientation of the ommatidia in the insect compound eye (see Chapter 11).

The mechanism for establishing planar polarity is not fully understood, especially how the axis of polarity across the tissue is determined. At the cellular level, however, some key proteins involved have been identified. The cell-surface receptor Frizzled (Fz) is one. In the fly abdominal epidermis, if Fz is absent then tissue polarity is disrupted and the hairs point in random directions (see top figure, panel b). But if the gene for Fz is turned off in just a small clone of epidermal cells, the unaffected cells immediately posterior to them reverse polarity, showing that polarity is influenced by intercellular—cell to cell—communication (see figure, panel c; red arrows indicate the reversed direction; the black arrow shows the normal direction of polarity from anterior (A) to posterior (P)). We met Fz earlier as the receptor for Wingless, but in the case of planar cell polarity it does not act through the canonical Wnt pathway shown in Box 1E (p. 26) and is unlikely to be stimulated by Wingless or other *Drosophila* Wnts.

Fz is thought to have at least three functions in planar cell polarity. The first is to receive long-range patterning information from upstream cues, which might involve the generation of a gradient of Fz activity across the whole axis of the tissue. Clonal repolarization experiments like those shown in the figure above suggest that the direction arrow of polarity points down the Fz gradient; cells point their hairs in the direction of a lower level of Fz and away from a higher level. Second, Fz is involved in the cell–cell communication that locally coordinates cell polarity. The third function of Fz is to provide an intracellular cue for the growth of the hairs.

It is not yet known how the direction of polarity is set across a tissue, but it is now clear that key interactions that determine and maintain the polarity of individual cells take place at the epidermal cell junctions. An asymmetry between the two sides of a junction is established by interactions between proteins in the adjacent cell membranes. In proximal to distal polarity in the wing, for example, Fz becomes associated with an atypical cadherin called Flamingo (Fmi, also known as Starry Night) at the distal end of each epidermal cell, while at the proximal end, Fmi is associated with the membrane protein Van Gogh (Vang, also

known as Strabismus). An asymmetric interaction across the junction between the Fz:Fmi and Fmi:Vang complexes seems to be central to establishing cell polarity—by some mechanism that is still to be determined. The cytoplasmic signaling proteins Dishevelled (Dsh) and Diego (Dgo) become associated with Fz:Fmi and the cytoplasmic protein Prickled (Pk) associates with Fmi:Vang (see lower figure). These cytoplasmic proteins seem not to be essential for establishing cell polarity *per se* but may act to convert the polarity signal into some of the manifestations of polarity within the cell.

Yet another set of proteins, which include the membrane protein Four-jointed, and the atypical cadherins Dachsous and Fat, have also been implicated in determining planar cell polarity. They form gradients along the axis of polarity in tissues, and one proposed role for these proteins has been as long-range graded extracellular signals that set the direction of polarity by interacting in some way with the proteins described above.

However, there is evidence that this system can operate independently in parallel with the Fz–Fmi–Vang system rather than in a strict linear relationship. A clone of cells overexpressing Dachsous or other proteins of that group can repolarize adjacent cells even when those cells lack both Fz and Fmi and so cannot form the junctional complexes. Unraveling planar cell polarity is still a fascinating work in progress.

Determining cell polarity is an important aspect of development in all animals. We shall see in Chapter 8, for example, how some of the same proteins regulate the changes in cell shape and polarity that help to reshape vertebrate embryos during gastrulation.

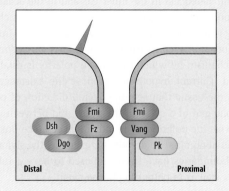

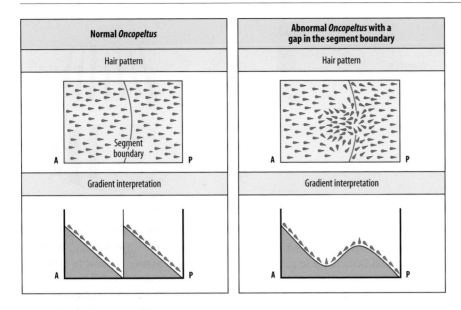

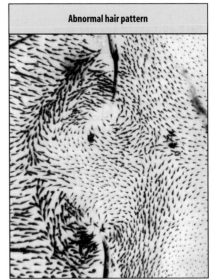

Fig 2.43 Gradients could specify polarity in the segments of *Oncopeltus*. The hairs on the cuticle point in a posterior direction and this may reflect an underlying gradient in a morphogen that is partly maintained by the segment boundary (left panels). When there is a gap in the boundary (center panels) the sharp discontinuity in morphogen concentration smoothes out, with a resulting local reversal of the gradient and of the direction in which the hairs point. This is illustrated in the photograph (right panel).

A natural case where loss of planar cell polarity can be observed is in adults of the plant-feeding bug *Oncopeltus*, which, like *Drosophila*, have hairs covering each segment, all pointing in a posterior direction (Fig. 2.43, left panel). In *Oncopeltus*, some individuals naturally have a gap in the segment boundary; this results in a rather precise pattern of hairs with altered orientation, many of the hairs near the gap pointing in the reverse direction (Fig. 2.43, center and right panels). If the slope of a gradient within each segment gives the hairs their polarity, a gap in a segment boundary would result in a local change in the gradient, the sharp change in concentration at the normal boundary being smoothed out and forming a local gradient running in the opposite direction. This would result in the hairs in this region pointing in the opposite direction. Mutations that cause similar reversals in polarity can be induced in clones of cells in the *Drosophila* abdominal epidermis, and can be similarly explained by the disruption of an informational gradient set up within a compartment. Genes essential for planar cell polarity have been identified in *Drosophila* and other organisms, together with some of the proteins involved in setting up the long-range gradient and in the signaling pathways that interpret it to set individual cell polarity (see Box 2F, p. 78).

2.26 Some insects use different mechanisms for patterning the body plan

Drosophila belongs to an evolutionarily advanced group of insects, in which all the segments are specified more or less at the same time in development, as shown both by the striped patterns of pair-rule gene expression in the acellular blastoderm and the appearance of all the segments shortly after gastrulation. This type of development is known as **long-germ development**, as the blastoderm corresponds to the whole of the future embryo. Many other insects, such as the flour beetle *Tribolium*, have a **short-germ development**. In short-germ development the blastoderm is short and forms only the anterior segments. The posterior segments are added by growth after completion of the blastoderm stage and gastrulation (Fig. 2.44). In spite of early differences between them, the mature germ band stages of both long- and short-germ

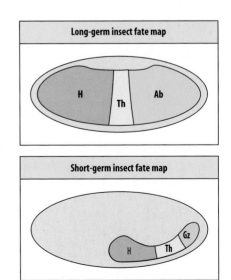

Fig 2.44 Differences in the development of long-germ and short-germ insects. Top panel: the general fate map of long-germ insects such as *Drosophila* shows that the whole of the body plan—head (H), thorax (Th), and abdomen (Ab)—is present at the time of initial germ-band formation. Bottom panel: in short-germ insects, only the anterior regions of the body plan are present at this embryonic stage. Most of the abdominal segments develop later, after gastrulation, from a posterior growth zone (Gz).

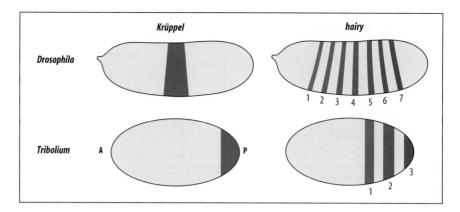

Fig 2.45 Gap and pair-rule gene expression in long-germ and short-germ insects at the time of germ-band formation. *Krüppel* (red) is a gap gene and *hairy* (green) is a pair-rule gene. The position of the *Krüppel* stripe in the short-germ embryo of *Tribolium* indicates that posterior regions of the body are not yet present at this stage. Similarly, only three *hairy* stripes, corresponding to the first three *hairy* stripes in a long-germ embryo, are present in *Tribolium*.

insect embryos look similar. This, therefore, is a common stage through which all insect embryos develop. In the long-germ embryos of the wasp *Nasonia*, gradients in the protein Orthodenticle pattern the anterior and posterior regions, in a manner similar to the action of Bicoid in *Drosophila*.

The question clearly arises as to what processes are common to the specification of the body plan in long- and short-germ insects. One clear difference is that whereas the patterning of the body plan in the long-germ band of *Drosophila* takes place before cell boundaries form, much of the body plan in short-germ insects is laid down at a later stage, when the posterior segments are generated during growth. At this stage the embryo is multicellular. So are the same genes involved?

There is very good evidence that the same genes and development processes are involved in the patterning of *Tribolium* and *Drosophila*. For example, the gap gene *Krüppel* is expressed at the posterior end of the *Tribolium* embryo at the blastoderm stage and not in the middle region, as in *Drosophila* (Fig. 2.45). It therefore seems to be specifying the same part of the body in the two insects. Similarly, only two repeats of the pair-rule stripes are present at the blastoderm stage, together with a posterior cap of pair-rule gene expression, in contrast to the seven in *Drosophila*. The genes *wingless* and *engrailed* are also expressed in a similar relation to that in *Drosophila*.

Although not many genes have been studied in detail in other insects, at least one, the gene *engrailed*, is known to be expressed in the posterior region of segments in a variety of insects. The pair-rule gene *even-skipped* (see Section 2.21), while present in the grasshopper (a short-germ insect), may not have a similar role in segmentation. It is, however, involved later in the development of the nervous system in the grasshopper, and is also expressed at the posterior end of the growing germ band.

Differences in early development are much more dramatic in some other insects. In certain parasitic wasps, the egg is small and undergoes cleavage to form a ball of cells, which then falls apart. Each of the resulting small clusters of cells—there can be as many as 400—can develop into a separate embryo. This wasp's development apparently does not depend on maternal information to specify the body axes, and in this respect resembles that of the early mammalian embryo.

Summary

The segmentation genes are involved in patterning the parasegments. One of the first genes to be activated is *engrailed*, which is expressed at the anterior margin of each parasegment, delineating a group of cells which becomes the posterior compartment of a segment. *engrailed* is also a selector gene in that it provides a group of cells with a long-term regional identity. *engrailed* expression shows lineage restriction: cells expressing *engrailed* define the posterior compartment of a segment, and cells never cross from the posterior into the anterior compartment. *engrailed* is turned on by the pair-rule genes, and its expression is maintained by the segmentation genes *wingless* and *hedgehog,* which stabilize the compartment boundary. The compartment boundary is involved in specifying pattern and cell polarity within the segment. Studies in other insects also suggest that a discrete gradient of positional information is set up in each segment, delimited by the boundaries, and this gradient patterns the segment. Adult epidermal cells are polarized. Hairs in the adult fly body and wing point along an axis, and this is due to a mechanism that gives rise to planar cell polarity. In contrast to *Drosophila,* some insects have a short-germ pattern of development in which posterior segments are added by growth after the cellular blastoderm stage.

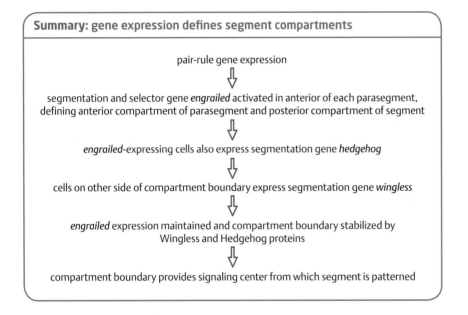

Summary: gene expression defines segment compartments

pair-rule gene expression

⇩

segmentation and selector gene *engrailed* activated in anterior of each parasegment, defining anterior compartment of parasegment and posterior compartment of segment

⇩

engrailed-expressing cells also express segmentation gene *hedgehog*

⇩

cells on other side of compartment boundary express segmentation gene *wingless*

⇩

engrailed expression maintained and compartment boundary stabilized by Wingless and Hedgehog proteins

⇩

compartment boundary provides signaling center from which segment is patterned

Specification of segment identity

Each segment has a unique identity, most easily seen in the larva in the characteristic pattern of denticles on the ventral surface. Since the same segmentation genes are turned on in each segment, what makes the segments different from each other? Their identity is specified by a class of master regulatory genes, known as **homeotic selector genes**, which set the future developmental pathway of each segment. A selector gene controls the activity of other genes and is required throughout development to maintain this pattern of gene expression. Homologs of the *Drosophila* selector genes that control segment identity have subsequently been discovered in virtually all other animals, where they broadly control identity along the antero-posterior axis, as we shall see in Chapter 5.

2.27 Segment identity in *Drosophila* is specified by Hox genes

The first evidence for the existence of genes that specify segment identity came from unusual and striking mutations in *Drosophila* that produced **homeotic transformations**—the conversion of one segment into another. Eventually the genes that produced these mutations were identified and the intricate way in which they specified segment identity was worked out. The discovery of these genes, now called the **Hox genes**, had an enormous impact on developmental biology. Homologous genes were found in other animals where they acted in a similar way, essentially specifying identity along the antero-posterior axis. Hox genes are now considered as one of the fundamental defining

Fig 2.46 The Antennapedia and bithorax homeotic selector gene complexes. The order of the genes from 3′ to 5′ in each complex reflects both the order of their spatial expression (anterior to posterior) and the timing of expression (3′ earliest).

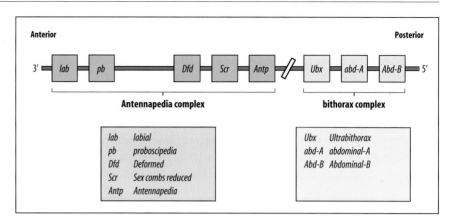

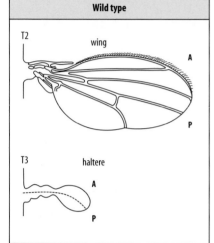

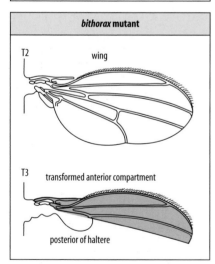

features of animals. In 1995, the American geneticist Edward Lewis shared the Nobel Prize in Physiology or Medicine for his pioneering work on the *Drosophila* homeotic gene complexes and how they function. In *Drosophila*, the Hox genes are organized into two gene clusters, which together make up the HOM-C complex (Fig. 2.46; see also Box 5E, p. 194). The Hox genes all encode transcription factors and take their name from the distinctive **homeobox** DNA motif that encodes part of the DNA-binding site of these proteins.

The two homeotic gene clusters in *Drosophila* are called the **bithorax complex** and the **Antennapedia complex** and are named after the mutations that first revealed their existence. Flies with the *bithorax* mutation have part of the haltere (the balancing organ on the third thoracic segment) transformed into part of a wing (Fig. 2.47), whereas flies with the dominant *Antennapedia* mutation have their antennae transformed into legs. Genes identified by such mutations are called **homeotic genes** because when mutated they result in **homeosis**—the transformation of a whole segment or structure into another related one, as in the transformation of antenna to leg. These bizarre transformations arise out of the homeotic selector genes' key role as identity specifiers. They control the activity of other genes in the segments, thus determining, for example, that a particular imaginal disc will develop as wing or haltere. The bithorax complex controls the development of parasegments 5–14, while the Antennapedia complex controls the identity of the more anterior parasegments. The action of the bithorax complex is the best understood and will be discussed first.

2.28 Homeotic selector genes of the bithorax complex are responsible for diversification of the posterior segments

The bithorax complex of *Drosophila* comprises three homeobox genes: *Ultrabithorax* (*Ubx*), *abdominal-A* (*abd-A*), and *Abdominal-B* (*Abd-B*). These genes are expressed in the parasegments in a combinatorial manner (Fig. 2.48, top panel). *Ubx* is expressed in all parasegments from 5 to 12, *abd-A* is expressed more posteriorly in parasegments 7–13, and *Abd-B* more posteriorly still from parasegment 10 onwards. Because

Fig 2.47 Homeotic transformation of the wing and haltere by mutations in the bithorax complex. Top panel: in the normal adult, both wing and haltere are divided into an anterior (A) and a posterior (P) compartment. Middle panel: the *bithorax* mutation transforms the anterior compartment of the haltere into an anterior wing region. The mutation *postbithorax* acts similarly on the posterior compartment, converting it into posterior wing (not illustrated). Bottom panel: if both mutations are present together, the effect is additive and the haltere is transformed into a complete wing, producing a four-winged fly.

Photograph courtesy of E. Lewis, from Bender, W., et al.: 1983. Illustration after Lawrence, P.: 1992.

the genes are active to varying extents in different parasegments, their combined activities define the character of each parasegment. *Abd-B* also suppresses *Ubx*, such that *Ubx* expression is very low by parasegment 14, as *Abd-B* expression increases. The pattern of activity of the bithorax complex genes is determined by gap and pair-rule genes.

The role of the bithorax complex was first indicated by classical genetic experiments. In larvae lacking the whole of the bithorax complex (see Fig. 2.48, second panel), every parasegment from 5 to 13 develops in the same way and resembles parasegment 4. The bithorax complex is therefore essential for the diversification of these segments, whose basic pattern is represented by parasegment 4. This parasegment can be considered as being a type of 'default' state, which is modified in all the parasegments posterior to it by the proteins encoded by the bithorax complex. It is because the genes of the bithorax complex are able to superimpose a new identity on the default state that they are called selector genes.

An indication of the role of each gene of the bithorax complex can be deduced by looking at embryos constructed so that the genes are put one at a time into embryos lacking the whole complex (see Fig. 2.48, bottom three panels). If only the *Ubx* gene is present, the resulting larva has one parasegment 4, one parasegment 5, and eight parasegments 6. Clearly, *Ubx* has some effect on all parasegments from 5 backward and can specify parasegments 5 and 6. If *abd-A* and *Ubx* are put into the embryo, then the larva has parasegments 4, 5, 6, 7, and 8, followed by five parasegments 9. So *abd-A* affects parasegments from 7 backward, and in combination, *Ubx* and *abd-A* can specify the character of parasegments 7, 8, and 9. Similar principles apply to *Abd-B*, whose domain of influence extends from parasegment 10 backward, and which is expressed most strongly in parasegment 14. Differences between individual segments may reflect differences in the spatial and temporal pattern of Hox gene expression.

These results illustrate an important principle, namely that the character of the parasegments is specified by the genes of the bithorax complex acting in a combinatorial manner. Their combinatorial effect can also been seen by taking the genes away one at a time, from the wild type. In the absence of *Ubx*, for example, parasegments 5 and 6 become converted to parasegment 4 (see Fig. 2.48, bottom panel). There is a further effect on the cuticle pattern in parasegments 7–14 in that structures characteristic of the thorax are now present in the abdomen, showing that *Ubx* exerts an effect in all these segments. Such abnormalities may result from expression of 'nonsense' combinations of the bithorax genes. For example, in such a mutant, abd-A protein is present in parasegments 7–9 without Ubx protein, and this is a combination never found normally. The site of appendages, such as legs, is determined by the Hox genes. For example, expression of the genes *Antp* and *Ubx* specifies the segments from which the second and third pair of legs, respectively, will emerge. The downstream targets of Hox proteins are beginning to be identified, and there may be a very large number of genes whose expression is affected.

Fig 2.48 The spatial pattern of expression of genes of the bithorax complex characterizes each parasegment. In the wild-type embryo (top panel) the expression of the genes *Ultrabithorax, abdominal-A*, and *Abdominal-B* are required to confer an identity on each parasegment. Mutations in the bithorax complex result in homeotic transformations in the character of the parasegments and the segments derived from them. When the bithorax complex is completely absent (second panel), parasegments 5–13 are converted into nine parasegments 4 (corresponding to segment T2 in the larva), as shown by the denticle and bristle patterns on the cuticle. The three lower panels show the transformations caused by the absence of different combinations of genes. When the *Ultrabithorax*. gene alone is absent (bottom panel), parasegments 5 and 6 are converted into 4. In each case the spatial extent of gene expression is detected by *in situ* hybridization (see Box 1D, p. 20). Note that the specification of parasegment 14 is relatively unaffected by the bithorax complex.

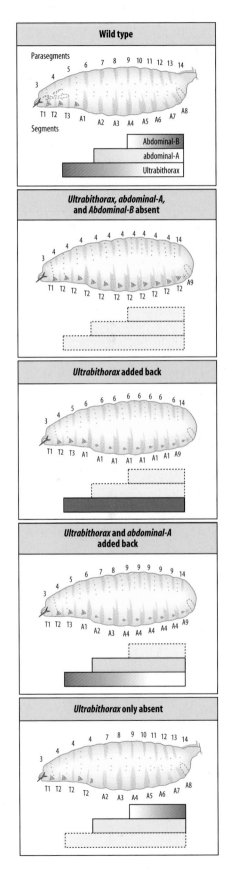

While both the gap and pair-rule proteins initially control the pattern of Hox gene expression, these proteins disappear within about 4 hours. The continued correct expression of the Hox genes involves two groups of genes—the Polycomb and Tritho-rax groups. The proteins of the Polycomb group maintain transcriptional repression of Hox genes where they are initially off, while the Trithorax group maintain expression in those cells where Hox genes have been turned on.

2.29 The Antennapedia complex controls specification of anterior regions

The Antennapedia complex comprises five homeobox genes (see Fig. 2.46), which control the behavior of the parasegments anterior to parasegment 5 in a manner similar to the bithorax complex. Since there are no novel principles involved, we just mention its role briefly. Several genes within the complex are critically involved in specifying particular parasegments. Mutations in the gene *Deformed* affect the ectodermally derived structures of parasegments 0 and 1, those in *Sex combs reduced* affect parasegments 2 and 3, and those in the *Antp* gene affect parasegments 4 and 5, the parasegments that produce the leg imaginal discs. The *Antennapedia* mutation that transforms antennae into legs in the adult fly results from the misexpression of the *Antp* gene in anterior segments in which it is not normally expressed.

2.30 The order of Hox gene expression corresponds to the order of genes along the chromosome

The bithorax and Antennapedia complexes possess some striking features of gene organization. Within both, the order of the genes in the complex is the same as the spatial and temporal order in which they are expressed along the antero-posterior axis during development. *Ubx*, for example, is on the 3′ side of *abd-A* on the chromosome, is more anterior in its pattern of expression, and is activated earlier. As we will see, the related Hox gene complexes of vertebrates, whose ancestors diverged from those of arthropods hundreds of millions of years ago, show the same correspondence between gene order and order of expression. This highly conserved temporal and spatial **co-linearity** must be related to the mechanisms that control the expression of these genes.

The complex yet subtle control to which the genes of the bithorax complex are subject is seen in experiments in which production of Ultrabithorax protein is forced in all segments. This targeted gene misexpression is achieved by linking the Ultra-bithorax protein-coding sequence to a heat-shock promoter—a promoter that is activated at 29°C—and introducing the novel DNA construct into the fly genome using a P element (see Box 2D, p. 61). When this transgenic embryo is given a heat shock for a few minutes, the extra *Ubx* gene is transcribed and the protein is made at high levels in all cells. This has no effect on posterior parasegments (in which Ubx protein is normally present), with the exception of parasegment 5 which, for unknown reasons, is transformed into parasegment 6 (this may reflect a quantitative effect of the Ubx protein). However, all parasegments anterior to 5 are also transformed into parasegment 6. While this seems a reasonably simple and expected result, consider what happens to parasegment 13.

Transcription of *Ubx* is normally suppressed in parasegment 13 in wild-type embryos, yet when the protein itself is produced in the parasegment as a result of heat shock, it has no effect. By some mechanism, the Ultrabithorax protein is rendered inactive in this parasegment. This phenomenon is quite common in the specification of the parasegments and is known as 'phenotypic suppression' or 'posterior dominance': it means that Hox gene products normally expressed in anterior regions are suppressed by more posterior products.

While the roles of the bithorax and Antennapedia complexes in controlling segment identity are well established, we do not yet know very much about how they interact with the genes acting next in the developmental pathway. These are the genes that actually specify the structures that give the segments their unique identities. What, for example, is the pathway that leads to a thoracic rather than an abdominal segment structure? Does each Hox gene activate just a few or many target genes? Do target genes with different functions act in concert to form a particular structure? Answers to these questions will help us understand how changes in a single gene can cause a homeotic transformation like antenna into leg. The role of the Hox genes in the evolution of different body patterns and in the evolution of limbs from fins will be considered in Chapter 15.

2.31 The *Drosophila* head region is specified by genes other than the Hox genes

As mentioned earlier in the chapter, the *Drosophila* larval head anterior to the mandibles is formed by the three most anterior parasegments. The segmental structure is also evident in the embryonic brain, which is composed of three segments (neuromeres). The specification of the anterior head and embryonic nervous system is not, however, under the control of either the pair-rule genes or the Hox genes. Instead, it appears to be determined at the blastoderm stage by the overlapping expression of the genes *orthodenticle*, *empty spiracles*, and *buttonhead*, which all encode gene-regulatory proteins and resemble the gap genes in their effect on phenotype; they are sometimes referred to as the 'head gap genes.' Unlike the gap genes regionalizing the trunk and abdomen, however, they do not regulate each other's activity. Nor is it known whether they work in a combinatorial way like the Hox genes to also confer specific identities on the head segments. *Orthodenticle* and *empty spiracles* have homologs that carry out similar functions in vertebrates, and it is clear that this head-specification system, like the Hox genes, is of ancient evolutionary origin in animals.

The genes *orthodenticle* and *empty spiracles* are examples of homeodomain transcription factors that lie outside the Hox gene clusters. The DNA-binding homeodomain is not confined to the proteins encoded by the Hox gene clusters, and similar domains are found in many other transcription factors involved in development in both *Drosophila* and vertebrates.

Summary

Segment identity is conferred by action of selector or homeotic genes, which direct the development of different parasegments so that each acquires a unique identity. Two clusters of selector genes, known generally as the Hox genes, are involved in specification of segment identity in *Drosophila*: the Antennapedia complex, which controls parasegment identity in the head and first thoracic segment, and the bithorax complex, which acts on the remaining parasegments. Segment identity seems to be determined by the combination of Hox genes active in the segment. The spatial pattern of expression of selector genes along the embryo body is initially determined by the preceding gap gene activity, but selector genes need to be switched on continuously throughout development to maintain the required phenotype. Mutations in genes of the Antennapedia and bithorax complexes can bring about homeotic transformations, in which one segment or structure is changed into a related one, for example an antenna into a leg. The Antennapedia and bithorax complexes are remarkable in that gene order on the chromosome corresponds to the spatial and temporal order of gene expression along the body.

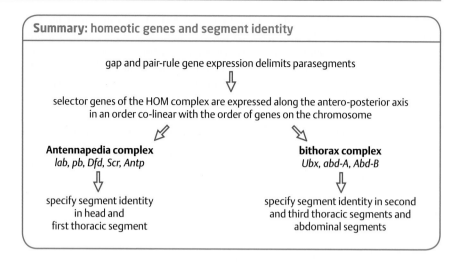

Summary: homeotic genes and segment identity

gap and pair-rule gene expression delimits parasegments

selector genes of the HOM complex are expressed along the antero-posterior axis
in an order co-linear with the order of genes on the chromosome

Antennapedia complex
lab, pb, Dfd, Scr, Antp

bithorax complex
Ubx, abd-A, Abd-B

specify segment identity
in head and
first thoracic segment

specify segment identity in second
and third thoracic segments and
abdominal segments

Summary to Chapter 2

The earliest stages of development in *Drosophila* take place when the embryo is a multinucleate syncytium. During oogenesis, the maternal gene products are deposited in the egg in a particular spatial pattern during oogenesis. This pattern will determine the main body axes and lay down a framework of positional information which, after fertilization, activates a cascade of zygotic gene activity that further patterns the body. The first zygotic genes to be activated along the antero-posterior axis are the gap genes, all encoding transcription factors, whose pattern of expression divides the embryo into a number of regions. The transition to a segmented body organization then starts with the activity of the pair-rule genes, whose sites of expression are specified by the gap gene proteins and which divide the axis into 14 parasegments. Along the dorso-ventral axis, zygotic gene expression also defines several regions, including the future mesoderm and future neural tissue. At the time of pair-rule gene expression, the embryo becomes cellularized and is no longer a syncytium. Segmentation genes pattern the segment and compartment boundaries are involved in patterning and polarizing segments. The identity of the segments is determined by two gene complexes which contain homeotic selector genes, known generally as the Hox genes. The spatial pattern of expression of these genes is largely determined by gap gene activity.

▨ End of chapter questions

Long answer (concept questions)

1. As we will see in future chapters, cleavage divides the eggs of vertebrates and many other animals into a cluster of separate cells. In contrast, the equivalent stage in the *Drosophila* embryo generates a syncytium. What is a syncytium, and how does this influence the early development of *Drosophila*?

2. Much attention has been paid to the process of segmentation in the *Drosophila* embryo. What are segments, and what is their significance to the strategy used by *Drosophila* to carry out pattern formation? Do humans have segments? Give examples.

3. Briefly summarize the mutation screen devised by Nüsslein–Volhard and Wieschaus to find developmental mutants. What role did denticle patterns play in their search for developmental mutants?

4. Developmental genes act hierarchically during pattern formation, first defining broad regions, which are secondarily refined to form a larger number of smaller regions (see Section 2.6). Summarize how this general principle is illustrated by early *Drosophila* embryogenesis.

5. The anterior end of the embryo is established by a locally high level of activity of the transcription factor Bicoid; the ventral side of the embryo is established by a locally high level of activity of the transcription factor Dorsal. Contrast the mechanisms by which these activities come to be localized in the required places.

6. Bicoid was the first morphogen whose mechanism of action was molecularly characterized; an important outcome of this work was the understanding of how a smoothly declining gradient of Bicoid can lead to a sharply delimited boundary of Hunchback expression. Use the Bicoid–Hunchback system to illustrate how an activation

Summary: main genes involved in specifying pattern in the early *Drosophila* embryo

	Gene	Maternal/zygotic	Nature of protein	Transcription factor (T), receptor (R), or signal protein (S)	Function (where known)
Antero-posterior system	*bicoid*	M	Homeodomain	T	Morphogen, provides positional information along AP axis
	hunchback	M/Z	Zinc fingers	T	Morphogen, provides positional information along AP axis
	nanos	M	RNA-binding protein		Helps to establish AP gradient of hunchback protein
	caudal	M	Homeodomain	T	Involved in specifying posterior region
	gurken	M	Secreted protein of TGF-α family	S	Posterior oocyte–follicle cell signaling
	oskar	M			Pole-cell determination
Terminal system	*torso*	M	Receptor tyrosine kinase	R	Terminal specification
	trunk	M		S	Ligand for torso
Gap genes	*hunchback*	Z	Zinc fingers	T	⎫
	Krüppel	Z	Zinc fingers	T	
	knirps	Z	Zinc fingers	T	Localize pair-rule gene expression
	giant	Z	Leucine fingers	T	
	tailless	Z	Zinc fingers	T	⎭
Pair-rule genes	*even-skipped*	Z	Homeodomain	T	Delimits odd-numbered parasegments
	fushi tarazu	Z	Homeodomain	T	Delimits even-numbered parasegments
	hairy	Z	Helix-loop-helix	T	
Segmentation genes	*engrailed*	Z	Homeodomain	T	Defines anterior region of parasegment and posterior region of segment
	hedgehog	Z	Membrane or secreted	S	⎫
	wingless	Z	Secreted	S	
	frizzled	Z	Membrane	R	Components of signaling pathways that pattern segments and stabilize compartment boundaries
	gooseberry	Z	Homeodomain	T	
	patched	Z	Membrane	R	
	smoothened	Z	G-protein-coupled	R	⎭
Selector genes bithorax complex	*Ultrabithorax*	Z	Homeodomain	T	⎫
	abdominal-A	Z	Homeodomain	T	Combinatorial activity confers identity on parasegments 5–13
	Abdominal-B	Z	Homeodomain	T	⎭
Antennapedia complex	*Deformed*	Z	Homeodomain	T	⎫
	Sex combs reduced	Z	Homeodomain	T	Combinatorial activity confers identity on parasegments anterior to 5
	Antennapedia	Z	Homeodomain	T	
	labial	Z	Homeodomain	T	⎭
Maintenance genes	Polycomb group	Z		T	⎫ Maintain state of homeotic genes
	Trithorax	Z		T	⎭
Dorso-ventral system					
Maternal genes	*Toll*	M	Membrane	R	Activation results in dorsal protein entering nucleus
	spätzle	M	Extracellular	S	Ligand for Toll protein
	dorsal	M		T	Morphogen, sets dorso-ventral polarity
	tube	M	Adaptor protein		⎫
	pelle	M	Protein kinase		Components of the Toll signaling pathway leading to Dorsal protein entering nuclei
	cactus	M	Cytoplasmic inhibitor		⎭
	gurken	M	Secreted protein of TGF-α family	S	Specifies oocyte axis
	pipe	M	Sulfotransferase	Enzyme	Part of pathway leading to Spätzle processing
Zygotic genes	*twist*	Z	Helix-loop-helix	T	⎫ Define mesoderm
	snail	Z	Zinc finger	T	⎭
	rhomboid	Z	Membrane protein	S	⎫
	zerknüllt	Z	Homeodomain	T	
	decapentaplegic	Z	Secreted protein of TGF-β family	S	Confer regional identity on dorso-ventral axis
	tolloid	Z	BMP-2 family	S	
	short gastrulation	Z		S	⎭

AP, antero-posterior; TGF, transforming growth factor.

threshold can translate a smooth gradient into a sharp developmental boundary. Be sure to incorporate the role of cooperative binding of Bicoid to the *hunchback* promoter in your answer.

7. The dorsal–ventral axis is patterned by a gradient of nuclear localization of the transcription factor Dorsal. Outline the steps that lead to this gradient of nuclear localization: include the Pipe enzyme, the Spätzle ligand, the Toll receptor, the Cactus protein, and finally Dorsal.

8. The specification of the 14 parasegments along the anterior–posterior axis depends on precise spatial expression of the pair-rule genes. Pair-rule gene expression depends in turn on the gap genes. Using the second stripe of *even-skipped* expression as a model, summarize the interactions that lead to proper expression of Bicoid, Hunchback, Krüppel, and Giant. In particular, describe how the anterior border of Krüppel is set, and how this boundary sets the posterior border of *eve* stripe 2 expression.

9. What is the special role of *engrailed* in the patterning of the *Drosophila* body plan?

10. *Drosophila* has played a central role in our understanding of insect development, but is not necessarily representative of all insects. How does the long-germ development of *Drosophila* differ from the short-germ band type of development seen in beetles, grasshoppers, and wasps?

11. What is meant by a homeotic transformation? Give an example.

12. Experiments demonstrate that the role of the Bithorax Complex is to drive the identity of the parasegments in the posterior portion of the animal away from the default identity of parasegment 4. Describe these experiments. How do these experiments illustrate the importance of the combinations of gene products present in specifying segment identity?

13. The generation of transgenic *Drosophila* using the P element transposon has played a complementary role to traditional mutation analysis in deciphering the genetic regulation of *Drosophila* development. Answer the following questions: Why is the P element important to this process? Why is the P element injected into the posterior of the egg, before any cellularization has occurred? Why are flies mutant for the *white* gene typically used as the hosts for these experiments? Last, what genetic crosses are necessary before the consequences of the P element injection can be observed?

Multiple choice (factual recall questions)

NB There is only one right answer to each question.

1. The 'spinal cord' of *Drosophila*—its 'nerve cord'

a) consists of two parallel structures that run laterally, along the sides of the animal.

b) is in no way analogous to the spinal cord of vertebrates, since *Drosophila* has no spine!

c) runs along the dorsal side, opposite the mouth as is the case in vertebrates such as humans

d) runs along the ventral surface, on the same side as the mouth

2. The process of segmentation will form how many parasegments?

a) 8 in the abdomen

b) 11 to form the thorax and abdomen

c) 14: 3 for the mouth, 3 in the thorax, and 8 in the abdomen

d) 3: head, thorax, and abdomen

3. When do the legs and wings of a fly form?

a) during embryogenesis, after imaginal disc cells are segregated away at the cellular blastoderm stage

b) during pupation, from the imaginal discs

c) during the larval instars

d) legs and wings are present during larval stages, but take their final form only after metamorphosis

4. Which protein (gene) functions to determine an anterior fate in the *Drosophila* embryo?

a) Bicoid

b) Caudal

c) Nanos

d) Torso

5. Maternal determinants are

a) genes that control differentiation as a female or male fruit fly

b) proteins and mRNAs packaged into the egg by the mother

c) proteins and mRNAs that determine the organizing axes of the ovary

d) substances present in the female that determine whether or not she will become a mother

6. Which is the correct hierarchy of gene activity in early *Drosophila* segmentation?

a) gap, segment polarity, pair-rule, maternal

b) maternal, gap, pair-rule, segment polarity

c) maternal, pair-rule, gap, segment-polarity

d) segment polarity, pair-rule, gap, maternal

7. A gap gene mutation could cause which of the following defects in the embryonic body plan?

a) every other segment is missing, resulting in T1, T3, A2, A4, etc., but no T2, A1, A3, and so on

b) patterning within each segment is abnormal, causing denticle belts to form across the entire segment

c) segments A2 through A6 are missing, but the rest of the pattern is essentially normal

d) the identity of one or more segments is transformed to that of a different segment

8. Segments are first positioned at the cell-by-cell level by the

a) gap genes

b) maternal genes

c) pair-rule genes

d) segment polarity genes

9. Engrailed expression defines the

a) anterior compartment of the segment

b) anterior margin of each segment

c) posterior compartment of the segment

d) posterior margin of each parasegment

10. Anterior patterning in the long-germ band wasp *Nasonia* is under the control of

a) Bicoid

b) Engrailed

c) Krüppel

d) Orthodenticle

Multiple choice answer key

1: d, 2: c, 3: b, 4: a, 5: b, 6: b, 7: c, 8: c, 9: c, 10: d.

■ General further reading

Adams, M.D., *et al.*: **The genome sequence of *Drosophila melanogaster*.** *Science* 2000, **267**: 2185–2195.

Ashburner, M.: Drosophila. *A Laboratory Handbook.* New York: Cold Spring Harbor Laboratory Press, 1989.

Lawrence, P.A.: *The Making of a Fly.* Oxford: Blackwell Scientific Publications, 1992.

Matthews, K.A., Kaufman, T.C., Gelbart, W.M.: **Research resources for *Drosophila*: the expanding universe.** *Nat Rev. Genet.* 2005, **6**: 179–193.

Stolc, V., *et al.*: **A gene expression map for the euchromatic genome of *Drosophila melanogaster*.** *Science* 2004, **306**: 655–660.

■ Section further reading

2.2 Cellularization is followed by gastrulation and segmentation

Stathopoulos, A., Levine, M.: **Whole-genome analysis of *Drosophila* gastrulation.** *Curr. Opin. Genet. Dev.* 2004, **14**: 477–484.

2.4 Many developmental mutations have been identified in *Drosophila* through induced large-scale genetic screening

Nusslein-Volhard. C., Wieschaus, E.: **Mutations affecting segment number and polarity in *Drosophila*.** *Nature* 1980, **287**: 795–801.

Rørth, P.: **A modular misexpression screen in *Drosophila* detecting tissue-specific phenotypes.** *Proc. Natl Acad. Sci. USA* 1996, **93**: 12418–12422.

2.7 Three classes of maternal genes specify the antero-posterior axis

St Johnston, D., Nüsslein-Volhard, C.: **The origin of pattern and polarity in the *Drosophila* embryo.** *Cell* 1992, **68**: 201–219.

2.8 Bicoid protein provides an antero-posterior gradient of a morphogen

Driever, W., Nüsslein-Volhard, C.: **The bicoid protein determines position in the *Drosophila* embryo in a concentration dependent manner.** *Cell* 1988, **54**: 95–104.

Ephrussi, A., St Johnston, D.: **Seeing is believing: the bicoid morphogen gradient matures.** *Cell* 2004, **116**: 143–152.

Gibson, M.C.: **Bicoid by the numbers: quantifying a morphogen gradient.** *Cell* 2007, **130**: 14–16.

Spirov, A., Fahmy, K., Schneider, M., Frei, E., Noll, M., Baumgartner, S.: **Formation of the bicoid morphogen gradient: an mRNA gradient dictates the protein gradient.** *Development* 2009, **136**: 605–614.

Teleman, A.A., Strigini, M., Cohen, S.M.: **Shaping morphogen gradients.** *Cell* 2001, **105**: 559–562.

2.9 The posterior pattern is controlled by the gradients of Nanos and Caudal proteins

Irish, V., Lehmann, R., Akam, M.: **The *Drosophila* posterior-group gene *nanos* functions by repressing *hunchback* activity.** *Nature* 1989, **338**: 646–648.

Murafta, Y., Wharton, R.P.: **Binding of pumilio to maternal *hunchback* mRNA is required for posterior patterning in *Drosophila* embryos.** *Cell* 1995, **80**: 747–756.

Rivera-Pomar, R., Lu, X., Perrimon, N., Taubert, H., Jackle, H.: **Activation of posterior gap gene expression in the *Drosophila* blastoderm.** *Nature* 1995, **376**: 253–256.

Struhl, G.: **Differing strategies for organizing anterior and posterior body pattern in *Drosophila* embryos.** *Nature* 1989, **338**: 741–744.

2.10 The anterior and posterior extremities of the embryo are specified by cell-surface receptor activation

Casali, A., Casanova, J.: **The spatial control of Torso RTK activation: a C-terminal fragment of the Trunk protein acts as a signal for Torso receptor in the *Drosophila* embryo.** *Development* 2001, **128**: 1709–1715.

Casanova, J., Struhl, G.: **Localized surface activity of torso, a receptor tyrosine kinase, specifies terminal body pattern in *Drosophila*.** *Genes Dev.* 1989, **3**: 2025–2038.

Coppey, M., Boettiger, A.N., Berezhkovskii, A.M., Shvartsman, S.Y.: **Nuclear trapping shapes the terminal gradient in the *Drosophila* embryo.** *Curr. Biol.* 2008, **18**: 915–919.

Furriols, M., Casanova, J.: **In and out of Torso RTK signalling.** *EMBO J.* 2003, **22**: 1947–1952.

Furriols, M., Ventura, G., Casanova, J.: **Two distinct but convergent groups of cells trigger Torso receptor tyrosine kinase activation by independently expressing torso-like.** *Proc. Natl Acad. Sci. USA* 2007, **104**: 11660–11665.

Li, W.X.: **Functions and mechanisms of receptor tyrosine kinase Torso signaling: lessons from *Drosophila* embryonic terminal development.** *Dev. Dyn.* 2005, **232**: 656–672.

2.11 The dorso-ventral polarity of the embryo is specified by localization of maternal proteins in the egg vitelline envelope

Anderson, K.V.: **Pinning down positional information: dorsal-ventral polarity in the *Drosophila* embryo.** *Cell* 1998, **95**: 439–442.

Sen, J., Goltz, J.S., Stevens, L., Stein, D.: **Spatially restricted expression of** *pipe* **in the** *Drosophila* **egg chamber defines embryonic dorsal-ventral polarity.** *Cell* 1998, **95**: 471–481.

2.12 Positional information along the dorso-ventral axis is provided by the Dorsal protein

Belvin, M.P., Anderson, K.V.: **A conserved signaling pathway: the** *Drosophila* **Toll-dorsal pathway.** *Annu. Rev. Cell Dev. Biol.* 1996, **12**: 393–416.

Imler, J.L., Hoffmann, J.A.: **Toll receptors in** *Drosophila*: **a family of molecules regulating development and immunity.** *Curr. Top. Microbiol. Immunol.* 2002, **270**: 63–79.

Roth, S., Stein, D., Nüsslein-Volhard, C.: **A gradient of nuclear localization of the dorsal protein determines dorso-ventral pattern in the** *Drosophila* **embryo.** *Cell* 1989, **59**: 1189–1202.

Steward, R., Govind, R.: **Dorsal-ventral polarity in the** *Drosophila* **embryo.** *Curr. Opin. Genet. Dev.* 1993, **3**: 556–561.

2.13 The antero-posterior axis of the *Drosophila* egg is specified by signals from the preceding egg chamber and by interactions of the oocyte with follicle cells

González-Reyes, A., St Johnston, D.: **The** *Drosophila* **AP axis is polarised by the cadherin-mediated positioning of the oocyte.** *Development* 1998, **125**: 3635–3644.

Huynh, J.R., St Johnston, D.: **The origin of asymmetry: early polarization of the** *Drosophila* **germline cyst and oocyte.** *Curr. Biol.* 2004, **14**: R438–R449.

Riechmann, V., Ephrussi, A.: **Axis formation during** *Drosophila* **oogenesis.** *Curr. Opin. Genet. Dev.* 2001, **11**: 374–383.

Torres, I.L., Lopez-Schier, H., St Johnston, D.: **A Notch/Delta-dependent relay mechanism establishes anterior-posterior polarity in** *Drosophila*. *Dev. Cell* 2005, **5**: 547–558.

Zimyanin, V.L., Belaya, K., Pecreaux, J., Gilchrist, M.J., Clark, A., Davis, I., St Johnston, D.: **In vivo imaging of** *oskar* **mRNA transport reveals the mechanism of posterior localization.** *Cell* 2008, **134**: 843–853.

2.14 Localization of maternal mRNAs to either end of the egg depends on the reorganization of the oocyte cytoskeleton

Becalska, A.N., Gavis, E.R.: **Lighting up mRNA localization in** *Drosophila* **oogenesis.** *Development* 2009, **136**: 2493–2503.

Martin, S.G., Leclerc, V., Smith-Litière, K., St Johnston, D.: **The identification of novel genes required for** *Drosophila* **antero-posterior axis formation in a germline clone screen using GFP-Staufen.** *Development* 2003, **130**: 4201–4215.

St Johnston, D.: **Moving messages: the intracellular localization of mRNAs.** *Nat Rev. Mol. Cell Biol.* 2005, **6**: 363–375.

2.15 The dorso-ventral axis of the egg is specified by movement of the oocyte nucleus followed by signaling between oocyte and follicle cells

González-Reyes, A., Elliott, H., St Johnston, D.: **Polarization of both major body axes in** *Drosophila* **by gurken-torpedo signaling.** *Nature* 1995, **375**: 654–658.

Jordan, K.C., Clegg, N.J., Blasi, J.A., Morimoto, A.M., Sen, J., Stein, D., McNeill, H., Deng, W.M., Tworoger, M., Ruohola-Baker, H.: **The homeobox gene** *mirror* **links EGF signalling to embryonic dorso-ventral axis formation through Notch activation.** *Nat. Genet.* 2000, **24**: 429–433.

Roth, S., Neuman-Silberberg, F.S., Barcelo, G., Schupbach, T.: **Cornichon and the EGF receptor signaling process are necessary for both anterior-posterior and dorsal-ventral pattern formation in** *Drosophila*. *Cell* 1995, **81**: 967–978.

van Eeden, F., St Johnston, D.: **The polarisation of the anterior-posterior and dorsal-ventral axes during** *Drosophila* **oogenesis.** *Curr. Opin. Genet. Dev.* 1999, **9**: 396–404.

2.16 The antero-posterior axis is divided up into broad regions by gap-gene expression

Hülskamp, M., Tautz, D.: **Gap genes and gradients—the logic behind the gaps.** *BioEssays* 1991, **13**: 261–268.

2.17 Bicoid protein provides a positional signal for the anterior expression of zygotic *hunchback*

Brand, A.H., Perrimon, N.: **Targeted gene expression as a means of altering cell fates and generating dominant phenotypes.** *Development* 1993, **118**: 401–415.

Gibson, M.C.: **Bicoid by the numbers: quantifying a morphogen gradient.** *Cell* 2007, **130**: 14–16.

Gregor, T., Tank, D.W., Wieschaus, E.F., Bialek W.: **Probing the limits to positional information.** *Cell* 2007, **130**: 153–164.

Ochoa-Espinosa, A., Yucel, G., Kaplan, L., Pare, A., Pura, N., Oberstein, A., Papatsenko, D., Small, S.: **The role of binding site cluster strength in Bicoid-dependent patterning in** *Drosophila*. *Proc. Natl Acad. Sci. USA* 2005, **102**: 4960–4965.

Simpson-Brose, M., Treisman, J., Desplan, C.: **Synergy between the Hunchback and Bicoid morphogens is required for anterior patterning in** *Drosophila*. *Cell* 1994, **78**: 855–865.

Struhl, G., Struhl, K., Macdonald, P.M.: **The gradient morphogen Bicoid is a concentration-dependent transcriptional activator.** *Cell* 1989, **57**: 1259–1273.

2.18 The gradient in Hunchback protein activates and represses other gap genes

Rivera-Pomar, R., Jäckle, H.: **From gradients to stripes in** *Drosophila* **embryogenesis: filling in the gaps.** *Trends Genet.* 1996, **12**: 478–483.

Struhl, G., Johnston, P., Lawrence, P.A.: **Control of** *Drosophila* **body pattern by the hunchback morphogen gradient.** *Cell* 1992, **69**: 237–249.

Wu, X., Vakani, R., Small, S.: **Two distinct mechanisms for differential positioning of gene expression borders involving the** *Drosophila* **gap protein giant.** *Development* 1998, **125**: 3765–3774.

2.19 The expression of zygotic genes along the dorso-ventral axis is controlled by Dorsal protein

Cowden, J., Levine, M.: **Ventral dominance governs segmental patterns of gene expression across the dorsal-ventral axis of the neurectoderm in the** *Drosophila* **embryo.** *Dev. Biol.* 2003, **262**: 335–349.

Harland, R.M.: **A twist on embryonic signalling.** *Nature* 2001, **410**: 423–424.

Markstein, M., Zinzen, R., Markstein, P., Yee, K.P., Erives, A., Stathopoulos, A., Levine, M.: **A regulatory code for neurogenic gene expression in the** *Drosophila* **embryo.** *Development* 2004, **131**: 2387–2394.

Rusch, J., Levine, M.: **Threshold responses to the dorsal regulatory gradient and the subdivision of primary tissue territories in the** *Drosophila* **embryo**. *Curr. Opin. Genet. Dev.* 1996, **6**: 416–423.

Stathopoulos, A., Levine, M.: **Dorsal gradient networks in the** *Drosophila* **embryo**. *Dev. Biol.* 2002, **246**: 57–67.

2.20 The Decapentaplegic protein acts as a morphogen to pattern the dorsal region

Affolter, M., Basler, K.: **The Decapentaplegic morphogen gradient: from pattern formation to growth regulation**. *Nat. Rev. Genet.* 2007, **8**: 663–674.

Ashe, H.L., Levine, M.: **Local inhibition and long-range enhancement of Dpp signal transduction by Sog**. *Nature* 1999, **398**: 427–431.

Mizutani, C.M., Nie, Q., Wan, F.Y., Zhang, Y.T., Vilmos, P., Sousa-Neves, R., Bier, E., Marsh, J.L., Lander, A.D.: **Formation of the BMP activity gradient in the** *Drosophila* **embryo**. *Dev. Cell* 2005, **8**: 915–924.

Shimmi, O., Umulis, D., Othmer, H., O'Connor, M.B.: **Facilitated transport of a Dpp/Scw heterodimer by Sog/Tsg leads to robust patterning of the** *Drosophila* **blastoderm embryo**. *Cell* 2005, **120**: 873–886.

Srinivasan, S., Rashka, K.E., Bier, E.: **Creation of a Sog morphogen gradient in the** *Drosophila* **embryo**. *Dev. Cell* 2002, **2**: 91–101.

Wang, Y.-C., Ferguson, E.L.: **Spatial bistability of Dpp-receptor interactions driving** *Drosophila* **dorsal-ventral patterning**. *Nature* 2005, **434**: 229–234.

Wharton, K.A., Ray, R.P., Gelbart, W.M.: **An activity gradient of decapentaplegic is necessary for the specification of dorsal pattern elements in the** *Drosophila* **embryo**. *Development* 1993, **117**: 807–822.

2.22 Gap-gene activity positions stripes of pair-rule gene expression

Clyde, D.E., Corado, M.S., Wu, X., Paré, A., Papatsenko, D., Small, S.: **A self-organizing system of repressor gradients establishes segmental complexity in** *Drosophila*. *Nature* 2003, **426**: 849–853.

Luengo Hendriks, C.L., Keränen, S.V., Fowlkes, C.C., Simirenko, L., Weber, G.H., DePace, A.H., Henriquez, C., Kaszuba, D.W., Hamann, B., Eisen, M.B., Malik, J., Sudar D., Biggin, M.D., Knowles, D.W.: **Three-dimensional morphology and gene expression in the** *Drosophila* **blastoderm at cellular resolution I: data acquisition pipeline**. *Genome Biol.* 2006, **7**: R123.

Small, S., Levine, M.: **The initiation of pair-rule stripes in the** *Drosophila* **blastoderm**. *Curr. Opin. Genet. Dev.* 1991, **1**: 255–260.

Yu, D., Small, D.: **Precise registration of gene expression boundaries by a repressive morphogen in** *Drosophila*. *Curr. Biol.* 2008, **18**: 888–876.

2.23 Expression of the *engrailed* gene delimits a cell-lineage boundary and defines a compartment

Dahmann, C., Basler, K.: **Compartment boundaries: at the edge of development**. *Trends Genet.* 1999, **15**: 320–326.

Gray, S., Cai, H., Barolo, S., Levine, M.: **Transcriptional repression in the** *Drosophila* **embryo**. *Phil. Trans. R. Soc. Lond.* 1995, **349**: 257–262.

Harrison, D.A., Perrimon, N.: **Simple and efficient generation of marked clones in** *Drosophila*. *Curr. Biol.* 1993, **3**: 424–433.

Vincent, J.P., O'Farrell, P.H.: **The state of** *engrailed* **expression is not clonally transmitted during early** *Drosophila* **development**. *Cell* 1992, **68**: 923–931.

2.24 Segmentation genes stabilize parasegment boundaries and set up a focus of signaling at the boundary that patterns the segment

Alexandre, C., Lecourtois, M., Vincent, J.-P.: **Wingless and hedgehog pattern** *Drosophila* **denticle belts by regulating the production of short-range signals**. *Development* 1999, **126**: 5689–5698.

Beckett, K., Franch-Marro, X., Vincent, J.P.: **Glypican-mediated endocytosis of hedgehog has opposite effects in flies and mice**. *Trends Cell Biol.* 2008, **18**: 360–363.

Bejsovec, A., Wieschaus, E.: **Segment polarity gene interactions modulate epidermal patterning in** *Drosophila* **embryos**. *Development* 1993, **119**: 501–517.

Briscoe, J., Therond, P.: **Hedgehog signaling: from the** *Drosophila* **cuticle to anti-cancer drugs**. *Dev. Cell* 2005, **8**: 143–151.

Gordon, M., Nusse, R.: **Wnt signaling: multiple pathways: multiple receptors, and multiple transcription factors**. *J. Biol. Chem.* 2006, **281**: 22429–22433.

Hooper, J.E., Scott, M.P.: **Communicating with Hedgehogs**. *Nat. Rev. Mol. Cell Biol.* 2005, **6**: 306–317.

Johnson, R.L., Scott, M.P.: **New players and puzzles in the Hedgehog signaling pathway**. *Curr. Opin. Genet. Dev.* 1998, **8**: 450–456.

Larsen, C.W., Hirst, E., Alexandre, C., Vincent, J.-P.: **Segment boundary formation in** *Drosophila* **embryos**. *Development* 2003, **130**: 5625–5635.

Lawrence, P.A., Casal, J., Struhl, G.: **Hedgehog and engrailed: pattern formation and polarity in the** *Drosophila* **abdomen**. *Development* 1999, **126**: 2431–2439.

Tolwinski, N.S., Wieschaus, E.: **Rethinking Wnt signaling**. *Trends Genet.* 2004, **20**: 177–181

von Dassow, G., Meir, E., Munro, E.M., Odell, G.M.: **The segment polarity network is a robust developmental module**. *Nature* 2000, **406**: 188–192.

Wehrli, M., Dougan, S.T., Caldwell, K., O'Keefe, L., Schwartz, S., Vaizel-Ohayon, D., Schejter, E., Tomlinson, A., DiNardo, S.: **arrow encodes an LDL-receptor-related protein essential for Wingless signalling**. *Nature* 2000, **407**: 527–530.

Zhu, A., Scott, M.: **Incredible journey: how do developmental signals travel through tissue?** *Genes Dev.* 2004, **18**: 2983–2997.

2.25 Insect epidermal cells become individually polarized in an antero-posterior direction in the plane of the epithelium & Box 2F Planar cell polarity in *Drosophila*

Lawrence, P.A., Casal, J., Struhl, G.: **Cell interactions and planar polarity in the abdominal epidermis of** *Drosophila*. *Development* 2004, **131**: 4651–4664.

Lawrence, P.A., Struhl, G., Casal, J.: **Planar polarity: one or two pathways**. *Nat. Rev. Genet.* 2008, **8**: 555–363.

Ma, D., Yang, C., McNeill, H., Simon, M.A., Axelrod, J.D.: **Fidelity in planar cell polarity signalling**. *Nature* 2003, **421**: 543–547.

Strutt, D.: **The planar polarity pathway**. *Curr. Biol.* 2008, **16**: R898–R902.

Strutt, D., Strutt, H.: **Differential activities of the core planar polarity proteins during *Drosophila* wing patterning**. *Dev. Biol.* 2007, **302**: 181–194.

Wu, J., Mlodzik, M.: **The Frizzled extracellular domain is a ligand for Van Gogh/Stm during nonautonomous planar cell polarity signaling**. *Dev. Cell* 2008, **15**: 462–469.

2.26 Some insects use different mechanisms for patterning the body plan

Akam, M., Dawes, R.: **More than one way to slice an egg**. *Curr. Biol.* 1992, **8**: 395–398.

Angelini, D.R., Liu, P.Z., Hughes, C.L., Kaufman, T.C.: **Hox gene function and interaction in the milkweed bug *Oncopeltus fasciatus* (Hemiptera)**. *Dev. Biol.* 2005, **287**: 440–455.

Brown, S.J., Parrish, J.K., Beeman, R.W., Denell, R.E.: **Molecular characterization and embryonic expression of the *even-skipped* ortholog of *Tribolium castaneum***. *Mech. Dev.* 1997, **61**: 165–173.

French, V.: **Segmentation (and *eve*) in very odd insect embryos**. *BioEssays* 1996, **18**: 435–438.

French, V.: **Insect segmentation: genes, stripes and segments in 'Hoppers'**. *Curr. Biol.* 2001, **11**: R910–R913.

Lynch, J.A., Brent, A.E., Leaf, D.S., Pultz, M.A., Desplan, C.: **Localized maternal *orthodenticle* patterns anterior and posterior in the long germ wasp *Nasonia***. *Nature* 2006, **439**: 728–732.

Sander, K.: **Pattern formation in the insect embryo**. In *Cell Patterning, Ciba Found. Symp. 29*. London: Ciba Foundation, 1975: 241–263.

Tautz, D., Sommer, R.J.: **Evolution of segmentation genes in insects**. *Trends Genet.* 1995, **11**: 23–27.

2.27 Segment identity in *Drosophila* is specified by Hox genes

Lewis E.B.: **A gene complex controlling segmentation in *Drosophila***. *Nature* 1978, **276**: 565–570.

Stark, A.: **A single Hox locus in *Drosophila* produces functional microRNAs from opposite DNA strands**. *Genes Dev.* 2008, **22**: 8–13.

2.28 Homeotic selector genes of the bithorax complex are responsible for diversification of the posterior segments

Castelli-Gair, J., Akam, M.: **How the Hox gene *Ultrabithorax* specifies two different segments: the significance of spatial and temporal regulation within metameres**. *Development* 1995, **121**: 2973–2982.

Duncan, I.: **How do single homeotic genes control multiple segment identities?** *BioEssays* 1996, **18**: 91–94.

Lawrence, P.A, Morata, G.: **Homeobox genes: their function in *Drosophila* segmentation and pattern formation**. *Cell* 1994, **78**: 181–189.

Liang, Z., Biggin, M.D.: **Eve and ftz regulate a wide array of genes in blastoderm embryos: the selector homeoproteins directly or indirectly regulate most genes in *Drosophila***. *Development* 1998, **125**: 4471–4482.

Mann, R.S., Morata, G.: **The developmental and molecular biology of genes that subdivide the body of *Drosophila***. *Annu. Rev. Cell Dev. Biol.* 2000, **16**: 143–271.

Mannervik, M.: **Target genes of homeodomain proteins**. *BioEssays* 1999, **4**: 267–270.

Simon, J.: **Locking in stable states of gene expression: transcriptional control during *Drosophila* development**. *Curr. Opin. Cell Biol.* 1995, **7**: 376–385.

2.30 The order of Hox gene expression corresponds to the order of genes along the chromosome

Morata, G.: **Homeotic genes of *Drosophila***. *Curr. Opin. Genet. Dev.* 1993, **3**: 606–614.

2.31 The *Drosophila* head region is specified by genes other than the Hox genes

Rogers B.T., Kaufman, T.C.: **Structure of the insect head in ontogeny and phylogeny: a view from *Drosophila***. *Int. Rev. Cytol.* 1997, **174**: 1–84.

Vertebrate development I: life cycles and experimental techniques

- Vertebrate life cycles and outlines of development
- Experimental approaches to studying vertebrate development

In this chapter we examine the similarities and differences in the outlines of development of four vertebrate model animals: amphibians represented by the frog Xenopus, fish by the zebrafish, birds by the chick, and mammals by the mouse. The second part of the chapter introduces some of the experimental approaches used to investigate vertebrate development.

We saw in Chapter 2 how early development in insects is largely under the control of maternal factors that interact with each other to specify broadly different regions of the body. This initial blueprint is then elaborated on by the embryo's own genes. We shall now look at how the same task of establishing the outline of the body plan is achieved in early vertebrate development. All vertebrates, despite their many outward differences, have a similar basic body plan. The defining structures are the segmented backbone or **vertebral column** surrounding the spinal cord, with the brain at the head end enclosed in a bony or cartilaginous skull (Fig. 3.1). These prominent structures mark the **antero-posterior axis**, the main body axis of vertebrates. The head is at the anterior end of this axis, followed by the trunk with its paired appendages—limbs in terrestrial vertebrates (with the exception of snakes) and fins in fish—and in many vertebrates this axis terminates at the posterior end in a post-anal tail. The vertebrate body also has a distinct **dorso-ventral axis** running from the back to the belly, with the spinal cord running along the dorsal side and the mouth defining the ventral side. The antero-posterior and dorso-ventral axes together define the left and right sides of the animal. Vertebrates have a general **bilateral symmetry** around the dorsal midline so that outwardly the right and left sides are mirror images of each other. Some internal organs, such as lungs, kidneys, and gonads, are also present as symmetrically paired structures, but single organs such as the heart and liver are arranged asymmetrically with respect to the dorsal midline with the heart on the left and the liver on the right.

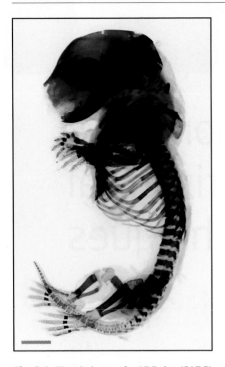

Fig. 3.1 The skeleton of a 17.5-day (E17.5) mouse embryo illustrates the vertebrate body plan. The skeletal elements in this embryo have been stained with Alcian blue (which stains cartilage) and Alizarin red (which stains bone). The vertebral column, which develops from blocks of somites, is divided up into cervical (neck), thoracic (chest), lumbar (lower back), and sacral (hip and lower) regions. The paired limbs can also be seen. Scale bar = 1 mm.

Photograph courtesy of M. Maden.

In this and the following two chapters we will look at four vertebrates whose early development has been particularly well studied—*Xenopus*, zebrafish, chicken, and mouse. We also touch on the early development of a human embryo. In this chapter, we will first outline the life cycles of these organisms and briefly describe the key features of their early development, providing the context for the later discussions. We will not at this stage consider the mechanisms of development, but focus on the changes in form that occur in early embryonic development to give the well-defined overall body plan of the vertebrate embryo, with its characteristic structures such as the notochord and neural tube (see Fig. 3.8). We will take the embryo from the earliest cell divisions of the fertilized egg, through the cell movements of gastrulation, the dynamic process in which the embryo is rearranged to form the three main germ layers in their correct positions in the body, to the end of neurulation, which forms the neural tube—the earliest appearance of the nervous system. The germ layers are the ectoderm, which gives rise to the nervous system and epidermis, the mesoderm, which gives rise to the skeleton, muscles, heart, blood, and some other internal organs and tissues, and the endoderm, which gives rise to the gut and associated glands and organs (see Box 1C, p. 15). In the second half of the chapter we will introduce some of the experimental techniques that are used to study vertebrate development.

Chapter 4 then looks in detail at the first two key events in development: the establishment of the antero-posterior and dorso-ventral axes, and the specification and early patterning of the germ layers. In particular, we will look at the setting up of the organizer regions (see Section 1.4) and the induction of the mesodermal germ layer and its patterning along the dorso-ventral axis. In Chapter 5 we continue the discussion of early development in vertebrates by looking at the role of the organizer region in directing development along the antero-posterior axis during and after the large-scale reorganization of tissues that occurs during gastrulation. This reorganization leads to the all-important induction of the prospective neural tissue from ectoderm—from which all the nervous system develops—and the formation of the notochord and somites from mesoderm, the somites subsequently developing into muscle and skeletal tissues (see Box 1B, p. 7). By this stage the embryo has become recognizably vertebrate. The development of individual structures and organs, such as limbs, eyes, heart, and the nervous system, is covered in later chapters. We shall take a more detailed look at the mechanics of vertebrate gastrulation and the underlying cell biology that allows such extensive remodeling of tissues in Chapter 8, along with gastrulation in simpler non-vertebrate animals such as the sea urchin.

Vertebrate life cycles and outlines of development

All vertebrate embryos pass through a broadly similar set of developmental stages, which are outlined for each organism in the life-cycle diagrams in this chapter (for example, the one for *Xenopus* in Fig. 3.3). After fertilization, the zygote undergoes cleavage. These are rapid cell divisions by which the embryo becomes divided into a number of smaller cells. This is followed by gastrulation, a set of cell movements that generates the three distinct cell layers: ectoderm, mesoderm, and endoderm (see Box 1C, p. 15). By the end of gastrulation, the ectoderm covers the embryo, and the mesoderm and endoderm have moved inside. The endoderm gives rise to the gut, and to its derivatives such as liver and lungs; the mesoderm forms the skeleton, muscles, connective tissues, kidneys, heart, and blood, as well as some other tissues; and the ectoderm gives rise to the epidermis and the nervous system.

Figure 3.2 shows the differences in form of the early embryos from the four model species and of a human embryo at the same stage. After gastrulation, all vertebrate embryos pass through a stage at which they all more or less resemble each other and show the specific embryonic features of the chordates, the phylum of animals to which vertebrates belong. This stage is called the **phylotypic stage**. At the phylotypic stage the head is distinct and the **neural tube**, the forerunner of the nervous system, runs along the dorsal midline of the antero-posterior (head to tail) axis. Immediately under the neural tube runs the **notochord**, a signature structure of chordates, flanked on either side by the **somites**, blocks of tissue from which the muscles and skeleton will derive. Features special to the different vertebrate groups, such as beaks, wings, and fins, appear later.

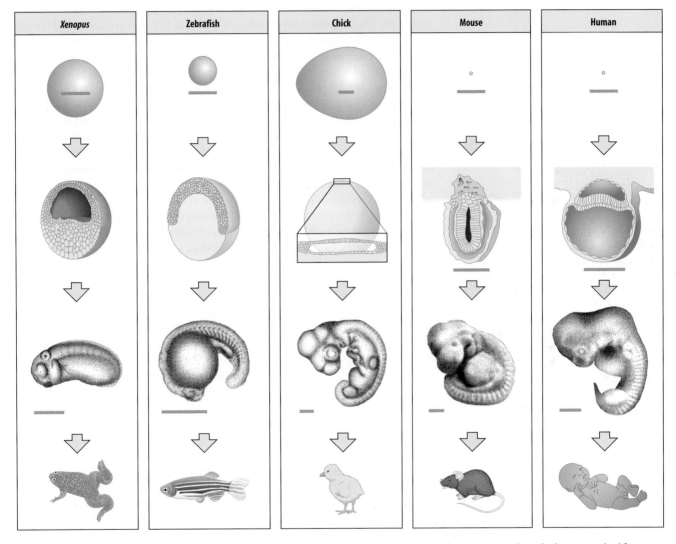

Fig. 3.2 **Vertebrate embryos show considerable differences in form before gastrulation but subsequently all go through a stage at which they look similar.** The eggs of the frog (*Xenopus*), zebrafish, chicken, and mouse are very different in size, while human eggs are about the same size as mouse eggs (top row). Scale bars in this row all represent 1 mm, except for the chicken egg, which represents 10 mm. Their early development (second row) is rather different. In this row, the embryos are shown in cross-section at the stage corresponding roughly to the *Xenopus* blastula (left panel) just before gastrulation commences. The main determinant of tissue arrangement is the amount of yolk (yellow) in the egg. The human and mouse embryos (on the right) at this stage have implanted into the uterine wall and

thus already developed some extra-embryonic tissues required for implantation. The mouse embryo proper is the small cup-shaped structure at the center, seen here in cross-section as a U-shaped layer of epithelium. The human embryo proper is a two-layered disc of cells. Scale bars for mouse and human embryos = 100 μm. After gastrulation and formation of the neural tube, vertebrate embryos pass through an embryonic stage at which they all look rather similar (third row), and which is known as the phylotypic stage. The body has developed, and neural tube, somites, notochord, and head structures are present. Scale bars = 1 mm. After this stage their development diverges again. Paired appendages, for example, develop into fins in fish, wings and legs in the chick and arms and legs in the human (bottom row).

The rod-shaped notochord is one of the earliest mesodermal structures that can be recognized in vertebrate embryos. This structure does not persist and its cells eventually become incorporated into the column of vertebrae that form the spine (see Fig. 3.1). The rest of the vertebral column, the skeleton of the trunk, and the muscles of the trunk and limbs develop from the somites, blocks of cells that form in an antero-posterior sequence from the mesoderm on either side of the notochord. Towards the end of gastrulation, the ectoderm overlying the notochord becomes specified as neural tissue and then rolls up to form the neural tube, from which the brain, spinal cord, and the rest of the nervous system derive (see Fig. 3.7). The overall similarity of the body plan in all vertebrates suggests that the developmental processes that establish it are broadly similar in the different animals. This is largely the case, although there are also considerable differences in development, especially at the earliest stages.

The differences in development in the model organisms particularly relate to how and when the axes are set up, and how the germ layers are established, as we shall see in Chapter 4, and are mainly due to the different modes of reproduction and the consequent form of the earliest embryo. Yolk provides all the nutrients for fish, amphibian, reptile, and bird embryonic development, and for the few egg-laying mammals such as the platypus. The eggs of most mammals, by contrast, are small and non-yolky, and the embryo is nourished for the first few days by fluids in the oviduct and uterus. Once implanted in the uterus wall, the mammalian embryo develops specialized **extra-embryonic membranes** that surround and protect the embryo and through which it receives nourishment from the mother via the **placenta**. Avian embryos also develop extra-embryonic membranes for obtaining nutrients from the yolk, for oxygen and carbon dioxide exchange through the permeable egg shell, and for waste disposal. Birds and mammals, both of which form an extra-embryonic membrane called an amnion, are known as **amniotes**, whereas amphibians and fish, which do not form extra-embryonic membranes, are known as **anamniotes**.

To study development, it is necessary to have a reliable way of identifying and referring to a particular stage of development. Simply measuring the time from fertilization is not satisfactory for most species, as there is considerable variation. Amphibians, for example, will develop quite normally over a range of temperatures, but the rate of development is quite different at different temperatures. Developmental biologists, therefore, divide the normal embryonic development of each species into a series of numbered stages, which are identified by their main features rather than by time after fertilization. A stage-10 *Xenopus* embryo, for example, refers to an embryo at a very early stage of gastrulation, whereas stage 54 is the fully developed tadpole with legs. Numbered stages have similarly been characterized for the chick embryo, and these provide a much finer staging system than measuring the time since the egg was laid. Even for mouse embryos, which develop in a much more constant environment inside the mother, there is considerable variation in developmental timing, even among embryos in the same litter. Therefore, although early mouse embryos are often staged according to days post-conception, with the morning on which the vaginal plug appears called 0.5 dpc, or E0.5 (embryonic day 0.5), this is only a guide, and once somites have formed, somite number can be used as a more accurate indication of developmental stage. Links to websites that describe all the stages for our four model vertebrates are given in General further reading. The techniques that can be used to study these four vertebrates are discussed in the later part of the chapter.

3.1 The frog *Xenopus laevis* is the model amphibian for developmental studies

The amphibian species most commonly used for developmental work is the African claw-toed frog, *Xenopus laevis*, which is completely aquatic and able to develop normally in tap water. *Xenopus laevis* is a tetraploid species, however, and the diploid species *X. tropicalis* is becoming increasingly used for genetic studies (see Section 1.6). Throughout this book, *Xenopus* will refer to *X. laevis*, unless otherwise specified. A great advantage of *Xenopus* is that its fertilized eggs are easy to obtain; females and males injected with the human hormone chorionic gonadotropin and put together overnight will mate, and the female will lay hundreds of eggs in the water, which are fertilized by sperm released by the male. Eggs can also be fertilized in a dish by adding sperm to eggs released after hormonal stimulation of the female. One advantage of artificial fertilization is that because the resulting embryos develop synchronously, many embryos, all at the same stage, can be obtained. The eggs are large (1.2–1.4 mm in diameter) and so are quite easily manipulated. The embryos of *Xenopus* are extremely hardy and are highly resistant to infection after microsurgery. It is also easy to culture fragments of early *Xenopus* embryos in a simple, chemically defined solution.

The *Xenopus* life cycle and main developmental stages are summarized in Fig. 3.3. The mature *Xenopus* egg has a distinct polarity, with a dark, pigmented **animal region** and a pale, yolky, and heavier **vegetal region** (Fig. 3.4). The axis running from the animal pole to the vegetal pole is known as the **animal–vegetal axis**. Before fertilization, the egg is enclosed in a protective **vitelline membrane**, which is embedded in a gelatinous coat. Meiosis is not yet complete: the first meiotic division has resulted in a small cell—a **polar body**—forming at the animal pole, but the second meiotic division is completed only after fertilization, when the second polar body also forms at the animal pole (see Box 9A, p. 338). At fertilization, one sperm enters the egg in the animal region. The egg completes meiosis and the egg and sperm nuclei fuse to form the diploid zygote nucleus.

Fig. 3.3 Life cycle of the African claw-toed frog *Xenopus laevis*. The numbered stages refer to standardized stages of *Xenopus* development. An illustrated list of all stages can be found on the Xenbase website listed in General further reading. The photographs show: an embryo at the blastula stage (top, scale bar = 0.5 mm); a tadpole at stage 41 (middle, scale bar = 1 mm); and an adult frog (bottom, scale bar = 1 cm).

Photographs courtesy of J. Slack (top, from Alberts, B., et al.: 1994) and J. Smith (middle and bottom).

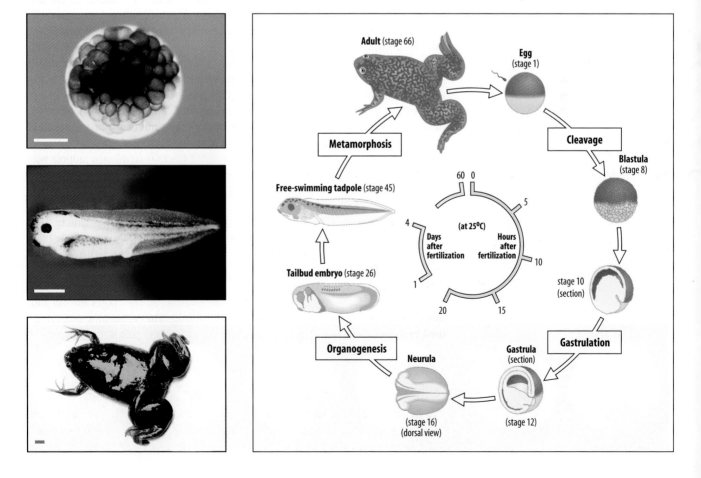

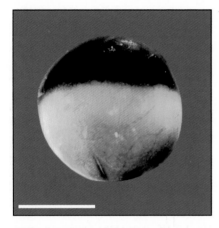

Fig. 3.4 A late-stage *Xenopus* oocyte.
The surface of the animal half (top) is pigmented and the paler, vegetal half of the egg is heavy with yolk. Scale bar = 1 mm.

Photograph courtesy of J. Smith.

Fig. 3.5 Cleavage of the *Xenopus* embryo. The upper panels are diagrams showing the fertilized egg and the first three cleavage divisions seen from the side. Polar bodies shown in the fertilized egg and the two-cell stage are attached to the animal pole. The photographs below show the cleaving *Xenopus* egg, but are taken from various angles (see also Fig. 1.14).

Photographs courtesy of R. Kessel, from Kessel, R.G., et al.: 1974.

The first cleavage division of the fertilized egg occurs about 90 minutes after fertilization and is along the plane of the animal–vegetal axis, dividing the embryo into equal left and right halves (Fig. 3.5). Further cleavages follow rapidly at intervals of about 20 minutes. The second cleavage is also along the animal–vegetal axis but at right angles to the first. The third cleavage is equatorial, at right angles to the first two, and divides the embryo into four animal cells and four larger vegetal cells. In *Xenopus* there is no cell growth between cell divisions at this early stage and so continued cleavage results in the formation of smaller and smaller cells; the cells deriving from cleavage divisions in animal embryos are often called **blastomeres**. Cleavage occurs synchronously and divisions occur in such a way that cells in the yolky vegetal half of the embryo are larger than those in the animal half. Inside this spherical mass of cells a fluid-filled cavity—the **blastocoel**—develops in the animal region, and the embryo is now called a **blastula**.

At the end of blastula formation, the *Xenopus* embryo has gone through about 12 cell divisions and is made up of several thousand cells. The mesoderm and endoderm, which will develop into internal structures, are located around the equator in the **marginal zone** and in the vegetal region, respectively, while the ectoderm, which will eventually cover the whole of the embryo, is still confined to the animal region (Fig. 3.6, first panel).

The next stage is gastrulation, which involves extensive movement and rearrangement of the germ layers specified in the blastula so that they become located in their proper positions in the body. Gastrulation involves changes in three dimensions, and so it can be difficult to visualize. The first external sign of gastrulation is a small slit-like infolding—the **blastopore**—that forms on the surface of the blastula on the future dorsal side (see Fig. 3.6, second panel). This region is of particular importance in development, as it is the site of the **embryonic organizer**, known as the **Spemann organizer** in amphibians, without which dorsal and axial development will not occur. The famous experiment by Hans Spemann and Hilde Mangold that demonstrated this is described in Fig. 1.9. Once gastrulation has started, the embryo is known as a **gastrula**. In *Xenopus*, the future endoderm and mesoderm in the marginal zone move inside the gastrula through the blastopore by rolling under the lip as coherent sheets of cells. This type of inward movement is known as **involution**. Once inside, the tissues converge towards the midline and extend along the antero-posterior axis beneath the dorsal ectoderm, elongating the embryo along the antero-posterior axis. At the same time, the ectoderm spreads downward

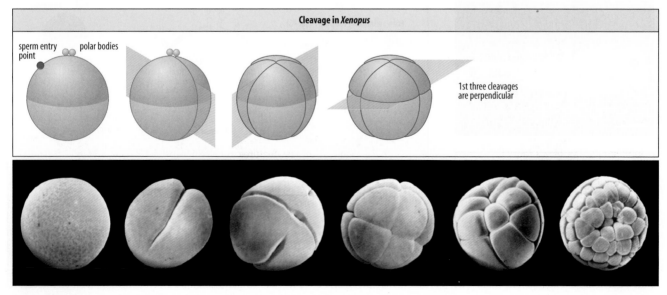

Cleavage in *Xenopus*

sperm entry point   polar bodies

1st three cleavages are perpendicular

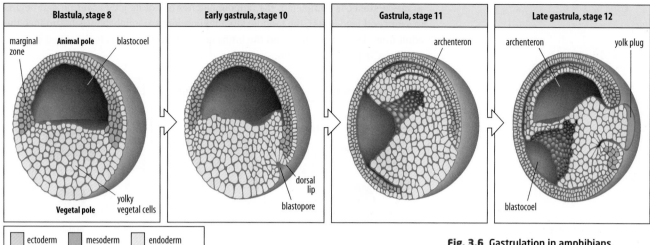

Fig. 3.6 Gastrulation in amphibians.
The blastula (first panel) contains several thousand cells and there is a fluid-filled cavity, the blastocoel, beneath the cells at the animal pole. Gastrulation begins (second panel) at the blastopore, which forms on the dorsal side of the embryo. Future mesoderm and endoderm of the marginal zone move inside at this site through the dorsal lip of the blastopore, the mesoderm ending up sandwiched between the endoderm and ectoderm in the animal region (third panel). The tissue movements create a new internal cavity–the archenteron–that will become the gut. Endoderm in the ventral region also moves inside through the ventral lip of the blastopore (fourth panel) and will eventually completely line the archenteron. At the end of gastrulation the blastocoel is considerably reduced in size.

After Balinsky, B.I.: 1975.

to cover the whole embryo by a process known as **epiboly**. The involuting layer of dorsal endoderm is closely applied to the mesoderm, and the space between it and the yolky vegetal cells is known as the **archenteron** (see Fig. 3.6, third panel). This is the precursor of the gut cavity.

The inward movement of endoderm and mesoderm begins dorsally and then spreads to form a complete circle of involuting cells around the blastopore (see Fig. 3.6, fourth panel). By the end of gastrulation, the blastopore has almost closed—the remaining slit will form the anus. The dorsal mesoderm now lies beneath the dorsal ectoderm, the lateral and ventral mesoderm is in its definitive position in the body plan, and the ectoderm has spread to cover the entire embryo. There is still a large amount of yolk present, which provides nutrients until the larva—the tadpole—starts feeding. During gastrulation, the mesoderm in the dorsal region starts to develop into the notochord and the somites, while the more lateral mesoderm—the **lateral plate mesoderm**—will later form mesoderm-derived internal organs such as the kidneys, with anterior lateral plate mesoderm giving rise to the heart. The cell and tissue movements that occur during gastrulation in *Xenopus* and other animals are discussed in more detail in Chapter 8.

Gastrulation is succeeded by **neurulation**—the formation of the neural tube, which is the early embryonic precursor of the central nervous system. The embryo is then called a **neurula**. The earliest visible sign of neurulation is the formation of the **neural folds**, which form on the edges of the **neural plate**, an area of columnar ectoderm cells overlying the notochord. The folds rise up, fold toward the midline and fuse together to form the neural tube, which sinks beneath the epidermis (Fig. 3.7). **Neural crest cells** detach from the top of the neural tube on either side of the site of fusion and migrate throughout the body to form various structures, as we describe in Chapter 8. The anterior neural tube gives rise to the brain; further back, the neural tube overlying the notochord will develop into the spinal cord. The formation of notochord, somites, and neural tube is described in Chapter 5. The embryo now begins to look something like a tadpole and we can recognize the main vertebrate features (Fig. 3.8). At the anterior end, the brain is already divided up into a number of regions, and the eyes and ears have begun to develop. There are also three **branchial arches** on each side, the most anterior of which will form the jaws. More posteriorly, the somites and notochord are well developed. The mouth breaks through at stage 40, about 2.5 days after fertilization. The post-anal tail of the tadpole is formed last. It develops from the **tailbud**, which gives rise to a continuation of notochord, somites, and neural tube in the tail. Further development gives rise to various organs and tissues such as blood and heart, kidneys, lungs, and liver. After organ formation, or

organogenesis, is completed, the tadpole hatches out of its jelly covering and begins to swim and feed. Later, the tadpole larva will undergo metamorphosis to produce the adult frog; the tail regresses and the limbs grow (metamorphosis is described in more detail in Chapter 13).

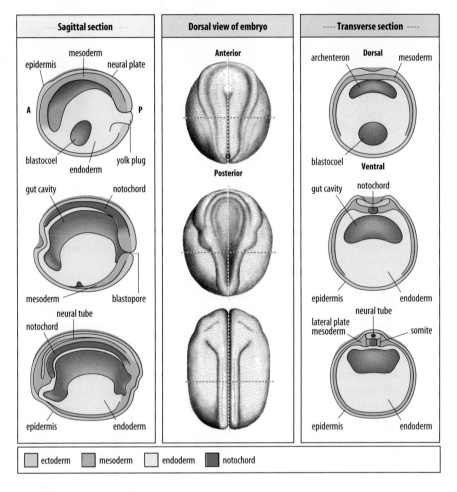

Fig. 3.7 Neurulation in amphibians. Top row: the neural plate develops neural folds just as the notochord begins to form in the midline (see middle row). Middle and bottom rows: the neural folds come together in the midline to form the neural tube, from which the brain and spinal cord will develop. During neurulation, the embryo elongates along the antero-posterior axis. The left panel shows sections through the embryo in the planes indicated by the yellow dashed lines in the center panel. The center panel shows dorsal surface views of the amphibian embryo. The right panel shows sections through the embryo in the planes indicated by the green dashed lines in the center panel. This diagram shows neurulation in a urodele amphibian embryo rather than *Xenopus*, as the neural folds in urodele embryos are more clearly defined.

Fig. 3.8 The early tailbud stage (stage 26) of a *Xenopus* embryo. At the anterior end, in the head region, the future eye is prominent and the ear vesicle (otic vesicle), which will develop into the ear, has formed. The brain is divided into forebrain, midbrain, and hindbrain. Just posterior to the site at which the mouth will form are the branchial arches, the first of which will form the lower jaw. More posteriorly, a succession of somites lies on either side of the notochord (which is stained brown in the photograph). The embryonic kidney (pronephros) is beginning to form from lateral mesoderm. Ventral to these structures is the gut (not visible in this picture). The tailbud will give rise to the tail of the tadpole, forming a continuation of somites, neural tube, and notochord. Scale bar = 1 mm.

Photograph courtesy of B. Herrmann.

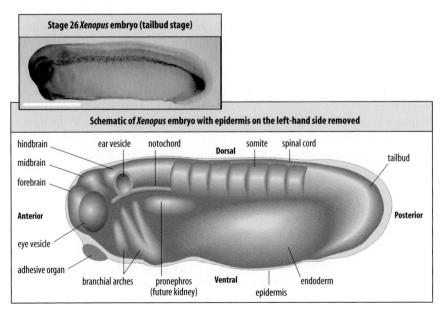

3.2 The zebrafish embryo develops around a large mass of yolk

The zebrafish has two great advantages as a vertebrate model for development: its short life-cycle of approximately 12 weeks makes genetic analysis, including large-scale genetic screening relatively easy (see Box 3C, p. 120), while the transparency of the embryo allows the fate and movements of individual cells during development to be observed. Because of the feasibility of genetic analysis in the zebrafish, it has turned out to be a useful model organism for studying some human diseases caused by genetic defects, in particular certain blood and cardiovascular disorders. The life cycle is shown in Fig. 3.9. The zebrafish egg is about 0.7 mm in diameter, with a clear animal–vegetal axis: the cytoplasm and nucleus at the animal pole sit upon a large mass of yolk. After fertilization, the zygote undergoes cleavage, but cleavage does not extend into the yolk and results in a mound of blastomeres perched above a large yolk cell. The first five cleavages are all vertical, and the first horizontal cleavage gives rise to the 64-cell stage about 2 hours after fertilization (Fig. 3.10).

Further cleavage leads to the **sphere stage**, in which the embryo is now in the form of a **blastoderm** of around 1000 cells lying on top of the yolk cell. The hemispherical blastoderm has an outer layer of flattened cells, one cell thick, known as the **outer enveloping layer**, and a **deep layer** of more rounded cells from which the embryo develops (Fig. 3.11). During the early blastoderm stage, blastomeres at the margin of the blastoderm merge and collapse into the yolk cell, forming a continuous layer of multinucleated non-yolky cytoplasm underlying the blastoderm that is called the **yolk syncytial layer**. The blastoderm, together with the yolk syncytial layer, spreads in a vegetal direction by epiboly and eventually covers the yolk cell.

Although the fish blastoderm and the amphibian blastula are different in shape, they are corresponding stages in development. In the fish, the endoderm is derived from the

Fig. 3.9 Life cycle of the zebrafish.
The zebrafish embryo develops as a mound-shaped blastoderm sitting on top of a large yolk cell. It develops rapidly and by 2 days after fertilization the tiny fish, still attached to the remains of its yolk, hatches out of the egg. The top photograph shows a zebrafish embryo at the sphere stage of development, with the embryo sitting on top of the large yolk cell (scale bar = 0.5 mm). The middle photograph shows an embryo at the 14-somite stage, showing developing organ systems. Its transparency is useful for observing cell behavior (scale bar = 0.5 mm). The bottom photograph shows an adult zebrafish (scale bar = 1 cm). An illustrated list of the numbered stages in zebrafish development can be found on the website listed in General further reading.

Photographs courtesy of C. Kimmel (top, from Kimmel, C.B., et al.: 1995), N. Holder (middle), and M. Westerfield (bottom).

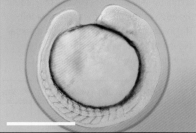

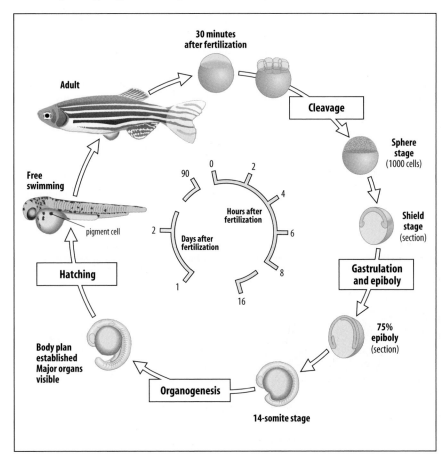

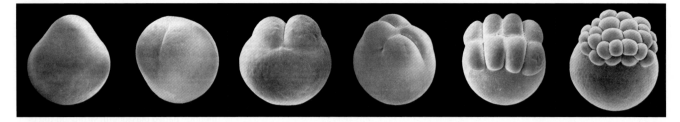

Fig. 3.10 Cleavage of the zebrafish embryo is initially confined to the animal (top) half of the embryo.

Photographs courtesy of R. Kessel, from Kessel, R.G., et al.: 1974.

Fig. 3.11 Epiboly and gastrulation in the zebrafish. At the end of the first stages of cleavage, around 4.3 hours after fertilization, the zebrafish embryo is composed of a mound of blastomeres sitting on top of the yolk, separated from the yolk by a multinucleate layer of cytoplasm called the yolk syncytial layer (first panel). With further cleavage and spreading out of the layers of cells (epiboly), the upper half of the yolk becomes covered by a blastoderm with a thickened edge, the germ ring, and a shield-shaped region is visible in the blastoderm on the dorsal side (second panel). Gastrulation occurs by involution of cells in a ring around the edge of the blastoderm (third panel). The involuting cells converge on the dorsal midline to form the body of the embryo encircling the yolk (fourth panel).

deep cells that lie right at the blastoderm margin. This narrow marginal layer is often known as **mesendoderm** because individual cells in this region can give rise to both endoderm and mesoderm. Deep cells located between four and six cell diameters away from the margin give rise exclusively to mesoderm. This overlap between prospective endoderm and mesoderm in the zebrafish differs from the arrangement in *Xenopus* in which the endoderm and mesoderm occupy more distinct locations. The ectoderm in the fish embryo, as in *Xenopus*, comes from cells in the animal region of the blastoderm.

By about 5.5 hours after fertilization the blastoderm has spread halfway to the vegetal pole, and the deep-layer cells accumulate to form a thickening around the blastoderm edge known as the embryonic germ ring (Fig. 3.11, second panel). At the same time, deep-layer cells within the germ ring converge towards the dorsal side of the embryo, eventually forming a compact shield-shaped region in the germ ring on the dorsal side that becomes visible at about 6 hours after fertilization, marking the **shield stage**. The shield region is analogous to the Spemann organizer of *Xenopus*. Gastrulation ensues and the mesendodermal cells and mesoderm cells roll under the blastoderm margin in the process of involution and move into the interior under the ectoderm. The earliest cells that involute become endoderm. As the mesendoderm is becoming internalized, the ectoderm continues to undergo epiboly, spreading in the vegetal direction until it covers the whole embryo, including the yolk (discussed in more detail in Chapter 8). Differences between gastrulation in the zebrafish and *Xenopus* are that in the fish inward movement of cells occurs all around the periphery of the blastoderm at about the same time, and that endoderm is derived predominantly from the dorsal and lateral margins of the blastoderm.

Once internalized into the gastrulating embryo, the mesendodermal cells migrate under the ectoderm towards the animal pole, the tissue converging to form the main axis of the embryo, and extending and elongating the embryo in an antero-posterior direction, as in *Xenopus*. The future mesodermal and endodermal cells are now beneath the ectoderm, and by the time that the blastoderm has spread about three-quarters of the way to the vegetal pole, a single layer of endoderm cells has formed in the embryo next to the yolk with the more superficial cells becoming mesoderm. By 9 hours the notochord has become distinct in the dorsal midline of the embryo, and involution

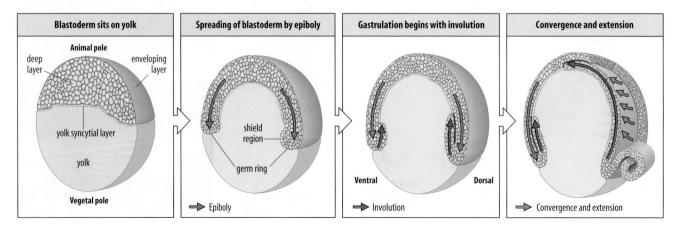

of cells around the blastoderm margin is complete by 10 hours. Somite formation, neurulation, and migration of neural crest cells then follow.

Over the next 12 hours the embryo elongates further, and rudiments of the primary organ systems become recognizable. Somites first appear at the anterior end at about 10.5 hours, and new ones are formed at intervals of initially 20 minutes and then 30 minutes; by 18 hours, 18 somites are present. Formation of the gut in zebrafish differs in several respects from that of *Xenopus*, chick, and mouse. In zebrafish, the gut starts to develop at a relatively late stage in gastrulation—at the 18-somite stage—and is formed by reorganization of cells within the internalized solid mass of endoderm to form a tubular structure. Neurulation in zebrafish begins, as in *Xenopus*, towards the end of gastrulation with the formation of the neural plate, an area of columnar ectoderm overlying the notochord. But unlike *Xenopus*, the whole of the neural plate in the zebrafish first forms a solid rod of cells that later becomes hollowed out internally to form the neural tube. The nervous system develops rapidly. Optic vesicles, which give rise to the eyes, can be distinguished at 12 hours as bulges from the brain, and by 18 hours the body starts to twitch. At 48 hours the embryo hatches, and the young fish begins to swim and feed.

3.3 Birds and mammals resemble each other and differ from *Xenopus* in some important features of early development

Before we describe the course of development in chick and mouse embryos separately, it is worth emphasizing some features of their development that differ from *Xenopus*. The first is the shape of the early embryo. The avian or mammalian structure corresponding to the spherical amphibian blastula just before gastrulation is not a hollow blastula but a layer of epithelium called the **epiblast**. The second difference is that at this stage there are no distinct contiguous regions of the epiblast that correspond to ectoderm, endoderm, and mesoderm. As we shall discuss in more detail in later chapters, the timing of specification of the germ layers is somewhat different from *Xenopus*. There is also considerable cell proliferation in the chick and mouse epiblasts before and during gastrulation, which causes cell mixing at this stage. The third difference, which is related to both the previous ones, is that the gastrulation process that leads to internalization of endoderm and mesoderm and the organization of the germ layers is rather different in appearance from that in *Xenopus*, with internalization of cells occurring in a straight furrow, rather than through a circular blastopore.

In mouse and chick, the equivalent to the amphibian blastopore is the **primitive streak**. The primitive streak is most easily visualized in the flat epiblast of the chick embryo and can be seen in the top photograph in Fig. 3.14. At gastrulation, epiblast cells converge on the primitive streak and pass through it as individual cells, spreading out underneath the surface and forming a bottom layer of endoderm and a middle layer of mesoderm (Fig. 3.12). Cells become specified as endoderm and mesoderm during their passage through the streak, with the cells remaining on the surface becoming ectoderm. Because of the essentially sheet-like form of the avian or mammalian epiblast at this stage, this initial phase of gastrulation does not directly form a gut cavity in the same way as in the spherical *Xenopus* gastrula. The gut is formed later, by a folding together of the lateral edges of the embryo, which results in a gut cavity entirely surrounded by layers of endoderm, mesoderm, and ectoderm. We shall now return to the chick to look at the course of its development in more detail.

3.4 The early chicken embryo develops as a flat disc of cells overlying a massive yolk

Avian embryos are very similar to those of mammals in the morphological complexity of the embryo and the general course of embryonic development, but are easier to obtain and observe. Many observations can be carried out simply by making a window in the

Fig. 3.12 Ingression of mesoderm and endoderm during gastrulation in the chick embryo. Gastrulation begins with the formation of the primitive streak, a region of proliferating and migrating cells, which elongates from the posterior marginal zone. Future mesodermal and endodermal cells migrate through the primitive streak into the interior of the blastoderm. During gastrulation, the primitive streak extends about halfway across the area pellucida (see Fig. 3.14). At its anterior end an aggregation of cells, known as Hensen's node, forms. As the streak extends, cells of the epiblast move toward the primitive streak (arrows), move through it, and then outward again underneath the surface to give rise internally to the mesoderm and endoderm internally, the latter displacing the endoblast.

Adapted from Balinsky, B.I., et al.: 1975.

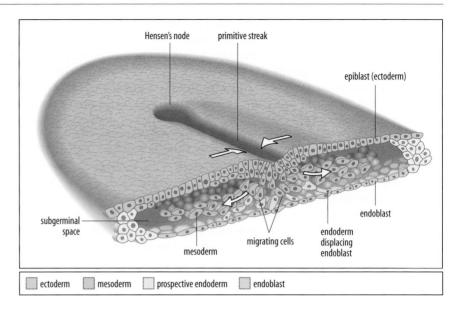

ectoderm   mesoderm   prospective endoderm   endoblast

egg shell, and the embryo can also be cultured outside the egg. This is particularly convenient for experimental microsurgical manipulations, tracing cell lineages by injecting marker dyes, and investigations of the effects of introduced genes and other treatments (see Section 3.6). Despite considerable differences in shape between chick and mouse embryos in the very early stages of development (see Fig. 3.2), gastrulation and later development are very similar in both, and studies of chick embryos complement those of mouse. The chick's flat early embryo is similar in shape to that of humans (see Fig. 3.2) and the topology of early human embryonic development is in some ways more easily compared with that of the chick than with the mouse, which has a cup-shaped embryo.

The large yolky egg cell is fertilized and begins to undergo cleavage while still in the hen's oviduct. Because of the massive yolk, cleavage is confined to a small patch of cytoplasm several millimeters in diameter, which contains the nucleus and lies on top of the yolk. Cleavage in the oviduct results in the formation of a disc of cells called the **blastoderm** or **blastodisc**. During the 20-hour passage down the oviduct, the egg becomes surrounded by extracellular albumen (egg white), the shell membranes, and the shell (Fig. 3.13). At the time of laying, the blastoderm, which is analogous to the early amphibian blastula, is composed of some 20,000–60,000 cells. The chick developmental cycle is shown in Fig. 3.14.

The early cleavage furrows extend downward from the surface of the cytoplasm but do not completely separate the cells, whose ventral faces initially remain open to the yolk. The central region of the blastoderm, under which a cavity called the **subgerminal space** develops, is translucent and is known as the **area pellucida**, in contrast to the outer region, which is the darker **area opaca** (Fig. 3.15). A layer of cells called the **hypoblast** develops over the yolk to form the floor of the cavity. The hypoblast eventually gives rise to extra-embryonic structures such as those that connect the embryo to its source of nutrients in the yolk. The embryo proper is formed from the remaining upper layer of the blastoderm, known as the epiblast.

The first morphological structure that presages the antero-posterior polarity of the embryo is a crescent-shaped ridge of small cells called **Koller's sickle**, located at the boundary between the area opaca and area pellucida at the posterior end of the embryo. Koller's sickle defines the position in which the streak will form and the region of the epiblast immediately adjacent to the sickle is known as the **posterior marginal zone**. The streak is first visible as a denser region that then gradually extends as a narrow stripe to just over half way across the area pellucida, eventually forming a furrow in the dorsal face of the epiblast (see Fig. 3.12).

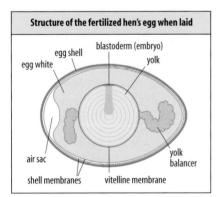

Structure of the fertilized hen's egg when laid

Fig. 3.13 The structure of a hen's egg at the time of laying. Cleavage begins after fertilization while the egg is still in the oviduct. The albumen (egg white) and shell are added during the egg's passage down the oviduct. At the time of laying, the embryo is a disc-shaped cellular blastoderm lying on top of a massive yolk, which is surrounded by the egg white and shell.

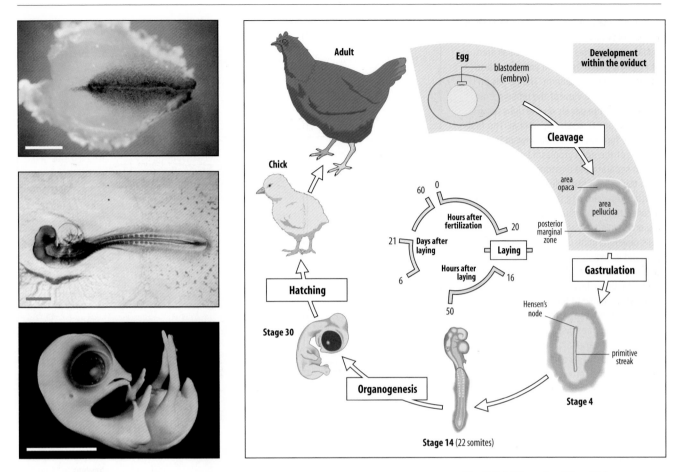

Fig. 3.14 Life cycle of the chicken.
The egg is fertilized in the hen and by the time it is laid cleavage is complete and a cellular blastoderm lies on the yolk. After gastrulation, the primitive streak forms. Regression of Hensen's node is associated with somite formation. The photographs show: the primitive streak (stained brown by staining with antibody against Brachyury protein) surrounded by the area pellucida (top, scale bar = 1 mm); a stage 14 embryo (50–53 hours after laying) with 22 somites (the head region is well defined and the transparent organ adjacent to it is the ventricular loop of the heart; middle, scale bar = 1 mm); a stage-35 embryo, about 8.5-9 days after laying, with a well-developed eye and beak (bottom, scale bar = 10 mm). A descriptive list of the numbered stages (Hamilton and Hamburger stages) in chick development can be found on the website listed in General further reading.

Top photograph courtesy of B. Herrmann, from Kispert, A., et al: 1995.

Unlike *Xenopus*, cell proliferation and growth of the chick embryo continues throughout gastrulation. Epiblast cells converge on the primitive streak, and as the streak moves forward from the posterior marginal zone, cells in the furrow move inward and spread out anteriorly and laterally beneath the upper layer, forming a layer of loosely connected cells, or **mesenchyme**, in the subgerminal space (see Figs 3.12 and 3.15). The primitive streak is thus similar to the blastopore region of amphibians, but cells move inwards individually, rather than as a coherent sheet, a type of inward movement known as **ingression**. The ingressing cells give rise to mesoderm and endoderm, whereas the cells that remain on the surface of the epiblast give rise to the ectoderm.

The primitive streak in the chick embryo is fully extended by 16 hours after laying. At the anterior end of the streak there is a condensation of cells known as **Hensen's node**, where cells are also moving inwards. Hensen's node is the major organizing center for the early chick embryo, equivalent to the Spemann organizer in amphibians, and is formed of cells derived from Koller's sickle as well as cells recruited from the epiblast.

After the primitive streak has elongated to its full length, some cells from Hensen's node start to migrate forwards along the midline under the epiblast, to give rise to the **prechordal plate mesoderm** and the **head process**. The prechordal plate mesoderm is a looser mass of cells anterior to the head process, the anterior part of the notochord. After the head process forms, the primitive streak begins to regress, with Hensen's node moving back towards the posterior end of the embryo (Fig. 3.16). As the node regresses, the notochord is laid down in its wake, lengthening the head process along the dorsal midline. As the notochord is laid down, the mesoderm immediately on each side of it begins to form the somites. These processes are discussed in detail in Chapters 4 and 5. The first pair of somites is formed at about 24 hours after laying and new ones are formed at intervals of 90 minutes. The rest of the mesoderm lateral

Fig. 3.15 Cleavage and epiblast formation in the chick embryo. By the time the egg is laid, cleavage has divided the small area of egg cytoplasm free from yolk into a disc-shaped cellular blastoderm. The panels on the left show the view of the embryo from above, while the panels on the right show cross-sections through the embryo. The first cleavage furrows extend downward from the surface of the egg cytoplasm and initially do not separate the blastoderm completely from the yolk. In the cellular blastoderm the central area overlying the subgerminal space is called the area pellucida and the marginal region is called the area opaca. A layer of cells develops immediately overlying the yolk and is known as the hypoblast. This will give rise to extra-embryonic structures, while the upper layers of the blastoderm—the epiblast—give rise to the embryo. The chick primitive streak starts to form when the hypoblast becomes displaced forward from the posterior marginal zone by a new layer of cells called the secondary hypoblast or endoblast, which grows out from the zone. The cells of the epiblast move toward the primitive streak (arrows), move through it, and then outward again underneath the surface to give rise to the mesoderm and endoderm internally, the latter displacing the endoblast.

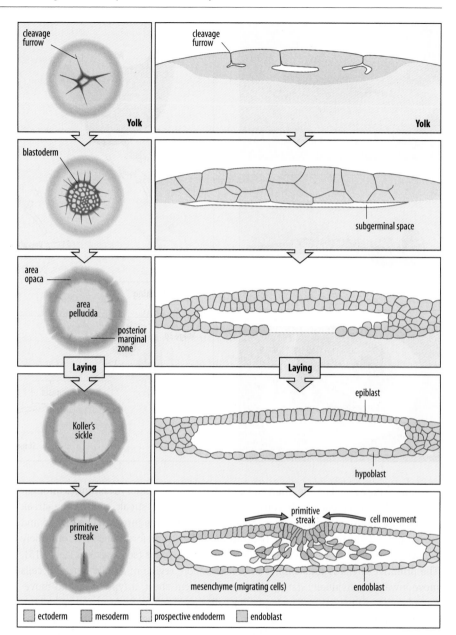

Fig. 3.16 Regression of Hensen's node. After extending about halfway across the blastoderm, the primitive streak begins to regress, with Hensen's node moving in a posterior direction as the head fold and neural plate begin to form anterior to it. As the node moves backward, the notochord develops in the area anterior to it and somites begin to form on either side of the notochord.

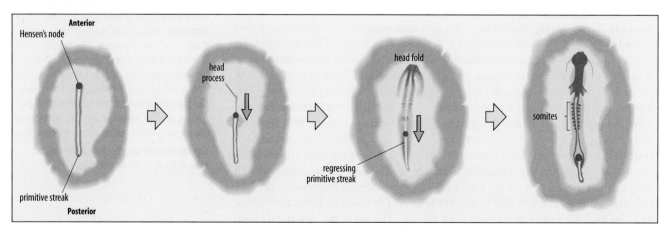

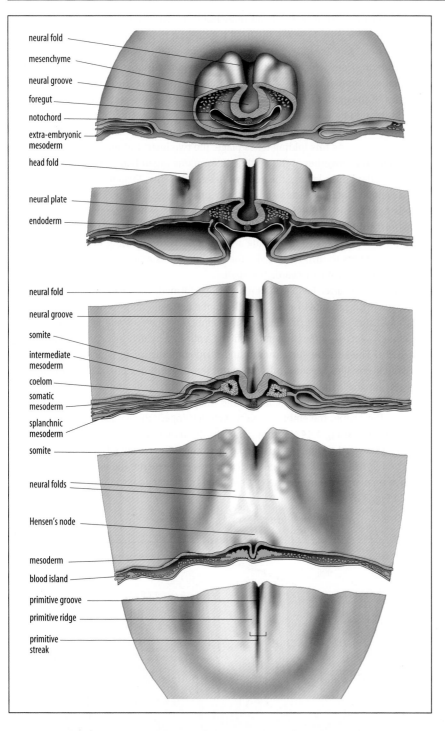

Fig. 3.17 Development of the neural tube and mesoderm in the chick embryo. Once the notochord has formed, neurulation begins in an anterior to posterior direction. The figure shows a series of sections along the antero-posterior axis of a chick embryo. Neural-tube formation is well advanced at the anterior end (top two sections), where the head fold has already separated the future head from the rest of the blastoderm and the ventral body fold has brought endoderm from both sides of the body together to form the gut. During neurulation, the neural plate changes shape: neural folds rise up on either side and form a tube when they meet in the midline. The mesenchyme in this region will give rise to head structures. Further back (middle sections), in the future trunk region of the embryo, notochord and somites have formed and neurulation is starting. At the posterior end, behind Hensen's node (bottom section), notochord formation, somite formation, and neurulation have not yet begun. The mesoderm internalized through the primitive streak starts to form structures appropriate to its position along the antero-posterior and dorso-ventral axes. For example, in the future trunk region, the intermediate mesoderm will form the mesodermal parts of the kidney, and the anterior splanchnic mesoderm will give rise to the heart. The body fold will continue down the length of the embryo, forming the gut and also bringing paired organ rudiments that initially form on each side of the midline (e.g. those of the heart and dorsal aorta) together to form the final organs lying ventral to the gut. Blood islands, from which the first blood cells are produced, form from the ventral-most part of the lateral mesoderm.

After Patten, B.M.: 1971.

Fig. 3.18 Scanning electron micrograph of chick showing early somites and neural tube. Blocks of somites can be seen adjacent to the neural tube, with the notochord lying beneath it. The lateral plate mesoderm flanks the somites. Scale bar = 0.1 mm.

Photograph courtesy of J. Wilting.

to the somites is the lateral plate mesoderm, and this will develop into organs such as heart, kidneys, and the vascular system and blood (Fig. 3.17). The node eventually forms a center of stem cells in the tailbud, which gives rise to the post-anal tail.

As the notochord forms, neural tissue begins to develop, first as the neural plate, which is first evident as a region of columnar ectoderm above the notochord. The neural plate then folds upwards and inwards so that its sides come together and eventually fuse in the dorsal midline to form the neural tube, initially leaving the anterior and posterior ends open. The fused neural tube becomes covered over by epidermis; Fig. 3.18 shows a section through the chick embryo showing the fused neural tube and the notochord beneath it.

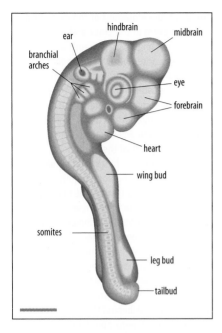

Fig. 3.19 The 40-somite stage of the chick embryo. Development of the head region and the heart are quite well advanced and the wing and leg buds are present as small protrusions.

Fig. 3.20 The extra-embryonic structures and circulation of the chick embryo.

A chick embryo at 4 days of incubation is depicted *in situ*. The embryo lies within the fluid-filled amniotic cavity enclosed by the amnion, which provides a protective chamber. The yolk is surrounded by the yolk sac membrane. The vitelline vein takes nutrients from the yolk sac to the embryo and the blood is returned to the yolk sac via the vitelline artery. The umbilical artery takes waste products to the allantois and the umbilical vein brings oxygen to the embryo. The arteries are shown in red and the veins in blue but this does not denote the oxygenation status of the blood. As the embryo grows, the amniotic cavity enlarges; the allantois also increases in size and its outer layer fuses with the chorion, the membrane under the shell, while at the same time the yolk sac shrinks. Note that in this diagram the allantois has been enlarged so that the umbilical vessels can be shown clearly.

After Patten, B.M.: 1951

Just after the head process (head notochord) appears and Hensen's node starts to regress, all three germ layers in the head region start to fold ventrally, to begin to generate the **head fold** (see Fig. 3.16, third image from left). This delimits a pouch lined by endoderm, from which the pharynx and foregut will arise (see top of Fig. 3.17). Eventually, a similar fold appears at the tail end to define the hindgut. In addition, the sides of the embryo fold together to form the rest of the gut, which only remains open in the middle (umbilical region) and subsequently the mesoderm and ectoderm grow over to form the ventral body wall. This key morphogenetic event is known as **ventral closure**. As this folding takes place, the two heart rudiments that start out on either side come together at the midline to form one organ lying ventral to the gut. By just over 2 days after laying, the embryo has reached the 22-somite stage, the head is well-developed, optic vesicles and auditory pits are present, and the heart and blood vessels have formed. Blood vessels and blood islands, where the first blood cells are being formed, have developed in the extra-embryonic tissues; the vessels connect up with those of the embryo to provide a circulation with a beating heart. At this stage the embryo starts to turn on its side; the head develops a flexure so that the right eye comes to lie uppermost towards the shell.

By 3–3.5 days after laying, 40 somites have formed, the head is now much more developed with prominent eyes, and the limbs are beginning to develop (Fig. 3.19). At 4 days after laying, extra-embryonic membranes have developed through which the embryo gets its nourishment from the yolk and which also provide protection (Fig. 3.20). The **amnion** surrounds a fluid-filled amniotic sac in which the embryo lies and which provides mechanical protection; the **chorion** lies outside the amnion just beneath the shell; the **allantois** receives excretory products and provides the site of oxygen and carbon dioxide exchange; and the **yolk sac** surrounds the yolk. By about 10 days after laying, the embryo is very well-developed; the wings, legs, and beak are now formed (see Fig 3.14, bottom panel. In the remaining time before hatching, the embryo grows in size, the internal organs become fully developed, and down feathers grow on the wings and body. The chick hatches 21 days after the egg is laid. A picture of a live quail embryo inside the egg obtained by magnetic resonance imaging can be seen in Fig. 3.27.

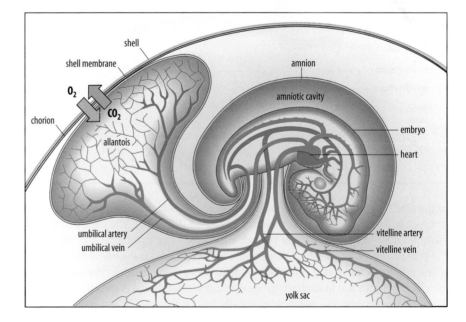

3.5 Early development in the mouse involves the allocation of cells to form the placenta and extra-embryonic membranes

The mouse has a life cycle of 9 weeks from fertilization to mature adult (Fig. 3.21), which is relatively short for a mammal. This makes genetic analysis relatively easy and is one reason the mouse has become the main model organism for mammalian development and was the first mammal after humans to have its complete genome sequenced. As in all our model organisms, knowledge of the genome sequence enables profiles of all the genes expressed at a particular stage in development to be determined (Box 3A, p. 115). Particular advantages of the mouse in this respect are the relative ease with which mice with a particular genetic constitution can be routinely produced using **transgenic** techniques (see Section 3.9), and the ability to produce mice in which particular genes are rendered completely inactive or 'knocked out'.

Mammalian eggs are much smaller than either chick or *Xenopus*, about 80–100 μm in diameter for both mouse and human, and they contain no yolk. The unfertilized egg is shed from the ovary into the oviduct and is surrounded by a protective external coat, the **zona pellucida**, which is composed of mucopolysaccharides and glycoproteins. Fertilization takes place internally in the oviduct; meiosis is then completed and the second polar body forms (see Box 9A, p. 338). Cleavage starts while the fertilized egg is still in the oviduct. Early cleavages are very slow compared with *Xenopus* and chick, the first occurring about 24 hours after fertilization, the second about 20 hours later, and subsequent cleavages at about 12-hour intervals. Cleavage produces a solid ball of

Fig. 3.21 The life cycle of the mouse. The egg is fertilized in the oviduct, where cleavage also takes place before implantation of the blastocyst in the uterine wall at 5 days after fertilization. Gastrulation and organogenesis then take place over a period of about 7 days and the remaining 6 days before birth are largely a time of overall growth. After gastrulation the mouse embryo undergoes a complicated movement known as turning, in which it becomes surrounded by its extra-embryonic membranes (not shown here). The photographs show (from top): a fertilized mouse egg just before the first cleavage (scale bar = 10 μm); the anterior view of a mouse embryo at E8 (scale bar = 0.1 mm); and a mouse embryo at E14 (scale bar = 1 mm). An illustrated list of the stages in mouse development can be found on the website listed in General further reading.

Photographs courtesy of: T. Bloom (top, from Bloom, T.L.: 1989); N. Brown (middle); and J. Wilting (bottom).

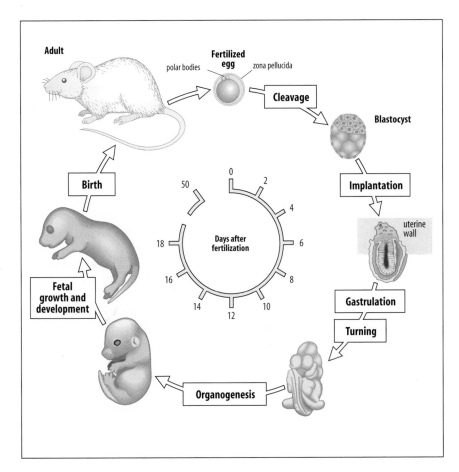

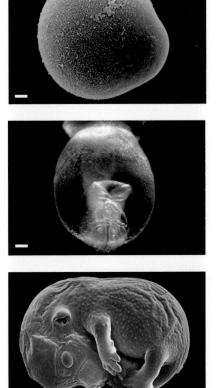

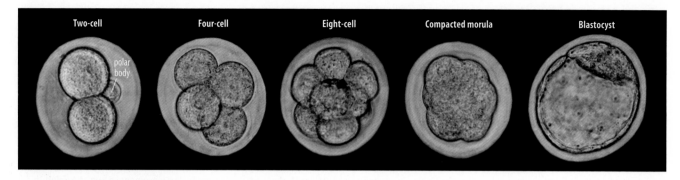

| Two-cell | Four-cell | Eight-cell | Compacted morula | Blastocyst |

Fig. 3.22 Cleavage in the mouse embryo. The photographs show the cleavage of a fertilized mouse egg from the two-cell stage through to the formation of the blastocyst. After the eight-cell stage, compaction occurs, forming a solid ball of cells called the morula, in which individual cell outlines can no longer be discerned. The internal cells of the morula give rise to the inner cell mass, which can be seen as the compact clump at the top of the blastocyst. It is from this that the embryo proper forms. The outer layer of the hollow blastocyst—the trophectoderm—gives rise to extra-embryonic structures.

Photographs courtesy of T. Fleming.

cells, or blastomeres, called a morula (Fig. 3.22). At the eight-cell stage the blastomeres increase the area of cell surface in contact with each other in a process called compaction. After compaction, the cells are polarized; their exterior surfaces carry microvilli, whereas their inner surfaces are smooth. Further cleavages are somewhat variable and are both radial and tangential, so that by the equivalent of the 32-cell stage the morula contains about 10 internal cells and more than 20 outer cells.

A special feature of mammalian development is that the early cleavages give rise to two distinct groups of cells—the trophectoderm and the inner cell mass. The internal cells of the morula give rise to the inner cell mass and the outer cells to the trophectoderm. The trophectoderm will give rise to extra-embryonic structures, such as the placenta, through which the embryo gains nourishment from the mother. The embryo proper develops from a subset of cells in the inner cell mass. At this stage (E3.5) the mammalian embryo is known as a blastocyst (see Fig. 3.21). Fluid is pumped by the trophectoderm into the interior of the blastocyst, which causes the trophectoderm to expand and form a fluid-filled cavity (the blastocoel) containing the inner cell mass at one end.

From E3.5 to E4.5 the inner cell mass becomes divided into two regions. The surface layer in contact with the blastocoel becomes the primitive endoderm, and will contribute to extra-embryonic membranes, whereas the remainder of the inner cell mass—the primitive ectoderm or epiblast—will develop into the embryo proper as well as giving rise to some extra-embryonic components. At this stage, about E4.5, the embryo is released from the zona pellucida and implants into the uterine wall.

The course of early post-implantation development of the mouse embryo from around E4.5 to E8.5 appears more complicated than that of the chick, partly because of the need to produce a larger variety of extra-embryonic membranes, and partly because the epiblast is distinctly cup-shaped in the early stages. This is a peculiarity of mouse and other rodent embryos. Human and rabbit embryos, for example, are flat blastodiscs, much more resembling that of the chick (see Fig. 3.2). Despite the different topology, however, gastrulation and the later development of the mouse embryo are in essence very similar to that of the chick.

The first 2 days of mouse post-implantation development are shown in Fig. 3.23. After the initial adhesion of the blastocyst to the uterine epithelium, the cells of the mural trophectoderm—the region not in contact with the inner cell mass—replicate their DNA without cell division, giving rise to primary trophoblast giant cells that invade the uterus wall and surround most of the conceptus, forming an interface with the maternal tissue. The uterus wall envelops the blastocyst, and the polar trophectoderm cells in contact with the inner cell mass continue to divide, forming the ectoplacental cone and the extra-embryonic ectoderm, which both contribute to the placenta. The outer cells of the ectoplacental cone differentiate into secondary trophoblast giant cells. Some cells from the primitive endoderm migrate to cover the whole inner surface of the mural trophectoderm. They become the parietal endoderm, which eventually becomes Reichert's membrane, a sticky layer of cells and

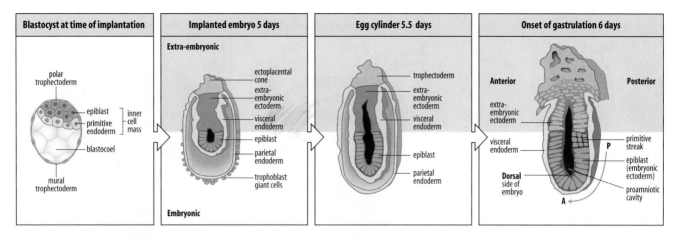

| Blastocyst at time of implantation | Implanted embryo 5 days | Egg cylinder 5.5 days | Onset of gastrulation 6 days |

Fig. 3.23 Early post-implantation development of the mouse embryo. First panel: before implantation, the fertilized egg has undergone cleavage to form a hollow blastocyst, in which a small group of cells, the inner cell mass, will give rise to the embryo, while the rest of the blastocyst forms the trophectoderm, which will develop into extra-embryonic structures. At the time of implantation the inner cell mass divides into two regions: the primitive ectoderm or epiblast, which will develop into the embryo proper, and the primitive endoderm, which will contribute to extra-embryonic structures. Second panel: the mural trophectoderm gives rise to trophoblast giant cells, which invade the uterine wall, helping to anchor the blastocyst to it. The blastocyst becomes surrounded by the uterine wall. The polar trophectoderm in contact with the epiblast proliferates and forms extra-embryonic tissues—the ectoplacental cone and extra-embryonic ectoderm—which contribute to the placenta. The epiblast elongates and develops an internal cavity (proamniotic cavity), which gives it a cup-shaped form. Third panel: the cylindrical structure containing both the epiblast and the extra-embryonic tissue derived from the polar trophectoderm is known as the egg cylinder. Fourth panel: the beginning of gastrulation is marked by the appearance of the primitive streak (brown) at the posterior of the epiblast (P). It starts to extend anteriorly (arrow) towards the bottom of the cylinder. For simplicity, the parietal endoderm and trophoblast giant cells are not shown in this panel or in subsequent figures.

extracellular matrix that has a barrier function. The remaining primitive endoderm cells form the **visceral endoderm**, which covers the elongating **egg cylinder** containing the epiblast.

By E5, an internal cavity has formed inside the epiblast, which then becomes cup-shaped—U-shaped when seen in cross-section (see Fig. 3.23, second and third panels). The epiblast is now a curved single layer of epithelium, which at this stage contains about 1000 cells. The embryo develops from this layer. The first easily visible sign that marks the future antero-posterior axis is the appearance of the primitive streak at about E6. The streak starts as a localized thickening at the edge of the cup on one side; this side will correspond to the future posterior end of the embryo. The initial development of the primitive streak in the mouse is similar to that in the chick. Over the next 12–24 hours it elongates until it reaches the bottom of the cup. A condensation of cells—the **node**—becomes distinguishable at the anterior end of the extended streak and corresponds to Hensen's node in the chick embryo. To make primitive streak formation easier to compare with that of the chick, imagine the epiblast cup spread out flat. In mouse gastrulation, as in chick, epiblast cells converge on the primitive streak, and proliferating cells migrate through it to spread out laterally and anteriorly between the ectoderm and the visceral endoderm to form a mesodermal layer (Fig. 3.24). Development from E7 is shown in Fig. 3.25.

Some epiblast-derived cells pass through the streak and enter the visceral endoderm, gradually displacing it to form the definitive **embryonic endoderm** on the outside of the cup, which is the future ventral side of the embryo. Cells migrating anteriorly from the node form the head process and notochord, while prospective mesodermal cells from the region surrounding the node migrate anteriorly to form the somites. Cell proliferation continues during gastrulation, and the embryo anterior to the node grows rapidly in size; as in the chick, the node eventually becomes a center of stem cells in the tailbud, giving rise to the post-anal tail.

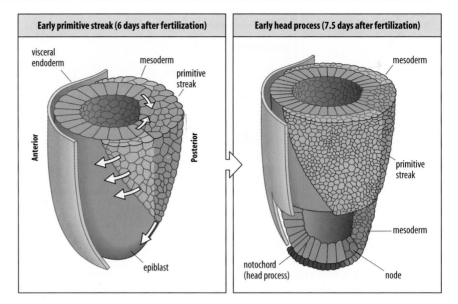

Fig. 3.24 Gastrulation in the mouse embryo. Left panel: as in the chick, gastrulation in the mouse embryo begins when epiblast cells converge on the posterior of the epiblast and move under the surface, forming the denser primitive streak (brown) where the cells are becoming internalized. Once inside, the proliferating cells spread out laterally between the epiblast surface and the visceral endoderm to give a layer of prospective mesoderm (light brown). Some of the internalized cells will eventually replace the visceral endoderm to give definitive endoderm (not shown on these diagrams for simplicity), which will form the gut. Right panel: as gastrulation proceeds, the primitive streak lengthens and reaches the bottom of the cup, with the node at the anterior end. The node gives rise to the notochord, which forms the structure known as the head process. Part of the visceral endoderm and the mesoderm has been cut away in this diagram to show the node and notochord. Note that, given the topology of the mouse embryo at this stage, the germ layers appear inverted (ectoderm on the inner surface of the cup, endoderm on the outer) if compared with the frog gastrula.

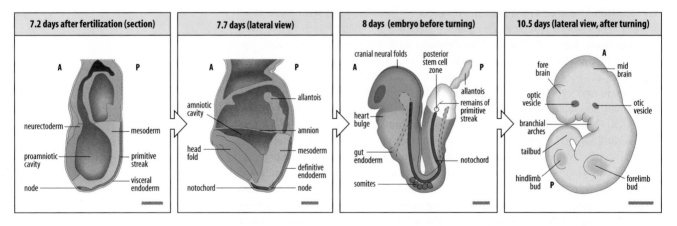

Fig. 3.25 Schematic views of the early development of the mouse embryo to the completion of gastrulation and neurulation. First panel: by 7.5 days, the primitive streak (brown) has extended to the bottom of the epiblast and the node has formed. The anterior ectoderm (blue) becomes prospective neurectoderm, which will give rise to the brain and spinal cord. Mesoderm is shown as light brown. Second panel: the anterior part of the embryo grows in size and the head fold appears. Definitive endoderm (green) replaces visceral endoderm (yellow) to form an outer layer on the ventral surface of the embryo. The notochord (red) begins to form. Third panel: by 8 days there has been further growth of the embryo anterior to the node, the head is distinct, the neural folds have formed, the foregut and hindgut have closed, and somites are beginning to form on either side of the notochord. The embryo is covered by a layer of ectoderm that will form the epidermis, which is not shown on this diagram. Fourth panel: by 10.5 days, gastrulation and neurulation are complete. The embryo has undergone a complicated turning process around day 9 that brings the dorsal and ventral sides into their final positions. Scale bars for the first three panels = 100 μm; fourth panel, scale bar = 75 μm.

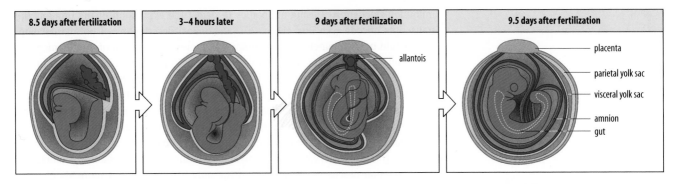

8.5 days after fertilization	3–4 hours later	9 days after fertilization	9.5 days after fertilization

Fig. 3.26 Turning in the mouse embryo. Between 8.5 and 9.5 days, the mouse embryo becomes entirely enclosed in the protective amnion and amniotic fluid. The visceral yolk sac, a major source of nutrition, surrounds the amnion and the allantois connects the embryo to the placenta.

After Kaufman, M.H.: 1992.

Somite formation and organogenesis start at the anterior end of the mouse embryo and proceed posteriorly. The first somite is formed at around E7.5 and new ones are formed at intervals of 120 minutes. At around E8, neural folds have also started to form at the anterior end on the dorsal side of the embryo and a head is apparent (see Fig. 3.25). The embryonic endoderm—initially exposed on the ventral surface of the embryo—becomes internalized to form the foregut and hindgut, the lateral surfaces eventually folding together to internalize the gut completely with subsequent overgrowth of mesoderm and ectoderm to form the ventral body wall in the process of ventral closure. The heart and liver move into their final positions relative to the gut, and the head becomes distinct. By about midway through embryonic development, gastrulation and neurulation are complete: the mouse embryo has a distinct head and the forelimb buds are starting to develop. A complicated 'turning' process has occurred around E9 that produces a more recognizable mouse embryo, surrounded by its extra-embryonic membranes (Fig. 3.26). As a result of turning, the initial cup-shaped epiblast has turned inside out so that the dorsal surface is now on the outside; the ventral surface, with the umbilical cord that connects it to the placenta, is facing inwards. (Turning is another developmental quirk peculiar to rodents; human embryos are surrounded by their extra-embryonic membranes from the beginning.) Organogenesis in the mouse proceeds very much as in the chick embryo. From fertilization to birth takes around 18 to 21 days, depending on the strain of mouse.

Experimental approaches to studying vertebrate development

This part of the chapter introduces some of the main techniques used to study the developmental biology of vertebrates. It is not intended to be comprehensive, but to indicate general experimental approaches and to describe in more detail a few techniques that are very commonly used. Rather than reading it straight through, you may wish to use it as a reference to come back to, to put into context some of the experiments described in the rest of the book. Many of the techniques described here are also applicable to non-vertebrates (see Chapters 2 and 6).

The earliest studies of vertebrate embryonic development involved careful and detailed study of whole and dissected embryos under the microscope, and simply watching, describing, and drawing the development of easily accessible embryos such as those of newts and frogs. From the seventeenth century onwards, observations of animal and human embryos by zoologists and physicians gradually yielded the comprehensive anatomical descriptions of embryonic development that are the foundation of modern developmental biology. Hensen's node and Koller's sickle (see Section 3.4), for example, are named after embryologists working in the early twentieth century. Careful observation is just as important in developmental biology today, even if the need to draw by hand the images you see in the microscope has been superseded by digital-imaging technology.

Once ways of detecting the expression of genes *in situ* were developed in the latter half of the twentieth century—both riboprobes to detect the gene transcripts and immunohistochemistry using antibodies to detect the expression of the proteins the transcripts encode (see Box 1D, p. 20), anatomical descriptions of normal development could be supplemented by mapping the expression patterns of developmentally important genes. Projects are currently under way to produce comprehensive online 'atlases' of the normal expression patterns of hundreds of genes at different developmental stages for the mouse and the chick (see General further reading). Techniques such as DNA microarray analysis can also detect and identify the expression of large numbers of gene at the same time and so can provide information about all the genes being expressed in a particular tissue or at a particular stage in development (Box 3A, p. 115). Observation on its own cannot unravel the mechanisms underlying developmental processes, however, and the only way to find out more about these is to disturb the developmental process in some specific way and see what happens. Techniques for interfering with development can be very broadly divided into two types, and many experiments in developmental biology will use a combination of these two approaches. Classical experimental embryological techniques manipulate the embryo by physical intervention—by removing or adding cells to cleavage-stage embryos or transplanting blocks of cells from one embryo to another, for example. The other class of techniques is based on genetics and genomics, and is used to disturb the expression of developmentally important genes by mutation, gene silencing, overexpression or misexpression—expressing the gene at a time or place where it is not normally expressed.

We shall start by taking a brief look at experimental manipulation in *Xenopus* and chick embryos, and then describe some of the techniques from molecular biology, genetics, and genomics that have revolutionized the study of developmental biology over the past few decades. Other technical advances that have benefited developmental biology over the past 20 years are the great improvement in computer-aided microscopic imaging techniques, the development of fluorescent labels in a vast range of colors that allow the imaging of living embryos, and the introduction of new forms of imaging into the field, such as magnetic resonance imaging (MRI) and optical projection tomography (OPT). Images of live avian embryos inside the egg can now be obtained by magnetic resonance imaging (Fig. 3.27).

3.6 Not all techniques are equally applicable to all vertebrates

You will soon notice as you read further in this book that experiments involving the microsurgical manipulation of vertebrate embryos largely feature *Xenopus* and the chick. Few spontaneous developmental mutations are known in these animals, and chick and amphibian embryos were being studied long before developmental genes had

Fig. 3.27 A live 9-day quail embryo inside its egg. The image on the left was obtained by magnetic resonance imaging (MRI).

Photograph courtesy of Suzanne Duce.

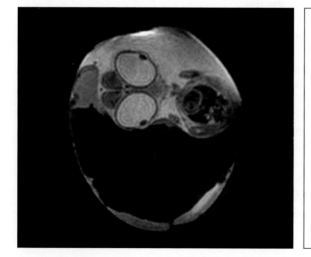
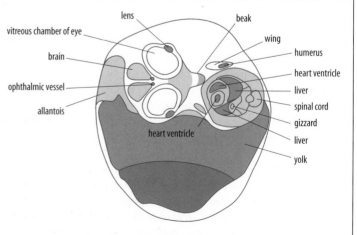

Box 3A Gene-expression profiling by DNA microarray

In this book we deal with the general principles underlying development so we do not describe every gene known so far to be involved in laying down the body plans of the animals we cover. However, now that the genomes of the model vertebrates have been sequenced, systematic genome-wide approaches are being used to identify all the genes involved in a particular developmental process. Genomic sequence information can also be used to identify the targets of the transcription factors that coordinate gene expression in space and time.

The identification of all the genes expressed in a particular tissue or at a particular stage of development can be accomplished by carrying out genome-wide screens for gene expression using **DNA microarrays**, often known as **DNA chips**. This technology enables the levels of RNA transcripts of thousands of genes to be measured simultaneously, and the main use of microarrays in developmental biology is to monitor the changes in gene expression that occur within a tissue or embryo at different stages of development or after experimental manipulation. Microarrays for studying gene expression come in various formats, but the most widely used are flat surfaces studded with a regular array of clusters or 'spots' of DNA fragments of known sequence, each cluster representing a DNA sequence found in a specific protein-coding gene. These DNA fragments are known as the 'probes,' and a microarray can contain thousands to hundreds of thousands of different probes. Microarrays composed of hundreds of thousands of oligonucleotide probes specially synthesized to cover the whole genome sequence of mouse and human are now commercially available. Alternatively, for more limited applications, probes can be made from cDNAs that have been made by reverse transcription of mRNA and so represent just the genes being expressed by a given tissue.

To determine which genes are being expressed, the total mRNA from the tissue of interest is extracted, then either

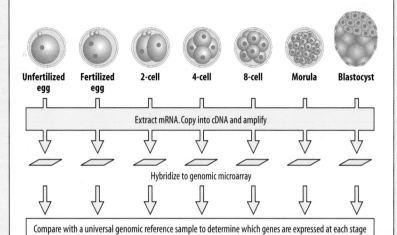

Unfertilized egg   **Fertilized egg**   **2-cell**   **4-cell**   **8-cell**   **Morula**   **Blastocyst**

Extract mRNA. Copy into cDNA and amplify

Hybridize to genomic microarray

Compare with a universal genomic reference sample to determine which genes are expressed at each stage

amplified as RNA or converted into cDNA and amplified by the polymerase chain reaction (PCR). The nucleic acid is then tagged with a fluorescent dye and hybridized to the microarray. A reference sample of mRNA, similarly treated and tagged with a different fluorescent dye, is also hybridized to the same microarray. The microarray is then scanned and the ratios of the signals from both dyes at each spot are recorded and converted into a relative expression level for the test sample.

One such experiment, for example, aimed to find all the genes expressed at particular stages in early mouse development. Genes expressed in the unfertilized egg, fertilized egg, two-cell stage, four-cell stage, eight-cell stage, morula, and blastocyst have been identified by extracting mRNA from 500 eggs or embryos at the appropriate stage, then labeling it and hybridizing it to miniaturized microarrays of unique oligonucleotides of 60 bases each, representing all of the mouse genes (see figure). The binding patterns of the labeled RNAs indicate which genes are expressed at the different stages. Figure 3.35 illustrates another technique that uses DNA microarrays to help reconstruct the gene regulatory networks involved in particular developmental processes.

been identified. *X. laevis* is also unsuitable for conventional genetic analysis because of its long generation time (1–2 years) and its tetraploid genome. Amphibian and avian embryos are, however, robust to surgical manipulation, and easily accessible to the experimenter at all stages in their development, unlike those of mammals (Fig. 3.28). Microsurgical manipulations in *Xenopus*, such as the removal of particular blastomeres, or the grafting of cells from one embryo into another, provided much of the information about the location and function of organizer regions in the blastula and their role in development; several experiments of this type are illustrated in Chapter 4. Blastomere removal identifies regions of the embryo that are essential for further development, while transplantation experiments are used to test the developmental potential of a

The suitability of different vertebrate embryos for classical experimental embryological manipulation				
	Frog (*Xenopus laevis*)	**Zebrafish**	**Chick**	**Mouse**
Are living embryos easily available and observable at all stages of their development?	Yes. Eggs can be obtained and externally fertilized and embryos develop to swimming tadpole stage in the aquarium. Large batches of eggs can be fertilized at the same time and develop synchronously to give large numbers of embryos at the same developmental stage.	Yes. As for *Xenopus*	For most of it. Eggs are fertilized internally and so the very early stages of development—to blastoderm stage—occur within the mother hen. Laid eggs are easily obtainable, can be incubated in the laboratory, and the development of the embryo observed at any time up to hatching. Early embryos can also be cultured up to about stage 10 out of the egg.	Only very early stages. *In vitro* fertilized eggs can be grown in culture up to the blastocyst stage, after which they have to be replaced in a surrogate mother mouse to continue development. Early isolated embryos can be cultured for a limited time (see Section 3.6).
Is the embryo amenable to experimental manipulation, e.g. blastomere removal or addition, or tissue grafting?	Embryo can be manipulated surgically up to the neurula stage (see Chapter 4). Embryos are comparatively large, and are highly resistant to infection after microsurgery	Amenable to surgical manipulation up to the neurula stage and to some extent in later stages. The embryo is transparent, which helps in observing developmental changes and the movement of labeled cells	Can be manipulated surgically up to quite late stages in embryonic development, including limb development (see Chapters 4, 5, 11, and 12).	Only very early embryos up to blastocyst stage can be manipulated if they are to be replaced in a surrogate mother for further development (see Figs 3.34 and 4.12). Isolated postimplantation embryos can be cultured for a limited time (see Section 3.6).
Is fate mapping and lineage tracing relatively easy?	Yes, in the early embryo. Single cells in the early blastula can be injected with a fluorescent non-toxic dye or RNA encoding fluorescent protein (GFP) (see Chapter 4)	Yes, as in *Xenopus*	Relatively easy and cells can be labeled at quite late embryonic stages by dye injection; the use of chick–quail chimeric embryos; grafts from GFP chickens (see below) or electroporation of DNA encoding a reporter protein or fluorescent protein (see Section 3.7).	In the past it was difficult. Now becoming easier owing to transgenic techniques for restricting reporter gene expression to particular cells (see Section 3.9 and Box 3B) and also inducible transgenes.

Fig. 3.28 The suitability of the different model vertebrates for classical experimental embryological manipulation.

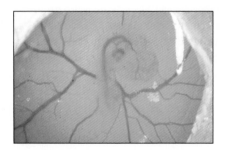

Fig. 3.29 A chick embryo can be observed through a window in the eggshell. The almost transparent 3.5-day embryo (H&H stage 21) can be just seen lying on top of the yolk. After an experimental procedure has been carried out, the window can be sealed with adhesive tape and the embryo left to develop further.

particular region of the embryo or to determine the time at which the fate of cells in a particular location becomes irreversibly determined (see Section 1.12). The cells of early amphibian embryos carry their own store of nutrients in the form of yolk platelets and so explants of embryonic tissue can be cultured in a simple salt medium for several days. This allows experiments that investigate the inducing effects of one tissue on another when placed together in culture, and experiments of this type in *Xenopus* are described in Chapters 4 and 5. Similar explant experiments have been carried out with chick embryonic tissue embedded in collagen gels and using cell culture medium.

The early chick embryo (blastoderm stage) can be removed from the egg and cultured on its vitelline membrane with thin albumen in a watch glass. The embryo will then develop for up to about 36 hours (Hamilton and Hamburger stage 10). Experiments in which pieces of one chick embryo are grafted into another at different stages have also provided much of the information on the organizer properties of different regions of the embryo and the generation of the primitive streak (discussed in Chapters 4 and 5). Older embryos can be accessed through a window cut in the egg shell and manipulated while remaining in the egg (Fig. 3.29). The window is then re-sealed with adhesive tape and the embryo allowed to continue development so that the effects of the procedures can be assessed. The accessibility of the chick embryo allows many types of experimental procedures: part of a developing structure such as a limb bud can be removed; tissue can be grafted to new sites; and the effects of chemicals such as growth factors on development can be investigated by implanting small inert beads soaked in these agents into a specific location in the embryo (examples of all these types of experiment in the study of chick limb development are described in Chapter 11).

After implantation, mouse development *in vivo* is hidden from view and can only be followed by isolating embryos at different stages (see Fig. 4.12) and cells can be injected into the inner cell mass of the blastocyst (see Section 3.9). Tissues from very early embryos can be cultured for a limited time. For example, inner cell masses isolated from 3.5-day embryos and cultured *in vitro* for several days will develop certain embryonic and extraembryonic tissues and some structures typical of a normal embryo at the corresponding stage. Mouse embryos between stages E6.5 and E12.5 have been cultured for 24–48 hours in roller tubes, which allows manipulations such as node transplantation. Experimental investigation of living embryonic tissues at later stages requires dissection of embryos and the growth of tissue or organ explants (such as limb buds, embryonic kidney, lung) *in vitro*.

3.7 Fate mapping and lineage tracing reveal which cells in the early embryo give rise to which adult structures

Fate maps like those that will be discussed in Chapter 4 show what a particular region of an embryo will develop into during normal development. For both biological and technical reasons, fate mapping in the early embryo is easiest in *Xenopus*. At the blastula stage, cells that will eventually form the three germ layers are arranged in distinct regions and are accessible from the surface of the embryo (see Fig. 3.6). Single blastomeres can be easily labeled with a non-toxic marker such as DiI (which fluoresces red) or fluorescein-linked dextran (see Fig. 4.18) or by expression of a fluorescent protein such as GFP by injecting GFP RNA.

Fate mapping by dye injection into individual cells is more difficult in the chick embryo as the cells are much smaller, but it has been used to trace the lineage of cells derived from Hensen's node. Chick–quail chimeras have also been used for cell-lineage tracing. Chick and quail embryos are very similar in their development and early embryos composed of a mixture of chick and quail cells will develop normally in culture. The cells of the two species can be distinguished by the different appearance of their nuclei; quail cells have a prominent nucleolus that stains red with an appropriate stain (see Fig. 5.6). Quail cells can also be distinguished from chick cells by immunohistochemical staining with labeled antibodies for quail proteins. For lineage tracing, cells are taken from a particular site in an early quail embryo and transplanted into the same site in a chick embryo of the same age. Later-stage embryos are then stained and sectioned to determine where the quail cells have ended up. Recently, transgenic chicken embryos expressing GFP in all cells have become available (see Section 3.9) and can be used for long-term fate mapping.

Defined small populations of cells, or even single cells, can also be conveniently marked in chick embryos in the egg (*in ovo*) or in culture by introducing a DNA encoding a fluorescent protein. Despite the small cell size in chick embryos compared with *Xenopus* early blastomeres, DNA constructs can be targeted to a relatively small number of cells by the technique of **electroporation**. The DNA is injected into the embryo at the desired site with a fine pipette and pulses of electric current applied using very fine electrodes. The current makes the membranes of nearby cells permeable to the injected DNA.

Cell-lineage tracing and fate mapping by dye injection or electroporation of nucleic acids is technically much more difficult in mammalian embryos than in *Xenopus* or the chick because the embryos are less accessible inside the mother. Nevertheless these same techniques have been used to follow cell fate in mouse embryos between E6.5 and E8.5, stages that can be maintained in culture (see Section 3.6). Researchers have found other ingenious ways around the problem, although some of the techniques used required a great deal of skill. For example, in experiments carried out in the early 2000s to study the migration of neurons in the developing brain, a DNA vector containing a GFP gene was injected into the lateral ventricles of the brains of mouse embryos *in utero* (at around stage E14), accompanied by electroporation, without harming the embryos. Sections of the brains of later-stage embryos and of mouse pups after birth were examined by fluorescence microscopy to reveal the locations of the GFP-labeled neurons, which originate in the layer of cells immediately under the ventricle surface (see Chapter 12).

Lineage tracing in the mouse is now done much more conveniently using transgenic embryos expressing a reporter gene in a particular set of cells under the control of the

Cre/*loxP* system (Box 3B) or a reporter gene linked to a suitable promoter (see Section 3.9). Techniques such as these are continually being refined to give better resolution, and have proved particularly useful for tracing later events in embryogenesis, such as organ development, the development of the nervous system, and cell behavior in epithelia in which cells are continually being replaced, such as skin and gut (see Section 10.7). Clonal cell-lineage analysis is now becoming possible using a transgenic mouse in which a drug-inducible Cre recombinase (linked to the reporter gene LacZ (see Box 1D, p. 20) or to GFP) is expressed in every cell; treatment of the pregnant mouse with an appropriate dose of the drug will lead to a low frequency of recombination and generate clones of marked cells in the embryos. This strategy led to the discovery of dorso-ventral compartments in the developing mouse limb (see Chapter 11).

Box 3B Insertional mutagenesis and gene knock-outs in mice: the Cre/*loxP* system

One very powerful and widely used way of targeting a gene knock-out to a specific tissue and/or a particular time in development is provided by the **Cre/*loxP* system**. The target gene is first 'floxed' by inserting a *loxP* sequence of 34 base pairs on either side of the gene (hence 'floxed', for 'flanked by *lox* '). This targeting is carried out using embryonic stem cells in culture (see Section 3.9) and transgenic mice are then produced by introducing the genetically modified cells into blastocysts (see Fig. 3.34) and deriving lines of transgenic mice. The line of mice with the floxed gene is then crossed with another line of transgenic mice carrying the gene for the recombinase Cre. *loxP* sequences are recognized by Cre, and the resulting recombination reaction excises all the DNA between the two *loxP* sites. In the offspring, if Cre is expressed in all cells, then all cells will excise the floxed target, causing a knock-out of the target gene in all cells. However, if the gene for Cre is under the control of a tissue-specific promoter, so that, for example, it is only expressed in heart tissue, the target gene will only be excised in heart tissue (see figure). If the *Cre* gene is linked to an inducible control region, it is possible to induce excision of the target gene at will by exposing the mice to the inducing stimulus.

A significant number of knock-outs of a single gene result in mice developing without any obvious abnormality, or with fewer and less severe abnormalities than might be expected from the normal pattern of gene activity. A striking example is that of *myoD*, a key gene in muscle differentiation, which is described in Chapter 10. In *myoD* knock-outs, the mice are anatomically normal, although they do have a reduced survival rate. The most likely explanation for this is that the process of muscle differentiation contains a certain amount of redundancy and that other genes can substitute for some of the functions of *myoD* (see Section 10.8).

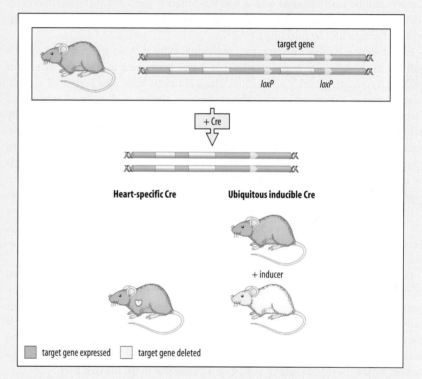

However, it is unlikely that any gene is without any value at all to an animal. It is much more likely that there is an altered phenotype in these apparently normal animals, which is too subtle to be detected under the artificial conditions of life in a laboratory. Redundancy is thus probably apparent rather than real. A further complication is the possibility that, under such circumstances, related genes with similar functions may increase their activity to compensate for the mutated gene.

As well as knocking out a coding sequence, the Cre/loxP method can be used to target promoters in such a way that the associated gene is permanently turned on. This technique is used in cell-lineage studies in mice to turn on a reporter gene like *GFP* or *lacZ* in particular cells at a particular stage in development. The progeny of the labeled cells can then be distinguished in later-stage embryos or after birth.

3.8 Developmental genes can be identified by spontaneous mutation and by large-scale mutagenesis screens

Rare spontaneous mutations were for a long time the only way that genetic disturbances of development could be studied, and such mutations have been identified in mice (see Fig. 1.12) and in chickens. Informative spontaneous mutations are rare, however, even in mice, and as they are only noticed if the animal survives birth, mutations in important developmental genes that lead to the death of a mouse embryo *in utero* (**embryonic lethal mutations**) are likely to be missed. A lethal developmental mutation in chicken was only discovered because of the reduced hatchability of the eggs. Many more developmental mutations have been produced by inducing random mutation in large numbers of organisms by chemical treatments or irradiation by X-rays, and then screening for mutants of developmental interest. The aim, where possible, is to treat a large enough population so that, in total, a mutation is induced in every gene in the genome. This sort of approach is best used in organisms that breed rapidly and that can conveniently be obtained and treated in very large numbers. Among vertebrates, this approach has been applied to both mice and zebrafish. Zebrafish in particular, are a potentially valuable vertebrate system for large-scale mutagenesis because large numbers can be handled, and the transparency and large size of the embryos makes it easier to identify developmental abnormalities (Box 3C, p. 120). However, unlike the genetic screening in *Drosophila* described earlier in the book (see Box 2A, p. 42), there are no genetic means so far of automatically eliminating unaffected individuals in zebrafish. This means that all the progeny of a cross have to be examined visually for any developmental abnormality.

Mutagenesis screens like those for zebrafish are also carried out in mice, but because of the mouse's larger size, longer generation time, and the fact that embryos cannot be screened so easily, they are more difficult and costly to set up. Nevertheless, many such screens have been carried out. Chemical mutagenesis generates both dominant mutations with varying degrees of gene function, which are particularly relevant to human conditions, and recessive lack-of-function mutations. Dominant mutations can be recognized in offspring in the F_1 generation (Fig. 3.30), but more complicated breeding programs like that described for zebrafish have to be used to reveal recessive mutations.

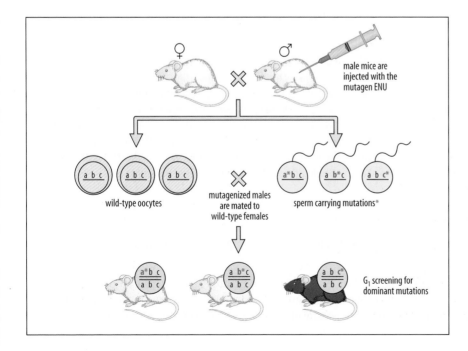

Fig. 3.30 A mutagenesis screen for dominant mutations in the mouse. Progeny with a dominant mutation, such as the dark brown coat color shown here, can be detected in the first generation (G_1). ENU, ethylnitrosurea.

Box 3C Large-scale mutagenesis in zebrafish

A screening program using zebrafish involves breeding for three generations (see figure). Male fish treated with a chemical mutagen are crossed with wild-type females; their F_1 male offspring are crossed again with wild-type females, and the female and male siblings from each of these crosses are themselves crossed. The offspring from each of these pairs are examined separately for homozygous mutant phenotypes. If the F_1 fish carry a mutation, then in 25% of the F_2 matings two heterozygotes will mate and 25% of their offspring will be homozygous for the mutation. Zebrafish can also be made to develop as haploids by fertilizing the egg with sperm heavily irradiated with UV light. This allows one to detect early-acting recessive mutations without having to breed the fish to obtain homozygous embryos.

Two new techniques can be used to identify or create mutations in zebrafish genes about which nothing may be known apart from their sequences. **TILLING** (targeting-induced local lesions in genomes) is used to identify mutations generated by chemical mutagenesis in particular genes. First developed for plants, it has also been applied to *Drosophila, Caenorhabditis*, and

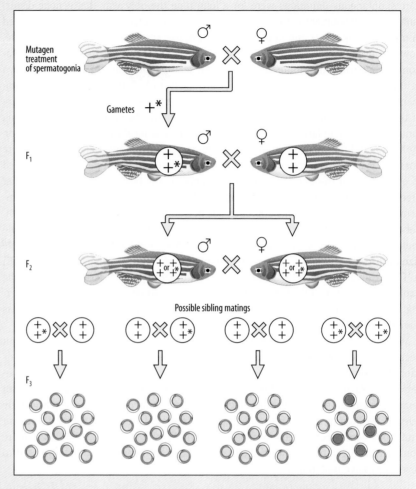

the rat. All that is necessary is knowledge of the DNA sequence of the gene. Male fish treated with a chemical mutagen are crossed with wild-type females. A sample of tissue from each F_1 male, which will be heterozygous for any point mutations generated, is collected and, at the same time, its sperm are frozen and stored. Mutations in genes of interest are detected by first making large amounts of the mutated DNA by the polymerase chain reaction (PCR) and then hybridizing the mutated DNA with unmutated DNA from the same gene. Mismatched bases will reveal the site of a mutation and the DNA is then sequenced to identify the precise change. Males and/or frozen sperm corresponding to each

mutation are used to generate families of fish with that mutation, which reveal any phenotypic change associated with it.

A relatively novel method for generating mutations relies on engineering targeting specificity into **zinc-finger nucleases** (**ZFNs**), which cleave DNA, and using them to cleave and disrupt chosen genes *in vivo*. This approach has been used in zebrafish to target the No-tail (*Ntl*) gene, the zebrafish version of *Brachyury*. Like *Brachyury* in the mouse (see Fig. 1.13), *No-tail* is required for tail formation. About 30% of zebrafish embryos injected at the one-cell stage with mRNA for *Ntl*-specific ZFN lacked tails, showing that *Ntl* had been rendered non-functional.

3.9 Transgenic techniques enable animals to be produced with mutations in specific genes

To study the function of a particular gene, it is more efficient to be able to specifically alter that gene in the animal rather than having to trust to random mutagenesis. Strains of mice with a particular mutant genetic constitution can now be produced

Fig. 3.31 Chicks transgenic for green fluorescent protein (GFP). Transgenic chickens can be obtained by injecting DNA carried by a self-inactivating lentiviral vector into the subgerminal space beneath the blastodisc in newly laid eggs. The chicks illustrated are second-generation offspring from an original transgenic bird, and carry the GFP transgene in the germline. The bird in the center is not transgenic.

Photograph from McGrew, J.M. et al.: 2004.

relatively routinely, and as long as the mutation is recessive, animals heterozygous for the mutation can be maintained as breeding stock, even for mutations that may be lethal to the embryo when homozygous. Animals into which an additional or altered gene has been introduced are known as **transgenic** animals. Transgenic techniques are most highly developed in mice, and we shall describe transgenesis here in relation to mice. Transgenic zebrafish can be obtained by injecting DNA into the one-cell-stage embryo and the first transgene to be successfully delivered into zebrafish was the delta-crystallin gene of the chicken. Transgenesis has quite recently been demonstrated in principle in chickens, by the injection of lentiviral vectors containing the gene encoding GFP under the blastoderm of a newly laid egg (Fig. 3.31). And although *X. laevis* is less amenable to germline transgenesis, the transgenic techniques that have been developed for its close relative *X. tropicalis*, which is diploid and has a much shorter generation time of 5–9 months, can also be used. For *Xenopus*, the strategy is to introduce the DNA transgene into sperm nuclei and then transplant these nuclei into unfertilized eggs. The relative suitability of our model vertebrates for genetics-based techniques is summarized in Fig. 3.32.

The suitability of the different model vertebrates for genetics-based techniques for studying development				
	Frog (*Xenopus laevis*)	**Zebrafish**	**Chick**	**Mouse**
Genome sequenced	The diploid *X. tropicalis* genome has been sequenced	Yes	Yes	Yes
Spontaneous mutations	No	Yes	Yes	Yes
Induced mutations (mutational screens)	Yes	Yes (see Box 3C)	No	Yes
Gene silencing or gene knock-out in somatic cells	Gene silencing by morpholino antisense RNAs (MOs) (Box 6A).		Gene silencing by MOs and RNA interference (RNAi) (Box 6A)	Gene knock-out by Cre/*loxP* system (requires making transgenic animals first, Box 3B)
Germline gene knock-out (transgenic animals)	No	Yes	Not yet	Yes
Gain-of-function (e.g. misexpression and ectopic expression) of specific genes in the whole embryo	By injection of mRNAs into the fertilized egg or very early embryo (see e.g. Fig. 4.4)		Not routinely possible so far but transgenic chicks have been produced (Fig. 3.31)	Transgenic mice with gain-of-function mutations can be made
Targeting of gene overexpression or misexpression in time and space	Expression can be targeted to some extent by injection of RNAs into specific blastomeres	No	Yes in somatic tissue	Transgenic mice with inducible or cell-type specific expression of a transgene can be made
Embryonic stem cells can be cultured and differentiated *in vivo*	No	No	Yes	Yes. Mutant transgenic mice can also be made using ES cells mutated by homologous recombination (see Section 3.9 and Chapter 10)

Fig. 3.32 The suitability of the different model vertebrates for genetics-based study of development.

Two main techniques for producing transgenic mice are currently in use. One is to inject a transgene composed of DNA encoding the required gene and any necessary regulatory regions directly into the male pronucleus of a fertilized egg, and then replace the egg in a surrogate mother. If the transgene becomes incorporated into the genome it will be present in all the cells of the embryo, including the germline, and will be expressed according to the promoter and other regulatory regions it contains. In this way, mice that overexpress a given gene in particular cells or at particular times during development can be produced, and the effect of misexpressing a gene in a tissue or at a time where it is not usually active can be investigated.

A newer technique for producing transgenic mice uses embryonic stem cells (ES cells) that have been mutated *in vitro*, and this is now the most commonly used transgenic technique for producing specific gene knock-outs—the permanent deletion of a gene function. ES cells are pluripotent cells derived from the inner cell mass of a mouse blastocyst; they can be maintained in culture indefinitely and grown in large numbers and are being investigated for their possible use in regenerative medicine (discussed in Chapter 10). ES cells injected into the cavity of an early blastocyst become incorporated into the inner cell mass and so will become part of all the embryo's tissues, even giving rise to germ cells and gametes. For example, if ES cells from a brown-pigmented mouse are introduced into the inner cell mass of a blastocyst from a white mouse, the mouse that develops from this embryo will be a chimera of 'brown' and 'white' cells. In the skin, this mosaicism is visible as patches of brown and white hairs.

To generate transgenic mice with a particular mutation, the ES cells are mutated while growing in culture and the mutated cells are introduced into a blastocyst, which is then placed in a surrogate mother to continue its development. Mutations are targeted to a particular gene by the technique of homologous recombination, in which specially tailored DNA constructs are introduced into the cell by transfection. A transfected DNA molecule will usually insert randomly in the genome, but by including some sequence that is homologous with the desired target gene, it is possible to make it insert at a predetermined site (Fig. 3.33). Homologous recombination between the transfected DNA and the target gene in the ES cell results in an insertion that renders

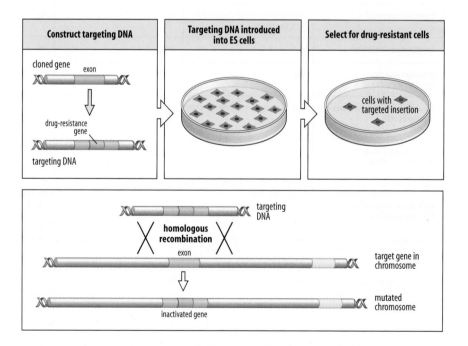

Fig. 3.33 Production of mutant mouse ES cells by homologous recombination. The gene to be targeted for gene knock-out is first cloned and a drug-resistance or other selectable marker is inserted into it, which also renders the gene non-functional. When the modified gene is introduced into cultured mouse ES cells, it will undergo homologous recombination with the corresponding gene in a small number of the ES cells. These can then be selected by their drug resistance.

the gene non-functional, and this is now the main way of producing gene knock-outs. Homologous recombination works particularly well in mice. The DNA to be introduced must contain enough sequence homology with its target to insert within it in at least a few cells in the ES cell culture. The introduced DNA usually carries a drug-resistance gene so that cells containing the insertion can be selected by adding the drug, which kills the other, unmodified, cells. The mutated ES cells are then introduced into the inner cell mass of a blastocyst, producing a transgenic mouse carrying a mutation in a known gene (Fig. 3.34). When one gene is replaced by another functional gene using these techniques, it is called a gene knock-in.

Because the animals produced by ES-cell transfer are chimeras composed of both mutant and normal cells, they may show few, if any, effects of the mutation. If the mutant gene has entered the germline, however, strains of mice heterozygous for the altered gene can be intercrossed to produce homozygotes that are viable or fail to develop, depending on the gene involved. We shall see examples of gene knock-outs in mice being used to study the functions of Hox genes in regionalizing the somitic mesoderm and the neural tube in Chapters 5 and 12.

3.10 Gene function can also be tested by transient transgenesis and gene silencing

Germline transgenesis is not the only way of altering gene expression, and producing transgenic mice is still a relatively expensive process. Techniques of transient transgenesis or somatic transgenesis are easier and cheaper to implement, and are particularly valuable in amphibians and the chick, in which germline transgenesis has not been a practical option. The early *Xenopus* embryo has large blastomeres and is very suitable for experiments in which the mRNA of a gene suspected of having a developmental role is injected into a specific blastomere to see the effect of its overexpression. mRNA injection is also used in *Xenopus* and other embryos to see if a specific gene can 'rescue' normal development after a region of the early embryo has been removed or the embryo's own gene has been silenced. Experiments of this type are described in Chapter 4.

Transient transgenesis for both short-term and long-term overexpression or misexpression of genes is also widely used to study gene function in the chick embryo. For short-term expression, genes are delivered into cells by electroporation (see Section 3.6). For longer-term overexpression, the gene is inserted into a replication-competent retroviral vector, which is then injected into the embryo. The infected cells express the gene as part of the viral replication cycle and, as the virus replicates, it spreads the gene to neighboring cells.

The transient counterpart to the gene knock-out is the technique of gene silencing or gene knockdown, in which a gene is not mutated or removed from the genome, but its expression is blocked by targeting the mRNA and preventing its translation. Techniques for doing this are discussed in Box 6A, p. 220. Because it is not possible to generate loss-of-function mutations in *Xenopus* and chick embryos, gene silencing by the injection of morpholino antisense RNAs is particularly useful and this technique is also widely used in zebrafish. Morpholinos are stable RNAs designed to be complementary to a specific mRNA. When introduced into the cells of an embryo, they hybridize with the target mRNA, thus preventing its translation into protein. Injection of a morpholino into a fertilized *Xenopus* egg will knock down gene expression in all the cells of the early embryo; electroporation of morpholinos into cells of early chick embryos will produce local knockdown. The technique of RNA interference, which is also described in Box 6A (p. 220) is also used in embryology. RNAi constructs encoding the interfering RNAs have been electroporated into the neural tube of chick embryos to knock down gene expression.

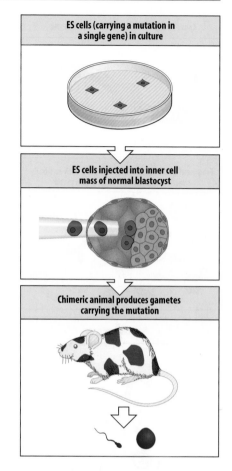

Fig.3.34 Production of transgenic mice by the introduction of mutated ES cells into the blastocyst. ES cells injected into a blastocyst will become part of the inner cell mass, and can give rise to all the cell types in the mouse, including the germline. The first generation of transgenic mice will be chimeras of the blastocyst genotype and the ES cell genotype, but if they carry the mutant gene in the germline, they can be interbred over several generations to produce strains of heterozygous or homozygous transgenic mice.

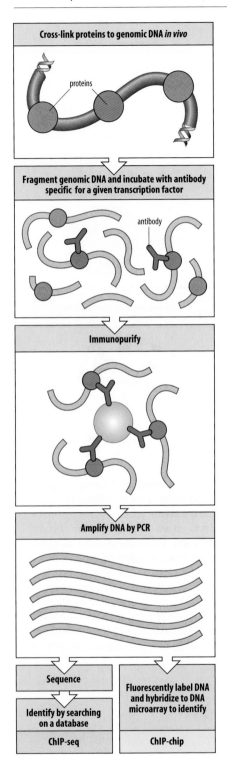

Cross-link proteins to genomic DNA *in vivo*

proteins

Fragment genomic DNA and incubate with antibody specific for a given transcription factor

antibody

Immunopurify

Amplify DNA by PCR

Sequence

Identify by searching on a database

ChIP-seq

Fluorescently label DNA and hybridize to DNA microarray to identify

ChIP-chip

Fig. 3.35 ChIP-chip and ChIP-seq. These two variants of chromatin immunoprecipitation are used to discover the binding sites in DNA for transcription factors and other DNA-binding proteins. After cross-linking any bound proteins to the DNA by chemical treatment, the DNA is fragmented and treated with an antibody specific for the protein of interest. The antibody complex is purified and the DNA extracted. It is then amplified by the polymerase chain reaction (PCR) and either analyzed by DNA microarray (ChIP-chip) or by DNA sequencing (ChIP-seq).

3.11 Gene regulatory networks in embryonic development can be revealed by chromatin immunoprecipitation techniques

One powerful new approach to investigating development relies on a combination of bioinformatics and experimental molecular and cell biology to describe the network of interactions between genes and their transcription factors, and so identify **gene regulatory networks**. Once the genome sequence of an animal is known, the technique of chromatin immunoprecipitation followed by DNA microarray analysis (**ChIP-chip**), can be used to identify the target genes that a given transcription factor binds to *in vivo* (Fig. 3.35). Embryos at a given embryonic stage are treated so that proteins bound to the *cis*-regulatory control elements in the DNA are chemically linked to the DNA. The DNA is then isolated and sheared into small segments and treated with an antibody specific for a given transcription factor. This will precipitate, or 'pull down,' all the DNA segments to which that transcription factor was bound in the embryo's cells. The identity of the DNA can then be determined by digesting away the protein, labeling the DNA segment and hybridizing it to DNA microarrays containing genomic sequences lying upstream of known genes (see Box 3A, p. 115). Alternatively, the protein-free DNA can be sequenced, when the technique is known as **ChIP-seq**.

The associated gene can then be identified by searching the databases of genome sequences. ChIP-seq has been used, for example, to identify the gene targets of the transcription factor Ntl (No-tail), the zebrafish equivalent of Brachyury (see Fig. 1.13), in mid-gastrula zebrafish embryos, and thus begin to uncover the gene regulatory network involved in mesoderm specification and patterning in this model animal (see Box 4E, p. 162).

Summary to Chapter 3

All vertebrates have a similar basic body plan. The defining vertebrate structures are the vertebral column, which surrounds the spinal cord, and the bony or cartilaginous skull that forms the head and encloses the brain. The vertebrate body has three main axes of symmetry—the antero-posterior axis running from head to tail, the dorso-ventral axis running from back to belly, and bilateral symmetry around the dorsal midline, so that outwardly the right and left sides are mirror images of each other. Some internal organs, such as lungs, kidneys, and gonads, are also present as symmetrically paired structures, but single organs, such as the heart and liver, are arranged asymmetrically with respect to the dorsal midline. Early vertebrate embryos pass through a set of developmental stages—cleavage, gastrulation, and neurulation—to form embryos that broadly resemble each other—a stage known as the phylotypic stage. There are considerable differences in the development of the different model organisms before the phylotypic stage, which relate to how and when the axes are set up, and how the germ layers are established. These differences are mainly due to the different modes of reproduction of the animals and the consequent form of the earliest embryo. Structures typical of the different vertebrate groups, such as fins, beaks, wings, and tails, develop after the phylotypic stage. The body pattern and the initial organ rudiments are laid down when the embryo is still very small. Growth in size takes place at later stages. The mechanisms underlying development can only be studied by disturbing the normal developmental process in specific ways and observing what happens. Techniques for interfering with development can be broadly divided into two types: experimental embryological techniques that manipulate the embryo by removing or adding cells to cleavage-stage embryos or transplanting blocks of cells from one embryo to another; and genetics-based techniques that disturb the expression of developmentally important genes by mutation, gene silencing, overexpression or misexpression.

■ End of chapter questions

Long answer (concept questions)

1. All vertebrates share certain features; what are those features? Referring to Fig. 3.2, identify those features in the row of illustrations that shows the phylotypic stage.

2. After fertilization, all embryos undergo cleavage. In the *Xenopus* embryo, what is the relationship of the animal-vegetal axis to the plane of the first three cleavages? Describe how cleavage in the chick embryo differs from cleavage in frogs and mice; why do these differences occur?

3. Compare the terms blastula, blastoderm, and blastocyst. To what organisms do they refer? What features do they share, and how do they differ?

4. Gastrulation in *Xenopus* leads to the establishment of the three germ layers through the process of involution. What are the three germ layers, and what is their relationship to each other in the gastrula (stage 11)? What is the archenteron, and what will it become?

5. Distinguish between the following regions in the chick embryo: epiblast and hypoblast; posterior marginal zone, Koller's sickle, and Hensen's node; endoblast and endoderm; mesenchyme and mesoderm.

6. Hensen's node of the gastrulating chick embryo is conceptually analogous to the blastopore of the *Xenopus* embryo, but despite this analogy, several differences between the two structures exist. Elaborate on the following differences: (a) the ingression of cells in the chick versus the involution of cells in the frog; (b) the movement of Hensen's node in the chick versus the lack of movement of the blastopore; (c) the formation of the notochord in two species.

7. During neurulation in the chick embryo, the precursors to several adult structures first form. What structures will form from: neural plate; somites; intermediate mesoderm; splanchnic mesoderm; notochord? (See Fig. 3.17.)

8. Distinguish between the inner cell mass and the trophectoderm in the mouse embryo. Outline the events that will lead to the formation of the epiblast and a primitive streak.

9. What are some of the features that would make an animal species a good 'model organism' for the study of development? For example, what features make a chicken an attractive model for development compared to frogs? In contrast, what features does the mouse offer over chickens?

10. Briefly summarize the production of transgenic mice and of knock-out mice. How do the two processes differ, both in technique and in outcome?

Multiple choice (factual recall questions)

NB There is only one right answer to each question.

1. In which portion of the frog's life cycle would a frog appear most similar to a mammal?

a) blastula

b) phylotypic stage

c) gastrulation

d) fertilized egg

2. The darkly pigmented end of a *Xenopus* egg is called

a) dorsal

b) the animal pole

c) the vegetal pole

d) yolk

3. Which of the following choices puts the early developmental stages of the frog in their correct order?

a) blastula, cleavage, gastrula, neurula

b) cleavage, gastrulation, neurulation, organogenesis

c) neurulation, gastrulation, cleavage, organogenesis

d) tadpole, embryo, neurula, gastrula

4. Cleavage in the zebrafish gives rise to the

a) 14-somite stage

b) gastrula

c) sphere stage

d) shield stage

5. The primitive streak of mouse and chick embryos is equivalent to which amphibian structure?

a) archenteron

b) blastocoel

c) blastopore

d) neural folds

6. At the blastocyst stage, the mammalian embryo is composed of

a) a blastocoel, yolk, and animal cap cells

b) an epiblast overlying a hypoblast

c) ectodermal, mesodermal, and endodermal cells

d) trophectoderm and inner cell mass

7. The identification of all the genes expressed in a particular tissue or at a particular stage of development can be done using

a) *in situ* hybridization

b) insertional mutagenesis

c) microarrays

d) zinc-finger nucleases

8. The advantage of using the Cre/loxP system for targeted mutagenesis is

a) a gene can be deleted in a particular tissue or at a desired time in development

b) the Cre/loxP system allows germline transgenesis in amphibians and chickens

c) the Cre/loxP system is more specific than homologous recombination

d) the deletion of a gene by the Cre/loxP system is reversible

9. The least favorable organism for traditional genetic analysis is

a) *Drosophila*

b) mouse

c) *Xenopus*

d) zebrafish

10. Electroporation is

a) fusion of adjacent cells using an electrical pulse

b) injection of a gene into the nucleus of a cell

c) the introduction of a fluorescent lineage maker into cells

d) use of brief pulses of electricity to cause cells to take up DNA

Multiple choice answer key

1: b, 2: b, 3: b, 4: c, 5: c, 6: d, 7: c, 8: a, 9: c, 10: d.

■ General further reading

Bard, J.B.L.: *Embryos. Color Atlas of Development*. London: Wolfe, 1994.

Carlson, B.M.: *Patten's Foundations of Embryology*. New York: McGraw-Hill, 1996.

Descriptions of developmental stages of frog, zebrafish, chick and mouse embryos

Xenopus: developmental stages: Xenbase – Nieuwkoop and Faber stage series [http://www.xenbase.org/xenbase/original/atlas/NF/NF-all.html] (accessed 13 May 2010); movies of cleavage and gastrulation [http://www.xenbase.org] (accessed 13 May 2010).

Zebrafish: developmental stages Karlstrom Lab [http://www.bio.umass.edu/biology/kunkel/fish/zebra/devstages.html] (accessed 13 May 2010); ZFIN Embryonic Developmental Stages [http://zfin.org/zf_info/zfbook/stages/stages.html] (accessed 13 May 2010); developmental movie: Karlstrom, R.O., Kane, D.A.: **A flipbook of zebrafish embryogenesis**. *Development* 1996, **123**: 1–461 [http://www.bio.umass.edu/biology/karlstrom/Karlstrom-Lab.html] (accessed 13 May 2010).

Chick: list and description of Hamilton & Hamburger developmental stages: UNSW Embryology – Chicken Developmental Stages [http://embryology.med.unsw.edu.au/Otheremb/chick1.htm] (accessed 13 May 2010).

Mouse: Edinburgh Mouse Atlas Project [http://genex.hgu.mrc.ac.uk] (accessed 13 May 2010).

Human: The Multidimensional Human Embryo [http://embryo.soad.umich.edu] (accessed 13 May 2010); UNSW Embryology: Carnegie staging of human embryos [http://embryology.med.unsw.edu.au] (accessed 13 May 2010).

■ Section further reading

3.1 The frog *Xenopus laevis* is the model amphibian for developmental studies

Hausen, P., Riebesell, H.: *The Early Development of* Xenopus laevis. Berlin: Springer-Verlag, 1991.

Nieuwkoop, P.D., Faber, J.: *Normal Tables of* Xenopus laevis. Amsterdam: North Holland, 1967.

3.2 The zebrafish embryo develops around a large mass of yolk

Kimmel, C.B., Ballard, W.W., Kimmel, S.R., Ullmann, B., Schilling, T.F.: **Stages of embryonic development of the zebrafish**. *Dev. Dyn.* 1995, **203**: 253–310.

Warga, R.M., Nusslein-Volhard, C.: **Origin and development of the zebrafish endoderm**. *Development* 1999, **26**: 827–838.

Westerfield, M. (Ed.): *The Zebrafish Book; A Guide for the Laboratory Use of Zebrafish* (Brachydanio rerio). Eugene, Oregon: University of Oregon Press, 1989.

3.4 The early chicken embryo develops as a flat disc of cells overlying a massive yolk

Bellairs, R., Osmond, M.: *An Atlas of Chick Development* (2nd edn). London: Academic Press, 2005.

Chuai, M., Weijer, C.J.: **The mechanisms underlying primitive streak formation in the chick embryo**. *Curr. Top. Dev. Biol.* 2008, **81**: 135–156.

Hamburger, V., Hamilton, H.L.: **A series of normal stages in the development of a chick**. *J. Morph.* 1951, **88**: 49–92.

Lillie, F.R.: *Development of the Chick: An Introduction to Embryology*. New York: Holt, 1952.

Patten, B.M.: *The Early Embryology of the Chick*. New York: McGraw-Hill, 1971.

Stern, C.D.: **Cleavage and gastrulation in avian embryos (version 3.0)**. *Encyclopedia of Life Sciences* 2009, http://www.els.net/ (13 May 2010).

3.5 Early development in the mouse involves the allocation of cells to form the placenta and extra-embryonic membranes

Cross, J.C., Werb, Z., Fisher, S.J.: **Implantation and the placenta: key pieces of the developmental puzzle**. *Science* 1994, **266**: 1508–1518.

Kaufman, M.H.: *The Atlas of Mouse Development* (2nd printing). London: Academic Press, 1994.

Kaufman, M.H., Bard, J.B.L.: *The Anatomical Basis of Mouse Development*. London: Academic Press, 1999.

Experimental approaches to studying vertebrate development

Hogan, H., Beddington, R., Costantini, F., Lacy, E.: *Manipulating the Mouse Embryo. A Laboratory Manual* (2nd edn). New York: Cold Spring Harbor Laboratory Press, 1994.

McGrew, M.J., Sherman, A., Ellard, F.M., Lillico, S.G., Gilhooley, H.J., Kingsman, A.J., Mitrophanous, K.A., Sang, H.: **Efficient production of germline transgenic chickens using lentiviral vectors**. *EMBO Rep.* 2004, **5**: 728–733.

Sharpe, P., Mason, I.: *Molecular Embryology Methods and Protocols* (2nd edn). New York: Humana Press; 2008.

Southall, T.D., Brand, A.H.: **Chromatin profiling in model organisms**. *Brief. Funct. Genomics Proteomics* 2007, **6**: 133–140.

Stern, C.D.: **The chick: a great model system just became even greater**. *Dev. Cell* 2004, **8**: 9–17.

Box 3A Gene-expression profiling by DNA microarray

Hamatani, T., Carter, M.G., Sharov, A.A., Ko, M.S.: **Dynamics of global gene expression changes during mouse preimplantation development**. *Dev. Cell* 2004, **6**: 117–131.

Butte, A.: **The use and analysis of microarray data**. *Nat. Rev. Microarray Collection* 2004 [http://www.nature.com/reviews/focus/microarrays/index.html] (accessed 13 May 2010).

Box 3B Insertional mutagenesis and gene knock-outs in mice: the Cre/*loxP* system

Yu, Y., Bradley, A.: **Engineering chromosomal rearrangements in mice**. *Nat. Rev. Genet.* 2001, **2**: 780–790.

Box 3C Large-scale mutagenesis in zebrafish

Doyon, Y., McCammon, J.M., Miller, J.C., Faraji, F., Ngo, C., Katibah, G.E., Amora, R., Hocking, T.D., Zhang, L., Rebar, E.J., Gregory, P.D., Urnov, F.D., Amacher, S.L.: **Heritable targeted gene disruption in zebrafish using designed zinc-finger nucleases**. *Nat. Biotechnol.* 2008, **26**: 702–708.

Stemple, D.: **Tilling—a high throughput harvest for functional genomics**. *Nat. Rev. Genet.* 2004, **5**: 1–6.

4

Vertebrate development II: axes and germ layers

- ■ Setting up the body axes
- ■ The origin and specification of the germ layers

In this chapter we examine the mechanisms that specify the body axes and germ layers of the early embryo in Xenopus, zebrafish, chick, and mouse. We will first look at axis formation and explore the relative roles of maternal factors in the egg, external factors, and the embryo's own developmental program. One of the key processes of early vertebrate development is the development of the three germ layers—ectoderm, endoderm, and mesoderm—from the fertilized egg, and we shall see how the different groups of animals achieve this in somewhat different ways. Despite its outward bilateral symmetry, the vertebrate body has a left-right internal asymmetry, as indicated by the positions of the heart and liver, and we shall see that the origins of this asymmetry are established very early in embryonic life.

Vertebrate development proceeds with respect to well-defined antero-posterior (head to tail) and dorso-ventral (back to belly) axes. In this chapter we will look in detail at the first two key events in vertebrate development. The first is the establishment of these axes, and the second is the specification and early patterning of the germ layers, in particular the induction of the mesodermal germ layer and its patterning along the dorso-ventral axis. In the next chapter (Chapter 5) we shall look at patterning of the vertebrate embryo along the antero-posterior axis and the all-important induction of the prospective nervous system from ectoderm.

For some vertebrates, the regions of the egg that will give rise to the endoderm and ectoderm seem already at least partly specified, but the mesodermal germ layer is not pre-specified in any vertebrate embryo, and its induction as part of the embryo's developmental program is one of the key events in early development. In *Xenopus*, for example, factors in the animal region of the egg specify that region as prospective ectoderm, while factors in the vegetal half of the egg specify that region as prospective endoderm. The mesoderm, in contrast, is produced after development has begun by signals from the vegetal-region cells acting on adjacent animal-region cells to specify a band of mesoderm around the equator of the blastula. As the blastula develops, the mesoderm becomes patterned along the dorso-ventral axis so that it becomes subdivided into regions that give rise to different structures (Fig. 4.1). The timing of mesoderm induction differs considerably between the different vertebrate groups, however, although many of the signaling molecules used are the same.

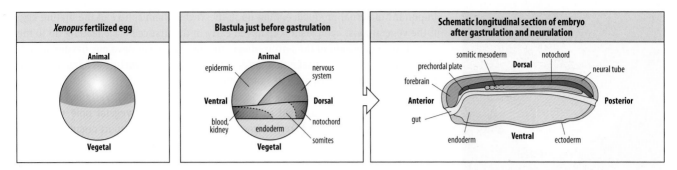

Xenopus fertilized egg	Blastula just before gastrulation	Schematic longitudinal section of embryo after gastrulation and neurulation

Setting up the body axes

We first consider how the antero-posterior and dorso-ventral axes are established in vertebrate embryos and whether they are already present in the egg or are specified later. In other words, to what extent is the embryo already patterned as a result of maternal factors laid down in the egg during its development in the ovary? The maternal contribution to early development differs considerably among our four model organisms. How the body axes will be established in amphibians and zebrafish is under the control of factors already present in the egg, whereas birds and mammals follow a different developmental strategy. Finally, we will briefly consider the intriguing question of how the left–right asymmetry, or 'handedness', of a number of internal organs might be determined.

4.1 The animal-vegetal axis is maternally determined in *Xenopus* and zebrafish

The very earliest stages in development of both *Xenopus* and the zebrafish are exclusively under the control of maternal factors present in the egg. As we saw in *Drosophila* (see Chapter 2), maternal factors are the mRNA and protein products of genes that are expressed by the mother during oogenesis and laid down in the egg while it is being formed. The *Xenopus* egg possesses a distinct polarity even before it is fertilized, and this polarity influences the pattern of cleavage. One end of the egg, the **animal pole**, which rotates to be uppermost when the egg is fertilized, has a heavily pigmented surface, whereas most of the yolk is located toward the opposite, unpigmented end, the **vegetal pole** (see Fig. 3.4). The animal half contains the egg nucleus, which is located close to the animal pole. These differences define the **animal–vegetal axis**. The pigment itself has no role in development but is a useful marker for real developmental differences that exist between the animal and vegetal halves of the egg. The planes of early cleavages that occur after fertilization are related to the animal–vegetal axis. The first cleavage is parallel with the axis and often defines a plane corresponding to the midline, dividing the embryo into left and right sides. The second cleavage is in the same plane at right angles to the first and divides the egg into four cells, while the third cleavage is at right angles to the axis and divides the embryo into animal and vegetal halves, each composed of four blastomeres (see Fig. 3.5).

Maternal factors are of key importance in the very early development of *Xenopus* and zebrafish because most of the embryo's own genes do not begin to be transcribed until after a stage called the mid-blastula transition, which will be discussed later in this chapter. The *Xenopus* egg contains large amounts of maternal mRNAs and there are also large amounts of stored proteins; there is, for example, sufficient histone protein for the assembly of more than 10,000 nuclei, quite enough to see the embryo through the first 12 cleavages until its own genes begin to be expressed. The egg also contains a number of maternal mRNAs that encode proteins with specifically developmental roles. These mRNAs are localized along the animal–vegetal axis of the egg during its development in the mother (oogenesis), with most of the developmentally

Fig. 4.1 Comparison of the axes and fate maps of the *Xenopus* egg and blastula with the structures in the tailbud embryo. The *Xenopus* egg has an initial axis of symmetry running from the animal to the vegetal pole—the animal-vegetal axis (first panel). This symmetry is broken by the specification of one side of the blastula as the dorsal side (second panel). After gastrulation and neurulation, the embryo has elongated and now has a distinct anterior (head) to posterior (tail) axis and a distinct dorso-ventral axis at right angles to it (third panel). In *Xenopus*, the endoderm and ectoderm are specified by maternal factors already present in the egg, but the mesoderm is induced from ectoderm by signals from the vegetal region during the blastula stage.

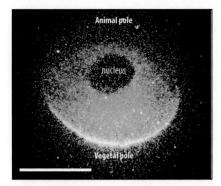

Fig. 4.2 Distribution of mRNA for the growth factor Vg-1 in the unfertilized amphibian egg. *In situ* hybridization with a radioactive probe for maternal Vg-1 mRNA in a *Xenopus* late oocyte shows its localization in the cortex of the vegetal region, as indicated by the yellow color (which represents silver grains viewed by special lighting). In this image, made before the use of fluorescent labels to detect nucleic acid hybridization was commonplace, the section through the egg was covered with a radiographic emulsion after hybridization of the probe, and the presence and density of the radioactive probe at any location was determined by the amount of silver grains deposited. Scale bar = 1 mm.

Photograph courtesy of D. Melton.

important maternal products ending up in the vegetal hemisphere of the mature egg. If the vegetal-most cytoplasm is removed early in the first cell cycle leading up to the first cleavage division of the fertilized egg, the embryo never develops beyond a radially symmetric mass of tissue that lacks any indication of structures characteristic of the antero-posterior axis, such as notochord or somites.

Localization of maternal mRNAs occurs by one of two routes. Early in oogenesis, some mRNAs are concentrated in the cortex at the future vegetal pole via transport involving a cytoplasmic region known as the message transport organizer region (METRO), which is associated with the mitochondrial cloud—a region of the oocyte cytoplasm where mitochondria are being generated. The mechanism underlying the selective trapping and transport of mRNAs by the METRO is not yet clear, but it may involve association of the mRNAs with the cytoskeleton and small vesicles of endoplasmic reticulum. Later in oogenesis, other mRNAs are transported to the vegetal cortex along microtubules by kinesin motor proteins, a similar mechanism to that of mRNA transport in the *Drosophila* oocyte (see Fig. 2.21).

The expression of the developmentally important maternal mRNAs in the *Xenopus* egg is regulated such that in most cases they are only translated after fertilization. The proteins encoded by several of these mRNAs are candidates for signals that specify early polarity and set further development in train. One maternal mRNA encodes the signaling protein Vg-1 (secreted signaling proteins used in vertebrate early development are listed in Box 4A, p. 131). *Vg-1* mRNA is synthesized during early oogenesis and becomes localized in the vegetal cortex of mature oocytes (Fig. 4.2), moving into the vegetal cytoplasm before fertilization. It is translated after fertilization and functions as an early signal for some aspects of mesoderm induction. Another vegetally localized mRNA encodes the protein Xwnt-11, which provides one of the early signals required for dorso-ventral axis specification. Xwnt-11 is a member of the Wnt family of signaling proteins related to the Wingless protein in *Drosophila*. Other components of the Wnt signaling pathway are also encoded by maternal mRNAs (see Box 1E, p. 26). The other main class of maternal factors with developmental roles comprises the transcription factors that switch on and regulate zygotic gene expression. Maternal mRNA encoding the T-box transcription factor VegT is localized to the vegetal hemisphere and is translated after fertilization. VegT has a key role in specifying both endoderm and mesoderm and its localization in the vegetal region is crucial to this function.

The animal–vegetal axis of the egg is in some sense related to the antero-posterior axis of the tadpole, as the head will form from the animal region, but the axes of the tadpole and the fertilized egg are not directly comparable (see Fig. 4.1). Which side of the animal region will form the head is not determined until the dorsal side of the embryo is specified, which does not happen until after fertilization. The precise position of the embryo's antero-posterior axis thus depends on the specification of the dorso-ventral axis.

In the zebrafish fertilized egg, maternal factors are also distributed along the animal–vegetal axis and key factors for axis formation are present in the vegetal region. As in *Xenopus*, removal of the vegetal-most yolk early in the first cell cycle of the fertilized egg results in radially symmetric embryos that lack structures characteristic of the antero-posterior axis.

4.2 Localized stabilization of the transcriptional regulator β-catenin specifies the future dorsal side and the location of the main embryonic organizer in *Xenopus* and zebrafish

The spherical unfertilized egg of *Xenopus* is radially symmetric about the animal–vegetal axis, and this symmetry is broken only when the egg is fertilized. Sperm entry sets in motion a series of events that defines the dorso-ventral axis of the gastrula

Box 4A Intercellular protein signals in vertebrate development

Proteins that are known to act as signals between cells during development belong to seven main families. Some of these families, such as the fibroblast growth factors (FGFs), were originally identified because they were essential for the survival and proliferation of mammalian cells in tissue culture. The members of the seven families are either secreted or are signaling proteins spanning the plasma membrane, and they provide intercellular signals at many stages of development. Other signal molecules, such as insulin and the insulin-like growth factors, and the neurotrophins, are used by embryos for particular tasks, but the seven families listed in the table are the main families involved throughout vertebrate development in vertebrates. Homologs of these proteins are also important in embryonic development in non-vertebrates (see Chapter 2, Chapter 6, and examples throughout the rest of the book).

The signaling molecules listed in the table all act by binding to receptors on the surface of a target cell. This produces a signal that is passed, or transduced, across the cell membrane by the receptor to the intracellular biochemical signaling pathways. Each type of signal protein has a corresponding receptor or set of receptors, and only cells with the appropriate receptors on their surface can respond. The intracellular pathways can be complex and involve numerous different proteins. We have already introduced the canonical Wnt intracellular signaling pathway in Box 1E (p. 26) and the Nodal signaling pathway is illustrated in Fig. 4.33 and Box 4B, p. 143. Figures illustrating the other signaling pathways are cross-referenced in the table. In the developmental context, the important responses of cells are either a change in gene expression, with specific genes being switched on or off, or a change in the cytoskeleton, leading to changes in cell shape and motility.

Some signaling proteins, such as those of the transforming growth factor-β (TGF-β) family, act as dimers; two molecules form a covalently linked complex that binds to and activates the receptor, bringing two receptor proteins together to form a dimer (see Fig. 4.33). In some cases, the active form of the protein signal is a heterodimer, made up of two different members of the same family. Secreted signals can diffuse or be transported short distances through tissue and set up concentration gradients. Some signal molecules, however, remain bound to the cell surface and thus can only interact with receptors on a cell that is in direct contact. One such is the protein Delta, which is membrane-bound and interacts with the receptor protein Notch on an adjacent cell (see Box 5C, p. 190 for the Notch signaling pathway).

For consistency in this book we have hyphenated the abbreviations for all secreted signaling proteins where the name allows—for example Vg-1, BMP-4, and TGF-β.

Common intercellular signaling proteins in vertebrate embryonic development

Signal molecule family	Receptors	Examples of roles in development
Fibroblast growth factor (FGF)		
Twenty-five FGF genes have been identified in vertebrates, not all of which are present in all vertebrates	FGF receptors (receptor tyrosine kinases)	Roles at all stages in development. Maintenance of mesoderm formation (this chapter). Induction of spinal cord (see Chapter 5 and Box 5B for signaling pathway). Signal from apical ectodermal ridge in limb bud (Chapter 11)
Epidermal growth factor (EGF)		
Epidermal growth factor	EGF receptors (receptor tyrosine kinases)	Cell proliferation and differentiation. Limb patterning (Chapter 11)
Transforming growth factor-β (TGF-β)		
A large protein family that includes activin, Vg-1, bone morphogenetic proteins (BMPs), Nodal, and Nodal-related proteins	Type I and Type II receptor subunits (receptor serine/threonine kinases) Receptors act as heterodimers	Roles at all stages in development. Signaling pathway in Fig. 4.33. Mesoderm induction and patterning (this chapter)
Hedgehog		
Sonic hedgehog and other members	Patched	Positional signaling in limb and neural tube (Chapters 11 and 12). Signaling pathway in Box 11C. Involved in determination of left–right asymmetry in chick embryos (this chapter)
Wingless (Wnt)		
Nineteen Wnt genes have been identified in vertebrates, not all of which are present in all vertebrates	Frizzled family of receptors (seven-span transmembrane proteins)	Roles at all stages in development. For canonical and non-canonical signaling pathways see Box 1E and Box 8C. Dorso-ventral axis specification in *Xenopus* (this chapter). Limb development (Chapter 11)
Delta and Serrate		
Transmembrane signaling proteins	Notch	Roles at all stages in development. Left–right asymmetry (this chapter). Somite development (see Chapter 5, signaling pathway in Box 5C)
Ephrins		
Transmembrane signaling proteins	Eph receptors (receptor tyrosine kinases)	Formation of rhombomeres in embryonic hindbrain (Chapter 5). Guidance of developing blood vessels (Chapter 11). Guidance of axon navigation in the nervous system (Chapter 12)

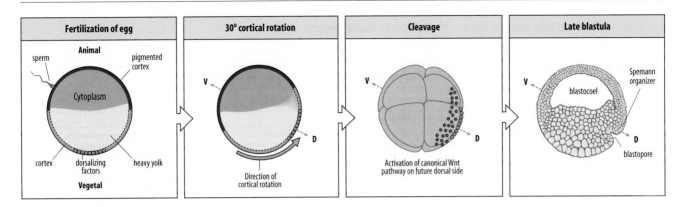

| Fertilization of egg | 30° cortical rotation | Cleavage | Late blastula |

Fig. 4.3 The future dorsal side of the amphibian embryo develops opposite the site of sperm entry as a result of the redistribution of maternal dorsalizing factors by cortical rotation. After fertilization (first panel), the cortical layer just under the cell membrane rotates by 30° towards the future dorsal region. This results in dorsalizing factors, such as *XWnt-11* mRNA and Dishevelled (Dsh) protein (a component of the Wnt signaling pathway)—indicated by green dots—being relocated from their initial position at the vegetal pole to a position approximately opposite to the site of sperm entry (second panel). The factors are transported further away from the vegetal pole along microtubules that lie at the interface of the cortex and the yolky cytoplasm. The Wnt pathway is activated where these factors end up (third panel, red dots indicate activation of the Wnt signaling pathway), which results in an accumulation of β-catenin in nuclei on the future dorsal side of the blastula (not shown here). In the blastula (fourth panel), gastrulation starts in the dorsal region, adjacent to the site of the Spemann organizer, which is formed in this region. V = ventral; D = dorsal.

(see Section 3.1 for a general outline of *Xenopus* development), with the dorsal side forming more or less opposite the sperm's entry point.

Sperm entry can occur anywhere in the animal hemisphere of the *Xenopus* egg and it causes the outer layer of cytoplasm, the cortex, to loosen from the inner dense cytoplasm so that it is able to move independently. The cortex is a gel-like layer rich in actin filaments and associated material. During the first cell-cycle, the cortex rotates about 30° against the underlying cytoplasm in a direction away from the site of sperm entry (Fig. 4.3). As a result of this **cortical rotation**, an array of sub-cortical microtubules becomes oriented with their plus ends pointing away from the site of sperm entry and they act as tracks for the directed movement of proteins and mRNAs.

Cortical rotation and transport along cortical microtubules leads to the relocation of maternal factors originally located at the vegetal pole, such as *XWnt-11* mRNA and Dishevelled protein (a Wnt signaling pathway component), to sites on the equator opposite the sperm entry point, creating asymmetry in the fertilized egg, such that when the second cleavage division occurs only two of the four blastomeres inherit these maternal factors. These factors specify their new location as the future dorsal side of the embryo, conferring a second axis of symmetry, the dorso-ventral axis, on the fertilized egg (see Fig. 4.3). For this reason, they are often called **dorsalizing factors**. Establishment of the future dorsal region is a key event in *Xenopus* development, as it sets up the conditions necessary for the formation of the main embryonic organizer, the Spemann organizer, in this region. As we saw in Chapter 1, the embryonic organizer was originally discovered as a small region of the embryo, active in the early stages of development, which, when grafted to another embryo, could induce the formation of a partial second embryo (see Fig. 1.9).

Any failure of cortical rotation and translocation of the dorsalizing factors blocks all further anterior and dorsal development, leading to a highly abnormal and non-viable embryo. Cortical rotation can be prevented by irradiating the ventral side of the egg with ultraviolet (UV) light, which disrupts the microtubules responsible for transporting the dorsalizing factors. Embryos developing from treated eggs are **ventralized**; that is, they are deficient in structures normally formed on the dorsal side and develop excessive amounts of blood-forming mesoderm, a tissue normally only present at the embryo's ventral midline. With increasing doses of radiation, both dorsal and anterior structures are lost, and the ventralized embryo appears as little more than a small distorted cylinder. The effects of UV irradiation are, however, 'rescued' by tipping the egg. Because of gravity, the dense yolk slides against the cortex, thus effectively restoring rotation and the localization of maternal factors.

The maternal dorsalizing factors act by stimulating a signaling pathway known as the **canonical Wnt/β-catenin pathway**. This pathway leads to the local stabilization of the protein β-catenin in the cells of the future dorsal region by preventing its degradation there. The canonical Wnt pathway (see Box 1E, p. 26) is one of several signaling pathways that can be stimulated by Wnt proteins and we have already seen it in action in *Drosophila* in response to Wingless. β-Catenin has a dual role as a gene-regulatory protein and as a protein that links cell-adhesion molecules to the cytoskeleton (see Box 8B, p. 293). In the specification of the dorsal region it is acting as a gene-regulatory protein. The dorsalizing effects of β-catenin can be seen when its mRNA is injected into ventral vegetal cells, which then form a second main body axis (Fig. 4.4).

Maternal β-catenin is initially present throughout the vegetal region, but when bound to a 'destruction complex' of proteins that includes the protein kinase GSK-3β, which phosphorylates it, the phosphorylated β-catenin is targeted for proteasomal degradation (see Box 1E, p. 26). Activation of the canonical Wnt signaling pathway by the dorsalizing factors blocks the activity of the destruction complex, thus preventing the degradation of β-catenin on the prospective dorsal side of the embryo. By the two-cell stage, β-catenin has started to accumulate in the cytoplasm of these cells, and by the end of the next two cleavage divisions it has entered the nuclei of the dorsal cells, where it helps switch on zygotic genes required for the next stage in development.

In *Xenopus*, maternal XWnt-11, translated in the dorsal region from its mRNA is a likely activator of the canonical Wnt signaling pathway at this stage. Further evidence that a Wnt protein is required for dorsalization comes from the finding that inhibiting expression of one of the *Xenopus* Wnt receptors, Xfz7 (*Xenopus* Frizzled 7), using antisense morpholino oligonucleotides (see Box 6A, p. 220) causes loss of dorsal structures. *XWnt-11* mRNA is present in the ventral region as well, and Wnt signaling might be limited to the dorsal region by maternal 'antagonists' or inhibitors of Wnt signaling present in the ventral region. As we shall see thoughout this chapter, inhibition of the activity of secreted signaling proteins is an important developmental strategy for restricting signaling to specific regions. The Wnt pathway might also be activated directly via pathway components such as Dishevelled, which is moved into the future dorsal region as a result of cortical rotation (see Fig. 4.3), or by inhibitory proteins acting on the GSK-3β complex. The key role played by inhibition of GSK-3β activity in establishing the future dorsal region explains the long-established finding that treatment of *Xenopus* embryos with lithium chloride at this stage dorsalizes them, promoting the formation of dorsal and anterior structures at the expense of ventral and posterior structures. Lithium is known to block GSK-3β activity.

The dorso-ventral axis is specified similarly in zebrafish, although in zebrafish the site of sperm entry is not involved in determining the future dorsal side and there is no cortical rotation. Specification of the dorsal region in the zebrafish involves activation of β-catenin by a maternal dorsalizing activity that moves in a microtubule-dependent manner from the vegetal region towards the animal pole and comes to lie under the cellular blastoderm. The stimulus for this movement is so far unknown. The maternal dorsalizing factors activate the β-catenin pathway in the dorsal yolk syncytial layer and dorsal margin blastomeres, and β-catenin accumulates in nuclei there (see Fig. 4.7).

4.3 Signaling centers develop on the dorsal side of *Xenopus* and zebrafish blastulas

The combined actions of β-catenin and other maternal factors present in the dorsal side of the vegetal region lead to the formation of a **signaling center** in this region. A signaling center is a region of an embryo that produces signaling molecules that act as morphogens to pattern the embryo and set cell fates. The first signaling center that develops in the dorsal–vegetal region in the *Xenopus* blastula is known as the **blastula organizer** or **Nieuwkoop center**. The center is named after the Dutch embryologist Pieter Nieuwkoop, who discovered the organizing properties of the vegetal region through experiments in which explants of animal cap tissue were combined with vegetal explants. Before those experiments, in the late 1960s, it had been thought that mesoderm, like endoderm and ectoderm, was intrinsically specified by a particular cytoplasmic region in the amphibian egg.

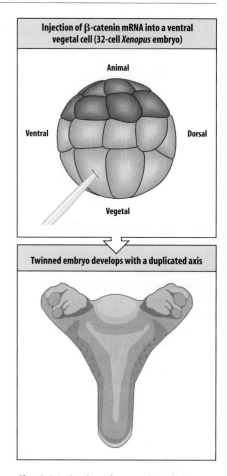

Injection of β-catenin mRNA into a ventral vegetal cell (32-cell *Xenopus* embryo)

Twinned embryo develops with a duplicated axis

Fig. 4.4 Induction of a new dorsal side by injection of β-catenin mRNA. Injection of mRNA encoding β-catenin into ventral vegetal cells can specify a new Nieuwkoop center (discussed in Section 4.3) at the site of injection, leading to a twinned embryo. Injection of some other vegetally localized maternal mRNAs, such as *Vg-1*, have similar effects.

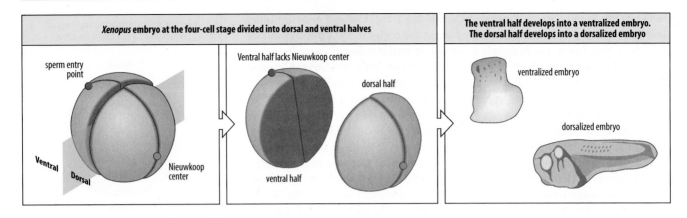

Fig. 4.5 The Nieuwkoop center is essential for normal development. If a *Xenopus* embryo is divided into a dorsal and a ventral half at the four-cell stage, the dorsal half containing the Nieuwkoop center develops as a dorsalized embryo lacking a gut, while the ventral half, which has no Nieuwkoop center, is ventralized and lacks both dorsal and anterior structures.

The Nieuwkoop center sets the initial dorso-ventral polarity in the blastula, and was originally thought to send a distinct signal required for establishment of the Spemann organizer. The Spemann organizer arises just above the Nieuwkoop center at the late blastula–early gastrula stage and is crucial for further development of both the antero-posterior and dorso-ventral axes of the embryo and the induction of the nervous system from the ectoderm. More recent work indicating molecular similarities between the Nieuwkoop center generated in tissue explants and the Spemann organizer as defined *in vivo* suggests, however, that it might be more appropriate to consider the Nieuwkoop center as a version of the Spemann organizer that manifests itself in explant experiments, or as an early stage in Spemann organizer formation.

Whatever the true nature of the Nieuwkoop center, the importance of the dorsal region as an organizer is shown by experiments that divide the embryo into two at the four-cell stage in such a way that one half contains the future dorsal region while the other does not (Fig. 4.5). The dorsal half will develop most structures of the embryo, although it will lack a gut, which develops from ventral blastomeres. The ventral half will develop much more abnormally, with little relevance to the cells' normal fates, producing a distorted, radially symmetric ventralized embryo that lacks all dorsal and anterior structures; it has developed in the same way as an embryo in which cortical rotation has been completely blocked by UV irradiation. This experiment shows that dorsal and ventral regions are established by the four-cell stage.

The influence of the Nieuwkoop center can also be seen in experiments in which cells from the dorsal–vegetal region of a 32-cell *Xenopus* embryo are grafted into the ventral side of another embryo. This gives rise to a twinned embryo with two dorsal sides (Fig. 4.6). Cells from the graft itself contribute to endodermal tissues but not to the mesodermal and neural tissues of the new axis. This shows that the grafted cells are able to induce these fates in the cells of the host embryo. In contrast, grafting ventral cells to the dorsal side has no effect. As we saw in Chapter 1, grafts of the

Fig. 4.6 The Nieuwkoop center can specify a new dorsal side. Grafting vegetal cells containing the Nieuwkoop center from the dorsal side of one 32-cell *Xenopus* blastula to the ventral side of another results in the formation of a second axis and the development of a twinned embryo. The grafted cells signal but do not contribute to the second axis.

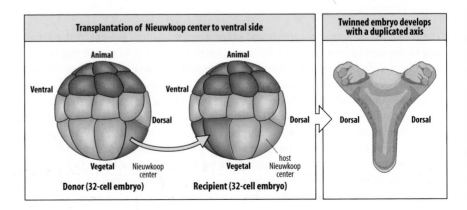

Spemann organizer also produce twinned embryos but in this case the grafted tissue forms the notochord of the secondary axis.

We can now understand the result of Roux's classic experiment (see Fig. 1.8) in which he destroyed one cell of a frog embryo at the two-cell stage and got a half-embryo rather than a half-sized whole embryo. The crucial feature of his experiment, it turns out, was that the killed cell remained attached, but the embryo did not 'know' it was dead. The remaining living cell could still develop a functional dorsal organizer but it developed as if the other half of the embryo was still there. If the killed cell had been separated from the living blastomere, the embryo could have regulated and developed into a small-sized whole embryo.

Although β-catenin becomes localized in nuclei in both the animal and vegetal regions and specifies a large region as dorsal, the future organizing centers—the Nieuwkoop center and the Spemann organizer—arise in the vegetal/equatorial region. These centers develop in this region because organizing function depends not only on β-catenin but also on the maternal transcription factor VegT, which is present throughout the vegetal region from fertilization onwards. By the 16- to 32-cell stage, β-catenin protein can be detected in nuclei on the dorsal side (Fig. 4.7), where in vegetal cells expressing VegT it activates zygotic genes such as *siamois*, which is involved in the induction of Spemann organizer function. VegT is also required to induce expression of the genes encoding Nodal-related proteins, signaling proteins that induce mesoderm

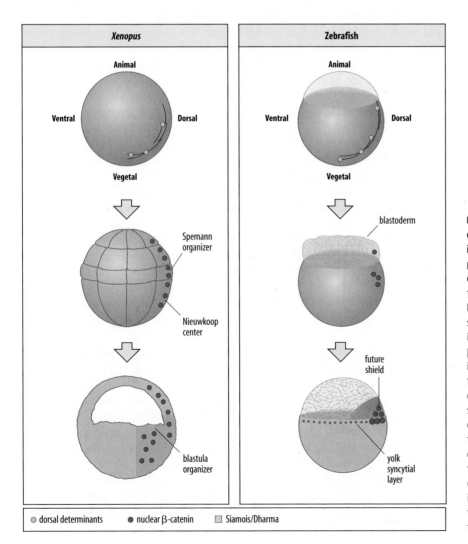

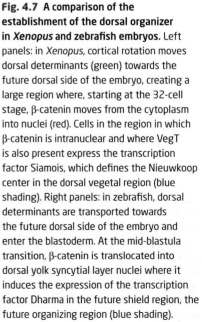

Fig. 4.7 A comparison of the establishment of the dorsal organizer in *Xenopus* and zebrafish embryos. Left panels: in *Xenopus*, cortical rotation moves dorsal determinants (green) towards the future dorsal side of the embryo, creating a large region where, starting at the 32-cell stage, β-catenin moves from the cytoplasm into nuclei (red). Cells in the region in which β-catenin is intranuclear and where VegT is also present express the transcription factor Siamois, which defines the Nieuwkoop center in the dorsal vegetal region (blue shading). Right panels: in zebrafish, dorsal determinants are transported towards the future dorsal side of the embryo and enter the blastoderm. At the mid-blastula transition, β-catenin is translocated into dorsal yolk syncytial layer nuclei where it induces the expression of the transcription factor Dharma in the future shield region, the future organizing region (blue shading).

(see Box 4A, p. 131). Named after the mouse protein **Nodal**, members of this protein family are key factors in mesoderm induction in all vertebrates, and in the frog they are known as the *Xenopus* Nodal-related factors (Xnrs). They are members of the TGF-β family of signaling proteins and we will discuss them further when we look at the induction and patterning of the mesoderm in the last part of this chapter.

In the zebrafish embryo, β-catenin becomes localized in nuclei of the dorsal yolk syncytial layer and the marginal blastomeres overlying it, which will become the shield region (see Section 3.2). The shield contains an organizing center analogous to the Spemann organizer. β-catenin induces the expression of a gene encoding a transcription factor, called Dharma (also known as Bozozok), which is essential for inducing the organizing function of the shield and the development of head and trunk regions (see Fig. 4.7).

4.4 The antero-posterior and dorso-ventral axes of the chick blastoderm are related to the primitive streak

The chick embryo develops from a circular blastoderm sitting on the top of the yolk (see Section 3.3). Like the amphibian blastula, the chick blastoderm is initially radially symmetric. This symmetry is broken when the posterior end of the embryo is specified. The future posterior end becomes evident soon after the egg is laid by the appearance of a ridge of cells known as Koller's sickle at one side of the blastoderm. As we saw in Section 3.3, this sickle lies adjacent to the **posterior marginal zone**, and marks the site at which the primitive streak will develop at gastrulation. Unlike *Xenopus* and the zebrafish, in the chick the embryo's own genes are expressed from the start of cleavage and it is unlikely that maternal determinants are involved in early development to any great extent.

The primitive streak indicates the direction of the antero-posterior axis of the embryo: the start of the streak marks the posterior end of the embryo and the streak elongates in an anterior direction. The location of the posterior marginal zone is therefore crucial to setting the body axes, and it turns out that this is determined by gravity. During the passage of the fertilized egg through the hen's uterus, which takes about 20 hours, the egg moves pointed end first and rotates slowly around its long axis, each revolution taking about 6 minutes. Cleavage has already started at this stage, and the blastoderm contains many thousands of cells by the time the egg is laid. As the egg rotates, the blastoderm remains tilted at an angle of about 45° to the vertical. The downward-pointing edge indicates the future head end of the embryo and the posterior marginal zone develops at the edge of the blastoderm that is uppermost (Fig. 4.8). It is still completely unknown how this effect of gravity breaks the blastoderm's initial radial symmetry. Whatever the answer, the axis is not irreversibly determined by these events. The chick blastoderm is highly regulative. Even at the stage when the blastoderm contains around 20,000 cells, it can be cut into many fragments, each of which will develop a complete embryonic axis.

The posterior marginal zone can be thought of as an organizing center analogous in some ways with the Nieuwkoop center in *Xenopus*, since it can induce the formation of a new body axis. If the posterior marginal zone from one blastoderm is grafted to another position on a second blastoderm of the same age it can induce a new primitive streak at this position (Fig. 4.9). The primitive streak is initiated by the concerted actions of various signaling proteins, which we shall return to later in the chapter. One of the earliest signals involved in streak initiation in the chick is the TGF-β family member Vg-1, which is expressed in the posterior marginal zone. When cells expressing Vg-1 are grafted to another part of the marginal zone in another blastoderm (essentially the same experiment as that shown in Fig. 4.9), they can induce a complete new streak. We have already encountered this signaling protein in *Xenopus*, where it is similarly able to induce a new body axis (see Fig. 4.4).

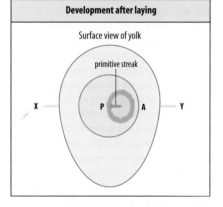

Rotation in oviduct

X   Y

X–Y cross-section during rotation

blastoderm

P

A

X   Yolk   Y

Albumen

Development after laying

Surface view of yolk

primitive streak

X   P A   Y

Fig. 4.8 Gravity defines the antero-posterior axis of the chick. Rotation of the egg in the oviduct of the mother results in the blastoderm being tilted in the direction of rotation, though it tends to remain uppermost. The posterior marginal zone (P) develops at that side of the blastoderm that was uppermost, and initiates the primitive streak. A = anterior.

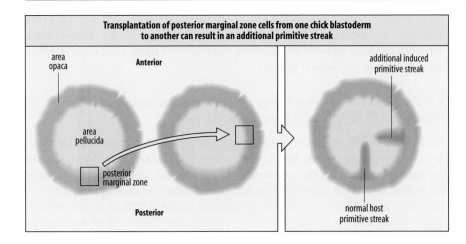

Transplantation of posterior marginal zone cells from one chick blastoderm to another can result in an additional primitive streak

Fig. 4.9 The posterior marginal zone of the chick blastodisc induces the primitive streak. The posterior marginal zone marks the posterior end of the antero-posterior axis. Grafting posterior marginal zone cells expressing the signaling protein Vg-1 to a lateral site in the marginal zone of another blastoderm can result in the formation of an extra primitive streak at that position. Two streaks often do not develop, however, as whichever forms first tends to inhibit the development of the other.

In the chick, however, often only one axis develops in the grafted embryos—either the host's normal axis or the one induced by the graft. This suggests that the more advanced of the two organizing centers inhibits streak formation elsewhere. It has been shown that Vg-1 induces an inhibitory signal that can travel across the embryo, a distance of around 3 mm in about 6 hours, to prevent another streak forming. Vg-1 is thus required for initiating streak formation but at the same time is involved in inhibiting the formation of a second streak.

Once the primitive streak has formed and starts to elongate, the antero-posterior axis is defined. After the node starts to regress, the head can be seen forming at the anterior end (Fig. 4.10). The streak also defines the future dorso-ventral axis of the embryo, with the most ventral structures (such as extra-embryonic mesoderm and lateral plate) being derived from the more posterior end of the streak and dorsal structures such as the notochord from the node (see Fig. 4.21). The relation of the streak to the dorso-ventral axis is the result of patterning of mesoderm cells while they are in the primitive streak, and will be explained in more detail later in the chapter.

Primitive-streak formation only starts when the hypoblast becomes displaced away from the posterior marginal region by growth of a layer of cells known as the endoblast (see Fig. 3.15). The hypoblast actively inhibits streak formation, as shown by experiments in which its complete removal leads to the formation of multiple streaks in random positions. This inhibition is due to a protein produced by the hypoblast

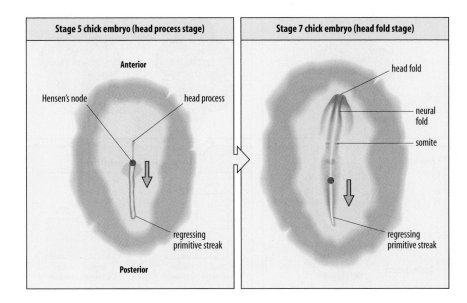

Fig. 4.10 The direction of the primitive streak marks the antero-posterior axis of the chick embryo. After extending to its full length, the primitive streak begins to regress, with Hensen's node moving in a posterior direction (first panel). As the node moves backward, the notochord (head process) develops anterior to it, from cells of the node that are left behind as it regresses. The head fold and neural plate form in the epiblast anterior to the furthest extent of the primitive streak (second panel). As the node regresses, somites begin to form on either side of the notochord.

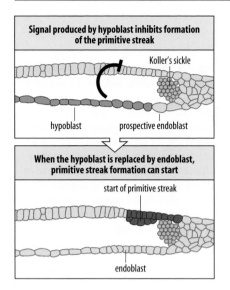

Fig. 4.11 The primitive streak begins to form once the hypoblast has been replaced by endoblast. Signaling in the posterior marginal zone that enables posterior streak formation is initially inhibited by a protein secreted by the hypoblast, a layer of cells that initially lies below the blastodisc over the underlying yolk (see Fig. 3.15). When the hypoblast has been displaced by endoblast, which grows in from the posterior marginal zone, this protein is no longer made and the primitive streak can begin to form.

Fig. 4.12 The specification of the blastomeres that will form the inner cell mass of a mouse embryo depends on whether they are on the outside or the inside of the embryo. Left panels: following cleavage the mouse embryo is a solid ball of cells (a morula). This develops into the hollow blastocyst, which consists of an inner cell mass surrounded by a layer of trophectoderm. Right panels: to determine whether the position of a blastomere in the morula determines whether it will develop into trophectoderm or inner cell mass, labeled blastomeres (blue) from a four-cell mouse embryo are separated and combined with unlabeled blastomeres (gray) from another embryo. By following the fate of the descendants of the labeled cells, it can be seen that blastomeres on the outside of the aggregate more often give rise to trophectoderm, 97% of the labeled cells ending up in that layer. The reverse is true for the origin of the inner cell mass, which derives predominantly from blastomeres on the inside.

which antagonizes the function of one of the signaling proteins required for streak formation, and which will be discussed in more detail later in the chapter. When hypoblast has been replaced by endoblast, this inhibitory signal ceases and primitive streak formation can begin (Fig. 4.11).

4.5 The definitive antero-posterior and dorso-ventral axes of the mouse embryo are not recognizable early in development

The mammalian egg differs considerably from the eggs of either *Xenopus* or the chick, as it contains no yolk and the early stages in development are concerned with generating extra-embryonic yolk sac membranes and the placenta connecting the embryo to the mother, so that the embryo can be nourished. There is no clear sign of polarity in the mouse egg, and any contribution of maternal factors is difficult to determine as zygotic gene activation occurs in the one-cell embryo and is essential for development beyond the two-cell stage. Several claims have been made that the first cleavage of the zygote is related to an axis of the embryo and that the first two cells (blastomeres) are biased towards different fates. This is an interesting possibility but is not yet clear, as different workers get different results. Furthermore, the highly regulative nature of early development in the mouse argues against the importance of maternal determinants. It has been known for a long time that both blastomeres of the two-cell mouse embryo can give rise to normal embryos if separated. In addition, individual blastomeres from mouse embryos up to the eight-cell stage can contribute extensively to both embryonic and placental tissues, and normal development can still occur even after cells are removed or added to a **preimplantation embryo**—the blastocyst before it implants into the wall of the uterus. This extensive regulative ability would not, however, preclude the possibility that, in normal development, individual blastomeres in early embryos have different developmental properties and fates and maternal determinants might be distributed differently between the blastomeres—the debate over this continues.

The early cleavages in mouse embryos do not follow a well-ordered pattern. Some are parallel to the egg's surface, so that the morula has distinct outer and inner cell populations. By the 32-cell stage, the morula has developed into a blastocyst, a hollow sphere of epithelium containing the inner cell mass—some 10–15 cells—attached to the epithelium at one end (Fig. 4.12, left panels). The epithelium of the blastocyst

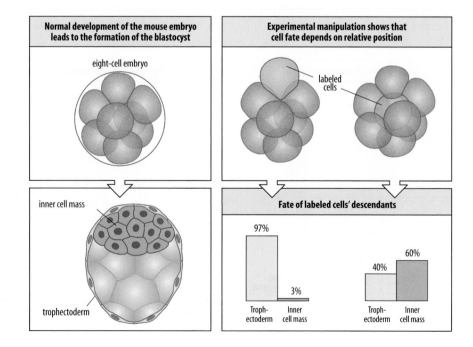

will form the trophectoderm, which gives rise to extra-embryonic structures associated with implantation and formation of the placenta, a structure unique to mammalian development. The cells of the inner cell mass are pluripotent, meaning that they can give rise to all the cell types in the embryo (see Section 1.17, p. 29), and they remain pluripotent up to stage E4.5.

Inner cell mass cells can be isolated and grown in culture to produce pluripotent **embryonic stem cells** (**ES cells**). As described in Chapter 10, these cells have been shown to give rise to all tissues of the embryo, including the germ cells, if introduced into a blastocyst, and this property is exploited to produce mice carrying particular mutations or particular transgenes. ES cells that have been mutated or genetically engineered in culture are introduced into a blastocyst, which is then placed in a surrogate mother (see Fig. 3.34). If the introduced ES cells contribute to the germ cells of the embryo, a line of mice can be bred that carry the mutation in all their cells. Human ES cells are also being investigated for their medical therapeutic potential (discussed in Chapter 10).

The specification of cells as either inner cell mass or trophectoderm in early mouse embryos depends on their relative positions in the cleaving embryo. Determination of their fate occurs after the 32-cell stage, and during earlier stages all the cells seem to be equivalent in their ability to give rise to either tissue. The most direct evidence for the effect of position comes from taking individual blastomeres from disaggregated eight-cell embryos, labeling them, and then combining the labeled blastomeres in different positions with respect to unlabeled blastomeres from another embryo. If the labeled cells are placed on the outside of a group of unlabeled cells they usually give rise to trophectoderm; if they are placed inside, so that they are surrounded by unlabeled cells, they more often give rise to inner cell mass (see Fig. 4.12, right panels). If a whole embryo is surrounded by blastomeres from another embryo, it can all become part of the inner cell mass of a giant embryo. Aggregates composed entirely of either 'outside' or 'inside' cells of early embryos can also develop into normal blastocysts, showing that there is no specification of these cells, other than by their position, at this stage.

The segregation of trophectoderm and the pluripotent inner cell mass lineages is the first differentiation event in mammalian embryonic development, and is essential for the formation of the placenta. The transcription factor Cdx2 is involved in the differentiation of trophectoderm, while the transcription factor Oct3/4 is involved in keeping the inner cell mass pluripotent. Oct3/4 is expressed from the two-cell stage and all blastomeres at the eight-cell stage co-express both transcription factors. As the morula forms, Cdx2 expression levels become slightly higher in the outer cells, possibly as a result of asymmetric cell divisions. Then, because of reciprocal negative feedback between Cdx2 and Oct3/4, Cdx expression levels increase and Oct3/4 expression levels decrease in the outer cells, whereas Oct3/4 levels of expression increase and Cdx expression levels decrease in the inner cells. By the blastocyst stage, Cdx2 expression is confined to the trophectoderm and Oct3/4 to the cells of the inner cell mass. In the absence of Cdx2, trophectoderm initially forms in the blastocyst but does not differentiate and is not maintained. An additional transcription factor, Tead4, is required for formation of the trophectoderm.

Before the blastocyst stage the embryo is a symmetric spheroidal ball of cells. The first asymmetry appears in the blastocyst, where the blastocoel develops asymmetrically, leaving the inner cell mass attached to one part of the trophectoderm (Fig. 4.13). The blastocyst now has one distinct axis running from the site where the inner cell mass is attached—the **embryonic pole**—to the opposite, **abembryonic**, pole, with the blastocoel occupying most of the abembryonic half. This axis is known as the **embryonic–abembryonic axis**. It corresponds in a geometric sense to the dorso-ventral axis of the epiblast, but is not related to specification of cell fate.

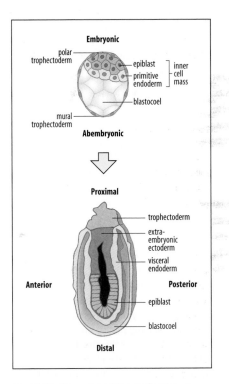
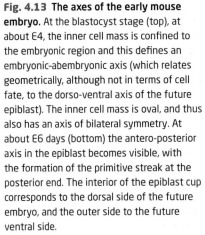

Fig. 4.13 The axes of the early mouse embryo. At the blastocyst stage (top), at about E4, the inner cell mass is confined to the embryonic region and this defines an embryonic-abembryonic axis (which relates geometrically, although not in terms of cell fate, to the dorso-ventral axis of the future epiblast). The inner cell mass is oval, and thus also has an axis of bilateral symmetry. At about E6 days (bottom) the antero-posterior axis in the epiblast becomes visible, with the formation of the primitive streak at the posterior end. The interior of the epiblast cup corresponds to the dorsal side of the future embryo, and the outer side to the future ventral side.

By about 4.5 days after fertilization, the inner cell mass has become differentiated into two tissues: **primary** or **primitive endoderm** (which will form extra-embryonic structures) forms its outer, or blastocoelic, surface, and this surrounds the **epiblast**, from which the embryo and some extra-embryonic structures will develop. The blastocyst implants in the uterine wall, and the trophectoderm at the embryonic pole proliferates to form the ectoplacental cone, producing extra-embryonic ectoderm that pushes the inner cell mass across the blastocoel. A cavity—the proamniotic cavity—is then formed within the epiblast as its cells proliferate. The mouse epiblast is now a monolayer and is formed into a cup shape, U-shaped in section, with a layer of visceral endoderm on the outside. At this stage, about E5.0–E5.25, the egg cylinder has a proximo-distal polarity in relation to the site of implantation with the ectoplacental cone at the proximal end, and the cup-shaped epiblast at the distal end (see Fig. 4.13). Extra-embryonic ectoderm forms a distinct layer of cells between the ectoplacental cone and the epiblast. Proximo-distal polarity within the epiblast is generated by Nodal signaling. A small region of visceral endoderm over the distal-most end of the epiblast cup is induced as **distal visceral endoderm** (**DVE**) by Nodal signaling from the epiblast. The DVE then starts to express genes encoding the homeodomain transcription factors Hex and Lim-1 (Lhx1), which are markers of anterior tissues, and also expresses genes encoding extracellular antagonists of Wnt and Nodal signaling.

4.6 Movement of the distal visceral endoderm indicates the definitive antero-posterior axis in the mouse embryo

The true antero-posterior axis of the mouse embryo is thought to be formed at E5.5–E6.0, about 12–24 hours before gastrulation begins, by a symmetry-breaking event in which the DVE moves rapidly to one side of the epiblast cup, displacing the more proximal visceral endoderm. This movement specifies the prospective anterior region of the embryo, and the endodermal cells derived from the DVE are now called the **anterior visceral endoderm** (**AVE**). The AVE specifies this side of the epiblast as anterior, in part by preventing the formation of the primitive streak here (Fig. 4.14). Active cell migration is the probable underlying mechanism for this asymmetric movement, although increased cell proliferation may also be involved. The AVE continues to express Hex and Lim1, together with a range of Wnt and Nodal antagonists. As a result, the epiblast lying next to it is specified as anterior ectoderm, from which the brain and other anterior structures will eventually develop.

Despite the differences in topology, this process is developmentally similar to that described earlier for the chick (see Section 4.4): the DVE/AVE is equivalent to the chick

Fig. 4.14 The symmetry-breaking event in the early mouse embryo is the specification of the distal visceral endoderm. At around E5.5, before primitive streak formation, visceral endoderm (yellow) at the distal-most end of the cup is specified as distal visceral endoderm (DVE; green) by signals from the epiblast (red arrows) and begins to proliferate and extend to one side of the epiblast, where it becomes the anterior visceral endoderm (AVE). The anterior fate is restricted to the distal region by inhibitory signals from the extra-embryonic ectoderm to the proximal visceral endoderm (barred blue lines). The extended AVE signals to the adjacent epiblast (green arrows) and induces anterior ectoderm (pale green), which includes neural ectoderm. At E6.5 the primitive streak (brown) has begun to form on the opposite side of the epiblast, thus marking the posterior end of the axis.

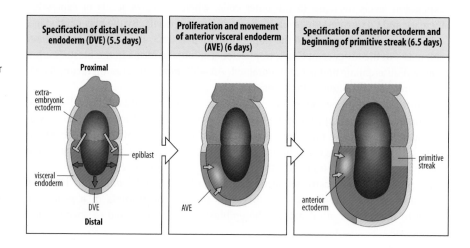

hypoblast, which also prevents primitive streak formation. In both cases, this influence is due to these cells producing proteins that counteract Nodal signaling: Cerberus in chick; Lefty-1 and Cerberus-like 1 in mouse. The movement of the DVE to the anterior side thus leads directly to the initiation of the primitive streak at around E6.25, and its positioning on the opposite side of the cup, which becomes the posterior end of the embryo. When a mouse node, which develops as part of the primitive streak, is transplanted to another part of the epiblast, a secondary axis is induced but this almost invariably lacks the most anterior regions, including the forebrain. This has been interpreted as evidence that determination of the anterior part of the embryo, by the AVE movement described earlier, is developmentally independent of determination of the rest of the body.

The extra-embryonic ectoderm in the E5.5 embryo seems to play a role in initially restricting the DVE to the most distal region of the visceral endoderm. Nodal is present throughout the epiblast and could, in principle, induce DVE and subsequently AVE fate throughout the visceral endoderm. If the extra-embryonic ectoderm is removed, the region of the visceral endoderm expressing anterior markers expands. Signaling by bone morphogenetic proteins (BMPs) (see Box 4A, p. 131) in proximal extra-embryonic ectoderm may contribute to restricting AVE fate and preventing the genes specifying anterior fate being expressed in proximal regions of the visceral endoderm. We shall return to this later in the chapter, when we discuss the formation of the primitive streak in more detail.

4.7 The bilateral symmetry of the early embryo is broken to produce left-right asymmetry of internal organs

Vertebrates are bilaterally symmetric about the midline of the body for many structures, such as eyes, ears, and limbs. But, while the vertebrate body is outwardly symmetric, most internal organs are in fact asymmetric with respect to the left and right sides. This is known as **left–right asymmetry**. In mice and humans, for example, the heart is on the left side, the right lung has more lobes than the left, the stomach and spleen lie towards the left, and the bulk of the liver towards the right. This handedness of organs is remarkably consistent, but there are rare individuals, about one in 10,000 in humans, who have the condition known as **situs inversus**, a complete mirror-image reversal of handedness. Such people are generally asymptomatic, even though all their organs are reversed. A similar condition can be produced in mice that carry the *iv* mutation, in which organ asymmetry is randomized (Fig. 4.15). In this context, 'randomized' means that some *iv* mutant mice will have the normal organ asymmetry and some the reverse.

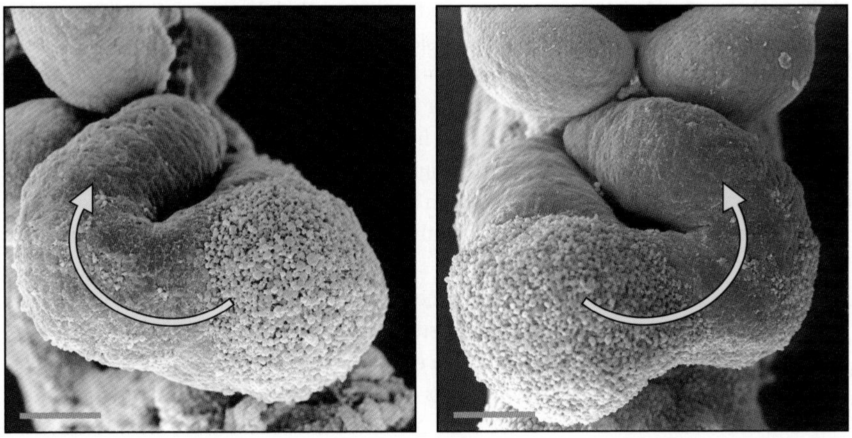

Fig. 4.15 Left-right asymmetry of the mouse heart is under genetic control. Each photograph shows a mouse heart viewed frontally after the loops have formed. The normal asymmetry of the heart results in it looping to the right, as indicated by the arrow (left panel). 50% of mice that are homozygous for the mutation in the *iv* gene have hearts that loop to the left (right panel). Scale bar = 0.1 mm.

Photographs courtesy of N. Brown.

Specification of left and right is fundamentally different from specifying the other axes of the embryo, as left and right have meaning only after the antero-posterior and dorso-ventral axes have been established. If one of these axes were reversed, then so too would be the left–right axis (this is the reason that handedness is reversed when you look in a mirror: your dorso-ventral axis is reversed, and so left becomes right and vice versa). One suggestion to explain the development of left–right asymmetry is that an initial asymmetry at the molecular level is converted to an asymmetry at the cellular and multicellular level. If that were so, the asymmetric molecules or molecular structure would need to be oriented with respect to both the antero-posterior and dorso-ventral axes.

Although the mechanisms by which left–right symmetry is initially broken are still not known, the subsequent cascade of events that leads to organ asymmetry is better understood. The stimulation of a 'leftward' flow of extracellular fluid across the embryonic midline by transient populations of ciliated cells has been shown to be critical in mouse, zebrafish and *Xenopus* embryos in inducing asymmetric expression of genes involved in establishing left versus right. (see Fig. 4.17). In the mouse, ciliated cells are found on the ventral surface of the node and create fluid flow in a pit formed below the node where there is no underlying endoderm. In the zebrafish, the ciliated epithelium lines a fluid-filled vesicle known as Kupffer's vesicle, which develops at the midline in the zebrafish embryo's tailbud at the end of gastrulation, while in *Xenopus*, a triangular region of ciliated mesoderm, the amphibian gastrocoel roof plate, straddles the midline in the roof of the archenteron. In chick embryos, however, motile cilia do not appear to be present on the ventral surface of the node and instead passive leftward movement of cells appears to be involved in establishing asymmetrical gene expression. Disrupting the action of the cilia in either the Kupffer's vesicle of zebrafish or the gastrocoel roof plate of *Xenopus*, for example by using antisense morpholinos (see Box 6A, p. 220) to knock down expression of ciliary dynein genes, leads to a breakdown in right–left asymmetry. In mouse embryos, mutations that affect the motor protein dynein and block the movement of cilia, as in *iv* mouse mutant and Kartagener's syndrome in humans, result in randomized asymmetry (see Fig. 4.15). Reversal of the leftward flow in the ventral node by an imposed fluid movement also reverses left–right asymmetry in mouse embryos.

One of the key proteins in establishing 'leftness,' and which comes to be expressed at high levels on the left side of the embryo, is Nodal. Nodal is initially expressed symmetrically on both sides of the node in both mouse and chick embryos, and adjacent to Kupffer's vesicle in zebrafish, and the gastrocoel roof plate in *Xenopus*. Initial small differences in Nodal expression on right and left sides induced by leftward flow of fluid flow across the midline in mouse, zebrafish and *Xenopus* then become amplified as a result of a positive feedback loop—as one effect of Nodal signaling is to activate the Nodal gene, and so produce more Nodal protein (Box 4B, p. 143). Asymmetric Nodal expression in the left side of the region is followed by widespread and stable activation of Nodal expression in the left lateral-plate mesoderm together with expression of the transcription factor Pitx2, another key determinant of leftness. This left-sided pattern of gene expression in the lateral plate mesoderm that gives rise to internal organs such as the heart is highly conserved, being found in the mouse, zebrafish, *Xenopus*, and chick. It has been shown in chick embryos that if the expression of either Nodal or Pitx2 in the lateral plate mesoderm is made symmetric, then organ asymmetry is randomized.

An early indication of left–right asymmetry in both *Xenopus* and chick embryos is the asymmetric activity of a proton–potassium pump (H^+/K^+-ATPase). This ATPase can be detected as early as the third cleavage division—the eight-cell stage—in *Xenopus*, and treatment of early embryos with drugs that inhibit the pump cause randomization

Box 4B Fine-tuning Nodal signaling

Mouse Nodal and the corresponding Nodal-related proteins in other vertebrates are secreted signaling proteins that are essential for the induction and patterning of the mesoderm and the establishment of left–right asymmetry in early vertebrate embryos. Signaling molecules that have such powerful effects on development must be tightly regulated to ensure that the right amounts are present in the right place at the right time, and the strength of Nodal signaling can be modulated and regulated at several different levels as indicated on the figure.

One is the production of a functional protein (1). Like many other growth factors, Nodal is synthesized and secreted as an inactive precursor protein that has to be converted to the active form by proteolytic cleavage; this is carried out by enzymes called proprotein convertases. If these are not present, then even if the *nodal* gene is expressed, no functional protein will be made.

A second level of control is due to extracellular inhibitors that bind to Nodal and/or its receptors (2) and prevent the signal from reaching the target cell, as discussed in this chapter. Like other TGF-β family members, Nodal binds to heterodimeric receptors known as type I/II receptors, and initiates an intracellular signaling cascade in which proteins called Smads are phosphorylated and enter the nucleus to regulate the expression of target genes (see Fig. 4.33). Extracellular antagonists of Nodal signaling include the proteins Cerberus and Lefty.

A third level of control of Nodal signaling is through regulation of its gene (3). The expression of *lefty*, *Cerberus* and the *nodal* gene itself can all be increased in response to Nodal signaling, and this can lead to either negative or positive feedback on the amount of Nodal present. A negative-feedback loop, in which Nodal increases the expression of *lefty*, fine tunes the level of Nodal signaling in the proximo-distal patterning of the mouse epiblast. Here, Nodal induces the distal visceral endoderm (DVE), which then secretes Cerberus-like 1 and Lefty-1, which in turn downregulate Nodal in an adjacent area of the epiblast, thus specifying this region as the anterior end of the antero-posterior axis. A positive feedback loop, in which Nodal signaling upregulates the expression of Nodal, operates in the establishment of left–right asymmetry and leads to high levels of Nodal on the left side of the node in mouse embryos (see Section 4.7). Yet another level of fine tuning of Nodal signaling discovered more recently involves the specific inhibition of translation of *nodal* and *lefty* mRNAs by microRNAs (4 on the figure above) (see Box 6B, p. 228 for how microRNAs act).

Surprisingly many of the molecules that play key roles in shaping the vertebrate body plan act as extracellular antagonists, or inhibitors, of intercellular signaling molecules and block or modulate signaling. Other widely used developmental signals such as Wnts and BMPs have antagonists that counteract their

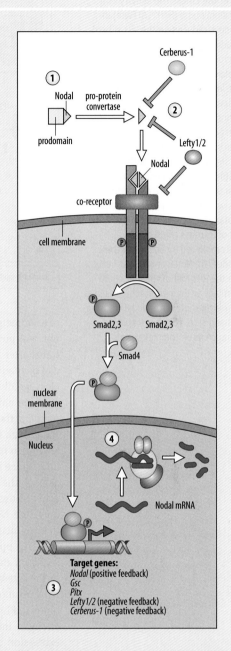

expression or activity at some level (see summary table, p. 167), and also make positive and negative feedback loops with these antagonists in a similar way to Nodal. In *Xenopus*, for example, the dorsalizing signals produced by Spemann's organizer are proteins, such as Noggin and Chordin, that block the action of the ventralizing signal, BMP-4 (see Section 4.17). In proximo-distal patterning of the epiblast in mouse embryos, the proximalizing signals Nodal and Wnt are blocked by the distally produced extracellular antagonists Cerberus-like 1 and Dickkopf-1, respectively.

Fig. 4.16 Determination of left–right asymmetry in the chick. The activity of a proton–potassium pump (H^+/K^+-ATPase) in Hensen's node is reduced on the left side of the node, leading to a membrane-potential difference across the node and an increased release of calcium (Ca^{2+}) into the extracellular space on the left side of the node (left panel). This leads to a greater activation of Notch signaling on the left side of the node, which in turn activates expression of the gene *nodal* in cells on the left side of the node. Nodal signaling together with Sonic hedgehog (Shh) then switches on *nodal* expression in the lateral plate mesoderm on the left side, which leads to expression of the transcription factor Pitx2, an important determinant of leftness (right panel). Shh activity is inhibited by activin on the right-hand side. The Nodal antagonist Lefty expressed in the left half of the notochord and the floorplate of the neural tube provides a midline barrier that prevents Nodal crossing to the right-hand side. This is a simplified version of a much more complex set of interactions.

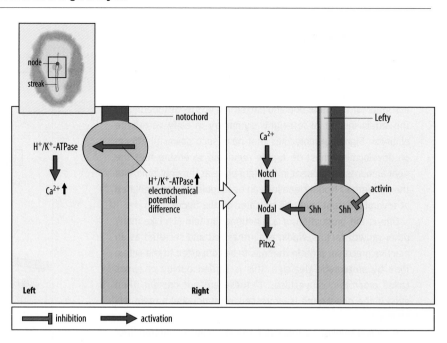

of asymmetry. How the asymmetry of the H^+/K^+-ATPase itself originates is not known, but one possibility is that the mRNA encoding the ion transporter and/or the ion transporter protein itself might be moved by motor proteins along oriented cytoskeletal elements in the fertilized egg. In *Xenopus*, disruption of cortical actin at the first cleavage division can affect left–right asymmetry.

In the chick, a detailed mechanism for left–right asymmetry based on the asymmetric activity of an H^+/K^+-ATPase in Hensen's node has been proposed (Fig. 4.16). It has been found that the activity of the pump is reduced on the left side of the node, leading to membrane depolarization and release of calcium from cells to the extracellular space on the left-hand side. The release of calcium in turn leads to the expression of the Notch ligands, Delta-like and Serrate, in a spatial pattern that could activate the transmembrane signaling protein Notch on the left side of the node only (the Notch signaling pathway is illustrated in Box 5C, p. 190). Notch activity, in conjunction with signaling by the secreted protein Sonic hedgehog (Shh) (see Box 11C, pp. 424–425, for the Shh signaling pathway), which is expressed at higher levels on the left side of the node, then leads to the left-sided expression of Nodal in the lateral-plate mesoderm. Left-sided expression of Shh may be due to passive leftward movement of Shh-expressing cells; Shh expression on the right side is also repressed by signaling by the TGF-β family member activin. The left-sided expression of Nodal induced by combined Notch and Shh signaling then in turn activates the expression of the transcription factor Pitx2 on the left side. If a pellet of cells secreting Shh is placed on the right side, then organ asymmetry is randomized. The right half of the embryo is protected from Nodal signaling by the protein Lefty secreted by the notochord and the floor of the neural tube. Lefty antagonizes Nodal signaling by binding to the Nodal protein and so prevents Nodal binding to and activating its receptor. Lefty might also interact with the receptor itself and so compete for Nodal binding (see Box 4B, p. 143).

Studies in mouse and zebrafish also implicate the release of calcium from cells on the left side of the node or Kupffer's vesicle, respectively, as part of the symmetry-breaking process. In the mouse, one hypothesis suggests that the leftward flow of fluid activates sensory cilia on the left side of the node, leading to calcium release and the upregulation of Nodal (Fig. 4.17). Other suggestions are that the leftward flow leads to the accumulation of a morphogen on the left of the node.

Summary

Setting up the body axes in vertebrates involves maternal factors, external influences, and cell–cell interactions. In the amphibian embryo, maternal factors determine the animal–vegetal axis, which corresponds approximately to the antero-posterior axis; the dorso-ventral axis is specified by the site of sperm entry and the resulting cortical rotation relocates maternal factors to the dorsal side, which leads to the establishment of the Nieuwkoop center. In the zebrafish, maternal factors also specify the dorsal side and the embryonic shield, the site of the embryonic organizer, although the symmetry-breaking event has not yet been identified in this case. In chick embryos, gravity determines the side of the blastoderm at which the posterior marginal zone, and thus the primitive streak, forms, marking the orientation of the antero-posterior axis. It is not clear yet whether specification of the axes in the mouse embryo involves any maternal component. The axes are definitively established later in the epiblast under the direction of signals from embryonic and extra-embryonic tissues. The symmetry-breaking event in the mouse appears to be the specification of the anterior end of the embryo, which is coupled to the determination of the position of the primitive streak at the posterior end of the epiblast. The generation of the consistent left-right organ asymmetry found in vertebrates is marked by the expression of the secreted protein Nodal and the transcription factor Pitx2 on the left side only. This asymmetry appears to be generated by various mechanisms, involving a proton-potassium pump asymmetry (in *Xenopus* and chick), the leftward movement of cilia (in mouse, humans, zebrafish, and *Xenopus*), and asymmetrically located calcium signaling, Notch pathway activation, Sonic hedgehog signaling and Nodal signaling.

Summary: vertebrate axis determination

	Dorso-ventral axis	Antero-posterior axis
Xenopus	sperm entry point and cortical rotation lead to specification of dorsal side opposite the point of sperm entry	specified in organizer
Zebrafish	initial symmetry-breaking event unknown. Localization of maternal Ndr1 and nuclear β-catenin specifies dorsal side and position of organizer	specified in organizer
Chick	posterior marginal zone specifies ventral end of dorso-ventral axis	gravity determines position of posterior marginal zone and thus the posterior end of the A–P axis
Mouse	interaction between inner cell mass and trophectoderm	specification and movement of dorsal visceral endoderm

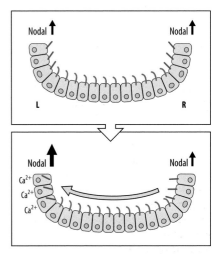

Fig. 4.17 **In the mouse embryo, cilia-directed leftward flow of extracellular fluid generates left-right asymmetry across the midline.** In the mouse, the initial event that breaks left-right symmetry is a short period of directed leftward flow of extracellular fluid caused by the movement of cilia on a pit on the ventral side of the node at around the late head-fold stage. Before this flow, there is a low level of expression of *nodal* mRNA on both sides of the node (top panel, indicated by short arrows). There are two types of cilia on the cells lining a pit on the ventral side of the node: central motile cilia (blue) that generate the flow and peripheral immotile cilia (red). Coordinated rotational beating of the central cilia generates a leftward flow of extracellular fluid. One hypothesis is that the immotile cilia act as mechanosensors for the direction of flow and their stimulation results in release of intracellular Ca^{2+} in the cells on the left side of the node, as illustrated in this figure. Propagation of the Ca^{2+} signal to nearby cells then leads to the upregulation of Nodal expression on the left side (bottom panel, larger arrow). Other hypotheses propose that the leftward flow leads to the accumulation of a chemical morphogen on the left-hand side of the node, and that this morphogen directly or indirectly causes the upregulation of Nodal.

The origin and specification of the germ layers

We have seen in the preceding sections how the main axes are laid down in various vertebrate embryos. We now focus on the earliest patterning of the embryo with respect to these axes: the specification of the three germ layers—endoderm, mesoderm, and ectoderm—and their further diversification (see Box 1C, p. 15).

All the tissues of the body are derived from these three germ layers. The mesoderm becomes subdivided into cells that give rise to notochord, muscle, heart and kidney, and blood-forming tissues, among others. The ectoderm becomes subdivided into cells that give rise to the epidermis of the skin and cells that develop into the nervous system. The endoderm gives rise to the gut and its associated organs such as the lungs, liver, and pancreas. We first look at the **fate maps** of early embryos of the different vertebrates, which tell us which tissues are generated from particular regions of the embryo (see Section 1.12). We then consider how the germ layers are specified and subdivided, with the main focus on *Xenopus*, in which these processes are best understood and where many of the genes and proteins involved have been identified. The fate maps and the genes involved in germ-layer specification are more similar in all our vertebrate models than are the mechanisms involved in early axis specification.

4.8 A fate map of the amphibian blastula is constructed by following the fate of labeled cells

The external appearance of the *Xenopus* blastula at the 32-cell stage gives no indication of how the different regions will develop, but individual cells can be identified with respect to the animal–vegetal axis (defined by pigmentation) and dorso-ventral axes (defined by the sperm entry point). By following the fate of individual cells, or groups of cells, we can make a map on the blastula surface showing the regions that will give rise to, for example, notochord, somites, nervous system, or gut. The fate map shows where the tissues of each germ layer normally come from, but it indicates neither the full potential of each region for development nor to what extent its fate is already specified or determined in the blastula. In other words, we know what each of the early cells will give rise to—but the embryo does not. Early vertebrate embryos have considerable capacity for regulation when pieces are removed or are transplanted to a different part of the same embryo (see Section 1.12). This implies considerable developmental plasticity at this early stage and also that the actual fate of cells is heavily dependent on the signals they receive from neighboring cells.

One way of making a fate map is to stain various parts of the surface of the early embryo with a lipophilic dye such as DiI, and observe where the labeled region ends up. Individual cells can also be labeled by injection of stable high-molecular-weight molecules, such as fluorescein-labeled dextran, which cannot pass through cell membranes and so are restricted to the injected cell and its progeny; as flourescein fluoresces green when excited by UV light, the fluorescein-labeled dextran can be easily detected under a fluorescence microscope. The fluorescent protein green fluorescent protein (GFP), which produces a green light when excited by UV, is now widely used for such purposes (see Box 1D, p. 20). Figure 4.18 shows a *Xenopus* embryo labeled with fluorescein-dextran-amine for fate mapping.

Fig. 4.18 Fate mapping of the early *Xenopus* embryo. Left panel: a single cell in the embryo, C3, is labeled by injection of fluorescein-dextran-amine, which fluoresces green under UV light. Right panel: a cross-section of the embryo, made at the tailbud stage, shows that the labeled cell has given rise to mesoderm cells on one side of the embryo. Scale bar = 0.5 mm.

Photograph courtesy of L. Dale.

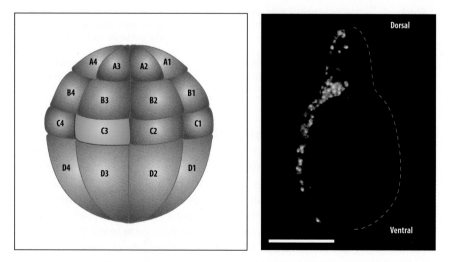

The fate map of the late *Xenopus* blastula (Fig. 4.19) shows that the yolky vegetal region, which occupies the lower third of the spherical blastula, gives rise to most of the endoderm. The yolk, which is present in all cells, provides the nutrition for the developing embryo, and is gradually used up as development proceeds. At the other pole, the animal hemisphere becomes ectoderm, which becomes further diversified into epidermis and the future nervous tissue. The future mesoderm forms from a belt-like region, the **marginal zone**, around the equator of the blastula. In *Xenopus*, but not in all amphibians, a thin outer layer of presumptive endoderm overlies the presumptive mesoderm in the marginal zone.

The fate map of the blastula makes clear the function of gastrulation. At the blastula stage, the endoderm, which gives rise to the gut, is on the outside and so must move inside. Similarly, the mesoderm, which will form internal tissues and organs, must move inwards. During gastrulation, the marginal zone moves into the interior through the dorsal lip of the blastopore, which lies above the Nieuwkoop center. The fate map of the mesoderm (see Fig. 4.19) shows that it is subdivided into different fates along the dorso-ventral axis of the blastula. The most dorsal mesoderm gives rise to the notochord, followed, going ventrally, by somites (which give rise to muscle tissue), lateral plate (which contains heart and kidney mesoderm), and blood islands (tissue where hematopoiesis first occurs in the embryo). There are also important differences between the future dorsal and ventral sides of the animal hemisphere: the epidermis comes mainly from the ventral side of the animal hemisphere, whereas the nervous system comes from the dorsal side. After neural tube formation the epidermis spreads to cover the whole of the embryo.

The terms dorsal and ventral in relation to the fate map can be somewhat confusing, because the fate map does not correspond exactly to a neat set of axes at right angles to each other. As a result of cell movements during gastrulation, cells from the dorsal side of the blastula give rise to some ventral parts of the anterior end of the embryo, such as the head, as well as to dorsal structures, and will also form some other ventral structures, such as the heart. The ventral region gives rise to ventral structures in the anterior part of the embryo, but will also form some dorsal structures posteriorly. This is why the dorsalized embryos described in Section 4.2 also have overdeveloped anterior structures and lack posterior regions. Indeed, a somewhat different interpretation of the standard fate map for *Xenopus* has been proposed in which the 'dorso-ventral' axis of the blastula is considered to be the future antero-posterior axis of the embryo.

4.9 The fate maps of vertebrates are variations on a basic plan

Fate maps of the early embryos of zebrafish, chick, and mouse have been prepared using techniques essentially similar to those used for *Xenopus*: cells in the early embryo are labeled and their fate followed. In the zebrafish embryo there is extensive cell-mixing during the transition from blastula to gastrula, and so it is not possible to construct a reproducible fate map at cleavage stages but only from gastrula stage. The zebrafish late blastula comprises a blastoderm of deep cells and a thin overlying enveloping layer of cells (Fig. 4.20). The overlying enveloping layer is largely protective and is eventually lost. At the beginning of gastrulation, the fate of deep-layer cells, from which all the cells of the embryo will come, is correlated with their position in respect to the animal pole. Cells at the dorsal and lateral margin of the blastoderm give rise to endoderm and mesoderm (mesendoderm), cells slightly further toward the animal pole give rise to mesoderm, while ectodermal cells come from the blastoderm nearest the animal pole (see Fig. 4.20). In general terms, the fate map of the zebrafish is rather similar to that of an amphibian, if one imagines the vegetal region of the amphibian blastula being replaced by one large yolk cell.

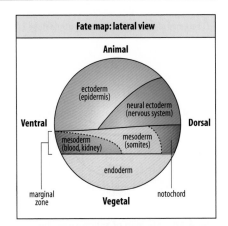

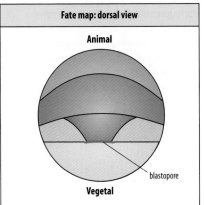

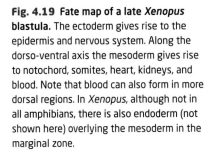

Fig. 4.19 Fate map of a late *Xenopus* blastula. The ectoderm gives rise to the epidermis and nervous system. Along the dorso-ventral axis the mesoderm gives rise to notochord, somites, heart, kidneys, and blood. Note that blood can also form in more dorsal regions. In *Xenopus*, although not in all amphibians, there is also endoderm (not shown here) overlying the mesoderm in the marginal zone.

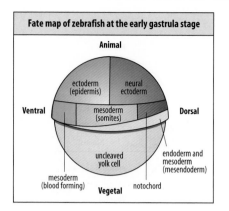

Fig. 4.20 Fate map of zebrafish at the early gastrula stage. The three germ layers come from the blastoderm, which sits on the lower hemisphere composed of an uncleaved yolk cell. The endoderm comes from the dorsal and lateral margins of the blastoderm, which also give rise to mesoderm (mesendoderm), and some has already moved inside.

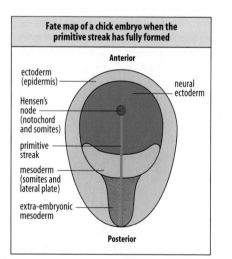

Fig. 4.21 Fate map of a chick embryo when the primitive streak has fully formed. The diagram shows a view of the dorsal surface of the embryo. Almost all the endoderm has already moved through the streak to form a lower layer, so is not shown.

It is difficult to produce a fate map of the chick embryo at the early blastoderm stage, the stage that roughly corresponds to the *Xenopus* blastula, because cells that will contribute to the different germ layers are mixed up in the epiblast. Roughly, cells adjacent to the posterior marginal zone at the early blastula stage are equivalent to the dorsal cells of the *Xenopus* blastula (future notochord). There are extensive cell movements both before and during the emergence of the primitive streak, but the picture becomes clearer once the primitive streak has formed and cells have started to move inside and be determined as mesoderm and endoderm.

At the stage shown in Fig. 4.21, the chick blastoderm has become a three-layered structure, with the endoderm forming the bottom layer (not visible in this dorsal view). Cells from the surface of the blastoderm have ingressed through the primitive streak into the interior and there formed the mesodermal and endodermal layers. Most of the cells in the outer surface of the blastoderm are now prospective ectoderm and will form neural tube and epidermis, but there are still regions of the outer blastoderm that will move through the streak and give rise to mesoderm. Hensen's node at the anterior end of the streak is prospective mesoderm; as the node regresses it leaves cells behind that will form the notochord and will also contribute to the somites. The mesoderm lying along the antero-posterior midline gives rise to somites and is flanked by cells that will form the lateral plate mesoderm (shown as a single region in Fig. 4.21) and structures such as the heart and kidney. In the lowest layer of the embryo, closest to the yolk, the presumptive endoderm (not visible in Fig. 4.21) is flanked by cells that will form extra-embryonic structures.

In the case of the mouse, most cells from the inner cell mass of an embryo less than 3.5 days old can give rise to many different embryonic tissues, as well as to some extra-embryonic structures, such as the visceral and parietal endoderm, and so at this stage a fate map cannot be constructed. At 6–7 days' gestation the mouse epiblast becomes transformed into the three germ layers by gastrulation. Gastrulation in the mouse is essentially very similar to gastrulation in the chick, but at this stage the mouse epiblast is folded into a cup, which makes the process of primitive streak and node formation more difficult to follow. Like the chick, there is extensive cell movement and cell proliferation in the epiblast. A detailed fate map of this stage has been established by tracing the descendants of single cells that have been labeled by injection with a dye. But descendants of a single cell can spread widely and give rise to cells of different germ layers, so that only about 50% of the labeled clones have progeny in only one germ layer.

The fate map of the mouse at the primitive-streak stage is basically similar to that of the chick (Fig. 4.22). As the mouse epiblast at this stage is cup-shaped, it may be helpful to look back at Figs 3.24 and 3.25, which show the cup-shaped embryo, to help interpret its fate map. Figure 4.22 shows the embryo as if the cup were opened out and flattened and viewed from the dorsal side, which is the interior of the cup. Apart from the topology, there are other differences from the chick fate map that are worth noting; one is that in the mouse the node gives rise exclusively to the notochord. Cells from the anterior and middle regions of the streak will move forward around the node to produce the other mesodermal tissues (see Fig. 3.26). The definitive gut endoderm and the midline axial mesoderm (at this stage called mesendoderm) derive from the extreme anterior end of the primitive streak, whereas the middle part of the streak gives rise mainly to somites and lateral-plate mesoderm. The posterior part of the streak will become the tailbud, and also provides the extra-embryonic mesoderm that forms the extra-embryonic membranes—the amnion, visceral yolk sac, and allantois.

We can see, therefore, that the fate maps of the different vertebrates are similar when we look at the relationship between the germ layers and the site of the inward movement of cells at gastrulation (Fig. 4.23). The differences are due mainly to the amount of yolk in the different eggs, which determines the patterns of cleavage and influences the shape of the early embryo. The similarity in relationship between the germ layers suggests that similar mechanisms will be involved in their specification. This is true to some extent, but as we shall see, there are still considerable differences.

4.10 Cells of early vertebrate embryos do not yet have their fates determined and regulation is possible

Early vertebrate embryos have considerable powers of regulation (see Section 1.3) when parts of the embryo are removed or rearranged. Experiments show that at the blastula stage, and even later, many cells are not yet specified or determined (see Section 1.12 and Fig. 1.22); their potential for development is greater than their position on the fate map suggests.

The state of determination of cells, or of small regions of an embryo, can be studied by transplanting them to a different region of a host embryo and seeing how they develop. If they are already determined, they will develop according to their original position. If they are not yet determined, they will develop in line with their new position. This can be shown experimentally by introducing a single labeled cell from a *Xenopus* blastula into the blastocoel of a later-stage embryo and following its fate. The transferred cell divides, and during gastrulation its progeny become distributed to different parts of the embryo.

In general, cells in transplants made from early blastulas are not yet determined; their progeny will differentiate according to the signals they receive at their new location. So cells from the vegetal pole, which would normally form endoderm, can contribute to a wide range of other tissues such as muscle or nervous system, when grafted at an early stage. Similarly, early animal pole cells, whose normal fate is epidermis or nervous tissue, can form endoderm or mesoderm. With time, cells gradually become determined, so that similar cells taken from later-stage blastulas and early gastrulas develop according to their fate at the time of transplantation.

Fragments of a fertilized *Xenopus* egg that are only one-fourth of the normal volume can develop into more-or-less normally proportioned but small embryos. There must therefore be a patterning mechanism involving cell interactions that can cope with such differences in size. There are, however, limits to the capacity for regulation. Isolated animal and vegetal halves of an eight-cell *Xenopus* embryo do not develop normally; and while dorsal halves of eight-cell embryos regulate to produce a reasonably normal embryo, the ventral halves do not. As we saw in Section 4.3, this is due to the development of the essential embryonic signaling centers in the dorsal half.

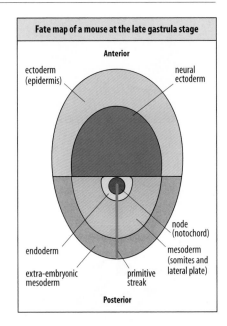

Fig. 4.22 Fate map of a mouse at the late gastrula stage. The embryo is depicted as if the 'cup' has been flattened and is viewed from the dorsal side. At this stage the primitive streak is at its full length.

Fig. 4.23 The fate maps of vertebrate embryos at comparable developmental stages. In spite of all the differences in early development, the fate maps of vertebrate embryos at stages equivalent to a late blastula or early gastrula show strong similarities. All maps are shown in a dorsal view. The future notochord mesoderm occupies a central dorsal position. The neural ectoderm lies adjacent to the notochord, with the rest of the ectoderm anterior to it. The mouse fate map depicts the late gastrula stage. The future epidermal ectoderm of the zebrafish is on its ventral side.

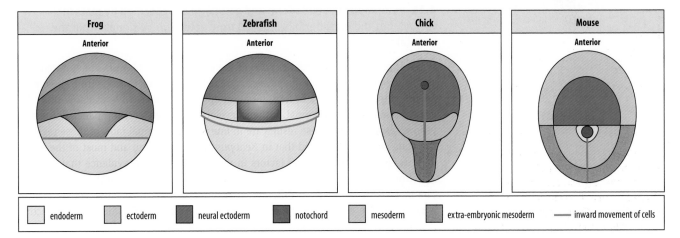

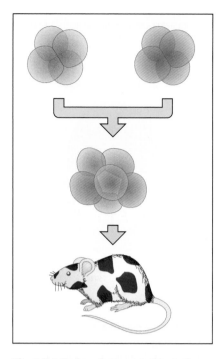

Fig. 4.24 Fusion of mouse embryos gives rise to a chimera. If a four-cell stage embryo of an unpigmented strain of mouse is fused with a similar embryo of a pigmented strain, the resulting embryo will give rise to a chimeric animal, with a mixture of pigmented and unpigmented cells. The distribution of the different cells in the skin gives this chimera a patchy coat.

The early chick embryo also has remarkable powers of regulation, and fragments of the blastoderm can give rise to whole embryos. The capacity to regulate persists right up to the time at which the primitive streak starts to form. In chick there is no mid-blastula transition, as zygotic genes are expressed from the start, and although Vg-1 is normally expressed in the posterior marginal zone (see Section 4.4), a new site of expression of Vg-1 appears when the embryo is cut into pieces.

Early mouse embryos can also regulate to achieve the correct size. Giant embryos formed by aggregation of several embryos in early cleavage stages can achieve normal size within about 6 days by reducing cell proliferation. The mouse embryo retains considerable capacity to regulate until late in gastrulation. Even at the primitive-streak stage, up to 80% of the cells of the epiblast can be destroyed by treatment with the cytotoxic drug mitomycin C, and the embryo can still recover and develop with relatively minor abnormalities.

Further evidence for regulation in mammalian embryos, including human, and in birds comes from twinning. As explained in Box 4C (p. 152), human twins usually result from division of the inner cell mass, but twinning also very rarely occurs as late as 9–15 days gestation, when the primitive streak is about to form. These powers of regulation of human embryos have enabled the development of reproductive technologies such as preimplantation genetic diagnosis (Box 4D, p. 153).

Because early vertebrate embryos show considerable capacity for regulation and many of the cells are not determined, this implies that cell–cell communication must determine cell fate. To study this question, one can create **chimeric** mice—that is, mice that are mosaics of cells with two different genetic constitutions—by fusing two embryos (Fig. 4.24). Chimeras made from a normal embryo and one that is genetically similar, but homozygous for a mutated version of a single gene, can be used to find out whether the effects of the gene are **cell-autonomous** or **non-cell-autonomous**. If only the mutant cells exhibit the mutant phenotype and are not 'rescued' by the normal cells, the gene is acting cell-autonomously. This means that the product of the gene is acting solely within the cell in which it is made, and is not influencing other cells. The cells of a black mouse, for example, remain black when put into a white mouse, and they do not change the white cells to black; they are thus autonomous with respect to pigmentation (see Fig. 4.24). In contrast, a gene is acting non-autonomously when either the mutant cells in the chimera appear to act normally or the normal cells start to show the mutant phenotype. Non-autonomous action is typically due to a gene product such as an intercellular signaling protein that is secreted by one cell and acts on others, whereas autonomous effects are due to proteins such as transcription factors that act only within the cell that makes them.

4.11 In *Xenopus* the endoderm and ectoderm are specified by maternal factors, but the mesoderm is induced from ectoderm by signals from the vegetal region

When explants from different regions of the early *Xenopus* blastula are cultured in a simple medium containing the necessary salts for ion balance, tissue from the region nearest the animal pole will form a ball of epidermal cells, while explants from the vegetal region are endodermal in their development (Fig. 4.25, top panels, blue and yellow, respectively). These results are in line with the normal fates of these regions. It is thus generally accepted that in *Xenopus* all the ectoderm and most of the endoderm is specified by maternal factors in the egg. There is no evidence that any signals from other regions of the embryo are necessary for their initial specification. The mesoderm, however, is different. It is induced from prospective ectoderm in a band round the equator of the blastula in response to signals from the prospective

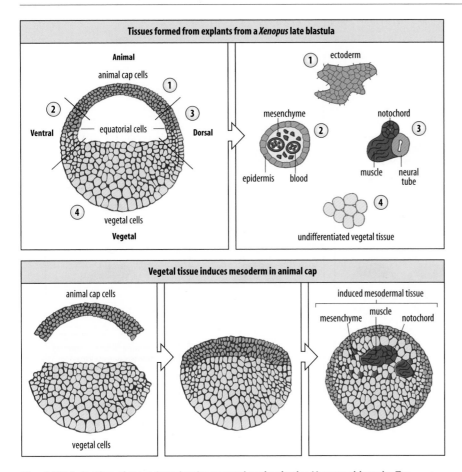

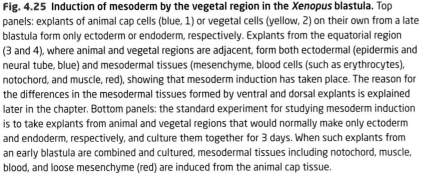

Fig. 4.25 **Induction of mesoderm by the vegetal region in the *Xenopus* blastula.** Top panels: explants of animal cap cells (blue, 1) or vegetal cells (yellow, 2) on their own from a late blastula form only ectoderm or endoderm, respectively. Explants from the equatorial region (3 and 4), where animal and vegetal regions are adjacent, form both ectodermal (epidermis and neural tube, blue) and mesodermal tissues (mesenchyme, blood cells (such as erythrocytes), notochord, and muscle, red), showing that mesoderm induction has taken place. The reason for the differences in the mesodermal tissues formed by ventral and dorsal explants is explained later in the chapter. Bottom panels: the standard experiment for studying mesoderm induction is to take explants from animal and vegetal regions that would normally make only ectoderm and endoderm, respectively, and culture them together for 3 days. When such explants from an early blastula are combined and cultured, mesodermal tissues including notochord, muscle, blood, and loose mesenchyme (red) are induced from the animal cap tissue.

endoderm. A small amount of anterior endoderm, which gives rise to the pharyngeal endoderm, is also induced from the ectoderm along with the mesoderm.

The formation of mesoderm can be observed by culturing explants from the blastula. When explants from the equatorial region of the late blastula, where animal and vegetal cells are adjacent, are cultured in isolation, they form ectodermal derivatives and mesoderm (see Fig. 4.25, top panels). When explants of animal and vegetal regions distant from the equator are cultured in contact with each other, mesodermal tissues are also formed (see Fig. 4.25, bottom panels). Confirmation that it is the **animal cap** cells, and not the vegetal cells, that are forming mesoderm was obtained by pre-labeling the animal region of the blastula with a cell-lineage marker and showing that the labeled cells form the mesoderm. Clearly, the vegetal region is producing a signal or signals that can induce mesoderm. Mesoderm can be distinguished by its

Box 4C Identical twins

The existence of identical, or mon-ozygotic, twins shows that early human embryos, like early-stage mouse embryos, can regulate. Identical twins are produced when an early embryo splits in half—unlike non-identical, or dizygotic, twins, which are the result of the fertilization of two separate eggs.

Monozygotic twinning can occur at various stages in early devel-opment and the stage at which it has occurred can be deduced by the arrangement of the extra-embryonic membranes and the pla-centa. Identical twins with separate amniotic and chorionic cavities and separate placentas are produced by the splitting of the morula during cleavage and before implantation, as shown here for twinning at the two-cell stage (figure, left). More commonly, twins have their own amniotic cavities but share a cho-rionic cavity and a placenta, which represents splitting of the very early inner cell mass inside the blastocyst (figure, center). A very small percentage of identical twins (about 4%) share both amniotic and chorionic cavities and a pla-centa, which means that twinning must have occurred quite late, with splitting of the inner cell mass at epiblast stages, between 9–15 days after fertilization (figure, right).
Figure adapted from Larsen, W.J.: 1997.

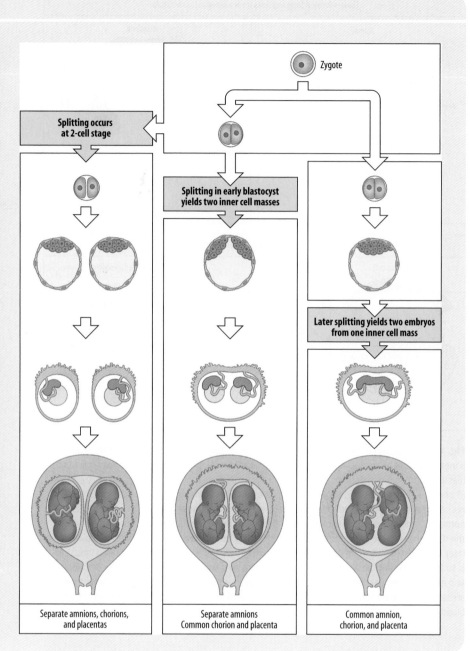

histology as, after a few days' culture, it can contain muscle, notochord, blood, and loose mesenchyme (connective tissue). It can also be identified by the typical proteins that cells of mesodermal origin produce, such as muscle-specific actin.

In this explant system, if the animal and vegetal explants are separated by a filter with pores too small to allow cell contacts to develop, mesoderm induction still takes place. This suggests that the mesoderm-inducing signals are in the form of secreted molecules that diffuse across the extracellular space, and do not pass directly from cell to cell via cell junctions (see Fig. 1.23). Differentiation of a mesodermal tissue such as muscle in the explant system appears to depend on a **community effect** in the responding cells. A few animal cap cells placed on vegetal tissue will not be induced to express muscle-specific genes. Even when a small number of individual cells are

Box 4D Preimplantation genetic diagnosis

We now take *in vitro* fertilization (IVF) of human eggs as a treatment for infertility almost for granted, even though it is only 30 years since the first IVF baby, Louise Brown, was born in the United Kingdom. More recently, it has become possible to determine the genotype of embryos produced by IVF before implantation without harming the embryo. This procedure is called **preimplantation genetic diagnosis (PGD)** and was developed in the late 1980s for use with fertile couples at risk of passing on a serious genetic disease to their children. The aim was to provide an alternative to prenatal diagnosis and the potential termination of an affected pregnancy.

Because of the capacity of human embryos to regulate, one blastomere can be removed from an IVF-produced embryo during its early cleavage *in vitro* without affecting subsequent development. The DNA from this blastomere can be amplified *in vitro* and tested for the presence or absence of mutations known to cause disease. Where there is a known high risk that the parents will pass on a particular genetic disease—for example, when the parents are both carriers for the cystic fibrosis gene—PGD can be used to ensure that an embryo that would develop the disease is not implanted into the mother.

The demand for PGD is likely to increase because it can be used not only to identify embryos with mutations that will inevitably cause a potentially fatal disease in infancy and childhood, but also to identify embryos with mutations in genes that predispose an individual to disease in later life. An example is the gene *BRCA1*, certain mutations in which predispose women to develop breast and ovarian cancer, and account for 80% of these tumors in women with a genetically inherited predisposition (5-10% of all breast and ovarian cancer). In males, mutations in *BRCA1* are

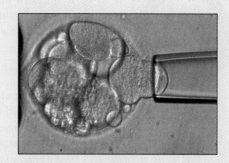

linked to an increased susceptibility to prostate cancer. By determining whether an embryo from a high-risk family carries a *BRCA1* disease allele the genetic susceptibility to these cancers could, in principle, be eliminated from a family. At least one baby has been born to such a high-risk couple after IVF and selection by PGD for the absence of the *BRCA1* disease allele. In the United Kingdom, PGD has been licensed by the Human Fertilisation and Embryology Authority (HFEA) for more than 60 genetic diseases.

There are practical and ethical questions in relation to PGD, such as which genetic conditions it should be applied to. An ethical question that arose recently concerned parents who wished to select an IVF embryo with the best HLA match to a sibling suffering from a rare blood disease, so that the IVF child could eventually donate stem cells to their sibling. The HFEA made a general ruling allowing this on a case-by-case basis in 2004. One UK family, who have a child with the rare condition Diamond–Blackfan anemia, have since had an HLA-matched daughter selected by PGD after IVF.

Photograph courtesy of Dr Malpani, MD, Malpani Infertility Clinic, Mumbai, India. www.drmalpani.com

placed between two groups of vegetal cells, induction does not occur. By contrast, larger aggregates of animal cap cells respond by strongly expressing muscle-specific genes (Fig. 4.26).

In contrast to mesoderm, the ectoderm and most of the endoderm are maternally specified in *Xenopus*. The transcription factor VegT has a key role in specifying the vegetal region as endoderm. VegT is translated from maternal mRNA present in the vegetal region of the egg and is inherited by the cells that develop from this region. Injection of *VegT* mRNA into animal cap cells induces expression of endoderm-specific markers, whereas blocking the translation of *VegT* mRNA in the vegetal region, by injecting oligonucleotides that are antisense to the mRNA (see Box 6A, p. 220), results in a loss of endoderm. VegT is also essential for mesoderm induction, as we shall see later. In VegT-depleted embryos, ectoderm forms throughout the embryo.

A strong candidate for a maternal determinant of ectoderm is the E3 ubiquitin-ligase Ectodermin, which is translated from maternal mRNA located at the animal pole of the fertilized egg and becomes localized in the nuclei of cells throughout the animal cap (Fig. 4.27). Ectodermin functions to counteract potential mesoderm-inducing signals, and in this way limits mesoderm formation to the equatorial zone. Ectodermin adds ubiquitin to Smad4, a gene regulatory protein activated as a result of BMP signaling (see Fig. 4.33), and this affects the interactions of Smad4 with other proteins.

Another determinant of ectoderm is the zygotic transcription factor Foxl1e, which is expressed in the cells of the animal half of the late blastula and is required to maintain their regional identity and their development as ectoderm (see Fig. 4.27). In the absence of Foxl1e, the animal cells mix with the cells of other germ layers and differentiate according to their new positions.

4.12 Mesoderm induction occurs during a limited period in the blastula stage

Mesoderm induction is one of the key events in vertebrate development, as the mesoderm itself is a source of essential signals for further development of the embryo. In *Xenopus*, mesoderm induction is largely completed before gastrulation begins. Explant experiments like those described above have shown that in *Xenopus* there is a period of about 7 hours at the blastula stage during which animal cells are competent to respond to a mesoderm-inducing signal; cells lose their competence to respond about 11 hours after fertilization. An exposure of about 2 hours to inducing signal is sufficient for the induction of at least some mesoderm.

One might expect that the timing of expression of mesoderm-specific genes, such as those of muscle, would be closely coupled to the time at which the mesoderm is induced—but it is not. Explant experiments have shown that, irrespective of when during the competent period the cells are exposed to the 2-hour induction, muscle-specific gene expression always starts at about 5 hours after the end of the competent period, a time that corresponds to mid-gastrula stage in the embryo (Fig. 4.28). Muscle gene expression can occur as early as 5 hours after induction, if induction occurs late in the period of competence, or as late as 9 hours after induction, if induction occurs early in the competent period. These results suggest that there is an independent

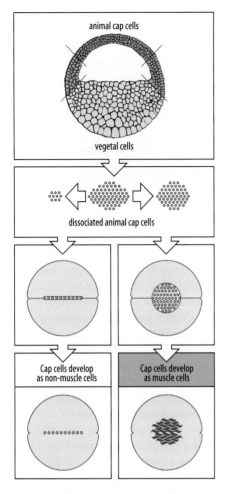

Fig. 4.26 The community effect. A single animal cap cell or a small number of animal cap cells in contact with vegetal tissue are not induced to become mesodermal cells and do not begin to express mesodermal markers such as muscle-specific proteins. A minimum of about 200 animal cap cells must be present for induction of muscle differentiation to occur.

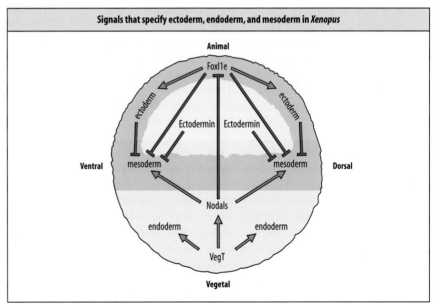

Fig. 4.27 Signals for germ-layer specification in *Xenopus*. In the vegetal half of the blastula, the maternal factor VegT is essential for both endoderm specification and, indirectly, for the production of Nodal proteins, which are one of the main signals that induce mesoderm from the prospective ectodermal cells of the animal cap. In the animal half of the blastula, the maternal factor Ectodermin counteracts potential mesoderm-inducing signals, and limits mesoderm formation to the equatorial zone. The zygotic transcription factor Foxl1e is expressed in the animal region and is necessary for the development of the animal-half cells as ectoderm and for preventing their mixing with more vegetal cells and developing as mesoderm or endoderm. In turn, Nodal signaling inhibits the expression of Foxl1e in the mesoderm.

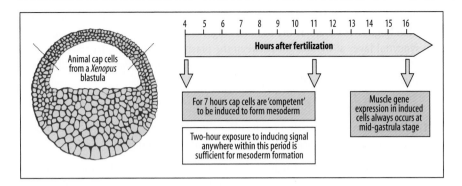

For 7 hours cap cells are 'competent' to be induced to form mesoderm

Two-hour exposure to inducing signal anywhere within this period is sufficient for mesoderm formation

Muscle gene expression in induced cells always occurs at mid-gastrula stage

Fig. 4.28 Timing of muscle gene expression is not linked to the time of mesoderm induction. Animal cap cells isolated from an early *Xenopus* blastula are competent to respond to mesoderm-inducing signals only for a period of about 7 hours, between 4 and 11 hours after fertilization. For expression of target genes to occur, exposure to inducer must be for at least 2 hours within this period. Irrespective of when the induction occurs within this competence period, muscle-gene activation occurs at the same time: 16 hours after fertilization.

timing mechanism by which the cells monitor the time elapsed since fertilization and then, provided that they have received the mesoderm-inducing signal, they express muscle-specific genes.

The extended period of competence gives the embryo latitude as to when mesoderm induction actually takes place, and means that the timing of the inductive signal does not have to be rigorously linked to a short time period. Given this latitude, an independent timing mechanism for mesodermal gene expression is one way to keep subsequent development coordinated, whatever the actual time of mesoderm induction.

4.13 Zygotic gene expression is turned on in *Xenopus* at the mid-blastula transition

Another event that appears to be under intrinsic temporal control, although not linked to mesoderm induction, is the beginning of transcription from the embryo's own genes. The *Xenopus* egg contains quite large amounts of maternal mRNAs, which are laid down during oogenesis. After fertilization, the rate of protein synthesis increases 1.5-fold, and a large number of new proteins are synthesized by translation of the maternal mRNA. There is, in fact, very little new mRNA synthesis until 12 cleavages have taken place and the embryo contains 4096 cells. At this stage, transcription from zygotic genes begins, cell cycles become asynchronous, and various other changes occur. This is known as the **mid-blastula transition**, although it does in fact occur in the late blastula, just before gastrulation starts. A few zygotic genes are expressed earlier, most notably two Nodal-related genes, *Xnr-5* and *Xnr-6*, which are expressed as early as the 256-cell stage.

How is the mid-blastula transition triggered? Suppression of cleavage but not DNA synthesis does not alter the timing of transcriptional activation, and so the timing is not linked directly to cell division. Nor are cell–cell interactions involved, as dissociated blastomeres undergo the transition at the same time as intact embryos. The key factor in triggering the mid-blastula transition seems to be the ratio of DNA to cytoplasm—the quantity of DNA present per unit mass of cytoplasm.

Direct evidence for this comes from increasing the amount of DNA artificially by allowing more than one sperm to enter the egg or by injecting extra DNA into the egg. In both cases, transcriptional activation occurs prematurely, suggesting that there may be some fixed amount of a general repressor of transcription present initially in the egg cytoplasm. As the egg cleaves, the amount of cytoplasm does not increase, but the amount of DNA does. The amount of repressor in relation to DNA gets smaller and smaller until there is insufficient to bind to all the available sites on the DNA and the repression is lifted. Timing of the mid-blastula transition thus seems to fit with an hourglass egg-timer model (Fig. 4.29). In such a model something has to accumulate, in this case DNA, until a threshold is reached. The threshold is determined by the initial concentration of the cytoplasmic factor, which does not increase.

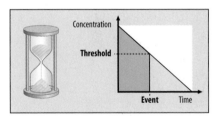

Fig. 4.29 A timing mechanism that could operate in development. A mechanism based on an analogy to an egg-timer could measure time to the mid-blastula transition. The decrease in the active concentration of some molecule, such as a transcriptional repressor, could occur with time, and the transition could occur when the repressor reaches a critically low threshold concentration. This would be equivalent to all the sand running into the bottom of the egg-timer. In the embryo a reduction in repressor activity per nucleus would in fact occur, because repressor concentration in the embryo as a whole remains constant but the number of nuclei increases as a result of cell division. The amount of repressor per nucleus thus gets progressively smaller over time.

Zebrafish embryos also go through a mid-blastula transition at which zygotic transcription begins. It occurs at the 512-cell stage, coincidentally with the formation of the syncytial layer at the margin between the blastoderm and yolk (see Section 3.2). Shortly after this, β-catenin activates zygotic genes such as *dharma*, *chordino*, and *nodal-related* in the future dorsal region (see Fig. 4.7).

4.14 Mesoderm-inducing and patterning signals in *Xenopus* are produced by the vegetal region, the organizer, and the ventral mesoderm

Once it had been established that mesoderm is induced by signals from the vegetal region, the next question was the nature of those signals and how the mesoderm becomes patterned along the dorso-ventral axis. In the *Xenopus* late blastula fate map, the mesoderm is divided into a number of regions along the dorso-ventral axis (see Fig. 4.19): the most dorsal region gives rise to the notochord; the next most dorsal to the somites (embryonic tissue that produces the skeletal muscles and the skeleton of the trunk); then comes mesoderm giving rise to organs such as the kidney; with the most ventral regions giving rise to blood only (although significant amounts of blood will also form from more dorsal mesoderm).

This patterning is not intrinsic to the mesoderm and has to be acquired. Experiments that combined different parts of the vegetal region with animal cap explants suggested differences in mesoderm-patterning activity along the dorso-ventral axis (Fig. 4.30). For example, dorsal vegetal tissue containing the Nieuwkoop center induces notochord and muscle from animal cap cells, whereas ventral vegetal tissue induces mainly blood-forming tissue and little muscle. Evidence for mesoderm-patterning signals from the organizer came from combining a fragment of dorsal marginal zone from a late blastula with a fragment of the ventral presumptive mesoderm. The ventral fragment will now form substantial amounts of muscle, whereas ventral mesoderm isolated from an early blastula and cultured on its own will produce mainly blood-forming tissue and mesenchyme.

These types of experiments, together with other evidence, indicated that the induction and initial patterning of the mesoderm in *Xenopus* is due to the concerted actions

Fig. 4.30 Differences in mesoderm induction by dorsal and ventral vegetal regions. The dorsal vegetal region of the *Xenopus* blastula, which contains the Nieuwkoop center, induces notochord and muscle from animal cap tissues, while ventral vegetal cells induce blood and associated tissues. This is good evidence for different inducing signals coming from the dorsal and ventral vegetal regions.

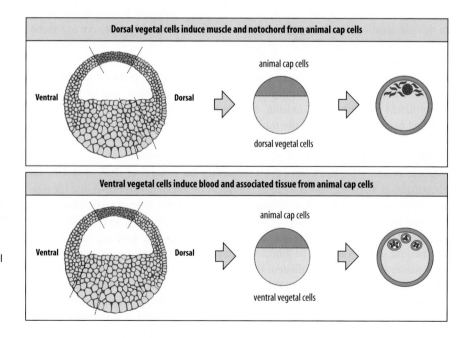

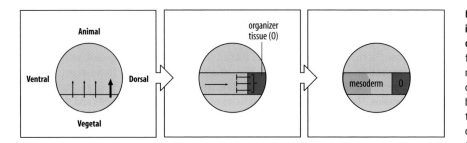

Fig. 4.31 Signals involved in mesoderm induction and patterning originate from different regions of the blastula. Signals from the vegetal region initially induce the mesoderm from the prospective ectoderm of the animal cap, with the organizer region being specified on the dorsal-most side where the inducing signal is strongest and of longest duration. Signals from the ventral side of the embryo pattern the mesoderm ventrally, and the extent of their influence is limited by signals emanating from the Spemann organizer region (0), which counteract them.

of signals emanating from different parts of the blastula (Fig. 4.31). The vegetal region initially produces signals that induce the mesoderm and establish the Spemann organizer on the dorsal side. Another set of signals gives the ventral mesoderm its character. At the same time the organizer emits yet another set of signals that counteract the ventral signals, limiting their range of action and so, for example, enabling the mesoderm adjacent to the organizer to give rise to somites. The outcome of all this activity is that by late blastula–early gastrula stage there is a band of prospective mesoderm around the equator of the embryo and this is becoming subdivided into regions with different fates.

The importance of the signals from the organizer in organizing both the dorso-ventral and the antero-posterior axis of the embryo is shown by a graft of the organizer into the ventral marginal zone of an early gastrula (Fig. 4.32). As discussed earlier, the organizer graft induces a complete new dorso-ventral axis. This, together with the induction of anterior and neural structures by the organizer, which we shall discuss in Chapter 5, results in a twinned embryo.

4.15 Members of the TGF-β family have been identified as mesoderm inducers

The explant experiments discussed in the previous sections suggested that secreted extracellular signaling proteins were involved in mesoderm induction. Two main approaches were then used to try and identify these factors in *Xenopus*. One was to apply candidate proteins directly to isolated animal caps in culture. The other was to inject mRNA encoding the suspected inducers into animal cells of the early blastula and then culture the cells. By itself, however, the ability to induce mesoderm in culture does not prove that a particular protein is a natural inducer in the embryo. Rigorous criteria must be met before such a conclusion can be reached. These include the presence of the protein signal and its receptor in the right concentration, place, and time in the embryo; the demonstration that the appropriate cells can respond to it; and the demonstration that blocking the response prevents induction. On all these criteria, growth factors of the TGF-β family (see Box 4A, p. 131) have been identified as the most likely candidates for mesoderm inducers. Members of this family known to be involved in early *Xenopus* development include the *Xenopus* Nodal-related proteins (Xnr-1, -2, -4, -5, -6), the protein Derrière, the bone morphogenetic proteins (BMPs), Vg-1, and activin.

Fig. 4.32 Transplantation of the Spemann organizer can induce a new axis in *Xenopus*. The Spemann organizer region produces signals required for mesoderm induction and patterning, and their effect can be seen by transplanting the Spemann organizer into the ventral region of another gastrula. The resulting embryo has two distinct heads, one of which was induced by the Spemann organizer. The organizer therefore produces signals that not only result in patterning the mesoderm dorso-ventrally, but also induce neural tissue and anterior structures. Scale bar = 1 mm.

Photograph courtesy of J. Smith.

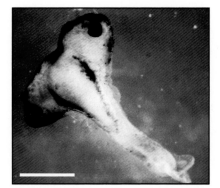

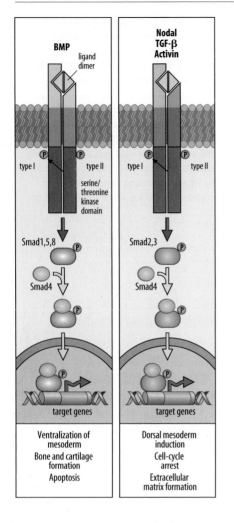

Fig. 4.33 Signaling by members of the TGF-β family of growth factors uses different combinations of receptor subunits and transcriptional regulatory proteins to activate different sets of target genes. Members of this family involved in early vertebrate development, such as Nodal and the Nodal-related proteins, bone morphogenetic proteins (BMPs), and activin, are dimeric ligands that act at cell-surface receptors. These receptors are heterodimers of two different subunits—type I and type II—with intracellular serine/threonine kinase domains. There are several different forms of type I and type II subunits and these combine to form distinctive receptors for different TGF-β family members. Binding of ligand causes subunit II to phosphorylate receptor subunit I, which in turn phosphorylates intracellular signaling proteins called Smads. These are named for the Sma and Mad proteins in *Drosophila*, in which they were discovered. The phosphorylated Smads themselves bind to another type of Smad, sometimes called co-Smads, forming a transcriptional regulatory complex that enters the nucleus to either activate or repress target genes. Different receptors use different Smads, thus leading to the activation of different sets of target genes. The identity of the type I subunit determines which Smads get activated. The biological response depends on the combination of the activated target genes and the particular cellular environment.

The signaling pathways stimulated by TGF-β growth factors are illustrated in Fig. 4.33. Confirmation of the role of TGF-β growth factors in mesoderm induction came initially from experiments that blocked activation of the receptors for these factors (Fig. 4.34). The activin type II receptor is a receptor for several TGF-β family growth factors and is uniformly distributed throughout the early *Xenopus* blastula. When mRNA for a mutant receptor subunit is injected into the early embryo and blocks TGF- β family signaling, mesoderm formation is prevented. Because different TGF-β growth factors can act through the same receptor, these experiments were not able to pinpoint the individual proteins responsible.

More recently, it has become possible to study mesoderm induction using more specific inhibitors of Nodal family proteins, and to detect the presence of signaling by using labeled antibodies to detect phosphorylated Smads and to distinguish their type. This latter technique enables the spatial patterns and dynamics of normal growth factor signaling to be visualized directly, and it is now widely used to detect signaling by TGF-β family members. For example, in *Xenopus*, signaling by BMPs can be distinguished from that of Nodal-related proteins or activin by looking for phosphorylated Smad1 or Smad2, respectively (see Fig. 4.33).

4.16 The zygotic expression of mesoderm-inducing and patterning signals in *Xenopus* is activated by the combined actions of maternal VegT and Wnt signaling

One maternal vegetal factor of great importance in the earliest stage of mesoderm induction is the transcription factor VegT (see Section 4.1). Directly and indirectly, VegT activates the zygotic expression of the Xnr proteins and Derrière. If VegT is severely depleted, expression of *Xnr* genes and *Derrière* is downregulated and almost no mesoderm develops, showing that VegT is crucial for its induction. Injection of mRNAs for the Xnrs or Derrière can rescue mesoderm formation in the VegT-depleted embryos, suggesting that these are most likely to be the direct mesoderm inducers. The rescue of head, trunk, and tail mesoderm by the mRNAs for the genes *Xnr-1*, *Xnr-4*, and *Xnr-5* indicates that they are involved in the general induction of all mesoderm, whereas *Derrière* mRNA rescues only trunk and tail, indicating that on its own it cannot induce the anterior-most mesoderm. However, Nodal-related proteins act as dimers (see Fig. 4.33), and assays of their activity in blastula explants have shown that they can heterodimerize and cooperate to induce mesoderm. FGF signaling (see Box 5B, p. 184) also appears to be necessary, but not sufficient by itself, for general mesoderm induction.

Fig. 4.34 A mutant activin receptor blocks mesoderm induction. Receptors for proteins of the TGF-β family function as dimers (see Fig. 4.33). Receptor function in a cell can be blocked by introducing mRNA encoding a mutant receptor subunit that lacks most of the cytoplasmic domain, and so cannot function. The mutant receptor subunit can bind ligand and forms heterodimers with normal receptor subunits, but the complex cannot signal. The mutant subunit thus acts in a dominant-negative fashion preventing receptor function. When mRNA encoding the mutant activin receptor subunit is injected into cells of the two-cell *Xenopus* embryo, subsequent mesoderm formation is blocked. No mesoderm or axial structures are formed except for the cement gland, the most anterior structure of the embryo.

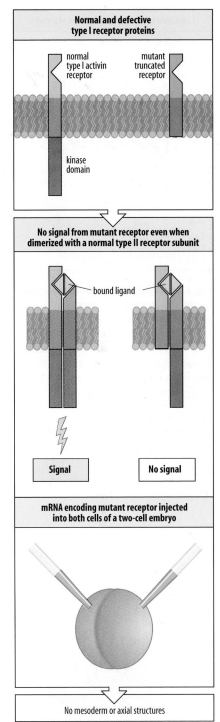

Numerous experiments have now shown that Xnrs are expressed in the *Xenopus* vegetal region in a graded fashion at the appropriate time for mesoderm induction, and that the difference between a dorsal and a ventral mesoderm fate is likely to be due to expression of a cocktail of different Xnrs or differences in Xnr levels along the dorso-ventral axis. Nuclear β-catenin stimulates *Xnr* gene transcription, and thus Xnr levels will be highest on the dorsal side, where the VegT signal, Xnrs, and the β-catenin signal overlap (Fig. 4.35). Signaling via Xnrs stimulates phosphorylation and activation of Smad2, and a wave of Smad2 activation, detected by anti-phospho-Smad2 antibodies, appears to sweep from the dorsal to the ventral side of the embryo from early to mid-gastrulation. All this evidence is consistent with the idea that the Spemann organizer is specified where Nodal-related signals emanating from the vegetal region are most intense and of longest duration (see Fig. 4.35).

4.17 Signals from the organizer pattern the mesoderm dorso-ventrally by antagonizing the effects of ventral signals

We can now consider the other signals that pattern the mesoderm along the dorso-ventral axis once it has been induced. Ventralization of the mesoderm is promoted by gradients of activity of two signaling molecules, BMP-4 and Xwnt-8, with their high

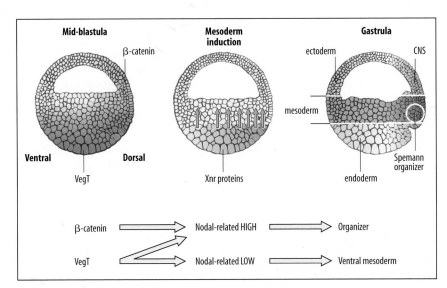

Fig. 4.35 A gradient in Nodal-related proteins may provide the initial signals in mesoderm induction. Maternal VegT in the vegetal region activates the transcription of Nodal-related genes (*Xnrs*). The presence of β-catenin on the dorsal side results in a dorsal-to-ventral gradient in the Xnr proteins. These induce mesoderm and, at high doses, specify the Spemann organizer on the dorsal side. This is a simplified version because other signals, such as Vg-1 and activin, also have roles. CNS, central nervous system.

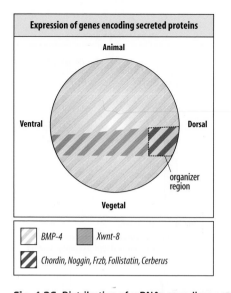

Expression of genes encoding secreted proteins

Animal

Ventral — Dorsal

organizer region

Vegetal

BMP-4   Xwnt-8

Chordin, Noggin, Frzb, Follistatin, Cerberus

Fig. 4.36 Distribution of mRNAs encoding secreted proteins in the *Xenopus* blastula. The secreted proteins that are made in the organizer block the action of BMP-4 and Xwnt-8, proteins that are made throughout the blastula and prospective mesoderm, respectively, where their genes are expressed.

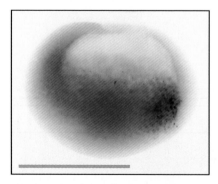

Fig. 4.37 Expression of *noggin* in the *Xenopus* blastula. Staining for expression of *noggin* mRNA locates it in the region of the Spemann organizer. Scale bar = 1 mm.

Photograph courtesy of R. Harland, from Smith, W.C., et al.: 1992.

points on the ventral side of the blastula. Initially, BMP-4 protein is made uniformly throughout the late *Xenopus* blastula, and zygotic Xwnt-8 protein is made in the prospective mesoderm (Fig. 4.36). As noted in Section 4.11, the potential mesoderm-inducing and patterning activity of BMP-4 is restricted to the equatorial zone by the action of Ectodermin. When the action of BMP-4 is blocked throughout the blastula by introducing a dominant-negative mutant receptor (see Fig. 4.34), the embryo is dorsalized, with ventral cells differentiating as both muscle and notochord. Conversely, overexpression of BMP-4 ventralizes the embryo.

The restriction of BMP and Wnt-8 signaling to the more ventral regions is due to dorsalizing signals that are secreted by the organizer in response to Nodal signaling and inhibit, or antagonize, BMP and Wnt-8 activity. The first of these antagonists to be discovered was the protein Noggin, which inhibits BMP-4 activity. The gene for Noggin was first discovered in a screen for factors that could rescue UV-irradiated ventralized *Xenopus* embryos, in which a dorsal region had not been specified by the usual activity of the β-catenin pathway. The *Noggin* gene is strongly expressed in the Spemann organizer (Fig. 4.37). Noggin protein cannot induce mesoderm in animal-cap explants but it can dorsalize explants of ventral marginal zone tissue, thus making it a good candidate for a signal that patterns the mesoderm along the dorso-ventral axis. The organizer was subsequently found to secrete a mixture of antagonists—the BMP antagonists Noggin, Chordin, and Follistatin, and the Wnt antagonist Frizzled-related protein (Frzb). The use of antagonists to restrict signaling to a particular region, or to modulate the level of signaling, is widespread in embryonic development, as noted earlier in relation to anterior specification and left–right asymmetry in chick and mouse (see Sections 4.6 and 4.7).

Noggin, Chordin, and Follistatin proteins interact with BMP-4 protein and prevent it from binding to its receptor. In this way, a functional gradient of BMP-4 activity is set up across the dorso-ventral axis with its highest point ventrally and little or no activity in the presumptive dorsal mesoderm. Frzb generates a similar ventral-to-dorsal gradient of Wnt activity by binding to Wnt proteins and preventing them acting dorsally. Yet another signaling antagonist is Cerberus, which inhibits the activity of BMPs, Nodal-related proteins, and Wnt proteins in prospective anterior tissues. In *Xenopus*, Cerberus is expressed in the organizer and the anterior endoderm, and it is involved in the suppression of mesoderm fate and induction of anterior structures, in particular the head, as discussed in more detail in Chapter 5. The summary table on p. 167 lists some of the main proteins so far identified as mesoderm inducers and mesoderm-patterning factors in *Xenopus*, while Fig. 4.38 summarizes which signaling systems are active and which are antagonized in the different germ layers at the blastula stage.

The antagonistic action of Chordin on BMP-4 mirrors that of the *Drosophila* Chordin homolog Sog on the BMP-4 homolog Decapentaplegic (Dpp) in the patterning of the dorso-ventral axis in the fly (see Section 2.20). In the fly, however, the dorso-ventral axis is inverted compared with that of vertebrates, and so Dpp specifies a dorsal fate whereas BMP-4 specifies a ventral fate (the inversion of the dorso-ventral axis during evolution is discussed in Section 15.6). Other components of the mechanism that forms the gradient of Dpp activity in *Drosophila* are also conserved in vertebrates. Xolloid, for example, is the *Xenopus* equivalent of the fly metalloproteinase Tolloid, and degrades Chordin. Xolloid is thought to act as a clearing agent for Chordin, reducing the extent of its long-range diffusion and helping to maintain a gradient of Chordin dorsalizing activity.

Mutagenesis screens have identified genes in zebrafish similar to those involved in mesoderm induction and dorso-ventral patterning in *Xenopus*, and they confirm the general outline of germ-layer specification. The zebrafish signaling protein Nodal-related 1 (Ndr1, also called Squint) is expressed in the dorsal yolk syncytial layer and overlying cells as a result of β-catenin signaling. Together with another Nodal-related protein, Ndr2 (Cyclops), Ndr1 is required for the specification of the marginal blasto-meres as mesendoderm: prospective endoderm and mesoderm. High levels of Nodal signaling are thought to specify endoderm and lower levels the mesoderm. Double mutants of *Ndr1* and *Ndr2* lack both head and trunk mesoderm, but some mesoderm still develops in the tail region. The Ndr proteins signal via Smad2. Nuclear accumu-lation of Smad2 and the formation of the complex between Smad2 and its co-Smad, Smad4, have been visualized in living transgenic zebrafish embryos carrying a Smad2 gene fused with a sequence encoding a fluorescent tag. This experiment revealed a graded distribution of Nodal-type signaling, most probably due to Ndr1, in the blas-toderm of these embryos, with highest activity at the margin of the blastoderm and grading off towards the animal pole, and confirmed that a gradient of Nodal signaling patterns the zebrafish blastoderm into endoderm, mesoderm, and ectoderm. Nodal signaling also appears to be highest and of longest duration in the dorsal marginal region of the blastoderm, suggesting that it may also help to pattern the mesoderm dorso-ventrally. Many of the other proteins involved in *Xenopus* mesoderm induc-tion and specification have been found to be involved, such as zebrafish versions of Chordin and BMPs.

4.18 Threshold responses to gradients of signaling proteins are likely to pattern the mesoderm

The mesoderm-inducing and patterning signals described in previous sections exert their developmental effects by inducing the expression of specific genes that are required for the further development of the mesoderm into, for example, notochord, muscle, or blood. Expression of the gene *Brachyury,* which encodes a T-box family transcription factor, is one of the earliest markers of mesoderm in all vertebrates, and Brachyury is thought to act as a key transcription factor in mesoderm specifica-tion and patterning. The *Xenopus* version of the *Brachyury* gene, *Xbra*, is expressed throughout the presumptive mesoderm in the late blastula and early gastrula from soon after the mid-blastula transition (Fig. 4.39), later becoming confined to the notochord (the dorsal-most derivative of the mesoderm), and to the tailbud (posterior mesoderm). *Brachyury* is essential for the development of posterior mesoderm. The gene was originally discovered in the mouse as a result of semi-dominant mutations that produced short tails; when homozygous, this mutation is lethal in the embryo as a result of a lack of development of posterior mesoderm (see Fig. 1.13). Loss of the zebrafish ortholog of *Brachyury*, called *No tail* (*Ntl*), similarly results in fish lacking a tail. *No tail* and other zebrafish genes involved in specifying the mesoderm have been linked up into a gene regulatory network using the techniques of chromatin immuno-precipitation followed by DNA sequencing (ChIP-seq) (Box 4E, p. 162).

One of the first zygotic genes to be expressed specifically in the *Xenopus* organizer, and in other vertebrate embryonic organizer regions, is *goosecoid,* which encodes a transcription factor somewhat similar to both the Gooseberry and Bicoid proteins of *Drosophila*—hence the name. In line with its presence in the organizer, microinjec-tion of *goosecoid* mRNA into the ventral region of a blastula mimics to some extent transplantation of the Spemann organizer (see Fig. 4.32), resulting in the formation of a secondary axis. Genes for other transcription factors are also specifically expressed in the organizer region (Fig. 4.40 and summary table, p. 167).

It is still not clear how the secreted signaling proteins present in the mesoderm turn on genes like *goosecoid* and *Brachyury* in the right place. For example, *Brachyury* can

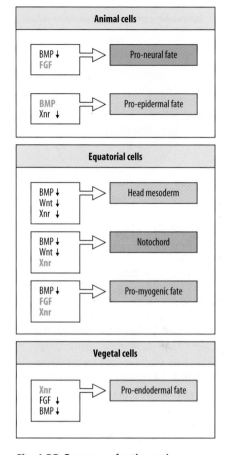

Fig. 4.38 Summary of active and antagonized signals in the different germ layers in a *Xenopus* blastula. Three main families of secreted signaling proteins are involved in germ-layer specification in *Xenopus*: FGFs, BMPs, and Nodals (Xnrs). Signals in green lettering are active. Those in black with a down arrow are inhibited.

Adapted from Heasman, J.: 2006.

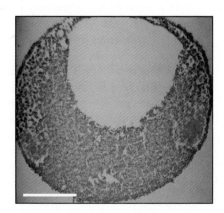

Fig. 4.39 Expression of *Brachyury* in the *Xenopus* blastula. A cross-section through the embryo along the animal–vegetal axis shows that *Brachyury* (red) is expressed in the prospective mesoderm. Scale bar = 0.5 mm.

Photograph courtesy of M. Sargent and L. Essex.

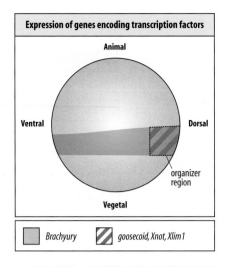

Expression of genes encoding transcription factors

Animal

Ventral — Dorsal

organizer region

Vegetal

Brachyury | goosecoid, Xnot, Xlim1

Fig. 4.40 Expression of transcription factor genes in a late *Xenopus* blastula. The expression domains of a number of zygotic genes that code for transcription factors correspond quite well to demarcations on the specification map, as judged by the distribution of their mRNAs. *Brachyury* is expressed in a ring around the embryo corresponding quite closely to the future mesoderm (see also Fig. 8.28). A number of transcription factors are specifically expressed in the region of the dorsal mesoderm that corresponds to the Spemann organizer and are required for its function. Xnot protein appears to have a role in the specification of notochord; Goosecoid and Xlim1 function are required for a transplanted Spemann organizer to be able to induce the formation of a new head.

be experimentally switched on in *Xenopus* animal caps by activin, but what induces it normally in the presumptive mesoderm *in vivo* is still unclear; a combination of Vg-1 and Xnrs is thought to be most likely, and its expression is probably maintained by FGF. As we have seen in the previous sections, there are gradients of activity of

Box 4E A zebrafish gene regulatory network

The zebrafish transcription factor No-tail (Ntl) is a homolog of the mouse mesodermal transcription factor Brachyury and the technique of chromatin immunoprecipitation followed by DNA sequencing (ChIP-seq, see Fig. 3.35) has been used to uncover the targets of Ntl in mesoderm specification, patterning and differentiation. Ntl activates a network of genes encoding transcription factors (for example, Snail and the T-box (Tbx) transcription factors), intercellular signals (for example, FGFs, Notch, and Wnt-11 in its later role as a signal directing convergent extension during gastrulation, see Box 8C, p. 310), and genes involved in the differentiation of specific cell types (for example, genes encoding the transcription factors Pax3, FoxD3, and Myf5 acting in muscle differentiation, see Chapter 10).

In the figure, genes are grouped to represent different aspects of

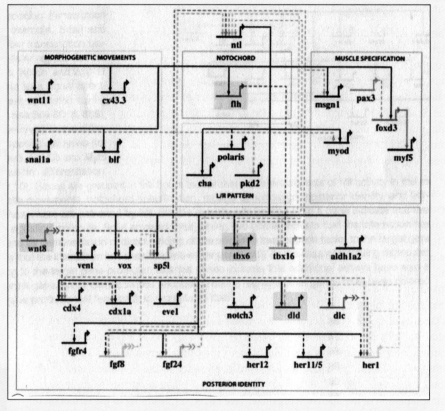

Ntl activity in the mesoderm: control of morphogenetic movements, notochord specification, muscle specification, posterior identity, and left-right patterning. Expression of all the genes shown here is activated by Ntl. Double arrowheads leading from a gene indicate that the gene product is an intercellular signaling molecule. Solid arrowed lines joining two genes indicate that the interaction has been verified both genetically and by demonstration of direct binding of the encoded transcription factor to the target gene promoter. Dashed lines indicate that the interaction

has been verified either genetically (in the case of signaling molecules) or by transcription factor binding to the target gene promoter. Shaded boxes indicate that additional assays have also shown direct regulation of the target gene by Ntl. Genes whose encoded proteins regulate other genes have been colored for clarity. Note the number of gene products that feed back to regulate expression of the *ntl* gene itself.

Figure from Morley, R.H.: 2009.

Fig. 4.41 Graded responses of early *Xenopus* tissue to increasing concentrations of activin. When animal cap cells are treated with increasing concentrations of activin, particular genes are activated at specific concentrations, as shown in the top panel. At intermediate concentrations of activin, *Brachyury* is induced, whereas *goosecoid,* which is typical of the organizer region, is only induced at high concentrations. If beads releasing a low concentration of activin are placed in the center of a mass of animal cap cells (lower left panel), expression of low-response genes such as *Brachyury* is induced immediately around the beads. With a high concentration of activin in the beads (lower right panel), *goosecoid* and other high-response genes are now expressed around the beads and the low-response genes farther away.

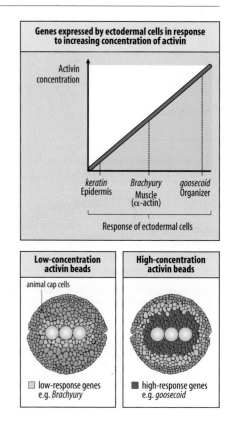

secreted signaling molecules throughout the mesoderm that could provide positional information for switching on developmental genes.

Experiments with the TGF-family member activin provide a beautiful example of how a diffusible protein could pattern a tissue by turning on particular genes at specific threshold concentrations. Although activin itself is not likely to be responsible for primary mesoderm patterning *in vivo*, it acts through the same receptor as the Xnrs, and animal cap cells from a *Xenopus* blastula respond to increasing doses of activin by activation of different mesodermal genes at different threshold concentrations. In this explant system, increasing concentrations of activin specify several distinct cell states that correspond to the different regions along the dorso-ventral axis. At the lowest concentrations of activin, only epidermal genes are expressed; no mesoderm is induced. Then, as the concentration increases, *Brachyury* is expressed, together with muscle-specific genes such as that for actin. With a further increase in activin concentration, *goosecoid* is expressed, corresponding to the dorsal-most region of the mesoderm—the Spemann organizer—which produces the notochord (Fig. 4.41). One can see, therefore how graded growth-factor signaling along the dorso-ventral axis could, in principle, activate transcription factors in particular regions and thus pattern the tissues. A 1.5-fold increase in activin concentration is sufficient to cause a change from formation of muscle to formation of notochord by explants of animal caps. Just how gradients are set up *in vivo* with the necessary precision is not clear, however. Simple diffusion of morphogen molecules may play a part, but more complex processes are likely to be involved, as we saw in the formation of the Dpp gradient in *Drosophila* (see Section 2.20). Some common cellular mechanisms that could be involved in gradient formation are indicated in Box 11A, p. 421.

How do cells distinguish between different concentrations of growth factor? Occupation of just 100 activin receptors per cell is required to activate expression of *Brachyury*, whereas 300 receptors must be occupied for *goosecoid* to be expressed. The link between the strength of the signal and gene expression may not be so simple, however. There are additional layers of intracellular regulation; cells expressing *goosecoid* at high activin concentrations also repress *Brachyury*, for example, and this involves the action of the Goosecoid protein itself, together with other proteins.

Excellent evidence for a secreted morphogen turning on mesodermal genes at specific threshold concentrations also comes from zebrafish. The putative morphogen here is the protein Ndr1, which is involved in patterning the mesoderm (see Section 4.17). Injection of *Ndr1* mRNA into a single cell of an early zebrafish embryo results in high-threshold Ndr1 target genes being activated in adjacent cells, whereas low-threshold genes are activated in more distant cells.

4.19 Mesoderm induction and patterning in the chick and mouse occurs during primitive-streak formation

In the chick, most mesoderm induction and patterning occurs at the primitive streak. Chick epiblast isolated before streak formation will form some mesoderm containing blood vessels, blood cells, and some muscle, but no dorsal mesodermal structures, such as notochord. Treatment of the isolated epiblast with activin, however, results in

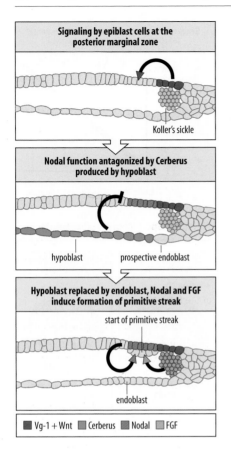

Fig. 4.42 Signals at the posterior marginal zone of the chick epiblast that initiate primitive streak formation. Epiblast cells in the posterior marginal zone overlying Koller's sickle secrete Vg-1 and Wnts. These signals induce expression of *nodal* in neighboring epiblast cells, but Nodal protein function is blocked by Cerberus, which is produced by the hypoblast. When the hypoblast has been displaced by endoblast, Nodal signaling in the epiblast and FGF signals from Koller's sickle induce the internalization of epiblast cells and formation of the primitive streak.

the additional appearance of notochord and more muscle, suggesting that TGF-β family members such as Nodal act as mesoderm-inducing and/or mesoderm-patterning signals in chick embryos as they do in *Xenopus*.

As we saw in Section 4.4, the posterior marginal zone in the chick embryo is crucial for the formation of the primitive streak, and the zone also supplies signals for mesoderm induction and patterning. The primitive streak is initiated in the posterior marginal zone by the actions of Wnt-8c, which is present throughout the marginal zone in a gradient with its high point at the posterior, and Vg-1, which is localized to the most posterior region of the marginal zone. Once the hypoblast and its inhibitory signal have been replaced by endoblast, Vg-1 and Wnt-8c induce the expression of Nodal in the posterior marginal zone and the streak (Fig. 4.42). Fibroblast growth factor (FGF) is expressed in Koller's sickle and together with Nodal induces the internalization of epiblast cells, by formation of the primitive streak and mesoderm. The protein Chordin is made in the cells at the tip of the streak. It antagonizes the signaling activity of bone morphogenetic proteins (BMPs), which are secreted by cells outside the streak and tend to inhibit the formation of axial mesoderm, as they do in *Xenopus*. All these signals are required for streak formation and a fully functional Hensen's node.

Mesodermal patterning also occurs in the streak, with different regions of the streak giving rise to different mesodermal tissues along the dorso-ventral axis: the tip of the streak, including Hensen's node, gives rise to the notochord and somites, and the posterior streak generates lateral mesoderm, including blood and blood vessels, and extra-embryonic mesoderm (Fig. 4.43). Thus, although the antero-posterior axis of the primitive streak marks the orientation of the future antero-posterior axis of the embryo, dorso-ventral patterning of the mesoderm also initially occurs along this axis. This occurs by the more posterior cells moving progressively forward and outwards to take up their positions along the dorso-ventral axis, which is orthogonal to the antero-posterior axis.

In the mouse, mesoderm induction occurs only in the primitive streak, which is initiated at around E6, starting from one small region at one side of the epiblast (see Fig. 3.24). This region is specified as a result of the determination of the anterior end of the embryo by the movement of the distal visceral endoderm, which then becomes the anterior visceral endoderm (AVE) (see Section 4.6). As the anterior ectoderm becomes determined, proximal epiblast cells that express genes characteristic of prospective mesoderm start to converge on the posterior proximal region and internalize at the point of convergence to initiate the primitive streak (Fig. 4.44, first and second panels). BMP-4 is expressed in the extra-embryonic ectoderm above the proximal rim of the epiblast and activates the expression of mesodermal markers in adjacent proximal epiblast cells. Nodal is required both to establish the posterior end of the primitive streak and for mesoderm formation, and is expressed in the primitive streak along with Wnts. In mutants lacking Nodal function, mesoderm does not form. Genes encoding secreted antagonists of Nodal and Wnt signaling, such as Cerberus-like 1 and Dickkopf, respectively, are expressed on the anterior side of the embryo and inhibit BMP-4, Nodal, and Wnt signaling in the AVE and its overlying ectoderm. This allows the ectoderm to develop as anterior tissues. Altogether, this indicates that, as in other vertebrates, graded Nodal signals pattern the mouse embryo, with the highest levels of Nodal becoming restricted to the proximal posterior region of

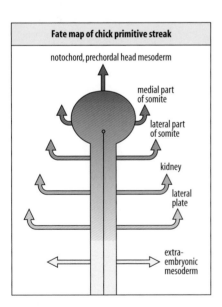

Fig. 4.43 Mesoderm is patterned within the chick primitive streak. Different parts of the primitive streak give rise to mesoderm with different fates. The arrows indicate the direction of movement of the mesodermal cells.

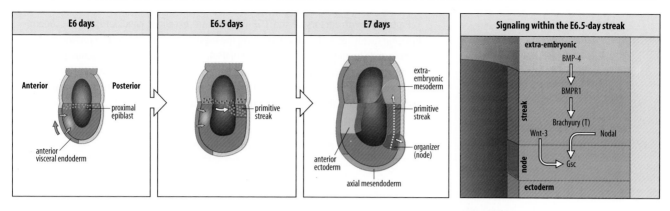

Fig. 4.44 Specification of the primitive streak in the mouse embryo. First panel: by E6 the anterior visceral endoderm (AVE) has induced anterior character in the underlying epiblast (small arrows) and early markers of the future primitive streak are restricted to the proximal rim of the epiblast. Second panel: BMP-4 (red dots) is transiently expressed in the adjacent extra-embryonic ectoderm, and the posterior movement (white arrow) of cells expressing primitive-streak markers (purple) results in the primitive streak forming opposite the AVE at 6.5 days. Third panel: by 7 days, extra-embryonic mesoderm (brown) is produced from the posterior end of the streak while the node (red) forms at the anterior end. The extreme anterior end of the streak gives rise to axial mesendoderm, which will form prechordal plate and the gut endoderm. Fourth panel: a possible scheme of the molecular interactions within the streak at 6.5 days. Gsc, goosecoid.

the epiblast and being required to induce primitive-streak formation. Dose-dependent Nodal signaling also patterns the mesendoderm and mesoderm, with definitive gut endoderm and prechordal plate being specified by high levels of Nodal signaling, the node by intermediate levels of signaling, and paraxial and lateral-plate mesoderm by low levels of Nodal signaling.

In the mouse, as in the chick, mesoderm patterning occurs in the primitive streak and the position along the primitive streak at which cells ingress determines their eventual fate. The earliest cells to ingress are those in the posterior streak which give rise to extra-embryonic tissues, while cells that ingress in more intermediate and anterior regions of the streak form lateral mesoderm, and cells ingressing at the anterior end of the streak give rise to anterior mesendoderm. Once inside the embryo, these cells from the anterior end of the streak move anteriorly along the midline, pattern the overlying neurecto-derm, and give rise to dorsal mesoderm such as the notochord. At the anterior-most tip of the streak, cells that will give rise to the definitive endoderm also ingress. How all the different cell lineages are segregated at the tip of the streak is not yet understood.

Having looked at the induction and initial patterning of the mesoderm, we are now in a position to consider the final emergence of the typical vertebrate body plan. In the next chapter, we will discuss patterning of the germ layers along the antero-posterior axis during gastrulation, the initial formation of the nervous system, and the development of notochord and somites from mesoderm.

Summary

There are strong similarities in the fate maps of the frog, zebrafish, chick, and mouse. Even though there is good evidence for the maternal specification of some regions such as the future endoderm in amphibians, the embryo can still undergo considerable regulation at the blastula stage. This implies that interactions between cells, rather than intrinsic factors, have a central role even in early amphibian development. This strategy is even more pronounced in the mouse and chick, where it appears to be position that determines cell fate.

In *Xenopus*, the mesoderm and a small amount of endoderm are induced from prospective ectoderm at the equator of the blastula by signals from the vegetal region. Patterning of the mesoderm along the dorso-ventral axis produces regions that, from ventral to dorsal, give rise to blood and blood vessels, internal organs such as kidneys, the somites, and the organizer region in the most dorsal mesoderm, which produces the notochord. Signals originating from the ventral side of the blastula, give the ventral mesoderm its character. These signals are counteracted by signals emanating from the organizer, which limit their influence.

Protein growth factors of the TGF-β family are excellent candidates for mesoderm-inducing factors as well as for patterning the mesoderm. Other secreted proteins, such as Noggin and Chordin, inhibit the action of BMP-4, and so are involved in specifying the dorsal mesoderm. *Brachyury* and *goosecoid* are early mesodermally expressed genes encoding transcription factors, and their pattern of expression may be specified by gradients of signaling proteins, the genes being turned on at particular threshold concentrations.

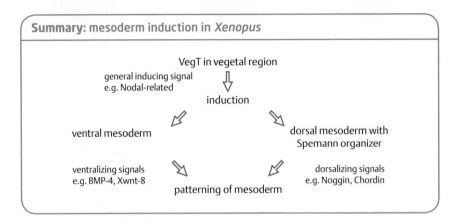

Summary: mesoderm induction in *Xenopus*

VegT in vegetal region

general inducing signal
e.g. Nodal-related

induction

ventral mesoderm

dorsal mesoderm with
Spemann organizer

ventralizing signals
e.g. BMP-4, Xwnt-8

patterning of mesoderm

dorsalizing signals
e.g. Noggin, Chordin

Summary to Chapter 4

All vertebrates have the same basic body plan. During early development, the antero-posterior and dorso-ventral axes of this body plan are set up. The mechanism appears to be different in frog, chick, zebrafish, and mouse but can involve localized maternal determinants, external signals, and cell-cell interactions. This early patterning also establishes bilateral asymmetry. It is possible to construct a fate map in the early embryo for the three germ layers—mesoderm, endoderm, and ectoderm. The fate maps of the different vertebrates have strong similarities. At this early stage the embryos are still capable of considerable regulation and this emphasizes the essential role of cell-cell interactions in development. In *Xenopus,* signals emanating from different regions of the blastula are involved in mesoderm induction and early patterning. Good candidates for these signals have been identified and include members of the TGF-β family and Wnt proteins. At particular concentrations, these signals activate mesoderm-specific genes such as *Brachyury* and so gradients could pattern the mesoderm. The summary table on p. 166 lists all genes considered in this chapter in relation to *Xenopus.*

■ End of chapter questions

Long answer (concept questions)

1. The organization of the *Xenopus* embryo begins with the establishment of the dorso-ventral axis. A critical component of this event is the accumulation of β-catenin in cells of the dorsal marginal zone. Describe the series of events that lead to this accumulation. Include the maternal deposition of factors into the oocyte, events accompanying sperm entry, cortical rotation, and the Wnt signaling pathway.

2. Give a molecular explanation for the ventralizing effects of: UV irradiation of the egg, and treatment of the embryo with lithium.

3. The blastula organizer of *Xenopus*, also known as the Nieuwkoop center, was defined by various experiments. Describe the results of the following experiments:

a) The two dorsal cells of a four-cell embryo are isolated from the two ventral cells. What does each half embryo develop into?

b) Cells from the animal region of a late blastula are cultured together with cells from the dorsal vegetal region.

c) Cells from the dorsal vegetal region of a 32-cell embryo are grafted into the ventral marginal zone of an early gastrula.

4. What is the result of the following experiment?
The dorsal lip of the blastopore from one early *Xenopus* gastrula is grafted into the ventral marginal zone of another early gastrula (see also Section 1.4). What is the explanation?

5. Describe the developmental roles of the VegT protein found in fertilized eggs and early embryos in *Xenopus* embryonic development. Include in your answer what kind of protein it is; whether it is maternally supplied, or is transcribed and translated from the

Summary: main genes involved in patterning of axes and germ layers in *Xenopus*

Gene	Maternal/ zygotic	Type of protein	Where expressed	Function of protein
Specification of germ layers and dorso-ventral axis				
VegT	M	Transcription factor	Vegetal region	Endoderm specification; activates expression of mesoderm inducers
Ectodermin	M	Ubiquitin ligase	Animal hemisphere	Ectoderm specification; inhibits mesoderm formation
Vg-1	M	TGF-β family	Vegetal region; RNA enriched dorsally after fertilization	Mesoderm induction
Xwnt-11	M	Wnt family	Vegetal region; RNA translocated to dorsal side after fertilization	Specification of dorsal structures and organizer formation
Dishevelled	M	Wnt pathway signaling protein	Protein associated with vesicles that move to dorsal side of embryo after fertilization	Specification of dorsal structures and organizer formation
β-catenin	M	Acts in Wnt pathway; regulates gene expression	Protein enriched in dorsal nuclei in response to Wnt signaling	Specification of dorsal structures and organizer formation
GSK-3	M	Protein kinase	Protein depleted on dorsal side	Suppression of dorsalizing signals
axin	M	Binds β-catenin	RNA throughout zygote; more protein dorsally than ventrally	Suppression of dorsalizing signals
Derrière	Z	TGF-β family	Vegetal hemisphere and marginal region	Posterior-mesoderm induction
Xnr-1,2,4,5,6	Z	TGF-β family	Vegetal hemisphere and marginal region; higher dorsally	Mesoderm induction
FGF (several types)	Z	Secreted signals	Vegetal hemisphere and marginal region	Posterior-mesoderm development
Mesoderm patterning				
Brachyury	Z	Transcription factor	Throughout prospective mesoderm	Posterior-mesoderm formation
Xwnt-8	Z	Wnt family	Ventral and lateral regions of prospective mesoderm	Mesoderm ventralization
BMP-4	Z	TGF-β family	Throughout late blastula; then excluded from organizer	Mesoderm ventralization
Activin	Z	TGF-β family	Throughout late blastula/ early gastrula	Mesoderm induction and patterning
Organizer function				
siamois	Z	Transcription factor	Nieuwkoop center	Induction of organizer
noggin	Z	Secreted signal	Spemann organizer	Mesoderm dorsalization by antagonizing BMP-4
chordin	Z	Secreted signal	Spemann organizer	Mesoderm dorsalization by antagonizing BMP-4
frizbee	Z	Secreted signal	Spemann organizer	Mesoderm dorsalization by antagonizing Xwnt-8
cerberus	Z	Secreted signal	Spemann organizer	Promotes head development by inhibiting Wnt, Nodal-related, and BMP signaling
goosecoid	Z	Transcription factor	Spemann organizer	Organizer function
Xlim-1	Z	Transcription factor	Spemann organizer	Gastrulation and head formation
Xnot	Z	Transcription factor	Spemann organizer	Notochord specification

embryo's own genes; where it is located in the fertilized egg; and what germ layers it is involved in specifying, whether alone or in combination with other proteins.

6. What is the posterior marginal zone of the chick embryo? What is its significance to chick development? Describe the external forces that lead to its formation and the signaling molecules associated with it.

7. The establishment of left–right asymmetry is associated with higher levels of the TGF-β family signaling molecule Nodal on the left side of the embryo, relative to the right. What are the influences that led to this asymmetry of Nodal concentration? Include positive feedback, Notch signaling, Sonic hedgehog signaling, and the Lefty protein in your answer.

8. A simple view of gastrulation in *Xenopus* would be that the blastula is like a soft ball, and the involution of cells during gastrulation is analogous to poking your finger into the ball at the blastopore, until you can touch the ventral surface. Examine the fate map of the *Xenopus* embryo (Fig. 4.19); what is wrong with this simple view? Provide a more accurate description of the process in *Xenopus* (refer to Fig. 3.6 for help). Where would the mesoderm have to be located in the blastula for the simple invagination method to work?

9. Specification, determination, and differentiation describe different states of development of a cell. Define each of these terms, and describe what experiments are used to distinguish these three different states of development (refer to Sections 1.8, 1.12, and 4.11 in constructing your answer)? Speculate on what sort of changes might have occurred at the molecular level in a cell between specification and determination.

10. Early amphibian development is driven largely by maternally supplied factors until the mid-blastula transition. What is the mid-blastula transition? What is the model that explains how it occurs? What experimental evidence supports this model?

11. 'Dorsal vegetal tissue containing the Nieuwkoop center induces notochord and muscle from animal cap cells, whereas ventral vegetal tissue induces mainly blood-forming tissue and little muscle.' (Section 4.14). What is the significance of these observations? What signaling events are required to explain these observations.

12. An early event in axis formation is the accumulation of the transcription factor β-catenin in the nuclei of cells on one side of the *Xenopus* embryo. Which side of the embryo does this specify? Describe briefly how the actions of β-catenin lead to specification of the Spemann organizer.

13. What is the developmental function of the TGF-β family member BMP-4 in the blastula? Since BMP-4 is initially present throughout the blastula, how are its effects restricted to particular regions (a) of the ectoderm, and (b) of the dorsal mesoderm?

14. The TGF-β signaling pathway is an example of the remarkable conservation of signaling pathways during evolution. Compare the activities of BMP-4 in *Xenopus* and Decapentaplegic in *Drosophila*. Are the antagonists of these signaling molecules in the two systems also homologs?

15. In this chapter, we have seen several examples of the importance of transcriptional control and cell–cell signaling in development.

If you had to choose between transcription factors, signals of the TGF-β family, or signals of the Wnt family for further study into the mechanisms that specify and pattern the mesoderm in vertebrates, which would you choose? Justify your answer.

Multiple choice (factual recall questions)

NB There is only one right answer to each question.

1. The nervous system, heart, and liver are derived from _____, _____, and _____, respectively.

a) all are derived from mesoderm

b) ectoderm, mesoderm, and endoderm

c) endoderm, mesoderm, and ectoderm

d) mesoderm, ectoderm, and endoderm

2. The positioning of the dorso-ventral axis in *Xenopus* is determined by _____, while the positioning of the antero-posterior axis in chicks is determined by _____.

a) gravity; point of sperm entry

b) maternal factors in both cases

c) position in the egg chamber; gravity

d) the point of sperm entry; gravity

3. The 'organizer' in *Xenopus* is responsible for

a) inducing a mesodermal fate in nearby cells

b) initiating involution and gastrulation

c) specifying the dorsal region of the embryo

d) all of these events

4. A key factor that specifies the organizer in *Xenopus* embryos is:

a) GSK-3β

b) Xfz7

c) XWnt-11

d) β-catenin

5. How would you produce a chimeric mouse?

a) Divide a morula into two separate clusters of blastomeres and let each cluster form an embryo.

b) Introduce cells from the inner cell mass of one blastocyst into the inner cell mass of another blastocyst of the same genetic constitution.

c) Divide a fertilized mouse egg into two halves and let each develop separately.

d) Introduce cells from the inner cell mass of one blastocyst into the inner cell mass of another blastocyst of a different genetic constitution.

6. If cells from the animal pole of a frog blastula (animal cap cells) are placed in direct contact with cells from the vegetal region, what is the result?

a) An embryo forms with only ectodermal and endodermal derivatives.

b) Animal cap cells are induced by the vegetal cells to form mesodermal derivatives.

c) The embryo regulates and forms a normal embryo.

d) Vegetal hemisphere cells are induced by the animal cap cells to form mesodermal derivatives.

7. Maternal factors are

a) the energy-rich yolk present in the eggs of many species

b) the genes contributed to the embryo by the maternal haploid genome

c) hormonal influences encountered by the mammalian embryo in the uterus

d) developmentally active mRNAs and proteins packaged into the egg by the mother

8. The activity of the protein Noggin is

a) to antagonize BMP-4, and block ventralization of the mesoderm

b) to signal through the Noggin pathway and establish the organizer

c) to specify a mesodermal fate

d) to turn on genes that determine the anterior-most cells in the embryo, which will form the head

9. Vg-1, Xnrs, BMPs and activin are all members of which family of signaling molecules?

a) FGF

b) Hedgehog

c) TGF-β

d) Wnt

Multiple choice answer key

1: b, 2: d, 3: b, 4: d, 5: d, 6: b, 7: d, 8: a, 9: c, 10: c.

■ Section further reading

4.1 The animal-vegetal axis is maternally determined in *Xenopus* and zebrafish

Heasman, J.: **Patterning the early *Xenopus* embryo**. *Development* 2006, **133**: 1205–1217.

Schier, A.F., Talbot, W.S.: **Molecular genetics of axis formation in zebrafish**. *Annu. Rev. Genet.* 2005, **39**: 561–613.

Weaver, C., Kimelman, D.: **Move it or lose it: axis specification in *Xenopus***. *Development* 2004, **131**: 3491–3499.

4.2 Localized stabilization of the transcriptional regulator β-catenin specifies the future dorsal side and the location of the main embryonic organizer in *Xenopus* and zebrafish

Dosch, R., Wagner, D.S., Mintzer. K.A., Runke, G., Wiemelt, A.P., Mullins, M.C.: **Maternal control of vertebrate development before the midblastula transition: mutants from the zebrafish I**. *Dev Cell* 2004, **6**: 771–780.

Gerhart, J., Danilchik, M., Doniach, T., Roberts, S., Browning, B., Stewart, R.: **Cortical rotation of the *Xenopus* egg: consequences for the antero-posterior pattern of embryonic dorsal development**. *Development* (**Suppl.**) 1989, 37–51.

Heasman, J.: **Maternal determinants of embryonic cell fate**. *Semin. Cell Dev. Biol.* 2006, **17**: 93–98.

Kodjabachian, L., Dawid, I.B., Toyama, R.: **Gastrulation in zebrafish: what mutants teach us**. *Dev. Biol.* 1999, **126**: 5309–5317.

Logan, C.Y., Nusse, R.: **The Wnt signalling pathway in development and disease**. *Annu. Rev. Cell Dev. Biol.* 2004, **20**: 781–801.

Pelegri, F.: **Maternal factors in zebrafish development**. *Dev. Dyn.* 2003, **228**: 535–554.

Sokol, S.Y.: **Wnt signaling and dorso-ventral axis specification in vertebrates**. *Curr. Opin. Genet. Dev.* 1999, **9**: 405–410.

Tao, Q., Yokota, C., Puck, H., Kofron, M., Birsoy, B., Yan, D., Asashima, M., Wylie, C.C., Lin, X., Heasman, J.: **Maternal wnt 11 activates the canonical wnt signaling pathway required for axis formation in *Xenopus* embryos**. *Cell* 2005, **120**: 857–871.

4.3 Signaling centers develop on the dorsal side of *Xenopus* and zebrafish blastulas

Smith, J.: **T-box genes: what they do and how they do it**. *Trends Genet.* 1999, **15**: 154–158.

Vonica, A., Gumbiner, B.M.: **The *Xenopus* Nieuwkoop center and Spemann-Mangold organizer share molecular components and a requirement for maternal Wnt activity**. *Dev. Biol.* 2007, **312**: 90–102.

4.4 The antero-posterior and dorso-ventral axes of the chick blastoderm are related to the primitive streak

Bertocchini, F., Skromne, I., Wolpert, L., Stern, C.D.: **Determination of embryonic polarity in a regulative system: evidence for endogenous inhibitors acting sequentially during primitive streak formation in the chick embryo**. *Development* 2004, **131**: 3381–3390.

Khaner, O., Eyal-Giladi, H.: **The chick's marginal zone and primitive streak formation. I. Coordinative effect of induction and inhibition**. *Dev. Biol.* 1989, **134**: 206–214.

Kochav, S., Eyal-Giladi, H.: **Bilateral symmetry in chick embryo determination by gravity**. *Science* 1971, **171**: 1027–1029.

Seleiro, E.A.P., Connolly, D.J., Cooke, J.: **Early developmental expression and experimental axis determination by the chicken Vg-1 gene**. *Curr. Biol.* 1996, **11**: 1476–1486.

Stern, C.D.: **Cleavage and gastrulation in avian embryos (version 3.0)**. *Encyclopedia of Life Sciences* 2009 http://www.els.net/ (13 May 2010).

4.5 The definitive antero-posterior and dorso-ventral axes of the mouse embryo are not recognizable early in development

Arnold, S.J., Robertson, E.J.: **Making a commitment: cell lineage allocation and axis patterning in the early mouse embryo**. *Nature Mol. Cell Biol. Rev.* 2009, **10**: 91–103.

Beddington, R.S.P., Robertson, E.J.: **Axis development and early asymmetry in mammals**. *Cell* 1999, **96**: 195–209.

Bischoff, M., Parfitt, D.E., Zernicka-Goetz, M.: **Formation of the embryonic-abembryonic axis of the mouse blastocyst: relationships between orientation of early cleavage divisions and pattern of symmetric/asymmetric divisions**. *Development* 2008, **135**: 953–962.

Deb, K., Sivaguru, M., Yul Yong, H., Roberts, M.: **Cdx2 gene expression and trophectoderm lineage specification in mouse embryos**. *Science* 2006, **311**: 992–996.

Hillman, N., Sherman, M.I., Graham, C.: **The effect of spatial arrangement on cell determination during mouse development**. *J. Embryol. Exp. Morph.* 1972, **28**: 263–278.

Perea-Gomez, A., Camus, A., Moreau, A., Grieve, K., Moneron, G., Dubois, A., Cibert, C., Collignon, J.: **Initiation of gastrulation in the mouse embryo is preceded by an apparent shift in the orientation of the antero-posterior axis**. *Curr. Biol.* 2004, **14**: 197–207.

Rivera-Perez, J.A.: **Axial specification in mice: ten years of advances and controversies**. *J. Cell Physiol.* 2007, **213**: 654–660.

Robertson, E.J., Norris, D.P., Brennan, J., Bikoff, E.K.: **Control of early anterior-posterior patterning in the mouse embryo by TGF-beta signalling**. *Phil. Trans. R Soc. Lond. B Biol. Sci.* 2003, **358**: 1351–1357.

Rodriguez, T.A., Srinivas, S., Clements, M.P., Smith, J.C., Beddington, R.S.: **Induction and migration of the anterior visceral endoderm is regulated by the extra-embryonic ectoderm**. *Development* 2005, **132**: 2513–2520.

Rossant, J., Tam, P.P.L.: **Blastocyst lineage formation, early embryonic asymmetries and axis patterning in the mouse**. *Development* 2009, **136**: 701–713.

Srinivas, S., Rodriguez, T., Clements, M., Smith, J.C., Beddington, R.S.P.: **Active cell migration drives the unilateral movements of the anterior visceral endoderm**. *Development* 2004, **131**: 1157–1164.

Takaoka, K., Yamamoto, M., Hamada, H.: **Origin of body axes in the mouse embryo**. *Curr. Opin. Genet. Dev.* 2007, **17**: 344–350.

Zernicka-Goetz, M.: **Developmental cell biology: cleavage pattern and emerging asymmetry of the mouse embryo**. *Nat. Rev. Mol. Cell Biol.* 2005, **6**: 919–928.

4.7 The bilateral symmetry of the early embryo is broken to produce left-right asymmetry of internal organs

Blum, M., Beyer, T., Weber, T., Vivk, P., Andre, P., Bitzer, E., Schweickert, A.: *Xenopus*, **an ideal model system to study vertebrate left-right asymmetry**. *Dev. Dyn.* 2009, **238**: 1215–1225.

Blum, M., Weber, T., Beyer, T., Vick, P.: **Evolution of leftward flow**. *Semin. Cell Dev. Biol.* 2008, **20**: 464–471.

Brennan, J., Norris, D.P., Robertson, E.J.: **Nodal activity in the node governs left-right asymmetry**. *Genes Dev.* 2002, **16**: 2339–2344.

Brown, N.A., Wolpert, L.: **The development of handedness in left/right asymmetry**. *Development* 1990, **109**: 1–9.

Gros, J., Feistel. K., Viebahn, C., Blum, M., Tabin, C.J.: **Cell movements at Hensen's node establish left/right asymmetric gene expression in the chick**. *Science* 2009, **324**: 941–944.

Levin, M., Palmer, A.P.: **Left-right patterning from the inside out: widespread evidence for intracellular control**. *BioEssays* 2007, **29**: 271–287.

McGrath, J., Somlo, S., Makova, S., Tian, X., Brueckner, M.: **Two populations of node monocilia initiate left-right asymmetry in the mouse**. *Cell* 2003, **114**: 61–73.

Rana, A.A., Barbera, J.P., Rodriguez, T.A., Lynch, D., Hirst, E., Smith, J.C., Beddington, R.S.P.: **Targeted deletion of the novel cytoplasmic dynein mD2LIC disrupts the embryonic organiser, formation of body axes and specification of ventral cell fates**. *Development* 2004, **131**: 4999–5007.

Raya, A., Izpisua Belmonte, J.C.: **Unveiling the establishment of left-right asymmetry in the chick embryo**. *Mech. Dev.* 2004, **121**: 1043–1054.

Raya, A., Izpisua Belmonte, J.C.: **Insights into the establishment of left-right asymmetries in vertebrates**. *Birth Defects Res C Embryo Today* 2008, **84**: 81–94.

Schlueter, J., Brand, T.: **Left-right axis development: examples of similar and divergent strategies to generate asymmetric morphogenesis in chick and mouse embryos**. *Cytogenet. Genome Res.* 2007, **117**: 256–267.

Shen, M.M.: **Nodal signaling: developmental roles and regulation**. *Development* 2007, **134**: 1023–1034.

4.8 A fate map of the amphibian blastula is constructed by following the fate of labeled cells

Dale, L., Slack, J.M.W.: **Fate map for the 32 cell stage of *Xenopus laevis***. *Development* 1987, **99**: 527–551.

Gerhart J.: **Changing the axis changes the perspective**. *Dev. Dyn.* 2002, **225**: 380–383.

Lane, M.C., Smith, W.C.: **The origins of primitive blood in *Xenopus*: implications for axial patterning**. *Development* 1999, **126**: 423–434.

4.9 The fate maps of vertebrates are variations on a basic plan

Beddington, R.S.P., Morgenstern, J., Land, H., Hogan, A.: **An *in situ* transgenic enzyme marker for the midgestation mouse embryo and the visualization of inner cell mass clones during early organogenesis**. *Development* 1989, **106**: 37–46.

Gardner, R.L., Rossant, J.: **Investigation of the fate of 4–5 day post-coitum mouse inner cell mass cells by blastocyst injection**. *J. Embryol. Exp. Morph.* 1979, **52**: 141–152.

Kimmel, C.B., Warga, R.M., Schilling, T.F.: **Origin and organization of the zebrafish fate map**. *Development* 1990, **108**: 581–594.

Lawson, K.A., Meneses, J.J., Pedersen, R.A.: **Clonal analysis of epiblast fate during germ layer formation in the mouse embryo**. *Development* 1991, **113**: 891–911.

Smith, J.L., Gesteland, G.M., Schoenwolf, G.C.: **Prospective fate map of the mouse primitive streak at 7.5 days of gestation**. *Dev. Dyn.* 1994, **201**: 279–289.

Stern, C.D.: **The marginal zone and its contribution to the hypoblast and primitive streak of the chick embryo**. *Development* 1990, **109**: 667–682.

Stern, C.D., Canning, D.R.: **Origin of cells giving rise to mesoderm and endoderm in chick embryo**. *Nature* 1990, **343**: 273–275.

4.10 Cells of early vertebrate embryos do not yet have their fates determined and regulation is possible

Lewis, N.E., Rossant, J.: **Mechanism of size regulation in mouse embryo aggregates**. *J. Embryol. Exp. Morph.* 1982, **72**: 169–181.

Snape, A., Wylie, C.C., Smith, J.C., Heasman, J.: **Changes in states of commitment of single animal pole blastomeres of *Xenopus laevis***. *Dev. Biol.* 1987, **119**: 503–510.

Tam, P.P., Rossant, J.: **Mouse embryonic chimeras: tools for studying mammalian development**. *Development* 2003, **130**: 6155–6163.

Wylie, C.C., Snape, A., Heasman, J., Smith, J.C.: **Vegetal pole cells and commitment to form endoderm in *Xenopus laevis*. *Dev. Biol.* 1987, **119**: 496–502.

4.11 In *Xenopus* the endoderm and ectoderm are specified by maternal factors, but the mesoderm is induced from ectoderm by signals from the vegetal region

Dale, L.: **Vertebrate development: multiple phases to endoderm formation.** *Curr. Biol.* 1999, **9**: R812–R815.

Dupont, S., Zacchigna, L., Cordenonsi, M., Soligo, S., Adorno, M., Rugge, M., Piccolo, S.: **Germ-layer specification and control of cell growth by Ectodermin, a Smad4 ubiquitin ligase.** *Cell* 2005, **121**: 87–99.

Mir, A., Kofron, M., Zorn, A.M., Bajzer, M., Haque, M., Heasman, J., Wylie, C.C.: **FoxI1e activates ectoderm formation and controls cell position in the *Xenopus* blastula.** *Development* 2007, **134**: 779–788.

White, J.A., Heasman, J.: **Maternal control of pattern formation in *Xenopus laevis*.** *J. Exp. Zool.* B 2008, **310**: 73–84.

Xanthos, J.B., Kofron, M., Wylie, C., Heasman, J.: **Maternal VegT is the initiator of a molecular network specifying endoderm in *Xenopus laevis*.** *Development* 2001, **128**: 167–180.

4.12 Mesoderm induction occurs during a limited period in the blastula stage

Gurdon, J.B., Lemaire, P., Kato, K.: **Community effects and related phenomena in development.** *Cell* 1993, **75**: 831–834.

4.13 Zygotic gene expression is turned on in *Xenopus* at the mid-blastula transition

Davidson, E.: *Gene Activity In Early Development.* New York: Academic Press, 1986.

O'Boyle, S., Bree, R.T., McLoughlin, S., Grealy, M., Byrnes, L.: **Identification of zygotic genes expressed at the midblastula transition in zebrafish.** *Biochem. Biophys. Res. Commun.* 2007, **358**: 462–468.

Yasuda, G.K., Schübiger, G.: **Temporal regulation in the early embryo: is MBT too good to be true?** *Trends Genet.* 1992, **8**: 124–127.

4.14 Mesoderm-inducing and patterning signals in *Xenopus* are produced by the vegetal region, the organizer, and the ventral mesoderm

Agius, E., Oelgeschläger, M., Wessely, O., Kemp, C., De Robertis, E.M.: **Endodermal Nodal-related signals and mesoderm induction in *Xenopus*.** *Development* 2000, **127**: 1173–1183.

Heasman, J.: **Patterning the early *Xenopus* embryo.** *Development* 2006, **133**: 1205–1217.

Kimelman, D.: **Mesoderm induction: from caps to chips.** *Nat. Rev. Genet.* 2006, **7**: 360–372.

4.15 Members of the TGF-β family have been identified as mesoderm inducers

Amaya, E., Musci, T.J., Kirschner, M.W.: **Expression of a dominant negative mutant of the FGF receptor disrupts mesoderm formation in *Xenopus* embryos.** *Cell* 1991, **66**: 257–270.

Birsoy, B., Kofron, M., Schaible, K., Wylie, C., Heasman, J.: **Vg1 is an essential signaling molecule in *Xenopus* development.** *Development* 2006, **133**: 15–20.

Massagué, J.: **How cells read TGF-β signals.** *Nat Rev. Mol. Cell Biol.* 2000, **1**: 169–178.

Schier, A.F.: **Nodal signaling in vertebrate development.** *Annu. Rev. Cell. Dev. Biol.* 2003, **19**: 589–621.

4.16 The zygotic expression of mesoderm-inducing and patterning signals in *Xenopus* is activated by the combined actions of maternal VegT and Wnt signaling

De Robertis, E.M., Larrain, J., Oelgeschläger, M., Wessely, O.: **The establishment of Spemann's organizer and patterning of the vertebrate embryo.** *Nat. Rev. Genet.* 2000, **1**: 171–181.

Fletcher, R.B., Harland, R.M.: **The role of FGF signaling in the establishment and maintenance of mesodermal gene expression in *Xenopus*.** *Dev. Dyn.* 2008, **237**: 1243–1254.

Harvey, S.A., Smith, J.C.: **Visualisation and quantification of morphogen gradient formation in the zebrafish.** *PLoS Biol.* 2009, **7**: e101.

Kofron, M., Demel, T., Xanthos, J., Lohr, J., Sun, B., Sive, H., Osada, S-I., Wright, C., Wylie, C., Heasman, J.: **Mesoderm induction in *Xenopus* is a zygotic event regulated by maternal VegT via TGFβ growth factors.** *Development* 1999, **126**: 5759–5770.

Niehrs, C.: **Regionally specific induction by the Spemann–Mangold organizer.** *Nat. Rev. Genet.* 2004, **5**: 425–434.

4.17 Signals from the organizer pattern the mesoderm dorso-ventrally by antagonizing the effects of ventral signals

De Robertis, E.M.: **Spemann's organizer and self-regulation in amphibian embryos.** *Nature Mol. Cell Biol. Rev.* 2006, **7**: 296–302.

Gonzalez, E.M., Fekany-Lee, K., Carmany-Rampey, A., Erter, C., Topczewski, J., Wright, C.V., Solnica-Krezel, L.: **Head and trunk in zebrafish arise via coinhibition of BMP signaling by bozozok and chordino.** *Genes Dev.* 2000, **14**: 3087–3092.

Oelgeschläger, M., Larrain, J., Geissert, D., De Robertis, E.M.: **The evolutionarily conserved BMP-binding protein Twisted gastrulation promotes BMP signalling.** *Nature* 2000, **405**: 757–763.

Piccolo, S., Sasai, Y., Lu, B., De Robertis, E.M.: **Dorsoventral patterning in *Xenopus*: inhibition of ventral signals by direct binding of chordin to BMP-4.** *Cell* 1996, **86**: 589–598.

Piepenburg, O., Grimmer, D., Williams, P.H., Smith, J.C.: **Activin redux: specification of mesodermal pattern in *Xenopus* by graded concentrations of endogenous activin B.** *Development* 2004, **131**: 4977–4986.

Schier, A.F.: **Axis formation and patterning in zebrafish.** *Curr. Opin. Genet. Dev.* 2001, **11**: 393–404.

Zimmerman, L.B., De Jesús-Escobar, J.M., Harland, R.M.: **The Spemann organizer signal noggin binds and inactivates bone morphogenetic protein 4.** *Cell* 1996, **86**: 599–606.

4.18 Threshold responses to gradients of signaling proteins are likely to pattern the mesoderm

Chen, Y., Schier, A.F.: **The zebrafish Nodal signal Squint functions as a morphogen.** *Nature* 2001, **411**: 607–609.

Green, J.B.A., New, H.V., Smith, J.C.: **Responses of embryonic *Xenopus* cells to activin and FGF are separated by multiple dose thresholds and correspond to distinct axes of the mesoderm**. *Cell* 1992, **71**: 731–739.

Gurdon, J.B., Standley, H., Dyson, S., Butler, K., Langon, T., Ryan, K., Stennard, F., Shimizu, K., Zorn, A.: **Single cells can sense their position in a morphogen gradient**. *Development* 1999, **126**: 5309–5317.

Jones, C.M., Armes, N., Smith, J.C.: **Signaling by TGF-β family members: short-range effects of Xnr-2 and BMP-4 contrast with the long-range effects of activin**. *Curr. Biol.* 1996, **6**: 1468–1475.

Papin, C., Smith, J.C.: **Gradual refinement of activin-induced thresholds requires protein synthesis**. *Dev. Biol.* 2000, **217**: 166–172.

Schulte-Merker, S., Smith, J.C.: **Mesoderm formation in response to *Brachyury* requires FGF signalling**. *Curr. Biol.* 1995, **5**: 62–67.

Box 4E A zebrafish gene regulatory network

Morley, R.H., Lachanib, K., Keefe, D., Gilchrist, M.J., Flicek, P., Smith, J.C., Wardle, F.C.: **A gene regulatory network directed by zebrafish No tail accounts for its roles in mesoderm formation**. *Proc. Natl Acad. Sci. USA* 2009, **106**: 3829–3834.

4.19 Mesoderm induction and patterning in the chick and mouse occurs during primitive-streak formation

Chapman, S.C., Matsumoto, K., Cai, Q., Schoenwolf, G.C.: **Specification of germ layer identity in the chick gastrula**. *BMC Dev. Biol.* 2007, **7**: 91.

Chu, G.C., Dunn, N.R., Anderson, D.C., Oxburgh, L., Robertson, E.J.: **Differential requirements for Smad4 in TGFbeta-dependent patterning of the early mouse embryo**. *Development* 2004, **131**: 3501–3512.

Dunn, N.R., Vincent, S.D., Oxburgh, L., Robertson, E.J., Bikoff, E.K.: **Combinatorial activities of Smad2 and Smad3 regulate mesoderm formation and patterning in the mouse embryo**. *Development* 2004, **131**: 1717–1728.

Lu, C.C., Robertson, E.J.: **Multiple roles for Nodal in the epiblast of the mouse embryo in the establishment of anterior-posterior patterning**. *Dev. Biol.* 2004, **273**: 149–159.

Vincent, S.D., Dunn, N.R., Hayashi, S., Norris, D.P., Robertson, E.J.: **Cell fate decisions within the mouse organizer are governed by graded Nodal signals**. *Genes Dev.* 2003, **17**: 1646–1662.

Zakin, L., Reversade, B., Kuroda, H., Lyons, K.M., De Robertis, E.M.: **Sirenomelia in Bmp7 and Tsg compound mutant mice: requirement for Bmp signaling in the development of ventral posterior mesoderm**. *Development* 2005, **132**: 2489–2499.

Vertebrate development III: patterning the early nervous system and the somites

- ■ The role of the organizer and neural induction
- ■ Somite formation and patterning
- ■ Initial antero-posterior patterning of the brain

During and after gastrulation, the vertebrate embryo becomes further patterned along the antero-posterior and dorso-ventral axes. This patterning is carried out by a combination of signals from various regions of the embryo and the interpretation of positional identity by cells. A key developmental event that accompanies gastrulation is the induction of the nervous system from the dorsal ectoderm by a complex cascade of signals from adjacent tissues, including the organizer. At around the same time, the main antero-posterior organization of the rest of the body starts to become evident, with the segmentation of mesoderm along the antero-posterior axis into blocks of tissue called somites—which give rise to the spinal column, ribs, and the skeletal muscles of the trunk. The expression of genes involved in encoding position along the antero-posterior axis is central to this patterning, as is the mechanism by which their spatial patterns of expression are initially specified.

In Chapter 4, we saw how the body axes are set up and how the three germ layers are initially specified in vertebrate embryos. Although amphibian, fish, chick, and mouse embryos share some features at these stages, there are some significant differences. As we approach the phylotypic stage—the embryonic stage common to all vertebrates (see Fig. 3.2)—the similarities between vertebrate embryos become greater.

During gastrulation, the germ layers—mesoderm, endoderm, and ectoderm—move to the positions in which they will develop into the structures of the larval or adult body. The antero-posterior body axis of the vertebrate embryo emerges clearly, with the head at one end and the future tail at the other (Fig. 5.1). In this chapter, we focus mainly on the patterning of ectoderm that will develop into the nervous system, and on the formation and patterning of the mesoderm-derived **somites**, blocks of tissue that give rise to the skeleton and skeletal muscles of the trunk, and to some of the dermis of the skin. The cell movements of gastrulation and the action of the organizer region are crucial to establishing the vertebrate body plan and will be discussed in this chapter in relation to their role in patterning. A detailed discussion of the mechanisms underlying the movements of cells and tissues during gastrulation is deferred to Chapter 8.

Fig. 5.1 The final body plan of the embryo emerges during gastrulation and neurulation. Left, fate map of a late *Xenopus* blastula. Right, a schematic of a sagittal section through a tailbud stage *Xenopus* embryo after gastrulation and neurulation. The mesoderm and endoderm have moved inside. The mesoderm gives rise to the prechordal plate (brown), notochord (red), somites (orange), and lateral mesoderm (not shown in this view). The endoderm (yellow) moves inside to line the gut. The neural tube (dark blue) forms from dorsal ectoderm and the ectoderm that will form the epidermis (light blue) covers the whole embryo. The antero-posterior axis has emerged, with the head at the anterior end.

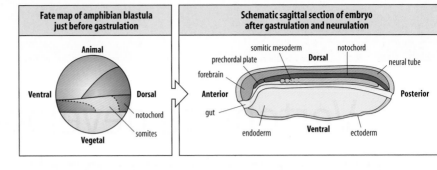

In *Xenopus*, the dorsal mesoderm (the organizer region) in the late blastula is internalized at gastrulation to form the prechordal plate mesoderm, which lies anterior to the notochord and gives rise to the ventral head mesoderm, and the rigid rod-like **notochord** that runs along the dorsal midline. The notochord is flanked immediately on each side by more ventral mesoderm, which by the end of gastrulation is beginning to segment to form blocks of somites, starting at the anterior end (see Fig. 5.1). The vertebrate notochord is a transient structure, and its cells eventually become incorporated into the vertebrae and intervening discs of the spinal column. As gastrulation proceeds, **neurulation**, the first stage of nervous system formation, begins. The ectoderm overlying the notochord rolls upwards on both sides of the dorsal midline to form the **neural folds**, which meet over the midline to form a tubular structure, the **neural tube** (see Figs. 3.7). The internal structure of the *Xenopus* embryo just after the end of neurulation is illustrated in Fig. 5.2. The main structures that can be recognized at this stage are the neural tube, the notochord, the somites, the lateral plate mesoderm, which lies ventral to the somites on either side, and the endoderm lining the gut.

Both the mesodermal structures along the antero-posterior axis of the trunk and the ectodermally derived nervous system have a distinct antero-posterior organization. An initial antero-posterior patterning of the germ layers occurs during mesoderm induction, as mesoderm becomes differentiated from endoderm and ectoderm, while some important aspects of later antero-posterior patterning in both the mesoderm and the nervous system are controlled by the Hox genes, which are the vertebrate equivalent of the Hox genes responsible for antero-posterior patterning in *Drosophila* (see Chapter 2).

Fig. 5.2 A cross-section through a stage 22 *Xenopus* embryo just after gastrulation and neurulation are completed. The germ layers are now all in place for future development and organogenesis. The most dorsal parts of the somites have already begun to differentiate into the dermomyotome, which will give rise to the trunk and limb muscles and the dermis, as described later in the chapter. Scale bar = 0.2 mm.

Photograph from Hausen, P. and Riebesell, M.: 1991.

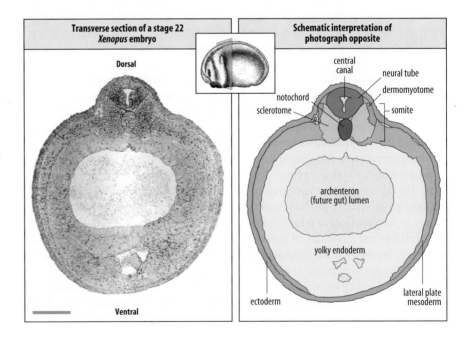

In the first part of this chapter we consider the function of the embryonic organizer in establishing the basic antero-posterior organization of vertebrate embryos, focusing on its role in the induction of the nervous system. In the second part of the chapter we describe the development of the somites and discuss the role of the Hox genes in the antero-posterior patterning of the somitic mesoderm, as shown by their effects on the somite-derived vertebrae of the spinal column. Finally, we will look at the antero-posterior patterning of the hindbrain, in which Hox genes are again involved.

The role of the organizer and neural induction

We shall first consider the role of the organizer region, which is crucially important both in neural induction and in organizing the antero-posterior axis. The Spemann organizer of amphibians, the shield in zebrafish, Hensen's node in the chick, and the equivalent node region in the mouse all have a similar global organizing function in vertebrate development. They can induce a complete body axis if transplanted to another embryo at an appropriate stage (see Chapter 4), and so are able to organize and coordinate both dorso-ventral and antero-posterior aspects of the body plan, as well as induce neural tissue from ectoderm. In the mouse, the anterior visceral endoderm is required in addition to the node for the induction of structures anterior to the end of the notochord, such as the head and forebrain (see Section 4.6).

During gastrulation, the ectoderm in the dorsal region of the embryo becomes specified as neuroectoderm, the **neural plate** (see Fig. 3.7). During the subsequent stage of neurulation, the neural plate forms the neural tube, which eventually differentiates into the central nervous system—the brain and the spinal cord—and the peripheral nervous system. The neural tube throws off neural crest cells, which migrate throughout the body to give rise to the sensory and autonomic parts of the peripheral nervous system. It also gives rise to some non-neural structures, most notably some of the tissues of the face and jaws, such as bone and cartilage, tissues that in other parts of the body are derived from mesoderm. The contribution of the neural crest cells to these tissues in the head is a rare example of cells from one germ layer giving rise to tissues characteristic of another. That particular example will be discussed later in this chapter, while the enormous differentiation potential of neural crest cells generally is discussed in Chapter 10, and details of their migration routes in the rest of the body are considered in Chapter 8.

The nervous system must develop in the correct relationship with other body structures, particularly the mesodermally derived structures that give rise to the skeleto-muscular system. Thus, patterning of the nervous system must be linked to that of the mesoderm, and this is coordinated through the organizer, which is involved in both. In this part of the chapter, we consider the induction of the neuroectoderm and formation of the neural tube, along with the specification of the prospective neural crest cells. The antero-posterior patterning of the brain and neural crest will be discussed in the last part of the chapter.

The function of the organizer has been best studied in amphibians, and we have already described its role in the dorso-ventral patterning of the mesoderm in *Xenopus* in Chapter 4. We now discuss its profound effects on the antero-posterior axis. The function of the equivalent region in the chick, Hensen's node, has also been extensively studied because chick embryos are particularly amenable to surgical procedures, such as transplantation (see Section 3.6).

5.1 The inductive capacity of the organizer changes during gastrulation

In amphibians, the action of the Spemann organizer is dramatically demonstrated in what is classically known as **primary embryonic induction**. The organizer is located

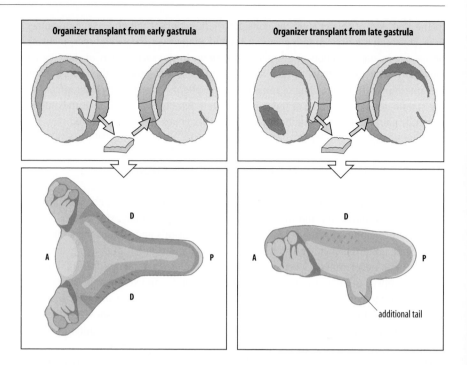

Fig. 5.3 The inductive properties of the organizer change during gastrulation. A graft of the organizer region, from the dorsal lip of the blastopore of an early frog gastrula to the ventral side of another early gastrula, results in the development of an additional anterior axis at the site of the graft (left panels). A graft from the dorsal lip region of a late gastrula to an early gastrula only induces formation of tail structures (right panels). A = anterior; P = posterior; D = dorsal; V = ventral.

in the dorsal lip of the blastopore and if the blastopore dorsal lip from a very early gastrula is grafted to the ventral side of the marginal zone of another early gastrula, it can induce a complete second embryo with a well defined head, a central nervous system, a trunk region, and a tail (Fig. 5.3, left panels, and see Fig. 4.32). This organizer function is often called the 'head organizer.' Various other treatments, such as grafting dorsal vegetal blastomeres containing the Nieuwkoop center to the ventral side, produce a similar result (see Fig. 4.6), but what all these treatments have in common is that, directly or indirectly, they result in the formation of a new Spemann-organizer region. Classic experiments also showed that a graft of the blastopore dorsal lip from a mid-gastrula to an early gastrula induces a trunk and tail but no head; this organizer function is often called the 'trunk organizer'. A grafted dorsal lip from a late gastrula induces only a tail (Fig. 5.3, right panels). These results were interpreted to mean that as gastrulation proceeds, the antero-posterior axis of the embryo becomes specified, and so the cells that make up the organizer at later stages of gastrulation induce only posterior structures. We now also know that both the quantity and the nature of the inducing signals produced by the organizer change as gastrulation proceeds.

The cells present at the dorsal lip of the blastopore in a *Xenopus* early gastrula themselves give rise during gastrulation to anterior endoderm, prechordal plate mesoderm, and the notochord (Fig. 5.4). In addition to providing cells for these axial structures, the organizer region has patterning and inductive properties: it helps to pattern the adjacent more ventral mesoderm, as we saw in Chapter 4, and it induces the neural plate in the adjacent dorsal ectoderm and gives it an initial antero-posterior pattern. The organizer of the early gastrula is therefore a complex signaling center with different parts expressing different genes, having different inductive capacities, and giving rise to different structures. Because of these complex properties, understanding how the organizer region controls the overall pattern of the antero-posterior axis is not straightforward. As gastrulation proceeds and cells move inwards, the cellular composition of the dorsal lip changes, and at one level this explains the different inductive properties displayed by the organizer over time in the experiments described above (see Fig. 5.3). For example, in the early *Xenopus* gastrula, the vegetal portion of the organizer that is fated to produce prechordal plate mesoderm (shown in brown in Fig. 5.4) expresses proteins, such as the transcription factor XOtx2, that are characteristic of

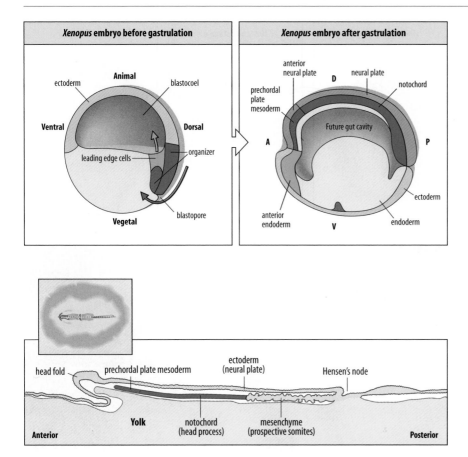

Fig. 5.4 Different parts of the *Xenopus* organizer region give rise to different tissues. In the early gastrula (left panel) the organizer is located in the dorsal lip of the blastopore. The cells of the leading edge (orange) are the first to internalize and give rise to anterior endoderm of the neurula stage (right panel). The deep cells (brown) are next to internalize and give rise to the prechordal plate, the mesoderm anterior to the notochord, which forms the ventral head mesoderm. The remainder of the organizer gives rise to notochord (red).

Adapted from Kiecker, C. and Niehrs, C: 2001.

Fig. 5.5 Notochord and head-fold formation in the chick embryo. This schematic diagram shows a sagittal section through the chick embryo (inset, dorsal view) at the stage of head-fold formation as Hensen's node starts to regress. As the node regresses, the notochord (sometimes called the head process at this stage) starts to form anterior to it with the prechordal plate mesoderm immediately anterior to the notochord. The undifferentiated mesenchyme on either side of the notochord will form somites. In the main picture the anterior somitic mesoderm (in which one somite has formed, see inset) has been cut away to show the notochord.

anterior structures. Experiments investigating the inductive capacity of different parts of the organizer show that the ability to induce heads—'head organizer' function—is also restricted to this vegetal region. The more dorsal part of the organizer (shown in red in Fig. 5.4) is characterized by expression of the transcription factor Xnot, and can induce trunk and tail structures but not heads.

The avian equivalent of the Spemann organizer is Hensen's node, the region at the anterior end of the primitive streak in the chick blastoderm. The node contributes cells to prechordal plate mesoderm, notochord, somites, and gut endoderm, as well as producing inducing signals. In the chick embryo, the notochord forms in the dorsal midline anterior to Hensen's node, mostly from the axial mesoderm left behind as the node and primitive streak regress (Fig. 5.5). The neural plate develops above it and somites are formed from mesoderm on either side of the notochord. Notochord and somite formation and neurulation proceed similarly in the mouse (see Figs 3.25 and 3.26).

The properties of the avian node have been investigated by transplanting quail nodes to chick embryos. Quail cells can be distinguished from chick cells by their distinctive nuclei, which can be detected in histological sections (Fig. 5.6) or by using a species-specific antibody. When, for example, a node from a gastrula-stage quail embryo is grafted beneath the lateral epiblast of a chick embryo at the same stage of development, the transplanted node can induce the formation of a complete additional axis with a head. When the node is taken from an older embryo at the head-process-stage, for example, then it only induces a trunk and not a head (Fig. 5.7). Thus the outcome of grafting nodes at different stages is similar to that seen with organizer grafts of different stages in *Xenopus.* In the mouse, the node precursors can induce a similar axis duplication on transplantation to the lateral epiblast of an early embryo, with the exception of the forebrain, which requires additional signals from the anterior visceral endoderm (see Fig. 4.14).

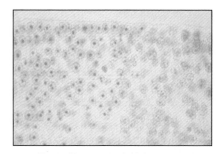

Fig. 5.6 Photograph of quail-chick chimeric tissue. The quail cells are on the left and the chick cells are on the right. Note the heavily staining nuclei of the quail cells.

Photograph courtesy of Nicole Le Douarin.

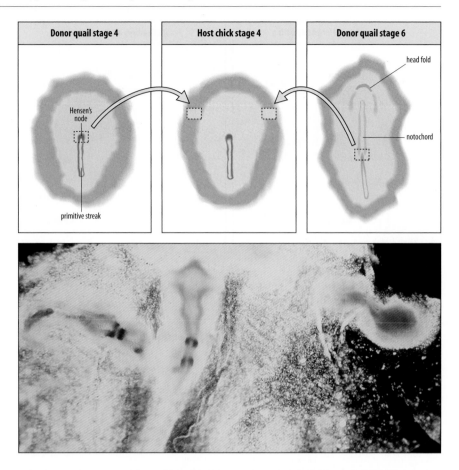

Fig. 5.7 Hensen's node can induce a new axis in avian embryos. When Hensen's node from a stage-4 quail embryo (left) is grafted to the extra-embryonic area opaca (dark gray), just outside the embryonic area pellucida (light gray), of a host chick embryo at the same stage of development (center), a complete new axis, including brain tissue, forms at the site of transplantation (photo, embryo on the left). The embryo in the center of the photograph has formed from the original primitive streak of the host chick embryo. The two stripes of dark purple staining in the left and center embryos indicate expression of *Krox20* mRNA, a marker for the hindbrain, as detected by *in situ* hybridization. Quail cells are distinguished from chick cells by staining reddish brown with anti-quail antibody. Although some of the tissue of the new axis (embryo on left) is formed from the quail graft, most has been induced from chick tissue that does not normally form an embryo. A node from a quail embryo at the head-process stage of development (stage 6, right) grafted to a stage-4 chick embryo produces a new axis with only a trunk (photo, right embryo) and most of the new tissue is derived from the quail graft, as indicated by the reddish-brown staining throughout. *Photograph from Stern, C.D.: 2005.*

	Xenopus gastrula	Mouse gastrula
	mesoderm / organizer	organizer
Genes in organizer region		
Genes encoding transcription factors	Brachyury	Brachyury
	goosecoid	goosecoid
	Xlim1	Lim1
Genes encoding secreted proteins	Xnr3	Nodal
	chordin, noggin	chordin, noggin
	Cerberus	Cerberus-related

Fig. 5.8 Genes expressed in the Spemann organizer region of the *Xenopus* gastrula, and in the node in the mouse gastrula. There is a similar pattern of gene activity in the two animals, with homologous genes being expressed. The expression of some of these genes, such as *Brachyury*, is not confined to the organizer.

A number of proteins are specifically expressed in the vertebrate organizer and are known to be required for its function. Many are common to all vertebrates (Fig. 5.8). Goosecoid, for example, is an early and reliable marker of the organizer that is expressed, in *Xenopus*, in the cells that will give rise to foregut, prechordal plate, and notochord, and which have internalized by the mid-gastrula stage. Although *goosecoid* expression is required for head formation in the normal course of development, *goosecoid* mRNA injected into ventral blastomeres of *Xenopus* embryos induces a secondary axis lacking a head. It is important to bear in mind, however, that although many of the same proteins are produced in and around the organizer in different vertebrates, they do not all appear to have precisely the same functions in all our model animals.

We saw in Chapter 4 that gradients of signaling by TGF-β family proteins (Nodal and BMPs) pattern the dorso-ventral axis of the *Xenopus* gastrula (see Sections 4.17

and 4.18). Now we see that a gradient of Wnt/β-catenin signaling patterns the antero-posterior axis. So any account of early vertebrate development needs to incorporate these orthogonal TGF-β gradients and Wnt gradients into a three-dimensional coordinate system of positional information that functions during axial patterning in the gastrula (Fig. 5.9). The essential ingredients for head formation in *Xenopus* are the inhibition in anterior tissues of Nodal, BMP and Wnt signals, which are present in the gastrula at this time (see Section 4.17): an extra head can be induced by the simultaneous inhibition of BMP and Wnt signaling, or of Nodal and BMP signaling in the ventral region of an early gastrula. As we saw in Chapter 4, antagonists of BMPs and Wnts are secreted by the *Xenopus* organizer: the proteins Chordin, Noggin, and Follistatin can antagonize BMP, and Dickkopf1antagonizes Wnt. The protein Cerberus, which is produced by the anterior endoderm, antagonizes Wnt, Nodal, and BMP signaling (see Chapter 4 summary figure, p. 167). The head will form where both BMP and Wnt signaling are low, whereas the tail forms where both are high. The important function of the Spemann organizer is to regulate these gradients by secreting growth-factor antagonists. Morphogen gradients at right angles to each other also operate to pattern the *Drosophila* wing (see Chapter 11), and represent an important way of patterning an embryonic tissue.

5.2 The neural plate is induced in the ectoderm

The induction of neural tissue from ectoderm was first indicated by the organizer-transplant experiment in frogs illustrated in Fig. 5.3; in the secondary embryo that forms at the site of transplantation, a nervous system develops from the host ectoderm that would normally have formed ventral epidermis. This suggested that neural tissue could be induced from as-yet-unspecified ectoderm by signals emanating from the organizer mesoderm. The requirement for induction was confirmed by experiments that replaced prospective neural plate ectoderm with prospective epidermis before gastrulation; the transplanted prospective epidermis developed into neural tissue (Fig. 5.10). This showed that the formation of the nervous system is dependent on an inductive signal.

In the chick embryo, neural tissue can similarly be induced in the epiblast—in both the area pellucida and the area opaca—by grafts from the primitive-streak mesoderm (see Fig. 5.7). Inducing activity is initially located in the anterior primitive streak and Hensen's node. Later, during regression of the node, inducing activity becomes

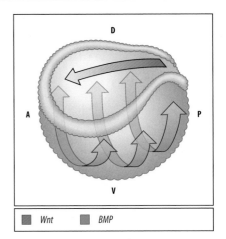

Fig. 5.9 Double-gradient model of embryonic axis formation. The model shows how gradients of Wnts (blue) and BMPs (green) perpendicular to each other regulate antero-posterior and dorso-ventral patterning in the amphibian embryo. The color scales of the arrows indicate the signaling gradients; arrows indicate the direction of spreading of the signals. For example, the formation of the tail requires a higher level of Wnt signaling than does the head. Patterning begins at gastrula stages, but for clarity is depicted here in an early neurula.
Adapted from Niehrs, C.: 2004.

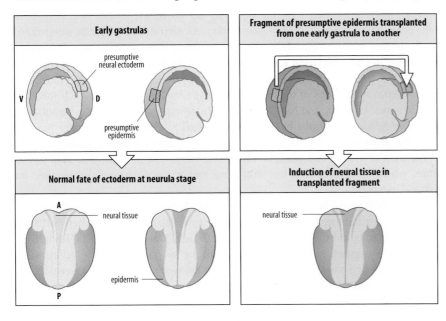

Fig. 5.10 The nervous system of *Xenopus* is induced during gastrulation. The left panels show the normal developmental fate of ectoderm at two different positions in the early gastrula. The right panels show the transplantation of a piece of ventral ectoderm, whose normal fate is to form epidermis, from the ventral side of an early gastrula to the dorsal side of another, where it replaces a piece of dorsal ectoderm whose normal fate is to form neural tissue. In its new location, the transplanted prospective epidermis develops not as epidermis but as neural tissue, and forms part of a normal nervous system. This shows that the ventral tissue has not yet been determined at the time of transplantation, and that neural tissue is induced during gastrulation.

much reduced and by the four-somite stage has completely disappeared. The competence of the ectoderm to respond to neural-inducing signals disappears at the time of head-process formation, suggesting that in normal development neural induction has ended by this time.

An enormous amount of effort was devoted in the 1930s and 1940s to trying to identify the signals involved in neural induction in amphibians. Researchers were encouraged by the finding that a dead organizer region could still induce neural tissue. It seemed to be merely a matter of hard work to isolate the chemicals responsible. The search proved fruitless, however, for it appeared that an enormous range of different substances were capable of varying degrees of neural induction. As it turned out, this was because salamander ectoderm, the main experimental material used, has a high propensity to develop into neural tissue on its own. This is not the case with *Xenopus* ectoderm, although simply dissociating *Xenopus* ectodermal cells for a few hours can result in their differentiation as neural cells after reaggregation. The molecules responsible for neural induction have still not been definitively identified, although there are now some strong candidates.

A key discovery in the study of neural induction was the finding that the BMP inhibitor Noggin, the first secreted protein isolated from the Spemann's organizer, could induce neural differentiation in ectoderm explants from *Xenopus* embryos. In the late *Xenopus* blastula, BMPs are expressed throughout the ectoderm (see Fig. 4.36), but their expression is subsequently lost in the neural plate. These results suggested that neural plate could only develop if BMP signaling is absent. BMP inhibitors such as Noggin and Chordin produced by the organizer therefore became attractive candidates for inhibiting BMP signaling in the prospective neural plate and allowing neural induction to proceed. As BMP signaling maintains expression of the BMP genes, antagonizing BMP signaling would also shut down BMP expression.

These observations led to the so-called 'default model' for neural induction in *Xenopus*. This proposed that the default state of the dorsal ectoderm is to develop as neural tissue, but that this pathway is blocked by the presence of BMPs, which promote the epidermal fate. The role of the organizer is to lift this block by producing proteins that inhibit BMP activity; the region of ectoderm that comes under the influence of the organizer will then develop as neural ectoderm. The organizer produces several antagonists including the BMP antagonists Chordin and Noggin, the protein Cerberus, which can antagonize BMP, Nodal, and Wnt signals, and the Wnt antagonist, Dickkopf1 (see Fig. 5.8, and Chapter 4 summary figure, p. 167). According to this model, the antagonists secreted from the organizer act on ectoderm that lies adjacent to the organizer at the beginning of gastrulation. As gastrulation proceeds, internalized cells derived from the organizer continue to secrete these proteins, which continue to act on the now-overlying ectoderm (see Fig. 5.4). Elimination of these organizer antagonists individually in amphibians has rather modest effects on neural induction. But when combinations of the BMP antagonists Chordin, Noggin, and Follistatin, or of Cerberus, Chordin, and Noggin were simultaneously depleted in the organizer of the frog *Xenopus tropicalis* by antisense morpholino oligonucleotides (see Box 6A, p. 220), there was a dramatic failure of neural and other dorsal development and an expansion of ventral and posterior fates. These experiments show that BMP inhibitors, which are produced at the right place and at the right time, are required for neural induction.

The default model is not the complete answer, however, as neural development in both *Xenopus* and the chick also requires the growth factor FGF, even when BMP inhibition is lifted by the presence of Noggin and Chordin. Precisely how FGF fits into the picture in *Xenopus* is still much debated, with the most recent work suggesting that BMP inhibition is most important for inducing anterior neural tissue, whereas FGF is required in addition to the inhibition of BMPs for giving a posterior character to the prospective neuroectoderm (Fig. 5.11).

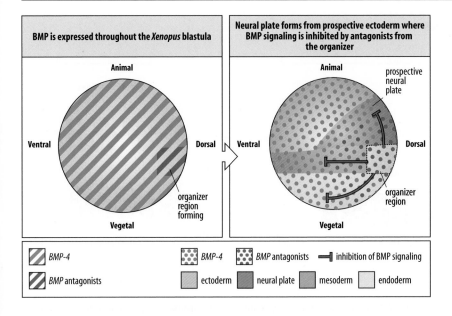

BMP is expressed throughout the *Xenopus* blastula	Neural plate forms from prospective ectoderm where BMP signaling is inhibited by antagonists from the organizer

Fig. 5.11 Inhibition of BMP signaling is required for induction of neural plate. BMP (green) is expressed throughout the early blastula. In the late blastula/early gastrula, BMP antagonists secreted from the organizer inhibit BMP signaling in adjacent regions. This negative action is not on its own enough to induce formation of neural tissue, which also requires a positive input from FGF signaling (not shown).

In the chick, the earliest steps in neural induction occur before gastrulation in the blastoderm, where FGF signaling is also involved in mesoderm induction (see Section 4.19). Blastoderm cells are thought to be prevented from taking on a neural-plate fate at that time because the chromosomal regions encoding neural-inducing genes are at that time in a 'silenced' state in which genes cannot be switched on and transcribed. Experiments in chick embryos suggest that the lifting of this inhibition during gastrulation is associated with the removal of repressor proteins from the regulatory regions of such genes and the remodeling of the chromatin into an active state by chromatin-remodeling complexes (Box 5A, p. 182). Blastoderm cells that receive FGF signals but have not formed mesoderm by this stage can now form neural plate.

One of the key genes activated by FGF signaling in neural induction in chick embryos is *Churchill*, which encodes a zinc finger transcription factor. When *Churchill* expression is downregulated using an antisense morpholino oligonucleotide (see Box 6A, p. 220), the neural plate does not form. The activation of *Churchill* as a result of FGF signaling leads indirectly to repression of genes characteristic of mesoderm and activation of the gene for the neural-specific transcription factor Sox2, the earliest definitive marker of neural plate in the chick.

Not all the effects of FGF in neural induction in chick or *Xenopus* embryos may result from a direct inducing effect. FGF activates the mitogen-activated protein kinase (MAPK) intracellular signaling pathway (Box 5B, p. 184) and studies in both *Xenopus* and mouse provide evidence that MAPK can interfere with BMP signaling by causing an inhibitory phosphorylation of Smad1, which is part of the intracellular signaling pathway stimulated by BMPs (see Fig. 4.33). Therefore, FGF signaling could also contribute to the inhibition of BMP signaling, and could help to allow neural induction this way (see Fig. 5.11). The MAPK signaling pathway is also activated when *Xenopus* ectodermal cells are dissociated, which might explain, as noted earlier, why this dissociation so easily induces a neural fate.

Neural induction is therefore a complex multistep process and, as discussed earlier, the very first stages are likely to occur in the blastula or blastoderm, even before a distinct organizer region becomes detectable. The individual signaling proteins have multiple roles in development, and their roles change over time, which makes it difficult to disentangle their real contributions to any particular process. Their developmental roles also differ between different vertebrates. An essential similarity in the mechanism of neural induction among vertebrates is likely, however, as Hensen's node from a chick embryo can induce neural gene expression in *Xenopus* ectoderm

Box 5A Chromatin-remodeling complexes

Whether a gene is expressed or not depends not only on whether appropriate gene-specific DNA-binding regulatory proteins are present, but also on the state of the **chromatin**—the complex of DNA and proteins of which the chromosomes are composed. Chromatin proteins include the histones, which package DNA into structures called **nucleosomes**, and other proteins that bind directly to DNA or to proteins bound to DNA, and affect chromatin structure and the availability of the DNA for tran-

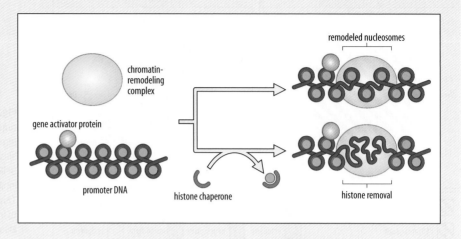

scription. In highly simplified terms, **euchromatin** is chromatin in which genes are available for transcription whereas in **heterochromatin** they are unavailable—or 'silenced'.

In the case of the chick embryo, the gene *Sox2* is an early marker for neural plate but is not expressed before gastrulation even though FGF, which is involved in neural-plate induction, is active early on. Experiments investigating the regulation of *Sox2* in cultured chick embryos have shown that, like many other genes, its expression requires the action of a chromatin-remodeling protein called Brm. Brm has ATPase activity and is an essential component of numerous multiprotein ATP-dependent **chromatin-remodeling complexes** that are active in cells. Such complexes can be recruited to promoter sites in the DNA (see Fig. 1.19) and they are thought to use the energy of ATP hydrolysis to loosen the attachment of DNA to histones and thus enable nucleosomes to be slid along DNA (see figure above, upper arrow) or the histone core to be removed from the DNA (see figure, lower arrow). This frees the promoter DNA to bind other gene regulatory proteins, RNA polymerase, and the rest of the transcriptional machinery, thus enabling the gene to be transcribed.

Chromatin-remodeling complexes work in conjunction with other gene regulatory proteins. To determine the effects of candidate regulatory proteins on *Sox2* expression, DNAs encoding

active or inactive forms of the proteins, in different combinations, were transfected into the area opaca of early chick epiblasts, an area outside the embryo that would never normally express *Sox2*, to see whether they could induce *Sox2* expression. From the results of these experiments, a series of steps leading from inactive *Sox2* in the early embryo to active *Sox2* in the neural plate could be proposed. The early exposure to FGF is proposed to prime *Sox2* for eventual expression by inducing an activator protein that binds to the *Sox2* control region along with Brm, but the gene is kept inactive by the binding of repressive heterochromatin-inducing proteins, which are ubiquitously expressed throughout the embryo. The inhibition is lifted by the specifically timed expression of a protein that disrupts the complex, removing the repressors and enabling the activator protein, in conjunction with the now active chromatin-remodeling complex, to switch on gene expression (see figure below). This general strategy for preventing the premature expression of a crucial gene in the presence of potential activating signals is likely to be widely deployed in development.

The recruitment of chromatin-modeling complexes and heterochromatin-inducing proteins to specific sites on the chromosome often depends on pre-existing chemical modifications to chromatin such as DNA methylation and histone methylation and acetylation, which are discussed in Box 10A, p. 374.

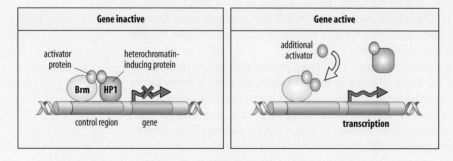

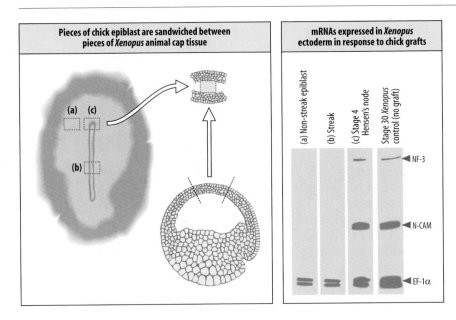

Fig. 5.12 Hensen's node from a chick embryo can induce gene expression characteristic of neural tissue in *Xenopus* ectoderm. Tissues from different parts of the primitive-streak stage of a chick epiblast are placed between two fragments of animal cap tissue (prospective ectoderm) from a *Xenopus* blastula. The induction in the *Xenopus* ectoderm of genes that characterize the nervous system is detected by looking for expression of mRNAs for neural cell adhesion molecule (N-CAM) and neurogenic factor-3 (NF-3), which are expressed specifically in neural tissue in stage-30 *Xenopus* embryos. Only transplants from Hensen's node induce the expression of these neural markers in the *Xenopus* ectoderm. (EF-1α is a ubiquitous transcription factor expressed in all cells and is used as a control.)

After Kintner, C.R., et al.: 1991.

(Fig. 5.12), which suggests that there has been an evolutionary conservation of inducing signals. Moreover, early nodes induce gene expression characteristic of anterior amphibian neural structures, whereas older nodes induce expression typical of posterior structures. These results are in line with the theory that the vertebrate node specifies different antero-posterior positional values at different times, and they confirm the essential similarity of Hensen's node and the Spemann organizer.

In *Xenopus*, the elongation of the posterior neural plate is part of the general elongation of the embryo along the antero-posterior axis and contributes to the generation of the spinal cord (see Fig. 3.7). This elongation process is discussed in more detail in Chapter 8. In chick and mouse, on the other hand, the whole spinal cord is generated from a small region of proliferating stem cells that develops in the ectoderm at the posterior end of the neural plate on both sides of the node and primitive streak (Fig. 5.13). This stem zone becomes distinct just before somite formation begins and moves posteriorly along with the node, leaving behind neural progenitor cells that produce the spinal cord.

In zebrafish, there appears to be a clear difference in the mechanism of induction of anterior neural tissue close to the shield in the dorsal region, and the induction of the posterior neural ectoderm that will give rise to the spinal cord. Whereas the organizer

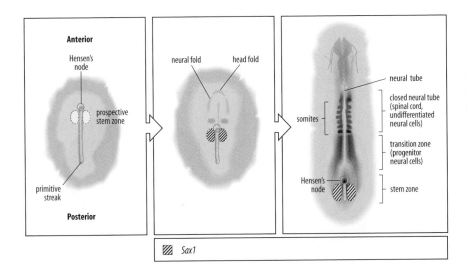

Fig. 5.13 The spinal cord is formed from a zone of stem cells that arises in the posterior neural plate. In the chick embryo, a restricted region of stem cells forms in the posterior neural plate alongside the node and anterior primitive streak just before the onset of somitogenesis (yellow). This stem-cell region comes to express Sax1, a gene encoding a homeodomain-containing protein, and moves posteriorly with the node as it regresses, leaving behind progenitor neural cells that will form the spinal cord. The complete spinal cord is formed from this stem-cell zone.

Adapted from Delfino-Machin, M., et al.: 2005.

Box 5B The FGF signaling pathway

The developmentally important growth factors FGF and EGF (epidermal growth factor), along with numerous other growth and differentiation factors, signal through transmembrane receptors with intracellular tyrosine kinase domains. Several different signaling pathways lead from these receptors, which control many aspects of cell behavior throughout an animal's life. The signaling pathway shown here is a simplified version of a pathway triggered by FGF that leads to changes in gene expression that promote cell survival, cell growth, cell division, or differentiation, depending on the cells involved and the developmental context. There are multiple different forms of FGF and FGF receptors, which are also used in different contexts. For example, in the chick, FGF-8 is involved in somite formation and in signaling from the organizing center in the brain (see Sections 5.5 and 5.10) whereas FGF-10 is crucial for limb development (discussed in Chapter 11).

The intracellular signaling pathway from these receptors illustrated here is often known as the Ras-MAPK pathway because of the key involvement of the small G protein Ras and the activation of a cascade of serine/threonine kinases which ends in the activation of a mitogen-activated protein kinase (MAPK). The name comes from the fact FGF and other growth factors can act as mitogens, agents that stimulate cell proliferation. In one variation or another, the central Ras-MAPK module occurs in the signaling pathways from many different tyrosine-kinase receptors.

Binding of extracellular FGF to its receptor results in dimerization of two ligand-bound receptor molecules, activating the intracellular tyrosine kinase domains which then phosphorylate each other. The phosphorylated receptor tails recruit adaptor proteins (Grb and Sos), which in turn recruit and activate Ras at the plasma membrane. This results in the binding and activation

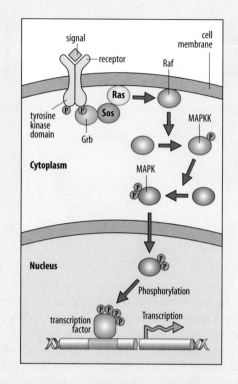

of the first serine/threonine kinase in the cascade, which in the mammalian pathway is called Raf. This phosphorylates and activates the next protein kinase, the MAPK kinase (MAPKK), which phosphorylates and activates MAPK. MAPK may then phosphorylate other kinases and can also enter the nucleus and phosphorylate transcription factors, thus activating gene expression. There are several different MAPKs in mammalian cells that are used in different pathways and target different transcription factors.

in zebrafish contributes to the induction of anterior neural tissue by inhibiting BMP signaling, the ectoderm that develops into the spinal cord is some distance away from the organizer on the ventral-vegetal side of the embryo, where organizer signals do not reach (Fig. 5.14). The initiator of neural development in this ventral-vegetal

Fig. 5.14 Prospective spinal cord in the zebrafish embryo is distant from the organizer. In the zebrafish, the ectoderm that will form the spinal cord is situated on the other side of the gastrula from the organizer, in the ventral–vegetal ectoderm, and is too far away to be influenced by signals from the organizer. FGF signals in the ventral–vegetal region induce this ectoderm as neuroectoderm and BMPs promote its development as posterior neural tissue (spinal cord). The epidermis is not shown here for simplicity. By this stage it covers the whole embryo.

Adapted from Kudoh, T., et al.: 2004.

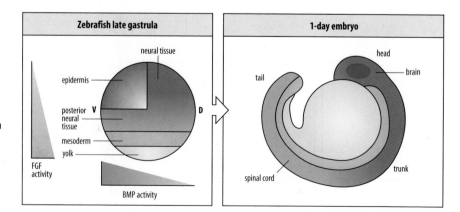

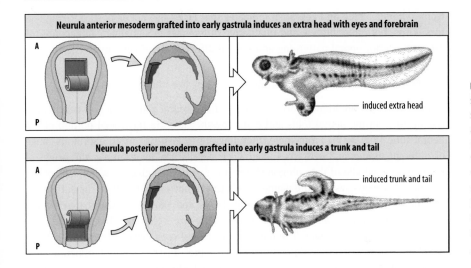

Fig. 5.15 Induction of the nervous system by the mesoderm is region specific. Mesoderm from different positions along the dorsal antero-posterior axis of early newt neurulas induces structures specific to its region of origin when transplanted to ventral regions of early gastrulas. Anterior mesoderm induces a head with a brain (top panels), whereas posterior mesoderm induces a posterior trunk with a spinal cord ending in a tail (bottom panels).

After Mangold, O.: 1933.

ectoderm is FGF, and BMP signals, which are high in the ventral region, in this case appear to push the neuroectoderm towards a posterior fate.

5.3 The nervous system is initially patterned by signals from the mesoderm

Whatever the mechanisms of neural induction eventually turn out to be, it is clear that the neural plate can be patterned by signals from the mesoderm. Pieces of mesoderm taken from different positions along the antero-posterior axis of a newt neurula and placed in the blastocoel of an early newt embryo induce neural structures at the site of transplantation. The structures that are formed correspond more or less to the original position of the transplanted mesoderm: pieces of anterior mesoderm induce a head with a brain, whereas posterior pieces induce a trunk with a spinal cord (Fig. 5.15). Another indication of positional specificity comes from the observation that pieces of neural plate induce similar regional neural structures in adjacent ectoderm when transplanted beneath the ectoderm of another gastrula.

Both qualitative and quantitative differences in signaling by the mesoderm can account for antero-posterior neural patterning. Qualitatively different inducers are secreted by the mesoderm at different positions along the antero-posterior axis. In *Xenopus*, Wnt antagonists such as Cerberus, Dickkopf1, and Frizzled-related protein are predominantly secreted from anterior endoderm and mesoderm beneath the future head, whereas BMP antagonists are released from mesoderm beneath both head and trunk. The combinatorial action of BMP and Wnt antagonists, therefore, could promote development of a head with a brain, whereas BMP antagonists alone induce a trunk with a spinal cord. Mouse embryos mutant for Dickkopf1 are headless, indicating that Wnt inhibition is necessary for head formation. However, when over-expressed in *Xenopus*, Dickkopf1 is only able to induce extra heads when in combination with BMP inhibitors. This shows that it is the combined inhibition of BMP and Wnt signaling that mimics the action of the anterior mesoderm.

Quantitative differences in Wnt/β-catenin signaling are present along the axis. Moving more posteriorly along the axis, the levels of Wnt/β-catenin signaling progressively increase, resulting in a gradient with the highest level of Wnt/β-catenin signaling at the posterior end of the embryo (see Fig. 5.9). There is evidence that high levels of Wnt signaling give prospective neural tissue a more posterior identity (Fig. 5.16). In *Xenopus* animal cap explants taken from embryos injected with mRNA for the BMP antagonist Noggin, and used to represent anterior neuralized tissue, increasing levels of Wnt signaling induce the expression of different posterior marker genes. Wnts have also been shown to posteriorize the neural plate in chick and zebrafish. Evidence

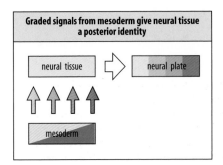

Fig. 5.16 Mesoderm patterns the neural plate. Quantitative differences in a signal along the axis could help pattern the neural plate. For example, the levels of Wnt/β-catenin signaling increase from anterior to posterior and the high levels at the posterior end could confer a more posterior identity on the neural plate.

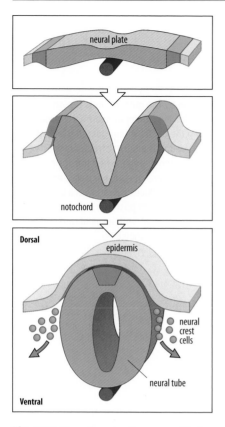

Fig. 5.17 Neural crest cells are specified at the lateral borders of the neural plate. First panel: prospective neural crest cells (green) are specified at the sides of the neural plate (blue) where levels of BMP are just high enough to prevent neural plate specification. Second panel: formation of the neural folds carries these cells to the dorsal crests of the neural tube, from which they soon migrate away (third panel).

from *Xenopus*, chick and zebrafish indicates that FGFs also act as posteriorizing factors. Thus, both quantitative and qualitative differences in signaling can account for antero-posterior neural patterning and several different signals are involved.

5.4 Neural crest cells arise from the borders of the neural plate

Neural crest cells are induced at the borders of the neural plate, which at neurulation rise up to form the neural folds and come together to form the dorsal portion of the neural tube (Fig. 5.17). From there, they migrate to give rise to an enormous range of different tissues and cell types. Neural crest cells are, for example, the origin of the neurons and glia of the autonomic nervous system, the glial Schwann cells of the peripheral nervous system, the pigment cells (melanocytes) of the skin, the adrenaline (epinephrine)-producing cells of the adrenal glands, and, most surprisingly, much of the bone, cartilage, muscle, and connective tissue of the face—cell types that are more commonly of mesodermal origin. The migration of neural crest cells is described in more detail in Chapter 8 and their differentiation in Chapter 10. The migration of neural crest from the hindbrain to the branchial arches, from which some of the structures and tissues of the vertebrate head are derived, is touched on briefly later in this chapter in relation to the antero-posterior patterning of the hindbrain and the cells derived from it.

The induction of the neural crest is a multi-step process that in *Xenopus* starts at the early gastrula stage and continues until neural tube closure. A current view of neural crest induction is that crest cells form in a band in the ectoderm along each lateral border of the neural plate, at a position where BMP signaling is just above the level that blocks neural plate formation (see Fig. 5.11). As might be expected from their role in neural plate induction, Wnt and FGF signals also appear to be involved in neural crest specification. Induction of neural crest is marked by expression of the transcription factors Sox9 and Sox10, which in turn activate the gene *snail*, an early marker of neural crest. This gene is the vertebrate version of the *Drosophila* gene *snail* (see Section 2.19). As we shall see in Chapter 8, the Snail protein is involved in enabling the morphological transition that an epithelial cell undergoes to become a migratory cell. Such epithelial-to-mesenchymal transitions occur in many situations during animal development, and neural crest migration is just one example.

Summary

Patterning of the early vertebrate embryo, along both the antero-posterior and dorso-ventral axes, is closely related to the action of the Spemann organizer and its morphogenesis during gastrulation. When grafted to the ventral side of an early gastrula, the Spemann organizer induces both a new dorso-ventral axis and a new antero-posterior axis, with the development of a second twinned embryo. The organizer secretes growth factor antagonists, which set up gradients of TGF-β and Wnt signaling to pattern the embryo. In the zebrafish, the embryonic shield is the organizer, while in chick development, Hensen's node serves a function similar to that of the Spemann organizer, and it too can specify a new antero-posterior axis. In the mouse, the node can specify a new axis apart from the most anterior forebrain, for which induction by the anterior visceral endoderm is also necessary.

The vertebrate nervous system, which forms from the neural plate, is induced both by early signals within the ectoderm and by signals from the mesoderm that comes to lie beneath prospective neural plate ectoderm during gastrulation. In *Xenopus*, inhibition of BMP signaling by proteins such as Noggin produced by the organizer is required for neural induction. Patterning of the neural plate involves both quantitative and qualitative differences in signaling by the mesoderm with high levels of Wnt and FGF signaling specifying the more posterior structures.

Somite formation and antero-posterior patterning

By the end of gastrulation, the vertebrate embryo has elongated along the antero-posterior axis, the neural tube is evident, the notochord has formed, and the rest of the mesoderm is located on either side of it, eventually completely surrounding the endodermally derived gut. The fate maps of vertebrates (see Figs 4.19–4.23) show that the notochord develops from the most dorsal region of the mesoderm, and somites from more ventro-lateral mesoderm on either side of the midline, which is known as the **paraxial mesoderm**. Lateral to the paraxial mesoderm lies the intermediate mesoderm that gives rise to the kidney, and more lateral still is the lateral plate mesoderm, which becomes divided into somatic mesoderm and splanchnic mesoderm by formation of the coelom (see Fig. 3.17, which shows these regions of mesoderm in sections of the chick embryo). The anterior lateral plate mesoderm gives rise to the heart while limb buds arise from lateral plate mesoderm at the appropriate levels along the antero-posterior axis of the body.

The somites give rise to the bone and cartilage of the trunk, including the spinal column, to the skeletal muscles, and to the dermis of the skin on the dorsal side of the body, and their patterning provides much of the body's antero-posterior organization. The vertebrae, for example, have characteristic shapes at different positions along the spine. We will first examine the development of the somites and how their different developmental identities along the antero-posterior axis are specified by Hox genes. We conclude our look at somite formation by discussing how individual somites are patterned and which structures they give rise to. We end this part of the chapter by revisiting the embryonic brain and looking at aspects of its early patterning along the antero-posterior axis.

5.5 Somites are formed in a well-defined order along the antero-posterior axis

Somite number can be highly variable between different vertebrates; birds and humans have around 50 whereas snakes have up to several hundred. Much of the work on somite formation has been done on the chick because of the ease with which the process can be observed, and so we will use the chick as our main model organism here. In the chick embryo, somite formation occurs on either side of the notochord in the paraxial mesoderm anterior to the regressing Hensen's node (Fig. 5.18). The first five somites in the chick contribute to the most posterior part of the skull and, anterior to this, the paraxial mesoderm does not segment. It will form some facial muscles and bones—the rest of the skull is of neural crest origin. Somite formation proceeds in an anterior to posterior direction in the unsegmented mesoderm—the **pre-somitic mesoderm**—lying between the most recently formed somite and the node. As the node regresses, the length of this region remains fairly constant. Changes in cell shape and intercellular contacts in the pre-somitic mesoderm result in the formation of distinct blocks of cells—the somites. Somites are formed in pairs, one on either side of the notochord, with the two somites in a pair forming simultaneously. A pair of somites forms every 90 minutes in the chick, every 120 minutes in the mouse, every 45 minutes in *Xenopus*, and every 30 minutes in zebrafish.

The sculpting of the somites involves tissue separation and cell movements, as well as integration of cells at the anterior and posterior somite borders. Somite formation in *Xenopus*, in particular, displays several unique features. A group of cells of the unsegmented mesoderm, which are oriented perpendicular to the notochord, separates from the rest of the mesoderm by formation of an intersomitic furrow and forms a distinct block—the prospective somite. The block then undergoes a 90° rotation. This rotation involves a series of cellular movements and reorientations that have not been well investigated, but which seem to depend on cells sensing their locations within the block and undergoing morphological changes and cellular rearrangements according to their positions.

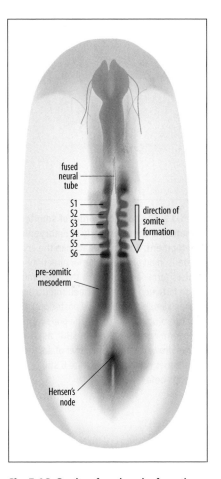

fused neural tube

S1
S2
S3
S4
S5
S6

direction of somite formation

pre-somitic mesoderm

Hensen's node

Fig. 5.18 Somites form in pairs from the paraxial mesoderm. Somites are blocks of tissue that form from the mesoderm lying immediately on either side of the notochord, the paraxial mesoderm. They are made in an anterior to posterior direction from unsegmented pre-somitic mesoderm which is produced from a pool of stem cells around the node. The first five somites give rise to the posterior part of the skull; the remaining somites produce the muscles and bones of the trunk and tail and the muscles of the limbs.

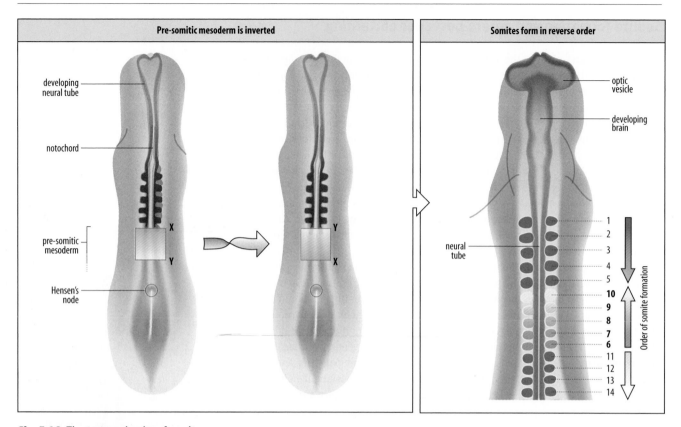

Fig. 5.19 The temporal order of somite formation is specified early in embryonic development. Somite formation in the chick proceeds in an antero-posterior direction. Somites form sequentially in the pre-somitic region between the last-formed somite and Hensen's node, which moves posteriorly. If the antero-posterior axis of the pre-somitic mesoderm is rotated through 180°, as shown by the arrow, the temporal order of somite formation is not altered—somite 6 still develops before somite 10.

In the chick, the cells that give rise to the somites originate in the epiblast on either side of the anterior primitive streak and move into it at gastrulation to form a population of somitogenic stem cells located around Hensen's node. The stem cells divide, and those that remain in the stem-cell region continue to be self-renewing stem cells, but those left behind as the node regresses form the pre-somitic mesoderm. As new cells are being added to the pre-somitic mesoderm at the posterior end of the chick embryo, somites are forming at the anterior end. In this way, the length of the pre-somitic mesoderm remains more or less constant, enough to make about 12 somites.

The sequence of somite formation in the unsegmented region is unaffected by transverse cuts in the plate of pre-somitic mesoderm, suggesting that somite formation is an autonomous process and that, at this time, no extracellular signal specifying antero-posterior position or timing is involved. Even if a piece of the unsegmented mesoderm is rotated through 180°, each somite still forms at the normal time, but with the sequence of formation running in the opposite direction to normal in the inverted tissue (Fig. 5.19). So, before somite formation begins, a molecular pattern that specifies the time of formation of each somite has already been laid down in the pre-somitic mesoderm; we shall return to this patterning process later. Given the existence of this pattern, the prospective identity of each somite must be related to the temporal order in which their cells entered the pre-somitic mesoderm.

Somite formation is largely determined by an internal 'clock' in the pre-somitic mesoderm. This clock is represented by periodic cycles of gene expression, such as that of the genes *c-hairy1* and *c-hairy2* in the chick embryo, whose expression sweeps from the posterior to the anterior end of the pre-somitic mesoderm with a period of 90 minutes, the time it takes for a pair of somites to form. In a newly formed somite, *c-hairy 1* expression becomes restricted to the posterior end of the somite, where it persists, while a new wave of *c-hairy 1* expression starts at the tail end of the pre-somitic mesoderm (Fig. 5.20). Oscillating gene expression has also been detected

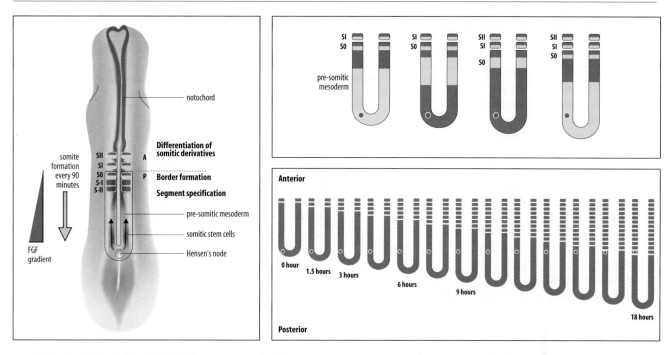

Fig. 5.20 Somite formation in the chick. As shown in the left-hand panel, the somites are generated successively from pre-somitic mesoderm, which is derived from somitic stem cells in the primitive streak. As pre-somitic cells are released into the posterior pre-somitic mesoderm, a new pair of somites buds from the anterior end every 90 minutes. SI, the most recently formed somite; SII, the last but one somite formed; S0, somite in the process of formation, whose boundaries are not yet set; S-I, S-II, blocks of pre-somitic cells that will form somites. The position in which the border is formed to generate each somite is specified by a threshold level of FGF-8 gradient (green dotted line). At formation, each somite acquires antero-posterior polarity, after which it can respond to the signals that pattern it along the antero-posterior and dorso-ventral axes. The somite has distinct anterior (gray) and posterior (yellow) halves. The top right-hand panel shows stages in one of the cycles of *c-hairy 1* expression (yellow) that sweep from posterior to anterior of the pre-somitic mesoderm every 90 minutes. During each cycle, a given pre-somitic cell experiences distinct phases when *c-hairy 1* is expressed and when it is not expressed. The lower right-hand panel shows the history of a pre-somitic cell (red dot) from the time it enters the pre-somitic mesoderm until it is incorporated into a somite. Somitic cells in the anterior somites will have experienced fewer cycles of *c-hairy 1* expression before they leave the node region for the pre-somitic mesoderm than will cells in posterior somites, and this could define a clock that is both linked to somite segmentation and 'tells' the somite its position along the antero-posterior axis.

in the mouse and zebrafish pre-somitic mesoderm. How these gene oscillations are related to regular somite segmentation is not yet clear, and the underlying mechanism for determining the segment boundary may differ in different vertebrates.

In the mouse and chick, another of the genes with cyclic expression encodes Lunatic fringe, an enzyme that modulates activity of the Notch–Delta signaling pathway (Box 5C, p. 190). This pathway is widely involved in determining cell fate and delimiting boundaries during embryonic development. As we shall see later, somites are functionally divided into anterior and posterior halves with different fates, and one function of Notch signaling in mice seems to be to establish this anterior–posterior boundary. Both Notch and Delta are transmembrane proteins, and this pathway is an example of the transmission of signals by direct cell–cell contact. In zebrafish, one proposed function of Notch activity during somitogenesis is the synchronization of oscillating gene expression between neighboring somitic stem cells. Similarly, mouse embryos in which Delta–Notch signaling is either made constitutively active or abolished entirely often do not form somites, and if they do, the somites vary in size and shape and are different on each side of the body.

The timing and position of somite formation seem to be determined by the interaction of the segmentation clock with the intercellular signaling protein FGF-8. In the chick and the mouse at this time there is an FGF-8 gradient in both mesoderm and

Box 5C The Notch signaling pathway

The family of Notch transmembrane signaling proteins are involved in numerous developmental decisions that set cell fate, ranging from neuronal precursors in *Drosophila* and vertebrates to the differentiation of epidermal cells in mammalian skin. The ligands for Notch are also membrane-bound proteins and so Notch signaling requires direct cell-cell contact. This means that Notch delivers a very spatially precise signal that can set the fate of a single cell out of many, as we shall see when we look at the development of the *Drosophila* eye in Chapter 11 and neuroblast specification in Chapter 12.

The Notch ligands are the members of the Delta and Serrate/Jagged1 families. Binding of ligand on one cell to Notch on an adjacent

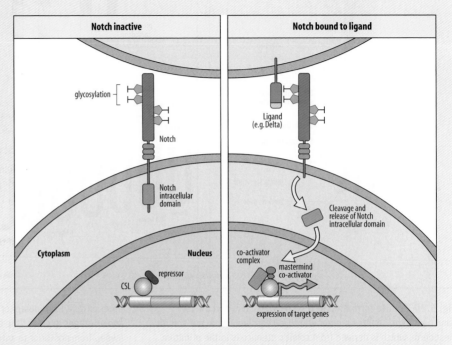

cell activates intracellular signaling from the receptor. The Notch receptor has a large extracellular domain with several glycosylation sites that are specifically modified by glycosyltransferases, including those of the Fringe family. Glycosylation modulates Notch signaling by affecting the ability of the extracellular domain to bind to different ligands, facilitating its binding to Delta ligands while reducing the ability to bind to Serrate ligands. Activation of the Notch receptor involves the enzymatic cleavage of the cytoplasmic tail of the receptor (the Notch intracellular domain) and its translocation to the nucleus to bind and activate a transcription factor of the CSL family (CSL is named after

the Notch-activated transcription factors CBF1/RBPkJ in mammals, Suppressor of Hairless (Su(H)) in *Drosophila* and LAG-1 in *C. elegans*). In the absence of Notch signaling, CSL is complexed with repressor proteins. Activation of the CSL complex by binding of the Notch intracellular domain involves release of the repressors and recruitment of the co-activator protein Mastermind and other co-activators. The outcome of the Notch signaling pathway is the transcription of specific genes. Notch signaling is very versatile. In different organisms and in different developmental circumstances, activation of Notch is used to switch a wide range of different genes on or off.

ectoderm, with its high point at the node and decreasing anteriorly. The gradient regresses in a posterior direction along with the node, probably because *FGF-8* mRNA is only made in cells in and around the node and is progressively degraded in the cells that are left behind when the node moves posteriorly. The result is a gradient in intracellular *FGF-8* mRNA, which after translation and secretion of FGF-8 gives an extracellular head-to-tail FGF-8 gradient with its high point eventually residing in the tailbud of the embryo (Fig. 5.21). Somite formation occurs where FGF-8 reaches a sufficiently low threshold level. As the gradient moves along with the node, therefore, it successively defines regions in the pre-somitic mesoderm where somites start to form. The small secreted signaling molecule retinoic acid (Box 5D, pp. 192–193) has an antagonistic effect on FGF action. It forms a gradient in the opposite direction to FGF, and by antagonizing FGF, prevents the pre-somitic region from continually getting longer. Retinoic acid is synthesized in the somites and diffuses both posteriorly and anteriorly. It is also thought to be involved in maintaining the bilateral symmetry of somite development by buffering the somites, in some way not yet understood, against the signals that are setting up left–right asymmetry in the lateral mesoderm at this stage in development (see Chapter 4).

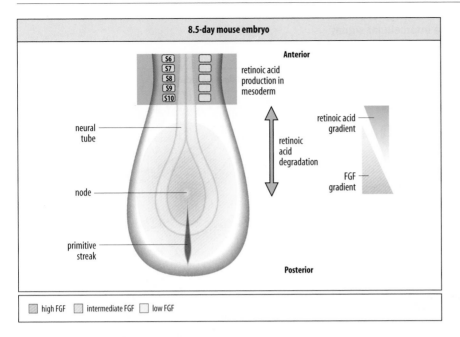

8.5-day mouse embryo

Anterior

retinoic acid
production in
mesoderm

retinoic acid
gradient

retinoic
acid
degradation

FGF
gradient

neural
tube

node

primitive
streak

Posterior

high FGF ☐ intermediate FGF ☐ low FGF

Fig. 5.21 FGF and retinoic acid gradients help to pattern the antero-posterior axis in the mouse embryo. An antero-posterior gradient of FGF develops in the embryo, with its high point at the node. This schematic diagram shows the dorsal view of the posterior half of an 8.5-day mouse embryo with 10 pairs of somites formed and the neural tube partly closed but still open at the posterior end of the future spinal cord. The FGF gradient is formed by cells in and around the node synthesizing *FGF* mRNA, which is then gradually degraded when cells leave the region anteriorly to form the pre-somitic mesoderm. This results in a gradient of translated and secreted FGF protein that is continuously moving posteriorly as the embryo elongates. The gradient in the mesoderm fades out at around the position of the last-formed somite, suggesting that somites are formed when the FGF level drops to some low threshold. Retinoic acid synthesized and secreted by the somites forms an opposing gradient that antagonizes the actions of FGF.

Adapted from Deschamps, J. and Van Nes, J.: 2005.

Somites differentiate into particular axial structures depending on their position along the antero-posterior axis. The easiest somite-derived structures to detect are the vertebrae. The anterior-most somites contribute to the most posterior region of the skull, those posterior to them form cervical vertebrae, and those more posterior still develop as thoracic vertebrae with ribs. Specification by position has occurred during gastrulation before somite formation itself begins; for example, if unsegmented somitic mesoderm from the presumptive thoracic region of the chick embryo is grafted to replace the presumptive mesoderm of the neck region, it will still form thoracic vertebrae with ribs (Fig. 5.22). How then is the pre-somitic mesoderm patterned so that somites acquire their identity and form particular vertebrae?

5.6 Identity of somites along the antero-posterior axis is specified by Hox gene expression

The antero-posterior patterning of the mesoderm is most clearly seen in the differences in the vertebrae, each individual vertebra having well defined anatomical characteristics

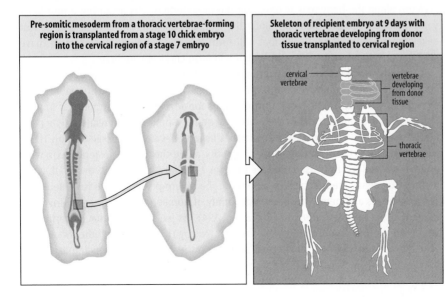

Pre-somitic mesoderm from a thoracic vertebrae-forming region is transplanted from a stage 10 chick embryo into the cervical region of a stage 7 embryo

Skeleton of recipient embryo at 9 days with thoracic vertebrae developing from donor tissue transplanted to cervical region

cervical
vertebrae

vertebrae
developing
from donor
tissue

thoracic
vertebrae

Fig. 5.22 The pre-somitic mesoderm has acquired a positional identity before somite formation. Pre-somitic mesoderm that will give rise to thoracic vertebrae is grafted to an anterior region of a younger embryo that would normally develop into cervical vertebrae. The grafted mesoderm develops according to its original position and forms ribs in the cervical region.

Box 5D Retinoic acid: a small-molecule intercellular signal

Retinoic acid is a small diffusible molecule derived from vitamin A (retinol). Animals obtain vitamin A through their diet, and it has been known for a long time through work with pigs that vitamin A deficiency can lead to developmental defects. It subsequently emerged that an excess of vitamin A also leads to developmental defects. Sadly, babies with craniofacial malformations have been born to women who had been treated for severe skin diseases with synthetic vitamin A derivatives, not knowing they were pregnant.

Retinol is present at high levels in chicken eggs and in mammalian embryos is supplied by the maternal circulation. Retinol binds to retinol-binding proteins in the extracellular space and is transferred into cells by the transport receptor Simulated by retinoic acid 6 (STRA6). Once inside the cell, retinol is bound by an intracellular retinol-binding protein (CRBP). Retinoic acid is produced in a two-step reaction (see figure, above), in which retinol is first metabolized to retinaldehyde by retinol dehydrogenases (mainly RDH10) or alcohol dehydrogenase (ADH); retinaldehyde is then metabolized to retinoic acid by retinaldehyde dehydrogenases (RALDH). There are three vertebrate retinaldehyde dehydrogenases, encoded by the genes *Raldh1, 2,* and *3,* and expression of these genes is tightly regulated both temporally and spatially in vertebrate embryos to ensure that retinoic acid is produced at very precise locations and specific times during development.

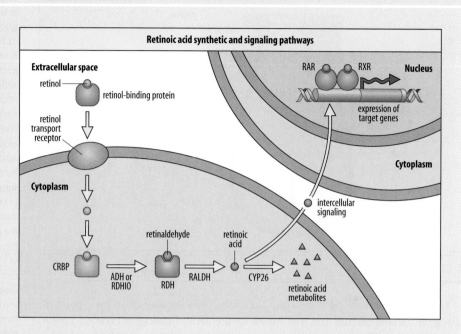

depending on its location along the axis. The most anterior vertebrae are specialized for attachment and articulation of the skull, while the cervical vertebrae of the neck are followed by the rib-bearing thoracic vertebrae and then those of the lumbar region, which do not bear ribs, and finally, those of the sacral and caudal regions. Patterning of the skeleton along the body axis is based on the mesodermal cells acquiring a positional value that reflects their position along the axis and so determines their subsequent development. Mesodermal cells that will form thoracic vertebrae, for example, have different positional values from those that will form cervical vertebrae.

Patterning along the antero-posterior axis in all animals involves the expression of genes that specify positional identity along the axis. The concept of positional identity, or positional value, has important implications for developmental strategy; it implies that a cell or a group of cells in the embryo acquires a unique state related to its position at a given time, and that this determines its later development (see Section 1.15). Homeotic genes that specify positional identity along the antero-posterior axis were originally identified in *Drosophila* (see Chapter 2) and it was later discovered that the related Hox genes are involved in patterning the antero-posterior axis in vertebrates. The Hox genes are members of the large family of homeobox genes that are involved in many aspects of development (Box 5E, p. 194). As we shall see later, patterning of vertebrate embryos along the antero-posterior axis by Hox genes and other homeobox genes is not confined to mesodermal structures; the hindbrain, for example, is also divided into distinct regions by characteristic patterns of Hox gene expression.

Retinoic acid is broken down by enzymes of the P450 cyto-chrome system, mainly by the subfamily of CYP26 enzymes which, like the RALDHs, are expressed in very precise patterns in the embryo. Intricate homeostatic buffering mechanisms involving regulation of expression of the genes encoding retinoic-acid-metabolizing enzymes and end-product feedback serve to maintain appropriate levels of retinoic acid. If retinoic acid levels increase, *Raldh* gene expression is reduced and expression of *Cyp26* genes is increased; when retinoic acid levels fall the reverse happens.

Retinoic acid acting as an intercellular signal is shown in the figure, although it probably also acts as a signal within the cells that produce it. Retinoic acid is lipid-soluble and so can pass through cell membranes. Inside a cell it is bound by pro-teins that transport it to the nucleus, where it binds receptors that act directly as transcriptional regulators. Mammals and birds have three retinoic acid receptor (RAR) proteins—α, β and γ—which act as dimers in combination with the retinoid X recep-tor proteins (RXR; also α, β and γ). In the absence of retinoic acid, the heterodimeric RAR:RXR receptors are bound to regula-tory DNA sequences known as retinoic-acid-response elements (RAREs) and repress the associated genes. Binding of retinoic acid to its nuclear receptor leads to activation of the gene asso-ciated with the response element. Gene activation by retinoic acid can involve not only the direct initiation of transcription but also the recruitment of co-activator proteins that induce chro-matin remodeling, which enables the gene to be expressed (see Box 5A, p. 182).

Among the genes with a retinoic-acid-response element is the gene for RARβ itself, and so one direct response to retinoic acid is a switch-ing on of the *RARβ* gene. Expres-sion of *RARβ* can therefore be used as a reporter of retinoic acid signal-ing and transgenic mice have been made in which the reporter gene *lacZ* (encoding β-galactosidase, see Box 1D, p. 20) is hooked up to the *RARβ* retinoic-acid-response element. β-Galactosidase can then be revealed in embryos and tissues by a histochemical reaction which gives a blue color, showing which cells were responding to retinoic acid signaling. The photograph above shows an E9.5 mouse embryo.

Other important developmental signals that act through intra-cellular receptors of the same superfamily as the RAR and RXR proteins are the steroid hormones (such as estrogen and tes-tosterone, discussed in Chapters 9 and 13) and the thyroid hor-mones (thyroxine and tri-iodothyroxine; discussed in Chapter 13). Like retinoic acid these molecules are lipid soluble; they diffuse through the plasma membrane unaided and bind to intracellu-lar receptor proteins. The ligand-receptor complex is then able to act as a transcriptional regulator, binding directly to specific response elements in the DNA to activate (or in some cases repress) transcription of a number of genes simultaneously.
Photograph courtesy of Pascal Dolle.

The Hox genes are the most striking example of a widespread conservation of devel-opmental genes in animals. Although it was widely believed that there are common mechanisms underlying the development of all animals, it seemed at first unlikely that genes involved in specifying the positional identity of the segments of *Drosophila* would specify positional identity in vertebrates. However, the strategy of comparing genes by sequence homology proved extremely successful in identifying vertebrate Hox genes related to the homeotic genes of *Drosophila*. Numerous other genes first identified in *Drosophila*, in which the genetic basis of development is far better under-stood than in any other animal, have also proved to have counterparts involved in development in vertebrates.

Most vertebrates have four separate clusters of Hox genes that are thought to have arisen by duplications of the clusters themselves during vertebrate evolution, and some of the Hox genes within each cluster may also have arisen by duplication (see Box 5E, p. 194). The zebrafish is unusual in having seven Hox gene clusters, as a result of further duplication. A particular feature of Hox gene expression in both insects and vertebrates is that the genes in the clusters are expressed in a temporal and spatial order that reflects their order on the chromosome. This is a unique feature in development, as it is the only known case where a spatial arrangement of genes on a chromosome corresponds to a spatial pattern in the embryo.

A simple idealized model illustrates the key features by which a Hox gene cluster could record positional identity by a combinatorial mechanism. Consider four genes—I, II, III, and IV—arranged along a chromosome in that order (Fig. 5.23). The genes are

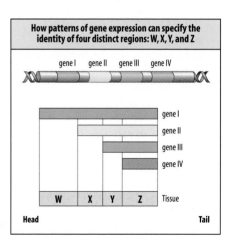

Fig. 5.23 Gene activity can provide positional values. The model shows how the pattern of gene expression along a tissue can specify the distinct regions W, X, Y, and Z. For example, only gene I is expressed in region W but all four genes are expressed in region Z.

Box 5E The Hox genes

The proteins encoded by the Hox genes of vertebrates belong to a large group of gene-regulatory proteins that all contain a similar DNA-binding region of around 60 amino acids. This region, known as the **homeodomain**, contains a helix-turn-helix DNA-binding motif that is characteristic of many DNA-binding proteins. The homeodomain is encoded by a DNA motif of around 180 base pairs termed the **homeobox**, a name that came originally from the fact that this gene family was discovered through mutations that produce a homeotic transformation—a mutation in which one structure replaces another.

Clusters of homeotic genes involved in specifying segment identity were first discovered in *Drosophila*. For example, in one homeotic mutant in *Drosophila*, a segment in the fly's body that does not normally bear wings is transformed to resemble the adjacent segment that does bear wings, resulting in a fly with four wings instead of two.

The Hox genes in *Drosophila* are organized into two distinct gene complexes, the Antennapedia complex and the bithorax complex (known collectively as the HOM-C cluster). In vertebrates, the Hox genes are organized into clusters, known as the Hox complexes, and the homeoboxes of the genes are related in DNA sequence to the homeoboxes of genes of the two gene complexes in *Drosophila*. In each Hox cluster the order of the genes from 3′ to 5′ in the DNA is the order in which they are expressed along the antero-posterior axis and specify positional identity. In the mouse, there are four unlinked Hox complexes, designated Hoxa, Hoxb, Hoxc, and Hoxd (originally called Hox1, Hox2, Hox3, and Hox4), located on chromosomes 6, 11, 15, and 2, respectively (see figure; on chromosomes 7, 17, 12, and 2 in the human genome). All Hox genes thus resemble each other to some extent; the homology is most marked within the homeobox and less marked in sequences outside it. Genes that have arisen by duplication and divergence within a species are known as **paralogs**, and the corresponding genes

in the different clusters (e.g. *Hoxa4*, *Hoxb4*, *Hoxc4*, *Hoxd4*) are known as a **paralogous subgroup**. In the mouse and human there are 13 paralogous subgroups.

The Hox genes and their role in development are of ancient origin. The mouse and frog genes are similar to each other and to those of *Drosophila*, both in their coding sequences and in their order on the chromosome. In both *Drosophila* and vertebrates, these homeotic genes are involved in specifying regional identity along the antero-posterior axis. Many other genes contain a homeobox but they do not, however, belong to a homeotic complex, nor are they involved in homeotic transformations. Other subfamilies of homeobox genes in vertebrates include the **Pax genes**, which contain a homeobox typical of the *Drosophila* gene *paired*. All these genes encode transcription factors with various functions in development and cell differentiation.

Illustration after Veraksa et al.: 2000.

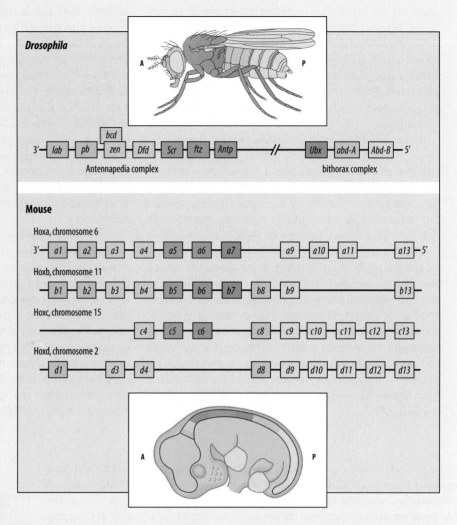

expressed in a corresponding order along the antero-posterior axis of a tissue. Thus, gene I is expressed throughout the tissue with its anterior boundary at the anterior end. Gene II has its anterior boundary in a more posterior position and expression continues posteriorly. The same principles apply to the two other genes. This pattern of expression defines four distinct regions, coded for by the expression of different combinations of genes. If the amount of gene product is varied within each expression domain, for example by interactions between the genes, many more regions can be specified.

The role of the Hox genes in vertebrate axial patterning has been best studied in the mouse, because it is possible to either knock out particular Hox genes or alter their expression in transgenic mice (see Section 3.9 and Box 3B, p. 118). As in all vertebrates, the Hox genes start to be expressed in mesoderm cells of mouse embryos at an early stage of gastrulation when these cells begin to move internally away from the primitive streak and towards the anterior. The 'anterior' Hox genes are expressed first. More posterior Hox genes are expressed as gastrulation proceeds and clearly defined patterns of Hox gene expression are most easily seen in the mesoderm after somite formation and the neural tube after neurulation.

Figure 5.24 shows the expression patterns of three Hox genes in mouse embryos at day E9.5. Typically, the expression of each gene is characterized by a sharp anterior border and extends posteriorly. Although there is considerable overlap in the spatial pattern of expression of Hox genes, particularly of neighboring genes in a cluster, almost every region in the mesoderm along the antero-posterior axis is characterized by a specific combination of expressed Hox genes, as shown by a summary picture of the anterior borders of expression of Hox genes in mouse embryonic mesoderm (Fig. 5.25). For example, the most anterior somites are characterized by expression of *Hoxa1* and *Hoxb1* and no other Hox genes are expressed in this region. By contrast, all Hox genes are expressed in the more posterior regions. The Hox genes can therefore provide a code for regional identity. The spatial and temporal order of expression is similar in the mesoderm and the ectodermally derived nervous system, but the positions of the boundaries between regions of gene expression in the two germ layers do not always correspond. The most anterior region of the body in which Hox genes are expressed is the hindbrain; other homeobox genes such as *Emx* and *Otx*, but not Hox genes, are expressed in more anterior regions of the vertebrate body—midbrain, forebrain, and the anterior part of the head.

If we focus on just the Hoxa complex of genes, we see that the most anterior border of expression in the mesoderm is that of *Hoxa1* in the posterior head mesoderm, while *Hoxa11*, a very posterior gene in the Hoxa complex, has its anterior border of expression in the sacral region (see Fig. 5.25). This exceptional correspondence, or **co-linearity**, between the order of the genes on the chromosome and their order of spatial and temporal expression along the antero-posterior axis is typical of all the Hox

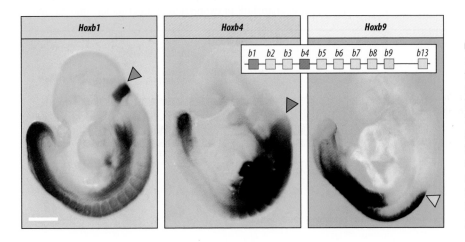

Fig. 5.24 Hox gene expression in the mouse embryo after neurulation. The three panels show lateral views of 9.5 dpc mouse embryos stained for expression of a LacZ reporter (blue) linked to promoters for the *Hoxb1*, *Hoxb4*, and *Hoxb9* genes. These genes are expressed in the neural tube and the mesoderm. The arrowheads indicate the anterior boundary of expression of each gene within the neural tube. The position of the three genes within the Hoxb gene complex is indicated (inset). Scale bar = 0.5 mm.
Photographs courtesy of A. Gould.

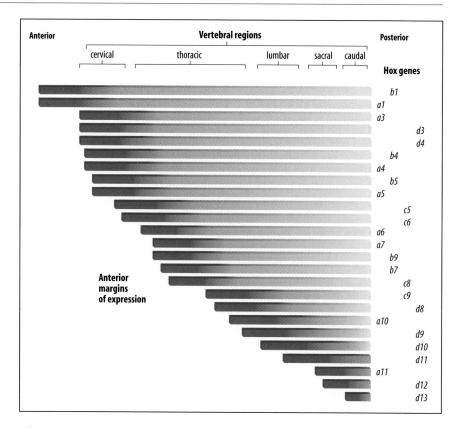

Fig. 5.25 A summary of Hox gene expression along the antero-posterior axis of the mouse mesoderm. The anterior border of expression of each gene is shown by the red blocks. Expression usually extends backwards some distance but the posterior margin of expression may be poorly defined. The pattern of *Hox* gene expression specifies the identity of the tissues at different positions. This figure represents an overall picture of Hox gene expression and is not a snapshot of gene expression at any particular time.

Fig. 5.26 Patterns of *Hox* gene expression in the mesoderm of chick and mouse embryos, and their relation to regionalization. The pattern of Hox gene expression varies along the antero-posterior axis. The vertebrae (squares) of the spinal column are derived from the embryonic somites (circles), 40 of which are shown here. The vertebrae have characteristic shapes in each of the five regions along the antero-posterior axis: cervical (C), thoracic (T), lumbar (L), sacral (S), and caudal (Ca). Which somites form which vertebrae differs in chick and mouse. For example, thoracic vertebrae start at somite 19 in the chick, but at somite 12 in the mouse. Despite this difference, the transition from one region to another corresponds to a similar pattern of *Hox* gene expression in chick and mouse, such that the anterior boundaries of expression of *Hoxc5* and *Hoxc6* lie on either side of the transition from cervical to thoracic vertebrae (light blue to mid blue) in both species. Similarly, the anterior boundaries of *Hoxd9* and *Hoxd10* are at the transition between the lumbar and sacral regions (green to dark blue) in both chick and mouse.

After Burke, A.C.: 1995.

clusters. The genes of each Hox complex are expressed in an orderly sequence, with the gene lying most 3′ in the cluster being expressed the earliest and in the most anterior position. The correct expression of the Hox genes is dependent on their position in the cluster, and anterior genes (3′) are expressed before more posterior (5′) genes.

Support for the idea that the Hox genes are involved in controlling regional identity comes from comparing their patterns of expression in mouse and chick with the well defined anatomical regions—cervical, thoracic, sacral, and lumbar (Fig. 5.26). Hox gene expression corresponds well with the different regions. For example, even though the number of cervical vertebrae in birds (14) is twice that of mammals, the anterior boundaries of *Hoxc5* and *Hoxc6* gene expression in both chick and mouse lie on either side of the cervical/thoracic boundary. A correspondence between Hox gene expression and region is also similarly conserved among vertebrates at other anatomical boundaries.

It must be emphasized that the summary picture of Hox gene expression given in Fig. 5.25 does not represent a 'snapshot' of expression at a particular time but rather

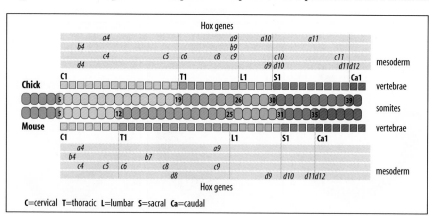

an integrated picture of the overall pattern of expression. Some genes are switched on early and expression is then reduced, whereas others are switched on considerably later; the most posterior Hox genes, such as *Hoxd12* and *Hoxd13*, are expressed in the post-anal tail. Moreover, this summary picture reflects only the general expression of the genes in any particular region; not all Hox genes expressed in a region are expressed in all the cells. Nevertheless, the overall pattern suggests that the combination of Hox genes expressed provides positional identity. In the cervical region, for example, each somite, and thus each vertebra, could be specified by a unique pattern of Hox gene expression or **Hox code**.

As we saw in Fig. 5.22, grafting experiments show that the character of the somites is already determined in the pre-somitic mesoderm and that somitic tissue transplanted to other levels along the axis retains its original identity. As would be expected if the Hox code specifies vertebral identity, the transplanted pre-somitic mesoderm also retains its original pattern of Hox gene expression.

5.7 Deletion or overexpression of Hox genes causes changes in axial patterning

If the Hox genes do provide positional values that determine a region's subsequent development, then one would expect morphological changes if their pattern of expression is altered. This is indeed the case. To investigate how Hox genes control patterning, their expression can be abolished by mutation, or they can be expressed in abnormal positions. Hox gene expression can be eliminated from the developing mouse embryo by gene knock-out (see Section 3.9 and Box 3B, p. 118). Experiments along these lines indicate that the absence of given Hox genes affects patterning in a way that agrees with the idea that they are normally providing cells with positional identity. In some cases, however, knocking out an individual Hox gene produces little effect, or affects only one tissue, and more substantial changes in patterning are only seen when two or more genes, particularly **paralogous genes** from different Hox complexes (see Box 5E, p. 194), are knocked out together.

Mice in which *Hoxd3* is deleted show structural defects in the first and second cervical vertebra, where this gene is normally strongly expressed (see Fig. 5.25), and in the sternum (breastbone). More posterior structures, where the inactivated gene is also normally expressed, show no evident defects, however. This observation illustrates a general principle of Hox gene expression, which is that more posteriorly expressed Hox genes tend to over-ride the function of more anterior Hox genes whenever they are coexpressed. This phenomenon is known as **posterior dominance** or **posterior prevalence**. Other Hox genes are expressed together with *Hoxd3* in the lumbar and sacral regions of the body and this explains why there are no posterior defects when *Hoxd3* is deleted. Another general point illustrated by the *Hoxd3* knock-out is that the effects of a Hox gene knock-out can be tissue specific, in that only certain tissues in which the Hox gene is normally expressed are affected, whereas other tissues at the same position along the antero-posterior axis appear normal. For example, although *Hoxd3* is expressed in neural tube, branchial arches, and paraxial mesoderm at the same antero-posterior level in the body, only the vertebrae are defective in a *Hoxd3* knock-out. The apparent absence of an effect in such cases may be due to redundancy, with paralogous genes from another complex being able to compensate. *Hoxa3* is expressed strongly in the same tissues as *Hoxd3* in the cervical region, but when *Hoxa3* alone is knocked out there are no defects in cervical vertebrae; instead other defects are observed, including reductions in cartilage elements derived from the second branchial arch. When both *Hoxd3* and *Hoxa3* are knocked out together, much more severe defects are seen in both vertebrae and branchial arch derivatives, suggesting that the two genes normally function together in these tissues.

Loss of Hox gene function often results in **homeotic transformation**—the conversion of one body part into another. This is the case in the *Hoxd3* knock-out mouse we

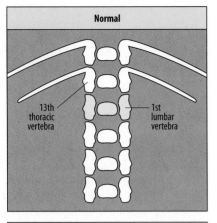

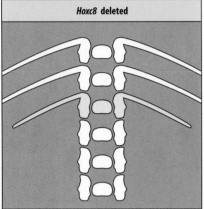

Fig. 5.27 Homeotic transformation of vertebrae due to deletion of *Hoxc8* in the mouse. In loss-of-function homozygous mutants of *Hoxc8,* the first lumbar vertebra (yellow) is transformed into a rib-bearing 'thoracic' vertebra. The mutation has resulted in the transformation of the lumbar vertebra into a more anterior structure.

have just considered. Detailed analysis of the cervical region in these mice revealed that the atlas vertebra (the first cervical vertebra) appears to be transformed into the basal occipital bone at the base of the skull, and that the axis vertebra (the second cervical vertebra) has acquired morphological features resembling the atlas. Thus, in the absence of *Hoxd3* expression, cells acquire a more anterior positional value and develop into more anterior structures. The double knock-out of *Hoxa3* and *Hoxd3* results in a deletion of the atlas. The complete absence of this bone in these double knock-outs suggests that one target of Hox gene action is the cell proliferation required to build such a structure from the somite cells. Unfortunately, very few direct gene targets transcriptionally regulated by Hox proteins have yet been identified.

Another example of homeotic transformation is seen in *Hoxc8*-knock-out mice. In normal embryos, *Hoxc8* is expressed in the thoracic and more posterior regions of the embryo from late gastrulation onward (see Fig. 5.25). Mice homozygous for deletion of *Hoxc8* die within a few days of birth, and have abnormalities in patterning between the seventh thoracic vertebra and the first lumbar vertebra. The most obvious homeotic transformations are the attachment of an eighth pair of ribs to the sternum and the development of a 14th pair of ribs on the first lumbar vertebra (Fig. 5.27). Thus, as with the knock-outs discussed above, the absence of *Hoxc8* has led to some of the cells that would normally express it acquiring a more anterior positional value, and developing accordingly.

Some homeotic transformations only occur if all the paralogous genes are knocked out together. In mice, in the absence of all the Hox10 paralogous subgroup genes, there are no lumbar vertebrae and there are ribs on all posterior vertebrae; in the absence of all the Hox11 paralogous subgroup genes several sacral vertebrae become lumbar. These homeotic transformations do not occur if only some members of the paralogous subgroup are mutated, again suggesting redundancy of Hox gene function.

In contrast to the effects of knocking out Hox genes, the ectopic expression of Hox genes in anterior regions that normally do not express them can result in transformations of anterior structures into structures that are normally more posterior. For example, when *Hoxa7*, whose normal anterior border of expression is in the thoracic region (see Fig. 5.25), is expressed throughout the whole antero-posterior axis, the basal occipital bone of the skull is transformed into a pro-atlas structure, normally the next more posterior skeletal structure.

5.8 Hox gene expression is activated in an anterior to posterior pattern

In all vertebrates, the Hox genes begin to be expressed at an early stage of gastrulation, when mesodermal cells begin their gastrulation movements. The genes expressed most anteriorly, which correspond to those at the 3′ end of a cluster, are expressed first. If an 'early' *Hoxd* gene is relocated to the 5′ end of the *Hoxd* complex, for example, its spatial and temporal expression pattern then resembles that of the neighboring *Hoxd13* (see Box 5E, p. 194). This shows that the structure of the Hox complex is crucial in determining the pattern of Hox gene expression.

Unlike the situation in *Drosophila*, where the activation of Hox genes clearly depends on factors unequally distributed along the antero-posterior axis (see Chapter 2), the mechanism of Hox gene activation in vertebrates is more complex and less well understood. One way the antero-posterior pattern of Hox gene expression might be established in the somitic mesoderm is through linking gene activation to the time spent by cells in the somite stem-cell region (see Fig. 5.20). In the chick and the mouse, the whole of the mesoderm alongside the dorsal midline derives from a small population of stem cells located in the anterior primitive streak and later in the tailbud. It has been shown that genes of the *Hoxb* cluster follow a strict temporal order of co-linear activation in this stem-cell region, the 3′ genes being activated before the 5′ ones. Expression of genes that are activated in the stem-cell region is maintained when the cells leave to form the pre-somitic mesoderm, and in this way the temporal pattern of Hox gene activation can be converted into positional information; the cells expressing different Hox genes are generated progressively along the antero-posterior axis as structures are laid down in an antero-posterior sequence. This hypothesis is similar to one proposed for specifying position along the proximo-distal axis of the vertebrate limb bud (see Chapter 11).

A gradient in retinoic acid from anterior to posterior along the trunk region of the mouse embryo (see Fig. 5.21) could also be involved in activating Hox genes in an antero-posterior pattern. Retinoic acid has been shown experimentally to alter the expression of Hox genes. More recently, genes encoding retinoic acid receptors and other proteins of the retinoid signaling pathway have been knocked out in mouse embryos, and this results in homeotic transformations of the axial skeleton. Mouse embryos lacking the gene *Cyp26a1*, which encodes an enzyme that degrades retinoic acid (see Box 5D, pp. 192–193), have many severe defects, including posterior homeotic transformations of cervical vertebrae into vertebrae carrying ribs. These transformations occur in regions where retinoic acid signaling is higher than normal and where anterior expansion of *Hoxb4* expression has also been seen. A role for retinoic acid in Hox gene activation is supported by studies in zebrafish. The zebrafish mutation *giraffe*, which affects the *Cyp26* gene, has anteriorly extended Hox gene expression and vertebral transformations. From vertebrate genome sequences, predicted genes for microRNAs (miRNAs) have been detected within Hox clusters and it has been proposed that the encoded miRNAs may be involved in post-transcriptional regulation of Hox protein expression (see Box 6B, p. 228, for how miRNAs work). More of the predicted target genes of a given miRNA tend to lie on the 3′ side of the miRNA gene rather than the 5′ side. Given this arrangement, the Hox miRNAs may help to repress the function of more anterior Hox genes, and this could explain the general principle of posterior prevalence—that more posteriorly expressed Hox genes tend to over-ride the function of more anterior Hox genes wherever they are coexpressed.

We have concentrated here on the expression of Hox genes in the mesoderm, but they are also expressed in a patterned way in the neural tube after its induction, and we shall look at this aspect of antero-posterior regionalization later in the chapter.

5.9 The fate of somite cells is determined by signals from the adjacent tissues

We shall now return to the individual somite and consider how it is patterned to give rise to different tissues. This patterning process is quite independent of the antero-posterior patterning of the whole pre-somitic mesoderm by Hox gene expression. Individual somites are subdivided into anterior and posterior halves with different properties, and this subdivision is linked to the segmentation clock. In the mouse, the transcription factor Mesp2 is expressed in a somite-wide stripe in the pre-somitic mesoderm that will give rise to the next somite (S0), and its expression then shrinks to just the anterior part of the somite (Fig. 5.28). Mesp2 activates a cascade of gene expression that leads to its expression being maintained in the anterior part of the somite. The restricted expression of *Mesp2* is thought to be a consequence of Notch–Delta signaling. The differences between the two halves of the somite are later used by neural crest cells and motor nerves as guidance cues to generate the periodic arrangement of spinal nerves and ganglia (see Chapter 8). The subdivision is also important for vertebra formation, as each vertebra derives from the anterior half of one somite and the posterior half of the preceding one, a process known as 'resegmentation'.

Individual somites are also patterned along their medio-lateral axis. But before we consider the individual somite, we will first look at how the mesoderm as a whole is patterned medio-laterally, such that paraxial mesoderm is formed on either side of the notochord (medially) and lateral plate mesoderm further away (laterally). In chick embryos, the mesoderm cells along the medio-lateral axis of the somite originate from different sites along the primitive streak and are brought together during gastrulation.

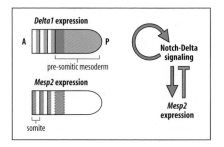

Fig. 5.28 Division of somites into anterior and posterior halves. Positive and negative feedback loops between Delta1 and Mesp2 (right) are essential for dividing the somite into an anterior half expressing Mesp2 and a posterior half expressing Delta1 (left), shown in dorsal views of the tail regions of E9.5 mouse embryos. Mesp2 is initially expressed in a somite-wide stripe in pre-somitic mesoderm and then as the somite forms Mesp expression is maintained in the anterior half.

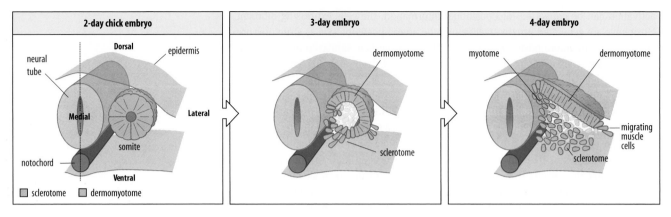

Fig. 5.29 **The fate map of a somite in the chick embryo.** The ventral medial quadrant (blue) gives rise to the sclerotome cells, which migrate to form the cartilage of the vertebrae. The rest of the somite—the dermomyotome—forms the myotome, which gives rise to all the trunk muscles, while the epithelial dermomyotome region gives rise to the dermis of the skin. The dermomyotome also gives rise to muscle cells that migrate into the limb buds.

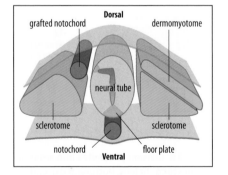

Fig. 5.30 **A signal from the notochord induces sclerotome formation.** A graft of an additional notochord to the dorsal region of a somite in a 10-somite chick embryo suppresses the formation of the dermomyotome from the dorsal portion of the somite, and instead induces the formation of sclerotome, which develops into cartilage. The graft also affects the shape of the neural tube and its dorso-ventral pattern.

The cells of the medial part of the somite come from the primitive streak close to Hensen's node, whereas the cells in the lateral part come from slightly more posterior in the primitive streak, and cells that form the lateral plate mesoderm originate even more posteriorly (see Fig. 4.43). The level of BMP-4 signaling mediates medio-lateral patterning of the mesoderm; it is high in the lateral plate mesoderm and lower in pre-somitic mesoderm, the decrease being due to the expression of the BMP antagonist Noggin in the pre-somitic mesoderm. The importance of low-level BMP signaling for somite formation is shown by the fact that inhibition of BMP by Noggin is sufficient to convert lateral plate mesoderm into somites.

The somites of the vertebrate embryo give rise to major axial structures: the cartilage cells of the embryonic axial skeleton—the vertebrae and ribs; all the skeletal muscles, including those of the limbs; and much of the dermis. The fate maps for particular somites have been made by grafting somites from a quail into a corresponding position in a chick embryo at a similar stage of development and following the fate of the quail cells. More recently, transgenic chickens expressing GFP have been created and grafting GFP-expressing chick tissues into normal chick embryos will allow long-term fate mapping experiments in the same avian species.

Cells located in the dorsal and lateral regions of a newly formed somite make up the **dermomyotome**, which expresses the gene *Pax3*, a homeobox-containing gene of the *paired* family (see Box 5E, p. 194). The dermomyotome gives rise to the **myotome**, from which muscle cells originate, and it later forms the epithelial sheet over the myotome from which the dermis originates. The dermomyotome also contains cells that contribute to the vasculature. Cells from the medial region of the myotome form mainly axial and back (epaxial) muscles, whereas lateral myotome cells migrate to give rise to abdominal and limb (hypaxial) muscles. All committed muscle precursors express muscle-specific transcription factors of the MyoD family. The ventral part of the medial somite contains **sclerotome** cells that express the *Pax1* gene and migrate ventrally to surround the notochord and develop into vertebrae and ribs (Fig. 5.29).

Which cells will form cartilage, muscle, or dermis is not yet determined at the time of somite formation. This is clearly shown by experiments in which the dorso-ventral orientation of newly formed somites is inverted but normal development nevertheless ensues. Specification of the fate of somite tissue requires signals from adjacent tissues. In the chick, determination of myotome occurs within hours of somite formation, whereas the future sclerotome is only determined later. Both the neural tube and the notochord produce signals that pattern the somite and are required for its future development. If the notochord and neural tube are removed, the cells in the somites undergo apoptosis; neither vertebrae nor axial muscles develop, although limb musculature still does.

The role of the notochord in specifying the fate of somitic cells has been shown by experiments in the chick, in which an extra notochord is implanted to one side of the neural tube, adjacent to the somite. This has a dramatic effect on somite differentiation, provided the operation is carried out on unsegmented pre-somitic mesoderm.

When the somite develops, there is an almost complete conversion to cartilage precursors (Fig. 5.30), suggesting that the notochord is an inducer of cartilage. In normal development, cartilage-inducing signals come from the notochord and the ventral region of the neural tube, the floor plate. Other signals come from the dorsal neural tube and the overlying ectoderm to specify the medial part of the dermomyotome, while signals from the lateral plate mesoderm are involved in specifying the lateral demomyotome (Fig. 5.31).

Some of these signals have been identified. In the chick, both the notochord and the floor plate of the neural tube express Sonic hedgehog (Shh), a secreted protein that is a key molecule for positional signaling in a number of developmental situations. We encountered Shh in Chapter 1 as a signal involved in craniofacial development (see Box 1F, p. 28) and also met it in Chapter 4 with respect to the asymmetry of structures about the midline. We shall meet it again in Chapter 11 (see Box 11C, pp. 424–425 for the Shh signaling pathway) in connection with limb development, and in Chapter 12 in the dorso-ventral patterning of the neural tube. In somite patterning, high levels of Shh signaling specify the ventral region of the somite and are required for sclerotome development. Low levels of Shh signaling, together with signals from the dorsal neural tube and from the overlying non-neural ectoderm, specify the dorsal region of the somite as dermomyotome (see Fig. 5.31). Wnt signals are good candidates for the dorsal signals. BMP-4 is the signal from the lateral plate mesoderm. Tendons arise from cells that come from the dorso-lateral domain of the sclerotome and specifically express the transcription factor Scleraxis. This tendon-progenitor region is induced at the boundary of the sclerotome and myotome.

Regulation of expression of the *Pax* homeobox genes in the somite by signals from the notochord and neural tube seems to be important in determining cell fate. *Pax3* is initially expressed in all cells that will form somites. Its expression is then modulated by signals from BMP-4 and Wnt proteins so that it becomes confined to muscle precursors. It is then further downregulated in cells that differentiate as the muscles of the back, but remains switched on in the migrating presumptive muscle cells that populate the limbs. Mice that lack a functional *Pax3* gene—*Splotch* mutants—lack limb muscles. In the chick, *Pax1* has been implicated in the formation of the scapula, a key element in the shoulder girdle, part of which is contributed by somites. All the scapula-forming cells express *Pax1*. But unlike the *Pax1*-expressing cells of the vertebrae, which are of sclerotomal origin, the blade of the scapula is formed from dermomyotome cells of chick somites 17–24, while the head of the scapula is derived from lateral plate mesoderm.

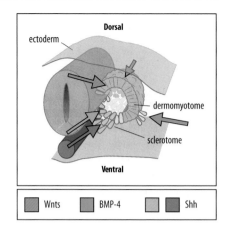

Fig. 5.31 The somite is patterned by signals secreted by adjacent tissues. The sclerotome is thought to be specified by a diffusible signal, Sonic hedgehog protein (Shh) (red and orange arrows), produced by the notochord (red) and the floor plate (orange) of the neural tube. Signals from the dorsal neural tube and ectoderm such as Wnts (blue arrows) would specify the dermomyotome, together with lateral signals such as BMP-4 (green arrow) from the lateral plate mesoderm.

After Johnson, R.L.: 1994.

Summary

Somites are blocks of mesodermal tissue that are formed after gastrulation. They form sequentially in pairs on either side of the notochord, starting adjacent to the posterior region of the hindbrain at the anterior end of the embryo. The somites give rise to the vertebrae, to the muscles of the trunk and limbs, and to most of the dermis of the skin. The pre-somitic mesoderm is patterned with respect to the antero-posterior axis while cells are in the node before the somites form. The first manifestation of this pattern is the expression of the Hox genes in the pre-somitic mesoderm. The positional identity of the somites is specified by the combinatorial expression of genes of the Hox complexes along the antero-posterior axis, from next to the hindbrain to the posterior end of the embryo, with the order of expression of these genes along the axis corresponding to their order along the chromosome. Mutation or overexpression of a Hox gene results, in general, in localized defects in the anterior parts of the regions in which the gene is expressed, and can cause homeotic transformations. We can think of Hox genes as providing positional information that specifies the identity of a region and its later development. They act on downstream gene targets about which we know relatively little.

Patterning of the mesoderm with respect to the medio-lateral axis and the formation of somites in the paraxial mesoderm depends on the inhibition of signals that would otherwise specify lateral plate mesoderm. Individual somites are also patterned by signals from the notochord, neural tube, and ectoderm, which induce particular regions of each somite to give rise to muscle, cartilage, or dermis.

The initial regionalization of the vertebrate brain

The developing vertebrate brain becomes divided into three main regions—**forebrain** (prosencephalon), **midbrain** (mesencephalon), and **hindbrain** (rhombencephalon)—illustrated here for 2-day and 4.5-day chick embryos (Fig. 5.32). Considerable lateral expansions develop from the embryonic forebrain and these are the rudiments of the optic vesicles, from which the eyes will form. The embryonic forebrain gives rise to the cerebral hemispheres (telencephalon), and the thalamus and the hypothalamus. The cortex of the cerebral hemispheres is where the highest-level processing centers for sensory information, motor control, learning, and memory are located, while the hypothalamus is the brain's link with the endocrine system, and controls the release of hormones from the pituitary. The embryonic midbrain gives rise to the tectum, which is the site of integration and relay centers for signals coming to and from the hindbrain and also for inputs from the sensory organs, such as the eye (see Chapter 12). The embryonic hindbrain is characterized by a series of seven or eight transient swellings, known as rhombomeres, and gives rise to the cerebellum, the pons and the medulla oblongata, which are concerned with the regulation of basic body activities that are independent of conscious control, such as regulation of muscle tone and posture (by the cerebellum), and heartbeat, breathing, and blood pressure.

Fig. 5.32 Local signaling centers in the developing vertebrate brain. Lateral view of the brain of a 2-day chick embryo (HH stage 13; left panel) and of a 4.5-day chick embryo (HH stage 24: right panel). The two signaling centers—the isthmus at the midbrain-hindbrain boundary (MHB) and the zona limitans intrathalamica (ZLI) in the forebrain—are indicated by dark blue lines. Bi-directional signaling from these centers (indicated by black arrows) is responsible for patterning the adjacent brain regions and involves FGF-8 in the MHB and Sonic hedgehog (Shh) in the ZLI. Wnt-1 is also expressed on the anterior side of the MHB. The boundary between *Otx2* expression in the midbrain (violet) and *Gbx2* expression in the anterior hindbrain (yellow) is linked to formation of the MHB signaling center.

Adapted from Kiecker, C., and Lumsden, A.: 2005.

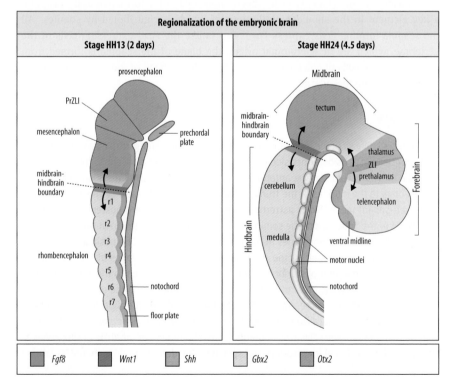

5.10 Local signaling centers pattern the brain along the antero-posterior axis

There are two local organizing regions in the developing midbrain and forebrain; one, known as the **isthmus**, is located at the midbrain–hindbrain boundary and regulates patterning of the posterior midbrain and the anterior hindbrain; the other, known as the **zona limitans intrathalamica**, is located in the forebrain and regulates the patterning of forebrain on both the anterior and posterior sides (see Fig. 5.32). The isthmus is already a major signaling center in the 2-day chick brain and is still active in brain of 4.5-day chick embryos, whereas the zona limitans intrathalamica is formed later and acts as a signaling center in the 4.5-day chick brain.

The signaling properties of the isthmus were first shown by grafting experiments in 2-day chick embryos. If the isthmus is grafted into the anterior midbrain or posterior forebrain, it converts that tissue into posterior midbrain, whereas if grafted to the posterior hindbrain, the isthmus converts that tissue into anterior hindbrain (which develops into the cerebellum). FGF-8 is expressed at the isthmus and is a good candidate for a bidirectional signal controlling the patterning of posterior midbrain and anterior hindbrain. Induced expression of FGF-8 in the chick anterior midbrain, for example, results in the expression of genes characteristic of the posterior midbrain. The zebrafish mutant *acerebellar* has a mutation that abolishes expression of FGF-8 almost completely, and as the name suggests, mutant fish lack the anterior hindbrain (r1, from which the cerebellum derives) and its associated organizer.

The midbrain–hindbrain organizer region develops at the border between the expression domains of transcription factors Otx2 (on the midbrain side) and Gbx2 (on the hindbrain side). Initial division into an Otx2-expressing anterior territory and a Gbx2-expressing posterior territory occurs in the neural plate, and is one of the earliest patterning events in brain development. Later, in the neural tube, FGF-8 is expressed on the hindbrain side of the border and Wnt-1 on the midbrain side (see Fig. 5.32). The other local signaling center, the zona limitans intrathalamica, signals bi-directionally via Shh, and is essential for formation of the pre-thalamus anteriorly and the thalamus posteriorly (see Fig. 5.32).

5.11 The hindbrain is segmented into rhombomeres by boundaries of cell-lineage restriction

Patterning of the posterior region of the vertebrate head and the hindbrain involves segmentation along the antero-posterior axis. The type of patterning of the nervous system does not occur along the spinal cord, where the pattern of dorsal root ganglia and ventral motor nerves at regular intervals—one pair per somite—is imposed by the somites. In the chick embryo, three segmented systems can be seen in the posterior head region by 3 days of development (Fig. 5.33). The mesoderm on either side of the notochord is subdivided into somites, the hindbrain (the rhombencephalon) is divided into eight **rhombomeres** (r1 to r8), and a series of branchial arches (b1 to b4) that are populated by neural crest cells migrating from the hindbrain have formed. The branchial arches give rise to the tissues of the head and throat other than the brain. A similar division of the hindbrain into rhombomeres occurs in the other model vertebrates.

Development of the hindbrain region of the head involves several interacting components. The neural tube produces both the segmentally arranged cranial nerves that innervate the face and neck, and also neural crest cells, which give rise both to peripheral nerves and most of the facial skeleton. In addition, the otic vesicle which develops opposite rhombomeres 5 and 6 gives rise to the ear. The main skeletal elements of the head in this region develop from the first three branchial arches. For example, the first arch gives rise to the jaws, while the second arch develops into

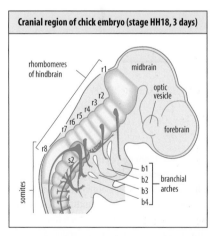

Cranial region of chick embryo (stage HH18, 3 days)

Fig. 5.33 The nervous system in a 3-day chick embryo (HH stage 18). At this stage in development the hindbrain is divided into eight rhombomeres (r1 to r8). The positions of three cranial nerves that arise from the hindbrain (V, VII, IX) are shown in green; b1 to b4 are the four branchial arches; b1 gives rise to the jaws; s = somite.

Adapted from Lumsden, A.: 1991.

the bony parts of the ear. This region of the head is a particularly valuable model for studying patterning along the antero-posterior axis because of the presence of the numerous different structures ordered along it.

Immediately after the neural tube of the chick embryo closes in this region, the future hindbrain becomes transversely constricted to define eight rhombomeres (see Fig. 5.33). The cellular basis for these constrictions is not entirely understood but is likely to involve differential cell division or changes in cell shape. The boundaries between rhombomeres subsequently become clearly differentiated from the adjacent neuroepithelium, with a marked increase in the extracellular space between cells. Cell-lineage tracing experiments show that once formed, individual rhombomeres are compartments of cell-lineage restriction; that is, cells and their descendants are confined within the rhombomere in which they originate and do not cross from one side of the boundary to the other. Tracking the descendants of individual labeled cells shows that before constrictions become visible, the descendants of a given cell can form a clone that spans a boundary. After the constrictions appear, however, clones of labeled cells are confined to a single rhombomere (Fig. 5.34). This lineage restriction suggests that each rhombomere is acting as an independent developmental unit. In this respect a rhombomere is behaving rather like a compartment, which is an important feature of insect development, as we saw in Chapter 2, but seems to be rare in vertebrates. The rhombomeres are no longer visible morphologically after about 4.5 days in the chick embryo (see Fig. 5.32).

Clues to how cells from adjacent rhombomeres might be prevented from mixing come from the discovery that the cells of odd-numbered rhombomeres seem to share some adhesive property that is not shared by even-numbered rhombomeres. This was originally revealed by experiments in which an odd-numbered and an even-numbered rhombomere from non-adjacent positions were placed together—and a new boundary formed between them. No new boundary formed when two odd-numbered rhombomeres were placed together, however, suggesting that their cells are compatible with each other. The boundary region itself is unlikely to be acting purely as a mechanical barrier, as cells from adjacent rhombomeres do not mix even when the boundary

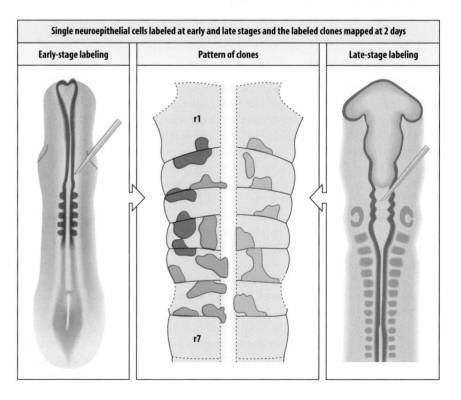

Fig. 5.34 Lineage restriction in rhombomeres of the embryonic chick hindbrain. Single cells are injected with a label (rhodamine-labeled dextran) at an early stage (left panel) or a later stage (right panel) of neurulation, and their descendants are mapped 2 days later. Cells injected before rhombomere boundaries form give rise to some clones that span two rhombomeres (dark red) as well as those that do not cross boundaries (red). Clones marked after rhombomere formation never cross the boundary of the rhombomere that they originate in (blue).

Adapted from Lumsden, A.: 1991.

between them is surgically removed. The finding that particular cell-surface proteins of the **ephrin** family and their **Eph receptors** are expressed separately in alternate rhombomeres provides one possible mechanism for preventing cell mixing. Ephrins and their receptors are transmembrane signaling proteins that interact with each other on adjacent cells and can both signal to their respective cells. They direct either repulsive or attractive cell–cell interactions (Box 5F, p. 205). In particular, the EphA4 receptor, which is strongly expressed in rhombomeres r3 and r5, and its ligand ephrin B2, which is expressed in r2, r4, and r6, could provide repulsive interactions at each rhombomere boundary and so prevent cell mixing (see Fig. 5.34). The expression of EphA4 in r3 and r5 is under the control of the transcription factor Krox20, which is itself expressed

---

Box 5F Eph receptors and their ephrin ligands

Eph receptors and their ephrin ligands are transmembrane proteins that can generate both repulsive and adhesive interactions between cells. Because they are both attached to a cell surface, they can only act through direct cell–cell contact, like Notch and Delta. They are best known as contact-dependent guidance molecules for growing axons in the nervous system, especially in the guidance of axons of retinal neurons to their correct locations in the brain's visual-processing centers; their role in axon guidance is discussed in Chapter 12. Among other roles, they act as guidance molecules for migrating neural crest cells, and appear to be involved in the final stages of somite formation—the separation of the new somite from the pre-somitic mesoderm. And according to work in mice, Ephs and ephrins are required to correctly set up the local neuronal circuits in the spinal cord that enable you to walk with alternate legs, rather than have to hop like a rabbit.

The 14 different Eph receptors in mammals fall into two structural classes: EphAs (EphA1–EphA8 and EphA10) and EphBs (EphB1–Eph B4 and EphB6). Ephrins similarly can be classified into ephrin As (ephrins A1–A5) and ephrin Bs (ephrins B1–B3). The A-type receptors typically bind to ephrin As, and the B-type receptors to ephrin Bs, but EphA4 is an exception, being able to bind both ephrin A and B ligands. Ephrins binding to their Eph receptors can generate a bidirectional signal, in which both the Eph and the ephrin send a signal to their own cell (see figure). Unlike most of the signaling receptors we have encountered so far, the main target of Eph-ephrin signaling is the actin cytoskeleton, whose rearrangement causes the changes in cell morphology and behavior that underlie repulsion.

Ephs are receptor tyrosine kinases, with an intracellular kinase domain, whereas ephrins signal by associating with and activating a cytoplasmic kinase. Ephrin-Eph contact is best known for producing repulsion between the interacting cells, as occurs at the interface between different rhombomeres (see Section 5.11), but Eph-ephrin interactions can also result in attraction and adhesion. Whether a particular Eph-ephrin pairing repels or attracts is not straightforward, and seems to depend on a range of different factors, including the cell types involved, the numbers of receptors present, and their degree of clustering.

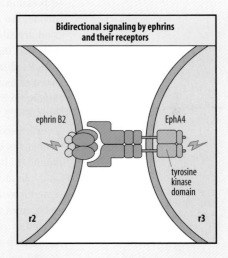

The repulsion commonly mediated by Eph-ephrin pairs is something of a paradox, given their inherent affinity for each other. One solution to the paradox was the discovery that repulsion by ephrin As involves cleavage of their ectodomain, which enables cell disengagement. Internalization of Eph-ephrin complexes by endocytosis is another possible solution. Repulsive interactions seem generally to require a relatively high level of signaling from the Eph tyrosine kinase. If this is lacking or at a low level, then the binding of Eph and ephrin will tend to promote cell adhesion. An example of this has been found in the closure of the neural tube in mice, in which a form of EphA7 lacking its kinase domain is produced by the cells of the neural folds as a result of alternative RNA splicing. The neural fold cells also express ephrin A5, and the adhesive interaction between EphA7 and ephrin A5 on opposing folds is essential for neural tube closure. Mouse mutants that lack ephrin A5 have neural tube defects resembling the human condition of anencephaly, in which the cranial neural tube fails to close, the forebrain does not develop, and the fetus is usually stillborn. In tissues other than the neural tube, in which EphA7 is produced with an intact tyrosine kinase domain, its interaction with ephrin A5 causes repulsion between the interacting cells.

in two stripes in the neural plate that become r3 and r5. In turn, separate sets of transcription factors have been shown to activate *Krox20* in r3 and r5, respectively.

The division of the hindbrain into rhombomeres has functional significance. A similar pattern of neuronal cell types develops along the antero-posterior axis of each rhombomere, and at the same time, each rhombomere acquires a unique identity in respect of the particular cranial nerves and other tissues it gives rise to. The delimitation and future development of individual rhombomeres is under the control of Hox and other transcription factors, and we shall next consider how the Hox genes provide positional information and identity both to the developing hindbrain itself and to the neural crest cells that migrate away from it.

5.12 Hox genes provide positional information in the developing hindbrain

Visible constriction of the hindbrain into rhombomeres is preceded by the patterned expression of transcription factor genes, including *Krox20* and Hox genes, that initially divides the neural tube into the prospective rhombomere segments. Hox gene expression provides at least part of molecular basis for the identities of both the rhombomeres and the neural crest at different positions in the hindbrain. No Hox genes are expressed in the most anterior part of the head or in r1, but Hox genes from four paralogous groups (1, 2, 3, and 4) are expressed in the rest of the mouse embryonic hindbrain in a well-defined pattern, which closely correlates with the segmental pattern (Fig. 5.35). The most anterior Hox gene expressed at this stage is *Hoxa2*, which is expressed in r2. In general, the different paralogous groups have different anterior margins of expression. For example, the anterior border of *Hoxb2* expression is at the boundary between r2 and r3, whereas the anterior margin of expression of *Hoxb3* is at the border between r4 and r5 (Fig. 5.36). Transplantation of rhombomeres from an anterior to a more posterior position alters the pattern of Hox gene expression so that it becomes the same as that normally expressed at the new location.

Studies at the molecular level have provided some indication to how Hox gene expression is controlled in the individual rhombomeres. For example, although the *Hoxb2* gene is expressed in the three contiguous rhombomeres r3, r4, and r5, its expression in r3 and r5 is controlled quite independently from that in r4. The regulatory region of the *Hoxb2* gene carries two separate *cis*-regulatory modules, one of which regulates its expression in r3 and r5, while expression in r4 is controlled through the other. In r3 and r5, *Hoxb2* is activated in part by the transcription factor Krox20, which is expressed in these rhombomeres but not in r4 (see Fig. 5.35 and Section 5.11), and there are binding sites for Krox20 in the corresponding regulatory

Fig. 5.35 Expression of Hox genes in the hindbrain. The expression of genes of three paralogous Hox complexes in the hindbrain (rhombomeres r1 to r8) is shown. *Hoxa1* and *Hoxd1* are not expressed at this stage. Note that there are no Hox genes expressed in r1. The Hox genes whose anterior expression starts in each rhombomere are listed underneath. With the exception of *Hoxb1*, which is only expressed in r4, all the Hox genes listed are also expressed in all rhombomeres posterior to their most anterior expression. The gene encoding the transcription factor Krox20 is expressed in rhombomeres 3 and 5.

After Krumlauf, R.: 1993.

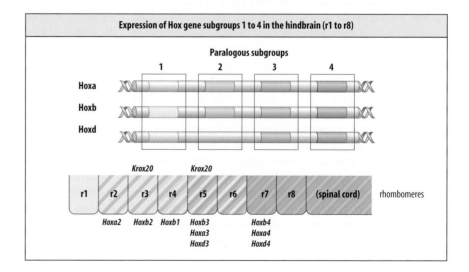

control region of *Hoxb2*. In the case of *Hoxb4*, whose anterior boundary of expression is at the boundary between r6 and r7, there is evidence that the gene is switched on by signals both from within the neural tube and from adjacent somites. One of these signals could be retinoic acid, which is produced by the somites and has been established to have a role in patterning the hindbrain. An absence of retinoic acid leads to loss of hindbrain rhombomeres, whereas an excess causes a transformation of cell fate from anterior to posterior.

Evidence that Hox genes determine the fate and future development of cells in the rhombomeres comes from misexpression experiments in which *Hoxb1* (normally expressed in r4) was expressed in r2. Each pair of rhombomeres produces motor axons that project to a single branchial arch: axons from rhombomeres r2/r3 project to the first branchial arch, whereas those from r4/r5 project to the second arch (see Fig. 5.33). *Hoxb1* is normally expressed in r4 but not in r2. If *Hoxb1* is misexpressed in an r2 rhombomere, however, this then sends axons to the second arch. This is yet another example of a homeotic transformation produced by misexpression of Hox genes.

Hox genes are not expressed in the anterior-most neural tissue—r1 of the hindbrain, the midbrain and forebrain. Instead, homeodomain transcription factors such as Otx and Emx are expressed anterior to the hindbrain and specify antero-posterior pattern in the anterior brain. As we saw earlier, the posterior limit of *Otx2* expression delineates the midbrain–hindbrain boundary in the chick (see Fig. 5.32). The *Drosophila orthodenticle* gene and the vertebrate *Otx* genes are homologous and provide a good example of the conservation of gene function during evolution. *orthodenticle* is expressed in the posterior region of the future *Drosophila* brain, and mutations in the gene result in a greatly reduced brain. In mice, *Otx1* and *Otx2* are expressed in overlapping domains in the developing forebrain and midbrain, and mutation in *Otx1* leads to brain abnormalities and epilepsy. Mice with a defective *Otx* gene can be partly rescued by replacing it with *orthodenticle*, even though the sequence similarity of the two proteins is confined to the homeodomain region. Human *Otx* can even rescue *orthodenticle* mutants in *Drosophila*.

5.13 Neural crest cells from the hindbrain migrate to populate the branchial arches

The pattern of Hox gene expression in the epidermal ectoderm of the branchial arches at any particular position along the antero-posterior axis is similar to that in the neural tube and neural crest. Neurons and neural crest cells obtain their positional values from the Hox genes they express as part of the neural tube and retain it when they migrate. Neural crest from the hindbrain migrates to populate the branchial arches, where it will contribute to mesodermal tissues of the face and skull. This migration is accomplished in about 10 hours between 1 and 2 days of incubation in the chick embryo. The ectoderm may acquire its positional value from the crest cells that come to lie beneath it, and thus come to express the same pattern of Hox genes.

The cranial neural crest that migrates from the rhombomeres of the dorsal region of the hindbrain has a segmental arrangement, correlating closely with the rhombomere from which the crest cells come. This has been revealed by labeling chick neural crest cells *in vivo* and following their migration pathways. Crest cells from rhombomeres 2, 4, and 6 populate the first, second, and third branchial arches, respectively (Fig. 5.37).

The crest cells have already acquired a positional value before they begin to migrate. When crest cells of rhombomere 4 are replaced by cells from rhombomere 2 taken from another embryo, these cells enter the second branchial arch but develop into structures characteristic of the first arch, to which they would normally have

Fig. 5.36 Gene expression in the hindbrain. The photograph shows a coronal section through the hindbrain of a E9.5 mouse embryo, which is transgenic for two reporter constructs. The first construct contains the *lacZ* gene under the control of an enhancer from *Hoxb2*, which directs expression in rhombomeres 3 and 5 (revealed as blue staining). The second construct contains an alkaline phosphatase gene under the control of an enhancer from *Hoxb1*, which directs expression in rhombomere 4 (revealed as greenish-brown staining). A similar enhancer directing expression in rhombomere 4 exists for *Hoxb2*. Anterior is to the left, and the positions of five of the rhombomeres are indicated (r2 to r6). Scale bar = 0.1 mm.

Photograph courtesy of J. Sharpe, from Lumsden, A. and Krumlauf, R.: 1996.

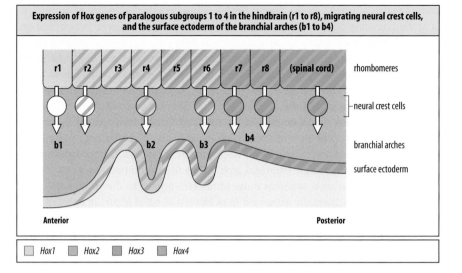

Fig. 5.37 Expression of Hox genes in the branchial region of the head. The expression of Hox genes in the hindbrain (rhombomeres r1 to r8), neural crest, branchial arches (b1 to b4), and surface ectoderm is shown. The arrows indicate the migration of neural crest cells into the branchial arches. Note the absence of neural crest migration from r3 and r5.

After Krumlauf, R.: 1993.

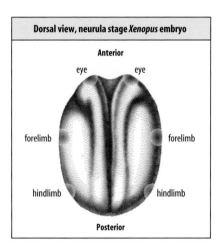

Fig. 5.38 The *Xenopus* embryo has become regionalized by the neurula stage. Various organs such as limbs, heart, and eyes will develop from specific regions (red) of the neurula after gastrulation is complete. Some of these regions, like the limb buds, are already determined at this stage and will not form any other structure. The boundaries of the regions are not sharply defined, however, and within each region or 'field' considerable regulation is still possible.

migrated. This can result in the development of an additional lower jaw in the chick embryo. However, neural crest cells have some developmental plasticity and their ultimate differentiation depends on signals from the tissues into which they migrate.

Gene knock-outs in mice have also shown that the Hox genes are involved in patterning the hindbrain region, though the results are not always easy to interpret; knock-out of a particular Hox gene can affect different populations of neural crest cells in the same animal, such as those that will form neurons and those that will form skeletal structures. Knock-out of the *Hoxa2* gene, for example, results in skeletal defects in that region of the head corresponding to the normal domain of expression of the gene, which extends from rhombomere 2 backwards. Segmentation itself is not affected, but the skeletal elements in the second branchial arch, all of which come from neural crest cells derived from rhombomere 4, are abnormal. The usual elements, such as the stapes of the inner ear are absent, but instead, some of the skeletal elements normally formed by the first branchial arch develop, such as Meckel's cartilage, a precursor of an element in the lower jaw. Thus, suppression of *Hoxa2* causes a partial homeotic transformation of one segment into another. The converse effect is seen when *Hoxa2* is misexpressed in a more anterior position. Expression of *Hoxa2* in the first branchial arch, in which it is not normally expressed leads to transformation of the first-arch cartilages, such as the quadrate and Meckel's cartilage, which is a precursor element of the lower jaw, into second-arch cartilages such as those of the tongue skeleton. In addition, if *Hox2a* is transfected into prospective anterior neural crest, the cells do not develop in their normal way and the skeletal structures they give rise to are abolished.

These observations, together with those described earlier in this chapter, show that during gastrulation the cells of vertebrates acquire positional values along the antero-posterior axis, and that this positional identity is encoded by the genes of the Hox complexes. Many of the anatomical differences between vertebrates are probably simply due to differences in the subsequent gene targets of Hox gene actions, which result in the emergence of different but homologous skeletal structures—the mammalian jaw or the bird's beak, for example.

5.14 The embryo is patterned by the neurula stage into organ-forming regions that can still regulate

At the neurula stage in *Xenopus*, the body plan has been established and the regions of the embryo that will form limbs, eyes, heart, and other organs have become determined (Fig. 5.38). This contrasts sharply with the blastula stage, at which time no such determination has occurred. The basic vertebrate phylotypic body plan is thus established during gastrulation. Although the positions of various organs are fixed, there is not yet any overt sign of differentiation. Nevertheless, numerous grafting experiments have shown that the potential to form a given organ is now confined to specific regions. Each of these regions has, however, considerable capacity for regulation, so that if part of the region is removed a normal structure can still form. For example, the region of the neurula that will form a forelimb will, when transplanted to a different region, still develop into a limb. If part of a future limb region is removed, the remaining part can still regulate to develop a normal limb. The development of limbs and other organs is discussed in Chapter 11.

Summary

The developing brain is divided into three regions—forebrain, midbrain, and hindbrain—along the antero-posterior axis. Signaling centers located at the midbrain-hindbrain boundary and within the forebrain produce signals that pattern adjacent regions of the hindbrain, midbrain, and forebrain, respectively. The hindbrain is segmented into rhombomeres, with the cells of each rhombomere respecting their boundaries. A Hox gene code provides positional values for the rhombomeres of the hindbrain, and the neural crest cells that derive from them, while other genes specify more anterior regions. Neural crest cells that migrate from the hindbrain populate the branchial arches in a position-dependent fashion. By the neurula stage, after gastrulation, the body plan has been established.

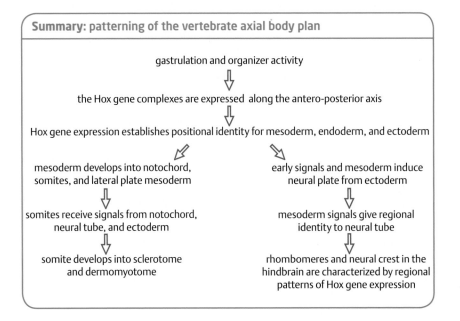

Summary: patterning of the vertebrate axial body plan

gastrulation and organizer activity

⇩

the Hox gene complexes are expressed along the antero-posterior axis

⇩

Hox gene expression establishes positional identity for mesoderm, endoderm, and ectoderm

↙ ↘

mesoderm develops into notochord, somites, and lateral plate mesoderm

early signals and mesoderm induce neural plate from ectoderm

⇩ ⇩

somites receive signals from notochord, neural tube, and ectoderm

mesoderm signals give regional identity to neural tube

⇩ ⇩

somite develops into sclerotome and dermomyotome

rhombomeres and neural crest in the hindbrain are characterized by regional patterns of Hox gene expression

Summary to Chapter 5

The germ layers specified during blastula formation become further patterned along the antero-posterior and dorso-ventral axes during gastrulation. The primary embryonic organizer is involved in the initial patterning that underlies the regionalization of the antero-posterior body axis. Positional identity of cells along the antero-posterior axis is encoded by the combinatorial expression of genes of the four Hox gene complexes. There is both spatial and temporal co-linearity between the order of Hox genes on the chromosomes and the order in which they are expressed along the axis from the hindbrain backwards. Inactivation or overexpression of Hox genes can lead both to localized abnormalities and to homeotic transformations of one 'segment' of the axis into another, indicating that these genes are crucial in specifying regional identity. At the end of gastrulation, the basic body plan has been laid down and the nervous system induced. Specific regions of each somite give rise to cartilage, muscle, and dermis, and these regions are specified by signals from the notochord, neural tube, lateral plate mesoderm, and ectoderm. Induction and patterning of the nervous system involves both signals in the early embryo and from the underlying mesoderm. Signaling centers pattern regions of the hindbrain, midbrain, and forebrain and in the hindbrain, Hox gene expression provides positional values for both neural tissue and neural crest cells.

■ End of chapter questions

Long answer (concept questions)

1. Distinguish between the notochord, the spinal cord, and the vertebral column.

2. Describe the relationships between the following structures: neuroectoderm, neural plate, neural folds, neural tube, and neural crest.

3. What experiment defines the phenomenon of primary embryonic induction? How do the results of this experiment change depending on the stage of the gastrula from which the dorsal lip of the blastopore is derived?

4. Review what we have learned about the specification of anteriormost mesoderm in the *Xenopus* gastrula including *goosecoid* activation in response to TGF-β family signaling, and the requirement for antagonists of ventral signals for head formation.

5. The 'default model' for neural induction in *Xenopus* proposes that the development of dorsal ectoderm as neural tissue is the default pathway. However, there are signals from both the vegetal cells and from ventral mesoderm, which could prevent dorsal ectoderm from forming. In light of this, outline the influences required for dorsal ectoderm to form. Include Ectodermin, Chordin (or noggin), and Cerberus.

6. The text states (Section 5.1): 'The important function of the Spemann organizer is to regulate these [BMP and Wnt] gradients by secreting growth-factor antagonists.' Discuss this statement.

7. Somite formation in chicks proceeds from anterior to posterior in response to a gradient of FGF-8. Describe this gradient: Where does the FGF-8 originate? Why does it form a gradient? How does the gradient trigger somite formation?

8. The Hox gene clusters of mice (Box 5E) illustrate several mechanisms that operate in genome evolution: genes can become duplicated to give rise to paralogs, the entire cluster can become duplicated to give rise to paralogous groups of genes, and within any given cluster, genes can be lost. Give an example of each of these three mechanisms.

9. The Hox genes are expressed in the vertebrate mesoderm both temporally and spatially in order, from one end of the Hox gene complex to the other. Refer to Fig. 5.26 and compare Hox gene expression in the mouse posterior cervical region to that in the anterior thoracic

region. How do the gene expression patterns illustrate the temporal gradient of expression along the Hox clusters? How do the gene expression patterns illustrate how different combinations of gene expression can produce spatial differences in cell identity along an axis?

10. Compare the consequences of *Hoxc8* gene deletion in the mouse (Fig. 5.27) with deletion of the *Ubx* gene in *Drosophila* (Fig. 2.48). Are the transformations seen toward anterior or posterior identities? What do these transformations say about the roles of Hox genes in identity specification? Do these transformations in mice qualify as homeotic transformations?

11. How does the zebrafish *giraffe* (*gir*) mutation support the hypothesis that a gradient of retinoic acid may be involved in the co-linearity of the vertebrate axis and Hox gene expression? Recalling that genes are typically named for the mutant phenotype produced, speculate on the phenotype of the *gir* mutation. If a deletion (a loss-of-function) of the *Cyp26a1* gene in mice causes cervical vertebrae to become more posterior rib-bearing vertebrae, would the *gir* mutation similarly be a loss-of-function, or, conversely, might it instead be a mutation that causes increased expression of the gene (gain of function)? Explain.

12. Describe the effects of the experiment shown in Fig. 5.30, in which a notochord from a donor embryo is grafted adjacent to the dorsal region of the neural tube in a host embryo. What signaling molecule is proposed to be responsible for the observed induction?

13. What kind of proteins are the Pax proteins? What is their relationship to Hox proteins? What is the role of Pax3 in the somite?

14. What are two signaling centers that form in the developing vertebrate brain? What signals do they emit? In what way do they qualify as true signal centers (or 'organizers')?

15. What are rhombomeres? What similarities can you list between segmentation of the vertebrate hindbrain into rhombomeres and segmentation in the embryo *Drosophila*?

16. What are ephrins? How do they differ from most of the signaling molecules that have been discussed? (Hint: are they secreted?) What is their role in the formation of rhombomeres?

17. Describe an experiment that demonstrates a homeotic transformation of rhombomeres. Outline the strategy used and compare this with other strategies that result in homeotic transformations.

Multiple choice (factual recall questions)

NB There is only one right answer to each question.

1. Somites will give rise to

a) heart and blood

b) internal organs of the abdomen

c) the vertebral column and skeletal muscles

d) the spinal cord

2. The equivalents of the *Xenopus* organizer in chicks and zebrafish are

a) Hensen's node and the embryonic shield

b) the hypoblast and the yolk syncytial layer

c) the notochord in both cases

d) the primitive streak and the mesendoderm

3. Chordin acts

a) as a TGF-β signaling molecule

b) to antagonize Wnt signaling

c) to bind DNA and activate transcription

d) to block the activity of BMP-4

4. Cerberus acts

a) as a TGF-β signaling molecule

b) to antagonize Wnt signaling

c) to bind DNA and activate transcription

d) to enhance the activity of BMP-4

5. Somite formation in chicks proceeds

a) from anterior to posterior, as Hensen's node retreats

b) from posterior to anterior, as the primitive streak elongates

c) from the dorsal midline outward, into lateral plate

d) simultaneously along the axis

6. In the early blastula, Wnt signaling and the subsequent nuclear accumulation of β-catenin specifies a dorsal fate, whereas after gastrulation, Wnt/β-catenin signaling specifies

a) a dorsal fate

b) a posterior fate

c) a ventral fate

d) an anterior fate

7. The sclerotome of the somite gives rise to _____, whereas the myotome gives rise to _____

a) muscles, vertebrae

b) peripheral nervous system, muscles

c) spinal cord, limb muscles

d) vertebrae, muscles of the back

8. Hox genes that are expressed more anteriorly comprise Hox genes

a) with all possible numbers and letters, since there is no systematic relationship between the gene designations and the expression patterns of the Hox genes

b) with higher numbers, such as *Hoxa13*

c) with letters, so that anterior genes are *Hoxa1–13*, and posterior genes are *Hoxp1–13*

d) with lower numbers, such as *Hoxa1*

9. Neural crest cells will contribute to which tissues or structures?

a) melanocytes in the skin

b) mesoderm of the face and skull

c) peripheral nervous system

d) all of these

Multiple choice answer key

1: c, 2: a, 3: d, 4: b, 5: a, 6: b, 7: d, 8: a, 9: d.

■ Section further reading

5.1 The inductive capacity of the organizer changes during gastrulation

Agathon, A., Thisse, C., Thisse, B.: **The molecular nature of the zebrafish tail organizer**. *Nature* 2003, **424**: 448–452.

Beddington, R.S.P., Robertson, E.H.: **Axis development and early asymmetry in mammals**. *Cell* 1999, **96**: 195–209.

Brickman, J.M., Jones, C.M., Clements, M., Smith, J.C., Beddington, R.S.P.: **Hex is a transcriptional repressor that contributes to anterior identity and suppresses Spemann organiser function**. *Development* 2000, **127**: 2303–2315.

Glinka, A., Wu, W., Delius, H., Monaghan, P., Blumenstock, C., Niehrs, C.: **Dickkopf-1 is a member of a new family of secreted proteins and functions in head induction**. *Nature* 1998, **391**: 357–362.

Glinka, A., Wu, W., Onichtchouk, D., Blumenstock, C., Niehrs, C.: **Head induction by simultaneous repression of Bmp- and Wnt-signalling in Xenopus**. *Nature* 1997, **389**: 517–519.

Griffin, K., Patient, R., Holder, N.: **Analysis of FGF function in normal and no tail zebrafish embryos reveals separate mechanisms for formation of the trunk and tail**. *Development* 1995, **121**: 2983–2994.

Jansen, H.J., Wacker, S.A., Bardine, N., Durston, A.J.: **The role of the Spemann organizer in anterior-posterior patterning of the trunk**. *Mech. Dev.* 2007, **124**: 668–681.

Jones, C.M., Broadbent, J., Thomas, P.Q., Smith, J.C., Beddington, R.S.P.: **An anterior signalling centre in *Xenopus* revealed by the homeobox gene XHex**. *Curr. Biol.* 1999, **9**: 946–954.

Joubin, K., Stern, C.D.: **Molecular interactions continuously define the organizer during the cell movements of gastrulation**. *Cell* 1999, **98**: 559–571.

Kiecker, C., Niehrs, C.: **The role of prechordal mesendoderm in neural patterning**. *Curr. Opin. Neurobiol.* 2001, **11**: 27–33.

Mukhopadhyay, M., Shtrom, S., Rodriguez-Esteban, C., Chen, L., Tsuku, T., Gomer, L., Dorward, D.W., Glinka, A., Grinberg, A., Huang, S.-P., Niehrs, C., Izpisúa Belmonte, J.-C., Westphal, H.: **Dickkopf1 is required for embryonic head induction and limb morphogenesis in the mouse**. *Dev. Cell* 2001, **1**: 423–434.

Niehrs, C.: **Regionally specific induction by the Spemann-Mangold organizer**. *Nat. Rev. Genet.* 2004, **5**: 425–434.

Piccolo, S., Agius, E., Leyns, L., Bhattacharya, S., Grunz, H., Bouwmeester, T., De Robertis, E.M.: **The head inducer Cerberus is a multifunctional antagonist of Nodal, BMP and Wnt signals.** *Nature* 1999, **397**: 707–710.

Schneider, V.A., Mercola, M.: **Spatially distinct head and heart inducers within the *Xenopus* organizer region.** *Curr. Biol.* 1999, **9**: 800–809.

5.2 The neural plate is induced in the ectoderm

Bachiller, D., Klingensmith, J., Kemp, C., Belo, J.A., Anderson, R.M., May, S.R., MacMahon, J.A., McMahon, A.P., Harland, R.M., Rossant, J., De Robertis, E.M.: **The organizer factors Chordin and Noggin are required for mouse forebrain development.** *Nature* 2000, **403**: 658–661.

Delfino-Machín, M., Lunn, J.S., Breitkreuz, D.N., Akai, J., Storey, K.G.: **Specification and maintenance of the spinal cord stem zone.** *Development* 2005, **132**: 4273–4283.

De Robertis, E.M.: **Spemann's organizer and self-regulation in amphibian embryos.** *Nat. Rev. Mol. Cell Biol.* 2006, **4**: 296–302.

De Robertis, E.M., Kuroda, H.: **Dorsal-ventral patterning and neural induction in *Xenopus* embryos.** *Annu. Rev. Dev. Biol.* 2004, **20**: 285–308.

Londin, E.R., Niemiec, J., Sirotkin, H.I.: **Chordin, FGF signaling, and mesodermal factors cooperate in zebrafish neural induction.** *Dev. Biol.* 2005, **279**: 1–19.

Marchal, L., Luxardi, G., Thomé, V., Kodjabachian L.: **BMP inhibition initiates neural induction via FGF signaling and Zic genes.** *Proc. Natl Acad. Sci. USA* 2009; **106**: 17437–17442.

Stern, C.: **Neural induction: old problems, new findings, yet more questions.** *Development* 2005, **132**: 2007–2021.

Streit, A., Berliner, A.J., Papanayotou, C., Sirulnik, A., Stern, C.D.: **Initiation of neural induction by FGF signalling before gastrulation.** *Nature* 2000, **406**: 74–78.

Wilson, S.L., Rydström, A., Trimborn, T., Willert, K., Nusse, R., Jessell, T.M., Edlund, T.: **The status of Wnt signaling regulates neural and epidermal fates in the chick embryo.** *Nature* 2001, **411**: 325–329.

Box 5A Chromatin-remodeling complexes

Ho, L., Crabtree, G.R.: **Chromatin remodelling during development.** *Nature* 2010, **463**: 474–484.

5.3 The nervous system is initially patterned by signals from the mesoderm

Ang, S.L., Rossant, J.: **HNF-3β is essential for node and notochord formation in mouse development.** *Cell* 1994, **78**: 561–574.

Aybar, M.J., Mayor, R.: **Early induction of neural crest cells: lessons learned from frog, fish and chick.** *Curr. Opin. Genet. Dev.* 2002, **12**: 452–458.

Doniach, T.: **Basic FGF as an inducer of antero-posterior neural pattern.** *Cell* 1995, **85**: 1067–1070.

Foley, A.C., Skromne, I., Stern, C.D.: **Reconciling different models of forebrain induction and patterning: a dual role for the hypoblast.** *Development* 2000, **127**: 3839–3854.

Kiecker, C., Niehrs, C.: **A morphogen gradient of Wnt/beta-catenin signaling regulates anteroposterior neural patterning in Xenopus.** *Development* 2001, **128**: 4189–4201.

Kudoh, T., Concha, M.L., Houart, C., Dawid, I.B., Wilson, S.W.: **Combinatorial Fgf and Bmp signalling patterns the gastrula ectoderm into prospective neural and epidermal domains.** *Development* 2004, **131**: 3581–3592.

Pera, E.M., Ikeda, A., Eivers, E., De Robertis, E.M.: **Integration of IGF, FGF, and anti-BMP signals via Smad1 phosphorylation in neural induction.** *Genes Dev.* 2003, **17**: 3023–3028.

Sasai, Y., De Robertis, E.M.: **Ectodermal patterning in vertebrate embryos.** *Dev. Biol.* 1997, **182**: 5–20.

Sharman, A.C., Brand, M.: **Evolution and homology of the nervous system: cross-phylum rescues of *otd/Otx* genes.** *Trends Genet.* 1998, **14**: 211–214.

Sheng, G., dos Reis, M., Stern, C.D.: **Churchill, a zinc finger transcriptional activator, regulates the transition between gastrulation and neurulation.** *Cell* 2003, **115**: 603–613.

Stern, C.D.: **Initial patterning of the central nervous system: how many organizers?** *Nat Rev. Neurosci.* 2001, **2**: 92–98.

Storey, K., Crossley, J.M., De Robertis, E.M., Norris, W.E., Stern, C.D.: **Neural induction and regionalization in the chick embryo.** *Development* 1992, **114**: 729–741.

5.4 Neural crest cells arise from the borders of the neural plate

Huang, X., Saint-Jeannet, J-P.: **Induction of the neural crest and the opportunities of life on the edge.** *Dev. Biol* 2004, **275**: 1–11.

5.5 Somites are formed in a well-defined order along the antero-posterior axis

Bray, S.J.: **Notch signalling: a simple pathway becomes complex.** *Nat. Rev. Mol. Cell Biol.* 2006, **7**: 678–689.

Dequeant, M.L., Pourquié, O.: **Segmental patterning of the vertebrate embryonic axis.** *Nat. Rev. Genet.* 2008, **9**: 370–382.

Dubrulle, J., Pourquié, O.: ***fgf8* mRNA decay establishes a gradient that couples axial elongation to patterning in the vertebrate embryo.** *Nature* 2004, **427**: 419–422.

Dubrulle, J., Pourquié, O.: **Coupling segmentation to axis formation.** *Development* 2004, **131**: 5783–5793.

Kawakami, Y., Raya, A., Raya, R.M., Rodriguez-Esteban, C., Izpisua Belmonte, J.C.: **Retinoic acid signalling links left-right asymmetric patterning and bilaterally symmetric somitogenesis in the zebrafish embryo.** *Nature* 2005, **435**: 165–171.

Kieny, M., Mauger, A., Sengel, P.: **Early regionalization of somitic mesoderm as studied by the development of the axial skeleton of the chick embryo.** *Dev Biol.* 1972, **28**: 142–161.

Lai, E.: **Notch signaling: control of cell communication and cell fate.** *Development* 2004, **131**: 965–973.

Lewis, J.: **Autoinhibition with transcriptional delay: a simple mechanism for the zebrafish somitogenesis oscillator.** *Curr. Biol* 2003, **13**: 1398–408.

Stern, C.D., Charité, J., Deschamps, J., Duboule, D., Durston, A.J., Kmita, M., Nicolas, J.-F., Palmeirim, I., Smith, J.C., Wolpert, L.: **Head–tail patterning of the vertebrate embryo: one, two or many unsolved problems?** *Int. J. Dev Biol.* 2006, **50**: 3–15.

Vermot, J., Pourquié, O.: **Retinoic acid coordinates somitogenesis and left-right patterning in vertebrate embryos.** *Nature* 2005, **435**: 215–220.

5.6 Identity of somites along the antero-posterior axis is specified by *Hox* gene expression

Burke, A.C., Nelson, C.E., Morgan, B.A., Tabin, C.: **Hox genes and the evolution of vertebrate axial morphology.** *Development* 1995, **121**: 333–346.

Godsave, S., Dekker, E.J., Holling, T., Pannese, M., Boncinelli, E., Durston, A.: **Expression patterns of Hoxb in the *Xenopus* embryo suggest roles in antero-posterior specification of the hindbrain and in dorso-ventral patterning of the mesoderm.** *Dev. Biol.* 1994, **166**: 465–476.

Kondo, T., Duboule, D.: **Breaking colinearity in the mouse HoxD complex.** *Cell* 1999, **97**: 407–417.

Krumlauf, R.: **Hox genes in vertebrate development.** *Cell* 1994, **78**: 191–201.

Nowicki, J.L., Burke, A.C.: **Hox genes and morphological identity: axial versus lateral patterning in the vertebrate mesoderm.** *Development* 2000, **127**: 4265–4275.

Box 5D Retinoic acid—a small-molecule intercellular signal

Niederreither, K., Dolle, P.: **Retinoic acid in development: towards an integrated view.** *Nat. Rev. Genet.* 2008, **9**: 541–553.

Rossant, J., Zirngibl, R., Cado, D., Shago, M., Giguère, V.: **Expression of a retinoic acid response element-hsplacZ transgene defines specific domains of transcriptional activity during mouse embryogenesis.** *Genes Dev.* 1991, **5**: 1333–1344.

5.7 Deletion or overexpression of Hox genes causes changes in axial patterning

Condie, B.G., Capecchi, M.R.: **Mice with targeted disruptions in the paralogous genes Hoxa3 and Hoxd3 reveal synergistic interactions.** *Nature* 1994, **370**: 304–307.

Duboule, D.: **Vertebrate Hox genes and proliferation: an alternative pathway to homeosis?** *Curr. Opin. Genet. Dev.* 1995, **5**: 525–528.

Favier, B., Le Meur, M., Chambon, P., Dollé, P.: **Axial skeleton homeosis and forelimb malformations in Hoxd11 mutant mice.** *Proc. Natl Acad. Sci. USA* 1995, **92**: 310–314.

Kessel, M., Gruss, P.: **Homeotic transformations of murine vertebrae and concomitant alteration of the codes induced by retinoic acid.** *Cell* 1991, **67**: 89–104.

Ruiz-i-Altaba, A., Jessell, T.: **Retinoic acid modifies mesodermal patterning in early *Xenopus* embryos.** *Genes Dev.* 1991, **5**: 175–187.

Wellik, D.M., Capecchi, M.R.: **Hox10 and Hox11 genes are required to globally pattern the mammalian skeleton.** *Science* 2003, **301**: 363–367.

5.8 Hox gene expression is activated in an anterior to posterior pattern

Duboule, D.: **Vertebrate Hox gene regulation: clustering and/or colinearity?** *Curr. Opin. Genet. Dev.* 1998, **8**: 514–518.

Sakai, Y., Meno, C., Fujii, H., Nishino, J., Shiratori, H., Saijoh, Y., Rossant, J., Hamada, H.: **The retinoic acid-inactivating enzyme CYP26 is essential for establishing an uneven distribution of retinoic acid along the anteroposterior axis within the mouse embryo.** *Genes Dev.* 2006, **15**: 213–225.

Vasiliauskas, D., Stern, C.D.: **Patterning the embryonic axis: FGF signaling and how vertebrate embryos measure time.** *Cell* 2001, **106**: 133–136.

Wacker, S.A., Janse, H.J., McNulty, C.L., Houtzager, E., Durston, A.J.: **Timed interactions between the Hox expressing non-organiser mesoderm and the Spemann organiser generate positional information during vertebrate gastrulation.** *Dev. Biol.* 2004, **268**: 207–219.

Yekta, S., Tabin, C., Bartel, D.P.: **MicroRNAs in the Hox network: an apparent link to posterior prevalence.** *Nat. Rev. Genet.* 2008, **9**: 789–796.

5.9 The fate of somite cells is determined by signals from the adjacent tissues

Brand-Saberi, B., Christ, B.: **Evolution and development of distinct cell lineages derived from somites.** *Curr. Topics Dev. Biol.* 2000, **48**: 1–42.

Brent, A.E., Braun, T., Tabin, C.J.: **Genetic analysis of interactions between the somitic muscle, cartilage and tendon cell lineages during mouse development.** *Development* 2005, **132**: 515–528.

Huang, R., Zhi, Q., Patel, K., Wilting, J., Christ, B.: **Dual origin and segmental organisation of the avian scapula.** *Development* 2000, **127**: 3789–3794.

Olivera-Martinez, I., Coltey, M., Dhouailly, D., Pourquié, O.: **Mediolateral somitic origin of ribs and dermis determined by quail-chick chimeras.** *Development* 2000, **127**: 4611–4617.

Pourquié, O., Fan, C.-M., Coltey, M., Hirsinger, E., Watanabe, Y., Bréant, C., Francis-West, P., Brickell, P., Tessier-Lavigne, M., Le Douarin, N.M.: **Lateral and axial signals involved in avian somite patterning: a role for BMP-4.** *Cell* 1996, **84**: 461–471.

Tonegawa, A., Takahashi, Y.: **Somitogenesis controlled by Noggin.** *Dev. Biol.* 1998, **202**: 172–182.

5.10 Local signaling centers pattern the brain along the antero-posterior axis

Brocolli, V., Boncinelli, E., Wurst, W.: **The caudal limit of Otx2 expression positions the isthmaic organizer.** *Nature* 1999, **401**: 164–168.

Kiecker, C., Lumsden, A.: **Compartments and their boundaries in vertebrate brain development.** *Nat. Rev. Neurosci.* 2005, **6**: 553–564.

Rhinn, M., Brand, M.: **The midbrain-hindbrain boundary organizer.** *Curr. Opin. Neurobiol.* 2001, **11**: 34–42.

5.11 The hindbrain is segmented into rhombomeres by boundaries of cell-lineage restriction

Cooke, J.E., Moens, C.B.: **Boundary formation in the hindbrain: *Eph* only it were simple.** *Trends Neurobiol.* 2002, **25**: 264–267.

Klein, R.: **Neural development: bidirectional signals establish boundaries.** *Curr. Biol.* 1999, **9**: R691–R694.

Lumsden, A.: **Segmentation and compartition in the early avian hindbrain.** *Mech. Dev.* 2004, **121**: 1081–1088.

Xu, Q., Mellitzer, G., Wilkinson, D.G.: **Roles of Eph receptors and ephrins in segmental patterning.** *Phil Trans. Roy. Soc. B* 2000, **355**: 993–1002.

Box 5F Eph receptors and their ephrin ligands

Klein, R.: **Eph/ephrin signaling in morphogenesis, neural development and plasticity.** *Curr. Opin. Cell Biol.* 2004, **16**: 580–589.

Kullander, K., Butt, S.J.B., Lebret, J.M., Lundefeld, L., Restrepo, E., Ryderström, A., Klein, Rudiger, Kiehn, O.: **Role of Eph4A and ephrin B3 in local neuronal circuits that control walking**. *Science* 2003, **299**: 1889–1892.

5.12 Hox genes provide positional information in the developing hindbrain

Bell, E., Wingate, R.J., Lumsden, A.: **Homeotic transformation of rhombomere identity after localized Hoxb1 misexpression**. *Science* 1999, **284**: 2168–2171.

Grammatopoulos, G.A., Bell, E., Toole, L., Lumsden, A., Tucker, A.S.: **Homeotic transformation of branchial arch identity after Hoxa2 overexpression**. *Development* 2000, **127**: 5355–5365.

Grapin-Botton, A., Bonnin, M-A., McNaughton, L.A., Krumlauf, R., Le Douarin, N.M.: **Plasticity of transposed rhombomeres: Hox gene induction is correlated with phenotypic modifications**. *Development* 1995, **121**: 2707–2721.

Hunt, P., Krumlauf, R.: **Hox codes and positional specification in vertebrate embryonic axes**. *Annu. Rev. Cell Biol.* 1992, **8**: 227–256.

Krumlauf, R.: **Hox genes and pattern formation in the branchial region of the vertebrate head**. *Trends Genet.* 1993, **9**: 106–112.

Nonchev, S., Maconochie, M., Vesque, C., Aparicio, S., Ariza-McNaughton, L., Manzanares, M., Maruthainar, K., Kuroiwa, A., Brenner, S., Charnay, P., Krumlauf, R.: **The conserved role of Krox-20 in directing Hox gene expression during vertebrate hindbrain segmentation**. *Proc. Natl Acad. Sci. USA* 1996, **93**: 9339–9345.

Rijli, F.M., Mark, M., Lakkaraju, S., Dierich, A., Dolle, P., Chambon, P.: **A homeotic transformation is generated in the rostral branchial region of the head by disruption of Hoxa2, which acts as a selector gene**. *Cell* 1993, **75**: 1333–1349.

Wassef, M.A., Chomette, D., Pouilhe, M., Stedman, A., Havis, E., Desmarquet-Trin Dinh, C., Schneider-Maunoury, S., Gilardi-Hebenstreit, P., Charnay, P., Ghislain, J.: **Rostral hindbrain patterning involves the direct activation of a Krox20 transcriptional enhancer by Hox/Pbx and Meis factors**. *Development* 2008, **135**: 3369–3378.

White, R.J., Schilling, T.F.: **How degrading: Cyp26s in hindbrain development**. *Dev. Dyn.* 2008, **237**: 2775–2790.

5.13 Neural crest cells from the hindbrain migrate to populate the branchial arches

Keynes, R., Lumsden, A.: **Segmentation and the origin of regional diversity in the vertebrate central nervous system**. *Neuron* 1990, **4**: 1–9.

Le Douarin, N.M., Creuzet, S., Couly, G., Dupin, E.: **Neural crest plasticity and its limits**. *Development* 2004, **131**: 4637–4650.

5.14 The embryo is patterned by the neurula stage into organ-forming regions that can still regulate

De Robertis, E.M., Morita, E.A., Cho, K.W.Y.: **Gradient fields and homeobox genes**. *Development* 1991, **112**: 669–678.

Development of nematodes, sea urchins, and ascidians

- Nematodes   : - Echinoderms   : - Ascidians

Having looked at early embryonic development in Drosophila *and vertebrates, we will now examine some aspects of the early development of three other model invertebrate organisms that introduce some different developmental mechanisms. The nematode* Caenorhabditis elegans *is a very important model for the specification of fate associated with asymmetric cell division, as much of its early development involves patterning on a cell-by-cell basis rather than by morphogens affecting groups of cells. Sea urchins, which represent the echinoderms, are models for highly regulative development, and so their development illustrates some principles common to that of vertebrates, which is also highly regulative, but in a simpler system. Complete gene circuits underlying early sea-urchin development have been identified. The ascidians are of interest as they are chordates, as are the vertebrates; comparison of ascidian and vertebrate development is therefore of great interest in understanding chordate evolution.*

This chapter considers aspects of body-plan development in three model invertebrate organisms—nematodes, sea urchins (representing echinoderms), and the ascidians (commonly known as sea squirts), which are of interest as they are chordates. The evolutionary relationships between the organisms discussed in this chapter are shown in Fig. 1.11. All conform to the general plan of animal development: cleavage leads to a blastula (or blastula-equivalent stage in *Caenorhabditis*), which then undergoes gastrulation with the emergence of a body plan.

There is an old, and now less fashionable, distinction sometimes made between so-called regulative and mosaic development—the former involving mainly cell–cell interactions whereas the latter is based on localized cytoplasmic factors and their distribution through **asymmetric cell divisions** (see Section 1.17). Nematodes and ascidians show many instances of mosaic-like development, while sea-urchin development is highly regulative. There are, however, elements of both types of development in most embryos.

A feature of nematodes and ascidians is that cell fate is often specified on a cell-by-cell basis, a typical characteristic of mosaic development, and in general does not rely on positional information established by gradients of morphogens. This contrasts with the specification of cell fate in groups of cells in *Drosophila*, vertebrates, and the sea urchin. The early embryos of many invertebrates contain far fewer cells than those of flies and vertebrates, with each cell acquiring a unique identity at an early stage of development. In the nematode, for example, there are only 28 cells when gastrulation starts, compared with thousands in *Drosophila*. Specification on a cell-by-cell basis often makes use of asymmetric cell division and the unequal distribution of cytoplasmic factors. Daughter cells resulting from asymmetric cell division often adopt different fates not as a result of extracellular signals, but autonomously, as a result of the unequal distribution of some factor between them. However, asymmetric cell division in the early stages of development does not mean that cell–cell interactions are absent or unimportant in these organisms. Differences in the fates of two daughter cells produced by asymmetric division can also be specified by extracellular factors and cell signaling.

We begin with the nematode *Caenorhabditis elegans*, which has been studied intensively and in which many key developmental genes and signaling pathways have been identified. In this animal, specification is largely on a cell-by-cell basis. We then consider sea urchins, which develop much more like vertebrates, in that their embryos rely heavily on intercellular interactions and are highly regulative, and patterning involves groups of cells. Finally we consider ascidians, with particular emphasis on the role of the localization of cytoplasmic determinants in their early development.

Nematodes

The free-living soil nematode *C. elegans*, whose life cycle is shown in Fig. 6.1, is an important model organism in developmental biology. Its advantages are its suitability for genetic analysis, the small number of cells and their invariant lineage, and the transparency of the embryo, which allows the formation of each cell to be observed. This led Sydney Brenner to choose *C. elegans* as a model organism for studying the genetic basis of development. Brenner and his co-workers, Robert Horvitz and

Fig. 6.1 Life cycle of the nematode *Caenorhabditis elegans*. After cleavage and embryogenesis there are four larval stages (L1–L4) before the sexually mature adult develops. Adults of *C. elegans* are usually hermaphrodite, although males can develop. The photographs show: the two-cell stage (top, scale bar = 10 µm); an embryo after gastrulation with the future larva curled up (middle, scale bar = 10 µm); and the four larval stages and adult (bottom, scale bar = 0.5 mm).

Photographs courtesy of J. Ahringer.

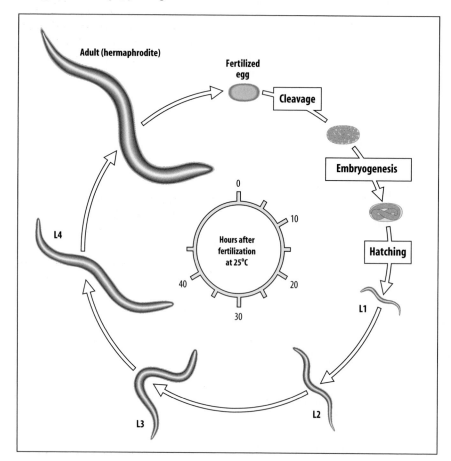

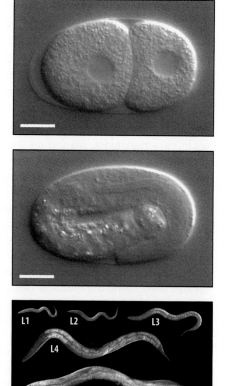

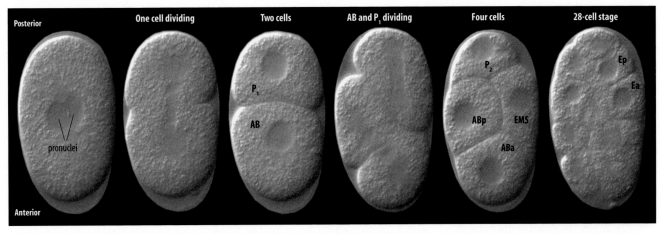

| Posterior | One cell dividing | Two cells | AB and P₁ dividing | Four cells | 28-cell stage |

pronuclei — P₁, AB — P₂, ABp, EMS, ABa — Ep, Ea

Anterior

John Sulston, were awarded the Nobel Prize in Physiology or Medicine in 2002 for discoveries concerning genetic control of organ development and apoptosis—cell suicide—in their studies on the worm.

C. elegans has a simple anatomy and the adults are about 1 mm long and just 70 μm in diameter. Nematodes can be grown on agar plates in large numbers and early larval stages can be stored frozen and later resuscitated. *C. elegans* adults are mainly **hermaphrodite**; in nematodes, these are females that make their own sperm for a short time and then switch to making oocytes. They are capable of self-fertilization. Small numbers of males exist and can be used in mating experiments. Embryonic development is rapid, the larva hatching after 15 hours at 20°C, although maturation through larval stages to adulthood takes about 50 hours.

The nematode egg is small, ovoid and only 50 μm long. Polar bodies are formed after fertilization. Before the male and female nuclei fuse, there is what appears to be an abortive cleavage, but after fusion of the nuclei, true cleavage begins (Fig. 6.2). The first cleavage is asymmetric and generates an anterior AB cell and a smaller posterior P₁ cell. At the second cleavage, AB divides to give ABa anteriorly and ABp posteriorly, while P₁ divides to give P₂ and EMS. At this stage the main axes can already be identified, as P₂ is posterior and ABp dorsal.

The AB cells and the P₂ cell will also divide in a well-defined pattern to give rise to the other tissues of the worm. Gastrulation starts at the 28-cell stage, when the descendants of the E cell (itself produced by division of the EMS cell) that will form the gut move inside. Not all cells that are formed during embryonic development survive; programmed death, or apoptosis, of specific cells is an integral feature of nematode development.

The newly hatched larva (Fig. 6.3), while similar in overall organization to the mature adult, is sexually immature and lacks a gonad and its associated structures,

Fig. 6.2 Cleavage of the *Caenorhabditis elegans* embryo. After fertilization, the pronuclei of the sperm and egg fuse. The egg then divides into a large anterior AB cell and a smaller, posterior P₁ cell. At the next cell division, AB divides into ABa and ABp, while P₁ divides into P₂ and EMS. Each of the cells will continue to divide in a well-defined pattern within these groups to give rise to particular cell types and tissues. Gastrulation begins at the 28-cell stage. By this stage, further division of the EMS cell has produced the E cells, Ea and Eb, which will give rise to the gut, and MS cells (not labeled here) which give rise to a variety of other cell types.

Photographs courtesy of J. Ahringer.

Fig. 6.3 *Caenorhabditis elegans* larva at the L1 stage (20 hours after fertilization). The vulva will form from the gonad primordium.

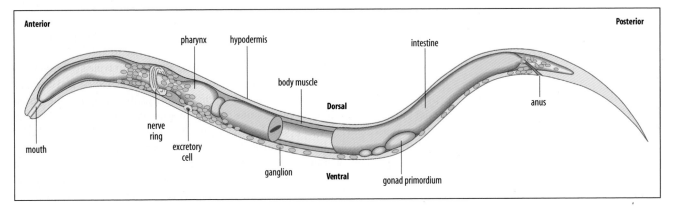

Anterior — Posterior

pharynx, hypodermis, intestine, body muscle, **Dorsal**, mouth, nerve ring, excretory cell, ganglion, **Ventral**, gonad primordium, anus

such as the vulva of hermaphrodites, which are required for reproduction. Post-embryonic development takes place during a series of four larval stages separated by molts. The additional cells in the adult are derived largely from precursor blast cells (P cells) that are distributed along the body axis. Each of these blast cells founds an invariant lineage involving between one and eight cell divisions. The vulva, for example, is derived from blast cells P5, P6, and P7. One can think of post-embryonic development in the nematode as the addition of adult structures to the basic larval plan.

It is a triumph of direct observation that, with the aid of Nomarski interference microscopy, the complete lineage of every cell in the nematode *C. elegans* has been worked out. The pattern of cell division is largely invariant—it is virtually the same in every embryo. The larva, when it hatches, is made up of 558 cells, and after four further molts this number has increased to 959, excluding the germ cells, which vary in number. This is not the total number of cells derived from the egg, as 131 cells die during development as a result of apoptosis, which is discussed in more detail in Chapter 10. As the fate of every cell at each stage is known, a fate map can be accurately drawn at any stage and thus has a precision not found in any vertebrate. As with any fate map, however, even where there is an invariant cell lineage, this precision in no way implies that the lineage must determine the fate or that the fate of the cells cannot be altered. As we shall see, interactions between cells have a major role in determining cell fate in the nematode.

The complete genome of *C. elegans* has been sequenced, and contains nearly 20,000 predicted genes. *C. elegans* does not have extensive alternative RNA splicing, so that most genes code for just one protein, which may be part of the reason the anatomically simple worm apparently needs more genes than the more complex fruit fly. Around 1700 genes have been identified as affecting development, two-thirds of which were found using the technique of RNA interference (RNAi; Box 6A, p. 220). Many nematode developmental genes are related to genes that control development in *Drosophila* and other animals; they include the Hox genes (see Box 5E, p. 194), and genes for signaling proteins of the TGF-β, Wnt, and Notch families (see Box 4A, p. 131) and their associated intracellular pathways. One exception is the Hedgehog signaling pathway (see Fig. 2.40), which is not present in *Caenorhabditis*, although it does have genes for proteins that can be considered to be related to Hedgehog. Although zygotic gene expression begins at the four-cell stage, maternal components control almost all of the development up to gastrulation at the 28-cell stage.

6.1 The antero-posterior axis in *Caenorhabditis elegans* is determined by asymmetric cell division

The first cleavage of the nematode egg is unequal, dividing the egg into a large anterior AB cell and a smaller posterior P_1 cell. This asymmetry defines the antero-posterior axis and is determined at fertilization. The P_1 cell behaves rather like a stem cell; at each further division it produces one P-type cell and one daughter cell that will embark on another developmental pathway. For the first three divisions, the P-cell daughters give rise to body cells, but after the fourth cleavage they only give rise to germ cells (Fig. 6.4). Division of the AB cell gives rise to anterior and posterior AB daughter cells. The anterior daughter, ABa, gives rise to typically ectodermal tissues, such as the epidermis (called the hypodermis in nematodes) and nervous system, but also to a small portion of the mesoderm of the pharynx. The posterior daughter, ABp, also makes neurons and hypodermis as well as some specialized cells. At second cleavage, P_1 divides asymmetrically to give P_2 and EMS, which will subsequently divide into MS and E. MS gives rise to many of the body's muscles and the posterior half of the pharynx, while E is the precursor of the 20 cells of the endoderm of the mid-gut. The C cell, which is derived from the P_2 cell at the third cleavage, forms hypodermis and body-wall muscle; D from the fourth cleavage of the

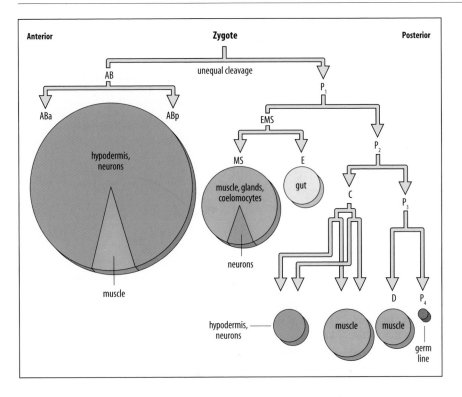

Fig. 6.4 Cell lineage and cell fate in the early *Caenorhabditis elegans* embryo. The cleavage pattern is invariant. The first cleavage divides the fertilized egg into a large anterior AB cell and a smaller posterior P_1 cell and the descendants of these cells have constant lineages and fates. AB divides into ABa, which produces neurons, hypodermis, and muscle (of the pharynx), and ABp, which produces neurons, hypodermis, and some specialized cells. P_1 divides to give EMS and P_2. The EMS cell then divides to give cells MS and E. MS gives rise to muscle, glands, and coelomocytes (free-floating spherical cells), and E produces the gut. Further divisions of the P lineage are rather like stem-cell divisions, with one daughter of each division (C and D) giving rise to a variety of tissues while the other (P_2 and P_3) continues to act as a stem cell. Eventually, P_4 gives rise to the germ cells.

P cell produces only muscle. All the cells undergo further invariant divisions, and at about 100 minutes after fertilization gastrulation begins.

Before fertilization, there is no evidence of any asymmetry in the nematode egg, but sperm entry sets up an antero-posterior polarity in the fertilized egg that determines the position of the first cleavage division and the future embryonic antero-posterior axis. This cleavage is both unequal and asymmetric, forming the large AB cell at the anterior end and the smaller P_1 cell at the posterior. The existence of polarity in the fertilized egg becomes evident before the first cleavage. A cap of actin microfilaments forms at the future anterior end, and a set of cytoplasmic granules—the so-called **P granules**, which contain maternal mRNAs and proteins required for development of the germline cells—become localized at the future posterior end. (The P granules are not the determinants of polarization; their redistribution simply reflects it.) P granules remain in the P daughters of cell division, eventually becoming localized to the P_4 cell, which gives rise to the germline (Fig. 6.5).

The initial polarization of the fertilized egg seems to be due to the centrosome that the sperm nucleus brings into the egg. The centrosome, or cell center, is a small structure composed of microtubules that organizes the microtubule cytoskeleton of the cell; it duplicates to form the two centrosomes that nucleate each pole of the microtubule spindle that organizes chromosome distribution in mitosis and meiosis (the mitotic cell cycle is discussed in Box 1B, p. 7). The egg cortex, the layer of cytoplasm immediately under the plasma membrane, contains a dense network of actin

Fig. 6.5 Localization of P granules after fertilization. The movement of P granules is shown during the development of a fertilized egg of *Caenorhabditis elegans* from sperm entry up to the 26-cell stage. In the left panel, the embryo is stained for DNA, to visualize the chromosomes, and in the right panel for P granules. (a) Fertilized egg with egg nucleus at anterior end and sperm nucleus at posterior end. At this stage the P granules are distributed throughout the egg. (b) Fusion of sperm and egg nucleus. The P granules have already moved to the posterior end. (c) Two-cell stage. The P granules are localized in the posterior cell. (d) 26-cell stage. All the P granules are in the P_4 cell, which will give rise to the germline only.
Photographs courtesy of W. Wood, from Strome, S., et al.: 1983.

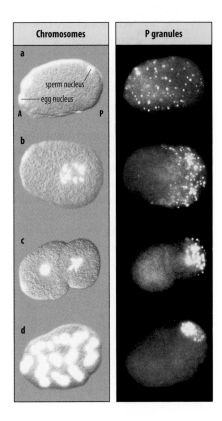

Box 6A Gene silencing by antisense RNA and RNA interference

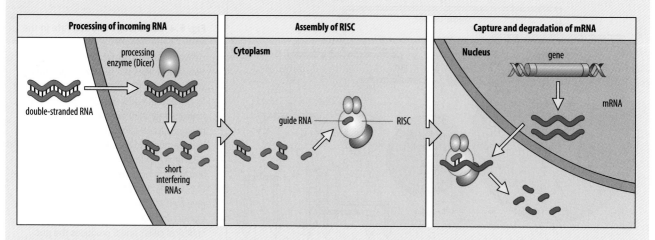

Gene knock-outs remove the function of a gene permanently by disrupting its DNA. Another set of techniques for suppressing gene function rely on destroying or inhibiting the mRNA. Because the gene itself is left untouched, and thus suppression is, in principle, reversible, this type of loss of function is generally known as **gene silencing**. All methods of gene silencing rely on the introduction into embryos or cultured cells of an RNA with a sequence complementary to that of the specific mRNA being targeted. Depending on the particular technique, the introduced RNA either binds to the mRNA and prevents it from being translated, and/or targets a nuclease to bind to the mRNA and degrade it.

The first technique of this type to be developed uses short synthetic **antisense RNAs**, which are usually chemically modified (for example, morpholino RNAs) to increase their stability within the cell. Morpholino oligonucleotides have been used extensively to block specific gene expression, especially in sea urchins, *Xenopus*, and zebrafish.

A more recent addition to the gene-silencing armoury is the technique of **RNA interference (RNAi)**, which recruits a natural RNA-degrading mechanism that is apparently ubiquitous in multicellular eukaryotes from plants to mammals. The phenomenon was first uncovered in plants during transgenic experiments, when it was found that introduction of a gene very similar to one of the plant's own genes blocked the expression of both the introduced gene and the endogenous gene. It turned out that suppression was occurring at the level of the mRNAs,

which appeared to be rapidly degraded. RNAi is now known to be caused by the degradation of double-stranded RNAs by a cellular enzyme called Dicer, which chops them into short lengths of around 21–23 nucleotides. This so-called **short interfering RNA (siRNA)** is unwound into single-stranded RNA, which becomes incorporated into a nuclease-containing protein complex known as RISC (for RNA-induced silencing complex). The RNA-binding component in RISC is a ribonuclease called Argonaute, also called Slicer in mammalian cells. The siRNA acts a 'guide RNA' to target RISC to any mRNA containing an exactly complementary sequence. The mRNA is then degraded by the nuclease in the RISC (see figure). RNAi is particularly effective and easy to carry out in *C. elegans*. Adult worms injected with double-stranded RNA will show suppression of the corresponding gene in the embryos they produce. Worms can also be soaked in the appropriate RNA or fed on bacteria expressing the required double-stranded RNA. The natural role of the cellular RNAi machinery is thought to be defense against transposons and viruses, many of which produce double-stranded RNAs in the course of replication.

In plants, fungi, nematodes, *Drosophila*, and most non-mammalian cells, RNAi is carried out by introducing an appropriate double-stranded RNA, which is then processed into siRNAs within the cells. In mammalian cells, however, double-stranded RNAs trigger another defensive response that interferes with the gene-silencing effect, and RNAi is instead achieved by introducing the siRNAs themselves, or by expressing artificial DNA constructs from which siRNAs are transcribed.

filaments associated with the motor protein myosin, forming contractile macromolecular assemblies of actomyosin. Interaction of the centrosome with the actin network causes the network to contract asymmetrically away from the point of sperm entry about 30 minutes after fertilization. This leads to a general flow of cortical components towards the future anterior end of the zygote, and cytoplasmic flow towards the future posterior end. Among the cortical proteins that are relocated by this movement are a group of maternal proteins known as **PAR (partitioning) proteins**. These are initially uniformly distributed, but become concentrated in anterior or posterior cortex after cortical movement and control the asymmetric divisions in the early embryo.

The PAR proteins were discovered in a very ingenious way. Vast numbers of embryos would have had to be screened to detect mutants defective in asymmetric development. The process was streamlined by using a mutant—*egl*—in which the fertilized eggs are not released and the larvae hatch out inside the parent, devouring it from the inside and killing it. Mutations that halt development early on spare the parent, and this is how the maternal *par* genes were discovered.

Partitioning-defective embryos have defects in the asymmetric divisions that divide the zygote into blastomeres. If the normal pattern of division is not precisely maintained, future development is affected. The *par* genes encode several different types of proteins, including protein kinases and other proteins involved in intracellular signaling pathways. PAR proteins similar to those in *C. elegans* are involved in establishing and maintaining cell polarity in many different situations, and are present in animals ranging from nematodes to mammals. In *C. elegans*, the correct localization of PAR proteins is required for the correct positioning and orientation of mitotic spindles in the early cleavages to ensure division at the required place in the cell and in the required plane.

The cortical flow resulting from sperm entry and actomyosin network contraction leads to the anterior localization of PAR-3 and PAR-6, and the posterior localization of PAR-1 and PAR-2. This asymmetric distribution then results, by mechanisms that are not yet completely understood, in the positioning of the first mitotic spindle posterior to the center of the cell, and thus to a first cleavage that is both unequal (producing cells of different sizes) and functionally asymmetric (producing cells containing different developmental determinants). The PAR proteins are thought to be involved in regulating the forces that pull on each pole of the spindle to position it in the cell. The mechanisms that control the first two cleavage divisions are extremely complex, as a large-scale RNAi experiment that individually suppressed around 98% of the nematode's genes showed that more than 600 gene products were involved in these two stages.

6.2 The dorso-ventral axis in *Caenorhabditis elegans* is determined by cell–cell interactions

Despite the highly determinate cell lineage in the nematode, cell–cell interactions are involved in specifying the dorso-ventral axis. At the time of the second cleavage, if the future anterior ABa cell is pushed and rotated with a glass needle, not only is the antero-posterior order of the daughter AB cells reversed, but the cleavage of P_1 is also affected, so that the position of the P_1 daughter cell EMS relative to the AB cells is inverted (Fig. 6.6). The manipulated embryo develops completely normally but 'upside down', with its dorso-ventral axis inverted within the eggshell. ABp develops as the anterior cell and ABa as the posterior cell of the pair, and the polarity of P_2 has been reversed compared to its polarity in the unmanipulated embryo. This shows that the dorso-ventral polarity of the embryo cannot already be fixed at this stage and that the usual developmental fate of these cells can be changed: this implies that their fate is specified by cell–cell interactions. The normal development after this manipulation also implies that the left–right axis of symmetry, which is defined in relation to the antero-posterior and dorso-ventral axes, is not yet determined. Indeed, the left–right axis can be reversed by manipulation at a slightly later stage, as we see next.

Adult worms have a well defined left–right asymmetry in their internal structures, and the pattern of development on the left and right sides of the embryo is strikingly different, more so than in many other animals. Not only do cell lineages on the left and right differ, but some cells move from one side to the other during embryonic development. The concept of left and right only has meaning once the antero-posterior

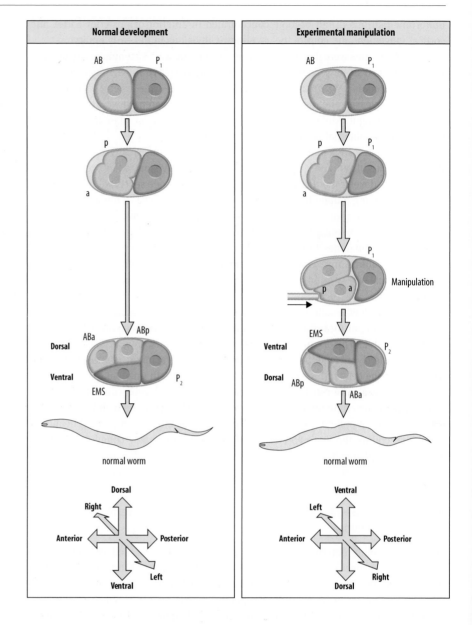

Fig. 6.6 Reversal of dorso-ventral polarity at the four-cell stage of the nematode. In normal embryos, the AB cell rotates at cleavage so that ABa is anterior. If the AB cell is mechanically rotated in the opposite direction at second cleavage, the ABp cell is now anterior. This manipulation also displaces the P_1 cell so that when it divides, the position of the EMS daughter cell is reversed with respect to the AB cells. Development is still normal, but the dorso-ventral axis is inverted (the worm develops 'upside down' in the egg case) a and p show the anterior/posterior orientation of the AB cell before and after manipulation.

After Sulston, J.E., et al.: 1983.

and dorso-ventral axes are defined (see Section 4.7). In *C. elegans*, specification of left and right normally occurs at the third cleavage, and handedness can be reversed by experimental manipulation at this stage. At third cleavage, ABa and ABp each divides to produce laterally disposed right and left daughter cells (for example, ABa produces ABal and ABar). The plane of cleavage is, however, slightly asymmetric, so that the left daughter cell lies just a little anterior to its right-hand sister. If the cells are manipulated with a glass rod during this cleavage, their positioning can be reversed so that the right-hand cell lies slightly anterior (Fig. 6.7). This small manipulation is sufficient to reverse the handedness of the animal.

While the molecular mechanisms specifying handedness are not known, mutation of the maternal gene *gpa-16* results in near-randomization of spindle orientation. Using temperature-sensitive mutant alleles of *gpa-16*, and raising the embryos at a temperature at which the first two cleavages are often normal, it proved possible to pinpoint the third cleavage as the one at which handedness is determined. In mutant worms that survived to maturity, handedness had been reversed from right to left in around 50%, and this could be traced back to a skewing of the mitotic spindle axis

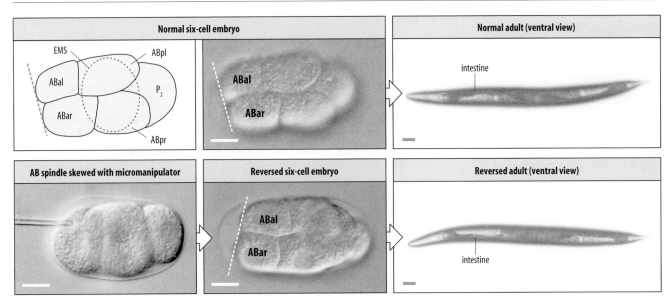

through 90° at third cleavage. Embryos that died had complete misorientation of spindle axes at the third cleavage. The protein GPA-16 is a G-protein α subunit which, together with other proteins, is involved in the positioning of centrosomes, and thus of the mitotic spindle.

There are ways of generating functional asymmetry other than unequal cell division: for example, at a much later stage in nematode development a functional difference is generated between two taste receptors, one on the left and one on the right, so that each expresses genes for different chemoreceptors. This asymmetry in gene expression is controlled by a microRNA, *lsy-6*.

6.3 Both asymmetric divisions and cell–cell interactions specify cell fate in the early nematode embryo

Although cell lineage in the nematode is invariant, experimental evidence, such as that described above, shows that cell–cell interactions are crucial in specifying cell fate at very early stages. Otherwise, the reversal of ABa and ABp by micromanipulation just as they are being formed (see Fig. 6.7) would not give rise to a normal worm. ABa and ABp must initially be equivalent, and their fate must be specified by interactions with adjacent cells. Evidence for such an interaction comes from removing P_1 at the first cleavage: the result is that the pharyngeal cells normally produced by ABa are not made.

What are the interactions that specify the non-equivalence of the two AB descendants? If ABp is prevented from contacting P_2 it develops as an ABa cell. Thus, the P_2 blastomere is responsible for specifying ABp. The induction of an ABp fate by P_2 involves the proteins GLP-1 in the ABp cell and APX-1 in P_2. These proteins are the nematode counterparts of the receptor protein Notch and its ligand Delta, respectively and, like Notch and Delta, are associated with the cell membrane. As we have seen in previous chapters, interactions between Notch and its various ligands set cell fate in many developmental situations throughout the animal kingdom.

GLP-1 is one of the earliest proteins to be spatially localized during nematode embryogenesis. The maternal *glp-1* mRNA is uniformly present throughout the embryo but its translation is repressed in the P-cell lineage; at the two-cell stage the GLP-1 protein is thus only present in the AB cell. This strategy is highly reminiscent of the localization of maternal Hunchback protein in *Drosophila* (see Section 2.9), and, as in that case, in *C. elegans* other maternal proteins binding to the 3′ untranslated region of *glp-1* mRNA suppress its translation.

Fig. 6.7 Reversal of handedness in *Caenorhabditis elegans*. At the six-cell stage in a normal embryo, the AB cell on the left side (ABal) is slightly anterior to that on the right (top left panels, scale bar = 10 μm). Manipulation that makes the AB cell on the right side (ABar) more anterior results in an animal with reversed handedness (bottom left panels, scale bars = 10 μm). The right panels show the resulting normal and reversed adults; scale bars = 50 μm.

Photographs courtesy of W. Wood, from Wood, W.B.: 1991.

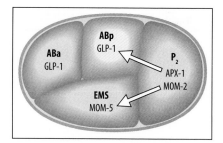

Fig. 6.8 The P$_2$ cell is the source of inductive signals that determine antero-posterior polarity and cell fate. ABp becomes different from ABa as a result of signaling from the adjacent P$_2$ cell by the Delta-like protein APX-1, which interacts with the Notch-like receptor GLP-1 on the ABp cell. P$_2$ also sends a Wnt-like inductive signal, MOM-2, to the EMS cell that interacts with a Frizzled-like receptor, MOM-5, on EMS to set up a posterior difference in the cell that determines the different fates of anterior and posterior daughter cells. These interactions depend on cell contact.

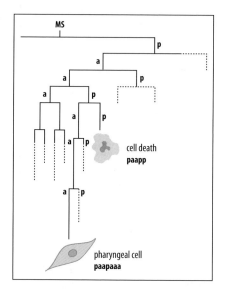

Fig. 6.9 Cell fate is linked to the pattern of cell divisions. The figure illustrates part of the cell lineages generated by the MS blastomere, which give rise to body muscle and the mesodermal cells of the pharynx. Each division produces an anterior (a) and a posterior (p) cell. The lineage paapp, for example, always gives rise to a cell that undergoes programmed cell death and dies by apoptosis. The lineage paapaaa, on the other hand, always results in a particular pharyngeal cell.

After second cleavage the ABa and ABp cells both contain GLP-1. The two cells are then directed to become different by a local inductive signal, sent to the ABp cell from the P$_2$ cell at the four-cell stage (Fig. 6.8). This signal depends on cell contact: it is delivered by APX-1 (produced by the P$_2$ cell), which acts as an activating ligand for GLP-1 in the ABp cell membrane. As a result of this induction, the descendants of ABa and ABp respond differently to later signals from descendants of the EMS lineage.

The EMS cell gives rise to daughter cells (MS and E) that give rise to mesoderm and endoderm, respectively (see Fig. 6.4). Specification of the EMS cell as a mesendo-dermal precursor also involves localized maternal determinants in the EMS cell and inductive signals from the adjacent P$_2$ cell (see Fig. 6.8). One maternal determinant involved in giving EMS its mesendodermal fate is the transcription factor SKN-1, the product of the *skin excess* (*skn-1*) gene. *skn-1* mRNA is uniformly distributed at the two-cell stage, but there are much higher levels of SKN-1 protein in the nucleus of P$_1$ than in AB. Maternal-effect mutations that abolish the function of *skn-1* result in the EMS blastomere adopting a mesectodermal fate; that is, it develops much like its sister cell P$_2$ (see Fig. 6.4), giving rise to mesoderm (body-wall muscle) and ectoderm (hypodermis—hence the name 'skin excess' for the *skn-1* mutant), but no pharyngeal cells and, in most mutants, no endoderm. SKN-1 acts by switching on the zygotic genes *med-1* and *med-2*, which encode transcription factors that are necessary for MS fate and which contribute to the E fate in parallel with other maternal inputs.

Formation of endoderm from the EMS cell also requires an inductive signal. If removed from the influence of its neighbors towards the end of the four-cell stage, an EMS cell can develop in isolation to produce gut structures, but if isolated at the beginning of that stage, it cannot. Removal of P$_2$ at the early four-cell stage results in no gut being formed, indicating that P$_2$ is the cell that delivers the inductive signal. This was confirmed by recombining isolated EMS and P$_2$ cells, which restores development of endoderm; recombining EMS with other cells from the four-cell embryo has no effect. The P$_2$ cell signal to the EMS cell is a Wnt protein (MOM-2), which interacts with a Frizzled receptor (MOM-5) on the EMS cell surface (see Fig. 6.8), the point of contact specifying the future posterior cell, which will be E. Signaling by P$_2$ thus serves to break two symmetries in the embryo: P$_2$ makes ABp different from ABa, and it also polarizes the EMS cell so that its division produces a posterior E cell and an anterior MS cell.

So we begin to see how a combination of cell–cell interactions and localized cytoplasmic determinants specify cell fate in the early nematode embryo. The result of killing individual cells by laser ablation at the 32-cell stage, when gastrulation begins, suggests that many of the cell lineages are now determined, because when a cell is destroyed at this stage it cannot be replaced and its normal descendants are absent. Cell–cell interactions are, however, required again later to effect cells' final differentiation.

Cell differentiation in the nematode is closely linked to the pattern of cell division. Each blastomere undergoes a unique and nearly invariant series of cleavages that successively divide cells into anterior and posterior daughter cells. Cell fate appears to be specified by whether the final differentiated cell is descended through the anterior (a) or posterior (p) cell at each division. In the lineage generated from the MS blastomere, for example, the cell resulting from the sequence p-a-a-p-p undergoes apoptosis (Fig. 6.9), whereas that resulting from p-a-a-p-a-a-a gives rise to a particular pharyngeal cell. What is the causal relation between the pattern of division and cell differentiation? One answer appears to lie in the anterior daughter cell having a higher intranuclear level of the maternal protein POP-1, a transcriptional regulatory protein, compared with its posterior sister. In the case of the EMS blastomere, for example, the anterior daughter cell, the MS blastomere, has a higher level of intranuclear POP-1 than the posterior daughter, the E blastomere, as a result of an asymmetric division of the EMS cell. In a maternal-effect *pop-1* mutant that does not have any POP-1 protein, MS adopts an E-like fate because the antero-posterior distinction is lost.

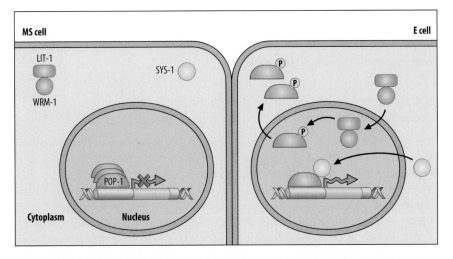

Fig. 6.10 The *Caenorhabditis* Wnt/β-catenin asymmetry pathway determines anterior and posterior cell fates in the MS and E cells. The Wnt (MOM-2) signal originally received by Frizzled receptors (MOM-5) at the posterior end of the EMS cell (not shown) sets up anterior-posterior differences in the EMS cell that become evident in its daughter MS (anterior) and E (posterior) cells. POP-1 is the *C. elegans* versions of the transcription factor TCF. SYS-1 and WRM-1 are nematode-specific β-catenins. In the E cell, SYS-1 is stabilized and accumulates in the nucleus. In addition, a complex of WRM-1 and the protein kinase LIT-1 also accumulates in the nucleus and phosphorylates POP-1. This leads to the export of POP-1 from the E-cell nucleus, reducing the amount to a level at which SYS-1 can act as co-activator for POP-1 and switch on endoderm-specifying target genes. In the MS cell, in contrast, neither SYS-1 nor the WRM-1–LIT-1 complex accumulate in the nucleus and so levels of POP-1 remain high and target genes remain suppressed. The interactions between the various pathway components are still being worked out.

The differential distribution of POP-1 in the EMS daughter cells is the ultimate result of the P_2 Wnt signal (MOM-2) acting on the receptor Frizzled (MOM-5) on the EMS cell (see Fig. 6.8). POP-1 is the *C. elegans* version of the vertebrate transcription factor TCF. In the 'canonical Wnt pathway' β-catenin enters the nucleus as a result of Wnt signaling, interacts with TCF and converts it from a transcriptional repressor to a transcriptional activator (see Box 1E, p. 26). In the context of the EMS asymmetric cell division, the Wnt signal acts via a different intracellular signaling pathway, the 'Wnt/β-catenin asymmetry pathway' that has not been found in *Drosophila* or vertebrates. This pathway has two branches that independently control the actions of two nematode-specific β-catenins, WRM-1 and SYS-1. The MOM-2 Wnt signal received at the posterior end of the EMS cell polarizes the cell, resulting in the differential localization of cytoplasmic factors controlling the pathway in the anterior (MS) and posterior (E) daughter cells after cell division. As a result of the localization, a complex of WRM-1 and a protein kinase (LIT-1) accumulates in the E-cell nucleus, which leads to the phosphorylation of POP-1 and its export from the nucleus (Fig. 6.10). The other β-catenin, SYS-1, also accumulates in the anterior nucleus, and is able to act as co-activator to the reduced amounts of POP-1 and help to switch on endoderm-specifying genes. In the MS cell, on the other hand, WRM-1–LIT-1 is exported from the nucleus, thus keeping POP-1 levels high there and suppressing the endodermal fate. The Wnt/β-catenin asymmetry pathway is thought to be a general mechanism of regulating anterior/posterior cell fate throughout nematode development, switching on different target genes in different situations.

At the 80-cell gastrula stage a fate map can be made for the nematode embryo (Fig. 6.11). At this stage, cells from different lineages that will contribute to the same organ, such as the pharynx, cluster together. The gut is made clonally from E derivatives only. Genes that appear to act as organ-identity genes may now be

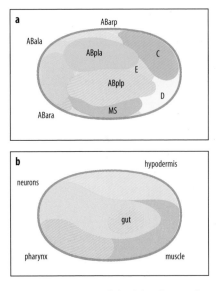

Fig. 6.11 Fate map of the 80-cell gastrula of *Caenorhabditis elegans*. In panel (a), the regions of the embryo are color-coded according to their blastomere origin. In panel (b) they are colored according to the organs or tissues they will ultimately give rise to. At this stage in development, cells from different lineages that contribute to the same tissue or organ have been brought together in the embryo. For clarity, the domains occupied by ABpra and ABprp are not shown in panel a. Anterior is to the left, ventral down.

After Labouesse, M. and Mango, S.E.: 1999.

expressed by the cells in the cluster. The *pha-4* gene, for example, seems to be an organ-identity gene for the pharynx. Mutations in *pha-4* result in loss of the pharynx, whereas ectopic expression of normal *pha-4* in all the cells of the embryo leads to all cells expressing pharyngeal cell markers.

6.4 Hox genes specify positional identity along the antero-posterior axis in *Caenorhabditis elegans*

The nematode body plan is very different from that of vertebrates or *Drosophila*, and there is no segmental pattern along the antero-posterior axis; nevertheless, as in other organisms, Hox genes are involved in specifying positional identity along this axis. The nematode contains a large number of homeobox genes, of which only six are orthologous with those in the Hox gene clusters of *Drosophila* or vertebrates. Four of these genes—*lin-39*, *ceh-13*, *mab-5*, and *egl-5*—are relatively closely linked, with the remaining two, *php-3* and *nob-1*, some distance away on the same chromosome. Comparisons with other nematode species indicate loss of Hox genes and disruption of the ancestral Hox gene cluster in the lineage leading to *C. elegans* and its close relatives.

Of the *C. elegans* Hox genes, only *ceh-13*, which is required for anterior organization, is essential for embryonic development, and it seems that the other Hox genes carry out their function of regional specification at the larval stage. With the exception of *ceh-13*, the order of the genes along the chromosome is reflected in their spatial pattern of expression, as shown for three Hox genes in Fig. 6.12. *lin-39* appears to control the fate of mid-body cells and is known to be involved in regulating the development of the vulva in hermaphrodites (discussed later in this chapter), whereas *mab-5* controls the development of a region slightly more posterior, and *egl-5* provides positional identity for posterior structures. *php-3* and *nob-1* are also posterior group genes. As in other organisms, mutations in Hox genes can cause cells in one region of the body to adopt fates characteristic of other body regions. For example, a mutation in *lin-39* can result in larval mid-body cells expressing fates characteristic of more anterior or more posterior body regions.

Despite the fact that nematode Hox gene expression occurs in a regional pattern, the pattern may not be lineage dependent. For example, in the larva, cells expressing the Hox gene *mab-5* all occur in the same region (see Fig. 6.12) but are quite unrelated by lineage. The expression of *mab-5* is due to extracellular positional signals.

6.5 The timing of events in nematode development is under genetic control that involves microRNAs

Embryonic development gives rise to a larva of 558 cells and there are then four larval stages, which produce the adult. Because each cell in the developing nematode can be identified by its lineage and position, genes that control the fates of individual cells at

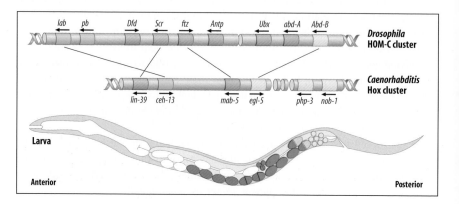

Fig. 6.12 The *Caenorhabditis elegans* Hox gene cluster and its relation to the HOM-C gene cluster of *Drosophila*. The nematode contains a cluster of six Hox genes, four of which show homologies with Hox genes of the fly. The pattern of expression of three of the genes in the larva is shown.

After Bürglin, T.R., et al.: 1993.

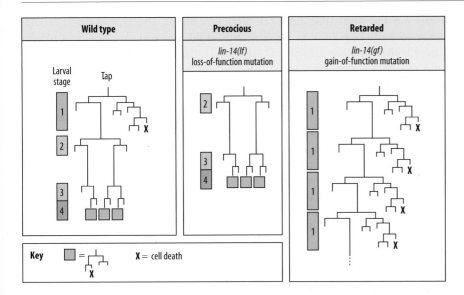

Fig. 6.13 Cell-lineage patterns in wild-type and heterochronic mutants of *Caenorhabditis elegans*. The lineage of the T-blast cell (T.ap) continues through four larval stages (left panel). Mutants in the gene *lin-14* show disturbance in the timing of cell division, resulting in changes in the patterns of cell lineage. Loss-of-function mutations result in a precocious lineage pattern, with the pattern of development of early stages being lost (center panel). Gain-of-function mutations result in retarded lineage patterns, with the patterns of the early larval stage being repeated (right panel).

specific times in development can also be identified. This enables the genetic control of timing during development to be studied particularly easily in *C. elegans*. The order of developmental processes is of central importance, as well as the time at which they occur. Genes must be expressed both in the right place and at the right time. One well-studied example of timing in nematode development is the generation of different patterns of cell division and differentiation in the four larval stages of *C. elegans*, which are easily distinguished in the developing cuticle.

Mutations in a small set of genes in *C. elegans* change the timing of cell divisions in many tissues and cell types. Mutations that alter the timing of developmental events are called **heterochronic**. The first two heterochronic genes to be discovered in *C. elegans* were *lin-4* and *lin-14*, and we shall use them to illustrate both the phenomenon of heterochrony and the control of this process by microRNAs. Different mutations in either of these genes can produce either 'retarded' or 'precocious' development.

Changes in developmental timing resulting from mutations in *lin-14* can be illustrated by what happens to the T-cell lineage (Fig. 6.13). In wild-type individuals, the T.ap cell gives rise to hypodermal cells, neurons, and their support cells by a pattern of stereotyped divisions in larval stages L1 and L2. During larval stages L3 and L4, some of the T-cell descendants divide further to give rise to other adult structures. Loss-of-function mutations in *lin-14* result in a precocious phenotype—for example, the pattern of cell divisions seen in L1 is lost and post-embryonic development starts with the cell divisions normally seen in L2. In contrast, gain-of-function mutations in *lin-14* result in retarded development. Post-embryonic development begins normally, but the cell-division patterns of the first or second larval stages are repeated.

It has been suggested that the genes that control timing of developmental events may do so by controlling the concentration of some substance, causing it to decrease with time (Fig. 6.14). This temporal gradient could control development in much the

Fig. 6.14 A model for the control of the temporal pattern of *Caenorhabditis elegans* larval development. Top panel: the stage-specific pattern of larval development is determined by a temporal gradient of the protein LIN-14, which decreases during larval development. A high concentration in the early stages specifies the pattern of normal early-stage development as shown in Fig. 6.13 (left panel). The reduction in LIN-14 at later stages is due to the inhibition of *lin-14* mRNA translation by *lin-4* RNA. Bottom panels: loss-of-function (lf) mutations in *lin-14* result in the absence of the first larval stage (L1) lineage, whereas gain-of-function (gf) mutations, which maintain a high level of LIN-14 throughout development, keep the lineage in an L1 phase. Loss-of-function mutations in *lin-4* result in a lifting of repression of *lin-14*, a continued high activity of LIN-14, and a repetition of the L1 lineage.

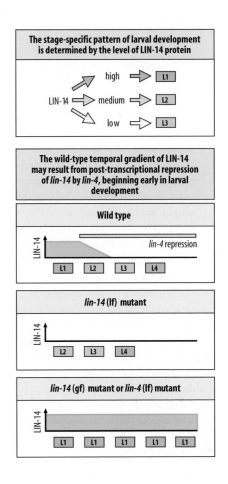

same way that a spatial gradient can control patterning. This type of timing mechanism seems to operate in *C. elegans*, because the concentration of LIN-14 protein drops tenfold between the first and later larval stages. Differences in the concentration of LIN-14 at different stages of development may specify the fates of cells, with high concentrations specifying early fates and low concentrations later fates. Thus, the decrease of LIN-14 during development could provide the basis of a precisely ordered temporal sequence of cell activities. Dominant gain-of-function mutations of *lin-14* keep LIN-14 protein levels high, and so in these mutants the cells continue to behave as if they are at an earlier larval stage. In contrast, loss-of-function mutations result in abnormally low concentrations of LIN-14, and the larvae therefore behave as if they were at a later larval stage.

Developmental timing in *C. elegans* was one of the first processes discovered in which gene expression was found to be controlled by microRNAs (Box 6B). The expression of *lin-14* is controlled post-transcriptionally by *lin-4*, which encodes a microRNA that represses the translation of *lin-14* mRNA and downregulates the

Box 6B Gene silencing by microRNAs

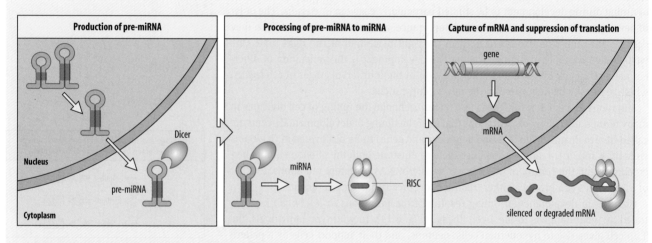

A new category of genes involved in development has been recognized quite recently, which codes not for proteins but for specialized short RNAs, called **microRNAs (miRNAs)**. These regulate gene expression by preventing specific mRNA transcripts from being translated. This natural method of gene silencing has similarities with RNA interference by small interfering RNAs (siRNAs) described in Box 6A (p. 220), but appears to be a distinct phenomenon, although there is some overlap in the processing machinery. MicroRNAs were first identified in *Caenorhabditis elegans* as the products of the genes *let-7* and *lin-4*, which control the timing of development (see text). Developmental roles are known for a few other miRNAs in nematodes, and for the miRNA *bantam* in *Drosophila*, which prevents cell death. There is now evidence that many human genes code for miRNAs and regulatory roles for human miRNAs are being revealed in cancer.

The primary miRNA transcript is several hundred nucleotides long and contains an inverted repeat of the mature miRNA

sequence. After initial processing in the nucleus, the transcript is folded back on itself to form a double-stranded RNA hairpin, the pre-miRNA, which is exported to the cytoplasm (see figure). Here it is cleaved by the enzyme Dicer to produce a short single-stranded miRNA about 22 nucleotides long. As in RNA interference, miRNAs become incorporated as guide RNAs in an RNA-induced silencing protein complex (RISC; see Box 6A, p. 220) and the complex is specifically targeted to an mRNA. Unlike siRNAs, animal miRNAs are typically not perfectly complementary to their target mRNA and have a few mismatched bases. Once bound, the protein complex renders the mRNA inactive and suppresses translation, and the mRNA may be degraded. miRNA can also repress transcription of genes.

Other small noncoding RNAs are known that have a role in regulating gene expression but work in a different way by physically binding to the chromosomes to silence expression. *Xist* RNA, which is involved in X-chromosome inactivation in mammals, is one example and is discussed in Chapter 9.

concentration of LIN-14 protein. LIN-14 itself is a transcription factor, but no target genes have yet been identified that can explain its heterochronic loss-of-function and gain-of-function phenotypes. Increasing synthesis of *lin-4* RNA during the later larval stages generates the temporal gradient in LIN-14, as indicated by the fact that loss-of-function mutations in *lin-4* have the same effect as gain-of-function mutations in *lin-14*. Expression of another gene involved in timing, *lin-41*, is also regulated post-transcriptionally by a microRNA, *let-7*, which represses its translation.

6.6 Vulval development is initiated by the induction of a small number of cells by short-range signals from a single inducing cell

The vulva forms the external genitalia of the adult hermaphrodite worm and connects with the uterus. It is an important example of a structure that is initially specified as a small number of cells—just four, one inducing and three responding. Unlike the other aspects of *Caenorhabditis* development discussed so far, the vulva is an adult structure that develops in the last larval stage. The mature vulva contains 22 cells, with a number of different cell types, and more than 40 genes are involved in its development. It is derived from ectodermal cells originating from the AB blastomere. Six of these precursor blast cells (P cells, not to be confused with the P (posterior) cells of the early embryo) persist in the larva in a row aligned antero-posteriorly on the ventral side of the larva, and posterior daughters of three of these—P5p, P6p, and P7p—give rise to the vulva, each having a well defined lineage. One of the functions of the Hox gene *lin-39* is to prevent the fusion of the six P cells to form hypodermis in the early larval stages, which is the fate of the rest of the ventral ectodermal cells of this lineage. In discussing the development of the vulva, three distinct cell fates are conventionally distinguished—primary (1°), secondary (2°), and tertiary (3°). The primary and secondary fates refer to different cell types in the vulva, whereas the tertiary fate is non-vulval. P6p normally gives rise to the primary lineage, whereas P5p and P7p give rise to secondary lineages. The other three P cells give rise to the tertiary lineage, divide once and then fuse to form parts of the hypodermis (Fig. 6.15). Initially, however, all six P cells are equivalent in their ability to develop as vulval cells. One of the questions we address here is how the final three cells become selected as the vulval precursor cells.

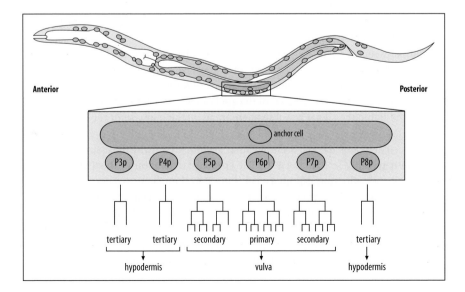

Fig. 6.15 Development of the nematode vulva. The vulva develops from three cells, P5p, P6p, and P7p, in the post-embryonic stages of nematode development. Under the influence of a fourth cell—the anchor cell—P6p undergoes a primary pathway of differentiation that gives rise to eight vulval cells. P6p is flanked by P5p and P7p, which each undergo a secondary pathway of differentiation that gives rise to seven cells of different vulval cell type. Three other P cells nearby adopt a tertiary fate and give rise to hypodermis.

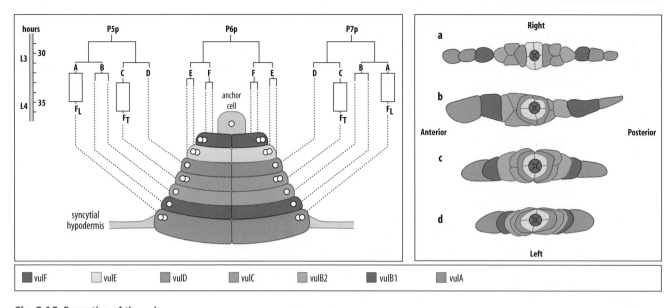

| ■ vulF | □ vulE | ■ vulD | ■ vulC | ■ vulB2 | ■ vulB1 | ▨ vulA |

Fig. 6.16 Formation of the vulva by migration and fusion of specified precursors. The left-hand diagram shows a lateral representation of the seven rings of the vulva at around 39 hours and the cell lineages that give rise to them. The daughter cells of the A cell fuse, as do the daughter cells of the C cell. The open circles represent nuclei. At a later stage there will be further fusion of cells in the rings. The blue bar on the left shows the time in hours and larval stages. The right panel shows a ventral view of the developing vulva. The sequence (a) to (d) shows morphogenesis of the vulva over a period of 5 hours, from 34 hours. Changes in the shape of the cells and cell fusion give rise to the conical structure seen in the left panel.

After Sharma-Kishore, R., et al.: 1999.

The fates of the three vulval precursor cells are specified by an inductive signal from a fourth cell, the gonadal anchor cell, which confers a primary fate on P6p, the cell nearest to it, and a secondary fate on P5p and P7p, the cells lying just beyond it (see Fig. 6.15). In addition, once induced, the primary cell inhibits its immediate neighbors from expressing a primary fate. P cells that do not receive the anchor cell signal adopt a tertiary fate.

After induction, the lineage of these P cells is fixed. If one of the daughter cells is destroyed, the other does not change its fate. There is no evidence for cell interactions in the further development of these three P-cell lineages, and cell fate is thus probably specified by asymmetric cell divisions. The vulva is formed from the 22 cells derived from the three lineages by a precise pattern of cell division, movement and cell fusion that forms a structure of seven concentric rings stacked on each other to form a conical structure (Fig. 6.16).

How are just three Pp cells specified, and how is the primary fate of the central cell made different from the secondary fate of its neighbors? The six Pp cells are initially equivalent, in that any of them can give rise to vulval tissue. The key determining signal is that provided by the anchor cell. The vital role of the anchor cell is shown by cell-ablation experiments; when the anchor cell is destroyed with a laser microbeam, the vulva does not develop.

Familiar signaling pathways are involved in vulval induction. The anchor cell's signal is the secreted product of the *lin-3* gene, which is the counterpart of the growth factor EGF in other animals. Mutations in *lin-3* result in the same abnormal development that follows removal of the anchor cell: no vulva is formed. The receptor for the inductive signal is a transmembrane tyrosine kinase of the EGF receptor (EGFR) family, encoded by the *let-23* gene (Fig. 6.17). The inductive signal from the anchor cell activates the EGFR intracellular signaling pathway most strongly in P6p, but also detectably in P5p and P7p. The strongest activation of this pathway leads to the P6p cell adopting a primary cell fate. P6p then sends a lateral signal to its two neighbors that both prevents them from adopting a primary fate and induces a secondary fate. This signal is composed of three proteins of the Delta family, which interact with the transmembrane receptor LIN-12, a member of the Notch family, on P5p and P7p. The canonical Wnt signaling pathway (see Box 1E, p. 26), involving a third *C. elegans* β-catenin (BAR-1), is involved in the adoption of cell fate in response to this signal.

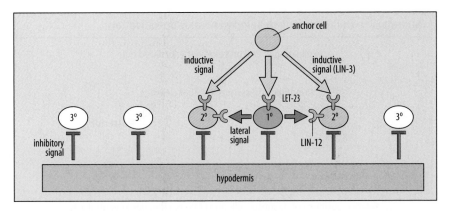

Fig. 6.17 Intercellular interactions in vulval development. The anchor cell produces a diffusible signal LIN-3, which induces a primary fate in the precursor cell closest to it by binding to the receptor LET-23. It also induces a secondary fate in the two P cells slightly further away, where the concentration of the signal is lower. The cell adopting a primary fate inhibits adjacent cells from adopting the same fate by a mechanism of lateral inhibition involving LIN-12, and also induces a secondary fate in these cells. A constitutive signal from the hypodermis inhibits the development of both primary and secondary fates, but is overruled by the initial inductive signal from the anchor cell.

Vulval development varies in many ways, both minor and major, among nematode species as a result of evolutionary divergence and convergence within this large group of organisms. Although the vulva has the same form in all species examined and is derived from much the same small set of P cells, some species have a vulva located posteriorly, whereas others, like *C. elegans*, have a centrally located vulva. The distinctive *C. elegans* mechanism of vulval induction, in which general vulval fate and a primary cell fate are induced at the same time via a signal from a single gondal anchor cell, is very recently evolved and probably arose within the genus *Caenorhabditis*. Other species of nematode have a variety of vulval induction mechanisms that differ from that of *C. elegans* to a greater or lesser extent.

Summary

Specification of cell fate in the nematode embryo is intimately linked to the pattern of cleavage, and provides an excellent example of the subtle relationships between maternally specified cytoplasmic differences and very local and immediate cell–cell interactions. The antero-posterior axis is specified at the first cleavage by the site of sperm entry and specification of the dorso-ventral and left-right axes involves local cell–cell interactions. Gene products are asymmetrically distributed during the early cleavage stages, but specification of cell fates in the early embryo is crucially dependent on local cell–cell interactions, which are mediated by Wnt and Delta–Notch signaling. The development of the gut, which is derived from a single cell, requires an inductive signal by an adjacent cell. A small cluster of homeobox genes provides positional identity along the antero-posterior axis of the larva. The timing of developmental events in the larva is controlled by changes in the levels of various LIN proteins that decrease over time as a result of post-transcriptional repression by microRNAs. Formation of the adult nematode vulva involves both a well defined cell lineage and inductive cell interactions involving the EGF and Notch signaling pathways. The detailed mechanism of vulval development varies among nematode species.

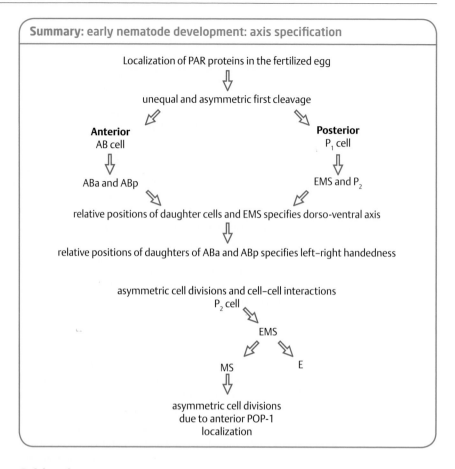

Summary: early nematode development: axis specification

Localization of PAR proteins in the fertilized egg

⇩

unequal and asymmetric first cleavage

Anterior
AB cell

⇩

ABa and ABp

Posterior
P₁ cell

⇩

EMS and P₂

relative positions of daughter cells and EMS specifies dorso-ventral axis

⇩

relative positions of daughters of ABa and ABp specifies left–right handedness

asymmetric cell divisions and cell–cell interactions
P₂ cell

EMS

MS   E

asymmetric cell divisions
due to anterior POP-1
localization

Echinoderms

Echinoderms include the sea urchins and starfish. Because of their transparency and ease of handling, sea-urchin embryos have long been used as a model developmental system. Another useful feature is that echinoderms are more closely related to vertebrates than are the other main model invertebrate organisms, *Drosophila* and *Caenorhabditis*. Echinoderms are **deuterostomes**, like vertebrates and ascidians, and so these three groups have certain basic developmental features in common. Deuterostomes are coelomate animals that have radial cleavage of the egg, and in which the primary invagination of the gut at gastrulation forms the anus, with the mouth developing independently. Arthropods (which include *Drosophila*) and nematodes, on the other hand, belong to the **protostomes**, in which cleavage of the zygote is not radial and in which gastrulation primarily forms the mouth.

For some years the echinoderms were somewhat neglected as a general developmental model as they were not amenable to investigation by conventional genetics or by the transgenic techniques that have proved so successful in other organisms. With the advent of new genetic techniques and the ability to identify genes by homology with genes from other organisms, the genetic and molecular basis of sea urchin development can now be studied. The genome sequence of the purple sea urchin *Strongylocentrotus purpuratus* has been completed, and the complete gene-regulatory network that governs early development is being worked out.

6.7 The sea-urchin embryo develops into a free-swimming larva

The fertilized sea-urchin egg is surrounded by a double membrane, inside which the embryo develops to blastula stage. The blastula hatches from the membrane,

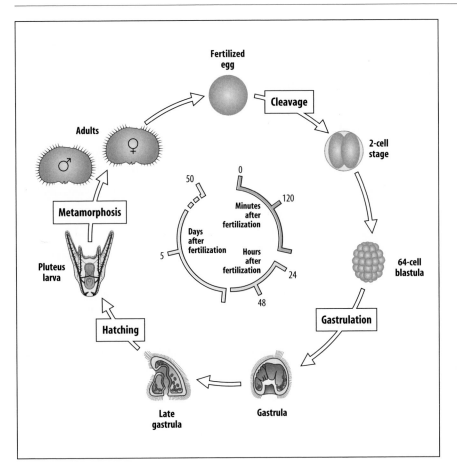

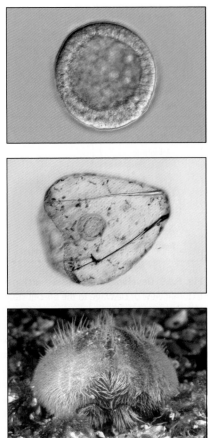

Fig. 6.18 Life cycle of the sea urchin *Strongylocentrotus purpuratus*. Eggs released by the female are fertilized externally by male sperm and develop into a blastula. After cleavage and gastrulation, the embryo hatches as a pluteus larva, which then undergoes metamorphosis into the mature radially symmetrical adult. The photographs show: the blastula (top); the pluteus larva (middle); and the adult (bottom).

Top and middle photographs courtesy of Jim Coffman. Bottom Photograph from Oxford Scientific Films.

undergoes gastrulation, and develops into a bilaterally symmetrical free-swimming larva called a **pluteus**. This eventually undergoes metamorphosis into the radially symmetrical adult (Fig. 6.18). Developmental studies are confined to the development of the embryo into the larva, as metamorphosis is a complex and poorly understood process. The sea-urchin embryo is classically regarded as a model of regulative development. It formed the basis for Driesch's ideas at the beginning of the twentieth century, that the position of the cells in an embryo determined their fate (see Section 1.3).

The sea-urchin egg divides by radial cleavage. The first three cleavages are symmetric but the fourth is asymmetric, producing four small cells at one pole of the egg, the vegetal pole, thus defining the animal–vegetal axis of the egg. The first two cleavages are at right angles to each other and divide the egg in the plane of the animal–vegetal axis (Fig. 6.19). The third cleavage, in contrast, is equatorial and divides the embryo into animal and vegetal halves. At the next cleavage the animal cells again divide in a plane parallel to the animal–vegetal axis, but the vegetal cells divide asymmetrically to produce four **macromeres**, which contain about 95% of the cytoplasm, and four much smaller **micromeres**. At fifth cleavage, the micromeres divide asymmetrically again, so that there are four small micromeres at the vegetal pole with four larger micromeres above them. Continued cleavage results in a hollow spherical blastula composed of about 1000 ciliated cells that form an epithelial sheet enclosing the blastocoel.

Gastrulation in sea-urchin embryos, the mechanism of which we consider in detail in Chapter 8, starts about 10 hours after fertilization, with the mesoderm and endoderm moving inside from the vegetal region. In *S. purpuratus*, the first event is

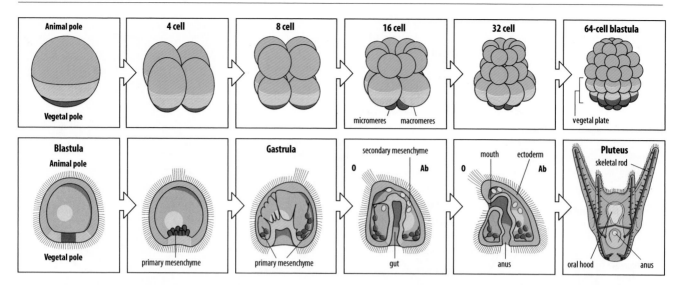

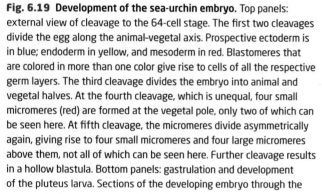

Fig. 6.19 Development of the sea-urchin embryo. Top panels: external view of cleavage to the 64-cell stage. The first two cleavages divide the egg along the animal-vegetal axis. Prospective ectoderm is in blue; endoderm in yellow, and mesoderm in red. Blastomeres that are colored in more than one color give rise to cells of all the respective germ layers. The third cleavage divides the embryo into animal and vegetal halves. At the fourth cleavage, which is unequal, four small micromeres (red) are formed at the vegetal pole, only two of which can be seen here. At fifth cleavage, the micromeres divide asymmetrically again, giving rise to four small micromeres and four large micromeres above them, not all of which can be seen here. Further cleavage results in a hollow blastula. Bottom panels: gastrulation and development of the pluteus larva. Sections of the developing embryo through the plane of the animal-vegetal axis are shown, starting from a hollow blastula. Gastrulation begins at the vegetal pole, with the entry of about 40 primary mesenchyme cells into the interior of the blastula. The gut invaginates from this site and fuses with the mouth, which invaginates from the opposite side of the embryo. The mouth defines one end of an oral (O)-aboral (Ab) axis, which will be the main axis of symmetry of the pluteus larva. The secondary mesenchyme comes from the tip of the invagination. During further development, growth of skeletal rods laid down by the primary mesenchyme results in the extension of the four 'arms' of the pluteus larva. The view of the pluteus is from the oral side. The oral hood narrows into the mouth and channels food into it.

the entry into the blastocoel of about 32 primary mesenchyme (mesodermal) cells at the vegetal pole (see Fig. 6.19). They migrate along the inner face of the blastula wall to form a ring in the vegetal region and lay down calcareous rods that will form the internal skeleton of the pluteus larva. The endoderm, together with the mesodermal secondary mesenchyme, then starts to **invaginate** at the vegetal pole. Invagination involves an inward movement of cells in a sheet, rather like pushing a finger into a balloon. In the embryo, the invagination is driven by local cellular forces, which are discussed in Chapter 8. The invagination eventually stretches right across the blastocoel, where it fuses with a small invagination in the future mouth region on the ventral side. Thus the mouth, gut, and anus are formed. Before the invaginating gut fuses with the mouth, secondary mesenchyme cells at the tip of the invagination migrate out as single cells and give rise to mesoderm, such as muscle and pigment cells. The embryo is then a feeding pluteus larva.

We will focus here on the initial patterning of the early sea-urchin embryo along its two main axes. We shall see that, despite the great differences in final form between sea urchins and vertebrates, some of the same basic developmental mechanisms and gene circuits are used in both. Because of its body form, the sea-urchin pluteus larva does not have antero-posterior and dorso-ventral axes like insect and vertebrate embryos. Instead, the two axes that are defined in the embryo are the **animal–vegetal axis** and an **oral–aboral axis**. The oral–aboral axis is defined by the position of the mouth and in turn defines the larva's plane of bilateral symmetry. An animal–vegetal asymmetry is already present in the unfertilized egg, but the oral–aboral axis is only established after fertilization and early cleavage. The sea urchin is also used as a simple model of gastrulation and to study fertilization, and it is discussed in relation to these topics in Chapters 8 and 9, respectively.

6.8 The sea-urchin egg is polarized along the animal–vegetal axis

The sea-urchin egg has a well-defined animal–vegetal polarity that appears to be related to the site of attachment of the egg in the ovary. Polarity is marked in some species by a fine canal at the animal pole, whereas in other species there is a band of pigment granules in the vegetal region. Early development is intimately linked to this egg axis. The first two planes of cleavage are always parallel to the animal–vegetal axis, and at the fourth cleavage, which is unequal (see Fig. 6.19), the micromeres are formed at the vegetal pole. The micromeres give rise to the primary mesenchyme, and both the micromeres and the primary mesenchyme are initially specified by cytoplasmic factors localized at the vegetal pole of the egg.

The animal–vegetal axis is stable and cannot be altered by centrifugation, which redistributes larger organelles such as mitochondria and yolk platelets. Isolated fragments of egg also retain their original polarity. Blastomeres isolated at the two-cell and four-cell stages, each of which contains a complete animal–vegetal axis, give rise to mostly normal but small pluteus larvae (Fig. 6.20, left panel). Similarly, eggs that are fused together with their animal–vegetal axes parallel to each other form giant, but otherwise normal, larvae.

There is, in contrast, a striking difference in the development of animal and vegetal halves if they are isolated at the eight-cell stage. An isolated animal half merely forms a hollow sphere of ciliated ectoderm, whereas a vegetal half develops into a larva that is variable in form but is usually vegetalized; that is, it has a large gut and skeletal rods, and a reduced ectoderm lacking the mouth region (see Fig. 6.20, right panel). On occasion, however, vegetal halves from the eight-cell stage can form normal pluteus larvae if the third cleavage is slightly displaced toward the animal pole.

Together, these observations show that there are maternally determined differences along the animal–vegetal axis that are necessary to specify either animal or vegetal fates. Despite this, it is clear that the sea-urchin embryo has considerable capacity for regulation, implying the occurrence of cell–cell interactions.

Fig. 6.20 Development of isolated sea urchin blastomeres. Left panel: if isolated at the four-cell stage, each blastomere develops into a small but normal larva. Right panel: an isolated animal half from the eight-cell stage forms a hollow sphere of ciliated ectoderm, whereas an isolated vegetal half usually develops into a highly abnormal embryo with a large gut, and some skeletal structures, but with a reduced ectoderm.

Isolation at four-cell stage gives four small larvae	Animal and vegetal halves develop differently when isolated
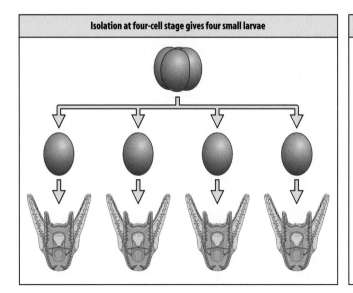	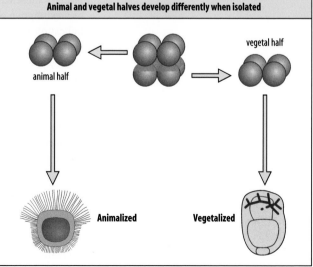

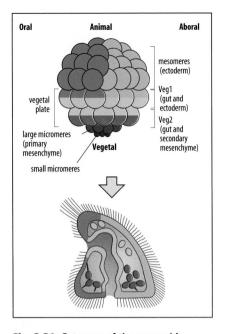

Fig. 6.21 Fate map of the sea-urchin embryo. Lineage analysis has shown four main regions at the 60-cell stage. The embryo is divided along the animal-vegetal axis into three bands: the micromeres, which give rise to the primary mesenchyme; the vegetal plate, which gives rise to the endoderm, secondary mesenchyme, and some ectoderm; and the mesomeres, which give rise to ectoderm. The ectoderm is divided into oral and aboral regions. Prospective mesoderm is colored in red; prospective endoderm in yellow; prospective neural tissue in dark blue; and prospective epidermis in light blue.

After Logan, C.Y. and McClay, D.R.: 1997.

6.9 The sea-urchin fate map is finely specified, yet considerable regulation is possible

At the 60-cell stage, five regions of cells can be distinguished along the animal–vegetal axis and a simplified fate map can be constructed (Fig. 6.21). This is composed of three bands of cells: the small and large micromeres at the vegetal pole, which will give rise to mesoderm (the primary mesenchyme that forms the skeleton and some adult structures); the vegetal plate, which comprises the next two tiers of cells (Veg1 and Veg2) and gives rise to endoderm, mesoderm (the secondary mesenchyme that forms muscle and connective tissues), and some ectoderm; and the remainder of the embryo, which gives rise to ectoderm and is divided into future oral and aboral regions. The ectoderm gives rise to the outer epithelium of the embryo and to neurogenic cells.

Using vital dyes as lineage tracers, the pattern of cleavage and of cell fates have both been mapped. The pattern of cleavage has been shown to be invariant. Unlike the nematode, however, this stereotypical cleavage pattern seems irrelevant to normal development. Compressing an early embryo by squashing it with a cover slip, and thus altering the pattern of cleavage, still results in a normal embryo after further development. Thus the stereotypical cleavage pattern does not, in most cases, predict an invariance of cell fate, and the embryo is highly regulative: lineage and cell fate are to a large extent separable. The regulative capacity is not uniform, however. The micromeres, for example, have a fixed fate once they are formed and this fate cannot be changed. They give rise only to primary mesenchyme, and no other fate has been observed, even if they are grafted to another site on the embryo. The fate of the vegetal macromeres is not fixed when they are formed. Vegetal macromeres labeled at the 16-cell stage give rise to endoderm, mesoderm, and ectoderm. If cells are labeled at the 60-cell stage, the Veg2 cells derived from these macromeres become either mesoderm (secondary mesenchyme) or endoderm, but never ectoderm, while the Veg1 cells give rise to endoderm and ectoderm but not mesoderm. This shows that endomesoderm specification has occurred by this stage, but that the boundary between endoderm and ectoderm has not yet been set. This occurs slightly later. Single Veg1 cell progeny labeled after the eighth cleavage give rise to either endodermal or ectodermal tissues, but not both.

When isolated and cultured, some regions of the sea-urchin embryo develop more or less in line with their normal fate. Micromeres isolated at the 16-cell stage will form primary mesenchyme cells and may even form skeletal rods. An isolated animal half of presumptive ectoderm forms an animalized ciliated epithelial ball, but there is no indication of mouth formation. Although these experiments suggest that localized cytoplasmic factors are involved in specifying cell fate, this does not tell the full story, as they do not reveal the role of inductive cell-cell interactions in the normal process of development. The ability of a vegetal half to form an almost complete embryo has already been noted. Another dramatic example of regulation is seen when the animal and vegetal blastomeres from half of an eight-cell embryo are combined with the animal blastomeres from another; this unlikely combination can develop into a completely normal embryo. How does this occur? The answer, as we see next, is the ability of the vegetal region to act as an organizer.

6.10 The vegetal region of the sea-urchin embryo acts as an organizer

The regulative capacities of the early sea-urchin embryo are due in large part to an organizing center that develops in the vegetal region and which, like the amphibian Spemann organizer (see Fig. 4.32), has the potential ability to induce the formation of an almost complete body axis. This center is initially set up by the activities of maternal factors and becomes located in the micromeres formed at the fourth

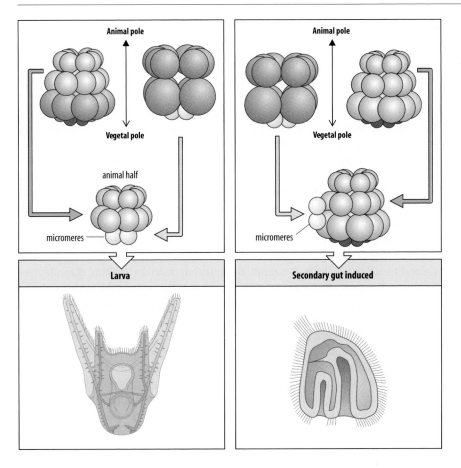

Animal pole

Vegetal pole

animal half

micromeres

Larva

Animal pole

Vegetal pole

micromeres

Secondary gut induced

Fig. 6.22 The inductive action of micromeres. Left panels: combining the four micromeres from a 16-cell sea-urchin embryo with an animal half from a 32-cell embryo results in a normal larva. An animal half cultured on its own merely forms ectoderm (not shown). Right panels: micromeres implanted into the side of another 32-cell embryo induce the formation of another gut at the implantation site.

cleavage division. It produces signals that induce the adjacent vegetal macromeres to adopt an **endomesodermal** fate. The endomesoderm gives rise to the sea-urchin spicules (skeletal rods), muscle, and gut. The signals from the organizer set in train a relay of events that subsequently specifies the endomesoderm more finely as mesoderm or endoderm, and helps set the boundary between endoderm and ectoderm. Ablation of the micromeres at the fourth cleavage results in abnormal development, but if they are removed at the sixth cleavage, after 2–3 hours of contact, gastrulation is delayed but normal larvae can develop.

Striking evidence for an organizer came from experiments that combined micromeres isolated from a 16-cell embryo with an isolated animal half from a 32-cell embryo. This combination can give rise to an almost normal larva (Fig. 6.22). The micromeres clearly induce some of the presumptive ectodermal cells in the animal half to form a gut, and the ectoderm becomes correctly patterned. Further evidence for the organizer-like properties of the micromeres comes from grafting them to the side of an intact embryo. Here, they induce endoderm in the presumptive ectoderm, which then invaginates to form a second gut. There is some evidence that the closer the graft is to the vegetal region, the greater the size of the invagination, implying a graded ability along the animal–vegetal axis to respond to a signal from micromeres.

The regulative capacity of the very early embryo can even compensate for loss of the micromeres. If micromeres are removed as soon as they are formed at the fourth cleavage (see Fig. 6.19), regulation occurs and a normal larva will still develop. The most vegetal region assumes the property of micromeres, giving rise to skeleton-forming cells and acquiring organizing properties.

6.11 The sea-urchin vegetal region is demarcated by the nuclear accumulation of β-catenin

The sea-urchin vegetal region and its organizing properties are specified by a mechanism very similar to the one that marks out the future dorsal region in the early *Xenopus* embryo (discussed in Chapter 4). In the sea urchin, the early blastomeres accumulate maternal β-catenin in their nuclei as a result of activation of part of the canonical Wnt pathway (see Box 1E, p. 26). This accumulation is not uniform, however; β-catenin concentrations are high in the micromere nuclei whereas it is almost absent from nuclei in the most animal region of the future ectoderm. This differential stabilization of intranuclear β-catenin along the animal–vegetal axis is at least partly controlled by the local activation in the vegetal region of maternal Dishevelled protein, which is an intracellular activator of the pathway by which β-catenin is stabilized. Within the nucleus, β-catenin acts as a co-activator of the transcription factor TCF to activate the expression of zygotic genes required to specify a vegetal fate.

Treatments that inhibit β-catenin degradation, such as lithium chloride (see Section 4.2) cause vegetalization, resulting in sea-urchin embryos with a reduced ectoderm and an enlarged gut. The main effect of lithium on whole embryos is to expand the area of intranuclear β-catenin accumulation beyond the normal vegetal region (the region receiving a Wnt signal), shifting the border between endoderm and ectoderm toward the animal pole. Similarly, overexpression of β-catenin also causes vegetalization and can induce endoderm in animal caps. In contrast, preventing the nuclear accumulation of β-catenin in vegetal blastomeres completely inhibits the development of both endoderm and mesoderm. Micromeres depleted of β-catenin, for example, are unable to induce endoderm when transplanted to animal regions.

Maternal β-catenin and the transcription factor Otx activate the gene *pmar1* in the micromeres at the fourth cleavage. *pmar1* encodes a transcription factor that is produced exclusively in micromeres during cleavage and early blastula stages and is required for the micromeres to express their organizer function and also for their development as primary skeletogenic mesenchyme. The outline of endomesoderm specification is illustrated in Fig. 6.23. Soon after they are formed, the micromeres produce a signal, called the early signal (ES), which induces the endomesoderm state in the Veg2 cells. A second signal, produced after the seventh cleavage, is the ligand Delta, which acts on the Notch receptor in adjacent Veg2 cells to specify them as pigment and secondary mesenchyme cells. At later stages Notch–Delta signaling is also involved in endoderm specification and in delimiting the endoderm–ectoderm boundary. The secreted signaling protein Wnt-8, first expressed by micromeres after the fourth cleavage, acts back on the micromeres that produce it to maintain the stabilization of β-catenin and thus maintain micromere function. Wnt-8 acting later in the Veg2 endoderm is involved in directing the specification of endoderm in adjacent Veg1 cells. In this way, the sea urchin mesoderm and endoderm is specified by a series of short-range signals that result in changes in gene expression.

6.12 The genetic control of the skeletogenic pathway is known in considerable detail

The pluteus larva has an internal skeleton of rods composed of calcium magnesium salts and matrix proteins, many of which are acidic glycoproteins. The skeletal material is secreted by skeleton-forming (skeletogenic) primary mesenchyme cells, which are derived from the micromeres. The gene-regulatory network underlying the development of the skeletogenic cells, from the events occurring in the early blastula to activation of genes for matrix proteins, has been worked out in some detail. The genes in this developmental pathway, and the interactions between them, were identified

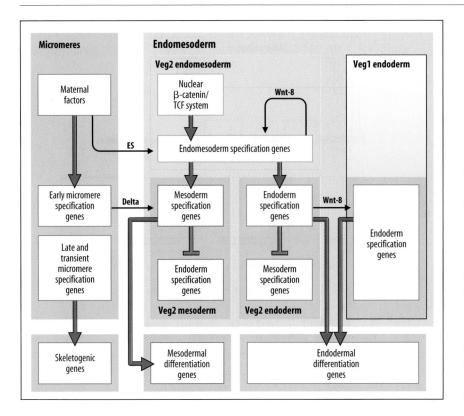

Fig. 6.23 An outline of endomesoderm specification in the sea urchin. The white boxes represent networks of genes that are reponsible for the developmental functions indicated. The background shading represents the embryonic territories involved. Peach, the large micromere precursors that will give rise to skeletal tissue; yellow, the early endomesoderm, which segregates at late cleavage into mesoderm (violet) and endoderm (blue). The red arrows and blue barred lines refer to gene interactions; the thin black arrows represent intercellular signals. ES, early signal.

Adapted from Oliveri, P. and Davidson, E.H.: 2004.

both by measuring gene activity directly and by using antisense morpholino RNAs (see Box 6A, p. 220) to silence specific gene expression. Every known regulatory gene that is expressed in the skeletogenic lineage has been incorporated into a genetic regulatory network; here we focus on the activation of genes that specify the primary mesenchyme as skeleton-forming tissue.

As described in Section 6.11, the zygotic gene *pmar1* is activated exclusively in the micromeres, and this is the first step in specifying the micromeres as future skeleton-forming tissue (Fig. 6.24). If *pmar1* mRNA is made to be present in all the cells, the whole embryo turns into skeleton-forming mesenchyme. *pmar1* encodes a transcriptional repressor for the gene *hesC*, which itself encodes a repressor. The HesC protein is repressing genes essential for the micromeres to develop as primary mesenchyme, and so the repression of *hesC* transcription by Pmar1 allows the activation of these genes (for example, *alx1*, *ets1* and *tbr*) in the micromeres only (see Fig. 6.24). This is another example of a common developmental strategy by which development is allowed to proceed by using repression to lift a pre-existing repression. *Alx1* and *ets1* encode transcription factors, which activate the next tier of genes, and so on, in conjunction with intercellular signaling, for example via the Notch–Delta pathway. The end result is the activation of genes that encode skeletal matrix proteins and other proteins involved in the laying down and mineralization of the skeleton. Some of these genes, such as those for matrix proteins Sm27 and Sm50, can be seen at the bottom of Fig. 6.24.

The network of interactions leading to skeletogenic mesoderm is just a small part of the whole network that specifies the endomesoderm and endoderm (Fig. 6.25). The complexity of these networks is hardly surprising given how complex the regulatory regions of developmental genes can be. As we saw in Chapter 2, the regulatory region of a *Drosophila* pair-rule gene, such as *even-skipped*, contains many binding sites for transcription factors that can both activate and repress gene activity (see Fig. 2.34). These target sites are grouped into discrete subregions—regulatory modules—which

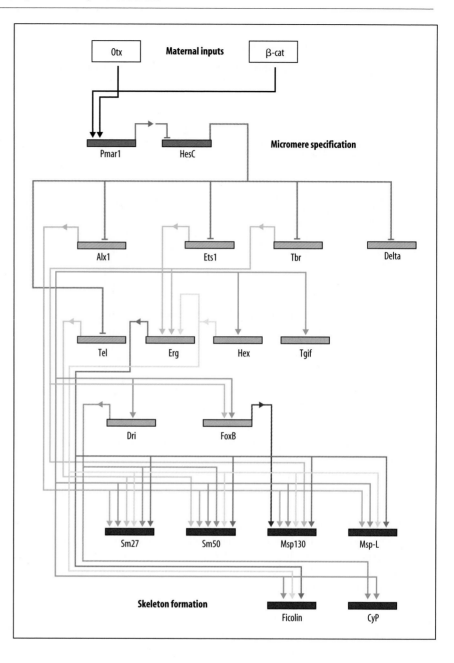

Fig. 6.24 Specification of the sea urchin micromeres and the activation of their organizing and skeletogenic function. The diagram shows a simplified version of the initial specification of the micromere lineage as discussed in the text. Arrows indicate that the gene product activates its target gene, a barred line indicates that it suppresses the gene. The maternal transcription factors β-catenin and Otx are localized in micromere nuclei and activate transcription of *pmar1* (red). This gene encodes a transcriptional repressor that suppresses expression of *hesC* (brown), the repressor of micromere fate, which is otherwise active throughout the embryo. Genes repressed by HesC protein are indicated in blue. When this repression is lifted in the micromeres, the products of these genes activate further sets of genes (orange and black) that carry out the organizing and skeletogenic properties of the micromeres. Genes at the bottom of the diagram colored in black encode skeletal matrix proteins (e.g. Sm27 and Sm50) and other proteins involved in producing the endoskeleton.

Adapted from Oliveri, P., et al.: 2008.

each control expression of the gene at a particular site in the embryo. Many sea urchin genes similarly have *cis*-regulatory control regions composed of a number of modules that govern gene expression at different times and in different parts of the embryo.

Although all the genes controlling sea-urchin development have yet to be identified, many whose expression varies with both space and time have been analyzed in detail. Their regulatory regions prove to be complex and modular, as illustrated by the regulatory region of the sea-urchin *Endo*-16 gene, a gene coding for a secreted glycoprotein of unknown function. *Endo*-16 is initially expressed in the vegetal region of the blastula, in the presumptive endoderm (Fig. 6.26). After gastrulation, *Endo*-16 expression increases and becomes confined to the middle region of the gut. The regulatory region controlling *Endo*-16 expression is about 2200 base pairs long and contains at least 30 target sites, to which 13 different transcription factors can bind. These regulatory sites seem to fall into several subregions, or modules. The

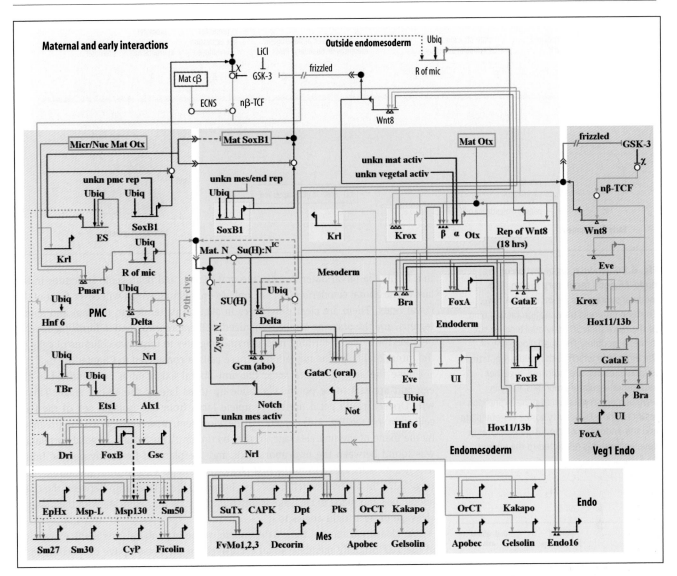

Fig. 6.25 The gene regulatory network for sea-urchin endomesoderm specification. The part of the network shaded in gray at the top corresponds to the nuclear localization of β-catenin under the control of maternal factors. The rest of the network refers to zygotic gene expression. PMC, primary mesenchyme (micromeres); Mes, secondary mesenchyme; Veg1 Endo, Veg1-derived endoderm; Endo, rest of endoderm.

Adapted from Olivieri, P. and Davidson, E.H.: 2004.

function of each module has been determined by attaching it to a reporter gene and injecting the recombinant DNA construct into a sea-urchin egg. In such experiments, module A, for example, is found to promote expression of its attached reporter gene in the presumptive endoderm, whereas modules D and C prevent expression in the primary mesenchyme. Midgut expression is under control of module B. Modules D, C, E, and F are activated by the action of added lithium and are involved in the vegetalizing action of *Endo-*16.

The modular nature of the *Endo-*16 control region is similar to that of the pair-rule genes in *Drosophila*, where different modules are responsible for expression in each of the seven stripes. The sea-urchin example illustrates yet again the importance of gene-control regions in integrating and interpreting developmental signals.

6.13 The oral–aboral axis in sea urchins is related to the plane of the first cleavage

The other main axis in the sea urchin-embryo is the oral–aboral axis, which defines the main axis of the pluteus larva and also defines the plane of larval bilateral symmetry. The oral ectoderm gives rise to the cells of the stomodeum (the mouth), the epithelium

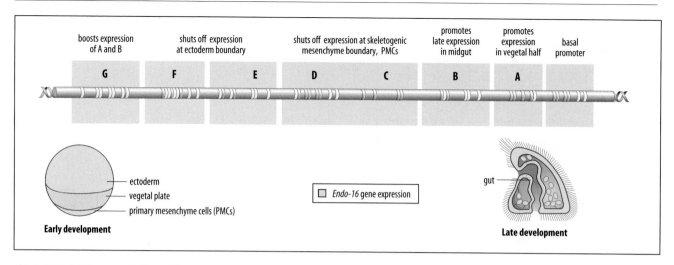

Fig. 6.26 Modular organization of the *endo-16* gene regulatory region. The modular subregions are indicated A to G. Thirteen different transcription factors bind to the more than 50 individual binding sites in these modules. The spatial domains of the embryo in which each module regulates gene expression are indicated, together with other regulatory modules. Early in development, regulation of gene expression restricts *Endo*-16 activity to the vegetal plate and the endoderm of the future gut; later in development it is restricted to the midgut.

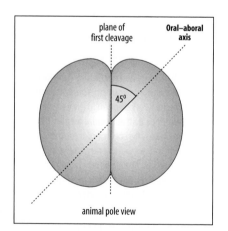

Fig. 6.27 Position of the oral-aboral axis of the sea urchin, *Strongylocentrotus purpuratus*, embryo in relation to first cleavage. When viewed from the animal pole, the oral-aboral axis is usually 45° clockwise from the plane of first cleavage. Note that at this stage the polarity of the axis is not determined, only the positions of the two ends.

that forms the larval outer layer in the oral region, and neurogenic structures in the larva. The aboral ectoderm gives rise only to epithelium, which will cover most of the larval body. There are clear differences in skeletal patterning in the oral and aboral regions, and the oral side can be recognized well before invagination forms the mouth, because the skeleton-forming mesenchyme migrates to form two columns of cells on the future oral face. The migration of these cells is considered in Chapter 8.

Unlike the animal–vegetal axis, the oral–aboral axis cannot be identified in the egg and appears to be potentially labile up to as late as the 16-cell stage. This is why Driesch could get a normal complete larva from one isolated cell of the two-cell embryo (see Fig. 1.8). In normal development, however, the axis is predicted by the plane of the first cleavage. In the sea urchin *S. purpuratus*, a good correlation was found between the oral–aboral axis and the plane of first cleavage: the future axis lies 45° clockwise from the first cleavage plane as viewed from the animal pole (Fig. 6.27). In other sea urchin species, however, the relationship between the oral–aboral axis and plane of cleavage is different; the axis may coincide with the plane of cleavage or be at right angles to it.

One clear indication of the establishment of the oral–aboral axis is the expression of the TGF-β family member Nodal exclusively in the presumptive oral ectoderm in the early blastula, at around the 60- to 120-cell stage. The initial event that breaks the radial symmetry of the fertilized egg and sets the oral end of the oral–aboral axis by initiating Nodal expression may be an underlying mitochondrial gradient. Oral–aboral polarity correlates with a mitochondrial gradient (highest in the oral region), which produces an intracellular gradient in hydrogen peroxide (H_2O_2) and reactive oxygen species. Experimental destruction of H_2O_2 in the two-cell embryo suppresses the initial expression of Nodal and pushes oral–aboral polarity towards aboral. It is likely that the stress-activated kinase p38, which is upregulated by H_2O_2, is an intermediary in this process.

6.14 The oral ectoderm acts as an organizing region for the oral-aboral axis

Nodal is one of the earliest zygotic genes to be expressed in the sea urchin, and Nodal signaling seems to be instrumental in organizing and patterning the oral–aboral axis. If Nodal activity is blocked, the embryo remains radially symmetrical and differentiation of ectoderm into oral and aboral ectoderm is blocked. Overexpression of Nodal also abolishes the oral–aboral axis, but in this case all the ectoderm is converted to an oral fate. But if Nodal expression is completely blocked up to the eight-cell stage and Nodal is then expressed in a single blastomere, even in one that would not normally express it, development can be 'rescued.' The treated blastulas develop oral–aboral polarity, with the Nodal-expressing cells becoming oral ectoderm, and will frequently even develop into normal pluteus larvae.

These experiments suggest that Nodal activity sets up a signaling center with organizing properties along the oral–aboral axis. High levels of Nodal signaling in the oral ectoderm result in the expression of the transcription factor Goosecoid (a characteristic protein of the amphibian Spemann's organizer) and the secreted signaling protein BMP-2/4. In the sea urchin, Goosecoid is thought to suppress aboral fate in the oral ectoderm, whereas BMP-2/4 is thought to diffuse from its source in the oral ectoderm and induce an aboral fate in the rest of the ectoderm. Another protein expressed in the prospective oral ectoderm as a consequence of Nodal activity is the TGF-β family member Lefty, which is an antagonist of Nodal signaling and restricts Nodal activity to the oral ectoderm. If the function of Lefty is blocked, most of the ectoderm develops into oral ectoderm. This situation is reminiscent of the interactions involved in mesoderm induction and patterning in vertebrates, showing how the same developmental gene circuits can be deployed for different purposes in different species. In the early sea-urchin embryo, Nodal appears to have no role in specifying and patterning the mesoderm, which is one of its primary roles in vertebrates (see Chapter 4).

For correct overall development, patterning along one axis must be spatially and temporally linked to what is happening along the other. In the sea urchin, the oral–aboral axis is linked to the animal–vegetal axis by an indirect effect of the intranuclear localization of β-catenin in the vegetal region (described in Section 6.11). In the early blastula, the transcription factor FoxQ2 is initially produced throughout the animal hemisphere and suppresses expression of the *nodal* gene there. As the wave of intranuclear β-catenin localization proceeds from the vegetal pole towards the animal hemisphere, the signals it generates restrict FoxQ2 expression to a small region at the animal pole known as the animal plate (which eventually gives rise to neurogenic ectoderm). This removal of FoxQ2 allows expression of Nodal in the prospective oral ectoderm.

There is a second role for Nodal signaling in the sea urchin—the establishment of internal left–right asymmetry (see Section 4.7). Inside the sea-urchin larva, a structure called the adult rudiment, from which the adult tissues will be produced during morphogenesis, is formed only on the left side. As in vertebrates, a regulatory circuit comprising Nodal, Lefty, and Pitx2 regulates the generation of this asymmetry, but the expression of these genes is reversed in the sea urchin compared with vertebrates. Nodal signals produced in the ectoderm on the right side of the larva inhibit rudiment formation on the right side.

Summary

Sea-urchin embryos have a remarkable capacity for regulation even though they also have an invariant early cell lineage. The sea-urchin egg has a stable animal-vegetal axis before fertilization, which appears to be specified by the localization of maternal cytoplasmic factors in the egg. After fertilization, maternal factors specify the formation of an organizer region in the most vegetal region of the egg, where the micromeres are formed. This organizer, like the Spemann organizer in amphibians, is able to specify an almost complete body axis. Localized activation of the Wnt pathway and nuclear accumulation of β-catenin specifies the vegetal region and the organizing properties of the micromeres. The oral-aboral axis is determined after fertilization. Signals from the micromeres induce and pattern the endomesoderm along the animal-vegetal axis, whereas the oral-aboral axis is specified and patterned by the activation of the Nodal signaling pathway in the future oral region. The site and time of expression of developmental genes are controlled by complex gene-regulatory networks, which have been worked out in some detail.

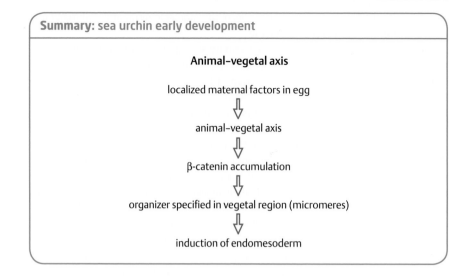

Summary: sea urchin early development

Animal–vegetal axis

localized maternal factors in egg

⇩

animal–vegetal axis

⇩

β-catenin accumulation

⇩

organizer specified in vegetal region (micromeres)

⇩

induction of endomesoderm

Ascidians

Adult ascidians, also known as sea squirts, are sessile marine animals. They are urochordates, which are included in the same phylum—the Chordata—as the vertebrates because their free-living, tadpole-like larvae possess a notochord, neural tube, and muscles, and are rather similar to the tailbud stage of a vertebrate embryo. The larva, which has about 2600 cells, undergoes metamorphosis into a sac-like sessile adult (Fig. 6.28). Although they are chordates, some aspects of ascidian development are very different from those of vertebrates. Ascidian embryos have an invariant cleavage pattern (Fig. 6.29), and localized cytoplasmic factors appear to have a much more important role in specifying cell fate. The genomes of the sea squirts *Ciona intestinalis* and *C. savignyi* have been sequenced, and *C. intestinalis* has an estimated 15,800 genes, which is comparable to the other model invertebrates. Knowing the genome sequences will enable the full panoply of genomic techniques to be applied to the study of ascidian development, and enable its similarities to, and differences from, the development of vertebrates and other chordates to be investigated at the genetic and molecular level.

Knowledge of the *C. intestinalis* genome has already been exploited in a large-scale screen for *cis*-regulatory control elements controlling Hox gene expression. Nine Hox genes have been identified in *Ciona*, seven of which are dispersed on the same chromosome, with the rest on another chromosome. Hox genes are expressed along the neural tube during development, and some are expressed in the endoderm that gives rise to the intestine. But, unlike flies and vertebrates, the timing of expression of the various Hox genes in *Ciona* seems not to be coordinated, and several are missing. Some 670 transcription factors have already been identified from the genome sequence, along with 119 major signal transduction proteins. All the major vertebrate developmental intercellular signaling pathways are present in ascidians, including the Hedgehog, Wnt, TGF-β, and Notch pathways.

Ascidian embryogenesis has long been regarded as a typical example of mosaic development, with cytoplasmic factors specifying cell fate during cleavage, and cell interactions playing only a relatively minor part. It is now clear, however, that cell interactions are more important than was previously thought.

6.15 Animal–vegetal and antero-posterior axes in the ascidian embryo are defined before first cleavage

The developmental axes in an ascidian embryo are related to the early pattern of cleavage. Like amphibian and sea-urchin eggs, the unfertilized ascidian egg is polarized

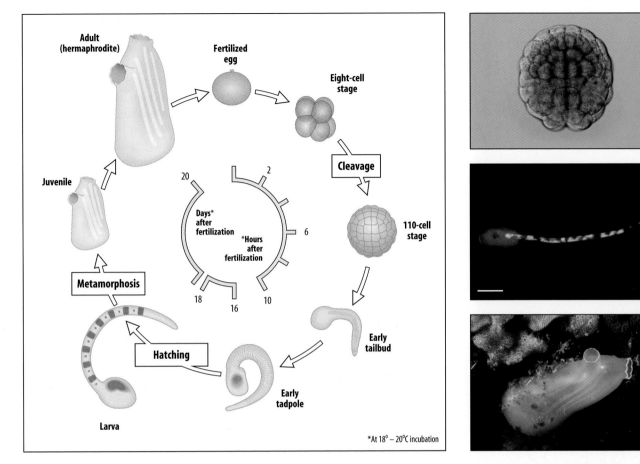

Fig. 6.28 Life-cycle diagram of the ascidian *Ciona intestinalis.* *Ciona* is hermaphrodite and eggs are fertilized externally. The fertilized egg takes about 18 hours to hatch into a larva, depending on the temperature of the water. The free-swimming larva undergoes metamorphosis into the sessile juvenile (about 2 cm tall) in around 20 days. Development of the juvenile into a sexually mature adult (which is about 5–8 cm tall) takes a further 2 months.

The photographs show: (top) a 110-cell embryo; (middle) a larva (with some notochord cells labeled green; scale bar = 0.1 mm); (bottom) adult *C. intestinalis.*

Top photograph courtesy of Shigeki Fujiwara and Naoki Shimozono; middle photograph courtesy of J. Corbo, from Corbo, J.C., et al.: 1997; bottom photograph courtesy of Andrew Martinez.

along an animal–vegetal axis, with the prospective territories of ectoderm, mesoderm, and endoderm lying along this axis from animal to vegetal. Another axis, which is conventionally called the antero-posterior axis, is established perpendicular to this. Note that this naming convention differs from that of the axes in amphibian embryos. This difference is largely semantic, and is due to the way in which ascidian embryos gastrulate and to a difference in the position of the developing embryo used as a reference point by ascidian researchers. The animal–vegetal axis of the egg is conventionally considered as the future ventro-dorsal axis of the tadpole, as the vegetal pole becomes covered with dorsal tissues at gastrulation.

Fig. 6.29 Cleavages in the egg of the ascidian *Halocynthia roretzi.* The first cleavage divides the fertilized egg along the dorso-ventral plane into left and right halves. From this stage onwards the development of the embryo is bilaterally symmetrical. Corresponding right- and left-half blastomeres are given the same designation, but left-half blastomeres and their descendants are conventionally underlined. The second cleavage divides the embryo along the antero-posterior axis. Note that the view of the embryo has been rotated by 90° towards the viewer, between the first and second cleavages, so that the antero-posterior axis can be shown clearly.

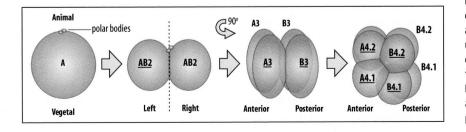

Important maternal factors are localized in the vegetal region of the fertilized egg, among them components of the Dishevelled–β-catenin signaling pathway. As in the sea urchin, localized activation of this pathway stabilizes maternal β-catenin and induces its preferential entry into nuclei in vegetal cells, where it is required to specify endoderm in the vegetal region. In contrast, the development of an epidermal fate in animal blastomeres requires suppression of β-catenin function.

The animal–vegetal and antero-posterior axes of the ascidian embryo are specified in the fertilized egg before first cleavage by movement of the cortical cytoplasm. The unfertilized egg is radially symmetrical along its animal–vegetal axis. As in *Xenopus*, sperm entry at fertilization triggers a rotation of the egg cortex that breaks this symmetry (see Section 4.2). In the ascidian, part of the cortical cytoplasm moves towards the vegetal pole, generating a characteristic bulge at the pole. This movement is associated with the cytoskeleton, principally cortical actin filaments and a deeper network of intermediate filaments. The cortical region that relocates is rich in mitochondria and is called the **myoplasm**. It contains maternal cytoplasmic determinants, such as the mRNA *macho-1*, that specify muscle, and it will eventually give rise to the muscle of the ascidian tadpole's tail. Its position after cortical movement specifies the posterior end of the antero-posterior axis, which is also where gastrulation will start.

The first cleavage is in the plane of the animal–vegetal axis and gives rise to two cells with similar developmental potential, each of which can regulate to grow into a half-size tadpole if separated. The second cleavage is in the same plane but at right angles to the first and delimits the anterior and posterior halves of the embryo. The third cleavage demarcates the animal and vegetal halves (see Fig. 6.29). The planes of further cleavages are strictly controlled; this stereotyped pattern of cleavage determines what cytoplasmic maternal determinants the resulting cells contain and thus how they will develop. An important influence on further unequal cleavages in the posterior region is the so-called chromosome-attracting body (CAB) in the myoplasm. The CAB contains PAR proteins, which, as in nematodes (see Section 6.1), help position the mitotic spindle so that the cell divides asymmetrically.

By the time the embryo reaches the 110-cell stage, at which gastrulation begins, the fate of each blastomere has become restricted by a combination of the maternal cytoplasmic determinants it contains and intercellular signaling, such that most of the cells give rise to a single cell type in the larva (Fig. 6.30). When individual blastomeres from this stage are isolated and cultured they develop into the cell types specified by the fate map.

Despite the prominent role of cytoplasmic determinants in specifying cell fate in ascidians, the development of some types of mesodermal tissue at least depends on inductive signals, as in vertebrates and echinoderms. Three distinct types of mesodermal tissue are formed in ascidian embryos—the muscles of the larval tail, the mesenchymal tissue that gives rise to internal tissues, and the notochord. Here we shall look at the ways in which the fate of these three tissues is determined.

6.16 In ascidians, muscle is specified by localized cytoplasmic factors

The fate of the myoplasm in the ascidian *Styela* is particularly easy to follow as it contains yellow pigment granules that are visible in the embryo, and this was the first indication, discovered at the beginning of the twentieth century, that a particular region of the ascidian egg cytoplasm could give rise to a particular tissue. A hundred years later we are at last beginning to find out how. The cells that acquire the yellow myoplasm during cleavage give rise to the muscle cells of the larval tail (Fig. 6.31). Before fertilization, the yellow granules are more or less uniformly distributed throughout the egg; following fertilization and cortical rotation, there is a dramatic rearrangement of the myoplasm to form the yellow crescent at the equator.

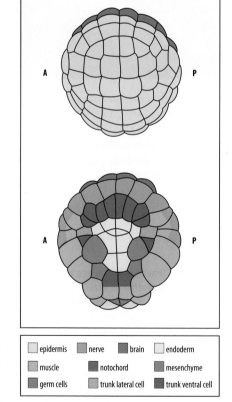

epidermis	nerve	brain	endoderm
muscle	notochord	mesenchyme	
germ cells	trunk lateral cell	trunk ventral cell	

Fig. 6.30 Fate map of the 110-cell ascidian embryo. Top panel: view looking down on the animal pole. Almost all the cells of the animal half become ectoderm. Bottom panel: view from the vegetal pole.

Adapted from Nishida, H.: 2005.

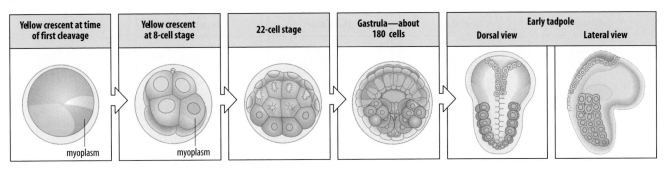

| Yellow crescent at time of first cleavage | Yellow crescent at 8-cell stage | 22-cell stage | Gastrula—about 180 cells | Early tadpole |
| | | | | Dorsal view — Lateral view |

myoplasm — myoplasm

By the eight-cell stage, the myoplasm is largely confined to the two vegetal posterior cells, with a small amount in adjacent cells (see Fig. 6.31). In the species *Halocynthia roretzi*, the cell lineage has been worked out in detail by the injection of a tracer into the early cells. The two posterior B4.1 cells (see Fig. 6.29) that contain myoplasm contribute to the primary muscle cells, which are 28 of the 42 muscle cells that lie on each side of the tail. The secondary muscle cells are derived from blastomeres adjacent to B4.1 at the eight-cell stage and end up at the top of the tail. The lineage is complex; for example, at the 128-cell stage one of the descendants of B4.1 is still an endomesodermal cell that will give rise to both muscle and endoderm. So, while there is a good correlation between muscle development and myoplasm, these observations on their own did not establish that it is something in the myoplasm that is causing differentiation into muscle.

Experiments that altered the distribution of the myoplasm also provided suggestive, but not conclusive, evidence that the myoplasm on its own can specify muscle cells. But the most persuasive evidence that the myoplasm specifies muscle is provided by more recent experiments that deplete it of a key maternal mRNA—*macho-1* mRNA. This is localized in the myoplasm in the egg and its depletion results in loss of the primary muscle cells in the tail. If the posterior-vegetal cytoplasm containing the *macho-1* mRNA is removed from the fertilized egg, the blastomeres that would normally develop as muscle develop as nerve cord. Conversely, injection of *macho-1* mRNA into non-muscle cell lineages causes ectopic muscle differentiation.

Fig. 6.31 Muscle development and cytoplasmic determinants in the ascidian *Styela*. Following fertilization, the myoplasm, which is colored with yellow granules, moves laterally and toward the equator. This movement forms a yellow crescent at the future posterior end of the embryo. Gastrulation starts at this site. The muscle of the ascidian tadpole's tail comes both from cells that contain the yellow myoplasm and the cells adjacent to them.

After Conklin, E.G.: 1905.

6.17 Notochord, neural precursors, and mesenchyme in ascidians require inducing signals from neighboring cells

When first visible in the early tadpole, the notochord of the *Ciona* larva consists of a single row of 40 cells aligned along the center of the tail, which undergo shape changes to eventually form a hollow tube. The notochord derives mainly from A lineage cells with a small contribution from the B lineage and requires induction by adjacent vegetal cells for its formation. Blastomeres that normally give rise to notochord do not do so if they are isolated at the 32-cell stage, unless they are combined with vegetal blastomeres. However, prospective notochord cells isolated at the 110-cell stage do develop into notochord.

The specification of the notochord precursor cells of the A lineage provides a good example of how a pattern of asymmetric cell division, together with signals from neighboring cells, can allocate daughter cells to different germ layers. Blastomeres A7.3 and A7.7 in the 64-cell embryo give rise to notochord only, whereas their sister cells, A7.4 and A7.8, respectively, give rise to spinal cord. (This applies also to the corresponding cells on the right-hand side of the embryo. We will follow events on just one side for simplicity.) The division that creates these cells with very different fates occurs between the 32-cell and 44-cell stage (Fig. 6.32). When A6.2 and A6.4 in the 32-cell embryo divide, the posterior daughter of each division, the one adjacent to the vegetal blastomeres contributes to the notochord. The anterior daughter, adjacent to animal cap cells, gives rise to neural tissue (spinal cord).

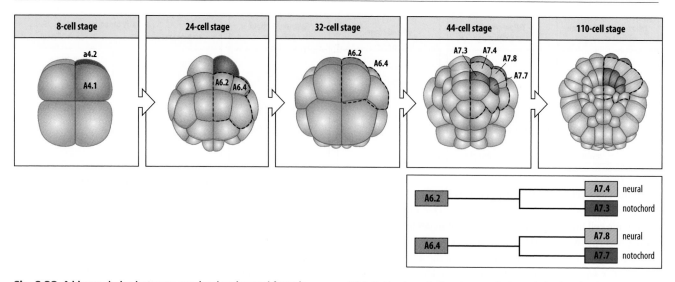

| 8-cell stage | 24-cell stage | 32-cell stage | 44-cell stage | 110-cell stage |

Fig. 6.32 A binary choice between notchord and neural fates in ascidian embryos. The embryo is viewed from the vegetal pole, in the same orientation as shown in the lower panel in Fig. 6.30. Anterior is up. All notochord cells are derived from blastomeres A4.1 and B4.1, formed at the eight-cell stage. In the A lineage, the notochord derives from an asymmetric division of the the A6.2 and A6.4 mother cells at the late 32-cell stage. These cells were formed from A4.1 by two successive rounds of cell division. At the 44-cell stage, A6.2 and A6.4 divide along their antero-posterior axis to give rise to one (posterior) notochord precursor each (red, A7.3 and A7.7) and one (anterior) neural precursor each (blue, A7.4 and A7.8). Each precursor cell then divides in the medio-lateral plane to generate four notochord precursors and four neural precursors in the 110-cell embryo.

After Picco, V., et al.: 2007.

As we saw with the asymmetric division of the *Caenorhabditis* EMS cell into an MS cell and an E cell (see Section 6.3), the assignment of notchord and neural fates to the daughters of A6.2 and A6.4 is the result of a polarization of the mother cells by intercellular signals followed by a functionally asymmetric division. The growth factor FGF can induce notochord *in vitro* and is probably the inducer of a notochord fate *in vivo*, as preventing FGF signaling in the early embryo prevents notochord formation. If the FGF signaling pathway is blocked, both the daughter cells of A6.2 and A6.4 are neural precursors. The acquisition of a neural fate requires contact with anterior animal cap cells and Eph–ephrin signaling, which suppresses the notochord fate. The involvement of Eph receptors and their ephrin ligands as signals that determine cell fate in the early embryo came as something of a surprise. In other chordates they are usually involved in adhesive cell–cell interactions, such as those that delimit rhombomere boundaries in the chick brain (see Section 5.11), and in the selective adhesive interactions that guide growing axons to their targets during nervous system development (see Chapter 12).

Although the early development of ascidians is very different from that of vertebrates, the presence of a notochord and its induction by vegetal cells is a striking parallel between the two groups. Moreover, the same genes are involved in notochord specification in both ascidians and vertebrates. The gene *Brachyury* is expressed in early mesoderm in vertebrates and then becomes confined to the notochord (see Section 4.18). Expression of the ascidian homolog of *Brachyury* is first detected at the 64-cell stage in the A-lineage precursors of the notochord, and this stage appears to correspond with the time at which induction is complete. Ectopic expression of the ascidian *Brachyury* gene can transform endoderm to notochord. Thus, there appear to be similar mechanisms involved in notochord formation in all chordates.

Specification of mesenchymal precursors (which will give rise to internal tissues) from a mesodermal precursor also requires signals from adjacent cells. The mesenchyme precursors are located adjacent to the endodermal cells at the vegetal pole of the 110-cell embryo (see Fig. 6.30). When blastomeres are isolated at the 32-cell stage, all those blastomeres whose normal fate is mesenchyme develop into muscle. It appears that a signal from the endoderm, probably FGF, normally suppresses muscle formation in these blastomeres. In the absence of such signals they develop into muscle, as a result of the presence of maternal muscle determinants, such as Macho-1, within them. The interplay between intercellular signals and intracellular determinants in setting cell fate is complicated. Embryos lacking Macho-1 form notochord in place of mesenchyme, whereas the overexpression of *macho-1* causes the usual notochord precursors to develop as mesenchyme.

Summary

In the chordate ascidians, there is evidence that localized cytoplasmic factors are involved in specifying cell fate, particularly of muscle, but that cell interactions are also involved. The notochord develops through a well-defined cell lineage but also requires induction. The ascidian homolog of the vertebrate gene *Brachyury*, which is involved in specifying the notochord in vertebrates, is expressed in the presumptive ascidian notochord after induction. The third type of mesodermal tissue in ascidian embryos, the mesenchyme, also requires inductive signals from neighboring cells.

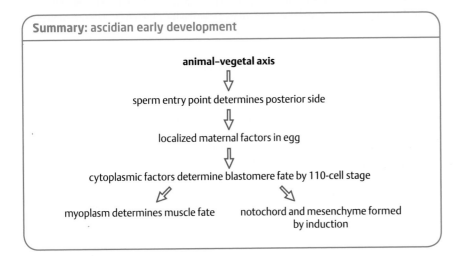

Summary: ascidian early development

animal–vegetal axis

⇓

sperm entry point determines posterior side

⇓

localized maternal factors in egg

⇓

cytoplasmic factors determine blastomere fate by 110-cell stage

⇙   ⇘

myoplasm determines muscle fate   notochord and mesenchyme formed by induction

Summary to Chapter 6

In many invertebrates, such as the nematode and the ascidian, cytoplasmic localization and cell lineage, together with local intercellular interactions, seem to be very important in specifying cell fate. However, there is very little evidence in any of these animals for key signaling regions corresponding to the vertebrate organizer, or for morphogen gradients as in *Drosophila*. This may reflect the predominance of mechanisms that specify cell fate on a cell-by-cell basis, rather than by groups of cells as in vertebrates and insects.

Although genetic evidence of the kind available in *Drosophila* is lacking in other invertebrates, except in the case of the nematode *Caenorhabditis elegans*, it is clear that maternal genes are crucial in specifying early patterning. There are numerous examples of the first cleavage in the embryo being related to the establishment of the embryo's axes and evidence that this plane of cleavage is maternally determined. Asymmetric cleavage associated with the unequal distribution of cytoplasmic determinants seems to be very important at early stages. Cytoplasmic movements that are involved in this localization of determinants often occur after fertilization, and in this respect they may resemble the cortical rotation of the amphibian egg.

There can be no doubt that there are cell interactions, even at early stages, in all of the organisms in this chapter; an example is the specification of left-right asymmetry in nematodes, where a small alteration in cell relationships can reverse their handedness. But in most of these embryos the interactions are highly local, only affecting an adjacent cell. This is unlike vertebrates, in which interactions may occur over several cell diameters and involve groups of cells.

By contrast, in sea-urchin development, which is highly regulative, cell interactions play a major role, and thus echinoderms resemble the vertebrates in this respect. The micromeres at the vegetal pole have properties similar to those of the vertebrate organizer, and cells are specified in groups rather than on a cell-by-cell basis. Ascidians are members of the chordates, the same phylum as vertebrates, and show a mixture of mosaic development and cell-cell interactions.

■ End of chapter questions

Long answer (concept questions)

1. What is meant by asymmetric cell division: does it necessarily mean that one daughter is bigger than the other? What other possibility qualifies a cell division event as being asymmetric? How are both types of asymmetric cell division illustrated by the first division in *Caenorhabditis elegans*?

2. Compare and contrast RNAi and miRNA gene silencing. Incorporate these hints into your answer: experimental versus natural; ~ 22 nucleotides; RISC; and degradation versus translation.

3. How is the antero-posterior axis of *C. elegans* determined? What is the role of the PAR proteins in axis determination?

4. During studies of cell lineage in *C. elegans*, a convention arose that allowed the unambiguous identification of each of its 558 cells (at hatching). Key to this system is the designation of the anterior daughter of each cell division with an 'a', and the posterior daughter with a 'p.' Refer to Fig. 6.9 and describe why the indicated pharyngeal cell is called 'MS paapaaa.' Similarly, why is the apoptotic cell called 'MS paapp'?

5. To what class of signaling proteins do the *C. elegans* proteins GLP-1 and APX-1 belong? What is the role of the GLP-1/APX-1 system in dorso-ventral cell fate specification? 'APX' stands for 'anterior pharynx in excess;' why was the mutant gene named 'APX' (refer to Section 6.1 and Fig. 6.4)?

6. What are heterochronic genes? What feature of *C. elegans* development makes this a favorable organism for the identification of such genes?

7. *lin-4* encodes a miRNA that represses translation of *lin-14* mRNA. Since *lin-14* maintains the Tap cell in a first larval (L1) state, this repression is required for the cell to transition to a L2 identity. Based on this mechanism, what is the phenotype of a *lin-14* loss-of-function mutant? What is the phenotype of a *lin-4* loss-of-function mutant? What would be the phenotype of a worm with loss-of-function mutations in both genes? Explain your answer.

8. The anchor cell of *C. elegans* is required for the induction of vulval development. To trace this induction, indicate which cells express the EGF, EGF receptor, Delta, and Notch proteins, and the consequences of these proteins in those cells.

9. What are the similarities between gastrulation in *Strongylocentrotus* and *Xenopus* that make them both deuterostomes?

10. Compare the first two cleavages in the frog and sea urchin. In what ways are they similar or different morphologically? Are the embryos similar or different molecularly; that is, have the determinative events that occur in *Xenopus* by the four-cell stage also occurred in sea urchins? Describe the differences in terms of 'regulation' during development.

11. The micromeres of sea urchins act as an organizer. Describe two experiments that support this statement (see Fig. 6.22).

12. *pmar1* is a transcription factor in the paired class of homeodomain factors, which acts as a switch in the specification of endomesoderm (versus ectoderm) in sea urchins. Outline the role of β-catenin and Disheveled in activation of *pmar1* in the micromeres (note that Wnt is not involved at this point). What signaling pathway is activated later in the micromeres in response to *pmar1* expression (see Fig. 6.24)?

13. Sea-urchin larvae have two overt axes, animal–vegetal and oral–aboral, and a more subtle internal left–right asymmetry. Compare the molecular bases of these three axes with the three axes of frogs (this may be most easily done in the form of a table, with the six types of axes on the side and the signaling molecules and transcription factors along the top, or in some other form of your own choosing).

14. What is the myoplasm of ascidian embryos? How is the myoplasm related to cortical rotation, and what tissue will it form?

15. Why are ascidians chordates, but not vertebrates? What similarities and differences exist between the formation of the notochord of ascidians and the notochord of amphibian or chick embryos?

Multiple choice (factual recall questions)

NB There is only one right answer to each question.

1. *C. elegans* is best known for

a) its formation of an inductive signaling center analogous to the organizer of *Xenopus*

b) its fully defined cell lineage during development

c) the large size and easy manipulation of its blastomeres

d) its mechanisms of metamorphosis

2. P granules of nematodes are

a) determinants of the antero-posterior axis

b) determinants of the germ line

c) precursors of TGF-β family signaling proteins

d) remnants of the sperm pronucleus

3. In flies, frogs, and chicks, gradients of morphogens determine the future antero-posterior and dorso-ventral axes of the developing embryo. How is the antero-posterior axis determined in *C. elegans*?

a) Bicoid protein is translated in the anterior of the fertilized egg, leading to a gradient that determines the antero-posterior axis.

b) Opposing gradients of chordin and BMP-4 establish the antero-posterior axis.

c) Sperm entry leads to a reorganization of the cytoskeleton and redistribution of maternally packaged PAR proteins, which in turn determine the antero-posterior axis.

d) β-catenin becomes localized to the nucleus in the future anterior cells after fertilization.

4. What are the *lin-4* and *lin-14* genes of *C. elegans*?

a) *lin-4* and *lin-14* are *C. elegans* versions of the Hox genes.

b) *lin-4* and *lin-14* are genes that are named for their control of the first division, and hence the lineage of the AB and P1 cells.

c) *lin-4* controls the timing of transcription of *lin-14*, and so is referred to a 'heterochronic' gene.

d) *lin-4* encodes a miRNA that represses *lin-14* translation, which in turn regulates timing during larval development.

5. In *C. elegans*, the only Hox gene utilized during embryogenesis is

a) *ceh-13*, the *C. elegans* ortholog of the *Drosophila labial* gene

b) *egl-5*, the *C. elegans* ortholog of the *Drosophila Abd-B* and the vertebrate *Hox9-13* genes

c) *lin-39*, a *C. elegans* gene related to the *Drosophila Deformed* and *Sex combs reduced* family of homeodomain proteins

d) *mab-5*, a *C. elegans* gene related to the *Drosophila Antennapedia*, *Ultrabithorax*, and *abdominal-A* family of homeodomain proteins

6. Which of these model organisms is closest evolutionarily to humans?

a) Fruit flies, because of their sophisticated nervous system and the presence of legs.

b) Plants, because their early divergence from animal lineages means they are closely related to animals, especially humans.

c) Sea urchins, despite their radial symmetry and lack of limbs.

d) Worms, despite their few cells, small genomes, and abbreviated complement of Hox genes.

7. Which of the following statements about regulation in sea-urchin embryos is true?

a) Any blastomere at the eight-cell stage can still regulate to form a full larva, but by the 60-cell stage, blastomeres will develop according to their presumptive fate.

b) Blastomeres at the four-cell stage each still possesses a portion of the original animal–vegetal axis, and if isolated will go on the form a fully formed, yet smaller, larva.

c) Blastomeres from the animal or vegetal poles of the embryo can be isolated and will go on to form an embryo, as long as either an animal or vegetal pole is included.

d) Up until the hatching of the embryo as a pluteus larva, any blastomere can be separated from the others and will regulate to go on and develop into a fully formed and full-sized larva.

8. In which ways is the organizer region of the sea urchin similar to that of frogs?

a) The organizer accumulates β-catenin, which in turn activates the expression of cell–cell signaling molecules.

b) The organizer forms near the dorsal lip of the blastopore.

c) The organizer induces the dorso-ventral axis.

d) The organizer will express high levels of Nodal.

9. In the ascidian *Styela* the yellow-pigmented myoplasm marks cells fated to become muscle cells in the tail; however, at the molecular level, the key determinant of tail muscle seems to be:

a) expression of the *Brachyury* gene

b) induction of muscle by FGF

c) localization of β-catenin to the nuclei in the vegetal portion of the embryo

d) the gene product of the *macho-1* gene

10. The role of the *Brachyury* gene in ascidians is similar to that in *Xenopus* in that

a) *Brachyury* is the master-switch gene for muscle development on both organisms

b) *Brachyury* sets up the animal–vegetal axis

c) signaling that leads to induction of the mesoderm results in expression of the *Brachyury* gene

d) the *Brachyury* gene is expressed specifically in those cells that will form somites, and hence muscle

Multiple choice answer key

1: b, 2: b, 3: c, 4: d, 5: a, 6: c, 7: b, 8: a, 9: d, 10: c.

■ Further reading

Nematodes

Sommer, R.J.: **Evolution of development in nematodes related to *C. elegans*** (December 14, 2005), *WormBook*, ed. The *C. elegans* Research Community, WormBook, doi/10.1895/wormbook.1.46.1, http://www.wormbook.org (date accessed 13 May 2010).

Sulston, J.E., Scherienberg, E., White, J.G., Thompson, J.N.: **The embryonic cell lineage of the nematode *Caenorhabditis elegans***. *Dev. Biol.* 1983, **100**: 64–119.

Sulston, J.: **Cell lineage**. In *The Nematode* Caenorhabditis elegans. Edited by Wood, WB. New York: Cold Spring Harbor Laboratory Press, 1988: 123–156.

Wood, W.B.: **Embryology**. In *The Nematode* Caenorhabditis elegans. Edited by Wood, WB. New York: Cold Spring Harbor Laboratory Press, 1988: 215–242.

6.1 The antero-posterior axis in *Caenorhabditis elegans* is determined by asymmetric cell division

Cheeks, R.J., Canman, J.C., Gabriel, W.N., Meyer, N., Strome, S., Goldstein, B.: **C. elegans PAR proteins function by mobilizing and stabilizing asymmetrically localized protein complexes.** *Curr. Biol* 2004, **14**: 851–862.

Cowan, C.R., Hyman, A.A.: **Acto-myosin reorganization and PAR polarity in C. elegans.** *Development* 2007, **134**: 1035–1043.

Goldstein, B., Macara, I.G.: **The PAR proteins: fundamental players in animal cell polarization.** *Dev. Cell* 2007, **13**: 609–622.

Munro, E., Nance, J., Priess, J.R.: **Cortical flows powered by asymmetrical contraction transport PAR proteins to establish and maintain anterior-posterior polarity in the early C. elegans embryo.** *Dev. Cell* 2004, **7**: 413–424.

Box 6A Gene silencing by antisense RNA and RNA interference

Brennecke, J., Malone, C.D., Aravin, A.A., Sachidanandam, R., Stark, A., Hannon, G.J.: **An epigenetic role for maternally inherited piRNAs in transposon silencing.** *Science* 2008, **322**: 1387–1392.

DasGupta, R., Nybakken, K., Booker, M., Mathey-Prevot, B., Gonsalves, F., Changkakoty, B., Perrimon, N.: **A case study of the reproducibility of transcriptional reporter cell-based RNAi screens in Drosophila.** *Genome Biol.* 2007, **8**: R203.

Dykxhoorn, D.M., Novind, C.D., Sharp, P.A.: **Killing the messenger: short RNAs that silence gene expression.** *Nat. Rev. Mol. Cell. Biol* 2003, **4**: 457–467.

Eisen, J.S., Smith, J.C.: **Controlling morpholino experiments: don't stop making antisense.** *Development* 2008, **135**: 1735–1743.

Kamath, R.S., Fraser, A.G., Dong, Y., Poulin, G., Durbin, R., Gotta, M., Kanapin, A., Le Bot, N., Moreno, S., Sohrmann, M., Welchman, D.P., Zipperlen, P., Ahringer, J.: **Systematic functional analysis of the Caenorhabditis elegans genome using RNAi.** *Nature* 2003, **421**: 231–237.

Nature Insight: **RNA interference.** *Nature* 2004, **431**: 337–370.

Sonnichsen, B., *et al.*: **Full-genome RNAi profiling of early embryogenesis in Caenorhabditis elegans.** *Nature* 2005, **434**: 462–469.

6.2 The dorso-ventral axis in *Caenorhabditis elegans* is determined by cell–cell interactions

Bergmann, D.C., Lee, M., Robertson, B., Tsou, M.F., Rose, L.S., Wood, W.B.: **Embryonic handedness choice in C. elegans involves the Galpha protein GPA-16.** *Development* 2003, **130**: 5731–5740.

Wood, W.B.: **Evidence from reversal of handedness in C. elegans embryos for early cell interactions determining cell fates.** *Nature* 1991, **349**: 536–538.

6.3 Both asymmetric divisions and cell–cell interactions specify cell fate in the early nematode embryo

Eisenmann, D.M.: **Wnt signaling** (June 25, 2005), *WormBook*, ed. The *C. elegans* Research Community, WormBook, doi/10.1895/wormbook.1.7.1, http://www.wormbook.org (date accessed 13 May 2010).

Evans, T.C., Crittenden, S.L., Kodoyianni, V., Kimble, J.: **Translational control of maternal glp-1 mRNA establishes an asymmetry in the C. elegans embryo.** *Cell* 1994, **77**: 183–194.

Kidd, A.R. 3rd, Miskowski, J.A., Siegfried, K.R., Sawa, H., Kimble, J.: **A beta-catenin identified by functional rather than sequence criteria and its role in Wnt/MAPK signaling.** *Cell* 2005, **121**: 761–772.

Lyczak, R., Gomes, J.E., Bowerman, B.: **Heads or tails: cell polarity and axis formation in the early Caenorhabditis elegans embryo.** *Dev. Cell* 2002, **3**: 157–166.

Maduro, M.F.: **Structure and evolution of the C. elegans embryonic emdomesoderm network.** *Biochim. Biophys. Acta* 2008, **1789**: 250–260.

Mello, G.C., Draper, B.W., Priess, J.R.: **The maternal genes apx-1 and glp-1 and the establishment of dorsal-ventral polarity in the early C. elegans embryo.** *Cell* 1994, **77**: 95–106.

Mizumoto, K., Sawa, H.: **Two βs or not two βs: regulation of asymmetric division by β-catenin.** *Trends Cell Biol.* 2007, **17**: 465–473.

Mizumoto, K., Sawa, H.: **Cortical β-catenin and APC regulate asymmetric nuclear β-catenin localization during asymmetric cell division in C. elegans.** *Dev. Cell* 2007, **12**: 287–299.

Park, F.D., Priess, J.R.: **Establishment of POP-1 asymmetry in early C. elegans embryos.** *Development* 2003 **130**: 3547–3556.

Park, F.D., Tenlen, J.R., Priess, J.R.: **C. elegans MOM-5/frizzled functions in MOM-2/Wnt-independent cell polarity and is localized asymmetrically prior to cell division.** *Curr. Biol.* 2004, **14**: 2252–2258.

Phillips, B.T., Kidd III, A.R., King, R., Hardin, J., Kimble, J.: **Reciprocal asymmetry of SYS-1/β-catenin and POP-1/TCF controls asymmetric divisions in Caenorhabditis elegans.** *Proc. Natl Acad. Sci. USA* 2007, **104**: 3231–3236.

6.4 Hox genes specify positional identity along the antero-posterior axis in *Caenorhabditis elegans*

Aboobaker, A.A., Blaxter, M.L.: **Hox gene loss during dynamic evolution of the nematode cluster.** *Curr. Biol.* 2003, **13**: 37–40.

Clark, S.G, Chisholm, A.D., Horvitz, H.R.: **Control of cell fates in the central body region of C. elegans by the homeobox gene lin-39.** *Cell* 1993, **74**: 43–55.

Cowing, D., Kenyon, C.: **Correct Hox gene expression established independently of position in Caenorhabditis elegans.** *Nature* 1996, **382**: 353–356.

Salser, S.J., Kenyon, C.: **A C. elegans Hox gene switches on, off, on and off again to regulate proliferation, differentiation and morphogenesis.** *Development* 1996, **122**: 1651–1661.

Van Auken, K., Weaver, D.C., Edgar, L.G, Wood, W.B.: **Caenorhabditis elegans embryonic axial patterning requires two recently discovered posterior-group Hox genes.** *Proc. Natl Acad. Sci. USA* 2000, **97**: 4499–4503.

6.5 The timing of events in nematode development is under genetic control that involves microRNAs

Ambros, V.: **Control of developmental timing in Caenorhabditis elegans.** *Curr. Opin. Genet. Dev.* 2000, **10**: 428–433.

Austin, J., Kenyon, C.: **Developmental timekeeping: marking time with antisense.** *Curr. Biol.* 1994, **4**: 366–396.

Grishok, A., Pasquinelli, A.E., Conte, D., Li, N, Parrish, S., Ha, I., Baillie, D.L., Fire, A., Ruvkun, G., Mello, C.C.: **Genes and** mechanisms related to RNA interference regulate expression of the small temporal RNAs that control *C. elegans* developmental timing. *Cell* 2001, **106**: 23–34.

Box 6B Gene silencing by microRNAs

Bartel, D.P.: **MicroRNAS: genomics, biogenesis, mechanism and function.** *Cell* 2004, **116**: 281–297.

He, L., Hannon, G.J.: **MicroRNAs: small RNAs with a big role in gene regulation.** *Nat Rev. Genet.* 2004, **5**: 522–531.

Liu, J.: **Control of protein synthesis and mRNA degradation by microRNAs.** *Curr. Opin. Cell Biol.* 2008, **20**: 214–221.

Murchison, E.P., Hannon, G.J.: **miRNAs on the move: miRNA biogenesis and the RNAi machinery.** *Curr. Opin. Cell Biol.* 2004, **16**: 223–229.

6.6 Vulval development is initiated by the induction of a small number of cells by short-range signals from a single inducing cell

Kenyon, C.: **A perfect vulva every time: gradients and signaling cascades in *C. elegans*.** *Cell* 1995, **82**: 171–174.

Sharma-Kishore, R., White, J.G., Southgate, E., Podbilewicz, B.: **Formation of the vulva in *Caenorhabditis* elegans: a paradigm for organogenesis.** *Development* 1999, **126**: 691–699.

Sundaram, M., Han, M.: **Control and integration of cell signaling pathways during *C. elegans* vulval development.** *BioEssays* 1996, **18**: 473–480.

Yoo, A.S., Bais, C., Greenwald, I.: **Crosstalk between the EGFR and LIN-12/Notch pathways in *C. elegans* vulval development.** *Science* 2004, **303**: 663–636.

Echinoderms

Horstadius, S.: *Experimental Embryology of Echinoderms*. Oxford: Clarendon Press, 1973.

6.7 The sea-urchin embryo develops into a free-swimming larva

Raff, R.A., Raff, E.C.: **Tinkering: new embryos from old—rapidly and cheaply.** *Novartis Found. Symp.* 2007, **284**: 35–45; discussion 45–54, 110–115.

Wilt, F.H.: **Determination and morphogenesis in the sea-urchin embryo.** *Development* 1987, **100**: 559–576.

Sea Urchin Genome Sequencing Consortium: **The genome of the sea urchin *Strongylocentrotus purpuratus*.** *Science* 2006, **314**: 941–952. Erratum in: *Science* 2007, **315**: 766.

Sea urchin interactive online poster: *Science* 2006 www.sciencemag. org/sciext/seaurchin (date accessed 13 May 2010).

6.8 The sea-urchin egg is polarized along the animal-vegetal axis

Davidson, E.H., Cameron, R.S., Ransick, A.: **Specification of cell fate in the sea-urchin embryo: summary and some proposed mechanisms.** *Development* 1998, **125**: 3269–3290.

Henry, J.J.: **The development of dorsoventral and bilateral axial properties in sea-urchin embryos.** *Semin. Cell Dev. Biol.* 1998, **9**: 43–52.

Logan, C.Y., McClay, D.R.: **The allocation of early blastomeres to the ectoderm and endoderm is variable in the sea-urchin embryo.** *Development* 1997, **124**: 2213–2223.

6.9 The sea urchin fate map is finely specified, yet considerable regulation is possible

Angerer, L.M., Angerer, R.C.: **Patterning the sea-urchin embryo: gene regulatory networks, signaling pathways, and cellular interactions.** *Curr. Top. Dev. Biol.* 2003, **53**: 159–198.

Angerer, L.M., Oleksyn, D.W., Logan, C.Y., McClay, D.R., Dale, L., Angerer, R.C.: **A BMP pathway regulates cell fate allocation along the sea urchin animal-vegetal embryonic axis.** *Development* 2000, **127**: 1105–1114.

Sweet, H.C., Hodor, P.G., Ettensohn, C.A.: **The role of micromere signaling in Notch activation and mesoderm specification during sea-urchin embryogenesis.** *Development* 1999, **126**: 5255–5265.

Wessel, G.M., Wikramanayake, A.: **How to grow a gut: ontogeny of the endoderm in the sea-urchin embryo.** *BioEssays* 1999, **21**: 459–471.

6.10 The vegetal region of the sea-urchin embryo acts as an organizer

Ransick, A., Davidson, E.H.: **A complete second gut induced by transplanted micromeres in the sea-urchin embryo.** *Science* 1993, **259**: 1134–1138.

6.11 The sea-urchin vegetal region is specified by the nuclear accumulation of β-catenin

Ettensohn, C.A.: **The emergence of pattern in embryogenesis: regulation of beta-catenin localization during early sea urchin development.** *Sci STKE* 2006, **14**: pe48.

Weitzel, H.E., Illies, M.R., Byrum, C.A., Xu, R., Wikramanayake, A.H., Ettensohn, C.A.: **Differential stability of beta-catenin along the animal-vegetal axis of the sea-urchin embryo mediated by dishevelled.** *Development* 2004, **131**: 2947–2955.

6.12 The genetic control of the skeletogenic pathway is known in considerable detail

Davidson, E.H.: **A view from the genome: spatial control of transcription in sea urchin development.** *Curr. Opin. Genet. Dev.* 1999, **9**: 530–541.

Ettensohn, C.A.: **Lessons from a gene regulatory network: echinoderm skeletogenesis provides insights into evolution, plasticity and morphogenesis.** *Development* 2009, **136**: 11–21.

Oliveri, P., Davidson, E.H.: **Gene regulatory network controlling embryonic specification in the sea urchin.** *Curr. Opin. Genet. Dev.* 2004, **14**: 351–360.

Oliveri, P., Tu, Q., Davidson, E.H.: **Global regulatory logic for specification of an embryonic cell lineage.** *Proc. Natl Acad. Sci. USA* 2008, **105**: 5955–5962.

Kirchnamer, C.V., Yuh, C.V., Davidson, E.H.: **Modular *cis*-regulatory organization of developmentally expressed genes: two genes transcribed territorially in the sea-urchin embryo, and additional examples.** *Proc. Natl Acad. Sci. USA* 1996, **93**: 9322–9328.

6.13 The oral-aboral axis in sea urchins is related to the plane of the first cleavage

Cameron, R.A., Fraser, S.E., Britten, R.J., Davidson, E.H.: **The oral-aboral axis of a sea-urchin embryo is specified by first cleavage.** *Development* 1989, **106**: 641–647.

6.14 The oral ectoderm acts as an organizing region for the oral-aboral axis

Coffman, J.A., Coluccio, A., Planchart, A., Robertson, A.J.: **Oral-aboral axis specification in the sea-urchin embryo III. Role of mitochondrial redox signaling via H$_2$O$_2$**. *Dev. Biol.* 2009, **330**: 123–130.

Duboc, V., Lepage, T.: **A conserved role for the nodal signaling pathway in the establishment of dorso-ventral and left–right axes in deuterostomes**. *J. Exp. Zool. (Mol. Dev. Evol.)* 2008, **310B**: 41–53.

Duboc, V., Lapraz, F., Besnardeau, L., Lepage, T.: **Lefty acts as an essential modulator of Nodal activity during sea urchin oral-aboral axis formation**. *Dev. Biol.* 2008, **320**: 49–59.

Duboc, V., Rottinger, E., Besnardeau, L., Lepage, T.: **Nodal and BMP2/4 signaling organizes the oral-aboral axis of the sea-urchin embryo**. *Dev. Cell* 2004, **6**: 397–410.

Yaguchi, S., Yaguchi, J., Angerer, R.C., Angerer, L.M.: **A Wnt-FoxQ2-nodal pathway links primary and secondary axis specification in sea-urchin embryos**. *Dev. Cell* 2008, **14**: 97–107.

Ascidians

Dehal, P., Satou, Y., Campbell, R.K., Chapman, J., Degnan, B., De Tomaso, A., Davidson, B., Di Gregoriao, A., Gelpke, M., Goodstein, D.M., *et al.*: **The draft genome of *Ciona intestinalis*: insights into chordate and vertebrate origins**. *Science* 2002, **298**: 2157–2167.

Keys, D.N., Lee, B.I., Di Gregorio, A., Harafuji, N., Detter, J.C., Wang, M., Kahsai, O., Ahn, S., Zhang, C, Doyle, S.A., Satoh, N., Satou, Y, Saiga, H., Christian, A.T., Rokhsar, D.S., Hawkins, T.L., Levine, M., Richardson, P.M.: **A saturation screen for cis-acting regulatory DNA in the Hox genes of *Ciona intestinalis***. *Proc. Natl Acad. Sci. USA* 2005, **102**: 679–683.

Lemaire, P., Smith, W.C., Nishida, H.: **Ascidians and the plasticity of the chordate developmental program**. *Curr. Biol.* 2008, **18**: R620–R631.

Passamaneck, Y.J., Di Gregorio, A.: ***Ciona intestinalis*: chordate development made simple**. *Dev. Dyn.* 2005, **233**: 1–19.

Satoh, N.: *The Developmental Biology of Ascidians*. Cambridge: Cambridge University Press, 1994.

Shoguchi, E., Hamaguchi, M., Satoh, N.: **Genome-wide network of regulatory genes for construction of a chordate embryo**. *Dev. Biol.* 2008, **316**: 498–509.

6.15 Animal-vegetal and antero-posterior axes in the ascidian embryo are defined before first cleavage

Nishida, H.: **Specification of embryonic axis and mosaic development in ascidians**. *Dev. Dyn.* 2005, **233**: 1177–1193.

Sardet, C., Paix, A., Prodon, F., Dru, P., Chenevert, J.: **From oocyte to 16-cell stage: cytoplasmic and cortical reorganizations that pattern the ascidian embryo**. *Dev. Dyn.* 2007, **236**: 1716–1731.

6.16 In ascidians, muscle is specified by localized cytoplasmic factors

Bates, W.R.: **Development of myoplasm-enriched ascidian embryos**. *Dev. Biol.* 1988, **129**: 241–252.

di Gregorio, A., Levine, M.: **Ascidian embryogenesis and the origins of the chordate body plan**. *Curr. Opin. Genet. Dev.* 1998, **7**: 457–463.

Kim, G.J., Nishida, H.: **Suppression of muscle fate by cellular interaction as required for mesenchyme formation during ascidian embryogenesis**. *Dev Biol.* 1999, **214**: 9–22.

Nishida, H., Sawada, K.: ***macho-1* encodes a localized mRNA in ascidian eggs that specifies muscle fate during embryogenesis**. *Nature* 2001, **409**: 724–728.

6.17 Notochord, neural precursors, and mesenchyme in ascidians require inducing signals from neighboring cells

Corbo, J.C., Levine, M., Zeller, R.W.: **Characterization of a notochord-specific enhancer from the *Brachyury* promoter region of the ascidian, *Ciona intestinalis***. *Development* 1997, **124**: 589–602.

Kim, G.J., Yamada, A., Nishida, H.: **An FGF signal from endoderm and localized factors in the posterior-vegetal egg cytoplasm pattern the mesodermal tissues in the ascidian embryo**. *Development* 2000, **127**: 2853–2862.

Nakatani, Y., Nishida, H.: **Induction of notochord during ascidian embryogenesis**. *Dev. Biol.* 1994, **166**: 289–299.

Nakatani, Y., Yasuo, H., Satoh, N., Nishida, H.: **Basic fibroblast growth factor induces notochord formation and the expression of As-T, a Brachyury homolog, during ascidian embryogenesis**. *Development* 1996, **122**: 2023–2031.

Picco, V., Hudson, C., Yasuo, H.: **Ephrin–Eph signalling drives the asymmetric division of notochord/neural precursors in *Ciona* embryos**. *Development* 2007, **134**: 1491–1497.

Plant development

- Embryonic development
- Meristems
- Flower development and control of flowering

Because of plant cells' rigid cell walls and lack of cell movement within tissues, a plant's architecture is very much the result of patterns of oriented cell divisions. Despite this apparent invariance, however, cell fate in development is largely determined by similar means as in animals—by a combination of positional signals and intercellular communication. As well as communicating by extracellular signals and cell-surface interactions, plant cells are interconnected by cytoplasmic channels, which allow movement of proteins such as transcription factors directly from cell to cell.

The plant kingdom is very large, ranging from the algae, many of which are unicellular, to the multicellular land plants, which exist in a prodigious variety of forms. Plants and animals probably evolved the process of multicellular development independently, their last common ancestor being a unicellular eukaryote some 1.6 billion years ago. Plant development is, therefore, of interest not simply for its own sake and for its agricultural importance. By looking at the similarities and differences between plant and animal development, a study of plant development can shed light on the way that developmental mechanisms in two groups of multicellular organisms have evolved independently and under different sets of developmental constraints.

Do plants and animals use the same developmental mechanisms? As we shall see in this chapter, the logic behind the spatial layouts of gene expression that pattern a developing flower is similar to that of Hox gene action in patterning the body axis in animals, but the genes involved are completely different. Such similarities between plant and animal development are due to the fact that the basic means of regulating gene expression are the same in both, and thus similar general mechanisms for patterning gene expression in a multicellular tissue are bound to arise. As we shall see, many of the general control mechanisms we have encountered in animal development, such as asymmetric cell division, the response to positional signals, lateral inhibition, and changes in gene expression in response to extracellular signals, are all present in plants. Differences between plants and animals in the way development is controlled arise from some different ways in which plant cells can communicate with each other compared with animal cells, from the existence of rigid cell walls and the lack of large-scale cell movements, to the fact that environment has a much greater impact on plant development than on that of animals.

One general difference between plant and animal development is that most of the development occurs not in the embryo but in the growing plant. Unlike an animal embryo, the mature plant embryo inside a seed is not simply a smaller version of the organism it will become. All the 'adult' structures of the plant—shoots, roots, stalks, leaves, and flowers—are produced after germination from localized groups of undifferentiated cells known as **meristems**. Two meristems are established in the embryo: one at the tip of the root and the other at the tip of the shoot. These persist in the adult plant, and almost all the other meristems, such as those in developing leaves and flower shoots, are derived from them. Cells within meristems can divide repeatedly and can potentially give rise to all plant tissues and organs. This means that developmental patterning within meristems to produce organs such as leaves and flowers continues throughout a plant's life.

Plant and animal cells share many common internal features and much basic bio-chemistry, but there are some crucial differences that have a bearing on plant development. One of the most important is that plant cells are surrounded by a framework of relatively rigid cell walls. There is therefore virtually no cell migration in plants, and major changes in the shape of the developing plant cannot be achieved by the movement and folding of sheets of cells. In plant development, form is largely generated by differences in rates of cell division and by division in different planes, followed by directed enlargement of the cells.

As in animal development, one of the main questions in plant development is how cell fate is determined. Many structures in plants normally develop by stereotyped patterns of cell division, but despite this observation, which implies the importance of lineage, cell fate is, in many cases, also known to be determined by factors such as position in the meristem and cell–cell signaling. The cell wall would seem to impose a barrier to the passage of large signaling molecules, such as proteins, although it is very thin in some regions, such as the meristems; but all known plant extracellu-lar signaling molecules—such as auxins, gibberellins, cytokinins, and ethylene—are small molecules that easily penetrate cell walls. Plant cells also communicate with each other through fine cytoplasmic channels known as **plasmodesmata**, which link neighboring plant cells through the cell wall, and they may be the channels by which some developmentally important gene-regulatory proteins, and even mRNAs, move directly between cells. The size of the channel varies and those in shoot cells have the largest diameter.

Another important difference between plant and animal cells is that a complete, fertile plant can develop from a single differentiated somatic cell and not just from a fertilized egg. This suggests that, unlike the differentiated cells of adult animals, some differentiated cells of the adult plant may retain **totipotency**. Perhaps they do not become fully determined in the sense that adult animal cells do, or perhaps they are able to escape from the determined state, although how this could be achieved is as yet unknown. In any case, this difference between plants and animals illus-trates the dangers of the wholesale application to plant development of concepts derived from animal development. Nevertheless, the genetic analysis of development in plants is turning up instances of genetic strategies for developmental patterning rather similar to those of animals.

The small cress-like weed *Arabidopsis thaliana* has become the model plant for genetic and developmental studies, and will provide many of the examples in this chapter. We will begin by describing the main features of its morphology, life cycle, and reproduction.

7.1 The model plant *Arabidopsis thaliana* has a short life-cycle and a small diploid genome

The equivalent of *Drosophila* in the study of plant development is the small crucifer *Arabidopsis thaliana*, commonly called thale cress, which is well-suited to genetic and developmental studies. It is a diploid (unlike many plants, which are polyploid)

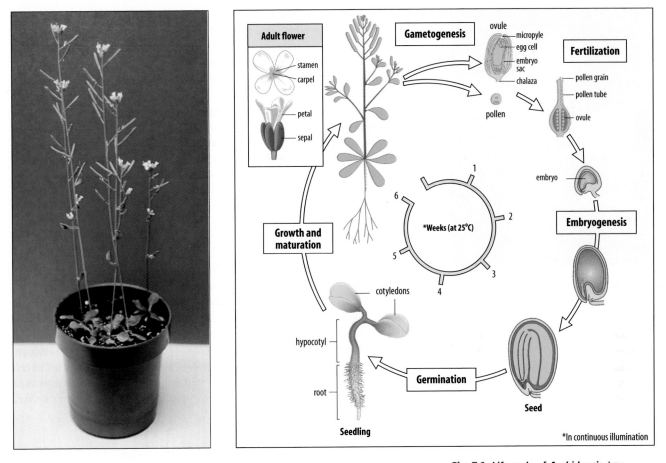

Fig. 7.1 Life cycle of *Arabidopsis*. In flowering plants, egg cells are contained separately in ovules inside the carpels. Fertilization of an egg cell by a male nucleus from a pollen grain takes place inside the ovary. The egg then develops into an embryo contained within the ovule coat, forming a seed. *Arabidopsis* is a dicotyledon, and the mature embryo has two wing-like cotyledons (storage organs) at the apical (shoot) end of the main axis—the hypocotyl—which contains a shoot meristem at one end and a root meristem at the other. Following germination, the seedling develops into a plant with roots, a stem, leaves, and flowers. The photograph shows five mature *Arabidopsis* plants.

and has a relatively small, compact genome that has been sequenced and which contains about 27,000 protein-coding genes. It is an annual, flowering in the first year of growth, and develops as a small ground-hugging rosette of leaves, from which a branched flowering stem is produced with a flowerhead, or **inflorescence**, at the end of each branch. It develops rapidly, with a total life cycle in laboratory conditions of some 6–8 weeks, and like all flowering plants, mutant strains and lines can easily be stored in large quantities in the form of seeds. The life cycle of *Arabidopsis* is shown in Fig. 7.1.

Each *Arabidopsis* flower (Fig. 7.2) consists of four sepals surrounding four white petals; inside the petals are six stamens, the male sex organs, which produce pollen containing the male gametes. Petals, sepals, and the other floral organs are thought to derive evolutionarily from modified leaves. At the center of the flower are the female sex organs, which consist of an ovary of two carpels, which contain the **ovules**. Each ovule contains an egg cell. Fertilization of an egg cell occurs when a pollen grain deposited on the carpel surface grows a tube that penetrates the carpel and delivers two haploid pollen nuclei to an ovule. One nucleus fertilizes the egg cell, while the other fuses with two other nuclei in the ovule. This forms a triploid cell that will develop into a specialized nutritive tissue—the **endosperm**—that surrounds the fertilized egg cells and provides the food source for embryonic development.

Following fertilization, the embryo develops inside the ovule, taking about 2 weeks to form a mature seed, which is shed from the plant. The seed will remain dormant until suitable external conditions trigger germination. The early stages of germination and seedling growth rely on food supplies stored in the **cotyledons** (seed leaves), which are storage organs developed by the embryo. *Arabidopsis* embryos have two

Fig. 7.2 An individual flower of
Arabidopsis. Scale bar = 1 mm.

cotyledons, and thus belong to the large group of plants known as dicotyledons (dicots for short). The other large group of flowering plants is the monocotyledons (monocots), which have embryos with one cotyledon; monocots typically have long, narrow leaves and include many important staple crops, such as wheat, rice, and maize. Agriculturally important dicots include potato, tomato, sugar beet, and most other vegetables.

At germination, the shoot and root elongate and emerge from the seed. Once the shoot emerges above ground it starts to photosynthesize and forms the first true leaves at the shoot apex. About 4 days after germination the seedling is a self-supporting plant. Flower buds are usually visible on the young plant 3–4 weeks after germination and will open within a week. Under ideal conditions, the complete *Arabidopsis* life cycle thus takes about 6–8 weeks (see Fig. 7.1).

Embryonic development

Formation of the embryo, or embryogenesis, occurs inside the ovule, and the end result is a mature, dormant embryo enclosed in a seed awaiting germination. During embryogenesis, the shoot–root polarity of the plant body, which is known as the apical–basal axis, is established, and the shoot and root meristems are formed. Plants also possess radial symmetry, as seen in the concentric arrangement of the different tissues in a plant stem, and this radial axis is also set up in the embryo. Development of the *Arabidopsis* embryo involves a rather invariant pattern of cell division (which is not the case in all plant embryos) and this enables structures in the *Arabidopsis* seedling to be traced back to groups of cells in the early embryo to provide a fate map (see Section 7.2).

7.2 Plant embryos develop through several distinct stages

Arabidopsis belongs to the angiosperms, or flowering plants, one of the two major groups of seed-bearing plants; the other is the gymnosperms, or conifers. The typical course of embryonic development in angiosperms is outlined in Box 7A. Like an animal zygote, the fertilized plant egg cell undergoes repeated cell divisions, cell growth, and differentiation to form a multicellular embryo. The first division of the zygote is at right angles to the long axis, dividing it into an apical cell and a basal cell, and establishing an initial polarity that is carried over into the apical–basal polarity of the embryo and into the apical–basal axis of the plant. In many species, the first zygotic division is unequal, with the basal cell larger than the apical cell. The basal cell divides to give rise to the suspensor, which may be several cells long (see Box 7A). This attaches the embryo to maternal tissue and is a source of nutrients. The apical cell divides vertically to form a two-celled proembryo, which will give rise to the rest of the embryo. In some species, the basal cell contributes little to the further development of the embryo, but in others, such as *Arabidopsis*, the topmost suspensor cell is recruited into the embryo, where it is known as the hypophysis, and contributes to the embryonic root meristem and root cap.

The next two divisions produce an eight-cell octant-stage embryo, which develops into a globular-stage embryo of around 32 cells (Fig. 7.3). The embryo elongates and the cotyledons start to develop as wing-like structures at one end, while an embryonic root forms at the other. This stage is known as the heart stage. Two groups of undifferentiated cells capable of continued division are located at each end of this axis; these are known as the apical meristems. The meristem lying between the cotyledons gives rise to the shoot, while the one at the opposite end of the axis, towards the end of the embryonic root, will drive root growth at germination. The region in between the embryonic root and the future shoot will become the seedling stem or

Box 7A Angiosperm embryogenesis

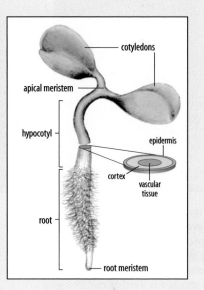

In flowering plants (angiosperms), the egg cell is contained within an ovule inside the ovary in the flower (right inset in the bottom panel). At fertilization, a pollen grain deposited on the surface of the stigma puts out a pollen tube, down which two male gametes migrate into the ovule. One male gamete fertilizes the egg cell while the other combines with another cell inside the ovule to form a specialized nutritive tissue, the endosperm, which surrounds, and provides the food source for, the developing embryo.

The small annual weed *Capsella bursa-pastoris* (shepherd's purse) is a typical dicotyledon. The first, asymmetric, cleavage divides the zygote transversely into an apical and a basal cell (bottom panel). The basal cell then divides several times to form a single row of cells—the suspensor—which, in many angiosperm embryos, takes no further part in embryonic development, but may have an absorptive function; in *Capsella*, however, it contributes to the root meristem. Most of the embryo is derived from the apical cell. This undergoes a series of stereotyped divisions, in which a precise pattern of cleavages in different planes gives rise to the heart-shaped embryonic stage typical of dicotyledons. This develops into a mature embryo that consists of a cylindrical body with a meristem at either end, and two cotyledons.

The early embryo becomes differentiated along the radial axis into three main tissues—the outer epidermis, the prospective vascular tissue, which runs through the center of the main axis and cotyledons, and the ground tissue (prospective cortex) that surrounds it.

The ovule containing the embryo matures into a seed (left inset in the bottom panel), which remains dormant until suitable external conditions trigger germination and growth of the seedling. A typical dicotyledon seedling (top right panel) comprises the shoot apical meristem, two cotyledons, the trunk of the seedling (the hypocotyl), and the root apical meristem. The seedling may be thought of as the phylotypic stage of flowering plants. The seedling body plan is simple. One axis—the apical–basal axis—defines the main polarity of the plant. The shoot forms at the apical pole and the root at the basal pole. The shoot meristem is, therefore, referred to in the seedling and adult plant simply as the apical meristem. A plant stem also has a radial axis, which is evident in the radial symmetry in the hypocotyl and is continued in the root and shoot. In the center is the vascular tissue, which is surrounded by cortex, and an outer covering of epidermis. At later stages, structures such as leaves and other organs have a dorso-ventral axis running from the upper surface to the lower surface.

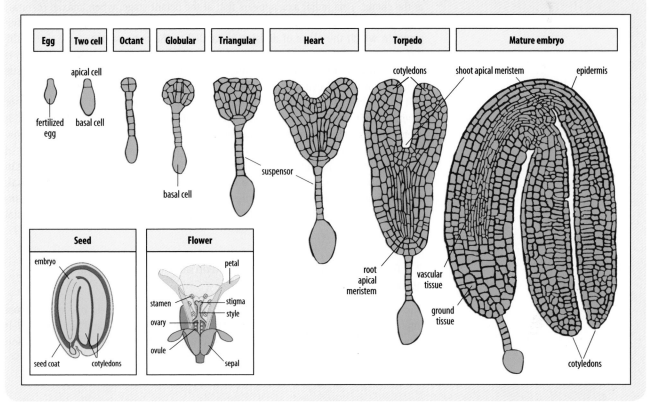

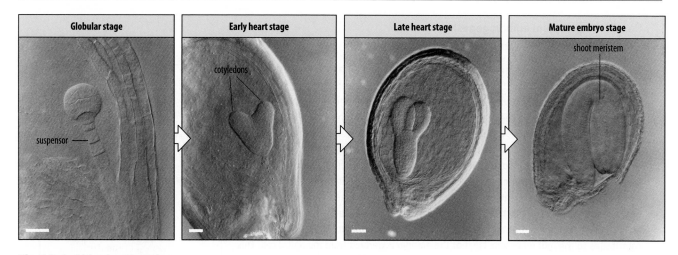

| Globular stage | Early heart stage | Late heart stage | Mature embryo stage |

suspensor

cotyledons

shoot meristem

Fig. 7.3 *Arabidopsis* embryonic development. Light micrographs (Nomarski optics) of cleared wild-type seeds of *A. thaliana*. The cotyledons can already be seen at the heart stage. The embryo proper is attached to the seed coat through a filamentous suspensor. Scale bar = 20 μm.

Photographs courtesy of D. Meinke, from Meinke, D.W.: 1994.

hypocotyl. Almost all above ground adult plant structures are derived from the apical meristems. The main exception is radial growth in the stem, which is most evident in woody plants, and which is produced by the **cambium**, a ring of secondary meristematic tissue in the stem. After the embryo is mature, the apical meristems remain quiescent until germination.

Seedling structures can be traced back to groups of cells in the early embryo to provide a fate map (Fig. 7.4). In *Arabidopsis*, patterns of cleavage up to the 16-cell stage are highly reproducible, and even at the octant stage it is possible to make a fate map for the major regions of the seedling along the apical–basal axis. The upper tier of cells gives rise to the cotyledons and the shoot meristem, the next tier is the origin of the hypocotyl, and the bottom tier together with the region of the suspensor where it joins the embryo will give rise to the root (see Section 7.3). At the heart stage, the fate map is clear.

The radial pattern in the embryo comprises three concentric rings of tissue: the outer **epidermis**, the ground tissue (**cortex** and **endodermis**), and the vascular tissue at the center. This radial axis appears first at the octant stage, when adaxial (central) and abaxial (peripheral) regions become established. Subsequently, **periclinal** divisions, in which the plane of division is parallel to the outer surface, give rise to the different rings of tissue, and **anticlinal** divisions, in which the plane of cell division is at right angles to the outer surface, increase the number of cells in each ring of tissue (Fig. 7.5). At the 16-cell stage the epidermal layer, or dermatogen, is established.

It is not known to what extent the fate of cells is determined at this stage or whether their fate is dependent on the early pattern of asymmetric cell divisions. It seems that cell lineage is not crucial, as mutations have been discovered that show that cell division can be uncoupled from pattern formation. The *fass* mutation in *Arabidopsis* disrupts the regular pattern of cell division by causing cells to divide in random orientation, and produces a much fatter and shorter seedling than normal. But although the *fass* seedling is misshapen, it has roots, shoots, and even eventually flowers in the correct places and the correct pattern of tissues along the radial axis is maintained.

7.3 Gradients of the signal molecule auxin establish the embryonic apical–basal axis

The small, organic molecule **auxin** (indoleacetic acid or IAA) is one of the most important and ubiquitous chemical signals in plant development and plant growth. It causes changes in gene expression by promoting the ubiquitination and degradation

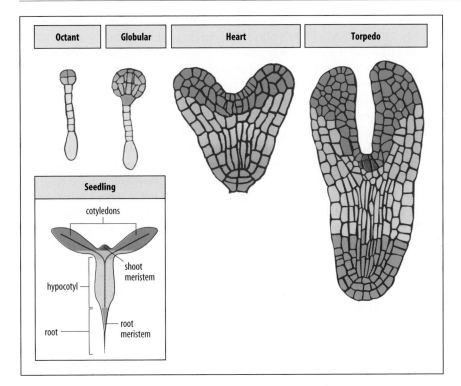

Fig. 7.4 Fate map of the *Arabidopsis* embryo. The stereotyped pattern of cell division in dicotyledon embryos means that at the globular stage it is already possible to map the three regions that will give rise to the cotyledons (dark green) and shoot meristem (red), the hypocotyl (yellow), and the root meristem (purple) in the seedling.

After Scheres, B., et al.: 1994.

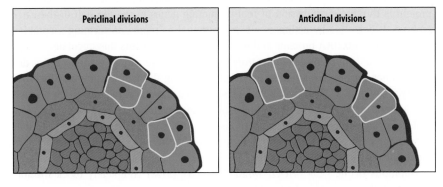

Fig. 7.5 Periclinal and anticlinal divisions. Periclinal cell divisions are parallel to the organ surface, whereas anticlinal divisions are at right angles to the surface.

of transcriptional repressors known as **Aux/IAA proteins**. In the absence of auxin, these bind to proteins called **auxin-response factors (ARFs)** to block the transcription of so-called auxin-responsive genes. Auxin-stimulated degradation of Aux/IAA proteins frees the auxin-response factors, which can then activate, or in some cases repress, these genes (Fig. 7.6). Auxin-responsive genes include genes involved in the regulation of cell division and cell expansion, as well as genes involved in specifying cell fate. In some cases auxin appears to be acting as a classical morphogen, forming a concentration gradient and specifying different fates according to a cell's position along the gradient.

The earliest known function of auxin in *Arabidopsis* is in the very first stage of embryogenesis, where it establishes the apical–basal axis. Immediately after the first division of the zygote, auxin is actively transported from the basal cell into the apical cell, where it accumulates. It is transported out of the basal cell by the auxin-efflux protein PIN7, which is localized in the apical plasma membrane of the basal cell. The auxin is required to specify the apical cell, which gives rise to all the apical embryonic structures such as shoot apical meristem and cotyledons. Through the subsequent cell divisions, transport of auxin continues up through the suspensor cells and into the

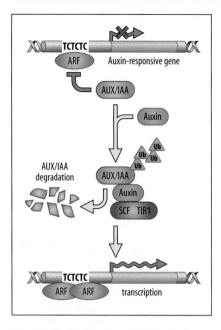

Fig. 7.6 Auxin signaling pathway. In the absence of auxin, AUX/IAA proteins bind to and repress the activity of transcription factors in the AUXIN RESPONSE FACTOR (ARF) family, which bind TGTCTC-containing DNA sequence elements in the promoters of auxin-responsive genes. Auxin targets the ubiquitin ligase complex SCF/TIR1 to the AUX/IAA protein, causing it to be ubiquitinated (Ub). This modification targets AUX/IAA for degradation and thus lifts the repression of the auxin-responsive gene, which can then be transcribed.

Adapted from Chapman, E.J. and Estelle, M.: 2009.

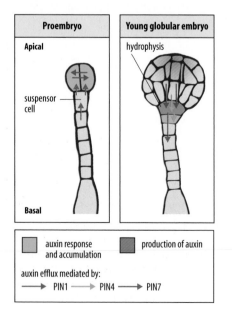

cell at the base of the developing embryo until the globular-stage embryo of about 32 cells. The apical cells of the embryo then start to produce auxin, and the direction of auxin transport is suddenly reversed. PIN7 proteins in the suspensor cells move to the basal faces of the cells. Other *PIN* genes are activated, and the concerted actions of the PIN proteins cause auxin from the apical region to be transported into the basal region of the globular embryo, from which will develop the hypocotyl, root meristem, and embryonic root (Fig. 7.7).

Expression of the homeobox genes *WOX2* and *WOX8* in the asymmetrically dividing zygote seems to be essential for the formation of the embryonic auxin gradient. Both genes are expressed in the zygote, but after division, *WOX2* expression is restricted to the apical cell and its descendants whereas *WOX8* is expressed only in the basal cell lineage. Expression of *WOX8* appears to be required for the continued expression of *WOX2* and for the establishment of an auxin gradient, most probably by effects of the WOX proteins on expression of the PIN proteins that direct auxin transport.

The importance of auxin in specifying the basal region of the embryo is illustrated by the effects of mutations in the cellular machinery for auxin transport or auxin response. The gene *MONOPTEROS* (*MP*), for example, encodes one of the auxin-response factors, ARF5. Embryos mutant for *MP* lack the hypocotyl and the root meristem, and abnormal cell divisions are observed in the regions of the octant-stage embryo that would normally give rise to these structures.

An early step in axis formation in the embryo is the specification of cells as potential shoot or root. The mutation *topless-1* causes a dramatic switching of cell fate, transforming the whole of the apical region into a second root region and abolishing development of cotyledons and shoot meristem. The normal function of the TOPLESS protein is to repress expression of root-promoting genes in the top half of the early embryo. It does this by acting as a transcriptional co-repressor, forming a complex together with Aux/IAA proteins and the auxin-response factors they regulate.

The *topless-1* mutation is temperature-sensitive, and so can be made to act at different times during development by simply changing the temperature at which the embryos are grown. Such manipulations show that the apical region can be respecified as root between the globular and heart stage, even after it has begun to show molecular signs of developing cotyledons and shoot meristem. This suggests that even though apical cell fate is normally specified by this stage of embryonic development, the decision is not irreversible.

Mutations in the gene *SHOOT MERISTEMLESS* (*STM*) completely block the formation of the shoot apical meristem, but have no effect on the root meristem or other parts of the embryo. *STM* encodes a transcription factor that is also required to maintain cells in the pluripotent state in the adult shoot meristem. The pattern of *STM* expression develops gradually, which is also typical of several other genes that characterize the shoot apical meristem. Expression is first detected in the globular stage in one or two cells and only later in the central region between the two cotyledons (Fig. 7.8).

Fig. 7.7 The role of auxin in patterning the early embryo. Left panel: auxin produced in the original basal cell accumulates in the two-celled proembryo (green) through transport in the basal to apical direction mediated by the protein PIN7, which is located in the apical membranes of the basal and suspensor cells (purple arrows). Another PIN protein, PIN1, transports auxin between the two cells of the proembryo (red arrows). Cell division distributes auxin into the developing embryo. Right panel: at the globular stage, free auxin starts to be produced at the apical pole and the direction of auxin transport is reversed. PIN7 becomes localized in basal membranes of the suspensor cells and the proteins PIN1 and PIN4 (orange arrows) transport auxin from the apical region into the most basal cell of the embryo, known as the hypophysis, where it accumulates.

7.4 Plant somatic cells can give rise to embryos and seedlings

As gardeners well know, plants have amazing powers of regeneration. A complete, new plant can develop from a small piece of stem or root, or even from the cut edge of a leaf. This reflects an important difference between the developmental potential of plant and animal cells. In animals, with few exceptions, cell determination and differentiation are irreversible. By contrast, many somatic plant cells remain totipotent. Cells from roots, leaves, stems, and even, for some species, a single, isolated protoplast—a cell from which the cell wall has been removed by enzymatic treatment—can be grown in culture and induced by treatment with the appropriate growth hormones to give rise to a new plant (Fig. 7.9). Plant cells proliferating in culture can give rise to cell clusters that pass through a stage resembling normal embryonic development, although the pattern of cell divisions are not the same. These 'embryoids' can then develop into seedlings. Plant cells can also form callus, an apparently disorganized mass of cells, and the callus can form new shoot or root meristems, and therefore new shoots and roots. The ability of plants to regenerate from somatic cells means that transgenic plants can be easily generated using the plant pathogenic bacterium *Agrobacterium tumefaciens* as the carrier of the gene to be transferred (Box 7B, p. 264).

The ability of single somatic cells to give rise to whole plants has two important implications for plant development. The first is that maternal determinants may be of little or no importance in plant embryogenesis, as it is unlikely that every somatic cell would still be carrying such determinants. Second, it suggests that many cells in the adult plant body are not fully determined with respect to their fate, but remain totipotent. Of course, this totipotency seems only to be expressed under special conditions, but it is, nevertheless, quite unlike the behavior of animal cells. It is as if such plant cells have no long-term developmental memory, or that such memory is easily reset.

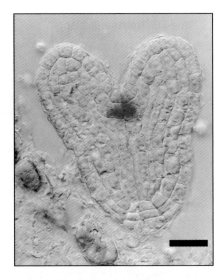

Fig. 7.8 Section through a late heart-stage *Arabidopsis* embryo showing expression of **SHOOT MERISTEMLESS (STM)**. At this stage, the STM RNA (stained red) is expressed in cells located between the cotyledons. Scale bar = 25 µm.

Photograph courtesy of K. Barton. From Long, J.A. and Barton, K.B.: 1998.

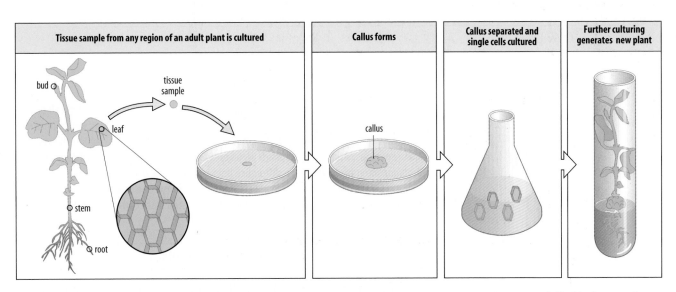

| Tissue sample from any region of an adult plant is cultured | Callus forms | Callus separated and single cells cultured | Further culturing generates new plant |

bud
tissue sample
leaf
stem
root
callus

Fig. 7.9 Cultured somatic cells from a mature plant can form an embryo and regenerate a new plant. The illustration shows the generation of a plant from single cells. If a small piece of tissue from a plant stem or leaf is placed on a solid agar medium containing the appropriate nutrients and growth hormones, the cells will start to divide to form a disorganized mass of cells known as a callus.

The callus cells are then separated and grown in liquid culture, again containing the appropriate growth hormones. In suspension culture, some of the callus cells divide to form small cell clusters. These cell clusters can resemble the globular stage of a dicotyledon embryo, and with further culture on solid medium, develop through the heart-shaped and later stages to regenerate a complete new plant.

Box 7B Transgenic plants

One of the most common ways of generating transgenic plants containing new and modified genes is through infection of plant tissue in culture with the bacterium *Agrobacterium tumefaciens*, the causal agent of crown gall tumors. *Agrobacterium* is a natural genetic engineer. It contains a **plasmid**—the Ti plasmid—that contains the genes required for the transformation and proliferation of infected cells to form a callus. During infection, a portion of this plasmid—the T-DNA (shown in red below)—is transferred into the genome of the plant cell, where it becomes stably integrated. Genes experimentally inserted into the T-DNA will therefore also be transferred into the plant cell chromosomes. Ti plasmids, modified so that they do not cause tumors but still retain the ability to transfer T-DNA, are widely used as vectors for gene transfer in dicotyledonous plants. The genetically modified plant cells of the callus can then be grown into a complete new transgenic plant that carries the introduced gene in all its cells and can transmit it to the next generation.

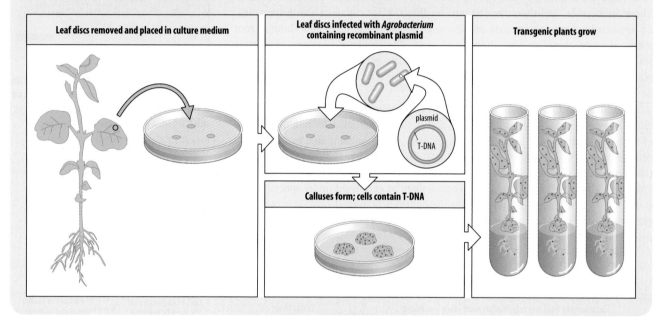

Leaf discs removed and placed in culture medium

Leaf discs infected with *Agrobacterium* containing recombinant plasmid

Transgenic plants grow

plasmid

T-DNA

Calluses form; cells contain T-DNA

Summary

Early embryonic development in many plants is characterized by asymmetric cell division of the zygote, which specifies apical and basal regions. Embryonic development in flowering plants establishes the shoot and root meristems from which the adult plant develops. Auxin gradients are involved in specifying the apical-basal axis of the *Arabidopsis* embryo. One major difference between plants and animals is that, in culture, a single somatic plant cell can develop through an embryo-like stage and regenerate a complete new plant, indicating that some differentiated plant cells retain totipotency.

Summary: early development in flowering plants

first asymmetric cell division and auxin signaling in embryo establishes apical–basal axis

⇩

embryonic cell fate is determined by position

⇩

shoot and root meristems of seedling give rise to all adult plant structures

Meristems

In plants, most of the adult structures are derived from just two regions of the embryo, the embryonic shoot and root meristems, which are maintained after germination. The embryonic shoot meristem, for example, becomes the shoot apical meristem of the growing plant, giving rise to all the stems, leaves, and flowers. As the shoot grows, lateral outgrowths from the meristem give rise to leaves and to side shoots. In flowering shoots, the vegetative meristem becomes converted into one capable of producing **floral meristems** that make flowers, not leaves. In *Arabidopsis*, for example, the shoot apical meristem changes from a vegetative meristem that makes leaves around it in a spiral pattern to an **inflorescence meristem** that then produces floral meristems, and thus flowers, around it in a spiral pattern. The first stages of future organs are known as **primordia** (singular **primordium**). Each primordium consists of a small number of **founder cells** that produce the new structure by cell division and cell enlargement, accompanied by differentiation.

There is usually a time delay between the initiation of two successive leaves in a shoot apical meristem, and this results in a plant shoot being composed of repeated modules. Each module consists of an **internode** (the cells produced by the meristem between successive leaf initiations), a **node** and its associated leaf, and an axillary bud (Fig. 7.10). The axillary bud itself contains a meristem, known as the **lateral shoot meristem**, which forms at the base of the leaf, and can produce a side shoot when the inhibitory influence of the main shoot tip is removed. Root growth is not so obviously modular, but similar considerations apply, as new lateral meristems initiated behind the root apical meristem give rise to lateral roots.

Shoot apical meristems and root apical meristems operate on the same principles, but there are some significant differences between them. We will use the shoot apical meristem to illustrate the basic principles of meristem structure and properties, and then discuss roots.

7.5 A meristem contains a small, central zone of self-renewing stem cells

Shoot apical meristems are rarely more than 250 µm in diameter in angiosperms and contain a few hundred relatively small, undifferentiated cells that are capable of cell division. Most of the cell divisions in normal plant development occur within meristems, or soon after a cell leaves a meristem, and much of a plant's growth in size is due to cell elongation and enlargement. Cells leave the periphery of the meristem to form organs such as leaves or flowers, and are replaced from a small central zone of slowly dividing, self-renewing stem cells or **initials** at the tip of the meristem (Fig. 7.11). In *Arabidopsis* this zone comprises around 12 to 20 cells. Initials behave in the same way as animal stem cells (see Chapter 10). They can divide to give one daughter that remains a stem cell and one that loses the stem-cell property. This daughter cell continues to divide and its descendants are displaced towards the peripheral zone of the meristem, where they become founder cells for a new organ or internode, leave the meristem, and differentiate. A very small number of long-term stem cells at the meristem center may persist for the whole life of the plant.

Meristem stem cells are maintained in the self-renewing state by cells underlying the central zone that form the **organizing center**. As we shall see, it is the

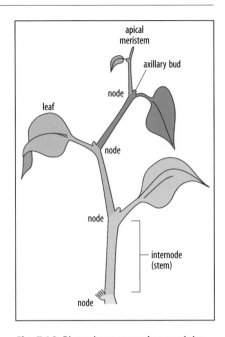

Fig. 7.10 Plant shoots grow in a modular fashion. The shoot apical meristem produces a repeated basic structural module. The vegetative shoot module typically consists of internode, node, leaf, and axillary bud (from which a side branch may develop). Successive modules are shown here in different shades of green. As the plant grows, the internodes behind the meristem lengthen and the leaves expand.

After Alberts, B., et al.: 2002.

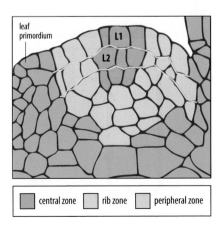

Fig. 7.11 Organization of the *Arabidopsis* shoot meristem. A longitudinal section is shown. The meristem has three main layers, L1, L2, and an inner layer, as indicated by the yellow lines, and is divided functionally into a central zone (red), a rib zone (yellow), and a peripheral zone (blue). The stem cells or initials lie in the central zone, while the peripheral zone contains proliferating cells that will give rise to leaves and side shoots. The rib zone gives rise to the central tissues of the plant stem.

microenvironment maintained by the organizing center that gives stem cells their identity.

The undetermined state of meristem stem cells is confirmed by the fact that meristems are capable of regulation. If, for example, a seedling shoot meristem is divided into two or four parts by vertical incision, each part becomes reorganized into a complete meristem, which gives rise to a normal shoot. Provided some subpopulation of organizing center cells plus overlying stem cells is present, a normal meristem will regenerate. If a shoot apical meristem is completely removed, no new apical meristem form, but the incipient meristem at the base of the leaf is now able to develop and form a new side shoot. In the presence of the original meristem, this prospective meristem remains inactive, as active meristems inhibit the development of other meristems nearby—as a result of auxin transport from the shoot apex, among other factors. This regulative behavior is in line with cell–cell interactions being a major determinant of cell fate in the meristem.

7.6 The size of the stem-cell area in the meristem is kept constant by a feedback loop to the organizing center

Numerous genes that control the behavior of the cells in the meristem are known. The gene *STM*, which is involved in specifying the shoot meristem in the development of the *Arabidopsis* embryo (see Section 7.3) is, for example, expressed throughout adult shoot meristems but is suppressed as soon as cells become part of an organ primordium. Its role seems to be to maintain meristematic cells in an undifferentiated state, as loss of *STM* function results in all the meristem cells being incorporated into organ primordia. In contrast, mutations in the *Arabidopsis CLAVATA* (*CLV*) genes increase the size of the meristem as a result of an increase in the number of stem cells. Normally, despite the continual exit of cells from the stem-cell pool, the number of stem cells in a meristem is kept roughly the same throughout a plant's life, by division of the remaining stem cells. The role of the *CLV* genes in regulating stem-cell numbers is understood in some detail, and involves feedback between the stem cells and the organizing center that underlies them.

In *Arabidopsis*, cells of the organizing center express a homeobox transcription factor, WUSCHEL (WUS). This is required to produce a signal (as yet unknown) that gives the overlying cells their stem-cell identity. Mutations in *WUS* result in termination of the shoot meristem and cessation of growth as a result of the loss of stem cells, while its overexpression increases stem-cell numbers. The stem cells express *CLAVATA3* (*CLV3*), which encodes a secreted protein that acts indirectly to repress *WUS*. This feedback loop could control *WUS* activity in the organizing center and suppress *WUS* activation in neighboring cells, thus limiting the extent of *WUS* expression (Fig. 7.12, top panel). In turn, this would regulate the extent of the stem-cell zone

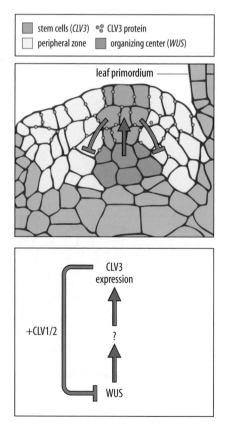

stem cells (*CLV3*) CLV3 protein
peripheral zone organizing center (*WUS*)

leaf primordium

CLV3 expression

+CLV1/2

?

WUS

Fig. 7.12 Regulation of the stem-cell population in a shoot meristem. Intercellular signals control the position and size of the stem-cell population in the *Arabidopsis* shoot meristem. Top panel: the organizing center (purple) expresses the transcription factor WUS, which induces the production of an as yet unknown intercellular signal (red arrow) that maintains the overlying stem cells (orange). The stem cells express and secrete the signal protein CLAVATA3 (CLV3; orange dots), which moves laterally and downwards and indirectly represses transcription of the *WUS* gene in the surrounding cells, acting through its cell-surface receptor proteins CLV1 and CLV2. CLV3 thus limits the extent of the area specified as stem cells. The descendants of the stem cells are continuously displaced into the peripheral zone of the meristem (pale yellow), where they are recruited into leaf primordia. Bottom panel: the negative-feedback loop whereby WUS expression is restricted by CLV3. Expression of WUS in the organizing center is responsible for the production of an unknown signal that induces the expression of CLV3. CLV3 in turn signals through CLV1/2 to suppress the expression of *WUS*.
Adapted from Brand, U., et al.: 2000.

above the organizing center. If stem-cell numbers temporarily fall, for example, less CLV3 is produced and *WUS* activity increases, with a consequent increase in stem-cell numbers. More CLV3 is then produced and limits the extent of *WUS* activity. Other CLAVATA proteins are involved in the feedback loop (Fig. 7.12, bottom panel).

Meristems can be induced elsewhere in the plant by the misexpression of genes involved in specifying stem-cell identity, yet another indication that stem-cell identity is conferred by cell–cell interactions and not by an embryonically specified cell lineage.

7.7 The fate of cells from different meristem layers can be changed by changing their position

Other evidence that the fate of a meristematic cell is determined by its position in the meristem, and thus the intercellular signals it is exposed to, comes from observing the fates of cells in the different meristem layers. As well as being organized into central and peripheral zones, the apical meristem of a dicotyledon, such as *Arabidopsis*, is composed of three distinct layers of cells (Fig. 7.13). The outermost layer, L1, is just one cell thick. Layer L2, just beneath L1, is also one cell thick. In both L1 and L2, cell divisions are anticlinal—that is, the new wall is in a plane perpendicular to the

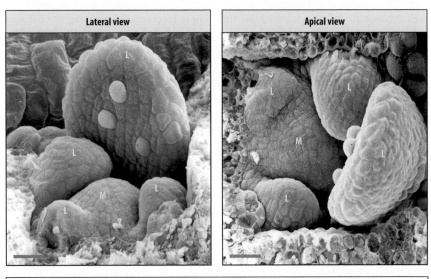

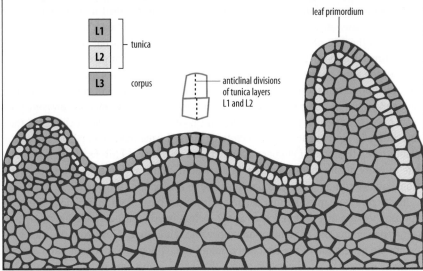

Fig. 7.13 Apical meristem of *Arabidopsis*. Top panels: scanning electron micrographs showing the organization of the meristem at the young vegetative apex of *Arabidopsis*. The plant is a *clavata1* mutant, which has a broadened apex that allows for a clearer visualization of the leaf primordia (L) and the meristem (M). Scale bar = 10 μm. Bottom panel: diagram of a vertical section through the apex of a shoot. The three-layered structure of the meristem is apparent in the most apical region. In layer 1 (L1) and layer 2 (L2), the plane of cell division is anticlinal; that is, at right angles to the surface of the shoot. Cells in layer 3 (L3, the corpus) can divide in any plane. A leaf primordium is shown forming at one side of the meristem.

Photographs courtesy of M. Griffiths.

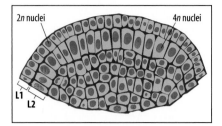

Fig. 7.14 A periclinal meristem chimera composed of cells of two different genotypes. In L1 the cells are diploid whereas the cells of L2 are tetraploid—that is, they have double the normal chromosome number—and are larger and easily recognized.

After Steeves, T.A., et al.: 1989.

layer—thus maintaining the two-layer organization. The innermost layer is L3, in which the cells can divide in any plane. L1 and L2 are often known as the tunica, and L3 as the corpus.

To find which tissues each layer can give rise to, the fates of cells in the different layers can be followed by marking one layer with a distinguishable mutation, such as a change in pigmentation or a polyploid nucleus. When a complete layer is genetically different from the others in this way, the organism is known as a **periclinal chimera** (Fig. 7.14) and the fate of cells from this layer can be traced. As we saw in relation to animal embryos, chimeras are organisms composed of cells of two different genotypes (see Section 3.9). Plant chimeras can be made by inducing mutation in the apical meristem of a seed or shoot tip by X-irradiation, or by treatment with chemicals such as colchicine that induce polyploidy.

Layer L1 gives rise to the epidermis that covers all structures produced by the shoot, while L2 and L3 both contribute to cortex and vascular structures. Leaves and flowers are produced mostly from L2; L3 contributes mainly to the stem. Although the three layers maintain their identity in the central region of the meristem over long periods of growth, cells in either L1 or L2 occasionally divide periclinally; new cell walls are formed parallel to the surface of the meristem, and thus one of the new cells invades an adjacent layer. This migrant cell now develops according to its new position, showing that cell fate is not necessarily determined by the meristem layer in which the cell originated, and that intercellular signaling is involved in specifying or changing its fate in its new position. The anticlinal pattern of cell division in L2 becomes disrupted when leaf primordia start to form, when the cells divide periclinally, as well as anticlinally.

The transcription factor Knotted-1 in maize is homologous to *Arabidopsis* STM and, like STM, it is expressed throughout the shoot meristem to keep cells in an undifferentiated state. It is one example of a transcription factor that moves directly from cell to cell. The *KNOTTED-1* gene is expressed in all layers except L1, but the protein is also found in L1, suggesting that it can move between cells, perhaps via plasmodesmata. In *Knotted-1* gain-of-function mutants, in which the gene is misexpressed in leaves, Knotted-1 protein fused to green fluorescent protein has also been observed to move from the inner layers of the leaf to the epidermis, but not in the opposite direction.

7.8 A fate map for the embryonic shoot meristem can be deduced using clonal analysis

Much of our knowledge about the general properties of meristems outlined in previous sections comes from studies some decades ago that determined how the embryonic shoot apical meristem is related to the development of a plant over its whole lifetime. What was the fate of individual 'embryonic initials', as the stem cells were known at the time? Did particular regions in the embryonic meristem give rise to particular parts of the adult plant? Individual initials can be marked by mutagenesis of seeds by X-irradiation or transposon activation, to give cells with a different color from the rest of the plant, for example. If the marked meristem cell and its immediate progeny populate only part of one layer of the meristem (in contrast to a periclinal chimera), then this area will give rise to visible sectors of marked cells in stems and organs as the plant grows; this type of chimera is known as a **mericlinal chimera** (Fig. 7.15). The fate of individually marked initials in mericlinal chimeras can be determined using clonal analysis in a manner similar to its use in *Drosophila* (see Box 2E, p. 73).

In maize, the marked sectors usually start at the base of an internode and extend apically, terminating within a leaf. Some sectors include just a single internode and leaf, representing the progeny of an embryonic initial that is lost from the meristem before the generation of the next leaf primordium. Others, on the other hand, extend through numerous internodes, showing that some embryonic initials remain in the

Fig. 7.15 Tobacco plant mericlinal chimera. This plant has grown from an embryonic shoot meristem in which an albino mutation has occurred in a cell of the L2 layer. The affected area occupies about a third of the total circumference of the shoot, suggesting that there are three apical initial cells in the embryonic shoot meristem.

Photograph courtesy of S. Poethig.

meristem for a long time, contributing to a succession of nodes and internodes. In sunflowers, marked clones have been observed to extend through several internodes up into the flower, showing that a single initial can contribute to both leaves and flowers.

From the analysis of hundreds of mericlinal chimeras, fate maps of the embryonic shoot meristems of several species were constructed, which shed light on the properties of the shoot meristem and how it behaves during normal development. These fate maps are probabilistic because it was not possible to know the location of the marked cell in the embryonic meristem, which is inaccessible inside the seed.

The probabilistic fate map for the maize embryonic shoot apical meristem indicates that the three most apical cells in L1 give rise to the male inflorescence (the tassel and spike; Fig. 7.16). The remainder of the maize meristem can be divided into five tiers of cells that produce internodes and leaves, and which form overlapping concentric domains on the fate map. The outermost domain contributes to the earliest internode–leaf modules, while the inner domains give rise to internodes and leaves successively higher up the stem. A fate map has been similarly constructed for the embryonic shoot apical meristem of *Arabidopsis* (Fig. 7.17). Most of the *Arabidopsis* embryonic meristem gives rise to the first six leaves, whereas the remainder of the shoot, including all the flowerheads, is derived from a very small number of embryonic cells at the center of the meristem. Unlike maize, the number of leaves in *Arabidopsis* is not fixed—growth is said to be **indeterminate**. There is no relation between particular

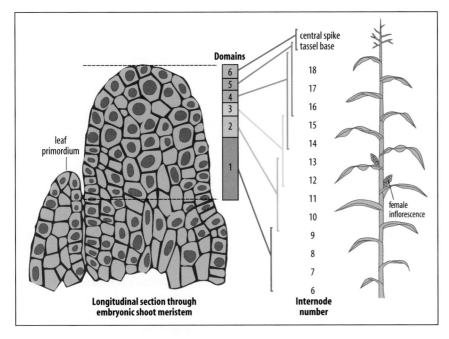

Fig. 7.16 Probabilistic fate map of the shoot apical meristem in the mature maize embryo. In maize, the primordia of the first six leaves are already present in the mature embryo and are excluded from the analysis. At the time the cells were marked, the meristem contained about 335 cells, which will give rise to 12 more leaves, the female inflorescences, and the terminal male inflorescence (the tassel and spike). A longitudinal section through the embryonic apical dome—the shoot meristem—is shown on the left. Clonal analysis shows that it can be divided into six vertically stacked domains, each comprising a set of initials that can give rise to a particular part of the plant. The number of initials in layers L1 and L2 of each domain at the embryonic stage can be estimated from the final extent of the corresponding marked sectors in the mature plant (see text). The fate of each domain, as estimated from the collective results of clonal analysis of many different plants, is shown on the right. Domain 6, comprising the three most apical L1 cells of the meristem, will give rise to the terminal male inflorescence. The fate of cells in the other domains is less circumscribed. Domain 5, for example, which consists of around eight L1 cells surrounding domain 6 and the underlying four or so L2 cells, can contribute to nodes 16-18; domain 4 to nodes 14-18; and domain 3 to nodes 12-15. The female inflorescences, which develop in the leaf axils, are derived from the corresponding domains.

After McDaniel, C.N., et al.: 1988.

Fig. 7.17 Probabilistic fate map of the embryonic shoot meristem of *Arabidopsis*. The L2 layer of the meristem is depicted as if flattened out and viewed from above. The numbers indicate the leaf, as shown on the plant below, to which each group of meristem cells contributes, and indicate the sequence in which the leaves are formed. The inflorescence shoot (i) is derived from a small number of cells in the center of the layer.

After Irish, V.E.: 1991.

cell lineages and particular structures, which indicates that position in the meristem is crucial in determining cell fate. One exception is that germ cells always arise from L2. The L2 layer is a clone, and so there is, in the case of germ cells, a relation between fate and a particular cell lineage.

The conclusions from the clonal analysis experiments are that the initials that contribute to a particular structure are simply those that happen to be in the appropriate region of the meristem at the time; they have not been prespecified in the embryo as, say, flower or leaf.

7.9 Meristem development is dependent on signals from other parts of the plant

To what extent does the behavior of a meristem depend on other parts of the plant? It seems to have some autonomy, because if a meristem is isolated from adjacent tissues by excision it will continue to develop, although often at a much slower rate. Excised shoot apical meristems of a variety of plants can be grown in culture, where they will develop into shoots complete with leaves if the growth hormones auxin and cytokinin are added. The behavior of the meristem *in situ*, however, is influenced in more subtle ways by interactions with the rest of the plant.

As we saw earlier, the apical meristem of maize gives rise to a succession of nodes, and terminates in the male flower. The number of nodes before flowering is usually between 16 and 22. This number is not controlled by the meristem alone, however. Evidence comes from culturing shoot tips consisting of the apical meristem and one or two leaf primordia. Meristems taken from plants that have already formed as many as 10 nodes still develop into normal maize plants with the full number of nodes. The isolated maize meristem has no memory of the number of nodes it has already formed, and repeats the process from the beginning. The meristem itself is therefore not determined in the embryo with respect to the number of nodes it will form. In the plant, control over the number of nodes that are formed must therefore involve signals from the rest of the plant to the meristem, finally directing the meristem to terminate node formation and form a tassel.

7.10 Gene activity patterns the proximo-distal and adaxial–abaxial axes of leaves developing from the shoot meristem

Leaves develop from groups of founder cells within the peripheral zone of the shoot apical meristem. The first indication of leaf initiation in the meristem is usually a swelling of a region to the side of the apex, which forms the **leaf primordium** (Fig. 7.18). This small protrusion is the result of increased localized cell multiplication and altered patterns of cell division. It also reflects changes in polarized cell expansion.

Two new axes that relate to the future leaf are established in a leaf primordium. These are the **proximo-distal axis** (leaf base to leaf tip) and the **adaxial–abaxial axis** (upper surface to lower surface, sometimes called dorsal to ventral). The latter is termed adaxial–abaxial as it is related to the radial axis of the shoot. The upper surface of the leaf derives from cells near the center of this axis (adaxial), while the lower surface derives from more peripheral cells (abaxial). The two leaf surfaces carry out different functions and have different structures, with the top surface being specialized for light capture and photosynthesis. In *Arabidopsis*, flattening of the leaf along the adaxial–abaxial axis occurs after leaf primordia begin developing, but in monocots like maize the leaf is flattened as it emerges. The establishment of adaxial–abaxial polarity is likely to make use of positional information along the radial axis of the meristem.

Arabidopsis leaf primordia emerge from the shoot meristem with distinct programs of development in the adaxial and abaxial halves. This asymmetry can be seen from

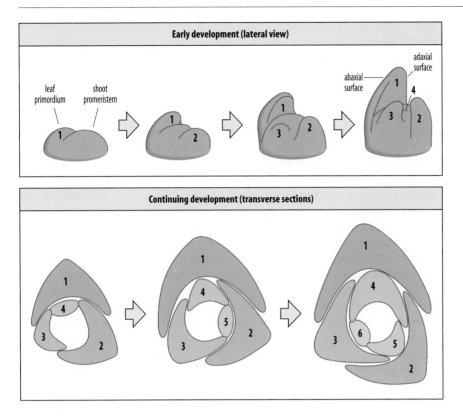

Early development (lateral view)

leaf primordium   shoot promeristem

abaxial surface   adaxial surface

Continuing development (transverse sections)

Fig. 7.18 Leaf phyllotaxis. In shoots where single leaves are arranged spirally up the stem, the leaf primordia arise sequentially in a mathematically regular pattern in the meristem. Leaf primordia arise around the sides of the apical dome, just outside the **promeristem** region. A new leaf primordium is formed slightly above and at a fixed radial angle from the previous leaf, often generating a helical arrangement of primordia visible at the apex. Top panel: lateral views of the shoot apex. Bottom panel: view looking down on cross-sections through the apex near the tip, at successive stages from the top panel.

After Poethig, R.S., et al.: 1985 (top panel); and Sachs, T.: 1994 (bottom panel).

the beginning, as the leaf primordium has a crescent shape in cross-section, with a convex outer (abaxial) side and a concave inner (adaxial) surface (see Fig. 7.18). Different genes are expressed in the future adaxial and abaxial sides. For example, the *Arabidopsis* gene *FILAMENTOUS FLOWER* (*FIL*) is normally expressed in the abaxial side of the leaf primordium, and specifies an abaxial cell fate. Its ectopic expression throughout the leaf primordium can cause all cells to adopt an abaxial cell fate, and the leaf develops as an arrested cylindrical structure.

The specification of adaxial cell fate in *Arabidopsis* involves the genes *PHABULOSA* (*PHAB*), *PHAVOLUTA* (*PHAV*), and *REVOLUTA* (*REV*). These encode transcription factors and are initially expressed in the shoot meristem and in the adaxial side of the primordium. Loss-of-function mutations in these genes result in radially symmetrical leaves with only abaxial cell types characteristic of the underside of the leaf. A microRNA (see Box 6B, p. 228) is involved in the restriction of *PHAB*, *PHAV*, and *REV* expression to the adaxial side, targeting and destroying their mRNAs on the abaxial side and thus limiting their activity to the adaxial side.

It has been suggested that, in normal plants, the interaction between adaxial and abaxial initial cells at the boundary between them initiates lateral growth, resulting in the formation of the leaf blade and the flattening of the leaf. The importance of boundaries in controlling pattern and form has already been seen in the parasegments of *Drosophila* (see Section 2.24) and further examples will be found in Chapter 11.

Development along the leaf proximo-distal axis also appears to be under genetic control. Like other grasses, a maize leaf primordium is composed of prospective leaf-sheath tissue proximal to the stem and prospective leaf-blade tissue distally. Mutations in certain genes result in distal cells taking on more proximal identities; for example, making sheath in place of blade. Similar proximo-distal shifts in pattern occur in *Arabidopsis* as a result of mutation. Positional identity along the proximo-distal axis may reflect the developmental age of the cells, distal cells adopting a different fate from proximal cells because they mature later.

7.11 The regular arrangement of leaves on a stem is generated by regulated auxin transport

As the shoot grows, leaves are generated within the meristem at regular intervals and with a particular spacing. Leaves are arranged along a shoot in a variety of ways in different plants, and the particular arrangement, known as phyllotaxy or **phyllotaxis**, is reflected in the arrangement of leaf primordia in the meristem. Leaves can occur singly at each node, in pairs, or in whorls of three or more. A common arrangement is the positioning of single leaves spirally up the stem, which can sometimes form a striking helical pattern in the shoot apex.

In plants in which leaves are borne spirally, a new leaf primordium forms at the center of the first available space outside the central region of the meristem and above the previous primordium (see Fig. 7.18). This pattern suggests a mechanism for leaf arrangement based on lateral inhibition (see Section 1.16), in which each leaf primordium inhibits the formation of a new leaf within a given distance. In this model, inhibitory signals emanating from recently initiated primordia prevent leaves from forming close to each other. There is some experimental evidence for this. In ferns, leaf primordia are widely spaced, allowing experimental microsurgical interference. Destruction of the site of the next primordium to be formed results in a shift toward that site by the future primordium, whose position is closest to it (Fig. 7.19). Recent studies indicate, however, that it is competition rather than inhibition that is responsible.

As we have seen in the determination of apical–basal polarity in the embryo, auxin is transported out of cells with the help of proteins such as PIN1 (see Section 7.3). In the shoot, auxin produced in the shoot tip below the meristem is transported upwards into the meristem, through the epidermis and the outermost meristem layer. The direction of auxin flow in the shoot apex is controlled by PIN1 and follows a simple rule: the side of the cell on which PIN1 is found is the side nearest the neighbor cell with the highest auxin concentration. Thus auxin transport is always towards a region of higher concentration. A high concentration of auxin is a primordium activator, and initially, auxin is pumped towards a new primordium that is developing at a site of high auxin concentration. This, however, depletes a zone of cells around the primordium of auxin, such that cells nearer the center of the meristem now have more auxin than the cells adaxial to the new primordium. This activates a feedback mechanism that causes PIN1 to move to the other side of these cells, and auxin now flows out of the new primordium towards the meristem, creating a new spot of high auxin concentration in the meristem farthest away from any new primordium (Fig. 7.20).

Fig. 7.19 Leaf primordia may be positioned by lateral inhibition or by competition. Leaf primordia on a fern shoot tip form in a regular order in positions 1 to 4. Primordia appear to form as far as possible from existing primordia, so 2 forms almost opposite 1. Normally, 4 will develop between 1 and 2, but if 1 is excised, 4 forms much further away from 2. This result can be interpreted either by the removal of lateral inhibition by primordium 1 or by the removal of the competitive effect of 1 for some primordium-inducing factor such as auxin.

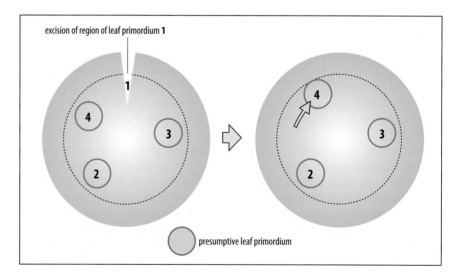

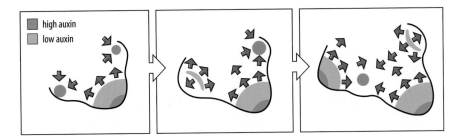

Fig. 7.20 Auxin-dependent mechanism of phyllotaxis in *Arabidopsis*. Auxin is transported through PIN proteins towards areas of high auxin concentration (dark green), at which an organ primordium will form. Cells around the developing primordium become depleted of auxin (light green), which causes the polarity of the PIN proteins to reverse so that auxin then flows away from the primordium. The red arrows indicate the direction of PIN polarity. It is proposed that this pattern of auxin circulation could set up the regular pattern of leaf and flower formation from meristems.

This leads to auxin peaks occurring sequentially, at the regular positions later occupied by new leaves.

7.12 Root tissues are produced from *Arabidopsis* root apical meristems by a highly stereotyped pattern of cell divisions

The organization of tissues in the *Arabidopsis* root tip is shown in Fig. 7.21. The radial pattern comprises single layers of epidermal, cortical, endodermal, and pericycle cells, with vascular tissue in the center (protophloem and protoxylem). Root apical meristems resemble shoot apical meristems in many ways and give rise to the root in a similar manner to shoot generation. But there are some important differences between the root and shoot meristems. The shoot meristem is at the extreme tip of the shoot, whereas the root meristem is covered by a root cap (which is itself derived from one of the layers of the meristem); also, there is no obvious segmental arrangement at the root tip resembling the node–internode–leaf module.

The root is set up early (see Section 7.2) and a well-organized embryonic root can be identified in the late heart-stage embryo (Fig. 7.22). An antagonistic interaction between auxin and cytokinin controls the establishment of the root stem-cell niche. Clonal analysis has shown that the seedling root meristem can be traced back to a set of embryonic initials that arise from a single tier of cells in the heart-stage embryo.

As in the shoot meristem, a root meristem is composed of an organizing center, called the **quiescent center** in roots, in which the cells divide only very rarely, and which is surrounded by stem-cell-like initials that give rise to the root tissue (see Fig. 7.22). The quiescent center is essential for meristem function. When parts of the meristem are removed by microsurgery, it can regenerate, but regeneration is always preceded by the formation of a new quiescent center. Laser destruction of individual quiescent-center cells shows that, as in the shoot meristem, a key function of the quiescent center is to maintain the immediately adjacent initials in the stem-cell state and prevent them from differentiating.

Each initial undergoes a stereotyped pattern of cell divisions to give rise to a number of columns, or **files**, of cells in the growing root (see Fig. 7.21); each file of cells in the root thus has its origin in a single initial. Some initials give rise to both endodermis and cortex, whereas others give rise to both epidermis and the root cap. Before it leaves the meristem, therefore, the undifferentiated progeny of an endodermis/cortex initial, for example, will divide asymmetrically to give one daughter that produces cortex and one that produces endodermis. The gene *SCARECROW* is necessary to confer this asymmetry on the dividing cell, and mutations in this gene give roots with no distinct endodermis or cortex but with a tissue layer with characteristics of both.

The normal pattern of cell divisions is not obligatory, however. As discussed earlier, *fass* mutants, which have disrupted cell divisions, still have relatively normal

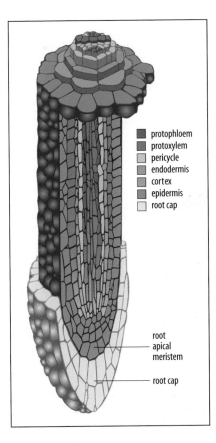

- ■ protophloem
- ■ protoxylem
- □ pericycle
- ■ endodermis
- □ cortex
- ■ epidermis
- □ root cap

root apical meristem

root cap

Fig. 7.21 The structure of the root tip in *Arabidopsis*. Roots have a radial organization. In the center of the growing root tip is the future vascular tissue (protoxylem and protophloem). This is surrounded by further tissue layers.

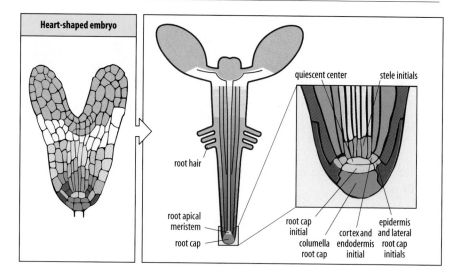

Fig. 7.22 Fate map of root regions in the heart-stage *Arabidopsis* embryo. The root grows by the division of a set of initial cells. The root meristem comes from a small number of cells in the heart-shaped embryo. Each tissue in the root is derived from the division of a particular initial cell. At the center of the root meristem is a quiescent center, which does not divide.

After Scheres, B., et al.: 1994.

patterning in the root. In addition, laser ablation of individual meristem cells does not lead to an abnormal root. The remaining initials undergo new patterns of cell division that replace the progeny of the cells that have been destroyed. Such observations show that, as in the shoot meristem, the fate of cells in the developing root meristem depends on their recognition of positional signals and not on their lineage.

As we saw in Section 7.3, auxin gradients play a major role in patterning the embryo and specifying the root region, and mutations that affect auxin localization lead to root defects. At the globular stage of embryonic development, the auxin-transport protein PIN1 is localized in cells in the future root region and the highest level of auxin is found adjacent to where the quiescent center will develop.

The role of auxin in root development continues into the adult plant. Auxin in the root is transported out of cells via the PIN proteins, and enters adjacent cells. Cells with raised auxin levels transport auxin better, probably due to an increase in the number of PIN proteins in the membrane as auxin prevents their endocytosis and recycling, and so there is a positive feedback loop that raises the local concentration.

Auxin plays a key role in patterning the growing root. Modeling of auxin movement, using the known distribution of PIN proteins in the membranes of different cell types, has shown that PIN-directed flow can explain the formation and maintenance of a stable auxin maximum at the quiescent center. Auxin is transported down the central vascular tissue of the root tip, forming a concentration maximum at the quiescent center, and then outwards and upwards through the outer layers of the root tip, forming gradients along both the basal-apical axis and laterally in the root. This modeling successfully simulates the effects of such auxin gradients on pattern formation, cellular differentiation and root growth in real time.

An idea of how the auxin gradients might be translated into effects on cell fate and cell behavior is provided by the graded distribution of the four PLETHORA-family transcription factors in the root. These transcription factors are required for proper root development and are expressed in a graded manner along the apical–basal axis, with expression maxima for all at the quiescent center—the region of maximum auxin concentration. Here they are essential for stem-cell maintenance and function. Lower concentrations of PLETHORA proteins correspond to the meristem region, where cells are proliferating, while even lower concentrations appear to be necessary for exit from the meristem and cell differentiation in the elongation zone. Although not yet proven experimentally, *PLETHORA* gene expression could provide a graded read-out of the auxin gradient that helps direct root patterning.

Other influences on root patterning are the interplay between auxin and the hormone cytokinin. Cytokinin helps regulate root meristem size and set the position of the transition zone—where cells stop dividing and start elongating and differentiating—by repressing both auxin transport and cellular responses to auxin in this region.

Auxin is also involved in the ability of plants to regenerate from a small piece of stem. In general, roots form from the end of the stem that was originally closest to the root, whereas shoots tend to develop from dormant buds at the end that was nearest to the shoot. This polarized regeneration is related to vascular differentiation and to the polarized transport of auxin. Transport of auxin from its source in the shoot tip toward the root leads to an accumulation of auxin at the 'root' end of the stem cutting, where it induces the formation of roots. One hypothesis suggests that polarity is both induced and expressed by the oriented flow of auxin.

One of the best examples of developmentally important transcription factor movement from one cell to another is found in roots. As noted earlier, expression of the gene *SCARECROW* is required for root cells to adopt an endodermal fate. This expression requires the transcriptional activator SHORT-ROOT (SHR). SHR is, however, not synthesized in the prospective endodermal cells, but in the adjacent cells on the inner side. SHR protein is transported from these cells outwards into the prospective endodermis, and this movement appears to be regulated and not simply due to diffusion.

7.13 Root hairs are specified by a combination of positional information and lateral inhibition

Root hairs are formed from epidermal cells at regular intervals around the root, and this regularity is thought to be achieved by a combination of responses to positional information and lateral inhibition by the movement of transcription factors between cells. Files of cells that will make root hairs alternate with files of non-hair-producing cells on the surface of the developing root. The importance of position is shown by the fact that if an epidermal cell overlies a junction between two cortical cells it forms a root hair, whereas if it contacts just one cortical cell it does not (Fig. 7.23). And if a cell changes its position in relation to the cortex, its fate will also change, from a potential hair-forming cell to a non-hair-forming cell and vice versa. Most cell divisions in the future epidermis are horizontal, increasing the number of cells per file, but occasionally a vertical anticlinal division occurs, pushing one of the daughter cells into an adjacent file. The daughter cell then assumes a fate corresponding to its new position in relation to the adjacent cortical cells.

The positional cues, as yet unknown, are thought to be detected by the epidermal cells through the SCRAMBLED protein, which is a receptor-like protein kinase. SCRAMBLED activity influences the activity of a network of transcription factors that control cell fate. Two groups of transcription factors have been identified by mutation experiments, one group promoting a root-hair fate and one suppressing it. A key transcription factor that seems to be regulated by SCRAMBLED activity is WEREWOLF, whose expression is suppressed in presumptive hair-forming cells, presumably in response to the positional signal. As well as promoting an atrichoblast fate, however, WEREWOLF is also required for the expression of transcription factors (CAPRICE, TRYPTYCHON, and ENHANCER OF TRYPTYCHON) that are needed to specify hair cells. After positional signaling, these proteins will only be produced in the presumptive atrichoblasts, but they are thought to move laterally into the adjacent epidermal cells and promote these cells' differentiation as trichoblasts by inhibiting genes that would otherwise give an atrichoblast fate.

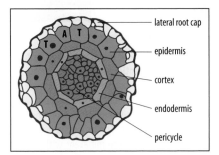

Fig. 7.23 Organization of cell types in the root epidermis. The epidermis is composed of two types of cells: trichoblasts (T), which will form root hairs, and atrichoblasts (A), which will not. Trichoblasts overlie the junction between two cortical cells and atrichoblasts are located over the outer tangential wall of cortical cells.
After Dolan, L., Scheres, B.: 1998.

Summary

Meristems are the growing points of a plant. The apical meristems, found at the tips of shoots and roots, give rise to all the plant organs—roots, stem, leaves, and flowers. They consist of small groups of a few hundred undifferentiated cells that are capable of repeated division. The center of the meristem is occupied by self-renewing stem cells, which replace the cells that are lost from the meristem when organs are formed. The fate of a cell in the shoot meristem depends upon its position in the meristem and interactions with its neighbors, as when a cell is displaced from one layer to another it adopts the fate of its new layer. Meristems can also regulate when parts are removed, in line with cell-cell interactions determining cell fate. Fate maps of embryonic shoot meristems show that they can be divided into domains, each of which normally contributes to the tissues of a particular region of the plant, but the fate of the embryonic initials is not fixed. The shoot meristem gives rise to leaves in species-specific patterns—phyllotaxy—which seem best accounted for by regulated transport of auxin. Lateral inhibition is involved in the regular spacing of hairs on root and leaf surfaces. In the root meristem, the cells are organized rather differently from those in the shoot meristem, and there is a much more stereotyped pattern of cell division. A set of initial cells maintains root structure by dividing along different planes.

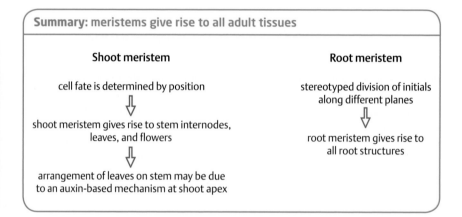

Summary: meristems give rise to all adult tissues

Shoot meristem	**Root meristem**
cell fate is determined by position	stereotyped division of initials along different planes
⇩	⇩
shoot meristem gives rise to stem internodes, leaves, and flowers	root meristem gives rise to all root structures
⇩	
arrangement of leaves on stem may be due to an auxin-based mechanism at shoot apex	

Fig. 7.24 Scanning electron micrograph of an *Arabidopsis* inflorescence meristem. The central inflorescence meristem (shoot apical meristem, SAM) is surrounded by a series of floral meristems (FM) of varying developmental ages. The inflorescence meristem grows indeterminately, with cell divisions providing new cells for the stem below, and new floral meristems on its flanks. The floral meristems (or floral primordia) arise one at a time in a spiral pattern. The most mature of the developing flowers is on the right (FM1), showing the initiation of sepal primordia surrounding a still-undifferentiated floral meristem. Eventually such a floral meristem will also form petal, stamen, and carpel primordia.

Photograph from Meyerowitz, E.M., et al.: 1991.

Flower development and control of flowering

Flowers contain the reproductive cells of higher plants and develop from the shoot meristem. In most plants, the transition from a vegetative shoot meristem to a floral meristem that produces a flower is largely, or absolutely, under environmental control, with daylength and temperature being important determining factors. In a plant such as *Arabidopsis*, in which each flowering shoot produces multiple flowers, the vegetative shoot meristem first becomes converted into an inflorescence meristem, which then forms floral meristems, each of which develops completely into a single flower (Fig. 7.24). Floral meristems are thus determinate, unlike the indeterminate shoot apical meristem. Flowers, with their arrangement of floral organs (sepals, petals, stamens, and carpels), are rather complex structures, and it is a major challenge to understand how they arise from the floral meristem.

The conversion of a vegetative shoot meristem into one that makes flowers involves the induction of so-called **meristem identity genes**. A key regulator of floral induction in *Arabidopsis* is the meristem identity gene *LEAFY* (*LFY*); a related gene in *Antirrhinum* is *FLORICAULA* (*FLO*). How environmental signals, such as daylength,

Fig. 7.25 Structure of an Arabidopsis flower. *Arabidopsis* flowers are radially symmetrical and have an outer ring of four identical green sepals, enclosing four identical white petals, within which is a ring of six stamens, with two carpels in the center. Bottom: floral diagram of the *Arabidopsis* flower representing a cross-section taken in the plane indicated in the top diagram. This is a conventional representation of the arrangement of the parts of the flower, showing the number of flower parts in each whorl and their arrangement relative to each other. *After Coen, E.S., et al.: 1991.*

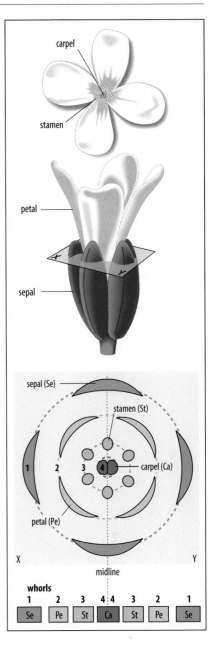

influence floral induction is discussed later. We will first consider the mechanisms that pattern the flower, in particular those that specify the identity of the floral organs.

7.14 Homeotic genes control organ identity in the flower

The individual parts of a flower each develop from a **floral organ primordium** produced by the floral meristem. Unlike leaf primordia, which are all identical, the floral organ primordia must each be given a correct identity and be patterned according to it. An *Arabidopsis* flower has four concentric whorls of structures (Fig. 7.25), which reflect the arrangement of the floral organ primordia in the meristem. The sepals (whorl 1) arise from the outermost ring of meristem tissue, and the petals (whorl 2) from a ring of tissue lying immediately inside it. An inner ring of tissue gives rise to the male reproductive organs—the stamens (whorl 3). The female reproductive organs—the carpels (whorl 4)—develop from the center of the meristem. In a floral meristem of *Arabidopsis*, there are 16 separate primordia, giving rise to a flower with four sepals, four petals, six stamens and a pistil made up of two carpels (see Fig. 7.25).

The primordia arise at specific positions within the meristem, where they develop into their characteristic structures. After the emergence of the primordia in *Antirrhinum*, cell lineages become restricted to particular whorls, rather like the lineage restriction to compartments in *Drosophila* (see Section 2.23). Lineage restriction occurs at the time when the pentagonal symmetry of the flower becomes visible and genes that give the different floral organs their identity are expressed. The lineage compartments within the floral meristem appear to be delineated by narrow bands of non-dividing cells.

Like the homeotic selector genes that specify segment identity in *Drosophila*, mutations in floral identity genes cause homeotic mutations in which one type of flower part is replaced by another. In the *Arabidopsis* mutant *apetala2*, for example, the sepals are replaced by carpels and the petals by stamens; in the *pistillata* mutant, petals are replaced by sepals and stamens by carpels. These mutations identified the floral organ identity genes, and have enabled their mode of action to be determined.

Homeotic floral mutations in *Arabidopsis* fall into three classes, each of which affects the organs of two adjacent whorls (Fig. 7.26). The first class of mutations, of

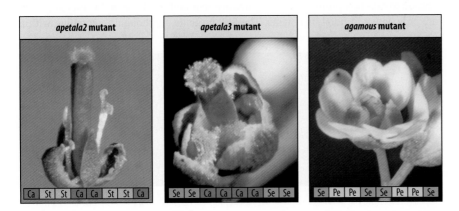

Fig. 7.26 Homeotic floral mutations in *Arabidopsis*. Left panel: an *apetala2* mutant has whorls of carpels and stamens in place of sepals and petals. Center panel: an *apetala3* mutant has two whorls of sepals and two of carpels. Right panel: *agamous* mutants have a whorl of petals and sepals in place of stamens and carpels. Transformations of whorls are shown inset, and can be compared to the wild-type arrangement, as shown in Fig. 7.25.

Photographs from Meyerowitz, E.M., et al.: 1991 (left panel), and Bowman, J.L., et al.: 1989 (center panel).

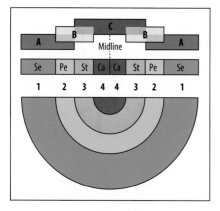

Fig. 7.27 The three overlapping regions of the *Arabidopsis* floral meristem that have been identified by the homeotic floral identity mutations. Region A corresponds to whorls 1 and 2, B to whorls 2 and 3, and C to whorls 3 and 4.

which *apetala2* is an example, affect whorls 1 and 2, giving carpels instead of sepals in whorl 1, and stamens instead of petals in whorl 2. The phenotype of the flower, going from the outside to the center, is therefore carpel, stamen, stamen, carpel. The second class of homeotic floral mutations affects whorls 2 and 3. In this class, *apetala3* and *pistillata* give sepals instead of petals in whorl 2 and carpels instead of stamens in whorl 3, with a phenotype sepal, sepal, carpel, carpel. The third class of mutations affects whorls 3 and 4, and gives petals instead of stamens in whorl 3 and sepals or variable structures in whorl 4. The mutant *agamous*, which belongs to this class, has an extra set of sepals and petals in the center instead of the reproductive organs.

These mutant phenotypes can be accounted for by an elegant model in which overlapping patterns of gene activity specify floral organ identity (Fig. 7.27) in a manner highly reminiscent of the way in which *Drosophila* homeotic genes specify segment identity along the insect's body. In detail, however, there are many differences, and quite different genes are involved. In this instance, plants and animals have, perhaps not surprisingly, independently evolved a similar approach to patterning a multicellular structure, but have recruited different proteins to carry it out.

In essence, the floral meristem is divided by the expression patterns of the homeotic genes into three concentric overlapping regions, A, B, and C, which partition the meristem into four non-overlapping regions corresponding to the four whorls. Each of the A, B, and C regions corresponds to the zone of action of one class of homeotic genes and the particular combinations of A, B, and C functions give each whorl a unique identity and so specify organ identity. Of the genes mentioned in Fig. 7.26, *APETALA1* (*AP1*) and *APETALA2* (*AP2*) are A-function genes, *APETALA3* (*AP3*) and *PISTILLATA* (*P1*) are B-function genes, and *AGAMOUS* (*AG*) is a C-function gene. The expression of *AP3* and *AG* in the developing flower is shown in Fig. 7.28. All the homeotic genes, also known as **floral organ identity genes**, encode transcription factors, and the B- and C-function proteins such as AP3 and AG contain a conserved DNA-binding sequence known as the MADS box. MADS-box genes are present in animals and yeast, but a role in development is known mainly in plants—although a MADS-box transcription factor, MEF2, is involved in muscle differentiation in animals. The original simple model for specifying floral organ identity is presented in more detail in Box 7C. Since the model was first proposed, more has become known about the activities and functions of the genes identified by the homeotic mutations, more genes controlling flower development have been discovered, and more 'functions' added.

One question that arose when the original floral organ identity model was investigated experimentally was why the ABC genes only showed their homeotic properties in the floral meristem and did not convert leaves into floral organs when artificially overexpressed in vegetative meristems, as might be expected for homeotic genes of this type. The answer came with the discovery of the *SEPALLATA* (*SEP*) genes, which also encode MADS-box proteins. These genes are required for the B and C functions and are only active in floral meristems. The SEP proteins are thought to combine with *B*

Fig. 7.28 Expression of *APETALA3* and *AGAMOUS* during flower development. *In situ* hybridization shows that *AGAMOUS* is expressed in the central whorls (left panel), whereas *APETALA3* is expressed in the outer whorls that give rise to petals and stamens (right panel).

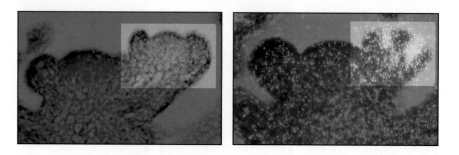

Fig. 7.29 Current status of the ABC model of floral organ identity. The regulatory genes *LEAFY*, *WUSCHEL* (*WUS*), and *UNUSUAL FLORAL ORGANS* (*UFO*) are expressed in specific domains in the floral meristem, which, together with repression of *APETALA1* by *AGAMOUS*, results in the pattern of ABC functions. ABC proteins and the co-factor SEP proteins assemble into complexes that specify the different organ identities.

Adapted from Lohmann, J.U., Weigel, D.: 2002

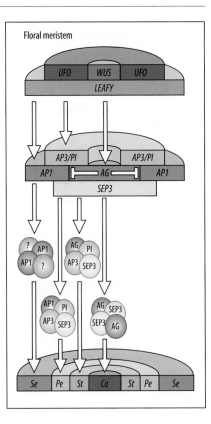

and *C* gene products to form active gene-regulatory complexes. A current view of the mechanism of specifying floral organ identity is shown in Fig. 7.29.

There is a better understanding of the functions and patterning of the floral homeotic genes than when the ABC model was first proposed. The MADS-box homeotic A-class gene *AP1* has been found to have a dual role: it acts early with other genes to specify general floral meristem identity and only later contributes to A function. It is induced by the meristem identity gene *LFY*, which is expressed throughout the meristem, and *AP1* is actively inhibited in the central regions of floral meristems by *AGAMOUS*. The expression of the A-function gene *APETALA2* (*AP2*) is translationally repressed by a microRNA, keeping the AP2 protein at a low level. *APETALA3* and *PISTILLATA1* are thought to be activated as a result of a meristem identity gene called *UNUSUAL FLORAL ORGANS* (*UFO*), which is expressed in the meristem in a pattern similar to that of B-function genes (see Fig. 7.29). *UFO* encodes a component of ubiquitin ligase, and is thought to exert its effects on flower development by targeting specific proteins for degradation. As we saw in animals, in relation to the control of β-catenin degradation (see Chapters 4 and 6), regulated degradation of proteins can be a powerful developmental mechanism. In the center of the floral meristem, the expression of *AGAMOUS* is partly controlled by *WUS*, which as we have seen earlier, is expressed in the organizing center of the vegetative shoot meristem and continues to be expressed in floral meristems.

Another group of genes that help pattern the floral organ primordia are genes that control cell division. The gene *SUPERMAN* is one example, controlling cell proliferation in stamen and carpel primordia, and in ovules. Plants with a mutation in this gene have stamens instead of carpels in the fourth whorl. *SUPERMAN* is expressed in the third whorl, and maintains the boundary between the third and fourth whorls.

Despite the enormous variation in the flowers of different species, the mechanisms underlying flower development seem to be very similar. For example, there are striking similarities between the genes controlling flower development of *Arabidopsis* and *Antirrhinum*, despite the quite different final morphology of the snapdragon flower. In developing *Arabidopsis* flowers, the patterns of activity of the corresponding genes fit well with the cell-lineage restriction to whorls seen in *Antirrhinum*.

7.15 The *Antirrhinum* flower is patterned dorso-ventrally as well as radially

Like *Arabidopsis* flowers, those of *Antirrhinum* consist of four whorls, but unlike *Arabidopsis*, they have five sepals, five petals, four stamens, and two united carpels (Fig. 7.30, left). Floral homeotic mutations similar to those in *Arabidopsis* occur in *Antirrhinum*, and floral organ identity is specified in the same way. Several of the *Antirrhinum* homeotic genes have extensive homology with those of *Arabidopsis*, the MADS box in particular being well conserved.

An extra element of patterning is required in the *Antirrhinum* flower, which has a bilateral symmetry imposed on the basic radial pattern common to all flowers. In whorl 2, the upper two petal lobes have a shape quite distinct from the lower three, giving the flower its characteristic snapdragon appearance. In whorl 3, the uppermost stamen is absent, as its development is aborted early on. *The Antirrhinum flower*

Box 7C The basic model for the patterning of the *Arabidopsis* flower

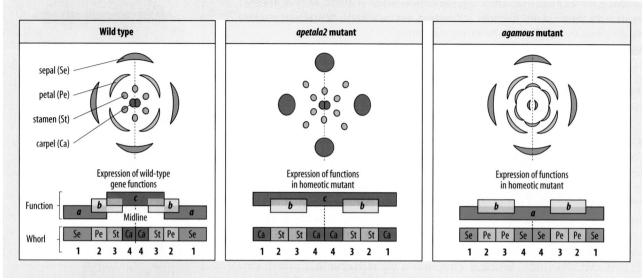

The floral meristem is divided into three overlapping regions, A, B, and C, each region corresponding to a class of homeotic mutations, as shown in Fig. 7.26 (see text). Three regulatory functions—*a*, *b*, and *c*—operate in regions A, B, and C, respectively, as shown in the panels above. In the wild-type flower (top left panel), it is assumed that *a* is expressed in whorls 1 and 2, *b* in 2 and 3, and *c* in whorls 3 and 4. In addition, *a* function inhibits *c* function in whorls 1 and 2 and *c* function inhibits *a* function in whorls 3 and 4—that is, *a* and *c* functions are mutually exclusive. *a* alone specifies sepals, *a* and *b* together specify petals, *b* and *c* stamens, and *c* alone carpels.

The homeotic mutations eliminate the functions of *a*, *b*, or *c*, and alter the regions within the meristem where the various functions are expressed. Mutations in *a*, such as *apetala2* (see center top panel), result in an absence of function *a*, and *c* spreads throughout the meristem, resulting in the half-flower pattern of carpel, stamen, stamen, carpel. Mutations in *b*, such as *apetala3* (see Fig. 7.26), result in only *a* functioning in whorls 1 and 2, and *c* in whorls 3 and 4, giving sepal, sepal, carpel, carpel. Mutations in *c* genes (such as *agamous*), result in *a* activity in all whorls, giving the phenotype sepal, petal, petal, sepal (see top right panel).

All the floral homeotic mutants discovered so far in *Arabidopsis* can be quite satisfactorily accounted for by this model (although there are small variations in gene numbers and expression patterns in other species that allow mutant phenotypes not seen in *Arabidopsis*), and particular genes can be assigned to each controlling function. Function *a* corresponds to the activity of genes such as *APETALA2*, *b* to *APETALA3* and *PISTILLATA*, and *c* to *AGAMOUS*. The model also accounts for the phenotype of double mutants, such as *apetala2* with *apetala3*, and *apetala3* with *pistillata*, as shown in the panels on the right.

This system emphasizes the similarity in function between the homeotic genes in animals and those controlling organ identity in flowers, although the genes themselves are completely different. The functional similarity with the Hox complex of *Drosophila* is further illustrated by the role of the *CURLY LEAF* gene of *Arabidopsis*, which is necessary for the stable maintenance of homeotic gene activity. *CURLY LEAF* is related to the *Polycomb* family of genes in *Drosophila* and is similarly required for stable repression of homeotic genes.

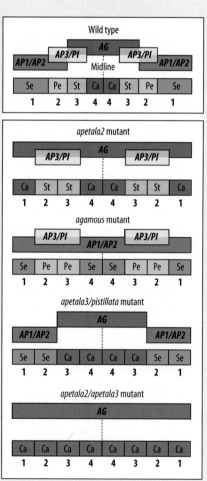

Dorso-ventral axis

Wild type　　　　　　**CYCLOIDEA mutant**

Fig. 7.30 Mutations in *CYCLOIDEA* make the *Antirrhinum* flower symmetrical. In the wild-type flower (left) the petal pattern is different along the dorso-ventral axis. In the mutant (right) the flower is symmetrical. All the petals are like the most ventral one in the wild type and are folded back.

Photograph courtesy of E. Coen, from Coen, E.S., et al.: 1991.

therefore has a distinct dorso-ventral axis. Another group of homeotic genes, different from those that govern floral organ identity, appear to act in this dorso-ventral patterning. For example, mutations in the gene *CYCLOIDEA*, which is expressed in the dorsal region, abolish dorso-ventral polarity and produce flowers that are more radially symmetrical (Fig. 7.30, right).

7.16 The internal meristem layer can specify floral meristem patterning

Although all three layers of a floral meristem (Fig. 7.31) are involved in organogenesis, the contribution of cells from each layer to a particular structure may be variable. Cells from one layer can become part of another layer without disrupting normal morphology, suggesting that a cell's position in the meristem is the main determinant of its future behavior. Some insight into positional signaling and patterning in the floral meristem can be obtained by making periclinal chimeras (see Section 7.7) from cells that have different genotypes and that give rise to different types of flower. From such chimeras, one can find out whether the cells develop autonomously according to their own genotype, or whether their behavior is controlled by signals from other cells.

As well as being produced by mutation, chimeras can also be generated by grafting between two plants of different genotypes. A new shoot meristem can form at the junction of the graft, and sometimes contains cells from both genotypes. Such chimeras can be made between wild-type tomato plants and tomato plants carrying the mutation *fasciated*, in which the flower has an increased number of floral organs per whorl. This phenotype is also found in chimeras in which only layer L3 contains *fasciated* cells (Fig. 7.32). The increased number of floral organs is associated with an overall increase in the size of the floral meristem, and in the chimeric plants this cannot be achieved unless the *fasciated* cells of layer L3 induce the wild-type L1 cells to divide more frequently than normal. The mechanism of intercellular signaling between L3 and L1 is not yet known. In *Antirrhinum*, the abnormal expression of *FLORICAULA* in only one meristem layer can result in flower development. These results illustrate the importance of signaling between layers in flower development.

7.17 The transition of a shoot meristem to a floral meristem is under environmental and genetic control

Flowering plants first grow vegetatively, during which time the apical meristem generates leaves. Then, triggered by environmental signals such as increasing daylength,

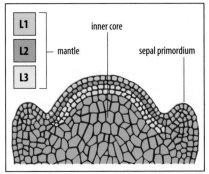

Fig. 7.31 Floral meristem. The meristem is composed of layers L1, L2, and L3. The inner core cells are derived from L3. The sepal primordia are just beginning to develop.

After Drews, G.N., et al.: 1989.

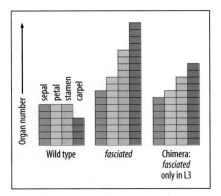

Fig. 7.32 Floral organ number in chimeras of wild-type and *fasciated* tomato plants. In the *fasciated* mutant there are more organs in the flower than in wild-type plants. In chimeras in which only layer L3 of the floral meristem contains *fasciated* mutant cells, the number of organs per flower is still increased, showing that L3 can control cell behavior in the outer layers of the meristem.

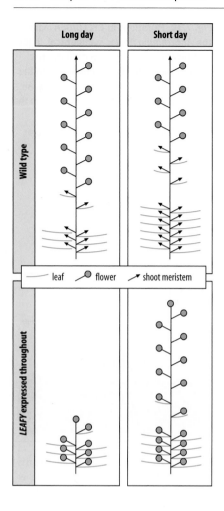

leaf —— | flower 🔵 | shoot meristem ↗

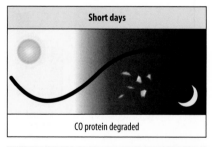

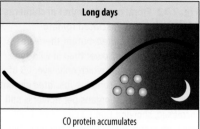

Fig. 7.33 Flowering can be controlled by daylength and LEAFY expression. As shown in the top panels, when wild-type *Arabidopsis* is grown under long-day conditions (left), few lateral shoots are formed before the apical shoot meristem begins to form floral meristems. When grown under short-day conditions, flowering is delayed and there are in consequence more lateral shoots. The gene *LEAFY* is normally expressed only in inflorescence and floral meristems, but if it is expressed throughout the plant (bottom panels), all shoot meristems produced are converted to floral meristems in both daylengths.

the plant switches to a reproductive phase and from then on the apical meristem gives rise only to flowers. There are two types of transition from vegetative growth to flowering. In the determinate type, the inflorescence meristem becomes a terminal flower, whereas in the indeterminate type the inflorescence meristem gives rise to a number of floral meristems. *Arabidopsis* is of the indeterminate type (Fig. 7.33). A primary response to floral inductive signals in *Arabidopsis* is the expression of floral meristem identity genes such as *LEAFY* and the dual-function *AP1* (see Section 7.13), which are necessary and sufficient for this transition. *LEAFY* potentially activates *AP1* throughout the meristem while also activating *AGAMOUS* in the center of the flower. *AGAMOUS* then represses the expression of *AP1* in the center, helping to restrict its floral organ identity function to region A (see Fig. 7.27). Mutations in floral meristem identity genes partly transform flowers into shoots. In a *leafy* mutant, which lacks LEAFY function, the flowers are transformed into spirally arranged sepal-like organs along the stem, whereas expression of *LEAFY* throughout a plant is sufficient to confer a floral fate on lateral shoot meristems and they develop as flowers (Fig. 7.33, bottom panels).

In *Arabidopsis*, flowering is promoted by increasing daylength, which predicts the end of winter and the onset of spring and summer (see Fig. 7.33). This behavior is called **photoperiodism**. In some strains, flowering is also accelerated after the plant has been exposed to a long period of cold temperature, a cue that winter has passed. This phenomenon is known as **vernalization**. Grafting experiments have shown that daylength is sensed not by the shoot meristem itself, but by the leaves. When the period of continuous light reaches a certain length, a diffusible flower-inducing signal is produced that is transmitted through the phloem to the shoot meristem. The pathway that triggers flowering involves the plant's **circadian clock**, the internal 24-hour timer that causes many metabolic and physiological processes, including the expression of some genes, to vary throughout the day. One of the genes regulated by the circadian clock is *CONSTANS (CO)*, which is a key gene in controlling the onset of flowering and provides the link between the plant's daylength-sensing mechanism and production of the flowering signal. The expression of *CO* oscillates on a 24-hour cycle under the control of the circadian clock, and its timing is such that the peak *CO* expression occurs towards the end of the afternoon. This means that in longer days, peak expression occurs in the light, whereas in short days, it will already be dark at this time. In the dark, the CO protein is degraded and so the circadian control ensures that CO only accumulates to high enough levels to trigger the flowering pathway when light conditions are favorable (Fig. 7.34).

CO is a transcription factor that activates a gene known as *FLOWERING LOCUS T (FT)*, producing the FT protein, which appears to act as the flowering signal. The FT protein is thought to travel from the leaf through the phloem to the shoot apical meristem, where it acts in a complex with the transcription factor FLOWERING LOCUS D (FD),

Fig. 7.34 The initiation of flowering is under the dual control of daylength and the circadian clock. The transcription factor CONSTANS (CO) is required for production of the flowering signal and is expressed in leaves under the control of the circadian clock. In short days, expression of the *CO* gene peaks in the dark and the protein is rapidly degraded. In long days, peak expression occurs in the light, and the CO protein accumulates.

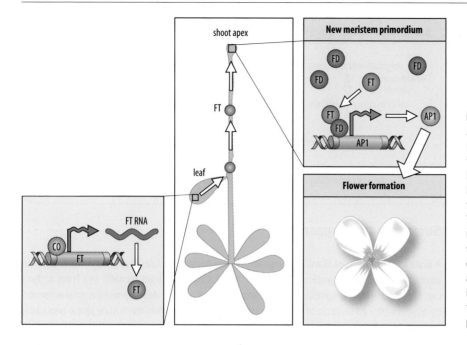

shoot apex

New meristem primordium

FD

FD

FD

FT

FT

FD

AP1

AP1

FT

leaf

Flower formation

FT RNA

CO

FT

FT

Fig. 7.35 Signals that initiate flowering in *Arabidopsis.* When the daylength becomes longer after winter, the transcription factor CO accumulates in the leaf, which in turn activates the transcription of *FT* in leaf phloem cells. The *FT* protein travels through the phloem to the shoot apex. This interacts with the transcription factor FD to form a complex, which acts together with the transcription factor LEAFY (whose own expression in upregulated in the shoot apex by FT) to activate key floral meristem identity genes such as *AP1*, which convert the vegetative meristem into one that will produce floral meristems.

which is expressed in the meristem, to turn on the expression of genes such as *AP1* that promote flowering (Fig. 7.35). If *FT* is activated in a single leaf, this is sufficient to induce flowering. Induction of flowering also requires the downregulation of a set of floral repressor genes such as *FLOWERING LOCUS C* (*FLC*). These suppress the transition from a vegetative to the flowering state until the positive signals to flower are received. FLC encodes a protein that binds to FT and suppresses its activity. After cold exposure, for example, FLC activity is low, and the repression of FT can be released.

Summary

Before flowering, which is triggered by environmental conditions such as daylength, the vegetative shoot apical meristem becomes converted into an inflorescence meristem, which either then becomes a flower or produces a series of floral meristems, each of which develops into a single flower. Genes involved in the initiation of flowering and patterning of the flower have been identified in both *Arabidopsis* and *Antirrhinum.* Flowering is induced by daylength acting together with the plant's natural circadian rhythms of gene expression to turn on a gene in the leaves that produces a flowering signal that is transported to the shoot meristem. This signal turns on the expression of meristem identity genes that are required for the transformation of the vegetative shoot meristem to an inflorescence meristem and the formation of floral meristems from the inflorescence meristem. Homeotic floral organ identity genes, which specify the organ types found in the flowers, have been identified from mutations that transform one flower part into another. On the basis of these mutations, a model has been proposed in which the floral meristem is divided into three concentric overlapping regions, in each of which certain floral identity genes act in a combinatorial manner to specify the organ type appropriate to each whorl. Studies with chimeric plants have shown that different meristem layers communicate with each other during flower development and that transcription factors can move between cells.

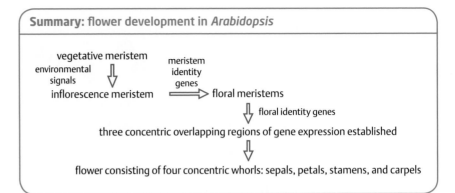

Summary: flower development in *Arabidopsis*

Summary to Chapter 7

A distinctive feature of plant development is the presence of relatively rigid walls and the absence of any cell migration. Another is that a single, isolated somatic cell from a plant can regenerate into a complete new plant. Early embryonic development is characterized by asymmetric cell division of the fertilized egg, which specifies the future apical and basal regions. During early development of flowering plants, both asymmetric cell division and cell–cell interactions are involved in patterning the body plan. During this process, the shoot and root meristems are specified and these meristems give rise to all the organs of the plant—stems, leaves, flowers, and roots. The shoot meristem gives rise to leaves in well-defined positions, a process involving regulated transport of a morphogen, auxin. The shoot meristem eventually becomes converted to an inflorescence meristem, which either becomes a floral meristem (in determinate inflorescences) or gives rise to a series of floral meristems, retaining its shoot meristem identity indefinitely (in indeterminate inflorescences). In floral meristems, each of which develops into a flower, homeotic floral organ identity genes act in combination to specify the floral organ types. Increasing daylength induces the synthesis of a flowering signal in the leaves that is transported to the shoot meristem where it induces flower formation.

■ End of chapter questions

Long answer (concept questions)

1. What features led to the adoption of *Arabidopsis thaliana* as the predominant model for plant development?

2. Distinguish between the following parts of a plant: shoot, root, node, leaf, meristem, sepal, petal, stamen, carpel.

3. What is the role of auxin before the 32-cell stage in *Arabidopsis* embryogenesis? What is the mechanism by which a differential in auxin concentrations is generated? What is the mechanism by which auxins influence gene expression?

4. Describe the process by which transgenic plants are produced. Include the role of the Ti plasmid, and of *Agrobacterium tumefaciens*.

5. What is a meristem? Describe the structure of the shoot meristem of *Arabidopsis*.

6. Contrast the roles of the homeobox genes *WUSCHEL* (*WUS*) and *SHOOT MERISTEMLESS* (*STM*) in formation and maintenance of shoot meristems.

7. What has the study of mericlinal chimeras revealed about cell specification during plant embryogenesis? Are cells specified in the embryo to form leaves and flowers, in a way analogous, for example, to the specification of cells as dorsal mesoderm in animal embryos?

8. What are the terms applied to the top and bottom surfaces of a leaf, in reference to the radial axis of the shoot? How are the transcription factors PHAB, PHAV, and REV restricted to the top surface?

9. How does auxin control the positioning of leaf primordia? Include the role of the PIN transporter and the concept of lateral inhibition in your answer.

10. What is SHORT-ROOT? How does it come to be present in endodermal cells? Might plasmodesmata be involved? Is there an analogous process involving the SHOOT MERISTEMLESS ortholog KNOTTED-1 of maize?

11. What is the nature of the homeotic mutations that can occur in *Arabidopsis*? What genes are involved?

12. What is the ABC model for flower development? How does it illustrate combinatorial control of cellular identity?

13. The MADS-box is named for the prototypic proteins in which it was found: MCM1 (*Saccharomyces*), AGAMOUS (*Arabidopsis*), DEFICIENS (*Antirrhinum*), and SRF (*Homo*). What is the function of the MADS-box in a protein? (Note that the MADS-box proteins are not related to SMADS; see Fig. 4.33.)

14. What modification of the ABC model derived from studies of *Arabidopsis* is required to explain flower development in *Antirrhinum*?

15. Through what mechanism is the photoperiod interpreted to trigger flower development in *Arabidopsis*?

Multiple choice (factual recall questions)

NB There is only one right answer to each question

1. How many genes are present in humans, *Drosophila melanogaster*, *Caenorhabditis elegans*, and *Arabidopsis*, respectively. (Although the number for humans has not yet been presented in the text, the number for the other organisms has been described.)

a) 19,000 – 27,000 – 14,000 – 19,000

b) 21,000 – 14,000 – 19,000 – 27,000

c) 27,000 – 21,000 – 19,000 – 14,000

d) 14,000 – 19,000 – 21,000 – 27,000

2. Embryogenesis in plants occurs

a) in the ovule, after the seed is fertilized and shed by the plant

b) in the seed after it germinates

c) in the seed, after the seed is fertilized

d) inside the ovule, before the seed is shed by the plant

3. Which statement is true about the totipotency of cells?

a) All animal and plant cells are totipotent.

b) In plants, many cells are totipotent, whereas in animals, only the fertilized egg is totipotent.

c) Mammalian embryonic stem cells are totipotent.

d) Only the stem cells of animals, and the meristem cells of plants are totipotent.

4. One of the earliest events in *Arabidopsis* development is formation of the _____ axis, in response to a gradient of _____.

a) adaxial–abaxial, cytokinins

b) apical–basal, auxin

c) apical–basal, Pin proteins

d) dorsal–ventral, miRNA

5. The fate map of the *Arabidopsis* embryo at the heart stage indicates that

a) although none of the adult structures have formed, the regions that will give rise to the meristems, which will in turn give rise to adult structures, can be identified

b) development in plants is so indeterminate that a true fate map cannot be drawn

c) the primordia of the leaves, stems, and roots have already formed

d) the three germ layers that will give rise to roots, stems, and leaves have formed

6. The *agamous* mutation causes the formation of

a) flowers with only petals and sepals

b) flowers with only sepals and carpels

c) flowers with only stamens and carpels

d) plants completely lacking flowers

7. Maintenance of the shoot meristems in adult *Arabidopsis* plants relies on which of the following mechanisms?

a) A homeobox transcription factor encoded by the WUSCHEL gene is expressed in the organizing center and initiates a signal to the overlying cells to behave as stem cells.

b) A transcription factor encoded by the SHOOT MERISTEMLESS gene is expressed in shoot meristem cells and maintains them in their undifferentiated state.

c) Shoot meristems cells secrete proteins encoded by the CLAVATA family that antagonize WUSCHEL expression, thereby restricting the size of the shoot meristems.

d) All of these are involved in specification and maintenance of shoot meristems.

8. What is meant by the word 'whorl' in discussing floral meristems?

a) Flowers consist of four different types of organs, which occur in concentric rings called 'whorls.'

b) The floral meristem as to rotate during flower formation, giving the process the name 'whorl.'

c) The flowers of *Arabidopsis* appear as the stem elongates in a pattern called a 'whorl.'

d) The six stamens in a dicot flower like that of *Arabidopsis* form a ring that is called the flower's 'whorl.'

9. In what way are the homeotic genes of flowering plants similar to those of *Drosophila* and other animals?

a) All homeotic genes encode transcription factors of the homeobox class.

b) Homeotic genes in both plants and animals encode transcription factors of the MADS-box type.

c) Mutations in the homeotic genes of flowers cause transformation of one organ into another.

d) The homeotic genes of flowers are derived during evolution from the same primordial genes used in animals.

10. How is the ABC model for floral identity in *Arabidopsis* reminiscent of the models for homeotic gene function derived from studies in *Drosophila*?

a) In both organisms, each homeobox gene specifies the identity of a different region of the adult.

b) In both organisms, one homeotic gene is expressed at the two ends, a second expressed in domains more central to that of the first, and a third expressed most centrally, thus contributing unambiguous identities to all regions of the organism.

c) In both organisms, it is often the combination of genes present that is critical in unambiguously specifying structures in the adult.

d) The key responsibility of the homeotic genes in both organisms is the patterning of antero-posterior identity.

Multiple choice answer key

1: b, 2: d, 3: b, 4: b, 5: a, 6: a, 7: d, 8: a, 9: c, 10: c.

General further reading

Meyerowitz, E.M.: *Arabidopsis—a useful weed*. *Cell* 1989, **56**: 263–264.

Meyerowitz, E.M.: Plants compared to animals: the broader comparative view of development. *Science* 2002, **295**: 1482–1485.

Section further reading

7.1 The model plant *Arabidopsis thaliana* has a short life-cycle and a small diploid genome & 7.2 Plant embryos develop through several distinct stages

Lloyd, C.: **Plant morphogenesis: life on a different plane**. *Curr. Biol.* 1995, **5**: 1085–1087.

Mayer, U., Jürgens, G.: **Pattern formation in plant embryogenesis: a reassessment**. *Semin. Cell Dev. Biol.* 1998, **9**: 187–193.

Meyerowitz, E.M.: **Genetic control of cell division patterns in developing plants**. *Cell* 1997, **88**: 299–308.

Torres-Ruiz, R.A., Jürgens, G.: **Mutations in the *FASS* gene uncouple pattern formation and morphogenesis in Arabidopsis development**. *Development* 1994, **120**: 2967–2978.

7.3 Gradients of the signal molecule auxin establish the embryonic apical-basal axis

Breuninger, H., Rikirsch, E., Hermann, M., Ueda, M., Laux, T.: **Differential expression of *WOX* genes mediates apical-basal axis formation in the *Arabidopsis* embryo**. *Dev. Cell* 2008, **14**: 867–876.

Friml, J., Vieten, A., Sauer, M., Weijers, D., Schwarz, H., Hamann, T., Offringa, R., Jürgens, G.: **Efflux-dependent auxin gradients establish the apical-basal axis of *Arabidopsis***. *Nature* 2003, **426**: 147–153.

Jenik, P.D., Barton, M.K.: **Surge and destroy: the role of auxin in plant embryogenesis**. *Development* 2005, **132**: 3577–3585.

Jürgens, G.: **Axis formation in plant embryogenesis: cues and clues**. *Cell* 1995, **81**: 467–470.

Long, J.A., Moan, E.I., Medford, J.I., Barton, M.K.: **A member of the knotted class of homeodomain proteins encoded by the *STM* gene of *Arabidopsis***. *Nature* 1995, **379**: 66–69.

Szemenyei, H., Hannon, M., Long, J.A.: **TOPLESS mediates auxin-dependent transcriptional repression during *Arabidopsis* embryogenesis**. *Science* 2008, **319**: 1384–1386.

7.4 Plant somatic cells can give rise to embryos and seedlings

Zimmerman, J.L.: **Somatic embryogenesis: a model for early development in higher plants**. *Plant Cell* 1993, **5**: 1411–1423.

7.5 A meristem contains a small central zone of self-renewing stem cells

Byrne, M.E., Kidner, C.A., Martienssen, R.A.: **Plant stem cells: divergent pathways and common themes in shoots and roots**. *Curr. Opin. Genet. Dev.* 2003, **13**: 551–557.

Großhardt, R., Laux, T.: **Stem cell regulation in the shoot meristem**. *J. Cell Sci.* 2003, **116**: 1659–1666.

Ma, H.: **Gene regulation: Better late than never?** *Curr. Biol.* 2000, **10**: R365–R368.

7.6 The size of the stem-cell area in the meristem is kept constant by a feedback loop to the organizing center

Brand, U., Fletcher, J.C., Hobe, M., Meyerowitz, E.M., Simon, R.: **Dependence of stem cell fate in *Arabidopsis* on a feedback loop regulated by *CLV3* activity**. *Science* 2000, **289**: 635–644.

Carles, C.C., Fletcher, J.C.: **Shoot apical meristem maintenance: the art of dynamical balance**. *Trends Plant Sci.* 2003, **8**: 394–401.

Clark, S.E.: **Cell signalling at the shoot meristem**. *Nat Rev. Mol. Cell Biol.* 2001, **2**: 277–284.

Lenhard, M., Laux, T.: **Stem cell homeostasis in the *Arabidopsis* shoot meristem is regulated by intercellular movement of CLAVATA3 and its sequestration by CLAVATA1**. *Development* 2003, **130**: 3163–3173.

Reddy, G.V., Meyerowitz, E.M.: **Stem-cell homeostasis and growth dynamics can be uncoupled in the *Arabidopsis* shoot apex**. *Science* 2005, **310**: 663–667.

Schoof, H., Lenhard, M., Haecker, A., Mayer, K.F.X., Jürgens, G., Laux, T.: **The stem cell population of *Arabidopsis* shoot meristems is maintained by a regulatory loop between the *CLAVATA* and *WUSCHEL* genes**. *Cell* 2000, **100**: 635–644.

Vernoux, T., Benfey, P.N.: **Signals that regulate stem cell activity during plant development**. *Curr. Opin. Genet. Dev.* 2005, **15**: 388–394.

7.7 The fate of cells from different meristem layers can be changed by changing their position

Castellano, M.M., Sablowski, R.: **Intercellular signalling in the transition from stem cells to organogenesis in meristems**. *Curr. Opin. Plant Biol.* 2005, **8**: 26–31.

Gallagher, K.L., Benfey, P.N.: **Not just another hole in the wall: understanding intercellular protein trafficking**. *Genes Dev.* 2005, 19: 189–195.

Laux, T., Mayer, K.F.X.: **Cell fate regulation in the shoot meristem**. *Semin. Cell Dev. Biol.* 1998, **9**: 195–200.

Sinha, N.R., Williams, R.E., Hake, S.: **Overexpression of the maize homeobox gene *knotted-1*, causes a switch from determinate to indeterminate cell fates**. *Genes Dev.* 1993, **7**: 787–795.

Turner, I.J., Pumfrey, J.E.: **Cell fate in the shoot apical meristem of *Arabidopsis thaliana***. *Development* 1992, **115**: 755–764.

Waites, R., Simon, R.: **Signaling cell fate in plant meristems: three clubs on one tousle**. *Cell* 2000, **103**: 835–838.

7.8 A fate map for the embryonic shoot meristem can be deduced using clonal analysis

Irish, V.F.: **Cell lineage in plant development**. *Curr. Opin. Genet. Dev.* 1991, **1**: 169–173.

7.9 Meristem development is dependent on signals from other parts of the plant

Doerner, P.: **Shoot meristems: intercellular signals keep the balance**. *Curr. Biol.* 1999, **9**: R377–R380.

Irish, E.E., Nelson, T.M.: **Development of maize plants from cultured shoot apices.** *Planta* 1988, **175**: 9–12.

Sachs, T.: *Pattern Formation in Plant Tissues.* Cambridge: Cambridge University Press, 1994.

7.10 Gene activity patterns the proximo-distal and adaxial-abaxial axes of leaves developing from the shoot meristem

Bowman, J.L.: **Axial patterning in leaves and other lateral organs.** *Curr. Opin. Genet. Dev.* 2000, **10**: 399–404.

Kidner, C.A., Martienssen, R.A.: **Spatially restricted microRNA directs leaf polarity through *ARGONAUTE1*.** *Nature* 2004, **428**: 81–84.

Waites, R., Selvadurai, H.R., Oliver, I.R., Hudson, A.: The *PHAN-TASTICA* **gene encodes a MYB transcription factor involved in growth and dorsoventrality of lateral organs** *in Antirrhinum.* *Cell* 1998, **93**: 779–789.

7.11 The regular arrangement of leaves on a stem is generated by regulated auxin transport

Berleth, T., Scarpella, E., Prusinkiewicz, P.: **Towards the systems biology of auxin-transport-mediated patterning.** *Trends Plant Sci.* 2007, **12**: 151–159.

Heisler, M.G., Ohno, C., Das, P., Sieber, P., Reddy, G.V., Long, J.A., Meyerowitz, E.M.: **Patterns of auxin transport and gene expression during primordium development revealed by live imaging of the *Arabidopsis* inflorescence meristem.** *Curr. Biol.* 2005, **15**: 1899–1911.

Jönsson, H., Heisler, M.G., Shapiro, B.E., Meyerowitz, E.M., Mjolsness, E.: **An auxin-driven polarized transport model for phyllotaxis.** *Proc. Natl. Acad. Sci. USA* 2006, **103**: 1633–1638.

Mitchison, G.J.: **Phyllotaxis and the Fibonacci series.** *Science* 1977, **196**: 270–275.

Reinhardt, D.: **Regulation of phyllotaxis.** *Int. J. Dev. Biol.* 2005, **49**: 539–546.

Scheres, B.: **Non-linear signaling for pattern formation.** *Curr. Opin. Plant Biol.* 2000, **3**: 412–417.

Schiefelbein, J.: **Cell-fate specification in the epidermis: a common patterning mechanism in the root and shoot.** *Curr. Opin. Plant Biol.* 2003, **6**: 74–78.

Smith, L.G., Hake, S.: **The initiation and determination of leaves.** *Plant Cell* 1992, **4**: 1017–1027.

7.12 Root tissues are produced from *Arabidopsis* root apical meristems by a highly stereotyped pattern of cell divisions

Costa, S., Dolan, L.: **Development of the root pole and patterning in *Arabidopsis* roots.** *Curr. Opin. Genet. Dev.* 2000, **10**: 405–409.

Dello Ioio, R., Nakamura, K., Moubayidin, L., Perilli, S., Taniguchi, M., Morita, M.T., Aoyama, T., Costantino, P., Sabatini, S.: **A genetic framework for the control of cell division and differentiation in the root meristem.** *Science* 2008, **322**: 1380–1384.

Dolan, L.: **Positional information and mobile transcriptional regulators determine cell pattern in *Arabidopsis* root epidermis.** *J. Exp. Bot.* 2006, **57**: 51–54.

Grieneisen, V.A., Xu, J., Marée, A.F.M., Hogeweg P., Scheres, B.: **Auxin transport is sufficient to generate a maximum and gradient guiding root growth.** *Nature* 2007, **449**: 1008–1013.

Sabatini, S., Beis, D., Wolkenfeldt, H., Murfett, J., Guilfoyle, T., Malamy, J., Benfey, P., Leyser, O., Bechtold, N., Weisbeek, P.,

Scheres, B.: **An auxin-dependent distal organizer of pattern and polarity in the *Arabidopsis* root.** *Cell* 1999, **99**: 463–472.

Scheres, B., McKhann, H.I., van den Berg, C.: **Roots redefined: anatomical and genetic analysis of root development.** *Plant Physiol.* 1996, **111**: 959–964.

van den Berg, C., Willemsen, V., Hendriks, G., Weisbeek, P., Scheres, B.: **Short-range control of cell differentiation in the *Arabidopsis* root meristem.** *Nature* 1997, **390**: 287–289.

Veit, B.: **Plumbing the pattern of roots.** *Nature* 2007, **449**: 991–992.

7.13 Root hairs are specified by a combination of positional information and lateral inhibition

Kwak, S.-H., Schiefelbein, J.: **The role of the SCRAMBLED receptor-like kinase in patterning the *Arabidopsis* root epidermis.** *Dev. Biol.* 2007, **302**: 118–131.

Schiefelbein, J., Kwak, S.-H., Wieckowski, Y., Barron, C., Bruex, A.: **The gene regulatory network for root epidermal cell-type pattern formation in *Arabidopsis*.** *J. Exp. Bot.* 2009, **60**: 1515–1521.

7.14 Homeotic genes control organ identity in the flower

Bowman, J.L., Sakai, H., Jack, T., Weigel, D., Mayer, U., Meyerowitz, E.M.: **SUPERMAN, a regulator of floral homeotic genes in *Arabidopsis*.** *Development* 1992, **114**: 599–615.

Breuil-Broyer, S., Morel, P., de Almeida-Engler, J., Coustham, V., Negrutiu, I., Trehin, C.: **High-resolution boundary analysis during *Arabidopsis thaliana* flower development.** *Plant J.* 2004, **38**: 182–192.

Coen, E.S., Meyerowitz, E.M.: **The war of the whorls: genetic interactions controlling flower development.** *Nature* 1991, **353**: 31–37.

Irish, V.F.: **Patterning the flower.** *Dev. Biol.* 1999, **209**: 211–220.

Krizek, B.A., Meyerowitz, E.M.: The *Arabidopsis* **homeotic genes *APETALA3* and *PISTILLATA* are sufficient to provide the B class organ identity function.** *Development* 1996, **122**: 11–22.

Krizek, B.A., Fletcher, J.C.: **Molecular mechanisms of flower development: an armchair guide.** *Nat Rev. Genet.* 2005, **6**: 688–698.

Lohmann, J.U., Weigel, D.: **Building beauty: the genetic control of floral patterning.** *Dev. Cell* 2002, **2**: 135–142.

Ma, H., dePamphilis, C.: **The ABCs of floral evolution.** *Cell* 2000, **101**: 5–8.

Meyerowitz, E.M., Bowman, J.L., Brockman, L.L., Drews, G.M., Jack, T., Sieburth, L.E., Weigel, D.: **A genetic and molecular model for flower development in *Arabidopsis thaliana*.** *Development Suppl.* 1991, **1**: 157–167.

Meyerowitz, E.M.: **The genetics of flower development.** *Sci. Am.* 1994, **271**: 40–47.

Sakai, H., Medrano, L.J., Meyerowitz, E.M.: **Role of *SUPERMAN* in maintaining *Arabidopsis* floral whorl boundaries.** *Nature* 1994, **378**: 199–203.

Vincent, C.A., Carpenter, R., Coen, E.S.: **Cell lineage patterns and homeotic gene activity during *Antirrhinum* flower development.** *Curr. Biol.* 1995, **5**: 1449–1458.

Wagner, D., Sablowski, R.W.M., Meyerowitz, E.M.: **Transcriptional activation of *APETALA 1*.** *Science* 1999, **285**: 582–584.

7.15 The *Antirrhinum* flower is patterned dorso-ventrally as well as radially

Coen, E.S.: **Floral symmetry**. *EMBO J.* 1996, **15**: 6777–6788.

Luo, D., Carpenter, R., Vincent, C., Copsey, L., Coen, E.: **Origin of floral asymmetry in *Antirrhinum***. *Nature* 1996, **383**: 794–799.

7.16 The internal meristem layer can specify floral meristem patterning

Szymkowiak, E.J., Sussex, I.M.: **The internal meristem layer (L3) determines floral meristem size and carpel number in tomato periclinal chimeras**. *Plant Cell* 1992, **4**: 1089–1100.

7.17 The transition of a shoot meristem to a floral meristem is under environmental and genetic control

An, H., Roussot, C., Suarez-Lopez, P., Corbesier, L., Vincent, C., Pineiro, M., Hepworth, S., Mouradov, A., Justin, S., Turnbull, C., Coupland, G.: **CONSTANS acts in the phloem to regulate a systemic signal that induces photoperiodic flowering of *Arabidopsis***. *Development* 2004, **131**: 3615–3626.

Becroft, P.W.: **Intercellular induction of homeotic gene expression in flower development**. *Trends Genet.* 1995, **11**: 253–255.

Blázquez, M.A.: **The right time and place for making flowers**. *Science* 2005, **309**: 1024–1025.

Corbesier, L., Vincent, C., Jang, S., Fornara, F., Fan, Q., Searle, I., Giakountis, A., Farrona, S., Gissot, L., Turnbull, C., Coupland, G.: **FT protein movement contributes to long-distance signaling in floral induction of *Arabidopsis***. *Science* 2007, **316**: 1030–1033.

Hake, S.: **Transcription factors on the move**. *Trends Genet.* 2001, **17**: 2–3.

Jaeger, K.E., Wigge, P.A.: **FT protein acts as a long-range signal in *Arabidopsis***. *Curr. Biol.* 2007, **17**: 1050–1054.

Kobayashi, Y., Weigel, D.: **Move on up, it's time for change - mobile signals controlling photoperiod-dependent flowering**. *Genes Dev.* 2007, **21**: 2371–2384.

Lohmann, J.U., Hong, R.L., Hobe, M., Busch, M.A., Parcy, F., Simon, R., Weigel, D.: **A molecular link between stem cell regulation and floral patterning in *Arabidopsis***. *Cell* 2001, **105**: 793–803.

Putterill, J., Laurie, R., Macknight, R.: **It's time to flower: the genetic control of flowering time**. *BioEssays* 2004, **26**: 363–373.

Sheldon, C.C., Hills, M.J., Lister, C., Dean, C., Dennis, E.S., Peacock, W.J.: **Resetting of *FLOWERING LOCUS C* expression after epigenetic repression by vernalization**. *Proc. Natl. Acad. Sci. USA* 2008, **105**: 2214–2219.

Valverde, F., Mouradov, A., Soppe, W., Ravenscroft, D., Samach, A., Coupland, G.: **Photoreceptor regulation of CONSTANS protein in photoperiodic flowering**. *Science* 2004, **303**: 1003–1006.

Morphogenesis: change in form in the early embryo

- Cell adhesion
- Cleavage and formation of the blastula
- Gastrulation movements
- Neural tube formation
- Cell migration
- Directed dilation

Changes in form in animal embryos are brought about by cellular forces that are generated in a variety of ways, including cell division, change in cell shape, rearrangement of cells within tissues, and the migration of individual cells from one part of the embryo to another. These cellular forces are most frequently generated within epithelial sheets by contraction of the cells' internal cytoskeleton; resisting these forces are the cells' own structure and the adhesive interactions that hold cells together in tissues. The generation of hydrostatic pressure within a structure, cell division, and cell migration also play a part in morphogenetic processes. The major morphogenetic event in early animal embryos is gastrulation, when a wholesale reorganization of the body plan occurs. Other examples of morphogenesis discussed in this chapter include notochord and neural tube formation, and the migration of neural crest cells to specific locations. Plants develop form through cell division and cell enlargement.

So far, we have discussed early development mainly from the viewpoint of developmental patterning and the assignment of cell fate. In this chapter, we look at embryonic development from a different perspective—the generation of form, or **morphogenesis**. All animal embryos undergo a dramatic change in shape during their early development. This occurs principally during gastrulation, the process that transforms a two-dimensional sheet of cells into the complex three-dimensional animal body. Gastrulation involves extensive rearrangements of cell layers and the directed movement of cells from one location to another.

If pattern formation can be likened to painting, morphogenesis is more akin to modeling a formless lump of clay into a recognizable shape. Change in form is largely a problem in cell mechanics: understanding it requires an appreciation of the forces involved in bringing about changes in cell shape and cell migration, the cellular machinery that carries them out, and how this is controlled during development.

Two key cellular properties involved in changes in animal embryonic form are **cell adhesiveness** and **cell motility**. Animal cells stick to one another, and to the extracellular matrix, through interactions involving cell-surface proteins. Changes in the adhesion proteins at the cell surface can therefore determine the strength of cell–cell adhesion and its specificity. These adhesive interactions affect the surface tension at the cell membrane, a property that contributes to the mechanics of the cell rearrangements that generate the three distinct germ layers during gastrulation. Different surface tensions are what keep two immiscible liquids such as water and oil separate, and embryologists had long noted the 'liquid-like' behavior of the tissues in gastrulating frog embryos, while observing how the separate germ layers formed and remained distinct from each other, slid over each other, spread out, and underwent internal rearrangement to shape the embryo. The modern **differential adhesion hypothesis** explains these liquid-like tissue dynamics as the consequences of surface tensions and interfacial tensions generated by cells that tend to stick to each other but are also able to change shape and to move.

The second key property—cell motility—encompasses both the migration of individual cells and the ability of cells to change shape while remaining part of a tissue. For example, folding of a cell sheet—a very common feature in embryonic development—is caused by changes in cell shape. Cells move and change shape by rearranging their internal cytoskeleton, particularly those elements of it that can cause contraction or constriction of part of the cell. Cellular contraction is generated by actin filaments complexed with the motor protein myosin, forming contractile structures similar to, but simpler than, those in muscle. In non-muscle cells, contractile actin–myosin complexes (actomyosin) are concentrated in the cell cortex—the region immediately underneath the cell membrane. Changes in shape that lead to cell migration involve the extension of actin-based structures, such as filopodia and lamellipodia, in the direction of movement (Box 8A). An additional force that operates during morphogenesis, particularly in plants but also in some aspects of animal embryogenesis, is hydrostatic pressure, which is generated by osmosis and fluid accumulation. In plants there is no cell movement during growth, and changes in form are generated by oriented cell division and cell expansion, as we saw in Chapter 7.

Changes in embryonic form are the final consequence of the precise spatio-temporal expression of proteins that control cell adhesion, cell motility, oriented cell division, and the generation of hydrostatic pressure. It is likely that changes in form are brought about by an earlier patterning process that determines which cells will express those proteins that are required to generate and harness the appropriate forces. An attractive hypothesis is that pattern-determining genes, such as the Hox genes, activate genes that control the expression of such proteins, and there is some evidence for this.

In this chapter we shall discuss mainly the morphological changes that occur during the development of the animal body plan. First, we will look at how cleavage of the zygote gives rise to the simple shape of the early embryo, of which the spherical blastocyst of the mouse and the blastulas of sea urchin and amphibian are good examples. We then consider the movements that occur during gastrulation and during neurulation—the formation of the neural tube in vertebrates—which involve folding of cell sheets and rearrangement of cell layers. In vertebrates, migration of cells from the neural crest after neurulation generates a variety of structures in the trunk and head, and we consider how these cells migrate to their correct sites. Finally, we look at directed dilation, in which hydrostatic pressure is the force driving changes in shape. Other morphogenetic mechanisms such as cell growth, cell proliferation, and cell death will be considered in relation to the development of particular organs and the growth of an organism as a whole (see Chapters 11, 12, and 13).

We begin by considering how cells adhere to each other, and how differences in adhesiveness and specificity of adhesion are involved in maintaining boundaries between tissues.

Box 8A Change in cell shape and cell movement

Cells actively undergo major changes in shape during development. The two main changes are associated with cell migration and with the infolding of epithelial sheets. Changes in shape are generated by the cytoskeleton, an intracellular protein framework that also controls changes in cell movement. There are three principal types of protein polymers in the cytoskeleton—**actin filaments (microfilaments)**, **microtubules**, and **intermediate filaments**—as well as many other proteins that interact with them. Actin filaments and microtubules are dynamic structures, polymerizing and depolymerizing according to the cell's requirements. Intermediate filaments are more stable, forming rope-like structures that transmit mechanical forces, spread mechanical stress, and provide mechanical stability to the cell. Microtubules play an important part in maintaining cell asymmetry and polarity, and provide tracks along which motor proteins convey other molecules and even organelles. They also form the spindles that segregate chromosomes at mitosis and meiosis. Polymerization and depolymerization of actin filaments at the leading edges of cells are involved in cell shape change and cell movements, and actin filaments also associate with the motor protein myosin to form contractile bundles of **actomyosin**, which are primarily responsible for force-generating contractions within cells.

Actin filaments are fine threads of protein about 7 nm in diameter and are polymers of the globular protein actin. They are organized into bundles and three-dimensional networks, which, in most cells, lie mostly just beneath the plasma membrane, forming the gel-like cell cortex. Numerous actin-binding proteins are associated with actin filaments, and are involved in bundling them together, forming networks, and aiding the polymerization and depolymerization of the actin subunits. Actin filaments can form rapidly by polymerization of actin subunits and can be equally rapidly depolymerized. This provides the cell with a highly versatile system for assembling actin filaments in a variety of different ways and in different locations, as required. The fungal drug cytochalasin D prevents actin polymerization and is a useful agent for investigating the role of actin networks. Actin filaments can also assemble with myosin into contractile structures, which act as miniature muscles. For example, contraction of the **contractile ring**, bundles of actomyosin arranged in a ring in the cell cortex, pinches animal cells in two at cell division (see figure, left panel). The contraction of a similar ring of actomyosin around the apical end of a cell leads to apical constriction and elongation of the cell (see figure, center panel).

Many embryonic cells can migrate over a solid substratum, such as the extracellular matrix. They move by extending a thin sheet-like layer of cytoplasm known as a lamellipodium (see figure, right panel), or long, fine cytoplasmic processes called filopodia. Both these temporary structures are pushed outwards from the cell by the assembly of actin filaments. Contraction of the actomyosin network at the front and the rear of the cell then draws the cell forward. To do this, the contractile system must be able to exert a force on the substratum and this occurs at focal contacts, points at which the advancing filopodia or lamellipodium are anchored to the surface over which the cell is moving. At focal contacts, integrins (see Box 8B, p. 293) both adhere to extracellular matrix molecules through their extracellular domains, and provide an anchor point for actin filaments through their cytoplasmic domains. Integrins transduce signals from the extracellular matrix across the plasma membrane at focal contacts, enabling the cell to sense the environment over which it is traveling and to adjust its movement accordingly.

The small GTP-binding proteins known as the Rho-family GTPases, which include Rho, Rac, and Cdc42, have a key role in regulating the actin cytoskeleton. Rac is required for extension of lamellipodia, for example, and Cdc42 is necessary to maintain cell polarity. Rho is involved in the planar cell polarity signaling pathway (see Box 8C, p. 310). Signals from the environment relayed to the cytoskeleton via the Rho-family proteins are thus able to control cell movement.

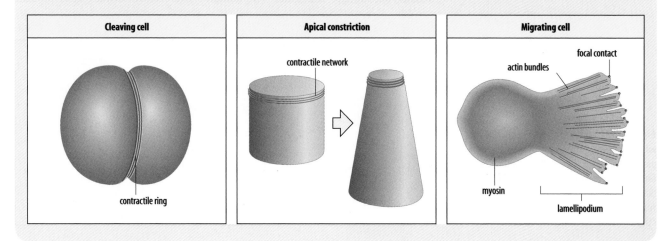

| Cleaving cell | Apical constriction | Migrating cell |

contractile ring

contractile network

focal contact

actin bundles

myosin

lamellipodium

Cell adhesion

The late embryo and the adult are composed of a variety of differentiated cell types, which are grouped together to form tissues such as skin, cartilage and muscle. The integrity of tissues is maintained by adhesive interactions between cells and between cells and the extracellular matrix; differences in cell adhesiveness also help maintain the boundaries between different tissues and structures. Cells stick to each other by means of **cell-adhesion molecules**, which are proteins on the cell surface that can bind strongly to proteins on other cell surfaces or in the extracellular matrix (Box 8B). The particular adhesion molecules expressed by a cell determine which cells it can adhere to, and changes in the adhesion molecules expressed are involved in many developmental phenomena. In epithelial tissues, with which we shall be most concerned here, adjacent cells are joined together by specialized structures called **adhesive cell junctions** that incorporate cell-adhesion molecules. The junctions discussed in this chapter are the **adherens junction**, in which the adhesion molecule is cadherin, which is linked intracellularly to the actin cytoskeleton, and the **tight junction**, which joins epithelial cells together with a tight 'seal' that forms a permeability barrier between the environments inside and outside the epithelium.

8.1 Sorting out of dissociated cells demonstrates differences in cell adhesiveness in different tissues

Differences in cell adhesiveness can be illustrated by experiments in which different tissues are confronted with one another in an artificial setting. Two pieces of early endoderm from an amphibian blastula will fuse to form a smooth sphere if placed in contact. In contrast, when a piece of early endoderm and a piece of early ectoderm are combined they initially fuse, but in time the endodermal and ectodermal cells separate, until only a narrow bridge connects the two types of tissue (Fig. 8.1).

In another type of experiment, cells from two different tissues are dissociated, mixed together, and then allowed to reaggregate. When cells of amphibian presumptive epidermis and presumptive neural plate are dissociated, mixed, and left to reaggregate, they sort out to reform the two different tissues (Fig. 8.2). The epidermal cells are eventually found on the outer face of the aggregate, surrounding a mass of neural cells—the same types of cell are now in contact with each other. Mixed ectodermal and mesodermal cells similarly sort themselves out to form a mass of cells with ectoderm on the outside and mesoderm on the inside.

This sorting out is the combined result of cell movement and differences in adhesiveness. Initially, cells move about randomly in the mixed aggregate, exchanging weaker for stronger adhesions. The adhesive interactions between cells produce

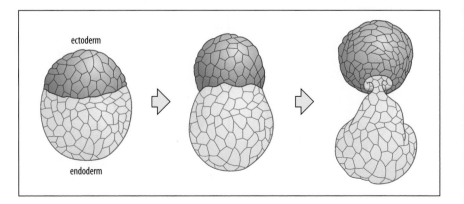

Fig. 8.1 Separation of embryonic tissues with different adhesive properties. When pieces of early ectoderm (blue) and early endoderm (yellow) from an amphibian blastula are placed together, they initially fuse, but then separate until only a narrow strip of tissue joins the two tissues.

Box 8B Cell-adhesion molecules and cell junctions

Three classes of adhesion molecules are particularly important in development (see figure). The **cadherins** are transmembrane proteins that, in the presence of calcium ions (Ca^{2+}), adhere to cadherins on the surface of another cell. Calcium-independent cell-cell adhesion involves a different structural class of proteins—members of the large **immunoglobulin superfamily**. The neural cell-adhesion molecule (N-CAM), which was first isolated from neural tissue, is a typical member of this family. Some immunoglobulin superfamily members, such as N-CAM, bind to similar molecules on other cells; others bind to a different class of adhesion molecule, the **integrins**. Integrins also act as receptors for molecules of the extracellular matrix, which they can use to mediate adhesion between a cell and its substratum.

About 30 different types of cadherins have been identified in vertebrates. Cadherins bind to each other through one or more binding sites located within the extracellular amino-terminal 100 amino acids. In general, a cadherin binds only to another cadherin of the same type, but they can also bind to some other molecules. Cadherins are the adhesive components in both **adherens junctions** (adhesive cell-cell junctions that are present in many tissues) and **desmosomes** (cell-cell junctions present mainly in epithelia). **Hemidesmosomes** are an integrin-based type of junction that rivets epithelial cells firmly to the extracellular matrix of the basement membrane.

As cells approach and touch one another, the cadherins cluster at the site of contact. They interact with the intracellular cytoskeleton through the connection of their cytoplasmic tails with **catenins** and other proteins, and thus can be involved in transmitting signals to the cytoskeleton. This adhesive and signaling role of α-, β-, and γ -catenins is separate from the role of β-catenin as a gene-regulatory protein (see, for example, Section 4.2). The interaction with the cytoskeleton is required

for normal strong cell-cell adhesion. In adherens junctions the connection is to cytoskeletal actin filaments, whereas in desmosomes and hemi-desmosomes it is to intermediate filaments such as keratin. Nectin, a member of the immunoglobulin superfamily of adhesion molecules, also clusters to adherens junctions in mammalian tissues and connects to the actin cytoskeleton.

Adhesion of a cell to the extracellular matrix, which contains proteins such as collagen, fibronectin, laminin, and tenascin, as well as proteoglycans, is by the binding of integrins to these matrix molecules. An integrin molecule is composed of two different subunits, an α subunit and a β subunit. Twenty-four different integrins are known so far in vertebrates, made up from eight β subunits and 18 α subunits. Many extracellular matrix molecules are recognized by more than one integrin.

Integrins not only bind to other molecules through their extracellular face, but they also associate with the actin filaments or intermediate filaments of the cell's cytoskeleton through complexes of proteins that are in contact with the integrin's cytoplasmic region. This association enables integrins to transmit information about the extracellular environment, such as extracellular matrix composition or the type of intercellular contact, to the cell. Integrins can, therefore, transmit signals from the matrix that affect cell shape, motility, metabolism, and cell differentiation. Integrins are also involved in adhesion between cells, binding either to adhesion molecules of the immunoglobulin superfamily or via a shared ligand to integrins on another cell surface.

A third type of cell junction present in epithelia is the **tight junction**, which forms a seal that prevents water and other molecules passing between the epithelial cells. Multiple lines of transmembrane proteins called claudins and occludins in adjoining cell membranes form a continuous seal around the apical end of the cell (see Fig. 8.12).

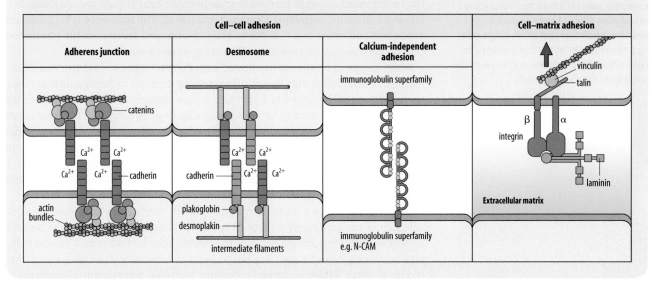

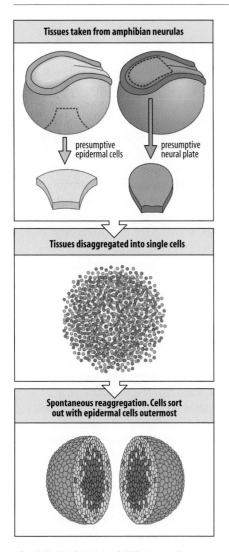

Fig. 8.2 Sorting out of different cell types. Ectoderm from the presumptive epidermis (gray) and from the presumptive neural plate (blue) of early amphibian neurulas are disaggregated into single cells by treatment with an alkaline solution. The cells, when mixed together, sort out with the epidermal cells on the outside.

different degrees of surface tension that is sufficient to generate the sorting-out behavior, just as two immiscible liquids, such as oil and water, separate out when mixed. As the constituent cells form stable contacts with each other, the tissues become reorganized and the strength of intercellular binding in the system as a whole is maximized. In general, if the adhesion between unlike cells is weaker than the average of the adhesions between like cells, the cells will segregate according to type, with the more cohesive tissue tending to be enveloped by the less cohesive one. In terms of the surface tension, those cells with the higher surface tension become surrounded by those with lower surface tension.

It is important to note that the above experiments on isolated cells *in vitro* simply indicate the different adhesiveness of different tissues; they do not define which tissue will be on the inside or the outside in the embryo itself. This is because in a real embryo, tissues are under pre-existing developmental constraints. In the embryo, the most cohesive tissue is the ectoderm, which is derived from the outermost layer of cells, forms a continuous sheet, and thus remains on the outer surface of the embryo, while the less cohesive mesoderm and endoderm become internalized. If ectodermal and mesendodermal cells from zebrafish are isolated and mixed *in vitro*, however, they will rearrange with the more cohesive ectodermal cells on the inside.

The *in vitro* experiments also show how differential cell adhesion can, in principle, stabilize boundaries between tissues. We have already seen an example of boundary formation by a somewhat different mechanism in Chapter 5, in the case of the rhombomere boundaries in the hindbrain. There, repulsive interactions between ephrins and their Eph receptors across a rhombomere boundary prevent cells from different rhombomeres intermingling, whereas cohesive Eph–ephrin interactions within a rhombomere promote cell adhesion (see Section 5.11).

8.2 Cadherins can provide adhesive specificity

Differential adhesiveness between cells is the result of differences in the types and numbers of adhesion molecules on cell surfaces. Evidence that the cadherins can provide adhesive specificity comes from studies in which cells with different cadherins on their surface are mixed together. Fibroblasts of the mouse L cell line do not express cadherins on their surface and do not adhere strongly to each other. But if the gene encoding E-cadherin is introduced (transfected) into L cells in a form that can be expressed, the cells produce that cadherin on their surface and stick together, forming a structure resembling a compact epithelium. The adherence is both calcium dependent, indicating that it is due to the cadherin, and specific, as the transfected cells do not adhere to untreated L cells lacking surface cadherins.

When groups of L cells are transfected with different types of cadherin and mixed together in suspension, only those cells expressing the same cadherin adhere strongly to each other: cells expressing E-cadherin adhere strongly to other cells expressing E-cadherin, but only weakly to cells expressing P- or N-cadherin. The amount of cadherin on the cell surface can also have an effect on adhesion. When cells expressing different amounts of the same cadherin are mixed together, those cells with more cadherin on their surface form an inner ball, surrounded by the cells with less surface cadherin (Fig. 8.3). Thus, quantitative differences in cell-adhesion molecules could maintain differential cell adhesion.

Cadherin molecules bind to each other through their extracellular domains, but adhesion is not exclusively controlled by these domains. The cytoplasmic domain associates with the actin filaments of the cytoskeleton by means of a protein complex containing catenins (see Box 8B, p. 293), and failure to make this association results in weak adhesion. In early *Xenopus* blastulas, for example, E-cadherin is

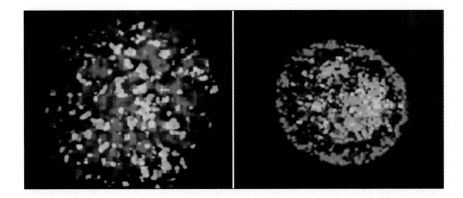

Fig. 8.3 **Sorting out of cells carrying different amounts of adhesion molecules on their surfaces.** When two cell lines with different amounts of N-cadherin on their surface are mixed, they reassort with the cells containing the most N-cadherin (green stained cells) ending up on the inside.
Photographs courtesy of M. Steinberg, from Foty, R.A., Steinberg, M.S.: 2005.

expressed in the ectoderm just before gastrulation and N-cadherin appears in the prospective neural plate. If E-cadherin lacking an extracellular domain is produced in the blastulas from injected mutated mRNA, the mutant cadherin will compete with the embryo's own intact cadherin molecules for the cytoskeletal association sites. The defective cadherin cannot influence cell–cell adhesion as it has no extracellular domain, but it blocks the intact cadherin's access to the cytoskeleton. The result is disruption of the ectoderm during gastrulation, indicating that a cadherin molecule must bind both to its partner on an adjacent cell and to the cytoskeleton of its own cell to create a stable adhesion. The initial binding of the extracellular cadherin domains transmits a signal to the cytoskeleton, which then stabilizes the interaction.

Summary

The adhesion of cells to each other and to the extracellular matrix maintains the integrity of tissues and the boundaries between them. The associations of cells with each other are determined by the cell-adhesion molecules they express on their surface: cells bearing different adhesion molecules, or different quantities of the same molecule, sort out into separate tissues. This is due to the effects of intercellular adhesiveness on the cortical cytosekeleton and thus on cell-surface tension and interfacial tension between tissues. Cell-cell adhesion is due mainly to two classes of surface proteins: the cadherins, which bind in a calcium-dependent manner to identical cadherins on another cell surface, and members of the immunoglobulin superfamily, some of which bind to similar molecules on other cells, while others bind to different molecules, such as integrins. The binding of this second class of surface proteins is calcium independent. Adhesion to the extracellular matrix is mediated by a third class of adhesion molecule—the integrins. Stable cell-cell adhesion via cadherins involves both physical contact between their external domains and the cytoplasmic domain of the cadherin molecule binding to the cell cytoskeleton via a protein complex containing catenins.

Cleavage and formation of the blastula

The first step in animal embryonic development is the division of the fertilized egg by cleavage into a number of smaller cells (blastomeres), leading in many animals to the formation of a hollow sphere of cells—the blastula (see Chapter 3). In many embryos, cleavage involves short mitotic cell cycles, in which cell division and mitosis succeed each other repeatedly without intervening periods of cell growth. During cleavage, therefore, the mass of the embryo does not increase. Early cleavage patterns can vary widely between different groups of animals (Fig. 8.4). In **radial cleavage**, cleavages occur at right angles to the egg surface and the first few cleavages produce tiers of blastomeres that sit directly over each other. This type of cleavage is characteristic of the deuterostomes, such as sea urchins and vertebrates. The eggs of molluscs and annelids, which are protostomes, illustrate another cleavage pattern, called **spiral cleavage**, in which successive divisions are at planes at slight angles to each other, producing a spiral arrangement of cells. In both radial and spiral cleavage, some divisions may be unequal. The first three cleavages in the sea urchin, for example, give rise to equal-sized blastomeres, whereas the later cleavage that forms the micromeres is unequal, one daughter cell being smaller than the other (see Fig. 6.19). In nematodes, the very first cleavage of the egg is unequal, producing two cells of unequal size (see Fig. 8.4). In the early *Drosophila* embryo, the nuclei undergo repeated divisions without cell division, forming a syncytium, and separation into individual cells only occurs later, with the growth of cell membranes between the nuclei (see Section 2.1).

The amount of yolk in the egg can influence the pattern of cleavage. In yolky eggs undergoing otherwise symmetric cleavage, a cleavage furrow develops in the least yolky region and gradually spreads across the egg, but its progress is slowed or even halted by the presence of the yolk. Cleavage may thus be incomplete for some time. This effect is most pronounced in the heavily yolked eggs of birds and zebrafish, where complete divisions are restricted to a region at one end of the egg, and the embryo is formed as a cap of cells sitting on top of the yolk (see Figs. 3.11 and 3.13). Even in moderately yolky eggs, such as those of amphibians, the presence of yolk can influence cleavage patterns. In frog eggs, for example, the later cleavages are unequal and asynchronous, resulting in an animal half composed of a mass of small cells and a yolky vegetal region composed of fewer and larger cells.

Two key questions arise in relation to early cleavage: how are the positions of the cleavage planes determined; and how does cleavage lead to a hollow blastula (or its equivalent), which has a clear inside–outside polarity?

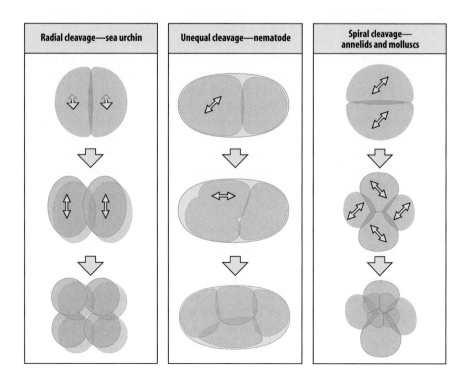

Fig. 8.4 Different patterns of early cleavage are found in different animal groups. Radial cleavage (for example, sea urchin) gives tiers of blastomeres sitting directly above each other. Unequal cleavage (for example, nematode) results in one daughter cell being larger than the other. In spiral cleavage (for example, molluscs and annelid worms), the mitotic apparatus is oriented at a slight angle to the long axis of the cell, resulting in a spiral arrangement of blastomeres.

8.3 The orientation of the mitotic spindle determines the plane of cleavage at cell division

The orientation of the plane of cleavage at cell division is of great importance in embryonic development. It not only determines the spatial arrangement of daughter cells relative to each other, and whether they will be of equal or unequal size, but it can also result in the apportionment of different cytoplasmic determinants to different daughter cells, thus determining their fate. We have already seen such effects in the unequal cleavages of early nematode development, which produce cells with different fates (see Chapter 6). The plane of cleavage can also be of great importance in later morphogenesis and growth. It determines, for example, whether an epithelial sheet remains as a single cell layer or becomes multilayered.

Cell division always occurs halfway between the two poles of the mitotic spindle. One possible explanation proposes that there is an overall increase in tension in the cell cortex due to cytoskeletal contraction, which causes the cell to round up. The asters at the poles of the spindle then interact with the cortex to relax the tension equally in the regions immediately adjacent to themselves, leaving the region between them to form a contractile ring and divide the cell into two. It has been shown experimentally that the position of a cleavage furrow can be specified by the asters alone, independently of the presence of a mitotic spindle. If a fertilized sea-urchin egg is deformed by inserting a glass bead that interrupts the first cleavage furrow, this creates a horseshoe-shaped cell in which the two nuclei generated by the first mitotic division are each segregated into an arm of the horseshoe (Fig. 8.5). At the next mitosis, spindles form normally in each arm of the horseshoe and each is bisected by a cleavage furrow, but a third furrow also forms between the two adjacent asters at the top of the horseshoe that are not connected by a spindle.

An aster is composed of microtubules that radiate out from a structure called a **centrosome**, which is an organizing center for microtubule growth in the cell. Before mitosis, the single centrosome in the cell becomes duplicated and the daughter centrosomes move to opposite sides of the nucleus and form the asters, whose interactions with cortical microtubules will anchor and help orient the mitotic spindle. The pattern of centrosome duplication and movement usually results in successive planes of cell division being at right angles to each other. An example of this is division of the nematode AB cell, one of the pair of cells formed by the first cleavage of the zygote, where the mitotic spindle in the dividing AB cell is oriented roughly at right angles to that of the zygote (Fig. 8.6). However, in the other product of the first cleavage, the P_1 cell, the centrosomes initially migrate to the same positions as in the AB cell but the nucleus and its associated centrosomes then rotates 90° towards the anterior of the cell, so that the eventual plane of cleavage of P_1 is the same as that of the zygote. The different planes of cleavage in the AB and P_1 cells are likely to be

Fig. 8.5 The positions of cleavage furrows are determined by the asters of the mitotic spindle. If the mitotic apparatus of a fertilized sea-urchin egg is displaced at the first cleavage by a glass bead, the cleavage furrow forms only on the side of the egg to which the mitotic apparatus has been moved. At the next cleavage, furrows bisect each mitotic spindle as expected, but an additional furrow, indicated by the blue triangle, forms between the two adjacent asters, even though there is no spindle between them. Chromosomes are not shown for simplicity.

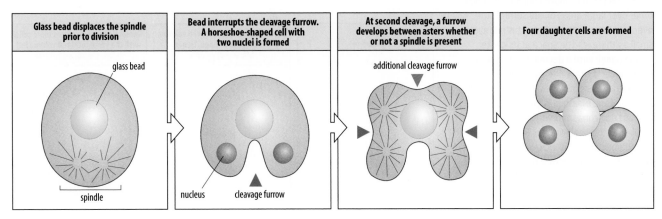

| Glass bead displaces the spindle prior to division | Bead interrupts the cleavage furrow. A horseshoe-shaped cell with two nuclei is formed | At second cleavage, a furrow develops between asters whether or not a spindle is present | Four daughter cells are formed |

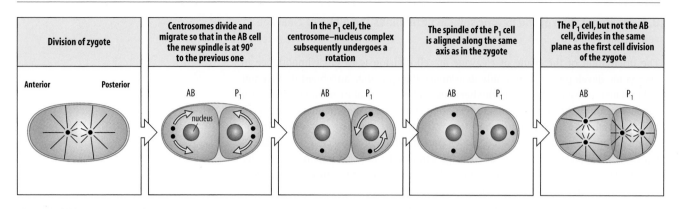

Fig. 8.6 Different planes of cleavage in different cells are determined by the behavior of the centrosomes. The first cleavage of the nematode zygote divides it into an anterior AB cell and a posterior P_1 cell. Before the next division, the duplicated centrosomes of the AB cell move apart in such a way that the plane of cleavage will be at right angles to the first division. The duplicated centrosomes in the P_1 cell initially move into a similar alignment, but before mitosis, the nucleus and its associated centrosomes rotate 90° so that cleavage is in the same plane as the first cleavage.

After Strome, S.: 1993.

under the ultimate control of cytoplasmic factors, such as the PAR proteins, that are differentially distributed at the first cleavage and have effects on spindle positioning (see Section 6.1).

In plants, the planes of successive cell divisions, along with the subsequent direction of cell elongation, are the main factors determining the forms of stems, roots, leaves, and flowers (see Chapter 7). When a plant cell divides, it is not pinched in half by a contractile actomyosin ring as in an animal cell. Instead, a new cell membrane and cell wall is formed *in situ* midway between the two poles of the mitotic spindle. In plants, the plane of division seems to be determined before mitosis begins, as a circumferential band of microtubules and actin filaments transiently forms in the cortex at the position of the future cell division.

A common form of early animal embryo is the hollow spherical blastula, composed of an epithelial sheet enclosing a fluid-filled interior. The development of such a structure from a fertilized egg depends both on particular patterns of cleavage and on changes in the way cells pack together, as shown in schematic form in Fig. 8.7. If the plane of cell division is always at right angles to the surface (radial), the cells will remain in a single layer. As cleavage proceeds individual cells get smaller, but the surface area of the cell sheet increases and a space—the blastocoel—forms in the interior, which increases in volume at each round of cell division (see Fig. 8.7). This is essentially how the sea urchin blastula is formed.

8.4 Cells become polarized in the sea-urchin blastula and the mouse morula

As a result of the pattern of cleavage, the surface of the sea-urchin egg, which is covered in microvilli, becomes the outer surface of the blastula. A layer of extracellular matrix called the hyaline layer, which contains the proteins hyalin and echinonectin, is secreted on to the external surface and the cells adhere to this sheet. Adhesive cell–cell junctions develop between adjacent blastomeres, forming them into an epithelium, and

Fig. 8.7 Cell packing can determine the volume of a blastula. Left panel: when adjacent cells in a spherical sheet, such as those of a blastula, make contact with each other over large areas of cell surface, the overall volume of the cell sheet is relatively small as there is little space at its center. Middle panel: with a decrease in the area of intercellular contact, the size of the internal space (the blastocoel), and thus the overall volume of the blastula, is greatly increased without any corresponding increase in cell number or total cell volume. Right panel: if the number of cells is increased by radial cleavage, and the packing remains the same, the blastocoel volume will increase further, again without any increase in total cell volume.

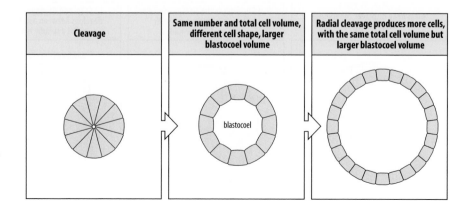

the cell contents become polarized in an apical–basal direction, with the Golgi apparatus oriented toward the apical (outer) surface. The cells are held together laterally by cadherin-based adherens junctions and by septate junctions, a type of cell junction found in invertebrates, but not chordates, and which is thought to provide the same type of permeability barrier as the vertebrate tight junction—preventing water, small molecules, and ions from permeating between the epithelial cells. On the inner (basal) surface of the epithelium, a basal lamina (an organized layer of extracellular matrix) is laid down. As the blastula develops, cilia are produced on the outer surface (Fig. 8.8).

In the mouse embryo, the earliest sign of structural differentiation is at the eight-cell stage, when the morula undergoes a process known as compaction (Fig. 8.9). Until then, the blastomeres form a loosely packed ball, with each cell surface uniformly covered by microvilli. At compaction, blastomeres flatten against each other, maximizing cell–cell contact, and cells become polarized, with microvilli becoming confined to the apical surfaces (Fig. 8.10). Radial cleavages produce two polarized cells, but some cleavages now occur tangentially (parallel to the surface), each producing one outer polarized cell and one inner nonpolarized cell. As development proceeds to form the blastocyst, the nonpolarized cells become the inner cell mass, which will give rise to

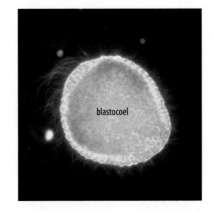

Fig. 8.8 Blastula of a sea-urchin embryo, Lytechinus pictus. A single layer of cells surrounds the hollow blastocoel. Scale bar = 10 μm.

Photograph from Raff, E.C. et al.: 2006.

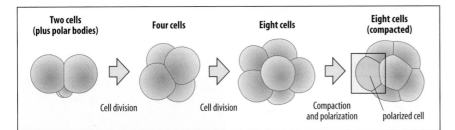

Fig. 8.9 Compaction of the mouse embryo. At the eight-cell stage the cells have relatively smooth surfaces and microvilli are distributed uniformly over the surface. At compaction, microvilli are confined to the outer surface and cells increase their area of contact with one another. Scale bar = 10 μm.

Photographs courtesy of T. Bloom, from Bloom, T.L: 1989.

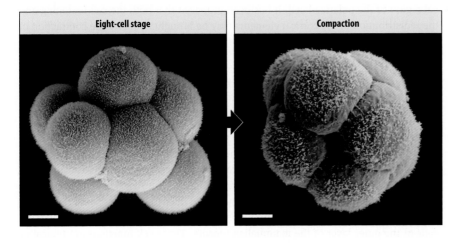

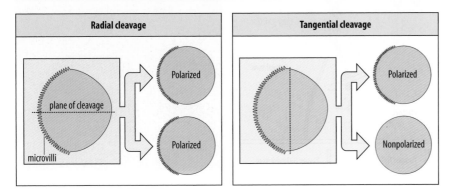

Fig. 8.10 Polarization of cells during cleavage of the mouse embryo. At the eight-cell stage, compaction takes place, with the cells forming extensive contacts with each other (top panel). The cells also become polarized. For example, microvilli, which were initially uniformly distributed over the cell surface, become confined to the outward-facing cell surface. Future cleavages can thus divide the cell either into two polarized cells (by a radial cleavage, bottom left panel) or into a polarized and a nonpolarized cell (by a tangential cleavage, bottom right panel). The tangential cleavages give rise to the inner cell mass.

the embryo proper, whereas the outer cells develop tight junctions and form the trophectoderm, the outer epithelium of the blastocyst (see Fig. 3.22, first panel).

The formation of cadherin-based adherens junctions (see Box 8B, p. 293) between blastomeres is a major feature of compaction. At the two-cell and four-cell stages, E-cadherin is uniformly distributed over blastomere surfaces and contact between cells is not extensive. At the eight-cell stage, however, E-cadherin becomes restricted to regions of intercellular contact, where it now acts for the first time as an adhesion molecule, forming adherens junctions. The change in the adhesive properties of E-cadherin results from its becoming linked to cytoskeletal elements in the cortex via β-catenin. Adheren-junction formation requires the action of protein kinase C-α. If this kinase is activated before the eight-cell stage, compaction occurs prematurely, whereas if it is inhibited, compaction fails to occur. It is likely that phosphorylation of junctional components such as β-catenin by the kinase is required for assembly of the adhesive form of the cadherin–catenin complex.

Compaction is associated with a radical remodeling of the cell cortex, which is presumably involved in the relocalization of E-cadherin and the redistribution of microvilli. The cytoskeletal proteins actin, spectrin, and myosin are cleared from the region of intercellular contact and become concentrated in a band around the apical region of the polarized cell. It has been proposed that both the cell flattening and the redistribution of the cortical elements may be brought about by the contraction of actin filaments, which draws the cortical elements to the apical pole.

8.5 Fluid accumulation as a result of tight-junction formation and ion transport forms the blastocoel of the mammalian blastocyst

Accumulation of fluid in the interior space, or blastocoel, of a blastocyst or blastula exerts an outward pressure on the blastula wall, and this hydrostatic pressure is one of the forces that forms and maintains the spherical shape. The blastocoel is formed in the preimplantation mammalian embryo by a cavitation process in which fluid accumulates in the interior of the developing blastocyst. The inflow of water is linked to the active transport of sodium ions and other electrolytes into the extracellular spaces (Fig. 8.11).

At the eight-cell stage, **tight junctions** start to be formed between the polarized cells of the outer layer of the mammalian morula. These cell junctions are typical of epithelia and form a seal that prevents water molecules, ions, and small molecules permeating in between the cells (Fig. 8.12). By the 32-cell stage this barrier is fully

Fig. 8.11 The blastocoel of the mammalian blastocyst is formed by the inflow of water creating an internal fluid-filled cavity. Left panel: active transport of sodium ions (Na$^+$) across the basolateral membranes of trophectoderm cells into the blastocoel cavity of the mouse blastocyst results in an increase in ion concentration in the extracellular fluid, which causes water to flow in from the surrounding isotonic solution. The blastocoel is sealed off from the outside medium by tight junctions between the trophectoderm cells at their outer edge, and so the inflow of water creates creates hydrostatic pressure that enlarges the blastocoel. Right panel: a surface view of a mouse blastocyst shows the tight junctions between epithelial cells.

Photograph from Eckert, J.J., Fleming, T.P.: 2008

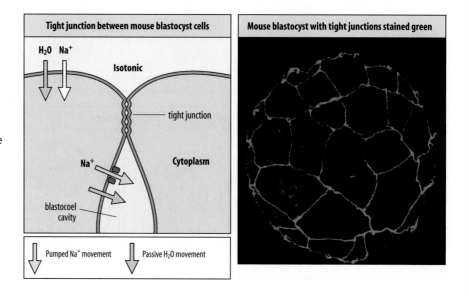

formed. At the same time, the Na$^+$/K$^+$-ATPase sodium pump, and other membrane transport proteins, become active in the basolateral membranes of the outer cells, transporting sodium and other electrolytes into the extracellular space. The cellular sodium is replaced by sodium from the fluid surrounding the blastocyst. As the ion concentration in the blastocoel fluid increases, water is drawn into the blastocoel through aquaporin water channels in the cell membranes, and the increase in hydrostatic pressure caused by the accumulating fluid stretches the surrounding epithelial layer. A similar mechanism based on ion transport seems to operate in drawing fluid into the blastocoel in the *Xenopus* blastula. In the case of *Xenopus*, a small blastocoel can be observed by electron microscopy as early as the two-cell stage.

8.6 Internal cavities can be created by cell death

Hollow structures can be created in several different ways during early embryonic development. In the case of the mammalian neural tube, for example, the tubular structure results from the folding of an epithelial sheet, as we shall see later in this chapter. We have just seen how the blastocoel is formed by a cavitation process in which water inflow enlarges an interior space. Another example of cavity formation, this time produced by cell death, occurs during the development of the mouse epithelial epiblast.

The epiblast, which gives rise to the mouse embryo, is derived from the inner cell mass (see Section 3.5). Initially, the epiblast is a solid cylindrical mass of cells, but it develops into an epithelial layer enclosing a fluid-filled cavity. Formation of this cavity results from programmed cell death—apoptosis—of the cells in the center of the epiblast (Fig. 8.13). It is likely that the surrounding cell layer—the visceral endoderm—sends a death signal to the epiblast, and that only the outer epiblast cells survive, as a result of their contact with the basement membrane. These survival signals are delivered by integrins in the cell membranes that interact with components of the basement membrane. The phenomenon of programmed cell death and its role in development is discussed further in later chapters.

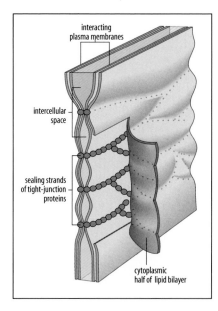

Fig. 8.12 The structure of the tight junction. The apposing membranes of adjacent epithelial cells are held closely together by the tight junction, which is located towards the apex of the cell and forms a tight seal between the extracellular and intracellular milieu. The membranes are held together by horizontal rows (sealing strands) of membrane proteins (occludins and claudins), which bind tightly to each other. *Adapted from Alberts, B., et al.: 2008.*

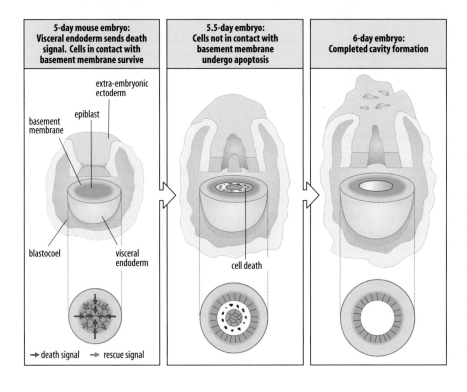

Fig. 8.13 Cavity formation in the epiblast of the mouse embryo. Left panel: the cells of the inner cell mass proliferate and form a solid epiblast (blue), which is surrounded by the visceral endoderm (yellow). Center and right panels: programmed cell death—apoptosis—in the epiblast results in the formation of a cavity (white). As shown at the bottom of the first panel, the signal for cell death die may come from the visceral endoderm (first panel). Only the epiblast cells directly attached to the basement membrane receive a rescue signal and survive (second and third panels). *After Coucouvanis, E., et al.: 1995.*

Summary

In many animals, the fertilized egg undergoes a cleavage stage that divides the egg into a number of small cells (blastomeres) and eventually gives rise to a hollow blastula. Various patterns of cleavage are found in different animal groups. In some animals, such as nematodes and molluscs, the plane of cleavage is of importance in determining the position of particular blastomeres in the embryo, and the distribution of cytoplasmic determinants. The plane of cleavage in animal cells is determined by the orientation of the mitotic spindle, which is in turn determined by the final positioning of the asters. In plant cells, the plane of cell division seems to be determined before mitosis begins. At the end of the cleavage period, the blastula essentially consists of a polarized epithelium surrounding a fluid-filled blastocoel. Fluid accumulation in the blastocoel may be partly due to active ion transport into the blastocoel drawing water in after it by osmosis. Cavity formation in the early mouse epiblast is caused by cell death.

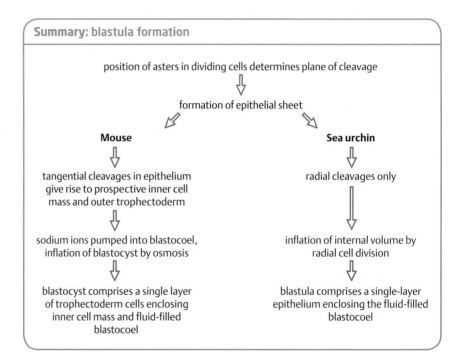

Summary: blastula formation

position of asters in dividing cells determines plane of cleavage

⇩

formation of epithelial sheet

Mouse	**Sea urchin**
⇩	⇩
tangential cleavages in epithelium give rise to prospective inner cell mass and outer trophectoderm	radial cleavages only
⇩	⇩
sodium ions pumped into blastocoel, inflation of blastocyst by osmosis	inflation of internal volume by radial cell division
⇩	⇩
blastocyst comprises a single layer of trophectoderm cells enclosing inner cell mass and fluid-filled blastocoel	blastula comprises a single-layer epithelium enclosing the fluid-filled blastocoel

Gastrulation movements

Movements of individual cells and cell sheets at gastrulation bring most of the tissues of the blastula (or its equivalent stage) into their appropriate position in relation to the body plan. The necessity for gastrulation is clear from fate maps of the blastula in, for example, sea urchins and amphibians, where the presumptive endoderm, mesoderm, and ectoderm can be identified as adjacent regions in an epithelial sheet, and the endoderm and mesoderm must therefore move from the surface to inside the early embryo (see Chapters 4–6). After gastrulation, these tissues become completely rearranged in relation to each other: for example, the endoderm develops into an internal gut and is separated from the outer ectoderm by a layer of mesoderm. Gastrulation thus involves dramatic changes in the overall structure of the embryo, converting it into a complex three-dimensional structure. During gastrulation, a program of cell activity involving changes in cell shape and adhesiveness remodels the embryo, so that the endoderm and mesoderm move inside and only ectoderm remains on the outside. The primary force for gastrulation is provided by changes in cell shape, as well as the ability of individual cells to migrate (see Box 8A, p. 291). In some embryos, all this remodeling occurs with little or no accompanying increase in cell number or total cell mass.

In this section, we first consider gastrulation mechanisms in sea urchins and insects, in which it is a relatively simple process. *Xenopus* serves as our model for the more complex gastrulation process in vertebrates.

8.7 Gastrulation in the sea urchin involves cell migration and invagination

Just before gastrulation begins, the late sea-urchin blastula consists of a single-layered ciliated epithelium surrounding a central fluid-filled blastocoel. The future mesoderm occupies the most vegetal region, with the future endoderm adjacent to it (Fig. 8.14). The rest of the embryo gives rise to ectoderm. The epithelium is polarized in an apical–basal direction: on its apical surface is the hyaline layer, while the basal surface facing the blastocoel is lined with a basal lamina. The epithelial cells are attached laterally to each other by septate junctions and adherens junctions.

Gastrulation begins with an **epithelial-to-mesenchymal transition**, with the most vegetal mesodermal cells becoming motile and mesenchymal in form. They become detached from each other and from the hyaline layer, and migrate into the blastocoel as single cells (Fig. 8.15) that have lost both their epithelial polarity and their cuboid shape. They are now known as primary mesenchyme (see Section 6.7). The transition to primary mesenchyme cells and entry into the blastocoel is foreshadowed by intense pulsatory activity on the inner face of these cells, while still part of the epithelium, and on occasion a small transitory infolding or invagination is seen in the surface of the blastula before migration begins properly. Cell internalization requires loss of cell–cell adhesion and is associated with repression of cadherin expression, the removal of cadherin from the cell surface by endocytosis, and the loss of α- and β-catenins.

The epithelial-to-mesenchyme transition is regulated in part by the sea-urchin version of the gene *snail*, which, like many other developmental genes, was first identified in *Drosophila* (see Section 2.19). Experiments using antisense morpholinos (see Box 6A, p. 220) to knock down *snail* expression in early sea-urchin embryos show that it is required for the repression of cadherin gene expression and for cadherin endocytosis in the epithelial-to-mesenchymal transition. The role of the Snail family of transcription factors in this type of cellular change is highly conserved across the animal kingdom, as we shall see in this chapter. The actions of Snail and a related

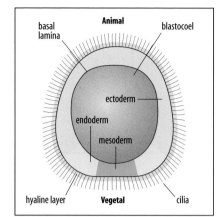

Fig. 8.14 The sea-urchin blastula before gastrulation. The prospective endoderm (yellow) and mesoderm (orange) are at the vegetal pole. The rest of the blastula is prospective ectoderm (gray). There is an extracellular hyaline layer and a basal lamina lines the blastocoel. The blastula surface is covered with cilia.

Fig. 8.15 Sea-urchin gastrulation. Cells of the vegetal mesoderm undergo the transition to primary mesenchyme cells and enter the blastocoel at the vegetal pole. This is followed by an invagination of the endoderm, which extends inside the blastocoel toward the animal pole, forming a clear archenteron. Filopodia extending from the secondary mesenchyme cells at the tip of the invaginating endoderm contact the blastocoel wall and draw the invagination to the site of the future mouth, with which it fuses, forming the gut. Scale bar = 50 μm.

Photographs courtesy of J. Morrill.

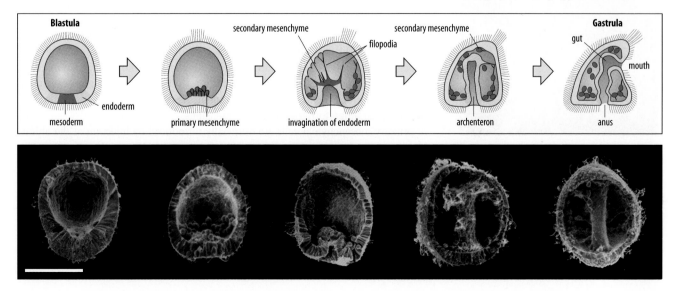

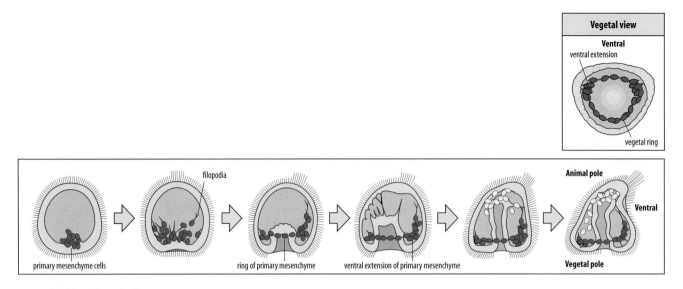

Fig. 8.16 Migration of primary mesenchyme in early sea-urchin development. The primary mesenchyme cells enter the blastocoel at the vegetal pole and migrate over the blastocoel wall by filopodial extension and contraction. Within a few hours they take up a well-defined ring-like pattern in the vegetal region with extensions along the ventral side.

transcription factor, Slug, are also of medical interest. A similar epithelial-to-mesenchymal transition takes place in cancer cells, enabling them to undergo metastasis and migrate from the original site of the tumor to set up secondary cancers elsewhere in the body. In these metastatic cells, the normal low expression of *snail* and *slug* has been increased.

After entry, the primary mesenchymal cells migrate within the blastocoel to form a characteristic pattern on the inner surface of the blastocoel wall. They first become arranged in a ring around the gut in the vegetal region at the ectoderm–endoderm border. Some then migrate to form two extensions towards the animal pole on the ventral (oral) side (Fig. 8.16). The migration path of individual cells varies considerably from embryo to embryo, but their final pattern of distribution is fairly constant. The primary mesenchyme cells later lay down the skeletal rods of the sea urchin endoskeleton by secretion of matrix proteins (see Section 6.12).

Primary mesenchymal cells move over the inner surface of the blastocoel wall by means of fine filopodia, which can be up to 40 μm long and can extend in several directions. Filopodia contain cross-linked bundles of actin filaments and extend forward by the rapid assembly of actin filaments that push out the leading edge of the filopodium. At any one time, each cell has on average six filopodia, most of which are branched. When filopodia make contact with, and adhere to, the basal lamina lining the blastocoel wall, they retract, drawing the cell body toward the point of contact. As each cell extends several filopodia (Fig. 8.17), some or all of which may retract on contact with the wall, there seems to be competition between the filopodia, the cell being drawn towards that region of the wall where the filopodia make the most stable contact. The movement of the primary mesenchymal cells therefore resembles a random search for the most stable attachment. As the cells migrate, their filopodia fuse with each other, forming cable-like extensions.

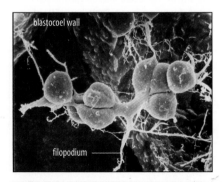

Fig. 8.17 Filopodia of sea-urchin mesenchyme cells. The scanning electron micrograph shows a group of primary mesenchyme cells, some of which have fused together, moving over the blastocoel wall by means of their numerous filopodia, which can extend and contract.

Photograph courtesy of J. Morrill, from Morrill, J.B., et al.: 1985.

If primary mesenchymal cells are artificially introduced by injection at the animal pole, they move in a directed manner to their normal positions in the vegetal region. This suggests that guidance cues, possibly graded, are distributed in the basal lamina. Even cells that have already migrated will migrate again to form a similar pattern when introduced into a younger embryo. The stability of contacts between filopodia and the blastocoel wall is one factor determining the pattern of cell migration. Analysis of movies of migrating cells suggests that the most stable contacts are made in the regions where the cells finally accumulate, namely the vegetal ring and the two ventro-lateral clusters (see Fig. 8.16). The signaling molecules fibroblast growth factor A (FGFA) and vascular epithelial growth factor (VEGF) have been implicated as possible guidance cues. In the blastula, FGFA is expressed in ectodermal regions to which the primary mesenchyme will migrate and the FGF receptor is expressed on the primary mesenchyme cells. If signaling by FGFA is inhibited, then so is cell migration and skeleton formation. Studies in vertebrates have shown that FGFs can attract migrating mesenchymal cells, and it is likely that they can also do so in the sea urchin. VEGF is expressed by the prospective ectoderm cells at the most ventral sites to which the mesenchyme cells migrate, while the VEGF receptor is again expressed exclusively in the prospective primary mesenchyme. As with FGF, if VEGF or its receptor are lacking, the cells do not reach their correct positions and the skeleton does not form.

The entry of the primary mesenchyme is followed by the invagination and extension of the endoderm to form the embryonic gut (the archenteron). The endoderm invaginates as a continuous sheet of cells (see Fig. 8.15). Formation of the gut occurs in two phases. During the initial phase, the endoderm invaginates to form a short, squat cylinder extending up to halfway across the blastocoel. There is then a short pause, after which extension continues. In this second phase, the cells at the tip of the invaginating gut, which will later detach as the secondary mesenchyme, form long filopodia, which make contact with the blastocoel wall. Filopodial extension and contraction pull the elongating gut across the blastocoel until it eventually comes in contact with and fuses with the mouth region, which forms a small invagination on the ventral side of the embryo (see Fig. 8.15). During this process the number of invaginating cells doubles, and this is in part due to cells around the site of invagination contributing to the hindgut.

How is invagination of the endoderm initiated? The simplest explanation is that a change in shape of the endodermal cells causes, and initially maintains, the change in curvature of the cell sheet (Fig. 8.18). At the site of invagination, the cuboid cells adopt a more elongated, wedge-shaped form, which is narrower at the outer (apical) face. This change in cell shape is the result of constriction of the cell at its apex by contraction of actomyosin in the cortex (see Box 8A, p. 291). Such a shape change would be sufficient to pull the outer surface of the cell sheet inward and to maintain invagination, as shown by a computer simulation in which apical constriction, modeled to spread over the vegetal pole, results in an invagination (Fig. 8.19).

The primary invagination only takes the gut about a third of the way to its final destination. The second phase of gastrulation involves two different mechanisms: contraction of filopodia in contact with the future mouth region, as described above, and a filopodia-independent process. Treatments that interfere with filopodial attachment to the blastocoel wall result in failure of the gut to elongate completely, but it still reaches about two-thirds of its complete length. Filopodia-independent extension is due to active rearrangement of cells within the endodermal sheet. If a sector of cells in the vegetal mesoderm is labeled with a fluorescent dye before gastrulation, it is seen to become a long, narrow strip as the gut extends (Fig. 8.20). The sheet of cells has elongated in one direction while becoming narrower in width. We will look at this type of cell rearrangement—known as convergent extension—in more detail later in this chapter, in relation to similar phenomena in amphibian and zebrafish gastrulation (see Box 8C, p. 310).

What guides the tip of the gut to the future mouth region? As the long filopodia at the tip of the gut initially explore the blastocoel wall, they make more stable contacts

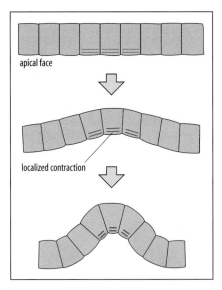

Fig. 8.18 Change in cell shape in a small number of cells can cause invagination of endodermal cells. Bundles of contractile filaments composed of actin and myosin contract at the outer edge of a small number of adjacent cells, making them wedge-shaped. As long as the cells remain mechanically linked to each other and to adjacent cells in the sheet, this local change in cell shape draws the sheet of cells inward at that point.

Fig. 8.19 Computer simulation of the role of apical constriction in invagination. Computer simulation of the spreading of apical constriction over a region of a cell sheet shows how this can lead to an invagination.

Illustration after Odell, G.M., et al: 1981.

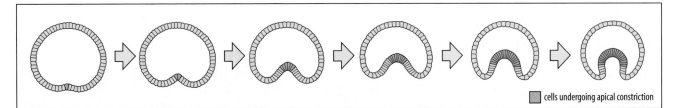

cells undergoing apical constriction

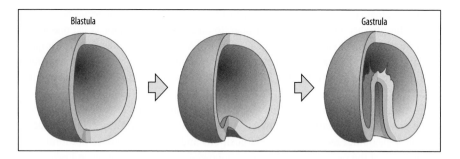

Fig. 8.20 Gut extension during sea-urchin gastrulation. Labeling of cells in the vegetal region of the blastula (green) shows that gut extension involves cell rearrangement within the endoderm, causing the labeled cells to become distributed in a long narrow strip.

at the animal pole and then in the region where the future mouth will form. Filopodia making contact there remain attached for 20–50 times longer than when attached to other sites on the blastocoel wall. Gastrulation in the sea urchin clearly shows how changes in cell shape, changes in cell adhesiveness, and cell migration all work together to cause a major change in embryonic form.

8.8 Mesoderm invagination in *Drosophila* is due to changes in cell shape that are controlled by genes that pattern the dorso-ventral axis

At the time gastrulation begins, the *Drosophila* embryo consists of a blastoderm of about 6000 cells, which forms a superficial single-celled layer (see Section 2.1). Gastrulation begins with the invagination of a longitudinal strip of future mesodermal cells (8–10 cells wide) on the ventral side of the embryo, to form a ventral furrow and then a tube inside the body. The tube breaks up into individual cells, which spread out to form a single layer of mesoderm on the interior face of the ectoderm (Fig. 8.21). The gut develops slightly later, by invaginations of prospective endoderm near the anterior and posterior ends of the embryo (see Fig. 2.3), which we shall not consider here.

Invagination of the mesoderm occurs in two phases. First, the central strip of cells develops flattened and smaller apical surfaces, and their nuclei move away from the apical surface. These changes in cell shape result in formation of the furrow and its folding into the interior of the embryo to form a tube, the central cells forming the tube proper while the peripheral cells form a 'stem' (see Fig. 8.21, center). This first phase takes about 30 minutes. During the second phase, which takes about an hour, the tube dissociates into individual cells, which proliferate and spread out laterally. The entry of cells into cell division is delayed during invagination and this delay is essential for the cell-shape changes to occur.

Mutant *Drosophila* embryos that have been dorsalized or ventralized (see Section 2.12) show that this behavior of the mesodermal cells is autonomous, as in the sea urchin (see Section 8.7), and is not affected by adjacent tissues. In dorsalized embryos, which have no mesoderm, no cells undergo nuclear migration or apical contraction. In ventralized embryos, in which most of the cells are mesoderm, these changes occur throughout the whole of the dorso-ventral axis.

Fig. 8.21 *Drosophila* gastrulation. Transverse sections of a *Drosophila* embryo undergoing gastrulation. The mesodermal cells (stained brown) are present in a longitudinal strip in the epithelium on the ventral side of the embryo. A change in their shape from cuboid to wedge-shaped, causes an invagination, which develops into a ventral furrow (left panel). Further apical contraction of the cells results in the mesoderm forming a tube on the inside of the embryo (center panel). The mesodermal cells then start to migrate individually to different sites (right panel). Scale bars = 50 μm.

Photographs courtesy of M. Leptin, from Leptin, M., et al.: 1992.

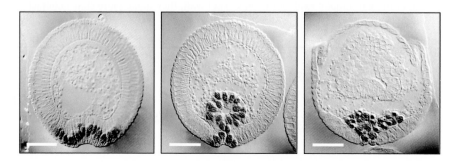

Gastrulation in *Drosophila* provided the first possibility of linking the action of patterning genes directly to morphogenesis, in this case by their involvement in causing changes in cell shape and cell adhesiveness. Mesoderm invagination is affected by mutations in the genes *twist* and *snail*, which are expressed in the prospective mesoderm before gastrulation as part of the patterning of the dorso-ventral axis (see Section 2.19). By visualizing the outlines of ventral cells by labeling cell membranes with green fluorescent protein, apical constriction of the prospective mesoderm can be seen to occur as a result of successive pulses of contraction. Labeling of myosin and actin showed that the constriction was due to pulsed contractions of an actomyosin network that formed under the apical cell surface. No contractile network formed at all in loss-of-function *snail* mutants, whereas in *twist* mutants, pulsed contractions still occurred but they did not produced apical constriction. Further investigations confirmed that while Snail was required to initiate the whole process, Twist was required to stabilize the state of contraction between pulses. In the absence of Twist function, the cell surface relaxes back to its original area after each contraction.

Twist and Snail also control the next stage of mesoderm internalization. When the central part of the mesoderm is fully internalized, its cells make contact with the ectoderm. Adherens junctions that had been established in the blastoderm are now lost—one of the results of Snail activity (see Section 8.7)—and the mesoderm undergoes an epithelial-to-mesenchymal transition, dissociating into individual cells that spread out to cover the ectoderm as a single-cell layer. Remarkably like the situation in the sea urchin, FGF signals from the ectoderm acting on an FGF receptor on the mesodermal cells are needed for the invaginating cells to spread. Failure to activate the FGF signaling pathway results in the mesodermal cells remaining near the site of invagination. Expression of the FGF receptor in the mesoderm is induced by Twist and Snail, while Snail also represses expression of FGF in the mesoderm, thus restricting FGF to the ectoderm.

Mesodermal cell spreading involves the cells switching from expression of ectodermal E-cadherin to N-cadherin, so that they no longer adhere to ectodermal cells. The change from E- to N-cadherin is also under the control of Snail and Twist. The actions of Snail suppress the production of E-cadherin in the mesoderm, while expression of Twist induces the production of N-cadherin.

8.9 Germ-band extension in *Drosophila* involves myosin-dependent remodeling of cell junctions and cell intercalation

Another dramatic change in shape that occurs in the *Drosophila* embryo is the extension of the germ band (see Fig. 2.4), which leads to an almost doubling in the length of the epithelial layer that forms the thorax and abdomen of the embryo. This extension is driven not by cell division or changes in cell shape, but by convergent extension of the ventral part of the epithelium. Adjacent cells intercalate between each other laterally—towards the midline—so that the tissue narrows, causing it to extend in the antero-posterior direction. This remodeling occurs over the whole of the ventral epithelium and is due to regulated changes in the cadherin-containing adherens junctions that normally hold the cells tightly together.

At the beginning of germ-band extension, the epidermal cells are packed in a regular hexagonal pattern, with their boundaries either parallel to the dorso-ventral axis or at 60° to it, as shown in Fig. 8.22. When extension starts, the adherens junctions on the faces parallel to the dorso-ventral axis shrink and disappear and the cells become diamond-shaped. New junctions parallel to the antero-posterior axis appear, making the boundaries hexagonal again and intercalating the cells along the dorso-ventral axis—which makes the tissue extend along the antero-posterior axis. This ability to undergo convergent extension along the antero-posterior axis reflects a prior antero-posterior patterning of cells in the epidermal sheet.

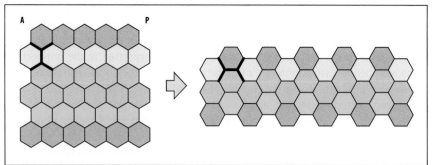

Fig. 8.22 Junction remodeling enables cells to undergo intercalation in *Drosophila* germ-band extension. The disassembly and reassembly of intercellular contacts in the germ-band epithelium drives the tissue to elongate along the antero-posterior axis and narrow along the dorso-ventral axis. The heavy black outline indicates how the four cells in contact with each other in the left-hand diagram have changed their intercellular contacts after junction remodeling (right-hand diagram).

Adapted from Bertet, C., et al.: 2004.

The mechanism of intercalation in the germ band involves the regulated localization and activity of myosin, which co-localizes with β-catenin–E-cadherin–actin complexes at the adherens junctions (see Box 8B, p. 293) and is enriched in shrinking junctions. The proposed mechanism is that regulated myosin contraction at the junctions prevents E-cadherin holding the cells together, enabling new contacts to be made that result in cell intercalation.

8.10 Dorsal closure in *Drosophila* and ventral closure in *Caenorhabditis elegans* are brought about by the action of filopodia

At about 11 hours after the beginning of *Drosophila* embryogenesis, when gastrulation is complete and the germ band has retracted again, the amnioserosa on the dorsal surface of the embryo is not yet covered by the epidermis. The epidermis now moves to close this gap, zipping-up the opening from both ends (Fig. 8.23). About 2 hours after this process starts, the two epithelial fronts have fused to form a seam along the dorsal midline, with the segment pattern exactly matched. The movement of the epidermis is due to the action of filopodia and lamellipodia, which extend and retract from the cells along the edges of the epithelial sheet. Filopodia are active along the whole epithelial front, extending up to 10 μm from the edge, and their behavior is similar to that seen in sea urchin gastrulation (see Section 8.7). Laser ablation of filopodial cells does not prevent closure, however, which is brought about by actomyosin-based contraction at the leading edges of the epidermal sheet and of the amnioserosa. The small GTPase Rac plays a key role in organizing actin filaments at the leading faces of the cells, and is also involved through the Jun N-terminal kinase (JNK) intracellular signaling pathway in controlling changes in cell shape and fusion. Dorsal closure occurs with remarkable accuracy, almost down to single-cell resolution. This appears to be achieved by cells at each edge that express the same genes identifying and fusing with each other. For example, the gene *engrailed* is expressed in a series of transverse stripes along the length of the embryo, corresponding to the posterior compartment of each segment (see Section 2.23). Cells expressing *engrailed* on one edge interact only with cells expressing *engrailed* on the other edge, and cells expressing the gene *patched*, which form a different set of stripes, fuse only with *patch*-expressing cells on the other edge, thus neatly matching up the stripes across the seam.

The *C. elegans* embryo undergoes a process similar to dorsal closure, but on the ventral surface. At the end of gastrulation, the epidermis only covers the dorsal region

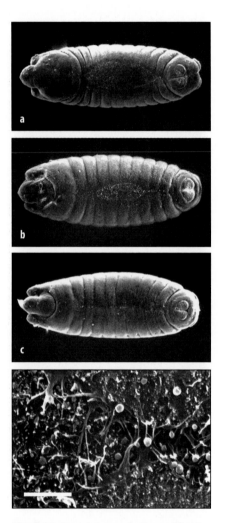

Fig. 8.23 Dorsal closure in *Drosophila*. The closure of the epidermis along the dorsal midline over a period of about 2 hours is shown in the micrographs in panels a–c. Starting at both ends of the opening, the edges of the epidermis come together over the amnioserosa (seen as the central oval area in a and b) and fuse to form the midline seam, matching the segment pattern on either side. The bottom panel shows filopodia acting as a zipper.

Photographs courtesy of A. Jacinto, from Jacinto, A., et al.: 2002.

and the ventral region is bare. The epidermis then spreads around the embryo until its edges finally meet along the ventral midline. Time-lapse studies show that the initial ventral migration is led by four cells, two on each side, which extend filopodia towards the ventral midline. Blocking filopodial activity by laser ablation or by cytochalasin D, which blocks the formation of actin filaments, prevents ventral closure, showing that the filopodia provide the driving force. Closure probably occurs by a zipper mechanism similar to that in the fly.

8.11 Vertebrate gastrulation involves several different types of tissue movement

Gastrulation in vertebrates involves a much more dramatic and complex rearrangement of tissues than in sea urchins, because of the need to produce a more complex body plan. In amphibians, fish, and birds, there is also the added complication of the presence of large amounts of yolk. But the outcome is the same: the transformation of a two-dimensional sheet of cells into a three-dimensional embryo, with ectoderm, mesoderm, and endoderm in the correct positions for further development of body structure. The main movements of gastrulation that we will discuss here are **involution**, which is the rolling-in of coherent sheets of endoderm and mesoderm at the blastopore, as occurs in *Xenopus* (Fig. 8.24); **epiboly**, which is the spreading and thinning of the ectodermal layers as the endoderm and mesoderm move inside and which can be observed in gastrulating frogs and zebrafish; and **convergent extension** (Box 8C, p. 310), which elongates the body axis. Gastrulation in mammals and birds occurs in the primitive streak (see Chapters 3 and 4) and involves the convergence of epiblast cells on the midline, the separation or **delamination** of cells individually from the epiblast, and their internalization, or **ingression**, followed by migration internally and convergent extension. We will begin with gastrulation in *Xenopus*.

In the late *Xenopus* blastula, the presumptive endoderm extends from the most vegetal region to cover the presumptive mesoderm. During gastrulation, the presumptive endoderm moves inside through the blastopore to line the gut, while all the cells in the equatorial band of mesoderm move inside to form a layer of mesoderm underlying the ectoderm, extending antero-posteriorly along the dorsal midline of the embryo (Fig. 8.25).

Gastrulation starts at a site on the dorsal side of the blastula, toward the vegetal pole. The first visible sign is the formation of bottle-shaped cells (bottle cells, see Fig. 8.25, second panel) by some of the presumptive mesodermal cells. The formation of the bottle shape is due to apical constriction of the cell (see Box 8A, p. 291). As in the sea urchin and *Drosophila*, this change in cell shape forms a groove in the blastula surface. In *Xenopus*, this starts as a small groove—the blastopore—whose dorsal lip is the site of the Spemann organizer. The layer of mesoderm and endoderm starts to involute around the blastopore, rolling into the interior of the blastula against its own under surface (see Fig. 8.25). As gastrulation proceeds, the region of involution

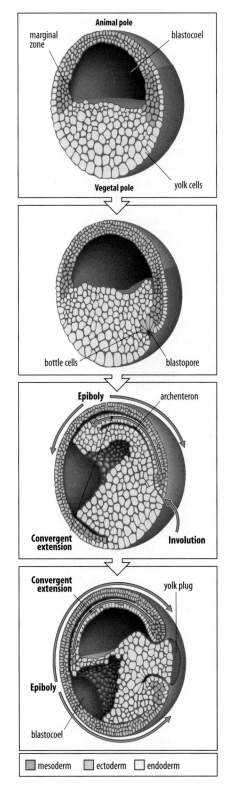

Fig. 8.24 Tissue movements during gastrulation of *Xenopus*. In the late blastula the future mesoderm (red) is in the marginal zone, overlaid by presumptive endoderm (yellow). Ectoderm is shown in blue. Gastrulation is initiated by the formation of bottle cells in the blastopore region, which is followed by the involution of mesoderm over the dorsal lip of the blastopore. Marginal zone endoderm and mesoderm move inside over the dorsal lip of the blastopore (events in this region are shown in more detail in Fig. 8.25). The marginal zone endoderm, which was on the surface of the blastula, now lies ventral to the mesoderm and forms the roof of the archenteron or future gut. At the same time, the ectoderm of the animal cap spreads downward. The mesoderm converges and extends along the antero-posterior axis. The region of involution spreads ventrally to include more endoderm, and forms a circle around a plug of yolky vegetal cells. The ectoderm spreads by epiboly.

After Balinsky, B.I.: 1975.

Box 8C Convergent extension

Convergent extension plays a key role in gastrulation and other morphogenetic processes. It is a mechanism for elongating a sheet of cells in one direction while narrowing its width, and occurs by rearrangement of cells within the sheet, rather than by cell migration or cell division (see top figure). It occurs, for example, in the extension of the mesoderm that elongates the antero-posterior axis in

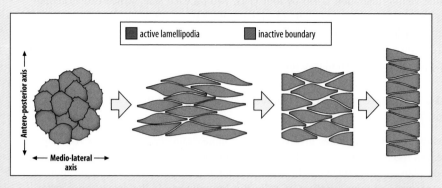

amphibian embryos (see Fig. 8.27). For convergent extension to take place, the axes along which the cells will intercalate and extend must already have been defined. The cells first become elongated in a direction at right angles to the antero-posterior axis—the medio-lateral direction. They also become aligned parallel to one another in a direction perpendicular to the direction of tissue extension (see top figure). Active movement is largely confined to each end of these elongated bipolar cells, which form active protrusions or filopodia,

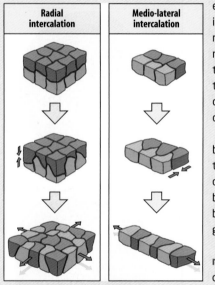

enabling them to exert traction on neighboring cells on each side and on the underlying substratum and to shuffle in between each other—or intercalate—always along the medio-lateral axis, with some cells moving medially and some laterally. This rearrangement is known as **medio-lateral interacalation**. There are some small protrusions on the anterior and posterior faces of the cells but these seem to hold apposed surfaces together rather than being involved in cell movement. At a tissue boundary, one end of the cell is captured and movement ceases, but cells can still exert traction on each other to draw the boundaries closer together.

The boundary of a tissue undergoing convergent extension is maintained by the behavior of cells at the boundary. When an active cell tip is at the boundary of the tissue, its movement ceases and the cell becomes monopolar (see top figure), with only the tip facing into the tissue still active. It is as if the cell tip that has reached the boundary has become fixed there. Precisely what defines a boundary is not clear, but boundaries between mesoderm, ectoderm, and endoderm seem to be specified before gastrulation commences.

A mechanically similar process called **radial intercalation** (see figure on left) causes the thinning of a multi-cell layer into a thinner sheet and its consequent extension around the edges, as seen during epiboly of the ectoderm during frog and zebrafish gastrulation (see, for example, Fig. 8.24). Radial intercalation occurs in the multilayered ectoderm of the animal cap, in which cells intercalate in a direction perpendicular to the surface, moving from one layer into that immediately above. This leads to an increase in the surface area of the cell sheet, as well as its thinning, and is in part the cause of epiboly.

The molecular mechanisms controlling convergent extension in vertebrates are related to those controlling planar cell polarity in *Drosophila* (see Box 2F, p. 78), and involve signaling by Wnts through Frizzled via the non-canonical Wnt planar cell polarity pathway (see figure on right). The much-simplified pathway shown here is a general compilation from different organisms. Signaling by Wnt through its receptor Frizzled (Fz) via Dishevelled activates RhoA and the Rho kinase Rok2, leading to changes in the cell's actin cytoskeleton. The transmembrane protein Van Gogh (Vang, also called Strabismus) and other proteins identified from *Drosophila* are also components of the vertebrate pathway, although how they all interact with each other is not yet clear. Vang is thought to interact with Dishevelled to stimulate the activation of another protein kinase, the Jun N-terminal kinase (JNK), which can activate the transcription factor AP1.

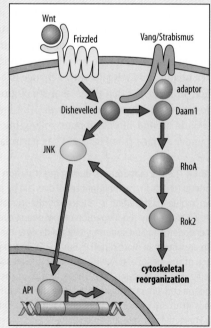

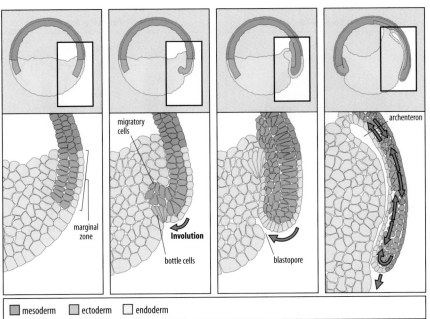
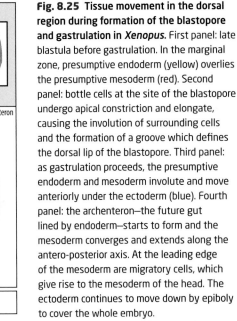

migratory cells

archenteron

marginal zone

Involution

bottle cells

blastopore

☐ mesoderm   ☐ ectoderm   ☐ endoderm

Fig. 8.25 Tissue movement in the dorsal region during formation of the blastopore and gastrulation in *Xenopus*. First panel: late blastula before gastrulation. In the marginal zone, presumptive endoderm (yellow) overlies the presumptive mesoderm (red). Second panel: bottle cells at the site of the blastopore undergo apical constriction and elongate, causing the involution of surrounding cells and the formation of a groove which defines the dorsal lip of the blastopore. Third panel: as gastrulation proceeds, the presumptive endoderm and mesoderm involute and move anteriorly under the ectoderm (blue). Fourth panel: the archenteron—the future gut lined by endoderm—starts to form and the mesoderm converges and extends along the antero-posterior axis. At the leading edge of the mesoderm are migratory cells, which give rise to the mesoderm of the head. The ectoderm continues to move down by epiboly to cover the whole embryo.

After Hardin, J.D., et al.: 1988.

spreads laterally and ventrally to involve the vegetal endoderm, and the blastopore eventually forms a circle around a plug of yolky cells (see Fig. 8.24). In time, the blastopore contracts, forcing the yolky cells into the interior where they form the floor of the gut.

The first mesodermal cells to involute migrate as individual cells over the blastocoel roof to give rise to very anterior mesodermal structures in the head. Behind them is mesoderm that enters, together with the overlying endoderm, as a single multilayered sheet. For this cell sheet, going through the narrow blastopore is rather like going through a funnel, and the cells become rearranged by convergent extension, which is a major feature of gastrulation. The mesoderm is initially in the form of an equatorial ring, but during gastrulation it converges and extends along the antero-posterior axis (Fig. 8.26). Thus, cells that are initially on opposite sides of the embryo come to lie next to each other (Fig. 8.27), whereas others become separated from each other along the long axis of the embryo. The migrating mesoderm cells are polarized in the direction of the animal pole and their migration depends on interaction with the fibronectin fibrils in the extracellular matrix lining the blastocoel roof. The assembly of fibronectin fibrils in the matrix is due to tension at the surface of the blastocoel roof epithelium, caused by cell–cell adhesion, being transmitted via integrins in the cell membrane that interact with fibronectin.

In *Xenopus*, convergent extension occurs in both the mesoderm and endoderm as they involute, and in the future neural ectoderm that gives rise to the spinal cord. Together with the elongation of the notochord, all these processes elongate the embryo in an antero-posterior direction. One can appreciate the dramatic nature of convergent extension by viewing the early gastrula from the blastopore. The future mesoderm can be identified by the expression of the gene *Brachyury*, and can be seen as a narrow ring around the blastopore (Fig. 8.28). As the *Brachyury*-expressing ring of tissue enters the gastrula it converges into a narrow band along the dorsal midline of the embryo, extending in the antero-posterior direction, and eventually only the prospective notochordal mesoderm along the midline itself continues to express *Brachyury*.

As gastrulation proceeds, the mesoderm comes to lie immediately beneath the ectoderm, while the endoderm immediately below it lines the roof of the archenteron, the future gut (see Fig. 8.25). Even though they are in contact, the mesoderm and ectoderm do not adhere to each other and so can continue to move independently of

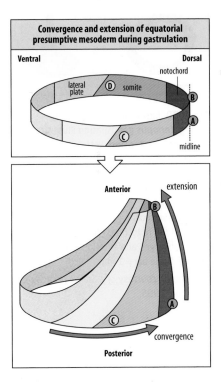

Convergence and extension of equatorial presumptive mesoderm during gastrulation

Ventral   Dorsal

notochord

lateral plate   D   somite   B

A

C   midline

Anterior   extension

B

A

C   convergence

Posterior

Fig. 8.26 Convergent extension of the mesoderm. Initially the mesoderm (red, orange and yellow) is in an equatorial ring, but during gastrulation it converges and extends along the antero-posterior axis. A–D are reference points, used to illustrate the extent of movement during convergent extension. In the bottom panel, D is hidden behind C.

Fig. 8.27 The rearrangement of mesoderm and endoderm during gastrulation in *Xenopus*. The presumptive mesoderm (red and orange) is present in a ring around the blastula underlying the endoderm (yellow), which is shown peeled back. During gastrulation, both these tissues move inside the embryo through the blastopore and completely change their shape, converging and extending along the antero-posterior axis, so that points A and B move apart. Thus, the two points C and D, originally on opposite sides of the blastula, come to lie next to each other. In the neural ectoderm (blue) there is also convergent extension, as shown by the movement of points E and F. Note that little cell division or cell growth has occurred in these stages, and all these changes have occurred by rearrangement of cells within the tissues.

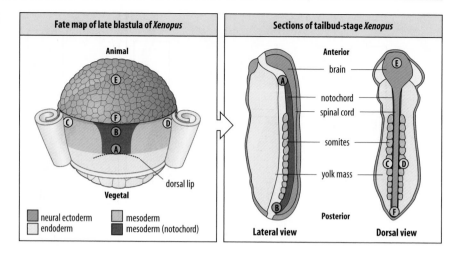

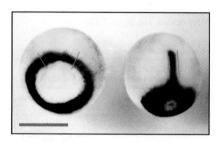

Fig. 8.28 Expression of *Brachyury* during *Xenopus* gastrulation illustrates convergent extension. Left: before gastrulation, the expression of *Brachyury* (dark stain) marks the future mesoderm, which is present as an equatorial ring when viewed from the vegetal pole. Right: as gastrulation proceeds, the mesoderm that will form the notochord (delimited by blue lines) converges and extends along the midline. Scale bar = 1 mm.

Photograph courtesy of J. Smith, from Smith, J.C., et al.: 1995.

each other. As the mesoderm and endoderm move inside, the ectoderm of the animal cap region spreads down over the vegetal region by epiboly and will eventually cover the whole embryo. Epiboly involves both stretching of the cells themselves and cell rearrangements, in which cells lower down in the ectodermal layers intercalate between the cells immediately above them—a process called **radial intercalation** (see Box 8C, p. 310). Together, these processes lead to thinning of the ectoderm layer and an increase in its surface area.

Gastrulation in the zebrafish has both similarities and differences to that of *Xenopus*. It begins by the blastoderm at the animal pole starting to spread out over the yolk cell and down towards the vegetal pole. As in *Xenopus*, this epibolic spreading is due to cells of the deep layer undergoing radial intercalation, leading to the layer becoming thinner and thus increasing its surface area (Fig. 8.29, left panel). By the time the blastoderm margin reaches about half-way down the yolk cell, the deep-layer cells have accumulated around the blastoderm edge to form a thickening known as the embryonic germ ring, and have formed two cell layers—prospective ectoderm overlying mesendoderm—under the superficial enveloping layer. The mesendodermal cells now move internally at the edge of the germ ring and then move upwards under the ectoderm towards the future anterior end of the embryo, forming a layer called the hypoblast (Fig. 8.29, center panel). The ectoderm, now sometimes called the epiblast, continues to undergo epiboly and move downwards. The next step is convergent extension of the mesendodermal and ectodermal layers towards the dorsal midline and along the future antero-posterior axis, which elongates the embryo (Fig. 8.29, right panel).

In amphibian and zebrafish gastrulation, cells undergoing convergent extension along the antero-posterior axis are polarized in the medio-lateral direction (see Box 8C, p. 310). The contractile forces generated at the tips of the elongated cells both move the cells towards the midline and extend the tissue. One can visualize the process of convergent extension in terms of lines of cells pulling on one another in a direction at right angles to the direction of extension. Because the cells at the boundary are trapped at one end, this tensile force causes the tissue to narrow and hence extend anteriorly. In fish embryos, convergent extension also involves an initial directed migration of individual mesodermal cells from lateral regions towards the midline, a movement known as **dorsal convergence**, followed by medio-lateral intercalation as mesodermal cells become incorporated into the future notochord at the midline (Fig. 8.30).

The presumptive notochord mesoderm is among the first of the tissues to internalize during gastrulation, and in amphibians it is the first of the dorsal structures to differentiate. Initially it can be distinguished from the adjacent somitic mesoderm by

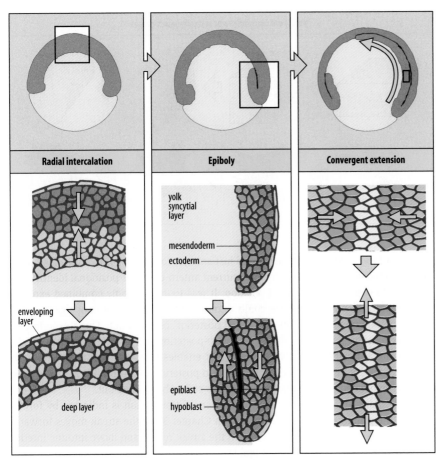

Radial intercalation | **Epiboly** | **Convergent extension**

yolk
syncytial
layer

mesendoderm

ectoderm

enveloping
layer

deep layer

epiblast

hypoblast

Fig. 8.29 Zebrafish gastrulation movements. Left panel: the first sign of gastrulation in the zebrafish embryo is spreading of the blastoderm over the surface of the yolk cell. This epiboly is caused by a thinning of the deep layer by a radial intercalation process. Center panel: when the blastoderm has reached about half the way down the yolk cell, separate layers of prospective mesendoderm (brown) and ectoderm (light blue) have formed and the mesendoderm starts to internalize. Ectoderm does not internalize and will continue to move down and spread out over the yolk cell. It will eventually cover the whole embryo. Right panel: convergent extension (blue arrow) of the mesendoderm towards the dorsal midline results in elongation of the embryo.

After Montero, J.A., et al.: 2004.

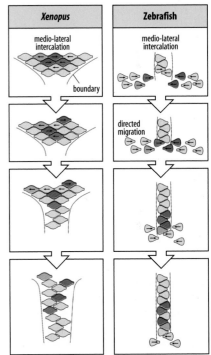

Xenopus | Zebrafish

medio-lateral intercalation | medio-lateral intercalation

boundary

directed migration

Fig. 8.30 Convergent extension in *Xenopus* and zebrafish. The left panels show convergent extension of mesoderm in *Xenopus* during the formation of the notochord, which occurs by medio-lateral intercalation of cells in a coherent tissue sheet. The right panels show formation of the notochord by convergent extension in the zebrafish, which occurs by directed movement of loosely packed mesenchymal mesodermal cells from the lateral regions towards the midline, insertion into the future notochord and medio-lateral intercalation within the notochord boundary.

After Wallingford, J.B., et al.: 2002.

the packing of its cells, and by the slight gap that forms the boundary between the two tissues, possibly reflecting differences in cell adhesiveness. The notochord elongates and develops into a stiff rod composed of a stack of thin flat cells shaped like pizza slices.

After an initial elongation by medio-lateral intercalation and convergent extension, the *Xenopus* notochord undergoes a further dramatic narrowing in width accompanied by an increase in height (Fig. 8.31). The cells become elongated at right angles to the main antero-posterior axis, foreshadowing the later pizza-slice arrangement of cells (see Fig. 8.31, bottom). The cells intercalate between their neighbors again, thus causing further convergent extension. A later stage in notochord development, which stiffens and further elongates the rod of mesoderm, involves a process called directed dilation and is discussed later in this chapter (Section 8.14).

The medio-lateral polarization of individual cells requires the vertebrate version of the Wnt-mediated planar cell polarity pathway (see Box 8C, p. 310). When this pathway is rendered non-functional, *Xenopus* mesodermal cells fail to elongate or to develop polarized lamellipodia. But another polarization system is also at work in the convergent extension of the vertebrate body axis, which maintains the correct direction of extension in this population of continually rearranging and moving cells. Before gastrulation, the organizer mesoderm is patterned along the antero-posterior axis as reflected, for example, in a gradient of Chordin expression in the prospective notochord mesoderm—the chordamesoderm—which helps to pattern the overlying neural ectoderm by antagonizing the activity of BMPs (see Section 5.2). Antero-posterior patterning seems to be a prerequisite for cell intercalation and convergent extension. This can be demonstrated by disaggregating explants of 'anterior' and 'posterior' chordamesoderm from a *Xenopus* early gastrula and then mixing the cells together.

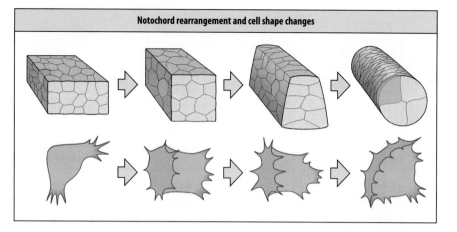

Fig. 8.31 Changes in cell arrangement and cell shape during notochord development in *Xenopus*. During convergent extension of the notochord, its height increases. At the same time, the cells become elongated. The bottom of the figure shows the eventual arrangement of cells, and their pizza-slice shape.

After Keller, R., et al.: 1989.

The chordamesoderm is situated at and just anterior to the upper lip of the blastopore. The cells recognize each other by their current antero-posterior positional identity and sort out to end up in the correct position. It was found that only combined explants of anterior and posterior tissue could undergo intercalation and convergent extension; combinations of 'all anterior' or 'all posterior' cells did not. These experiments suggest that the patterning along the antero-posterior axis is required to set up the medio-lateral polarity in individual cells that enables them to intercalate. How cells integrate the cues they receive from the antero-posterior polarization system and from the Wnt-mediated planar polarity system has still to be worked out.

In the chick and the mouse (and in humans), gastrulation is indicated by formation of the primitive streak, as described in Chapter 3. As the streak moves forward, epiblast cells separate from each other in the streak region and move into the interior as individual cells. The epithelial epiblast—the future ectoderm—gives rise to both the mesoderm and endoderm, as well as the ectoderm. Epiblast cells become specified as mesodermal and endodermal cells in the streak and move into the interior.

The mechanisms of streak formation, movement of epiblast cells, and delamination are not as clearly understood as are the involution movements in *Xenopus*. One view is that convergence at the posterior end of the streak provides much of the force for streak extension, and that the contraction and delamination of cells in the streak draws lateral epiblast cells towards and into the streak. In the chick epiblast, primitive-streak formation is preceded by extensive bilaterally symmetric movements of anterior epiblast cells outwards from the midline and backwards towards the posterior edge where the streak will begin (Fig. 8.32). These move-

Fig. 8.32 Polonaise movements in the chick epiblast during primitive-streak formation. Left panel: cell movements in the epiblast before laying start to bring cells towards the site at which the primitive streak will form. Second panel: around 6–7 hours after laying (stage 2 embryo) cells in the epiblast undergo characteristic extensive 'polonaise' movements outwards from the midline and back towards the start of the primitive streak; the region forming the streak starts to converge and extend by medio-lateral intercalation (see Fig. 8.30 and Box 8C, p. 310). Third panel: at around 12–13 hours after laying (stage 3) the streak has extended to about half its full length; as it extends, the epiblast cells in front of it move forward.

Adapted from Voiculescu, O., et al.: 2007.

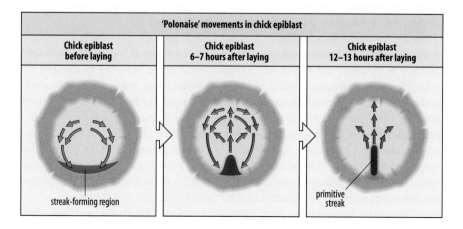

ments are known as 'polonaise' movements, as they resemble a formal eighteenth-century dance of that name. High-resolution time-lapse videos of the chick epiblast show that convergent extension is the most likely mechanism generating the cell movements. The intercalation of cells close to the streak generates forces in the epithelium that cause the movement of cells elsewhere in the epiblast. As with axis elongation in amphibians and zebrafish, this example of convergent extension in the chick involves the Wnt planar cell polarity pathway, and inhibition of this pathway blocks streak formation.

Summary

At the end of cleavage the animal embryo is essentially a closed sheet of cells, which is often in the form of a sphere enclosing a fluid-filled interior. Gastrulation converts this sheet into a solid three-dimensional embryonic animal body. During gastrulation, cells move from the exterior into the interior of the embryo, and the endodermal and mesodermal germ layers take up their appropriate positions inside the embryo. Gastrulation results from a well-defined spatio-temporal pattern of changes in cell shape, cell movements, and changes in cell adhesiveness, and the main forces underlying the morphological changes are generated by changes in cell motility and localized cellular contractions. In the sea urchin, gastrulation occurs in two phases. In the first, changes in shape and adhesion of a small region of prospective mesoderm cells in the blastula wall result in an epithelial-to-mesenchyme transition and migration of these cells into the interior and along the inner face of the blastula wall, followed by invagination of the adjacent part of the blastula epithelium, the prospective endoderm, to form the gut. Filopodia on the migrating primary mesenchyme cells explore the environment, drawing the cells to those regions of the wall where the filopodia make the most stable attachment; in this way guidance cues in the extracellular matrix lead the migrating cells to their correct locations. In the second phase, extension of the gut to reach the prospective oral (mouth) region on the opposite side of the embryo occurs by cell rearrangement within the endoderm, which causes the tissue to narrow and lengthen (convergent extension), and by traction of filopodia extending from the gut tip against the blastocoel wall. Invagination of the mesoderm in *Drosophila* occurs by a similar mechanism to that in sea urchin, whereas germ-band extension is due to myosin-driven remodeling of cell junctions and intercalation of cells, leading to convergent extension. Dorsal closure in *Drosophila* is driven by filopodial extension and retraction. Gastrulation in vertebrates involves more complex movements of cells and cell sheets and also results in elongation of the embryo along the antero-posterior axis. In *Xenopus* it involves three processes: involution, in which a double-layered sheet of endoderm and mesoderm rolls into the interior over the lip of the blastopore; convergent extension of endoderm and mesoderm in the antero-posterior direction to form the roof of the gut, and the notochord and somitic mesoderm, respectively; and epiboly—the spreading of the ectoderm from the animal cap region to cover the whole outer surface of the embryo. Both convergent extension and epiboly are due to cell intercalation, the cells becoming rearranged with respect to their neighbors. Convergent extension of the antero-posterior axis at gastrulation involves both the vertebrate version of the Wnt planar cell polarity signaling pathway and signals delivered by the pre-existing antero-posterior patterning in the mesoderm. Gastrulation in the chick and the mouse involves delamination of individual cells from the epiblast and their movement through the primitive streak into the interior of the embryo, where they form mesoderm and endoderm.

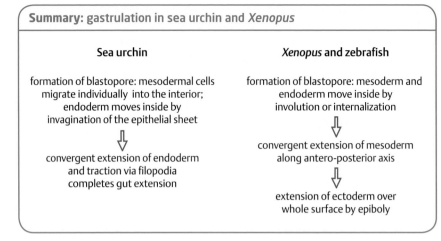

Summary: gastrulation in sea urchin and *Xenopus*

Sea urchin	***Xenopus* and zebrafish**
formation of blastopore: mesodermal cells migrate individually into the interior; endoderm moves inside by invagination of the epithelial sheet	formation of blastopore: mesoderm and endoderm move inside by involution or internalization
⬇	⬇
convergent extension of endoderm and traction via filopodia completes gut extension	convergent extension of mesoderm along antero-posterior axis
	⬇
	extension of ectoderm over whole surface by epiboly

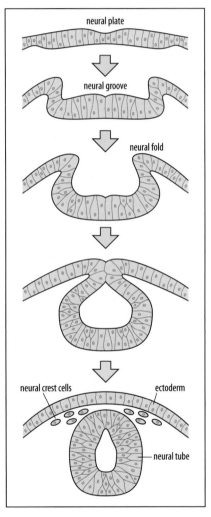

Fig. 8.33 Neural tube formation results from formation and fusion of neural folds.
Bending of the neural plate inwards at both sides creates the neural groove, with the neural folds at the edges. The neural folds come together and fuse to form a tube of epithelium—the neural tube. This detaches from the rest of the ectoderm, which becomes the epidermis. Neural crest cells detach from the dorsal neural tube and migrate away from it.

Neural tube formation

Neurulation in vertebrates results in the formation of the neural tube—a tube of epithelium derived from the dorsal ectoderm—which develops into the brain and spinal cord (see Section 5.10). Following induction by the mesoderm during gastrulation, the ectoderm that will give rise to the neural tube initially appears as a thickened plate of tissue—the neural plate—in which the cells have become more columnar than the cells of the adjacent ectoderm.

The vertebrate neural tube is formed by two different mechanisms in different regions of the body. The anterior neural tube forms the brain and the central nervous system in the trunk, and is essentially formed by the folding of the neural plate into a tube. At neurulation, the edges of the neural plate become raised above the surface, forming two parallel neural folds with a depression—the neural groove—between them (Fig. 8.33). The neural folds eventually come together along the dorsal midline of the embryo and fuse at their edges to form the neural tube, which then separates from the adjacent ectoderm. The surface layer of ectoderm becomes epidermis. The posterior neural tube in the lumbar and tail region, in contrast, develops from a solid rod of cells which develops an interior cavity or lumen. There are, however, variations in different vertebrates; in zebrafish and other fishes, for example, the whole of the neural plate first forms a solid rod, which later becomes hollow.

8.12 Neural tube formation is driven by changes in cell shape and convergent extension

The curvature of the neural plate and formation of the neural folds are associated with changes in cell shape. The cells at the edge of the plate are constricted at the apical surface, making them wedge-shaped (Fig. 8.34). This change in cell shape is, in principle, sufficient to draw the edges of the neural plate up into folds. Like the axial mesoderm, the neural tube undergoes convergent extension during its formation, which both extends it in the antero-posterior direction, and brings the neural folds close together, enabling them to fuse. The cells of the neural plate seem to become oriented for convergent extension in response to signals from midline structures such as the underlying notochord. Like convergent extension in the mesoderm, convergent extension and closure of the neural tube require a functional Wnt planar polarity pathway (see Box 8C, p. 310). If this pathway is defective, closure is incomplete, as observed in mouse embryos lacking Dishevelled, Flamingo, or the vertebrate Vang proteins, Vang-like 1 and 2, in which the neural tube remains open from the midbrain to the lower spine, a condition known as craniorachischisis.

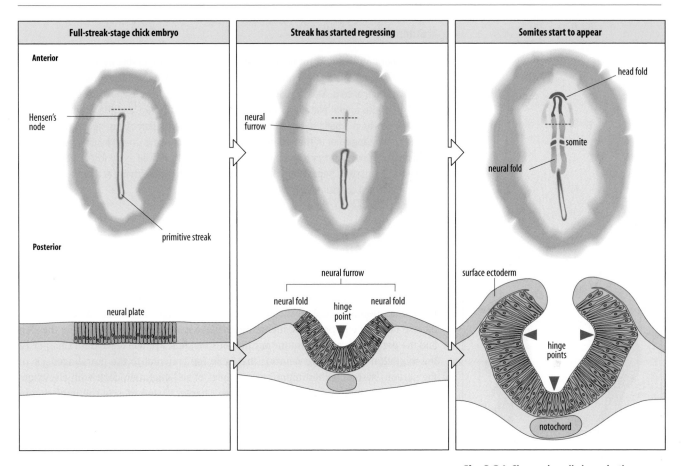

| Full-streak-stage chick embryo | Streak has started regressing | Somites start to appear |

Anterior

Hensen's node

primitive streak

Posterior

neural plate

neural furrow

neural furrow

neural fold   hinge point   neural fold

head fold

somite

neural fold

surface ectoderm

hinge points

notochord

Fig. 8.34 Change in cell shape in the neural plate during chick neurulation.
Top: surface views of the chick epiblast. Bottom: cross-sections taken through the epiblast at the sites indicated by the dashed lines in the top diagrams. Cells anterior to Hensen's node become elongated to form the neural plate. Cells in the center of the neural plate become wedge-shaped, defining the hinge point at which the neural plate bends. Additional hinge points with wedge-shaped cells also form on the sides of the furrow.
After Schoenwolf, G.C., et al.: 1990.

Neurulation does not occur along the whole of the neural plate at the same time, but starts in the region of the midbrain, and proceeds anteriorly and posteriorly. Changes in shape of neural furrow cells are observed in chick embryos, but whether they are a cause or a result of folding is not yet clear. The detailed cellular mechanism underlying the shape changes is not yet established, and in the chick there is some evidence that tissues lateral to the neural plate exert forces that help to throw up the neural folds. Cells in the midline of the chick neural furrow, the so-called hinge point, become wedge-shaped. Later, wedge-shaped cells are seen at additional hinge points on the sides of the furrow, where it curves round further to form a tube (see Fig. 8.34). When the folds meet they zip together so closely that the lumen almost disappears; it only opens up again after the folds fuse. As in many other situations, apical constriction of cells in the neural furrow and hinge points is thought to be due to the purse-string mechanism of actomyosin contraction illustrated in Box 8A, p. 291.

The neural tube, which is initially part of the ectoderm, separates from the presumptive epidermis after its formation as a result of changes in cell adhesiveness. In the chick, neural plate cells, like the rest of the ectoderm, initially have the adhesion molecule L-CAM on their surface. However, as the neural folds develop, the neural plate ectoderm begins to express both N-cadherin and N-CAM, whereas the adjacent ectoderm expresses E-cadherin. These differences may enable the neural tube to separate from the surrounding ectoderm and allow it to sink beneath the surface, the rest of the ectoderm reforming over it in a continuous layer. Changes in adhesiveness could even provide part of the mechanism for neural tube formation, in a manner similar to that described in Section 8.1 for the sorting out of cells.

Summary

In vertebrates, induction of neural tissue by the mesoderm during gastrulation is followed by neurulation—the formation of the neural tube, which will eventually form the brain and spinal cord. Neurulation in mammals, birds, and amphibians involves both the folding of the neural plate into a tube (in the head and trunk), and the formation of a central cavity in a solid rod of cells (in the tail). The formation of the neural folds and their coming together in the midline to form the neural tube is driven by changes in cell shape and convergent extension within the tube itself, as well as by forces generated by adjacent tissues. Separation of the neural tube from the overlying ectoderm requires changes in cell adhesiveness.

Cell migration

Cell migration is a major feature of animal morphogenesis, with cells moving over relatively long distances from one site to another. The example we shall discuss here is the migration of neural crest cells in vertebrate embryos to produce an enormously diverse range of end products, including cartilage in the face, melanocytes in the skin and the peripheral and the autonomic nervous systems. We will leave until Chapter 12 the migration of immature neurons that is so fundamental to the morphogenesis of the nervous system. Other important examples of cell migration dealt with elsewhere in the book include germ-cell migration (see Chapter 9) and muscle-cell migration in vertebrate limbs (see Chapter 11).

8.13 Neural crest migration is controlled by environmental cues

Neural crest cells of vertebrates have their origin at the edges of the neural plate and first become recognizable during neurulation (see Section 5.4). At the time of closure of the neural tube in vertebrate embryos, neural crest cells can be seen on each side of the midline (Fig. 8.35). They undergo an epithelial-to-mesenchymal transition and leave the midline, migrating away from it on either side. As we saw earlier, epithelial-to-mesenchymal transitions involve the transcription factors Slug and Snail, which control processes that transform nonmotile epithelial cells into migrating cells (see Section 8.7).

Neural crest cells give rise to a wide variety of different cell types, including cartilage in the head, pigment cells in the dermis, the medullary cells of the adrenal gland, glial Schwann cells, and to neurons of both the somato-sensory and the autonomic peripheral nervous systems. The differentiation of neural crest cells is discussed in Chapter 10. Here we focus on the migration pathways of the crest cells in the trunk region of the chick embryo. We have already discussed the migration of neural crest cells from the hindbrain region of the neural tube to form the branchial arches (see Section 5.13).

Various strategies have been used to follow the migration of neural crest cells. For example, because quail cells have a nuclear marker that distinguishes them from chick cells, grafting a neural tube from a quail embryo into a chick embryo allows the subsequent migration pathways of the quail neural crest cells in the chick embryo to be followed (see Fig. 8.35). It is also possible to identify migrating chick neural crest cells by tagging them with labeled monoclonal antibodies, by labeling them with a fluorescent dye such as DiI, or by expressing a photoactivatable green fluorescent protein (pa-GFP) in the neural tube. pa-GFP can then be made to fluoresce when required by a precisely focused beam of light, and the migration patterns of a small group of neural crest cells can be followed. Migrating neural crest cells in the zebrafish embryo can be seen in Box 1D (p. 20).

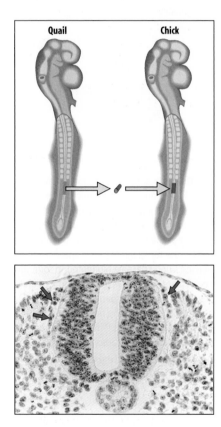

Fig. 8.35 Following cell-migration pathways by grafting a piece of quail neural tube to a chick host. A piece of neural tube from a quail embryo is grafted to a similar position in a chick host. The photograph shows migration of the quail neural crest cells (red arrows). Their migration can be tracked as quail cells have a nuclear marker that distinguishes them from chick cells.

Photograph courtesy of N. Le Douarin.

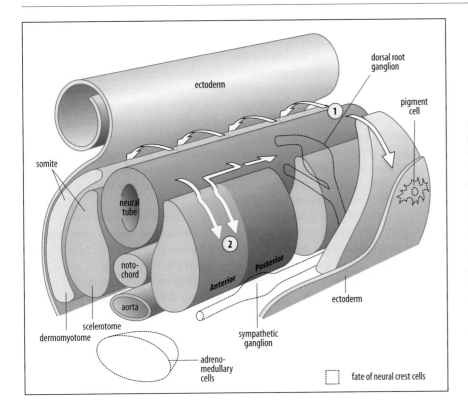

Fig. 8.36 Neural crest migration in the trunk of the chick embryo. One group of cells (1) migrates under the ectoderm to give rise to pigment cells (shown in dotted outline). The other group of cells (2) migrates over the neural tube and then through the anterior half of the somite; the cells do not migrate through the posterior half of the somite. Those that migrate along this pathway give rise to dorsal root ganglia, sympathetic ganglia, and cells of the adrenal cortex, and their future sites are also shown in dotted outline. Those cells opposite the posterior regions of a somite migrate in both directions along the neural tube until they come to the anterior region of a somite. This results in a segmental pattern of migration and is responsible for the segmental arrangement of ganglia.

There are two main migratory pathways for neural crest cells in the trunk of the chick embryo (Fig. 8.36). One goes dorso-laterally under the ectoderm and over the somites; cells that migrate this way mainly give rise to pigment cells, which populate the skin and feathers. The other pathway is more ventral, primarily giving rise to sympathetic and sensory ganglion cells. Some crest cells move into the somites to form dorsal root ganglia; others migrate through the somites to form sympathetic ganglia and adrenal medulla, but appear to avoid the region around the notochord. Trunk neural crest selectively migrates through the anterior half of the somite and not through the posterior half. Within each somite, neural crest cells are found only in the anterior half, even when they originate in neural crest adjacent to the posterior half of the somite. This behavior is unlike that of the neural crest cells taking the dorsal pathway, which migrate over the whole dorso-lateral surface of the somite. Neural crest migration to the anterior of each somite results in the distinct segmental arrangement of spinal ganglia in vertebrates, with one pair of ganglia for each pair of somites—one segment—in the embryo (Fig. 8.37). The segmental pattern of migration is due to different guidance cues on the two halves of each somite. If the somites are rotated through 180° so that their antero-posterior axis is reversed, the crest cells still migrate through the original anterior halves only.

Interactions between ephrins and Ephs are likely to contribute to the expulsion of neural crest cells from the posterior halves of the somites. In the chick, ephrin B1 is localized to the posterior halves of somites and acts as a repulsive guidance cue, interacting with EphB3 on the neural crest cells, whereas the anterior halves of the somites, through which the neural crest cells do travel, express EphB3. Membrane proteins called **semaphorins** are also involved in repelling neural crest cells from the posterior half of the somite. Interactions such as these provide a molecular basis for the segmental arrangement of the spinal ganglia.

The neural tube and notochord also both influence neural crest migration. If the early neural tube is inverted through 180°, before neural crest migration starts, so that the dorsal surface is now the ventral surface, one might think that the cells that normally

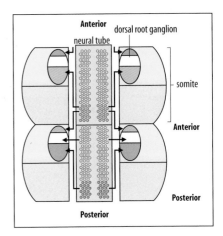

Fig. 8.37 Segmental arrangement of dorsal root ganglia is due to the migration of crest cells through the anterior half of the somite only. Neural crest cells cannot migrate through the posterior (gray) half of a somite but can migrate in either direction along the neural tube and through the anterior half of somites (yellow). The dorsal root ganglion in a given segment is thus made up of crest cells from the posterior region of the segment anterior to it, neural crest cells immediately adjacent to it (white), and crest cells from the posterior of its own segment.

migrate ventrally, now being nearer to their destination, would move ventrally. But this is not the case, and many of the crest cells move upward through the sclerotome in a ventral-to-dorsal direction, staying confined to the anterior half of each somite. This suggests that the neural tube somehow influences the direction of migration of neural crest cells. The notochord also exerts an influence, inhibiting neural crest cell migration over a distance of about 50 μm, and thus preventing the cells from approaching it.

The direction of neural crest migration does involve some chemotaxis towards specific chemical cues. The migration of streams of neural crest cells in particular directions mainly results from the cells becoming polarized in the direction of migration, probably due to the actions of the Wnt-mediated planar cell polarity pathway (see Box 8C, p. 310). Inhibition of components of this pathway, such as Wnts and Frizzled, inhibit neural crest migration.

Neural crest cells are guided by interactions with the extracellular matrix over which they are moving, as well as by cell–cell interactions. Neural crest cells can interact with extracellular matrix molecules by means of their cell-surface integrins (see Box 8B, p. 293). Avian neural crest cells cultured *in vitro* adhere to, and migrate efficiently on, fibronectin, laminin, and various collagens. Blocking adhesion of neural crest cells to fibronectin or laminin by blocking the integrin β_1 subunit *in vivo* causes severe defects in the head region but not in the trunk, suggesting that the crest cells in these two regions adhere by different mechanisms, probably involving other integrins. It is striking how neural crest cells in culture will preferentially migrate along a track of fibronectin, although the role of this extracellular matrix molecule in guiding the cells in the embryo is still unclear.

Summary

Streams of neural crest cells migrate from both sides of the dorsal neural tube to form a wide variety of tissues in different parts of the body. Their migration is guided by signals from other cells and by interactions with the extracellular matrix. In neural crest cells migrating in the trunk, repulsive interactions between ephrins in the posterior halves of the somites and Eph receptors on the neural crest cells prevent their migration over or through the posterior halves. Thus, neural crest cells that will give rise to the dorsal root ganglia of the peripheral nervous system accumulate adjacent to the anterior half of each somite, resulting in a segmental arrangement of paired ganglia along the antero-posterior axis.

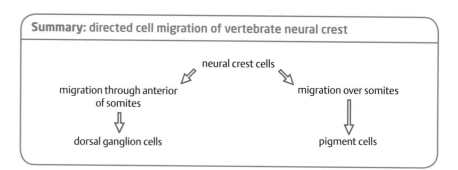

Summary: directed cell migration of vertebrate neural crest

neural crest cells

migration through anterior of somites → dorsal ganglion cells

migration over somites → pigment cells

Directed dilation

Hydrostatic pressure can provide the force for morphogenesis in a variety of situations. We have already seen how an increase in hydrostatic pressure inside the mammalian blastocyst causes an increase in volume and maintains the blastocyst in a roughly spherical shape (see Section 8.5). Here, we consider examples of **directed dilation**, where an increase in internal hydrostatic pressure causes a distinct asymmetric change in shape as a result of the particular properties of the structure. For example, if the circumferential resistance of a stretchable tube to internal pressure is much greater than the resistance lengthways, then an increase in the internal pressure will cause an increase in length (Fig. 8.38).

8.14 Later extension and stiffening of the notochord occurs by directed dilation

After the *Xenopus* notochord has formed, its volume increases threefold, and there is considerable further lengthening as it straightens and becomes stiffer. At this stage, the notochord has become surrounded by a sheath of extracellular material, which restricts circumferential expansion but does allow expansion in the antero-posterior direction. The cells within the notochord develop fluid-filled vacuoles and expand in volume as a result. They thus exert hydrostatic pressure on the notochord sheath, resulting in directed dilation: circumferential expansion of the notochord is prevented by the resistance of the sheath, and this ensures that the increase in volume (dilation) is directed along the notochord's long axis.

The vacuoles in the notochord cells are filled with glycosaminoglycans which, because of their high carbohydrate content, tend to attract water into the vacuoles by osmosis. It is this that produces the hydrostatic pressure that causes the increase in cell volume, and the consequent stiffening and straightening of the notochord. Changes in the structure of the sheath during the period of notochord elongation fit well with the proposed hydrostatic mechanism. The sheath contains both glycosaminoglycans, which have little tensile strength, and the fibrous protein collagen, which has a high tensile strength; during notochord dilation the density of collagen fibers increases, providing resistance to circumferential expansion. The crucial role of the sheath in dilation and elongation is shown by the fact that if it is digested away, the notochord buckles and folds and the notochord cells, instead of being flat, become rounded.

8.15 Circumferential contraction of hypodermal cells elongates the nematode embryo

During the early development of the nematode there is little change in body shape from the ovoid form of the fertilized egg, even during gastrulation. After gastrulation, about 5 hours after fertilization, the embryo begins to elongate rapidly along its antero-posterior axis. Elongation takes about 2 hours, during which time the embryo decreases in circumference about threefold and undergoes a fourfold increase in length.

This elongation is brought about by a change in shape of the hypodermal (epidermal) cells that make up the outermost layer of the embryo; their destruction by laser ablation prevents elongation. During embryo elongation, these cells change shape so that, instead of being elongated in the circumferential direction, they become elongated along the antero-posterior axis (Fig. 8.39). Throughout this elongation the hypodermal cells remain attached to each other by cell junctions. The junctions are also linked within the cells by actin-containing fibers that run circumferentially, and these fibers appear to shorten as the cells elongate. The disruption of actin filaments by cytochalasin D treatment blocks elongation, and so it is very likely that their contraction brings about the change in cell shape. Circumferential contraction of the hypodermal cells causes an increase in hydrostatic pressure within the embryo, forcing an extension in an antero-posterior direction. Circumferentially oriented microtubules may also have a mechanical role in constraining the expansion, in the same way

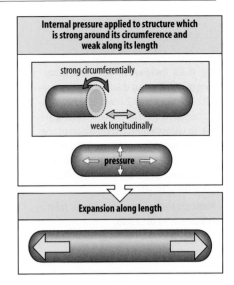

Fig. 8.38 Directed dilation. Hydrostatic pressure inside a constraining sheath or membrane can lead to elongation of the structure. If the circumferential resistance is much greater than the longitudinal resistance, as it is in the notochord sheath, the rod of cells inside the sheath lengthens.

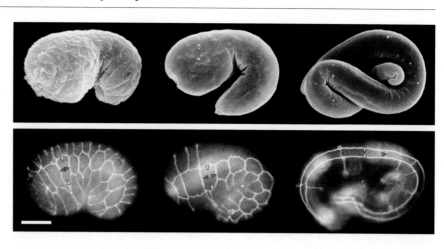

Fig. 8.39 Increase in nematode body length by directed dilation. The change in body shape over 2 hours is illustrated in the top panel. The increase in length is due to circumferential contraction of the hypodermal cells, as shown in the bottom panel. The change in shape of a single cell can be seen in the cell marked with arrows. Scale bar = 10 μm.

Photographs courtesy of J. Priess, from Priess, J.R., et al.: 1986.

as the sheath of the *Xenopus* notochord described earlier. Increase in nematode body length is thus another example of directed dilation.

8.16 The direction of cell enlargement can determine the form of a plant leaf

Cell enlargement is a major process in plant growth and morphogenesis, providing up to a 50-fold increase in the volume of a tissue. The driving force for expansion is the hydrostatic pressure—turgor pressure—exerted on the cell wall as the protoplast swells as a result of the entry of water into cell vacuoles by osmosis (Fig. 8.40). Plant-cell expansion involves synthesis and deposition of new cell-wall material, and is an example of directed dilation. The direction of cell growth is determined by the orientation of the cellulose fibrils in the cell wall. Enlargement occurs primarily in a direction at right angles to the fibrils, where the wall is weakest. The orientation of cellulose fibrils in the cell wall is thought to be determined by the microtubules of the cell's cytoskeleton, which are responsible for positioning the enzyme assemblies that synthesize cellulose at the cell wall. Plant growth hormones, such as ethylene and gibberellins alter the orientation in which the fibrils are laid down and so can alter the direction of expansion. Auxin aids expansion by loosening the structure of the cell wall.

The development of a leaf involves a complex pattern of cell division and cell elongation, with cell elongation playing a central part in the expansion of the leaf blade. Two mutations that affect the shape of the blade by affecting the direction of cell elongation have been identified. Leaves of the *Arabidopsis* mutant *angustifolia* are similar in length to the wild type but are much narrower (Fig. 8.41). In contrast, *rotundifolia* mutations reduce the length of the leaf relative to its width. Neither of these mutations affects the number of cells in the leaf. Examination of the cells in the developing leaf shows that these mutations are affecting the direction of elongation of the enlarging cells.

Fig. 8.40 Enlargement of a plant cell. Plant cells expand as water enters the cell vacuoles and thus causes an increase in intracellular hydrostatic pressure. The cell elongates in a direction perpendicular to the orientation of the cellulose fibrils in its cell wall.

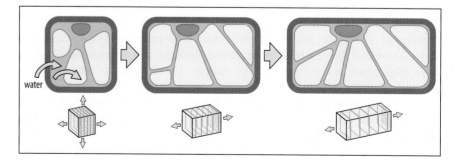

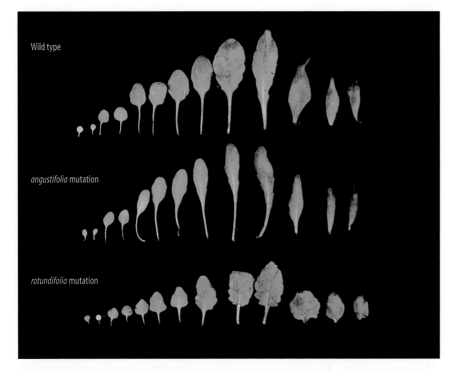

Fig. 8.41 The shape of the leaves of ***Arabidopsis*** **is affected by mutations affecting cell elongation.** The *angustifolia* mutation results in cells with narrow leaves, while *rotundifolia* mutations cause short, fat leaves to develop.

Photograph courtesy of H. Tsukaya, from Tsuge, T., et al.: 1996.

Summary

Directed dilation results from an increase in hydrostatic pressure inside a structure, and unequal peripheral resistance to this pressure. Extension of the notochord is brought about by directed dilation, in which the notochord interior increases in volume while its circumferential expansion is constrained by the notochord sheath, forcing it to elongate. Similarly, the nematode embryo elongates after gastrulation as a result of a circumferential contraction of the outer hypodermal cells that generates pressure on the internal cells, forcing the embryo to extend in an antero-posterior direction. In plants, the shape of a leaf is determined by the direction of cell enlargement, which is dictated by the distribution of cellulose fibrils in the cell wall. Directed dilation resulting from internal pressure elongates the cell where the cell wall is weakest.

Summary: directed dilation

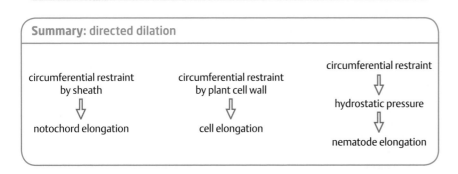

circumferential restraint
by sheath
⇩
notochord elongation

circumferential restraint
by plant cell wall
⇩
cell elongation

circumferential restraint
⇩
hydrostatic pressure
⇩
nematode elongation

Summary to Chapter 8

Changes in the shape of animal embryos as they develop are mainly due to forces generated by changes in cell shape, cell motility, and differences in cell adhesiveness. Formation of a hollow, spherical blastula or blastocyst is the result of particular patterns of cell division, cell packing, and cell polarization, and the spherical shape is maintained by flow of fluid into the interior and the hydrostatic pressure generated. Gastrulation involves major movements of cells, such that the future endoderm and mesoderm move inside the embryo to their appropriate positions in relation to the main body plan. Convergent extension has a key role in many developmental contexts, such as elongation of the antero-posterior body axis during vertebrate gastrulation, lengthening and closure of the neural tube, and germ-band extension in *Drosophila*. Formation of the vertebrate neural tube from the neural plate involves cell shape changes that cause the edges of the plate to curve upwards and finally fuse. The role of cell migration in morphogenesis is represented in this chapter by the neural crest cells, which leave the neural tube to give rise to a wide variety of tissues throughout the body. The phenomenon of directed dilation, involving hydrostatic pressure, is responsible for the later elongation of the notochord, the elongation of the nematode embryo and the direction of enlargement of plant cells.

■ End of chapter questions

Long answer (concept questions)

1. Name the four types of adhesive interactions between cells and between cells and the extracellular matrix. What are the transmembrane proteins that characterize each of these adhesive interactions? What are the particular features of an adherens junction: what cell adhesion molecule is used, what cytoskeletal component is involved, and what links the adhesion molecule to the cytoskeleton?

2. Describe the results of an experiment in which epidermis and neural plate from a neurula stage amphibian embryo are first disaggregated into single cells, then mixed together and reaggregated. How is this experiment interpreted in terms of the relative strength of 'self' adhesiveness in these two tissues? How does this result reflect the natural organization of these tissues in the embryo?

3. Define 'radial cleavage.' In the radial cleavage of an embryo such as the sea urchin, how are the centrosomes positioned at the first cleavage, with respect to the animal–vegetal axis? The second cleavage? The third cleavage?

4. What is compaction of the mouse embryo? How is compaction related to two important subsequent events in mouse embryogenesis: the formation of the blastocoel, and the formation of the inner cell mass?

5. Contrast the properties of an epithelial tissue with those of a mesenchymal tissue. How are cadherins related to the transition of epithelium to mesenchyme; how is this transition regulated, or triggered (think: '*snail*')?

6. Refer to the fate maps of sea urchins (Fig. 8.14) and *Xenopus* (Fig. 8.24): in sea urchins, mesoderm is most vegetal and endoderm lies between mesoderm and ectoderm; in *Xenopus*, endoderm is most vegetal and mesoderm lies between the endoderm and the ectoderm. Can you briefly describe the differences in gastrulation in these two organisms that explain these differing arrangements?

7. Contrast epiboly and convergent extension. How are these two processes used during *Xenopus* gastrulation?

8. If cells in a flat sheet constrict one surface (the apical surface), the sheet will develop a depression in that region. Give examples of how this mechanism is used to initiate gastrulation in sea urchins, *Drosophila*, and *Xenopus*.

9. How do changes in cell shape and changes in cell adhesiveness collaborate in the formation of the neural tube?

10. Describe the migration route of the neural crest cells that will form the dorsal root ganglia. What role do Ephs and ephrins play in this migratory pathway?

11. What is the influence of directed dilation on formation of the notochord? What role do sugars (glycosaminoglycans) play in this process?

12. In your own words, describe the cell movements that occur during gastrulation, from the first appearance of the blastopore, to the first appearance of the neural folds. Do this first for *Xenopus*, then describe the major differences seen in chicks and zebrafish.

Multiple choice (factual recall questions)

NB There is only one right answer to each question.

1. Adherens junctions play in important role in development; they are composed of

a) cadherins linked to other cadherins extracellularly, and to actin filaments intracellularly

b) cadherins linked to other cadherins extracellularly, and to intermediate filaments intracellularly

c) immunoglobulin superfamily (IgSF) molecules linked to other IgSF molecules extracellularly, and to the cytoskeleton intracellularly

d) integrins linked to extracellular matrix molecules extracellularly, and to the cytoskeleton intracellularly

2. If embryos are disaggregated with chemicals or proteases, and the individual cells are mixed together in culture, what will happen?

a) The cells will associate with one another randomly.

b) The cells will often sort themselves so that like cells are together.

c) The cells will reaggregate to form a normal embryo capable of continuing development.

d) The cells will regulate to form one or more normal embryos.

3. What type of cleavage is used in frog embryos?

a) radial

b) spiral

c) superficial

d) unequal

4. How do the endodermal cells of the sea urchin embryo carry out gut formation during gastrulation?

a) Changes in cell shape initiate the invagination; convergent extension extends the sheet of cells into the blastocoel, and finally, filopodia make contact with the future mouth region and pull the tip of the gut to that point.

b) They crawl inside the blastocoel and form a solid rod, which is then hollowed out to form the tube of the gut.

c) They move inside the blastocoel, and form the skeletal structures that will support the adult animal.

d) They move into the blastocoel as mesenchyme and crawl ventrally, to mark the location of the future mouth.

5. During gastrulation in the frog, the very first cells to move into the interior of the embryo through the blastopore come from the surface layer of cells in the marginal zone; they will become:

a) ectoderm

b) endoderm

c) mesoderm

d) yolk

6. Gastrulation in sea urchins, *Drosophila*, and *Xenopus* all begin with a change in cell shape, in which the apical surface of an epithelial sheet contracts. This process is called

a) convergent extension

b) epiboly

c) invagination

d) involution

7. In *Xenopus*, the mesoderm moves in through the blastopore by rolling around the dorsal lip in a process called

a) convergent extension

b) epiboly

c) invagination

d) involution

8. In *Xenopus*, the elongation of the mesoderm toward the anterior results from the intercalation of cells during a process called

a) convergent extension

b) epiboly

c) invagination

d) involution

9. In chicks, the formation of the neural tube relies on what cell biological processes?

a) Changes in adhesiveness between the prospective neural tube cells and the prospective epidermis in the ectoderm are the only changes required for neural tube formation.

b) Changes in cell shape in the neural plate and changes in adhesion molecule expression in the neural tube are required during formation of the neural tube.

c) Changes in cell shape in the neural plate are the only process required for neural tube formation.

d) Neural tube formation is a passive by-product of convergent extension in the notochord.

10. Neural crest cells taking the dorsal-lateral route will become

a) adrenal medulla

b) dorsal root ganglia

c) melanocytes

d) sympathetic ganglia

Multiple choice answer key

1: a, 2: b, 3: a, 4: a, 5: b, 6: c, 7: d, 8: a, 9: b, 10: c.

■ General further reading

Alberts, B. *et al.*: *Molecular Biology of the Cell* (5th edn). New York: Garland Science, 2008.

Aman, A., Piotrowski, T.: **Cell migration during morphogenesis**. *Dev. Biol.* 2010, **341**: 20–33.

Engler, A.J., Humbert, P.O., Wehrle-Haller, B., Weaver, V.M.: **Multiscale modeling of form and function**. *Science* 2009, **324**: 208–212.

Mattila, P.K., Lappalainen, P.: **Filopodia: molecular architecture and cellular functions**. *Nat. Rev. Mol. Cell Biol.* 2008, **9**: 446–454.

Stern, C. *et al.*: *Gastrulation: From Cells to Embryos*. New York: Cold Spring Harbor Laboratory Press; 2004.

■ Section further reading

8.1 Sorting out of dissociated cells demonstrates differences in cell adhesiveness in different tissues

Davis, G.S., Phillips, H.M., Steinberg, M.S.: **Germ-layer surface tensions and 'tissue affinities' in *Rana pipiens* gastrulae: quantitative measurements**. *Dev. Biol.* 1997, **192**: 630–644.

Steinberg, M.S.: **Differential adhesion in morphogenesis: a modern view**. *Curr. Opin. Genet. Dev.* 2007, **17**: 281–286.

Box 8B Cell-adhesion molecules and cell junctions

Hynes, R.O: **Integrins: bidirectional allosteric signaling mechanisms**. *Cell* 2002, **110**: 673–689.

Tepass, U., Truong, K., Godt, D., Ikura, M., Peifer, M.: **Cadherins in embryonic and neural morphogenesis**. *Nat Rev. Mol. Cell Biol.* 2000, **1**: 91–100.

Thiery, J.P.: **Cell adhesion in development: a complex signaling network**. *Curr. Opin. Genet. Dev.* 2003, **13**: 365–371.

8.2 Cadherins can provide adhesive specificity

Duguay, D., Foty, R.A., Steinberg, M.S.: **Cadherin-mediated cell adhesion and tissue segregation: qualitative and quantitative determinants**. *Dev. Biol.* 2003, **253**: 309–323.

Levine, E., Lee, C.H., Kintner, C., Gumbiner, B.M.: **Selective disruption of E-cadherin function in early *Xenopus* embryos by a dominant negative mutant**. *Development* 1994, **120**: 901–909.

Takeichi, M., Nakagawa, S., Aono, S., Usui, T., Uemura, T.: **Patterning of cell assemblies regulated by adhesion receptors of the cadherin superfamily**. *Proc. Roy. Soc. Lond. B* 2000, **355**: 885–896.

8.3 The orientation of the mitotic spindle determines the plane of cleavage at cell division

Backues, S.K., Konopka, C.A., McMichael, C.M., Bednarek, S.Y.: **Bridging the divide between cytokinesis and cell expansion**. *Curr. Opin. Plant. Biol.* 2007, **10**: 607–615.

Galli, M., van den Heuvel, S.: **Determination of the cleavage plane in early *C. elegans* embryos**. *Annu. Rev. Genet.* 2008, **42**: 389–411.

Glotzer, M.: **Cleavage furrow positioning**. *J. Cell Biol.* 2004, **164**: 347–351.

Gönczy, P., Rose, L.S.: **Asymmetric cell division and axis formation in the embryo**. In *WormBook* (ed. The C. elegans Research Community) (October 15, 2005). doi/10.1895/wormbook.1.30.1, http://www.wormbook.org (date accessed 21 May 2010).

8.4 Cells become polarized in the sea-urchin blastula and the mouse morula

Pauken, C.M., Capco, D.G.: **Regulation of cell adhesion during embryonic compaction of mammalian embryos: roles for PKC and beta-catenin**. *Mol. Reprod. Dev.* 1999, **54**: 135–144.

Sutherland, A.E., Speed, T.P., Calarco, P.G.: **Inner cell allocation in the mouse morula: the role of oriented division during fourth cleavage**. *Dev. Biol.* 1990, **137**: 13–25.

8.5 Fluid accumulation as a result of tight-junction formation and ion transport forms the blastocoel of the mammalian blastocyst

Barcroft, L.C., Offenberg, H., Thomsen, P., Watson, A.J.: **Aquaporin proteins in murine trophectoderm mediate transepithelial water movements during cavitation**. *Dev. Biol.* 2003, **256**: 342–354.

Eckert, J.J., Fleming, T.P.: **Tight junction biogenesis during early development**. *Biochim. Biophys. Acta* 2008, **1778**: 717–728.

Fleming, T.P., Sheth, B., Fesenko, I.: **Cell adhesion in the preimplantation mammalian embryo and its role in trophectoderm differentiation and blastocyst morphogenesis**. *Front. Biosci.* 2001, **6**: 1000–1007.

Watson, A.J., Natale, D.R., Barcroft, L.C.: **Molecular regulation of blastocyst formation**. *Anim. Reprod. Sci.* 2004, **82–83**: 583–592.

8.6 Internal cavities can be created by cell death

Coucouvanis, E., Martin, G.R.: **Signals for death and survival: a two-step mechanism for cavitation in the vertebrate embryo**. *Cell* 1995, **83**: 279–287.

8.7 Gastrulation in the sea urchin involves cell migration and invagination

Davidson, L.A., Oster, G.F., Keller, R.E., Koehl, M.A.R.: **Measurements of mechanical properties of the blastula wall reveal which hypothesized mechanisms of primary invagination are physically plausible in the sea urchin *Stronglyocentrotus purpuratus***. *Dev. Biol.* 1998, **204**: 235–250.

Duloquin, L., Lhomond, G., Gache, C.: **Localized VEGF signaling from ectoderm to mesenchyme cells controls morphogenesis of the sea urchin embryo skeleton**. *Development* 2007, **134**: 2293–2302.

Ettensohn, C.A.: **Cell movements in the sea urchin embryo**. *Curr. Opin. Genet. Dev.* 1999, **9**: 461–465.

Gustafson, T., Wolpert, L.: **Studies on the cellular basis of morphogenesis in the sea urchin embryo. Directed movements of primary mesenchyme cells in normal and vegetalized larvae**. *Exp. Cell Res.* 1999, **253**: 288–295.

Mattila, P.K., Lappalainen, P.: **Filopodia: molecular architecture and cellular functions**. *Nat. Rev. Mol. Cell Biol.* 2008, **9**: 446–454.

Miller, J.R., McClay, D.R.: **Changes in the pattern of adherens junction-associated β-catenin accompany morphogenesis in the sea urchin embryo**. *Dev. Biol.* 1997, **192**: 310–322.

Odell, G.M., Oster, G., Alberch, P., Burnside, B.: **The mechanical basis of morphogenesis. I. Epithelial folding and invagination**. *Dev. Biol.* 1981, **85**: 446–462.

Raftopoulou, M., Hall, A.: **Cell migration: Rho GTPases lead the way**. *Dev Biol.* 2004, **265**: 23–32.

Röttinger, E., Saudemont, A., Duboc, V., Besnardeau, L., McClay, D., Lepage, T.: **FGF signals guide migration of mesenchymal cells, control skeletal morphogenesis and regulate gastrulation during sea urchin development**. *Development* 2008, **135**: 353–365.

8.8 Mesoderm invagination in *Drosophila* is due to changes in cell shape that are controlled by genes that pattern the dorso-ventral axis

Martin, A.C., Kaschube, M., Wieschaus, E.F.: **Pulsed contractions of an actin–myosin network drives apical contraction**. *Nature* 2009, **457**: 495–499.

8.9 Germ-band extension in *Drosophila* involves myosin-dependent remodeling of cell junctions and cell intercalation

Bertet, C., Sulak, L., Lecuit, T.: **Myosin-dependent junction remodelling controls planar cell intercalation and axis elongation**. *Nature* 2004, **429**: 667–671.

8.10 Dorsal closure in *Drosophila* and ventral closure in *Caenorhabditis elegans* are brought about by the action of filopodia

Chin-Sang, I.D., Chisholm, A.D.: **Form of the worm: genetics of epidermal morphogenesis in *C. elegans***. *Trends Genet.* 2000, **16**: 544–551.

Jacinto, A., Woolner, S., Martin, P.: **Dynamic analysis of dorsal closure in *Drosophila*: from genetics to cell biology**. *Dev. Cell* 2002, **3**: 9–19.

Peralta, X.G., Toyama, Y., Hutson, M.S., Montague, R., Venakides, S., Kiehart, D.P., Edwards, G.S.: **Upregulation of forces and morphogenic asymmetries in dorsal closure during *Drosophila* development**. *Biophys. J.* 2007, **92**: 2583–2596.

Woolner, S., Jacinto, A., Martin, P.: **The small GTPase Rac plays multiple roles in epithelial sheet fusion—dynamic studies of *Drosophila* dorsal closure.** *Dev. Biol.* 2005, **282**: 163–173.

8.11 Vertebrate gastrulation involves several different types of tissue movement

Adams, D.S., Keller, R., Koehl, M.A.: **The mechanics of notochord elongation, straightening, and stiffening in the embryo of *Xenopus laevis*.** *Development* 1990, **100**: 115–130.

Dzamba, B.J., Jakab, K.R., Marsden, M., Schwartz, M.A., DeSimone, D.W.: **Cadherin adhesion, tissue tension, and noncanonical Wnt signaling regulate fibronectin matrix organization.** *Dev. Cell.* 2009, **16**: 421–432.

Goto, T., Davidson, L., Asashima, M., Keller, R.: **Planar cell polarity genes regulate polarized extracellular matrix deposition during frog gastrulation.** *Curr. Biol.* 2005, **15**: 787–793.

Heisenberg, C.P., Solnica-Krezel, L.: **Back and forth between cell fate specification and movement during vertebrate gastrulation.** *Curr. Opin. Genet. Dev.* 2008, **18**: 311–316.

Keller, R.: **Cell migration during gastrulation.** *Curr. Opin. Cell Biol.* 2005, **17**: 533–541.

Keller, R., Cooper, M.S., D'Anilchik, M., Tibbetts, P., Wilson, P.A.: **Cell intercalation during notochord development in *Xenopus laevis*.** *J. Exp. Zool.* 1989, **251**: 134–154.

Keller, R., Davidson, L.A., Shook, D.R.: **How we are shaped: the biomechanics of gastrulation.** *Differentiation* 2003, **71**: 171–205.

Montero, J.A., Carvalho, L., Wilsch-Brauninger, M., Kilian, B., Mustafa, C., Heisenberg, C.P.: **Shield formation at the onset of zebrafish gastrulation.** *Development* 2005, **132**: 1187–1198.

Montero, J.A., Heisenberg, C.P.: **Gastrulation dynamics: cells move into focus.** *Trends Cell Biol.* 2004, **14**: 620–627.

Ninomiya, H., Winklbauer, R.: **Epithelial coating controls mesenchymal shape change through tissue-positioning effects and reduction of surface-minimizing tension.** *Nat. Cell Biol.* 2008, **10**: 61–69.

Shih, J., Keller, R.: **Gastrulation in *Xenopus laevis*: involution—a current view.** *Dev. Biol.* 1994, **5**: 85–90.

Voiculescu, O., Bertocchini, F., Wolpert, L., Keller, R.E., Stern, C.D.: **The amniote primitive streak is defined by epithelial cell intercalation before gastrulation.** *Nature* 2007, **449**: 1049–1052.

Wacker, S., Grimm, K., Joos, T., Winklbauer, R.: **Development and control of tissue separation at gastrulation in *Xenopus*.** *Dev. Biol.* 2000, **224**: 428–439.

Wallingford, J.B., Fraser, S.E., Harland, R.M.: **Convergent extension: the molecular control of polarized cell movement during embryonic development.** *Dev. Cell* 2002, **2**: 695–706.

Box 8C Convergent extension

Keller, R., Davidson, L., Edlund, A., Elul, T., Ezin, M., Shook, D., Skoglund, P.: **Mechanisms of convergence and extension by cell intercalation.** *Proc R. Soc. Lond. B* 2000, **355**: 897–922.

Ninomiya, H., Elinson, R.P., Winklbauer, R.: **Antero-posterior tissue polarity links mesoderm convergent extension to axial patterning.** *Nature* 2004, **430**: 364–367.

Simons, M., Mlodzik, M.: **Planar cell polarity signaling: from fly development to human disease.** *Annu. Rev. Genet.* 2008, **42**: 517–540.

Torban, E., Kor, C., Gros, P.: **Van Gogh-like2 (Strabismus) and its role in planar cell polarity and convergent extension in vertebrates.** *Trends Genet.* **20**: 570–577.

Zallen, J.A.: **Planar polarity and tissue morphogenesis.** *Cell* 2007, **129**: 1051–1063.

8.12 Neural tube formation is driven by changes in cell shape and convergent extension

Colas, J.F., Schoenwolf, G.C.: **Towards a cellular and molecular understanding of neurulation.** *Dev. Dyn.* 2001, **221**: 117–145.

Davidson, L.A., Keller, R.E.: **Neural tube closure in *Xenopus laevis* involves medial migration, directed protrusive activity, cell intercalation and convergent extension.** *Development* 1999, **126**: 4547–4556.

Greene, N.D.E., Stanier, P., Copp, A.J.: **Genetics of human neural tube defects.** *Hum. Mol. Genet.* 2009, **18**: R113–R129.

Haigo, S.L., Hildebrand, J.D., Harland, R.M., Wallingford, J.B.: **Shroom induces apical constriction and is required for hinge-point formation during neural tube closure.** *Curr. Biol.* 2003, **13**: 2125–2137.

Kibar, Z., Capra, V., Gros, P.: **Toward understanding the genetic basis of neural tube defects.** *Clin. Genet.* 2007, **71**: 295–310.

Rolo, A., Skoglund, P., Keller, R.: **Morphogenetic movements during neural tube closure in *Xenopus* require myosin IIB.** *Dev. Biol.* 2009, **327**: 327–338.

Torban, E., Patenaude, A.M., Leclerc, S., Rakowiecki, S., Gauthier, S., Andelfinger, G., Epstein, D.J., Gros, P.: **Genetic interaction between members of the Vang1 family causes neural tube defects in mice.** *Proc. Natl. Acad. Sci. USA* 2008, **105**: 3449–3454.

Wallingford, J.B., Harland, R.M.: **Neural tube closure requires Dishevelled-dependent convergent extension of the midline.** *Development* 2002, **129**: 5815–5825.

8.13 Neural crest migration is controlled by environmental cues

Bronner-Fraser, M.: **Mechanisms of neural crest migration.** *BioEssays* 1993, **15**: 221–230.

Holder, N., Klein, R.: **Eph receptors and ephrins: effectors of morphogenesis.** *Development* 1999, **126**: 2033–2044.

Kuan, C.Y., Tannahill, D., Cook, G.M., Keynes, R.J.: **Somite polarity and segmental patterning of the peripheral nervous system.** *Mech. Dev.* 2004, **121**: 1055–1068.

Kuriyama, S., Mayor, R.: **Molecular analysis of neural crest migration.** *Philos. Trans. R. Soc. Lond. B Biol. Sci.* 2008, **363**: 1349–1362.

Nagawa, S., Takeichi, M.: **Neural crest emigration from the neural tube depends on regulated cadherin expression.** *Development* 1998, **125**: 2963–2971.

Poliakoff, A., Cotrina, M., Wilkinson, D.G.: **Diverse roles of Eph receptors and ephrins in the regulation of cell migration and tissue assembly**. *Dev. Cell* 2004, **7**: 465–480.

Thevenean, E., Marchant, L., Kuriyama, S., Gull, M., Moeppo, B., Parsons, M., Mayor, R.: **Collective chemotaxis requires contact-dependent cell polarity.** *Dev. Cell* 2010, **19**: 39–53.

Tucker, R.P.: **Neural crest cells: a model for invasive behavior**. *Int. J. Biochem. Cell Biol.* 2004, **36**: 173–177.

Xu, Q., Mellitzer, G., Wilkinson, D.G.: **Roles of Eph receptors and ephrins in segmental patterning**. *Proc. Roy. Soc. Lond. B* 2000, **353**: 993–1002.

8.14 Later extension and stiffening of the notochord occurs by directed dilation

Adams, D.S., Keller, R., Koehl, M.A.: **The mechanics of notochord elongation, straightening and stiffening in the embryo of *Xenopus laevis***. *Development* 1990, **110**: 115–130.

8.15 Circumferential contraction of hypodermal cells elongates the nematode embryo

Priess, J.R., Hirsh, D.I.: ***Caenorhabditis elegans* morphogenesis: the role of the cytoskeleton in elongation of the embryo**. *Dev. Biol.* 1986, **117**: 156–173.

8.16 The direction of cell enlargement can determine the form of a plant leaf

Jackson, D.: **Designing leaves. Plant morphogenesis**. *Curr. Biol.* 1996, **6**: 917–919.

Tsuge, T., Tsukaya, H., Uchimiya, H.: **Two independent and polarized processes of cell elongation regulate leaf blade expansion in *Arabidopsis thaliana* (L.) Heynh**. *Development* 1996, **122**: 1589–1600.

Germ cells, fertilization, and sex

- ■ The development of germ cells
- ■ Fertilization
- ■ Determination of the sexual phenotype

Animal embryos develop from a single cell, the fertilized egg or zygote, which is the product of the fusion of an egg and a sperm. In previous chapters we have considered mainly how the basic body plan of the animal embryo is generated by the development of the somatic cells and their assignment to the three germ layers. The zygote also gives rise to the germline cells—or germ cells—that develop into eggs and sperm, and through which an organism's genes are passed on to the next generation. As we shall discuss in this chapter, in animals, germline cells are specified and set apart in the early embryo, although functional mature eggs and sperm are only produced by the adult. An important property of germ cells is that they remain totipotent—able to give rise to all the different types of cells in the body. Nevertheless, eggs and sperm in mammals have certain genes differentially silenced during germ-cell development by a process known as genomic imprinting, and we will look at the implications of this for development. Embryonic development is initiated by the process of fertilization, and we shall discuss mechanisms in sea urchins and mammals that ensure that only one sperm enters the egg. In the animals we cover in this chapter, sex is determined by the chromosomal constitution, and we shall see that mechanisms of sex determination and of compensating for the unequal complement of sex chromosomes between males and females differ considerably in different species.

In sexually reproducing organisms, there is a fundamental distinction between the **germline cells (germ cells)** and the somatic or body cells (see Section 1.2): the former give rise to the **gametes**—eggs and sperm in animals—whereas somatic cells make no genetic contribution to the next generation. Germ cells have three key functions: the preservation of the genetic integrity of the germline, such as the prevention of aging; the generation of genetic diversity; and the transmission of genetic information to the next generation.

So far in this book, we have looked at the somatic development of our model organisms. Not surprisingly, a great deal of the biology of animals and plants is devoted to reproduction and sex, and in this chapter, we look at germ-cell formation, fertilization of the egg by the sperm, and sex determination, mainly in the mouse, *Drosophila*, and *Caenorhabditis*. As germ cells are the cells that will give rise to the next generation, their development is crucial, and in animals, germ cells are usually specified and set aside early in embryonic development. Plants, although reproducing sexually, differ from most animals in that their germ cells are not specified early in embryonic development, but during the development of the flowers (see Chapter 7). It is worth noting that some simple animals, such as the coelenterate *Hydra*, can reproduce asexually, by budding, and that even in some vertebrates, such as turtles, the eggs can develop without being fertilized.

We will begin this chapter by considering how the germ cells are specified in the early animal embryo and how they differentiate into eggs and sperm. Then, we will look at the **fertilization** and activation of the egg by the sperm—the vital step that initiates development. Finally, we turn to **sex determination**—why males and females are different from each other. In all the animals we look at here, sex is determined genetically, by the number and type of specialized chromosomes known as the **sex chromosomes**. Male and female embryos initially look the same, with sexual differences only emerging as a result of the activity of sex-determining genes located on the sex chromosomes.

The development of germ cells

In all but the simplest animals, the cells of the germline are the only cells that can give rise to a new organism. So, unlike somatic cells, which eventually all die, germ cells in a sense outlive the bodies that produced them. They are, therefore, very special cells. The outcome of germ-cell development is either a male gamete (the sperm in animals) or a female gamete (the egg). The egg is a particularly remarkable cell, as it ultimately gives rise to all the cells in an organism. In species whose embryos receive no nutrition from the mother after fertilization, the egg must also provide everything necessary for development, as the sperm contributes virtually nothing to the organism other than its chromosomes and a centrosome.

Animal germ cells typically differ from somatic cells by dividing less often during early embryogenesis. Later, they are the only cells to undergo the type of nuclear division called **meiosis**, by which gametes with half the number of chromosomes of the germ-cell precursor are produced. The halving of chromosome number at meiosis means that when two gametes come together at fertilization to form the zygote, the diploid number of chromosomes is restored. If there were no reduction division in germ-cell formation, the number of chromosomes in the somatic cells would double in each generation.

Germ cells undergo meiosis and differentiate into eggs and sperm within specialized reproductive organs called **gonads**—the **ovary** in females and the **testis** in males. But in many animals, including *Drosophila* and vertebrates, the germ cells are first specified in a region quite distant from where the gonads will eventually form, and migrate from their site of origin into the forming gonads. From the time they are first specified as germline cells until the time that they enter the gonads, precursor germ cells are known as **primordial germ cells**. In many animals, primordial germ cells are formed in locations that seem to protect them from the inductive signals that determine the fate of somatic cells, and instead shut down the somatic developmental program. In *Caenorhabditis elegans*, for example, there is evidence of a general repression of transcription in the primordial germ cells. In some germlines, other mechanisms that help suppress mutation and maintain genetic integrity are present. In *Drosophila*, for example, the movement of transposons in the genome, which can cause insertion mutations, is prevented by an RNA silencing pathway involving the Piwi-interacting small RNAs (piRNAs).

Germ cells are specified very early in some animals, although not in mammals, by cytoplasmic determinants present in the egg. We therefore start our discussion of germ-cell development by looking at the specification of primordial germ cells by special cytoplasm—the **germ plasm**—which is already present in these eggs.

9.1 Germ-cell fate is specified in some embryos by a distinct germ plasm in the egg

In flies, nematodes, fish, and frogs, molecules localized in specialized cytoplasm in the egg are involved in specifying the germ cells. The clearest example of this is in *Drosophila*, where primordial germ cells known as **pole cells** become distinct at the posterior pole of the egg about 90 minutes after fertilization, more than an hour before the cellularization of the rest of the embryo (see Fig. 2.2). The cytoplasm at the posterior pole is called the **pole plasm** and is distinguished by large organelles, the polar granules, which contain both proteins and RNAs. Two key experiments demonstrate that there is something special about the posterior cytoplasm. First, if the posterior end of the egg is irradiated with ultraviolet light, which destroys the pole plasm activity, no germ cells develop, although the somatic cells in that region do develop. Second, if pole plasm from an egg is transferred to the anterior pole of another embryo, the nuclei that become surrounded by the pole plasm are specified as germ-cell nuclei (Fig. 9.1). If the anterior cells containing pole plasm are transplanted into the posterior pole region of a third embryo, they develop as functional germ cells.

In Chapter 2, we saw how the main axes of the *Drosophila* egg are specified by the follicle cells in the ovary, and how the mRNAs for proteins such as Bicoid and Nanos become localized in the egg (see Section 2.14). The pole plasm also becomes localized at the posterior end of the egg under the influence of the follicle cells. Several maternal genes are involved in pole-plasm formation in *Drosophila*. Mutations in any of at least eight genes result in the affected homozygous individual being 'grandchildless.' Its offspring lack a proper pole plasm, and although they may develop normally in other ways, they lack germ cells and are therefore sterile. One of these eight genes is *oskar*, which plays a central role in the organization and assembly of the pole plasm; of the genes involved in pole-plasm formation, *oskar* is the only one to have its mRNA localized at the posterior pole. The signal for localization is contained in the 3' untranslated region of the mRNA. Staufen protein is required for the localization of *oskar* mRNA and may act by linking the mRNA to the microtubule system that is polarized along the antero-posterior axis (see Section 2.14). If the region of the *oskar* gene that codes for the 3' localization signal is replaced by the *bicoid* 3' localization signal, flies made transgenic with this DNA construct have *oskar* mRNA localized at the anterior end of the egg (Fig. 9.2).

In the nematode, a germ-cell lineage is set up at the end of the fourth cleavage division, with all the germ cells being derived from the P_4 blastomere (see Section 6.1). The P_4 cell is derived from three stem-cell-like divisions of the P_1 cell. At each of these divisions, one daughter produces somatic cells, whereas the other divides again to produce a somatic cell progenitor and a P cell. The egg contains P granules in its cytoplasm that become asymmetrically distributed before each cleavage division, and are subsequently confined to the P-cell lineage (Fig. 9.3). The association of germ-cell formation with the P granules suggested that they might have a role in germ-cell specification, and at least one P-granule component, the product of the *pgl-1* gene, has been shown to be necessary for germ-cell development. PGL-1 protein may specify germ cells by regulating some aspect of mRNA metabolism.

In both the fly and the nematode, repression of transcription is necessary to specify a germ-cell fate. In the nematode, the gene *pie-1* is involved in maintaining the stem-cell property of the P blastomeres. It encodes a nuclear protein that is expressed maternally and is not a component of P granules; the PIE-1 protein is present only in the germline blastomeres, and represses new transcription of zygotic genes in these blastomeres until PIE-1 disappears at around the 100-cell stage. This general repression may protect the germ cells from the actions of transcription factors that promote development into somatic cells. In the absence of zygotic transcription, maternal RNAs encode the germ-cell components.

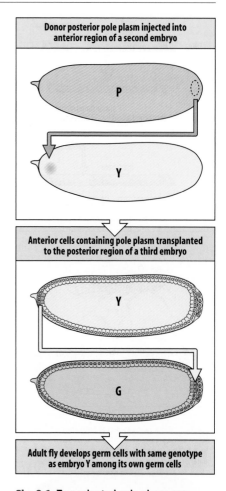

Fig. 9.1 Transplanted pole plasm can induce germ-cell formation in *Drosophila*. Pole plasm from a fertilized egg of genotype P (pink) is transferred to the anterior end of an early cleavage-stage embryo of genotype Y (yellow). After cellularization, cells containing pole plasm induced at the anterior end of embryo Y are transferred to the posterior end (a site from which germ cells can migrate into the gonad) of another embryo, of genotype G (green). The adult fly that develops from embryo G contains germ cells of genotype Y as well as those of G.

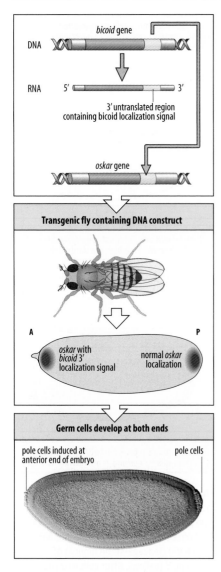

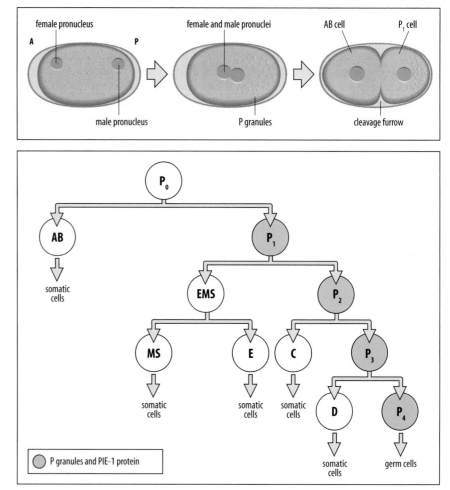

Fig. 9.2 The gene *oskar* is involved in specifying the germ plasm in *Drosophila*.
In normal eggs, *oskar* mRNA is localized at the posterior end of the embryo, whereas *bicoid* mRNA is at the anterior end. The localization signals for both *bicoid* and *oskar* mRNA are in their 3′ untranslated regions. By manipulating the *Drosophila* DNA, the localization signal of *oskar* can be replaced by that of *bicoid* (top panel). A transgenic fly is made containing the modified DNA. In its egg, *oskar* becomes localized at the anterior end (middle panel). The egg therefore has *oskar* mRNA at both ends, and germ cells develop at both ends of the embryo, as shown in the photograph (bottom panel). Thus, *oskar* alone is sufficient to initiate the specification of germ cells.

Photograph courtesy of R. Lehmann.

Fig. 9.3 P granules and PIE-1 protein become asymmetrically distributed to germline cells during cleavage of the nematode egg. Before fertilization, P granules are distributed throughout the egg. After fertilization, P granules become localized at the posterior end of the egg. At the first cleavage, they are only included in the P_1 cell (top panel), and thus become confined to the P cell lineage. The PIE-1 protein is only present in P cells. All germ cells are derived from P_4, which is formed at the fourth cleavage.

Xenopus and zebrafish also have distinct germ plasm. In *Xenopus*, distinct yolk-free patches of cytoplasm aggregate at the yolky vegetal pole after fertilization. When the blastomeres at the vegetal pole cleave, this cytoplasm is distributed asymmetrically, so that it is retained only in the most vegetal daughter cells, from which the germ cells are derived. Ultraviolet irradiation of the vegetal cytoplasm abolishes the formation of germ cells, and transplantation of fresh vegetal cytoplasm into an irradiated egg restores germ-cell formation. At gastrulation, the germ plasm is located in cells in the floor of the blastocoel cavity, among the cells that give rise to the endoderm. Cells containing the germ plasm are, however, not yet determined as germ cells and can contribute to all three germ layers if transplanted to other sites. At the end of gastrulation, the primordial germ cells are determined, and migrate out of the presumptive endoderm and into the **genital ridge**, which develops from mesoderm lining the abdominal cavity and will form the gonads. However, there are quite big differences in germ-cell origin in different groups of amphibians. In the Urodeles—the tailed amphibians that include the newts—there is no evidence for cytoplasmic localization of germ plasm, and the germ cells arise from lateral plate mesoderm. The evolutionary significance of this difference is not known.

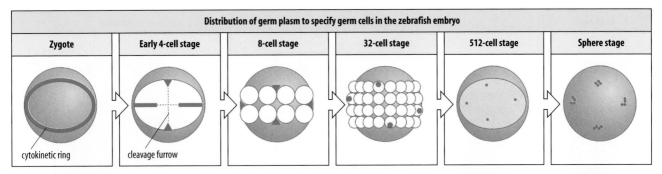

Fig. 9.4 Distribution of germ plasm in zebrafish embryo during cleavage. In this bird's-eye view of the animal pole of the egg, maternal *vasa* mRNA (blue) is a marker for the germ plasm. The *vasa* mRNA becomes localized to the cleavage furrows at first and second divisions. As a furrow completes division the *vasa* mRNA forms tight clumps at the distal ends, at or just outside the cell boundary but under the yolk cell membrane. At the 32-cell stage, the aggregates ingress into four adjacent cells and segregate asymmetrically at subsequent divisions, ending up in four widely separated cells at the 512-cell stage. At the sphere stage (cleavage cycle 13), the *vasa* mRNA becomes evenly distributed throughout the cells, and subsequent cell divisions produce a total of around 30 primordial germ cells in the four locations.

Adapted from: Pelegri, F.: 2003.

In the zebrafish, germ plasm containing a number of maternal mRNAs is present in several locations in the fertilized egg, and becomes localized to the distal ends of the cleavage furrows in the first few cleavages. Its fate can be followed during early development by tracking one of its components, such as maternal *vasa* mRNA (Fig. 9.4). By the 32-cell stage, germ plasm has been segregated into four cells. These cells then continue to divide asymmetrically, with the germ plasm segregating to one daughter cell at each division, so that even by the 512-cell stage there are still only four prospective germline cells in the embryo. At the sphere stage of the embryo (about 3.8 hours post-fertilization), the germ plasm expands to fill the cells, and equal divisions begin, eventually producing a total of around 30 primordial germ cells, which will migrate to the prospective ovaries or testes.

9.2 In mammals germ cells are induced by cell–cell interactions during development

There is no evidence for germ plasm in the mouse or other mammals or in the chick. Indeed, it seems that, although germ plasm is present in several of our developmental model organisms, it is the less prevalent mode of germline specification in animals generally. Germ-cell specification in the mouse involves cell–cell interactions, as cultured embryonic stem cells can, when injected into the inner cell mass, give rise to both germ cells and somatic cells (see Fig. 3.34). Mammalian germ cells were previously identified mainly by their high concentration of the enzyme alkaline phosphatase, which provides a convenient means of detecting them histochemically. Now that genes involved in germline specification have been identified, it is possible to identify primordial germ cells at a much earlier stage by detecting the expression of such genes.

The earliest detectable primordial germ cells can be identified in the mouse proximal epiblast just before the beginning of gastrulation. They form a cluster of six to eight cells expressing the transcriptional repressor protein Blimp1, which is thought to specifically repress genes associated with the somatic developmental program. The Blimp1-expressing cells are first found in the proximal epiblast at E6.25 days, in a region of prospective extra-embryonic mesoderm that will, after cell movement, be adjacent to the beginning of the primitive streak (Fig. 9.5). The Blimp1-positive cells proliferate and express the transmembrane protein Fragilis, which may help to demarcate germ cells from their somatic neighbors, and the protein Stella, which is exclusive to the germ-cell lineage.

By E7.5 days there are around 40 Stella-positive cells in the primitive streak and these represent the full complement of primordial germ cells that will eventually migrate to the mouse gonads (Fig. 9.5, third panel). Unlike somatic cells at this stage, the Hox genes are repressed in the primordial germ cells, which may explain their escape from a somatic cell fate and their retention of totipotency. Primordial germ cells also express Oct4, which as we shall see in Chapter 10 is a transcription factor associated with the maintenance of pluripotency in stem cells. The earliest events in establishing human germ cells remain unknown, but migratory germ cells have been reported at the fourth week of human embryonic development.

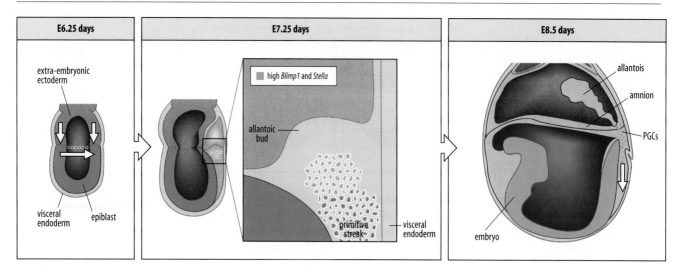

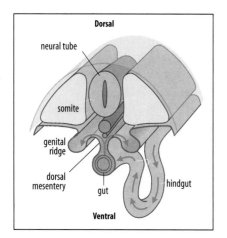

Fig. 9.5 Germ-cell formation in the mouse. First panel: a small number of primordial germ cells (PGCs) (white) expressing Blimp1 are first detectable in the proximal epiblast at 6.25 days after fertilization (E6.25). Second panel: during gastrulation, these cells and their surrounding prospective extra-embryonic mesoderm move to the posterior end of the embryo above the primitive streak, where the PGCs start also to express the germ-cell lineage gene *stella*. By E7.25, around 40 PGCs (orange) are present in the primitive streak. Third panel: at around E8.5, PGCs start to migrate to the gonads.

Fig. 9.6 Pathway of primordial germ-cell migration in the mouse embryo. In the final stage of migration, the cells move from the gut tube into the genital ridge, via the dorsal mesentery.

After Wylie, C.C., et al.: 1993.

9.3 Germ cells migrate from their site of origin to the gonad

In many animals, the primordial germ cells develop at some distance from the gonads, and only later migrate to them, where they differentiate into eggs or sperm. The reason for this separation of site of origin from final destination is not known, but it may be a mechanism for excluding germ cells from the general developmental upheaval involved in laying down the body plan, or a mechanism for selecting the healthiest germ cells, namely those that survive migration. The migration path of primordial germ cells is controlled by their environment; in *Xenopus*, for example, primordial germ cells transplanted to the wrong place in the blastula do not end up in the gonad.

The vertebrate gonad develops from mesoderm lining the abdominal cavity, which is known as the **genital ridge**. In *Xenopus*, primordial germ cells originating in the floor of the blastocoel migrate to the future gonad along a cell sheet that joins the gut to the genital ridge. Only a small number of cells start this journey, dividing about three times before arrival, so that about 30 germ cells colonize the gonad—about the same number as in the zebrafish. In contrast, the number of primordial germ cells that arrive at the genital ridge of the mouse is about 8000, starting from a population of around 40 at the posterior end of the primitive streak (see Fig. 9.5). In the mouse, the primordial germ cells enter the hindgut, and migrate along the dorsal wall until they reach the genital ridges (Fig. 9.6).

In chick embryos, the pattern of migration is different again: the germ cells originate in the epiblast and then migrate to the head end of the embryo. Most arrive at their final destination in the gonads by circulating in the blood, leaving the bloodstream at the hindgut and then migrating along an epithelial sheet to the gonad.

In zebrafish, small clusters of primordial germ cells are specified in four separate positions relative to the embryonic axis (see Fig. 9.4). These cells migrate to form bilateral lines of cells in the trunk at the level of somites 1–3 at around 6 hours post-fertilization, and then reach their final position at the level of somites 8–10 at around 12 hours, where they and the surrounding somatic cells form the gonads (Fig. 9.7). Zebrafish primordial germ cells face a challenging journey, as they originate in four separate locations, and it may be no coincidence that the journey seems to be broken up into a number of distinct stages. Germ cells taken from the margin of a mid-blastula embryo can even find their way to the gonad after transplantation to the head region (the animal pole) of a similar-stage embryo.

9.4 Germ cells are guided to their final destination by chemical signals

All migrating germ cells are continuously receiving signals for guidance, survival, and proliferation from the tissues through which they migrate. Some of these signals have

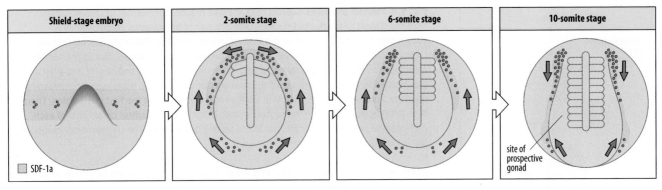

| Shield-stage embryo | 2-somite stage | 6-somite stage | 10-somite stage |

SDF-1a

site of prospective gonad

Fig. 9.7 Primordial germ cells in zebrafish migrate to the gonads in two stages. The diagrams show a dorsal view of the embryo. The chemoattractant SDF-1a (yellow) is expressed initially in a broad domain that includes the sites at which the primordial germ cells (PGCs, blue) are formed (see Fig. 9.4). PGCs migrate towards the region of highest SDF-1a, which is initially at the level of the first somite. Alterations in the pattern of SDF-1a expression then cause the PGCs to migrate posteriorly to the level of the eighth and tenth somites, where the gonads develop.

been identified; if they are lacking, germ cells are absent from the gonads or their numbers are greatly reduced. The main guidance cue in both zebrafish and in mice seems to be the chemoattractant protein SDF-1. In zebrafish, SDF-1 is secreted into the extracellular matrix by tissue cells in the regions through which the germ cells migrate and is thought to be the long-range guidance cue for the clusters of primordial germ cells to reach their final positions from their different starting points (see Fig. 9.7). Experiments that change the spatial expression of SDF-1 change the germ-cell migration patterns correspondingly. Also, knockdown by antisense RNA of either SDF-1 or its receptor CXCR4, which is expressed on the surface of primordial germ cells, disrupted migration.

In the mouse, the primordial germ cells in the primitive streak migrate anteriorly and become incorporated into the hindgut, which they eventually leave to migrate into the genital ridge (see Fig. 9.6). The mouse version of SDF-1 seems to act as a guidance cue for this final part of the journey at least, as in animals lacking either SDF-1 or its receptor, most germ cells get no further than the hindgut. The cell-surface receptor Kit, which is also produced in migrating germ cells, and its ligand the Steel protein, produced by the tissue cells over which the germ cells migrate, seem to be required for germ-cell survival and proliferation rather than for guidance. SDF-1 and CXCR4 are not only of interest to embryos; they are also expressed in some invasive human tumors, where they contribute to the ability of the tumor cells to spread to other sites in the body.

In *Drosophila*, primordial germ cells are formed next to the posterior midgut, and during gastrulation they are carried along the dorsal side of the embryo with the midgut primordium. They then migrate across the dorsal midgut epithelium to the adjacent mesoderm to join with gonadal precursor cells in forming the gonad. Screening for mutations that affect the direction of germ-cell migration has identified genes that keep the cells on track. The gene *wunen* is involved in repelling germ cells from the rest of the gut, and so preventing them from dispersing before they reach the gonadal mesoderm, whereas expression of the enzyme HMGCoA reductase in the prospective gonad is needed for the germ cells to be attracted to the prospective gonad. Whether this enzyme is involved in producing a chemoattractant, or has some other role to play, is so far not known.

9.5 Germ-cell differentiation involves a halving of chromosome number by meiosis

The sperm and egg are **haploid**—that is, they contain just one copy of each chromosome in contrast to the two copies present in the **diploid** somatic cells. At fertilization, therefore, the diploid chromosome number is reinstated. The primordial germ cells described in the previous sections are diploid, and reduction from the diploid to the haploid state during germ-cell development occurs during the specialized nuclear division known as meiosis, which does not occur until after germ cells have reached the gonads (Fig. 9.8). Meiosis comprises two cell divisions, in which the chromosomes are replicated before the first division, but not before the second, so that their

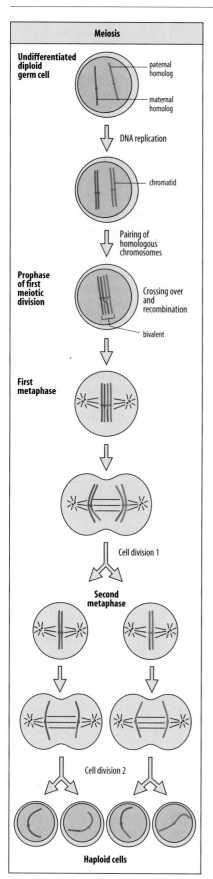

Meiosis

Undifferentiated diploid germ cell
— paternal homolog
— maternal homolog

DNA replication

— chromatid

Pairing of homologous chromosomes

Prophase of first meiotic division

Crossing over and recombination

bivalent

First metaphase

Cell division 1

Second metaphase

Cell division 2

Haploid cells

Fig. 9.8 Meiosis produces haploid cells. Meiosis reduces the number of chromosomes from the diploid to the haploid number. Only one pair of homologous chromosomes is shown here for simplicity. Before the first meiotic division the DNA replicates, so that each chromosome entering meiosis is composed of two identical chromatids. The paired homologous chromosomes (known as a bivalent) undergo crossing over and recombination, and align on the meiotic spindle at the metaphase of the first meiotic division. The homologous chromosomes separate, and each is segregated into a different daughter cell at the first cell division. There is no DNA replication before the second meiotic division. The daughter chromatids of each chromosome separate and segregate at the second cell division. The chromosome number of the resulting daughter cells is thus halved.

number is reduced by half. The cell divisions during meiosis can be unequal and give rise to a small structure called a **polar body** (Box 9A, p. 338).

During prophase of the first meiotic division, replicated homologous chromosomes pair up and undergo **recombination**, in which corresponding DNA sequences are exchanged between the homologs (see Fig. 9.8). Homologous chromosomes usually each carry different versions, or **alleles**, of many of the genes, and so meiotic recombination generates chromosomes with new combinations of alleles. Meiosis, therefore, results in gametes whose chromosomes carry different combinations of alleles compared with the parent. This means that when two gametes come together at fertilization, the resulting animal will differ in genetic constitution from either of its parents. So although we might resemble our parents, we never look exactly like them. Meiotic recombination is the main source of the genetic diversity present within populations of sexually reproducing organisms, including humans and other vertebrates.

Development of eggs and sperm follow different courses, even though they both involve meiosis, and the course is determined by the sex of the embryo. The development of the egg is known as **oogenesis**, and the main stages of mammalian oogenesis are shown in Fig. 9.9 (left panel). In mammals, germ cells undergo a small number of proliferative mitotic cell divisions as they migrate to the gonad. In the case of developing eggs, the diploid germline cells, now known as **oogonia**, continue to divide mitotically for a short time in the ovary. After entry into meiosis, they are called **primary oocytes**. Within the ovary, individual oocytes are each enclosed in a sheath of somatic ovarian cells called the **follicle**. Mammalian primary oocytes enter meiosis in the embryo but become arrested in the prophase of the first meiotic division, the stage at which homologous chromosomes pair up and recombination occurs (see Fig. 9.8). Arrest depends on high levels of cyclic AMP within the oocyte, which are generated as a result of signaling from G-protein-coupled receptors. The first meiotic division will not be completed until after ovulation in the adult, and the second meiotic division not until after fertilization.

Oocytes never proliferate again after entry into meiosis; thus, the number of oocytes at this embryonic stage is generally considered to be the maximum number of eggs a female mammal can ever have. In humans, most oocytes degenerate before puberty, leaving about 400,000 out of an original 6–7 million to last a lifetime. This number declines with age, with the decline becoming steeper after the mid-30 s until menopause, typically in the 50 s (Fig. 9.10). In mammals and many other vertebrates, oocyte development is held in suspension after birth for months (mice) or years (humans). When the female becomes sexually mature, the oocytes start to undergo further development, or maturation, as the result of hormonal stimuli. During maturation they grow up to 10-fold in size in mammals, and very much larger in some animals, such as frogs. In some vertebrates, such as frogs and fish, large numbers of oocytes become mature and are released, or ovulated, simultaneously at the end point of each reproductive cycle. In mammals, only a relatively small number of oocytes mature during each cycle and are released under the influence of luteinizing hormone. In mammals, meiosis resumes at the time of ovulation; the released oocyte completes the first meiotic division, with the production of one polar body, and proceeds as far as the metaphase of the second meiotic division. Here meiosis is arrested again, and is only completed

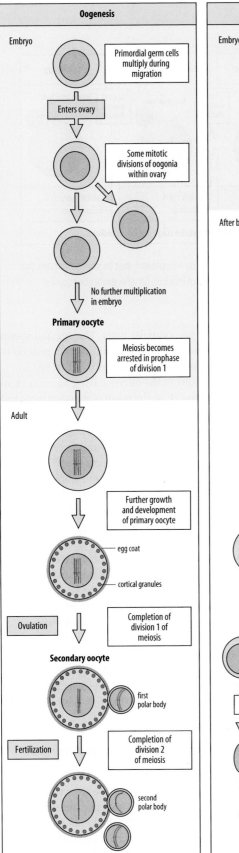

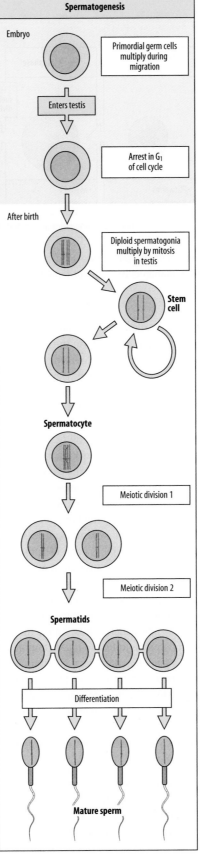

Fig. 9.9 Oogenesis and spermatogenesis in mammals. Left panel: after the germ cells that form oocytes enter the embryonic ovary, they divide mitotically a few times and then enter the prophase of the first meiotic division. No further cell multiplication occurs. Further development occurs in the sexually mature adult female. This includes a 10-fold increase in mass, the formation of external cell coats, and the development of a layer of cortical granules located under the oocyte plasma membrane. In each cycle, a group of follicles starts to grow, oocyte growth and maturation follows; a few eggs are ovulated but most degenerate. Eggs continue to mature in the ovary under hormonal influences, but become blocked in the second metaphase of meiosis, which is only completed after fertilization. Polar bodies are formed at meiosis (see Box 9A, p. 338). Right panel: germ cells that develop into sperm enter the embryonic testis and become arrested at the G_1 stage of the cell cycle. After birth, they begin to divide mitotically again, forming a population of stem cells (spermatogonia). These give off cells that then undergo meiosis and differentiate into sperm. Sperm can therefore be produced indefinitely.

Box 9A Polar bodies

Polar bodies are small cells formed by meiosis during the development of an oocyte into an egg. In this highly schematic illustration, the segregation of only one pair of chromosomes is shown for simplicity. There are two cell divisions associated with meiosis, and one daughter from each division is almost always very small compared with the other, which becomes the egg—hence the term polar bodies for these smaller cells.

The timing of meiosis in relation to the development of the oocyte varies in different animals, and in some species meiosis is completed and the second polar body formed only after fertilization (see figure). In general, polar-body formation is of little importance for later development, but in some animals the site of formation is a useful marker for the embryonic axes.

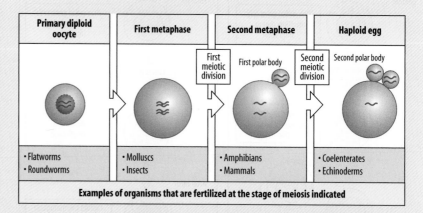

Primary diploid oocyte	First metaphase	Second metaphase	Haploid egg
• Flatworms • Roundworms	• Molluscs • Insects	• Amphibians • Mammals	• Coelenterates • Echinoderms

Examples of organisms that are fertilized at the stage of meiosis indicated

after fertilization, with the production of the second polar body. The egg and sperm pronuclei then fuse to form the zygote nucleus. The time of fertilization in relation to the stage of oocyte development differs in different animal groups (see Box 9A).

A major error in human oocyte meiosis results in the oocyte having an extra chromosome 21, known as a **trisomy** of chromosome 21, as the result of non-disjunction of homologous chromosomes 21 at the first meiotic division. This trisomy is the cause of Down syndrome, and is one of the most common genetic causes of congenital malformations and learning disability. Trisomies and other errors in chromosome segregation are relatively common in human eggs, far more so than in human sperm, and their incidence rises exponentially as oocytes age. The majority of human embryos with too many or too few chromosomes are not viable, because of the imbalance in the expression of large numbers of genes. They undergo developmental arrest at the preimplantation stage, or fail to implant, or are spontaneously aborted during gestation.

The strategy for **spermatogenesis**—the production of sperm—is quite different from oogenesis. Diploid germ cells that give rise to sperm do not enter meiosis in the embryo, but become arrested at an early stage of the mitotic cell cycle in the embryonic testis. They resume mitotic proliferation after birth. Later, in the sexually mature animal, spermatogonial stem cells give rise to differentiating spermatocytes, which undergo meiosis, each forming four haploid spermatids that mature into sperm (see Fig. 9.9, right panel). Thus, unlike the fixed number of oocytes in female mammals, sperm continue to be produced throughout the life of the organism.

In *Drosophila*, there is a continuous production of both eggs and sperm from a population of stem cells in females and males, respectively. Oogenesis begins with the division of a stem cell (see Fig. 2.16), and so there is no intrinsic limitation to the number of eggs that a female fly can produce.

9.6 Oocyte development can involve gene amplification and contributions from other cells

Eggs vary enormously in size among different animals, but they are always larger than the somatic cells. A typical mammalian egg cell is about 0.1 mm in diameter, a frog's egg about 1 mm, and a hen's egg about 3 cm (i.e. the yolk; the white of the egg is extracellular material) (see Fig. 3.2). To achieve growth to such a large size, a variety of mechanisms have evolved, some organisms using several of them together. One

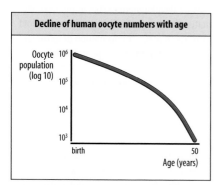

Decline of human oocyte numbers with age

Oocyte population (log 10)

10^6
10^5
10^4
10^3

birth   50
Age (years)

Fig. 9.10 Decline in human oocyte numbers with age. The graph shows the changes in numbers of oocytes in the human ovary with age, peaking at about 6-7 million (in early fetal development) and then decreasing by apoptotic cell death. At puberty there remain about 400,000 oocytes and only a very small number of these, between 400 and 500, are released throughout a woman's lifetime.

strategy is to increase the overall number of gene copies in the developing oocyte, as this proportionately increases the amount of mRNA that can be transcribed, and thus the amount of protein that can be synthesized. Vertebrate oocytes, for example, are arrested in prophase of the first meiotic division and so have double the normal diploid number of genes. Transcription and oocyte growth continue while meiosis is halted.

In addition, insects and amphibians produce many extra copies of selected genes whose products are needed in very large quantities in the egg. During amphibian oocyte development, the rRNA genes are amplified from hundreds of copies to millions; their subsequent translation produces enough ribosomal proteins to ensure sufficient ribosomes for protein synthesis during oocyte growth and early embryonic development. In insects, the genes that encode the proteins of the egg membrane, the chorion, become amplified in the surrounding follicle cells, which produce the egg membrane.

Yet another strategy is for the oocyte to rely on the synthetic activities of other cells. In insects, the nurse cells adjacent to the oocyte make many mRNAs and proteins and deliver them to the oocyte (see Fig. 2.16). Yolk proteins in birds and amphibians are made by liver cells, and are carried by the blood to the ovary, where the proteins enter the oocyte by endocytosis and become packaged into yolk platelets. In *C. elegans*, yolk proteins are made in intestinal cells, and in *Drosophila* in the fat body, from where they are transported to the oocytes. In amphibian eggs, the oocyte is polarized from an early stage onwards, and the yolk platelets accumulate at the vegetal pole. In animals that depend on maternal determinants for their early development, these mRNAs are produced by the mother and delivered into the egg, where they become localized to the appropriate position by microtubule-based transport.

9.7 Factors in the cytoplasm maintain the totipotent potential of the egg

Like all other cells in the body, mature gametes are specialized cells that have undergone a program of differentiation. As we shall consider in more detail in Chapter 10, except in rare instances, cell differentiation does not involve any alteration in the sequence or amount of DNA, but instead involves **epigenetic** changes. These are chemical modifications to DNA and to the chromosomal proteins associated with it that change the structure of the chromosome locally so that some genes are selectively silenced while others can still be expressed. Some somatic cell types, such as muscle cells and nerve cells, do not divide further once differentiation is complete. In cell types that retain the ability to divide, such as liver cells (hepatocytes) or connective tissue fibroblasts, the epigenetic changes associated with that cell type are passed on to the daughter cells, so that they are also hepatocytes or fibroblasts, respectively. Once differentiation is under way, none of these somatic cell types can in normal circumstances give rise to a completely different type of differentiated cell, let alone to the whole range of cell types needed to build a new organism.

Unlike differentiated somatic cells, mature gametes must retain totipotent potential, so that after fertilization their combined genomes can direct the development of a complete individual. In the egg, factors present in the cytoplasm are responsible for maintaining the egg genome in a totipotent state. This is demonstrated most dramatically by the ability of the cytoplasm of an unfertilized egg to 'reprogram' the nucleus from a differentiated somatic cell so that it reverts to a totipotent state and is able to direct the development of a new organism. This forms the basis of animal cloning by somatic cell nuclear transfer, which has been achieved in amphibians and some species of mammals, including mice, and is described in Chapter 10.

9.8 In mammals some genes controlling embryonic growth are 'imprinted'

Cloning mammals turned out to be much more difficult to achieve than some other animals, and this is most probably due to the phenomenon of **genomic imprinting**—by which certain genes are switched off in either the egg or the sperm during their

Fig. 9.11 Paternal and maternal genomes are both required for normal mouse development. A normal biparental embryo has contributions from both the paternal and maternal nuclei in the zygote after fertilization (left panel). Using nuclear transplantation, an egg can be constructed with two paternal or two maternal nuclei from an inbred strain. Embryos that develop from an egg with two maternal genomes—gynogenetic embryos (center panel)—have underdeveloped extra-embryonic structures. This results in development being blocked, although the embryo itself is relatively normal and well developed. Embryos that develop from eggs with two paternal genomes—androgenetic embryos (right panel)—have normal extra-embryonic structures, but the embryo itself only develops to a stage where a few somites have formed.

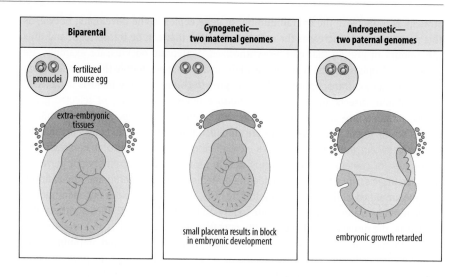

development and remain silenced in the genome of the early embryo. Evidence for imprinting in mammals came initially from the demonstration that the maternal and paternal genomes make different contributions to embryonic development.

Mouse eggs can be manipulated by nuclear transplantation to have either two paternal genomes or two maternal genomes, and can be reimplanted into a mouse for further development. The embryos that result are known as **androgenetic** and **gynogenetic** embryos, respectively. Although both kinds of embryo have a diploid number of chromosomes, their development is abnormal. The embryos with two paternal genomes have well-developed extra-embryonic tissues, but the embryo itself is abnormal, and does not proceed beyond a stage at which several somites are present. By contrast, the embryos with diploid maternal genomes have relatively well-developed embryos, but the extra-embryonic tissues—placenta and yolk sac—are poorly developed (Fig. 9.11). These results clearly show that both the maternal and the paternal genome are necessary for normal mammalian development: the two parental genomes make different contributions, and both are required for the normal development of the embryo and the placenta. This is the apparent reason that mammals cannot be naturally produced **parthenogenetically**, by activation of an unfertilized egg.

Such observations suggested that the paternal and maternal genomes must be epigenetically modified during germ-cell differentiation. The paternal and maternal genomes contain the same set of genes, but the imprinting process turns on or off certain genes in either the sperm or the egg, so that they are, or are not, expressed during development. Imprinting implies that the affected genes carry a 'memory' of being in a sperm or an egg.

Imprinting is a reversible process. The reversibility is important, because in the next generation any of the chromosomes could end up in male or female germ cells. Inherited imprinting is probably erased during early germ-cell development, and imprinting is later established afresh during germ-cell differentiation. When mammals are cloned using donor nuclei from differentiated somatic cells, these nuclei will not have gone through the normal reprogramming and imprinting process that occurs during germ-cell formation, and this could account for the large number of failures and abnormalities in the cloned embryos.

The imprinted genes affect not only early development but also the later growth of the embryo. Further evidence that imprinted genes are involved in embryonic growth comes from studies on chimeras made between normal embryos and androgenetic or gynogenetic embryos. When inner cell mass cells from gynogenetic embryos are injected into normal embryos, growth is retarded by as much as 50%. But when androgenetic inner cell mass cells are introduced into normal embryos, the chimera's growth is increased by up to 50%. The imprinted genes on the male genome thus significantly increase the growth of the embryo.

At least 80 imprinted genes have been identified in mammals, some of which encode non-coding RNAs (ncRNAs). Some imprinted genes are involved in growth control. The insulin-like growth factor IGF-2 is required for embryonic growth; its gene, *Igf2*, is imprinted in the maternal genome—that is, it is turned off, so that only the paternal gene is active (Fig. 9.12). Direct evidence for the *Igf2* gene being imprinted comes from the observation that when sperm carrying a mutated, and thus defective, *Igf2* gene fertilize a normal egg, small offspring result. This is because only very low levels of IGF-2 are produced from the imprinted maternal gene and this is not enough to make up for the loss of IGF-2 expression from the paternal genome. In contrast, if the non-functional mutant gene is carried by the egg's genome, development is normal, with the required IGF-2 activity being provided from the normal paternal gene.

Closely linked to the *Igf2* gene on mouse chromosome 7 is the gene *H19* (of unknown function), which is imprinted in the opposite direction. It is expressed from the maternal chromosome but not from the paternal one (see Fig. 9.12). By manipulating the chromosomes of an egg it was possible to make a diploid egg with two maternal genomes but with one genome having the normal female imprinting of *Igf2* and *H19* while in the other *Igf2* was not imprinted, as in the male germ cell (*H19* was absent in this genome). This egg could develop normally, unlike an embryo with two female genomes.

A possible evolutionary explanation for the reciprocal imprinting of genes that control growth invokes parental-conflict theory: the theory that the reproductive strategies of the father and mother are different. Paternal imprinting promotes embryonic growth, whereas maternal imprinting reduces it, for example. The father wants to have maximal growth for his own offspring, so that his genes have a good chance of surviving and being carried on. This can be achieved by having a large placenta, as a result of producing growth hormone, whose production is stimulated by IGF-2. The mother, who may mate with different males, benefits more by spreading her resources over all her offspring, and so wishes to prevent too much growth in any one embryo. Thus, a gene that promotes embryonic growth is turned off in the mother. Paternal genes expressed in the offspring could be selected to extract more resources from mothers, because an offspring's paternal genes are less likely to be present in the mother's other children. There are, however, many effects of imprinted genes other than on growth.

Imprinting occurs during germ-cell differentiation, and so a mechanism is required both for maintaining the imprinted condition throughout development and for wiping it out during the next cycle of germ-cell development. One mechanism for maintaining imprinting is DNA methylation, an epigenetic modification in which methyl groups are attached to cytosines in DNA. As we shall see in Chapter 10, DNA methylation is associated with gene silencing. Evidence that DNA methylation is required for imprinting comes from mice in which the methylation process is aberrant. In these mice, the *Igf2* gene is no longer imprinted and is expressed from the maternal chromosome as well as the paternal one. In addition to DNA methylation, other factors that have been implicated in imprinting are non-coding regulatory RNAs (ncRNAs), repressor proteins of the Polycomb group, and the chemical modification of histone proteins, the proteins that help package DNA in the chromosomes. Some of these ncRNAs are involved in imprinting adjacent genes. The ncRNA *Air*, for example, is encoded in the Igfr cluster of imprinted genes and is necessary for their paternal imprinting.

A number of developmental disorders in humans are associated with imprinted genes. One is Prader–Willi syndrome, which is linked to a loss of expression of the paternal copy of a gene on chromosome 15, usually as a result of a deletion of a small region of the chromosome containing that gene. Infants fail to thrive and later can become extremely obese; they also show mental retardation and mental disturbances such as obsessional-compulsive behavior. Angelman syndrome results from the loss

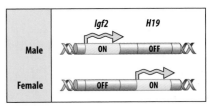

Fig. 9.12 Imprinting of genes controlling embryonic growth. In mouse embryos, the paternal gene for insulin-like growth factor 2 (*Igf2*) is on, but the gene on the maternal chromosome is off. In contrast, the closely linked gene *H19* is switched on in the maternal genome but silenced in the paternal genome.

of the same region of the maternal chromosome 15. The effect in this case is severe motor and mental retardation. Beckwith–Wiedemann syndrome is due to a generalized disruption of imprinting on a region of chromosome 7. There is excessive fetal overgrowth and an increased predisposition to cancer.

Summary

In many animals, the germline cells are specified by localized cytoplasmic determinants in the egg, whose localization is controlled in the mother by cells surrounding the oocyte. In contrast, germ cells in mammals are not specified by maternal determinants, but by intercellular interactions in the embryo. Once the germ cells are determined, they migrate from their site of origin to the gonads, where further development and differentiation takes place. In the gonads, the diploid germ-cell precursors undergo meiosis, eventually producing the haploid germ cells—eggs and sperm. The number of oocytes in a female mammal is fixed before birth, whereas sperm production in male mammals is continuous throughout adult life. Eggs are always larger than somatic cells, and in some animal groups are very large indeed. In order to achieve their increased size, specialized somatic cells surrounding the developing oocyte may provide some of the constituents, such as yolk; in addition, some genes producing materials required in large amounts may be amplified in the oocyte.

Both maternal and paternal genomes are necessary for normal mammalian development. Embryos with diploid maternal or paternal genomes develop abnormally. Certain genes in eggs and sperm are imprinted, so that the activity of the same gene is different depending on whether it is of maternal or paternal origin. Several imprinted genes are involved in growth control of the embryo. Improper imprinting can lead to developmental abnormalities in humans.

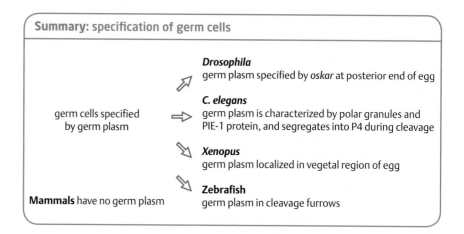

Summary: specification of germ cells

germ cells specified by germ plasm

Drosophila
germ plasm specified by *oskar* at posterior end of egg

C. elegans
germ plasm is characterized by polar granules and PIE-1 protein, and segregates into P4 during cleavage

Xenopus
germ plasm localized in vegetal region of egg

Zebrafish
germ plasm in cleavage furrows

Mammals have no germ plasm

Fertilization

Fertilization—the fusion of egg and sperm—is the trigger that initiates development. In many marine organisms, such as sea urchins, sperm released into the water by the males are attracted to eggs by chemotaxis—the sperm swim up a gradient of chemical released by the egg. The membranes of the egg and sperm fuse, and the sperm nucleus enters the egg cytoplasm, becoming the sperm **pronucleus**. In mammals and many other animals, fertilization triggers the completion of meiosis in the egg. The set of maternal chromosomes retained in the egg becomes the egg pronucleus, while the other set is segregated into the second polar body (see Box 9A, p. 338). The sperm and egg pronuclei (see Fig. 1.1) fuse to form the zygotic nucleus, and the egg starts dividing and embarks on its developmental program. Fertilization can be either external, as in frogs, or internal, as in *Drosophila*, mammals, and birds. Of all the sperm released by the male, only one fertilizes each egg. In many animals, including mammals, sperm penetration activates a blocking mechanism in the egg that prevents any further sperm entering. This is necessary because if more than one sperm nucleus enters the egg there will be additional sets of chromosomes and centrosomes, resulting in abnormal development. In humans, embryos with such abnormalities fail to develop. As we shall see in this chapter, there are multiple overlapping mechanisms for ensuring that only one sperm nucleus contributes to the zygote.

Both eggs and sperm are structurally specialized for fertilization. The specializations of the egg are directed to preventing fertilization by more than one sperm, and then to start development, while those of the sperm are directed to penetrating the egg. The unfertilized egg is usually surrounded by several protective layers outside the plasma membrane, and the eggs of many organisms have a layer of **cortical granules** just beneath the plasma membrane, whose contents are released on fertilization and help to block the entry of more sperm. These barriers together serve to make the egg impenetrable to more than a small number of sperm.

9.9 Fertilization involves cell-surface interactions between egg and sperm

Sperm are motile cells, typically designed for activating the egg and at the same time delivering their nucleus into the egg cytoplasm. They essentially consist of a nucleus, mitochondria to provide an energy source, and a flagellum for movement (Fig. 9.13). The anterior end is highly specialized to aid penetration. The sperm of *C. elegans* and some other invertebrates are unusual as they more closely resemble normal cells and move by ameboid motion.

After sperm have been deposited in the mammalian female reproductive tract, they undergo a process known as **capacitation**, which facilitates fertilization by removing certain inhibitory factors. There are very few mature eggs—usually one or two in humans and about 10 in mice—waiting to be fertilized, and fewer than a hundred of the millions of sperms deposited actually reach these eggs.

The sperm has to penetrate several physical barriers to enter the egg (Fig. 9.14). In mammalian eggs, the first is a sticky layer of hyaluronic acid and embedded somatic follicle cells, which are shed with the egg when it is released. These cells are called **cumulus cells**, as in low-power microsopic images of a released egg it appears to be surrounded by a 'cloud' of material. The first cloned mouse was called Cumulina as she was produced using the nucleus of a cumulus cell (see Chapter 10). Hyaluronidase activity on the surface of the sperm head helps it to penetrate this layer. The sperm next encounters the **zona pellucida**, a layer of fibrous glycoproteins secreted by the oocyte. This also acts as a physical barrier, but sperm are helped to penetrate it by the **acrosomal reaction**—the release of enzymes contained in the **acrosomal vesicle or acrosome** located in the sperm head (see Fig. 9.13).

Mammalian sperm carry various cell-surface proteins that are involved in binding to, and penetrating, the zona pellucida. One of these is the adhesive protein SED1.

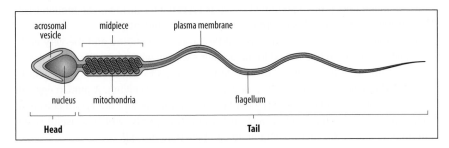

Fig. 9.13 A human sperm. The acrosomal vesicle at the anterior end of the sperm contains enzymes that are used to digest the protective coats around the egg. The plasma membrane on the head of the sperm contains various specialized proteins that bind to the egg coats and facilitate entry. The sperm moves by its single flagellum, which is powered by mitochondria. The overall length from head to tail is about 60 μm.

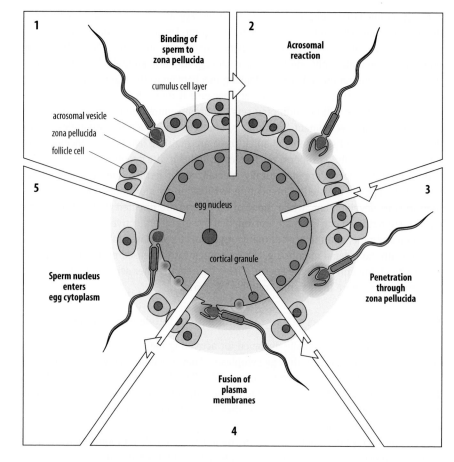

Fig. 9.14 Fertilization of a mammalian egg. After penetrating the follicle-derived cumulus cell layer, the sperm binds to the zona pellucida (1). This triggers the acrosomal reaction (2), in which enzymes are released from the acrosomal vesicle and break down the zona pellucida. This enables the sperm to penetrate the zona pellucida (3), and bind to the egg plasma membrane. The plasma membrane of the sperm head fuses with the egg plasma membrane (4). This activates the egg, causing a release of cortical granules and the sperm nucleus then enters the egg along with its mitochondria (5).

After Alberts, B., et al.: 1989.

As they progress through the epididymis of the testis, sperm become coated with this protein, which is secreted by the epididymal epithelium. Sperm lacking SED1 fail to bind the zona pellucida. After initial adhesion due to SED1 and other proteins, the acrosomal reaction is triggered when proteins in the plasma membrane of the sperm head bind to the zona pellucida glycoprotein ZP3. Binding to ZP3 triggers a signaling pathway in the sperm that results in release of the contents of the acrosome by exocytosis. The acrosomal enzymes break down the oligosaccharide side chains on the zona pellucida glycoproteins, making a hole in the zona pellucida that enables the sperm to approach the egg plasma membrane. In the sperm of many invertebrates, such as the sea urchin, the acrosomal reaction also results in the extension of a rod-like acrosomal process. This forms by polymerization of actin in the sperm cytoplasm and facilitates contact with the egg membrane.

The acrosomal reaction also exposes proteins on the sperm surface that can bind to the egg membrane and are involved in the fusion of sperm and egg membranes. One of these proteins, called Izumo, has been shown to be essential for sperm–egg fusion in mice. The ligand for Izumo on the egg plasma membrane is not yet known. The only egg membrane protein identified so far as essential for fusion is the protein CD9, but the sperm ligand for this receptor is similarly still unidentified.

Human and other mammalian eggs can be fertilized in culture and the very early pre-blastocyst embryo transferred to the mother's womb, where it implants and develops normally. This procedure of *in vitro* fertilization (IVF) has been of great help to couples who have, for a variety of reasons, difficulty in conceiving. A human egg can even be fertilized by injecting a single intact sperm directly into the egg in culture, a technique known as intracytoplasmic sperm injection (ICSI), which is useful when infertility is due to the sperm being unable to penetrate the egg.

9.10 Changes in the egg envelope at fertilization block polyspermy

Although many sperm attach to the coats around the egg, and even reach the plasma membrane, it is important that only one sperm nucleus fuses with the egg nucleus as the unbalanced chromosomal constitution resulting from the fusion of more than one sperm nucleus will be ultimately lethal to the embryo. Different organisms have different ways of ensuring fertilization by only one sperm. In birds, for example, many sperm penetrate the egg but only one sperm nucleus fuses with the egg nucleus; the other sperm nuclei are destroyed in the cytoplasm. It is likely in this case that the position at which the sperm enters the egg is important, with the sperm that enters immediately into the germinal vesicle—the region containing the egg chromosomes— acting as the fertilizing sperm. The DNA of sperm entering the cytoplasm outside this region is probably degraded by cytoplasmic DNases, whose presence has been detected in quail eggs. In sea urchins and mammals, on the other hand, the entry of more than one sperm is prevented. This general strategy is called the **block to polyspermy**. In many species, although not in mammals, the block to polyspermy consists of two stages—an initial rapid reaction followed by a slower one.

The eggs of sea urchins and *Xenopus* have a two-stage block to polyspermy. They are both externally fertilized and so are more likely than mammalian eggs to be bombarded simultaneously by large numbers of sperm. An immediate means of signaling that one sperm has fused with the egg membrane is therefore desirable. The **rapid block to polyspermy** in sea urchin is triggered within seconds by a transient depolarization of the egg plasma membrane that occurs on sperm–egg fusion. The electrical membrane potential across the plasma membrane goes from −70 mV to +20 mV within a few seconds of sperm entry (Fig. 9.15). The membrane potential then slowly returns to its original level. If depolarization is prevented, polyspermy occurs, but how depolarization blocks polyspermy is not yet clear. A similar rapid electrical block to polyspermy occurs in *Xenopus*.

As the membrane repolarizes, an impenetrable membrane called the **fertilization membrane** is formed around the egg. This is the result of the **slow block to polyspermy** and is triggered by a wave of calcium release in the egg that is due to sperm entry and leads to the cortical granules releasing their contents to the outside of the plasma membrane by exocytosis. This type of reaction occurs in mammalian eggs on fertilization, as well as in sea urchin and *Xenopus*. The wave of calcium also activates development, as we shall see in the next section. The unfertilized sea-urchin egg is surrounded by a **vitelline membrane**, which corresponds to the zona pellucida of mammalian oocytes. Release of the granule contents into the space between the plasma membrane and the vitelline layer causes this layer to lift off the plasma membrane. The cortical granule contents cross-link molecules in the vitelline membrane to produce a 'hardened' **fertilization membrane**, and also provide an additional jelly-like hyaline layer between it and the egg's plasma membrane (Fig. 9.16). Some of the released enzymes also cleave sperm-binding proteins. Together, these changes prevent

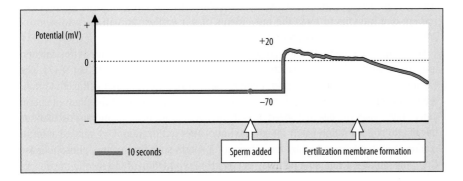

Fig. 9.15 Depolarization of the sea-urchin egg plasma membrane at fertilization. The resting membrane potential of the unfertilized sea-urchin egg is −70 mV. At fertilization, it changes rapidly to +20 mV, and then slowly returns to the original value. This depolarization may provide a fast block to polyspermy.

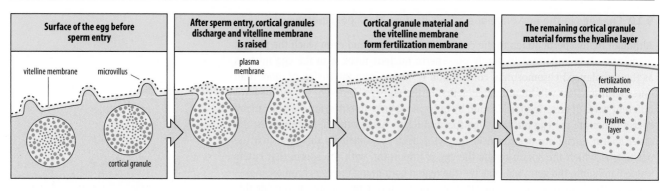

| Surface of the egg before sperm entry | After sperm entry, cortical granules discharge and vitelline membrane is raised | Cortical granule material and the vitelline membrane form fertilization membrane | The remaining cortical granule material forms the hyaline layer |

Fig. 9.16 The cortical reaction at fertilization in the sea urchin. The egg is surrounded by a vitelline membrane, which lies outside the plasma membrane. Membrane-bound cortical granules lie just beneath the egg plasma membrane. At fertilization, the cortical granules fuse with the plasma membrane, and some of the contents are extruded by exocytosis. These join with the vitelline membrane to form a tough fertilization membrane, which then lifts off the egg surface and prevents further sperm entry. Other cortical granule constituents give rise to a hyaline layer, which surrounds the egg under the fertilization membrane.

additional sperm from gaining access to the egg. The sea-urchin fertilization membrane dissolves at the blastula stage, and the blastula 'hatches' out.

Polyspermy is prevented in mammals both by mechanisms that regulate the numbers of sperm reaching the egg and by changes that occur to the zona pellucida and egg membrane once the first sperm has entered the egg. There is no rapid electrical block to polyspermy in mammalian eggs, but there is a slow block very similar to that in the sea urchin, which develops within 30 minutes to an hour after fertilization. After the first sperm fuses with the plasma membrane, the granules in the egg cortex are released by exocytosis and their contents form a layer immediately outside the egg plasma membrane. This material modifies the chemical structure of the zona pellucida so that sperm can no longer bind to it. Mammalian eggs also have a membrane block to polyspermy that develops over a similar timescale to the zona pellucida reaction. The membrane block has been demonstrated in mouse eggs whose zona pellucida has been removed before fertilization *in vitro*. Challenge with a second dose of sperm after the initial fertilization results in very few, if any, of these sperm entering the egg. The basis for the membrane block to polyspermy in mammalian eggs is currently not known.

9.11 Sperm–egg fusion causes a calcium wave that results in egg activation

The activation of the egg at fertilization initiates a series of events that result in the start of development. In the sea-urchin egg, for example, there is a several-fold increase in protein synthesis, and there are often changes in egg structure, such as the cortical rotation that occurs in amphibian eggs (see Section 4.2). Amphibian and mammalian eggs, which are arrested in second meiotic metaphase (see Box 9A, p. 338), now complete meiosis, after which the egg and sperm pronuclei fuse to form the diploid zygotic genome, and the fertilized egg enters mitosis. In mice and humans, the pronuclear membranes disappear before the pronuclei come together. In mammals, the sperm mitochondria are destroyed, and so all mitochondria in the fertilized egg are of maternal origin. The mitochondrial DNA genome is, therefore, inherited unchanged down the female line apart from the occasional change in DNA sequence as a result of mutation. This makes mitochondrial DNA an excellent material for studies that use the rare DNA sequence changes accumulated in the mitochondrial genomes of present-day humans to trace the movements of our earliest human ancestors out of Africa and their spread to different parts of the world.

Fertilization and egg activation are associated with an explosive release of free calcium ions (Ca^{2+}) within the egg, producing a wave of Ca^{2+} that travels across it (Fig. 9.17). The calcium release is triggered by sperm entry, and is both necessary and sufficient to initiate the onset of development. In sea-urchin eggs, the wave starts at the point of sperm entry and crosses the egg at a speed of 5–10 μm per second. In all mammals, oscillations in calcium concentration occur for several hours after fertilization. Ca^{2+} release at fertilization is probably triggered by the activity of a sperm-specific enzyme, phospholipase C-ζ. This initiates a signaling pathway that leads to production of the second messenger

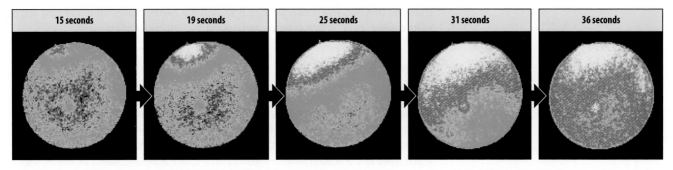

inositol 1,4,5-trisphosphate, which acts on receptors in intracellular membranes to release calcium from stores in organelles such as the endoplasmic reticulum.

The sharp increase in free Ca^{2+} is crucial for egg activation. The eggs of many animals can be activated if the Ca^{2+} concentration in the egg cytosol is artificially increased, for example, by direct injection of Ca^{2+}. Conversely, preventing calcium increase by injecting agents that bind it, such as the calcium chelator EGTA, blocks activation. It has been known for many years that *Xenopus* eggs can be activated simply by prodding them with a glass needle; this is due to a local influx in calcium at the site of insertion that triggers a calcium wave.

Calcium initiates the completion of meiosis in the fertilized egg by acting on proteins that control the cell cycle. The unfertilized *Xenopus* egg is maintained in the metaphase of the second meiotic division by the presence of high levels of a protein complex called maturation-promoting factor (MPF), which is a complex of a cyclin-dependent kinase (Cdk) and its partner cyclin. The effects of MPF are due to phosphorylation of a variety of protein targets by the kinase. For the egg to complete meiosis, the level of MPF activity must be reduced (Fig. 9.18). Similar Cdk–cyclin complexes control the mitotic cell cycle (see Section 13.2). The calcium wave results in the activation of the enzyme calmodulin-dependent protein kinase II. The activity of this kinase indirectly results in the degradation of the cyclin component of MPF, which allows the egg to

Fig. 9.17 Calcium wave at fertilization. A series of images showing an intracellular calcium wave at fertilization in a sea-urchin egg. The fertilizing sperm has fused just to the left of the top of the egg, and triggered the wave. Calcium ion concentration is monitored with a calcium-sensitive fluorescent dye, using confocal fluorescence microscopy. Calcium concentration is shown in false color: red is the highest concentration, then yellow, green, and blue. Times shown are seconds after sperm entry.

Photographs courtesy of M. Whitaker.

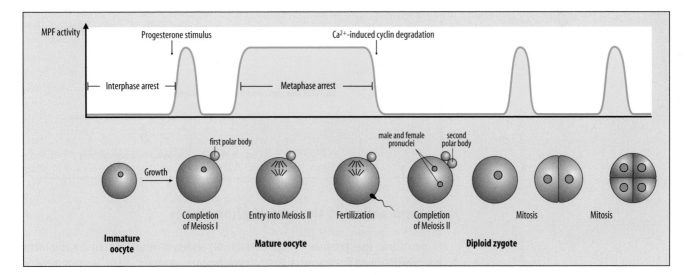

Fig. 9.18 Profile of maturation-promoting factor (MPF) activity in early *Xenopus* development. The immature *Xenopus* oocyte cell cycle is arrested. On receipt of a hormonal progesterone stimulus it enters, and completes, the first meiotic division, with the formation of the first polar body. It enters the second meiotic division, but becomes arrested again in metaphase. The egg is laid at this point.

At fertilization, the calcium wave leads to completion of meiosis, and the second polar body is formed. The zygote starts to cleave rapidly by mitotic divisions. Maturation-promoting factor (MPF) rises sharply just before each division of the meiotic and mitotic cell cycles, remains high during mitosis, and then decreases abruptly and remains low between successive mitoses.

complete meiosis. The pronuclei then fuse, and the zygote moves on to the next stage of its development, which is entry into the mitotic cell cycles of cleavage.

Summary

The fusion of sperm and egg at fertilization stimulates the egg to start dividing and developing. Both sperm and egg have specialized structures relating to fertilization. The initial binding of sperm to the mammalian egg is mediated by molecules in the zona pellucida (in mammals) or the corresponding vitelline membrane (in sea urchins), and leads to the release of the contents of the sperm acrosome, which facilitates the penetration of the sperm through the layers surrounding the egg and allows it to reach the egg plasma membrane. A block to polyspermy allows only one sperm to fuse with the egg and deliver its nucleus into the egg cytoplasm. In sea urchins, the first block is rapid and partial, and the second results from the release of egg cortical granule contents to the exterior to form an impenetrable fertilization membrane. A key role in egg activation after fertilization is played by the release of free calcium ions into the cytosol, which spread in a wave from the site of sperm fusion. In mammals, and most other vertebrates, fertilization triggers the completion of the second meiotic division; the sperm and egg haploid pronuclei give rise to the zygote nucleus, and the egg divides.

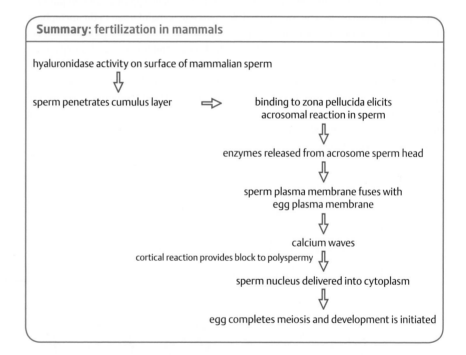

Summary: fertilization in mammals

hyaluronidase activity on surface of mammalian sperm

sperm penetrates cumulus layer ⟹ binding to zona pellucida elicits acrosomal reaction in sperm

enzymes released from acrosome sperm head

sperm plasma membrane fuses with egg plasma membrane

calcium waves

cortical reaction provides block to polyspermy

sperm nucleus delivered into cytoplasm

egg completes meiosis and development is initiated

Determination of the sexual phenotype

In organisms that produce two phenotypically different sexes, sexual development involves the modification of a basic developmental program. Early development is similar in both male and female embryos, with sexual differences only appearing at later stages. In the organisms considered here, somatic sexual phenotype—that is, the development of the individual as either male or female—is genetically fixed at fertilization by the chromosomal content of the gametes that fuse to form the fertilized egg. In mammals, for example, sex is determined by the Y chromosome: males are XY, females are XX.

Even among vertebrates, however, sex is not always determined by which chromosomes are present; in alligators, it is determined by the environmental temperature during incubation of the embryo, and some fish can switch sex as adults in response to environmental conditions. In insects, there is a wide range of different sex-determining mechanisms. Intriguing though these are, we focus here on those organisms in which the genetic and molecular basis of sex determination is best understood—mammals, *Drosophila*, and the nematode—in all of which sex is determined by chromosomal content, although by quite different mechanisms.

We first consider the determination of the somatic sexual phenotype. We then deal with the determination of the sex of the germ cells—whether they become eggs or sperm—and finally consider how the embryo compensates for the difference in chromosomal composition between males and females.

9.12 The primary sex-determining gene in mammals is on the Y chromosome

The genetic sex of a mammal is established at the moment of conception, when the sperm introduces either an X or a Y chromosome into the egg (Fig. 9.19). Eggs contain one X chromosome; if the sperm introduces another X the embryo will be female, if a Y it will be male. The presence of a Y chromosome causes testes to develop, and the hormones they produce switch the development of all somatic tissues to a characteristic male pathway and suppress female development. In the absence of a Y chromosome, the development of somatic tissues is along the female pathway. Specification of a gonad as a testis is controlled by a single gene on the Y chromosome, the **sex-determining region of the Y chromosome** (*SRY* in humans and *Sry* in mice), which was formerly known as testis-determining factor.

Evidence that a region on the Y chromosome actively determines maleness first came from two unusual human syndromes: Klinefelter syndrome, in which individuals have two X chromosomes and one Y (XXY), but are still males; and Turner syndrome, in which individuals have just one X chromosome (XO) and are female. Both these types of individual have some abnormalities; those with Klinefelter syndrome are infertile males with small testes, whereas females with Turner syndrome do not produce eggs. There are also rare cases of XY individuals who are female, and XX individuals who are phenotypically male. This is due to part of the Y chromosome being lost (in XY females) or to part of the Y chromosome being transferred to the X chromosome (in XX males). This can happen during meiosis in the male germ cells as the X and Y chromosomes are able to pair up, and crossing over can occur between them. Very rarely, this crossing over transfers the *SRY* gene from the Y chromosome onto the X (Fig. 9.20), thus leading to sex reversal.

The sex-determining region alone is sufficient to specify maleness, as shown by an experiment in which the mouse equivalent of the *SRY* gene (*Sry*) was introduced into the eggs of XX mice. These transgenic embryos developed as males, even though they lacked all the other genes on the Y chromosome. The presence of the *Sry* gene, which encodes a transcription factor, resulted in these XX embryos developing testes instead of ovaries. In these embryos, as in normal males, *Sry* was expressed in the developing gonad just before it started to differentiate and drove testis development by triggering the differentiation of precursor cells into testis-specific Sertoli cells rather than ovarian follicle cells. However, the *Sry* transgenic mice were not completely normal males, as other genes on the Y chromosome are necessary for the development of the sperm. These XX+*Sry* males were therefore infertile.

9.13 Mammalian sexual phenotype is regulated by gonadal hormones

All mammals, whatever their genetic sex, start off as embryos along a sexually neutral developmental pathway. The sex of the gonads is genetically determined, as the

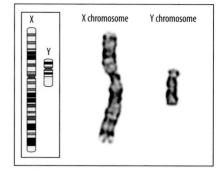

Fig. 9.19 The sex chromosomes in humans. If two X chromosomes (XX) are present, a female develops, whereas the presence of a Y chromosome (XY) leads to development of a male. The inset shows a diagrammatic representation of the banding in the chromosomes, which represent regions of increased chromatin condensation.

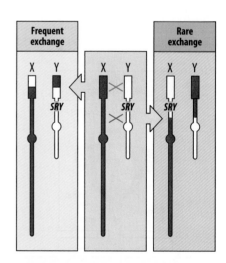

Fig. 9.20 Sex reversal in humans due to chromosomal exchange. At meiosis in male germ cells, the X and Y chromosomes pair up (center panel) and there is crossing over of the distal region (blue cross), which does not affect sexual development (left panel). On rare occasions, crossing over involves a larger segment that includes the *SRY* gene (red cross), so that the X chromosome now carries this male-determining gene (right panel). *After Goodfellow, P.N., et al.: 1993.*

presence of a Y chromosome results in the somatic cells of the embryo's gonads developing into testes rather than into ovaries. The testes secrete Müllerian-inhibiting substance, which suppresses female development by causing the embryonic precursor of the female reproductive organs to regress. It also induces cells to become Leydig cells, which secrete the male hormone testosterone, which stimulates the development of male reproductive organs. In XX individuals, the absence of a Y means that ovaries develop and all subsequent development is female.

The role of hormones in mammalian sexual development means that although the sex of the gonads is genetically determined, all the other cells in the mammalian body are neutral, irrespective of their chromosomal sex. It does not matter if they are XX or XY, as any future sex-specific development they undergo is controlled by hormones. The primary role of the testis in directing male development was originally demonstrated by removing the prospective gonadal tissue from early rabbit embryos. All the embryos developed as females, irrespective of their chromosomal constitution. To develop as a male, therefore, a testis has to be present. The testis exerts its effect on sexual differentiation of somatic tissues primarily by secreting the hormone testosterone.

The gonads in mammals develop in close association with the **mesonephros**; this is an embryonic kidney that contributes to both the male and female reproductive organs. Associated with the mesonephros on each side of the body are the **Wolffian ducts**, which run down the body to the cloaca, an undifferentiated opening. Another pair of ducts, the **Müllerian ducts**, run parallel to the Wolffian ducts and also open into the cloaca. In early mammalian development, before gonadal differentiation, both sets of ducts are present (Fig. 9.21). In females, in the absence of the testes, the Müllerian ducts develop into the **oviducts** (Fallopian tubes), which transport eggs from the ovaries to the uterus, while the Wolffian ducts degenerate.

In males, the expression of *Sry* results in the differentiation of Sertoli cells, which are the somatic cells of the testis and are essential for testis formation and for spermatogenesis, as they retain the germ cells that migrate into the gonad. Sertoli cell development involves upregulation of the transcription factor Sox9, which in turn induces production and secretion of **Müllerian-inhibiting substance**. This protein hormone in turn induces regression of the Müllerian duct, largely by apoptosis. The interstitial cell lineage in the testis then differentiates into Leydig cells, which produce testosterone. Under the influence of testosterone, the Wolffian duct develops into the **vas deferens**, the duct that carries sperm to the penis. There is evidence that the extracellular signaling molecule Wnt-4 represses testosterone production in the undifferentiated gonad, and that one of the functions of the *Sry* gene is to cause the downregulation of *Wnt-4* expression, which occurs when Sertoli cells differentiate. FGF-9 is also required for Sertoli cell differentiation and male mice lacking the gene *Fgf-9* develop as females.

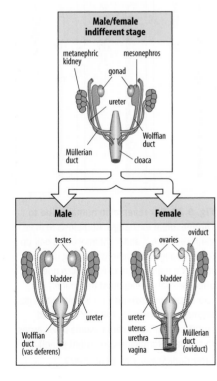

Fig. 9.21 Development of the gonads and related structures in mammals. Top panel: early in development, there is no difference between males and females in the structures that give rise to the gonads and related organs. The future gonads lie adjacent to the mesonephros, which are embryonic kidneys that are not functional in adult mammals. (The true kidney develops from the metanephros, from which the ureter carries urine to the bladder.) Two sets of ducts are present; the Wolffian ducts, which are associated with the mesonephros, and the Müllerian ducts. Both ducts enter the cloaca. Bottom left panel: after testes develop in the male, their secretion of Müllerian-inhibiting substance results in degeneration of the Müllerian duct by programmed cell death, whereas the Wolffian duct becomes the vas deferens, carrying sperm from the testis. Bottom right panel: in females, the Wolffian duct disappears, also by programmed cell death, and the Müllerian duct becomes the oviduct. The uterus forms at the end of the Müllerian ducts.

After Higgins, S.J., et al.: 1989.

The main secondary sexual characters that distinguish males and females are the reduced size of mammary glands in males, and the development of a penis and a scrotum in males instead of the clitoris and labia of females (Fig. 9.22). This is due to the action of the hormone testosterone. At early stages of embryonic development, the genital regions of males and females are indistinguishable. Differences only arise after gonad development, as a result of the action of testosterone in males. For example, in humans, the phallus gives rise to the clitoris in females and the end of the penis in males.

The role of hormones in sexual development is illustrated by rare cases of abnormal sexual development. Certain XY males develop as phenotypic females in external appearance, even though they have testes and secrete testosterone. They have a mutation that renders them insensitive to testosterone because they lack the testosterone receptor, which is present throughout the body. Conversely, genetic females with a completely normal XX constitution can develop as phenotypic males in external appearance, if they are exposed to male hormones during their embryonic development.

Sex-specific behavior is also affected by the hormonal environment as a result of the effects of hormones on the brain. For example, male rats castrated after birth develop the sexual behavioral characteristics of genetic females.

9.14 The primary sex-determining signal in *Drosophila* is the number of X chromosomes, and is cell autonomous

The external sexual differences between *Drosophila* males and females are mainly in the genital structures, although there are also some differences in bristle patterns and pigmentation, and male flies have a sex comb on the first pair of legs. In flies, sex determination of the somatic cells is cell autonomous—that is, it is specified on a cell-by-cell basis—and there is no process resembling the control of somatic sexual differentiation by hormones. Somatic sexual development is the result of a series of gene interactions that are initiated by the primary sex signal, and which act on a binary male/female genetic switch. The end result is the expression of just a few effector genes, whose activity controls the subsequent male or female differentiation of the somatic cells.

Like mammals, fruit flies have two unequally sized sex chromosomes, X and Y, and males are XY and females XX. But these similarities are misleading. In flies, sex is not determined by the presence of a Y chromosome, but by the number of X chromosomes. Thus, XXY flies are female and X flies are male. The chromosomal composition of each somatic cell determines its sexual development. This is beautifully illustrated by the creation of genetic mosaics in which the left side of the animal is XX and the right side X: the two halves develop as female and male, respectively (Fig. 9.23).

In flies, the presence of two X chromosomes results in the production of the protein Sex-lethal (Sxl), whose gene is located on the X chromosome. This leads to female development through a cascade of gene activation that first determines the sexual state and then produces the sexual phenotype. At the end of the sex-determination pathway is the *transformer* (*tra*) gene, which determines how the mRNA of the *doublesex* (*dsx*) gene is spliced; *dsx* encodes a transcription factor whose activity ultimately produces most aspects of somatic sex. The *dsx* gene is active in both males and females, but different protein products are produced in the two sexes as a result of the sex-specific RNA splicing (Fig. 9.24). Males and females thus express similar but distinct Doublesex (Dsx) proteins, which act in somatic cells to induce expression of sex-specific genes, as well as to repress characteristics of the opposite sex. Production of the

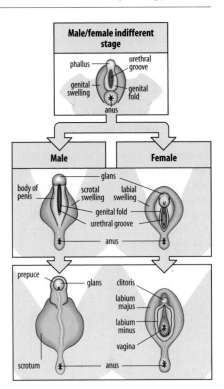

Fig. 9.22 Development of the genitalia in humans. At an early embryonic stage, the genitalia are the same in males and females (top panel). After testis formation in males, the phallus and the genital fold give rise to the penis, whereas in females they give rise to the clitoris and the labia minus. The genital swelling forms the scrotum in males and the labia majus in females.

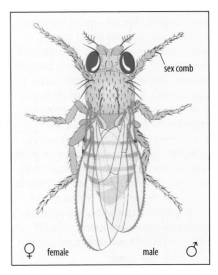

Fig. 9.23 A *Drosophila* female/male genetic mosaic. The left side of the fly is composed of XX cells and develops as a female, whereas the right side is composed of X cells and develops as a male. The male fly has smaller wings, a special structure, the sex comb, on its first pair of legs, and different genitalia at the end of the abdomen (not shown).

Fig. 9.24 Outline of the sex-determination pathway in *Drosophila*. The number of X chromosomes is the primary sex-determining signal, and in females the presence of two X chromosomes activates the gene *Sex-lethal* (*Sxl*). This produces Sex-lethal protein, whereas no Sex-lethal protein is made in males, which have only one X chromosome. The activity of *Sex-lethal* is transduced via the *transformer* gene (*tra*) and causes sex-specific splicing of *doublesex* RNA (*dsx^f*), such that the cells follow a female developmental pathway. In the absence of Sex-lethal protein, the splicing of *doublesex* RNA to give *dsx^m* RNA leads to male development.

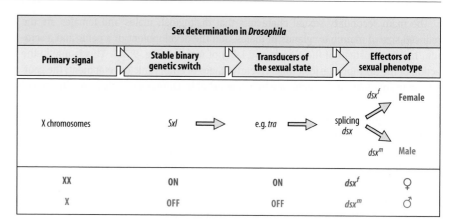

male form of the protein is the default pathway, but in the presence of Sxl, *tra* mRNA undergoes RNA splicing, and this, together with the actions of the Transformer-2 protein, leads to the female form of the Dsx protein being made, resulting in development as a female.

The *Sex-lethal* (*Sxl*) gene is only turned on in females with two X chromosomes. In the absence of early *Sxl* expression, male development occurs. Once *Sxl* is activated in females, it remains activated through an autoregulatory mechanism, which results in the Sxl protein being synthesized throughout female development. Early expression of *Sxl* in females occurs through activation of a promoter, P_e, at about the time of syncytial blastoderm formation. Sxl protein is synthesized and accumulates in the blastoderm of female embryos. At the cellular blastoderm stage, another promoter for *Sxl*, P_m, becomes active in both males and females and P_e is shut off, but the sex is already determined. Splicing of the RNA transcribed from P_m into functional *Sxl* mRNA needs some Sxl protein to be present already, and so can only happen in females (Fig. 9.25).

How does the number of X chromosomes control these key sex-determining genes? In *Drosophila*, the mechanism involves interactions between the products of so-called 'numerator' genes on the X chromosome and of genes on the autosomes, as well as maternally specified factors. Essentially, in females, the double dose of numerator proteins activates *Sxl* by binding to sites in the P_e promoter (see Fig. 9.25), overcoming the repression of Sxl by autosomally encoded proteins that would occur in males.

Fig. 9.25 Production of Sex-lethal protein in *Drosophila* sex determination. When two X chromosomes are present, the early establishment promoter (P_e) of the *Sex-lethal* (*Sxl*) gene is activated at the syncytial blastoderm stage in future females, but not in males. This results in the production of Sxl protein. Later, at the blastoderm stage, the maintenance promoter (P_m) of *Sxl* becomes active in both females and males, and P_e is turned off. The *Sxl* RNA is only correctly spliced if Sxl protein is already present, which is only in females. A positive feedback loop for Sxl protein production is thus established in females. The continued presence of Sxl protein initiates a cascade of gene activity leading to female development. If no Sxl protein is present, male development ensues.
After Cline, T.W.: 1993.

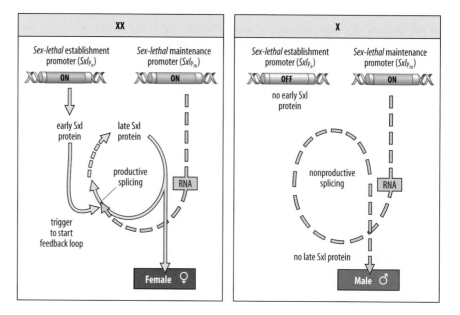

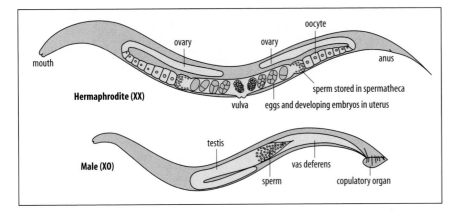

Fig. 9.26 Hermaphrodite and male *Caenorhabditis elegans.* The hermaphrodite has a 'two-armed' gonad and initially makes sperm, which is then stored in the spermotheca. It then switches to make eggs. The eggs are fertilized internally. The male makes sperm only.

The pathway outlined in Fig. 9.24 is an oversimplification, as *dsx* does not control all aspects of somatic sexual differentiation in *Drosophila*. There is an additional branch of the sex-differentiation pathway downstream of *tra*, which controls sexually dimorphic aspects of the nervous system and sexual behavior. This branch includes the gene *fruitless*, whose activity has been shown to be necessary for male sexual behavior.

In other dipteran insects, the same general strategy for sex determination is used, but there are marked differences at the molecular level. Only *dsx* has been found in dipterans distantly related to *Drosophila*; *Sxl* has been found in other dipterans but is not involved in sex determination there.

9.15 Somatic sexual development in *Caenorhabditis* is determined by the number of X chromosomes

In the nematode *C. elegans*, the two sexes are self-fertilizing hermaphrodite (essentially a modified female) and male (Fig. 9.26), although in other nematodes they are male and female. Hermaphrodites produce a limited amount of sperm early in development, with the remainder of the germ cells developing into oocytes. Sex in *C. elegans* (and other nematodes) is determined by the number of X chromosomes: the hermaphrodite (XX) has two X chromosomes, whereas the presence of just one X chromosome leads to development as a male (XO). One of the primary sex signals for hermaphrodite development is the SEX-1 protein, which is encoded on the X chromosome and is a key element in 'counting' the number of X chromosomes present. SEX-1 is a nuclear hormone receptor that represses the sex-determining gene *XO lethal* (*xol-1*), also present on the X chromosome. In the presence of a double dose of SEX-1 produced from the two X chromosomes, *xol-1* expression is inhibited, resulting in the development of a hermaphrodite. When only one X chromosome is present, *xol-1* is expressed at a high level and the embryo develops as a male.

A cascade of gene activity converts the level of *xol-1* expression into the somatic sexual phenotype (Fig. 9.27). Unlike *Drosophila*, sex determination in *C. elegans* requires intercellular interactions, as at least one of the genes involved encodes a secreted protein. At the end of the cascade is the gene *transformer-1* (*tra-1*), which encodes a transcription factor. Expression of TRA-1 protein is both necessary and sufficient to direct all aspects of hermaphrodite (XX) somatic cell development, as a gain-of-function mutation in *tra-1* leads to hermaphrodite development in an XO animal, irrespective of the state of any of the regulatory genes that normally control its activity. Mutations that inactivate *tra-1* lead to complete masculinization of XX hermaphrodites.

Before considering how the sex of the germ cells is determined and how the imbalance of X-linked genes between the sexes are dealt with in animals, we will digress briefly to touch upon sex determination in flowering plants.

Fig. 9.27 Outline of the somatic sex determination pathway in *Caenorhabditis elegans*. The primary signal for sex determination is set by the number of X chromosomes. When two X chromosomes are present, the expression of the gene *XO lethal* (*xol-1*) is low, leading to hermaphrodite development, whereas *xol-1* is expressed at a high level in males. There is a cascade of gene expression starting from *xol-1* that leads to the gene *transformer-1* (*tra-1*), which codes for a transcription factor. If *tra-1* is active, development as a hermaphrodite occurs, but if it is expressed at low levels, males develop. The product of the *hermaphrodite-1* (*her-1*) gene is a secreted protein, which probably binds to a receptor encoded by *transformer-2* (*tra-2*), inhibiting its function.

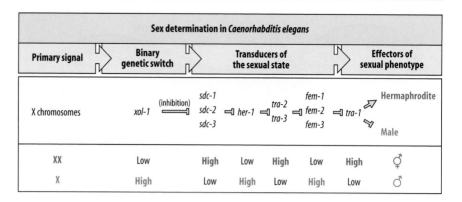

9.16 Most flowering plants are hermaphrodites, but some produce unisexual flowers

Unlike animals, plants do not set aside germ cells in the embryo, and germ cells are only specified when a flower develops. Any meristem cell can, in principle, give rise to a germ cell of either sex, and there are no sex chromosomes. The great majority of flowering plants (angiosperms) give rise to flowers that contain both male and female sexual organs, in which meiosis occurs. The male sexual organs are the stamens; they produce pollen, which contains the male gamete nuclei corresponding to the sperm of animals. The female reproductive structures are the carpels, which are either free or are fused to form a compound ovary (see Box 7A, p. 259). Carpels are the site of ovule formation, and each ovule produces an egg cell.

Angiosperms, such as *Arabidopsis*, undergo so-called 'double fertilization.' Each pollen grain contains two sperm nuclei, which are delivered into the ovule by the growth of a vegetative pollen tube. One nucleus fuses with the haploid egg cell, forming the zygote that develops into the embryo. The second nucleus fuses with the diploid 'central cell' in the ovule, forming a triploid cell that proliferates and develops into a storage tissue—the endosperm. In some plants, such as *Arabidopsis*, the endosperm nourishes the growth of the embryo, while in others it is broken down at germination to produce nutrients for the seedling.

Most flowering plants are hermaphrodite, bearing flowers with functional male and female sexual organs, as in *Arabidopsis*. In about 10% of flowering plant species, flowers of just one sex are produced. Flowers of different sexes may occur on the same plant, or be confined to different plants. The development of male or female flowers usually involves the selective resorption of either the stamens or pistil after they have been specified and have started to grow. In maize, for example, male and female flowers develop at particular sites on the shoot. The tassel at the tip of the main stem only bears flowers with stamens; the 'ears' at the ends of the lateral branches bear female flowers containing pistils. Sex determination becomes visible when the flower is still small, with the stamen primordia being larger in males and the pistil longer in females. The smaller organs eventually degenerate. The plant hormone gibberellic acid may be involved in sex determination, as differences in gibberellin concentration are associated with the different sexual organs. In the maize tassel, gibberellin concentration is 100-fold lower than in the developing ears. If the concentration of gibberellic acid is increased in the tassel, pistils can develop.

Genomic imprinting (see Section 9.7) occurs in flowering plants, but has only been detected in the endosperm and not in the embryo. The imprinting events take place in the central-cell genome and the sperm genome. In plants, as in mammals, imprinting involves gene silencing due to DNA methylation, histone modifications, Polycomb proteins and non-coding RNAs. Unlike mammals, however, the imprinting does not have to be removed to produce a new generation, as the endosperm is a temporary tissue that does not contribute cells to the embryo. As well as the silencing of paternal genes in sperm by DNA methylation and histone modification, a feature of angiosperm imprinting is the DNA demethylation of previously silenced maternal genes in the central cell, which then enables their expression. The activated genes include members of the Polycomb group, which then maintain the repression of incoming silenced paternal alleles and silence the maternal alleles of certain other genes. Imprinting in angiosperms could have evolved together with double fertilization as a means of preventing endosperm proliferation in the absence of fertilization.

9.17 Determination of germ-cell sex depends on both genetic constitution and intercellular signals

Determination of the sex of animal germ cells—that is, whether they will develop into eggs or sperm—is strongly influenced by the signals they receive when they become part of a gonad. In the mouse, for example, their future development is determined largely by the sex of the gonad in which they reside, and not by their own chromosomal constitution. In reality, chromosomal constitution and gonadal signals will almost always coincide, but it has been shown that germ cells from male mouse embryos can develop into oocytes rather than sperm if grafted into female embryonic gonads and vice versa.

There is a distinct difference in the timing of meiosis in male and female mammals. In male mouse embryos, the diploid germ cells stop dividing when they enter the gonad, becoming arrested in the G_1 phase of the mitotic cell cycle. They start dividing mitotically again after birth and enter meiosis some 7 to 8 days after birth. In female mouse embryos, the primordial diploid germ cells continue to proliferate for a few days after entering the genital ridge (see Sections 9.4 and 9.5). Diploid germ cells in the gonad first undergo a few rounds of mitotic division and then enter prophase of the first meiotic division; they then arrest at this primary oocyte stage until the mouse becomes a sexually mature female, about 6 weeks after birth, when at each reproductive cycle, selected oocytes complete the first meiotic division and start the second meiotic division (Fig. 9.28). Meiosis is only completed after fertilization.

Although the development of germ cells is normally heavily influenced by the environment in which they find themselves, if the usual external cues are absent, germ cells seem to follow an intrinsic pathway of differentiation. All mouse germ cells that enter meiosis before birth develop as eggs, whereas those not entering meiosis until after birth develop as sperm. Germ cells, whether XX or XY, that fail to enter the genital ridge and instead end up in adjacent tissues such as the embryonic adrenal gland or mesonephros, enter meiosis and begin developing as oocytes in both male and female embryos; thus, the default germ-cell sex appears to be female. XX/XY chimeric mouse embryos can be made by combining male and female four-cell embryos and, because of the presence of one Y chromosome, can develop testes. In these XX/XY embryos, XX germ cells that are surrounded by testis cells start to develop along the spermatogenesis pathway, but the later development of the gametes is abnormal. In a particular strain of Y/XX mice that develops ovaries rather than testes, XY germ cells appear to develop quite normally as oocytes but fail to develop further after fertilization as the spindle does not assembly properly for the second meiotic division. Experiments transferring XY oocyte nuclei into normal XX oocytes produced healthy offspring, indicating that the fault lies in the XY cytoplasm.

In *Drosophila*, the difference in the behavior of XY and XX germ cells depends initially on the number of X chromosomes, as in somatic cells; the *Sxl* gene again plays an important role, although most other elements in the sex-determination pathway may differ from those in somatic cells. Both chromosomal constitution and cell interactions are involved in the development of germ-cell sexual phenotype. Transplantation of genetically marked pole cells (see Section 9.1) into a *Drosophila* embryo of the opposite sex shows that male XY germ cells in a female XX embryo become integrated into the ovary and begin to develop as sperm; that is, their behavior is autonomous

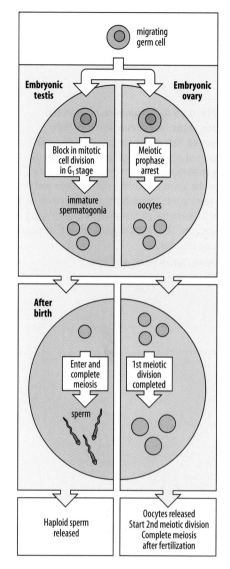

Fig. 9.28 The timing of meiosis in the germ cells differs considerably between males and females in mammals. Top panel: Migrating germ cells, whether XX or XY, enter meiotic prophase and start developing as oocytes unless they enter a testis. In the testis, the germ cells receive an inhibitory signal that blocks mitotic division and prevents them entering meiotic prophase. Bottom panel: In male mice after birth, immature diploid spermatogonia in the testis enter and complete meiosis in the testis to produce haploid germ cells which eventually mature into sperm (left). In female mice after birth, oocytes complete their first meiotic division in the ovary, but do not enter the second meiotic division meiosis until after they are released from the ovary. The egg only completes meiosis after fertilization.

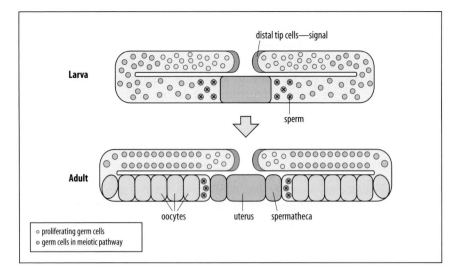

Fig. 9.29 Determination of germ-cell sex in the hermaphrodite nematode gonad. Top: during the larval stage, germ cells in a zone close to the distal tips of the gonad multiply; when they leave this zone in the larval stage they enter meiosis and develop into sperm. Bottom: in the adult, cells that leave the proliferative zone develop into oocytes. The eggs are fertilized as they pass into the uterus.

After Clifford, R., et al.: 1994.

with respect to their genetic constitution. By contrast, XX germ cells in a testis attempt to develop as sperm, showing a role for environmental signals. In neither case, however, are functional sperm produced.

The hermaphrodite of *C. elegans* provides a particularly interesting example of germ-cell differentiation, as both sperm and eggs develop within the same gonad. Unlike the somatic cells, which have a fixed lineage and number (see Section 6.1), the number of germ cells in an adult nematode is indeterminate, with about 1000 germ cells in each 'arm' of the gonad. At hatching of the first-stage larva, there are just two founder germ cells, which proliferate to produce the germ cells. The germ cells are flanked on each side by cells called distal tip cells, and their proliferation is controlled by a signal from the distal tip cells. This signal is the protein LAG-2, which is homologous to the Notch ligand Delta. The receptor for LAG-2 on the germ cells is GLP-1, which is similar both to nematode LIN-12, which is involved in vulva formation (see Section 6.6) and to Notch. We have already met GLP-1 acting to determine cell fate in the early embryo (see Section 6.3).

In *C. elegans*, entry of germ cells into meiosis from the third larval stage onward is controlled by the distal tip signal. In the presence of this signal, the cells proliferate, but as they move away from it, they enter meiosis and develop as sperm (Fig. 9.29, top). In the hermaphrodite gonad, all the cells that are initially outside the range of the distal tip signal develop as sperm, but cells that later leave the proliferative zone and enter meiosis develop as oocytes (see Fig. 9.29, bottom). The eggs are fertilized by stored sperm as they pass into the uterus. The male gonad has similar proliferative meiotic regions, but all the germ cells develop as sperm.

Sex determination of the nematode germ cells is somewhat similar to that of the somatic cells, in that the chromosomal complement is the primary sex-determining factor and many of the same genes are involved in the subsequent cascades of gene expression. The terminal regulator genes required for spermatogenesis are called *fem* and *fog*. In hermaphrodites, there must be a mechanism for activating the *fem* genes in some of the XX germ cells, so allowing them to develop as sperm.

9.18 Various strategies are used for dosage compensation of X-linked genes

In all the animals we have considered in this chapter there is an imbalance of X-linked genes between the sexes. One sex has two X chromosomes, whereas the other has one. This imbalance has to be corrected to ensure that the level of expression of genes carried on the X chromosome is the same in both sexes. The mechanism by which the imbalance in X-linked genes is dealt with is known as **dosage compensation**. Failure

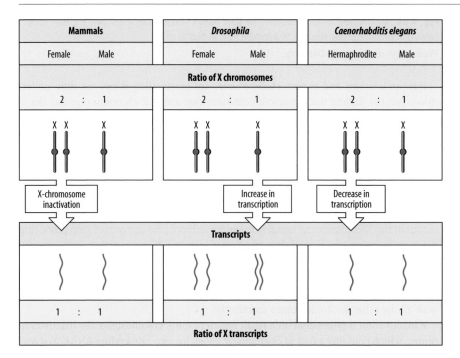

Mammals			Drosophila			Caenorhabditis elegans		
Female		Male	Female		Male	Hermaphrodite		Male
Ratio of X chromosomes								
2	:	1	2	:	1	2	:	1
X X		X	X X		X	X X		X
X-chromosome inactivation			Increase in transcription			Decrease in transcription		
Transcripts								
1	:	1	1	:	1	1	:	1
Ratio of X transcripts								

Fig. 9.30 Mechanisms of dosage compensation. In mammals, *Drosophila,* and *Caenorhabditis elegans* there are two X chromosomes in one sex and only one in the other. Mammals inactivate one of the X chromosomes in females; in *Drosophila* males there is an increase in transcription from the single X chromosome; and in *C. elegans* there is a decrease in transcription from the X chromosomes in hermaphrodites. The result of these different dosage compensation mechanisms is that the level of X chromosome transcripts is approximately the same in males and females.

to correct the imbalance leads to abnormalities and arrested development. Different animals deal with the problem of dosage compensation in different ways (Fig. 9.30).

Mammals, such as mice and humans, achieve dosage compensation in females by inactivating one X chromosome in each cell after the blastocyst has implanted in the uterine wall. Once an X chromosome has been inactivated in an embryonic cell, this chromosome is maintained in the inactive state in all the resulting somatic cells, and inactivation persists throughout the life of the organism (Fig. 9.31). The inactive X chromosome is replicated at each cell division but remains transcriptionally inactive and in a different physical state from the other chromosomes during the rest of the cell cycle. At mitosis all chromosomes become highly condensed. During the interphase of the cell cycle—the period between successive mitoses—all the other chromosomes decondense into extended threads of **chromatin**, the complex of DNA and proteins of which chromosomes are made, and are no longer visible under the light microscope. The inactive X chromosome, however, remains condensed and is visible in human cells as the Barr body (Fig. 9.32). Which X chromosome in a cell is inactivated seems to be random, and so female mammals are a mosaic of cells with different X chromosomes inactivated.

The mosaic effect of X inactivation is sometimes visible in the coats of female mammals. Female mice heterozygous for a coat pigment gene carried on the X chromosome have patches of color on their coat, produced by clones of epidermal cells that express the X chromosome carrying a functional pigment gene. The rest of the epidermis is composed of cells in which that X chromosome has been inactivated. The coat pattern of tortoiseshell cats is also due to X-linked mosaicism. They carry two co-dominant alleles, *orange* (X^O) and *black* (X^B), of the main coat color gene, which is located on the X chromosome. These alleles produce orange and black pigments, respectively. Pigment genes are expressed in melanocytes that derive from neural crest cells that migrate to the skin. Cells in which the chromosome carrying X^O is inactivated express the X^B allele and vice versa. In bi-colored tortoiseshell cats (Fig. 9.33), the two cell types are intermingled, producing the characteristic brindled appearance. In calico cats, which have distinct patches of white, orange, and black fur, there has been less intermingling and larger clones of cells of one color develop, along with white patches with no pigment.

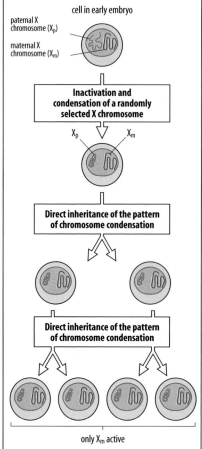

cell in early embryo

paternal X chromosome (X_p)

maternal X chromosome (X_m)

Inactivation and condensation of a randomly selected X chromosome

X_p X_m

Direct inheritance of the pattern of chromosome condensation

Direct inheritance of the pattern of chromosome condensation

only X_m active

Fig. 9.31 Inheritance of an inactivated X chromosome. In early mammalian female embryos one of the two X chromosomes, either the paternal X (X_p) or the maternal X (X_m), is randomly inactivated. In the figure, X_p is inactivated and this inactivation is maintained through many cell divisions. The inactivated chromosome becomes highly condensed.

After Alberts, B., et al.: 2002.

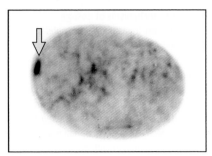

Fig. 9.32 Inactivated X chromosome (the Barr body). The photograph shows the Barr body (arrow) in the interphase nucleus of a female human buccal cell.

Photograph courtesy of J. Delhanty.

How cells count and choose chromosomes for inactivation is not yet fully understood. But the finding that tetraploid cells have two active X chromosomes has led to the following proposal, which depends on the levels of a signal produced by the X chromosomes themselves and invokes a negative-feedback mechanism between autosomes and the X chromosomes. Before inactivation, the X chromosomes are producing a signal (whether a protein or an RNA is not yet known) that binds to the other chromosomes in the cell. When all the binding sites are occupied, the autosomal chromosomes in turn produce a signal that can bind to and inactivate X chromosomes. The first X chromosome to acquire a critical mass of bound signal initiates its own inactivation. The output of the X-chromosome signal is therefore reduced, the autosomes stop producing the X-inactivating signal, with the result that just one X chromosome is inactivated per diploid genome. This type of mechanism can also accommodate the long-standing observation that in diploid female cells with an abnormal sex-chromosome constitution, such as XXY or XXXY, there is only one active X chromosome per somatic cell, with all the other X chromosomes inactivated.

X inactivation is dependent on a small region of the X chromosome, the inactivating center, which contains a pair of overlapping genes for two non-coding RNAs, *Xist* and *Tsix*, whose competing activities regulate inactivation. Before inactivation, both RNAs are transcribed at low levels from both X chromosomes. *Xist* expression then dramatically increases on one or other of the X chromosomes and ceases on the other. What causes this to happen is still unknown, but a candidate might be the still-hypothetical inactivating signal from the autosomes noted above. The *Xist* transcripts spread from their site of synthesis to coat the whole X chromosome, blocking transcription, including that of *Tsix*, and inactivating the chromosome. On the chromosome that is not inactivated, *Tsix* remains active for a little time but soon after X inactivation is complete, both *Tsix* and *Xist* are switched off on the active chromosome. *Xist* RNA continues to keep the inactivated chromosome inactive through subsequent cell divisions. The inactive X chromosome has a pattern of DNA methylation that differs from that of the

Fig. 9.33 The coloring of tortoiseshell cats is due to mosaicism of X-linked alleles due to X inactivation.

Foxfire, photograph courtesy of Bruce Goatly.

active X, and this methylation is likely to help keep it inactive. DNA methylation is one of the mechanisms used in mammals for long-term silencing of genes, and we have already seen its effects in the differential imprinting of maternal and paternal genes in the germline (see Section 9.8). DNA methylation and other epigenetic mechanisms for controlling gene expression are discussed in more detail in Box 10A (p. 374).

Dosage compensation in *Drosophila* works in a different way to that in mice and humans (see Fig. 9.30). Instead of repression of the 'extra' X activity in females, transcription of the X chromosome in males is increased nearly twofold. A set of male-specific genes, the MSL complex, controls most dosage compensation, and these are repressed in females by Sxl protein, thereby preventing excessive transcription of the X chromosome. The increased activity in males is regulated by the primary sex-determining signal, which results in the dosage compensation mechanism operating when *Sxl* is 'off'. In females, where *Sxl* is 'on', it turns off the dosage-compensation mechanism. As in the mouse, this regulatory mechanism involves non-coding RNA.

In *C. elegans*, dosage compensation is achieved by reducing the level of X chromosome expression in XX animals to that of the single X chromosome in XO males (see Fig. 9.30). The number of X chromosomes is communicated by a set of X-linked genes that can repress the master gene *xol-1*. A key event in initiating nematode dosage compensation is the expression of the protein SDC-2, which occurs only in hermaphrodites. It forms a complex specifically with the X chromosome and triggers assembly of a protein complex called the dosage compensation complex, which binds to a specific region on the X chromosome and reduces transcription.

Summary

The development of early embryos of both sexes is very similar. A primary sex-determining signal sets off development toward one or the other sex, and in mammals, *Drosophila*, and *C. elegans*, this signal is determined by the chromosomal complement of the fertilized egg. In mammals, the *Sry* gene on the Y chromosome is responsible for the embryonic gonad developing into a testis and producing hormones that determine male sexual characteristics. The sexual phenotype of the somatic cells is determined by the gonadal hormones. In *C. elegans* and *Drosophila*, the primary sex-determining signal is the number of X chromosomes. In *Drosophila*, the gene *Sex-lethal* is turned on in females but not in males, in response to this signal. In both cases, this results in further gene activity in which sex-specific RNA splicing is involved. In *C. elegans*, the gene *XO lethal* is turned off in hermaphrodites and on in males, eventually leading to sex-specific expression of the gene *transformer-1*, which determines the sexual phenotype. Somatic sexual differentiation in *Drosophila* is cell autonomous and is controlled by the number of X chromosomes; in *C. elegans*, cell–cell interactions are also involved. In mammals, signals from the gonads determine whether the germ cells develop into oocytes or sperm. Male germ cells in *Drosophila* develop along the sperm pathway even in an ovary, but female germ cells develop along the sperm pathway when placed in a testis. Most *C. elegans* adults are hermaphrodites, and produce both sperm and eggs from the same gonad.

Various strategies of dosage compensation are used to correct the imbalance of X chromosomes between males and females. In female mammals, one of the X chromosomes is inactivated; in *Drosophila* males the activity of the single X chromosome is upregulated; and in *C. elegans* the activity of the X chromosomes in XX hermaphrodites is downregulated to match that from the single X chromosome in males. Plants do not have a germline as any cell can, in principle, give rise to ova and sperm. Most plants produce flowers with hermaphrodite flowers, but some produce unisexual flowers, either on the same plant or on two separate plants. Plants also have a form of genomic imprinting involving sperm and endosperm cells.

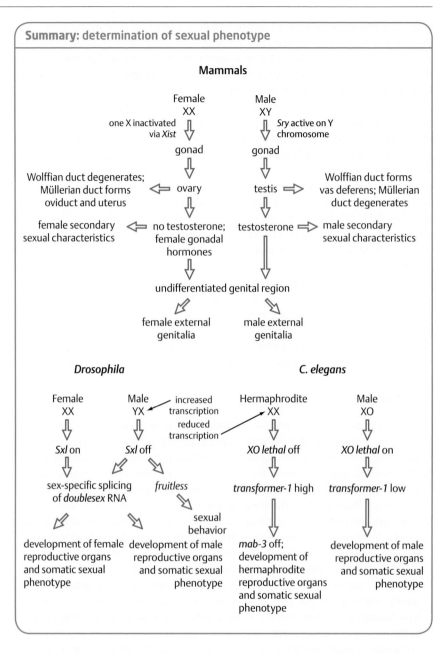

Summary to Chapter 9

In many animals, the future germ cells are specified by localized cytoplasmic determinants in the egg. Germ cells in mammals are unusual, as they are specified entirely by cell–cell interactions. In animals, the development of germ cells into sperm or egg depends both on chromosomal constitution and interactions with the cells of the gonad. At fertilization, fusion of sperm and egg initiates development, and there are mechanisms to ensure that only one sperm enters an egg. Both maternal and paternal genomes are required for normal mammalian development, as some genes are imprinted; for such genes, whether they are expressed or not during development depends on whether they are derived from the sperm or the egg. In many animals, the chromosomal constitution of the embryo determines which sex will develop. In mammals, the Y chromosome is male-determining; it specifies the development of a testis, and the hormones produced by the testis cause the development of male sexual characteristics. In the absence of a Y chromosome, the embryo develops as a female. In *Drosophila* and *C. elegans*, sexual development is initially determined by the number of X chromosomes, which sets in train a cascade of gene activity. In nematodes, the somatic sexual phenotype is determined by cell–cell interactions; in flies, somatic cell sexual differentiation is cell autonomous. Animals use a variety of strategies of dosage compensation to correct the imbalance in the number of X chromosomes in males and females.

■ End of chapter questions

Long answer (concept questions)

1. Distinguish between the germline, gametes, and somatic cells. What key functions are the responsibility of the germ cells?

2. Summarize the evidence for the existence of a special germ plasm in *Drosophila* and *Xenopus*. Is the presence of germ plasm a universal feature of animal development?

3. Draw a schematic of two generations of *Drosophila*: start with a male and a female, each of which is heterozygous for a loss-of-function recessive *oskar* mutation. Mate them together to get homozygous *oskar* males and females. Mate these *oskar* mutants to normal (wild-type) flies. How do these matings illustrate why such mutations are referred to as 'grandchildless'. Does it matter whether the homozygous *oskar* mutant is a male or a female? Explain.

4. Contrast the roles of SDF-1/CXCR4 and STEEL/KIT in germ cell migration.

5. Discuss the following aspects of meiosis: (1) Why is a 'reduction division' (meiosis I) important? (2) What is the significance of recombination in sexual reproduction? (3) Why do four sperm cells result from each cell undergoing meiosis during spermatogenesis, yet only one egg cell results during oogenesis—where do the other 'eggs' go?

6. When in human development do primary oocytes form? What happens to a primary oocyte between this time and its maturation into an egg ready to be fertilized? At what point in meiosis is the oocyte when fertilization occurs?

7. What is meant by an 'epigenetic' mechanism? What is meant by 'genomic imprinting'? What is the connection between the two?

8. Describe the process by which the sperm nucleus enters the egg. Include SED1, ZP3, and acrosomal enzymes in your answer.

9. Fertilization of a sea-urchin egg leads to a wave of calcium that sweeps across the egg. Is this calcium intracellular or extracellular in origin? What are two consequences of this increase in calcium?

10. What is MPF? Summarize the role of MPF-like complexes in the control of the various stages of the cell cycle.

11. Explain how chromosomal abnormalities can be used as evidence for the following statements. (1) Determination of the male sex in humans is controlled by the presence of the Y chromosome, not two X chromosomes. (2) Determination of female sex in *Drosophila* is controlled by the presence of two X chromosomes and not the presence of a Y chromosome. (3) One X chromosome in humans is sufficient to determine the female sex in humans. (4) One X chromosome is sufficient to determine the male sex in *Drosophila*. (5) *SRY* is the gene responsible for male sex determination in humans.

12. How does *SRY* trigger male development in mammals? Include: Sox9, Sertoli cells, Leydig cells, Müllerian inhibiting substance and testosterone in your answer.

13. Summarize the development of the gonads and related structures in mammals, focusing on the Wolffian and Müllerian ducts.

14. Summarize the information presented in Figs 9.24 and 9.25 on how cell-autonomous sexual differentiation occurs in *Drosophila*. Do not just copy the two figures, but instead incorporate the information into one comprehensive schematic.

15. What is different about the sexes in *Caenorhabditis elegans* compared with the sexes in our other model organisms? How does this affect its method of reproduction?

16. Briefly compare and contrast the sex-determination pathways in *Drosophila* and *C. elegans*. Are there similarities in the strategies used? Are there differences in the mechanisms involved?

17. Say how X-chromosome inactivation will achieve dosage compensation in humans with the chromosomal compositions: XY, XX, XXY, or XXX?

18. How does *Xist* control X-chromosome inactivation in mammals? Speculate why the other gene involved in this process, *Tsix*, might have been given this name.

Multiple choice (factual recall questions)

NB There is only one correct answer to each question.

1. The germ plasm in *Drosophila* is specified by

a) *bicoid*

b) Hox genes

c) *oskar*

d) the point of sperm entry

2. In mice, germ cells can be first identified

a) in prospective mesoderm of the proximal epiblast, as Blimp1-expressing cells

b) in the mesoderm that will form gonadal tissue

c) in the posterior tip of the embryo, as cells containing pole plasm

d) in the genital ridge, as Oct-4-expressing cells

3. Which feature of meiosis, as compared with mitosis, is most important for the production of gametes?

a) Homologous chromosomes separate in meiosis I, whereas sister chromatids separate in mitosis.

b) Meiosis leads to the production of haploid products, whereas mitosis leads to diploid daughter cells.

c) Meiosis provides an opportunity for recombination, whereas mitosis does not.

d) Typically, three of the four products do not become oocytes during meiosis in female organisms, whereas mitosis will produce four functional products after two rounds of cell division.

4. In what stage of which type of cell division, is the mammalian oocyte at the birth of the animal?

a) The oocyte is in G1 of the mitotic cell cycle.

b) The oocyte is in metaphase of meiosis II.

c) The oocyte is in prophase of meiosis I.

d) The oocyte is in prophase of mitosis.

5. What is the zona pellucida of the mammalian egg?

a) a hardened membrane that forms a physical block to polyspermy, formed from the vitelline membrane and the contents of the cortical granules

b) a layer of follicle-derived cells called cumulus cells

c) an extracellular layer of glycoproteins

d) the plasma membrane of the egg

6. The mammalian oviduct will form from the

a) mesonephros

b) Müllerian duct

c) ureter

d) Wolffian duct

7. The molecular activity of the Sxl protein of *Drosophila* is

a) as a 'numerator', counting the number of X chromosomes

b) as a transcription factor

c) to control RNA splicing

d) to signal to the Tra receptor

8. The human congenital disorder Angelman syndrome is due to the inheritance of a maternal chromosome 15 that has a small deletion of a specific region. Why does this deletion not behave as a recessive allele; that is, why is its loss not made up for by the intact region on the unmutated paternal chromosome 15?

a) Chromosomes with deletions do not go through mitosis correctly, so cell divisions in the embryo result in cells with abnormal numbers of chromosomes, and these cells do not contribute properly to the development of the organism.

b) The father's copy of chromosome 15 has genes in the region of the deletion that are imprinted, and thus inactive; in the absence of any active copies of these genes, development cannot proceed normally.

c) The genes in this portion of chromosome 15 are special in that they are required in two copies for normal development, and so the loss of one set does not allow normal development.

d) Two copies of every gene in the genome are required for development, so loss of one of the copies from this region perturbs development.

9. What is the cortical reaction, and why is it important?

a) The cortical reaction is the depolarization of the plasma membrane after sperm entry, which helps to block polyspermy.

b) The cortical reaction is the entry of Ca^{2+} ions into the egg via the cortex, which initiates development.

c) The cortical reaction is the fusion of the egg cortex with the egg plasma membrane, which allows the sperm to enter.

d) The cortical reaction is the release of the cortical granules after sperm entry, which converts the vitelline membrane into the fertilization membrane, which blocks polyspermy.

10. The *C. elegans* GLP-1 protein is similar to which mammalian signaling protein?

a) BMPs

b) Delta

c) Notch

d) testosterone receptor

Multiple choice answer key

1: c, 2: a, 3: b, 4: c, 5: c, 6: b, 7: c, 8: b, 9: d, 10: c.

■ General further reading

Chadwick, D., Goode, J.: *The Genetics and Biology of Sex Determination 2002*. Novartis Foundation Symposium 244. New York: John Wiley, 2002.

Cinalli, R.M., Rangan, P., Lehman, R.: **Germ cells are forever**. *Cell* 2008, **132**: 559–562.

Crews, D.: **Animal sexuality**. *Sci. Am.* 1994, **270**: 109–114.

Zarkower, D.: **Establishing sexual dimorphism: conservation amidst diversity?** *Nat. Rev. Genet.* 2001, **2**: 175–185.

■ Section further reading

9.1 Germ-cell fate is specified in some embryos by a distinct germ plasm in the egg

Extavour, C.G., Akam, M.: **Mechanisms of germ cell specification across the metazoans: epigenesis and preformation**. *Development* 2003, **130**: 5869–5884.

Matova, N., Cooley, L.: **Comparative aspects of animal oogenesis**. *Dev. Biol.* 2001, **231**: 291–320.

Mello, C.C., Schubert, C., Draper, B., Zhang, W., Lobel, R., Priess, J.R.: **The PIE-1 protein and germline specification in *C. elegans* embryos**. *Nature* 1996, **382**: 710–712.

Micklem, D.R., Adams, J., Grunert, S., St. Johnston, D.: **Distinct roles of two conserved Staufen domains in *oskar* mRNA localisation and translation**. *EMBO J.* 2000, **19**: 1366–1377.

Ray, E.: **Primordial germ-cell development: the zebrafish perspective**. *Nat. Rev. Genet.* 2003, **4**: 690–700.

Williamson, A., Lehmann, R.: **Germ cell development in *Drosophila***. *Annu. Rev. Cell Dev. Biol.* 1996, **12**: 365–391.

9.2 In mammals germ cells are induced by cell-cell interactions during development

Kurimoto, K., Yamaji, M., Seki, Y., Saitou, M.: **Specification of the germ cell lineage in mice: a process orchestrated by the PR-domain proteins, Blimp1 and Prdm14**. *Cell Cycle* 2008, **7**: 3514–3518.

McLaren, A.: **Primordial germ cells in the mouse**. *Dev. Biol.* 2003, **262**: 1–15.

Ohinata, Y., Payer, B., O'Carroll, D., Ancelin, K., Ono, Y., Sano, M., Barton, S.C., Obukhanych, T., Nussenzweig, M., Tarakhovsky, A., Saitou, M., Surani, M.A.: **Blimp1 is a critical determinant of the germ cell lineage in mice.** *Nature* 2005, **436**: 207–213.

Saitou, M., Barton, S.C., Surani, M.A.: **A molecular programme for the specification of germ cell fate in mice.** *Nature* 2002, **418**: 293–300.

Saitou, M., Payer, B., Lange, U.C., Erhardt, S., Barton, S.C., Surani, M.A.: **Specification of germ cell fate in mice.** *Philos. Trans. R. Soc. Lond. B Biol. Sci.* 2003, **358**: 1363–1370.

9.3 Germ cells migrate from their site of origin to the gonad & 9.4 Germ cells are guided to their final destination by chemical signals

Deshpande, G., Godishala, A., Schedl, P.: **Gγ1, a downstream target for the *hmgcr*-isoprenoid biosynthetic pathway, is required for releasing the Hedgehog ligand and directing germ cell migration.** *PLoS Genet.* 2009, **5**: e1000333.

Doitsidou, M., Reichman-Fried, M., Stebler, J., Koprunner, M., Dorries, J., Meyer, D., Esguerra, C.V., Leung, T., Raz, E.: **Guidance of primordial germ cell migration by the chemokine SDF-1.** *Cell* 2002, **111**: 647–659.

Knaut, H., Werz, C., Geisler, R., Nusslein-Volhard, C., Tubingen 2000 Screen Consortium: **A zebrafish homologue of the chemokine receptor Cxcr4 is a germ-cell guidance receptor.** *Nature* 2003, **421**: 279–282.

Molyneaux, K., Wylie, C.: **Primordial germ cell migration.** *Int. J. Dev. Biol.* 2004, **48**: 537–544.

Pelegri F.: **Maternal factors in zebrafish development.** *Dev. Dyn.* 2003, **228**: 535–554.

Raz, E.: **Guidance of primordial germ cell migration.** *Curr. Opin. Cell Biol.* 2004, **16**:169–173.

Santos, A.C., Lehmann, R.: **Germ cell specification and migration in *Drosophila* and beyond.** *Curr. Biol.* 2004, **14**: R578–R589.

Weidinger, G., Wolke, U., Köprunner, M., Thisse, C., Thisse, B., Raz, E.: **Regulation of zebrafish primordial germ cell migration by attraction towards an intermediate target.** *Development* 2002, **129**: 25–36.

Wylie, C.: **Germ cells.** *Cell* 1999, **96**: 165–174.

9.5 Germ-cell differentiation involves a halving of chromosome number by meiosis

De Rooij, D.G., Grootegoed, J.A.: **Spermatogonial stem cells.** *Curr. Opin. Cell Biol.* 1998, **10**: 694–701.

Hultén, M.A., Patel, S.D., Tankimanova, M., Westgren, M., Papadogiannakis, N., Jonsson, A.M., Iwarsson, E.: **The origins of trisomy 21 Down syndrome.** *Mol. Cytogenet.* 2008, **1**: 21–31.

Mehlmann, L.M.: **Stops and starts in mammalian oocytes: recent advances in understanding the regulation of meiotic arrest and oocyte maturation.** *Reproduction* 2005, **130**: 791–799.

Pacchierottia, F., Adler, I.-D., Eichenlaub-Ritter, U., Mailhes, J.B.: **Gender effects on the incidence of aneuploidy in mammalian germ cells.** *Environ. Res.* 2007, **104**: 46–69.

Vogta, E., Kirsch-Volders, M., Parry, J., Eichenlaub-Rittera, U.: **Spindle formation, chromosome segregation and the spindle checkpoint in mammalian oocytes and susceptibility to meiotic error.** *Mutat. Res.* 2008, **651**: 14–29.

9.6 Oocyte development can involve gene amplification and contributions from other cells

Browder, L.W.: *Oogenesis*. New York: Plenum Press, 1985.

Choo, S., Heinrich, B., Betley, J.N., Chen, A., Deshler, J.O.: **Evidence for common machinery utilized by the early and late RNA localization pathways in *Xenopus* oocytes.** *Dev. Biol.* 2004, **278**: 103–117.

de Rooij, D.G, Grootegoed, J.A.: **Spermatogonial stem cells.** *Curr. Opin. Cell Biol.* 1998, **10**: 694–701.

9.7 Factors in the cytoplasm maintain the totipotent potential of the egg

Gurdon, J.B.: **Nuclear transplantation in eggs and oocytes.** *J. Cell Sci. Suppl.* 1986, **4**: 287–318.

9.8 In mammals some genes controlling embryonic growth are 'imprinted'

Ideraabdullah, F.Y., Vigneau, S., Bartolomei, M.S.: **Genomic imprinting mechanisms in mammals.** *Mutat. Res.* 2008, **647**: 77–85.

Morison, I.M., Ramsay, J.P., Spencer, H.G.: **A census of mammalian imprinting.** *Trends Genet.* 2005, **21**: 457–465.

Reik, W., Walter, J.: **Genomic imprinting: parental influence on the genome.** *Nat. Rev. Genet.* 2001, **2**: 21–32.

Wood, A.J., Oakey, R.J.: **Genomic imprinting in mammals: emerging themes and established theories.** *PLoS Genet.* 2006, **2**: e147.

Wu, H-A., Bernstein, E.: **Partners in imprinting: noncoding RNA and Polycomb group proteins.** *Dev. Cell* 2008, **15**: 637–638.

9.9 Fertilization involves cell-surface interactions between egg and sperm

Ensslin, M.A., Lyng, R., Raymond, A., Copland, S., Shur, B.D.: **Novel gamete receptors that facilitate sperm adhesion to the egg coat.** *Soc. Reprod. Fertil. Suppl.* 2007, **63**: 367–383.

Jungnickel, M.K., Sutton, K.A., Florman, H.M.: **In the beginning: lessons from fertilization in mice and worms.** *Cell* 2003, **114**: 401–404.

Ohelndieck, K., Lennarz, W.J.: **Role of the sea-urchin egg receptor for sperm in gamete interactions.** *Trends Biochem. Sci.* 1995, **20**: 29–33.

Rubinstein, E., Ziyyat, A., Wolf, J.P., Le Naour, F., Boucheix, C.: **The molecular players of sperm-egg fusion in mammals.** *Semin. Cell Dev. Biol.* 2006, **17**: 254–263.

Shur, B.D., Ensslin, M.A., Rodeheffer, C.: **SED1 function during mammalian sperm-egg adhesion.** *Curr. Opin. Cell Biol.* 2004, **16**: 477–485.

Singson, A.: **Every sperm is sacred: fertilization in *Caenorhabditis elegans*.** *Dev. Biol.* 2001, **230**: 101–109.

9.10 Changes in the egg envelope at fertilization block polyspermy

Gardner, A.J., Evans, J.P.: **Mammalian membrane block to polyspermy: new insights into how mammalian eggs prevent fertilization by multiple sperm.** *Reprod. Fertil. Dev.* 2005, **18**: 53–61.

Hedrick, J.L.: **A comparative analysis of molecular mechanisms for blocking polyspermy: identification of a lectin-ligand** binding reaction in mammalian eggs. *Soc. Reprod. Fertil. Suppl.* 2007, **63**: 409–419.

Tian, J., Gong, H., Thomsen, G.H., Lennarz, W.J.: **Xenopus laevis sperm-egg adhesion is regulated by modifications in the sperm receptor and the egg vitelline envelope**. *Dev. Biol.* 1997, **187**: 143–153.

Wong, J.L., Wessel, G.M.: **Defending the zygote: search for the ancestral animal block to polyspermy**. *Curr. Top. Dev. Biol.* 2006, **72**: 1–151.

9.11 Sperm–egg fusion causes a calcium wave that results in egg activation

Knott, J.G., Gardner, A.J., Madgwick, S., Jones, K.T., Williams, C.J., Schultz, R.M.: **Calmodulin-dependent protein kinase II triggers mouse egg activation and embryo development in the absence of Ca^{2+} oscillations**. *Dev. Biol.* 2006, **296**: 388–395.

Swann, K., Yu, Y.: **The dynamics of calcium oscillations that activate mammalian eggs**. *Int. J. Dev. Biol.* 2008, **52**: 585–594.

Whitaker, M.: **Calcium at fertilization and in early development**. *Physiol. Rev.* 2006, **86**: 25–88.

Yu, Y., Halet, G., Lai, F.A., Swann, K.: **Regulation of diacylglycerol production and protein kinase C stimulation during sperm- and PLCzeta-mediated mouse egg activation**. *Biol. Cell* 2008, **100**: 633–643.

9.12 The primary sex-determining gene in mammals is on the Y chromosome

Capel, B.: **The battle of the sexes**. *Mech. Dev.* 2000, **92**: 89–103.

Koopman, P.: **The genetics and biology of vertebrate sex determination**. *Cell* 2001, **105**: 843–847.

Schafer, A.J., Goodfellow, P.N.: **Sex determination in humans**. *BioEssays* 1996, **18**: 955–963.

9.13 Mammalian sexual phenotype is regulated by gonadal hormones

Swain, A., Lovell-Badge, R.: **Mammalian sex determination: a molecular drama**. *Genes Dev.* 1999, **13**: 755–767.

Vainio, S., Heikkila, M., Kispert, A., Chin, N., McMahon, A.P.: **Female development in mammals is regulated by Wnt-4 signalling**. *Nature* 1999, **397**: 405–409.

9.14 The primary sex-determining signal in *Drosophila* is the number of X chromosomes, and is cell autonomous

Brennan, J., Capel, B.: **One tissue, two fates: molecular genetic events that underlie testis versus ovary development**. *Nat. Rev. Genet.* 2004, **5**: 509–520.

Hodgkin, J.: **Sex determination compared in *Drosophila* and *Caenorhabditis***. *Nature* 1990, **344**: 721–728.

MacLaughlin, D.T., Donahoe, M.D.: **Sex determination and differentiation**. *New Engl. J. Med.* 2004, **350**: 367–378.

9.15 Somatic sexual development in *Caenorhabditis* is determined by the number of X chromosomes

Carmi, I., Meyer, B.J.: **The primary sex determination signal of *Caenorhabditis elegans***. *Genetics* 152: 999–1015.

Meyer, B. J.: **X-chromosome dosage compensation**. In WormBook (June 25, 2005) (edited by The C. elegans Research Community). doi/10.1895/wormbook.1.8.1, http://www.wormbook.org (date accessed 21 May 2010).

Raymond, C.S., Shamu, C.E., Shen, M.M., Seifert, K.J., Hirsch, B., Hodgkin, J., Zarkower, D.: **Evidence for evolutionary conservation of sex-determining genes**. *Nature* 1998, **391**: 691–695.

9.16 Most flowering plants are hermaphrodite, but some produce unisexual flowers

Irisa, E.N.: **Regulation of sex determination in maize**. *BioEssays* 1996, **18**: 363–369.

Huh, J.H., Bauer, M.J., Hsieh, T.-F., Fischer, R.L.: **Cellular programming of plant gene imprinting**. *Cell* 2008, **132**: 735–744.

9.17 Determination of germ-cell sex depends on both genetic constitution and intercellular signals

Childs, A.J., Saunders, P.T.K., Anderson, R.A.: **Modelling germ cell development *in vitro***. *Mol. Hum. Reprod.* 2008, **14**: 501–511.

McLaren, A.: **Signaling for germ cells**. *Genes Dev.* 1999, **13**: 373–376.

Obata, Y., Villemure, M., Kono, T., Taketo, T.: **Transmission of Y chromosomes from XY female mice was made possible by the replacement of cytoplasm during oocyte maturation**. *Proc. Natl Acad. Sci. USA* 2008, **105**: 13918–13923.

Seydoux, G., Strome, S.: **Launching the germline in *Caenorhabditis* elegans: regulation of gene expression in early germ cells**. *Development* 1999, **126**: 3275–3283.

9.18 Various strategies are used for dosage compensation of X-linked genes

Akhtar, A.: **Dosage compensation: an intertwined world of RNA and chromatin remodeling**. *Curr. Opin. Genet. Dev.* 2003, **13**: 161–169.

Avner, P., Heard, E.: **X-chromosomes inactivation: counting, choice and initiation**. *Nat. Rev. Genet.* 2001, **2**: 59–67.

Csankovszki, G., McDonel, P., Meyer, B.J.: **Recruitment and spreading of the C. elegans dosage compensation complex along X chromosomes**. *Science* 2004, **303**: 1182–1185.

Latham, K.E.: **X chromosome imprinting and inactivation in preimplantation mammalian embryos**. *Trends Genet.* 2005, **21**: 120–127.

Panning, B.: **X-chromosome inactivation: the molecular basis of silencing**. *J. Biol.* 2008, **7**: 30.

Starmer, J., Magnuson, T.: **New model for random X chromosome inactivation**. *Development* 2009, **136**: 1–10.

Cell differentiation and stem cells

The differentiation of unspecialized cells into many different cell types occurs first in the developing embryo and continues after birth and throughout adulthood. The character of specialized cells, such as blood or muscle cells, is the result of a particular pattern of gene activity, which determines which proteins are synthesized. Thus, how this particular pattern of gene activity develops is a central question in cell differentiation, and is discussed in this chapter in relation to some well-studied model systems, including muscle, blood, and neural crest cell differentiation. We shall see that gene expression is under a complex set of controls that include the actions of transcription factors, chemical modification of DNA, and post-translational modifications to chromatin proteins. External signals play a key role in differentiation by triggering intracellular signaling pathways that affect gene expression. Another key question in development is the plasticity of differentiated cells—that is, their capacity to change their state of differentiation. As we shall see, although one cell type rarely changes into another under normal circumstances, such changes are possible. It is well-established that a nucleus from a differentiated cell can be reprogrammed when transferred into an enucleated egg, and become able to direct embryonic development, the process of cloning. Pluripotent embryonic stem cells and multipotent adult stem cells provide insights into the mechanisms that underlie the flexibility and reversibility of a cell's differentiated state, and have potential uses in regenerative medicine.

Embryonic cells start out morphologically similar to each other but eventually become different, acquiring distinct identities and specialized functions. As we have seen in previous chapters, many developmental processes, such as the early specification of the germ layers, involve transient changes in cell form, patterns of gene activity, and the proteins synthesized. **Cell differentiation**, in contrast, involves the gradual emergence of cell types that have a clear-cut identity in the adult, such as nerve cells, red blood cells, and fat cells (Fig. 10.1). There are more than 200 clearly recognizable differentiated cell types in mammals. Initially, embryonic cells fated to become different cell types differ primarily from each other only in their pattern of gene activity, and thus in the proteins present. Differentiation occurs over successive cell generations, the cells gradually acquiring new structural features while their potential fates become more and more restricted.

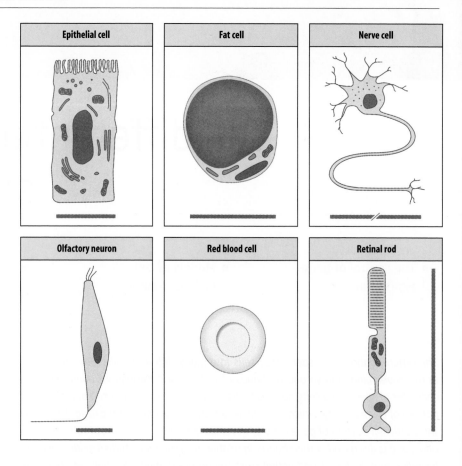

Fig. 10.1 Differentiated cell types.
Mammalian cell types come in various shapes and sizes. Scale bars: epithelial cell, 15 μm; fat cell, 100 μm; nerve cell, 100 μm–1 m; olfactory neuron, 8 μm; red blood cell, 8 μm; retinal rod, 20 μm.

Differentiated cells serve specialized functions and have achieved a terminal and usually stable state, in contrast to many of the transitory differences characteristic of earlier stages in development. The precursors of cartilage and muscle cells have no obvious structural differences from each other and so look the same and might be described as undifferentiated; nevertheless, they will differentiate as cartilage and muscle, respectively, when cultured under appropriate conditions. Similarly, at an early stage in their differentiation, the precursors of white blood cells are structurally indistinguishable from those of red blood cells, but are distinct in the proteins they express. The differences between the precursor cells at these early stages reflect differences in gene activity and thus the proteins they contain, which control their future development.

As with earlier processes in development, the central feature of cell differentiation is a change in gene expression. The genes expressed in a differentiated cell include not only those for a wide range of 'housekeeping' proteins, such as the glycolytic enzymes involved in energy metabolism, but also genes encoding cell-specific proteins that characterize a fully differentiated cell: hemoglobin in red blood cells, keratin in skin epidermal cells, and muscle-specific actins and myosins. It is important to realize how many different genes are active in any given cell in the embryo at any one time—it is of the order of several thousand, only a small number of which may be involved in specifying cell fate or differentiation. All the genes that are being expressed in a particular tissue or at a particular stage of development can be detected by DNA microarray techniques (see Box 3A, p. 115).

A differentiated cell is characterized by the proteins it contains. The presence of different proteins can result in considerable structural changes; mature mammalian red blood cells lose their nucleus and become biconcave discs stuffed with hemoglobin, whereas neutrophils, a type of white blood cell, develop a multi-lobed nucleus and a cytoplasm filled with secretory granules. Yet both have a similar early lineage. Expression of a single protein can change a cell's differentiated state. If the gene *myoD* is introduced into fibroblasts, for example, they will develop into muscle cells, as *myoD* encodes a master transcriptional regulator of muscle differentiation. As we shall see later in this chapter, a very dramatic effect is obtained when genes encoding four of the transcription factors expressed in embryonic stem cells (ES cells) are introduced into fibroblasts. The fibroblasts become pluripotent, which means that, like ES cells, they can be induced to differentiate into cells belonging to tissues derived from all three germ layers.

In the earliest stages of differentiation, differences between cells are not easily detected and probably consist of subtle alterations caused by a change in activity of a few genes. At this early stage, cells become **determined**, or committed, with respect to their developmental potential (see Section 1.12). The mesoderm of somites, for example, gives rise during normal development to muscle, cartilage, dermis, and vascular tissue, but not to other cell types. Cells that are determined with respect to their eventual fate will form only those appropriate cell types when grafted to a different site in the embryo; they retain their identity (see Fig. 1.21). Once determined for a particular fate, cells pass on that determined state to all their progeny.

Cell differentiation is known to be controlled by a wide range of external signals, from cell-surface proteins to secreted polypeptide cytokines and molecules of the extracellular matrix. Examples of all of these signals will be encountered in this chapter, but it is important to remember that, while the external signals that stimulate differentiation are often referred to as being 'instructive,' they are in general 'selective,' in the sense that, in practice, the number of developmental options open to a cell at any given time is limited. These options are set by the cell's internal state, which in turn reflects its developmental history. External signals cannot, for example, convert an endodermal cell into a muscle or nerve cell. Nevertheless the same external signals, such as Wnts, FGFs, and members of the TGF-β family, can be used again and again in different circumstances without upsetting previous development, as we have seen in early vertebrate development (see Box 4A, p. 131).

There can be a conflict between cell division and cell differentiation, with a cessation of cell proliferation being necessary for full cell-differentiation to occur. Cell proliferation is most evident before the terminal stage of differentiation, and most terminally differentiated cells divide only rarely. Some cells, such as skeletal muscle and nerve cells, do not divide at all after they have become fully differentiated. In cells that do divide after differentiation, the differentiated state, like the determined state, is passed on through all subsequent cell divisions. As the pattern of gene activity is the key feature in cell differentiation, this raises the question of how a particular pattern of gene activity is first established, and then how it is passed on to daughter cells.

We start this chapter with a consideration of the general mechanisms by which a pattern of gene activity can be established, maintained, and inherited at cell division. We then turn to a major question, namely the molecular basis of the specificity of cell differentiation, with muscle cells, blood cells, and neural crest cells providing our main examples. Finally, we look at the reversibility and plasticity of the differentiated state, particularly the special properties of stem cells, and the reprogramming of differentiated nuclei in oocytes that makes it possible to clone some animals— that is, to make an animal that is genetically identical to the animal from which the nucleus came. We shall also discuss the exciting recent discovery that pluripotency can be induced in fully differentiated cells taken from adults. This chapter concentrates on differentiation in animal cells. As we have seen in Chapter 7, plant cells do not have a permanently determined state, and a single somatic cell can give rise to a whole plant.

The control of gene expression

Every nucleus in the body of a multicellular organism is derived from the single zygotic nucleus in the fertilized egg. But patterns of gene activity in differentiated cells vary enormously from one cell type to another. The egg itself has a pattern of gene activity that is different from that of the embryonic cells to which it gives rise. This raises the question of how the particular pattern of gene activity in a differentiated cell is specified and how it is inherited.

To understand the molecular basis of cell differentiation, we first need to know how a gene can be expressed in a cell-specific manner. Why does a certain gene get switched on in one cell and not in another? We focus here on the regulation of **transcription**, which is the first (and generally the most significant) step in the expression of a gene (see Section 1.10). Control of protein synthesis can, however, also occur after transcription; for example, at the **translation** of mRNA to produce protein. We have seen an example of translational regulation in early *Drosophila* development, where the Nanos protein controls the production of Hunchback protein by preventing translation of maternal *hunchback* RNA (see Section 2.9), and in *Caenorhabditis elegans*, where a microRNA prevents the translation of *lin-14* RNA to give LIN-14 protein (see Section 6.5). In *Xenopus*, large numbers of maternal mRNAs are laid down in the egg but many are not translated until after the egg is fertilized (see Section 4.1).

10.1 Control of transcription involves both general and tissue-specific transcriptional regulators

Most of the key genes in development are initially in an inactive state and require activating transcription factors, or **activators**, to turn them on. These activators bind to specific *cis*-regulatory control regions in the DNA surrounding the gene, which are often called **enhancers**. The importance of these regions in tissue-specific gene expression was clearly demonstrated by experiments in which the control region of one tissue-specific gene was replaced by the control region of another (Fig. 10.2). We have already seen how such replacements are routinely used in experimental developmental biology to force the expression or misexpression of a gene in a particular tissue, and to mark cells expressing a particular gene in order to investigate its developmental function (see Box 1D, p. 20 and Box 2D, p. 61).

In eukaryotic cells, most regulated protein-coding genes are transcribed by RNA polymerase II. The binding of RNA polymerase to the promoter region of a gene in order to start transcription requires the cooperation of a set of so-called **general transcription factors**, which form an initiation complex with the polymerase at the promoter and help to activate the polymerase to start transcription at the correct place. For highly regulated genes, such as those involved in development, this complex cannot be formed and activated without additional help. The binding of activator proteins to sites in the *cis*-regulatory control regions of the gene is needed to help attract

Fig. 10.2 Tissue-specific gene expression is controlled by regulatory regions. The control region of the mouse elastase I gene is joined to a DNA sequence coding for human growth hormone. This DNA construct is injected into the nucleus of a fertilized mouse egg, where it becomes integrated into the genome. When the mouse develops, human growth hormone is made in the pancreas. A 213-base-pair stretch of DNA that includes the elastase I promoter and other control sites is sufficient to drive growth hormone expression in the pancreas. Normally, growth hormone is made only in the pituitary, and elastase I only in the pancreas.

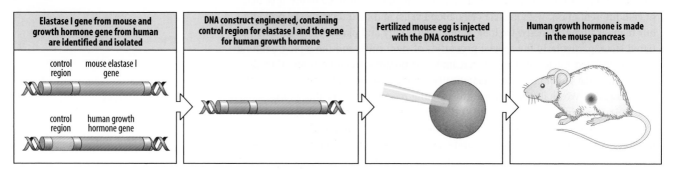

| Elastase I gene from mouse and growth hormone gene from human are identified and isolated | DNA construct engineered, containing control region for elastase I and the gene for human growth hormone | Fertilized mouse egg is injected with the DNA construct | Human growth hormone is made in the mouse pancreas |

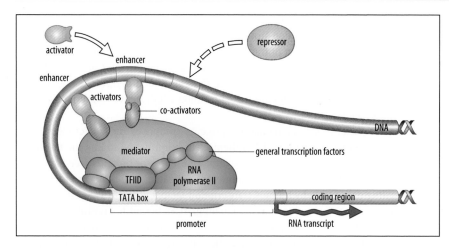

Fig. 10.3 Gene expression is regulated by the coordinated action of gene-regulatory proteins that bind to control regions in DNA. Most protein-coding genes in eukaryotic cells, including all highly regulated developmental genes, are transcribed by RNA polymerase II. The transcription machinery itself, composed of the polymerase and its associated general transcription factors (the core transcriptional machinery), binds to the promoter, which contains some binding sites, such as the TATA box, that are common to many genes transcribed by RNA polymerase II. The TATA box is recognized by the TATA-box binding protein (TBP), a component of the general transcription factor TFIID. Other control sites known as enhancers must be occupied by specific activator proteins and their associated co-activators before transcription can be initiated. These sites are located adjacent to the promoter and also further upstream, and sometimes downstream, of the coding region. The control regions of all highly regulated genes also include sites for repressor proteins, whose binding prevents transcription, for example, in the wrong tissue or at the wrong time during development. Control sites can be many kilobases away from the start point of transcription. The proteins bound to these distant sites are brought into contact with the transcription machinery by looping of the DNA. The Mediator protein complex is a large multi-protein complex that seems to be present in many RNA polymerase II initiation complexes and its role is to link the gene-regulatory proteins to the core transcriptional machinery. The fully operational initiation complex activates the RNA polymerase and releases it to start transcription.

the general transcription factors and polymerase and position them correctly on the promoter (Fig. 10.3).

For any given gene, the specificity of its activation is due to particular combinations of gene-regulatory proteins binding to individual sites in the control regions. These sites are typically 7–10 nucleotides long. At least 1000 different transcription factors are encoded in the genomes of *Drosophila* and *C. elegans*, and as many as 3000 in the human genome. On average, around five different transcription factors act together at a control region, and in some cases considerably more. This combinatorial action of gene-regulatory proteins is a key principle in the control of gene expression and is fundamental to the exquisite control and complexity of the gene expression that drives development. As well as sites that bind activators, control regions may contain sites that bind **repressors**, proteins that inhibit gene expression; these prevent a gene being expressed at the wrong time or in the wrong place. Examples of such complex control regions that regulate the temporal and spatial expression of a gene during embryonic development are shown in Fig. 2.34 for the *Drosophila even-skipped* (*eve*) pair-rule gene and in Fig. 6.26 for the sea urchin *Endo-16* gene, which encodes a calcium-binding extracellular matrix protein. The spatially delimited expression of *eve* in the second of seven transverse stripes in the embryo involves four different gene-regulatory proteins, which include activators and repressors, acting at 11 distinct sites. The correct expression of *Endo-16* throughout sea-urchin embryo development requires the participation of 13 gene-regulatory proteins acting at a potential 56 regulatory motifs within a control region that spans 2.3 kilobases.

Some control sites are located within the promoter region and are present in similar positions in most protein-coding genes. These sites, such as the TATA box, bind the general transcription factors and the RNA polymerase (see Fig. 10.3). Tissue-specific or developmental-stage expression is controlled by sites outside the immediate promoter region, such as those at which tissue-specific gene-regulatory proteins act. These sites vary greatly in type and position from gene to gene, and may be thousands of base pairs away from the start point of transcription. These distant sites are able to control gene activity because DNA can form loops, thus bringing the sites and their bound proteins into close proximity to the promoter. Proteins bound at distant sites can thus make contact with proteins at the promoter, forming a fully active transcription initiation complex (see Fig. 10.3).

Another important class of regulatory proteins comprises the **co-activators** and **co-repressors**, which do not bind DNA themselves but function to link the transcriptional machinery to the DNA-binding activator and repressor proteins (see Fig. 10.3). One co-activator we have already encountered is β-catenin, the end product of the Wnt signaling pathway, which binds transcription factors of the TCF/LEF family (see Box 1E, p. 26). In the absence of Wnt signaling, TCFs bound to control elements in genes that are targets of Wnt signaling are bound by a co-repressor, called Groucho in *Drosophila* and TLE in humans, and transcription is repressed. β-Catenin accumulating in the nucleus as the result of Wnt signaling displaces Groucho and binds to TCF, turning it into a transcriptional activator (see Box 1E, p. 26).

Transcriptional regulators fall into two main groups: those that are required for the transcription of a wide range of genes, and which are found in many different cell types; and those that are required for a particular gene or set of genes whose expression is tissue restricted, or **tissue-specific**, and which are only found in one or a very few cell types. We shall encounter both types of transcription factor in the following sections, when we look at the expression of globin genes in red blood cell precursors and the regulation of muscle genes in muscle precursor cells. In general, it can be assumed that activation of each gene involves a unique combination of transcription factors. It should also be noted that while differentiation may involve a relatively small number of genes, the number is not easy to establish, as each cell has several thousand active genes, most of which are probably concerned with housekeeping functions.

10.2 External signals can activate gene expression

Most of the molecules that act as signals between cells during development are proteins, and in the case of developmentally important signals their effect is usually to induce a change in gene expression. Protein signals bind to receptors in the cell membrane and the signal is relayed to the cell nucleus by intracellular signaling pathways such as the Wnt, Hedgehog, TGF-β, and FGF pathways we have already encountered (see Box 1E, p. 26, Fig. 2.40, Fig. 4.33, and Box 5B, p. 184). Non-protein, lipid-soluble signal molecules, such as retinoic acid, cross the plasma membrane unaided and also stimulate an intracellular signal pathway (see Box 5D, pp. 192–193). Once inside the cell, retinoic acid binds to receptor proteins, and the complex of retinoic acid and receptor is then able to act as a transcriptional regulator, binding directly to control sites in the DNA, known as retinoic acid response elements, to activate (or in some cases repress) transcription. Other lipid-soluble signal molecules that act through similar intracellular receptors include the steroid hormones, such as testosterone and estrogen in mammals. As we saw in Chapter 9, testosterone produced in the testis is responsible for the secondary sexual characteristics that make male mammals different from females. In insects, the steroid hormone ecdysone is responsible for metamorphosis (see Chapter 13) and induces the differentiation of a wide range of cell types. In all these cases, the hormone turns a whole set of genes on or off by acting, via its intracellular receptors, at a control element that is present in all these genes.

Estrogen is also involved in a classic example of tissue-specific gene expression, the induction of chick oviduct cells to produce the protein ovalbumin, a major component of egg white. The continued presence of estrogen is required for transcription of the ovalbumin gene; when estrogen is withdrawn, ovalbumin mRNA and protein disappear. The circulating estrogen activates expression of the ovalbumin gene only in the cells of the oviduct, whereas the ovalbumin gene in other cells, such as those of the liver, is unaffected. Indeed, the expression of the ovalbumin gene is among the most restricted known, but the precise mechanisms that lead to such a tissue-specific response are still elusive after decades of research. There is evidence that estrogen affects the structure of the chromosomal region containing the ovalbumin gene in oviduct cells but not others, thus allowing the hormone–receptor complexes access to the regulatory DNA of the gene. This tissue-specific effect of estrogen implies that other tissue-specific activators, repressors, and co-regulators exist that allow ovalbumin gene expression in the oviduct when stimulated with estrogen and prevent its expression in all other tissues.

10.3 The maintenance and inheritance of patterns of gene activity depend on chemical and structural modifications to chromatin, as well as on gene-regulatory proteins

A central feature of development in general, and cell differentiation in particular, is that some genes are maintained in an active state, while others are repressed and inactive. Many differentiated cells, including fibroblasts, liver cells, and myoblasts (cells that give rise to muscle cells), continue dividing after differentiation, and the particular pattern of gene activity characterizing the differentiated state is reliably transmitted through many cell divisions. Developmental genes generally have complex control regions, containing binding sites for a range of activating or inhibitory transcription factors, and whether or not a gene is activated depends on the precise combinations and levels of these transcription factors and their associated co-activators and co-repressors. A given gene-regulatory protein can also affect the activity of several, often many, other genes, activating some and repressing others, which provides an efficient means of coordinating the expression of large numbers of genes (Fig. 10.4).

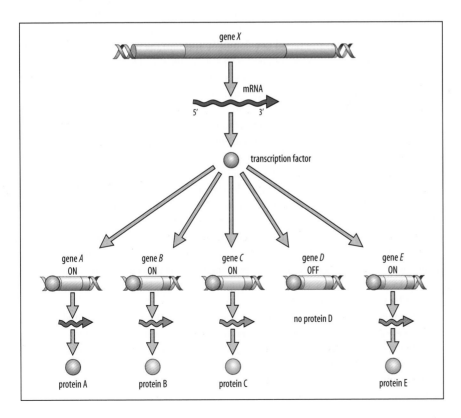

Fig. 10.4 A gene can regulate the activity of other genes when it codes for a transcription factor. Transcription factors can activate or repress the activity of genes by binding to their control regions. In the illustration, the activation of gene *X*, encoding a transcription factor, eventually leads to the production of four new proteins (A, B, C, and E), and the repression of protein D production.

After Alberts, B., et al.: 2002.

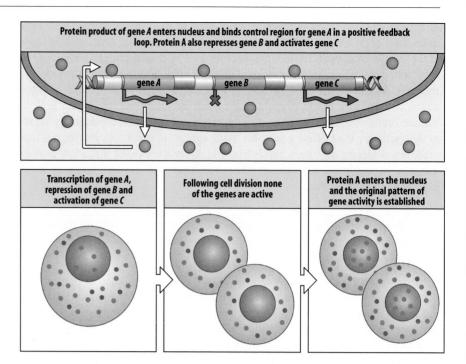

Fig. 10.5 Continued expression of gene regulatory proteins could maintain a pattern of differentiated gene activity. Top panel: transcription factor A is produced by gene *A*, and acts as a positive gene-regulatory protein for its own control region. Once activated, therefore, gene *A* remains switched on and the cell always contains A. Transcription factor A also acts on the control regions of genes *B* and *C* to repress and activate them, respectively, setting up a cell-specific pattern of gene expression. Bottom panels: after cell division, the cytoplasm of both daughter cells contains sufficient amounts of protein A to reactivate gene *A*, and thus maintain the pattern of expression of genes *B* and *C*.

One way of maintaining a particular pattern of gene activity is to ensure the continued production of these specific gene-regulatory proteins in the differentiated cell and its progeny. If the gene product itself acts as the positive regulatory protein, all that is then required for the pattern of gene activity to be maintained is the initial event that first activates the gene; once switched on the gene remains active (Fig. 10.5). This type of positive-feedback control of gene expression occurs in muscle-cell differentiation, where the MyoD protein acts as an activator of *myoD* gene expression. In *Drosophila*, the selector gene *engrailed* remains active in the posterior compartment of segments throughout embryonic and larval development and into the adult (see Chapter 2). Its expression is at least partly maintained by a similar positive feedback of the Engrailed transcription factor on its own gene, long after the pair-rule transcription factors that initiated expression of *engrailed* in the embryo have disappeared.

Another mechanism for maintaining and inheriting a pattern of active and inactive genes relies on chemical and structural changes in chromatin—the complex of DNA, histones, and other proteins of which chromosomes are made. Whether a stretch of chromosome can be transcribed depends on whether or not it is packaged into a chromatin structure that is accessible to transcription factors, RNA polymerase, and other necessary proteins. The compaction of chromatin that occurs at mitosis, for example, when the condensed chromosomes become visible under the light microscope, renders the chromosomes transcriptionally inactive. Localized structural alterations of this sort that persist throughout the cell cycle and are reinstated after cell division provide a means of shutting down, or silencing, genes in the long term. Indeed, this is one way in which expression of the *engrailed* gene is prevented in those regions of the fly in which it is not required. Chromatin that is packaged into a structure that cannot be transcribed is called heterochromatin.

Evidence that changes in the packing state of chromatin can be inherited through many cell divisions and can keep genes inactive over a long period was first provided by the phenomenon of X-chromosome inactivation in female mammals. One of the two X chromosomes in each somatic cell in females becomes highly condensed and inactive early in embryogenesis and is maintained in this state throughout the individual's lifetime. X inactivation is a dosage-compensation mechanism to equalize the level of gene expression from the X chromosome in males and females. The silencing mechanism involves the production of a non-coding RNA that coats the chromosome and is described in Section 9.17. Inactivation is not irreversible, however, as the inactive X is reactivated in the germline during the formation of egg cells, which each end up containing a single X chromosome.

Localized changes in chromatin structure and protein composition in chromosomes are thought to provide a general mechanism for long-term gene inactivation. A reflection of the difference in chromatin structure between active and inactive genes is that some active genes show a heightened sensitivity to digestion with the enzyme DNase I at sites known as hypersensitive sites. This is just one indication that in actively transcribed genes, the chromatin is packaged into a more 'open' structure that allows enzymes and other proteins, including the transcription apparatus, access to the DNA. When the chromatin is packed into heterochromatin, on the other hand, the DNA is inaccessible to DNase I and is protected from digestion.

Underlying these changes in the transcriptional properties of chromatin are chemical modifications to the DNA itself and to the associated histone proteins (Box 10A, p. 374). In vertebrates, methylation of cytosine at certain sites in the DNA is correlated with the absence of transcription in those regions. Moreover, the pattern of methylation can be faithfully inherited when the DNA replicates, thus providing a mechanism for passing on a pattern of gene activity to daughter cells (Fig. 10.6). After DNA replication, DNA methylase recognizes the presence of a methyl group on a cytosine and methylates the corresponding group on the other strand. DNA methylation is one way in which certain genes are rendered inactive in the maternal genome but not in the paternal genome in the phenomenon of genomic imprinting (see Section 9.8). The inactive X chromosome also has a pattern of DNA methylation that differs from that of the active X, and this methylation is likely to play a part in keeping it inactive.

The post-translational modification of histone proteins is also closely linked to regulation of gene activity. The histones in chromatin can undergo modifications such as methylation, acetylation, or phosphorylation on specific amino-acid residues, and this can change the properties of the chromatin (see Box 10A, p. 374).

Fig. 10.6 Inheritance of the pattern of methylation on DNA. Many cytosine bases occurring in cytosine-guanine (CG) pairs in vertebrate DNA are methylated (a CH_3 group is added). When the DNA is replicated, the new complementary strand lacks methylation but the pattern can be reinstated on the new strand by methylase enzymes that recognize the methylated cytosine of the CG pair on the old DNA strand and methylate the cytosine of the corresponding CG pair on the new strand. *After Alberts, B., et al.: 2002.*

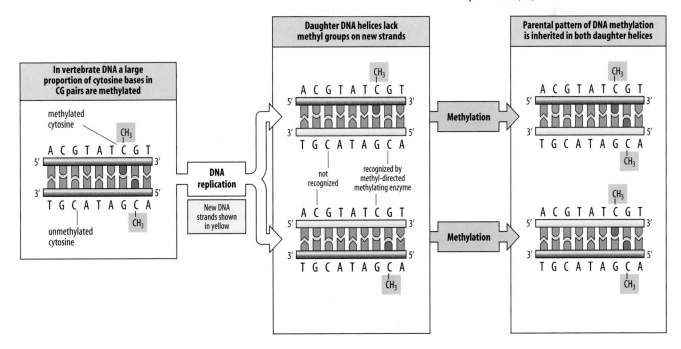

Box 10A Histones and Hox genes

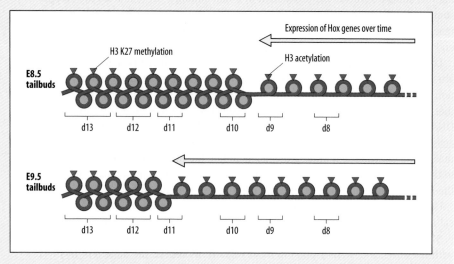

The post-translational modification of histone proteins is very closely linked to regulation of gene activity. The histones in chromatin can undergo modifications, such as methylation, acetylation, or phosphorylation, on specific amino-acid residues, particularly lysine (K), and this can change the transcriptional properties of the chromatin. Histone acetylation, for example, tends to be associated with regions of chromatin that are being transcribed, whereas methylation of particular lysine residues is more characteristic of transcriptionally silent heterochromatin, whereas methylation at

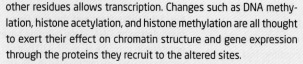

other residues allows transcription. Changes such as DNA methylation, histone acetylation, and histone methylation are all thought to exert their effect on chromatin structure and gene expression through the proteins they recruit to the altered sites.

Changes in chromatin status are often self-propagating as histone modifications, such as acetylation/deacetylation or methylation/demethylation, recruit the corresponding chromatin-modifying enzymes to the chromatin, causing a 'domino effect' that automatically propagates the change in chromatin structure along the chromosome.

This type of effect is likely to underlie the typical sequential activation of Hox genes in the order they occur in the gene cluster (see Box 5E, p. 194 and Fig. 5.24). The relationship between histone modification and expression of the genes of the Hoxd

cluster has been followed in excised mouse tail buds at successive stages of development using chromatin immunoprecipitation (ChIP) to detected the modified histone proteins and ChIP-chip to detect the target genes (see Section 3.11). At stage E8.5, the genes *Hoxd1-9* are already expressed and their chromatin is heavily acetylated, but there are few acetylation marks on *Hoxd10-13*, which are not being expressed at this stage, although acetylation is starting to appear on *Hoxd10*. Twelve hours later, *Hoxd10* is more heavily acetylated and is expressed. At stage E9.5, histone acetylation has spread over the remaining Hoxd genes, which are all now starting to be expressed. The opposite is seen when the progress of repressive methylation on H3 lysine 27 is tracked, with less methylation on active genes and strong methylation on those not yet expressed.

Summary

Transcriptional control is a key feature of cell differentiation. Tissue-specific expression of a eukaryotic gene depends on regulatory sites located in the *cis*-regulatory regions flanking the gene. These sites comprise both the promoter sequences adjacent to the start site of transcription, which bind RNA polymerase, and more distant sites that control the tissue-specific or developmental-stage—specific expression of the gene. The combination of different regulatory proteins bound to sites in the *cis*-regulatory regions determines whether a gene is active or not, and small differences in regulatory regions between genes can have significant effects on their pattern of expression. In some cases, cell-specific expression is due to the correct combination of regulatory proteins only being present in that cell type; in other cases, gene expression may be prevented by the gene being packed away in a state in which it is inaccessible to transcription factors and RNA polymerase. Gene-regulatory proteins interact not only with DNA but also with each other, forming multi-component transcription complexes that are responsible for initiating transcription by RNA polymerase. Extracellular signals, such as protein growth factors, do not themselves enter the cell but can specifically influence gene expression by acting at cell-surface receptors to stimulate intracellular signal transduction pathways. In contrast, signal molecules that can diffuse through cell membranes, such as retinoic acid or steroid hormones, form a complex with an intracellular receptor protein. The hormone-receptor complex then binds to the DNA at specific regulatory sites and acts as a transcription factor to influence gene expression.

Once established, the pattern of gene activity in a differentiated cell can be maintained over long periods of time and is transmitted to the cell's progeny. Mechanisms that maintain the pattern include the continued action and production of gene regulatory proteins, and long-term localized changes in the structure and properties of chromatin that are due to chemical modifications to DNA and to histones.

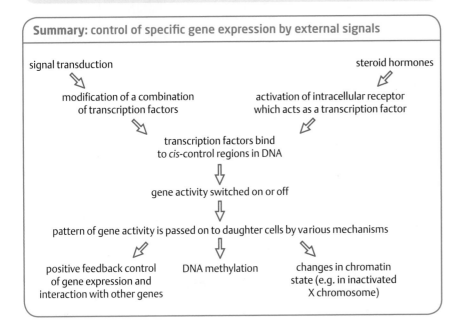

Summary: control of specific gene expression by external signals

signal transduction → modification of a combination of transcription factors

steroid hormones → activation of intracellular receptor which acts as a transcription factor

→ transcription factors bind to *cis*-control regions in DNA

→ gene activity switched on or off

→ pattern of gene activity is passed on to daughter cells by various mechanisms

→ positive feedback control of gene expression and interaction with other genes

→ DNA methylation

→ changes in chromatin state (e.g. in inactivated X chromosome)

Models of cell differentiation

Cellular differentiation is by no means confined to the embryo. A number of adult tissues, such as blood, skin, and the gut epithelial lining, are continually being renewed from undifferentiated stem cells, and are much easier to investigate than embryos. They are therefore among the best-studied examples of cell differentiation. We shall also consider skeletal muscle, which, although mainly differentiating in the embryo, can be renewed if necessary from stem cells in the adult. Embryonic neural crest cells, which we discussed in Chapter 8 in regard to their migration, give rise to a wide range of different structures, and we shall look here at some aspects of their differentiation. Finally, we consider apoptosis, or programmed cell death, which is the fate of very large numbers of developing cells, both in embryonic development and in the adult.

10.4 All blood cells are derived from multipotent stem cells

Hematopoiesis, or blood formation, is a particularly well-studied example of differentiation, partly because cells at all stages are relatively accessible in adult animals and can be grown in culture, and also because of its medical importance. In hematopoiesis, we are looking at the differentiation of a **multipotent** stem cell—a stem cell that can give rise to a limited range of differentiated cell types (see Section 1.17). We first look at the general process of hematopoiesis and at the external signals that influence

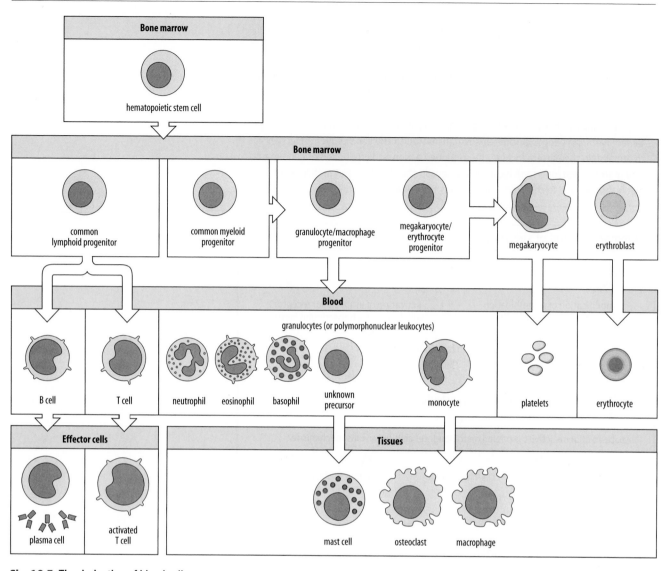

Fig. 10.7 The derivation of blood cells from multipotent stem cells. Multipotent stem cells in the bone marrow give rise to all of the blood cells, as well as to tissue macrophages, mast cells, and osteoclasts in bone. As well as renewing themselves, the stem cells are thought to give rise to uncommitted progenitors, which then split into more committed myeloid, erythroid, and lymphoid precursors. These then give rise to the different blood cell lineages (not all final differentiated cell types are shown). Hematopoietic growth factors influence the proliferation and differentiation of the different lineages.

After Janeway, C.A., et al.: 2005.

it, and then consider the control of gene expression in one cell type—the developing red blood cell.

All the blood cells in an adult mammal originate from hematopoietic stem cells located in the bone marrow. Hematopoietic stem cells originate from mesoderm in the yolk-sac blood islands of the early embryo; later, self-renewing hematopoietic stem cells are found in a region of the aorta and then in a number of definitive blood-forming sites, including the fetal liver and adult bone marrow. The stem cells are self-renewing and are the precursors of progenitor cells that become committed to one of the hematopoietic lineages at a later stage. Thus, hematopoiesis is, in effect, a complete developmental system in miniature, in which a single cell—the multipotent stem cell—gives rise to numerous different blood cell types. There is a continual turnover of blood cells, so that hematopoiesis must continue throughout life. As a measure of the complexity of hematopoiesis, hematopoietic stem cells from fetal liver have been found to express at least 200 transcription factors, a similar number of membrane-associated proteins, and around 150 signaling molecules.

Mammalian blood cells comprise numerous fully differentiated cell types, together with immature cells at various stages of differentiation (Fig. 10.7). The cell types are derived from three main lineages—the erythroid, lymphoid, and myeloid lineages. The first yields the red blood cells, or erythrocytes, and the megakaryocytes, which give rise to blood platelets. The lymphoid lineage produces the lymphocytes, which include the two antigen-specific cell types of the immune system, the B and T lymphocytes. In mammals, B lymphocytes develop in the bone marrow, whereas T lymphocytes develop in the thymus, from precursors that originate from the multipotent stem cells in the bone marrow and migrate from the bloodstream into the thymus. Both B and T lymphocytes undergo a further terminal differentiation after they encounter antigen. Terminally differentiated B lymphocytes become antibody-secreting plasma cells, while T cells undergo terminal differentiation into at least three functionally distinct effector T cells. The B and T lymphocytes of the immune system are the only known examples in vertebrates in which the DNA in the differentiated cell becomes irreversibly altered. The myeloid lineage gives rise to the rest of the white blood cells, or leukocytes. These are the eosinophils, neutrophils, and basophils (collectively known as granulocytes or polymorphonuclear leukocytes), mast cells, and monocytes. Monocytes differentiate into macrophages after they have left the bone marrow and entered tissues. Mast cells also reside in tissues.

Within the bone marrow, the different types of blood cells and their precursors are intimately mixed up with connective-tissue cells—the bone marrow stromal cells. The ability of stem cells both to self-renew and to differentiate into different cell types is tightly regulated by the cells and molecules in their immediate environment—the 'niche' that they inhabit. In the case of hematopoiesis, the stem-cell niche is provided by the bone marrow stromal cells. Secreted signal proteins of the Wnt and BMP families (see Box 4A, p. 131), together with retinoic acid, prostaglandin E_2 and Notch signaling, have been implicated in maintaining stem-cell proliferation and self-renewal. The existence of the multipotent stem cell can be inferred from the ability of bone marrow to reconstitute a complete blood and immune system when transplanted into individuals whose own bone marrow has been destroyed, a property that is exploited therapeutically in the use of bone marrow transplantation to treat diseases of the blood and immune system. The key experiment that showed this property was the transfusion of a suspension of bone marrow cells into a mouse that had received a potentially lethal dose of X-irradiation. The mouse would normally have died as a result of lack of blood cells, which, like other proliferating cells, are particularly sensitive to radiation, but transfusion of bone marrow cells reconstituted the hematopoietic system and the mouse recovered. Multipotent hematopoietic stem cells can be grown in culture, where they also proliferate extensively and differentiate into blood cells.

The multipotent stem cell generates progenitor cells that become irreversibly committed to one or other of the lineages leading to the different blood cell types. There is early commitment to either the myeloid or lymphoid lineage, for example, followed by later commitment to the individual lineages leading to the differentiated cell types. The hematopoietic system can thus be considered as a hierarchical system with the multipotent hematopoietic stem cell at the top (see Fig. 10.7). This stem cell generates uncommitted precursors that then become committed to the main lineages, after which they undergo further rounds of proliferation and further commitment, finally undergoing overt differentiation into the different cell types.

All this activity occurs in the microenvironment of the bone marrow stroma, and is regulated by external signals that are provided by the stromal cells in the form of hematopoietic growth factors and other cytokines. In this way, the numbers of the different types of blood cells can be regulated according to the individual's physiological need. Loss of blood results in an increase in red cell production, for example, whereas infection leads to an increase in the production of lymphocytes and other white blood cells.

A central question in considering the behavior of stem cells is how a single stem cell can divide to produce two daughter cells, one of which remains a stem cell, while the other gives rise to a lineage of differentiating cells. One possibility is that there is an intrinsic difference between the two daughter cells because the stem-cell division is an asymmetric division that results in the two cells acquiring a different complement of proteins. A clear example of this mechanism is neuroblast development in *Drosophila*, which is considered in Chapter 12. The second possibility is that external signals make the daughter cells different: a daughter cell that remains in the stem-cell niche continues to renew itself, whereas one that ends up outside the niche differentiates. In the case of hematopoiesis, both mechanisms may operate. On the one hand, four proteins have been found to segregate asymmetrically at cell division when human hematopoietic cells are cultured without the bone marrow stroma that provides the stem-cell niche *in vivo*. On the other hand, factors produced by bone marrow stromal cells that control stem-cell self-renewal have also been identified.

10.5 Colony-stimulating factors and intrinsic changes control differentiation of the hematopoietic lineages

Hematopoiesis can be characterized by the activities of a hierarchy of transcription factors, whose overlapping patterns of expression specify the various cell lineages. Some are only expressed in immature cells and are not lineage specific—the proto-oncogene product c-Myb falls into this class. Somewhat suprisingly, the hematopoietic stem cells express low levels of some of the genes that will be expressed later in the myeloid and erythroid lineages. Thus, gene repression, as well as activation, must occur when a cell becomes committed to a particular lineage. Some of the transcription-factor interactions that occur in hematopoietic stem-cell differentiation into the erythroid and myeloid lineages are illustrated in Fig. 10.8, and include activation, inhibition, and more complex interactions.

In blood-cell precursors, the transcription factor GATA1, for example, is necessary for the differentiation of the eythroid lineage, whereas another key transcription factor, PU.1, drives differentiation of the myeloid lineage. GATA1 and PU.1 bind to each other, thus inhibiting each other's ability to regulate the transcription of lineage-specific genes. The relative concentration of the two proteins determines cell fate. When the concentration of GATA1 is low, PU.1 will activate the myeloid program, and

Fig. 10.8 Some transcription factors of importance in blood cell differentiation. Some of the gene-regulatory interactions involved in the differentiation of the erythroid and myeloid lineages of blood cells. Macrophages, eosinophils, basophils, monocytes, and neutrophils belong to the myeloid lineage; the erythrocyte belongs to the erythroid lineage. Arrows indicate that the transcription factor stimulates differentiation of that lineage, whereas barred lines indicate inhibition. For example, the differentiation of neutrophils requires activation of the transcription factors CEBPE, PU.1, and CEBPA, and inhibition of CUTL1. c-Jun and the CEPBs are ubiquitous transcription factors, whereas PU.1 is more restricted to white-blood-cell differentiation.

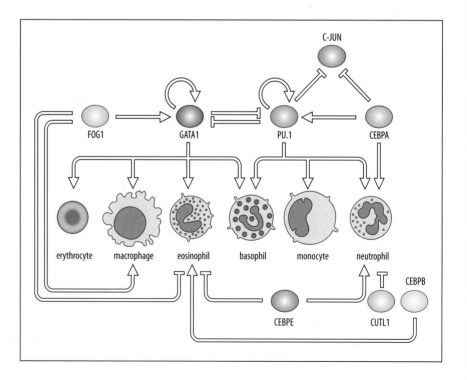

Some hematopoietic growth factors and their target cells	
Type of growth factor	Responding hematopoietic cells
Erythropoietin (EPO)	Erythroid progenitors
Granulocyte colony-stimulating factor (G-CSF)	Granulocytes, neutrophils
Granulocyte-macrophage colony-stimulating factor (GM-CSF)	Granulocytes, macrophages
Interleukin-3	Multipotent precursor cells
Stem cell factor (SCF)	Stem cells
Macrophage colony-stimulating factor (M-CSF)	Macrophages, granulocytes

Fig. 10.9 Hematopoietic factors and their target cells.

when the concentration of PU.1 is low, GATA1 will activate the erythroid program. As with all cell types, it is the combination of transcription factors, and not any particular one, that is responsible for the specific pattern of gene expression that leads to differentiation into a given cell type.

How then is the activity of all these factors controlled? Signals delivered by extracellular protein growth factors and differentiation factors have a key role. Studies on blood-cell differentiation in culture have identified at least 20 extracellular proteins, known generally as **colony-stimulating factors** or **hematopoietic growth factors**, that can affect cell proliferation and cell differentiation at various points in hematopoiesis (Fig. 10.9). Both stimulators and inhibitors have been identified. Not all the factors controlling blood-cell production are made by blood cells or stromal cells. The protein erythropoietin, for example, which induces the differentiation of committed red-blood-cell precursors, is mainly produced by the kidney in response to physiological signals that indicate red-blood-cell depletion.

Despite the well-established role of these factors in blood-cell proliferation and differentiation, there is evidence that the commitment of a cell to one or other pathway in the myeloid lineage might be a chance event, with the growth factors simply promoting the survival of particular cell lineages. Evidence for this comes from experiments where the two daughters of a single early progenitor cell (which can give rise to a range of different blood-cell types) are cultured separately but under the same conditions. Both daughter cells usually give rise to the same combination of cell types, but in about 20% of cases they give rise to dissimilar combinations.

Among the plethora of growth factors, it is difficult to distinguish those that might be exerting specific effects on differentiation from those that may be required for the survival and proliferation of a lineage or group of lineages. However, the contrasting functions of three growth factors—granulocyte-macrophage colony-stimulating factor (GM-CSF), macrophage colony-stimulating factor (M-CSF), and granulocyte colony-stimulating factor (G-CSF)—are well established. GM-CSF is generally required for the development of most myeloid cells from the earliest progenitors that can be identified. In combination with G-CSF, however, it tends to stimulate only granulocyte (principally neutrophil) formation from the common granulocyte-macrophage progenitor. This is in contrast to M-CSF, which, in combination with GM-CSF, tends to stimulate differentiation of monocytes (macrophages) from the same progenitors (Fig. 10.10).

These growth and differentiation factors have no strict specificity of activity, in the sense of each factor exerting a specific effect on one type of target cell only; rather, they act in different combinations on target cells to produce different results, the outcome also depending on the developmental history of their target.

10.6 Developmentally regulated globin gene expression is controlled by regulatory sequences far distant from the coding regions

We now focus on one type of blood cell, the red blood cell or erythrocyte, to consider the transcription factors that are active in these cells during their differentiation, and that control expression of their cell-specific genes. The main feature of

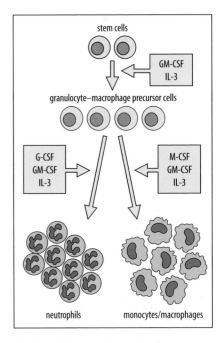

Fig. 10.10 Colony-stimulating factors can direct the differentiation of neutrophils and macrophages. Neutrophils and macrophages are derived from a common granulocyte-macrophage precursor cell. The choice of differentiation pathway can be determined by for example, the growth factors G-CSF and M-CSF. The growth factors GM-CSF and IL-3 are also required, in combination, to promote the survival and proliferation of cells of the myeloid lineage generally.

After Metcalf, D.: 1991.

red-blood-cell differentiation is the synthesis of large amounts of the oxygen-carrying protein hemoglobin, which involves the coordinated regulation of two different sets of globin genes. All the hemoglobin contained in a fully differentiated red blood cell is produced before its terminal differentiation. Terminally differentiated mammalian red blood cells lose their nucleus, whereas those of birds retain the nucleus but it becomes transcriptionally inactive. The program of erythropoiesis is primarily driven by the DNA-binding transcription factors GATA2 and GATA1, which bind to sites in the globin genes and recruit other gene-regulatory proteins.

Vertebrate hemoglobin is a tetramer of two identical α-type and two identical β-type globin chains. The α-globin and β-globin genes belong to different multigene families. Each family consists of a cluster of genes and the two families are located on different chromosomes. In mammals, different members of each family are expressed at various stages of development so that distinct hemoglobins are produced during embryonic, fetal, and adult life. Hemoglobin switching is a mammalian adaptation to differing requirements for oxygen transport at different stages of life; human fetal hemoglobin, for example, has a higher affinity for oxygen than adult hemoglobin, and so is able to efficiently take up the oxygen that has been released by maternal hemoglobin in the placenta. In turn, the oxygen affinity of adult hemoglobin is sufficient for it to pick up oxygen in the relatively oxygen-rich environment of the lungs. Here, we look at the control of expression of the β-globin genes, as examples of genes that are not only uniquely cell specific but are also developmentally regulated.

The human β-globin gene cluster contains five genes—ε, γ_G, γ_A, δ, and β (Fig. 10.11). These are expressed at different times during development: is expressed in the early embryo in the embryonic yolk sac; the two γ genes, which encode proteins differing by only one amino acid, are expressed in the fetal liver; and the δ and β genes are expressed in erythroid precursors in adult bone marrow. The protein products of all these genes combine with globins encoded by the α-globin complex to form physiologically different hemoglobins at each of these three stages of development.

Mutations in the globin genes are the cause of several relatively common inherited blood diseases. One of these, sickle-cell anemia, is caused by a point mutation in the gene coding for β-globin. In people homozygous for the mutation, the abnormal hemoglobin molecules aggregate into fibers, forcing the cells into the characteristic sickle shape. These cells cannot pass easily through fine blood vessels and tend to block them, causing many of the symptoms of the disease. They also have a much shorter lifetime than normal blood cells. Together these effects cause anemia—a lack

Fig. 10.11 The human globin genes. Human hemoglobin is made up of two identical α-type globin subunits and two identical β-type globin subunits, which are encoded by α-family and β-family genes on chromosomes 16 and 11, respectively. The composition of human hemoglobin changes during development. In the embryo it is made up of ζ (α-type) and ε (β-type) subunits; in the fetal liver, of α and γ subunits; and in the adult most of the hemoglobin is made up of α and β subunits, but a small proportion is made up of α and δ subunits. The regulatory regions α-LCR and β-LCR are locus control regions involved in switching on the hemoglobin genes at different stages.

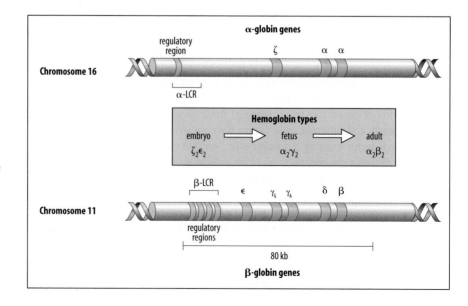

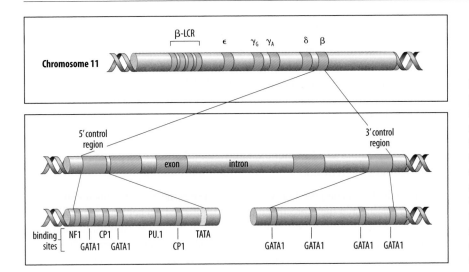

Fig. 10.12 The control regions of the β-globin gene. The β-globin gene is part of a complex of other β-type globin genes and is only expressed in the adult. Binding sites for the relatively erythroid-specific transcription factor GATA1, as well as for other non-tissue-specific transcription factors, such as NF1 and CP1, are located in the control regions. Upstream of the whole β-globin cluster, in the locus control region (LCR), are additional control regions required for high-level expression and full developmental regulation of the β-globin genes.

of functional red blood cells. Sickle-cell anemia is one of the few genetic diseases where the link between a mutation and the subsequent developmental effects on health is fully understood.

The control regions that regulate expression of the β-globin gene cluster are complex and extensive. Each gene has a promoter and control sites immediately upstream (to the 5′ side) of the transcription starting point, and there is also an enhancer downstream (to the 3′ side) of the β-globin gene, which is the last gene in the cluster (Fig. 10.12). But these local control sequences, which contain binding sites for transcription factors specific for erythroid cells, as well as for other more widespread transcriptional activators, are not sufficient to provide properly regulated expression of the β-globin genes.

The successive switching of different globin genes on and off during development is an intriguing feature of globin gene expression. This developmentally regulated expression of the β-globin gene cluster depends on a region a considerable distance upstream from the ε gene. This is the **locus control region** (LCR; see Fig. 10.12), which stretches over some 10,000 base pairs at around 5000 base pairs from the 5′ end of the ε gene. The LCR confers high levels of expression on any β-family gene linked to it, and is also involved in directing the developmentally correct sequence of expression of the whole β-globin gene cluster, even though the β-globin gene itself, for example, is around 50,000 base pairs away from the extreme 5′ end of the LCR. A similar control region has been found upstream of the α-globin gene cluster.

An attractive model for the control of globin-gene switching envisages an interaction of LCR-bound proteins with proteins bound to the promoters of successive globin genes (Fig. 10.13). The DNA between the LCR region and the globin genes is thought

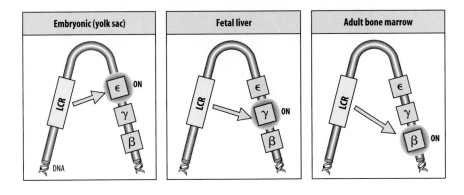

Fig. 10.13 A possible mechanism for the successive activation of β-family globin genes by the LCR during development. The LCR is thought to make contact with the promoter of each gene in succession, at different stages of development, thus controlling their temporal expression.

After Crossley, M., et al.: 1993.

to loop in such a way that proteins binding to the LCR can physically interact with proteins bound to the globin gene promoters. Thus, in erythrocyte precursors in the embryonic yolk sac the LCR would interact with the ε promoter, in fetal liver it would interact with the two γ promoters, and in the adult bone marrow with the β-gene promoter. There is some evidence that, for the β-globin locus at least, the LCR acts by promoting the transition from initiation of transcription to transcriptional elongation by RNA polymerase.

10.7 The epithelia of adult mammalian skin and gut are continually replaced by derivatives of stem cells

Mammalian skin is composed of two cellular layers—the **dermis**, which mainly contains fibroblasts, and the protective outer **epidermis**, which contains mainly keratin-filled cells called **keratinocytes**. A **basal lamina** or **basement membrane** composed of extracellular matrix separates the epithelial epidermis from the dermis. The epidermis, which we consider here, comprises a multilayered epithelium of keratinocytes (Fig. 10.14) and associated hair follicles and sebaceous and sweat glands. The epidermis between the hairs is known as the interfollicular epidermis. Because of the protective function of the epidermis, cells are continuously lost from its outer surface and must be replaced—every four weeks we have a brand new epidermis—and it is maintained throughout life by stem cells that reside in the basal layer of the epidermis in contact with the basement membrane, and at the base of the follicles.

Basal layer cells depend on stimuli received from fibroblasts in the dermis and the extracellular molecules of the basement membrane to remain proliferative. Once a cell leaves this stem-cell niche, it becomes committed to differentiate. Several hypotheses have been proposed to explain how asymmetrical divisions of basal cells could lead to one daughter remaining in the basal layer and the other becoming committed to differentiation (Fig. 10.15). After becoming committed, the cells differentiate further after leaving the basal layer. One terminal differentiation pathway culminates in the production of keratinocytes, which form the outermost, cornified epidermal layers that provide the skin's protective covering. In the interfollicular epidermis, differentiating

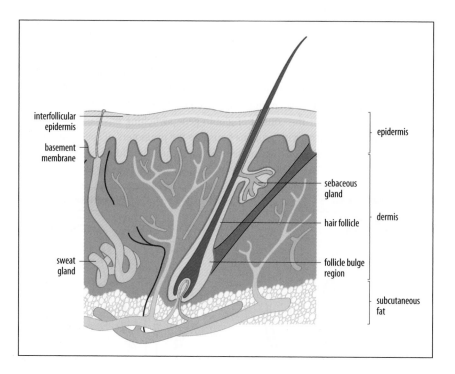

Fig. 10.14 Section through human skin showing epidermal structures. The skin is composed of an outer epidermis (pale orange) separated from the underlying dermis by a basement membrane. Structures that are epidermal in origin are shown in color here; dermal structures are in gray.

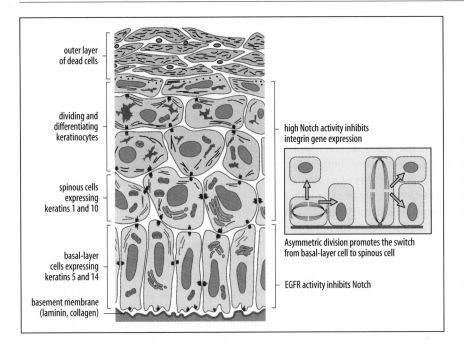

outer layer
of dead cells

dividing and
differentiating
keratinocytes

spinous cells
expressing
keratins 1 and 10

basal-layer
cells expressing
keratins 5 and 14

basement membrane
(laminin, collagen)

high Notch activity inhibits
integrin gene expression

Asymmetric division promotes the switch
from basal-layer cell to spinous cell

EGFR activity inhibits Notch

Fig. 10.15 Differentiation of keratinocytes in human epidermis. Descendants of basal-layer cells, which will become the keratinocytes of the epidermis, detach from the basement membrane, divide several times, and leave the basal layer before beginning to differentiate. Basal-layer cells express the intermediate filament proteins keratin 14 and keratin 5. Differentiation into keratinocytes involves a change in the keratin genes expressed and the production of large amounts of keratin 1 and keratin 10. In the intermediate layers, the cells are still large and metabolically active, whereas in the outer layers of the epidermis, the cells lose their nuclei, become filled with keratin filaments, and their membranes become insoluble owing to deposition of the protein involucrin. The dead cells are eventually shed from the skin surface.

keratinocytes move through the skin as they mature, until by the time they reach the outermost layer they are completely filled with the fibrous protein keratin and their membranes are strengthened with the protein involucrin. The dead cells eventually flake off the surface (see Fig. 10.15). The terminally differentiated cells of the sebaceous glands, by contrast, are lipid-filled sebocytes that eventually burst, releasing their contents onto the surface of the skin. Differentiation of a hair follicle is even more complex, comprising eight different cell lineages.

In living epidermal cells, keratins are an essential cytoskeletal component, forming the intermediate filaments through which epidermal cells are attached to each other, and giving skin its resilience and elasticity. Basal layer cells express keratin types 5 and 14 and mutations in the genes encoding these keratins are found in most patients with epidermolysis bullosa, a disease in which the epidermis is very fragile and can easily blister. Spinous cells expressing keratins 1 and 10 are located in the suprabasal layer of the epidermis and the switch from expression of keratin types 5 and 14 to expression of keratins 1 and 10 is a reliable indicator that cells have become committed to terminal differentiation. Several signals are now known to regulate this switch. Activation of Notch signaling promotes differentiation and is inhibited in basal-layer cells by various mechanisms, including signaling through the epidermal growth factor receptor (EGFR), which, as a consequence, contributes to maintaining basal cell fate (see Fig. 10.15). Cell-adhesion molecules also play a key role. The basal cell layer is attached to the basement membrane by specialized cell junctions that include integrin-containing hemidesmosomes and focal contacts (see Box 8B, p. 293). Basal cells express the integrin $\alpha_6\beta_4$, which is required for hemidesmosome formation and interacts with collagen and laminin in the extracellular matrix, and also express high levels of integrin $\alpha_3\beta_1$, which is a receptor for laminin. Mutations in the collagen associated with epidermal hemidesmosomes have been found in a subset of patients with epidermolysis bullosa blistering disease. Detachment of a cell from the basement membrane is accompanied by decreased amounts of integrins on the cell surface as a result of activation of Notch signaling.

One of the unresolved questions in epidermal cell biology is the location of the stem cells in the basal layer and whether all the cells within the basal layer are stem cells. Basal epidermal cells are heterogeneous, with some cells proliferating rapidly and others dividing far less frequently. It has been suggested that the slowly cycling cells are stem cells and that the surrounding proliferating basal cells, known as **transit-amplifying cells**, are committed to terminal differentiation. Other investigations tracking the fate of individual marked basal epidermal cells suggest, however, that all basal cells could be equivalent in their developmental potential. In the hair follicle, multipotent stem cells have been located in a region known as the bulge. Lineage-tracing experiments in transgenic mice in which, for example, green fluorescent protein (GFP) is expressed under the control of the keratin 5 promoter, showed that the cells that give rise to new hair follicles originate in the bulge, and under certain conditions, bulge cells can also contribute to sebaceous gland and epidermis. The descendants of single follicle bulge cells from mouse epidermis cultured in the laboratory can give rise to hair follicles, sebaceous glands, and epidermis when combined with dermal fibroblasts and used as skin grafts.

The epithelial cells lining the gut are also continuously replaced from stem cells. In the small intestine the epithelial cells form a single-layered epithelium that is folded to form villi that project into the gut lumen and crypts that penetrate the underlying connective tissue. Cells are continuously shed from the tips of the villi, while the stem cells that produce their replacements lie in the base of the crypts. The new cells generated by the stem cells move upward toward the tips of the villi. *En route*, they multiply in the lower half of the crypt and constitute a transit-amplifying population (Fig. 10.16). The turnover of cells in the mammalian small intestine is about 4 days.

A single crypt in the small intestine of the mouse contains about 250 cells comprising four main differentiated cell types. Paneth cells, goblet cells and enteroendrocrine cells are secretory cells, while the enterocyte, the predominant cell type in the epithelium of the small intestine, absorbs nutrients from the gut contents. About 150 of the crypt cells are proliferative, dividing about twice a day so that the crypt produces 300 new cells each day; about 12 cells move from the crypt into the villus each hour.

Canonical Wnt signaling leading to activation of the transcription factor TCF by its co-activator β-catenin (see Box 1E, p. 26) is essential for stem-cell proliferation, proliferation of the transit-amplifying population, and for the terminal differentiation of Paneth cells. Two types of stem cell have been identified in the mouse small intestine in the stem-cell niche at the base of each crypt. One type, originally identified by its slow rate of division, is located a short distance above the bottom of the crypt, at the edge of the stem-cell niche. The other stem cells, which are considerably more proliferative, are slender columnar cells—between four to six cells per crypt—that are interspersed with Paneth cells at the bottom of the crypt (see Fig. 10.16). Both sets of stem cells have distinctive patterns of gene expression. The slowly dividing cells express high levels of Bmi1, a component of the Polycomb complex, and long-term lineage tracing of cells expressing Bmi1 in transgenic mice using LacZ as a marker resulted in the labeling of entire crypts and villi after 12 months. The other population of stem cells expresses a Wnt target, the gene encoding Lgr5, a G-protein-coupled receptor whose ligand is still unknown. Experiments in which cells expressing *Lgr5*

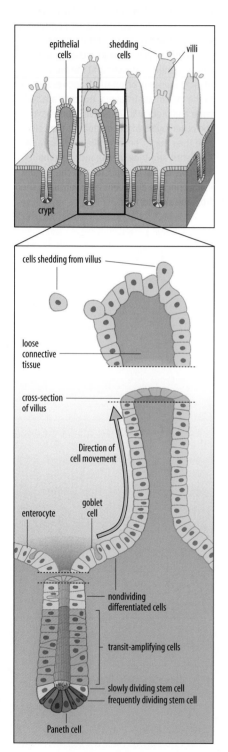

Fig. 10.16 Epithelial cells lining the mammalian gut are continually replaced. The upper panel shows the villi—epithelial extensions covering the wall of the small intestine. The lower panel is a detail of a crypt and villus. The stem cells from which the intestinal epithelium is renewed are found in the base of the crypts. One type of stem cell (yellow) is characterized by a slow rate of division, and is found at some distance from the bottom of the crypt, while a second type of stem cell (red) is interspersed between the Paneth cells in the bases of the crypts. These stem cells divide more frequently than the first type and are the ones mostly involved in routinely renewing the epithelium. The stem cells give rise to cells that proliferate and move upward. By the time they have left the crypt they have differentiated into epithelial cell types. They continue to move upwards on the projecting villus, and are finally shed from the tip. The total time of transit is about 4 days.

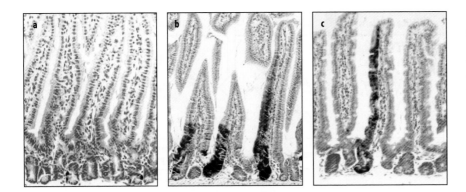

Fig. 10.17 Cell lineage tracing of Lgr5-expressing cells in the mouse small intestine. Transgenic mice which expressed LacZ from the Lgr5 promoter were used to trace the origin and movement of Lgr-expressing cells. The photos show histological analysis of LacZ activity (blue-staining cells) in small intestine: (a) 1 day after induction; (b) 5 days after induction; and (c) 60 days after induction.
From Barker, N., et al.: 2007.

and their progeny were marked irreversibly by expression of LacZ showed that these columnar cells could give rise to entire crypts and to ribbons of cells on adjacent villi after 2 months (Fig. 10.17). Thus, like the slowly cycling cells, the *Lgr5*-expressing columnar cells are indeed multipotent stem cells. It is not yet clear whether these two stem-cell populations serve different functions: one could perhaps be involved in normal tissue homeostasis, while the other provides a reserve of stem cells for regeneration.

The multipotency of the *Lgr5*-expressing stem cells was confirmed by isolating them and growing them in culture. When cultured under the right conditions, a single *Lgr5*-expressing cell can give rise to organized intestinal villus-crypt tissue containing all four differentiated cell types. Another epithelial structure, the mammary gland, has also been shown to be generated from stem cells. A complete mouse mammary gland, capable of producing milk, has been grown from a single stem cell.

Specification as either absorptive or secretory cells in the crypts of the small intestine also depends on Notch signaling. Under conditions in which Notch signaling is reduced, proliferative crypt cells are completely converted to goblet cells, whereas when Notch signaling is activated, there is increased proliferation of stem cells and a severe reduction in differentiated secretory cells (goblet cells, enteroendocrine and Paneth cells). Details of some of the transcription factors that operate in each of the four cell lineages are beginning to be revealed.

10.8 The MyoD family of genes determines differentiation into muscle

In Chapter 4 we considered the signals involved in specifying vertebrate muscle cells in the somite. Here we focus on the differentiation of cells that are already specified to become skeletal muscle; differentiation of cardiac muscle involves different transcription factors. Differentiation of vertebrate skeletal striated muscle can be studied in cell culture and provides a valuable model system for cell differentiation. Skeletal muscle cells derive from the myotome of the somites. **Myoblasts**, cells that are committed to forming muscle, can be isolated from chick or mouse embryos and mouse cells can be cloned to produce cell cultures derived from a single myoblast. These cells will proliferate until growth factors are removed from the culture medium, whereupon proliferation ceases and overt differentiation into muscle cells begins. The cells then begin to synthesize muscle-specific proteins such as actin, myosin II, and tropomyosin, which are part of the contractile apparatus, and muscle-specific enzymes, such as creatine phosphate kinase. Myoblasts also undergo structural change during differentiation. They first become bipolar in shape as a result of reorganization of the microtubule cytoskeleton, and then fuse to form multinucleate **myotubes** (Fig. 10.18). Fusion *in vivo* requires the presence of integrin β_1 on the cell membrane. Within about 20 hours of withdrawal of growth factors, typical striated **muscle fibers** can be seen.

The gene *myoD* is a member of a family of basic helix-loop-helix transcription-factor genes that are expressed only in muscle precursors and muscle cells. Genes

Fig. 10.18 Differentiation of striated muscle in culture. Myoblasts are cells that are committed to becoming muscle but have yet to show signs of differentiation. In the presence of growth factors they continue to multiply but do not differentiate. When growth factors are removed the myoblasts stop dividing; they align and fuse into multinucleate myotubes, which develop into myofibers that contract spontaneously.

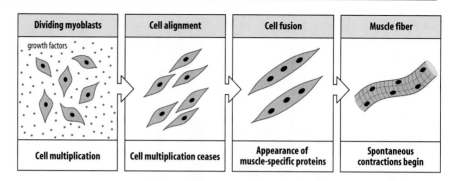

Dividing myoblasts	Cell alignment	Cell fusion	Muscle fiber
Cell multiplication	Cell multiplication ceases	Appearance of muscle-specific proteins	Spontaneous contractions begin

in this family can be considered as key controlling genes for muscle differentiation. Activation of these genes leads to the switching on of muscle-specific genes and differentiation. *myoD* will even induce differentiation into muscle if transfected into fibroblasts, which do not normally express either the MyoD protein or the muscle-specific structural proteins and enzymes. As well as *myoD*, the family includes three other genes—*mrf4*, *myf5*, and *myogenin*—which can all induce muscle differentiation in fibroblasts and other non-muscle cells. *mrf4*, *myf5*, and *myoD* are the first genes to be switched on in muscle precursors in mammals and they act as muscle-determination genes (Fig. 10.19). They are also activated by the transcription factor Pax3, which becomes restricted to muscle-cell precursors (see Section 5.9). Once turned on, these factors maintain their own expression by means of positive-feedback loops. Mrf4, Myf5, and MyoD in turn activate the *myogenin* gene, which is involved in muscle differentiation and maturation. *myoD* and *myf5* are expressed in proliferating, undifferentiated myogenic cells, whereas *myogenin* is only expressed in non-dividing differentiating cells; *mrf4* also has a second role in myofiber maturation.

Despite the demonstrated muscle-determining effects of MyoD and Myf5, gene knock-out experiments in transgenic mice (see Section 3.9 and Box 3B, p. 118) show that mice lacking MyoD make apparently normal skeletal muscle if *myf5* is active, suggesting redundancy in the system. Apparent redundancy is not uncommon in development, and has probably evolved to serve as a back-up mechanism to ensure the continuation of development in the face of fluctuations in gene expression (see Section 1.19). In mice lacking *myoD*, the level of *myf5* expression is raised, implying that the expression of *myoD* normally inhibits *myf5* expression, and that Myf5 protein can compensate for the absence of MyoD. Mice lacking Myf5 also make skeletal muscle, although they have other defects; the most obvious abnormality is a shortening of the ribs. Mice in which both Myf5 and MyoD are lacking, however, die before birth, although if such embryos still express *mrf4*, they make some muscle. Such experiments show that these proteins are acting as myogenic determination factors. Indeed, in embryos in which

Fig. 10.19 Key features of the differentiation of vertebrate skeletal muscle. External signals initiate muscle differentiation by activating the genes *mrf4*, *myoD* and *myf5*. One of these genes tends to be expressed preferentially (depending on the species) and their activity is mutually inhibitory and self-sustaining. The transcription factors encoded by *mrf4*, *myf5*, and *myoD* activate muscle-differentiation genes such as *myogenin*, which in turn activate expression of muscle-specific proteins, such as the proteins that make up the contractile apparatus in muscle cells.

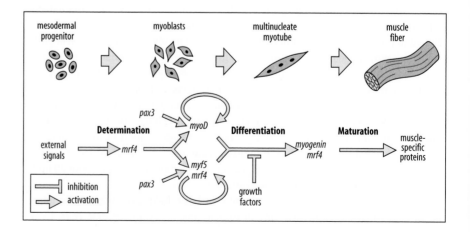

mutant *myf5* and *myoD* genes have been activated but which do not make functional Myf5 or MyoD protein, prospective muscle cells mislocate, and may be integrated into other differentiation pathways, such as cartilage and bone. Mice in which the *myogenin* gene has been knocked out, on the other hand, lack most skeletal muscle but precursor myoblasts are present. Myogenin is thus considered to be a myogenic differentiation factor rather than a protein essential for muscle determination.

The MyoD family of transcription factors activates transcription of muscle-specific genes by binding to a nucleotide sequence called the E-box, which is present in the control regions or promoter regions of these genes. MyoD forms a heterodimer with products of the E2.2 gene such as E_{12}, and this binds to the E-box. As the activity of MyoD and its relatives is essential for muscle-cell differentiation, we must now look at how the activity of MyoD could be regulated.

10.9 The differentiation of muscle cells involves withdrawal from the cell cycle, but is reversible

The proliferation and differentiation of myoblasts are mutually exclusive activities. Skeletal myoblasts that are multiplying in culture do not differentiate. Only when proliferation ceases does differentiation begin. In the presence of protein growth factors, myoblasts that are expressing both MyoD and Myf5 proteins continue to proliferate and do not differentiate into muscle. This means that the presence of MyoD and Myf5 is not in itself sufficient for muscle differentiation and that an additional signal or signals are required. In culture, this stimulus to differentiate can be supplied by the removal of growth factors from the medium. The myoblasts then withdraw from the cell cycle, cell fusion occurs, and differentiation takes place.

There is an intimate relationship between the proteins that control progression through the cell cycle and the muscle determination and differentiation factors. MyoD and Myf5 are phosphorylated by cyclin-dependent protein kinases, which become activated at key points in the cell cycle (see Fig. 13.2). Phosphorylation affects the activity of MyoD and Myf5 by making them more likely to be degraded. Thus, in actively proliferating cells, the levels of the myogenic factors will be kept low. A number of other proteins also interact with myogenic factors or with their active partners such as E_{12}, and inhibit their activity. One interacting protein is the transcription factor Id, which is present at high concentrations in proliferating cells, and which probably helps to prevent premature activation of muscle-specific genes.

The myogenic factors themselves actively interfere with the cell cycle to cause cell-cycle arrest. Myogenin and MyoD activate transcription of members of the p21 family of proteins, which block cell-cycle progression, causing the cell to withdraw from the cell cycle and to differentiate. Another key protein in preventing progression through the cell cycle is the retinoblastoma protein RB. RB is phosphorylated in proliferating cells by the cyclin-dependent kinase 4 (Cdk4) and dephosphorylation of RB is essential both for withdrawal from the cell cycle and for muscle differentiation. MyoD can interact strongly and specifically with Cdk4 to inhibit RB phosphorylation, which would also help to lead to cell-cycle arrest.

A more general mechanism for shutting down genes involved in cell proliferation has also been found to operate in muscle cells. Rather unexpectedly, this involves changing components of the core transcriptional machinery. One of the general transcription factors in the basic promoter–recognition complex described in Fig. 10.3 is TFIID, which is a multisubunit protein that recognizes the TATA box. TFIID is used in the promoter-recognition complex that controls the expression of genes involved in progression through the cell cycle and cell division. When the muscle cells undergo differentiation, however, TFIID is dismantled and is replaced by a different protein complex, which controls the expression of differentiation genes such as *myogenin*, among others (Fig. 10.20). This switch selectively turns on one transcriptional program

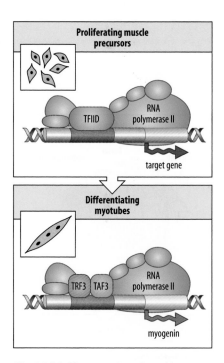

Fig. 10.20 The cessation of myoblast proliferation and the onset of muscle cell differentiation is accompanied by a change in the basal transcriptional complex. In myoblasts, as in most other cells studied, the basal transcription complex contains the general transcription factor TFIID, which recognizes the TATA box in gene promoters. In differentiating myotubes, TFIID is dismantled and is thought to be replaced by a complex containing the proteins TRF3 and TAF3. Along with specific activators, this complex recognizes the *myogenin* gene promoter and initiates transcription.

in the cell while effectively shutting down another, and could be a general mechanism for shutting down cell proliferation and initiating differentiation.

The opposition between cell proliferation and terminal differentiation makes developmental sense. In those tissues where fully differentiated cells do not divide further, it is essential that sufficient cells are produced to make a functional structure, such as a muscle, before differentiation begins. The reliance on external signals to induce differentiation, or to permit a cell already poised for differentiation to start differentiating, is a way of ensuring that differentiation only occurs in the appropriate conditions.

Despite the apparently terminal nature of muscle-cell differentiation, muscle cells provide an example of a differentiated cell that can undergo **dedifferentiation**—the loss of differentiated characteristics—re-enter the cell cycle and even give rise to other cell types when manipulated in culture—a phenomeon called **transdifferentiation**. The gene *Msx1* encodes a homeodomain-containing transcriptional repressor, which is expressed in determined but undifferentiated myoblasts during limb development and in other undifferentiated cells at the tip of the limb bud. When mouse muscle myotubes in culture were transfected with *Msx1*, 5–10% of the myotubes dedifferentiated, cleaving to form both smaller multinucleate myotubes and, more significantly, mononuclear cells that can undergo cell division. When dividing mononuclear cells were cultured in the appropriate conditions, some progeny of a single myotube expressed markers for other cell types, such as cartilage or fat cells. It was previously thought that a similar transdifferentiation of muscle cells to produce other cell types might occur in the regeneration of amphibian limbs, but this has been shown not to be the case in the axolotl (see Chapter 14).

10.10 Skeletal muscle and neural cells can be renewed from stem cells in adults

Once formed, skeletal muscle cells can enlarge by cell growth but do not divide. They can, however, be replaced. In adult mammalian muscle there are cells called **satellite cells** that can proliferate and differentiate into new muscle cells if the muscle is damaged. When satellite cells are activated, for example, by injury to the muscle, they proliferate and develop into myoblasts.

Satellite cells are the muscle stem cells. They lie between the basement membrane and the plasma membrane of mature muscle fibers and can be recognized by their expression of characteristic marker proteins, such as the cell-surface protein CD34 and the transcription factor Pax7 (Fig. 10.21). Under favorable conditions, the satellite cells associated with one myofiber (about 7 satellite cells per myofiber), grafted into an irradiated muscle in a mouse, can generate more than 100 new myofibers and will also undergo self-renewal. The irradiation kills the mouse's own satellite cells, thus making it impossible that the muscle could have regenerated on its own. Even a single muscle stem cell transplanted into irradiated mouse muscle can produce both muscle fibers and more satellite cells.

The function of Pax7 seems to overlap with that of the closely related Pax3. As discussed earlier, Pax3 is essential for early determination of muscle-cell precursors. In contrast, Pax7 is required for the maintenance of the satellite-cell population at the time of birth and is expressed in adult muscle. If, however, *Pax7* is deleted in mice 2–3 weeks after birth, limb skeletal muscle growth and regeneration is, surprisingly, unaffected and the satellite cells still make new muscle even though they no longer express the gene.

We will see in Chapter 12 how neurons and other nervous system cells develop from embryonic neural stem cells. Once fully differentiated, however, neurons in the mammalian brain do not divide, and for many years it was thought that no new brain neurons at all were produced in the adult mammal. Relatively recently, however, the production of new neurons, or **neurogenesis**, has been demonstrated as a normal

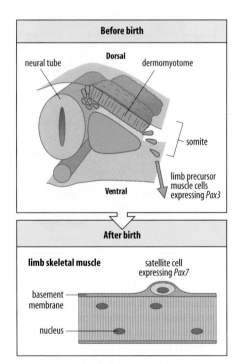

Fig. 10.21 Muscle satellite cells. During embryonic development, *Pax3*-expressing muscle cell precursors leave the dermomyotome and migrate into the limbs. In the limb muscle after birth, muscle stem cells remain associated with the muscle as satellite cells but now express *Pax7*.

occurrence in the adult mammalian brain, and neural stem cells have been identified in adult mammals that can generate neurons, astrocytes, and oligodendrocytes in culture. Neurogenesis in adult mammals occurs in just two regions: the subventricular zone of the lateral ventricle, and the subgranular zone of the dentate gyrus in the hippocampus. New neurons generated in the subventricular zone migrate to give rise to interneurons in the olfactory bulb, whereas the stem cells in the hippocampus give rise to new neurons of the dentate granule layer of the hippocampus. Slowly proliferating radial glial cells that express astrocyte-specific genes act as the adult neural stem cells. In the subventricular zone of the lateral ventricle, these stem cells give rise to proliferating transit-amplifying cells before maturing into neural progenitors (Fig. 10.22), whereas in the hippocampus, radial glial cells can give rise to neurons directly. Other adjacent glial cells in these regions provide a stem-cell niche, producing signals that control neural stem-cell behavior. Wnt signaling through the β-catenin pathway in neural stem cells has been shown to play an important role in regulating adult neurogenesis. Notch signaling determines the number of stem cells and regulates Sonic hedgehog signaling, which promotes their survival. Other sources of signals in the adult brain include the ependymal cells lining the ventricle, the blood vessels, and blood components, and some of the signaling molecules involved have been identified.

10.11 Embryonic neural crest cells differentiate into a wide range of different cell types

We now turn to the embryonic neural crest, which gives rise to a remarkable number of different cell types. As described in Chapter 5, neural crest cells arise from the dorsal neural tube and are recognizable as discrete cells when they emerge from the neural epithelium. In the cranial region of mammals, neural crest cells begin to emerge from the elevating neural folds before they fuse. The specification of the neural crest from uncommitted precursor cells at the margin of the neural plate is the result of inductive interactions between the neural plate and adjacent tissues—the non-neural ectoderm and the underlying mesenchyme. The classical model was that neural crest induction could be explained by the response of the ectoderm to intermediate levels of BMP signaling (see Chapter 5). It is now clear from work in many species that, in addition, Wnt and FGF signaling is critical for neural crest induction. Once neural crest cells have detached from the neural tube, they migrate to many sites, as discussed in Chapter 8.

Here we shall discuss their potential for differentiation. Neural crest cells give rise to the neurons and glia of the peripheral nervous system (which includes the sensory and autonomic nervous systems), endocrine cells, such as the chromaffin cells of the adrenal medulla, and melanocytes, which are pigmented cells found in the skin and other tissues (Fig. 10.23). In the head of mammalian and avian embryos, the neural

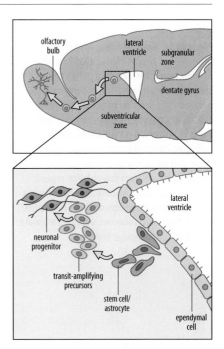

Fig. 10.22 Sites of neurogenesis in the adult mammalian brain. Neurogenesis occurs mainly in the subgranular layer of the dentate gyrus in the hippocampus and the anterior part of the subventricular zone, which is shown in more detail in this figure. Astrocyte-like neural stem cells give rise to rapidly dividing transit-amplifying cells, which differentiate into neurons. New neuronal cells generated in the dentate gyrus migrate to the granular layer, while new neuronal cells generated in the subventricular zone migrate to the olfactory bulb and differentiate there into interneurons. Endothelial cells of nearby blood vessels also provide signals for stem-cell differentiation (not shown).

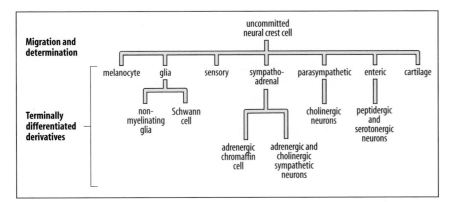

Fig. 10.23 Derivatives of the neural crest. Neural crest cells give rise to a wide range of cell types that include melanocytes, cartilage, glia, and various types of neurons distinguished by their functional specializations and the neurotransmitters they produce. Cholinergic neurons use acetylcholine as their neurotransmitter, adrenergic neurons principally use norepinephrine (noradrenaline), and peptidergic and serotonergic neurons produce peptide neurotransmitters and serotonin, respectively.

Fig. 10.24 Fate and developmental potential of neural crest cells. The fate map of the neural crest (bottom panel) shows that there is a general correspondence between the position of the neural crest cells along the antero-posterior axis and the position of the structure to which these cells give rise (top panel). For example, the crest that gives rise to the mesectoderm of the head comes from the anterior region. The developmental potential of neural crest cells, with the exception of the presumptive mesectoderm, is, however, very much greater than their presumptive fate.

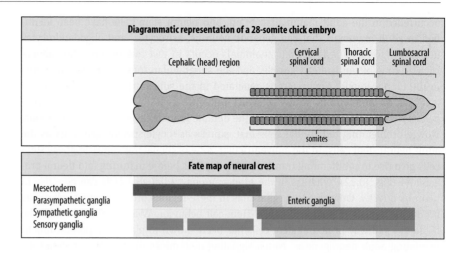

crest also gives rise to mesodermal types of tissues in the form of the facial bones, cartilage, and dermis. Neural crest with potential to form mesodermal as well as ectodermal derivatives is called the **mesectoderm**.

The tremendous differentiation potential of the neural crest was originally discovered by removing the neural folds from amphibian embryos and noting which structures failed to develop. This was later followed by an extensive series of experiments using quail neural crest transplanted into chick embryos. Quail cells contain a distinctive nucleus that distinguishes them from chick cells (see Fig. 5.6). It is therefore possible to track the transplanted quail cells and see which tissues they contribute to and what cell types they differentiate into. To construct a fate map of the neural crest, small regions of quail dorsal neural tube were grafted to equivalent positions in chick embryos of the same age, at a stage before neural crest migration. The fate map shows a general correspondence between the position of crest cells along the antero-posterior axis and the axial position of the cells and tissues they will give rise to (Fig. 10.24). For example, the neural crest cells that give rise to tissues in the facial and pharyngeal regions are derived from crest anterior to somite 5, whereas ganglion cells of the autonomic sympathetic system come from crest posterior to somite 5.

To what extent is the fate of neural crest cells fixed before migration? Some crest cells are unquestionably multipotent. Single neural crest cells injected with a tracer shortly after they have left the neural tube can be seen to give rise to a number of different cell types, both neuronal and non-neuronal. Also, by changing the position of neural crest before cells start to migrate, it has been shown that the developmental potential of these cells is greater than their normal fate would suggest. Multipotent stem cells have been isolated from neural crest that can give rise to neurons, glia, and smooth muscle. Similar multipotent cells have even been isolated from the peripheral nerves of mammalian embryos days after neural crest migration. Almost all of the cell types to which the neural crest can give rise will differentiate in tissue culture, and the developmental potential of single neural crest cells has been studied in this way. Most of the clones that develop from neural crest cells taken at the time of migration contain more than one cell type, showing that the cell was multipotent at the time it was isolated. As neural crest cells migrate, however, their potential decreases progressively: the size of cultured clones gets smaller and the diversity of cell types they contain is less. Thus, shortly after the beginning of migration, the neural crest is a mixed population of multipotent cells and cells whose potential is already restricted.

Numerous extracellular signaling proteins that affect the differentiation and proliferation of neural crest cells have been identified from mutations in mice and experiments with cultured cells. For example, melanocyte survival and differentiation is stimulated by Wnts and by the cytokines endothelin and stem-cell factor (SCF), gliogenesis by neuregulin (also called glial growth factor), while BMPs are required for the development of autonomic neurons. Activation of the canonical Wnt/β-catenin pathway in the neural tube promotes the formation of sensory neurons by neural crest at the expense of virtually all other derivatives, and brain-derived neurotrophic factor (BDNF) produced by the neural tube has been implicated in the survival of sensory neurons. A number of different transcription factors have been identified that are expressed at the neural plate border and in pre-migratory and early migrating neural crest progenitors. These include members of the SoxE family of transcription factors, such as Sox9, which has been shown to be important for neural crest survival in both mouse and zebrafish embryos, and Sox10, which is expressed in migrating neural crest cells (see Box 1D, p. 20) and then regulates the terminal differentiation of neurons and melanocytes.

Melanocytes are derived from neural crest cells that follow a dorsal migration route under the ectoderm. Mice homozygous for either the *white spotting* (*W*) or *Steel* mutations (genes originally identified by unusual coat color) lack melanocytes in the skin and other tissues. Although mutations in the two different genes apparently have the same gross effect, the genes code for two quite different proteins, but which nevertheless function in the same pathway. *white spotting* codes for the cell-surface protein Kit, which is the receptor for SCF. It is expressed on developing melanoblasts, which are the undifferentiated precursors of skin melanocytes. *Steel*, in contrast, codes for SCF, which is produced by the fibroblasts and keratinocytes that surround the melanoblast. Mutations in either gene block melanocyte differentiation, showing that the signal generated by the interaction of SCF and Kit is essential for differentiation (Fig. 10.25). There is a very rare human condition known as piebaldism in which melanocytes fail to develop in very localized regions of the body, including the scalp and forehead—giving rise to a white forelock—and the ventral trunk and limbs. It is now known that this condition is due to mutations in the human *KIT* gene.

Neural crest cells give rise to both the sensory dorsal root ganglia and the autonomic ganglia of the peripheral nervous system. These ganglia contain both neurons and glia, the neurons differentiating first. So what determines whether a neural crest cell gives rise to a neuron or to glia? There is evidence that activation of Notch in neural crest stem cells inhibits neuronal formation and promotes differentiation into glia. There is also evidence that Notch ligands, such as Delta, are expressed on neuroblasts, the precursors of neurons. Thus, once a neuroblast has embarked on the neurogenic differentiation pathway under the influence of neurogenic signals such as BMP-2, it may provide a feedback signal that prevents the surrounding cells from developing further as neuroblasts and promotes formation of glia instead. Wnt signaling biases neural crest cells towards the sensory neuron lineage, whereas BMP signals influence them to acquire properties of sympathetic neurons.

In mammal and bird embryos, the neural crest in the head, which migrates to the branchial arches (see Fig. 5.37), appears to give rise not only to the usual ectodermal derivatives, but also to cells that form facial bones, cartilage, and dermis. Recently, however, the presence of non-neural cells has been shown in the neural folds immediately adjacent to the neural crest in the heads of mouse embryos. A non-neural lateral epithelial domain characterized by the expression of various proteins normally associated with connective tissues can be distinguished from the adjacent neural crest region. It has therefore been suggested that this lateral epithelium might be the source of mesectodermal cells and that these are distinct from the neural crest. However, this intriguing idea has not yet been confirmed by the demonstration that these cells migrate to the branchial arches.

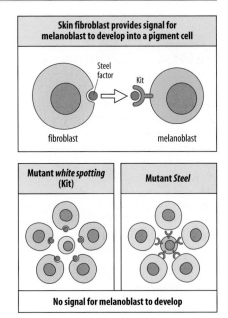

Fig. 10.25 **The receptor Kit and its ligand the Steel factor are involved in melanoblast differentiation.** When either the gene encoding Kit (the gene *white spotting,* or *W*) or the *Steel* gene is mutant, melanoblasts fail to differentiate into melanocytes.

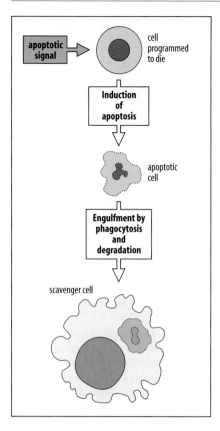

Fig. 10.26 Programmed cell death. During development many cells undergo programmed cell death or apoptosis. Apoptosis may be triggered by an external signal (as shown here) or cells may be programmed to die unless they receive a specific survival signal.

Hox gene expression in the cranial neural crest (see Section 5.13) is a determinant of crest potential. For example, no Hox genes are expressed in the anterior region of the head neural folds, which give rise to the cells that form the bones of the face, whereas anterior Hox genes are expressed in more posterior parts of the head neural crest. If the genes *Hoxa2, Hoxa3,* or *Hoxb4* are ectopically expressed in the anterior head neural crest of chick embryos, no facial skeletal structures differentiate. And in transgenic mouse embryos in which the entire *Hoxa* gene cluster has been inactivated altogether in the posterior cranial neural crest, these cells now give rise to anterior structures normally produced by cells that do not express Hox genes.

10.12 Programmed cell death is under genetic control

Selective cell death is a normal part of development. While not strictly cell differentiation, it is convenient to consider it here in these terms. It is involved, for example, in the morphogenesis of the vertebrate limb, where interdigital cell death is essential for separating the digits (see Chapter 11), and the development of the vertebrate nervous system involves the death of large numbers of neurons. Programmed cell death is particularly important in the development of the nematode (see Chapter 6): 959 somatic cells come from the egg and 131 die during development. In all these cases, the dying cell undergoes a type of cell suicide known as **apoptosis**, which requires both RNA and protein synthesis and is quite distinct from the death that occurs as a result of pathological damage. In apoptosis, the cytosolic calcium concentration rises and leads to the activation of an endonuclease, which fragments the chromatin. The cell contents remain bounded by a cell membrane throughout, even though the cell breaks up into fragments. The dying cell is finally phagocytosed by scavenger cells (Fig. 10.26). These features distinguish apoptosis from cell death due to damage, where the whole cell tends to swell and eventually burst open (lyze); this type of cell death is known as **necrosis**.

Studies of apoptosis in the nematode have shown that, although many different types of cells die, their death is initiated through a common mechanism centered on the actions of CED-3, which is a type of protease called a **caspase**. CED-3 triggers the cellular changes that lead to apoptotic cell death. The nematode caspase-activation pathway is shown in Fig. 10.27 (left panel). CED-4 in this pathway is an adaptor protein that activates CED-3. In nematodes with mutations that inactivate either *ced-3* or *ced-4*, none of the 131 cells that normally die do so, but appear instead to differentiate in the same way as their sister cells. Such worms appear to survive well and die at the usual age of several weeks.

The nematode's cell-death program is under the control of another protein, CED-9, which acts as a brake on the process by preventing the activation of CED-3 by CED-4. If *ced-9* is inactivated by mutation, many cells that normally do not die will die. And if the *ced-9* gene is made abnormally active by mutation, no cell deaths occur. Apoptosis occurs when the actions of the protein EGL-1 remove inhibition by CED-9. EGL-1 is produced in response to pro-apoptotic signals. A similar pathway is present in *Drosophila* and mammals (Fig. 10.27, middle and right panels). The mammalian homologs of *ced-9* are the Bcl-2 family of genes (Fig. 10.27, right panel). The sequence homologies between the gene *Bcl-2* and *ced-9* are so strong that *Bcl-2* introduced into nematodes will function in place of *ced-9*. Mammals also possess caspases homologous to CED-3, which initiate the proteolytic cascade that leads to cell death.

Programmed cell death in animals is far from being a rare phenomenon. The cells in all tissues are intrinsically programmed to undergo cell death, and are only prevented from dying by positive control signals from neighboring cells. Cell death also plays a key role both in controlling growth and preventing cancer (see Chapter 13).

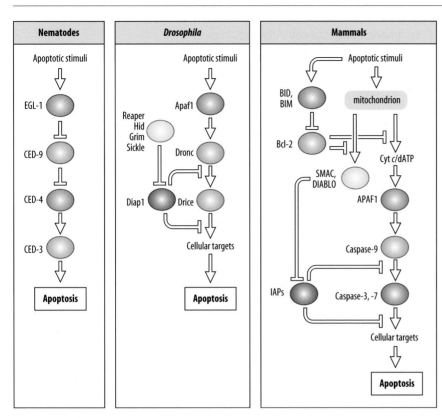

Fig. 10.27 Apoptotic pathways in nematodes, *Drosophila*, and mammals. Left panel: in nematodes, programmed cell death during development requires the activation of a pathway that ends in the activation of the caspase CED-3, a type of protease, by the protein CED-4. CED-3 triggers the cellular changes that lead to cell death. The actions of CED-3 and CED-4 are inhibited by CED-9. If CED-9 is active, the cell will not undergo apoptosis. Inhibition by CED-9 is lifted by the actions of EGL-1, which is expressed in response to pro-apoptotic signals. Middle panel: a similar pathway is present in *Drosophila*, with the addition of a second caspase (Drice). Diap1 is a negative regulator of the caspases that is not present in nematodes. Its inhibition is lifted by the actions of proteins such as Reaper and Sickle. Right panel: the apoptotic pathway in mammals is even more complex, but uses the same main components. Apoptotic signals are transduced through the EGL-1 homologs BIM and BID to cause the release of pro-apoptotic factors from mitochondria. One of these is cytochrome *c*, which triggers the caspase-activation pathway. In the absence of apoptotic signals, cell death is inhibited by BCL-2, which is homologous to CED-9. In all three diagrams, the proteins are color-coded to show homology.

Summary

Several adult tissues, such as the blood and the skin, are continuously being replaced by the differentiation of new cells from stem cells. Hematopoiesis illustrates how a single progenitor cell, the multipotent stem cell, can be both self-renewing and generate a range of different cell lineages. The differentiation of the various lineages of blood cells seems to depend on external signals, although there is some evidence for an intrinsic tendency to diversification. External factors are, however, essential for the survival and proliferation of the specific cell types. The epithelial tissues of adult mammalian skin and gut are also replaced from self-renewing stem cells. The derivation of blood cells from the multipotent stem cell involves a successive restriction of developmental potential as cells become committed to the different lineages. Similar processes occur in the diversification of the embryonic neural crest. Before they leave the crest, some crest cells still have a broad developmental potential, and environmental signals both direct the pathways of neural crest differentiation and promote the survival of particular cell types.

In general, cell differentiation is controlled by complex combinations of transcription factors, whose expression and activity are influenced by external signals. Some transcription factors are common to many cell types, but others have a very restricted pattern of expression. Muscle differentiation, for example, can be brought about by the expression of transcription factors specific to the muscle lineage, such as MyoD, which initiate the differentiation program and bind to the control regions of muscle-specific genes. Programmed cell death is a common fate for cells, especially during development, and there is evidence that positively acting signals are generally required to prevent programmed cell death and allow cells to survive.

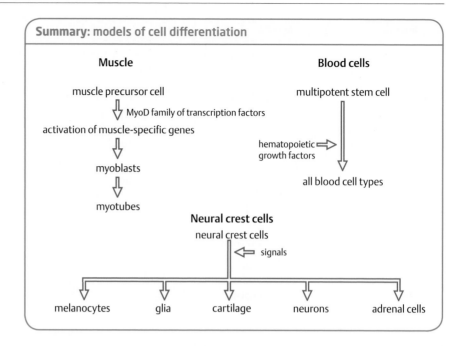

Summary: models of cell differentiation

The plasticity of gene expression

In earlier parts of this chapter we have looked at examples of genes being turned on and off during development and have considered some of the mechanisms involved. These chiefly involve the binding of different combinations of gene-regulatory proteins to the control regions of genes. How reversible are these changes, and thus how reversible is cell differentiation? We have already seen several examples of mutipotent stem cells that both self-renew and give rise to a range of different cell types. If such stem cells could be produced reliably and in sufficient numbers, it might be possible to use them to replace cells that have been damaged or lost by disease or injury. This is one of the main aims of the field of **regenerative medicine**. The therapeutic use of stem cells will depend on understanding precisely how gene activity can be controlled in stem cells to give the desired cell type, and just how plastic stem cells are. Could, for example, blood stem cells give rise to neurons under appropriate conditions? How easy is it to alter the pattern of gene activity? Is it possible to make pluripotent stem cells from differentiated cells?

We will start the discussion of cell plasticity by examining the extent to which the pattern of gene activity in differentiated cells can revert to that found in the fertilized egg. One way of finding out whether it can be reversed in practice is to place the nucleus of a differentiated cell in a different cytoplasmic environment, one that contains a different set of gene-regulatory proteins.

10.13 Nuclei of differentiated cells can support development

The most dramatic experiments addressing the reversibility of differentiation have investigated the ability of diploid nuclei from cells at different stages of development to replace the nucleus of an egg and to support normal development. If they can do this, it would indicate that no irreversible changes have occurred to the genome during differentiation. It would also show that a particular pattern of nuclear gene activity is determined by whatever transcription factors and other regulatory proteins are being synthesized in the cytoplasm of the cell. Such experiments were first carried out using the eggs of amphibians, which are particularly robust for experimental manipulation.

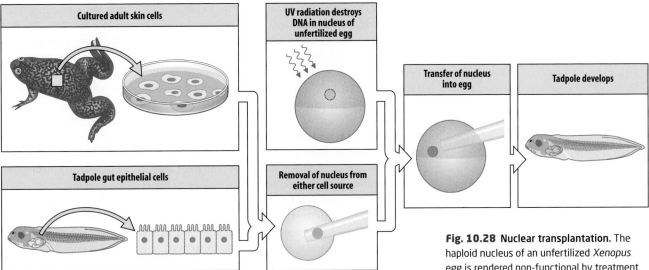

Fig. 10.28 Nuclear transplantation. The haploid nucleus of an unfertilized *Xenopus* egg is rendered non-functional by treatment with ultraviolet radiation. A diploid nucleus taken from an epithelial cell lining the tadpole gut or from a cultured adult skin cell is transferred into the enucleated egg, and can support the development of an embryo until at least the tadpole stage.

In the unfertilized eggs of *Xenopus*, the nucleus lies directly below the surface at the animal pole. A dose of ultraviolet radiation directed at the animal pole destroys the DNA within the nucleus, and thus effectively removes all nuclear function. These enucleated eggs can then be injected with a diploid nucleus taken from a cell at a later stage of development to see whether it can function in place of the inactivated nucleus. The results are striking: nuclei from early embryos and from some types of differentiated cells of larvae and adults, such as gut and skin epithelial cells, can replace the egg nucleus and support the development of an embryo up to the tadpole stage (Fig. 10.28), and in a small number of cases even into an adult. The organisms that result are a **clone** of the animal from which the somatic cell nucleus was taken, as they have the identical genetic constitution. The process by which they are produced is called **cloning**.

Nuclei from adult skin, kidney, heart, and lung cells, as well as from the intestinal cells and myotome of *Xenopus* tadpoles, can support development when transplanted into enucleated eggs. This technique is known as **somatic cell nuclear transfer**. However, the success rate with nuclei from somatic cells of adults is very low, with only a small percentage of nuclear transplants developing past the cleavage stage. In general, the later the developmental stage the nuclei come from, the less likely they are to be able to support development. Transplantation of nuclei taken from blastula cells is much more successful; by transplanting nuclei from the same blastula into several enucleated eggs, clones of genetically identical frogs can be obtained (Fig. 10.29). An even greater success rate is achieved if the process is repeated with nuclei from the blastula stage of a cloned embryo. These results show that, at least in those differentiated cell types tested, the genes required for development are not irreversibly altered. More importantly, when exposed to the egg's cytoplasmic factors, they behave as the genes in the zygotic nucleus of a fertilized egg would. So in this sense at least, many of the nuclei of the embryo and the adult are equivalent, and their behavior is determined entirely by factors present in the cell.

What about organisms other than *Xenopus*? Similar results have been obtained in insects, where nuclei from the blastoderm stage can, when returned to the egg, participate in the formation of a wide range of tissue types. In ascidians, the genetic equivalence of nuclei at different developmental stages can also be demonstrated. In

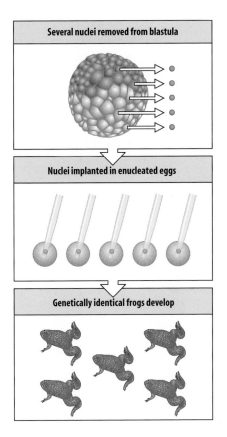

Fig. 10.29 Cloning by nuclear transfer. Nuclei from cells of the same blastula are transferred into enucleated, unfertilized *Xenopus* eggs. The frogs that develop are clones as they all have the same genetic constitution.

plants, single somatic cells can give rise to fertile adult plants, demonstrating complete reversibility of the differentiated state at the cellular level (see Fig. 7.9). The first mammal to be cloned by somatic cell nuclear transfer was a lamb, the famous Dolly. In this case, the nucleus was taken from a cell line derived from the udder. Mice have been similarly cloned using, for example, nuclei from the somatic cumulus cells that surround the recently ovulated egg, and several generations of these mice have been produced by repeating the procedure.

In general, the success rate of cloning by somatic cell nuclear transfer in mammals is extremely low, and the reasons for this are not yet well understood. Most cloned mammals derived from nuclear transplantation die before birth, and those that survive are usually abnormal in some way. The most likely cause of failure is incomplete reprogramming of the donor nucleus to remove all the epigenetic modifications, such as DNA methylation and histone modification, that are involved in determining and maintaining the differentiated cell state, as discussed earlier, and thus a failure to restore the DNA to a state resembling that of a newly fertilized oocyte. A related cause of abnormality may be that the reprogrammed genes have not gone through the normal imprinting process that occurs during germ-cell development, where different genes are silenced in the male and female parents (see Section 9.8). The abnormalities in adults that do develop from cloned embryos include early death, limb deformities and hypertension in cattle, and immune impairment in mice, and all these defects are thought to be due to abnormalities of gene expression that arise from the cloning process. Dolly the sheep developed arthritis at 5 years old and was humanely put down at the age of 6 years (relatively young for sheep) after developing a progressive lung disease, although there is no evidence that cloning was a factor in Dolly contracting the disease. Studies have shown that some 5% of the genes in cloned mice are not correctly expressed and that almost half of the imprinted genes are incorrectly expressed.

The generation of cloned blastocysts following the transfer of an adult somatic cell nucleus has more recently been achieved in primates and humans by modifying the protocols used for nuclear transfer in other animals. In both cases, nuclei from adult skin cells have been transferred into enucleated oocytes, many of which then underwent cleavage, with a small number eventually forming blastocysts. One of the reasons for attempting to clone mice and other mammals is to generate **embryonic stem cells (ES cells)** from the cells of the inner cell mass of a cloned blastocyst (see Section 3.9). ES cells have been successfully derived from cloned monkey blastocysts but this has not yet been achieved for human blastocysts. Cloning of an adult primate by nuclear transfer has not yet been accomplished and cloning a human being by these means, even if it could be achieved, is banned in most countries on both practical and ethical grounds. Despite reports in the media that humans have been cloned, none of these reports has been verified.

10.14 Patterns of gene activity in differentiated cells can be changed by cell fusion

Nuclear transplantation into eggs, particularly frog eggs, is facilitated by their size and the large amount of cytoplasm. With other cell types, particularly differentiated cells, injection of nuclei into foreign cytoplasm is not possible. It is possible, however, to expose the nucleus of one cell type to the cytoplasm of another type by fusing the two cells together. In this process, which can be brought about by treatment with certain chemicals or viruses, the plasma membranes fuse and nuclei from different cells come to share a common cytoplasm. A cell-division inhibitor is then used to ensure that the two nuclei remain separate.

A striking example of the reversibility of gene activity after cell fusion is provided by the fusion of chick red blood cells with human cancer cells in culture. Unlike mammalian red blood cells, mature chick red blood cells have a nucleus. Gene activity in the nucleus is, however, completely turned off and it thus produces no mRNA. When a chick red blood cell is fused with a cell from a human cancer cell line, gene expression in the red blood cell nucleus is reactivated, and chick-specific proteins are expressed. This expression of chick-specific proteins shows that human cells contain cytoplasmic factors capable of initiating transcription in chick nuclei.

Fusion of differentiated cells with striated muscle cells from a different species provides further evidence for reversibility of gene expression in differentiated cells (Fig. 10.30). Multinucleate striated muscle cells are good partners for cell-fusion studies as they are large, and muscle-specific proteins can easily be identified. Differentiated human cells representative of each of the three germ layers have been fused with mouse multinucleate muscle cells, so that the human cell nuclei are exposed to the mouse muscle cytoplasm. This results in muscle-specific gene expression being switched on in the human nuclei. For example, human liver cell nuclei in mouse muscle cytoplasm no longer express liver-specific genes; instead, their muscle-specific genes are activated and human muscle proteins are made. Furthermore, expression of genes such as *MyoD*, which as we have seen can initiate muscle differentiation, is activated, thus showing that reprogramming of the liver-cell nuclei in mouse muscle cytoplasm appears to proceed through the same steps as those taken when a muscle cell differentiates.

These results clearly show that patterns of gene expression in differentiated cells can be changed, and that gene expression can be controlled by factors present in the cytoplasm. At least some of these factors are transcription factors, as we will see below, and this leads to the conclusion that the differentiated state may be at least partly maintained by the continuous action of transcription factors on their target genes (see Section 10.3).

10.15 The differentiated state of a cell can change by transdifferentiation

A fully differentiated cell is generally stable, which is essential if it is to serve a particular function in the mature animal. If it has retained the ability to divide, the cell passes its differentiated state on to all its descendants. In some long-lived cells, such as neurons, which do not divide once they are differentiated, the differentiated state must remain stable for many years. Plant cells maintain their differentiated state while in the plant, but do not maintain it when placed in cell culture (see Chapter 7). Under certain conditions differentiated animal cells are not stable, and this provides yet another demonstration of the potential reversibility of patterns of gene activity. The change of one differentiated cell type into another is known as **transdifferentiation**. The allied phenomenon in which there is a switch of committed but not yet differentiated progenitor cells into a different lineage is known as **transdetermination**. The best studied examples of transdetermination occur in regenerating *Drosophila* imaginal discs, where rare transdetermination events result in the homeotic transformation of one type of adult structure into another.

Transdifferentiation occurs very rarely as part of normal development, and in vertebrates is more commonly seen in regeneration (discussed in Chapter 14) and certain pathological conditions. There are some examples of transdifferentation in normal invertebrate development, however. In *Caenorhabitis elegans*, in which the fate of every cell can be followed, a few cells appear to undergo transdifferentiation. An epithelial cell (known as Y), which forms part of the rectum, is generated in the embryo and has the morphological and molecular hallmarks of a differentiated epithelial cell. In the second larval stage, this cell leaves the rectum epithelium and migrates to form a motor neuron (known as PDA), which now possesses an axonal process and makes synaptic connections. Thus a fully differentiated epithelial cell with no detectable neuronal features undergoes transdifferentiation into a neuron with no residual

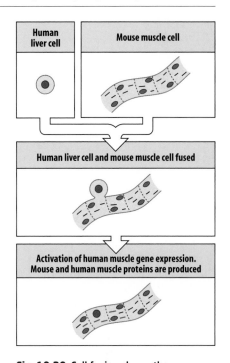

Fig. 10.30 Cell fusion shows the reversibility of gene inactivation during differentiation. A human liver cell is fused with a mouse muscle cell. The exposure of the human nucleus (red) to the mouse muscle cytoplasm results in the activation of muscle-specific genes and repression of liver-specific genes in the human nucleus. Human muscle proteins are made together with mouse muscle proteins. This shows that the inactivation of muscle-specific genes in human liver cells is not irreversible.

epithelial characteristics. It is not yet clear whether the Y cell dedifferentiates and then redifferentiates as the PDA cell or whether dedifferentiation and redifferentiation occur at the same time.

A role for the extracellular matrix in determining the state of cell differentiation is suggested by a particularly interesting example of transdifferentiation in jellyfish striated muscle. This can be made to transdifferentiate into two different types in succession. When a small patch of striated muscle and its associated extracellular matrix (mesogloea) is cultured, the striated muscle state is maintained. However, if the cultured tissue is treated with enzymes that degrade the extracellular matrix, the cells form an aggregate, and within 1–2 days some have transdifferentiated into smooth muscle cells, which have a different cellular morphology. This is followed by the appearance of a second cell type—nerve cells. This example indicates a role for the extracellular matrix in maintaining the differentiated state of the striated muscle.

The classic example of transdifferentiation in vertebrates is lens regeneration in the adult newt eye (Fig. 10.31). When the lens is completely removed by surgery (lentectomy), a new lens vesicle starts to form around 10 days post-lentectomy from the dorsal region of the pigmented epithelium of the iris. The iris cells first dedifferentiate—they lose their pigmentation and alter their shape from a flat to a columnar epithelium—and start to proliferate. They then redifferentiate to form a new lens. The lens of the adult mammalian or avian eye cannot regenerate *in vivo*, but the pigmented epithelial cells of the embryonic chick retina can be induced to transdifferentiate in culture. A single pigmented cell from the embryonic retina can be grown in culture to produce a monolayer of pigmented cells. On further culture in the presence of hyaluronidase, serum, and phenylthiourea, the cells lose their pigment and retinal cell characteristics. If cultured at a high density with ascorbic acid, the cells then start to take on the structural characteristics of lens cells, and to produce the lens-specific protein crystallin. It should be noted that, in both the newt and the chick, transdifferentiation occurs to a developmentally related cell type. Retinal and iris pigment cells and lens cells are all derived from the ectoderm (vertebrate eye development is discussed in Chapter 11).

Transdifferentiation could also underlie well-documented examples in humans, usually associated with tissue damage and repair, in which rare differentiated cells of one

Fig. 10.31 Lens regeneration in adult newt. Top panels: Removal of the lens from the eye of a newt results in regeneration of a new lens from the dorsal pigmented epithelium of the iris. Bottom panels: sections through an intact and a regenerating newt eye viewed by scanning electron microscopy. (a) A section through an intact eye. Note that the aqueous chamber (a) is clear, whereas the vitreous chamber (v) is dominated by compacted layers of extracellular matrix (pink). Lens is gray, cornea is shown as light blue, the retina as purple and the ligaments and matrix associated with the attachment to the lens is colored brown. (b) A section through an eye 5 days post-lentectomy. The aqueous chamber has been invaded by extracellular matrix and the iris has become thickened. (c) A section through an eye 25 days post-lentectomy. The aqueous chamber is gradually clearing, the lens, which developed from the dorsal rim of the iris, is now attached to the ventral iris (arrow), and extracellular matrix in the vitreous chamber is becoming reorganized. Magnification × 50.

Micrographs from Tsonis, P.A., et al.: 2004.

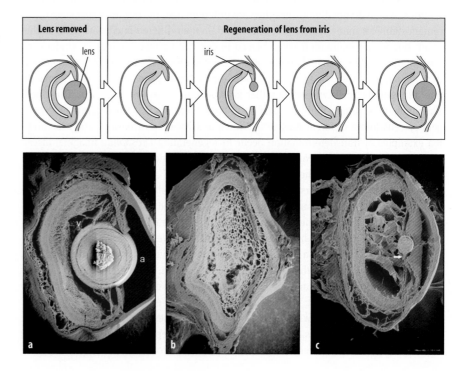

type appear in an inappropriate location in another tissue. Such examples include the appearance of hepatic cells in the pancreas and vice versa, and gastric-type epithelium in the esophagus in place of stratified squamous epithelium. In the latter example, known as Barrett's metaplasia, the change predisposes the patient to developing esophageal cancer, and illustrates the clinical importance of transdifferentiation. More optimistically, it might be possible to harness the process of transdifferentiation to produce selected differentiated cell types for regenerative medicine, as discussed later.

10.16 Embryonic stem cells can proliferate and differentiate into many cell types in culture

The multipotent stem cells that are responsible for tissue renewal in adults are limited in the range of different cell types they give rise to, and require a specific stem-cell niche to maintain their stem-cell state, such as the bone marrow stroma that supports hematopoietic stem cells (see Section 10.4). In contrast, pluripotent early embryonic cells can differentiate into cell types of all three germ layers and do not require a special niche.

The primary examples of pluripotent stem cells in mammals are the embryonic stem cells (ES cells) derived from the inner cell mass of the blastocyst. Mouse ES cells have been studied intensively. They can be maintained in culture for long periods, apparently indefinitely, but if injected into a blastocyst that is then returned to the uterus, they can contribute to all the types of cells in that embryo, but not to extraembryonic structures. One way of testing the pluripotent nature of ES cells is to return them to a blastocyst formed of tetraploid cells, which themselves can only form placental tissues (Box 10B). The ES cells then form the entire embryo.

Box 10B Testing ES cell potential in tetraploid blastocysts

One reason for cloning mice is to generate embryonic stem cells (ES cells) from the cells of the inner cell mass of a cloned blastocyst. These cells can be grown in culture to provide large numbers of pluripotent stem cells of a known genetic constitution for further experiments; for example, in mouse models of human diseases. To determine whether cultured ES cells were truly pluripotent and could support the complete development of a fetus, they were injected into tetraploid host blastocysts derived from the fusion of blastomeres from two-cell diploid embryos in culture (see figure). The chimeric blastocyst was then replaced in the uterus of a surrogate mother mouse. The tetraploid cells are only able to form the placenta, and therefore injecting ES cells into such embryos provides a way of confirming and studying their full developmental potential. Many of the embryos developed into normal mice, and all the tissues in the subsequent embryos came from the injected ES cells.

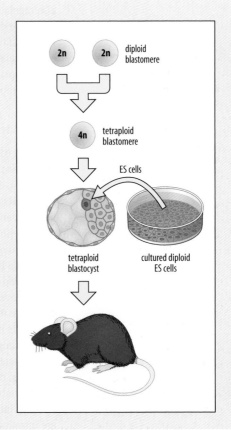

To maintain mouse ES cells in a pluripotent state in cell culture, the cells must express a particular combination of transcription factors (most notably Oct 3/4, Sox2, c-Myc and Kfl4), whose expression together is restricted to pluripotent stem cells. It was found that transcription of these genes could be maintained when the cytokines leukemia inhibitory factor (LIF) and BMP-4 were added to the culture medium. More recently it has been shown that small-molecule drugs that block differentiation by inhibiting mitogen-activated kinase (MAPK) signaling pathways, and which promote survival by inhibiting GSK3 signaling, can be used to derive and propagate ES cells from mouse and rat blastocysts. This may turn out to be a universal approach to deriving pluripotent ES cells.

The first human ES cells were produced in 1998. They express the same transcription factors as do mouse ES cells, but the signaling pathways that support self-renewal in cell culture are different. FGF and activin/Nodal signaling but not LIF are required for maintaining human ES cells, while BMP signaling promotes differentiation, the opposite effect to that in mouse ES cells. Several hundred human ES cell lines have now been established around the world.

ES cells can be made to differentiate into a particular cell type by manipulating the culture conditions, particularly in respect of the growth factors present. If grown in suspension, ES cells spontaneously form aggregates called embryoid bodies, which can differentiate into many cell types. When cultured in the absence of serum, which contains various growth and differentiation factors, many of which are still unknown, ES cells differentiate into neural cells. Under other treatments, ES cells can differentiate into heart muscle, blood islands, neurons, pigmented cells, epithelia, fat cells, macrophages, and even germ cells. For example, aggregates of ES cells successively treated with particular combinations of FGF-2, EGF and platelet-derived growth factor continue to proliferate as long as these growth factors are present. But when they are withdrawn, the cells differentiate into either of two glial cell types—astrocytes or oligodendrocytes. Cell shape can also regulate stem-cell commitment. Human mesenchymal stem cells allowed to flatten in culture differentiate as osteocytes (bone cells), whereas if they remain rounded they become adipocytes (fat cells). In regard to the potential use of neural stem cells, it is encouraging that mouse ES cells made to differentiate specifically into neural progenitors by treatment with retinoic acid can, when transplanted into the neural tube of a mouse embryo, develop into neurons appropriate to their new location.

10.17 Stem cells could be a key to regenerative medicine

The goal of regenerative medicine is to restore the structure and function of damaged or diseased tissues. As stem cells can proliferate and differentiate into a wide range of cell types, they are strong candidates for use in **cell-replacement therapy**, the restoration of tissue function by the introduction of new healthy cells. This type of therapy might eventually offer an alternative to conventional organ transplantation from a donor, with its attendant problems of rejection and shortage of organs, and might also be able to restore function to tissues such as brain and nerves. Both ES cells and adult stem cells have been studied with cell-replacement therapy in mind. Transdifferentiation is also emerging as an alternative route to generating selected cell types, but the recent development of **induced pluripotent stem cells** (**iPS cells**) (Box 10C, p. 401) offers the most exciting new opportunities.

ES cells have the advantage over adult stem cells in that they can differentiate into a wide range of different cell types and could, in theory, be used to repair any tissue. But stem cells from an unrelated donor embryo will still cause immune rejection reactions, and because of the proliferative capacity and pluripotency of ES cells, the risk of embryonal tumors—tumors derived from embryonic cells—is high. Cultured ES cells introduced into another embryo will develop normally, but when ES cells are introduced under the skin of a genetically identical adult mouse they give rise to

Box 10C Induced pluripotent stem cells

The most dramatic example of reprogramming of adult differentiated cells is the production of embryonic stem (ES) cell-like cells, known as induced pluripotent stem cells (iPS cells) by the introduction of the genes for four transcription factors associated with pluripotency (see figure). It had previously been shown that when adult differentiated somatic cells are fused with ES cells they are transformed into pluripotent cells, suggesting that cytoplasmic factors in the ES cells might be able to confer ES cell-like properties on differentiated cells. Nevertheless it was surprising that so few factors are needed to induce pluripotency.

The first iPS cells were produced from skin fibroblasts isolated from transgenic mice that carried a drug-selectable marker linked to expression of a gene associated with pluripotency, either *Oct3/4* or *Nanog*. The genes encoding four transcription factors associated with pluripotency in ES cells—*Oct3/4*, *Sox2*, *Kfl4*, and *c-Myc*—were introduced into the adult fibroblasts using retroviral vectors. The transfected cells were then treated with the appropriate drugs to select for the rare iPS cells. Most transfected cells did not express either *Oct3/4* or *Nanog* and therefore had no drug resistance and died. The remaining colonies of cells that survived resembled ES cells and could be isolated and expanded in culture. The inserted genes in the iPS cells shut themselves off once the reprogramming process was completed, and the cells reactivated their own pluripotency genes to maintain the pluripotent state. The same set of transcription factor genes was very soon found to be able reprogram adult human fibroblasts, but the conditions needed to expand human iPS cell populations are different, with mouse cells requiring LIF and human cells FGF (see Section 10.16).

The pluripotency of mouse iPS cells was tested by injecting them into normal mouse blastocysts, which were then replaced in a surrogate mother. The iPS cells contributed to all cell types in the resulting embryos, including the germline. When these iPS chimeric mice were crossed with normal mice, iPS-derived gametes successfully participated in fertilization and live mice were born. More recently, fertile adult mice generated entirely from iPS cells have been derived from tetraploid blastocyts into which fibroblast-derived iPS cells were injected (see Box 10B, p. 399, for a description of this definitive procedure for testing pluripotency). The efficiency of the generation of healthy mice from iPS cells by this procedure is still very low, however, and many of the pups produced died soon after birth.

Advances have been made over the past few years in inducing pluripotency without the need to introduce a persisting viral vector. This will be a prerequisite for any medical use for iPS cells in cell replacement. Transposon-based vectors have been produced that will snip themselves and the genes they carry cleanly out of the genome after the transformation to pluripotency is complete. Other approaches involve searching for small-molecule drugs that will act as co-activators for a cell's own pluripotency genes.

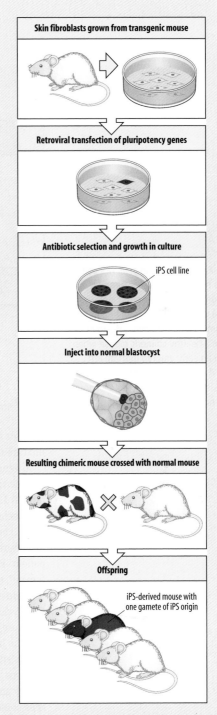

Skin fibroblasts grown from transgenic mouse

Retroviral transfection of pluripotency genes

Antibiotic selection and growth in culture

iPS cell line

Inject into normal blastocyst

Resulting chimeric mouse crossed with normal mouse

Offspring

iPS-derived mouse with one gamete of iPS origin

An equal effort is going into reducing the number of genes that need to be introduced. Adult neural stem cells have been reprogrammed by introducing just one of the transcription factors, Oct4. Adult neural stem cells already express high levels of Sox2 and c-Myc, however, and this is why introduction of Oct4 alone is sufficient to induce pluripotency.

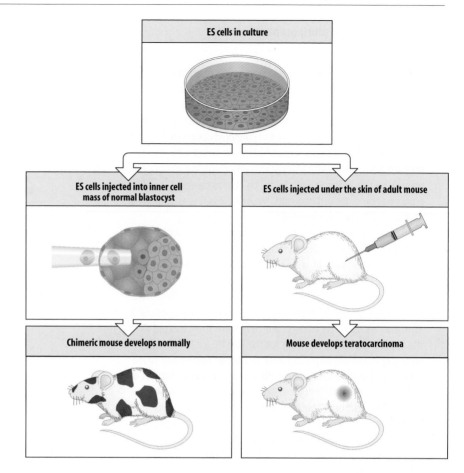

Fig. 10.32 Embryonic stem cells (ES cells) can develop normally or into a tumor, depending on the environmental signals they receive. Cultured ES cells originally obtained from an inner cell mass of a mouse contribute to a healthy chimeric mouse when injected into an early embryo (bottom left panels), but if injected under the skin of an adult mouse, the same cells will develop into a teratocarcinoma (bottom right panels).

a tumor known as a **teratocarcinoma** (Fig. 10.32). These unusual tumors contain a mixture of differentiated cells.

To generate ES cells of the patient's own tissue type, it would be a great advantage to be able to make use of **therapeutic cloning** by somatic cell nuclear transfer (see Section 10.13). A nucleus from a somatic cell from the patient would be introduced into an enucleated human oocyte which would then be allowed to develop to the blastocyst stage. The blastocyst could then be used as a source of ES cells that could be maintained in culture, and which would not cause immune rejection when implanted into the patient. But although human blastocysts have now been made by somatic cell nuclear transfer, the success rate is very low and no ES cells have yet been derived from them.

There are claimed to be ethical issues associated with the use of stem cells and therapeutic cloning. To make human ES cells, the blastocyst must be destroyed, and there are those who believe that this is a destruction of a human life. There is good evidence that the blastocyst does not necessarily represent an individual at this very early stage, as it is still capable of giving rise to twins at a later stage. And in practice, many early embryos are lost during assisted reproduction involving *in vitro* fertilization (IVF), a widely accepted medical intervention. Acceptance of IVF and rejection of the use of ES cells could be seen as a contradiction. New alternative approaches, such as transdifferentiation and the use of induced pluripotent stem cells, could produce cells of a patient's own tissue type but avoid the ethical issues associated with therapeutic cloning.

Adult stem cells have also been considered for use in cell-replacement therapy. Adult hematopoietic stem-cell transplants in the form of bone marrow transplants from donors have been successfully used for many years as therapy for certain immunological diseases and cancers, despite associated problems such as graft-versus-host rejection. The advantage of adult stem cells in certain other conditions is that they could be taken from the patient's own undamaged tissues and re-implanted where needed, without any problems of tissue rejection. This is already done in the case of skin grafts, which use the patient's own skin. But the range of cell types and tissues that could be repaired using a patient's own stem cells is limited. The isolation of stem cells from the brain, for example, would not be feasible. There have been reports over the years that adult hematopoietic stem cells will differentiate into cell types quite different from their normal range when transplanted into another type of tissue. However, it was subsequently shown that hematopoietic stem cells had not been able to give rise to cells outside their normal repertoire but instead had fused with existing cells.

10.18 Various approaches can be used to generate differentiated cells for cell-replacement therapies

One of the challenges facing cell-replacement therapy is devising differentiation protocols that will generate the particular cell types needed for repair, whether from ES cells, adult stem cells, or by transdifferentiation of pre-existing cell types. The generation of insulin-producing pancreatic β cells to replace those destroyed in type 1 diabetes is a prime medical target, and we shall use it to illustrate the experimental approaches that are being developed for cell-replacement therapy generally.

In type 1 diabetes, the insulin-producing β cells of the endocrine pancreas (the pancreatic islets) are destroyed by an autoimmune reaction, resulting in non-production of insulin and the potentially lethal build-up of sugar in the blood. Unlike blood or muscle, the pancreas does not contain dedicated stem cells from which it can regenerate, and people with type 1 diabetes face a lifetime of dependence on insulin injections. Pancreas or islet-cell transplants from cadaver donors have had some success, but tissue for transplantation is scarce and treatment with powerful immunosuppressant drugs is necessary to prevent rejection. Transplantation of insulin-producing pancreatic cells derived from the patient's own cells, along with the induction of immune tolerance to these cells, might eventually provide one long-term solution.

Treatments that direct the differentiation of ES cells towards making endoderm derivatives such as pancreatic cells have been particularly difficult to find. And to be useful for cell replacement, the insulin-producing cells generated must also be responsive to the signals, such as glucose, that switch on insulin production as required. Nevertheless, using knowledge of the signals that induce endoderm and pancreas development in mouse embryos, progress has been made in devising methods for differentiating human ES cells into pancreatic progenitor cells. Human ES cells put through a 9-day laboratory protocol involving stepwise treatments with different signal molecules differentiate into endoderm, expressing the endoderm-specific transcription factor FoxA2, but only a few of the cells in these cultures (around 5%) express Pdx1, a transcription factor essential for pancreas development. The next step was therefore to find ways of boosting the differentiation of pancreatic progenitor cells. A library of 5000 organic compounds was screened and one called indolactam V was found to stimulate the differentiation of Pdx1-positive cells. The percentage of Pdx1-expressing cells increased to 45% when endoderm cultures derived from the human ES cells were treated with indolactam V together with FGF-10, one of the signals involved in embryonic pancreatic development. When the Pdx1-expressing cells were transplanted under the kidney capsule in a mouse, a site that allows continued differentiation, they differentiated into insulin-secreting cells (Fig. 10.33).

Induced pluripotent stem cells (iPS cells) offer the greatest opportunities for generating pluripotent cells of a patient's own tissue type for cell replacement. As described in Box 10C (p. 401), iPS cells were first derived from mouse fibroblasts by introducing and expressing genes for four transcription factors, Oct 3/4, Sox2, c-Myc and Kfl4, which are associated with pluripotency in ES cells. In regard to diabetes, iPS cells have been derived from fibroblasts taken from patients with type 1 diabetes.

Fig. 10.33 Three experimental routes for producing insulin-producing cells by nuclear reprogramming. Two routes start with a skin fibroblast. In somatic cell nuclear transfer, the nucleus of the fibroblast is introduced into an unfertilized egg and embryonic stem cells (ES cells) are isolated from the resulting blastocyst. Alternatively, cultured fibroblasts can be transfected with the genes for the pluripotency transcription factors (Oct4, Sox2, c-Myc, and Kfl4) to produce induced pluripotent cells (iPS cells). The ES cells and iPS cells can then be differentiated into insulin-producing cells *in vitro*. A third possible route is the transdifferentiation of adult exocrine pancreatic cells or liver cells into insulin-producing cells *in vivo* by the targeted transfection of these organs with genes (including that for the transcription factor Pdx1) required for differentiation of endocrine pancreatic cells. Steps in red have been accomplished for human cells. Note that insulin-producing glucose-responsive cells can also be made starting with human ES cells derived from a normal blastocyst and differentiated *in vitro*.

Adapted from Gurdon, J.B., Melton, D.A.: 2008.

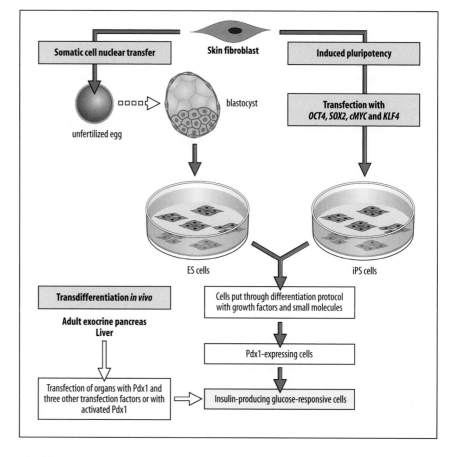

The fibroblasts were reprogrammed using the human genes *OCT4*, *SOX2* and *KLF4*, and were subsequently redifferentiated into insulin-producing cells (see Fig. 10.33). In the immediate future, such cells will be used to investigate the cellular and molecular defects underlying type 1 diabetes, which are still largely unknown.

Transdifferentiation may be yet another way of generating replacement cells, although this approach is likely to be limited to developmentally related cell types. The feasibility of this approach for generating insulin-producing endocrine cells from liver cells and from pancreatic exocrine cells has been demonstrated experimentally. The liver and the pancreas both derive from endoderm, and arise from adjacent regions of endoderm in the embryo, but the transcription factor Pdx1, which is essential for pancreatic development, is not expressed in the liver. When an activated form of Pdx1 is expressed transiently in the liver in transgenic *Xenopus* tadpoles under the control of a liver-specific promoter, however, part or all of the liver is converted to both exocrine and endocrine pancreatic tissue, while proteins characteristic of differentiated liver cells disappear from the cells in the converted regions. In another example of transdifferentiation, progenitor liver cells in adult mice have also been induced to develop into insulin-secreting cells resembling pancreatic islet cells by infecting the liver with viral vectors expressing the transcription factor neurogenin, which is another pancreatic transcription factor.

Transdifferentiation of pancreatic exocrine cells into endocrine β cells has also been achieved in adult mice. A combination of three transcription factors required for endocrine pancreatic cell differentiation, including Pdx1, were delivered into the pancreas using genetically modified adenoviruses. Endocrine cells were induced and, although they were not found in islets, these induced cells closely resembled their normal counterparts and had the ability to secrete insulin. Transdifferentiation was efficient (around 20%) and relatively fast, with the first insulin-positive cells appearing on day 3 after treatment.

There are, however, many obstacles to overcome before the potential therapeutic benefits of technologies based on ES cells, iPS cells, or transdifferentiation can be realized. The most effective current procedures for making iPS cells or causing transdifferentiation depend on using viral vectors to introduce the pluripotency genes. Viral vectors can integrate randomly into the genome and might activate genes that cause cancer, and so alternative ways of inducing expression of the necessary transcription factors will need to be found before medical applications are possible. Fortunately, it is possible to remove the vectors along with the introduced genes after successful conversion to pluripotency and still maintain a stable iPS cell phenotype. This is because once pluripotency is induced by the introduced genes, the cell's own pluripotency genes become activated. In addition, there has been some success in obtaining iPS cells by introducing the four pluripotency proteins into mouse embryonic fibroblasts rather than the genes.

But even if non-viral methods of inducing pluripotency are perfected, it will not eliminate the risk of tumor induction in patients undergoing cell-replacement therapy with ES cells or iPS cells. If undifferentiated pluripotent cells remain after the differentiation process and are introduced into the patient they could cause tumors. Only stringent selection procedures that ensure no undifferentiated cells are present in the transplanted cell population will overcome this problem. And it is not yet clear how stable differentiated ES cells and iPS cells will be in the long term. A final, but not insignificant, obstacle to be overcome is finding ways of generating enough cells for an effective treatment. At present, ES cells are easier to expand to large numbers in culture than are adult stem cells or even iPS cells.

We have focused here on just one medical target, the generation of pancreatic insulin-producing cells to treat type 1 diabetes. Similar strategies could also be used in other diseases. The neurodegenerative Parkinson's disease is another medical target and iPS cells have successfully been made from people with this disease. In the near future, iPS cells derived from patients will probably be most useful for investigating disease processes in the context of human cells, as animal models do not always mimic human disease completely. For example, iPS cells have been derived from skin fibroblasts from a child with genetically determined spinal muscular atrophy, and motor neurons have been produced in culture from these cells. Neuron production in the child's iPS cell cultures differed from neuron production in cultures of iPS cells made from the child's healthy mother, with the number and size of motor neurons produced decreasing over time in cultures derived from the child. Such cultures could be used to test the effects of drugs, which could lead to new clinical treatments. More generally, human iPS cells could be useful for screening compounds for toxicity and for teratogenicity in drug development.

Summary

Transplantation of nuclei from differentiated animal cells into fertilized eggs, and cell-fusion studies, show that the pattern of gene expression in the nucleus of a differentiated cell can often be reversed, implying that it is determined by factors supplied by the cytoplasm and that no genetic material has been lost. Although the differentiated state of an animal cell *in vivo* is usually extremely stable, some cases of differentiation are reversible. Transdifferentiation of one differentiated cell type into another has been shown to occur during regeneration and in cultured cells. Both amphibians and some mammals can be cloned by nuclear transplantation from an adult somatic cell into an enucleated egg. Stem cells are self-renewing and can differentiate into a wide range of cell types. They are being investigated with a view to being used in cell-replacement therapy. It has been found that the introduction of four transcription factors expressed in embryonic stem cells into adult differentiated cells induces these cells to become pluripotent. These induced pluripotent stem cells together with transdifferentiation provide alternative approaches for producing cells for cell-replacement therapy.

Summary to Chapter 10

Cell differentiation leads to distinguishable cell types whose specialized characters and properties are determined by their pattern of gene activity and thus the proteins they produce. The pattern of gene activity in a differentiated cell can be maintained over long periods of time and can be transmitted to the cell's progeny. Maintenance and inheritance of a pattern of gene activity probably involves a combination of several mechanisms, including the continued action of gene-regulatory proteins, changes in the packing state of chromatin, and chemical modifications to DNA.

Differentiation reflects the establishment of a particular pattern of gene expression rather than the loss of genetic material. Lymphocytes are an exception as their DNA has undergone irreversible changes during differentiation. The differentiated state of an animal cell *in vivo* is usually extremely stable, but in some cases transdifferentiation of one differentiated cell type into another can occur, for example, during tissue regeneration. The cloning of amphibians and some mammals by somatic cell nuclear transfer into enucleated eggs also shows that the nuclei of differentiated cells can be reprogrammed and restored to a state in which they are able to direct embryonic development. Adult somatic cells can also be induced to develop into pluripotent cells by expression of four transcription factors associated with pluripotency in embryonic stem cells. Pluripotent cells that can proliferate and differentiate into cell types from all three germ layers could be the key to regenerative medicine.

Apart from the use of universal mechanisms of gene regulation, there may be little similarity between the detailed mechanisms of differentiation in different cell types. The understanding of any given differentiation pathway lies in the detail of the external signals, internal signal transduction pathways, gene-regulatory proteins, and gene products at work in each particular case. Induction of cell differentiation by signaling molecules such as growth factors requires the execution of a complex program of intracellular events. Different stimuli, or even the same stimulus at different developmental stages, can activate the same internal signaling pathway, yet activate different genes in different cells, because of the cells' own developmental histories. Signal transduction at the cell membrane sets in motion a cascade of events within the cell, which results, for example, in the phosphorylation and activation of transcription factors and the consequent switching on and off of gene expression. Programmed cell death is a common fate of cells during development, and there is evidence that positively acting signals are generally required to prevent it and to allow cells to survive.

▇ End of chapter questions

Long answer (concept questions)

1. Use the Glossary to review the terms fate, specification, and determination. How do these cellular states differ from the topic of this chapter, differentiation?

2. Contrast the concept of 'housekeeping' genes (and proteins) from the cell-specific genes (and proteins) that characterize, and define, a differentiated cell type. Propose a microarray experiment that would distinguish the tissue-specific gene expression from housekeeping gene expression in muscle cells and neurons. (The principles behind a DNA microarray analysis are described in Box 3A, p.115.)

3. Gene function is often studied by misexpression of the protein in an abnormal site in the embryo or abnormal tissue in the adult body to see what effects the protein has. What are the essential features of the transcriptional apparatus that make misexpression experimentally possible (include the role of both *cis*-regulatory DNA sequences and transcription factors)?

4. Create two lists: one of peptide signaling molecules, and one of lipid-soluble signaling molecules. How do their mechanisms of action differ? What generalization(s) can you make regarding the way they activate a transcription factor?

5. Cell division entails the separation of the two strands of the DNA during DNA replication, which disrupts patterns of gene expression that have been previously established in the cell. Nonetheless, the two daughter cells must somehow 'remember' the differentiated state of the parental cell. Discuss how the differentiated state of a cell is remembered, using transcription factors, chromatin structure, and DNA methylation.

6. Hematopoiesis involves the development of lymphocytes (B and T cells), red blood cells, macrophages, and osteoclasts (among other blood-cell types) from cells in the bone marrow. Outline the general stages that lead to each of the cell types named, giving the name of the cell from which they all derive.

7. What is meant by 'stem-cell niche' and what is its role? For one of the stem-cell niches described in the text give examples of the signaling molecules that enable it to carry out this role.

8. Differentiation of the erythroid lineage requires signaling by erythropoietin (Epo), a peptide signaling molecule made by the kidney in response to red blood cell deficiency. One target of Epo is the GATA-1 transcription factor, and Epo is itself a target of GATA-1. β-globin is also a target of GATA-1. Using these facts, propose a pathway from a physiological need for more red blood cells, through the multipotent stem cell, to the globin-expressing red blood cell.

9. Describe the choice between differentiation as a neutrophil and differentiation as a macrophage at the level of the transcription factors involved. Describe the same choice at the level of the signaling molecules involved.

10. The mouse gene encoding M-CSF is named *op*, as mice mutant for this gene develop the bone disease osteopetrosis, in which the bones become abnormally hard and brittle. Review the lineages of the blood cell types and explain why this mutation would affect bone.

11. Why are different globins required at different stages of development? How is the switching between globin types thought to be controlled?

12. Describe the process by which epithelial cells of the intestine are replaced from a stem-cell population. What is the role of Wnt signaling in this process?

13. Despite its ability to act as a master regulator of muscle differentiation, mouse knockouts of *myoD* are viable. Refer to Fig. 10.19, and propose a reason that *myoD* is not essential. In contrast, knock-out of both *myoD* and *myf5* is lethal; knock-out of *mrf4* is also lethal; how would you interpret this result?

14. The differentiation of muscle cells requires withdrawal from the cell cycle. Elaborate on the following points regarding growth versus differentiation of myoblasts: (1) Removal of growth factors causes muscle precursor cells to differentiate in culture. (2) MyoD activates expression of p21. (3) Loss of Rb will allow myoblasts to re-enter the cell cycle.

15. Referring to Figure 10.23, fill in the extracellular signaling proteins that promote the various fates of neural crest cells, much as was done for blood cells in Fig. 10.9.

16. What is stem-cell factor (SCF)? Why would defects in SCF signaling by loss of its receptor, Kit, lead to a mouse phenotype called 'white spotting'?

17. Contrast apoptosis with necrosis. What is the role of caspases in apoptosis?

18. Describe the process referred to as cloning by nuclear transplantation. How does the success of this process illustrate the importance of the transcription factor complement present in any given cell? In contrast, how does the difficulty of the process illustrate the importance of epigenetic mechanisms in development?

19. ES cells and iPS cells are both pluripotent stem cells but are obtained in different ways. Describe how each is obtained.

20. What are the potential advantages of iPS cells over ES cells for therapeutic purposes? What are the disadvantages? What other approaches for obtaining stem-like cells are possible, and what problems does each of them circumvent?

Multiple choice (factual recall questions)

NB There is only one correct answer to each question.

1. The many different cell types in the blood of an adult mammal are derived from

a) single type of multipotent stem cell found in the bone marrow

b) differentiated cells that first developed in the fetal liver, but now reside in the bone marrow

c) division of differentiated cell types while circulating in the blood

d) stem cells that became committed to each specific blood cell type during embryonic development

2. What might be the result of producing a mouse knock-out for G-CSF?

a) The mouse would be unable to make granulocytes.

b) The mouse would be unable to make macrophages.

c) The mouse would lack red blood cells.

d) The mouse would lose its hematopoietic stem cell population completely.

3. The stem cells responsible for renewal of the keratinocytes of the skin are found in which layer of the epidermis?

a) basal lamina

b) basal layer

c) dermis

d) keratinocyte layer

4. The introduction of *myoD* into fibroblasts will cause

a) apoptosis

b) commitment to the myeloid lineage

c) differentiation into muscle cells

d) formation of red blood cells

5. Differentiation, as opposed to determination, of muscle cells is dependent on

a) Mrf4

b) Myf5

c) MyoD

d) myogenin

6. What are satellite cells?

a) Satellite cells are circulating cells 'orbiting' in the blood.

b) Satellite cells are muscle stem cells.

c) Satellite cells are muscle-associated support cells, much as glial cells are neuron-associated support cells.

d) Satellite cells are neuronal stem cells.

7. Neural crest cells can give rise to

a) cartilage of the face

b) parasympathetic ganglia

c) sensory ganglia

d) all of these

8. The human anti-apoptotic protein that is homologous to Ced-9 from *C. elegans* is

a) Bcl-2

b) Caspase-9

c) Ced-3

d) Diablo

9. If the lens of the eye is removed in the adult, which of the following organisms can produce a new lens by transdifferentiation of iris cells?

a) chicken

b) human

c) mouse

d) newt

10. ES cells are

a) cells isolated from teratocarcinomas

b) embryonic stem cells derived from the inner cell mass of the mammalian embryo

c) epidermal stem cells found in the basal layer of the epidermis

d) human stem cells generated by treatment of mature cell types with a specific combination of transcription factors

Multiple choice answer key

1: a, 2: a, 3: b, 4: c, 5: d, 6: b, 7: d, 8: a, 9: d, 10: b.

■ Section further reading

10.1 Control of transcription involves both general and tissue-specific transcriptional regulators

Cheung, W.L., Briggs, S.D., Allis, C.D.: **Acetylation and chromosomal functions**. *Curr. Opin. Cell Biol.* 2000, **12**: 328–333.

Levine, M., Tjian, R.: **Transcription regulation and animal diversity**. *Nature* 2003, **424**: 147–151.

Mannervik, M., Nibur, Y., Zhang, H., Levine, M.: **Transcriptional coregulators in development**. *Science*, 1999, **284**: 606–609.

10.2 External signals can activate genes

Brivanlou, A.H., Darnell, J.E.: **Signal transduction and control of gene expression**. *Science* 2002, **295**: 813–818.

Dougherty, D.C., Park, H.M., Sanders, M.M.: **Interferon regulatory factors (IRFs) repress transcription of the chicken ovalbumin gene**. *Gene* 2009, **439**: 63–70.

10.3 The maintenance and inheritance of patterns of gene activity depend on chemical and structural modifications to chromatin, as well as on gene-regulatory proteins

de Laat, W., Grosveld, F.: **Spatial organization of gene expression: the active chromatin hub**. *Chromosome Res.* 2003, **11**: 447–459.

Ho, L., Crabtree, G.R.: **Chromatin remodelling during development**. *Nature* 2010, **463**: 474–484.

10.4 All blood cells are derived from multipotent stem cells

Ema, H., Nakauchi, H.: **Self-renewal and lineage restriction of hematopoietic stem cells**. *Curr. Opin. Genet. Dev.* 2003, **13**: 508–512.

Phillips, R.L., Ernst, R.E., Brunk, B., Ivanova, N., Mahan, M.A., Deanehan, J.K., Moore, K.A., Overton, G.C., Lemischka, I.R.: **The genetic program of hematopoietic stem cells**. *Science* 2000, **288**: 1635–1640.

Zon, L.I.: **Intrinsic and extrinsic control of haematopoietic stem-cell renewal**. *Nature* 2008, **453**: 306–313.

10.5 Colony-stimulating factors and intrinsic changes control differentiation of the hematopoietic lineages

Anguita, E., Hughes, J., Heyworth, C., Blobel, G.A., Wood, W.G., Higgs, D.R.: **Globin gene activation during haemopoiesis is driven by protein complexes nucleated by GATA-1 and GATA-2**. *EMBO J.* 2004, **23**: 2841–2852.

Kluger, Y., Lian, Z., Zhang, X., Newburger, P.E., Weissman, S.M.: **A panorama of lineage-specific transcription in hematopoiesis**. *BioEssays* 2004, **26**: 1276–1287.

Metcalf, D.: **Control of granulocytes and macrophages: molecular, cellular, and clinical aspects**. *Science* 1991, **254**: 529–533.

Orkin, S.H.: **Diversification of haematopoietic stem cells to specific lineages**. *Nat. Rev. Genet.* 2000, **1**: 57–64.

10.6 Developmentally regulated globin gene expression is controlled by regulatory sequences far distant from the coding regions

Engel, J.D., Tanimoto, K.: **Looping, linking, and chromatin activity: new insights into β-globin locus regulation**. *Cell* 2000, **100**: 499–502.

Tolhuis, B., Palstra, R.J., Splinter, E., Grosveld, F., de Laat, W.: **Looping and interaction between hypersensitive sites in the active beta-globin locus**. *Mol. Cell* 2002, **10**: 1453–1465.

10.7 The epithelia of adult mammalian skin and gut are continually replaced by derivatives of stem cells

Barker, N., van Es, J.H., Kuipers, J., Kujala, P., van den Born, M., Cozijnsen, M., Haegebarth, Andrea, Korving, J., Begthel, H., Peters, P.J., Clevers, H.: **Identification of stem cells in small intestine and colon by marker gene Lgr5**. *Nature* 2007, **449**: 1003–1008.

Coulombe, P.A., Kerns, M. L., Fuchs, E.: **Epidermolysis bullosa simplex: a paradigm for disorders of tissue fragility**. *J. Clin. Invest.* 2009, **119**: 1784–1793.

Fuchs, E.: **The tortoise and the hair: slow-cycling cells in the stem cell race**. *Cell* 2009, **137**: 811–819.

Fuchs, E.: **Finding one's niche in the skin**. *Cell Stem Cell* 2009, **4**: 499–502.

Fuchs, E., Nowak, J.A.: **Building epithelial tissues from skin stem cells**. *Cold Spring Harb. Symp. Quant. Biol.* 2008, **73**: 333–350.

Sato, T., Vries, R.G., Snippert, H.J., van de Wetering, M., Barker, N., Stange, D.E., van Es, J.H., Abo, A., Kujala, P., Peters, P.J., Clevers, H.: **Single Lgr5 stem cells build crypt villus structures *in vitro* without a mesenchymal cell niche**. *Nature* 2009, **459**: 262–265.

Shackleton, M., Vaillant, F., Simpson, K.J., Stingl, J., Smyth, G.K., Asselin-Labat, M-L., Wu, L., Lindeman, G.J., Visvader, J.E.: **Generation of a functional mammary gland from a single stem cell**. *Nature* 2006, **439**: 84–88.

Van der Flier, L.G., Clevers, H.: **Stem cells self-renewal and differentiation in the intestinal epithelium**. *Annu. Rev. Physiol.* 2009, **71**: 241–260.

10.8 The MyoD family of genes determines differentiation into muscle

Kassar-Duchossoy, L., Gayraud-Morel, B., Gomes, D., Rocancourt, D., Buckingham, M., Shinin, V., Tajbakhsh, S.: **Mrf4 determines skeletal muscle identity in Myf5: Myod double-mutant mice**. *Nature* 2004, **431**: 466–471.

Tapscott, S.J.: **The circuitry of a master switch: Myod and the regulation of skeletal muscle transcription**. *Development* 2005, **132**: 2685–2695.

10.9 The differentiation of muscle cells involves withdrawal from the cell cycle, but is reversible

Deato, M.D., Tjian, R.: **An unexpected role of TAFs and TRFs in skeletal muscle differentiation; switching of core promoter complexes**. *Cold Spring Harb. Symp.* 2008, **73**: 217–225.

Novitch, B.G., Mulligan, G.J., Jacks, T., Lassar, A.B.: **Skeletal muscle cells lacking the retinoblastoma protein display defects in muscle gene expression and accumulate in S and G_2 phases of the cell cycle**. *J. Cell Biol.* 1996, **135**: 441–456.

Odelberg, S.J., Kolhoff, A., Keating, M.: **Dedifferentiation of mammalian myotubes induced by *msx1***. *Cell* 2000, **103**: 1099–1109.

10.10 Skeletal muscle and neural cells can be renewed from stem cells in adults

Collins, C.A., Partridge T.A.: **Self-renewal of the adult skeletal muscle satellite cell**. *Cell Cycle* 2005, **4**: 1338–1341.

Dhawan, J., Rando, T.A.: **Stem cells in postnatal myogenesis: molecular mechanisms of satellite cell quiescence, activation and replenishment**. *Trends Cell Biol.* 2005, **15**: 663–673.

Lepper, C., Conway, S.J., Fan, C.M.: **Adult satellite cells and embryonic muscle progenitors have distinct genetic requirements**. *Nature* 2009, **460**: 627–631.

Lie, D-C., Colamarino, S.A., Song, H-J., Desire, L., Mira, H., Consiglio, A., Lein, E.S., Jessberger, S., Lansford, H., Dearier, A.R., Gage, F.H.: **Wnt signalling regulates adult hippocampal neurogenesis**. *Nature* 2005, **437**: 1370–1375.

Ninkovic, J., Gotz, M.: **Signaling in adult neurogenesis: from stem cell niche to neuronal networks**. *Curr. Opin. Neurobiol.* 2007, **17**: 338–344.

Sacco, A., Doyonnas, R., Kraft, P., Vitorovic, S., Blau, H.M.: **Self-renewal and expansion of single transplanted muscle stem cells**. *Nature* 2008, **456**: 502–506.

Relaix, F., Rocancourt, D., Mansouri, A., Buckingham, M.A.: **pax3/pax7 dependent population of skeletal muscle progenitor cells**. *Nature* 2005, **435**: 948–953.

10.11 Embryonic neural crest cells differentiate into a wide range of different cell types

Anderson, D.J.: **Genes, lineages and the neural crest**. *Proc. R. Soc. Lond. B* 2000, **355**: 953–964.

Breau M.A., Pietri, T., Stemmler, M.P., Thiery, J.P., Weston, J.A.: **A non-neural epithelial domain**. *Proc. Natl Acad. Sci. USA* 2008, **105**: 7750–7757.

Bronner-Fraser, M.: **Making sense of the sensory lineage**. *Science* 2004, **303**: 966–968.

Dorsky, R.I., Moon, R.T., Raible, D.W.: **Environmental signals and cell fate specification in premigratory neural crest**. *BioEssays* 2000, **22**: 708–716.

Le Douarin, N.M., Creuzet, S., Couly, G., Dupin, E.: **Neural crest cell plasticity and its limits**. *Development* 2004, **131**: 4637–4650.

Lee, H.Y., Kleber, M., Hari, L., Brault, V., Suter, U., Taketo, M.M., Kemler, R., Sommer, L.: **Instructive role of Wnt/beta-catenin in sensory fate specification in neural crest stem cells**. *Science* 2004, **303**: 1020–1023.

Meier, P., Finch, A., Evan, G.: **Apoptosis in development**. *Nature* 2000, **407**: 796–801.

Minoux, M., Antonarakis, G.S., Kmita, M., Duboule, D., Rijli, F.M.: **Rostral and caudal pharyngeal arches share a common neural crest ground pattern**. *Development* 2009, **136**: 637-634.

Morrison, S.J., White, P.M., Zock, C., Anderson, D.J.: **Prospective identification, isolation by flow cytometry, and *in vivo* self-renewal of multipotent mammalian neural crest stem cells**. *Cell* 1999, **96**: 737–749.

Morrison, S.J., Perez, S.E., Qiao, Z., Verdi, J.M., Hicks, C., Weinmaster, G., Anderson, D.J.: **Transient Notch activation initiates an irreversible switch from neurogenesis to gliogenesis by neural crest stem cells**. *Cell* 2000, **101**: 499–510.

Sauka-Spengler, T., Bronner-Fraser, M.: **A gene regulatory network orchestrates neural crest formation**. *Nat. Rev. Mol. Cell Biol.* 2008, **9**: 557–568.

10.12 Programmed cell death is under genetic control

Jacobson, M.D., Weil, M., Raff, M.C.: **Programmed cell death in animal development**. *Cell* 1997, **88**: 347–354.

Raff, M.: **Cell suicide for beginners**. *Nature* 1998, **396**: 119–122.

Riedl, S.J., Shi, Y: **Molecular mechanisms of caspase regulation during apoptosis**. *Nat. Rev. Mol. Cell Biol.* 2004, **5**: 897–907.

Sancho, E., Batlle, E., Clevers, H.: **Live and let die in the intestinal epithelium**. *Curr. Opin. Cell Biol.* 2003, **15**: 763–770.

Vaux, D.L., Korsmeyer, S.J.: **Cell death in development**. *Cell* 1999, **96**: 245–254.

10.13 Nuclei of differentiated cells can support development

Eggan, K., Baldwin, K., Tackett, M., Osborne, J., Gogos, J., Chess, A., Axel, R., Jaenisch, R.: **Mice cloned from olfactory sensory neurons**. *Nature* 2004, **428**: 44–49.

Gurdon, J.B.: **Nuclear transplantation in eggs and oocytes**. *J. Cell Sci. Suppl.* 1986, **4**: 287–318.

Gurdon, J.B., Melton, D.A.: **Nuclear reprogramming in cells**. *Science* 2008, **322**: 1811–1815.

Humpherys, D., Eggan, K., Akutsu, H., Friedman, A., Hochedlinger, K., Yanagimachi, R., Lander, E.S., Golub, T.R., Jaenisch, R.: **Abnormal gene expression in cloned mice derived from embryonic stem cell and cumulus cell nuclei**. *Proc. Natl Acad. Sci. USA* 2002, **99**: 12889–12894.

Rhind, S.M., Taylor, J.E., De Sousa, P.A., King, T.J., McGarry, M., Wilmut, I.: **Human cloning: can it be made safe?** *Nat. Rev. Genet.* 2003, **4**: 855–864.

Wilmut, I., Taylor, J.: **Primates join the club**. *Nature* 2007, **450**: 485–486.

10.14 Patterns of gene activity in differentiated cells can be changed by cell fusion

Blau, H.M.: **How fixed is the differentiated state? Lessons from heterokaryons**. *Trends Genet.* 1989, **5**: 268–272.

Blau, H.M., Baltimore, D.: **Differentiation requires continuous regulation**. *J. Cell Biol.* 1991, **112**: 781–783.

Blau, H.M., Blakely, B.T.: **Plasticity of cell fates: insights from heterokaryons**. *Cell Dev. Biol.* 1999, **10**: 267–272.

Pomerantz, J.H., Mukherjee, S., Palermo, A.T., Blau, H.M.: **Reprogramming to a muscle fate by fusion recapitulates differentiation**. *J. Cell Sci.* 2009, **122**: 1045–1053.

10.15 The differentiated state of a cell can change by transdifferentiation

Horb, M.E., Shen, C.N., Tosh, D., Slack, J.M.: **Experimental conversion of liver to pancreas**. *Curr. Biol.* 2003, **13**: 105–115.

Jarriault, S., Shwab, Y., Greenwald, I.A.: ***Caenorhabitis elegans* model for epithelial-neuronal transdifferentiation**. *Proc. Natl Acad. Sci. USA* 2008, **105**: 3790–3795.

Slack, J.M.W.: **Metaplasia and transdifferentiation from pure biology to the clinic**. *Nat. Rev. Mol. Cell Biol.* 2007, **8**: 369–378.

Tsonis, P.A., Madhavan, M., Tancous, E.E., Del Rio-Tsonis, K.: **A newt's eye view of lens regeneration**. *Int. J. Dev. Biol.* 2004, **48**: 975–980.

10.16 Embryonic stem cells can proliferate and differentiate into many cell types in culture

Ho, A.D.: **Kinetics and symmetry of divisions of hematopoietic stem cells**. *Exp. Hematol.* 2005, **33**: 1–8.

Loebel, D.A., Watson, C.M., De Young, R.A., Tam, P.P.: **Lineage choice and differentiation in mouse embryos and embryonic stem cells**. *Dev. Biol.* 2003, **264**: 1–14.

McBeath, R., Pirone, D.M., Nelson, C.M., Bhadriraju, K., Chen, C.S.: **Cell shape, cytoskeletal tension, and RhoA regulate stem cell lineage commitment**. *Dev. Cell* 2004, **6**: 483–495.

Molofsky, A.V., Pardal, R., Morrison, S.J.: **Diverse mechanisms regulate stem cell self-renewal**. *Curr. Opin. Cell Biol.* 2004, **16**: 700–707.

Plachta, N., Bibel, M., Tucker, K.L., Barde, Y.A.: **Developmental potential of defined neural progenitors derived from mouse embryonic stem cells**. *Development* 2004, **131**: 5449–5456.

West, J.A., Daley, G.Q.: ***In vitro* gametogenesis from embryonic stem cells**. *Curr. Opin. Cell Biol.* 2004, **16**: 688–692.

Ying, Q.L., Nichols, J., Chambers, I., Smith, A.: **BMP induction of Id proteins suppresses differentiation and sustains embryonic stem cell self-renewal in collaboration with STAT3**. *Cell* 2003, **115**: 281–292.

10.17 Stem cells could be a key to regenerative medicine

Jaenisch, R.: **Human cloning—the science and ethics of nuclear transplantation**. *N. Engl. J. Med.* 2004, **351**: 2787–2792.

McClaren, A.: **Ethical and social considerations of stem cell research**. *Nature* 2001, **414**: 129–131.

Okita, I., Ichisaka, T., Yamanka, S.: **Generation of germline-competent induced pluripotent stem cells**. *Nature* 2007, **448**: 313–317.

Pera, M.F., Trounson, A.O.: **Human embryonic stem cells: prospects for development**. *Development* 2004, **131**: 5515–5525.

Pomerantz, J., Blau, H.M.: **Nuclear reprogramming: a key to stem cell function in regenerative medicine**. *Nat. Cell Biol.* 2004, **6**: 810–816.

Rolletschek, A., Wobus, A.M.: **Induced human pluripotent stem cells: promises and open questions**. *Biol. Chem.* 2009, **390**: 845–849.

Wurmer, A.E., Palmer, T.D., Gage, F.H.: **Cellular interactions in the stem cell niche**. *Science* 2004, **304**: 1253–1255.

Yechoor, V., Liu, V., Espiritu, C., Paul, A, Oka, K., Kojima, H., Chan, L.: **Neurogenin3 is sufficient for transdetermination of hepatic progenitor cells into neo-islets in vivo but not transdifferentiation of hepatocytes**. *Dev. Cell* 2009, **16**: 358–373.

10.18 Various approaches can be used to generate differentiated cells for cell-replacement therapies

Chen, S., Borowiak, M., Fox, J.L., Maehr, R., Osafune, K., Davidow, L., Lam, K., Peng, L.F., Schreiber, S.L., Rubin, L.L., Melton, D.: **A small molecule that directs differentiation of human ESCs into the pancreatic lineage**. *Nat. Chem. Biol.* 2009, **5**: 258–265.

Maehr, R., Chen, S., Snitow, M., Ludwig, T., Yagasaki, L., Goland, R., Leibel, R.L., Melton, D.A.: **Generation of pluripotent stem cells from patients with type 1 diabetes**. *Proc. Natl Acad. Sci. USA* 2009, **106**: 15768–15773.

Zhou, Q., Brown, J., Kanarek, A., Rajagopal, J., Melton, D.A.: ***In vivo* reprogramming of adult pancreatic exocrine cells to beta-cells**. *Nature* 2008, **455**: 627–632.

11

Organogenesis

- The vertebrate limb
- Insect wings and legs

- Vertebrate and insect eyes

- Internal organs: tracheal system, lungs, kidneys, blood vessels, heart, and teeth

Once the basic animal body plan has been laid down, the development of organs as varied as insect wings and vertebrate eyes begins. One question is whether quite new developmental mechanisms and principles are brought into play, or whether organogenesis depends on the same basic mechanisms as those used in early development. Organ development involves very large numbers of genes and, because of this complexity, general principles can be quite difficult to distinguish. Nevertheless, many of the mechanisms used in organogenesis, such as positional information, are similar to those of earlier development, and certain signals are used again and again.

So far, we have concentrated almost entirely on the aspects of development involved in laying down the basic body plan in various organisms, and on early morphogenesis and cell differentiation. We now turn to the development of specific organs and structures—**organogenesis**—which is a crucial phase of development that will ultimately lead to the embryo at last becoming a fully functioning organism, capable of independent survival.

The development of certain organs has been studied in great detail and they provide excellent models for looking at developmental processes, such as pattern formation, the specification of positional information, induction, change in form, and cellular differentiation. In this chapter, we first consider the development of some of these classic model systems—the vertebrate limb, the legs and wings of *Drosophila*, and vertebrate and insect eyes. We then consider some internal organs. The structure of many internal organs is based on tubules—for example, the lungs, the vascular system, and the kidneys—and we look at the different ways in which tubules can be formed. We then look briefly at the initial patterning and regionalization of the heart, which develops from a tube of mesoderm, and finally at the induction and patterning of teeth. The development of other major organs, such as the gut, liver, and pancreas, involves no new principles and we will not consider them here. The liver is interesting as an example of a mammalian organ that can regenerate, and some aspects of liver growth and regeneration are discussed in Chapters 13 and 14.

The cellular mechanisms involved in organogenesis are essentially similar to those encountered in earlier stages of development; they are merely employed in different spatial and temporal patterns. Many of the genes and signaling molecules involved will be familiar from earlier chapters. We shall, however, see that the mechanisms involved are much more complex and that, while one can identify some general principles, such

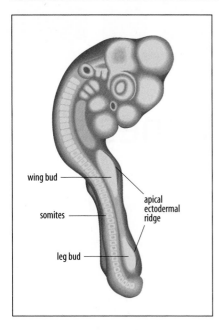

Fig. 11.1 The limb buds of the chick embryo. Limb buds appear on the flanks of the embryo on the third day of incubation (only the limb buds on the right side are shown here). They are composed of mesoderm, with an outer covering of ectoderm. Along the tip of each runs a thickened ridge of ectoderm, the apical ectodermal ridge.

as the use of positional information, there is much additional detail for which there are no unifying principles—it just works that way in that particular organ.

The vertebrate limb

The vertebrate embryonic limb is a particularly good system in which to study pattern formation, as the basic pattern is initially quite simple, and limbs have distinct elements of pattern along all three axes that can be easily recognized. The limb is also a good model for studying cellular interactions within a structure containing a large number of cells, and for elucidating the role of intercellular signaling in development. Mice are used to study some aspects of limb development, mainly through spontaneous mutants and artificial gene knock-outs, but the basic principles of limb patterning have been most extensively investigated in chick embryos, because here the developing limbs themselves are easily accessible for microsurgical manipulation. A window is made in the egg shell and the limb buds manipulated while the embryo remains in the egg (see Fig. 3.29). After surgery, the window can be re-sealed with adhesive tape and the embryo allowed to continue development, so that the effects of the procedures on limb development can be assessed.

In chick embryos, the first signs of limb development can be seen around the third day after the egg is laid, when the structures of the main body axis are already well-established. Small protrusions—the **limb buds**—arise from the body wall of the embryo (Fig. 11.1). By 10 days, the main features of the limbs are well-developed. Figure 11.2 shows the pattern of the elements that make up the skeleton; they are first formed as cartilage and are later replaced by bone (how bones grow is described in Section 13.8). Structures in the surface epithelium, such as feather buds, can also be seen. The limb at this stage also has muscles and tendons. The limb has three developmental axes: the **proximo-distal axis** runs from the base of the limb to the tip; the **antero-posterior** axis runs parallel with the body axis (in the human hand it goes from the thumb (anterior) to the little finger (posterior), and in the chick wing from digit 2 to digit 4); the **dorso-ventral** axis is the third axis—in the human hand it runs from the back of the hand to the palm.

11.1 The vertebrate limb develops from a limb bud

The early limb bud has two major components—a core of loose mesenchymal cells derived from the lateral plate mesoderm, and an outer layer of ectodermal epithelial

Fig. 11.2 The embryonic chick wing. The photograph shows a stained whole mount of the wing of a chick embryo 10 days after the egg has been laid. By this time, the main cartilaginous elements (e.g. humerus, radius, and ulna) have been laid down. They later become ossified to form bone. The muscles and tendons are also well developed at this stage but cannot be seen in this type of preparation. Feather buds can be seen particularly along the posterior margin of the limb. The three developmental axes of the limb are proximo-distal, antero-posterior, and dorso-ventral, as shown in the top panel. Note that the chick wing has only three digits, which have been called 2, 3, and 4. Scale bar = 1 mm.

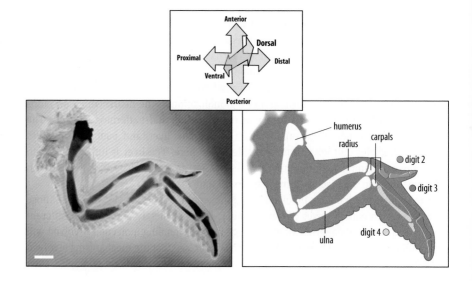

cells (Fig. 11.3). The skeletal elements and connective tissues of the limb develop from these mesenchyme cells, but the mesodermal cells that give rise to the limb muscles have a separate lineage. They derive from the somites and migrate into the limb bud (see Section 5.9). The limb vasculature is, at least in part, also formed by cells derived from the somites. At the very tip of the limb bud is a thickening in the ectoderm—the **apical ectodermal ridge** or **apical ridge**, which runs along the dorso-ventral boundary (Fig. 11.4). Directly beneath the apical ectodermal ridge lies a region that is composed of rapidly proliferating undifferentiated mesenchymal cells. It is only when cells leave this zone that they begin to differentiate. As the bud grows, the cells start to differentiate and cartilaginous structures begin to appear in the mesenchyme. The proximal part of the limb—that is, the part nearest to the body—is the first to differentiate, and differentiation proceeds distally as the limb extends. Of the structures found in the developing limb, the patterning of the cartilage has been the best studied, as cartilage can be stained and seen easily in whole mounts of the embryonic limb (see Fig. 11.2). The disposition of muscles and tendons is more intricate, and although this can be studied in whole mounts by staining with antibodies for tissue-specific proteins, histological examination of serial sections through the limb may be needed.

The first clear sign of cartilage differentiation is the increased packing of groups of cells, a process known as **condensation**. The cartilage elements are laid down in a proximo-distal sequence in the chick wing—the humerus, radius, and ulna, the wrist elements (carpals), and then three easily distinguishable digits, 2, 3, and 4 (see Fig. 11.2). Figure 11.5 compares this sequence in the development of a chick wing with the similar sequence in the development of a mouse forelimb.

The chick limb bud at 3 days is about 1 mm wide by 1 mm long, but by 10 days it has grown around 10-fold, mostly in length. The basic pattern has been laid down well before then. But even at 10 days, the limb is still small compared with the size of the limb when the chick hatches. The apical ridge disappears as soon as all the basic elements of the limb are in place, and growth occupies most of the subsequent development of the limb, both before and after hatching. During the later growth phase, the cartilaginous elements are largely replaced by bone. Nerves only enter the limb after the cartilage has been laid down, at around 4.5 days, and we shall discuss this in Chapter 12. The problem of limb patterning is to understand how the basic pattern of cartilage, muscle, and tendons is formed in the right places, and how they make the right connections with each other. The first consideration, however, is how limb buds develop at the appropriate positions on the body.

11.2 Genes expressed in the lateral plate mesoderm are involved in specifying the position and type of limb

The forelimbs and hindlimbs of vertebrates arise at precise positions along the antero-posterior axis of the body. Transplantation experiments have shown that the lateral plate mesoderm that gives rise to the limb-bud mesenchyme becomes determined to form limbs in these exact positions, long before limb buds are visible. Like the pre-somitic mesoderm discussed in Chapter 5, the lateral plate mesoderm of the trunk becomes regionalized by Hox gene expression along the antero-posterior axis, and so it is possible that Hox gene coding could determine prospective forelimb and hindlimb regions, thus specifying the position and type of a limb bud. This is analogous to the way in which the different thoracic segments and their appendages are specified in insect development (see Chapter 2). The homeodomain transcription factor Pitx1 (which is related to the Otx homeodomain proteins) is expressed in the hindlimb region and plays a key role in determining the differences between hindlimb and forelimb.

Proteins of the fibroblast growth factor (FGF) family appear to be key signaling molecules in initiating both forelimb and hindlimb development in vertebrates. Localized application of FGF-4 to the flank of a chick embryo, between the wing and leg buds,

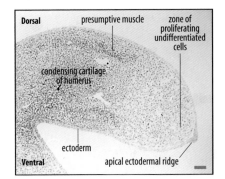

Fig. 11.3 Cross-section through a developing limb bud. The thickened apical ectodermal ridge is at the tip. Beneath the ridge is a region of proliferating undifferentiated cells. Proximal to this region, mesenchyme cells condense and differentiate into cartilage. Presumptive muscle cells migrate into the limb from the adjacent somites and form dorsal and ventral muscle masses. Scale bar = 0.1 mm.

Fig. 11.4 Scanning electron micrograph of a chick limb bud at 4.5 days after laying, showing the apical ectodermal ridge. Scale bar = 0.1 mm.

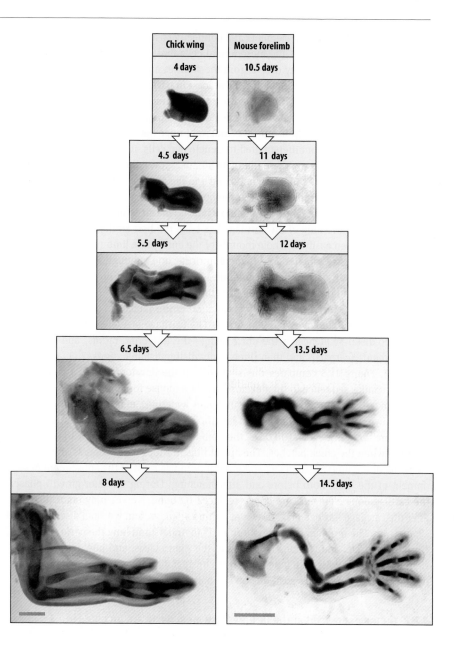

Fig. 11.5 The development of the chick wing and mouse forelimb is similar. The skeletal elements are laid down, as cartilage, in a proximo-distal sequence as the limb bud grows outward. The cartilage of the humerus is laid down first, followed by the radius and ulna, wrist elements, and digits. Scale bar = 1 mm.

induces an ectopic limb bud. Wing buds develop from application of FGF to the anterior flank, whereas leg buds develop from application to the posterior flank (Fig. 11.6). The T-box transcription factors, Tbx5 and Tbx4, which are related to the mesodermal marker Brachyury (see Section 4.18), are expressed in forelimbs and hindlimbs, respectively, and are required for limb initiation via FGF. In humans, mutations in the related T-box gene *Tbx5* are associated with the Holt–Oram syndrome, which is most commonly characterized by abnormalities of the thumbs and heart defects.

FGF is involved in establishing and maintaining the two main organizing regions of a limb. These are the apical ectodermal ridge, which is essential for limb outgrowth and for correct patterning along the proximo-distal axis of the limb, and a mesodermal polarizing region, which is situated on the posterior side of the limb bud and is crucial for determining pattern along the antero-posterior limb axis, as discussed later in the chapter. Cells of the polarizing region express the signaling protein Sonic hedgehog (Shh), and staining for Shh mRNA has been used to reveal the polarizing regions in the normal and the ectopic limb buds in the embryo shown in Fig. 11.6. The gene *Fgf10* is expressed in limb-forming regions, and in the mouse embryo, in which, unlike chick, transgenic knock-outs are possible, knock-out of either *Fgf10* or the gene encoding its receptor results in embryos lacking limb buds.

In chick embryos, Wnt proteins produced and secreted by the mesoderm play a key role in determining where FGFs are expressed and maintained. A Hox gene coding in the mesoderm could therefore determine where limbs develop by specifying where Wnts, and thus FGFs, are expressed, and so where an apical ridge and a polarizing region develop. Evidence for the combined actions of *Hox* genes in determining the position of the forelimb bud along the antero-posterior axis comes from knock-out mice lacking *Hoxb5* expression, in which the forelimbs develop at a more anterior level. The different patterns of *Hox* gene expression at different positions along the antero-posterior axis of the body could also be involved in specifying whether the limb will be a forelimb or a hindlimb.

11.3 The apical ectodermal ridge is required for limb outgrowth

We shall next look at how the limb buds start to grow and develop. One crucial signaling region in the limb bud is the apical ectodermal ridge, whose presence is essential for limbs to grow and develop. The apical ectodermal ridge consists of closely packed columnar epithelial cells, which are directly connected to each other by gap junctions. Their tight packing gives the ridge a mechanical strength that probably is what keeps the limb flattened along its dorso-ventral axis. The initial outgrowth of the limb bud depends on proliferating undifferentiated mesenchymal cells lying beneath the apical ectodermal ridge. Perhaps surprisingly, the establishment of the early bud does not reflect a local increase in cell division in the limb-bud region, but rather a decrease from a previously high rate of cell proliferation along the rest of the flank of the embryo.

The experiment that showed the importance of the apical ectodermal ridge involved removing the ridge from a chick limb bud by microsurgery. This resulted in a significant reduction in outgrowth and the limb was anatomically truncated, with distal parts missing. The proximo-distal level at which the limb is truncated depends on the time at which the ridge is removed (Fig. 11.7). The earlier the ridge is removed, the greater the effect; removal at a late stage only results in loss of the ends of the digits, whereas when the ridge is removed at an early stage most of the limb is missing. Following apical-ridge removal, proliferation of cells at the tip of the limb bud is greatly reduced and there is also cell death in that region. The importance of signaling by the

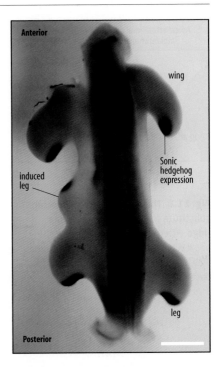

Fig. 11.6 Result of local application of FGF-4 to the flank of a chick embryo. FGF-4 is applied to a site close to the leg bud, where it induces outgrowth of an additional leg bud. The dark regions indicate *Sonic hedgehog* transcripts. *Sonic hedgehog* is expressed at the anterior of the additional leg bud and the leg that subsequently develops has an opposite polarity to that of the normal limbs. It is not known why *Sonic hedgehog* is expressed in this way in the additional leg bud. Scale bar = 1 mm.

Photograph courtesy of J-C. Izpisúa-Belmonte, from Cohn, M.J., et al.: 1995.

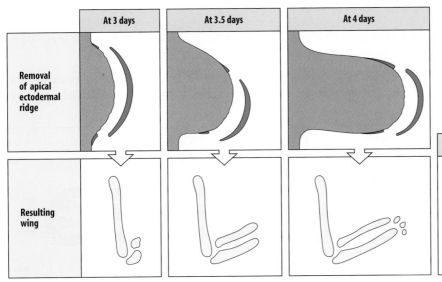

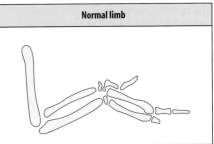

Fig. 11.7 The apical ectodermal ridge is required for proximo-distal development. Limbs develop in a proximo-distal sequence. Removal of the ridge from a developing limb bud leads to truncation of the limb; the later the ridge is removed, the more complete the resulting limb.

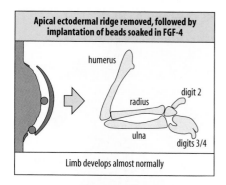

Apical ectodermal ridge removed, followed by implantation of beads soaked in FGF-4

humerus

radius

digit 2

ulna

digits 3/4

Limb develops almost normally

Fig. 11.8 The growth factor FGF-4 can substitute for the apical ectodermal ridge. After the ridge is removed, grafting heparin beads that release FGF-4 into the limb bud results in almost normal limb development.

apical ectodermal ridge in promoting outgrowth has also been shown by grafting an isolated ridge to the dorsal surface of an early chick wing bud. The result is an ectopic outgrowth, which may even develop cartilaginous elements and digits.

A key signal from the ridge is provided by FGFs. The gene for FGF-8 is expressed throughout the ridge and genes for FGF-4 and two other FGFs are expressed in the posterior region. Experiments in chick wing buds showed that FGF-8 can act as a functional substitute for an apical ridge. If the ridge is removed and beads that release the growth factor are grafted into the tip of the wing bud in its place, more-or-less normal outgrowth of the limb continues. If sufficient FGF-8 is provided to the outgrowing cells, a fairly normal limb develops. Similar results are obtained with FGF-4 (Fig. 11.8). Experiments in transgenic mice in which *Fgf* genes are knocked-out specifically in the apical ectodermal ridge, also show that FGF signaling is required for limb-bud outgrowth and normal development. It is tricky to delete FGF signaling completely using transgenic approaches because four different *Fgf* genes are expressed in the apical ectodermal ridge. Complete limb truncations are produced only when three of these *Fgf* genes are knocked out. All this evidence together confirmed that apical-ridge signaling is essential for normal vertebrate limb development.

11.4 Patterning of the limb bud involves positional information

We have just seen how the apical ridge controls the outgrowth of the limb, but how are the skeletal and other elements of the limb laid down in the correct positions? Some aspects of vertebrate limb development fit very well with a model of pattern formation based on positional information. The developing chick limb bud behaves as if its future development is determined by the positions of its cells with respect to the main limb axes while the cells are in the undifferentiated region at the tip of the limb bud (Fig. 11.9).

Central to the idea of positional information is the distinction between positional specification and interpretation. Cells acquire positional information first and then interpret these positional values according to their developmental history. It is the difference in developmental history that makes wings and legs different, for positional information is signaled in wing and leg buds in the same way. An attractive hypothesis for limb patterning is that a single three-dimensional positional field controls the development of the cells that give rise to all the mesodermal limb elements—cartilage, muscle, and tendons. The specification of the three axes—proximo-distal, antero-posterior, and dorso-ventral—is, as we shall see, linked by molecular signals emanating from different parts of the bud.

11.5 How position along the proximo-distal axis of the limb bud is specified is still a matter of debate

We will start by discussing patterning along the limb's all-important proximo-distal axis, which runs from the shoulder to the tips of the digits. The cartilage elements

Fig. 11.9 Cells acquire positional values along the antero-posterior and proximo-distal axes. In the early bud there are two signaling regions—the polarizing region at the posterior margin (red), and the apical ectodermal ridge (blue). The polarizing region specifies position along the antero-posterior axis. How position is specified along the proximo-distal axis is still under debate (see Section 11.5). Cartilage begins to differentiate first in the most proximal region, and the cartilaginous elements are laid down in a proximo-distal sequence (carpals not shown).

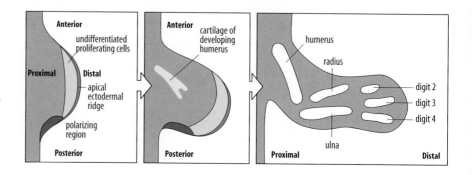

are laid down in a proximal to distal sequence, with the proximal elements beginning to differentiate first. The mechanisms that pattern the proximo-distal axis are, however, still a matter of some debate, with several different hypotheses currently being proposed. For simplicity, in this section we will discuss mainly the longest-standing model, which derives from work on chick embryo limbs and proposes that proximo-distal patterning is specified by the length of time cells spend in the zone of undifferentiated cells at the tip of the limb. This region has therefore been called the **progress zone**, and we shall refer to this model as the 'timing model.' An alternative mechanism for proximo-distal patterning has been proposed more recently, based on newer experiments in chick and mouse, and we shall refer to this as the 'two-signal' model. There is evidence in support of both models, but neither is entirely consistent with all the known evidence, so it is clear that there is more to be learned about how the limb is patterned in the proximo-distal direction.

As the limb bud grows, cells are continually leaving the zone of undifferentiated cells at the distal end. Because the limb bud extends from proximal to distal, the earliest cells leaving the zone develop into the humerus and those leaving last form the tips of the digits. The timing model proposes that, if cells could measure the time they spend in the undifferentiated zone, by counting cell divisions, for example, this would give them their positional value along the proximo-distal axis (Fig. 11.10, upper panel). The length of time cells spend in the somite stem-cell region has similarly been suggested to account for the specification of the position of a somite along the antero-posterior axis (see Section 5.8). A timing mechanism of this sort for the limb is consistent with the experimental observation that removal of the apical ridge results in a distally truncated limb (see Fig. 11.7). according to the timing theory, removal of

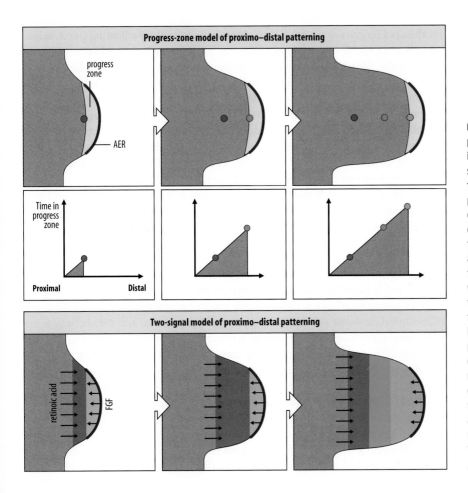

Fig. 11.10 A cell's proximo-distal positional value may depend on the time it spends in the 'progress zone' or be specified by opposing signal gradients in the early limb bud. The upper two sets of panels show the 'timing' model. Cells are continually leaving the progress zone. If the cells could measure how long they spend in the zone, this could specify their position along the proximo-distal axis. Cells that leave the zone early (red) form proximal structures whereas cells that leave it last (blue) form the tips of the digits. The lower panel shows the 'two-signal' model. Cells in the early limb bud acquire positional values from the presence of two opposing signal gradients, one of retinoic acid from the most proximal region (gray) specifying proximal (red) and one of FGF from the apical ectodermal ridge specifying distal (turquoise and later blue). As the limb bud grows, cells that are out of range of both signaling sources acquire an 'intermediate' positional value (green).
Adapted from Zeller, R., et al.: 2009.

the apical ridge means that cells in the progress zone will no longer proliferate, and so the more distal structures cannot form.

Another line of evidence put forward in support of a timing mechanism is that killing or blocking the proliferation of cells in the progress zone of a chick wing bud at an early stage (by X-irradiation, for example), results in the absence of proximal structures, whereas distal ones are present and can be almost normal. Because many cells in the irradiated progress zone do not divide, fewer cells than usual will leave the zone during each unit of time. As the bud grows out, the progress zone becomes repopulated by the proliferation of surviving cells, which would spend a longer time than usual in the progress zone and therefore acquire a distal positional value and form normal distal structures, such as digits. According to other models, proximal structures suffer most because, although the few cells that left soon after irradiation will be molecularly specified as proximal, there will not be enough of them to form a cartilage element. Similar arguments could be made to account for the absence of proximal limb structures in patients born in the late 1950s and early 60s whose mothers took the drug thalidomide to ease morning sickness. Thalidomide is known to interfere with the development of blood vessels, and this could have resulted in extensive cell death throughout the early limb bud. The effects of thalidomide may, it has been suggested, be explained by their simply eliminating cartilage precursor cells at a time when proximal regions are forming but distal differentiation has not yet commenced.

We have already seen that FGFs produced by the apical ridge are key signals in limb development, but what specific part do they play in proximo-distal patterning? According to the timing model, the function of FGF signaling is to maintain the progress zone, and cells might measure time by the length of time they are exposed to FGF signals. There is, however, also evidence that retinoic acid, which is synthesized by cells at the base of the limb buds and in the adjoining trunk, might contribute to specifying proximal identity in the limb. Retinoic acid targets include genes of the *Meis* family, which encode homeodomain transcription factors, and misexpression of *Meis* genes in the chick wing leads to distal abnormalities or truncations. Retinoic acid signaling and *Meis* expression are normally restricted to the proximal region of the limb bud by FGF signaling from the apical ridge, while FGF signaling is restricted to the distal region of the limb bud. This interplay between FGF and retinoic acid signaling is not unique to the developing limb. Opposing gradients of retinoic acid and FGF have been proposed to operate during growth and antero-posterior patterning of the main body axis and to control differentiation of the spinal cord (see Section 5.5). For the limb, it has been proposed that these two signals may specify a complete proximo-distal axis at an early stage, rather than the axis being constructed progressively over time as proposed by the timing model, and this newer model is known as the 'two-signal' model (see Fig. 11.10, lower panel).

Recent genetic data from the mouse has been interpreted in terms of the two-signal model, and is based on limb patterns that result when FGF signaling in the apical ridge of mouse limb buds is incrementally reduced by genetically deleting *Fgf* genes in the apical ridge one by one. The proposal is that proximal and distal components of the limb pattern are first specified by retinoic acid and FGF signaling, respectively, and that subsequent outgrowth of the limb, controlled by FGFs produced by the apical ridge, leads to the emergence of a region of cells outside the range of both distal and proximal signals. As a consequence, this region adopts an 'intermediate' fate—and forms the radius and ulna. It is not clear, however, whether this suggested mechanism for generating intermediate fates is universally applicable. When the tip of an old chick wing bud, which will form distal limb components, is grafted on to a stump of a young wing bud, which will form proximal limb components, subsequent outgrowth and development generates a wing that lacks radius and ulna. This failure of such composite chick wing buds to form intermediate structures is difficult to reconcile with the model.

11.6 The polarizing region specifies position along the limb's antero-posterior axis

The specification of pattern along the antero-posterior axis of the limb bud is better understood and the subject of less contention. Patterning along this axis is what specifies the individual digits and gives them their identities. In this and the next two sections we shall look at how a standard set of digits is specified, the signaling molecules that operate in antero-posterior patterning, and, finally, how each digit is given its unique identity. The organizing region for the antero-posterior axis is a mesodermal region at the posterior margin of the limb bud, known as the **polarizing region** or the **zone of polarizing activity (ZPA)** (Fig. 11.11).

The polarizing region of a vertebrate limb bud has organizing properties that are almost as striking as those of the Spemann organizer in amphibians. When the polarizing region from an early chick wing bud is grafted to the anterior margin of another early chick wing bud, a wing with a mirror-image pattern develops: instead of the normal pattern of digits—2 3 4—the pattern 4 3 2 2 3 4 develops (Fig. 11.12, center panels). The pattern of muscles and tendons in the limb shows similar mirror-image changes. A polarizing-region graft can also specify another ulna anteriorly, showing that the polarizing region influences antero-posterior pattern throughout the region distal to the elbow or the knee.

The additional digits come from the host limb bud and not from the graft, showing that the grafted polarizing region has altered the developmental fate of the host cells in the anterior region of the limb bud. The limb bud widens in response to the polarizing graft, which enables the additional digits to be accommodated. Widening of a limb bud is associated with an increase in the extent of the apical ectodermal ridge.

One way that the polarizing region could specify position along the antero-posterior axis is by producing a morphogen that forms a posterior to anterior gradient (Box 11A, p. 421). The concentration of morphogen could specify the position of cells along the antero-posterior axis with respect to the polarizing region located at the posterior margin of the limb. Cells could then interpret their positional values by developing specific structures at particular threshold concentrations of morphogen. Digit 4, for example, would develop at a high concentration, digit 3 at a lower one, and digit 2 at an even lower one (see Fig. 11.12, top panels). According to this model, a graft of an additional polarizing region to the anterior margin would set up a mirror-image gradient of morphogen, which would result in the 4 3 2 2 3 4 pattern of digits that is observed (see Fig. 11.12, center panels).

If the action of the polarizing region in specifying the character of a digit is due to the level of the signal, then when the signal is weakened, the pattern of digits should be altered in a predictable manner. Grafting small numbers of polarizing region cells to a limb bud results in the development of only an additional digit 2 (see Fig. 11.12, bottom panels). An analogous result can be obtained by leaving the polarizing region graft in place for a short time and then removing it. When left for 15 hours, an additional digit 2 develops, whereas the graft has to be in place for up to 24 hours for a digit 3 to develop.

The limb buds of a number of other vertebrates, including mouse, pig, ferret, and turtles, and even human limb buds, have been shown to have a polarizing region. When the posterior margin of a limb bud from an embryo of these species is cut out and grafted to the anterior margin of a chick wing bud, additional digits are produced. The additional digits induced are chick wing digits, which shows that, although the polarizing-region signal is conserved among vertebrates, its interpretation depends on the responding cells.

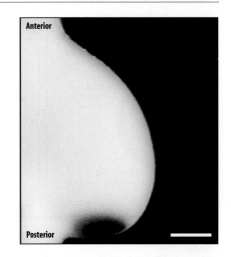

Fig. 11.11 The polarizing region or zone of polarizing activity (ZPA) of a chick limb bud is located at the posterior margin of the bud. The polarizing region can be detected by its expression of *Sonic hedgehog* (*Shh*) RNA (blue stain). Scale bar = 0.1 mm. *Photograph courtesy of C. Tabin.*

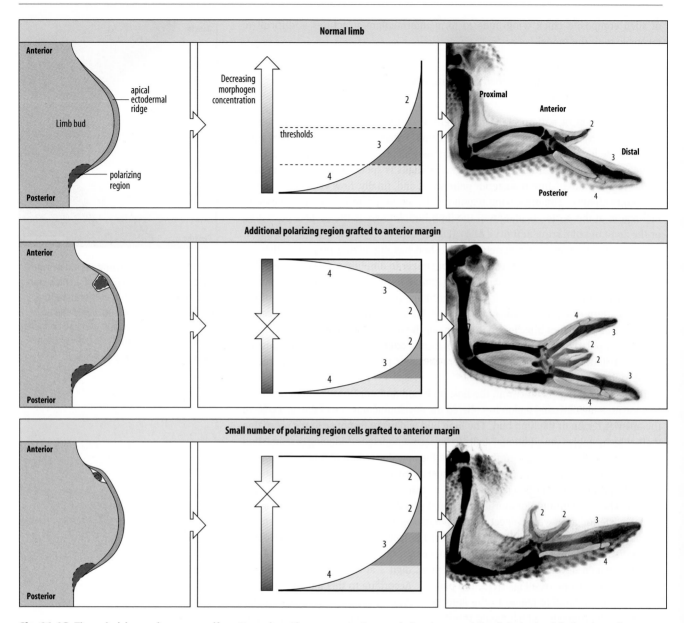

Fig. 11.12 The polarizing region can specify pattern along the antero-posterior axis. If the polarizing region is the source of a graded morphogen, the different digits could be specified at different threshold concentrations of signal, as shown for a normal limb bud in the top panels panels, with digit 4 developing where morphogen concentration is high and digit 2 where it is lowest. Threshold concentrations of morphogen that give rise to the different digits are indicated in the center panel of each row and the digits in the right-hand panels are color-coded to match. Grafting an additional polarizing region to the anterior margin of a limb bud (center row of panels) would result in a mirror-image gradient of signal, and thus the observed mirror-image duplication of digits. A small number of polarizing region cells grafted to the anterior margin of the limb bud produces only a weak signal and so only an extra digit 2 is produced (bottom row of panels).

11.7 Sonic hedgehog produced by the polarizing region is likely to be the primary morphogen patterning the antero-posterior axis of the limb

A persuasive case can be made for the secreted protein Sonic hedgehog (Shh) acting as a morphogen and a gradient of Shh protein being a key component of the natural polarizing signal. The *Sonic hedgehog* gene (*Shh*) is expressed in the polarizing region of the limb buds of both chick and mouse embryos and the embryos of many other vertebrates that have been studied (see Fig. 11.11), and by the use of anti-Shh antibodies, a graded distribution of Shh protein has been detected in the posterior region of the limb bud at a distance from the cells that produce it. Shh protein is involved in numerous patterning processes elsewhere in vertebrate development; for example, in somite patterning (see Section 5.9), in the establishment of internal left–right asymmetry (see Section 4.7), and in the patterning of the neural tube (discussed in Chapter 12). The *Drosophila* Hedgehog protein is a key signaling molecule in the patterning of the segments in the *Drosophila* embryo (see Section 2.24), as well as of *Drosophila* legs and wings, as we shall see later in this chapter.

Cultured chick fibroblasts transfected with a retrovirus containing the *Shh* gene acquire the properties of a polarizing region; when grafted to the anterior margin of a wing bud they cause the development of a mirror-image wing. A bead soaked in Shh protein will give the same result. The bead must be left in place for 16 to 24 hours for extra digits to be specified, and the pattern of digits depends on the concentration

Box 11A Positional information and morphogen gradients

Pattern formation in development can be specified by positional information, which is in turn specified by a gradient in some property. The basic idea, related to the French Flag problem (see Fig. 1.25), is that a specialized group of cells at one boundary of the area to be patterned secrete a molecule whose concentration decreases with distance from the source, thus forming a gradient of information. Any cell along the gradient then 'reads' the concentration at that particular point, and interprets this to respond in a manner appropriate to its position by switching on a particular pattern of gene expression, for example. Molecules that can induce such changes in cell fate are known as morphogens. Gradients of positional information are used for various tasks during development: regionalization of the antero-posterior and dorso-ventral axes and patterning of segments and imaginal discs in insects; mesoderm patterning in vertebrates; vertebrate limb patterning; and patterning along the dorso-ventral axis of the vertebrate neural tube, among others.

It is still unclear how gradients of positional information are formed, particularly the relative roles of morphogen diffusion and cell–cell interactions, although there is evidence for gradients of morphogen molecules in many cases. There are several questions about positional gradients still to be solved. How is positional information specified over distances of up to 20 cell diameters? Is it the extracellular or intracellular concentration of a morphogen that specifies positional identity? Does endocytosis play a key role?

Morphogens could travel by simple diffusion through extracellular spaces or, as described for the Decapentaplegic protein in *Drosophila* dorso-ventral patterning (see Section 2.20), a gradient can be formed by the interaction of the morphogen with various other proteins, both secreted extracellular proteins and cell-surface receptors. Morphogens may also be transported through cells by endocytic vesicles (see figure). In other cases, receptor-mediated endocytosis followed rapidly by degradation has been proposed as a way of producing a sharp gradient in protein activity. It is most likely that a cell measures a particular morphogen concentration according to how many receptors for the morphogen are occupied, and thus by the strength of the signal that is transmitted from the receptors. Despite years of work on the topic, however, it is still not clear whether mechanisms based on secreted diffusible

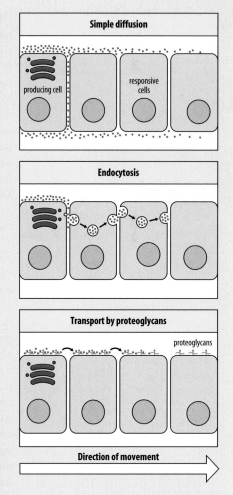

morphogens will be sufficiently reliable for pattern formation. Moreover, the concentration of a diffusible morphogen in the extracellular space is continually changing over time. An alternative to simple diffusion would be that positional information is specified by a mechanism involving the generation of a gradient of activity based on short-range interactions between cells, such as occur in the generation of planar cell polarity (see Box 8C, p. 310).

of Shh protein in the bead. A higher concentration is required for a digit 4 to develop than for a digit 2, for example. It has also been shown that Shh has a direct effect on mesenchymal cell proliferation by controlling expression of genes encoding cell-cycle regulators, thus revealing a mechanism that contributes to the widening of the limb bud following a polarizing graft.

The signaling molecule retinoic acid (see Box 5D, p. 192–193) was the first defined chemical that was found to mimic the signaling of the polarizing region of the chick wing and can induce mirror-image duplications of the digits when applied to the anterior margin of a chick wing bud. But it was later shown that retinoic acid induces *Shh* expression in anterior tissue and this explains the digit duplications. We have already discussed the evidence that retinoic acid signaling may pattern the proximal part of the limb but how this is related to its ability to induce *Shh* expression is not clear.

Mutations in mice and other animals, including humans, have been discovered that cause **preaxial polydactyly**. 'Polydactyly' denotes limbs with extra digits and 'pre-axial' that the extra digits form anteriorly, characteristically forming an extra 'thumb.' These mutations affect the polarizing region and provide more evidence for the crucial role of Shh in specifying the pattern of digits (Box 11B, p. 423).

The next question is how Shh exerts its effects on limb development. The intracellular signaling pathway that is activated in response to Shh involves the Gli transcription factors Gli1, Gli2, and Gli3, which are homologous to Ci in the Hedgehog signaling pathway in *Drosophila* (see Fig. 2.40). The vertebrate Shh signaling pathway is shown in Box 11C, pp. 424–425. In the presence of Shh signaling, the Gli proteins function as transcriptional activators, while in its absence, Gli2 and Gli3 are processed to short forms and Gli3 functions as a transcriptional repressor. Analysis of mice mutant for the *Gli* genes has shown that Gli3 is the key regulator of limb patterning, and several mutations in human Gli3 have also been found to be associated with limb abnormalities. *Gli3* knock-out mice have polydactyly and the digits are all of the same, unpatterned, type. This indicates that in normal circumstances the repressor form of Gli3 is preventing digit formation in the anterior part of the limb bud, where there is no Shh protein, and that in the posterior part of a normal limb bud, a Shh protein gradient will produce different ratios of Gli3 activator to Gli3 repressor along the antero-posterior axis and in this way will determine digit pattern.

In the limb buds of *Shh* knock-out mice, in contrast, high levels of repressor Gli3 are produced throughout the bud and prevent digit formation altogether. But when *Gli3* is knocked out as well as *Shh*, the limbs are similar to those of the *Gli3* knock-out mouse embryos. Taken together, these experiments show that the main function of Shh signaling in the limb bud is to suppress repressor Gli3 function in the posterior part of the bud. Some mouse mutants with reduced levels of repressor Gli and polydactyly do not have defects in the genes encoding Shh or components of the signaling pathway, but instead have mutations in genes essential for the formation of a cell structure called the **primary cilium**. This immotile microtubule-based structure is essential for Shh signaling and the processing of Gli3, as described in Box 11C, pp. 424–425. However, the fact that a limb with many digits, but no antero-posterior pattern, can develop when both Shh and Gli3 function are lacking implies the existence of some self-organizing mechanism in the limb bud that is capable of setting up a regular set of digit-like elements but cannot give them distinct identities. Such a mechanism could, for example, be based on the principle of reaction–diffusion (see Box 11D, p. 432), which has been proposed to be responsible for patterns such as the stripes on angelfish.

11.8 Transcription factors might specify digit identity

Several transcription factors have been implicated in digit formation and in conferring digit identity, including *Hoxd13* and *Hoxa13* (as we shall see later in the chapter),

Box 11B Too many fingers: mutations that affect antero-posterior patterning can cause polydactyly

Mutations have been discovered in mice that affect the function of the polarizing region and cause **preaxial polydactyly**. 'Polydactyly' denotes limbs with extra digits and 'preaxial' the fact that the extra digits form anteriorly. These mouse mutants, such as the Sasquatch mutant, provide more evidence for the crucial role of Shh in specifying the pattern of digits, as it has been shown that formation of the extra anterior digit in Sasquatch is due to additional anterior expression of the *Shh* gene in the limb buds. The Sasquatch mutation maps to a long-range *cis*-regulatory region in the mouse genome, which has become known as the ZRS (zone of polarizing activity regulatory sequence). The ZRS governs limb-specific *Shh* expression even though it is located 1 Mb away from the *Shh* gene. This example of long-range control is similar to the developmental control of β-globin expression in red blood cells by the locus control region (discussed in Section 10.6).

Mutations in the ZRS have also been found in polydactylous cats, with extra anterior digits, including the celebrated tribe of cats that inhabit the novelist Ernest Hemingway's old home in Key West, Florida (see photo). The tribe is descended from a polydactylous cat given to Hemingway by a sea captain, and all the polydactylous individuals have the same single-nucleotide substitution in the ZRS. There are also inherited human conditions based on mutations and/or duplications of the ZRS. Human patients with preaxial polydactyly also have mutations in the ZRS upstream of the human *SHH* gene, and several other polydactylous conditions in human patients are associated with microduplications within this regulatory sequence.

In contrast to the extra digits that develop from limb buds with additional anterior *Shh* gene expression, limbs of mouse embryos in which the *Shh* gene has been knocked-out lack distal structures, although proximal structures are still present. At best, one very small digit develops in the hindlimb. *Shh* knock-out mouse embryos also have other defects, including craniofacial defects such as holoprosencephaly (see Box 1F, p. 28) and patterning defects in the neural tube. Complete deletion of the ZRS in mouse embryos also leads to loss of posterior *Shh* expression in the limb buds and results in limb truncations resembling those in *Shh* knock-out embryos. In embryos with the ZRS deletion, however, no other structures are affected because Shh function is deleted just in the limb bud. Loss of distal limb structures is also seen in the very rare inherited human condition, **acheiropodia**, in which structures below the elbow or knee fail to develop but no other defects are present. Acheiropodia has been shown to be due to a chromosomal deletion near the ZRS, suggesting that there may be a *cis*-regulatory region in this position in the human genome that regulates *SHH* expression in the polarizing region.

Just because a morphogen gradient model is consistent with many experimental results in the chick wing, however, this is not sufficient to prove that it is the mechanism that specifies all of the digits in the limbs of other vertebrates. In the mouse limb, for example, cell-lineage analysis of cells that express Shh show that the two most posterior digits arise from cells of the polarizing region itself, and it is suggested that the identity of these two digits is determined by the length of time these cells spend in the polarizing region.

Photograph from Lettice, L.A., et al.: 2008.

and *Sall* and *Tbx* genes. Each of the four toes of the chick leg has a different number of bones (the phalanges) and so the different digits can be easily distinguished. This makes it an excellent system for studying how digit identity is specified. In the chick leg, expression of *Tbx2* and *Tbx3* in the most posterior interdigital region may contribute to specifying the identity of digit 4 (the most posterior toe), while expression of *Tbx3* alone in the next interdigital region may specify the identity of digit 3. This conclusion comes from experiments in which *Tbx2* and *Tbx3* were misexpressed in chick leg buds and changed one digit into the other. Limb defects are seen in human syndromes associated with mutations in *TBX3* (mammary ulnar syndrome) and *SALL1* (Townes–Brockes syndrome). The *Tbx* and *Sall* genes are homologs of the *Omb* and *Spalt* genes, respectively, which are involved in antero-posterior patterning of the *Drosophila* wing (discussed later in this chapter).

Box 11C Sonic hedgehog signaling and the primary cilium

The intercellular signaling protein Sonic hedgehog is a vertebrate homolog of the Hedgehog protein that plays a major role in the patterning of the *Drosophila* embryo (see Chapter 2). Mice and humans have three genes for hedgehog proteins—Sonic hedgehog (Shh), Indian hedgehog (Ihh), and Desert hedgehog (Dhh); zebrafish have a fourth hedgehog gene, Tiggywinkle, as a result of genome duplication. The signaling pathway stimulated by Shh and the other vertebrate hedgehog proteins (see figure) has many similarities to the Hedgehog signaling pathway in *Drosophila* (see Fig. 2.40). There are vertebrate homologs of the membrane proteins Patched and Smoothened, which sense the presence or absence of a hedgehog signal, and the vertebrate counterparts of the *Drosophila* transcription factor Cubitus interruptus (Ci) are the Gli transcription factors Gli1, Gli2, and Gli3. The name Gli comes from the initial identification of one of these genes as being amplified and highly expressed in a human glioma. Like Ci, Gli3 and Gli2 can be processed so that they act as transcriptional activators in the presence of a hedgehog signal and a transcriptional repressor in its absence. Proteins homologous with Cos2 and Su(fu) in *Drosophila* (see Fig. 2.40) form complexes with the Gli proteins and keep them in an inactive state.

A major difference between a vertebrate hedgehog signaling pathway and Hedgehog signaling in *Drosophila*, is that in vertebrates Gli processing takes place at the tip of a cell structure called the primary cilium. Most vertebrate cells have a primary cilium, which, like all cilia, is based on microtubules and extends from a basal body derived from a centrosome. Unlike most other types of cilia, the primary cilium is non-motile.

In the absence of a hedgehog ligand, such as Shh, Patched represses Smoothened activity and moves into the cilium membrane whereas Smoothened does not. The Gli3 and Gli2 proteins are phosphorylated, which targets them for cleavage when they encounter the proteasome at the base of the cilium. Cleavage produces short forms that act as transcriptional repressors.

The repressor Gli3 and Gli2 then enter the nucleus and prevent the transcription of Shh target genes. When Shh is present, it binds to Patched and activates Smoothened. Hedgehog ligand bound to Patched is internalized in membrane vesicles and plays no further part in signaling. Membrane vesicles containing the activated Smoothened are targeted to the cilium membrane, fuse with it, and deliver Smoothened into the membrane. At the tip of the cilium, Smoothened antagonizes the inhibition of Gli proteins by their binding proteins and prevents phosphorylation of Gli3 and Gli2. The activated Gli proteins are then transported along microtubules down the cilium to the cell body, where they enter the nucleus and activate transcription of target genes.

The first clues to the importance of the primary cilium in Shh signaling came from a genetic screen in mice for embryonic patterning mutants following mutagenesis with ethylnitrosourea (ENU) (see Fig. 3.30). Two mutants were identified with defects in Shh signaling in the neural tube, one of which survived long enough for limb polydactyly to be observed. Most unexpectedly, a search against the databases revealed that the genes mutated in these embryos encoded proteins similar to those that transport material within flagella (a longer version of cilia), and which were known to be essential for the growth and maintenance of flagella in the unicellular alga *Chlamydomonas*. Subsequently, mutations in genes encoding subunits of intraflagellar dynein and kinesin motor proteins were also found to result in defective Shh signaling, providing confirmation of the essential role of cilia in vertebrate hedgehog signaling.

All known mutations that prevent formation of the primary cilium cause patterning defects in the limb and the neural tube. In the absence of the primary cilium, both Gli repressor and Gli activator functions fail. Because of the importance of Gli repressor function in antero-posterior limb patterning, this leads to polydactyly, whereas the neural tube is dorsalized, which is due to the importance of Gli activator function in its dorso-ventral patterning (see Section 12.6).

Sal1, *Tbx2*, and *Hoxd13* are among the genes recently identified by a genome-wide analysis of gene expression in mouse limb buds that looked for genes regulated by Gli3 repressor. A Gli3 repressor tagged with a Flag label (which can be detected by specific antibodies) was expressed in the developing limbs of transgenic mice and the genomic sites occupied by the tagged Gli3 repressor were identified by chromatin immunoprecipitation (see Section 3.11). These results, together with transcriptional profiling by DNA microarrays of anterior and posterior regions of normal developing mouse limbs and limbs from mouse mutants with defective Shh signaling, identified more than 200 genes that are putative targets for the Gli3 repressor during limb development. From these data, a gene regulatory network mediated by Gli3 in the developing limb is being constructed (for other examples of gene regulatory networks see Box 4E, p. 162, and Section 6.12). So far, this network for the mouse limb has linked up about 10 genes, but many more will eventually be added. This is just the

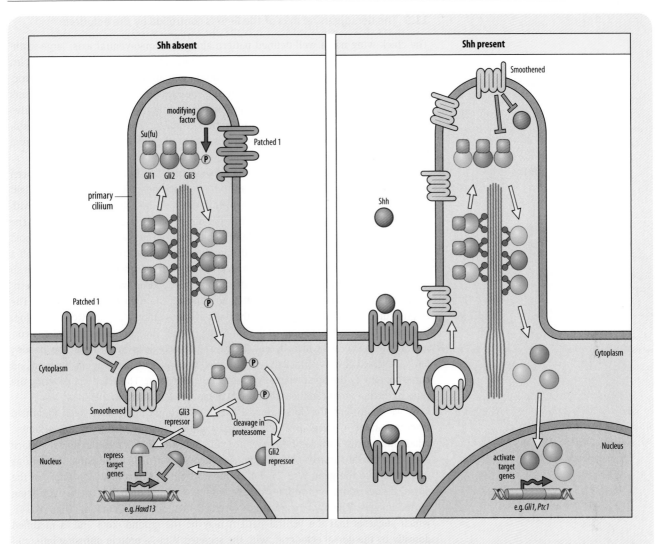

Defects in the formation of cilia and basal bodies have been linked to several dozen human congenital disorders, collectively known as **ciliopathies**. For example, patients with Bardet-Beidl syndrome (BBS), in which the primary cila are affected, have a range of defects that can include polydactyly, polycystic kidney disease, and hearing loss. Mutations in any of 11 genes can cause BBS, and some of these genes encode proteins that are part of the basal body and/or the centrosome, thus providing an explanation for defects in ciliogenesis.

beginning of unraveling the molecular basis for digit patterning downstream of Shh signaling.

Another Gli-regulated gene identified by microarray profiling was *Bmp2*, which encodes a secreted signaling protein of the TGF-β family and is expressed in the posterior of the early limb bud. BMPs are also expressed later in limb development in **interdigital regions**, the regions between the forming digits. Experiments on chick embryo leg indicate that BMP signaling in interdigital tissue might regulate digit identity even at late stages in digit formation. Removal of interdigital tissue or the implantation into interdigital regions of beads coated with the BMP antagonist Noggin (see Section 4.17), induce changes in the number of phalanges in the adjacent digit that are consistent with a more anterior identity. Each forming digit is characterized by a specific level of activity of Smad transcription factors, which provide the read-out of the levels of BMP signaling (see Fig. 4.33) and would activate genes involved in the subsequent stages of digit patterning.

11.9 The dorso-ventral axis of the limb is controlled by the ectoderm

The chick wing has a well-defined pattern along the dorso-ventral axis: large feathers are only present on the dorsal surface, and muscles and tendons have a complex dorso-ventral organization. Dorsal and ventral compartments of cell-lineage restriction (see Section 2.23 for an explanation of compartments) are specified in the trunk ectoderm of early chick embryos, including the ectoderm overlying the presumptive limb-bud regions, before any buds are visible; the apical ridge develops at the boundary between the compartments. Similar dorso-ventral compartments in the ectoderm have also been found in mouse limb buds.

The development of pattern along the dorso-ventral axis in the mesoderm has been studied by recombining ectoderm taken from left limb buds with mesoderm from right limb buds in such a way that the dorso-ventral axis of the ectoderm is reversed with respect to the underlying mesoderm, but the antero-posterior axis is unchanged. The right and left wing buds are removed from a chick embryo and, after treatment with cold trypsin, the ectoderm can be pulled off from the mesodermal core just like taking off a glove. Ectoderm from a left bud is then recombined with mesoderm from a right bud so that the dorso-ventral axis of the ectoderm runs the opposite way to that of the mesoderm—the equivalent of putting a left-hand glove on the right hand (Fig. 11.13). The reconstituted limb buds are grafted to the flank of a host embryo and allowed to develop. In general, the proximal regions of the limbs that develop have dorso-ventral polarity corresponding to that of the mesoderm, whereas the distal regions, in particular the 'hands', have a reversed dorso-ventral axis, with the pattern of muscles and tendons reversed and corresponding to the dorso-ventral polarity of the ectoderm. The ectoderm covering the sides of the limb bud can therefore specify dorso-ventral pattern in the limb.

Genes controlling the dorso-ventral axis in vertebrate limbs have been identified from mutations in mice, as a dorso-ventral pattern is easily visible in mouse limbs, where the ventral surface of the paw normally has no fur, whereas the dorsal surface does. The dorsal surfaces of the digits also have claws (nails in humans). Mutations that inactivate the gene *Wnt-7a*, for example, result in limbs in which many of the dorsal tissues adopt ventral fates to give a double ventral limb, the two halves being mirror images. The *Wnt-7a* gene is expressed in the dorsal ectoderm (Fig. 11.14), which suggests that the ventral pattern may be the ground state and is modified dorsally by the dorsal ectoderm, with the secreted Wnt-7a protein diffusing into, and playing a key role in, patterning the dorsal mesoderm. The ventral ectoderm is characterized by expression of the gene *Engrailed-1* (*En-1*), which encodes a homeodomain transcription factor (see Section 2.23). Mutations that destroy *En-1* function result in *Wnt-7a* being expressed ventrally as well as dorsally, giving a double dorsal limb. *En-1* expression is induced by BMP signaling in the ventral ectoderm.

One function of Wnt-7a is to induce expression of the gene encoding the LIM homeobox transcription factor Lmx1b in the mesenchyme underlying the dorsal

Fig. 11.13 Reversing the dorso-ventral axis of limb bud ectoderm. The only way a left-hand glove (representing the ectoderm from a left-hand limb bud) can fit on a right hand (the mesoderm of a right hand) is by turning it over so that the former ventral side (patterned) is now uppermost. The dorso-ventral polarity of the glove is now reversed in relation to the dorso-ventral axis of the hand. The antero-posterior relationship remains the same.

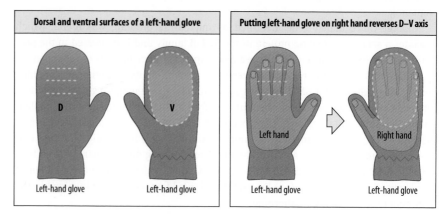

ectoderm (see Fig. 11.14). Clonal analysis of marked cells in mouse limb buds has recently revealed dorso-ventral compartments in the mesenchyme, with the progeny of dorsal cells being restricted to the dorsal half of the limb bud and the progeny of ventral cells to the ventral half. This is the first time that cell-lineage restrictions have been shown in mesenchymal tissue; all the other compartments discovered so far in both vertebrates and invertebrates are in epithelia—for example, the *Drosophila* epidermis (see Chapter 2) and the rhombomeres of the vertebrate hindbrain (see Chapter 5). The compartmentalization of the mesenchyme may serve to control gene expression; *Lmx1b*, for example, is expressed exactly in the dorsal compartment. In chick embryos, Lmx1b has been shown to specify a dorsal pattern in the mesoderm. Ectopic expression of Lmx1b in the ventral mesoderm of the chick limb results in the cells adopting a dorsal fate, resulting in a mirror-image dorsal limb. Loss-of-function mutations in human *LMX1B* cause an inherited disease called nail-patella syndrome, in which these dorsal structures (nails and kneecaps) are malformed or absent.

11.10 Development of the limb is integrated by interactions between signaling centers

It is crucial that patterning along all three limb axes is integrated so that the correct anatomy of the limb is generated. Patterning is integrated by interactions between the signals; Wnt-7a from dorsal ectoderm, FGFs from the apical ectodermal ridge, and Shh from the polarizing region. In many mice mutant for loss of *Wnt-7a* function, posterior digits are missing, suggesting that *Wnt-7a* is also required for normal antero-posterior patterning, and Shh expression in the polarizing region is also reduced in these mice. Similar results are observed in chick embryos when the dorsal ectoderm of the wing bud is removed.

A positive-feedback loop has been discovered between the polarizing region and the apical ectodermal ridge that links development along the proximo-distal and antero-posterior axes. Shh signaling maintains the level of *Fgf* gene expression in the ridge, while FGF signaling in turn maintains *Shh* gene expression in the polarizing region. This feedback loop also involves BMP, another of the proteins required to pattern the antero-posterior axis (see Section 11.7). A minimum level of BMP signaling is required to maintain outgrowth of the limb bud, but if the level rises too high the apical ridge regresses. The level of BMP signaling in the early limb bud is controlled by the BMP antagonist, Gremlin. *Gremlin* expression is first induced in the limb bud mesoderm by BMP-4 signaling and a negative-feedback loop is rapidly set up in which Gremlin protein antagonizes BMP activity, keeping it low. *Fgf* expression is consequently maintained in the apical ridge and, in turn, maintains *Shh* expression in the polarizing region in the earliest stages of limb-bud development. Shh signaling subsequently maintains *Gremlin* expression and thus low levels of BMP activity in the limb-bud mesoderm, reinforcing the positive-feedback loop between apical ridge FGF signaling and polarizing region Shh signaling. The epithelial–mesenchymal positive-feedback loop takes about 12 hours to complete, and is linked to the rapid negative-feedback loop between BMP-4 and Gremlin. This self-regulating system is an excellent example of the feedback control mechanisms that make development so robust and reliable (see Section 1.19).

11.11 Different interpretations of the same positional signals give different limbs

The signals controlling limb-bud outgrowth and patterning, such as FGF and Shh, are the same in chick wing and leg, but they are interpreted differently. A polarizing region from a wing bud, for example, will specify additional mirror-image digits if grafted into the anterior margin of a leg bud. However, these digits develop as toes, not wing digits, as the signal is being interpreted by leg cells. In a similar manner,

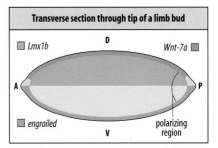

Fig. 11.14 The ectoderm controls dorso-ventral pattern in the developing limb. The gene for the secreted signal Wnt-7a is expressed in the dorsal ectoderm and the gene *engrailed* is expressed in the ventral ectoderm. Expression of the gene for the transcription factor Lmx1b is induced in the dorsal mesoderm compartment by Wnt-7a signaling and Lmx1b is involved in specifying dorsal structures.

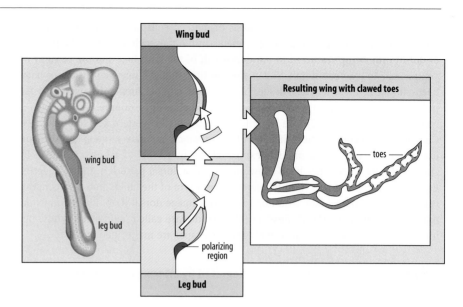

Fig. 11.15 Chick proximal leg bud cells grafted to a distal position in a wing bud acquire distal positional values. Proximal tissue from a leg bud that would normally develop into a thigh is grafted to the tip of a wing bud underneath the apical ectodermal ridge. It acquires more distal positional values and interprets these in terms of leg structures, forming clawed toes at the tip of the wing.

signals are conserved between different vertebrates; for example, a mouse apical ectodermal ridge grafted in place of a chick apical ectodermal ridge can provide the appropriate signal for early chick limb development. Signals are, however, interpreted according to the origin of the responding cells: when grafted into the anterior margin of a chick wing bud, a polarizing region from a mouse limb bud specifies additional wing structures, not mouse structures (see Section 11.6). Thus, the signals from the different polarizing regions are the same, as are the signals from the different apical ridges; the difference in the structures formed by the limbs is a consequence of how the signals are interpreted and depends on the genetic constitution and developmental history of the responding cells in the limb bud.

A further demonstration of the interchangeability of positional signals comes from grafting limb bud tissue to different positions along the proximo-distal axis of the bud. If tissue that would normally give rise to the thigh is grafted from the proximal part of an early chick leg bud to the tip of an early wing bud, it develops into toes with claws (Fig. 11.15). The tissue has acquired a more distal positional value after transplantation, but interprets this value according to its own developmental program, which is to make leg structures. One limb-specific protein is the transcription factor Pitx1. It is expressed in the leg bud only and its ectopic expression in the chick wing results in the development of leg-like distal structures, such as toes and claws.

11.12 Hox genes establish the polarizing region and also provide a code for limb patterning

Hox genes specify position along the antero-posterior axis of the vertebrate body (see Section 5.6). In the early limb bud, Hox genes establish the polarizing region and later on in development they provide positional values for the limb pattern. At least 23 different Hox genes are expressed during mouse and chick limb development. Attention has been largely focused on the genes of the *Hoxa* and *Hoxd* gene clusters, which are related to the *Drosophila Abdominal-B* gene (see Box 5E, p. 194) and genes in the 5′ regions of these clusters are expressed in both forelimbs and hindlimbs.

The expression of *Hoxa* and *Hoxd* genes during limb development is very dynamic—that is, it changes as the limb develops. In later limb buds, the pattern of expression changes in response to signaling by the polarizing region, in concert with signaling by the apical ectodermal ridge. But early on, *Hoxd9* and *Hoxd10* genes are expressed in the lateral plate mesoderm as it begins to thicken to become a limb bud. Expression of *Hoxd11*, *Hoxd12*, and *Hoxd13* genes is then initiated in rapid succession in progressively more restricted domains centered on the posterior distal region of both forelimbs and hindlimbs, and these domains then extend along the limb bud as it grows out, as shown for the chick wing bud in Fig. 11.16 (upper panel). Expression of *Hoxa9–13* is also initiated around the same time and gives rise to similar overlapping domains of expression along the proximo-distal axis, but without a posterior bias, and these domains also become more extended along the limb bud as it grows out. At the stage shown in Fig. 11.16 (lower panel), *Hoxa9* expression covers the complete region in which the *Hoxa* genes are expressed, and successive genes are expressed in more and more distal regions. At later stages of limb development, *Hoxa11* expression becomes restricted to the intermediate region of the limb and *Hoxd11* is also expressed there; *Hoxd13* and *Hoxa13* are expressed throughout the distal region of the limb, where the digits are beginning to form, while *Hoxd10–12* are also expressed distally except in the most anterior part of the bud.

Sophisticated genetic experiments in mice have succeeded in identifying *cis*-regulatory regions that drive *Hoxd* gene expression in the limb. An enhancer lying 3' (downstream) to the cluster drives the nested pattern of *Hoxd* gene expression in early limb buds and another enhancer, lying 100 kb 5' (upstream) of the cluster, drives *Hoxd* gene expression in the tissue that will form digits. The particularly strong activation of *Hoxd13* throughout the digital region (compared to the other *Hoxd* genes) may be because it is the first gene that this distant control region comes into contact with.

The early phase of *Hox* gene expression in the limb is involved in establishing the polarizing region. Genetic deletion of all *Hoxa* and *Hoxd* genes in the mouse forelimb bud results in the absence of *Shh* expression and severe limb truncations. Conversely, genetic manipulations leading to inversions of the *Hoxd* cluster or deletions of large parts of the downstream control region result in *Hoxd* genes in the 5' part of the cluster being expressed in the patterns of 3' *Hoxd* genes, so that *Hoxd13* is expressed throughout the early bud instead of being restricted to the posterior. The unrestricted expression of *Hoxd13* results in *Shh* expression both anteriorly and posteriorly, and the development of duplicated limb patterns. *Hoxb8*, one of the few Hox genes outside the *Hoxa* and *Hoxd* clusters that is expressed in the forelimb, also seems to be involved in establishing *Shh* expression.

Hox gene expression is maintained and elaborated as the limb bud develops in response to combined Shh and FGF signaling. These later phases of Hox gene expression are involved in recording positional information in the limb and, in some cases, are also involved in the growth of the cartilaginous elements once they have formed. If the Hox genes are involved in recording positional information, then experimental manipulations that lead to changes in the pattern of the limb skeletal elements should cause a corresponding change in domains of Hox expression. When a polarizing region is grafted to the anterior margin of a wing bud, there is indeed a change to a mirror-image pattern of the expression of Hoxd genes (Fig. 11.17). The change occurs within 24 hours of grafting, which is about the time required for the polarizing region to exert its effect (see Section 11.6).

As outlined above, the proximo-distal sequence of expression of *Hoxa* and *Hoxd* genes eventually comes to correspond generally with the three main proximo-distal regions of the limb. *Hoxa9* and *Hoxd9* are expressed in the upper part of the forelimb, where the humerus forms. *Hoxa11* and *Hoxd11* are expressed in the lower limb, where the radius and ulna (or tibia and fibula) develop, and *Hoxa13* and *Hoxd13* are expressed in the wrist (ankle) and digit-forming regions. There are some differences in the pattern of expression of the Hox paralogous subgroups 9 to 13 in the forelimb and hindlimb. Hox9 paralogs are not expressed in the hindlimb, but there are only relatively small variations in the pattern of expression of the other groups.

Gene function can only be inferred from expression patterns, and mutational experiments are needed to prove function. We need to answer the question of whether

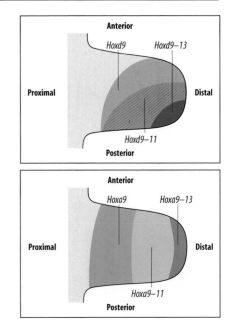

Fig. 11.16 Pattern of *Hox* gene expression in the chick wing bud. The *Hoxd* genes (upper panel) are expressed in a nested pattern centered on the posterior tip of the wing bud, *Hoxd13* being expressed most posteriorly and distally. The *Hoxa* genes (lower panel) are similarly expressed in a nested pattern along the proximo-distal axis, *Hoxa13* being expressed most distally.

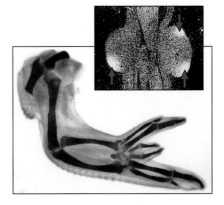

Fig. 11.17 Change in expression of Hoxd genes following a graft of the polarizing region. A polarizing region grafted to the anterior margin of a limb bud results in mirror-image expression of the *Hoxd13* genes and a mirror-image duplication of the digits (main panel). The inset shows the expression of *Hoxd13* (arrows) at the posterior tip in a normal limb bud (inset, left). In the operated limb (inset, right) *Hoxd13* is expressed at both the anterior and posterior tips.

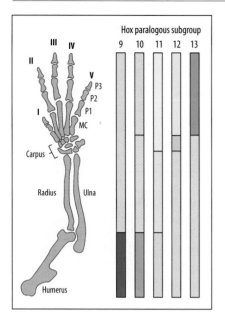

Fig. 11.18 Functional domains of Hox gene expression in the patterning of the mouse forelimb. Colored regions denote regions of Hox gene influence on patterning as determined from gene knock-out experiments. Hox9 (red) and Hox10 (orange) paralogs function together to pattern the upper forelimb (humerus), Hox10 paralogs also affect the middle forelimb (radius and ulna), as represented by the lighter orange shading. Hox11 paralogs (yellow) are mostly involved in patterning this middle region, with some influence on pattern in the wrist and 'hand' region (lighter yellow shading). Hox12 paralogs (green) predominantly pattern the wrist, while Hox13 paralogs (purple) predominantly pattern the hand. MC = metacarpal, P = phalanx.
Adapted from Wellik, D.M. and Capecchi, M.R.: 2003.

altering Hox gene expression in the limb leads to homeotic changes similar to those seen in vertebrae after Hox gene knock-out (see Section 5.7). The results of Hox gene knock-outs in mice are quite difficult to interpret in relation to the limb, but they do show that Hox genes clearly influence the patterning of different regions of the mouse limb along the proximo-distal axis (Fig. 11.18). This conclusion is based on the phenotypes of mice mutant for genes in the same paralogous subgroup. For example, paralogous subgroup Hox10 controls the development of proximal elements, whereas Hox11 patterns the radius and ulna in the forelimb (tibia and fibula in the hindlimb). Inactivation of genes of the Hox13 subgroup disrupts digit development, as one might expect from the expression of *Hoxd13* and *Hoxa13* in the distal region of the limb. Because of the dynamic nature of Hox gene expression as the limbs develop, it is difficult to interpret these results in terms of either the progress zone model or the two-signal model for proximo-distal patterning of the limb (see Section 11.5). The fact that the domain of *Hoxd13* at the tip of the limb is maintained by FGF signaling as the limb grows out could be seen as consistent with the two-signal model, but it is not clear that the cells that express *Hoxd13* at early stages have already been specified to form digits.

Which target genes the Hox proteins subsequently act on to produce the pattern along the proximo-distal axis is still largely unknown, but expression of the ephrin receptors EphA7 and EphA3 has been shown to be regulated by *Hoxd13* and *Hoxa13*. Ephrins and their receptors are involved in various cell-biological processes, such as the formation of regional boundaries (see Section 5.11), cell sorting, and guidance of cell migration (see Chapter 12).

Mutations in Hox genes are known to underlie human conditions in which limbs are abnormal. Mutations in *Hoxd13* lead to fusion of digits and some types of polydactyly, while mutations in *Hoxa13* lead to hand-foot-genital syndrome, but the precise phenotypes seen in patients are difficult to explain given our current knowledge.

11.13 Self-organization may be involved in the development of the limb bud

Signaling from the polarizing region might not be the only means by which a series of cartilaginous elements can develop along the antero-posterior axis of the limb, as it seems that the limb bud has some capacity for self-organization. Cells from a disaggregated limb bud can be reaggregated and develop digits, even in the absence of a polarizing region. Mesenchymal cells disaggregated from early chick limb buds are thoroughly mixed to disperse the polarizing region, or the mesenchymal cells are taken from just the anterior half of a limb bud, which lacks a polarizing region. The disaggregated cells are placed in an ectodermal jacket (prepared as described in Section 11.9), and grafted to a site where they can acquire a blood supply, but which is 'neutral' in terms of providing polarizing signals, such as the dorsal surface of an older limb. Limb-like structures develop from the grafted reaggregated limb bud cells, even though they have developed in the absence of a polarizing region. Several long cartilaginous elements may form in the more proximal regions of these abnormal limbs, although none of the proximal elements can be easily identified with normal structures. More distally, however, reaggregated hindlimb bud cells develop

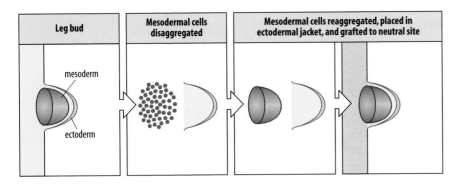

Leg bud	Mesodermal cells disaggregated	Mesodermal cells reaggregated, placed in ectodermal jacket, and grafted to neutral site

mesoderm

ectoderm

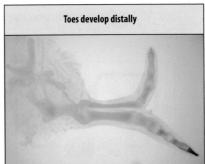

Toes develop distally

identifiable toes (Fig. 11.19). The fact that well-formed cartilaginous elements can develop at all in the absence of a discrete polarizing region shows that the bud has considerable capacity for self-organization.

These results suggest that the limb bud might be able to generate a basic **prepattern** consisting of equivalent cartilaginous elements. A prepattern is a basic organization that is generated automatically in the embryo and can be identified in advance of the later development of a similar pattern of structures. The pattern that eventually develops does not have to follow the prepattern exactly but may be a modified form of the prepattern. The initially equivalent cartilaginous elements of the limb prepattern could then be given their identities and further refined in response to positional information involving signals such as Shh and the activation of Hox genes and their downstream targets (see Sections 11.7 and 11.12). The mechanism for generating the prepattern could be based on a **reaction–diffusion mechanism** (Box 11D, p. 432). In the wing, for example, a reaction–diffusion or related mechanism could result in a single peak in some morphogen forming in the proximal region of the limb, which would specify a prepattern for the humerus. More distally, alterations in the reaction–diffusion conditions, due to changes in proximo-distal positional information, could give rise to three peaks of the morphogen, giving the cartilaginous elements of the three wing digits. These prepatterns could then be modified by the signals that specify antero-posterior and dorso-ventral positional information.

In this reaction–diffusion model, polydactyly could simply result from a chance widening of the limb bud. If a reaction–diffusion mechanism generates a periodic pattern of cartilage elements across the limb as the digits are forming, merely widening the limb bud by some small developmental accident would enable a further digit to develop.

Fig. 11.19 Reaggregated limb bud cells form digits in the absence of a localized polarizing region. Mesodermal cells from a chick leg bud are disaggregated, mixed to disperse all the cells, including the polarizing region, and then reaggregated, placed in an ectodermal jacket, and grafted to a neutral site. Well-formed toes develop distally.

11.14 Limb muscle is patterned by the connective tissue

Limb muscle cells have a different origin from that of the limb connective tissue cells, including muscle-associated connective tissue cells. If quail somites are grafted into a chick embryo at a site opposite where the wing bud will develop, the wing that forms will have muscle cells of quail origin, but all its other cells will be of chick origin. The cells that give rise to limb muscles migrate into the limb bud from the somites as myoblasts at a very early stage (see Section 5.9). After migration, the myoblasts multiply and initially form dorsal and ventral blocks of presumptive muscle (Fig. 11.20). These blocks undergo a series of divisions to give the individual muscles. The initial migration of myoblasts into the limb is restricted so that they enter either the dorsal or the ventral region; cells do not cross the dorsal–ventral compartment boundary, which is characterized by the dorsal expression of the transcription factor Lmx1b (see Fig. 11.14). Presumptive muscle cells, at least initially, do not acquire positional values in the same way as do cartilage and connective tissue cells, and are all equivalent. However, *Hoxa11* is expressed in the myoblasts when they enter the limb and is

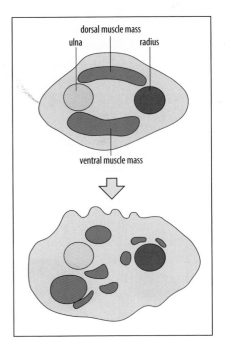

dorsal muscle mass

ulna   radius

ventral muscle mass

Fig. 11.20 Development of muscle in the chick limb. A cross-section through the chick limb in the region of the radius and ulna shortly after cartilage formation shows presumptive muscle cells present as two blocks—the dorsal muscle mass and the ventral muscle mass (upper panel). These blocks then undergo a series of divisions to give rise to individual muscles (lower panel).

Box 11D Reaction–diffusion mechanisms

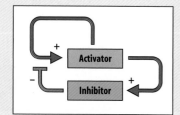

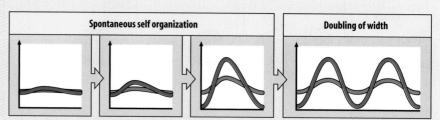

Spontaneous self organization

Doubling of width

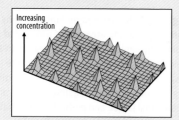

Increasing concentration

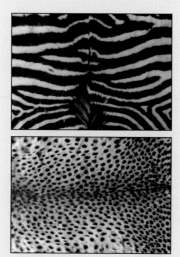

There are some self-organizing chemical systems that spontaneously generate spatial patterns of concentration of some of their molecular components. The initial distribution of the molecules is uniform, but over time the system forms wave-like patterns (see figure). The essential feature of such a self-organizing system is the presence of two or more types of diffusible molecules that interact with one another. For that reason, such a system is known as a **reaction–diffusion system**. For example, if the system contains an activator molecule that stimulates both its own synthesis and that of an inhibitor molecule, which in turn inhibits synthesis of the activator, a type of lateral inhibition will occur so that synthesis of activator is confined to one region (see upper diagram, top left).

Under appropriate conditions, which are determined by the reaction rates and diffusion constants of the components, a closed system of a certain size can spontaneously develop a spatial pattern of activator with a single concentration peak. If the size of the system is increased, two peaks will eventually develop, and so on. Such a mechanism could thus in theory generate periodic patterns such as the arrangement of digits or the sepals or petals of flowers. When the chemicals can diffuse in two dimensions, the system can give rise to a number of peaks that are somewhat irregularly spaced (see lower diagram, top left).

Such a system may underlie some of the patterns of pigmentation that are common throughout the animal kingdom, such as the patterns of spots and stripes seen in the zebra and cheetah (see photos, left). How these patterns are generated is not yet known, but one possibility is a reaction–diffusion mechanism. Assuming that pigment is synthesized in response to some activator, and that synthesis only occurs at high activator concentrations, some animal color patterns can be mimicked by a reaction–diffusion system in computer simulations.

A characteristic feature of reaction–diffusion patterns is that new intermediate peaks appear as the system grows in size. The angelfish *Pomocanthus semicirculatus* provides a remarkable example of striping that could be generated by a reaction-diffusion mechanism (see below). Juvenile *P. semicirculatus*, less than 2 cm long, have only three dorso-ventral stripes. As the fish grow, the intervals between the stripes increase until the fish is around 4 cm long. New stripes then appear between the original stripes and the stripe intervals revert to those present at the 2 cm stage. As the fish

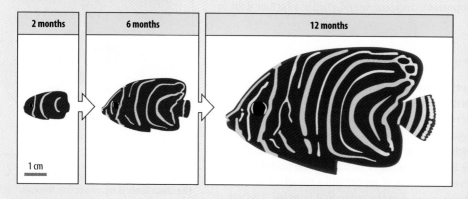

2 months   6 months   12 months

1 cm

grows larger, the process is repeated. This type of dynamic patterning is what would be expected of a reaction–diffusion mechanism. Computer simulations of reaction–diffusion mechanisms can also generate the patterns seen on a wide variety of mollusc shells. Nevertheless, there is as yet no direct evidence for a reaction–diffusion system patterning any developing organism.

Top figure after Meinhardt, H., et al.: 1974.

probably induced by the surrounding mesenchyme. Cells in the posterior region of the dorsal and ventral masses express *Hoxa13*. In both cases, the pattern of Hox gene expression in the myoblasts is different from that of the surrounding mesenchyme.

Evidence for the early equivalence of myoblasts comes from experiments where somites are grafted from the future neck region of the early chick embryo to replace the normal wing somites. The myoblasts that enter the wing therefore come from neck somites, but a normal pattern of limb muscles still develops. This shows that the muscle pattern is determined by the connective tissue into which the myoblasts migrate, rather than by the myoblasts themselves. One way this could be achieved is if the prospective muscle-associated connective tissue had surface or adhesive properties that the myoblasts recognized and were attracted to. If the pattern of adhesiveness in the connective tissue changed over time, this could result in associated changes in the migration of presumptive muscle cells, and this could account for the splitting of the muscle masses.

The myoblasts proliferating in the limb express Pax3. When *Pax3* is downregulated, cell proliferation stops and the cells differentiate (see Section 10.10). Signals from the overlying ectoderm, including BMP-4, prevent premature differentiation. The development of tendons is marked by the expression of the transcription factor Scleraxis (see Section 5.9).

11.15 The initial development of cartilage, muscles, and tendons is autonomous

As we have seen, a graft of the polarizing region causes the development of a mirror-image pattern of all the elements of the limb, not only the skeletal elements but also muscles and tendons. This indicates that the patterns of cartilage, tendon, and muscle in the limb are all specified by the same set of signals and there is little or no communication between them—that is, the development of each element is autonomous. If, for example, just the tip of an early chick wing bud is removed and grafted to the flank of a host embryo, it initially develops into normal distal structures with a wrist and three digits. The long tendon that normally runs along the ventral surface of digit 3 starts to develop, even though both its proximal end and the muscle to which it attaches are absent. The tendon does not continue to develop, however, because it does not make the necessary connection to a muscle, and so is not put under tension. The mechanism whereby the correct connections between tendons, muscles, and cartilage are established has still to be determined. It is clear, however, that there is little or no specificity involved in making such connections; if the tip of a developing limb is inverted dorso-ventrally, dorsal and ventral tendons can join up with inappropriate muscles and tendons. They simply make connections with those muscles and tendons nearest to their free ends—they are promiscuous.

11.16 Joint formation involves secreted signals and mechanical stimuli

The first event in joint formation is the formation of an 'interzone' at the site of the prospective joint, in which the cartilage-producing cells become fibroblast-like and cease producing the cartilage matrix, thus separating one cartilage element from another. At this time, Wnt-14 and the BMP-related protein GDF-5 are expressed in the presumptive joint region. Ectopic expression of Wnt-14 can induce the early steps in joint formation, and mice lacking GDF-5 have some of their joints missing. A little later, the musculature begins to function and muscle contraction occurs. At this point, high levels of hyaluronan, a component of the proteoglycans that lubricate joint surfaces, are made by the cells of the interzone; hyaluronan is secreted in vesicles that coalesce to form the joint cavity, which separates the articulating surfaces of the joint from each other. Synthesis of hyaluronan by the cells lining these surfaces depends on the mechanical stimulation that results from muscle contraction, and in the absence

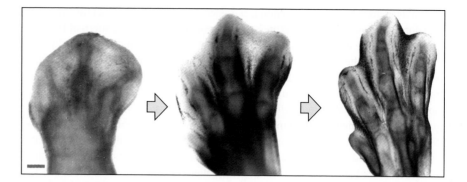

Fig. 11.21 Cell death during leg development in the chick. Programmed cell death in the interdigital region results in separation of the toes. Scale bar = 1 μm.

Photographs courtesy of V. Garcia-Martinez, from Garcia-Martinez, V., et al.: 1993.

of this activity, the joints fuse. Cells isolated from the interzone region and grown in culture respond to mechanical stimulation by increased secretion of hyaluronan when the substratum to which they are attached is stretched.

11.17 Separation of the digits is the result of programmed cell death

Programmed cell death by apoptosis plays a key role in molding the form of chick and mammalian limbs, especially the digits. The region where the digits form is initially plate-like, as the limb is flattened along the dorso-ventral axis. The cartilaginous elements of the digits develop from the mesenchyme at specific positions within this plate. Separation of the digits then depends on the death of the cells between these cartilaginous elements (Fig. 11.21). BMPs may be involved in signaling this cell death: if the functioning of BMP receptors is blocked in the developing chick leg, cell death does not occur and the digits are webbed.

Programmed cell death is a normal part of patterning and cell differentiation (see Chapter 10). The fact that ducks and other waterfowl have webbed feet, whereas chickens do not, is simply the result of less cell death between the digits of waterfowl. If chick-limb mesoderm is replaced with duck-limb mesoderm, cell death between the digits is reduced and the chick limb develops 'webbed' feet (Fig. 11.22). It is the mesoderm that determines the patterns of cell death both within the mesoderm and the overlying ectoderm. In amphibians, digit separation is not due to cell death, but results from the digits growing more than the interdigital region.

Programmed cell death also occurs in other regions of the developing limb, such as the anterior margin of the limb bud, between the radius and ulna, and in the wing polarizing region at a later stage. Indeed, it was the investigation of cell death in this region of a chick wing bud by transplanting it to the anterior margin of another chick wing bud that led to the discovery that it acts as a polarizing region. One suggestion for the role of cell death in the polarizing region is that it could control the number of cells expressing Shh.

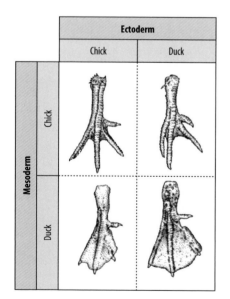

Ectoderm	
Chick	Duck

(Mesoderm: Chick / Duck rows)

Fig. 11.22 The mesoderm determines the pattern of cell death. In the developing webbed feet of ducks and other water birds, there is less cell death between the digits than in birds without webbed feet. When the mesoderm and ectoderm of embryonic chick and duck limb buds are exchanged, webbing develops only when duck mesoderm is present.

Summary

Positioning and patterning of the vertebrate limb is largely carried out by intercellular interactions that provide the cells with positional information. The position of the limb buds along the antero-posterior axis of the body is probably related to Hox gene expression. There are two key signaling regions within the limb bud. One is the apical ectodermal ridge, which is required for limb bud outgrowth; the second is the polarizing region at the posterior margin of the limb, which specifies pattern along the antero-posterior axis. The signals from the apical ridge are fibroblast growth factors (FGFs) and are essential for limb outgrowth. Shh protein is produced by the polarizing region and provides a graded positional signal across the antero-posterior axis of the limb bud that is pivotal for determining digit number and pattern. Pattern across the dorso-ventral axis is specified by the ectoderm. Hox genes are expressed in a well-defined spatio-temporal pattern within the limb bud and may provide the molecular basis of positional identity. The limb muscle cells are not generated within the limb but migrate in from the somites and are patterned by the limb connective tissue. Patterning of the cartilaginous elements may involve a self-organizing mechanism. In birds and mammals, separation of the digits is achieved by programmed cell death.

Summary: vertebrate limb development

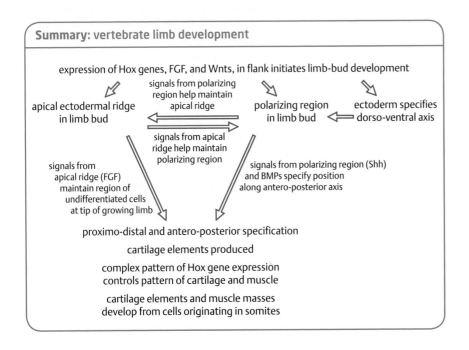

expression of Hox genes, FGF, and Wnts, in flank initiates limb-bud development

signals from polarizing region help maintain apical ridge

apical ectodermal ridge in limb bud

polarizing region in limb bud

ectoderm specifies dorso-ventral axis

signals from apical ridge help maintain polarizing region

signals from apical ridge (FGF) maintain region of undifferentiated cells at tip of growing limb

signals from polarizing region (Shh) and BMPs specify position along antero-posterior axis

proximo-distal and antero-posterior specification

cartilage elements produced

complex pattern of Hox gene expression controls pattern of cartilage and muscle

cartilage elements and muscle masses develop from cells originating in somites

Insect wings and legs

The adult organs and appendages of *Drosophila*, such as the wings, legs, eyes, and antennae, develop from imaginal discs, which are excellent experimental systems for analyzing pattern formation. The discs invaginate from the embryonic ectoderm as simple pouches of epithelium during embryonic development and remain as such until metamorphosis (see Fig. 2.3). Although all imaginal discs superficially look rather similar, they develop according to the segment in which they are located. The specification and basic patterning of the imaginal discs in the thoracic segments as wing or leg occur within the embryonic epithelium, at the time when the segments are being patterned and given their identity. The wing and leg discs are initially specified as clusters of 20–40 cells, and during larval development they grow about 1000-fold. The adult *Drosophila* compound eyes develop from imaginal discs at the anterior end of the embryo. A feature of imaginal discs is that if cells are removed or destroyed, the disc will regrow to the correct size and develop normally. The mechanism underlying this is not yet known: one proposal invokes the gradient mechanism for determining organ size discussed in Box 13A (p. 514).

Although insect wings and legs differ so greatly in appearance, they are partially homologous structures, and the strategy of their patterning is similar. Moreover, the mechanisms by which they are patterned, and even the genes involved, show some amazing similarities to the patterning of the vertebrate limb. We will start by discussing the patterning of the wing disc and its development into the adult wing and then compare the patterning of the leg disc. The development of the *Drosophila* eye is discussed later in the chapter.

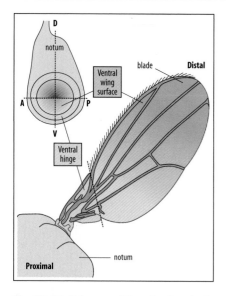

Fig. 11.23 Fate map of the wing imaginal disc of *Drosophila*. The wing disc before metamorphosis is an ovoid epithelial sheet bisected by the boundary between the anterior (A) and the posterior (P) compartments. There is also a compartment boundary between the future ventral (V) and dorsal (D) surfaces of the wing. At metamorphosis, the ventral surface will fold under the dorsal surface (see Fig. 11.24). The disc also contains part of the dorsal region of thoracic segment 2 in the region of wing attachment.

After French, V., et al.: 1994.

11.18 Positional signals from compartment boundaries pattern the wing imaginal disc

Imaginal discs are essentially epithelial sheets, which, in the case of wing and leg discs, will give rise to epidermis in the adult. The epidermis of body segments, and of the wings and legs that derive from them, is divided into anterior and posterior compartments—regions of cell lineage restriction (see Section 2.23). These correspond to the compartment boundaries in the segments in which they develop. In the wing disc epithelium, a second compartment boundary between the dorsal and ventral regions develops during the second larval instar (Fig. 11.23). The proximal parts of the wing and leg discs give rise to parts of the thoracic body wall.

The adult wing is a largely epidermal structure in which two epidermal layers—the dorsal and ventral surfaces—are close together. At metamorphosis, by which time patterning of the imaginal discs is largely complete, wing and leg discs undergo a series of profound anatomical changes. Essentially, the invaginated epithelial pouch is turned inside out, as its cells differentiate and change shape. In the case of the wing, the epithelial pouch extends and folds so that one surface comes to lie beneath the other to form the double-layered wing structure (Fig. 11.24). The zinc-finger transcription factors Elbows and No-ocelli act in both the wing and leg imaginal discs to repress genes specifying the body wall, and thus permit the appendages to develop.

In the patterning of the wing disc, signaling regions are set up along the compartment boundaries (Fig. 11.25). We shall look first at the antero-posterior patterning produced from the signaling center at the anterior–posterior compartment boundary. Cells at this boundary form a signaling region that specifies pattern along the antero-posterior axis of the wing. A cascade of events sets up the antero-posterior signaling center. It begins with the expression of the *engrailed* gene in the posterior compartment of the disc (see Fig. 11.25), which reflects the pattern of gene expression in the embryonic parasegment from which the disc derives. We shall return to the question of what gives wing and leg discs their segmental identity later in this chapter.

As we saw in relation to the setting up of compartments in the *Drosophila* embryo, the cells that express *engrailed* also express the gene *hedgehog* (see Section 2.24). The maintenance of compartment boundaries in the wing depends partly on communication between compartments. At the antero-posterior compartment boundary, the secreted Hedgehog protein acts as a short-range morphogen over about 10 cell diameters and induces adjacent cells in the anterior compartment to express the *decapentaplegic (dpp)* gene (see Fig. 11.25). As noted in Box 11A (p. 421), Hedgehog protein may not move by simple diffusion or endocytosis. The Hedgehog signal acts on the anterior cells through a receptor complex involving the proteins Patched and Smoothened (see Fig. 2.40).

Decapentaplegic protein is secreted at the antero-posterior compartment boundary (see Fig. 11.25) and is probably the positional signal for patterning both the anterior and posterior compartments along the antero-posterior axis. Dpp appears to form a

Fig. 11.24 Schematic representation of the development of the wing blade from the imaginal disc. Initially, the dorsal and ventral surfaces are in the same plane. At metamorphosis, the invaginated pouch turns inside out and extends, so the dorsal and ventral surfaces of the wing come together.

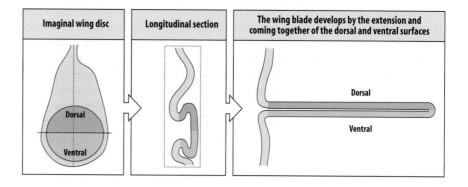

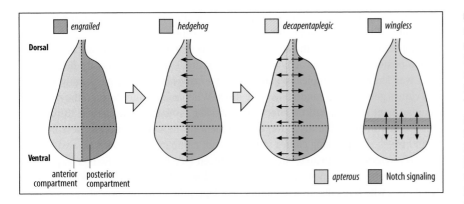

Fig. 11.25 Establishment of signaling regions in the wing disc at the compartment boundaries. The disc is divided into anterior and posterior compartments and, later in development, into dorsal and ventral compartments. The gene *engrailed* is expressed in the posterior compartment, where the cells also express the gene *hedgehog* and secrete the Hedgehog protein. Where Hedgehog protein interacts with anterior compartment cells, the gene *decapentaplegic* is activated, and Decapentaplegic protein is secreted into both compartments, as indicated by the black arrows. At the dorso-ventral compartment boundary, the *wingless* gene is activated and the signaling molecule is the protein Wingless.

concentration gradient across the wing disc that provides a long-range signal controlling the localized expression of the genes for the transcription factors Spalt (Sal), Omb and Brinker in the wing disc (Fig. 11.26), and they provide a further level of patterning in the wing. Decapentaplegic acts as a morphogen, specifying the expression of *spalt* and *omb* when particular threshold concentrations are reached (Fig. 11.27); low levels induce *omb*, whereas higher levels are required to induce *spalt*. As noted earlier, *Omb* and *Spalt* are related to the *Tbx* and *Sall* genes, respectively, which are involved in vertebrate limb patterning (see Section 11.8).

Decapentaplegic regulates its target genes by binding to the receptor protein, Thick veins, and activating an intracellular signaling pathway that results in changes in gene expression. One of the key genes regulated by Dpp in regard to wing patterning is *brinker*, which encodes a transcription factor. Dpp signaling downregulates *brinker* expression, thus forming a gradient of intranuclear Brinker protein that is the inverse of the Dpp gradient (see Fig. 11.26). Brinker represses some of the target genes of Dpp, such as *omb*, and so repression of *brinker* by Dpp is required for Dpp to be able to activate some of its target genes. These genes are then expressed in patterns that reflect their sensitivity to activation by Dpp and repression by Brinker (see Fig. 11.26). In addition, *brinker* reduces cell growth, while Dpp promotes it.

The evidence that Dpp is a morphogen in the wing disc comes from several types of experiments. First, clones of cells unable to respond to a Dpp signal do not express *spalt* or *omb*. Second, ectopic expression of the *dpp* gene in a region that does not normally express *spalt* and *omb* leads to the localized activation of these genes around the *dpp*-expressing cells. Formation of the long-range Dpp gradient in the wing disc is a complex process and is not yet fully understood. It appears to involve receptor-mediated endocytosis, as blocking the action of dynamin, an intracellular protein required for the internalization of endocytic vesicles, blocks gradient formation. The

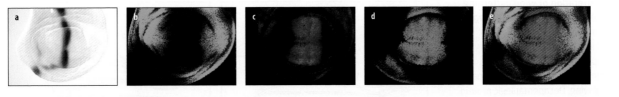

Fig. 11.26 Domains of gene expression initiated by Dpp signaling pattern the *Drosophila* wing disc. Panel (a) shows the expression of *decapentaplegic* (blue) at the antero-posterior compartment border of the wing blade. Dpp signaling represses expression of the gene *brinker* (green, panel (b)), which is therefore expressed only in two regions on either side of the zone of *dpp* expression. In turn *brinker* represses the expression of *sal* (red, panel (c)) and *omb* (blue, panel (d)), which are

therefore expressed in a wide strip over the antero-posterior border, where brinker is not expressed. Panel (e) shows a composite overlay of (b), (c), and (d).

(a) courtesy of K. Basler, from Nellen, D., et al.: 1996. (b–e) from Moser, M., Campbell, G.: 2005.

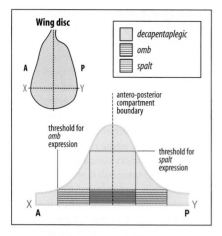

Fig. 11.27 A model for patterning the antero-posterior axis of the wing disc. The decapentaplegic protein is produced at the compartment boundary. The schematic cross-section of the wing disc (X-Y) along the antero-posterior axis shows how the presumed gradient in decapentaplegic in both anterior and posterior compartments activates the genes *spalt* and *omb* at their threshold concentrations.

glypican Dally, present on the cell surface, is also involved in both shaping the Dpp gradient and influencing signaling via the Dpp receptor Thick veins.

The pattern of veins on the adult wing blade is one of the ultimate outcomes of the antero-posterior patterning activity of Hedgehog and Dpp in the wing disc (Fig. 11.28). This was shown by ectopically expressing the *hedgehog* gene in genetically marked cell clones generated at random in wing discs. When such clones form in the posterior compartment they have little effect (as *hedgehog* is normally expressed throughout this compartment in the disc) and development of the wing is more or less normal. With *hedgehog*-expressing clones in the anterior compartment, however, a wing with a mirror-symmetric repeated pattern is produced (see Fig. 11.28). Ectopic expression of *hedgehog* has set up new sites of *dpp* expression in the anterior compartment of the wing disc, which results in new gradients of Dpp protein being formed.

There is no single or simple process by which the morphogens Hedgehog and Dpp specify the vein pattern: rather, the position of each vein is likely to be specified by a unique combination of factors, including the level of Dpp and Hedgehog signaling, which compartment the cells are in, and the induction or repression of particular transcription factors. For example, the lateral vein L5 in the posterior compartment develops at the border between *omb* and *brinker* expression domains. It seems, therefore, that Dpp and Hedgehog do not directly pattern the veins, but set up a series of cell–cell interactions that specify positional information across the wing disc, perhaps even to the level of single cells. It is striking how precise the pattern of veins is if one compares the wings on both sides of a fly.

11.19 A signaling center at the boundary between dorsal and ventral compartments patterns the *Drosophila* wing along the dorso-ventral axis

The wing disc epithelium is also subdivided into dorsal and ventral compartments (see Fig. 11.25), which correspond to the dorsal and ventral surfaces of the adult wing (see Fig. 11.23). These compartments develop after the wing disc is formed—during the second larval instar. The dorsal and ventral compartments were originally identified by cell-lineage studies, but they are also defined by expression of the homeotic selector gene *apterous*, which is confined to the dorsal compartment, and which defines the dorsal state. The Apterous protein is a structural relative of Lmx1b from the mouse, which specifies a dorsal pattern in the mesoderm of the mouse limb bud (see Section 11.9).

Like the anterior–posterior compartment boundary, the boundary between dorsal and ventral compartments acts as an organizing center, and Wingless is the signaling molecule at this boundary (see Fig. 11.25). Wingless is one member of the large Wnt family of secreted signal proteins, and we have already seen it in action in *Drosophila* embryonic development (see Section 2.24). Reduction of *wingless* function results

Fig. 11.28 Alteration of wing patterning due to ectopic expression of *hedgehog* and *decapentaplegic*. When *hedgehog* is expressed in a clone of cells in the anterior compartment, a new source of Decapentaplegic protein is set up. Left panel: in the normal wing, Decapentaplegic protein is made at the compartment boundary, where Hedgehog is expressed. Right panel: the ectopic expression of *hedgehog* in the anterior compartment results in a new source of Decapentaplegic protein and a new wing pattern develops in relation to this new source.

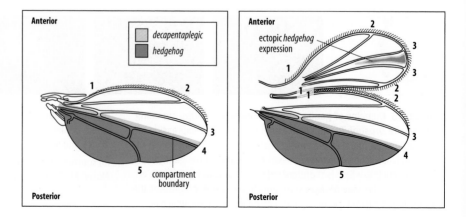

in loss of wing structures; the gene is named after this adult phenotype. Apterous induces the expression of Serrate, a Notch ligand, in dorsal cells, and restricts the expression of the other Notch ligand, Delta, to ventral cells. Notch signaling at the compartment boundary induces expression of Wingless. Wingless in turn induces expression Delta and Serrate at the boundary, thus maintaining its own expression. At the early stages of wing-disc development, Apterous also induces expression of the glycosyltransferase Fringe (see Box 5C, p. 190) in dorsal cells. Fringe promotes Notch signaling in response to Delta, but inhibits it in response to Serrate, which helps to localize Notch signaling, and thus Wingless expression, to the compartment boundary (see Fig. 11.25). Apterous is also thought to induce the expression of cell-surface proteins that, in conjunction with Fringe-modulated Notch signaling, maintain the boundary of lineage restriction once Apterous itself is no longer expressed. Wingless serves a role in the dorso-ventral patterning system analogous to Dpp in the antero-posterior patterning system, and activates expression of various genes at specific thresholds, which further pattern the wing and are responsible for its outgrowth. One of these genes is *vestigial*, which encodes a transcriptional co-activator essential for wing development and outgrowth and which is induced in a broad stripe centered on the dorso-ventral boundary (Fig. 11.29).

How the proximo-distal axis of the wing blade is patterned remains unclear in *Drosophila*, but there is evidence for both short- and long-range activation of genes in the wing blade by Wingless. We shall see later in the chapter how the colorful pattern on butterfly wings is a manifestation of proximo-distal patterning.

While both Dpp and Wingless may act as morphogens, whose graded concentrations provide cells with positional information, the formation of these gradients is not due to simple diffusion. The gradient of activity can also be modified in other ways, as both Dpp and Wingless can regulate the expression of their receptors. Dpp, for example, represses the expression of one of its receptors, Thick veins, so that fewer receptors are present where Dpp levels are high. Wing disc cells also have thin actin-based extensions called cytonemes that project to the antero-posterior signaling boundary. These projections may also be involved in long-range cell–cell communication.

As well as the general asymmetry of wing structure, which is due to patterning along the various axes, the individual cells in the adult *Drosophila* wing epidermis become polarized in a proximo-distal direction; this is reflected by the pattern of hairs (trichomes) on the wing surface. Each of the approximately 30,000 cells in the wing epidermis produces a single hair, which points distally (see Fig. 11.28, which shows only those hairs that would be visible on the edge of the wing). This is an example of planar cell polarity and what is known of the underlying mechanism is outlined in Box 2F, p. 78.

11.20 The leg disc is patterned in a similar manner to the wing disc, except for the proximo-distal axis

Insect legs are essentially jointed tubes of epidermis and thus have a quite different structure from those of vertebrates. The epidermal cells secrete the hard outer cuticle that forms the exoskeleton. Internally, there are muscles, nerves, and connective tissues. The easiest way of relating the leg imaginal disc to the adult leg is to think of it as a collapsed cone. Looking down on the disc, one can imagine it as a series of concentric rings, each of which will form a proximo-distal segment of leg. A change in the shape of the epithelial cells is responsible for the outward extension of the leg at metamorphosis. The process is rather like turning a sock inside out, with the result that the center of the disc ends up as the distal end, or tip, of the leg (Fig. 11.30). The outermost ring gives rise to the base of the leg, which is attached to the body, and the rings nearer the center give rise to the more distal structures (Fig. 11.31).

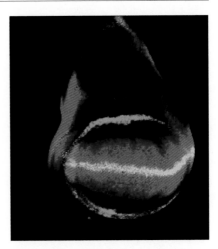

Fig. 11.29 Expression of *wingless* and *vestigial* in the wing disc. *wingless* expression is shown in green, while *vestigial* is red. Both *wingless* and *vestigial* are expressed at the dorso-ventral boundary, as indicated by the yellow stripe.

Photograph courtesy of K. Basler, from Zecca, M., et al.: 1996.

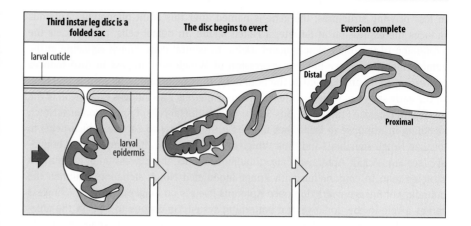

Fig. 11.30 *Drosophila* leg disc extension at metamorphosis. The disc epithelium, which is an extension of the epithelium of the body wall, is initially folded internally. At metamorphosis it extends outward, rather like turning a sock inside out. The red arrow in the first panel is the viewpoint that produces the concentric rings shown in Fig. 11.31.

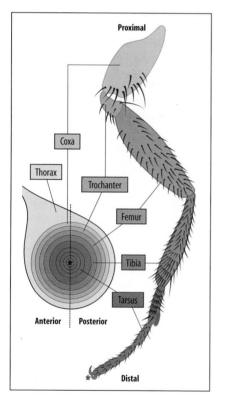

Fig. 11.31 Fate map of the leg imaginal disc of *Drosophila*. The disc is a roughly circular epithelial sheet, which becomes transformed into a tubular leg at metamorphosis. The center of the disc becomes the distal tip of the leg and the circumference gives rise to the base of the leg—this defines the proximo-distal axis. The tarsus is divided into five tarsal segments. The presumptive regions of the adult leg, such as the tibia, are thus arranged as a series of circles with the future tip at the center. A compartment boundary divides the disc into anterior and posterior regions.

After Bryant, P.J.: 1993.

The leg disc contains some 30 cells at its initial formation but grows to over 10,000 cells by the third instar. The first steps in the patterning of the leg disc along its antero-posterior axis are the same as in the wing. The *engrailed* gene is expressed in the posterior compartment and induces expression of *hedgehog*. The Hedgehog protein induces a signaling region at the antero-posterior compartment border. In the dorsal region of the leg disc, the expression of *decapentaplegic* is induced, as it is in the wing. In the ventral region, however, Hedgehog induces expression of *wingless* instead of *decapentaplegic* along the border, and Wingless protein acts as the positional signal (Fig. 11.32).

The patterning of the proximo-distal axis is far better understood in the leg disc than in the wing. It also involves interactions between Wingless and Decapentaplegic. The distal end of the proximo-distal axis is the point at which Wingless and Decapentaplegic expression meet, and is marked by expression of the homeodomain transcription factor Distal-less, which marks what will be the distal tip of the leg. Distal-less is not used in *Drosophila* wing development.

Patterning along the proximo-distal axis takes place sequentially and produces a two-dimensional pattern in the leg disc that will be converted at metamorphosis into the three-dimensional tubular leg. Early expression of *distal-less* is governed by a particular module in the gene's control region and lasts for just a few hours. Later

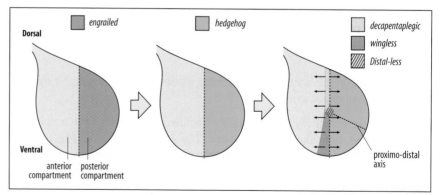

Fig. 11.32 Establishment of signaling centers in the antero-posterior compartment of the leg disc, and the specification of the distal tip. The gene *engrailed* is expressed in the posterior compartment and induces expression of *hedgehog*. Where Hedgehog protein meets and signals to anterior compartment cells, the gene *decapentaplegic* is expressed in dorsal regions and the gene *wingless* is expressed in ventral regions. Both of these genes encode secreted proteins. Expression of the gene *Distal-less*, which specifies the proximo-distal axis, is activated where the Wingless and Decapentaplegic proteins meet.

Fig. 11.33 Regional subdivision of the *Drosophila* leg along the proximo-distal axis. The pattern of gene expression in the leg is shown on the right. The stages of gene expression are shown in the two columns on the left, and are viewed as if looking down on the disc. Because of the way in which the leg extends from the disc, the center corresponds to the future tip of the leg, and the more proximal regions to successive rings around it. *decapentaplegic* (*dpp*) and *wingless* (*wg*) are initially expressed in a graded manner along the antero-posterior compartment boundary and together induce *Distal-less* (*Dll*) in the center and repress *homothorax* (*hth*), which is expressed in the outer region. They then induce *dachshund* (*dac*) in a ring between *Dll* and *hth*. Further signaling leads to these domains overlapping. While *hth* expression corresponds to proximal regions and *Dll* to distal regions, there is no simple relation between the other genes and the leg segments.

After Milan, M. and Cohen, S.M.: 2000.

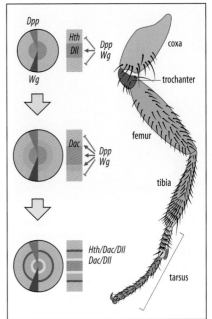

expression of *distal-less* in the future distal region is directed by other *cis*-regulatory modules. Another homeodomain transcription factor, Homothorax, is expressed in the peripheral region surrounding Distal-less, which is the future proximal region (Fig. 11.33). Their actions lead to the expression of the gene *dachshund*, which also encodes a transcription factor, in a ring between *Distal-less* and *homothorax*, which eventually leads to some overlap of the expression of *dachshund* with *Distal-less* and *homothorax*. Each of these genes is required for the formation of particular leg regions, but the expression domains do not correspond precisely with leg segments. Expression of *dachshund*, for example, corresponds to femur, tibia, and proximal tarsus. There is no evidence for a proximo-distal compartment boundary; cells expressing *homothorax* and *Distal-less* are, however, prevented from mixing at the interface between their two territories.

Additional patterning of the leg involves a gradient of EGF receptor activity from distal to proximal. At the third instar larval stage, the EGF receptor ligand Vein and the signaling pathway component Rhomboid are expressed at the central point of the disc. Genes coding for transcription factors Bric-a-brac and Bar are expressed at different levels in the tarsal segments and may determine their identity. The level of activity of these genes may be determined by the gradient of EGF receptor activity. Leg joint formation requires Notch signaling. Delta and Serrate, ligands for Notch, are expressed as a ring in each leg segment and their activation of Notch results in specification of the joint-forming cells.

11.21 Butterfly wing markings are organized by additional positional fields

The variety of color markings on butterfly wings is remarkable: more than 17,000 species can be distinguished. Many of these patterns are variations on a basic 'ground plan' consisting of bands and concentric eyespots (Fig. 11.34). The wings are covered with overlapping cuticular scales, which are colored by pigment synthesized and deposited by the underlying epidermal cells. How are these patterns specified? Butterfly wings develop from imaginal discs in the caterpillar in a similar way to *Drosophila* wings. Surgical manipulation has shown that the eyespot is specified at a late stage in the development of the wing disc, and that the pattern is dependent on a signal emanating from the center of the spot. A number of the genes that control wing development in *Drosophila*, such as *apterous*, are expressed in the butterfly in a spatial and temporal pattern similar to that in the fly. Thus, as in *Drosophila*, the shape and structure of the butterfly wing is patterned by a field of positional information. The patterning of the pigmentation, however, involves the establishment of additional fields of positional information.

Unlike *Drosophila* wings, development of butterfly wings does involve expression of *Distal-less*, and it is this that provides the extra dimension of pattern. The expression pattern of *Distal-less* in butterfly wing discs suggests that the mechanism used

Fig. 11.34 Butterfly wing pattern. Ventral view of a female African butterfly *Bicyclus anynana*, showing wing color pattern and prominent eyespots. Scale bar = 5 mm.

Photograph courtesy of V. French and P. Brakefield.

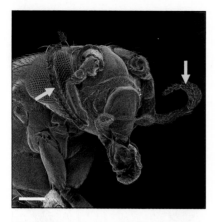

Fig. 11.35 Scanning electron micrograph of *Drosophila* carrying the *Antennapedia* mutation. Flies with this mutation have the antennae converted into legs (arrows). Scale bar = 0.1 mm.

Photograph by D. Scharfe, from Science Photo Library.

to delineate the color pattern on the butterfly wing is similar to the mechanism that specifies positional information along the proximo-distal axis of the insect leg. In the butterfly wing, *Distal-less* is expressed in the center of the eyespot, whereas in the *Drosophila* leg disc it is expressed in the central region corresponding to the future tip of the limb. Thus, it is possible that eyespot development and distal leg patterning involve similar mechanisms, although in the butterfly wing Notch expression precedes that of Distal-less. The eyespot may be thought of as a proximo-distal pattern superimposed on the two-dimensional wing surface. The center of the eyespot may represent the distal-most positional value, with the surrounding rings representing progressively more proximal positions, as in the leg. The site of the eyespot may be specified with reference to the primary wing pattern (that is, the anterior, posterior, dorsal, and ventral compartments); a secondary coordinate system centered on the eyespot would then be established, with expression of the *Distal-less* gene defining the central focus.

11.22 Different imaginal discs can have the same positional values

Patterning of legs and wings involves similar signals, such as Decapentaplegic protein, yet the actual pattern that develops is very different. This implies that the wing and leg discs interpret positional signals in different ways. This interpretation is under the control of the Hox genes and can be illustrated in regard to the leg and antenna. If the Hox gene *Antennapedia*, which is normally expressed in parasegments 4 and 5 (see Section 2.27) and specifies the discs for the second pair of legs, is expressed in the head region, the antennae develop as legs (Fig. 11.35). No homeotic genes are normally expressed in the antennal disc. Using the technique of mitotic recombination (see Box 2E, p. 73), it is possible to generate a clone of *Antennapedia*-expressing cells in a normal antennal disc. These cells develop as leg cells, but exactly which type of leg cell depends on their position along the proximo-distal axis; if, for example, they are at the tip, they form a claw. It is as if the positional values of the cells in the antenna and leg are similar, and the difference between the two structures lies in the interpretation of these values, which is governed by the expression or non-expression of the *Antennapedia* gene. This can be illustrated with respect to the French flag and the Union Jack (Fig. 11.36)—cells develop according to both their position and their developmental history, which determines the genes that they are expressing at any given time. This principle applies also to wing and haltere imaginal discs. Thus, we find a similarity in developmental strategy between insects and vertebrates; both use the same positional information in appendages like legs and wings, and interpret it differently. Still to be resolved is the question of how the expression of a single transcription factor like Antennapedia can transform an antenna into a leg. This requires an understanding of its downstream targets. So far, there is evidence that Antennapedia acts as a repressor of antennal identity in the leg by, for example, preventing the coexpression of *homothorax* and *Distal-less* in the femur region, which occurs in the antenna but not in the leg disc, where these two genes are expressed in adjacent and non-overlapping domains (see Fig. 11.33). In evolutionary terms, the antennal state may be the ground state for limb development.

The character of a disc and how positional information is interpreted is therefore determined by the pattern of Hox gene expression (or the absence of Hox expression in the case of the antennal disc). Insect legs develop only on the three thoracic segments and not on the abdominal segments and, in *Drosophila*, wings develop only on the second thoracic segment. These adult structures are segment-specific because the particular type of imaginal disc that gives rise to them is only formed by certain parasegments. There are no appendages on abdominal segments in *Drosophila* because the genes required for formation of leg and wing discs are suppressed in the

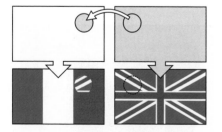

Fig. 11.36 Cells interpret their position according to their developmental history and genetic make-up. For example, if there are two flags that use the same positional information to produce a different pattern, then a graft from one to the other would result in the graft developing according to its intrinsic pattern, but with the pattern appropriate to its new position. This is what happens in imaginal discs.

abdomen. The type of disc formed in a particular thoracic segment is typically specified by the action of one of the Hox genes expressed in the segment. For example, expression of the genes *Antennapedia* and *Ultrabithorax* specifies the second and third pair of legs, respectively.

The leg imaginal discs arise from small clusters of ectodermal cells in parasegments 3–6, which contribute to the future thoracic segments of the embryo (Fig. 11.37). In the embryo, each disc initially contains around 25 cells and is formed during growth of the blastoderm. The discs arise at the parasegment boundaries, with anterior and posterior compartments of adjacent parasegments contributing to each disc. In the future second thoracic segment, the leg disc splits off a second disc early in its development, which becomes the wing disc. Similarly, an additional disc, which develops into the haltere, a balancing organ, is formed in the future third thoracic segment.

Because imaginal discs are formed across parasegment boundaries, mutations in Hox genes in *Drosophila* embryos can cause compartment-specific homeotic transformations of halteres into wings. In normal *Drosophila* adults, the wing is on the second thoracic segment and the haltere on the third thoracic segment; they arise from imaginal discs originating at the boundaries of parasegments 4/5 and 5/6, respectively (see Fig. 11.37). In the normal embryo, the *Ultrabithorax* gene, one of the genes of the bithorax complex (see Fig. 2.46), is expressed in parasegments 5 and 6 and is involved in specifying their identity. The *bithorax* mutation (*bx*), which causes the *Ultrabithorax* gene to be misexpressed, can transform the anterior compartment of the third thoracic segment, and thus of the haltere, into the corresponding anterior compartment of the second segment: the anterior half of the haltere thus becomes a wing (see Fig. 2.47). The *postbithorax* mutation (*pbx*), which affects a regulatory region of the *Ultrabithorax* gene, transforms the posterior compartment of the haltere into a wing (Fig. 11.38). If both mutations are present in the same fly, the effect is additive and the result is a fly that has four wings but cannot fly. Another mutation, *Haltere mimic*, causes a homeotic transformation in the opposite direction: the wing is transformed into a haltere.

As with the antenna and leg, it is possible to generate a mosaic haltere with a small clone of cells containing an *Ultrabithorax* mutation (such as *bithorax*) in the disc; the cells in the clone make wing structures, which correspond exactly to those that would form in a similar position in a wing. It is as if the positional values in haltere and wing discs are identical and all that has been altered in the mutant is how this positional information is interpreted (see Fig. 11.38). In fact, other imaginal discs seem to have similar positional fields.

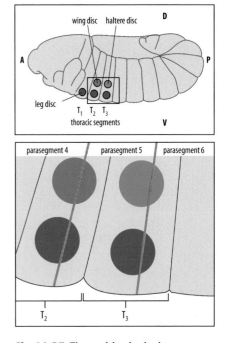

Fig. 11.37 The position in the late *Drosophila* embryo of the imaginal discs that give rise to the adult thoracic appendages. The imaginal discs for the legs, wings, and halteres lie across the parasegment boundaries in the thoracic segments as shown for the wing and haltere discs in the lower panel.

Summary

The legs and wings of *Drosophila* develop from epithelial sheets—imaginal discs—that are set aside in the embryo. Hox genes acting in the parasegments specify which sort of appendage will form and can direct the interpretation of positional information. The leg and wing imaginal discs are divided at an early stage into anterior and posterior compartments. The boundary between the compartments is a pattern-organizing center and a source of signals that pattern the disc. In the wing disc, expression of *decapentaplegic* is activated by the Hedgehog protein at the anteroposterior compartment boundary and the Decapentaplegic and Hedgehog proteins act as antero-posterior patterning signals. The wing disc also has dorsal and ventral compartments, with Wingless produced at the dorso-ventral boundary acting as a patterning signal. In the leg disc, the antero-posterior compartment boundary is established in a very similar way, except that Hedgehog activates *wingless* instead of *decapentaplegic* in the ventral region and the Wingless protein acts as the patterning signal in this region. There is no evidence of dorsal and ventral

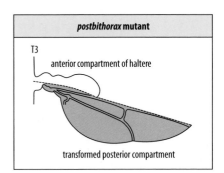

Fig. 11.38 Effects of the *postbithorax* mutation in *Drosophila*. The mutation *postbithorax* acts on the posterior compartment of a haltere, converting it into the posterior half of a wing.

compartments in the leg disc. The proximo-distal axis of the leg is specified by the interaction between Decapentaplegic and Wingless proteins, which activates genes such as *dachshund* and *homothorax* in concentric domains corresponding to regions along the leg. The colorful eyespots on butterfly wings may be patterned by a mechanism similar to that used to organize the proximo-distal axis of the insect leg.

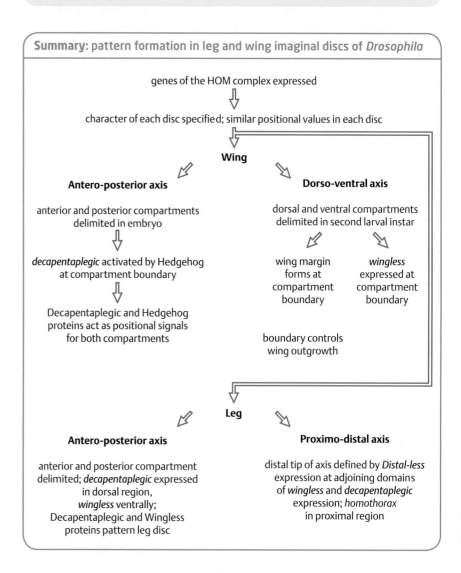

Summary: pattern formation in leg and wing imaginal discs of *Drosophila*

genes of the HOM complex expressed

⇩

character of each disc specified; similar positional values in each disc

Wing

Antero-posterior axis

anterior and posterior compartments delimited in embryo

⇩

decapentaplegic activated by Hedgehog at compartment boundary

⇩

Decapentaplegic and Hedgehog proteins act as positional signals for both compartments

Dorso-ventral axis

dorsal and ventral compartments delimited in second larval instar

wing margin forms at compartment boundary

wingless expressed at compartment boundary

boundary controls wing outgrowth

Leg

Antero-posterior axis

anterior and posterior compartment delimited; *decapentaplegic* expressed in dorsal region, *wingless* ventrally; Decapentaplegic and Wingless proteins pattern leg disc

Proximo-distal axis

distal tip of axis defined by *Distal-less* expression at adjoining domains of *wingless* and *decapentaplegic* expression; *homothorax* in proximal region

Vertebrate and insect eyes

Structures as complex as the compound eyes of insects and the camera eyes of vertebrates are a remarkable achievement of evolution (Fig. 11.39). Camera eyes and compound eyes do share some basic similarities. They all contain a lens to focus the light, a retina composed of light-sensing photoreceptor cells, and a pigmented layer that absorbs stray light and prevents it interfering with photoreceptor signaling. And despite the great difference in the anatomy of the final eye, some of the same transcription factors specify eye formation in insects and vertebrates. We shall begin here with the development of the vertebrate camera eye, which is essentially an offshoot of the forebrain. In the vertebrate eye, light enters through the pupil at the front of the eye and passes through a convex transparent lens, which focuses the light on

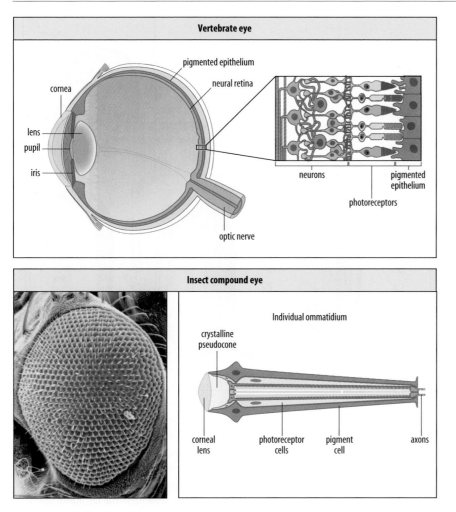

Vertebrate eye

pigmented epithelium

neural retina

cornea

lens

pupil

iris

neurons

pigmented epithelium

photoreceptors

optic nerve

Insect compound eye

Individual ommatidium

crystalline pseudocone

corneal lens

photoreceptor cells

pigment cell

axons

Fig. 11.39 Vertebrate and insect eyes. Vertebrate eyes (upper panel) are 'camera' eyes with a single lens that focuses light on photoreceptor cells in the neural retina lining the back of the eyeball. The photoreceptor cells connect to retinal neurons, whose axons form the optic nerve, which connects the eye to the brain. The *Drosophila* compound eye (lower panel) is composed of numerous individual light-sensing structures called ommatidia; a section through one is shown on the right. Each ommatidium has a lens, through which light passes to activate the photoreceptor cells, which are surrounded by pigment cells.

the photosensitive retina lining the back of the eyeball (see Fig. 11.39). Photoreceptor cells in the retina—the rods and cones—register the incoming photons and pass signals on to nerve cells, which transmit them via the optic nerve to the brain, where they are decoded.

11.23 The vertebrate eye develops from the neural tube and the ectoderm of the head

In developmental terms, the vertebrate eye is essentially an extension of the forebrain, together with a contribution from the overlying ectoderm and migrating neural crest cells. The development of an eye starts at E8.5 in the mouse and around 22 days in human embryos with the formation of a bulge, or evagination, in the epithelial wall of the posterior forebrain, in the region called the diencephalon. This evagination is called the optic vesicle; an optic vesicle is formed on either side of the head and extends to meet the surface ectoderm (Fig. 11.40). The optic vesicle interacts with the ectoderm to induce formation of the lens placode, which is a thickened region of ectoderm from which the lens will develop. The lens placode is part of a larger region of head ectoderm that gives rise to the epithelial placodes of some other sensory organs, including the placode that gives rise to the semi-circular canals, cochlea, and endolymphatic duct of the ear, and the placode that gives rise to the olfactory epithelium in the nose.

After induction of the lens placode, the tip of the optic vesicle invaginates to form a two-layered cup, the optic cup, adjacent to the placode. The inner epithelial layer of

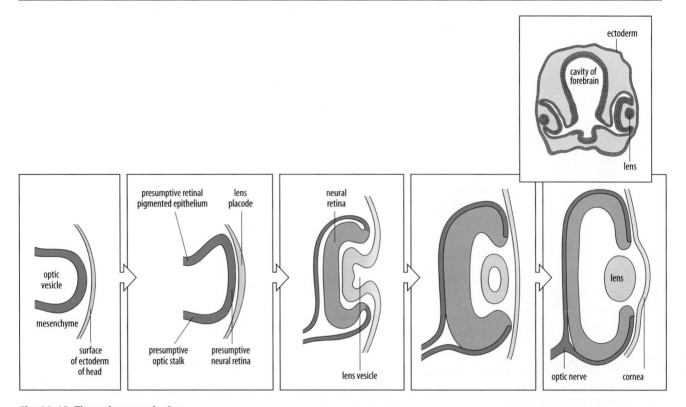

Fig. 11.40 The main stages in the development of the vertebrate eye. The optic vesicle (blue) develops as an outgrowth of the forebrain and induces the lens placode in the surface ectoderm of the head. The optic vesicle invaginates to form a two-layered cup, the optic cup, around the developing lens. The inner layer of the optic cup forms the neural retina and the outer layer the pigmented epithelium underlying the retina. The lens vesicle detaches from the surface ectoderm to form the lens and the remaining overlying ectoderm forms the cornea. The iris develops from the rim of the optic cup.

Adapted from Adler, R. and Valeria Canto-Soler, M.: 2007.

the optic vesicle will form the neural retina, while the outer layer will form the retinal pigment epithelium. The lens placode then invaginates and detaches from the surface ectoderm to form a small hollow sphere of epithelium that will develop into the lens. The lens is formed by proliferation of cells of the epithelium on the anterior side of the lens vesicle, with the new cells moving into the center of the lens, where they start to manufacture crystallin proteins. The cells eventually lose their nuclei, mitochondria, and internal membranes to become completely transparent lens fibers filled with crystallin. Renewal of lens fibers continues after birth, but proceeds much more slowly than in the embryo. In the adult chicken, the transformation of an epithelial cell into a crystallin-filled fiber takes two years.

The cornea is a transparent epithelium that seals the front of the eye. It is composed of inner and outer layers with different developmental origins. The inner layer is formed by mesenchymal neural crest cells that migrate into the anterior eye chamber to form a thin layer overlying the lens. Neural crest cells also contribute to other structures in the anterior part of the eye. The outer layer of the cornea is derived from the surface ectoderm adjacent to the eye. Most of the iris develops from the rim of the optic cup, with neural crest contributing to the anterior iris.

The neural retina in the vertebrate eye develops distinct layers of cells, with the photoreceptor cells forming the innermost layer, underneath layers of ganglion and bipolar cells. This is in contrast to the camera eyes of octopus and squid in which the photoreceptor cells are on the surface of the neural retina. Visual signals are transmitted from the eye to the brain via the optic nerve, which is formed by the axons of the neural ganglion cells. In Chapter 12 we will see how the neurons of the optic nerve connect with precise positions in the visual-processing centers in the brain in order to produce a 'map' of the retina.

Although evagination of the optic vesicles does not occur until neural tube closure is almost completed, the specification of cells as prospective eye cells occurs much earlier, in the neural plate. Key conserved eye-specifying transcription factors, such as Pax6, Six3, and Otx2, are expressed in anterior neural plate at late gastrula stages as part of the initial antero-posterior patterning of the neural plate. They continue to be expressed in the optic vesicle epithelium and the lens placode, as well as in other precursors of sensory organs such as the olfactory placode. Retinal precursor cells are first specified as a single 'eye field' across the center of the anterior neural plate. The eye field eventually becomes separated into two lateral regions of cells that will give rise to the optic vesicles after neural tube formation. Separation is achieved by downregulating expression of Pax6 and the other eye-specifying transcription factors along the midline. Failure of this separation could be one of the causes of cyclopia in infants, in which a single abnormal central eye is produced. The normal downregulation of the expression of Pax6 and other transcription factors in the forebrain midline is likely to be mediated by Shh signaling, which is known to be required for the correct specification of ventral midline structures. As we saw in Chapter 1 (see Box 1F, p. 28), a failure of Shh signaling can result in the condition of holoprosencephaly, which in its most severe form results in failure of the forebrain to divide into right and left hemispheres, and loss of midline facial structures, resulting in cyclopia.

The same set of transcription factors is essential for eye formation throughout the animal kingdom. The classic example of a gene with a conserved basic function is *Pax6*, which is required for the development of light-sensing structures in all bilaterian animals (animals with bilateral symmetry)—from the simple light-sensing organs of planarians to the compound eyes of insects (described later in this chapter), and the camera eyes of vertebrates and cephalopods. *Pax6* was initially identified from the genetic analysis of mutations causing abnormal eye development in mice and humans and is homologous with the *Drosophila* gene *eyeless*. Mouse embryos with defective *Pax6* function have smaller eyes than normal, or no eyes at all. People heterozygous for mutations in *Pax6* have a variety of eye malformations collectively known as **aniridia**, because of the partial or complete absence of the iris; these patients also have cognitive defects, as *Pax6* has a number of roles in brain development other than eye formation. In *Xenopus*, injection of *Pax6* mRNA into an animal pole blastomere at the 16-cell stage leads to formation of ectopic eye-like structures, with well-formed lenses and epithelial optic cups, in the tadpole head. Even more amazingly, the ectopic expression of *Pax6* from mouse, *Xenopus*, ascidians, or squid in *Drosophila* imaginal discs causes the development of *Drosophila*-type compound eye ommatidia on adult structures such as antennae (Fig. 11.41).

Formation of the lens is a crucial step in vertebrate eye development. Among much other evidence, this is nicely demonstrated by experiments in the teleost fish *Astyanax mexicanus*, which has both surface-dwelling sighted forms and cave-dwelling blind forms—called cavefish (see Fig. 15.18). A small optic primordium with an optic cup and a lens is formed in cavefish embryos but the lens then undergoes massive apoptosis. An expansion of Shh signaling along the embryonic midline in the cavefish has been

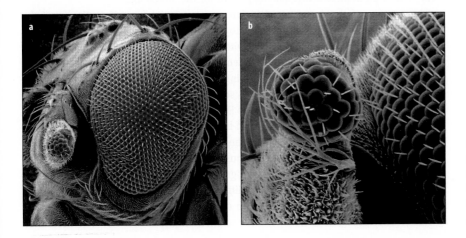

Fig. 11.41 *Pax6* is a master gene for eye development. The ectopic expression of mouse *Pax6* in *Drosophila* antennal disc results in compound eye structures developing on the antenna.

Photographs from Gehring, W.J.: 2005.

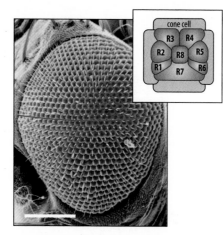

Fig. 11.42 Scanning electron micrograph of the compound eye of adult *Drosophila*. Each unit is an ommatidium. At the third larval instar, the ommatidial cluster (inset) is made up of eight photoreceptor neurons (R1–R8) and four cone cells. Red line indicates the equator of eye. Scale bar = 50 μm.

implicated in lens apoptosis. As a consequence of lens non-development, the cornea, the iris, and other eye structures at the anterior of the eye, fail to develop. Transplantation of a lens from the surface-dwelling *A. mexicanus* into a cavefish optic cup during embryogenesis reverses lens degeneration, indicating that the absence of a lens is at least partly responsible for the non-development of the cavefish eye. The effect of expanded Shh signaling in eye development was investigated by increasing Shh expression in embryos of surface-dwelling *Astyanax*, which normally develop into sighted adults. When *Shh* mRNA was injected into one side of a cleavage-stage embryo, expression of the eye gene *Pax6* was downregulated unilaterally in the corresponding developing eye region, and the larvae that developed lacked an eye on that side of the head.

11.24 Patterning of the *Drosophila* eye involves cell–cell interactions

The *Drosophila* compound eye is quite different in structure from the vertebrate eye. It is composed of about 800 identical photoreceptor organs called **ommatidia** (singular **ommatidium**) arranged in a hexagonal array of crystalline regularity (Fig. 11.42). Each fully developed ommatidium is made up 20 cells: eight photoreceptor neurons (R1–R8), together with four overlying cone cells (which secrete the lens), and eight additional pigment cells (not shown in the inset in Fig. 11.42). The genetic analysis of ommatidium development has provided one of the best model systems for studying the patterning of a small group of cells. An important early finding from lineage analysis was that the pattern of each ommatidium is specified by cell–cell interactions and is not based on cell lineages.

The eye develops from the single-layered epithelial sheet of the eye imaginal disc, located at the anterior end of the larva. Specification and patterning of the cells of the ommatidia begins in the middle of the third larval instar. Patterning starts at the posterior of the eye disc and progresses anteriorly, taking about 2 days, during which time the disc grows eight times larger. One of the earliest events in eye differentiation is the formation of a groove, the **morphogenetic furrow**, which sweeps across the disc from posterior to anterior in response to a wave of signals that initiate development of ommatidia from the eye disc cells. As the furrow moves across the epithelium from posterior to anterior, clusters of cells that will give rise to the ommatidia appear behind it, spaced in a hexagonal array (Fig. 11.43). The furrow moves slowly across the disc, at a rate of 2 hours per row of ommatidial clusters. As the morphogenetic furrow moves forward, the cells behind it start to differentiate to form regularly spaced ommatidia. The ommatidia are arranged in rows, with each row half an ommatidium out of register with the previous one. This gives the characteristic hexagonal packing arrangement. The first cells to differentiate are the R8 photoreceptor neurons. These appear as regularly spaced cells in each row, separated from each other by about eight cells. This separation sets the spacing pattern for the ommatidia.

The passage of the morphogenetic furrow is essential for differentiation of the ommatidia, as mutations that block its progress also block the differentiation of new rows of ommatidia, resulting in a fly with abnormally small eyes. Although there are no anterior and posterior compartments in the eye discs, the cells just behind the furrow can be regarded as resembling posterior cells, similar to the situation in the wing-disc posterior compartment (see Section 11.18). They secrete Hedgehog protein, which triggers the expression of Decapentaplegic, which in turn causes the cells to become competent to form neural tissue. The gene Wingless also plays a part in eye disc patterning. It is expressed at the lateral edges of the eye disc, and prevents the furrow starting in these regions. We thus see that, even though imaginal discs give rise to very diverse structures, the key signals involved in patterning the leg, wing, and eye are similar, although they have different roles in each case.

Once specified, each R8 cell initiates a cascade of signals that eventually recruits a cluster of 20 cells that form the mature ommatidium. The first cells to be recruited

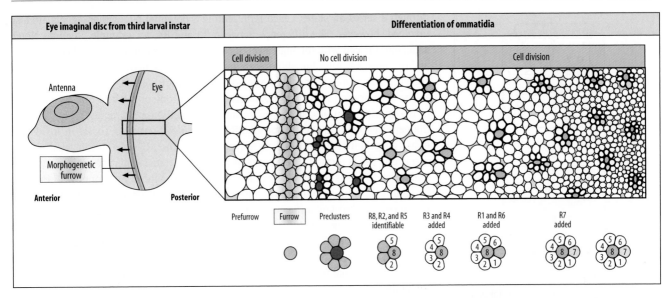

Eye imaginal disc from third larval instar	Differentiation of ommatidia

Fig. 11.43 Development of ommatidia in the *Drosophila* compound eye. The compound eye develops from the eye imaginal disc, which is part of a larger disc that also gives rise to an antenna. During the third larval instar, the morphogenetic furrow develops in the eye disc and moves across it in a posterior to anterior direction. Ommatidia develop behind the furrow, the photoreceptor neurons being specified in the order shown, R8 developing first and R7 last. The individual ommatidia are regularly spaced in a hexagonal grid pattern.

After Lawrence, P.: 1992.

are the prospective photoreceptor cells, which are sensory neurons. R2 and R5 differentiate on either side of R8, to form two functionally identical neurons. R3 and R4, which are a slightly different type of photoreceptor, are specified next. All these cells become arranged in a semi-circle with R8 at the center. R1 and R6 differentiate next and almost complete the circle, which is finally closed by the differentiation of R7 adjacent to R8 (see Fig. 11.43).

The clusters rotate 90°, so that R7 comes to be closest to the equator of the disc and R3 furthest away; rotation is in the opposite direction in the dorsal and ventral halves of the eye. This means that the dorsal and ventral regions of the eye have distinct and different polarities, with ommatidia on the dorsal and ventral sides of the equator having mirror-image symmetry (see Fig. 11.42). The polarization of the ommatidia is yet another example of planar cell polarity (see Box 2F, p. 78) and involves a higher level of Frizzled signaling in R3 than in R4.

The determination of the equator of the eye results from the specification of the dorsal region of the eye disc by members of the *Iroquois* gene complex, which is followed by specification of the equator itself, involving the actions of Notch, its ligands Serrate and Delta, and the secreted protein Fringe, in a similar way to the specification of the dorso-ventral compartment boundary in the wing (see Section 11.19).

The regular spacing of the ommatidia within the eye involves a **lateral inhibition** mechanism that spaces the R8 cells. All cells in the eye disc initially have the capacity to differentiate as R8 cells, and as the morphogenetic furrow passes they start to do so. But some inevitably gain a lead and are thus able to inhibit the differentiation of another R8 cell over a range of around three cell diameters. Cells that will give rise to R8 express the gene *atonal*. Inhibitors of *atonal* that space the R8 cells are the secreted Scabrous protein and Notch.

In the eye, cell fate is specified and determined cell by cell, not in groups of cells. Two proteins crucial in the patterning of an individual ommatidium are the EGF receptor and its ligand Spitz. One model for patterning the ommatidium is based on both EGF receptor activation and the age of the cells (Fig. 11.44). Spitz is produced by the three earliest-specified and most centrally located cells—R8, R2, and R5. It activates the EGF receptor on neighboring cells, and this recruits R3, R4, R1, R6, and R7 to a photoreceptor fate. The responding cells also secrete the protein Argos. This diffuses away and inhibits more distant cells from being activated by Spitz, so that no more cells in the prospective ommatidial cluster develop as photoreceptors. The actual character of each photoreceptor may be determined by the age of the cell, with cells passing

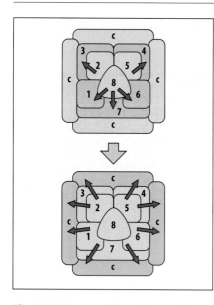

Fig. 11.44 Sequential recruitment of photoreceptors and cone cells during ommatidial development. The three earliest-specified and most centrally located photoreceptor cells—R8, R2, and R5—produce the protein Spitz. It activates the *Drosophila* EGF receptor DER on neighboring cells, which recruits R3, R4, R1, R6, and R7 to a photoreceptor fate (top). These then also produce Spitz, which interacts with DER on the prospective cone cells (c) to recruit them to the ommatidium. Once cells are determined they secrete the protein Argos. This diffuses away and inhibits more distant cells from being activated by Spitz, so that no more cells in the prospective ommatidial cluster develop as photoreceptors.

Adapted from Freeman, M.: 1997.

through a series of 'states', each representing a potential fate. Other signals are also involved, however. The difference between R3 and R4, for example, involves Notch signaling; the cell with the high level of Notch activity is inhibited from becoming R3, and becomes R4. After the photoreceptors have differentiated, the four lens-producing cone cells develop, and finally the surrounding ring of accessory cells.

The specification of R7 as a photoreceptor cell is one of the best understood cases of the specification of cell fate on an individual cell basis. It requires expression of the gene *sevenless* in the prospective R7 and *bride-of-sevenless* (*boss*) in R8. When either gene is inactivated, the phenotype is the same: R7 does not develop and an extra cone cell is formed. The Sevenless protein is a transmembrane receptor tyrosine kinase, and Boss is its ligand. Sevenless protein is produced not only by R7 but also by other cells in the ommatidium, including lens cells. Thus, expression of Sevenless is a necessary, but not sufficient, condition for R7 specification. Using genetic mosaics it can be shown that for R7 to develop, only R8 need express Boss protein and that this is the signal by which R8 induces R7. Boss is also an integral membrane protein; it is present on the apical surface of the R8 cell, where it makes contact with R7. A second signal for R7 is provided by the R1/R6 pair, which must activate Notch in the R7 cell.

11.25 Eye development in *Drosophila* is initiated by the actions of the same transcription factors that specify eye-precursor cells in vertebrates

The *Drosophila* version of *Pax6* is called *eyeless*, as mutations in this gene result in the reduction or complete absence of the compound eye. *eyeless* is expressed in the region of the eye disc anterior to the morphogenetic furrow (see Fig. 11.43). Ectopic expression of the *eyeless* gene in other imaginal discs results in the development of ectopic eyes, and these have been induced in this way in wings, legs, antennae, and halteres. The fine structure of these ectopic eyes is remarkably normal, and distinct ommatidia are present, although they are not connected to the nervous system. It is estimated that some 2000 genes are eventually activated as a result of *eyeless* activity, and all of them are required for eye morphogenesis. The mode of action of *eyeless* in *Drosophila* may be similar to that of the Hox genes, as it appears to change the interpretation of positional information in discs in which it is ectopically expressed.

As we saw earlier, the remarkable conservation of *Pax6/eyeless* function is demonstrated by the fact that *Pax6* from the mouse can substitute for *eyeless* and induce ectopic eye structures in *Drosophila* if expressed in the imaginal disc for the antenna (see Fig. 11.41). Whatever the source of the *Pax6* gene, the eye structures that form are always *Drosophila* ommatidia, not eyes like those of the animal from which the *Pax6* gene came. Pax6 is acting as a master gene that turns on *Drosophila*'s own eye development program.

The discovery of this underlying conservation of function now suggests to many biologists that the very different eyes found in different groups of animals are examples of parallel evolution that has adapted the same genetic circuitry and certain cell specializations that were present in the ancestors of all bilaterian animals, or even earlier. Animals are thought to have evolved from unicellular cells, probably most resembling flagellate protozoa (discussed in Section 15.2), which are proposed to have had a light-sensitive organelle similar to the 'eyespot' in some modern flagellates. The eyespot transmits signals to the flagella and enables the cell to move towards light. The evolution of multicellularity and of differentiated cell types eventually led to simple types of photoreceptor cells, such as those in present-day jellyfish larva, which have individual photosensitive cells in the ectoderm. Jellyfish photoreceptor cells not only use opsins as the photoreceptor proteins, as do all animals and some unicellular organisms, but also contain the pigment melanin, which is the pigment in the retinal pigmented epithelium of the vertebrate eye. The function of the pigment is to absorb any light that passes through the photoreceptor layer and so prevent it reflecting back onto the photoreceptors and producing an imprecise signal.

The next step in the evolution of the eye would have been the differentiation of unicellular light-sensing cells into two cell types: a sensory photoreceptor cell and a pigment cell. The idea that the origin of the eye involved two such cells goes back to Charles Darwin, who considered the evolution of the vertebrate eye one of the most difficult problems in evolution. Simple two-celled 'pigment cup' light-sensitive organs are present in the larvae of many bilaterian animals, while the more complex frontal eye of the larval amphioxus (a cephalochordate) consists of a pigment cup and two rows of photoreceptor cells connected by neurons to the central nervous system, foreshadowing the eyes of vertebrates. When and how Pax6 became such a crucial factor in eye development is still unknown.

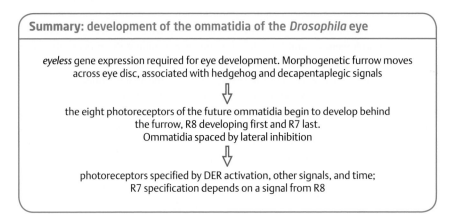

Summary: development of the ommatidia of the *Drosophila* eye

eyeless gene expression required for eye development. Morphogenetic furrow moves across eye disc, associated with hedgehog and decapentaplegic signals

⇩

the eight photoreceptors of the future ommatidia begin to develop behind the furrow, R8 developing first and R7 last.
Ommatidia spaced by lateral inhibition

⇩

photoreceptors specified by DER activation, other signals, and time;
R7 specification depends on a signal from R8

Summary

The vertebrate eye develops as an extension of the forebrain. Evagination of the forebrain lateral wall produces the optic vesicle, which in turn induces formation of the lens from the surface ectoderm. Invagination of the tip of the optic vesicle produces a two-layered optic cup, which surrounds the lens and develops into the eyeball, with the inner epithelial layer forming the neural retina, and the outer epithelium developing into the pigmented epithelium at the back of the retina. Development of the lens is essential for the eye to develop further. The transcription factor Pax6 is essential for eye development in animals as diverse as vertebrates and *Drosophila*. The *Drosophila* compound eye contains around 800 individual ommatidia, arranged in a regular hexagonal pattern, and develops from an imaginal disc. The regular spacing of the ommatidia is achieved by lateral inhibition by the photoreceptor neuron R8. The patterning of the eight photoreceptor neurons in each ommatidium is due to local cell–cell interactions, in which the photoreceptor cells are specified and differentiate in a strict order, initiated by R8.

Internal organs: insect tracheal system, vertebrate lungs, kidneys, blood vessels, heart, and teeth

All the structures discussed so far are external, and this has made them relatively easy to study. We now consider some, mainly vertebrate, internal organs. Their development illustrates developmental phenomena we have not yet encountered, such as the formation and branching of tubular epithelia and the organization of loosely organized mesenchyme into epithelium.

Epithelia are the most common type of tissue organization in animals, and many organs, such as the lungs, kidneys, and blood vessels, are predominantly composed of functionally specialized epithelia. Epithelial cells adhere tightly together to form a cell sheet, which can be single layered, as in the **endothelium** of capillaries and kidney tubules, or multilayered, as in skin. A common feature of the organs we shall discuss first is that the epithelium forms branching tubes.

11.26 The *Drosophila* tracheal system is a model for branching morphogenesis

The bodies of multicellular animals contain a multitude of tubes and tubules composed of epithelia, including blood vessels, kidney tubules, and the branched airways in the mammalian lung. Many of these tubular systems undergo extensive branching during their development, a phenomenon known as **branching morphogenesis**. The development of the *Drosophila* tracheal system provides an excellent model for branching morphogenesis and has identified genes controlling the process that also act in the morphogenesis of the vertebrate lung.

Air enters the tracheal system of the *Drosophila* larva through openings in the body wall called spiracles, and oxygen is delivered to the tissues by some 10,000 or so fine tubules, which develop during embryogenesis from 20 ectodermal placodes (10 on each side). Each placode produces the tracheal system for a metamere (one lateral half of a complete segment) in the larva. At the start of germ-band retraction, towards the end of embryonic development (see Section 2.2), the placode ectoderm invaginates to form a hollow sac of around 80 cells, which gives rise through successive branching to hundreds of fine terminal branches. Remarkably, the extension of the sacs to form branched tubes does not involve any further cell proliferation but is all achieved by directed cell migration, cell rearrangement by intercalation, and changes in cell shape. As development proceeds, branches from different placodes fuse to form a body-wide network of interconnected tubes.

The initial invagination of a tracheal placode involves constriction of apical cell surfaces and changes in cell shape, very similar to the internalization of the *Drosophila* mesoderm (see Fig. 8.21). The first tubules are formed when some cells in the wall of the sac develop filopodia, which enable them to migrate toward a source of chemoattractant, drawing an elongating tube of cells behind them (Fig. 11.45). These tubes then branch by a combination of cell intercalation and remodeling of intercellular junctions to produce secondary tubules composed of single cells joined end to end, with each cell wrapped around itself to form the tubule lumen. The chemoattractant guiding tube extension is the protein Branchless, which is the *Drosophila* version of mammalian FGF. It is expressed in clusters of mesenchymal cells in a pattern determined by the previous antero-posterior and dorso-ventral patterning of the embryo. Branchless acts via the *Drosophila* version of an FGF receptor called Breathless, which is expressed in the tracheal cells and which is also involved in initiating branching. Genes such as *breathless* and *branchless* were identified by mutations that disrupted tracheal morphology—hence their names. Another such gene is *sprouty*, which is expressed by the tracheal cells. Sprouty protein prevents excess branching by antagonizing Breathless signaling, which prevents the more proximal cells from forming branches. In the absence of Sprouty, many more secondary branches form than normal.

As the tubules extend into the larval body, low oxygen levels lead to local expression of Branchless in the mesenchyme. Breathless signaling in response induces cells at the tips of the secondary tubules to sprout to form the fine terminal branches of the tracheal system. Each tip cell forms a much-branched unicellular sprout, with the lumen of the tubules formed within the cell itself (see Fig. 11.45). Excessive terminal

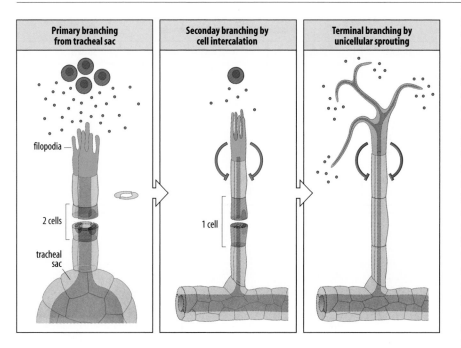

Primary branching from tracheal sac	Secondary branching by cell intercalation	Terminal branching by unicellular sprouting

filopodia

2 cells

tracheal sac

1 cell

Fig. 11.45 Branching in the *Drosophila* tracheal system is solely the result of cell rearrangement and remodeling. In the first stage of tracheal development, localized secretion of Branchless protein (blue) causes cells of the tracheal sac epithelium to form filopodia and move towards the source of Branchless, creating a primary branch (left panel). A secondary branch is formed by cells intercalating between each other and wrapping round themselves to form a tube of single cells placed end to end, with a lumen running through the center (center panel). At the tip of the secondary branch, the gene *sprouty* is induced, and the Sprouty protein (green) inhibits branching further away from the source of Branchless. Tip cells that will undergo terminal branching are then specified by Notch-Delta signaling. A tip cell puts out fine cytoplasmic extensions that develop a lumen and branch extensively (right panel). Cells behind the tip are inhibited by the Notch signaling (red barred lines) from adopting a tip-cell fate.

branching is prevented by Notch–Delta signaling, which assigns tip-cell fate and prevents the more proximal cells becoming tip cells and being able to form terminal branches.

11.27 The vertebrate lung also develops by branching of epithelial tubes

The paired lungs of vertebrates develop from two buds on each side of the end of the embryonic trachea (the windpipe) that grow out to form the bronchi. The lung is one of the internal organs that has left–right asymmetry (see Section 4.7) and this becomes evident very early in its development. Each bud grows out as a tubular epithelium, which will become a bronchus, and in humans it then makes one further branch on the right and three on the left to form the bronchioles (Fig. 11.46). These in turn make finer and finer branches, which terminate in thin-walled sacs of epithelium, the alveoli, where gas exchange occurs—oxygen is taken up by the blood and carbon dioxide released.

Unlike the tracheae of *Drosophila*, the outgrowth and branching of tubules in the vertebrate lung is the result of cell proliferation at the tip of the advancing tube, rather than cell migration. Nevertheless, sprouting and outgrowth of tubules from the main tube depends, as in the fly, on interaction of the tubular epithelium with signals from the surrounding mesenchyme, and many of these are the same as in the fly. FGF-10 is secreted by the splanchnic mesoderm cells adjacent to the lung epithelium, and interacts with the receptor FGFR2b, which is related to *Drosophila* Breathless, expressed by the lung epithelial cells. As in the fly, activation of FGFR2b induces expression of *Sprouty* in the lung cells, which acts as negative feedback to prevent the more proximal cells in the main tube from forming branches.

Another signaling protein essential for lung development is Shh, which is expressed in cells at the tips of the extending tubes. In mice with a homozygous loss of Shh function, the trachea and esophagus fail to separate properly and the primary lung buds fail to grow and branch, resulting in rudimentary sac-like lungs. Wnt signaling via the planar polarity pathway (see Box 8C, p. 310) is involved in regulating the proximo-distal polarity of the developing tubules. Mice lacking Wnt-5a have a truncated trachea which branches excessively.

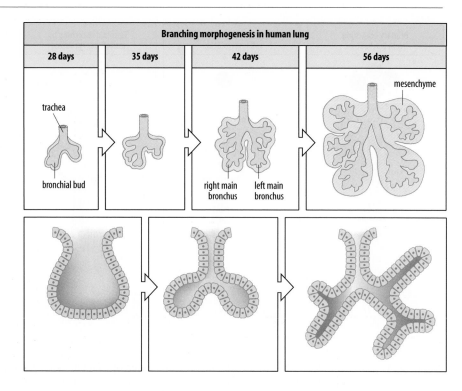

Fig. 11.46 Outgrowth and branching of the bronchial buds of the mammalian lung involve cell proliferation as well as deformation of an epithelial layer. Top panels: the lung epithelium develops by successive branching of the bronchial buds in response to signals from the surrounding mesenchyme. Bottom panels: deformation of a small section of an epithelial sheet accompanied by cell proliferation initially forms a pouch. The process is repeated to form successive branches. Branching in the developing mammalian lung is accompanied by cell proliferation.

11.28 The development of kidney tubules involves reciprocal induction by the ureteric bud and surrounding mesenchyme

The mammalian kidney is mainly composed of a set of convoluted tubules that receive fluid and salts from the blood, balance its composition, and then convey the final product—urine—to the ureter, which carries it out of the body. One end of each tubule ends in a glomerulus; this is the site at which blood capillaries release fluid and salts that are then taken up by the tubule. The other end of the tubule connects to the ureter. The kidney is formed from two regions of mesoderm, the ureteric bud and the metanephric mesenchyme (another name for the mammalian kidney is the metanephros). The ureteric bud branches off the Wolffian duct, which develops from a primitive form of kidney, the pronephros, which functions as the adult kidney in amphibians and fish. In mammals, the Wolffian ducts develop into the vas deferens in males (see Section 9.13).

The bud is induced to form at the posterior region of the Wolffian duct by the adjacent metanephric mesenchyme. Under the influence of the mesenchyme, the ureteric bud branches to form the proximal portions of the urine-carrying tubules, which are known as the collecting ducts. The ureteric bud induces the mesenchymal cells to condense around it and form epithelial structures that develop into renal tubules or nephrons. Each tubule elongates to form a glomerulus at one end, where filtration of the blood will occur, while the other end fuses to the ureteric collecting duct connecting to the ureter (Fig. 11.47). The intermediate portion of the tubule is convoluted and its epithelial cells become specialized for the resorption of ions from urine. The mammalian kidney is a good developmental system to study as it can develop in organ culture; explanted metanephric mesenchyme will form many glomeruli and tubules over a period of about 6 days, even though a blood supply is absent.

The development of the ureteric bud and the mesenchyme depends on mutual inductive interactions, neither being able to develop in the absence of the other. The mesenchyme induces the ureteric bud to grow out by secreting glial-cell-derived neurotrophic factor (GDNF), as well as two members of the FGF family. The site of

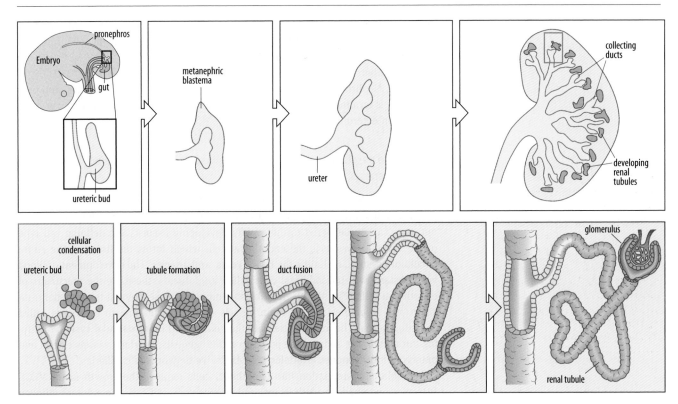

Fig. 11.47 Kidney development is an example of mesenchymal-to-epithelial transition. The kidney develops from a loose mass of mesenchyme, the metanephric blastema, which is induced to form tubules by the ureteric bud. The ureteric bud is itself induced by the mesenchyme to grow and branch to form the collecting ducts of the kidney that connect to the ureter. The mesenchyme cells form cellular condensations that become epithelial tubules, which open into the collecting system formed from the ureteric bud. Each tubule develops a glomerulus, through which waste products are filtered out of the blood.

expression of GDNF is critical for positioning ureteric bud development and restricting it to a single site; positioning of GDNF expression is under the control of the proteins Slit2 and its receptor Robo2. As we shall see in Chapter 12, these proteins are also involved in the guidance of axons in nervous-system development. GDNF in the mesenchyme is the main controller of ureteric bud branching, and its receptor, Ret, is present in the bud. The cellular basis for branching is similar to that described earlier for lung development (see Section 11.27). Activation of Ret activates the expression of *sprouty*, which, as in the lung, inhibits branching in the region behind the tip of the tubule. Knock-out of the gene for either GDNF or Ret results in the absence of ureteric bud outgrowth.

As the bud branches it induces the overlying mesenchyme at the tips to condense and to form epithelium that then forms the nephrons. Among the growth factors secreted by the bud and involved in this induction are FGF-2, leukemia inhibitory factor (LIF), and TGF-β2. A key protein that must be expressed in the mesenchyme for tubule formation to occur is the zinc-finger transcription factor WT1. Loss-of-function mutations in the *WT1* gene are associated with Wilm's tumor, a cancer of the kidney in children, and the gene is therefore known as the Wilm's tumor suppressor gene.

A very early response to induction is the expression of the transcription factor Pax2 in the mesenchyme, which is essential for subsequent tubule development, probably through its stimulation of the expression of GDNF. The expression of *WT1* then increases, the loose mass of mesenchyme cells condenses, and the cells start to express a matrix glycoprotein, syndecan, on their surface. This condensing stage is followed by the formation of distinct cellular aggregates, in which the mesenchymal cells become polarized and acquire an epithelial character and secrete the signaling protein Wnt-4. Each aggregate then forms an S-shaped tube, which elongates and differentiates to form the functional unit of a renal tubule and glomerulus. During this transition, the composition of the extracellular matrix secreted by the cells changes: mesenchymal collagen I is replaced by basal lamina proteins, such as collagen IV and laminin, that are typically secreted by epithelial cells. The adhesion molecules (see

Box 8B, p. 293) expressed by the cells also change; for example, the N-CAM expressed by the mesenchymal cells is replaced by E-cadherin in the epithelial cells. Integrins are also involved in the epithelial–mesenchyme interactions.

The number of nephrons per kidney varies widely, from around 230,000 to 1,800,000. Nephron number is determined by the number of bud branches, and how branching is terminated is not well understood. Many different proteins appear to be involved in determining nephron number.

11.29 The vascular system develops by vasculogenesis followed by angiogenesis

Not surprisingly, the vascular system, including blood vessels and blood cells, is among the first organ systems to develop in vertebrate embryos, so that oxygen and nutrients can be delivered to the rapidly developing tissues. The defining cell type of the vascular system is the endothelial cell, which forms the lining of the entire circulatory system, including the heart, veins, and arteries. The development of blood vessels starts with mesodermal cells called **angioblasts**; these are the precursors of endothelial cells (Fig. 11.48). Angioblasts undergo a mesenchymal-to-epithelial transition and assemble into the main vessels of the vasculature, in a process called **vasculogenesis**. The initial vessels are then elaborated into a vascular system ramifying throughout the body by the process of **angiogenesis**, in which vessels extend and branch to form venules, arterioles, and extensive networks of capillaries.

Angioblast differentiation into endothelial cells requires the growth factor VEGF (vascular endothelial growth factor) and its receptors. VEGF is also a potent mitogen for endothelial cells, stimulating their proliferation. The primary blood vessels, such as the dorsal aorta and the main veins, arise from angioblasts located in the lateral mesoderm. VEGF is secreted by axial structures, such as somites, and its expression is driven by Shh signaling in the notochord. Expression of the *Vegf* gene is induced by lack of oxygen or hypoxia, and thus an active organ using up oxygen promotes its own vascularization.

Fig. 11.48 Vasculogenesis and angiogenesis. First panel: angioblasts assemble into simple tubes of endothelium during vasculogenesis. Second and third panels: the tubes are then elongated during angiogenesis, which involves the development of filopodia on endothelial cells (inset third panel), which enable the cell to move towards a source of a chemomattractant (blue circles) resulting in vessel extension, branching, and growth.

New blood capillaries are formed by sprouting from pre-existing blood vessels (see Fig. 11.48). The capillary grows by degradation of the extracellular matrix and proliferation of cells at the tip of the sprout. Cells at the tip extend filopodia-like processes that guide and extend the sprout. The response of the tip cells to VEGF is to express the Notch ligand Delta-like 4, which then activates Notch in adjacent proximal cells. Notch signaling blocks expression of the VEGF receptor and is one of the mechanisms that

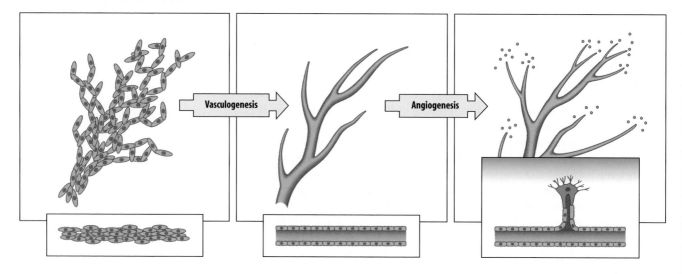

confines outgrowth to the tip of the tube. This is reminiscent of the Delta–Notch feedback mechanism deployed in *Drosophila* tracheal development (see Section 11.26).

During their development, blood vessels navigate along specific paths towards their targets, with the filopodia at the leading edge responding to both attractant and repellent cues on other cells and in the extracellular matrix. Cues in the extracellular environment are provided by proteins of the netrin and semaphorin families, which can act on the tips of growing blood vessels to block filopodial activity. Netrins and semaphorins are also involved in guidance of axons (discussed in Chapter 12), and there is a striking similarity in the signals and mechanisms that guide developing blood vessels and neurons. Blood-vessel morphogenesis also requires modulation of adhesive interactions between endothelial cells and the extracellular matrix and between the endothelial cells themselves. Changes in integrin-mediated cell adhesion to the extracellular matrix through focal contacts are particularly important in this context.

VEGF stimulates endothelial cells to degrade the surrounding basement membrane matrix and to migrate and proliferate. In the skin of mice, embryonic nerves form a template that directs the growth of arteries. The nerves secrete VEGF, which may both attract blood vessels and specify them as arteries. Many solid tumors produce VEGF and other growth factors that stimulate angiogenesis, and blocking new vessel formation is a means of reducing tumor growth. The monoclonal antibody bevacizumab (Avastin), an anti-angiogenic agent that targets VEGF, has been approved for clinical use in combination chemotherapy for colorectal cancer.

The first tubular structures of the vasculature are formed by endothelial cells and these vessels are then covered by pericytes and smooth muscle cells. Arteries and veins are defined by the direction of blood flow, as well as by structural and functional differences. Evidence from lineage tracing shows that angioblasts are specified as arterial or venous before they form blood vessels; their identities are still undetermined, however, and they can switch identity. Arterial and venous capillaries join each other in capillary beds, which form the interchange sites between the arterial and venous blood systems. The cell sorting and guidance molecule ephrin B2 is expressed in arterial blood vessels, while its receptor EphB4 is expressed in venous vessels, and interaction between them in primitive capillary networks may be required for endothelial cell sorting out into distinct arterial and venous vessels. Lymphatic vessels are also developmentally part of the vascular system and originate by budding from veins. The expression of the homeobox gene Prox1 marks an early stage of commitment to the lymphatic lineage.

Angiogenesis is not just a property of the embryo. It can occur throughout life and, properly regulated, is the means of repairing damaged blood vessels. Excessive or abnormal angiogenesis is, however, the hallmark of many human diseases. These include cancer, obesity, psoriasis, atherosclerosis, and arthritis. Human disorders relating to the nervous system that involve abnormalities in the vasculature include Alzheimer's disease, where blockage of the vasculature is a major cause of disease symptoms, and motor neuron disease.

11.30 The development of the vertebrate heart involves specification of a mesodermal tube that is patterned along its long axis

The heart is one of the first large organs to form in the embryo. It is of mesodermal origin and is first established as a single tube consisting of two epithelial layers—the inner **endocardium**, which is sheet of endothelium, and the outer **myocardium**, which will become the contractile cardiac muscle cells. During development, this tube becomes divided longitudinally into two chambers: the atrial and ventricular chambers. A two-chambered heart is the basic adult form in fish, but in higher vertebrates, such as birds and mammals, looping and further partitioning give rise to the four-chambered heart. In humans, about 1 in 100 live-born infants have some congenital heart malformation, while *in utero*, heart malformation leading to death of the embryo occurs in between 5 and 10% of conceptions (the different numbers reflect different studies).

The lateral plate mesoderm is the major source of heart mesoderm. In all vertebrates, precursor heart cells come to lie in two patches of lateral plate mesoderm on either side of the midline. They then undergo a complex morphogenesis, move towards the midline and fuse to form the heart tube (Fig. 11.49). A number of mutations in zebrafish have been found that disrupt this process and result in two laterally positioned hearts—a condition known as cardia bifida. One of the mutated genes is called *miles apart*, and codes for a receptor that binds lysosphingolipids; sphingosine 1-phosphate is the likely ligand in this case. *Miles apart* is not expressed in the migrating heart cells themselves, but in cells on either side of the midline, and thus may be involved in directing the migration of the presumptive heart cells.

As we saw in Chapter 4, heart mesoderm in *Xenopus* is initially specified by signals from the organizer during the general process of mesoderm induction and dorso-ventral patterning. In the chick and mouse, cells that will eventually become heart cells ingress through the primitive streak and become part of the lateral plate mesoderm. In the chick, presumptive heart cells ingress in a roughly similar antero-posterior order to their eventual position in the heart, but at this point they are not yet determined as heart cells nor is the antero-posterior patterning of the presumptive heart mesoderm fixed. A few hours after ingression, the cells have become committed to be heart cells and will differentiate into cardiac muscle cells if isolated. The antero-posterior patterning of the prospective heart mesoderm appears to become set soon afterwards, just before the formation of the heart tube.

There is some evidence that the heart develops as a modular organ in which each anatomical region is controlled by a distinct transcriptional program. The regions are the segments along the heart tube—atria, left ventricle, right ventricle, and ventricular outflow tract in the mammalian heart (Fig. 11.50). The precursor cells for these regions appear to acquire specific patterns of gene expression according to their position along the antero-posterior axis. The homeodomain transcription factor Nkx2.5, for example, is required for heart development and is one of the first markers of early heart cells; mutations in the *Nkx2.5* gene in mice and humans result in heart abnormalities. *Nkx2.5* is expressed in the developing heart in a complex pattern. Seven different activating regions and three repressor regions have been identified in the control region of the *Nkx2.5* gene. Activating regions include those that lead to its expression in the entire heart tube, or just in the right ventricle and outflow tract. Retinoic acid is thought to act as a signal for antero-posterior positional values in heart development.

In mice, experiments that traced the lineage of cells in the early heart region revealed the existence of two distinct lineages of heart precursor cells that derived

Fig. 11.49 Early heart development in the mouse. Left panel: schematic of a transverse section through a mouse embryo at the level of the heart at around 7.5 days. Paired heart rudiments have formed from lateral plate mesoderm on either side of the ventral midline. Gut endoderm is shown in yellow. Right panel: by 8.5 days, the ventral surface of the embryo has folded to form the foregut and the body cavity, and the two heart rudiments have fused to form a single tubular structure composed of the myocardium, which will form heart muscle, and the epithelial endocardium lining the internal surface. The heart is surrounded by the pericardial cavity.

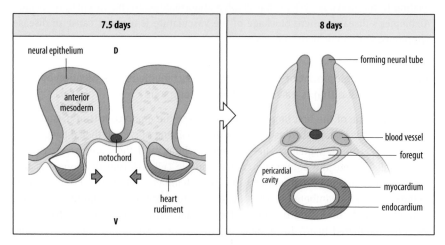

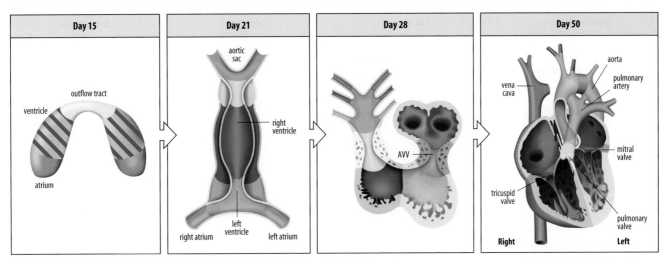

| Day 15 | Day 21 | Day 28 | Day 50 |

Day 15: ventricle, outflow tract, atrium

Day 21: aortic sac, right ventricle, left ventricle, right atrium, left atrium

Day 28: AVV

Day 50: vena cava, aorta, pulmonary artery, mitral valve, tricuspid valve, pulmonary valve, Right, Left

Fig. 11.50 Schematic of human heart development. By day 15 of human embryonic development, cardiogenic precursors have formed a crescent, as shown in the first panel, in which the main regions of the heart are already specified. The two arms of the crescent fuse along the midline to give a linear heart tube, which is patterned along the antero-posterior axis with the regions and chambers of the mature heart (second panel). After looping (third panel) these regions are disposed approximately in their eventual positions. Later development results in further patterning (fourth panel), and the formation of valves between, for example, the atria and ventricles. AVV, atrioventricular valve region.

After Srivastava, D. and Olson, E.N.: 2000.

from a common precursor very early in development. One lineage of cells gives rise to the primitive left ventricle and outflow tract; all other heart regions are colonized by both lineages.

The later development of the heart involves asymmetric looping of the heart tube. This is not well understood but is related to the left–right asymmetry of the embryo (see Section 4.7). The extracellular matrix molecule plectin is expressed earlier on the left side compared with the right and this may be as the result of the expression of Pitx2 on the left side. The heart tube then forms the separate chambers—four in the case of the mammalian heart. In addition, cells of neural crest origin contribute to the outflow tracts; they are essential, for example, for the formation of the pulmonary artery and aorta from the single embryonic outflow tract (see Fig. 11.50). Defects in these regions account for 30% of congenital heart defects in humans, some of which are the result of developmental perturbations in neural crest cell specification and migration.

There are remarkable similarities—although we should no longer be surprised—between the genes involved in heart development in *Drosophila* and in vertebrates. The homeobox gene *tinman* is required for heart formation in *Drosophila*, and is a homolog of vertebrate *Nkx2.5*. When vertebrate *tinman*-like genes are expressed in *Drosophila* they can substitute for *Drosophila tinman* and rescue some of the abnormalities caused by lack of normal *tinman* functions. In *Drosophila*, *dpp* maintains *tinman* expression in the dorsal mesoderm, while in vertebrates, BMPs (relatives of Decapentaplegic) have been shown to induce *Nkx2.5* and are also able to induce cardiac cell differentiation. Nkx2.5 or Tinman act together with GATA transcription factors to turn on cardiac-specific gene expression, although GATA transcription factors are not expressed exclusively in cardiac precursors.

11.31 A homeobox gene code specifies tooth identity

Teeth develop from a well-defined series of interactions between the oral epithelium and the mesenchyme of the jaws, which is of neural crest origin. Teeth develop from areas of thickened oral epithelium called the tooth placodes. The epithelium invaginates and the surrounding mesenchyme condenses, forming a tooth primordium or tooth germ (Fig. 11.51). In each primordium, a specialized group of cells—the enamel knot—acts as a signaling center for later development, such as cusp formation. The mesenchyme forms the dental papilla, which gives rise to the pulp and dentine, while the enamel is secreted by the epithelial cells. Tooth primordia that give rise to different

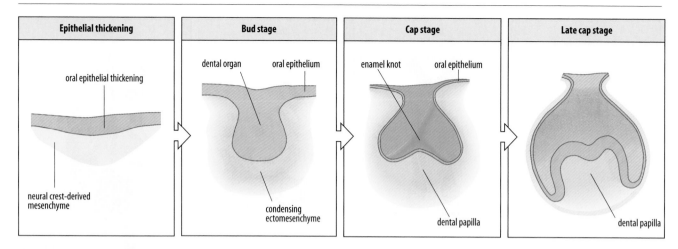

| Epithelial thickening | Bud stage | Cap stage | Late cap stage |

Fig. 11.51 Schematic diagram of the stages of tooth development.

types of teeth, such as incisors and molars, are specified at precise positions, and it is this positioning that we will discuss here.

Early tooth development is controlled by reciprocal interactions, by way of secreted signals, between the oral epithelium and the underlying mesenchyme. The sites of tooth formation are determined initially by signals produced by the epithelium. BMP-4 is expressed in the epithelium at all sites of tooth formation, and induces, or maintains, expression of the homeodomain transcription factor Msx1 in the underlying mesenchyme. FGF signaling is also involved in a similar pattern. Shh is expressed in the epithelium at sites of tooth formation throughout tooth development, whereas Wnt-7b is expressed elsewhere in the oral epithelium except at sites of tooth formation. It is likely that the formation of supernumerary teeth is the result of increased Shh signaling, which prevents the death of any supernumerary tooth buds that form, and increased Wnt signaling, which increases the number of tooth buds formed from the oral epithelium.

The reciprocal expression of these latter two signaling molecules establishes the boundaries between the oral epithelium and the dental epithelium that invaginates to form the tooth. As tooth-bud development proceeds, the direction of signaling changes so that signals secreted by the mesenchyme direct the development of the epithelium; for example, the development of the enamel knot.

In rodents, it has been established that the type of tooth that develops—incisor or molar—is determined by the expression of homeobox genes in the mesenchyme. These homeobox genes, which include *Barx1, Dlx1, Dlx2, Msx1,* and *Msx2* among others, provide a spatial code in the facial mesoderm (Fig. 11.52), analogous to the Hox

Fig. 11.52 The expression domains of four homeobox genes in the mesoderm of the mandible before initiation of tooth primordia. The positions at which the incisors and molars will develop are indicated in black.

After Ferguson, C.A., et al.: 2000.

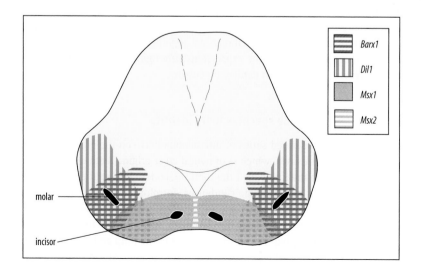

gene code along the antero-posterior body axis. Thus, *Dlx2* and *Barx1* are expressed in mesenchymal cells that will form molars, while *Msx1* and *Msx2* are expressed in cells that will form incisors. In mouse mutants lacking both *Dlx2* and *Dlx1* (a similar gene that can substitute for *Dlx2* in development), molars do not develop, but ectopic cartilage structures form at the molar sites. The homeodomain transcription factor Islet1, which we shall meet again in neural development, is expressed exclusively in developing incisors. It has a positive regulatory interaction with BMP-4, which induces *Msx1* expression in the incisor region.

Summary

Many internal organs, such as lungs, kidneys, and blood vessels, are formed of tubular epithelial structures. Tubular structures can be formed *de novo* by two basic mechanisms: invagination and remodeling of a pre-existing epithelium, as in the *Drosophila* trachea and the vertebrate lung, or by the aggregation of mesenchymal cells to form a tubular epithelium, as in the initial formation of kidney tubules and blood vessels. Branching of the *Drosophila* tracheal system and the airways of the vertebrate lung are controlled by conserved signals produced by the surrounding mesenchyme. The vascular system develops from mesenchymal endothelial cells that assemble during vasculogenesis into tubular epithelial structures that are the precursors of the large blood vessels. The tubules then extend and branch into finer vessels in the process of angiogenesis. The development of the mammalian kidney involves both mesenchymal-to-epithelial transitions and reciprocal induction. The ureteric bud is induced by the prospective kidney mesenchyme to grow and branch, while the ureteric bud induces mesenchymal cells to condense together and form tubules. The vertebrate heart initially forms as a linear tube made up of an inner endocardium and an outer myocardium. Its later development may be modular and is controlled by genes similar to those that regulate heart formation in *Drosophila*. Teeth develop from an invagination of the oral epithelium and the underlying mesenchyme and their positioning and further development is controlled by reciprocal signaling between epithelium and mesenchyme. A homeobox gene code in the mesenchyme specifies tooth identity.

Summary to Chapter 11

The development of organs involves similar processes and, in some cases, the same genes, as those used earlier in development. The pattern of the vertebrate limb is specified along three axes, at right angles to each other. Signaling molecules provide positional information, and interpretation of this information involves Hox genes. The development *of Drosophila* wings and legs from imaginal discs depends on positional signals generated at compartment boundaries, and some of these signal proteins are similar to those used in vertebrate limb patterning. The development of internal vertebrate organs such as the vascular system, lungs, and teeth involves epithelial morphogenesis and, in some cases, branching of tubular structures.

▨ End of chapter questions

Long answer (concept questions)

1. Describe an experiment that showed that FGFs play a critical role in the initiation of limb bud formation. What causes the expression of FGFs? Propose a linear gene expression pathway from Hox gene expression to limb bud formation based on these facts.

2. Describe the location and extent of the apical ectodermal ridge with respect to the three main axes of the limb.

3. What is the consequence of removing the ridge at different times during limb formation? How do you interpret these results? Cite evidence for your interpretation.

4. Describe the proposed relationship between FGF and retinoic acid signaling in specifying identity along the proximo-distal axis in the developing limb.

5. Where in the limb bud is the polarizing region (also known as the zone of polarizing activity or ZPA)? What properties of this region make it an organizing region (see Fig. 11.12)? Through which signaling molecule does it exert its effects? What is the evidence for the importance of this molecule?

6. What is Gli3? What is its role in specifying digit identity, and through what downstream genes does it appear to exert this role?

7. What is the consequence of grafting the ectoderm from a left limb bud onto a right mesodermal core, such that the dorso-ventral axis is reversed without altering the antero-posterior axis? What conclusion is drawn from this experiment? How are Wnt and Engrailed involved in this patterning, and how do we know?

8. What are the signals from the dorsal ectoderm and the apical ectodermal ridge that are required to maintain Shh expression in the polarizing region? In turn, how does Shh maintain signaling from the apical ectodermal ridge?

9. What is the origin of the muscle cells in a limb? What is the evidence that the limb mesoderm controls the pattern of muscles that forms, so that the musculature is appropriate to the limb?

10. How is the identity of the different *Drosophila* imaginal discs established in the embryo—for example, how is it determined whether a disc will form a wing, a leg, or some other structure? How does misexpression of *Antennapedia* in the head region illustrate this?

11. The gene *engrailed* is expressed in the posterior compartment of the wing disc in the larva and in the posterior compartment of segments in the embryo. Compare and contrast the signals downstream of posterior compartment *engrailed* expression in the wing disc and in segmentation (refer back to Chapter 2 for information about segmentation).

12. The *Drosophila* genes *wingless*, *apterous* (Pteron is Greek for wing), and *vestigial* can all reduce or eliminate the formation of the wing when mutated. Based on the roles of these genes in the wing imaginal disc, propose mechanisms by which these mutations cause the wing defects.

13. What type of protein does the gene *Pax6* encode and discuss its role in eye development? What human conditions result from defects in the normal function or expression of *Pax6*?

14. Describe the developmental events leading to formation of the lens of the eye in vertebrate embryos. What have experiments in the teleost genus *Astyanax* contributed to our understanding of the importance of the lens in eye development?

15. What is the role of signaling through the EGF receptor, DER, in development of the *Drosophila* ommatidial cell cluster?

16. Summarize the general processes involved in blood-vessel formation in vertebrates. What is the role of VEGF in blood-vessel formation?

Multiple choice (factual recall questions)

NB There is only one correct answer to each question.

1. Cartilage (bone-forming tissue) and connective tissue of the vertebrate limb originate from

a) mesodermal cells that migrate into the limb bud from the somites

b) mesodermal mesenchyme of the limb bud

c) the apical ectodermal ridge

d) the progress zone

2. Removal of the apical ectodermal ridge of a limb bud leads to

a) continued development of the proximal structures, but no formation of new distal structures

b) degeneration of the limb bud

c) regeneration of a new apical ridge from adjacent epidermal tissue

d) regeneration of the entire limb bud from underlying mesoderm

3. Grafting of a second polarizing region into the anterior region of a limb bud results in

a) formation of additional digits in mirror symmetry with the normal digits

b) formation of a second limb bud at the site of the graft

c) formation of new digits that lack any specific identity

d) formation of new digits, but only those of anterior identity

4. The programmed cell death that separates the digits is dependent upon which signaling pathway?

a) BMP

b) FGF

c) Shh

d) Wnt

5. The *Drosophila apterous* gene is related to the mouse *Lmx1b* gene, and both are involved in

a) controlling segmentation along the antero-posterior axis

b) determining whether or not wings are made

c) specifying dorsal identity in their respective appendages

d) specifying what structure is made from a given segment

6. The formation of the eyespot on butterfly wings is dependent on the expression of _____ in the disc, which is not expressed in the *Drosophila* wing disc.

a) *Distal-less*

b) *Engrailed*

c) *Hedgehog*

d) *Wingless*

7. The engineered expression of the mouse *Pax6* gene in the *Drosophila* leg disc

a) has no effect, since mouse genes cannot work in insects

b) can cause formation of mouse-type eye structures on the leg

c) will transform the leg disc into a wing disc

d) can cause formation of *Drosophila* eye structures on the leg

8. Drugs that block VEGF are used for

a) correcting birth defects

b) preventing limb development in mice

c) studying eye development in *Drosophila*

d) treating cancer

9. The heart develops from

a) foregut endoderm

b) invaginations of the ectodermal epithelium

c) lateral plate mesoderm

d) mesodermal angioblasts

10. The type of tooth that will form in different regions of the jaw is controlled by

a) BMP expression

b) homeobox gene expression

c) lateral inhibition

d) previous patterning in the mesoderm

Multiple choice answer key

1: b, 2: a, 3: a, 4: a, 5: c, 6: a, 7: d, 8: d, 9: c, 10: b.

▨ Section further reading

11.1 The vertebrate limb develops from a limb bud

Delaurier, A., Burton, N., Bennett, M., Baldock, R., Davidson, D., Mohun, T.J., Logan, M.P.: **The Mouse Limb Anatomy Atlas: an interactive 3D tool for studying embryonic limb patterning.** *BMC Dev. Biol.* 2008, **8**: 83–89.

Tickle, C.: **The contribution of chicken embryology to the understanding of vertebrate limb development.** *Mech. Dev.* 2004, **121**: 1019–1029.

11.2 Genes expressed in the lateral plate mesoderm are involved in specifying the position and type of limb

Kawakami, Y., Capdevila, J., Buscher, D., Itoh, T., Rodriguez Esteban, C., Izpisua Belmonte, J.C.: **WNT signals control FGF-dependent limb initiation and AER induction in the chick embryo.** *Cell* 2001, **104**: 891–900.

Minguillon, C., Buono, J.D., Logan, M.P.: ***Tbx5* and *Tbx4* are not sufficient to determine limb-specific morphologies but have common roles in initiating limb outgrowth.** *Dev. Cell* 2005, **8**: 75–84.

11.3 The apical ectodermal ridge is required for limb outgrowth

Fernandez-Teran, M., Ros, M.A.: **The apical ectodermal ridge: morphological aspects and signaling pathways.** *Int. J. Dev. Biol.* 2008, **52**: 857–871.

Mariani, F.V., Ahn, C.P., Martin, G.R.: **Genetic evidence that FGFs have an instructive role in limb proximal-distal patterning.** *Nature* 2008, **453**: 401–405.

Niswander, L., Tickle, C., Vogel, A., Booth, I., Martin, G.R.: **FGF-4 replaces the apical ectodermal ridge and directs outgrowth and patterning of the limb.** *Cell* 1993, **75**: 579–587.

11.4 Patterning of the limb involves positional information

Niswander, L.: **Pattern formation: old models out on a limb.** *Nat. Rev. Genet.* 2003, **4**: 133–143.

Towers, M., Tickle, C.: **Growing models of vertebrate limb development.** *Development* 2009, **136**: 179–190.

11.5 Position along the proximo-distal axis of the limb bud may be specified is still a matter of debate

Galloway, J.L., Delgado, I., Ros, M.A., Tabin, C.J.: **A reevaluation of X-irradiation-induced phocomelia and proximodistal limb patterning.** *Nature* 2009, **460**: 400–404.

Summerbell, D., Lewis, J.H., Wolpert, L.: **Positional information in chick limb morphogenesis.** *Nature* 1973, **244**: 492–496.

Tabin, C., Wolpert, L.: **Rethinking the proximodistal axis of the vertebrate limb in the molecular era.** *Genes Dev.* 2007, **21**: 1433–1442.

Therapontos, C., Erskine, L., Gardner, E.R., Figg, W.D., Vargesson, N.: **Thalidomide induces limb defects by preventing angiogenic outgrowth during early limb formation.** *Proc. Natl Acad. Sci. USA* 2009, **106**: 8573–8578.

Wolpert, L., Tickle, C., Sampford, M.: **The effect of cell killing by X-irradiation on pattern formation in the chick limb.** *J. Embryol. Exp. Morph.* 1979, **50**: 175–193.

Zeller, R., Lopez-Rios, J., Zuniga, A.: **Vertebrate limb bud development: moving towards integrative analysis of organogenesis.** *Nat. Rev. Genet.* 2009, **10**: 845–858.

11.6 The polarizing region specifies position along the limb's antero-posterior axis

Litingtung, Y., Dahn, R.D., Li, Y., Fallon, J.F., Chiang, C.: **Shh and Gli3 are dispensable for limb skeleton formation but regulate digit number and identity.** *Nature* 2002, **418**: 979–983.

Riddle, R.D., Johnson, R.L., Laufer, E., Tabin, C.: *Sonic hedgehog* **mediates polarizing activity of the ZPA.** *Cell* 1993, **75**: 1401–1416.

te Welscher, P., Zuniga, A., Kuijper, S., Drenth, T., Goedemans, H.J., Meijlink, F., Zeller, R.: **Progression of vertebrate limb development through SHH-mediated counteraction of GLI3.** *Science* 2002, **298**: 827–830.

Tickle, C.: **Making digit patterns in the vertebrate limb.** *Nat. Rev. Mol. Cell Biol.* 2006, **7**: 1–9.

Towers, M., Mahood, R., Yin, Y., Tickle, C.: **Integration of growth and specification in chick wing digit-patterning.** *Nature* 2008, **452**: 882–886.

Box 11A Positional information and morphogen gradients

Kerszberg, M., Wolpert, L.: **Specifying positional information in the embryo: looking beyond morphogens.** *Cell* 2007, **130**: 205–209.

11.7 Sonic hedgehog produced by the polarizing region is likely to be the primary morphogen patterning the antero-posterior axis of the limb

Riddle, R.D., Johnson, R.L., Laufer, E., Tabin, C.: **Sonic hedgehog mediates the polarizing activity of the ZPA.** *Cell* 1993, **75**: 1401–1416.

Yang, Y., Drossopoulou, G., Chuang, P.T., Duprez, D., Marti, E., Bumcrot, D., Vargesson, N., Clarke, J., Niswander, L., McMahon, A., Tickle, C.: **Relationship between dose, distance and time** in Sonic Hedgehog-mediated regulation of anteroposterior polarity in the chick limb. *Development* 1997, **124**: 4393–4404.

Box 11B Too many fingers: mutations that affect antero-posterior patterning can cause polydactyly

Ahn, S., Joyner, A.L.: **Dynamic changes in the response of cells to positive hedgehog signaling during mouse limb patterning.** *Cell* 2004, **118**: 505–516.

Harfe, B.D., Scherz, P.J., Nissim, S., Tian, H., McMahon, A.P., Tabin, C.J.: **Evidence for an expansion-based temporal Shh gradient in specifying vertebrate digit identities.** *Cell* 2004, **118**: 517–528.

Hill, R.E., Heaney, S.J., Lettice, L.A.: **Shh: restricted expression and limb dysmorphologies.** *J. Anat.* 2003, **202**: 13–20.

Lettice, L.A., Hill, A.E., Devenney, P.S., Hill, R.E.: **Point mutations in a distant Shh cis-regulator generate a variable regulatory output responsible for preaxial polydactyly.** *Hum. Mol. Genet.* 2008, **17**: 978–985.

Box 11C Sonic hedgehog signaling and the primary cilium

Eggenschwiler, J.T., Anderson, K.V.: **Cilia and developmental signaling.** *Annu. Rev. Cell Dev. Biol.* 2007, **23**: 345–373.

Huangfu, D., Anderson, K.V.: **Cilia and Hedgehog responsiveness in the mouse.** *Proc. Natl. Acad. Sci. USA* 2005, **102**: 11325–11330.

Huangfu, D., Liu, A., Rakeman, A.S., Murcia, N.S., Niswander, L., Anderson, K.V.: **Hedgehog signalling in the mouse requires intraflagellar transport proteins.** *Nature* 2003, **426**: 83–87.

11.8 Transcription factors might specify digit identity

Dahn, R.D., Fallon, J.F.: **Interdigital regulation of digit identity and homeotic transformation by modulated BMP signaling.** *Science* 2000, **289**: 438–441.

Suzuki, T., Takeuchi, J., Koshiba-Takeuchi, K., Ogura, T.: ***Tbx* genes specify posterior digit identity through Shh and BMP signaling.** *Dev. Cell* 2004, **6**: 43–53.

Suzuki, T., Hasso, S.M., Fallon, J.F.: **Unique SMAD1/5/8 activity at the phalanx-forming region determines digit identity.** *Proc. Natl. Acad. Sci. USA* 2008, **105**: 4185–4190.

Vokes, S.A., Ji, H., Wong, W.H., McMahon, A.P.: **A genome-scale analysis of the *cis*-regulatory circuitry underlying Shh-mediated patterning of the mammalian limb.** *Genes Dev.* 2008, **22**: 2651–2663.

11.9 The dorso-ventral axis of the limb bud is controlled by the ectoderm

Altabef, M., Clarke, J.D.W., Tickle, C: **Dorso-ventral ectodermal compartments and origin of apical ectodermal ridge in developing chick limb.** *Development* 1997, **124**: 4547–4556.

Arques, C.G., Doohan, R., Sharpe, J., Torres, M.: **Cell tracing reveals a dorsoventral lineage restriction plane in the mouse limb bud mesenchyme.** *Development* 2007, **134**: 3713–3722.

Geduspan, J.S., MacCabe, J.A.: **The ectodermal control of mesodermal patterns of differentiation in the developing chick wing.** *Dev. Biol.* 1987, **124**: 398–408.

Riddle, R.D., Ensini, M., Nelson, C., Tsuchida, T., Jessell, T.M., Tabin, C.: **Induction of the LIM homeobox gene *Lmx1* by *Wnt-7a* establishes dorsoventral pattern in the vertebrate limb.** *Cell* 1995, **83**: 631–640.

11.10 Development of the limb is integrated by interactions between signaling centers

Bénazet, J.D., Bischofberger, M., Tiecke, E., Gonçalves, A., Martin, J.F., Zuniga, A., Naef, F., Zeller, R.: **A self-regulatory system of interlinked signaling feedback loops controls mouse limb patterning.** *Science* 2009, **323**: 1050–1053.

Niswander, L., Jeffrey, S., Martin, G.R., Tickle, C.: **A positive feedback loop coordinates growth and patterning in the vertebrate limb.** *Nature* 1994, **371**: 609–612.

Pizette, S., Niswander, L.: **BMPs negatively regulate structure and function of the limb apical ectodermal ridge.** *Development* 1999, **126**: 883–894.

Zeller, R., Lopz-Rios, J., Zuniga, A.: **Vertebrate limb development: moving towards integrative analysis of organogenesis.** *Nature Rev. Genet.* 2009, **10**: 845–855.

11.11 Different interpretations of the same positional signals give different limbs

Krabbenhoft, K.M., Fallon, J.F.: **The formation of leg or wing specific structures by leg bud cells grafted to the wing bud is influenced by proximity to the apical ridge.** *Dev. Biol.* 1989, **131**: 373–382.

Logan, M.: **Finger or toe: the molecular basis of limb identity.** *Development* 2003, **130**: 6401–6410.

Logan, M., Tabin, C.J.: **Role of Pitx1 upstream of Tbx4 in specification of hindlimb identity.** *Science* 1999, **283**: 1736–1739.

11.12 Hox genes establish the polarizing region and also provide a code values for limb patterning

Charité, J., De Graaff, W., Shen, S., Deschamps, J.: **Ectopic expression of *Hoxb-8* causes duplication of the ZPA in the forelimb and homeotic transformation of axial structures.** *Cell* 1994, **78**: 589–601.

Goodman, F.R.: **Limb malformations and the human HOX genes.** *Am. J. Med. Genet.* 2002, **112**: 256–265

Kmita, M., Tarchini, B., Zàkàny, J., Logan, M., Tabin, C.J., Duboule, D.: **Early developmental arrest of mammalian limbs lacking *HoxA/HoxD* gene function.** *Nature* 2005, **435**: 1113–1116.

Nelson, C.E., Morgan, B.A., Burke, A.C., Laufer, E., DiMambro, E., Muytaugh, L.C., Gonzales, E., Tessarollo, L., Parada, L.F., Tabin, C.: **Analysis of Hox gene expression in the chick limb bud.** *Development* 1996, **122**: 1449–1466.

Wellik, D.M., Capecchi, M.R.: **Hox10 and Hox11 genes are required to globally pattern the mammalian skeleton.** *Science* 2003, **301**: 363–367.

Zakany, J., Kmita, M., Duboule, D.: **A dual role for Hox genes in limb anterior-posterior asymmetry.** *Science* 2004, **304**: 1669–1672.

Zakany, J., Duboule, D.: **The role of Hox genes during vertebrate limb development.** *Curr. Opin. Genet. Dev.* 2007, **17**: 359–366.

11.13 Self-organization may be involved in the development of the limb bud

Hardy, A., Richardson, M.K., Francis-West, P.N., Rodriguez, C., Izpisúa-Belmonte, J.C., Duprez, D., Wolpert, L.: **Gene expression, polarising activity and skeletal patterning in reaggregated hind limb mesenchyme.** *Development* 1995, **121**: 4329–4337.

Box 11D Reaction-diffusion mechanisms

Kondo, S., Asai, R.: **A reaction–diffusion wave on the skin of the marine angelfish** *Pomocanthus*. *Nature* 1995, **376**: 765–768.

Meinhardt, H., Gierer, A.: **Pattern formation by local self-activation and lateral inhibition.** *BioEssays* 2000, **22**: 753–760.

Murray, J.D.: **How the leopard gets its spots.** *Sci. Am.* 1988, **258**: 80–87.

11.14 Limb muscle is patterned by the connective tissue

Amthor, H., Christ, B., Patel, K.: **A molecular mechanism enabling continuous embryonic muscle growth—a balance between proliferation and differentiation.** *Development* 1999, **126**: 1041–1053.

Hashimoto, K., Yokouchi, Y., Yamamoto, M., Kuroiwa, A.: **Distinct signaling molecules control** *Hoxa-11* **and** *Hoxa-13* **expression in the muscle precursor and mesenchyme of the chick limb bud.** *Development* 1999, **126**: 2771–2783.

Robson, L.G., Kara, T., Crawley, A., Tickle, C.: **Tissue and cellular patterning of the musculature in chick wings.** *Development* 1994, **120**: 1265–1276.

Schweiger, H., Johnson, R.L., Brand-Sabin, B.: **Characterization of migration behaviour of myogenic precursor cells in the limb bud with respect to Lmx1b expression.** *Anat. Embryol.* 2004, **208**: 7–18.

11.15 The initial development of cartilage, muscles, and tendons is autonomous

Kardon, G.: **Muscle and tendon morphogenesis in the avian hind limb.** *Development* 1998, **125**: 4019–4032.

Ros, M.A., Rivero, F.B., Hinchliffe, J.R., Hurle, J.M.: **Immunohistological and ultrastructural study of the developing tendons of the avian foot.** *Anat. Embryol.* 1995, **192**: 483–496.

11.16 Joint formation involves secreted signals and mechanical stimuli

Khan, I.M., Redman, S.N., Williams, R., Dowthwaite, G.P., Oldfield, S.F., Archer, C.W.: **The development of synovial joints.** *Curr. Top. Dev. Biol.* 2007, **79**: 1–36.

Spitz, F., Duboule, D.: **The art of making a joint.** *Science* 2001, **291**: 1713–1714.

11.17 Separation of the digits is the result of programmed cell death

Garcia-Martinez, V., Macias, D., Gañan, Y., Garcia-Lobo, J.M., Francia, M.V., Fernandez-Teran, M.A., Hurle, J.M.: **Internucleosomal DNA fragmentation and programmed cell death (apoptosis) in the interdigital tissue of the embryonic chick leg bud.** *J. Cell. Sci.* 1993, **106**: 201–208.

Zuzarte-Luis, V., Hurle, J.M.: **Programmed cell death in the embryonic vertebrate limb.** *Semin. Cell Dev. Biol.* 2005, **16**: 261–269.

11.18 Positional signals from compartment boundaries pattern the wing imaginal disc & 11.19 A signaling center at the boundary between dorsal and ventral compartments patterns the *Drosophila* wing along the dorso-ventral axis

Baeg, G.H., Selva, E.M., Goodman, R.M., Dasgupta, R., Perrimon, N.: **The Wingless morphogen gradient is established by the cooperative action of Frizzled and heparan sulfate proteoglycan receptors.** *Dev. Biol.* 2004, **276**: 89–100.

Blair, S.S.: **Notch signaling: Fringe really is a glycosyltransferase.** *Curr. Biol.* 2000, **10**: R608–R612.

Crozatier, M., Glise, B., Vincent, A.: **Patterns in evolution: veins of the** *Drosophila* **wing.** *Trends Genet.* 2004, **20**: 498–505.

Entchev, E.V., Schwabedissen, A., González-Gaitán, M.: **Gradient formation of the TGF-β homolog Dpp.** *Cell* 2000, **103**: 981–991.

Fujise, M., Takeo, S., Kamimura, K., Matsuo, T., Aigaki, T., Izumi, S., Nakato, H.: **Dally regulates Dpp morphogen gradient formation in the** *Drosophila* **wing.** *Development* 2003, **130**: 1515–1522.

Han, C., Belenkaya, T.Y., Wang, B., Lin, X.: *Drosophila* **glypicans control the cell-to-cell movement of Hedgehog by a dynamin-independent process.** *Development* 2004, **131**: 601–611.

Kruse, K., Pantazis, P., Bollenbach, T., Julicher, F., González-Gáitan, M.: **Dpp gradient formation by dynamin-dependent endocytosis: receptor trafficking and the diffusion model.** *Development* 2004, **131**: 4843–4856.

Lawrence, P.: **Morphogens: how big is the picture?** *Nat. Cell Biol* 2001, **3**: E151–E154.

Lawrence, P.A., Struhl G.: **Morphogens, compartments, and pattern: lessons from** *Drosophila?* *Cell* 1996, **85**: 951–961.

Matakatsu, H., Blair, S.S.: **Interactions between Fat and Daschous and the regulation of planar cell polarity in the** *Drosophila* **wing.** *Development* 2004, **131**: 3785–3794.

Milán, M., Cohen, S.M.: **A re-evaluation of the contributions of Apterous and Notch to the dorsoventral lineage restriction boundary in the** *Drosophila* **wing.** *Development* 2003, **130**: 553–562.

Morata, G.: **How** *Drosophila* **appendages develop.** *Nature Rev. Mol. Cell Biol.* 2001, **2**: 89–97.

Moser, M., Campbell, G.: **Generating and interpreting the Brinker gradient in the** *Drosophila* **wing.** *Dev. Biol.* 2005, **286**: 647–658.

Muller, B., Hartmann, B., Pyrowolakis, G., Affolter, M., Basler, K.: **Conversion of an extracellular Dpp/BMP morphogen gradient into an inverse transcriptional gradient.** *Cell* 2003, **113**: 221–233.

Piddini, E., Vincent, J.P.: **Interpretation of the Wingless gradient requires signaling-induced self-inhibition.** *Cell* 2009, **136**: 296–307.

Rauskolb, C., Correia, T., Irvine, K.D.: **Fringe-dependent separation of dorsal and ventral cells in the** *Drosophila* **wing.** *Nature* 1999, **401**: 476–480.

Strutt, D.I.: **Asymmetric localization of Frizzled and the establishment of cell polarity in the** *Drosophila* **wing.** *Mol. Cell* 2001, **7**: 367–375.

Tabata, T.: **Genetics of morphogen gradients.** *Nature Rev. Genet.* 2001, **2**: 620–630.

Teleman, A.A., Strigini, M., Cohen, S.M.: **Shaping morphogen gradients.** *Cell* 2001, **105**: 559–562.

Zecca, M., Basler, K., Struhl, G.: **Sequential organizing activities of engrailed, hedgehog and decapentaplegic in the** *Drosophila* **wing.** *Development* 1995, **121**: 2265–2278.

11.20 The leg disc is patterned in a similar manner to the wing disc, except for the proximo-distal axis

Emerald, B.S., Cohen, S.M.: **Spatial and temporal regulation of the homeotic selector gene** *Antennapedia* **is required for the establishment of leg identity in** *Drosophila*. *Dev. Biol.* 2004, **267**: 462–472.

Kojima, T.: **The mechanism of *Drosophila* leg development along the proximodistal axis.** *Dev. Growth Differ.* 2004, **46**: 115–129.

McKay, D.J., Estella, C., Mann, R.S.: **The origins of the *Drosophila* leg revealed by the *cis*-regulatory architecture of the *Distalless* gene.** *Development* 2009, **136**: 61–71.

11.21 Butterfly wing markings are organized by additional positional fields

Brakefield, P.M., French, V.: **Butterfly wings: the evolution of development of colour patterns.** *BioEssays* 1999, **21**: 391–401.

French, V., Brakefield, P.M.: **Pattern formation: a focus on Notch in butterfly spots.** *Curr. Biol.* 2004, **14**: R663–R665.

11.22 Different imaginal discs can have the same positional values

Carroll, S.B.: **Homeotic genes and the evolution of arthropods and chordates.** *Nature* 1995, **376**: 479–485.

Morata, G.: **How *Drosophila* appendages develop.** *Nature Rev. Mol. Cell Biol.* 2001, **2**: 89–97.

Si Dong, P.D., Chu, J., Panganiban, G.: **Coexpression of the homeobox genes *Distal-less* and *homothorax* determines *Drosophila* antennal identity.** *Development* 2000, **127**: 209–216.

11.23 The vertebrate eye develops from the neural tube and the ectoderm of the head

Chow, R.L., Lang, R.A.: **Early eye development in vertebrates.** *Annu. Rev. Cell Dev. Biol.* 2001, **17**: 255–296.

Donner, A.L., Lachke, S.A., Maas, R.L.: **Lens induction in vertebrates: Variations on a conserved theme of signaling events.** *Semin. Cell Dev. Biol.* 2006, **17**: 676–685.

Gehring, W.J.: **New perspectives on eye development and the evolution of eyes and photoreceptors.** *J. Hered.* 2005, **96**: 171–184.

Streit, A.: **The preplacodal region: an ectodermal domain with multipotential progenitors that contribute to sense organs and cranial sensory ganglia.** *Int. J. Dev. Biol.* 2007, **51**: 447–461.

Takahashi, S., Asashima, M., Kurata, S., Gehring, W.J.: **Conservation of *Pax 6* function and upstream activation by Notch signaling in eye development of frogs and flies.** *Proc. Natl Acad. Sci. USA* 2002, **99**: 2020–2025.

11.24 Patterning of the *Drosophila* eye involves cell-cell interactions

Baonza, A., Casci, T., Freeman, M.: **A primary role for the epidermal growth factor receptor in ommatidial spacing in the *Drosophila* eye.** *Curr. Biol.* 2001, **11**: 396–404.

Bonini, N.M., Choi, K.W.: **Early decisions in *Drosophila* eye morphogenesis.** *Curr. Opin. Genet. Dev.* 1995, **5**: 507–515.

Chou, W.H., Huber, A., Bentrop, J., Schulz, S., Schwab, K., Chadwell, L.V., Paulsen, R., Britt, S.G.: **Patterning of the R7 and R8 photoreceptor cells of *Drosophila*: evidence for induced and default cell-fate specification.** *Development* 1999, **126**: 607–616.

Frankfort, B.J., Mardon, G.: **R8 development in the *Drosophila* eye: a paradigm for neural selection and differentiation.** *Development* 2002, **129**: 1295–1306.

Freeman, M.: **Cell determination strategies in the *Drosophila* eye.** *Development* 1997, **124**: 261–270.

Strutt, H., Strutt, D.: **Polarity determination in the *Drosophila* eye.** *Curr. Opin. Genet. Dev.* 1999, **9**: 442–446.

Tomlinson, A., Struhl, G.: **Decoding vectorial information from a gradient: sequential roles of the receptors Frizzled and Notch in establishing planar polarity in the Drosophila eye.** *Development* 1999, **126**: 5725–5738.

Tomlinson, A., Struhl, G.: **Delta/Notch and Boss/Sevenless signals act combinatorially to specify the *Drosophila* R7 photoreceptor.** *Mol. Cell* 2001, **7**: 487–495.

11.25 Eye development in *Drosophila* is initiated by the actions of the same transcription factors that specify eye-precursor cells in vertebrates

Gehring, W.J., Kazuho, I.: **Mastering eye morphogenesis and eye evolution.** *Trends Genet.* 1999, **15**: 371–377.

Halder, G., Callaerts, P., Gehring, W.J.: **Induction of ectopic eyes by targeted expression of the eyeless gene in *Drosophila*.** *Science* 1995, **267**: 1788–1792.

11.26 The *Drosophila* tracheal system is a model for branching morphogenesis

Affolter, M., Caussinus, E.: **Branching morphogenesis in *Drosophila*: new insights into cell behaviour and organ architecture.** *Development* 2008, **135**: 2055–2064.

Ribeiro, C., Neumann, M., Affolter, M.: **Genetic control of cell intercalation during tracheal morphogenesis in *Drosophila*.** *Cell* 2004, **14**: 2197–2207.

11.27 The vertebrate lung also develops by branching of epithelial tubes

Chuang, P-T., McMahon, A.P.: **Branching morphogenesis of the lung: new molecular insights into an old problem.** *Trends Cell Biol.* 2003, **13**: 86–91.

Hogan, B.L.M., Kolodziej, P.A.: **Molecular mechanisms of tubulogenesis.** *Nat Rev. Genet.* 2002, **3**: 513–523.

Horowitz, A., Simons, M.: **Branching morphogenesis.** *Circ. Res.* 2008, **103**: 784–795.

Pepicelli, C.V., Lewis, P.M., McMahon, A.P.: **Sonic hedgehog regulates branching morphogenesis in the mammalian lung.** *Curr. Biol.* 1998, **8**: 1083–1086.

11.28 The development of kidney tubules involves reciprocal induction by the ureteric bud and surrounding mesenchyme

Barasch, J., Yang, J., Ware, C.B., Taga, T., Yoshida, K., Erdjument-Bromage, H., Tempst, P., Parravicini, E., Malach, S., Aranoff, T., Oliver, J.A.: **Mesenchymal to epithelial conversion in rat metanephros is induced by LIF.** *Cell* 1999, **99**: 377–386.

Gao, X., Chen, X., Taglienti, M., Rumballe, B., Little, M.H., Kreidberg, J.A.: **Angioblast-mesenchyme induction of early kidney development is mediated by Wt1 and Vegfa.** *Development* 2005, **132**: 5437–5449.

Grieshammer, U., Le, Ma, Plump, A.S., Wang, F., Tessier-Lavigne, M., Martin, G.R.: **SLIT2-mediated ROBO2 signaling restricts kidney induction to a single site.** *Dev. Cell* 2004, **6**: 709–717.

Kuure, S., Vuolteenaho, R., Vainio, S.: **Kidney morphogenesis: cellular and molecular regulation.** *Mech. Dev.* 2000, **94**: 47–56.

Schedl, A., Hastie, N.D.: **Cross-talk in kidney development.** *Curr. Opin. Genet. Dev.* 2000, **10**: 543–549.

Shah, M.M., Sampogna, R.V., Sakurai, H., Bush, K.T., Nigam, S.K.: **Branching morphogenesis and kidney disease**. *Development* 2004, **131**: 1449–1462.

Shakya, R., Watanabe, T., Costantini, F.: **The role of GDNF/Ret signaling in ureteric bud cell fate and branching morphogenesis**. *Dev. Cell* 2005, **8**: 65–74.

Vainio, S., Muller, U.: **Inductive tissue interactions, cell signaling, and the control of kidney organogenesis**. *Cell* 1997, **90**: 975–978.

11.29 The vascular system develops by vasculogenesis followed by angiogenesis

Carmeliet, P.: **Angiogenesis in health and disease**. *Nat Med.* 2003, **9**: 653–660.

Carmeliet, P., Tessier-Lavigne, M.: **Common mechanisms of nerve and blood vessel wiring**. *Nature* 2005, **436**: 193–200.

Coultas, L., Chawengsaksophak, K., Rossant, J.: **Endothelial cells and VEGF in vascular development**. *Nature* 2005, **438**: 937–945.

Harvey, N.L., Oliver, G.: **Choose your fate: artery, vein or lymphatic vessel?** *Curr. Opin. Genet. Dev.* 2004, **14**: 499–505.

Hogan, B.L.M., Kolodziej, P.A.: **Molecular mechanisms of tubulogenesis**. *Nat Rev. Genet.* 2002, **3**: 513–523.

Jain, R.K.: **Molecular recognition of vessel maturation**. *Nat Med.* 2003, **9**: 685–692.

Larrivée, B., Freitas, C., Suchting, S., Brunet, I., Eichmann, A.: **Guidance of vascular development: lessons from the nervous system**. *Circ. Res.* 2009,**104**: 428–441.

Lu, X., Le Noble, F., Yuan, L., Jiang, Q., De Lafarge, B., Sugiyama, D., Breant, C., Claes, F., De Smet, F., Thomas, J.L., Autiero, M., Carmeliet, P., Tessier-Lavigne, M., Eichmann, A.: **The netrin receptor UNC5B mediates guidance events controlling morphogenesis of the vascular system**. *Nature* 2004, **432**: 179–186.

Nelson, K.S., Beitel, G.J.: **More than a pipe dream: uncovering mechanisms of vascular lumen formation**. *Dev. Cell* **17**: 435–436.

Rafii, S., Lyden, D: **Therapeutic stem and progenitor cell transplantation for organ vascularization and regeneration**. *Nat Med.* 2003, **9**: 702–712.

Reese, D.E., Hall, C.E., Mikawa, T.: **Negative regulation of midline vascular development by the notochord**. *Dev. Cell* 2004, **6**: 699–708.

Roman, B.L., Weinstein, B.M.: **Building the vertebrate vasculature: research is going swimmingly**. *BioEssays* 2000, **22**: 882–893.

Rossant, J., Hirashima, M.: **Vascular development and patterning: making the right choices**. *Curr. Opin. Genet. Dev.* 2003, **13**: 408–412.

Strilić, B., Kucera, T., Eglinger, J., Hughes, M.R., McNagny, K.M., Tsukita, S., Dejana, E., Ferrara, N., Lammert, E.: **The molecular bases of vascular lumen formation in the developing mouse aorta**. *Dev. Cell* 2009, **17**: 505–515.

Wang, H.U., Chen, Z-F., Anderson, D.J.: **Molecular distinction and angiogenic interaction between embryonic arteries and veins revealed by ephrin-B2 and its receptor Eph-B4**. *Cell* 1998, **93**: 741–753.

11.30 The development of the vertebrate heart involves specification of a mesodermal tube that is patterned along its long axis

Brand, T.: **Heart development: molecular insights into cardiac specification and early morphogenesis**. *Dev. Biol.* 2003, **288**: 1–19.

Carmeliet, P.: **Angiogenesis in health and disease**. *Nat Med.* 2003, **9**: 653–660.

Driever, W.: **Bringing two hearts together**. *Nature* 2000, **406**: 141–142.

Harvey, R.P.: **Patterning the vertebrate heart**. *Nat Rev. Genet.* 2002, **3**: 544–556.

Kelly, R.G., Buckingham, M.E.: **The anterior heart-forming field: voyage to the arterial pole of the heart**. *Trends Genet.* 2002, **18**: 210–216.

Linask, K.K., Yu, X., Chen, Y., Han, M.D.: **Directionality of heart looping: effects of Pitx2c misexpression on flectin asymmetry and midline structures**. *Dev. Biol.* 2002, **246**: 407–417.

Meilhac, S.M., Esner, M., Kelly, R.G., Nicolas, J.F., Buckingham, M.E.: **The clonal origin of myocardial cells in different regions of the embryonic mouse heart**. *Dev. Cell* 2004, **6**: 685–698.

Moorman, A.F., Christoffels, V.M.: **Cardiac chamber formation: development, genes and evolution**. *Physiol Rev.* 2003, **83**: 1223–1267.

Rosenthal, N., Xavier-Neto, J.: **From the bottom of the heart: anteroposterior decisions in cardiac muscle differentiation**. *Curr. Opin. Cell Biol.* 2000, **12**: 742–746.

Srivastava, D., Olson, E.N.: **A genetic blueprint for cardiac development**. *Nature* 2000, **407**: 221–226.

Stainier, D.Y.R.: **Zebrafish genetics and vertebrate heart formation**. *Nature Rev. Genet.* 2001, **2**: 39–48.

11.31 A homeobox gene code specifies tooth identity

Cobourne, M.T., Sharpe, P.T.: **Making up the numbers: the molecular control of mammalian dental formula**. *Semin. Cell Dev. Biol.* 2010, **21**: 314–324.

Ferguson, C.A., Hardcastle, Z., Sharpe, P.T.: **Development and patterning of the dentition**. *Linn. Soc. Symp.* 2000, **20**: 188–201.

Mitsiadis, T.A., Angeli, I., James, C., Lendahl, U., Sharpe, P.T.: **Role of Islet1 in the patterning of murine dentition**. *Development* 2003, **130**: 4451–4460.

Sarkar, L., Cobourne, M., Naylor, S., Smalley, M., Dale, T., Sharpe, P.T.: **Wnt/Shh interactions regulate ectodermal boundary formation during mammalian tooth development**. *Proc. Natl. Acad. Sci. USA* 2000, **97**: 4520–4524.

Tucker, A., Sharpe, P.: **The cutting-edge of mammalian development: how the embryo makes teeth**. *Nat. Rev. Genet.* 2004, **5**: 499–508.

Development of the nervous system

The nervous system is the most complex of all the organ systems in the animal embryo. In mammals, for example, billions of nerve cells, or neurons, develop a highly organized pattern of connections, creating the neuronal network that makes up the functioning brain and the rest of the nervous system. There are many hundreds of different types of neurons, differing in identity and the connections they make, even though they may look quite similar. Understanding the developmental processes that shape the nervous system is not simple, as they involve neuronal cell differentiation, morphogenesis, and migration. Neurons send out long processes from the cell body and these must be guided to find their targets. The picture is further complicated by the activity of the nerve cells, which can refine the pattern of connections they make.

All nervous systems, vertebrate and invertebrate, provide a system of communication through a network of electrically excitable cells—the nerve cells or **neurons**—of varying sizes, shapes, and functions (Fig. 12.1). The nervous system also contains non-neuronal cells known collectively as **glia**, which have a variety of supporting roles. The Schwann cells in the peripheral nervous system are **glial cells**, as are the oligodendrocytes and astrocytes of the central nervous system. Neurons connect with each other and with other target cells, such as muscle, at specialized junctions known as **synapses**. A neuron receives input from other neurons through its highly branched **dendrites** and, if the signals are strong enough to activate the neuron, it generates a new electrical signal—a nerve impulse, or action potential—at the **cell body**. This electrical signal is then conducted along the **axon** to the **axon terminal**, or nerve ending, which makes a synapse with the dendrites or cell body of another neuron or with the surface of a muscle cell. The dendrites and axon terminals of individual neurons can be extensively branched, and a single neuron in the central nervous system can receive as many as 100,000 different inputs. At a synapse, the electrical signal is converted into a chemical signal, in the form of a chemical neurotransmitter, which is released from the nerve ending and acts on receptors in the membrane of the opposing target cell to generate or suppress a new electrical signal. The nervous system can only function properly if the neurons are correctly connected to one another, and thus a central question surrounding nervous-system development is how the connections between neurons develop with the appropriate specificity. The number of neurons in the human brain seems generally to be estimated at around 100 billion. How many of them have unique or similar identities is not known.

In this chapter, we shall use examples from both invertebrate and vertebrate central nervous systems and this raises the question: do the central nervous systems of vertebrates and invertebrates trace back to a common precursor or are they of independent evolutionary origin? Anatomically, a central nervous system is a spatially delimited nervous tissue composed of neurons with specialized functions that are interconnected by highly organized tracts of axons. The central nervous system receives sensory inputs from the body, processes them, and integrates them centrally, and produces outputs that guide behavior, such as movement. Such an organization broadly defines the central nervous systems of animals as far apart evolutionarily as annelid worms and mammals, and is in contrast to the diffuse nervous system, or nerve net, of cnidarians (*Hydra* and sea anemones), which receives sensory input and processes output without central integration.

One important aspect of central nervous-system development in both vertebrates and invertebrates is the early segregation of the ectoderm into a 'non-neural' epidermal portion and a 'neural' portion—the **neuroectoderm** (see Section 5.2). In the Bilateria, that is, all animals with bilateral symmetry, the neuroectoderm is located at the anterior end of the embryo, where the brain and associated sensory organs develop. Similarities can be traced between the embryonic central nervous systems of simple invertebrates and those of vertebrates. In the annelid worm *Platynereis*, for example, the neuroectoderm is subdivided longitudinally into partially overlapping domains of progenitor neurons that correspond to similar domains in the vertebrate neural tube and give rise to similar cell types as those of vertebrates. In vertebrates, as discussed later in this chapter, the genes that pattern the central nervous system are sensitive to signaling by members of the bone morphogenetic protein (BMP) family, and this is also the case in *Platynereis*. All this evidence suggests that a central nervous system architecture was present in the last common ancestor of the bilaterians, and supports a common origin for the nervous system in Bilateria.

For all its complexity, the nervous system is the product of the same kind of cellular and developmental processes as those involved in the development of other organs. The overall process of nervous-system development can be divided into four major stages: the specification of neural cell (neuron or glial cell) identity; the migration of neurons and the outgrowth of axons to their targets; the formation of synapses with target cells, which can be other neurons, muscle or gland cells; and the refinement of synaptic connections through the elimination of axon branches and cell death.

We have already considered some aspects of the development of the vertebrate nervous system, including neural induction (Chapter 5), the formation of the neural tube (Chapter 8), the segmental arrangement of sensory dorsal root ganglia (Chapter 8), and the differentiation of neurons from neural crest cells (Chapter 10). We will revisit some of these events when we consider the specification of neural-cell identity. A short chapter, such as this one, cannot cover all aspects of nervous-system development, and the general focus here is on the mechanisms that control the identity of cells in the nervous system and that pattern their synaptic connections.

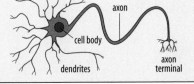

Fig. 12.1 Neurons come in many shapes and sizes. Upper panel: a montage of the various types of neurons found in the optic lobe of the avian brain. In the brain there are many more nervous connections that are not shown here. Lower panel: a single neuron consists of a small, rounded cell body, which contains the nucleus, and from which extend a single axon and a 'tree' of much-branched dendrites. Signals from other neurons are received at the dendrites, processed, and integrated in the cell body, and an outgoing signal—a nerve impulse—is generated and sent down the axon. We shall consider in more detail how neurons signal and communicate with each other in the last part of the chapter.

Specification of cell identity in the nervous system

Prospective neural tissue—the neuroectoderm—is distinguished from the rest of the ectoderm as part of the early patterning of invertebrate and vertebrate embryos, as discussed in previous chapters (Chapters 2 and 5). We will start this chapter by

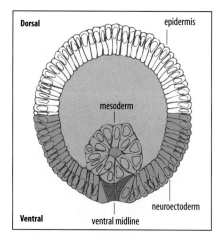

Fig. 12.2 Cross-section of a *Drosophila* early gastrula. At around 3 hours after fertilization, the neuroectoderm (blue) is located on either side of the ventral midline. The mesoderm (red), which originally lies along the ventral midline, has already been internalized.

considering the next stage in the process of generating a nervous system: how individual cells in the neuroectoderm become specified first as neural stem cells, or **neuroblasts**, and subsequently give rise to neurons and glia (see Chapter 10). We will first consider *Drosophila*, which has revealed some key developmental processes in **neurogenesis**, or the formation of neurons. We then turn to the more complex process of neurogenesis in the vertebrate neural tube. As the epithelial cells of the neural tube proliferate and differentiate into neurons, the wall of the tube develops a multi-layered structure, with the neurons migrating outwards to form the different layers. We shall look at the formation of this layered structure in relation to the mammalian cortex, which develops from the forebrain region of the neural tube.

The nervous systems of both invertebrates and vertebrates also contain enormous numbers of glial cells as well as neurons. These also develop from the neural stem cells of the neuroectoderm, but we shall not consider the specification and development of glial cells in any detail here.

12.1 Neurons in *Drosophila* arise from proneural clusters

In *Drosophila*, the cells that will give rise to the central nervous system of the larva are specified at an early stage of embryonic development, as part of the patterning that divides the embryo into different regions along the dorso-ventral and antero-posterior axes (see Section 2.19). In insects, the main nerve cord runs ventrally, rather than dorsally as in vertebrates, and the future central nervous system is specified as two longitudinal regions of neuroectoderm in the ventral half of the embryo, just above the mesoderm. The neuroectoderm, or neurogenic zone, comprises ectodermal cells that are fated to form either neural cells (neurons or glia) or epidermis. After gastrulation and internalization of the mesoderm, the neuroectoderm remains on the outside of the embryo on either side of the ventral midline (Fig. 12.2). Once individual cells are specified as neuroblasts in the neuroectoderm they move from the surface into the interior of the embryo, where they divide further and differentiate into neurons and glia, forming two tracts of axons running longitudinally on either side of the ventral midline, connected at intervals by neurons that cross the midline.

Despite very great differences between the insect central nervous system and that of vertebrates, there are intriguing parallels between the genes that pattern both systems at a very early stage. *Drosophila* neuroblasts form three longitudinal columns of cells on each side of the ventral midline, and the identity of these columns is specified while the cells are still in the neuroectoderm (Fig. 12.3). The homeodomain transcription factor genes *msh*, *ind*, and *vnd* are expressed in dorsal to ventral order in the *Drosophila* neuroectoderm, in response to the actions of earlier dorso-ventral patterning genes, such as *rhomboid* (see Section 2.19), and the activation of the *Drosophila* EGF signaling pathway. The developing vertebrate neural plate also has three longitudinal domains of prospective neural cells (see Fig. 12.3), which express genes homologous to those in the fly—*Msx*, *Gsh*, and *Nkx2*, respectively. This is a striking example of the evolutionary conservation of a mechanism for regional specification.

In response to previous patterning, the *Drosophila* neuroectoderm becomes subdivided along the antero-posterior and dorso-ventral axes into a precise orthogonal pattern of **proneural clusters**, each composed of three to five cells, which can be distinguished by the expression of genes known as **proneural genes** (Fig. 12.4). Initially, all the cells within a proneural cluster are capable of becoming neural precursor cells, or neuroblasts, but one cell, through an apparently random event, produces a signal that promotes its own development as a neuroblast and prevents the adjacent cells from following that fate. This cell will become the single neuroblast in the cluster, and can be distinguished, for example, by a high level of expression of a proneural gene such as *achaete*, whereas the surrounding cells no longer express *achaete* and will develop as epidermis (see Fig. 12.4).

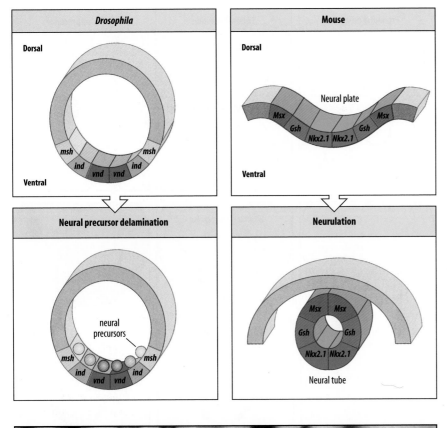

Fig. 12.3 The neuroepthelium of both *Drosophila* and vertebrates is organized into three columns of neural precursors on either side of the midline. The developing *Drosophila* embryonic central nervous system (left) and the vertebrate embryonic neural plate (right) are both organized into three longitudinal domains (columns) of gene expression on either side of the midline. The cells of each column express a specific homeobox gene and, as is evident before neurulation (upper panels), similar genes are expressed in the same medial-lateral order in *Drosophila* and in vertebrates, with *vnd/ Nkx2.1* expressed most medially (nearest the midline), *ind/Gsh* in an intermediate position, and *msh/Msx* expressed most laterally.

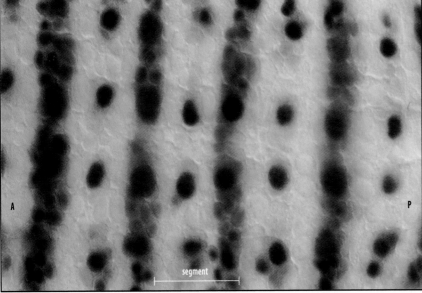

Fig. 12.4 Proneural clusters in the neuroectoderm of the *Drosophila* embryo. At the stage of development shown here, there are eight proneural clusters within each future segment, marked here by expression of the proneural gene *achaete* (black). Half of the clusters are formed in the region of engrailed expression (purple) at the posterior end of the segment, and half in a more anterior region. Some of the clusters have resolved into a single neural cell, the prospective neuroblast. The embryo is viewed here from the ventral surface; the ventral midline runs horizontally along the middle of the photo.

Photograph from Skeath, J.B., et al.: 1996.

The singling out of the prospective neuroblast is an example of lateral inhibition (Fig. 12.5) and is due to signaling between the prospective neuroblast and its neighbors via the transmembrane protein Notch and its transmembrane ligand Delta (see Box 5C, p. 190 for the Notch–Delta signaling pathway). All the cells in a proneural cluster initially can express both Notch and Delta. One cell will, however, start to express Delta sooner or more strongly than the others. Through the interaction of Delta with the Notch receptors on its less-advanced neighbors, this cell inhibits their further development as neural cells, and also suppresses their expression of Delta

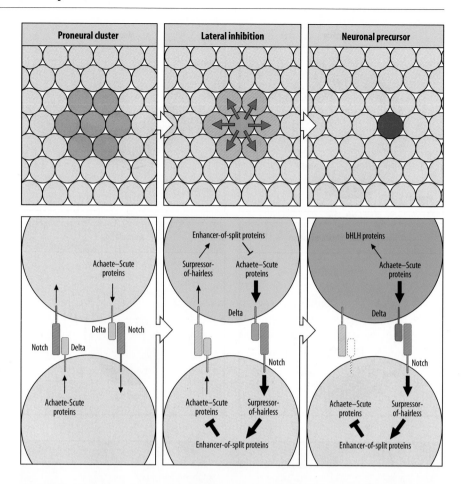

Fig. 12.5 The role of Notch signaling in lateral inhibition in neural development in *Drosophila*. As shown in the top row, the proneural cluster gives rise to a single neuronal precursor cell, the neuroblast or sensory organ precursor, by means of lateral inhibition. The rest of the cells in the cluster become epidermal or support cells. As shown in the bottom panels, Notch signaling between cells of the proneural cluster is initially similar and keeps the cells in the proneural state. An imbalance develops when one cell begins to express higher levels of Delta, the ligand for Notch, and thus activates Notch to a higher level in neighboring cells. This initial imbalance is rapidly amplified by a feedback pathway involving two transcription factors, Suppressor-of-hairless and Enhancer-of-split. Their activities repress the production of Achaete and Scute proteins and Delta protein in the affected cells. This prevents them from proceeding along the pathway of neuronal development and from delivering inhibitory Delta signals to the prospective neuroblast. bHLH, basic helix-loop-helix.

After Kandel, E.R., et al.: 2000.

(see Fig. 12.5, bottom row). In this way, the initially equivalent cells of the proneural cluster are resolved by lateral inhibition into one neuroblast per cluster. The other cells of the proneural cluster cease expression of Delta and of the proneural genes and go on to become epidermis. In the *Drosophila* ectoderm, if either Notch or Delta function is inactivated, the neuroectoderm makes many more neuroblasts and fewer epidermal cells than normal.

After specification, neuroblasts enlarge, leave the epithelium and move into the interior of the embryo, where they divide repeatedly and give rise to neurons and glia, as described in the next section. In *Drosophila*, neuroblasts are specified and exit the neuroectoderm in five waves over a period of around 90 minutes, to form an invariant pattern of approximately 30 neuroblasts in each hemisegment (a lateral half-segment) arranged in three longitudinal columns. Each neuroblast has a unique identity based on the particular embryonic segment it is part of, its position within the segment, and the time of its formation, and gives rise to a distinct set of neurons and glial cells in the larva.

12.2 The development of neurons in *Drosophila* involves asymmetric cell divisions and timed changes in gene expression

Before it leaves the neuroectoderm, a neuroblast becomes polarized in an apical–basal direction, with the apical and basal ends marked by the accumulation of specific sets of cytoplasmic determinants. The determinants at the apical end help set the plane of cell division by influencing the orientation of the mitotic spindle, and also direct the correct positioning of the basal determinants, which are crucial for determining neural cell fate. Once it has left the surface of the embryo, the neuroblast behaves

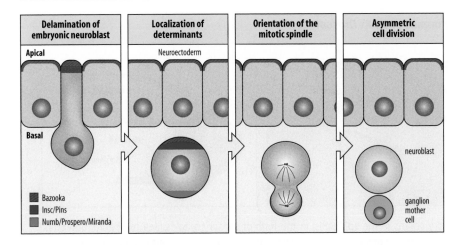

Delamination of embryonic neuroblast	Localization of determinants	Orientation of the mitotic spindle	Asymmetric cell division

■ Bazooka
■ Insc/Pins
■ Numb/Prospero/Miranda

Fig. 12.6 The formation of neuronal cells from neuroblasts in *Drosophila* by asymmetric cell division. Generation of ganglion mother cells of the central nervous system by the asymmetric division of neuroblasts in the *Drosophila* embryo. Once specified, neuroblasts move out of the neuroectodermal epithelium and then behave as neural stem cells. Each neuroblast divides asymmetrically to give an apical cell, which remains a neuroblast stem cell, and a smaller basal cell—the ganglion mother cell. The orientation of cell division and cell fate are specified by localized protein determinants. Numb protein is localized to one end of the neuroblast before division and the daughter that receives it becomes the ganglion mother cell that will give rise to neurons. Prospero and Miranda are required for the proper localization of Numb, and Inscuteable (Insc) and Pins are required for the correct orientation of cell division.

as a stem cell, dividing asymmetrically to give a larger apical cell, which remains a neural stem cell, and a smaller basal cell, the **ganglion mother cell**, which will divide once more and whose daughter cells will differentiate into neurons or glia (Fig. 12.6). Among the basal cytoplasmic determinants distributed to the ganglion mother cell by this asymmetric division is the transcription factor Prospero, which has been shown by DNA microarray profiling to regulate hundreds of target genes involved in suppressing neural stem cell behavior and promoting differentiation of the neuroblast.

The neuroblasts at the different locations along the body continue to divide according to the fly's developmental program, giving rise to a ganglion mother cell and a neural stem cell at each division. As a neuroblast generates more ganglion mother cells, the older cells are pushed further into the embryo, giving the final nerve cord a layered structure. As neurogenesis proceeds, the neuroblast undergoes successive changes in the expression of a particular set of transcription factors, which give the neurons produced at different times their different identities.

Because neurons with different functions must develop in the correct positions within the ventral nerve cord, the precise timing of the transition from expression of one transcription factor to the next in the neuroblasts is crucial for the correct formation of the nerve cord. The timing of the first transition seems to be linked to neuroblast division, whereas later transitions are linked to an intrinsic timing mechanism in the neuroblast that is independent of cell division.

The embryonic phase of neuroblast division and neurogenesis provides a functional central nervous system for the first instar *Drosophila* larva, but these neurons comprise only about 10% of the neurons that are eventually required by the adult fly. A further extended period of neurogenesis, involving most of the original embryonic neurons, occurs before metamorphosis and produces the remaining 90%.

Another classical system of neurogenesis in *Drosophila* is the generation of sensory bristles in the adult fly from neural cells that are specified in the imaginal discs (Box 12A, p. 474).

12.3 Specification of vertebrate neuronal precursors also involves lateral inhibition

The mechanisms involved in early neural specification in vertebrates are in many ways remarkably similar to those operating in *Drosophila*. As we saw in Chapter 5, the cells of the vertebrate nervous system derive from the neural plate, a region of columnar epithelium induced from the ectoderm on the dorsal surface of the embryo during gastrulation, and from sensory placodes in the head region that contribute to the cranial sensory ganglia and produce the sensory receptor cells of the eyes, ears, and nose (see Chapter 11). Towards the end of gastrulation, the neural plate begins

Box 12A Specification of the sensory organs of adult *Drosophila*

The sensory organs of adult *Drosophila* have provided a classic system for investigating both the patterning of the epithelia in which they arise, and the role of asymmetric localization of cytoplasmic determinants and asymmetric cell division to produce a structure composed of a neuron and non-neuronal cells. Adult flies have two

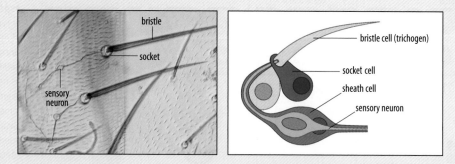

main types of sensory organ: external sensory organs in the epidermis, in the form of sensory bristles (see top figure) and other structures that act as mechanoreceptors and chemoreceptors, and internal chordotonal organs that monitor stretching of tissues. Each sensory organ is made up of four cells, one of which is a sensory neuron, and each organ arises from a single **sensory organ precursor** (**SOP**) cell that is initially specified in the epidermis and in the larval imaginal discs in a constant pattern from fly to fly.

Unlike the neuroblasts in the neuroectoderm (see Fig. 12.6), division of an SOP is oriented in such a way that both daughter cells stay within the epidermis (see figure on right), and this is reflected in the location of the Pins protein, which is involved in orienting the plane of cell division (compare Fig. 12.6).

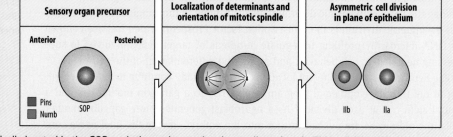

The Numb protein is also asymmetrically located in the SOP, such that only one daughter cell receives it. This cell (IIb) will give rise to a sensory neuron and a sheath cell at the next division, with the daughter cell that receives Numb protein becoming the neuron. The SOP daughter cell that did not receive Numb protein (IIa) gives rise at the next division to a bristle cell and a socket cell, which are non-neuronal cells. Notch signaling is also required for neuronal differentiation; if it is absent, a sheath cell develops instead of the neuron.

The pattern of sensory bristles in the wing epidermis is already present in the imaginal disc in the form of a pattern of sensory organ precursors, each of which will give rise to a sensory organ. As in the embryonic neuroectoderm (see Section 12.1), proneural clusters are first selected by expression of proneural genes–in this case genes of the *achaete–scute* complex–in a precise spatial pattern. Exhaustive genetic analysis revealed that this pattern is controlled by the use of different *cis*-regulatory modules in the *achaete–scute* complex DNA to determine gene expression at specific locations in the imaginal disc. The *achaete–scute* complex is an excellent example of how a spatial pattern of gene expression can be precisely controlled by highly complex control regions (see figure below). The top panel of the figure shows the position of various control modules that determine the position-specific expression of the *achaete* and *scute* genes. These modules are located both upstream and downstream of the *achaete* and *scute* genes themselves

(shown in black). The bottom panel shows the location of the proneural clusters in the imaginal disc (left) and of the corresponding sensory bristles in the adult wing and adjacent thoracic segments (right), with colors corresponding to the enhancer responsible for the site-specific expression of the genes.

Photograph courtesy of Y. Jan.

Diagram after Campuzano, S., et al.: 1992

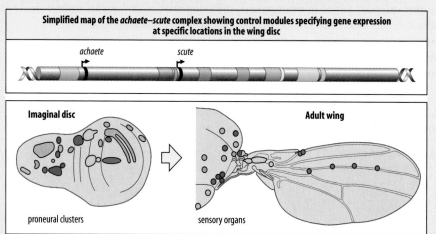

to fold and form the neural tube (see Fig. 5.17). Neural crest cells migrate away from the dorsal part of the tube and give rise to both the sensory neurons of the peripheral nervous system and to the autonomic nervous system, and to the glial Schwann cells of the peripheral nervous system (neural crest migration is discussed in Chapter 8). The cells that remain within the neural tube give rise to the brain and spinal cord, which together comprise the central nervous system.

Neuronal precursors are not generated simultaneously throughout the neural plate and its successor neural tube, but, as in *Drosophila*, are initially confined to three longitudinal stripes in the neural plate on each side of the dorsal midline (see Fig. 12.3). In the *Xenopus* spinal cord, the normal fate of the medial stripe (the one nearest the midline), which becomes the ventral region of the spinal cord, is to give rise to motor neurons. The more lateral stripes, which become the lateral and dorsal regions of the spinal cord, give rise to interneurons and sensory neurons.

The initial specification of the stripes in the neural plate ectoderm involves lateral inhibition, in which Delta–Notch signaling between adjacent cells restricts the expression of neuroblast-specifying proneural genes in the same way as it does in *Drosophila* (see Fig. 12.5). Prospective neuroblasts in the three stripes come to express Delta more strongly than adjacent cells, and the feedback loop set up by Notch activation in the adjacent cells shuts down proneural gene expression, leaving only the prospective neuroblasts expressing proneural genes. Vertebrate proneural proteins, such as the transcription factor **neurogenin** are homologs of *Drosophila* proneural transcription factors, while the proteins that inhibit specification as neural cells are vertebrate homologs of the Notch-activated transcription factor Su(H) and its target gene Enhancer of split (see Fig. 12.5). Cells that express neurogenin go on to express genes characteristic of differentiation into neurons, such as *NeuroD*. Confirmation of the role of Notch and Delta in vertebrates comes from the transgenic overexpression of Delta in the neural plate or expression of a mutant Notch that signals continuously, which both result in a reduction in the number of neurons formed. Inhibition of Delta, on the other hand, leads to an increase in the number of neurons, as does overexpression of neurogenin.

The neural tube is formed from the whole of the neural plate and will contain both committed neural precursors expressing neurogenin and cells in which the *neurogenin* gene is repressed. The latter form a pool of undifferentiated potential neural precursors in the neural tube, which can be brought into play later, as the first wave of neuroblasts to be formed stop dividing and start to differentiate. This ensures that neurogenesis can continue over a relatively long period of time.

12.4 Neurons are formed in the proliferative zone of the vertebrate neural tube and migrate outwards

In vertebrates, all the neurons and glia of the brain and spinal cord arise from a proliferative layer of epithelial cells that lines the lumen of the neural tube and is called the **ventricular zone**. As we saw in Chapters 5 and 8, the neural tube is initially made up of a single layer of epithelium. As the cells of the neural tube divide, the ventricular zone develops as a zone of rounded-up dividing cells on the inner surface of the tube. Once formed, neurons migrate outward from this zone and build up the neural tissue in concentric layers. In this way, the hollow neural tube eventually develops into the hollow spinal cord, and into the brain with its fluid-filled ventricles that represent the remains of the neural tube lumen.

The neural progenitor cells in the ventricular zone are multipotent neural stem cells, which give rise to many different types of neurons and to glia. In the ventricular zone of the dorsal forebrain, for example, cells known as **radial glial cells** can act as neural stem cells, generating both radial glia, whose role is described below, and some types of neurons. As with all stem cells, neural stem cells must generate both neural precursors and more stem cells. Two ways of achieving this seem to operate in the development of neurons in the developing nervous system. The stem cell can either divide 'vertically' and give rise to two similar cells—both stem cells or both neurons—or it can divide horizontally, parallel to the ventricular surface, with one daughter remaining a stem cell and the other developing into a neuron.

Once specified as neuroblasts, the cells migrate from the inner surface of the neural tube towards the outer surface. Here we shall look at the process linking neuron production and neuron migration with reference to the mammalian cerebral cortex, which develops from the dorsal part of the embryonic forebrain. The cortex is organized into six layers (I–VI), numbered from the cortical surface inward, and each layer contains neurons with distinctive shapes and connections. For example, large pyramidal cells are concentrated in layer V, and smaller stellate neurons predominate in layer IV. All of these neurons have their origin in the ventricular zone and migrate out to their final positions along radial glial cells—greatly elongated cells that extend across the developing neural tube (Fig. 12.7). The three-dimensional representation of radial fibers and migrating neurons was painstakingly reconstructed in the 1960s and 70s from electron-microscopic images of successive sections of the monkey fetal cortex, and the reconstruction drawn by hand in great detail.

The layer to which neurons migrate after their birth in the proliferative layer is related to the time when the neuron is born, and the identity of a cortical neuron is

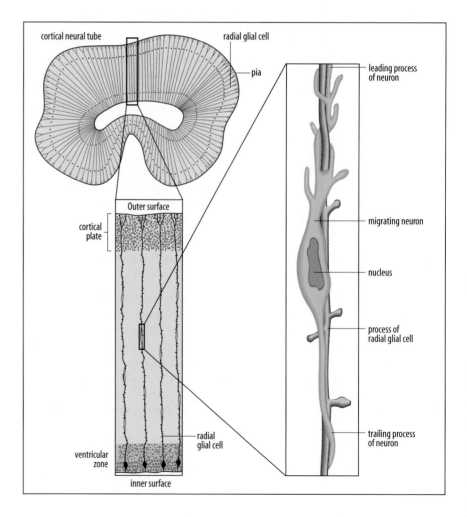

Fig. 12.7 Cortical neurons migrate along radial glial cells. Neurons generated in the ventricular zone migrate to their final locations in the cortex along radial glial cells that extend right across the wall of the neural tube.

After Rakic, P.: 1972.

thought to be specified before it begins to migrate. The neurons of the mammalian central nervous system do not divide once they have become specified, which is why newly formed neurons are often called **post-mitotic neurons**. A neuron's time of birth is defined by the last mitotic division that its progenitor cell underwent, and the experiments that originally established the relation between the birth time of a neuron and the layer in which it ended up are described in Box 12B (p. 478). The newly formed neuron is still immature; later it extends an axon and dendrites (see Fig. 12.1), and assumes the morphology of a mature neuron.

Neurons born at early stages of cortical development migrate to layers closest to their site of birth, whereas those born later end up further away, in more superficial layers (Fig. 12.8). The younger neurons must migrate past the older ones on their way to their correct position, giving rise to layers and columns of cell bodies. The exception is the first-formed layer, which remains as the outer layer. Mouse mutations that disrupt cortical neuronal migration have provided considerable insight into the process. The *reeler* mouse, as its name implies, has very poor motor coordination, which is hardly surprising as its cortical layers are arranged in reverse order. The *reeler* gene encodes an extracellular matrix molecule, reelin, that is normally expressed in the first-formed, outermost layer of the cortex. Mutations in *reeler* that cause a loss of reelin protein disrupt radial neuronal migration, with the successive waves of neurons being unable to bypass their predecessors. As one might expect, the migration of neurons in the developing brain is in fact much more complicated than described here; in addition to radial migration, brain neurons also migrate tangentially within the wall of the neural tube in a process that is independent of radial glial fibers.

The inside-to-outside migration of neurons in the developing mammalian cerebral cortex, means that the cortex has most of its cell bodies (the gray matter of the brain) located at the outer surface, while the axons (the white matter) extend through the intermediate zone. Compared with a strict outside-to-inside stacking up of successively formed neurons into serial layers of cell bodies and axons, this arrangement gives the mammalian cortex great potential for expansion during evolution.

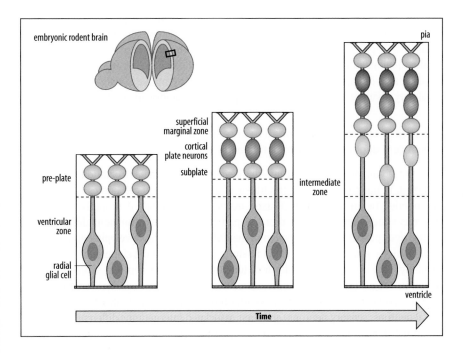

Fig. 12.8 The generation of neuronal layers in the mammalian cortex. Left: development of the cortex begins with the migration of the first wave of post-mitotic neurons (green) towards the outer surface of the neural tube to form the pre-plate above the ventricular zone. Middle: the second wave of newly formed neurons (blue) migrates past the first layer of the pre-plate to form the first layer of so-called cortical plate neurons. The upper layer of the pre-plate neurons becomes the superficial marginal zone and will become cortical layer I (cortex layers are numbered from the outside inwards). Right: a third wave of neurons (purple) migrates to form a second layer of cortical plate neurons above the first. Subsequent waves of neurons will each form a more superficial layer of cortical plate. The intermediate zone contains migrating neurons (light blue).

Adapted from Honda, T., et al.: 2003.

Box 12B Timing the birth of cortical neurons

The time at which a neuron is born determines what cortical layer it becomes part of. A neuron's time of birth is determined as the time that the precursor cell stops dividing and leaves the mitotic cell cycle. Some of the earliest experiments, done in the 1960s and 1970s, used radiolabeled isotopes to label cells so that their migration could be followed. If neuronal precursors are given a short pulse of a labeled compound that can be incorporated into DNA, those cells that undergo their last round of replication immediately after taking up the compound will have the

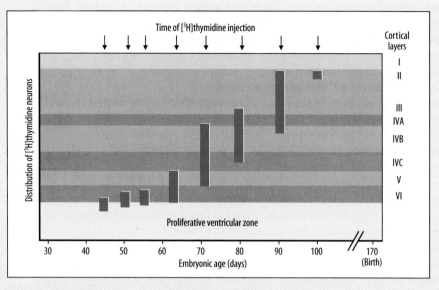

most heavily labeled DNA. If they continue to divide and replicate their DNA, the label becomes diluted among their progeny. Cells that stop dividing immediately after taking up the label can therefore be distinguished and their final destinations detected. The figure shows the results of an experiment of this type that originally determined the order in which neurons reached their final layers in the monkey visual cortex. Radioactively labeled thymidine ([³H]thymidine), which becomes incorporated into replicating DNA, was injected into the developing neural tube at different times during development. The red bars represent the

distribution of heavily labeled neurons found after each injection. Neurons born earliest remain closest to their site of birth, the proliferative ventricular zone. Neurons born later become part of successively higher cortical zones. Cortical zones are numbered from the surface of the neural tube inward; that is, layer V1 is the layer nearest to the ventricular zone. For this type of experiment nowadays, radioisotopic labeling has been superseded by the labeling of cells with, for example, replication-incompetent retroviruses carrying a marker gene such as *lacZ*, whose product can be detected histochemically.

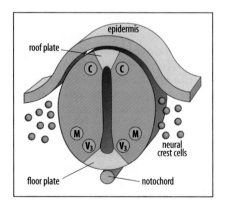

Fig. 12.9 Dorso-ventral organization in embryonic spinal neural tube. In neural tube that develops into spinal cord, a floor plate of non-neuronal cells develops along the ventral midline, and a roof plate of non-neuronal cells along the dorsal midline. Commissural neurons (C) differentiate in the dorsal region, near the roof plate. Motor neurons (M) and V3 interneurons differentiate close to the floor plate (see Fig. 12.12).

For many years it was thought that no new neurons could be generated in the adult mammalian brain, although adult neurogenesis was known to occur in the brains of songbirds, reptiles, amphibians and fish. Neural stem cells are now known to exist in limited regions of the adult mammalian brain, and can give rise to new neurons and glial cells, as described in Section 10.10. These adult neural stem cells are a population of slowly dividing radial glial cells, like those in the embryonic cortex.

12.5 The pattern of differentiation of cells along the dorso-ventral axis of the spinal cord depends on ventral and dorsal signals

Like the brain, the spinal cord initially develops by the radial migration of neurons, although in this case the final structure is of a central core of cell bodies (gray matter) surrounded by an outer layer of axon tracts (white matter). A properly functioning nervous system depends on neurons of particular types and functions differentiating in the right positions in the developing neural tube, and the developing spinal cord is a particularly good system in which to study this aspect of development. It has a particularly clear dorso-ventral pattern of functionally distinct neurons, which is similar along its length (Fig. 12.9). Future motor neurons and their associated interneurons are located ventrally, and form the ventral roots of the spinal cord, whereas commissural neurons, sensory neurons, and their associated interneurons differentiate primarily in the dorsal region. Each cell type is present symmetrically on either side of the midline. While motor and sensory neurons differentiating in the right and left halves

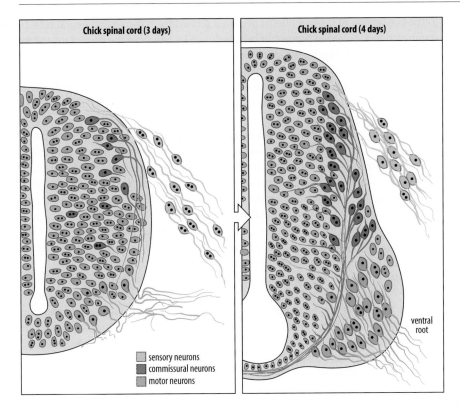

Chick spinal cord (3 days)	Chick spinal cord (4 days)

ventral
root

☐ sensory neurons
☐ commissural neurons
☐ motor neurons

Fig. 12.10 Formation of sensory and motor neurons in the chick spinal cord. Three days after an egg has been laid, motor neurons are beginning to form within the chick embryo neural tube. A day later, the motor neurons send at axons to form the ventral root. The sensory neurons that migrate from the dorsal part of the spinal cord are derived from neural crest cells and form the segmental dorsal root ganglia (see Section 10.11). The commissural neurons send axons towards and across the ventral midline.

of the spinal cord will remain on the side of the body in which they originated, the axons of commissural neurons will cross the midline of the spinal cord, providing a link between the two sides of the body at the level of the spinal cord (Fig. 12.10). The neurons of the sensory nervous system develop from neural crest cells that are located laterally and dorsally in the neural tube, and migrate away from the surface of the tube to form the dorsal root ganglia that lie at regular intervals on each side of the spinal cord (see Fig. 12.10 and Figs 8.36 and 8.37). As in the brain, the neurons and glia of the spinal cord are generated in a proliferative zone at the inner face of the neural tube.

We shall now consider how the dorso-ventral organization of the spinal cord is produced by signals received by the early neural tube, which induce the differentiation of particular types of neurons in the correct positions. In additional to the future nerve-cell precursors, a group of non-neural cells forms a region called the **floor plate** in the ventral midline of the neural tube, while another group of non-neural cells forms the **roof plate** in the dorsal midline. These two regions produce signals that pattern the neural tube in a dorso-ventral direction.

The floor plate is induced by signals secreted by the mesodermal notochord, which lies immediately below the neural tube. The patterning activity of the notochord on the neural tube can be shown by grafting a segment of notochord to a site lateral or dorsal to the neural tube (see Fig. 5.30, where the experimental set-up is described in regard to its effect on somite patterning). A second floor-plate is induced in the neural tube in direct contact with the transplanted notochord. Molecular markers of dorsal-cell differentiation are suppressed in the nearby neural cells, and additional motor neurons, which normally only develop in the ventral region, are generated. *In situ* staining for gene expression and misexpression experiments identified the inductive signal secreted by the notochord as the protein Sonic hedgehog (Shh). Shh induces the floor-plate cells themselves to produce Shh, which then forms a gradient of activity from ventral to dorsal in the neural tube, and acts as the ventral patterning signal (Fig. 12.11).

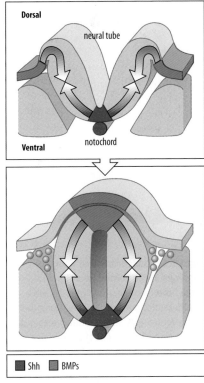

Dorsal

neural tube

Ventral notochord

☐ Shh ☐ BMPs

Fig. 12.11 Dorso-ventral patterning of the spinal cord involves both dorsal and ventral signals. As the neural tube forms (top panel), dorsal and ventral signaling occurs. The ventral signal from the floor plate is the Shh protein (red), whereas the dorsal signal (blue) includes BMP-4, which is induced during neural tube formation by the adjacent ectoderm. Ventral and dorsal signals oppose each other's action.

The conversion of neuroepithelial neuronal precursors in the ventral neural tube to committed motor-neuron progenitor cells is driven by Shh from the floor plate and by retinoic acid secreted by the adjacent mesoderm. Control of the proliferation of neural progenitor cells by mitogenic Wnt signals is also important in regulating the number of neurons produced. In line with the role of Shh and retinoic acid in neural differentiation *in vivo*, they can direct the development of embryonic stem cells in culture into motor neurons and interneurons.

The roof plate is initially specified by signals from the dorsal epidermal ectoderm, and signals from the roof plate pattern the dorsal half of the neural tube. From genetic knock-outs and cell-fate mapping, it has been shown that selective removal of the roof plate from the neural tube of the mouse embryo results in the loss of the most dorsal interneurons. The dorsalization signal includes BMPs: BMPs from the ectoderm (see Chapter 5) specify the roof plate, and BMPs produced in the roof plate then pattern the dorsal region of the neural tube. The BMP signals apparently ensure dorsal differentiation by opposing the action of signals from the ventral region, thus together patterning the dorso-ventral axis of the neural tube (see Fig. 12.11).

The spatial pattern of expression of BMPs in the dorsal neural tube is strongly influenced by the ventral signals, and this system bears a strong similarity to the mechanism for patterning the dorso-ventral body axis in early *Drosophila* development. In that case, the expression of the gene *decapentaplegic* is confined to dorsal regions because its expression is repressed in ventral regions by the transcription factor Dorsal (see Section 2.12). Like BMPs, Decapentaplegic is a member of the TGF-β family of secreted signals. Similarly, in the spinal cord, Shh and BMPs provide positional signals with opposing actions emanating from the two ends of the dorso-ventral axis.

12.6 Neuronal subtypes in the ventral spinal cord are specified by the ventral to dorsal gradient of Shh

The ventral spinal cord produces five different classes of neurons—motor neurons and four classes of interneurons, each distinguished by the expression of specific genes. Experiments *in vitro* have shown that these neuronal subtypes can be specified in a dose-dependent manner by Shh, the signal produced by the notochord and floor plate; the different neuronal subtypes can be generated from chick neural-plate explants in response to a two- to threefold change in concentration of Shh. The incremental changes in Shh concentration are mimicked by smaller changes in the activity of the transcription factor Gli3, which is activated by Shh signaling (see Box 11C, pp. 424–425), and which provides the key intracellular signal in a graded form (Fig. 12.12).

Fig. 12.12 Specification of neuronal subtypes in the ventral neural tube. Different neural subtypes (shown as different colors) are specified along the dorso-ventral axis of the ventral neural tube in response to a gradient of Sonic hedgehog (Shh) with its source in the floor plate. The Shh gradient results in an intracellular gradient of active transcription factor Gli3. Shh signaling mediates the repression of the so-called Class I homeodomain protein genes (*Pax7, Pax6, Dbx1, Dbx2,* and *Irx3*), whereas Class II homeodomain protein genes (*Nkx2.2* and *Nkx6.1*), are activated. Interactions between Class I and Class II genes pattern this expression further. Five neuronal subtypes are generated from the five domains. MN, motor neurons.

Adapted from Jessell, T.M.: 2000.

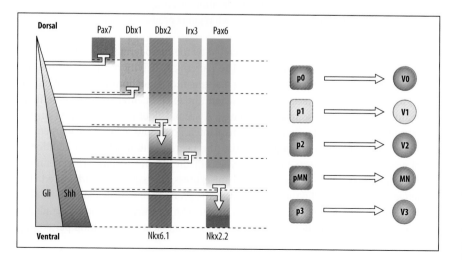

A number of homeobox transcription factor genes are regulated in ventral progenitor cells in response to Shh signaling, and these genes can be divided into two classes—one (class I) containing genes that are generally repressed by Shh, and one (class II) containing genes that are activated (see Fig. 12.12). The threshold levels of Shh that allow expression are different for each gene, so that overall, the responses of cells to the Shh gradient divides the dorso-ventral axis of the spinal cord into at least five regions of different types of progenitor neurons. Boundaries between each region are sharpened by interactions between the class I and class II proteins. The detailed mechanism by which Shh specifies the Gli gradient in the progenitor neurons *in vivo* is not yet completely worked out. As discussed elsewhere for morphogens in general (see Box 11A, p. 421), processes other than simple diffusion must be involved in forming the Shh gradient. Diffusion alone would not be a reliable way of setting up such a gradient, as the extracellular concentration of a diffusing molecule at any point changes over time, and its movement is affected by other extracellular molecules.

12.7 Spinal cord motor neurons at different dorso-ventral positions project to different trunk and limb muscles

The motor neurons developing in the ventral region of the spinal cord can be further classified on the basis of the position of their cell bodies along the cord dorso-ventral axis, and the muscles that their axons innervate. In the chick embryo, motor neurons are subdivided into longitudinal columns on each side of the midline—a median column nearest the midline, which runs the whole length of the spinal cord, and a lateral motor column (LMC), which is only present in the brachial and lumbar regions, where the forelimbs and hindlimbs, respectively, develop (Fig. 12.13). Motor neurons in the median motor column send axons to axial and body-wall muscles, whereas neurons of the LMC innervate the limb. The LMC is divided into two divisions:lateral and medial.

Once designated, each motor neuron subtype expresses a different combination of transcription factors of the LIM homeodomain family, which provides a neuron with its positional identity within the spinal cord. In the LMC, for example, neurons expressing Lim1 settle in the lateral division, whereas those expressing Isl1 settle in the medial division. Experiments tracing the outgrowth of axons from LMC motor neurons show that those in the lateral LMC send axons into the dorsal muscle masses

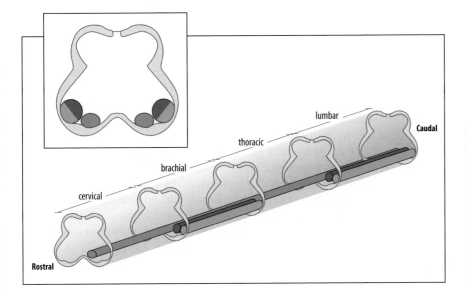

Fig. 12.13 The organization of the motor columns in the developing chick spinal cord. The organization of motor columns along the chick embryo spinal cord at stages 28-29. The median motor column is in blue, the medial division of the lateral column in red and the lateral division of the lateral motor column in green. The lateral motor columns are only present in the brachial and lumbar regions, in register with the position of limb formation.

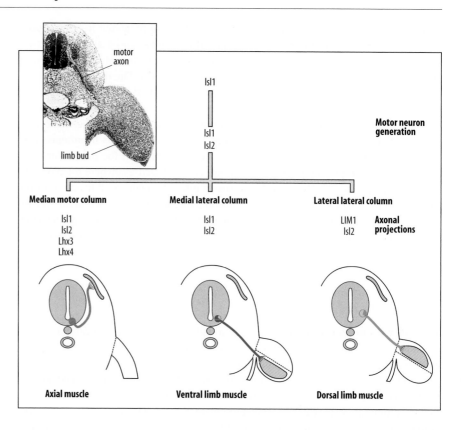

Fig. 12.14 Different motor neurons in the chick spinal cord express different LIM homeodomain proteins. The photograph (inset) shows a transverse section through a chick embryo at the level of the developing wing. Motor neurons can be seen entering the limb bud. Initially, all developing motor neurons express the LIM proteins Isl1 and Isl2. At the time of axon extension, motor neurons that innervate different muscle regions express different combinations of Isl1, Isl2, Lim1, Lhx3, and Lhx4.

Micrograph courtesy of K. Tosney.

in the limb bud, whereas medial LMC neurons send their axons ventrally into the limb bud (Fig. 12.14). We shall look in more detail at the mechanisms that guide the growing axon in the next part of the chapter.

12.8 Antero-posterior pattern in the spinal cord is determined in response to secreted signals from the node and adjacent mesoderm

As well as being organized along the dorso-ventral axis, neurons at different positions along the antero-posterior axis of the spinal cord become specified to serve different functions. Antero-posterior specification of neuronal function in the spinal cord was dramatically illustrated some 40 years ago by experiments in which a section of the spinal cord, which would normally innervate wing muscles, was transplanted from one chick embryo into the lumbar region, which normally serves the legs, of another embryo. Chicks developing from the grafted embryos spontaneously activated both legs together as though they were trying to flap their wings, rather than activating each leg alternately as if walking. These studies showed that motor neurons generated at a given antero-posterior level in the spinal cord had intrinsic properties characteristic of that position.

Neuronal fate along the antero-posterior axis of the spinal cord is determined initially by signal molecules secreted by the organizer (or node) and later by the adjacent paraxial mesoderm (see Chapter 5). This was first shown by experiments in which quail paraxial mesoderm was grafted to the thoracic level in chick embryos at the time of neural-tube closure. The result was that presumptive chick thoracic neurons became specified as brachial neurons. The signals from the node and the mesoderm form gradients along the antero-posterior axis of the embryo that activate patterns of Hox gene expression in post-mitotic neurons, giving groups of neurons a positional identity that determines their fate. The signals are composed of FGF, growth/differentiation factor (GDF)—a member of the TGF-β family—and retinoic acid (see Fig. 5.21). FGF is graded from posterior to anterior, and there is a shallow retinoic acid gradient at the anterior end of the spinal cord and a similar one of GDF at the posterior end.

The spinal cord becomes demarcated into different regions along the antero-posterior axis by combinations of expressed Hox genes. For example, *Hoxc6* is expressed in the motor neurons in the brachial region—adjacent to the forelimb—while *Hoxc9* is expressed in thoracic motor neurons. In some neuronal subtypes, different concentrations of the transcription factor FoxP1 act together with the Hox proteins to specify different neuronal identities. This is in addition to the gene expression specifying dorso-ventral identity described in the previous section.

A typical vertebrate limb contains more than 50 muscle groups with which neurons must connect in a precise pattern. Individual neurons express particular combinations of Hox genes, which determines which muscle they will innervate. As well as being organized into divisions within the LMC with ventral and dorsal trajectories into the limb (see Fig. 12.14), motor neurons serving the limbs are further subdivided into so-called 'pools' along the dorso-ventral and antero-posterior axes. The number of pools of motor neurons corresponds to the number of muscles in the limb, with the neurons in each pool directing their axons to a particular target muscle. The identity of the motor neurons in each pool is specified by the combinations of Hox genes expressed. For example, in the chick, the posterior limit of the region containing brachial motor neurons maps to the posterior boundary of the expression of Hox6 genes, while the anterior boundary is defined by the anterior limit of expression of *Hoxc6*. Interactions between genes of the Hox8, Hox7, Hox6, Hox5, and Hox4 paralogous subgroups specify the identities of pools of motor neurons that innervate particular muscles in the chick wing (Fig. 12.15).

So, all together, expression of genes resulting from dorso-ventral position, together with those resulting from antero-posterior position, confer a virtually unique identity on functionally distinct sets of neurons in the spinal cord.

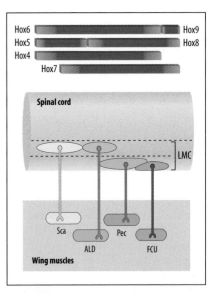

Fig. 12.15 Pools of motor neurons in the developing spinal cord make connections with specific limb muscles. The lateral columns of motor neurons in the brachial region of the developing chick spinal cord are organized into a number of groups, or pools, that each make connections to a specific muscle group in the developing wing. The identity of each pool of neurons is specified by interactions between the Hox genes expressed in the spinal cord in this region. LMC, lateral motor column (see Fig. 12.13). Muscles: Sca, scapulohumeralis posterior; Pec, pectoralis; ALD, anterior latissimus dorsi; FCU, flexor carpi ulnaris.

Adapted from Dasen, J.S., et al.: 2005.

Summary

In *Drosophila*, prospective neural tissue is specified early in development as a band of neuroectoderm along the dorso-ventral axis. Within the neuroectoderm, expression of proneural genes distinguishes clusters of neuroectoderm cells with the potential to form neural cells. As a result of lateral inhibition involving Notch and Delta, only one cell of the cluster finally gives rise to a neuroblast. Neuroblasts that give rise to the central nervous system behave as stem cells, undergoing repeated asymmetric division, each giving rise to a neuroblast and a daughter cell that can give rise to neurons.

The vertebrate nervous system is derived from the neural plate, which is specified during gastrulation. It contains the cells that form the brain and spinal cord, as well as the neural crest cells that contribute to the peripheral nervous system. Prospective neural cells are initially specified within the neural plate by a mechanism of lateral inhibition, similar to that in *Drosophila*. Neurons are born in the proliferative zone on the inner surface of the neural tube and then migrate to different locations within the developing brain and spinal cord. Patterning of neuronal cell types along the dorso-ventral axis of the spinal cord involves signals from both the ventral and dorsal regions. Secretion of Shh by the notochord and floor plate provides a graded signal that leads to specification of different neuronal subtypes in the ventral region of the spinal cord. Neuronal identity along the antero-posterior axis is specified by the combinatorial expression of Hox genes.

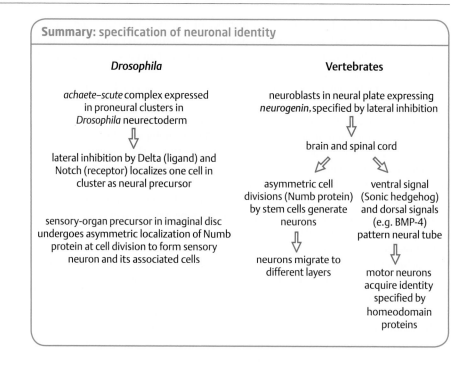

Summary: specification of neuronal identity

Drosophila	Vertebrates
achaete–scute complex expressed in proneural clusters in *Drosophila* neurectoderm	neuroblasts in neural plate expressing *neurogenin*, specified by lateral inhibition
⇩	⇩
lateral inhibition by Delta (ligand) and Notch (receptor) localizes one cell in cluster as neural precursor	brain and spinal cord
	↙ ↘
sensory-organ precursor in imaginal disc undergoes asymmetric localization of Numb protein at cell division to form sensory neuron and its associated cells	asymmetric cell divisions (Numb protein) by stem cells generate neurons ventral signal (Sonic hedgehog) and dorsal signals (e.g. BMP-4) pattern neural tube
	⇩ ⇩
	neurons migrate to different layers motor neurons acquire identity specified by homeodomain proteins

Axon navigation

We shall now look at a feature of development that is unique to the nervous system—the outgrowth and guidance of axons to their final targets. This stage of neuronal differentiation occurs mainly after immature post-mitotic neurons have migrated to their eventual locations. Each neuronal cell body extends an axon and dendrites, through which it will, respectively, send and receive signals. The working of the nervous system depends on the formation of discrete neuronal circuits, in which neurons make numerous and precise connections with each other (Fig. 12.16). In the rest of this chapter we will look at how these connections can be set up in various situations. What controls the growth of axons and the contacts they make with other cells, and how are such precise connections achieved? In this part of the chapter we will first consider axon outgrowth and the evidence that the growing axon is guided specifically to its target.

Fig. 12.16 Neurons make precise connections with their targets. Neurons (green) and their target cells usually develop in different locations. Connections between them are established by axonal outgrowth, guided by the movement of the axon tip (growth cone). Often, the initial set of relatively nonspecific synaptic connections is then refined to produce a more precise pattern of connectivity.

After Alberts, B., et al.: 2002.

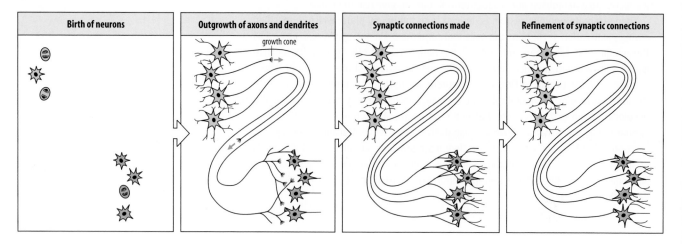

Birth of neurons	**Outgrowth of axons and dendrites**	**Synaptic connections made**	**Refinement of synaptic connections**

12.9 The growth cone controls the path taken by a growing axon

An early event in the differentiation of a neuron is the extension of its axon by the growth and migration of the **growth cone** located at the tip of the axon (Fig. 12.17). The growth cone is specialized for both movement and for sensing its environment for guidance cues. As in other cells capable of migration, such as the primary mesenchyme of the sea-urchin embryo (see Section 8.7), the growth cone can continually extend and retract filopodia at its leading edge, making and breaking connections with the underlying substratum to pull the axon tip forward. Between the filopodia, the edge of the growth cone forms thin ruffles—lamellipodia—similar to those on a moving fibroblast (see Box 8A, p. 291). Indeed, in its ultrastructure and mechanism of movement, the growth cone closely resembles the leading edge of a fibroblast crawling over a surface. Unlike a moving fibroblast, however, the extending axon also grows in size, with an accompanying increase in the total surface area of the neuron's plasma membrane. The additional membrane is provided by intracellular vesicles, which fuse with the plasma membrane.

The growth cone guides axon outgrowth, and is influenced by the contacts the filopodia make with other cells and with the extracellular matrix. In general, the growth cone moves in the direction in which its filopodia make the most stable contacts. In addition, its direction of migration can be influenced by diffusible extracellular signal molecules that bind to receptors on the growth-cone surface. Some of these extracellular cues promote axon extension, whereas others inhibit it by causing the retraction of the filopodia and the 'collapse' of the growth cone. The extension and retraction of filopodia and lamellipodia involve the assembly and disassembly of the actin cytoskeleton, and many of the extracellular signals that affect axon guidance are known to have effects on the actin cytoskeleton through the Ras-related GTPase family of intracellular signaling proteins. But precisely how growth cones transduce extracellular signals so as to extend or collapse filopodia is not yet fully understood.

Axon growth cones are guided by two main types of cue—attractive and repulsive. In addition, cues can act either at long or short range, thus giving four ways in which the growth cone can be guided (Fig. 12.18). Long-range attraction involves diffusible **chemoattractant** molecules released from the target cells. In contrast, **chemorepellents** are molecules that repel migrating cells or axons. Both chemoattractant and chemorepellent proteins have been identified in the developing nervous system. Short-range guidance is mediated by contact-dependent mechanisms involving molecules bound to other cells or to the extracellular matrix; again, such interactions can be either attractive or repulsive. Some signals can function as either an attractant or a repellent, depending on the cellular and developmental context.

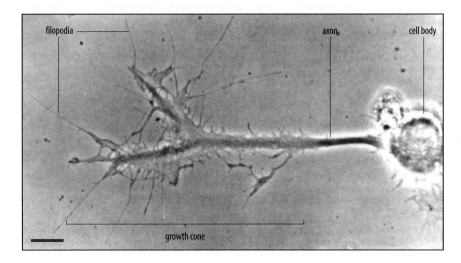

Fig. 12.17 A developing axon and growth cone. The axon growing out from a neuron's cell body ends in a motile structure called the growth cone. Many filopodia are continually extended and retracted from the growth cone to explore the surrounding environment. Scale bar = 10 μm.

Photograph courtesy of P. Gordon-Weeks.

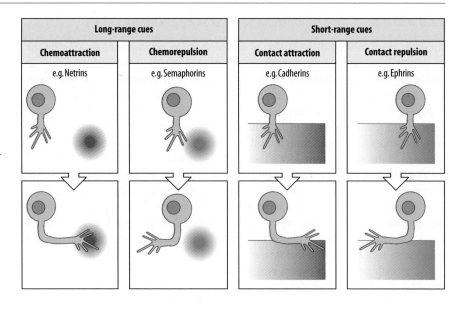

Fig. 12.18 Axon-guidance mechanisms. Four general types of mechanism can contribute to controlling the direction in which a growth cone moves: long-range attraction, long-range repulsion, short-range attraction, and short-range repulsion. Long-range cues are provided by secreted molecules (such as netrins and semaphorins) that form gradients in the extracellular matrix. Short-range cues are often provided by transmembrane proteins (such as the ephrins and their Eph receptors, and the cadherins, which bind to cadherins).

A number of different extracellular molecules that guide axons have been identified. The **semaphorins** are among the most prominent of the conserved families of axon-guidance molecules. There are seven different subfamilies, three of which are secreted, whereas the others are associated with the cell membrane. There are two classes of neuronal receptors for semaphorins—the plexins and neuropilins. Axon repulsion is a major theme of semaphorin action. Some, however, can either attract or repel growth cones; which they do will depend on the nature of the neuron and on the type of semaphorin receptor present in the neuron membrane. Different regions of the same neuron can respond in opposite ways to the same signal—the apical dendrites of pyramidal neurons in the cortex grow towards a source of semaphorin 3A, whereas their axons are repelled, which enables the receiving end of the neuron and the transmitting end of the neuron to be oriented in opposite directions by the same signal. This difference in response appears to be due to the presence of guanylate cyclase, a component of the pathway that transduces the semaphorin signal, in the dendrites but not in the axon.

Other classes of neuronal-guidance proteins, such as the secreted **netrins**, are also bifunctional, in that they attract some neurons and repel others. The attractant function of netrins is mediated by the DCC (deleted in colorectal cancer) family of receptors on growth cones. The **Slit proteins** are secreted glycoproteins with a repellent function. They repel a variety of axon classes by acting on receptors of the Robo family on neurons, but can also stimulate elongation and branching in sensory axons.

Guidance molecules that act by short-range cell–cell contact include the transmembrane **ephrins** and their **Eph receptors** (see Box 5F, p. 205), which in the nervous system mediate mainly repulsive interactions, and the **cadherins**, which bind to cadherins on adjacent cells (see Box 8B, p. 293), and which can act as short-range attractors of axons. As we shall see in the next section, ephrins are involved in guiding the axons of spinal cord motor neurons to their correct target muscles in the limb.

12.10 Motor neuron axons in the chick limb are guided by ephrin–Eph interactions

An animal's muscular system enables it to carry out an enormous variety of movements, which all depend on motor neurons from the spinal cord having taken the correct path and made the correct connections with muscles. The pattern of innervation of the chick wing, for example, is precise and always the same (Fig. 12.19). As we

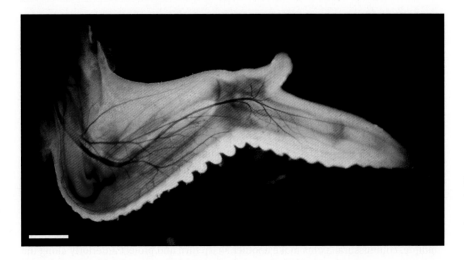

Fig. 12.19 Innervation of the embryonic chick wing. The nerves are stained brown and include both sensory and motor neurons. Scale bar = 1 mm.

Photograph courtesy of J. Lewis.

saw in Sections 12.7 and 12.8, particular 'pools' of neurons that will innervate specific muscles in the limb are already distinguished by the expression of particular combinations of LIM and Hox genes. To reach their correct targets, however, the growing axons require local guidance cues and these are provided by the limb tissue.

In the chick embryo, the motor neuron axons enter the developing limb bud at around 4.5 days, after cartilage elements, and dorsal and ventral muscle masses, have started to form. When the motor axons growing out from the spinal cord first reach the base of the limb bud, they are all mixed up in a single bundle. At the base of the limb bud, however, the axons separate out and form new nerves containing only those axons that will make connections with particular muscles.

Evidence that local cues in the limb are responsible for this sorting-out comes from experiments in chick embryos in which a section of the spinal cord whose motor neurons will innervate the hindlimb was inverted before the axons grew out. After inversion along the antero-posterior axis, motor axons from lumbo-sacral segments 1–3, which normally enter anterior parts of the limb, initially grew into posterior limb regions, but then took novel paths to innervate the correct muscles. So, even when the axon bundles entered in reverse order, the correct relationship between motor neurons and muscles was achieved. Motor axons can thus clearly find their appropriate muscles when their displacement relative to the limb is small. But with a large displacement, this is not the case. A complete dorso-ventral inversion of the limb bud resulted in the axons failing to find their appropriate targets. In such cases, the axons followed paths normally taken by other neurons. The mesoderm of the limb thus provides local cues for the motor axons and the axons themselves have an identity that allows them to choose the correct pathway.

These local cues include signals such as the transmembrane ephrins. Motor neurons in the medial division of the LMC (see Fig. 12.14) innervate ventral muscle masses in the limb, while those in the lateral division innervate dorsal limb muscles. EphA4 signaling in the lateral set of motor axons is a major determinant of the dorsal route they take into the limb. EphA4 is thought to mediate a repulsive interaction with its ligand ephrin A, which is restricted to the ventral limb mesenchyme. Migration of medial motor axons into the ventral mesenchyme appears to be regulated by another set of receptors that similarly respond to repulsive signals from the dorsal limb mesenchyme. The expression of EphA4 is determined by the LIM homeodomain proteins (see Section 12.7), with Lim1 expression in the lateral LMC motor neurons leading to raised levels of EphA4, whereas Isl1 expression in the medial LMC axons leads to a reduction in EphA4. The concentration of ephrin A in the dorsal limb mesenchyme is apparently kept at a low level by the expression there of the LIM protein Lmx1b (see Fig. 11.14).

12.11 Axons crossing the midline are both attracted and repelled

Although the development of the nervous system is largely symmetrical about the midline of the body, and proceeds independently on either side, one half of the body needs to know what the other half is doing. So, although some neurons must remain entirely within the lateral half of the body in which they were formed, many axons must cross the ventral midline of the embryo. Chemoattractant and chemorepellent molecules are involved in these decisions.

In the vertebrate spinal cord, commissural neurons provide the connections between the two sides of the cord. The cell bodies of commissural neurons are located in the dorsal region but they send their axons ventrally along the lateral margin of the cord. Signals from the roof plate prevent the commissural axons extending dorsally. About halfway along the margin of the spinal cord, the axons make an abrupt turn and project to the floor plate in the ventral midline, bypassing the motor neurons (Fig. 12.20; also see Fig. 12.10). After crossing the floor plate to the other side of the midline, commissural axons make another sharp turn and project anteriorly along the spinal cord towards the head; they do not re-cross the midline. Chemoattractant and chemorepellent molecules present in the ventral midline are involved in both enabling the commissural neurons to cross the midline and in preventing them from crossing back. Netrin-1 is a key attractant produced in the floor plate and in the midline, and knock-out of the mouse *netrin-1* gene, or the gene for its receptor DCC, results in abnormal commissural axon pathways (Fig. 12.21).

Axons approaching the midline of the spinal cord provide good examples of the long-range attraction and repulsion of growth cones. The axons of vertebrate commissural neurons are attracted to the floor plate at the ventral midline by the chemoattractant Netrin-1, which signals attraction via the netrin-1 receptor DCC. In both *Drosophila* and vertebrates, the mechanism that repels axons from the midline involves the extracellular protein Slit and its Robo receptors on the axons' growth cones. Slit is present at the midline and, when it binds to receptors called Robo1 and Robo2 on the growth cone, it exerts a chemorepellent effect. For vertebrate commissural axons approaching the midline, however, the chemorepellent effect of Slit is overcome by the Robo receptor-related protein Robo3 expressed on the growth cone, which interferes with Slit signaling through Robo1 and Robo2 (Fig. 12.22). Once the commissural axons cross the midline, Robo3 ceases to be expressed and the growth cone can be repelled from the midline again by Slit proteins, which prevents it crossing back. Mammalian commissural neurons make a sharp anterior turn after crossing the midline, which is thought to be due to a gradient of Wnt proteins in the floor plate with the high point anterior.

Fig. 12.20 Chemotaxis can guide commissural axons in the spinal cord. In the vertebrate spinal cord, the axons of spinal cord commissural neurons extend ventrally, and then toward the floor-plate cells. They cross the midline at the floor plate and then grow anteriorly along the spinal cord. The photograph shows a section through a rat spinal cord with the axons of the commissural neurons (those inside the spinal cord) stained green and the cell bodies red.

Photograph from Samantha Butler.

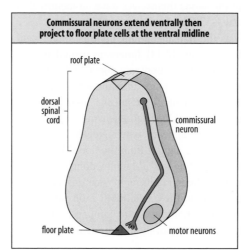

Commissural neurons extend ventrally then project to floor plate cells at the ventral midline

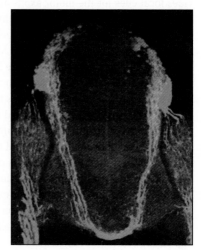

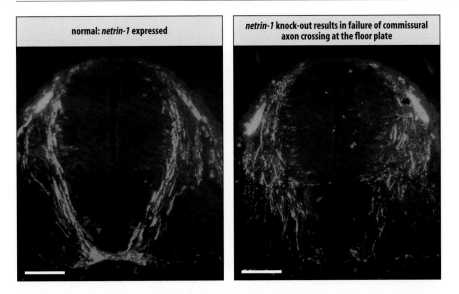

| normal: *netrin-1* expressed | *netrin-1* knock-out results in failure of commissural axon crossing at the floor plate |

Fig. 12.21 Effect of *netrin-1* gene knock-out in mice. In mice lacking *netrin-1*, the commissural axons do not migrate toward the floor plate (compare photograph in Fig. 12.20). Scale bar = 0.1 mm.

Photographs courtesy of M. Tessier-Lavigne, from Serafini, T., et al.: 1996.

12.12 Neurons from the retina make ordered connections with visual centers in the brain

An arguably even more complex task for the developing nervous system is to link up the sensory receptors that receive signals from the outside world with the neuronal targets in the brain that enable us to make sense of these signals. A characteristic feature of the vertebrate brain is the presence of topographic maps. That is, the neurons from one region of the nervous system project in an ordered manner to one region of the brain, so that nearest-neighbor relations are maintained. The vertebrate visual system has been the subject of intense investigation for many years, and the highly organized projection of neurons from the eye via the optic nerve to the brain is one of the best models we have to show how topographic neural projections are made. There are around 126 million individual photoreceptor cells in a human retina, and some animals with excellent night vision, such as owls, have many more. Each of those photoreceptor cells is continuously recording a minute part of the eye's visual field and the signals must be sent to the brain in an orderly manner so that they can be assembled into a coherent picture. Small sets of adjacent photoreceptor cells indirectly activate individual neurons—the retinal ganglion cells—whose axons are bundled together and exit the eye as the optic nerve (see Section 11.23 for a brief overview of the development of the eye itself). The optic nerve from each human eye is made up of over a million axons.

In birds and amphibians, the optic nerve connects the retina to a region of the midbrain called the **optic tectum**, where visual signals are processed. In birds and in larval amphibians, the optic nerve from the right eye makes connections with the left optic tectum, while the nerve from the left eye makes connections to the right-side of

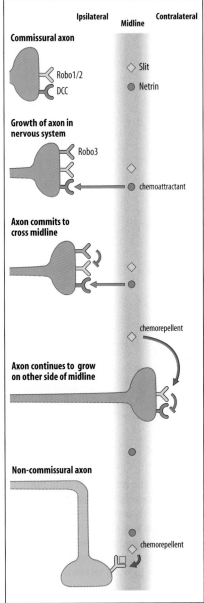

Fig. 12.22 Competing chemoattractant and chemorepellent signals both enable commissural axons to cross the midline and prevent them crossing back. Commissural axons are attracted to the midline by the chemoattractant protein Netrin (green circle) acting through its receptor DCC (green). As they grow toward the midline, the level of Robo1/2 (yellow), the receptors for the chemorepellent protein Slit (yellow diamond), are kept low. Commissural axon growth cones also start to express the cell-surface receptor Robo3 (orange) as they approach the midline, which inhibits the chemorepellent effect of Slit acting at Robo1/2. Once commissural axons reach and cross the midline, alternative splicing of Robo3 produces a cell-surface receptor that is now responsive to Slit and does not inhibit Robo1/2, and so repulsion by Slit can now prevent the axons from re-crossing the midline. Increased expression of Robo1/2 occurs after crossing and helps block attraction to the midline exerted by Netrin. Non-commissural axons do not express Robo3 and so are repelled from the midline by Slit.

Fig. 12.23 Comparison of amphibian and mammalian visual systems. Left panel: in the frog tadpole, neurons from the retina project directly to the optic tectum, neurons from the left retina connecting to the right optic tectum and those from the right retina connecting to the left tectum, crossing over at the optic chiasm. Right panel: in mammals with good binocular and stereoscopic vision, the main visual pathway is to the lateral geniculate nucleus (LGN), from which neurons project to the visual cortex. Neurons from one half of the retina project to the LGN on the same side as the eye, and those from the other half cross over to the opposite LGN at the optic chiasm. For the sake of clarity, the neurons from only one half of the retina are shown. N, nasal; T, temporal, L, lateral, M, medial. A, anterior, P, posterior.

After Goodman, C.S. and Shatz, C.J.: 1993.

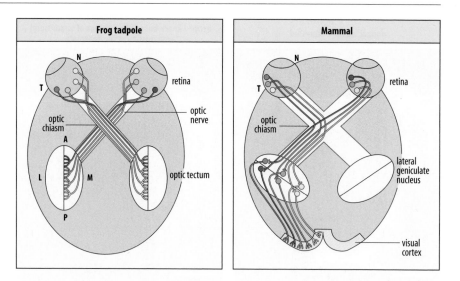

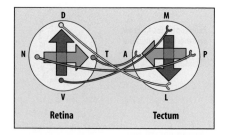

Fig. 12.24 The retina maps on to the tectum. In the amphibian tectum dorsal (D) retinal neurons connect to the lateral (L) side of the tectum and ventral (V) retinal neurons connect to the medial (M) tectum. Similarly, temporal (T) (or posterior) retinal neurons connect to anterior (A) tectum and nasal (N) (or anterior) retinal neurons connect to the posterior (P) tectum.

the brain, the axons crossing over each other in a structure called the **optic chiasm** (Fig. 12.23, left panel). In mammals, the main target for retinal axons are the paired **lateral geniculate nuclei** (**LGN**) in the forebrain, from which other neurons then convey the signals to the **visual cortex**, in which most visual processing occurs. A minor pathway in mammals goes from the retina to the **superior colliculus** in the midbrain (which is analogous to the avian and amphibian optic tectum), which is involved mainly in aiding the orientation of the eyes and head to a variety of sensory stimuli. As this pathway provides a simpler map than that to the lateral geniculate nucleus, it is often studied as the mammalian counterpart to the optic tetum.

In most mammals (and in frogs after metamorphosis), retinal axons from one eye do not all go to the same LGN. Instead, axons from the ventro-temporal region of the retina project to the LGN on the same side as the eye (**ipsilateral**), while the remaining retinal axons project to the LGN on the opposite side (**contralateral**) (Fig. 12.23, right panel). This separation of outputs occurs at the optic chiasm and enables excellent binocular vision in mammals with forward-facing eyes, such as primates and humans, by providing continuity of representation in the visual-processing structures in the brain. Birds, many of which also have excellent binocular vision, integrate the visual signals from each eye in other ways.

At the optic chiasm, retinal ganglion cell axons must make a decision to either avoid or traverse the midline of the brain. How this is achieved is not yet clear but it is likely that short-range guidance molecules, such as ephrins and their Eph receptors, are involved. In mice, for example, the ipsilateral retinal projection arises from neurons in the peripheral ventro-temporal crescent of the retina (see Fig. 12.23, right panel). These neurons express the guidance receptor EphB1 (see Box 5F, p. 205), which interacts with ephrin B2 on radial glia cells at the optic chiasm to repel the axons away from the midline and into the ipsilateral optic tract.

Retinal neurons map in a highly ordered manner on to their target structures, with a point-to-point correspondence between a position on the retina and one on the tectum. Figure 12.23 (left panel) shows how neurons in different positions along the naso-temporal (antero-posterior) axis of the tadpole retina, map in a reverse direction to points along the antero-posterior axis of the optic tectum. A similar reversed polarity is seen in the mapping of neurons along the dorso-ventral axis of the retina: dorsal retinal neurons project to the lateral region of the tectum and neurons from the ventral region of the retina project to the medial region of the tectum (Fig. 12.24).

Remarkably, in some lower vertebrates, such as fish and amphibians, the pattern of connections can be re-established with precision when the optic nerve is cut. The ends of the axons distal to the cut die, new growth cones form, and axon outgrowth reforms

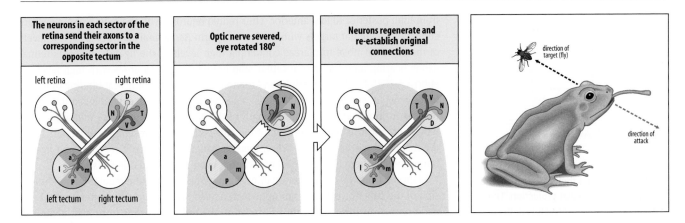

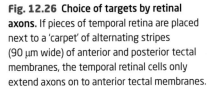

Fig. 12.25 Retino-tectal connections in amphibians are re-established in the original arrangement after severance of the optic nerve and rotation of the eye. First panel: neurons in the optic nerve from the left eye mainly connect to right optic tectum and those from the right eye to the left optic tectum. There is a point-to-point correspondence between neurons from different regions in the retina (nasal (N), temporal (T), dorsal (D), and ventral (V)) and their connections in the tectum (posterior (p), anterior (a), lateral (l), and medial (m), respectively). Center and right panels: if one optic nerve of a frog is cut and the eye rotated dorso-ventrally through 180°, the severed ends of the axons degenerate. When the neurons regenerate, they make connections with their original sites of contact in the tectum. However, because the eye has been rotated, the image falling on the tectum is upside down compared with normal. When the frog sees a fly above its head, it thinks the fly is below it, and moves its head downward to try to catch it (fourth panel).

connections to the tectum. In frogs, even if the eye is inverted through 180°, the axons still find their way back to their original sites of contact (Fig. 12.25). However, the animals subsequently behave as if their visual world has been turned upside down: if a visual stimulus, such as a fly, is presented above the inverted eye, the frog moves its head downward instead of upward, and can never learn to correct this error.

From such experiments it was suggested that each retinal neuron carries a chemical label that enables it to connect reliably with an appropriately chemically labeled cell in the tectum. This is known as the **chemoaffinity hypothesis** of connectivity. In the retino-tectal projection, it is thought that graded spatial distributions of a relatively small number of factors on the tectum provide positional information, which can be detected by the retinal axons. The spatially graded expression of another set of factors on the retinal axons would provide them with their own positional information. The development of the retino-tectal projection could thus, in principle, result from the interaction between these two gradients.

Such gradients were first found in the developing visual system of the chick embryo. An axon-guiding activity based on repulsion has been detected along the antero-posterior axis of the chick tectum. Normally, the temporal (posterior) half of the retina projects to the anterior part of the tectum and the nasal (anterior) retina projects to the posterior tectum. When offered a choice between growing on posterior or anterior tectal cells, temporal axons from an explanted chick retina show a preference for anterior tectal cells (Fig. 12.26). This choice is mediated by repulsion of the axons—as shown by the collapse of their growth cones—by a factor located on the surface of posterior tectal cells. Ephrins and their receptors are the likely molecules mediating this repulsion as they are expressed in reciprocal gradients in retina and tectum.

In the mouse, the role of ephrins and Ephs in creating a map has been investigated in the retino-collicular pathway, and repulsive interactions between the EphA family of receptors and their ephrin A ligands have been found to pattern this map in the antero-posterior dimension. Mapping of the retina on to the superior colliculus is similar to that from the retina to the optic tectum: retinal neurons from the nasal region map to the posterior colliculus and neurons from the temporal retina map to the anterior colliculus. EphAs are expressed in a gradient in the mouse retina, low in the nasal region, high in the temporal region, whereas ephrin As are graded in the superior colliculus

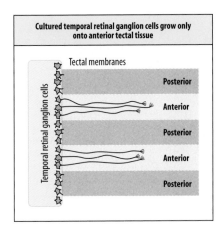

Fig. 12.26 Choice of targets by retinal axons. If pieces of temporal retina are placed next to a 'carpet' of alternating stripes (90 μm wide) of anterior and posterior tectal membranes, the temporal retinal cells only extend axons on to anterior tectal membranes.

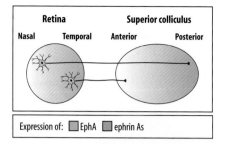

Expression of: ■ EphA ■ ephrin As

Fig. 12.27 Complementary expression of the ephrins and their receptors in the mouse retino-collicular projection. The superior colliculus is a structure in the mammalian midbrain to which some retinal neurons project and which is analogous to the tectum of amphibians and birds. The receptor EphA is a receptor tyrosine kinase and ephrin As are its ligands. Temporal retinal neurons, which express high levels of EphA on their surface, are repulsed by the posterior tectum, where high levels of ephrin A are present, but can make connections with the anterior tectum. Nasal retinal neurons make the best contacts in the posterior tectum, as they express low levels of EphA and thus are repulsed only where the levels of ephrins are highest.

from high posterior to low anterior. Thus retinal neurons with the most EphA receptors connect to collicular targets with the fewest ephrin As. Retinal axons bearing low levels of the receptor move into areas containing high levels of the ligands, whereas retinal axons bearing high levels of receptor stop before they reach areas with high levels of ephrin A (Fig. 12.27). A possible explanation for this behavior is that binding of ligand to the receptor sends a repulsive signal, and the axon ceases to extend when this signal reaches a threshold value. The strength of the signal received by the migrating axon will be proportional to the product of the concentrations of receptor and ligand. Thus, the threshold will be reached when there is a high level of both receptor and ligand, but not when there is a low level of receptor, even when ligand level is high.

Mapping from the dorso-ventral dimension of the retina to the medial-lateral dimension of the superior colliculus or optic tectum also involves graded expression of Ephs and ephrins. In this case, EphB is graded in the retinal neurons from high-ventral to low-dorsal and ephrin B in the tectum from high-medial to low-lateral. EphB and ephrin B mediate adhesive and attractive interactions, rather than repulsion, and this poses a conceptual problem. If their interaction were the only factor at work, all the retinal neurons would be attracted to the area of highest ephrin B, rather than mapping across the whole tectum. In this case, the necessary repulsive interactions in these dimensions appear to be provided by gradients of the signaling protein Wnt3 in the tectum and a receptor, Ryk, on the retinal axons.

Summary

Growth cones at the tip of the extending axon guide it to its destination. Filopodial activity at the growth cone is influenced by environmental factors, such as contact with the substratum and with other cells, and can also be guided by chemotaxis. Guidance involves both attraction and repulsion. In the development of motor neurons that innervate vertebrate limb muscles, the growth cones guide the axons so as to make the correct muscle-specific connections, even when their normal site of entry into the limb is disturbed. Combinations of transcription factors give each motor neuron an identity that determines its pathway. Attraction and repulsion control axons crossing the midline in both *Drosophila* and vertebrates. Gradients in diffusible molecules are probably responsible for the directional growth of commissural axons in the spinal cord. Neuronal guidance of axons from the retina to make the correct connections with the optic tectum involves gradients in cell-surface molecules both on the tectal neurons and on the retinal axons that can promote or repel growth cone approach, and there is competition between neurons for sites.

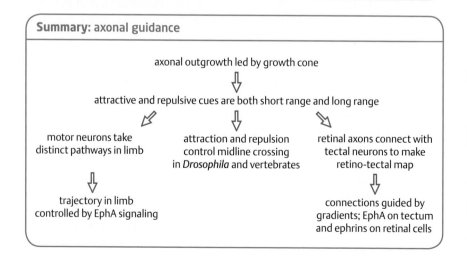

Summary: axonal guidance

axonal outgrowth led by growth cone

attractive and repulsive cues are both short range and long range

motor neurons take distinct pathways in limb

attraction and repulsion control midline crossing in *Drosophila* and vertebrates

retinal axons connect with tectal neurons to make retino-tectal map

trajectory in limb controlled by EphA signaling

connections guided by gradients; EphA on tectum and ephrins on retinal cells

Synapse formation and refinement

Having seen the outcome of axonal navigation, we now focus on the specialized connections—the **synapses**—that the axons make with their targets and which are essential for signaling between neurons and their target cells. Formation of synapses in the correct pattern is a basic requirement of any developing nervous system. The connections may be made with other nerve cells, with muscles, and also with certain glandular tissues. Here, we focus mainly on the development and stabilization of synapses at the junctions between motor neurons and muscle cells in vertebrates, which are called **neuromuscular junctions**. There are very many neuronal cell-types—hundreds if not thousands—and how they match up to make synapses reliably is a central problem. The molecular mechanisms are only beginning to be known. Studies on the vertebrate neuromuscular junction have guided much of the understanding of synapse formation.

Setting up the organization of a complex nervous system in vertebrates involves refining an initially rather imprecise organization by extensive programmed cell death (see Section 10.12). The establishment of a connection between a neuron and its target appears to be essential not only for the functioning of the nervous system, but also for the very survival of many neurons. Neuronal death is very common in the developing vertebrate nervous system; too many neurons are produced initially and only those that make appropriate connections survive. Survival depends on the neuron receiving neurotrophic factors, such as nerve growth factor, which are produced by the target tissue and for which neurons compete.

A special feature of nervous-system development is that fine-tuning of synaptic connections depends on the interaction of the organism with its environment and the consequent neuronal activity. This is particularly true of the vertebrate visual system, where sensory input from the retina in a period immediately after birth modifies synaptic connections so that the animal can perceive fine detail. Again, this refinement seems to involve competition for neurotrophic factors. We return to this topic after first considering synapse formation.

12.13 Synapse formation involves reciprocal interactions

We will start our look at synapse formation with the neuromuscular junction, as this is one of the most intensively studied and best-understood types of synapse. The mature neuromuscular junction is a complex structure involving extensive modification of the nerve ending and the muscle cell membrane (Fig. 12.28). Just before the junction, the axon terminal branches into a network-like arrangement. Each branch ends in a swelling, which is in contact with a special endplate region on the muscle fiber. The axon's plasma membrane is separated from the muscle-cell's plasma membrane by a narrow cleft (the **synaptic cleft**) filled with extracellular material secreted by both the neuron and the muscle cell. A basal lamina of extracellular material forms around the muscle-cell surface. The structure comprising axon plasma membrane, the opposing muscle-cell plasma membrane, and the cleft between them, is a synapse.

Electrical signals cannot pass across the synaptic cleft, and for the neuron to signal to the muscle, the electrical impulse propagated down the axon is converted at the terminal into a chemical signal. This is a chemical neurotransmitter that is released into the synaptic cleft from **synaptic vesicles** in the axon terminal. Molecules of the neurotransmitter diffuse across the cleft and interact with receptors on the muscle-cell membrane, causing the muscle fiber to contract. The neurotransmitter used by motor neurons connecting with skeletal muscles is acetylcholine. Because the signal travels from nerve to muscle, the axon terminal is called the **pre-synaptic** part of the junction and the muscle cell is the **post-synaptic** partner (see Fig. 12.28).

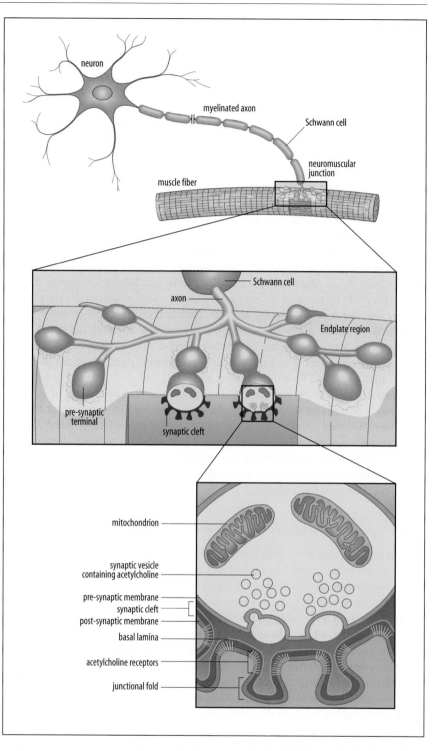

Fig. 12.28 Structure of the vertebrate neuromuscular junction. The motor neuron axon, which is covered in a myelin sheath produced by the glial Schwann cell, innervates a muscle fiber. At the neuromuscular junction, the axon branches in the endplate region, and makes synaptic connections with the muscle membrane. Communication between nerve and muscle is by release of the neurotransmitter acetylcholine from synaptic vesicles into the synaptic cleft. Acetylcholine diffuses across the cleft and binds to acetylcholine receptors on the muscle cell membrane.

After Kandel, E.R., et al.: 1991.

Development of a neuromuscular junction is progressive. Before the arrival of the axon there is already some pre-patterning on the muscle, with acetylcholine receptors concentrated in the central region of the muscle fiber. When the motor neuron terminal arrives, it releases the proteoglycan Agrin into the basal lamina, which binds to, and activates, the transmembrane muscle-specific kinase (MuSK), whose activity is required to maintain the clustering of acetylcholine receptors and the specialization of the muscle-cell surface at the site (Fig. 12.29). Signals from the muscle in turn induce differentiation of the pre-synaptic zone on the axon terminal

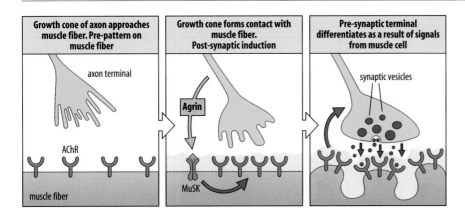

| Growth cone of axon approaches muscle fiber. Pre-pattern on muscle fiber | Growth cone forms contact with muscle fiber. **Post-synaptic induction** | **Pre-synaptic terminal differentiates as a result of signals from muscle cell** |

Fig. 12.29 Development of the neuromuscular junction. Before contact by an axon, acetylcholine receptors (AChR) are gathered in a central region of the muscle fiber. When a motor axon growth cone contacts a muscle fiber it releases the proteoglycan Agrin (blue arrow) into the extracellular material. Agrin binds to and activates the muscle-cell kinase MuSK, which in turn maintains clustering of AChR in the muscle-cell membrane and specialization of the post-synaptic surface. A signal (as yet unknown) from the muscle cell (purple arrow) then helps to induce the differentiation of the pre-synaptic terminal, with the formation of synaptic vesicles. Synaptic signaling by acetylcholine (small red circles) released from the synaptic vesicles can then begin.

and align this zone with the post-synaptic area on the muscle. These signals are not yet clearly identified, but the glycoprotein laminin β_2 in the basal lamina is required for correct alignment and full pre-synaptic differentiation. The growth factor neuregulin-1, which is related to epidermal growth factor and acts through the ErbB receptors, is required for the formation of some synapses, including neuromuscular junctions, where it helps mediate the expression of acetylcholine receptors on the muscle cell.

Neuregulin-1 is also involved in the development of glial cells, including the Schwann cells of the peripheral nervous system. In vertebrates, nerve impulses often have to travel long distances along axons in peripheral nerves—up to several meters in a large animal—and rapid propagation of nerve impulses over such distances is achieved by myelination of axons by glial cells. Schwann cells in the peripheral nervous system and oligodendrocytes in the central nervous system wrap themselves around axons so that their cell membranes form a fatty, insulating layer of myelin. Depolarization of the neuron membrane then only occurs at the small areas of bare neuronal membrane between one insulating cell and the next, and the nerve impulse is able to 'jump' over these areas without loss of signal. The progressive loss of neural function in the human disease multiple sclerosis is due to the destruction of myelin, and eventually leads to severe muscle weakness due to the reduction in signals from nerve to muscle.

The neuromuscular junction is a highly specialized type of synapse, but synapse formation within the vast network of neurons that form the central nervous system follows similar principles. Synapses are made between the axon terminal of one neuron and a dendrite, cell body, or more rarely the axon, of another neuron. An individual motor neuron in the spinal cord, for example, will receive input from thousands of other neurons on its dendrites and cell body. As with neuromuscular junctions, interneuronal synapse-formation is likely to involve reciprocal signaling between axon and dendrite. Synapses between neurons can form quite rapidly, on a time scale of about an hour, and observations suggest that filopodia on the dendrite initiate synapse formation by reaching out to the axon (Fig. 12.30). Signals from the axon, including release of neurotransmitter, may initially guide the dendrite filopodia. Many other factors have been shown to promote synapse formation, including cell-adhesion molecules and Wnt signaling. Cell-adhesion molecules implicated in synapse formation include the **neuroligins**, which are post-synaptic membrane proteins that bind to proteins called β-**neurexins** on the pre-synaptic axon surface, causing the neurexins to cluster, which induces differentiation in the pre-synaptic terminal. Neuroligins promote the function of excitatory synapses rather than inhibitory synapses. Cadherins, protocadherins, and members of the immunoglobulin superfamily (see Box 8B, p. 293) have also been found localized to synapses.

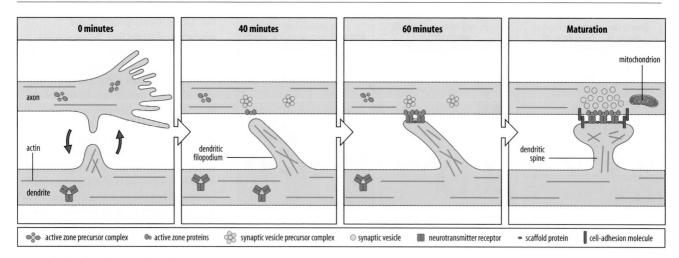

| 0 minutes | 40 minutes | 60 minutes | Maturation |

Legend: active zone precursor complex · active zone proteins · synaptic vesicle precursor complex · synaptic vesicle · neurotransmitter receptor · scaffold protein · cell-adhesion molecule

Fig. 12.30 Interneuronal synapse formation. A dendrite sends out filopodia that contact the axon growth cone. Reciprocal signaling (red arrows) initiates synapse formation aided by cell-adhesion molecules. In the axon, proteins that compose the pre-synaptic active zone gather at the site of contact along with synaptic vesicle precursors. In the dendrite, neurotransmitter receptors are synthesized and inserted into the post-synaptic membrane along with post-synaptic scaffold proteins. The post-synaptic connection develops into typical dendritic spines.

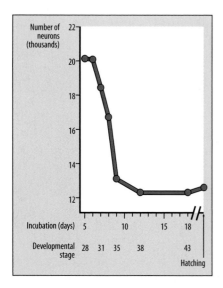

Fig. 12.31 Motor neuron death is a normal part of development in the chick spinal cord. The number of spinal cord motor neurons innervating a chick limb decreases by about half before hatching, as a result of programmed cell death during development. Most of the neurons die over a period of 4 days.

In *Drosophila*, the cell-adhesion molecule Dscam, a member of the immunoglobulin superfamily, is required both for axon guidance in the embryonic nervous system and in specifying particular neuronal connections. It initiates a signaling pathway that leads to activation of Rho-family GTPases that affect axonal migration by causing changes in the actin cytoskeleton. The Dscam gene is exceptional in that alternative splicing could produce an estimated 38,000 distinct forms of cell-surface receptor, thus raising the possibility that different forms of the protein might be produced in different neurons.

12.14 Many motor neurons die during normal development

During development, some 20,000 motor neurons are formed in the segment of spinal cord that provides innervation to a chick leg, but about half of them die soon after they are formed (Fig. 12.31). Cell death occurs after the axons have grown out from the cell bodies and entered the limb, at about the time that the axon terminals are reaching their potential targets—the skeletal muscles of the limb. The role of the target muscles in preventing cell death is suggested by two experiments. If the leg bud is removed, the number of surviving motor neurons decreases sharply. When an additional limb bud is grafted at the same level as the leg, providing additional targets for the axons, the number of surviving motor neurons increases.

Survival of a motor neuron depends on its establishing contacts with a muscle cell. Once a contact is established, the neuron can activate the muscle, and this is followed by the death of a proportion of the other motor neurons that are approaching the muscle cell. Muscle activation by neurons can be blocked by the drug curare, which prevents neuromuscular signal transmission, and this block results in a large increase in the number of motor neurons that survive.

Even after neuromuscular connections have been made, some are subsequently eliminated. At early stages of development, single muscle fibers are innervated by axon terminals from several different motor neurons. With time, most of these connections are eliminated, until each muscle fiber is innervated by the axon terminals from just one motor neuron (Fig. 12.32). This is due to competition between the synapses, with the most powerful input to the target cell destabilizing the less powerful inputs to the same target. This elimination process and the maturation of a stable neuromuscular junction take up to 3 weeks in the rat.

The well-established matching of the number of motor neurons to the number of appropriate targets by these mechanisms suggests that a general mechanism for the development of nervous-system connectivity in all parts of the vertebrate nervous system is that excess neurons are generated and only those that make the required connections are selected for survival. This mechanism is well-suited to regulating cell numbers by matching the size of the neuronal population to its targets. We now look in more detail at the neurotrophic factors that promote neuronal survival, and then consider the role of neural activity in the elimination of neuromuscular synapses.

12.15 Neuronal cell death and survival involve both intrinsic and extrinsic factors

Neuronal death during development takes place as a result of the activation of the apoptotic pathway (see Section 10.12). Apoptosis can be triggered or prevented either by an intrinsic developmental program or by external factors. In the development of the nematode, for example, particular neurons and other cells are intrinsically programmed to die (see Chapter 6), and the specificity of cell death can be shown to be under genetic control.

As we saw in Section 10.12, members of the Bcl-2 protein family control the apoptotic pathway in vertebrates—some acting to promote apoptosis and some to prevent it. A key family member promoting neuronal apoptosis during development is *Bax*; mice mutant for *Bax* have increased numbers of motor neurons in the face. A protein called Survivin is known to inhibit apoptosis during early neurogenesis. Mutant mice in which Survivin is deleted from embryonic day 12.5 onwards are born with much smaller brains than normal and numerous foci of apoptosis throughout the brain and spinal cord and die soon after birth.

In many developmental situations it seems that cells are programmed to die unless they receive specific survival signals. Extracellular factors play a key role in neuronal survival in vertebrates. The first one identified was nerve growth factor (NGF), whose discovery was due to the serendipitous observation that a mouse tumor implanted into a chick embryo evoked extensive growth of nerve fibers towards the tumor. This suggested that the tumor was producing a factor that promoted axonal outgrowth. The factor was eventually identified as a protein, NGF, using axon outgrowth in culture as an assay. NGF is necessary for the survival of a number of types of neurons, particularly those of the sensory and sympathetic nervous systems.

NGF is a member of a family of proteins known as the **neurotrophins**. Different types of neuron require different neurotrophins for their survival, and the requirement for certain neurotrophins also changes during development. The neurotrophin GDNF, for example, prevents the death of facial motor neurons, and is also expressed in developing limb buds. There is increasing evidence that individual neurotrophins act on a variety of neural cell types.

12.16 The map from eye to brain is refined by neural activity

The development and function of the nervous system depends not only on synapse formation, but also on the regulated disassembly of previously functional synaptic connections. We have already considered how axons from the retina make connections with the tectum so that a retino-tectal map is established. This map is initially rather coarse-grained, in that axons from neighboring cells in the retina make contacts over a large area of the tectum. This area is much larger early on than at later stages of development, when the retino-tectal map is more finely tuned. Fine-tuning of the map results, as in muscle, from the withdrawal of axon terminals from most of the initial contacts, and requires neural activity. This requirement is seen particularly clearly in the development of visual connections in mammals. If a mammal, including a human, is deprived of vision during a critical period, then it will suffer from poor visual acuity and will not be able to resolve fine detail.

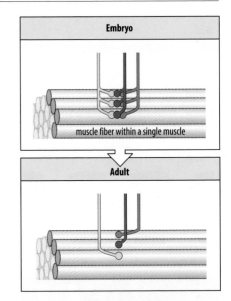

Fig. 12.32 Refinement of muscle innervation by neural activity. Initially, several motor neurons innervate the same muscle fiber. Elimination of synapses means that each fiber is eventually innervated by only one neuron.

After Goodman, C.S. and Shatz, C.J.: 1993.

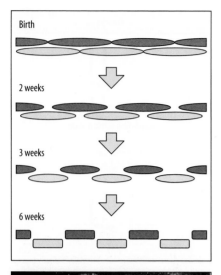

Birth

2 weeks

3 weeks

6 weeks

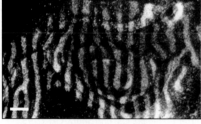

Fig. 12.33 Visualization of ocular dominance columns in the monkey visual cortex. A radioactive tracer is injected into one eye, from where it is transported to the visual cortex through the neurons. Tracer injected at birth is broadly distributed in the cortex. Tracer injected at later times becomes confined to alternating columns of cortical cells (brighter stripes), representing the ocular dominance columns for that eye, as seen in the photograph. Scale = 1 mm.

After Kandel, E.R., et al.: 1995.

In mammals with good binocular and stereoscopic vision, such as monkeys and humans, visual-system axons from the retina first connect to the lateral geniculate nucleus, on to which they map in an ordered manner. Input from half of each eye goes to the opposite side of the brain, whereas input from the other half of each eye goes to the same side of the brain (see Fig. 12.23, right panel).

In the lateral geniculate nucleus, as in the cortex, the neurons are arranged in layers. Each layer receives input from retinal axons from either the right or the left eye, but not both. Thus, inputs from left and right eyes are kept separate. Neurons from the lateral geniculate nucleus then send axons to the visual cortex (see Fig. 12.23, right panel). When there is a visual stimulus, the inputs from the retinal axons activate neurons in the lateral geniculate nucleus, which then activate neurons in the corresponding region of the visual cortex. The adult visual cortex consists of six cell layers, but we need only focus here on layer 4, which is where many axons from the lateral geniculate nucleus make connections. Layers in the lateral geniculate nucleus that receive left or right inputs both make connections with layer 4 of the visual cortex. This means that input from corresponding positions in both eyes, that is from the same part of the visual field, finally arrives at the same location in the cortex (see Fig. 12.23, right panel).

At birth, the nerve endings from the two eyes overlap and are mixed at their common final location, but with time the inputs from left and right eyes become separated into blocks of cortical cells about 0.5 mm wide, which are known as **ocular dominance columns** (Fig. 12.33). Adjacent columns respond to the same stimulus in the visual field; one column responding to signals from the left eye, and the next to signals from the right eye. This arrangement enables good binocular and stereoscopic vision. The columns can be detected and mapped by making electrophysiological recordings. They can also be directly observed by injecting a tracer, such as radioactive proline, into one eye. The tracer is taken up by retinal neurons, transported by the optic nerve to the lateral geniculate, and from there to the visual cortex, where its pattern can be detected by autoradiography. This reveals a striking array of stripes representing the input from one eye (see Fig. 12.33).

Neural activity and visual input are essential for the development and maintenance of the ocular dominance columns. While sensory input is important, spontaneous activity also plays a key role. Initial formation of the stripes in non-human primates occurs before visual experience, and involves spontaneously generated waves of action potentials in the mammalian retina. These waves may cause their effects through the release of neurotrophins that can remodel synaptic connections. If neural activity is blocked during development by the injection of tetrodotoxin, ocular dominance columns do not develop, and the inputs from the two eyes into the visual cortex remain mixed. If the input from one eye is blocked, the territory in the visual cortex occupied by the other eye's input expands at the expense of the blocked eye. The release of the inhibitory neurotransmitter γ-aminobutyric acid (GABA) is required to initiate the critical period.

The favored explanation for the formation of ocular dominance columns is based on competition between incoming axons (Fig. 12.34). Because of the initial connections that establish the map, individual cortical neurons can initially receive input from both eyes. Within a particular region, there will, therefore, be overlap of stimuli originating from the two eyes, and this overlap has to be resolved. Neighboring cells carrying input from the same eye tend to fire simultaneously in response to a visual stimulus; if they both innervate the same target cell, they can thus cooperate to excite it. As in muscle, stimulation of electrical activity in the target cell tends to strengthen the active synapses and suppress those that are not active at the time—cells that fire together, wire together. As there is competition between neurons for targets, this could generate discrete regions of cortical cells that respond only to one eye or the other, and so form the ocular dominance columns. Such a mechanism explains why experimental exposure of animals to continuous strobe lighting after birth, which causes simultaneous firing of neurons in both eyes, prevents the formation of ocular dominance columns.

In frogs the optic tecta do not normally receive input from both eyes until after morphogenesis and ocular dominance columns do not develop in the tecta under normal circumstances. Tecta that receive inputs from two eyes throughout development can, however, be produced by implanting a third eye primordium into the embryo. Retinal axons from the two eyes then segregate within the tectum and form alternating stripes that resemble the pattern of ocular dominance columns in cats and primates. In these tecta, the dendrites of some types of tectal neurons seem to respond to the stripe boundaries, as evidenced by marked changes in their normal direction of migration or by their termination at a stripe border. The behavior of these neurons in this artificial situation is further evidence that in all vertebrates, not simply mammals, highly localized interactions between incoming axon terminals and the post-synaptic dendrites with which they connect are critical in producing the patterning of a segregated, topographically organized cortex.

One possible mechanism for refining connections in response to neuronal activity involves local release of neurotrophins. A certain level of activation, or activation by two axons simultaneously, could induce the release of neurotrophins from the target cells, and only those axons that have been recently active might be able to respond to them. When neural activity in developing chick or amphibian embryos is blocked with drugs that prevent neuronal firing, no fine-grained retino-tectal map develops.

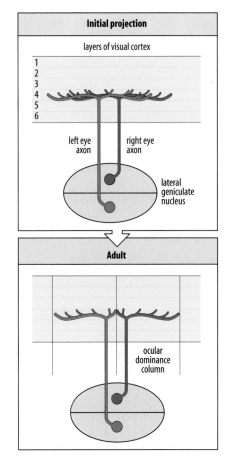

Fig. 12.34 Development of ocular dominance columns. Initially, neurons from the lateral geniculate nucleus, representing projections from both eyes and stimulated by the same visual stimulus, project to the same region of the visual cortex. (Only projection to layer 4 is shown here.) With visual stimulation, the neuronal connections separate out into columns, each representing innervation from only one eye. If stimulation by vision is blocked, ocular dominance columns do not form.

After Goodman, C.S. and Shatz, C.J.: 1993.

Summary

Neurons communicate with each other and with target cells, such as muscles, by means of specialized junctions called synapses. The formation of a neuromuscular junction involves changes in both pre-synaptic (neuronal) and post-synaptic (muscle cell) membranes once contact is made, and depends on reciprocal signaling between muscle and nerve cell. Reciprocal interactions also occur in the formation of synapses between axons and dendrites, and cell-adhesion molecules are likely to be involved. Initially, most mammalian muscle fibers are innervated by two or more motor axons, but nervous activity results in competition between synapses so that a single fiber is eventually innervated by only one motor neuron. Many neurons produced in the developing nervous system die. About half of the motor neurons that initially innervate the vertebrate limb undergo cell death; those that survive do so because they make functional connections with muscles. Many neurons depend on neurotrophins, such as nerve growth factor, for their survival, with different classes of neurons requiring different neurotrophins. Brain function is based not only on the assembly of synapses, but also on their regulated disassembly. Neural activity has a major role in refining the connections between the eye and the brain. In mammals, input from the left and right eyes is required for the development of ocular dominance columns in the visual cortex. These are adjacent columns of cells responding to the same stimulus from left and right eyes, respectively. Formation of the columns, which are essential for binocular vision, is a result of competition for cortical targets by axons carrying visual input from different eyes.

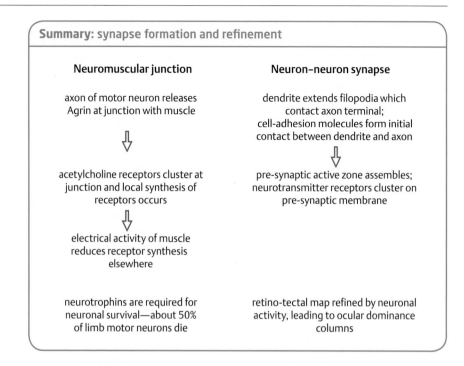

Summary: synapse formation and refinement

Neuromuscular junction	Neuron–neuron synapse
axon of motor neuron releases Agrin at junction with muscle	dendrite extends filopodia which contact axon terminal; cell-adhesion molecules form initial contact between dendrite and axon
⇩	⇩
acetylcholine receptors cluster at junction and local synthesis of receptors occurs	pre-synaptic active zone assembles; neurotransmitter receptors cluster on pre-synaptic membrane
⇩	
electrical activity of muscle reduces receptor synthesis elsewhere	
neurotrophins are required for neuronal survival—about 50% of limb motor neurons die	retino-tectal map refined by neuronal activity, leading to ocular dominance columns

Summary to Chapter 12

The processes involved in the development of the nervous system, with its numerous connections, are similar to those found in other developmental systems. Presumptive neural tissue is specified early in development–during gastrulation in vertebrates, and when the dorso-ventral axis is patterned in *Drosophila*. Within the neuroectodermal tissue, the specification of cells that give rise to neural cells involves lateral inhibition. The further development of neurons from neuronal precursors involves both asymmetric cell divisions and cell–cell signaling. The patterning of different types of neurons within the vertebrate spinal cord is due to both ventral and dorsal signals. As they develop, neurons extend axons and dendrites. The axons are guided to their destination by growth cones at their tips. Guidance is due to the growth cone's response to attractive and repulsive signals, which may be diffusible or bound to the substratum. Gradients in such molecules can guide the axons to their destination, as in the retino-tectal system of vertebrates. The functioning of the nervous system depends on the establishment of specific synapses between axons and their targets. Specificity appears to be achieved by an initial overproduction of neurons that compete for targets, with many neurons dying during development. Refinement of synaptic connections involves further competition. Neural activity plays a major role in refining connections, such as those between the eye and the brain.

▪ End of chapter questions

Long answer (concept questions)

1. In *Drosophila*, the central nervous system is ventral, whereas in vertebrates, the central nervous system is dorsal. Despite these differences, what similarities exist that would indicate a common evolutionary origin?

2. Outline the mechanism of lateral inhibition in the proneural cluster of *Drosophila*, in which the expression of Delta in one cell leads to loss of Delta in a neighboring cell. Compare the *Drosophila* pathway to that in vertebrates.

3. How do Numb and Pins proteins collaborate to cause one daughter cell to become a neuron after (a) division of the neuroblast in the *Drosophila* embryo? (b) After division of the sensory organ precursor in the *Drosophila* adult?

4. What are radial glial cells? Include in your answer: the embryonic structure in which they are found, their location in that structure, and their roles in development of the nervous system (name two roles).

5. Although the adult mammalian brain cannot in general produce new neurons, the wall of the lateral ventricle is an exception. Relate this observation to the embryological origin of neurons in the neural tube.

6. The formation of motor neurons in the developing spinal cord depends on a gradient of Sonic hedgehog (Shh) protein. What is the source of this Shh gradient? What Class I and Class II homeodomain protein genes are active in response to the precise level of this gradient required for motor neuron progenitors (pMN) to become specified as motor neurons (MN)? What role do the LIM-homeodomain proteins play in motor neuron identity?

7. Compare and contrast netrins, semaphorins, cadherins, and ephrins: in what way are netrins and cadherins similar in effect, and in what way are they different? In what way are semaphorins and ephrins similar, and in what way do they differ?

8. What is meant by the phrase 'the retina maps onto the tectum' (see Fig. 12.26)? How does the chemoaffinity hypothesis attempt to explain this phenomenon? How do the ephrins and their receptors fit into this hypothesis?

9. Describe the results of the experiment in which the optic nerve to a frog's eye is cut and the eye rotated by 180°. How is the subsequent response of the frog to a visual stimulus explained?

10. What is the role of neurotrophic factors, such as nerve growth factor, in refining the connections between neurons and their targets? Give a specific example of such action.

11. Draw a schematic diagram of a synapse, labeling the presynaptic cell and the post-synaptic cell and indicating the direction of signal transmission. Label the synaptic cleft and say why it prevents neurons from passing electrical signals directly to each other? How do neurons communicate across the synaptic cleft?

12. Describe the steps involved in the formation of the neuromuscular junction, starting with the agrin signal from the neuronal growth cone.

13. Cell death (apoptosis) in the developing nervous system is under the control of both intrinsic and extrinsic influences. Give examples of intrinsic (i.e. intracellular) and extrinsic (i.e. cell–cell signaling) proteins that control apoptosis in the nervous system during development.

14. Contrast the retinal inputs to the lateral geniculate nucleus in the mammalian brain with those to the tectum in the amphibian brain. What are ocular dominance columns, and why are they important? What is meant by 'cells that fire together, wire together'?

Multiple choice (factual recall questions)

NB There is only one correct answer to each question.

1. The genes of the achaete–scute complex encode proteins of which type?

a) cell–cell adhesion molecules

b) cytoskeletal proteins

c) membrane transport proteins

d) transcription factors

2. The vertebrate central nervous system (brain and spinal cord) is derived from

a) cells of the neural tube

b) neural crest cells that migrate along the dorso-lateral route

c) neural crest cells that migrate through the anterior half of the somites

d) somitic mesoderm

3. What would be the consequence of reducing Delta expression in the developing vertebrate neural tube?

a) Excess neurons are produced, because reduced Delta prevents the inhibition of neurogenesis in neighboring cells.

b) Excess neurons are produced, because the reduction in Delta expression increases neurogenin expression.

c) Fewer neurons are produced, because Delta signaling is required for neurogenesis.

d) There is no change in neurogenesis unless Delta expression is eliminated completely.

4. Which of the following is a determined neuronal cell type?

a) premigratory neural crest cells

b) floor plate cells

c) commissural cells

d) roof plate cells

5. Cortical neurons migrate to their final positions in the cortex along

a) astrocytes

b) commissural neurons

c) radial glia

d) roof-plate cells

6. Axons of spinal cord neurons from the lateral motor column in the chick embryo innervate

a) muscles of the body wall

b) muscles of the limb buds

c) muscles attached to the spine

d) muscles in the gut wall

7. Which of these is a cell-surface adhesion molecule that mainly attracts axons during their outgrowth?

a) neurotrophins

b) cadherins

c) netrins

d) semaphorins

8. Knock-outs of the mouse *netrin-1* gene have a phenotype similar to that of knock-outs of what other gene?

a) the DCC receptor

b) the Robo1 receptor

c) the semaphorin signaling molecule

d) the Slit signaling molecule

9. Neurons in the anterior (nasal) region of the right retina will project axons to

a) the anterior portion of the left tectum

b) the anterior portion of the right tectum

c) the posterior portion of the left tectum

d) the posterior portion of the right tectum

10. The synapses formed between motor neurons and muscles are called

a) myotomes

b) neuromuscular junctions

c) Nieuwkoop centers

d) retino-tectal projections

11. Schwann cells

a) are glial cells

b) are responsible for myelination of axons in the peripheral nervous system

c) are related to the oligodendrocytes of the central nervous system

d) are all of the above

12. The neurotransmitter used in the communication between motor neurons and muscles is

a) serotonin

b) dopamine

c) epinephrine

d) acetylcholine

Multiple choice answer key

1: d, 2: a, 3: a, 4: c, 5: c, 6: b, 7: b, 8: a, 9: c, 10: b, 11: d, 12: d.

▪ General further reading

Denes, A.S., Jékely, G., Steinmetz, P.R., Raible, F., Snyman, H., Prud'homme, B., Ferrier, D.E., Balavoine, G., Arendt, D.: **Molecular architecture of annelid nerve cord supports common origin of nervous system centralization in bilateria**. *Cell* 2007, **129**: 277–288.

Kandel, E.R., Schwartz, J.H., Jessell, T.H.: *Principles of Neural Science* (4th edn). New York: McGraw–Hill, 2000.

Kerszberg, M.: **Genes, neurons and codes: remarks on biological communication**. *BioEssays* 2003, **25**: 699–708.

▪ Section further reading

12.1 Neurons in *Drosophila* arise from proneural clusters

Cornell, R.A., Ohlen, T.V.: **Vnd/nkx, ind/gsh, and msh/msx: conserved regulators of dorsoventral neural patterning?** *Curr. Opin. Neurobiol.* 2000, **10**: 63–71.

Gibert, J.M., Simpson, P.: **Evolution of *cis*-regulation of the proneural genes**. *Int. J. Dev. Biol.* 2003, **47**: 643–651.

Skeath, J.B.: **At the nexus between pattern formation and cell-type specification: the generation of individual neuroblast fates in the *Drosophila* embryonic central nervous system**. *BioEssays* 1999, **21**: 922–931.

Weiss, J.B., Von Ohlen, T., Mellerick, D.M., Dressler, G., Doe, C.Q., Scott, M.P.: **Dorsoventral patterning in the *Drosophila* central nervous system: the intermediate neuroblasts defective homeobox gene specifies intermediate column identity**. *Genes Dev.* 1998, **12**: 3591–3602.

12.2 The development of neurons in *Drosophila* involves asymmetric cell divisions and timed changes in gene expression

Brody, T., Odenwald, W.: **Regulation of temporal identities during *Drosophila* neuroblast lineage development**. *Curr. Opin. Cell Biol.* 2005, **17**: 672–675.

Carmena, A.: **Signaling networks during development: the case of asymmetric cell division in the *Drosophila* nervous system**. *Dev. Biol.* 2008, **321**: 1–17.

Grosskortenhaus, R., Pearson, B.J., Marusich, A., Doe, C.Q.: **Regulation of temporal identity transitions in *Drosophila* neuroblasts**. *Dev. Cell* 2005, **8**: 193–202.

Jan, Y-N., Jan, L.Y.: **Polarity in cell division: what frames thy fearful asymmetry?** *Cell* 2000, **100**: 599–602.

Karcavich, R.E.: **Generating neuronal diversity in the *Drosophila* central nervous system: a view from the ganglion mother cells**. *Dev. Dyn.* 2005, **232**: 609–616.

Box 12A Specification of the sensory organs of adult *Drosophila*

Gómez-Skarmeta, J.L., Campuzano, S., Modolell, J.: **Half a century of neural prepatterning: the story of a few bristles and many genes**. *Nat. Rev. Neurosci.* 2003, **4**: 587–598.

Knoblich, J.A.: **Asymmetric cell division during animal development**. *Nat. Rev. Mol. Cell. Biol.* 2001, **2**: 11–20.

Modolell, J., Campuzano, S.: **The achaete-scute complex as an integrating device**. *Int. J. Dev. Biol.* 1998, **42**: 275–282.

12.3 Specification of vertebrate neuronal precursors also involves lateral inhibition

Chitins, A., Henrique, D., Lewis, J., Ish-Horowitcz, D., Kintner, C.: **Primary neurogenesis in *Xenopus* embryos regulated by a homologue of the *Drosophila* neurogenic gene *Delta***. *Nature* 1995, **375**: 761–766.

Ma, Q., Kintner, C., Anderson, D.J.: **Identification of *neurogenin*, a vertebrate neuronal determination gene**. *Cell* 1996, **87**: 43–52.

12.4 Neurons are formed in the proliferative zone of the neural tube and migrate outwards

D'Arcangelo, G., Curran, T.: ***Reeler*: new tales on an old mutant mouse**. *BioEssays* 1998, **20**: 235–244.

Gage, F.H.: **Mammalian neural stem cells**. *Science* 2000, **287**: 1433–1438.

Kriegstein, A., Alvarez-Buylla, A.: **The glial nature of embryonic and adult neural stem cells**. *Annu. Rev. Neurosci.* 2009, **32**: 149–184.

Lee, S.K., Lee, B., Ruiz, E.C., Pfaff, S.L.: **Olig2 and Ngn2 function in opposition to modulate gene expression in motor neuron progenitor cells**. *Genes Dev.* 2005, **19**: 282–294.

Lyuksyutova, A.I., Lu, C.C., Milanesio, N., King, L.A., Guo, N., Wang, Y., Nathans, J., Tessier-Lavigne, M., Zou, Y.: **Anterior-posterior guidance of commissural axons by Wnt-frizzled signaling**. *Science* 2003, **302**: 1984–1988.

Maricich, S.M., Gilmore, E.C., Herrup, K.: **The role of tangential migration in the establishment of the mammalian cortex.** *Neuron* 2001, **31**: 175–178.

Sauvageot, C.M., Stiles, C.D.: **Molecular mechanisms controlling cortical gliogenesis.** *Curr. Opin. Neurobiol.* 2002, **12**: 244–249.

Wichterle, H., Lieberam, I., Porter, J.A., Jessell, T.M.: **Directed differentiation of embryonic stem cells into motor neurons.** *Cell* 2002, **110**: 385–397.

Box 12B Timing the birth of cortical neurons

Rakic, P.: **Neurons in rhesus monkey visual cortex: systematic relation between time of origin and eventual disposition.** *Science* 1974, **183**: 425–427.

12.5 The pattern of differentiation of cells along the dorso-ventral axis of the spinal cord depends on ventral and dorsal signals

Jacob, J., Hacker, A., Guthrie, S.: **Mechanisms and molecules in motor neuron specification and axon pathfinding.** *BioEssays* 2001, **23**: 582–595.

Jessell, T.M.: **Neuronal specification in the spinal cord: inductive signals and transcriptional controls.** *Nat. Rev. Genet.* 2000, **1**: 20–29.

Megason, S.G., McMahon, A.P.: **A mitogen gradient of dorsal midline Wnts organizes growth in the CNS.** *Development* 2002, **129**: 2087–2098.

12.6 Neuronal subtypes in the ventral spinal cord are specified by the ventral to dorsal gradient of Shh

Dessaud, E., McMahon, A.P., Briscoe, J.: **Pattern formation in the vertebrate neural tube: a sonic hedgehog morphogen-regulated transcriptional network.** *Development* 2008, **135**: 2489–2503.

Stamataki, D., Ulloa, F., Tsoni, S.V., Mynett, A., Briscoe, J.: **A gradient of Gli activity mediates graded Sonic Hedgehog signaling in the neural tube.** *Genes Dev.* 2005, **19**: 626–641.

12.7 Spinal cord motor neurons at different dorso-ventral positions project to different trunk and limb muscles

Kania, A., Jessell, T.M.: **Topographic motor projections in the limb imposed by LIM homeodomain protein regulation of ephrin-A: EphA interactions.** *Neuron* 2003, **38**: 581–596.

Kania, A., Johnson, R.L., Jessell, T.M.: **Coordinate roles for LIM homeobox genes in directing the dorsoventral trajectory of motor axons in the vertebrate limb.** *Cell* 2000, **102**: 161–173.

12.8 Antero-posterior pattern in the spinal cord is determined in response to secreted signals from the node and adjacent mesoderm

Dasen, J.S., Jessell, T.M.: **Hox networks and the origins of motor neuron diversity.** *Curr. Top. Dev. Biol.* 2009, **88**: 169–200.

Ensini M., Tsuchida T.N., Belting, H.G., Jessell, T.M.: **The control of rostrocaudal pattern in the developing spinal cord: specification of motor neuron subtype identity is initiated by signals from paraxial mesoderm.** *Development* 1998, **125**: 969–982.

Sockanathan, S., Perlmann, T., Jessell, T.M.: **Retinoid receptor signaling in postmitotic motor neurons regulates rostrocaudal positional identity and axonal projection pattern.** *Neuron* 2003, **40**: 97–111.

12.9 The growth cone controls the path taken by the growing axon

Carmeliet, P., Tessier-Lavigne, M.: **Common mechanisms of nerve and blood vessel wiring.** *Nature* 2005, **436**: 193–200.

Lowery, L.A., Van Vactor, D.: **The trip of the tip: understanding the growth cone machinery.** *Nat. Rev. Mol. Cell. Biol.* 2009, **10**: 332–343.

Tear, G.: **Neuronal guidance: a genetic perspective.** *Trends Genet.* 1999, **15**: 113–118.

Tessier-Lavigne, M., Goodman, C.S.: **The molecular biology of axon guidance.** *Science* 1996, **274**: 1123–1133.

12.10 Motor neuron axons in the chick limb are guided by ephrin-Eph interactions

Dasen, J.S., Tice, B.C., Brenner-Morton, S., Jessell, T.M.: **A Hox regulatory network establishes motor neuron pool identity and target-muscle connectivity.** *Cell* 2005, **123**: 477–491.

Lance-Jones, C., Landmesser, L.: **Pathway selection by embryonic chick motoneurons in an experimentally altered environment.** *Proc. R. Soc. Lond. B* 1981, **214**: 19–52.

Tosney, K.W, Hotary, K.B., Lance-Jones, C.: **Specifying the target identity of motoneurons.** *BioEssays* 1995, **17**: 379–382.

12.11 Axons crossing the midline are both attracted and repelled

Giger, R.J., Kolodkin, A.L.: **Silencing the siren: guidance cue hierarchies at the CNS midline.** *Cell* 2001, **105**: 1–4.

Long, H., Sabatier, C., Ma, L., Plump, A., Yuan, W., Ornitz, D.M., Tamada, A., Murakami, F., Goodman, C.S., Tessier-Lavigne, M.: **Conserved roles for Slit and Robo proteins in midline commissural axon guidance.** *Neuron* 2004, **42**: 213–223.

Simpson, J.H., Bland, K.S., Fetter, R.D., Goodman, C.S.: **Short-range and long-range guidance by Slit and its Robo receptors: a combinatorial code of Robo receptors controls lateral position.** *Cell* 2000, **103**: 1019–1032.

Williams, S.E., Mason, C.A., Herrera, E.: **The optic chiasm as a midline choice point.** *Curr. Opin. Neurobiol.* 2004, **14**: 51–60.

Woods, C.G.: **Crossing the midline.** *Science* 2004, **304**: 1455–1456.

Zou, Y., Lyuksyutova, A.I.: **Morphogens as conserved axon guidance cues.** *Curr. Opin. Neurobiol.* 2007, **17**: 22–28.

12.12 Neurons from the retina make ordered connections with visual centers in the brain

Feldheim, D.A., Kim, Y-I., Bergemann, A.D., Frisen, J., Barbacid, M., Flanagan, J.G.: **Genetic analysis of ephrin A2 and ephrin A5 shows their requirement in multiple aspects of retinocollicular mapping.** *Neuron* 2000, **25**: 563–574.

Hansen, M.J., Dallal, G.E., Flanagan, J.G.: **Retinoic axon response to ephrin-As shows a graded concentration-dependent transition from growth promotion to inhibition.** *Neuron* 2004, **42**: 707–730.

Klein, R.: **Eph/ephrin signaling in morphogenesis, neural development and plasticity.** *Curr. Opin. Cell Biol.* 2004, **16**: 580–589.

Löschinger, J., Weth, F., Bonhoeffer, F.: **Reading of concentration gradients by axonal growth cones.** *Phil. Trans. R. Soc. Lond. B* 2000, **355**: 971–982.

McLaughlin, T., Hindges, R., O'Leary, D.D.: **Regulation of axial patterning of the retina and its topographic mapping in the brain.** *Curr. Opin. Neurobiol.* 2003, **13**: 57–69.

Petros, T.J., Shrestha, B.R., Mason, C.: **Specificity and sufficiency of EphB1 in driving the ipsilateral retinal projection**. *J. Neurosci.* 2009, **29**: 3463–3474.

Reber, M., Bursold, P., Lemke, G.: **A relative signalling model for the formation of a topographic neural map**. *Nature* 2004, **431**: 847–853.

Schmitt, A.M., Shi, J., Wolf, A.M., Lu, C-C., King, L.A., Zou, Y.: **Wnt-Ryk signalling mediates medial-lateral retinotectal topographic mapping**. *Nature* 2006, **439**: 31–37.

Wilkinson, D.G.: **Topographic mapping: organising by repulsion and competition?** *Curr. Biol.* 2000, **10**: R447–R451.

12.13 Synapse formation involves reciprocal interactions

Buffelli, M., Burgess, R.W., Feng, G., Lobe, C.G., Lichtman, J.W., Sanes, J.R.: **Genetic evidence that relative synaptic efficacy biases the outcome of synaptic competition**. *Nature* 2003, **424**: 430–434.

Chess, A.: **Monoallelic expression of protocadherin genes**. *Nat. Genet.* 2005, **37**: 120–121.

Goda, Y., Davis, G.W.: **Mechanisms of synapse assembly and disassembly**. *Neuron* 2003, **40**: 243–264.

Hua, J.Y, Smith, S.J.: **Neural activity and the dynamics of central nervous-system development**. *Nat. Neurosci.* 2004, **7**: 327–332.

Jan, Y-N., Jan, L.Y.: **The control of dendritic development**. *Neuron* 2003, **40**: 229–242.

Katz, L.C., Constantine-Paton, M.: **Relationships between segregated afferents and postsynaptic neurones in the optic tectum of three-eyed frogs**. *J. Neurosci.* 1988, **8**: 3160–3180.

Kummer, T.T., Misgled, T., Sanes, J.R.: **Assembly of the postsynaptic membrane at the neuromuscular junction**. *Curr. Opin. Neurobiol.* 2006, **16**: 74–82.

Levinson, J.N., El-Husseini, A.: **Building excitatory and inhibitory synapses: balancing neuroligin partnerships**. *Neuron* 2005, **48**: 171–174.

Li, Z., Sheng, M.: **Some assembly required: the development of neuronal synapses**. *Nat. Rev. Mol. Cell Biol.* 2003, **4**: 833–841.

Schmucker, D., Clemens, J.C., Shu, H., Worby, C.A., Xiao, J., Muda, M., Dixon, J.E., Zipursky, S.L.: ***Drosophila* Dscam is an axon guidance receptor exhibiting extraordinary molecular diversity**. *Cell* 2000, **101**: 671–684.

12.14 Many motor neurons die during normal development

Oppenheim, R.W.: **Cell death during development of the nervous system**. *Annu. Rev. Neurosci.* 1991, **14**: 453–501.

12.15 Neuronal cell death and survival involve both intrinsic and extrinsic factors

Birling, M.C., Price, J.: **Influence of growth factors on neuronal differentiation**. *Curr. Opin. Cell Biol.* 1995, **7**: 878–884.

Burden, S.J.: **Wnts as retrograde signals for axon and growth cone differentiation**. *Cell* 2000, **100**: 495–497.

Davies, A.M.: **Neurotrophic factors. Switching neurotrophin dependence**. *Curr. Biol.* 1994, **4**: 273–276.

Jiang, Y., de Bruin, A., Caldas, H., Fangusaro, J., Hayes, J., Conway, E.M., Robinson, M.L., Altura, R.A.: **Essential role for survivin in early brain development**. *J. Neurosci.* 2005, **25**: 6962–6970.

Pettmann, B., Henderson, C.E.: **Neuronal cell death**. *Neuron* 1998, **20**: 633–647.

Serafini, T.: **Finding a partner in a crowd: neuronal diversity and synaptogenesis**. *Cell* 1999, **98**: 133–136.

Šestan, N., Artavanis-Tsakonas, S., Rakic, P.: **Contact-dependent inhibition of cortical neurite growth mediated by Notch signaling**. *Science* 1999, **286**: 741–746.

12.16 The map from eye to brain is refined by neural activity

Del Rio, T., Feller, M.B.: **Early retinal activity and visual circuit development**. *Neuron* 2006, **52**: 221–222.

Horton, J.C., Adams, D.L.: **The cortical column: a structure without a function**. *Philos. Trans. R. Soc. Lond. B Biol. Sci.* 2005, **360**: 837–862.

Katz, L.C., Crowley, J.C.: **Development of cortical circuits: lessons from ocular dominance columns**. *Nat. Rev. Neurosci.* 2002, **3**: 34–42.

Katz, L.C., Shatz, C.J.: **Synaptic activity and the construction of cortical circuits**. *Science* 1996, **274**: 1133–1138.

Growth and post-embryonic development

- Growth
- Molting and metamorphosis
- Aging and senescence

Patterning of the embryo occurs on a small scale and is followed by growth. The control of growth, and therefore size, is a key problem in development, involving at its core the control of cell proliferation. Such control is also crucial in adult life in preventing cancer—the result of uncontrolled cell proliferation. In this chapter we look at the balance between intrinsic programs of cell proliferation during embryonic development, and growth that is stimulated by circulating factors such as hormones. Another aspect of post-embryonic development in many invertebrates is metamorphosis, in which the form of the animal is completely changed. Finally, the individual ages, a process that is not programmed genetically but which is due to cellular damage.

Development does not stop once the embryonic phase is complete. Most, but by no means all, of the growth in animals and plants occurs in the post-embryonic period, when the basic form and pattern of the organism has already been established. The basic patterning is done on a small scale, over dimensions of less than a millimeter. In many animals, the embryonic phase is immediately succeeded by a free-living larval or immature adult stage. In others, such as mammals, considerable growth occurs during a late embryonic or fetal period, while the embryo is still dependent on maternal resources. Growth then continues after birth. Growth is a central aspect of all developing systems, determining the final size and shape of the organism and its parts. Animals with a larval stage not only grow in size, but may undergo **metamorphosis**, in which the larva is transformed into the adult form. Metamorphosis often involves a radical change in form and the development of new organs.

We first consider the roles of intrinsic growth programs and of factors such as growth hormones in controlling both embryonic and post-embryonic growth. We shall also touch on the disturbance of growth control that leads to cancer. This is followed by a discussion of metamorphosis in insects and amphibians. Finally, we look at what might be considered an abnormal aspect of post-embryonic development—aging.

Growth

Growth is defined as an increase in the mass or overall size of a tissue or organism; this increase may result from cell proliferation, cell enlargement without division, or

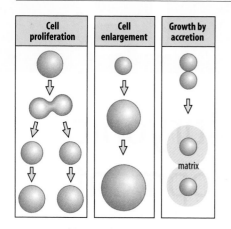

Fig. 13.1 The three main strategies for growth in vertebrates. The most common mechanism is cell proliferation—cell growth followed by division. A second strategy is cell enlargement, in which cells increase their size without dividing. The third strategy is to increase size by accretionary growth, such as secretion of a matrix.

by accretion of extracellular material, such as bone matrix or even water (Fig. 13.1). In animals, the basic body pattern is laid down when the embryo is still small. The growth program—that is, how much an organism or an individual organ grows and responds to factors like hormones—may also be specified at an early stage in development. Overall growth of the organism mainly occurs in the period after the basic pattern of the embryo has been established; there are, however, many examples where earlier organogenesis involves localized growth, as in the vertebrate limb bud (see Chapter 11), and the developing nervous system (see Chapter 12). Different rates of growth in different parts of the body, or at different times, during early development profoundly affect the shape of organs and the organism.

Unlike the situation in animals, where the embryo is essentially a miniature version of the free-living larva or adult, plant embryos bear little resemblance to the mature plant. Most of the adult plant structures are generated after germination by the shoot or root meristems, which have a capacity for continual growth (see Chapter 7). In woody plants, the cambial layer in the trunk, branches, and roots also retains proliferative capacity. It can give rise to the main tissues of the plant axis, enabling trees to increase in girth year after year.

13.1 Tissues can grow by cell proliferation, cell enlargement, or accretion

Although growth often occurs through **cell proliferation**, which increases the number of cells by mitotic cell division, this is only one of three main ways by which growth in size of an organ or body structure can occur (see Fig. 13.1). There is also a significant amount of programmed cell death in many growing tissues, and the overall growth rate is determined by the rates of cell death and cell proliferation.

A second strategy is growth by **cell enlargement**—that is, by individual cells increasing their mass and getting bigger. This is the case in *Drosophila* larval growth, for example, except for the imaginal discs, in which cells proliferate during the larval stages. And differences in size between closely related species of *Drosophila* are partly the result of the larger species having larger cells. In mammals, skeletal and heart muscle cells and neurons never divide again once differentiated, although they do increase in size. Neurons grow by the extension and growth of axons and dendrites, whereas muscle growth involves an increase in mass, as well as the fusion of satellite cells to pre-existing muscle fibers, providing additional nuclei to support the large cell mass. Much growth is through a combination of cell proliferation and cell enlargement. For example, the cells in the lens of the eye are produced by cell division from a proliferative zone for an extended period, whereas their differentiation involves considerable enlargement. Finally, cell enlargement is one of the main means by which plants grow in size.

The third growth strategy, **accretionary growth**, involves an increase in the volume of the extracellular space, which is achieved by secretion of large quantities of extracellular matrix by cells. This occurs in both cartilage and bone, where most of the tissue mass is extracellular.

Certain vertebrate tissues, including the blood (see Section 10.4) and epithelia (see Section 10.7), are continually renewed throughout an animal's lifetime by cell division and differentiation from a population of stem cells. The end products of this type of proliferative system, such as mature red blood cells and keratinocytes, are themselves incapable of division and eventually die.

13.2 Cell proliferation is controlled by regulating entry into the cell cycle

When a eukaryotic cell duplicates itself by mitosis it goes through a fixed sequence of stages called the **cell cycle**, which were described in Box 1B (p. 7). The cell grows in size, the DNA is replicated (S phase), and the replicated chromosomes undergo mitosis and become segregated into two daughter nuclei. The cell then divides (cytokinesis) to give two daughter cells. The cell cycle shown in Box 1B is the standard cell cycle of a dividing somatic animal cell. At different stages of development, or in specialized cell types, cell cycles vary in the presence or absence of particular stages, or the length of time taken to complete them. During the cleavage of the fertilized amphibian egg, for example, the growth phases G_1 and G_2 are virtually absent and so cells do not increase in size between each division (see Chapter 4). They therefore get smaller at each division. In *Drosophila* larval salivary glands there is no M phase, as the DNA replicates repeatedly without mitosis or cell division, leading to the formation of giant chromosomes. These are known as **polytene chromosomes**, as they are composed of large numbers of identical DNA molecules aligned side by side.

The timing of events in the cell cycle is controlled by a set of 'central' timing mechanisms (Fig. 13.2). Proteins known as **cyclins** control the passage through key transition points in the cell cycle. Cyclin concentrations oscillate during the cell cycle, and these oscillations correlate with transitions from one phase of the cycle to the next. Cyclins act by forming complexes with, and helping to activate, protein kinases known as **cyclin-dependent kinases (Cdks)**. These kinases phosphorylate proteins that trigger the events of each phase, such as DNA replication in S phase or mitosis in M phase.

Once a cell has entered the cycle and progressed beyond a point commonly known as 'Start', which occurs in mid- to late-G_1, it will continue through and complete the cycle without needing any further external signal. Transitions into successive phases are marked by **cell-cycle checkpoints** at which the cell monitors progress to ensure, for example, that an appropriate size has been reached, that DNA replication is complete, and that any DNA damage has been repaired. If such criteria are not met, progress into the next stage is delayed until all the necessary processes have been completed. If the cell has suffered some damage that cannot be repaired, the cell cycle will be arrested and the cell will usually undergo apoptosis. These mechanisms are intrinsic to all normal eukaryotic cells and, for example, prevent them from continuing to divide with damaged DNA.

Fig. 13.2 Progression through the cell cycle is regulated by the levels of cyclins. Different cyclins (G1/S, S and M) regulate progression through different phases of the cell cycle. The expression of each cyclin rises and falls throughout the cell cycle. As cyclin protein levels increase, the cyclins bind to their corresponding inactive cyclin-dependent kinases (Cdks) (not shown) forming active enzymes. These active kinases are ultimately responsible for driving cell-cycle events such as the initiation of DNA replication (S phase) and the entry into mitosis (M phase). There are a number of different checkpoints in the cell cycle (red bars) which ensure that the cell does not progress to the next stage until the previous stage has been completed successfully. At the end of mitosis, existing cyclins are targeted to the proteasome and destroyed, so that the new cells formed at cell division are re-set to early G1 phase.

Adapted from Morgan, D.O.: 2007.

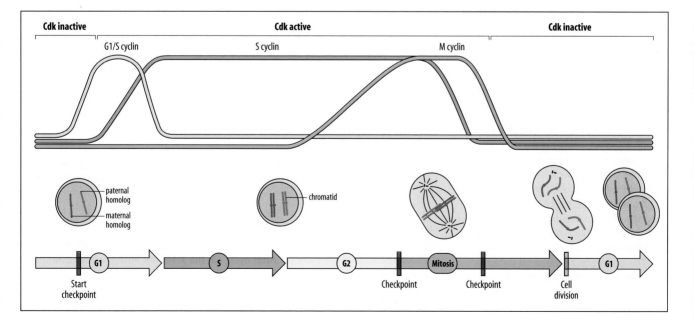

Extracellular factors control cell proliferation by stimulating entry into the cell cycle. Studies of animal cells in culture show that growth factors are essential for cells to multiply, with the particular growth factor or factors required depending on the cell type. When somatic cells are not stimulated to proliferate they are usually in a state known as G_0, into which they withdraw after mitosis (see Box 1B, p. 7). Growth factors enable the cell to proceed out of G_0 and progress through the cell cycle. Most cells in adult organisms are not actively proliferating and so need an external signal before they can re-enter the cell cycle. Numerous extracellular signaling proteins that can stimulate or inhibit cell proliferation have been discovered. Many of the extracellular signal molecules we met in earlier chapters in their roles as patterning agents, such as fibroblast growth factors (FGF) and members of the TGF-β family, were initially discovered through their ability to stimulate cell proliferation. They can also act in this way during embryonic and fetal growth, and in the control of cell proliferation in the adult. Other growth factors have specialized roles in directing the proliferation of particular tissues. One example is erythropoietin, which promotes the proliferation of red blood cell precursors (see Section 10.5).

A striking discovery over the past few decades has been the recognition that cells must receive signals, such as growth factors, not only for them to divide, but also simply to survive. In the absence of all growth factors, cells commit suicide by apoptosis, as a result of activation of an internal cell death program (see Section 10.12). There is a significant amount of cell death in all growing tissues, so that overall growth rate depends on the rates of both cell death and cell proliferation.

13.3 Cell division in early development can be controlled by an intrinsic developmental program

Compared with cells in adult organisms, embryonic cells proliferate much more freely, and in many organisms the pattern of cell division in the earliest embryonic stages is controlled by a cell-autonomous developmental program that does not depend on stimuli such as growth factors. One well-understood example comes from *Drosophila*, where the early cell-cycles are controlled by the proteins that pattern the embryo; these proteins exert their effects by interacting with components of the cell-cycle control system.

The first cell-cycles in the *Drosophila* embryo are represented by rapid and synchronous nuclear divisions, without any accompanying cell divisions, which create the syncytial blastoderm (see Fig. 2.2). There are virtually no G phases, just alternations of DNA synthesis (S phase) and mitosis. But at cycle 14, there is a major transition to a different type of cell cycle, a transition similar to the mid-blastula transition in frogs (see Section 4.13). A well-defined G_2 phase is seen in cycle 14, and the blastoderm becomes cellularized. A G_1 phase is seen at the 16th cycle. After the 17th or 18th cycle, cells in the epidermis and mesoderm stop dividing and differentiate. The cessation of cell proliferation is caused by the exhaustion of maternal cyclin E reserves originally laid down in the egg.

At the 14th cell cycle, distinct spatial domains with different cell-cycle times can be seen in the *Drosophila* blastoderm (Fig. 13.3). This patterning of cell cycles is produced by a change in the synthesis and distribution of a protein phosphatase called String, which exerts control on the cell cycle by dephosphorylating and activating a cyclin-dependent kinase. In the fertilized egg, the String protein is of maternal origin and is uniformly distributed. It therefore produces a synchronized pattern of nuclear division throughout the embryo. After cycle 13, the maternal String protein disappears and zygotic String protein becomes the controlling factor.

Zygotic *string* gene transcription occurs in a complex spatial and temporal pattern. Only cells in which the *string* gene is expressed enter mitosis. This results in variation in the rate of cell division in different parts of the blastoderm, which ensures that the

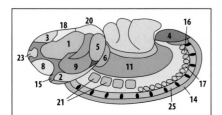

Fig. 13.3 Domains of mitosis in the *Drosophila* blastoderm. Areas composed of cells that divide at the same time are indicated by the various colors, and the numbers indicate the order in which these domains undergo mitosis at the 14th cell cycle, when zygotic String protein is first expressed. The schematic illustrates a lateral view of the embryo corresponding to a stage after mesoderm has been internalized and segmentation has begun (segments are indicated by the black marks along the lower surface). Anterior is to the left and dorsal is up. The gray region on the dorsal surface is the amnioserosa. The internalized mesoderm and some other domains are not visible in this view.

After Edgar, B.A., et al.: 1994.

correct number of cells is generated in different tissues. The pattern of zygotic *string* gene expression is controlled by transcription factors encoded by the early patterning genes, such as the gap and pair-rule genes and those patterning the dorso-ventral axis. One exception to the rule that expression of *string* leads to cell proliferation is the presumptive mesoderm, which is the first domain in which *string* is expressed but the tenth to start cell division. The delay in cell division is due to the expression of the gene *tribbles* in this region, as Tribbles protein degrades String. The delay is necessary to allow ventral furrow formation and thus mesoderm invagination (see Section 8.8), as cell division inhibits ventral furrow formation. The prospective mesodermal cells start to proliferate after they have been internalized. Thus, the early cell cycles in *Drosophila* development provide a good example of how a pattern of cell division can be under genetic control. In contrast to this intrinsic embryonic program, the cell proliferation and growth that occurs in the imaginal discs is modulated by extracellular signals produced by the previous patterning process, and involves the coordination of cell growth, cell proliferation, and cell death.

In early mammalian embryos, cell division times vary with developmental time and place. The first two cleavage cycles in the mouse last about 24 hours, and subsequent cycles take about 10 hours each. After implantation, the cells of the epiblast proliferate rapidly; cells in front of the primitive streak have a cycle time of just 3 hours.

13.4 Organ size can be controlled by both intrinsic growth programs and extracellular signals

Organ size in vertebrates is determined by both intrinsic developmental programs and extracellular factors that stimulate or inhibit growth, but the relative importance of these two mechanisms in different organs varies a good deal. The liver, for example, which is the main detoxifying organ in mammals, has excellent powers of regeneration in both embryo and adult, whereas the pancreas does not. Experiments to investigate these properties in the embryo make use of the fact that organ precursor cells can be specifically destroyed in embryonic mice using transgenic techniques. A strain of mice is generated that carries an introduced gene for diphtheria toxin under the control of a tetracycline-repressible promoter. Part of this promoter comes from a gene that is only expressed in the particular organ precursor cells targeted, and so only directs expression of the toxin in those cells. If a pregnant mouse of this strain is continuously fed tetracycline, the toxin is not expressed at all in the embryos she carries. If tetracycline is withdrawn, the toxin is produced in the targeted precursor cells, killing them but leaving other tissues in the body untouched.

To investigate the regenerative powers of the liver, tetracycline is withdrawn for a short period at the appropriate time during gestation, with the result that a proportion of the liver precursor cells in the embryos are destroyed. The longer the period without tetracycline, the greater the number of precursor cells lost. After such treatment, the embryonic liver grows back to a normal size, indicating that it does not arise from a fixed number of progenitor cells. Even in adult mammals, the liver will regrow to regain its normal mass after as much as two-thirds of it has been surgically removed.

In contrast, if some of the progenitor cells of the pancreas in a mouse embryo are destroyed (by the technique described above) once the pancreatic 'bud' has formed, a smaller than normal pancreas develops. The size of the embryonic pancreas seems, therefore, to be largely under intrinsic control. Another organ with intrinsic growth control is the thymus gland. If multiple fetal thymus glands are transplanted into a developing mouse embryo, each one grows to full size. If the same experiment is done with spleens, however, each spleen grows much smaller than normal, so that the final total mass of the spleens is equivalent to one normal spleen.

The liver and the spleen regulate their final size by a negative feedback mechanism. The developing or regenerating liver secretes some factors that stimulate cell proliferation and others that potentially inhibit growth, and these have been studied in detail in regard to the regenerating adult liver. Cell division is relatively rare in a normal adult liver but if a substantial part of it is surgically removed, a process known as resection, 95% of the remaining cells re-enter the cell cycle. In humans, as much as 70% of the liver can be removed, to cut out a tumor for example, and the liver will grow back to a normal size; however, if more than 75–80% is removed, it cannot regenerate.

In the regenerating human liver, hepatocytes start to divide a day after surgery, and the release of growth factors and cytokines by the remaining liver tissue is required for regrowth. The key growth factors that initiate regeneration and promote growth include hepatocyte growth factor (HGF) and epidermal growth factor (EGF), and other growth-promoting cytokines include tumor necrosis factor α (TNF-α) and interleukin-6. The serotonin produced by blood platelets seems also to stimulate liver regrowth by in turn stimulating the production of HGF by liver cells. Transforming growth factor β (TGF-β) acts as an inhibitor of liver growth, although it seems to be required for the proper organization of liver tissue towards the end of the regeneration period. The expression of TGF-β receptors by liver tissue temporarily decreases immediately after liver resection, preventing inhibition by TGF-β during this crucial period. As the liver regrows, TGF-β eventually reaches a level that blocks further growth.

When the liver gets to a certain size, the concentration of inhibitory factors in the circulation is sufficient to stop further growth—an example of negative feedback determining the size of an organ. The size of the spleen is determined in a similar way. In the experiment described above, the total mass of spleen tissue is being monitored by the embryo according to the level of growth-inhibitory proteins produced by the spleen tissue and circulating in the blood.

In muscle, the growth inhibitor myostatin, which is produced by the myoblasts (immature muscle cells) themselves, helps to regulate muscle size and cell numbers. Mutation in the *myostatin* gene in mice leads to a significant increase in muscle mass; the number of muscle fibers and their size are both increased. There is now some evidence for a similar negative signal controlling cell numbers in neural tissue. And in butterflies, growth-controlling interactions occur between imaginal discs; the removal of hindwing discs results in larger forewings and forelegs.

Even when they are capable of division, the cells of many adult tissues do not divide, or divide infrequently, under normal circumstances. Adult cells can, however, be induced to divide by injury or other stimuli, such as a reduction in the mass of the tissue, as in the case of the liver. Removal of a kidney leads to an increase in size of the remaining kidney, but in this case the growth is mainly the result of cell enlargement rather than cell proliferation. And in Chapter 14 we shall see how some adult amphibians and insects can regenerate whole limbs after amputation, complete with all the different types of tissue in their correct place.

Patterning of the embryo occurs while the organs are still very small. For example, human limbs have their basic pattern established when they are less than 1 cm long. Over the years, the limb grows to be at least one hundred times longer. How is this growth controlled? It appears that each of the cartilaginous elements in the embryonic limb has its own growth program. In the chick embryonic wing, the cartilaginous elements representing the long bones—the humerus and the ulna—are initially similar in size to the elements in the wrist (Fig. 13.4). Yet, with growth, the humerus and ulna increase many times in length compared with the wrist bones. These growth programs are specified when the elements are initially patterned and involve both cell multiplication and matrix secretion (accretion). Each skeletal element follows its own growth program even when grafted to a developmentally neutral site, provided that a good blood supply is established.

A classic illustration of an intrinsic growth program comes from grafting limb buds between large and small species of salamanders of the genus *Ambystoma*. A limb

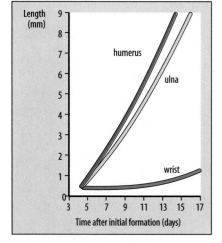

Fig. 13.4 Comparative growth of the cartilaginous elements in the embryonic chick wing. When first laid down, the cartilaginous elements of the humerus, ulna, and wrist are the same size, but the humerus and ulna then grow much more than the wrist element.

bud from the larger species grafted to the smaller species initially grows slowly, but eventually ends up at its normal size, which is much larger than any of the limbs of the host (Fig. 13.5). This indicates that whatever the circulatory factors, such as hormones, that influence growth, the intrinsic response of a tissue to them is crucial. Even though circulating homones are required for growth in many cases, such intrinsically patterned differential growth can affect the overall form of an organism considerably, as can be seen in Fig. 13.6. This figure shows that growth of the different parts of the body is not uniform, and different organs grow at different rates. At 9 weeks of development, the head of a human embryo is more than a third of the length of the whole embryo, whereas at birth it is only about a quarter. After birth, the rest of the body grows much more than the head, which is only about an eighth of the body length in the adult (see Fig. 13.6).

13.5 The amount of nourishment an embryo receives can have profound effects in later life

Whatever the type of growth program, no animal will reach its full potential size if inadequately nourished as an embryo and during the post-natal growth period. In mammals, inadequate nutrition or poor nutrition of the embryo not only has direct effects on embryonic and fetal growth, but can have serious, and in some ways unexpected, effects in adult life. Population studies in developed countries, such as the United Kingdom, have associated smaller size or relative thinness at birth (due to either maternal undernutrition or premature birth) and during early infancy with an increased risk of developing coronary heart disease, stroke, or type 2 diabetes in adult life. When undernutrition during early development is followed by improved nutrition later, whether later in gestation or in the early post-natal period, the period of 'catch-up' growth may carry a cost, as the body's resources are diverted from 'repair and maintenance' to growth. Premature birth itself, independent of size for gestational age, has been associated with insulin intolerance and glucose intolerance in pre-pubertal children that may continue into young adulthood and may be accompanied by high blood pressure. Catch-up growth seems to predispose to overweight and even obesity, and this may partly explain the effects on health in later life. One proposal is that in response to undernutrition *in utero* the fetus lays down more fat cells as a

Fig. 13.5 The size of limbs is genetically programmed in salamanders. An embryonic limb bud from a large species of salamander, *Ambystoma tigrinum*, grafted to the embryo of a smaller species, *Ambystoma punctatum*, grows much larger than the host limbs—to the size it would have grown in *Ambystoma tigrinum*.

Photograph from Harrison, R.G.: 1969.

Fig. 13.6 Different parts of the human body grow at different rates. At 9 weeks of development the head is relatively large but, with time, other parts of the body grow much more than the head.

After Gray, H.: 1995.

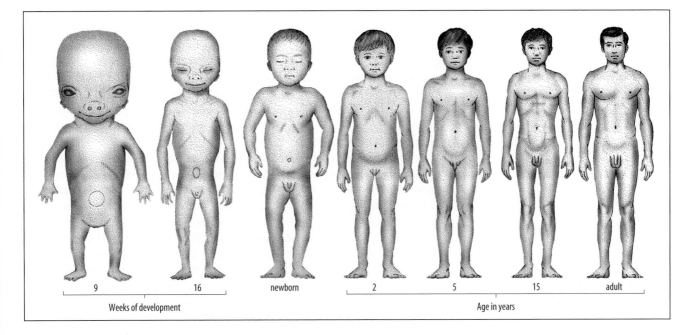

precaution. That is a good strategy if conditions after birth are indeed hard and food is short, but in conditions of plentiful food, this leads to undesirable consequences.

Experiments in animals support the effects of undernutrition and an unbalanced maternal diet observed in humans. The embryos of pregnant rats fed a low-protein, but otherwise calorie-sufficient, diet during the preimplantation period (0 to 4.5 days) showed altered development in multiple organ systems. If the pregnancy was allowed to go to full term, the offspring had low birth weight, increased post-natal growth, or adult-onset high blood pressure.

Obesity is associated with numerous diseases in later life, including type 2 diabetes and heart disease. Although much obesity in children and adults is due to overeating and lack of exercise, early developmental nutritional experience and genetic background can contribute. Human fatty tissue comprises some 40 billion adipose cells, with most of the bulk stashed under the skin, and obesity represents both greater numbers of adipose cells compared with lean people and excessive deposition of fat in these cells, which increases cell size. Humans are born with a certain number of adipose cells, with females generally having more than males. The number of adipose cells increases throughout late childhood and early puberty, and after this remains fairly constant. However, the number of adipose cells increases more rapidly in genuinely obese children than in lean children, meaning that they end up with more adipose cells. Once fat cells develop in the body, they remain there for life and they seldom die, although there is some turnover. Each month about 1% of adipose cells in the human body die and are replaced. Thus adult obesity is often linked to childhood obesity. Overweight people with extra fat cells can shrink the size of the cells and lose weight by burning the excess fat by careful dieting and exercise, but the fat cells themselves do not disappear, and are only too ready to start accumulating excess fat again.

13.6 Determination of organ size involves coordination of cell growth, cell division, and cell death

How do cells know to stop growing and dividing when an organ or body structure reaches its final size? Evidence that organ size and body size can be determined by monitoring overall dimensions, rather than the absolute number of cells or cell divisions, comes from a number of different sources. Species with either fewer or more than the diploid number of chromosomes—that is, haploid and polyploid organisms, respectively—provide one line of evidence. For a given cell type, cell size is usually proportional to ploidy, so haploid cells are half the volume of diploid cells and tetraploid cells twice the volume. Some salamander species are naturally tetraploid; they grow to the same size as their diploid relatives but have only half the number of cells. Artificially produced tetraploid mouse embryos also compensate for having larger cells by having fewer of them, but usually die before birth. In *Drosophila*, the final size of animals that are a mosaic of haploid and diploid cells is normal, with the haploid regions containing smaller but more numerous cells. Polyploid species of plants are, in fact, generally larger overall than their diploid relatives, but in plant mosaics of diploid and polyploid cells it is possible to find the same sort of size compensation as seen in animals.

The *Drosophila* wing imaginal disc has proved an interesting model system for studying how organ size might be determined. At its formation, the wing disc is initially composed of about 40 cells, and normally grows in the larva to about 50,000 cells. Cell division occurs throughout the disc, and then ceases uniformly when the correct size is reached. The adult wing is produced at metamorphosis by changes in cell shape without further cell division (see Fig. 11.24). Early experiments investigating wing growth showed that the final size of the wing does not depend on the imaginal disc undergoing a fixed number of cell divisions, or attaining a particular number of cells. Instead, final size seems to be controlled by some mechanism that monitors the overall size of the developing wing disc and adjusts cell division and cell size accordingly. Experiments show that there is no restriction on how much of the wing a given cell can make; the clonal descendants of a single cell can contribute from a tenth to as much as a half of the wing. Wing growth can also be experimentally decoupled from cell proliferation, as demonstrated by the fact that if cell division is blocked in either the anterior or posterior compartment in the larval wing disc, the final size of the wing is normal but the individual cells are larger.

In the same way, the size of a region or organ need not be determined by the rate of cell division. This can be shown by making mosaics of wild-type cells and slower-growing *Minute* cells (see Box 2E, p. 73). Within a compartment, the faster-dividing wild-type cells contribute more cells than the slow-growing *Minute* cells, but the size of the compartment remains the same as if all the cells were wild type.

In normal growth, the final size of the wing is likely to be achieved by a balance between cell proliferation and apoptosis. The recently discovered Hippo signaling pathway (Fig. 13.7), is likely to be an important coordinator of cell growth, cell division, and apoptosis in *Drosophila* and in other animals, suppressing cell proliferation and promoting apoptosis of extra, unwanted cells. The Hippo signaling network was discovered in *Drosophila* and has since been found in mammals and other vertebrates. It seems to be involved generally in controlling and coordinating cell proliferation and apoptosis in growing tissues, and components of the pathway have been found mutated in human cancers—yet another example of a pathway first identified in *Drosophila* development proving of importance to human disease. The Hippo pathway involves a cascade of protein kinases, of which Hippo is one, and the outcome of activation of the pathway is the inactivation of a transcriptional co-activator called Yorkie (Yki), which has a direct effect on cell proliferation (see Fig. 13.7). Loss-of-function mutations in Hippo or overexpression of Yki in *Drosophila* wing discs results in a dramatic increase in cell proliferation and a decrease in apoptosis, resulting in wing discs growing to almost eight times the normal size. When the pathway is activated, Yki is directly inhibited by phosphorylation by the protein kinase Warts (Wts), which is itself activated by phosphorylation by Hippo. The inactivation of Yki by the Hippo signaling network is therefore a mechanism for suppressing growth, and the Hippo pathway is a strong candidate for one means of terminating growth when an organ reaches its required size.

One target of the Hippo pathway in *Drosophila* is a microRNA (miRNA) called *bantam*, which is known both to promote cell proliferation and suppress apoptosis by suppressing production of the pro-apoptotic protein Hid. The Hippo pathway downregulates the production of *bantam* miRNA, and this could be one of the mechanisms by which the pathway promotes apoptosis and suppresses cell proliferation.

Because the Hippo pathway suppresses growth, Hippo, Warts, and associated components of the pathway are characterized as tumor-suppressor proteins. Tumor-suppressor genes are a class of genes defined by the fact that their absence or inactivation predisposes to uncontrolled cell proliferation and the development of cancer (see Section 13.10). In contrast, *Yki* is classed as an oncogene, as its overexpression promotes uncontrolled growth. In *Drosophila*, the Hippo pathway can be activated by signals from the atypical cadherin Fat. Whether it is activated by extracellular signals in mammalian tissues, what those might be, and how the pathway might be linked to the determination of mammalian organ size, are still unknown. In *Drosophila*, the Hippo pathway has been linked to a possible gradient mechanism for determining adult organ size (see Box 13A, p. 514).

13.7 Body size is also controlled by the neuroendocrine system in both insects and mammals

Body size depends not only on the rate of growth, but on how long growth continues. This is particularly evident in insects that undergo metamorphosis, in which growth

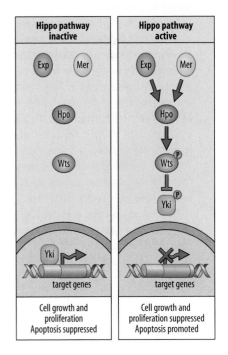

Fig. 13.7 The Hippo signaling pathway. The outcome of activation of the Hippo pathway is the inactivation of the transcriptional co-activator Yorkie (Yki). In the absence of signaling (left panel) Yki helps activate genes that promote cell growth and proliferation and suppress apoptosis. When the pathway is activated, the adaptor proteins Expanded (Exp) and Merlin (Mer) activate the protein kinase Hippo (Hpo). Hippo phosphorylates and activates the protein kinase Warts (Wts), which in turn phosphorylates and inactivates Yki, and so the growth-promoting and apoptosis-suppressing genes are not expressed.

Box 13A Gradients of signaling molecules could determine organ size

Evidence from various sources has suggested for some time that the final size of *Drosophila* imaginal discs, and thus of the adult organs, might be determined by a molecular gradient across the disc that could be formed as a result of previous patterning. The basic idea is that when the disc is small, the gradient is steep and the steepness of the gradient in some way promotes growth. As the organ grows, the gradient flattens out (see figure). Growth declines and finally terminates. The discovery of the Hippo pathway as a mechanism for regulating growth, and the discovery that it could be regulated by the cadherin Fat, turned attention to how the Hippo pathway might interact with molecular gradients of Fat-associated proteins present along the proximodistal axes of imaginal discs. The gradients are those of the atypical cadherin Dachsous (Ds) and the transmembrane serine/threonine kinase Four-jointed (Fj), which modulates the interaction of Ds with Fat.

Experiments in which Ds and Fj were overexpressed in distinct clones of cells in otherwise wild-type imaginal discs showed that the Hippo pathway was suppressed (which would allow cell proliferation) only where there was a significant difference in the levels of Ds/Fj expressed by neighboring cells. This occurs in cells on the boundary of the clone, where Ds/Fj-overexpressing cells are adjacent to wild-type cells. If a patch of cells all expresses the same level of Ds/Fj, no matter whether it is a low or a high level, the Hippo pathway is activated in these cells (which would suppress cell proliferation). Suppression or activation of the pathway was detected by looking at whether target genes of the pathway were transcribed or not.

The argument then goes that an initial steep gradient of Ds/Fj across a disc would allow it to grow, as there would be sharp discontinuities in the level of Ds/Fj between adjacent cells all across the disc. Growth will presumably stop as the

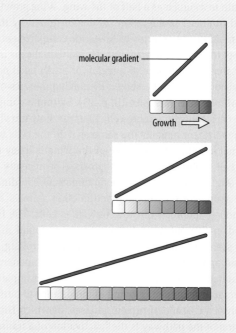

molecular gradient

Growth ⟹

disc reaches the required size, because the gradient has flattened out and the difference in Ds/Fj levels between cells is no longer sufficient to suppress the Hippo pathway. It is thought that these effects are being mediated in some way through the interaction of Ds/Fj with Fat, but precisely how is not yet known. Ds/Fj cannot simply be acting as conventional ligands that either activate or suppress Fat signaling. This proposed mechanism is reminiscent of the one for planar cell polarity in *Drosophila* (see Box 2F, p. 78), which also involves discontinuities in levels of Ds/Fj between cells. It will be fascinating to see how the same gradient of information can be read out to give instructions as to both organ size and cell directionality.

occurs only in the larval stages and the size of the adult is determined by the size the larva reaches before pupation. In normal circumstances the larva stops feeding and pupates when it reaches a size typical for the species, but if the larval stage is prolonged or shortened experimentally, adults that are, respectively, larger or smaller than usual can be produced. As discussed in more detail later in this chapter, pupation and metamorphosis in insects are initiated by a pulse of the steroid hormone **ecdysone**, which is released by the prothoracic gland under the stimulation of the neurohormone prothoracicotropic hormone (PTTH).

How body size is monitored by insect larvae and how the timing of the ecdysone signal is determined are still not entirely understood, and the mechanisms are different in different groups of insects. In *Drosophila*, for example, the larva must attain a minimum critical size, as otherwise it will not survive metamorphosis. In normal circumstances this minimum size is attained about half-way through the last instar and the larva continues to feed and grow until pupation (Fig. 13.8). If the larva is starved after it reaches the critical size, it still undergoes metamorphosis but produces smaller adults than normal. Experiments in *Drosophila* indicate that signaling via the insulin and target of rapamycin (TOR) pathways in the prothoracic

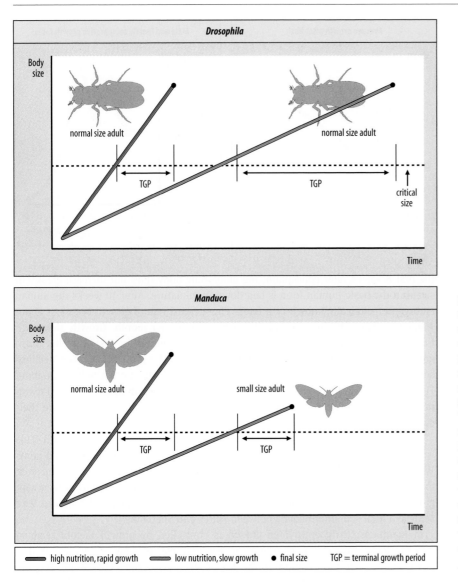

Fig. 13.8 Nutrition-dependent regulation of insect body size. The level of nutrition available regulates adult body size in different ways in different insects. Upper panel: in *Drosophila*, once the larva has reached a critical size, if nutrients are scarce the subsequent time the larva spends feeding (the terminal growth period, TGP) is simply prolonged until the larva reaches a normal size. It then undergoes metamorphosis, producing a normal-sized adult. Lower panel: in the tobacco hornworm (in which the adult is a moth), the larval terminal growth period (after reaching a critical size) remains the same whether nutrients are plentiful or scarce. This means that when nutrients are restricted a smaller adult is produced.

Adapted from Nijhout, H.F.: 2008.

gland in response to nutritional status plays an important part in determining the level of ecdysone production and thus the timing of metamorphosis. In larvae put on a low-nutrient diet after they had reached the critical size, pupation was delayed, enabling the larvae to grow to nearly normal size. This delay was accompanied by downregulation of the TOR pathway in the prothoracic gland, which had the effect of keeping ecdysone production at a low level. As the larvae did eventually pupate and metamorphose, it seems that in *Drosophila*, exposure to low levels of ecdysone over a period can have the same effect as the distinct pulse of ecdysone that is the usual signal for pupation.

Larval growth rate is governed ultimately by the level and quality of nutrients available, and, as in other animals, insulin-like peptide growth factors are involved in the direct control of cell growth. The ecdysone and PTTH mentioned earlier have no direct effect on growth rate—they act by affecting the length of the feeding period. Insulin-like peptides are also thought to be involved in coordinating growth with the availability of nutrients and with the hormonal signals that terminate the feeding period.

Human growth during the embryonic, fetal, and post-natal periods is typical of the different phases of growth in mammals. The human embryo increases in length from 150 mm at implantation to about 50 cm over the 9 months of gestation. During

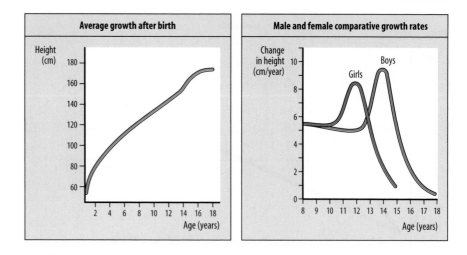

Fig. 13.9 Normal human growth. Left panel: an average growth curve for a human male after birth. Right panel: comparative growth rates of boys and girls. There is a growth spurt at puberty in both sexes, which occurs earlier in girls.

the first 8 weeks after conception, the embryonic body does not increase greatly in size, but the basic human form is laid down in miniature. After 10 weeks the human embryo is technically known as a fetus. The greatest rate of growth occurs at about 4 months, when the fetus grows as much as 10 cm per month. Growth after birth follows a well-defined pattern (Fig. 13.9, left panel). During the first year after birth, growth occurs at a rate of about 2 cm per month. The growth rate then declines steadily until the start of a characteristic adolescent growth spurt at puberty at about 11 years in girls and 13 years in boys (Fig. 13.9, right panel). In pygmies, sexual maturation at puberty is not accompanied by this adolescent growth spurt, hence their characteristic short stature.

The maternal environment plays an important role in controlling fetal growth. This is well illustrated by crossing a large shire horse with a much smaller Shetland pony. When the mother is a shire mare, the newborn foal is similar in size to a normal shire foal, but when the mother is a Shetland, the newborn is much smaller. However, with growth after birth, the offspring of both these crosses become similar in size, and achieve a final size intermediate between shires and Shetlands.

In mammals, early embryonic growth is controlled by growth factors secreted into the local extracellular environment. For example, we saw in Chapter 11 that FGFs control cell proliferation at the tip of the developing chick limb, and are required for proliferation and outgrowth of the bud. Evidence for the roles of individual growth factors in embryonic growth comes from the technique of gene knockout (see Box 3B, p. 118). For example, the **insulin-like growth factors 1 and 2 (IGF-1 and IGF-2)** have a key role in growth during embryonic development. They are single-chain protein growth factors that closely resemble insulin and each other in their amino-acid sequence and produce their effects on cells through a shared insulin-signaling pathway. Newborn mice lacking a functional *Igf2* gene develop relatively normally, but weigh only 60% of the normal newborn body weight. Mice in which the *Igf1* gene has been inactivated are also growth retarded. Both the factors and their receptors can be detected as early as the eight-cell stage of mouse development. *Igf2* is one of the genes that are imprinted in mammals (see Section 9.8); it is inactivated in maternal germ cells and is only expressed in the embryo from the paternal genome. Like many growth factors, the IGFs and FGFs also have important roles in post-natal growth and in controlling cell proliferation in adult mammals.

Growth hormone is a circulating protein hormone produced by the pituitary gland that is essential for the growth of humans and other mammals after birth. It is also synthesized in embryonic and fetal tissues, where it is thought to act directly on cells as an extracellular growth factor. Within the first year of birth, the pituitary gland begins to secrete growth hormone. A child with insufficient growth hormone grows less than normal, but if growth hormone is given regularly, normal growth is restored. In this case, there is a catch-up phenomenon, with a rapid initial response that tends to restore the growth curve to its original trajectory.

Production of growth hormone in the pituitary is under the control of two hormones produced in the hypothalamus: **growth hormone-releasing hormone**, which promotes growth hormone synthesis and secretion, and **somatostatin**, which inhibits its production and release. Growth hormone produces many of its effects by inducing the synthesis of IGF-1 (Fig. 13.10) and, to a lesser extent, IGF-2, which act directly on cells to promote proliferation. Post-natal growth, as well as embryonic growth, is largely due to the actions of the insulin-like growth factors, and complex hormonal regulatory circuits control their production.

Puberty is initiated by the activity of the hypothalamic neuronal network, which governs the intermittent release of **gonadotropin-releasing hormone (GnRH)** by hypothalamic neurons. The mechanism determining this timing is not known. One of the results of a pulse of GnRH is a sharp increase in the secretion of gonadotropins (luteinizing hormone and follicle-stimulating hormone) by the pituitary; these cause increased production of the steroid sex hormones—the estrogens and androgens. These in turn stimulate the production of pulses of growth hormone, which are responsible for the growth spurt at puberty.

13.8 Growth of the long bones occurs in the growth plates

An important aspect of post-embryonic vertebrate growth is the growth of the long bones of the limbs (humerus, femur, radius, ulna, tibia, and fibula). The long bones are initially laid down as cartilaginous elements (see Section 11.1) and then become ossified. The early growth of these elements involves both cell proliferation and matrix secretion in a well-defined pattern. The end of a long bone is known as the epiphysis and the central region as the diaphysis. In both fetal and post-natal growth, the cartilage is replaced by bone in a process known as **endochondral ossification**, in which ossification starts in the centers of the long bones and spreads outward (Fig. 13.11). Secondary ossification centers then develop at each end of the bone. The adult long bones thus have a bony shaft with cartilage confined to the articulating surfaces at each end, and to two internal regions near each end—the **growth plates**—in which growth occurs. In the growth plates, the cartilage cells, or **chondrocytes**, are usually arranged in columns, and various zones can be identified. Just next to the bony epiphysis is a narrow germinal zone, which contains stem cells. Next is a proliferative zone of cell division, followed by a zone of maturation, and a hypertrophic zone, in which the cartilage cells increase in size. Finally, there is a zone in which the cartilage cells die and are replaced by bone laid down by cells called **osteoblasts**, which differentiate from cells that form the perichondrium surrounding the cartilage. This stage involves Wnt signaling. There is a strong similarity to the development of skin, where basal stem cells give rise to dividing cells, which differentiate into keratinocytes and finally die (see Section 10.7).

The proliferation of chondrocytes at the ends of the long bones, and later in the growth plate, is controlled so that at a given distance from the end of the bone they stop dividing and enlarge to form the scaffolding for bone, as appropriate. In mice, the proliferation of chondrocytes is controlled by the secreted signaling proteins parathyroid-hormone-related protein (PHRP) and Indian hedgehog. Indian hedgehog belongs to the large family of Hedgehog signaling proteins. PHRP is secreted by the chondrocytes and perichondrial cells at the ends of the prospective bones and stimulates chondrocytes to proliferate, which prevents them from expressing Indian hedgehog. Once chondrocytes move out of the zone of influence of PHRP they stop proliferating, start to express Indian hedgehog, and become hypertrophic (Fig. 13.12). Indian hedgehog diffuses back into the pool of proliferating chondrocytes, where it increases

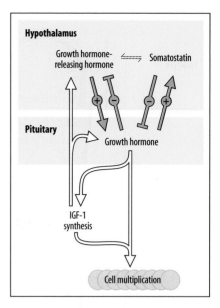

Fig. 13.10 Growth hormone production is under the control of the hypothalamic hormones. Growth hormone is made in the pituitary gland and is secreted. Growth hormone-releasing hormone from the hypothalamus promotes growth hormone synthesis, while somatostatin inhibits it. Growth hormone controls its own release by negative feedback signals to the hypothalamus. Growth hormone causes the synthesis of the insulin-like growth factor IGF-1 and this promotes the production of growth hormone.

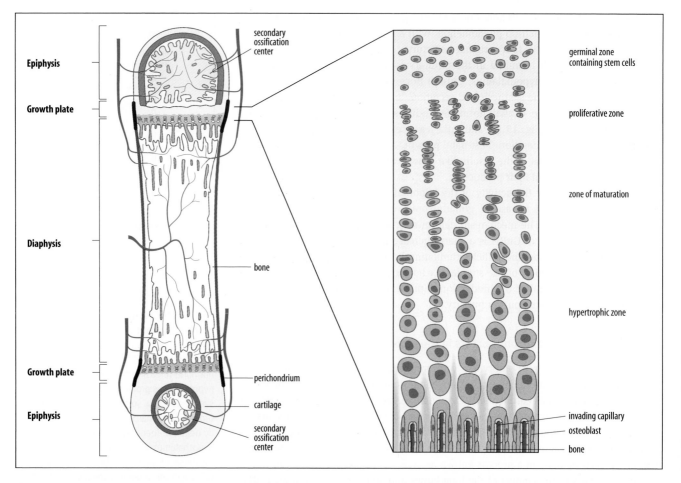

Fig. 13.11 Growth plates and endochondral ossification in the long bone of a vertebrate. The long bones of vertebrate limbs increase in length by growth from cartilaginous growth plates. The growth plates are cartilaginous regions that lie between the epiphysis of the future joint and the central region of the bone, the diaphysis. In the figure, bone has already replaced cartilage in the diaphysis, and more bone is being added at the growth plates. Within the growth plates, cartilage cells multiply in the proliferative zone, then mature and undergo hypertrophy (cell enlargement). They are then replaced by bone, which is laid down by specialized cells called osteoblasts. These derive both from perichondral cells and from cells that invade the bone along with the blood vessels. Secondary sites of ossification are located within the epiphyses. *After Wallis, G.A.: 1993.*

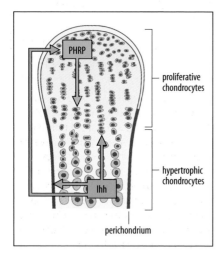

their rate of proliferation, and it also, by some mechanism as yet unknown, stimulates production of PHRP by the cells at the ends of the bone. It also acts on the adjacent perichondral cells to form bone-producing osteoblasts. In the absence of Indian hedgehog in mice there is an accelerated rate of hypertrophy and differentiation of chondrocytes to osteoblasts, resulting in short, stubby limbs. Cell proliferation is also

Fig. 13.12 Indian hedgehog (Ihh) and parathyroid-hormone-related protein (PHRP) form feedback loops that maintain chondrocyte proliferation and growth at the ends of developing bones. PHRP secreted by the chondrocytes at the ends of developing long bones acts on proliferating chondrocytes (blue) to maintain proliferation and thus prevents the expression of Indian hedgehog. Chondrocytes farther away from the end of the bone (orange) escape the influence of PHRP and express Ihh. This acts on adjacent chondrocytes to increase the rate of proliferation and also acts on perichondral cells to form osteoblasts. In some way that is not yet understood, production of Ihh also stimulates PHRP synthesis by the chondrocytes at the end of the bone, thus forming a positive feedback loop that maintains PHRP production. *Adapted from Kronenberg, H.M.: 2003.*

suppressed, and Indian hedgehog production increased, by FGFs that are produced in the perichondrium.

Growth hormone affects bone growth by acting on the growth plates. The cells in the germinal zone have receptors for growth hormone, and growth hormone is probably directly responsible for stimulating these stem cells to proliferate. Further growth, however, is probably mediated by IGF-1, whose production in the growth plate is stimulated by growth hormone. Thyroid hormones are also necessary for optimal bone growth; they act both by increasing the secretion of growth hormone and IGF-1, and by stimulating hypertrophy of the cartilage cells. FGF is also important for bone growth; the genetic defect that gives rise to achondroplasia (short-limbed dwarfism) is a dominant mutation in the FGF receptor-3, whose normal function is to limit, rather than promote, bone formation. When it is mutated, it limits growth abnormally in response to FGF.

The rate of increase in the length of a long bone is equal to the rate of new cell production per column multiplied by the mean height of an enlarged cell. The rate of new cell production depends both on the time cells take to complete a cycle in the proliferative zone, and the size of this zone. Different bones grow at different rates, and this can reflect the size of the proliferative zone, the rate of proliferation, and the degree of cell enlargement in the growth plate. If bone growth is prevented, for example by heavy loading, then there is some catch-up growth when the load is removed.

In view of the complexity of the growth plate, it is remarkable that human bones in limbs on opposite sides of the body can grow for some 15 years independently of each other, and yet eventually match to an accuracy of about 0.2%. This may be achieved by having many columns of cells in each plate, so that growth variation between cells is averaged out. When growth of a bone ceases the growth plate ossifies, and this occurs at different times for different bones. Ossification of growth plates occurs in a strict order in different bones and can therefore be used to provide a measure of physiological age. The timing of growth cessation in the growth plate appears to be instrinsic to the plate itself rather than to hormonal influence. The cessation of growth is due to the cessation of cell proliferation and this timing may be programmed in the cells. Cell senescence, and thus growth cessation, may be due to the chondrocyte stem cells having only a finite potential for division.

13.9 Growth of vertebrate striated muscle is dependent on tension

The number of striated (skeletal) muscle fibers in vertebrates is determined during embryonic development. Once differentiated, striated muscle cells lose the ability to divide. Post-embryonic growth of muscle tissue results from an increase in individual fiber size, both in length and girth, during which the number of myofibrils within the enlarged muscle fiber can increase more than 10-fold. Additional nuclei to support the functioning of the much-enlarged cell are provided by the fusion of satellite cells with the fiber. Satellite cells, which are undifferentiated cells lying adjacent to the differentiated muscle, also act as a reserve population of stem cells that can replace damaged muscle (see Section 10.10).

The increase in the length of a muscle fiber is associated with an increase in the number of sarcomeres—the functional contractile units—it contains. For example, in the soleus muscle of the mouse leg, as the muscle increases in length, the number of sarcomeres increases from 700 to 2300 at 3 weeks after birth. This increase in number seems to depend on the growth of the long bones putting tension on the muscle through its tendons. If the soleus muscle is immobilized by placing the leg in a plaster cast at birth, sarcomere number increases slowly over the next 8 weeks, but then increases rapidly when the cast is removed (Fig. 13.13). One can, therefore, see how bone and muscle growth are mechanically coordinated.

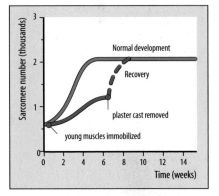

Fig. 13.13 The growth in length of the muscles attached to the long bones of the mouse leg depends on the tension provided by long bone growth. The length of a muscle is related to the number of sarcomeres—the basic contractile unit. If the limb is immobilized by a plaster cast in a position in which there is no tension on the muscle, there is little increase in muscle length until the cast is removed. There is then a rapid increase in length.

13.10 Cancer can result from mutations in genes that control cell multiplication and differentiation

Cancer can be regarded as a major perturbation of normal cellular behavior that results from certain mutations in somatic cells. Creating and maintaining tissue organization requires strict controls on cell division, differentiation, and growth. In cancer, cells escape from these normal controls, and proceed along a path of uncontrolled growth and migration that can kill the organism. There is usually a progression from a benign localized growth to malignancy in which the cells **metastasize**—migrate to many parts of the body where they continue to grow. Most cancers derive from a single abnormal cell that has acquired a number of mutations. Progression of a mutant cell to becoming a tumor-producing cell, a process known as tumor progression, is an evolutionary process, involving both further mutations and selection of those cells best able to proliferate.

The cells most likely to give rise to cancer are those that are undergoing continual division, such as stem cells. Because they replicate their DNA frequently, they are more likely than other cells to accumulate mutations that arise from errors in DNA replication. In almost all cancers, the cancer cells are found to have a mutation in one or more, and usually many, genes. An average of 63 mutations are found in a pancreatic cancer cell. There is a major rewiring of genetic circuits in cancer cells. Particular genes in which mutation can contribute to cancer formation have been identified in humans and other mammals. These genes are known as **proto-oncogenes**; when such a gene undergoes mutation it becomes an **oncogene**. In some cases, the presence of a single oncogene is capable of making a cell cancerous. At least 70 proto-oncogenes have been identified in mammals.

There is another group of genes in which mutations can also lead to cancer. These are the **tumor-suppressor genes**, in which inactivation or deletion of both copies of the gene is required for a cell to become cancerous. We have already seen why some of the components of the Hippo pathway are considered as tumor suppressors (see Section 13.6). The classic human example of a tumor caused by the loss of such a gene is the childhood tumor retinoblastoma, which is a tumor of retinal cells. Although retinoblastoma is normally very rare, some families have an inherited predisposition to it. This predisposition is determined by a single gene, and the affected families were the means of identifying this gene. The inherited defect in some of the families turned out to be a deletion of a particular region on one of the two copies of chromosome 13. This on its own does not cause the cells to be cancerous. However, if a retinal cell also acquires a deletion of the same region on the other copy of chromosome 13, a retinal tumor develops. The gene in this region that is responsible for susceptibility to retinoblastoma is known as the *retinoblastoma* (*RB*) gene. Both copies of the *RB* gene must be lost or inactivated for a cell to become cancerous (Fig. 13.14), and hence *RB* is regarded as a tumor-suppressor gene. The *RB* gene encodes a protein, RB, which is involved in the regulation of the cell cycle.

The tumor-suppressor gene *p53* plays a key role in many cancers; about half of all human tumors contain a mutated form of *p53*. This gene is not required for development *per se*, but when cells are exposed to agents that damage DNA, then *p53* is activated and arrests the cell cycle, giving the cell time to repair the DNA. The p53 protein thus prevents the cell from replicating damaged DNA and giving rise to mutant cells. Instead, p53 will cause the cell to die by apoptosis if the damage is too severe to be repaired. The mutant forms of *p53*, found in many cancers, do not promote apoptosis, and so the affected cells are more likely to accumulate mutations.

A major feature of cancer is the failure of tumor cells to differentiate properly. The majority of cancers—over 85%—occur in epithelia. This is not surprising when it is recalled that many epithelia (such as the epidermis and the lining of the gut) are constantly being renewed by division and differentiation of stem cells (see Section 10.7). In normal epithelia, cells generated by stem cells continue to divide for a little time

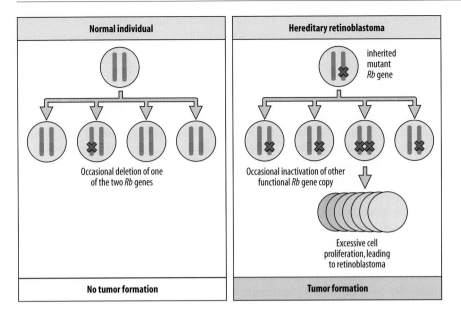

Fig. 13.14 The *retinoblastoma* (*RB*) gene is a tumor-suppressor gene. If only one copy of the *RB* gene is lost or inactivated, no tumor develops (left panel). In individuals already carrying an inherited mutant *RB* gene, if the other copy of the gene is lost or inactivated in a cell, that cell will generate a retinal tumor (right panel). Such individuals are thus at a much greater risk of developing retinoblastoma, and usually do so at a young age.

until they undergo differentiation, when they stop dividing. By contrast, cancerous epithelial cells continue to divide, although not necessarily more rapidly, and usually fail to differentiate. Another feature of cancer cells, unlike developing cells, is that when they divide, they are genetically unstable; the gain or loss of chromosomes is common in solid tumors.

This failure of cancer cells to differentiate is also clearly seen in certain leukemias—cancers of white blood cells. All blood cells are continually renewed from a multipotent stem cell in the bone marrow, by a process in which steps in differentiation are interspersed with phases of cell proliferation. The pathway eventually culminates in cell differentiation and a cessation of cell division (see Section 10.4). Several types of leukemia are caused by cells continuing to proliferate instead of differentiating. These cells become stuck at a particular immature stage in their normal development, a stage that can be identified by the molecules expressed on their cell surfaces.

A number of developmental genes that we have considered elsewhere are also involved in cancer. One view is that many cancers arise from tissue stem cells and that the Hedgehog- and Wnt-signaling pathways may promote cancer growth by promoting stem-cell renewal. For example, the first known mammalian member of the Wnt family was an oncogene (mouse *Int-1*, which is involved in brain development), and abnormal expression of Wnts can block cell differentiation. The Wnt pathway is crucial to the proliferation of intestinal cells, and overactivation of this pathway is known to be involved in colorectal cancer, most commonly through mutations in the gene for APC (adenomatous polyposis coli). Normally, APC helps to keep the pathway inhibited in the absence of Wnt signaling (see Box 1E, p. 26) and thus acts as a tumor suppressor. Mutations that inactivate APC lead to the constitutive activation of the pathway and thus to unregulated cell proliferation. The Hedgehog-signaling pathway is thought to control Wnt signaling in normal intestinal epithelia, confining Wnt activity to stem and progenitor cells, and mutations in Hedgehog pathway components that disrupt this function can lead to tumors. A link between Hedgehog signaling and cancer was first recognized by the finding that mutations in the Hedgehog receptor Patched (see Fig. 2.40) were a cause of the rare, inherited Gorlin's syndrome, one feature of which is multiple cancers of epidermal basal cells. Abnormal activation of the Hedgehog pathway is also present in nearly all pancreatic cancers, where it is thought to maintain cancer stem cells and to be involved in cancer progression. Mutations in the Notch pathway can also lead to a block in differentiation, and can therefore also result in a cancer, whereas members of the TGF-β family are involved in tumor suppression. Mutations in microRNAs (miRNAs) (which act as gene-regulatory RNAs) have also recently been shown to be involved in some human cancers. Some of the miRNAs involved can act as either tumor suppressors or oncogenes in different circumstances.

Most cancer-related deaths result from tumors that have spread from their site of origin to other tissues, the process known as **metastasis**. As most cancers have their origin in epithelial tissue, the ability of tumor cells to undergo an epithelial-to-mesenchymal transition is central to metastasis, as the tumor cells must lose their tissue organization and migrate as mesenchymal cells. Metastasis involves the loss of the E-cadherin-mediated cell–cell adhesion present in epithelia, allowing the cancer cells to migrate out of the epithelium and invade the underlying tissue. If the migrating cells enter the blood stream, they can be carried a long way from their origin. Epithelial-to-mesenchymal transitions are a feature of many stages of embryonic development and the control of this transition is discussed in more detail in Section 8.7.

Rarely, cancers can develop without any detectable alteration in the cell's genetic material. The clearest examples are **teratocarcinomas**, solid tumors that arise spontaneously from germ cells. Teratocarcinomas are unusual tumors, in that they can contain a bizarre mixture of differentiated cell types. Spontaneous teratocarcinomas usually occur in the ovary or testis, and are derived from the germ cells. In the mouse ovary, accidental activation of an unfertilized egg results in its development *in situ* to the stage of epiblast formation; the epiblast then gives rise to a tumor. Similarly, if the epiblast of an early mouse embryo is transplanted to any site in the body of an adult mouse in which it receives a good blood supply, it gives rise to a teratocarcinoma.

Various lines of evidence indicate that teratocarcinomas are not caused by genetic alterations. A mouse inner cell mass placed in culture gives rise to embryonic stem cells (ES cells), which can grow indefinitely in culture (discussed in Chapter 10). When put back into the inner cell mass of another embryo, ES cells contribute normally to many different tissues, including the germline, to produce a chimeric mouse. However, when the same ES cells are placed under the skin of an adult mouse, they develop into a teratocarcinoma (see Fig. 10.32). Transgenic mice containing tissues derived from ES cells do not have an increased probability of forming tumors, yet those same cells consistently form tumors when grafted into adult mice. This indicates that the teratocarcinoma must be the result of the ES cells receiving the wrong developmental signals, and not of a genetic change, and raises a warning in relation to the therapeutic use of ES cells, as discussed in Chapter 10.

13.11 Hormones control many features of plant growth

Plant growth is achieved by cell division in meristems and organ primordia, followed by irreversible cell enlargement, which achieves most of the increase in size (see Chapter 7 and Section 8.16). Unlike the protein growth hormones of animals, plant hormones are typically small organic molecules. Auxin (indole-3-acetic acid) is one of the main regulators of plant growth, and is implicated in a large number of developmental processes, including embryonic development (see Section 7.3), growth towards the light, tissue polarity, vascular tissue differentiation, and **apical dominance**, which is the suppression of the growth of lateral buds immediately below the apical bud. Apical dominance is caused by a diffusible inhibitor of bud outgrowth produced by the shoot apex, as shown by putting an excised apex in contact with an agar block, which can then, on its own, inhibit lateral growth. This inhibitor is auxin. It is produced by the apical bud, transported down the stem, and suppresses the outgrowth of buds that fall within its sphere of influence. If the apical bud is removed, apical dominance is also removed, and lateral buds start to grow (Fig. 13.15). Application of auxin to the cut tip replaces the suppressive effect of the apical bud.

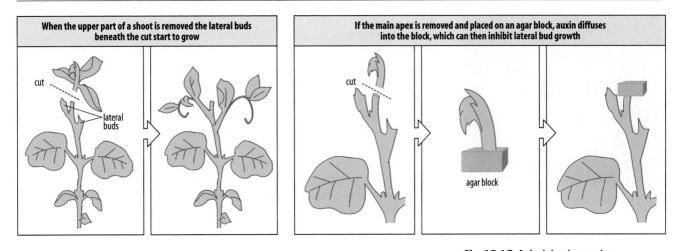

| When the upper part of a shoot is removed the lateral buds beneath the cut start to grow | If the main apex is removed and placed on an agar block, auxin diffuses into the block, which can then inhibit lateral bud growth |

Fig. 13.15 Apical dominance in plants. Left panels: the growth of lateral buds is inhibited by the apical meristem above them. If the upper part of the stem is removed, lateral buds start growing. Right panels: an experiment to show that apical dominance is due to inhibition of lateral bud outgrowth by a substance secreted by the apical region. This substance is the plant hormone auxin.

Another family of plant hormones, the gibberellins, regulate stem elongation and have some similar effects to auxin. Cytokinins, which can stimulate cell proliferation in culture, are derivatives of adenine. The chemical nature of plant hormones is very different from those of animals; nevertheless, they are thought to act through specific hormone-binding receptors and to stimulate intracellular signal transduction.

Plant growth is affected by a variety of environmental factors, such as temperature, humidity, and light. A seedling grown in the dark takes on a characteristic etiolated form in which chloroplasts do not develop, there is extensive elongation of internodes, and the leaves do not expand. The effect of light on plant growth (photomorphogenesis) is mediated through a family of intracellular receptor proteins called phytochromes, which respond to red light and regulate many aspects of plant development and growth.

Summary

Growth in both animals and plants mainly occurs after the basic body plan has been laid down and the organs are still very small. Final organ size can be controlled both by external signals and by intrinsic growth programs. In some cases, organ size can be determined by monitoring the dimensions of the growing organ rather than being set by cell number or cell size. In animals, growth can occur by cell multiplication, cell enlargement, and secretion of large amounts of extracellular matrix. A plant's growth in size is by cell division usually followed by considerable cell enlargement. In mammals, insulin-like growth factors are required for normal embryonic growth, and they also mediate the effects of growth hormone after birth. Human post-natal growth is largely controlled by growth hormone, which is made in the pituitary gland. Growth in the long bones occurs in response to growth hormone stimulation of the cartilaginous growth plates at either end of the bone. Cancer is the result of the loss of growth control and differentiation. Plant growth is due to cell division in meristems followed by cell enlargement, and is dependent on auxins, gibberellins, and other growth hormones.

Molting and metamorphosis

Many animals do not develop directly from an embryo into an 'adult' form, but into a larva from which the adult eventually develops by metamorphosis. The changes that occur at metamorphosis can be rapid and dramatic, the classic examples being the metamorphosis of a caterpillar into a butterfly, a maggot into a fly, and a tadpole into a frog. Another striking example of metamorphosis is the transformation of the pluteus

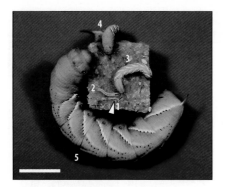

Fig. 13.16 Growth and molting of the caterpillar of the tobacco hawkmoth (*Manduca sexta*). The caterpillar, known as the tobacco hornworm, goes through a series of molts. The tiny hatchling (1)—indicated by the arrow—molts to become a caterpillar (2), and then undergoes three further molts (3, 4, and 5). The increase in size between molts is about twofold. The caterpillars are sitting on a lump of caterpillar food. Scale bar = 1 cm.

Photograph courtesy of S.E. Reynolds.

Fig. 13.17 Molting and growth of the epidermis in arthropods. The cuticle is secreted by the epidermis. At the start of molting, the cuticle separates from the epidermis—apolysis—and a fluid is secreted between them. The epidermis grows, becomes folded, and begins to secrete a new cuticle. Enzymes weaken the old cuticle, which is shed.

larva of the sea urchin into the adult (see Fig. 6.18). In some cases it is hard to see any resemblance between the animal before and after metamorphosis. The adult fly does not resemble the larva at all, because adult structures develop from the imaginal discs and so are completely absent from the larval stages (see Chapters 2 and 11). In frogs, the most obvious external changes at metamorphosis are the regression of the tadpole's tail and the development of limbs, although many other structural changes occur. In some insects, the entire body plan is transformed, with most larval tissues undergoing cell death as the adult tissues develop from the imaginal discs and histoblasts. In arthropods and nematodes, increase in size in larval and pre-adult stages requires shedding of the external cuticle, which is rigid, a process known as **molting**.

A number of features distinguish early embryogenesis from molting, metamorphosis, and other aspects of post-embryonic development. Whereas the signal molecules in early development act over a short range and are typically protein growth factors, many signals in post-embryonic development are produced by specialized endocrine cells, and include both protein and non-protein hormones. The synthesis of these hormones is orchestrated by the central nervous system in response to environmental cues, and there is complex feedback between the endocrine glands and their secretions.

13.12 Arthropods have to molt in order to grow

Arthropods have a rigid outer skeleton, the cuticle, which is secreted by the epidermis. This makes it impossible for the animals to increase in size gradually. Instead, increase in body size takes place in steps, associated with the loss of the old outer skeleton and the deposition of a new larger one. This process is known as **ecdysis** or molting. The stages between molts are known as instars. *Drosophila* larvae have three instars and molts. The increase in overall size between molts can be striking, as illustrated in Fig. 13.16 for the tobacco hornworm (*Manduca*).

At the start of a molt, the epidermis separates from the cuticle in a process known as apolysis, and a fluid (molting fluid) is secreted into the space between the two (Fig. 13.17). The epidermis then increases in area by cell multiplication or cell enlargement, and becomes folded. It begins to secrete a new cuticle, and the old cuticle is partly digested away, eventually splits, and is shed.

Molting is under hormonal control. Stretch receptors that monitor body size are activated as the animal grows, and this results in the brain secreting **prothoracicotropic hormone** (**PTTH**). This activates the prothoracic gland to release the steroid hormone ecdysone, which is the hormone that causes molting. A similar hormonal circuit controls metamorphosis, and is described more fully in the next section.

13.13 Metamorphosis is under environmental and hormonal control

When an insect larva has reached a particular stage, it does not grow and molt any further but undergoes a more radical metamorphosis into the adult form. Metamorphosis occurs in many animal groups other than arthropods, including amphibians. In both insects and amphibians, environmental cues, such as nutrition, temperature, and

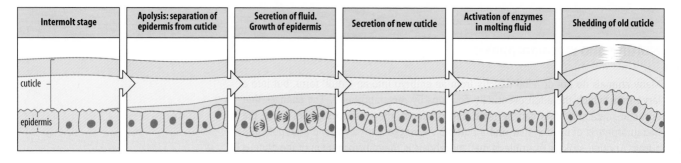

Intermolt stage	Apolysis: separation of epidermis from cuticle	Secretion of fluid. Growth of epidermis	Secretion of new cuticle	Activation of enzymes in molting fluid	Shedding of old cuticle

cuticle

epidermis

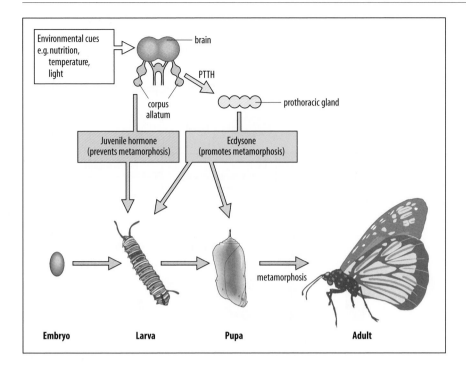

Fig. 13.18 Insect metamorphosis. The corpus allatum of a butterfly larva secretes juvenile hormone, which inhibits metamorphosis. In response to environmental changes, such as an increase in light and temperature, the corpus allatum of the final instar larva begins to secrete prothoracicotropic hormone (PTTH). This acts on the prothoracic gland to stimulate the secretion of ecdysone, the hormone that overcomes the inhibition by juvenile hormone and causes metamorphosis.

After Tata, J.R.: 1998.

light, as well as the animal's internal developmental program, control metamorphosis through their effects on neurosecretory cells in the brain. There are two groups of hormone-producing cells, one of which promotes metamorphosis, and the other inhibits it. Metamorphosis occurs when the inhibition, which is predominant in the larval stage, is overcome in response to environmental cues. The signals produced by the two sets of endocrine cells control the development of all the cells involved in metamorphosis. Larval tissues, such as gut, salivary glands, and certain muscles, undergo programmed cell death. The imaginal discs now develop into rudimentary adult appendages, such as wings, legs, and antennae. The nervous system is also remodeled.

In insects, temperature and light cues stimulate neurosecretory cells in the larva's central nervous system to release signals that act on a neurosecretory release site behind the brain, which then secretes PTTH. This acts on the prothoracic gland to stimulate the production of the steroid hormone ecdysone. It is ecdysone that promotes metamorphosis, as demonstrated by its ability to induce premature metamorphosis in *Drosophila* larvae. The action of ecdysone is counteracted by another hormone—juvenile hormone—produced by the corpus allatum, an endocrine gland located just behind the brain. As its name implies, juvenile hormone maintains the larval state. In butterflies, a pulse of ecdysone in the final instar larva triggers the beginning of pupa formation, and another pulse some days later initiates the later stages of metamorphosis (Fig. 13.18).

Ecdysone crosses the plasma membrane, where it interacts with intracellular ecdysone receptors that belong to the steroid hormone receptor superfamily. These receptors are gene-regulatory proteins, which are activated by binding their hormone ligand (see Box 5D, pp. 192–193). The hormone–receptor complex binds to the regulatory regions of a number of different genes, inducing a new pattern of gene activity characteristic of metamorphosis.

In amphibians, in response to nutritional status and environmental cues, such as temperature and light, the neurosecretory cells of the hypothalamus release corticotropin-releasing hormone, which acts on the pituitary gland, causing it to release thyroid-stimulating hormone (thyrotropin). (This action of corticotropin-releasing hormone is peculiar to non-mammalian vertebrates, and in *Xenopus* only occurs at the tadpole

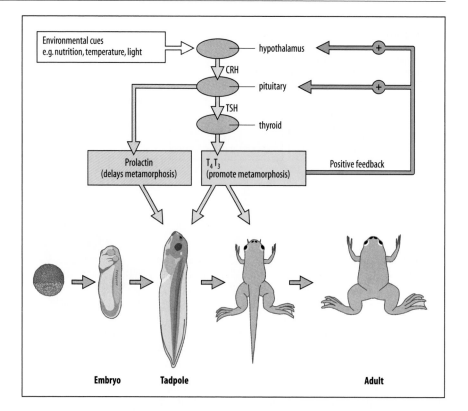

Fig. 13.19 Amphibian metamorphosis. Changes in the environment, such as an increase in nutritional levels, cause the secretion of corticotropin-releasing hormone (CRH) from the larval hypothalamus, which acts on the pituitary to release thyroid-stimulating hormone (TSH). This in turn acts on the thyroid glands to stimulate secretion of the thyroid hormones thyroxine (T_4) and tri-iodothyronine (T_3), which cause metamorphosis. The thyroid hormones also act on the hypothalamus and pituitary to maintain synthesis of CRH and TSH.

After Tata, J.R.: 1998.

stage; in adult frogs, as in mammals, the release of thyroid-stimulating hormone is due to thyrotropin-releasing hormone.) Thyroid-stimulating hormone in turn acts on the thyroid gland to stimulate the secretion of the **thyroid hormones** that bring about metamorphosis (Fig. 13.19). The thyroid hormones are the iodo-amino acids thyroxine (T_4), and tri-iodothyronine (T_3). They are signaling molecules of ancient origin, occurring even in plants. Although very different in chemical structure to ecdysone, they too pass through the plasma membrane and interact with intracellular receptors that belong to the same superfamily as the receptors for retinoic acid and steroid hormones (see Box 5D, pp. 192–193). The pituitary also produces the protein hormone prolactin, which was originally thought to be an inhibitor of metamorphosis; however, overexpression of prolactin does not prolong tadpole life but reduces tail resorption.

A striking feature of the hormones that stimulate metamorphosis is that, as well as affecting a wide variety of tissues, they affect different tissues in different ways, their effects varying from the subtle to the gross. In the tadpole limb, for example, thyroid hormones promote development and growth, whereas they cause cell death and degeneration in the tail. Fast muscles are the first to go, and later the notochord collapses. Yet in all these cases, the hormone produces these very different effects by binding to the same intracellular receptor. The difference in outcomes is due to the hormone–receptor complex switching on or off different sets of genes in different tissues, as a result of the different developmental histories of the regions. Each tissue has its own response to the hormones that cause metamorphosis, and some of these effects can be reproduced in culture. When excised *Xenopus* tadpole tails are exposed to thyroid hormones in culture, for example, they cause cell death and complete tissue regression. Metamorphosis also leads to changes in the responsiveness of cells to other signals; for example, in *Xenopus*, estrogen can only induce the synthesis of vitellogenin, a protein required for the yolk of the egg, after metamorphosis and the attainment of sexual maturity.

There are alterations in the expression of many genes—several hundred at least—during *Drosophila* metamorphosis. In *Drosophila*, because of the special characteristics of polytene chromosomes, changes in gene activity that occur during metamorphosis are actually visible. Cells in some larval tissues (such as the salivary glands) grow and repeatedly pass through S phase without undergoing mitosis and cell division. The cells become very large and can have several thousand times the normal complement of DNA. In salivary gland cells, many copies of each chromosome are packed side by side to form giant polytene chromosomes. When a gene is active, the chromosome at that site expands into a large localized 'puff', which is easily visible (Fig. 13.20). The puff represents the unfolding of chromatin and the associated transcriptional activity. When the gene is no longer active, the puff disappears. During the last days of larval life, a large number of puffs are formed in a precise sequence, a pattern that is under the direct influence of ecdysone.

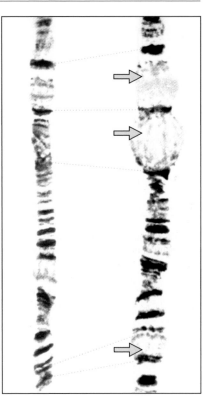

Fig. 13.20 Gene activity seen as puffs on the polytene chromosomes of _Drosophila_. A region of a chromosome is shown from a young third instar (left), and from an older larva (right), after ecdysone has induced puffs at three loci (arrowed).

Photograph courtesy of M. Ashburner.

Summary

Arthropod larvae grow by undergoing a series of molts in which the rigid cuticle is shed. Metamorphosis during the post-embryonic period can result in a dramatic change in the form of an organism. In insects, it is hard to see any resemblance between the animal before and after metamorphosis, whereas in amphibians, the change is somewhat less dramatic. Environmental and hormonal factors control metamorphosis. In both insects and amphibians there are two sets of hormonal signals, one promoting, the other delaying, metamorphosis. Thyroid hormones cause metamorphosis in amphibians, and ecdysone does the same in insects. In _Drosophila_, gene activity during metamorphosis can be monitored by localized puffing on the giant polytene chromosome.

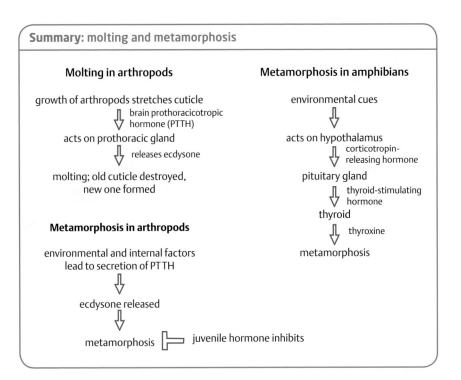

Summary: molting and metamorphosis

Molting in arthropods

growth of arthropods stretches cuticle
⇩ brain prothoracicotropic hormone (PTTH)
acts on prothoracic gland
⇩ releases ecdysone
molting; old cuticle destroyed, new one formed

Metamorphosis in arthropods

environmental and internal factors lead to secretion of PTTH
⇩
ecdysone released
⇩
metamorphosis ⊢ juvenile hormone inhibits

Metamorphosis in amphibians

environmental cues
⇩
acts on hypothalamus
⇩ corticotropin-releasing hormone
pituitary gland
⇩ thyroid-stimulating hormone
thyroid
⇩ thyroxine
metamorphosis

Aging and senescence

Organisms are not immortal, even if they escape disease or accidents. With aging—the passage of time—comes an increasing impairment of physiological functions, which reduces the body's ability to deal with a variety of stresses, and an increased susceptibility to disease. This age-related decline in function is known as **senescence**.

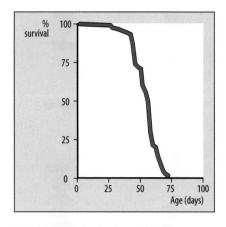

Fig. 13.21 Aging in *Drosophila*. The probability of dying increases rapidly at older ages.

Aging is observed in most multicellular animals, but there are notable exceptions, such as cnidarians (sea anemones, *Hydra*, and their relations) and planarians. All sexually reproducing animals age, whereas most asexual animals do not. *Hydra*, whose capacity for regeneration is discussed in the next chapter, does not age unless it undergoes sexual differentiation. The phenomenon of senescence raises many questions as to the underlying mechanisms, and these are still largely unanswered, but we can at least consider some general questions, such as whether senescence is part of an organism's post-embryonic developmental program or whether it is simply the result of wear and tear. Germ cells do not age; if they did, the species would die out.

Although individuals may vary in the time at which particular aspects of aging appear, the overall effect is summed up as an increased probability of dying in most animals, including humans, with increased age. This life pattern, which is illustrated in relation to *Drosophila* (Fig. 13.21), is typical of many animals, but there is little evidence that aging contributes to mortality in the wild; more than 90% of wild mice die during their first year. There are, however, exceptions, such as the Pacific salmon, in which death does not come after a process of gradual aging, but is linked to a certain stage in the life cycle, in this case to spawning.

One view of aging is that it is due to an accumulation of damage that eventually outstrips the ability of the body to repair itself, and so leads to the loss of essential functions. For example, some old elephants die of starvation because their teeth have worn out. The nematode *Caenorhabditis elegans* lives for an average or around 20 days and the major cellular change that occurs with age is the progressive deterioration of muscle. This deterioration has a random effect on the life span, which varies from 10 to 30 days. Nevertheless, there is clear evidence that senescence is under genetic control, as different species age at vastly different rates, as shown by their different life spans (Fig. 13.22). An elephant, for example, is born after 21 months' embryonic development, and at that point shows few, if any, signs of aging, whereas a 21-month-old mouse is already well into middle age and beginning to show signs of senescence.

The genetic control of aging can be understood in terms of the 'disposable soma' theory, which puts it into the context of evolution. The disposable soma theory

Longevity and time to attain reproductive maturity at puberty for various mammals			
	Maximum lifespan (months)	Length of gestation (months)	Age at puberty (months)
Human	1440	9	144
Finback whale	960	12	–
Indian elephant	840	21	156
Horse	744	11	12
Chimpanzee	534	8	120
Brown bear	442	7	72
Dog	408	2	7
Cattle	360	9	6
Rhesus monkey	348	5.5	36
Cat	336	2	15
Pig	324	4	4
Squirrel monkey	252	5	36
Sheep	240	5	7
Gray squirrel	180	1.5	12
European rabbit	156	1	12
Guinea-pig	90	2	2
House rat	56	0.7	2
Golden hamster	48	0.5	2
Mouse	42	0.7	1.5

Fig. 13.22 Table showing life span, length of gestation, and age at puberty for various mammals.

proposes that natural selection tunes the life history of the organism so that sufficient resources are invested in maintaining the repair mechanisms that prevent aging, at least until the organism has reproduced and cared for its young. Thus mice, which start reproducing when just a few months old, need to maintain their repair mechanisms for much less time than do elephants, which only start to reproduce when around 13 years old. In most species of animals in the wild, few individuals live long enough to show obvious signs of senescence and senescence need only be delayed until reproduction is complete. Cells have numerous mechanisms to delay aging, which are quite similar to the mechanisms used to prevent malignant transformation. These cellular mechanisms protect the cell from internal damage by reactive chemicals and routinely repair damage to DNA, which is occurring continually in living cells even when they are not actively dividing. They are particularly active in germ cells.

13.14 Genes can alter the timing of senescence

The maximum recorded life spans of animals show dramatic differences (see Fig. 13.22). Humans can live as long as 120 years, some owls 68 years, cats 28 years, *Xenopus* 15 years, mice 3.5 years, and the nematode about 25 days. Mutations in genes that affect life span have been identified in *C. elegans*, *Drosophila*, mice, and humans, and may give clues to the mechanisms involved; the ability to resist damage to DNA and the effects of oxygen radicals are important.

A *C. elegans* worm that hatches as a first instar larva in an uncrowded environment with ample food grows to adulthood and can survive for 25 days. In crowded conditions and when food is short, however, the animal enters a quiescent third-instar larval state known as the **dauer** state, where it neither eats nor grows until food becomes available again. When conditions become favorable, the dauer larva molts and becomes a fourth instar larva. The dauer state can last for 60 days and has no effect on the post-dauer life span: it is therefore considered to be a state in which the larva does not age. An insulin/IGF-1 signaling system plays an important role both in controlling entry into the dauer state in response to stress and in the overall control of fertility, life span, and metabolism in *C. elegans* and in *Drosophila*. Reduction in insulin/IGF-I signaling increases longevity (Fig. 13.23).

In *C. elegans*, mutations that cause a strong reduction in the expression of *daf-2*, which encodes a receptor in the insulin-signaling pathway, arrest development in the dauer state. The normal role of DAF-2 is to antagonize the activity of DAF-16, a transcription factor whose activity lengthens life span and increases resistance to some types of stress. A partial loss of DAF-2 function results in longer adult life-span after the dauer state but with reduced fertility and viability of the progeny, while a further reduction in DAF-2 at the larval stage by RNA interference increases life span even more, without shortening the dauer state. The presence of a germline seems to have a negative effect on life-span extension. Removal of germline precursor cells in *daf-2* mutant embryos resulted in their having a mean life-span of 125 days and remaining quite healthy—in humans, this would equate to a life span of 500 years. But when *daf-2* mutants were cultured alongside wild-type worms, the mutants became extinct in just a few generations, partly due to their reduced fertility. Microarray analysis of the effects of DAF-16 show that it activates stress-response and antimicrobial genes. In laboratory conditions it seems that aging animals in which DAF-16 is inhibited are killed by the bacteria on which they feed.

A similar system regulates aging in *Drosophila*. Mutations disabling the insulin/IGF-1 pathway almost double the life span in this animal, and calorie restriction also extends life span. The mutant flies, rather like the dauer larvae, enter a state of reproductive diapause or quiescence. These effects on life span may be due to resistance to oxidative stress. Aging in mice can be retarded by reduction in pituitary activity, and

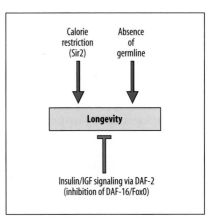

Fig. 13.23 Potential inputs to aging in *Caenorhabditis elegans* and *Drosophila*. Calorie restriction and the absence of a germline promote an extended life-span in both animals. In *Drosophila*, the calorie-restriction effect has been shown to require the histone deacetylase Sir2, which is likely to act on gene expression through its ability to modifiy chromatin (see Box 10A, p. 374). In *C. elegans*, insulin/IGF signaling has been shown to regulate life span via its inhibitory effect on expression of the transcription factor DAF-16 (a member of the FoxO family of transcription factors).

there is also evidence that female mice with a mutation in the IGF-1 receptor (which is the receptor activated by IGF-1 and IGF-2) can live 33% longer than usual.

There is evidence that oxidative damage accelerates aging and that reactive oxygen radicals are key players in causing cell damage. Reduction in food intake increases life span in other animals; rats on a minimal diet live about 40% longer than rats allowed to eat as much as they like. This is thought to be partly due to a reduced exposure to free radicals, which are formed during the oxidative breakdown of food. Free radicals are highly reactive, and can damage both DNA and proteins. A long-lived rodent species generates less reactive oxygen than the laboratory mouse. Oxidative stress also exerts its effects by damaging mitochondria.

Humans who are homozygous for the recessive gene defect known as Werner syndrome show striking effects of premature aging. There is growth retardation at puberty, and by their early twenties those affected by this syndrome have gray hair and suffer from a variety of illnesses, such as heart disease, that are typical of old age. Most people affected die before the age of 50. The gene affected in Werner syndrome has been identified, and is thought to encode a protein involved in unwinding DNA. Such unwinding is required for DNA replication, DNA repair, and gene expression. The inability to carry out DNA repair properly in Werner-syndrome patients could subject the genetic material to a much higher level of damage than normal. The link between Werner syndrome and DNA thus fits with the possibility that aging is linked to the accumulation of damage in DNA, but it may also be related to cell senescence. Hutchinson–Gilford progeria syndrome (progeria means premature aging) is even more severe in its effects. The gene affected in this syndrome encodes the intranuclear protein lamin A, and the primary cellular defect is an instability in the structure of the nuclear envelope, which means cells are more likely to die prematurely.

13.15 Cell senescence blocks cell multiplication

Fig. 13.24 Vertebrate fibroblasts can only go through a limited number of divisions in culture. Fibroblasts placed in culture are subcultured until they stop growing (top panels). The number of cell doublings in culture before they stop dividing is related to their maximum age, as indicated by the figures in brackets on the graph (bottom panel).

One might think that when cells are isolated from an animal, placed in culture, and provided with adequate medium and growth factors, they would continue to proliferate almost indefinitely. But this is not the case. For example, mammalian fibroblasts—connective tissue cells—will only go through a limited number of cell doublings in culture; the cells then stop dividing, however long they are cultured (Fig. 13.24). For normal fibroblasts, the number of cell doublings depends both on the species and the age of the animal from which they are taken. Fibroblasts taken from a human fetus

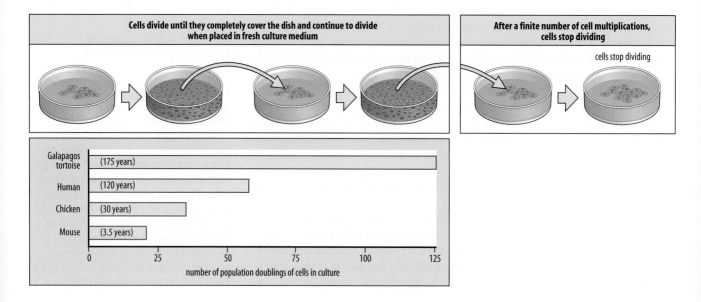

go through about 60 doublings, those from an 80-year-old about 30, and those from an adult mouse about 12–15 doublings. When the cells stop dividing, they appear to be healthy, but are stuck at some point in the cell cycle, often G_0; this phenomenon is known as **cell senescence**. Cells taken from patients with Werner's syndrome, who show an acceleration of many features of normal aging (see Section 13.14), make significantly fewer divisions in culture than normal cells. However, it is far from clear how this behavior of cells in culture is significant for the aging of the organism and to what extent it reflects the way the cells are cultured.

A feature shared by senescent cells in culture and *in vivo* is shortening of the **telomeres**. Telomeres are the repetitive DNA sequences at the ends of the chromosomes that preserve chromosome integrity and ensure that chromosomes replicate themselves completely without loss of information-encoding DNA at the ends. The length of the telomeres is reduced in older human cells. Telomere length is found to decrease at each DNA replication, suggesting that they are not completely replicated at each cell division and this may be related to senescence. If the enzyme telomerase, which maintains telomere length and is normally absent from cells in culture, is expressed in those cells, senescence in culture does not occur. However, it is not yet clear whether telomere shortening is a major cause of aging in somatic cells, as certain rodent cells such as Schwann cells can, under appropriate conditions, proliferate indefinitely and their telomeres do not control replication. ES cells also express telomerase and proliferate indefinitely.

Senescent cells that have stopped dividing have changes in the expression of proteins that are involved in cell-cycle control. The proteins p21 and p16 are often expressed; they are part of the tumor-suppressor pathways governed by p53 and RB in response to DNA damage (see Section 13.10). They act as inhibitors of cyclin-dependent kinases and prevent a cell with damaged DNA from entering the cell cycle. This helps to prevent senescent cells, with their DNA damage, from becoming cancerous.

Summary

Aging is largely caused by damage to cells, particularly by reactive oxygen, but is also under genetic control. Many normal cells in culture can only undergo a limited number of cell divisions, which correlate with their age at isolation and the normal life-span of the animal from which they came. Genes that can increase life span have been identified in *C. elegans* and *Drosophila* and may act by increasing the animals' resistance to oxidative stress.

Summary to Chapter 13

The form of many animals is laid down in miniature during embryonic development, and they then grow in size, keeping the basic body form, although different regions grow at different rates. Growth may involve cell multiplication, cell enlargement, and the laying down of extracellular material. In vertebrates, structures have an intrinsic growth program, which is under hormonal control. Arthropod larvae grow by molting and both they and other animals, such as frogs, undergo metamorphosis, which is under hormonal control. Cancer can be viewed as an aberration of growth, as it usually results from mutations that lead to excessive cell proliferation and failure of cells to differentiate. The symptoms of aging appear mainly to be caused by damage to cells that accumulates over time, and aging is also under genetic control. In plants, all adult structures are derived by growth from specialized regions called meristems.

■ End of chapter questions

Long answer (concept questions)

1. Define 'growth.' What are the main mechanisms of growth? How does cell death fit into a discussion of growth?

2. Why is it inaccurate to say that the cell cycle starts when cytokinesis produces two cells after M phase? According to molecular analysis of the cell cycle, each round of the cycle 'starts' at what point in the cycle? What are the molecular events that define this start point? (The cell-cycle diagram in Box 1B might also be helpful.)

3. Most cell types in adult vertebrates are not actively dividing; where in the cell cycle are they considered to be? What is required for them to re-enter the cell cycle? Name some adult differentiated cell types that are able to divide and some that are incapable of cell division.

4. In what way is the standard cell cycle modified during the first 13 pre-cellular divisions in *Drosophila* embryogenesis, to allow very rapid cell divisions? What is the maternally supplied protein that drives these early divisions; and what is this protein's function?

5. The final size of some organs, such as liver and spleen, is regulated by signaling between cells, whereas the size of other organs, such as pancreas and thymus, is under the control of a cell-intrinsic developmental program. Describe the experiments that reveal these two different mechanisms of growth control.

6. What are some of the consequences of poor maternal nutrition on the offspring?

7. In *Drosophila*, the Hippo protein and the *bantam* microRNA have opposing functions, as their names imply. What are the functions of these two regulators of growth? Outline the pathway from the cell-surface protein that initiates signaling to Hippo, and signalling to the *bantam* gene.

8. Address the following questions about human growth hormone: What is growth hormone? Where is it produced? How is its synthesis and secretion controlled? What other growth factor is a major mediator of growth hormone's effect on cellular proliferation?

9. Contrast the roles of chondrocytes and osteoblasts in growth of the long bones of vertebrates.

10. Summarize the effects of each of the following signaling molecules on chondrocytes: PHRP, Indian hedgehog, thyroid hormones, and FGF.

11. Review the various signaling molecules and signaling pathways discussed in the chapter, and indicate which could be classified as oncogenes, and which as tumor suppressors.

12. Elucidate the many connections found between Wnt, APC, β-catenin, and cadherins in the context of the progression of normal cells to metastatic cancer cells.

13. In one experiment, a frog tadpoleis exposed to thyroxine in the water in which it is being reared. In another experiment, the thyroid gland is removed from a tadpole. What results would you predict in each case. Justify your predictions.

14. Discuss the evidence for and against the statement: 'The changes that occur with aging are an inevitable part of an animal's genetically determined postembryonic developmental program.'

Multiple choice (factual recall questions)

NB There is only one correct answer to each question.

1. The maternally supplied *Drosophila* protein *string* is a
a) kinase
b) phosphatase
c) signaling molecule
d) transcription factor

2. What will be the result of grafting a limb bud from a large species of the salamander *Ambystoma* onto a smaller species?
a) The grafted bud will be unable to grow in a smaller animal, and will be lost.
b) The grafted bud will grow to a size appropriate to its host, and will thus be of the same size as the other limbs.
c) The grafted bud will grow to become a larger limb than the others, reflecting its origin in the larger species.
d) The grafted bud will initially grow larger than the others, but with time will regulate and become a limb of the same size as the other limbs of its host.

3. Growth is stimulated by which signaling molecules?
a) insulin
b) IGF-1 and -2
c) FGF
d) all of these may stimulate growth

4. Growth in the length of long bones in vertebrates occurs in the
a) diaphysis
b) epiphysis
c) growth plate
d) ossification centers

5. Normal postembryonic growth of striated muscle fibers occurs through
a) an increase in muscle fiber size, but not the number of muscle fibers
b) an increase in the number of muscle fibers
c) differentiation of stem cells to form new muscle fibers
d) division of the muscle cells to increase the total number of muscle cells

6. Which of the following statements is true?
a) The genes involved in cancer are all classed as oncogenes.
b) All genes involved in cancer are of a class called tumor suppressor genes.
c) Both oncogenes and tumor suppressor genes are involved in cancer.
d) No unifying classifications can be applied to the genes involved in cancer.

7. The majority of cancers occur in epithelia because
a) Epithelial tissues are more likely than other tissue types to be constantly renewed by division and differentiation of stem cells.

b) Growth regulatory pathways are more common in epithelia than in other tissue types, and hence are more likely to be deregulated in epithelia than in other tissues.

c) The cell adhesion contacts between cells in epithelia predispose these cells to abnormal growth.

d) The differentiation of epithelia is complex, and more likely to go wrong than other tissues, resulting in cancer.

8. Molting in insects and other arthropods is triggered by

a) auxin

b) ecdysone

c) Hedgehog

d) juvenile hormone

9. Which signaling pathway seems to play a role in aging in several different animals, from worms to mice?

a) ecdysone

b) FGF

c) Hippo

d) insulin/IGF-1

10. Which of the following is consistent with a model for aging in which stresses leading to DNA damage cause senescence?

a) Werner's syndrome is a premature aging illness, possibly caused by a defect in DNA repair.

b) Dietary restriction in mammals reduces the production of DNA-damaging free radicals in the mitochondria.

c) Senescent cells in culture that have stopped dividing often express p21 and p16, proteins that are normally activated in response to DNA damage.

d) All of these facts are consistent with a model for aging based on DNA damage.

Multiple choice answer key

1: b, 2: c, 3: d, 4: c, 5: a, 6: c, 7: a, 8: b, 9: d, 10: d.

▦ General further reading

Kirkwood, T.B.L.: **Understanding the odd science of aging**. *Cell* 2005, **120**: 437–447.

Nature Insight: **Cell division and cancer**. *Nature* 2004, **432**: 293–341.

Partridge, L.: **The new biology of ageing**. *Phil. Trans R. Soc. B* 2010, **365**: 147–154.

▦ Section further reading

13.1 Tissues can grow by cell proliferation, cell enlargement, or accretion

Goss, R.J.: *The Physiology of Growth*. New York: Academic Press, 1978.

13.2 Cell proliferation is controlled by regulating entry into the cell cycle

Morgan, D.O.: *The Cell Cycle: Principles of Control*. Oxford University Press 2007.

13.3 Cell division in early development can be controlled by an intrinsic developmental program

Edgar, B., Lehner, C.F.: **Developmental control of cell cycle regulators: a fly's perspective**. *Science* 1996, **274**: 1646–1652.

Follette, P.J., O'Farrell, P.H.: **Connecting cell behavior to patterning: lessons from the cell cycle**. *Cell* 1997, **88**: 309–314.

13.4 Organ size can be controlled by both intrinsic growth programs and extracellular signals

Amthor, H., Huang, R., McKinnell, I., Christ, B., Kambadur, R., Sharma, M., Patel, K.: **The regulation and action of myostatin as a negative regulator of muscle development during avian embryogenesis**. *Dev. Biol.* 2002, **251**: 241–257.

Shingleton, A.W.: **Body-size regulation: combining genetics and physiology**. *Curr. Biol.* 2005, **15**: R825–R827.

Stanger, B.Z.: **The biology of organ size determination**. *Diabetes Obes. Metab.* 2008, **10**: 16–26.

Stanger, B.Z., Tanaka, A.J., Melton, D.A.: **Organ size is limited by the number of embryonic progenitor cells in the pancreas but not in the liver**. *Nature* 2007, **445**: 886–891.

Taub, R.: **Liver regeneration: from myth to mechanism**. *Nat. Rev. Mol. Cell Biol.* 2004, **5**: 836–847.

13.5 The amount of nourishment an embryo receives can have profound effects in later life

Barker, D.J.: The Wellcome Foundation Lecture. 1994: **The fetal origins of adult disease**. *Proc. R. Soc. Lond.* 1995, **262**: 37–43.

Gluckman, P.D., Hanson, M.A., Cooper, C., Thornburg, K.L.: **Effect of *in utero* and early-life conditions on adult health and disease**. *N. Engl. J. Med.* 2008, **359**: 61–73.

13.6 Determination of organ size involves coordination of cell growth, cell division, and cell death

de la Cova, C., Abril, M., Bellosta, P., Gallant, P., Johnston, L.A.: ***Drosophila* myc regulates organ size by inducing cell competition**. *Cell* 2004, **117**: 107–116.

Martin, F.A., Herrera, S.C., Morata, G.: **Cell competition, growth and size control in the Drosophila wing imaginal disc**. *Development* 2009, **136**: 3747–3756.

Pan, D.: **Hippo signaling in organ size control**. *Genes Dev.* 2007, **21**: 886–897.

Box 13A Gradients of signaling molecules could determine organ size

Lawrence, P.A., Struhl, G., Casal, J.: **Do the protocadherins Fat and Daschous link up to determine both planar cell polarity and the dimensions of organs?** *Nat. Cell Biol.* 2008, **10**: 1379–1382.

Rogulja, D., Irvine, K.D.: **Regulation of cell proliferation by a morphogen gradient**. *Cell* 2005, **123**: 449–461.

Willecke, M., Hamaratoglu, F., Sansores-Garcia, L., Tao, C., Halder, G.: **Boundaries of Dachsous cadherin activity modulate the Hippo signaling pathway to induce cell proliferation**. *Proc. Natl Acad. Sci USA* 2008, **105**: 14897–14902.

13.7 Body size is also controlled by the neuroendocrine system in both insects and mammals

Mirth, C.K., Riddiford, L.M.: **Size assessment and growth control: how adult size is determined in insects**. *BioEssays* 2007, **29**: 344–355.

Nijhout, H.F.: **Size matters (but so does time), and it's OK to be different**. *Dev. Cell* 2008, **15**: 491–492.

Sanders, E.J., Harvey S.: **Growth hormone as an early embryonic growth and differentiation factor**. *Anat. Embryol.* 2004, **209**: 1–9.

Stern, D.: **Body-size control: how an insect knows it has grown enough**. *Curr. Biol.* 2003, **13**: R267–R269.

13.8 Growth of the long bones occurs in the growth plates

Kember, N.F.: **Cell kinetics and the control of bone growth**. *Acta Paediatr. Suppl.* 1993, **391**: 61–65.

Kronenberg, H.M.: **Developmental regulation of the growth plate**. *Nature* 2003, **423**: 332–336.

Nilsson, O., Baron, J.: **Fundamental limits on longitudinal bone growth: growth plate senescence and epiphyseal function**. *Trends Endocrinol. Metab.* 2004, **8**: 370–374.

Roush, W.: **Putting the brakes on bone growth**. *Science* 1996, **273**: 579.

13.9 Growth of vertebrate striated muscle is dependent on tension

Schultz, E.: **Satellite cell proliferative compartments in growing skeletal muscles**. *Dev. Biol.* 1996, **175**: 84–94.

Williams, P.E., Goldspink, G.: **Changes in sarcomere length and physiological properties in immobilized muscle**. *J. Anat.* 1978, **127**: 450–468.

13.10 Cancer can result from mutations in genes that control cell multiplication and differentiation

Beachy, P.A., Karhadkar, S.S., Berman, D.M.: **Tissue repair and stem cell renewal in carcinogenesis**. *Nature* 2004, **432**: 324–331.

Hunter, T.: **Oncoprotein networks**. *Cell* 1997, **88**: 333–346.

Jones, S., Zhang, X., Parsons, D.W., Lin, J.C., Leary, R.J., Angenendt, P., Mankoo, P., Carter, H., Kamiyama, H., Jimeno, A., *et al.*: **Core signaling pathways in human pancreatic cancers revealed by global genomic analyses**. *Science* 2008, **321**: 1801–1806.

Shilo, B.Z.: **Tumor suppressors. Dispatches from patched**. *Nature* 1996, **382**: 115–116.

Taipale, J., Beachy, P.A.: **The hedgehog and Wnt signalling pathways in cancer**. *Nature* 2001, **422**: 349–353.

Van Dyke, T.: **p53 and tumor suppression**. *N. Engl. J. Med.* 2007, **356**: 79–92.

Vernon, A.E., LaBonne, C.: **Tumor metastasis: a new Twist on epithelial-mesenchymal transitions**. *Curr. Biol.* 2004, **14**: R719–R721.

13.11 Hormones control many features of plant growth

Cosgrove, D.J.: **Plant cell enlargement and the action of expansins**. *BioEssays* 1996, **18**: 533–540.

Hauser, M.T., Morikami, A., Benfey, P.N.: **Conditional root expansion mutants of *Arabidopsis***. *Development* 1995, **121**: 1237–1252.

Raven, P.H., Evert, R.F., Eichhorn, S.E.: *The Biology of Plants* (7th edn). New York: W. H. Freeman, 2005.

Sablowski, R.: **Root development: the embryo within**. *Curr. Biol.* 2004, **14**: R1054–R1055.

13.12 Arthropods have to molt in order to grow & 13.13 Metamorphosis is under environmental and hormonal control

Brown, D.D., Cai, L.: **Amphibian metamorphosis**. *Dev. Biol.* 2007, **306**: 20–33.

De Loof, A.: **Ecdysteroids, juvenile hormone and insect neuropeptides: Recent successes and remaining major challenges**. *Gen. Comp. Endocrinol.* 2008, **155**: 3–13.

Huang, H., Brown, D.D.: **Prolactin is not juvenile hormone in *Xenopus laevis* metamorphosis**. *Proc. Natl Acad. Sci. USA* 2000, **97**: 195–199.

Nijhout, H.F.: **Size matters (but so does time), and it's OK to be different**. *Dev. Cell* 2008, **15**: 491–492.

Tata, J.R.: *Hormonal Signaling and Postembryonic Development*. Heidelberg: Springer, 1998.

Thummel, C.S.: **Flies on steroids—*Drosophila* metamorphosis and the mechanisms of steroid hormone action**. *Trends Genet.* 1996, **12**: 306–310.

White, K.P., Rifkin, S.A., Hurban, P., Hogness, D.S.: **Microarray analysis of *Drosophila* development during metamorphosis**. *Science* 1999, **286**: 2179–2184.

Wu, H.H., Ivkovic, S., Murray, R.C., Jaramillo, S., Lyons, K.M., Johnson, J.E., Calof, A.L.: **Autoregulation of neurogenesis by GDF11**. *Neuron* 2003, **37**: 197–207.

13.14 Genes can alter the timing of senescence

Arantes-Oliviera, N., Berman, J.R., Kenyon, C.: **Healthy animals with extreme longevity**. *Science* 2003, **302**: 611.

Campisi, J., d'Adda di Fagagna, F.: **Cellular senescence: when bad things happen to good cells**. *Nat. Rev. Mol. Cell Biol.* 2007, **8**: 729–740.

Finkel, T., Serrano, M., Blasco, M.A.: **The common biology of cancer and ageing**. *Nature* 2007, **448**: 767–774.

Harper, M.E., Bevilacqua, L., Hagopian, K., Weindruch, R., Ramsey, J.J.: **Ageing, oxidative stress, and mitochondrial uncoupling**. *Acta Physiol. Scand.* 2004, **182**: 321–331.

Kenyon, C.: **The plasticity of aging: insights from long-lived mutants**. *Cell* 2005, **120**: 449–460.

Kipling, D., Davis, T., Ostler, E.L., Faragher, R.G.: **What can progeroid syndromes tell us about human aging?** *Science* 2004, **305**: 1426–1431.

Kudlow, B.A., Kennedy, B.K., Monnat, R.J.: **Werner and Hutchinson-Gilford progeria syndromes: mechanistic basis of progerial diseases**. *Nat. Rev. Mol. Cell Biol.* 2007, **8**: 394–404.

Martinez, D.E.: **Mortality patterns suggest lack of senescence in hydra**. *Exp. Gerontol.* 1998, **33**: 217–225.

Murphy, C.T., Partridge, L., Gems, D.: **Mechanisms of ageing: public or private**. *Nat Rev. Genet.* 2002, **3**: 165–175.

Partridge, L.: **Some highlights of research on aging with invertebrates**. *Aging Cell* 2008, **7**: 605–608.

Tatar, M., Bartke, A., Antebi, A.: **The endocrine regulation of aging by insulin-like signals**. *Science* 2003, **299**: 1346–1351.

Weindruch, R.: **Caloric restriction and aging**. *Science* 1996, **274**: 46–52.

Yoshida, K., Fujisawa, T., Hwang, J.S., Ikeo, K., Gojobori, T.: **Degeneration after sexual differentiation in hydra and its relevance to the evolution of aging**. *Gene* 2006, **385**: 64–70.

13.15 Cell senescence blocks cell multiplication

Shay, J.W., Wright, W.E.: **When do telomeres matter?** *Science* 2001, **291**: 839–840.

Sherr, C.J., DePinho, R.A.: **Cellular senescence: mitotic clock or culture shock**. *Cell* 2000, **102**: 407–410.

Regeneration

- Limb and organ regeneration
- Regeneration in *Hydra*

Some animals, such as newts, can regenerate limbs, tails, and the eye lens. In the limb, this involves dedifferentiation of differentiated cells to progenitor cells and subsequent distal regrowth, with correct development guided by the generation of new distal positional values. Hydra, a small cnidarian, regenerates its head without growth and this involves gradients in inhibition and positional information set up by the head.

Many of the cells in the adult body, such as heart muscle and most neurons of the central nervous system, are the same cells that were originally generated during embryonic development. But some tissues, such as blood and epithelia, are continually being replaced by stem cells, and others, such as skeletal muscle, can be regenerated from quiescent stem cells if they are damaged (see, for example, Section 10.10). We have also seen many examples of the capacity of the embryo to self-regulate when parts of it are removed or rearranged (see, for example, Section 4.10). Here we look at the related phenomenon of **regeneration** in adult organisms. Regeneration is the ability of the fully developed organism to replace tissues, organs, and appendages by growth or repatterning of somatic tissue. Plants have remarkable powers of regeneration: a single somatic plant cell can give rise to a complete new plant (see Section 7.4). Some animals also show great ability to regenerate: small fragments of animals such as starfish, planarians (flatworms), and *Hydra* can give rise to a whole animal (Fig. 14.1). The ability of these animals to regenerate may

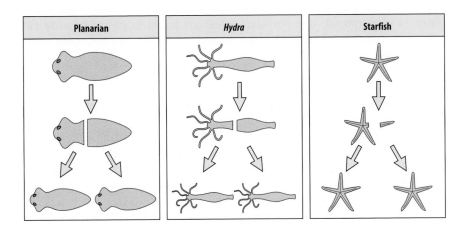

Planarian	*Hydra*	Starfish

Fig. 14.1 Regeneration in some invertebrate animals. A planarian, *Hydra*, and a starfish all show remarkable powers of regeneration. When parts are removed or a small fragment isolated, a whole animal can be regenerated.

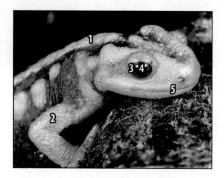

Fig. 14.2 The capacity for regeneration in urodele amphibians. The emperor newt can regenerate its dorsal crest (1), limbs (2), retina and lens (3 and 4), jaw (5), and tail (not shown).

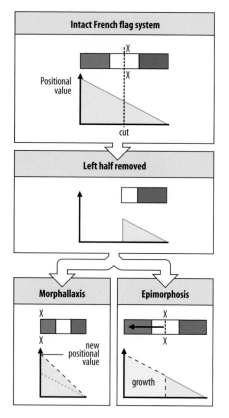

Fig. 14.3 Morphallaxis and epimorphosis. A pattern such as the French flag may be specified by a gradient in positional value (see Fig. 1.25). If the system is cut in half, it can regenerate in one of two ways. In regeneration by morphallaxis, a new boundary is established at the cut and the positional values are changed throughout. In regeneration by epimorphosis, new positional values are linked to growth from the cut surface.

be related to their ability to reproduce asexually; that is, to produce a complete new individual by budding or fission.

Some insects and other arthropods can regenerate lost appendages, such as legs, and among vertebrates, newts and other urodele amphibians show a remarkable capacity for regeneration, being able to regenerate complete new tails and limbs, as well as some internal tissues (Fig. 14.2). The lens in the newt eye, for example, regenerates from the pigmented epithelium of the iris, an example of transdifferentiation (see Fig. 10.31). Another striking case of regeneration in vertebrates is the zebrafish, which can regenerate the heart after removal of part of the ventricle. The regenerative powers of mammals are much more restricted. Mammals cannot regenerate the heart, for example, and so this ability in the genetically tractable zebrafish has implications for medical research. The mammalian liver can regrow if a part of it is removed, as liver cells (hepatocytes) retain the ability to divide (see Section 13.4), the antlers of male deer regrow each year, and fractured bones mend by a regenerative process. But mammals cannot regenerate lost limbs, although they do have a limited capacity to replace the ends of digits.

The issue of regeneration raises several major questions. Why are some animals able to regenerate and others not? What is the origin of the cells that give rise to the regenerated structures? What mechanisms pattern the regenerated tissue and how are these related to the patterning processes that occur in embryonic development? The lack of regenerative powers is not necessarily associated with increasing complexity: animals with no powers of regeneration at all include nematodes and rotifers. We will mainly focus in this chapter on two systems in which regeneration has been intensively studied: regeneration of limbs in amphibians (newts and axolotls) and insects; and regeneration of a whole animal—the freshwater cnidarian *Hydra*. Understanding regeneration in these systems could lead to progress in stimulating it in mammalian tissues and so help the development of medical ways of repairing tissues such as the heart and the spinal cord. There is also the question of whether stem cells are involved in regeneration, and so whether stem cells might be used to help repair damaged tissues (see Chapter 10). We will also look briefly at regeneration of heart muscle in zebrafish and the regeneration of peripheral nerves in mammals.

Regeneration requires, at the minimum, the production of a population of cells whose growth, differentiation, and patterning is regulated. A distinction has been drawn between two types of regeneration. In one—**morphallaxis**—there is little new cell division and growth, and regeneration of structure occurs mainly by the repatterning of existing tissue and the re-establishment of boundaries. Regeneration in *Hydra* is a good example of morphallaxis. By contrast, regeneration of a newt limb depends on the growth of completely new, correctly patterned structures, and this is known as **epimorphosis**. Both types of regeneration can be illustrated with reference to the French flag pattern (Fig. 14.3). In morphallaxis, new boundary regions are first established and new positional values are specified in relation to them; in epimorphosis, new positional values are linked to growth from the cut surface. In both cases, the origin of the progenitor cells for the regenerated tissue is a key issue.

Limb and organ regeneration

Urodele amphibians, such as newts and axolotls, show a remarkable capacity for regenerating body structures such as tails, limbs, jaws, and the lens of the eye (see Fig. 14.2). Regeneration of these structures involves cell proliferation and new growth, and is therefore of the epimorphic type. Regeneration of a structure such as an adult vertebrate limb, which contains a variety of fully differentiated cell types in a highly organized arrangement, raises the question of the origin of the cells that give rise to the regenerated structure. Do the differentiated cells dedifferentiate and start dividing? Are there special reserve cells? Do existing cells change their character after dedifferentiation? Damaged insect legs and other appendages can also regenerate. Insect regeneration has been studied mainly in the larval cockroach, which has relatively large legs that are easy to manipulate, and we shall also briefly discuss this system.

14.1 Amphibian limb regeneration involves cell dedifferentiation and new growth

Amputation of a newt limb is followed by a rapid migration of epidermal cells from the edges of the wound to form a covering over the wound surface. This process is essential for subsequent regeneration; if it is prevented by suturing the edges of the stump together, there is no regeneration. A mass of undifferentiated cells called the blastema then forms under the epidermal cap and this is what gives rise to the regenerated limb (Fig. 14.4). The blastema is formed from cells beneath the wound

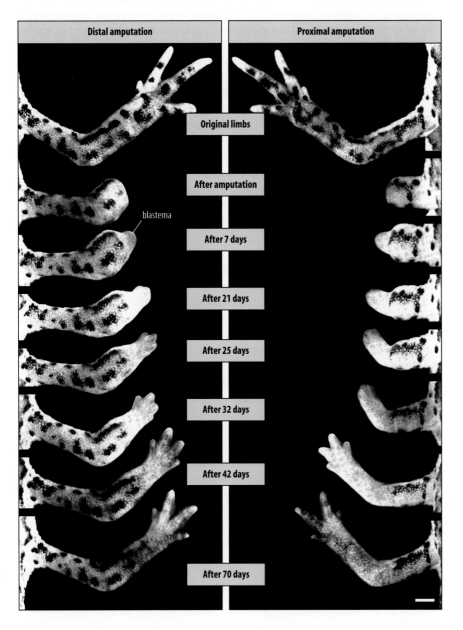

Distal amputation **Proximal amputation**

Original limbs

After amputation

blastema

After 7 days

After 21 days

After 25 days

After 32 days

After 42 days

After 70 days

Fig. 14.4 Regeneration of the forelimb in the red-spotted newt *Notophthalmus viridescens.* The left panel shows the regeneration of a forelimb after amputation at a distal (mid-radius/ulna) site. The right panel shows regeneration after amputation at a proximal (mid-humerus) site. At the top, the limbs are shown before amputation. Successive photographs were taken at the times shown after amputation. Note that the blastema gives rise to structures distal to the cut. Scale bar = 1 mm.

epidermis that lose their differentiated character and start to divide, eventually forming an elongated cone. As the limb regenerates over a period of weeks, the blastema cells differentiate into cartilage, muscle, and connective tissue. The blastema cells are derived locally from the mesenchymal tissues of the stump, close to the site of amputation. They come in particular from the dermis, but have also been shown to derive from cartilage and muscle.

The ability of newt skeletal muscle cells to dedifferentiate and return to a proliferating state is particularly intriguing, as vertebrate skeletal muscle cells do not normally divide after becoming fully differentiated. A general feature of vertebrate muscle differentiation is the withdrawal of muscle precursor cells from the cell cycle after myoblast fusion to produce myotubes. This withdrawal involves the dephosphorylation of the cell-cycle control protein Rb (retinoblastoma protein) (see Section 10.9), and cultured mouse muscle cells lacking Rb can re-enter the cell cycle. In the former muscle cells in the regenerating newt limb, Rb protein is inactivated by phosphorylation, allowing the cells to re-enter the cell cycle and divide. Re-entry into the cell cycle is also associated with the activation of thrombin in the blastema. Indeed, thrombin could be a key factor in regeneration in a variety of systems, as the regeneration of the newt lens from the dorsal margin of the iris also correlates with thrombin activity in that region.

Thrombin is a proteolytic enzyme that is more familiar as part of the blood-clotting cascade, but appears also to be involved in providing an environment for dedifferentiation. Multinucleate post-mitotic newt muscle cells can be converted to dividing mononuclear cells in culture in the presence of thrombin in the culture medium. Dedifferentiation of the muscle cells also involves renewed expression of the homeodomain transcription factor Msx1. This multifunctional transcriptional factor is known to prevent myogenic differentiation in mammalian cells (see Section 10.9), and its expression is characteristic of undifferentiated mesenchymal cells that can undergo regeneration. The question of the origin of regenerated muscle in newts was further complicated when cells functionally similar to the satellite stem cells that regenerate skeletal muscle fibers in adult mammals (see Section 10.10) were identified in adult newts, which had previously been thought to lack such cells.

The next question is whether the cells differentiating into cartilage and muscle in the regenerating newt limb are remaining true to type, or whether muscle satellite cells or dedifferentiated muscle cells can re-differentiate into another type of cell entirely—the process of transdifferentiation (see Section 10.15). Can dedifferentiated skeletal muscle cells in the stump re-differentiate as cartilage, for example? This question was originally addressed by experiments in which cultured newt limb muscle myotubes, which are multinucleate and have stopped dividing, were labeled in culture with a retrovirus expressing alkaline phosphatase, and then introduced into the blastema of a regenerating limb. Strongly labeled mononuclear cells were observed in the blastema a week after, and the majority of the introduced myotubes gave rise to mononuclear cells. After proliferating, the labeled mononuclear cells did appear to give rise to cartilage, as well as to new muscle, suggesting that transdifferentiation had occurred. Altogether, these observations on regenerating newt limbs suggested that the blastema cells were likely to have reverted to a homogeneous multipotent stem-cell state, from which they re-differentiated anew.

However, more recent experiments in the regenerating limb of the axolotl *Ambystoma mexicanum* give a very different picture. These experiments showed that the blastema cells did not revert to a multipotent state but retained a restricted developmental potential related to their origin. The fate of individual tissue types in a regenerating limb was traced by an ingenious technique. The first step was to produce transgenic axolotls that expressed green fluorescent protein (GFP) in all their cells. Next, a patch of tissue of a particular type, such as muscle or epidermis, was transplanted

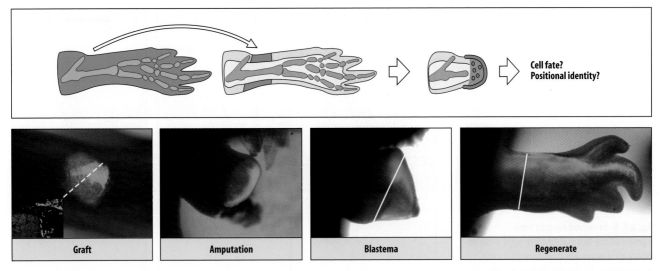

Fig. 14.5 Cells that regenerate the axolotl limb have restricted developmental potential. The experimental procedure is shown in the top row and photographs of the limb at various stages are shown in the bottom row. Cells of a particular tissue type (e.g. dermis as shown here) and position in the limb of a juvenile axolotl transgenic for green fluorescent protein, were transplanted into the limb of a non-transgenic animal and the limb amputated across the graft. In the regenerating limb that grew from the blastema, the transplanted cells and their progeny can be detected by their green fluorescence. These cells could then be tested for their tissue type. In the case illustrated, the transplanted dermis cells could give rise to new dermis and to the cartilage skeleton, but not muscle. When muscle cells were transplanted, they could only give rise to muscle. Cells taken from a proximal position in the limb could give rise to distal structures (as seen here by the labeled cells in the digits), but transplanted cartilage cells would not give rise to more-proximal structures.

Photographs from Kragl, M., et al.: 2009.

from these animals into the forelimbs of non-transgenic juveniles of the same species. The forelimb of the juvenile axolotl was amputated across the site of the transplant, and the fate of the glowing green transplanted cells could then be traced as the limb regenerated (Fig. 14.5). The tissue type of the fluorescent cells after re-differentiation was then identified by testing for tissue-specific markers.

These experiments showed that transplanted muscle tissue produced blastema cells that could only give rise to muscle. The experimental technique could not, however, resolve whether the regenerated muscle derived from satellite cells or from dedifferentiation of muscle cells or from both, as both cell types would have been present in the transplanted tissue. Cartilage and epidermal cells were also restricted in their further development, as were the glial Schwann cells, which provide the myelin coating of peripheral nerve fibers. One exception was the cells of the dermis, which could contribute to both new dermis and new cartilage skeleton, and the potential of fibroblasts to re-differentiate remains unclear. This investigation also found that cartilage-derived blastema cells retained a 'memory' of their original positional identity along the proximo-distal axis of the limb. Cartilage cells transplanted from a distal position in the original limb would not give rise to more proximal regions in the regenerating limb.

Growth of the blastema is dependent both on its nerve supply (Fig. 14.6) and on the overlying wound epidermis, which may play a role similar to the apical ectodermal ridge in limb development (see Section 11.3). In amphibian limbs in which the nerves have been cut before amputation, a blastema forms but fails to grow. The nerves have no influence on the character or pattern of the regenerated structure; it is the amount of innervation, not the type of nerve that matters. In regenerating adult salamander limbs, the nerves have been shown to provide an essential growth factor called anterior gradient protein (nAG), which is secreted first by the glial Schwann cells of the incoming nerves and later by the glandular cells of the wound epidermis. Treatment of the blastema of a denervated limb with nAG is sufficient to enable it to regenerate completely.

A striking instance of the influence of nerves on regeneration of amphibian limbs is that if a major peripheral nerve, such as the sciatic nerve, is cut and the branch inserted into a wound on a limb or on the surface of the adjacent flank, a supernumerary limb develops at that site. This experimental system provides an opportunity to study limb regeneration in the absence of the considerable cell damage caused by amputation. It was found, for example, that expression of the zinc finger transcription factor Sp9 by epidermal keratinocytes in the region of the wound was necessary for their dedifferentiation into an epithelium that could support limb regeneration.

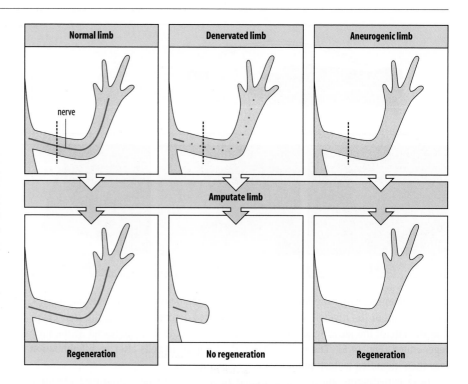

Fig. 14.6 Innervation and limb regeneration. Normal limbs require a nerve supply to regenerate (left panels). Limbs denervated prior to amputation will not regenerate (center panels). However, limbs that have never been innervated, because the nerve was removed during development, can regenerate normally in the absence of innervation (right panels).

An interesting phenomenon, as yet unexplained, is that if embryonic newt limbs are denervated very early in their development, and so do not become exposed to the influence of nerves, they can regenerate in the complete absence of any nerve supply (see Fig. 14.6, right panel). If an aneurogenic limb is subsequently innervated, however, it rapidly becomes dependent on nerves for regeneration. This suggests that dependence on innervation is imposed on the limb only after the nerves grow into it.

The *Xenopus laevis* tadpole can regenerate amputated limbs but, after metamorphosis, the young frog cannot, and if a limb is amputated, just a spike of tissue develops. Experiments investigating the epigenetic regulation of key developmental genes in the regenerating tissue have found that the ZRS (zone of polarizing activity regulatory sequence) of the Sonic hedgehog gene (*Shh*) is highly methylated in the froglet but not in the tadpole (see Box 11B, p. 423). The DNA methylation presumably leads to silencing of *Shh* expression via this enhancer in the froglet, and points to Shh as a key player in vertebrate limb regeneration, as it is in normal limb development (see Chapter 11).

Although mammals cannot regenerate whole limbs, many mammals, including young children, can regenerate the tips of their digits, provided the nail-generating tissue—the nail organ—is still present. In mice and children, the level from which digits are able to regenerate is limited to the base of the claw or nail, respectively. This probably reflects the presence of connective tissue cells under the nail organ that express the transcription factor Msx1, which regulates BMP-4 expression and is associated with undifferentiated mesenchymal cells, as noted earlier. In mammalian limb development, Msx1 is expressed at the tip of the embryonic limb bud, and in mice it continues to be expressed in the tips of the digits even after birth. The region in which digit regeneration in mice can occur corresponds with the expression of Msx1.

14.2 The limb blastema gives rise to structures with positional values distal to the site of amputation

Although proteins active in normal limb development also play a part in limb regeneration, regeneration is unlikely to occur by precisely the same mechanisms as in embryonic development. Regeneration always proceeds in a direction distal to the cut surface,

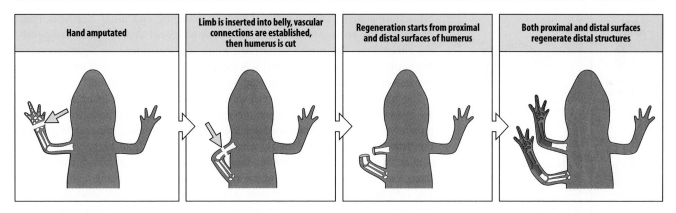

Hand amputated	Limb is inserted into belly, vascular connections are established, then humerus is cut	Regeneration starts from proximal and distal surfaces of humerus	Both proximal and distal surfaces regenerate distal structures

enabling replacement of the lost part. If the hand is amputated at the wrist, only the carpals and digits are regenerated, whereas if the limb is amputated through the middle of the humerus, everything distal to the cut (including the distal humerus) is regenerated. Positional value along the proximo-distal axis is therefore of great importance, and is at least partly retained in the blastema (see Section 14.1). The blastema has considerable morphogenetic autonomy. If it is transplanted to a neutral location that permits growth, such as the dorsal crest of a newt larva or even the anterior chamber of the eye, it gives rise to a regenerated structure appropriate to the position from which it was taken.

The growth of the blastema and the nature of the structures it gives rise to depend on the site of the amputation and not on the nature of the more proximal tissues. The limb is not, however, simply 'trying' to replace missing parts. This was shown in a classic experiment in which the distal end of a newt limb that had been amputated at the wrist was inserted into the belly of the same animal, so as to establish a blood supply to it. The limb was then cut mid-humerus. Both surfaces regenerated distally, even though the part attached to the belly already had a radius and ulna (Fig. 14.7).

The blastema is much larger than the embryonic limb bud—it can be ten times larger in terms of cell numbers. This size makes it most unlikely that there are signals that diffuse right across the blastema. Instead, regeneration can best be understood in terms of an adult limb having a set of positional values along its proximo-distal axis, which have been set up during embryonic development (see Section 11.5). The regenerating limb in some way reads the positional value at the site of the amputation and then regenerates all positional values distal to it. Epimorphic regeneration also involves the retention of embryonic processes, such as the ability to specify new positional values.

The ability of cells to recognize a discontinuity in positional values is illustrated by grafting a distal blastema to a proximal stump. In this experiment, the forelimb's stump and blastema have different positional values, corresponding to shoulder and wrist, respectively. The result is a normal limb in which structures between the shoulder and wrist have been generated by **intercalary growth**, predominantly from the proximal stump, while the cells from the wrist blastema mostly give rise to the hand (Fig. 14.8).

A fundamental question in any discussion of pattern formation is the molecular basis of the proposed positional information. A major advance on this front has been the identification of a cell-surface protein that is expressed in a graded manner along the proximo-distal axis of the newt limb. Proximo-distal identity in the blastema can be altered by treatment with retinoic acid, as discussed in more detail in the next section, and this property was used to compare gene expression in blastemas whose proximo-distal identity had been altered. An almost twofold difference in the level of expression of the gene for a cell-surface protein called Prod1 (the newt equivalent of the mammalian

Fig. 14.7 Limb regeneration is always in the distal direction. The distal end of a limb is amputated and the limb inserted into the belly. Once vascular connections are established, a cut is made through the humerus. Both cut surfaces regenerate the same distal structures even though, in the case of one of the regenerating limbs, distal structures are already present.

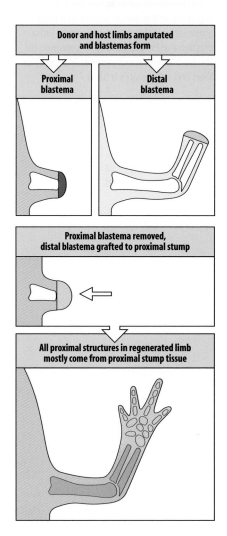

Donor and host limbs amputated and blastemas form

Proximal blastema	Distal blastema

Proximal blastema removed, distal blastema grafted to proximal stump

All proximal structures in regenerated limb mostly come from proximal stump tissue

Fig. 14.8 Proximo-distal intercalation in limb regeneration. A distal blastema grafted to a proximal stump results in intercalation of all the structures proximal to the distal blastema. Almost all of the intercalated region comes from the proximal stump.

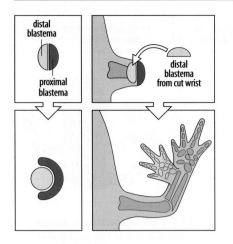

Fig. 14.9 Cell-surface properties vary along the proximo-distal axis. Left panels: when mesenchyme from distal and proximal blastemas is placed in contact in culture, the proximal mesenchyme engulfs the distal mesenchyme, which has greater adhesion between its cells. Right panels: if a distal blastema (in this case from a cut wrist) is grafted to the dorsal surface of a more proximal blastema, the regenerating wrist blastema will move distally to a position on the host limb that corresponds to its original level and regenerates a hand.

cell-surface protein CD59), was found between 'proximal' and 'distal' blastemas. Prod1 might therefore be used by the cells to determine positional identity by short-range cell–cell interactions. The growth factor nAG, which is essential for limb regeneration (see Section 14.1), is a ligand for Prod1. It is thought that nAG has no input into generating proximo-distal pattern, but acts through Prod1 to promote cell division.

The gradient in Prod1 fits very well with the observation that cell-surface properties are involved in regeneration. When mesenchyme from two blastemas from different proximo-distal sites are confronted in culture, the more proximal mesenchyme engulfs the distal (Fig. 14.9, left panels), whereas mesenchyme from two blastemas from similar sites maintains a stable boundary. This behavior is suggestive of a graded difference in cell adhesiveness along the axis, with adhesiveness being highest distally. The cells in the distal explant remain more tightly bound to each other than do the cells from the proximal blastema, which therefore spread to a greater extent. A difference in adhesiveness is also suggested by the behavior of a distal blastema when grafted to the dorsal surface of a proximal blastema so that their mesenchymal cells are in contact. Under these conditions, the distal blastema moves during limb regeneration to end up at the site from which it originated (see Fig. 14.9, right panels). This suggests that its cells adhere more strongly to the regenerated wrist region than they do to the proximal region to which it was transplanted. Transplantation of a shoulder-level blastema to a shoulder stump does not mobilize the stump tissue, but leads to a normal distal outgrowth from the shoulder blastema. All these experiments suggest that proximo-distal positional values in urodele limb regeneration are encoded as a graded property, probably in part at the cell surface, and that cell behavior—growth, movement, and adhesion—relevant to axial specification is a function of the expression of this property, relative to neighboring cells.

Maintaining the continuity of positional values by intercalation is a fundamental property of regenerating epimorphic systems and later in the chapter we will consider it in relation to the cockroach leg. Even normal regeneration by outgrowth from a blastema could be considered the result of intercalation between the cells at the level of amputation and those with the most distal positional values, as specified by the wound epidermis. It is not clear to what extent all blastemal cells inherit a particular positional value from their differentiated precursors, and to what extent they are subject to signals that induce the appropriate expression of positional value. In the axolotl experiment outlined in Fig. 14.5, cartilage cells retained their original positional identity whereas Schwann cells did not.

The precise relationship between Hox gene expression and positional identity is not understood, either for limb embryonic development or regeneration. During limb development in the embryo, the Hoxa genes are expressed with temporal and spatial co-linearity along the proximo-distal axis, as we saw in Chapter 11. In the regenerating axolotl limb, however, two Hoxa genes from the 3′ and 5′ ends of the complex are expressed in the stump cells at the same time—24 to 48 hours after amputation. This suggests that the most distal region of the blastema is specified first and that regeneration involves intercalation of positional values between this distal region and the stump. Grafting experiments also show that distal identity is established early, and that by 4 days the blastema is already divided into distinct zones that will give rise to the proximo-distal regions of the limb.

14.3 Retinoic acid can change proximo-distal positional values in regenerating limbs

We have already seen that retinoic acid is present in developing vertebrate limbs and how experimental treatment with retinoic acid can alter positional values in the developing chick limb (see Section 11.7). It also has striking effects on regenerating amphibian limbs.

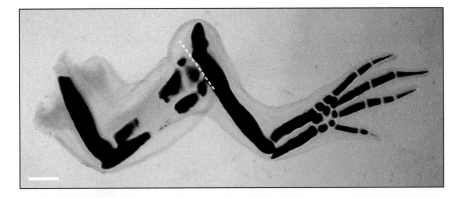

Fig. 14.10 Retinoic acid can proximalize positional values. A forelimb amputated at the level of the hand, as indicated by the dotted line, and then treated with retinoic acid, regenerates structures corresponding to a cut at the proximal end of the humerus. Scale bar = 1 mm.

Exposing a regenerating limb to retinoic acid results in the blastema becoming proximalized; that is, the limb regenerates as if it had originally been amputated at a more proximal site. For example, if a limb is amputated through the radius and ulna, treatment with retinoic acid will result not only in the regeneration of the elements distal to the cut, but also in the production of an extra complete radius and ulna. The effect of retinoic acid is dose-dependent, and with a high dose it is possible to regenerate a whole extra limb, including part of the shoulder girdle, on a limb from which only the hand has been amputated (Fig. 14.10). Retinoic acid, therefore, alters the proximo-distal positional value of the blastema, making it more proximal. It probably does this by affecting various pathways for specifying positional information, in particular increasing the expression of Prod1 (see Section 14.2), as overexpression of Prod1 in distal blastema cells also causes them to translocate to more proximal positions. Retinoic acid also appears to shift positional values in the blastema in a proximal direction through the activation of *meis* homeobox genes, which are involved in specifying proximal identity and are normally repressed in distal regions. Retinoic acid can also, under some experimental conditions, shift positional values along the antero-posterior axis in a posterior direction.

Retinoic acid can act through a variety of nuclear receptors, but only one (RAR δ2) is involved in changes in proximo-distal positional value in the limb. By constructing a chimeric receptor from a RAR δ2 DNA-binding portion and a thyroxine receptor hormone-binding portion, the retinoic acid receptor can be selectively activated by thyroxine, which enables the effects of its activation to be studied experimentally (Fig. 14.11). Cells in the distal blastema are first transfected with the chimeric receptor gene and then grafted to a proximal stump and treated with thyroxine. The transfected

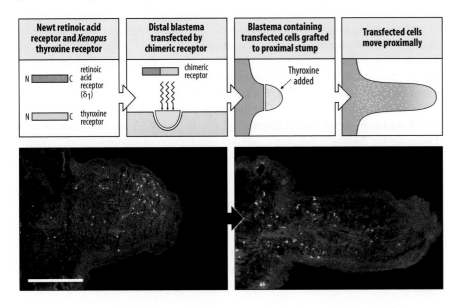

Fig. 14.11 Retinoic acid proximalizes the positional value of individual cells. Some of the cells of a newt distal blastema are transfected by a chimeric receptor, through which retinoic acid receptor function can be activated by thyroxine. This blastema is grafted to a proximal stump and treated with thyroxine. During intercalary growth, the transfected cells, which have been labeled, move proximally because their positional values have been proximalized by the activation of retinoic acid receptor function. The photographs illustrate proximalization of the transfected cells. Scale bar = 0.5 mm.

Photographs from Pecorino, L.T., et al.: 1996.

Fig. 14.12 Retinoic acid can induce additional limbs in the regenerating tail of a frog tadpole. After amputation, the regenerating tail stump of a tadpole of *Rana temporaria* is treated with retinoic acid at the time when hindlimbs are developing. This results in the appearance of additional hindlimbs in place of a new tail. Scale bar = 5 mm.

Photograph courtesy of M. Maden.

normal limb

cells behave as if they have been treated with retinoic acid. The result is a movement of the transfected blastemal cells to more proximal regions in the intercalating regenerating blastema, which shows that activation of the retinoic-acid pathway can proximalize the positional values of the cells, which then respond by movement to more proximal sites. Prod1 is a very strong candidate for mediating this change, as it too can cause cells to move proximally when its concentration is increased.

Another remarkable effect of retinoic acid is its ability to bring about a homeotic transformation of tails into limbs in tadpoles of the frog *Rana temporaria*. If the tail of a tadpole is removed, it will regenerate. But treatment of regenerating tails with retinoic acid at the same time as the hindlimbs are developing results in the appearance of additional hindlimbs in place of a regenerated tail (Fig. 14.12). There is, as yet, no satisfactory explanation for this result, but it has been speculated that the retinoic acid alters the antero-posterior positional value of the regenerating tail blastema to that of the site along the antero-posterior axis where hindlimbs would normally develop.

14.4 Insect limbs intercalate positional values by both proximo-distal and circumferential growth

The legs of some insects, such as the cockroach and cricket, can regenerate. Unlike *Drosophila*, these insects do not undergo a complete metamorphosis, and the juvenile stages (nymphs) are more similar to miniature adults. When the tibia of a third-instar nymph of the cricket *Gryllus bimaculatus* is amputated, it takes about 40 days to leg restore the adult. Wingless and Decapentaplegic signaling activate epidermal growth factor receptor (EGFR) signaling in a distal-to-proximal gradient in the cricket blastema, which patterns the regenerating distal leg.

In *Drosophila*, as we saw in Chapter 13, signaling through the cadherins Fat and Dachsous and the associated Hippo pathway coordinates cell proliferation and apoptosis in imaginal discs, determining their final size and thus the size of adult wings and legs (see Section 13.6). Experiments using RNA interference to investigate the role of Fat, Dachsous, and the Hippo pathway in cricket limb regeneration provide evidence that this seems also to be the case for an insect whose legs develop directly. Knockdown of either Fat or Dachsous resulted in a shorter than normal regenerated limb, whereas knockdown of another set of genes in the Hippo pathway produced a longer leg than normal. It is proposed that a Dachsous/Fat gradient exists along the limb and that the steepness of the gradient could control cell proliferation, reminiscent of the gradient model for determining organ size suggested for *Drosophila* (see Box 13A, p. 514).

Regenerating insect legs follow an epimorphic process of blastema formation and outgrowth that intercalates missing positional values, as described for the proximo-distal axis of the amphibian limb (see Section 14.2), even though the structure of the insect leg is different in many ways from those of vertebrates, especially in having an ectodermally derived exoskeleton. Intercalation of positional values, therefore, seems to be a general property of epimorphically regenerating systems. When cells with disparate positional values are placed next to one another, intercalary growth occurs in order to regenerate the missing positional values. Intercalation is particularly clearly illustrated by limb regeneration in the cockroach, and also occurs in regenerating *Drosophila* leg and wing imaginal discs after amputation.

A cockroach leg is made up of a number of distinct segments, arranged along the proximo-distal axis in the order: coxa, femur, tibia, tarsus. Each segment seems to contain a similar set of proximo-distal and circumferential positional values, and will intercalate missing positional values during regrowth. When a distally amputated tibia is grafted onto a host tibia that has been cut at a more proximal site, localized growth occurs at the junction between graft and host, and the missing central regions of the tibia are intercalated (Fig. 14.13, left panels). In contrast to amphibian regeneration, there is a predominant contribution from the distal piece. As in the amphibian, however, regeneration is a local phenomenon and the cells are indifferent to the overall pattern of the tibia. Thus, when a proximally cut tibia is grafted onto a more distal site, making an abnormally long tibia, regenerative intercalation again restores the missing positional values, making the tibia even longer (Fig. 14.13, right panels). The regenerated portion is in the reverse orientation to the rest of the limb, as indicated by the direction in which the bristles point, suggesting that the gradient in positional values also specifies cell polarity, as in insect body segments (see Chapter 2). These results also show that when

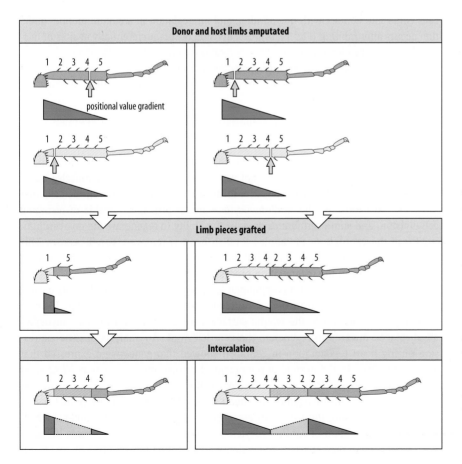

Fig. 14.13 Intercalation of positional values by growth in the regenerating cockroach leg. Left panels: when a distally amputated tibia (5) is grafted to a proximally amputated host (1), intercalation of the positional values 2–4 occurs, irrespective of the proximo-distal orientation of the grafts, and a normal tibia is regenerated. Right panels: when a proximally amputated tibia (1) is grafted to a distally amputated host (4), however, the regenerated tibia is longer than normal and the regenerated portion is in the reverse orientation to normal, as judged by the orientation of surface bristles. The reversed orientation of regeneration is due to the reversal in positional value gradient. The proposed gradient in positional value is shown under each figure.

After French, V., et al.: 1976.

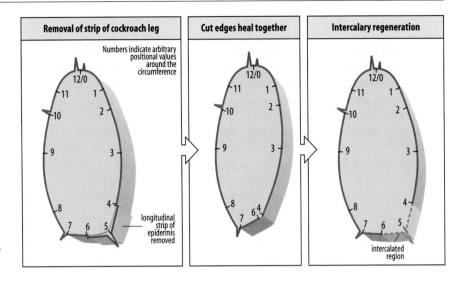

Fig. 14.14 Circumferential intercalation in the cockroach leg. The leg is seen in transverse section. When a piece of cockroach ventral epidermis is removed (left panel), the cut edges heal together (center panel). When the insect molts and the cuticle regrows, circumferential positional values are intercalated (right panel). The positional values are arranged around the circumference of the leg, rather like the hours on a clock face.
After French, V., et al.: 1976.

cells with non-adjacent positional values are placed next to each other, the missing values are intercalated by growth to provide a set of continuous positional values.

A similar set of positional values is present in each segment of the limb. Thus, a mid-tibia amputation, when grafted to the mid-femur of a host, will heal without intercalation. But grafting a distally amputated femur onto a proximally amputated host tibia results in intercalation, largely femur in type. There must be other factors making each segment different, rather like the segments of the insect larva.

Intercalary regeneration also occurs in a circumferential direction. When a longitudinal strip of epidermis is removed from the leg of a cockroach, normally non-adjacent cells come into contact with one another, and intercalation in a circumferential direction occurs after molting (Fig. 14.14). Cell division occurs preferentially at sites of mismatch around the circumference. One can treat positional values in the circumferential direction as a clock face, with values going continuously 12, 1, 2, 3 … 6 … 9 … 11. As in the proximo-distal axis, there is intercalation of the missing positional values.

14.5 Heart regeneration in the zebrafish involves the resumption of cell division by cardiomyocytes

One tissue that does not renew itself in mammals is the cardiac muscle or myocardium. Thus, damaged hearts do not repair themselves like other muscles do. If mammalian heart muscle is damaged, it forms a fibrous scar. However, the cells that make up heart muscle—the cardiomyocytes—are, in principle, capable of resuming cell division, and some vertebrates do have the ability to regenerate the myocardium. If 20% of the ventricle of the adult zebrafish heart is removed, it will regrow. This phenomenon is also seen in newts, but is of particular interest in the zebrafish as in this model organism the regeneration process can be studied genetically.

Unlike blastema formation in amphibian limb regeneration, zebrafish heart muscle regeneration seems not to involve dedifferentiation of the cardiomyocytes. Instead, cardiomyocytes nearest to the wound surface start to divide and the new myocytes are displaced inwards to regenerate the ventricle wall. An analysis of gene expression during heart regeneration in zebrafish showed that genes were expressed that are also involved in the normal development of cardiac muscle, such as *Nkx2.5*, an early marker of prospective heart cells (see Section 11.30). This gene was expressed only at a low level in the regenerating heart muscle, however, suggesting that regeneration was not using the same processes as embryonic heart development, which involves differentiation of cardiomyocytes from undifferentiated precursors. In addition, expression of the zebrafish homeodomain-containing transcription factors MsxB and MsxC (related to Msx1) was increased during regeneration, whereas these factors are not expressed during embryonic heart development.

14.6 The mammalian peripheral nervous system can regenerate

Complete regeneration of major organs is more limited in adult mammals than in fish or amphibians. One organ that can regrow functional tissue is the liver, as we saw in Chapter 13. Another mammalian system with considerable powers of regeneration in adults is the peripheral nervous system. But, unlike in the liver, the regeneration of peripheral nerves involves the regrowth of axons, and not the replacement of the cells themselves by cell division. If a neuronal cell body is destroyed, the neuron cannot regrow and cannot be replaced.

Axons of peripheral neurons in adult vertebrates, such as the motor and sensory axons running between the spinal cord and the ends of the limbs, can be up to hundreds of centimeters long. When such an axon is severed, a new growth cone forms at the cut surface and grows down the pathway of the original nerve trunk to make functional connections, leading to an almost complete recovery of function. In the case of motor neurons, the axon terminal finds the site of the original synapse on the muscle cell by recognizing the basal lamina that fills the synaptic cleft. Axon regrowth is promoted by the presence of the glial Schwann cells, which form the myelin covering peripheral neurons.

In contrast, the central nervous system of adult birds and mammals is not able to regenerate correct axonal and dendritic connections after damage to neurons. This is not due to an intrinsic deficit in the neurons themselves but rather is due to the damaged environment, which does not support, and even actively prevents, axon regeneration. Although the remaining cell body produces new outgrowths, these do not develop further. It seems that the failure of axons to regenerate is partly the result of inhibition by the glial cells of the central nervous system. Culturing immature neurons with oligodendrocytes, glial cells that are the source of myelin in the central nervous system, can induce growth-cone collapse, and astrocytes, another type of central nervous system glial cell, can be equally inhibitory in damaged grey matter. By contrast, when Schwann cells, the myelin-producing glial cells of the peripheral nervous system, are implanted into the central nervous system, they can still promote axon regrowth, with the regenerating axons migrating along the Schwann cells. Another reason for the lack of regrowth in a damaged central nervous system is that debris is cleared much more slowly than in the peripheral nervous system. It is not difficult to think of reasons why new axonal sprouting in the mammalian central nervous system, with its highly complex architecture, is generally inhibited. If axons were allowed to sprout unchecked, that would be very likely to disrupt the elaborate pattern of neuronal connections and destroy brain function.

Stem cells have great promise as sources for introducing new neurons and glia into damaged regions of the central nervous system. Embryonic stem cells (see Section 10.16), for example, can differentiate into neurons when transplanted into the mammalian brain. While such experiments are encouraging, it still needs to be shown that this approach can restore normal function after damage.

Summary

Urodele amphibians can regenerate amputated limbs and tails. Stump tissue at the site of amputation first dedifferentiates to form a blastema, which then grows and gives rise to a regenerated structure. The dedifferentiated cells of the blastema are the progenitors of the various cell types within the regenerated limbs. Regeneration is usually dependent on the presence of nerves, but limbs that have never been innervated can regenerate. Regeneration always gives rise to structures with positional values more distal than those at the site of amputation. When a blastema is grafted to a stump with different positional values, proximo-distal intercalation of the missing positional values occurs. Positional values may be related to a proximo-distal gradient in a cell-surface protein. Retinoic acid changes the positional values of the cells of the blastema, giving them more proximal values. Insect limbs can also regenerate, with intercalation of missing positional values occurring in both proximo-distal and circumferential directions. Among vertebrate internal organs, liver regenerates in mammals but heart muscle does not, although it can do so in amphibians and fish. Central nervous system neurons do not regenerate in adult mammals and birds after damage, but peripheral nerves can regrow axons and reform connections, although if the cell body is damaged the neuron cannot be replaced.

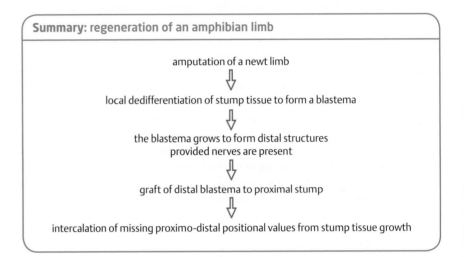

Summary: regeneration of an amphibian limb

amputation of a newt limb
⇩
local dedifferentiation of stump tissue to form a blastema
⇩
the blastema grows to form distal structures
provided nerves are present
⇩
graft of distal blastema to proximal stump
⇩
intercalation of missing proximo-distal positional values from stump tissue growth

Regeneration in *Hydra*

Hydra is a freshwater cnidarian consisting of a hollow tubular body about 0.5 cm long, with a head region at one end (the distal end) and a basal region at the other (proximal) end, by which it sticks to a surface (Fig. 14.15). The head consists of a small conical hypostome where the mouth opens, surrounded by a set of tentacles that are used for catching the small animals on which *Hydra* feeds. Unlike most of the animals discussed in this book, which have three germ layers, *Hydra* has only two. The body wall is composed of an outer epithelium, which corresponds to the ectoderm, and an inner epithelium lining the gut cavity, which corresponds to the endoderm. These two layers are separated by a gel-like basement membrane or mesoglea. There are about 20 different cell types in *Hydra*, which include highly specialized types such as neurons, secretory cells, and the nematocytes that are used to capture prey. Stem cells called interstitial cells give rise to these specialized cells in normal growth. *Hydra* has no central nervous system but has a network of neurons distributed throughout the body. The reason for looking at patterning and regeneration in *Hydra* is that it provides insight into an organizer region and developmental gradients that arose early in the evolution of animal development. It is likely that the more complex body patterns of other animals evolved from a simple body plan like that of *Hydra*.

14.7 *Hydra* grows continuously but regeneration does not require growth

Well-fed *Hydra* are in a dynamic state of continuous growth and pattern formation. Asexual reproduction is by budding, but in hard times *Hydra* can reproduce sexually. The cells of both epithelial layers proliferate steadily and, as the tissues grow, cells are displaced

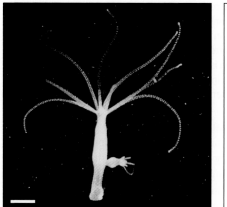

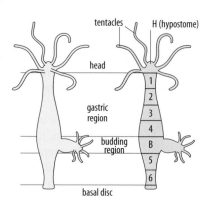

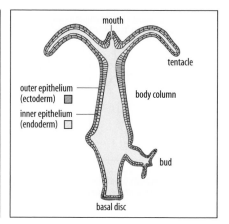

along the body column toward the head or foot. In order for an adult *Hydra* to maintain a constant size, excess cells must be continually lost. Cell loss occurs at the tips of the tentacles and at the basal disc of the foot bud. Most of the excess cell production is taken up by the asexual budding of new *Hydra* from the body column. Budding occurs about two-thirds of the way down the body (see Fig. 14.15); the body wall evaginates by a morphogenetic change in cell shape, generated locally by the cells in the bud region, to form a new column that develops a head at the end and then detaches as a small new *Hydra*.

Continuous growth in *Hydra* means that cells are continually changing their relative positions and are forming new structures as they move up or down the body column. There must, therefore, be mechanisms for repatterning cells during this dynamic process. It is these mechanisms that give *Hydra* its remarkable capacity for regeneration. If the body column of a *Hydra* is cut transversely, the lower piece will regenerate a head and the upper piece will regenerate a foot. Thus, what structure the cells regenerate at a cut surface depends on their relative position within the regenerating piece. The cut surface nearest the original head end forms a head—this shows that *Hydra* has a well-defined overall polarity. This polarity is maintained even in small pieces of the body and this can be seen when a short piece is cut out of the body column: the distal end regenerates a head while the proximal end becomes the basal disc.

Regeneration in *Hydra* does not require cell division and new growth and is therefore an example of morphallactic regeneration. When a short fragment of the column regenerates, there is no initial increase in size and the regenerated animal will be a small *Hydra*. Only after feeding will the animal return to a normal size. The lack of a growth requirement for regeneration is shown in heavily irradiated *Hydra*; no cell divisions occur in these animals but they can still regenerate more or less normally. *Hydra* lacking interstitial cells also regenerate normally. Even more dramatically, a mixture of dissociated *Hydra* cells will form an animal.

The basis for the enormous regenerative capacity of *Hydra* may reside in the stem-cell-like nature of its epithelial cells. These cells are continually self-renewing and do not seem to age, and they are the only cells that can stimulate regeneration. The epithelial cells also give rise to various cell types, such as those of the basal disc and the gut lining, with the type of cell produced being determined by positional values along the body axis.

14.8 The head region of *Hydra* acts both as an organizing region and as an inhibitor of inappropriate head formation

At the beginning of the twentieth century, it was found that grafting a small fragment of the hypostome region of a *Hydra* into the gastric region of another *Hydra* induced a new head, complete with tentacles, and a body axis (Fig. 14.16). Similarly,

Fig. 14.15 *Hydra*. This freshwater cnidarian has a head with tentacles and mouth at one end and a sticky foot at the other (photograph, left panel). It can reproduce by budding. For the purposes of grafting experiments, the body column is divided into a series of regions, as indicated in the center panel. The body wall is made up of two epithelial layers, corresponding to the ectoderm and endoderm found in other organisms featured in this book (right panel). Scale bar = 1 mm.

Photograph courtesy of W. Müller from Müller, W.A.: 1989.

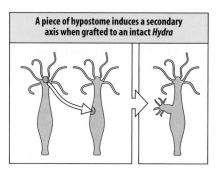

Fig. 14.16 The hypostome can induce a new head and body in *Hydra*. When an excised fragment of hypostome is grafted into the gastric region of another, intact *Hydra*, it can induce the formation of a complete new secondary axis with a head and tentacles.

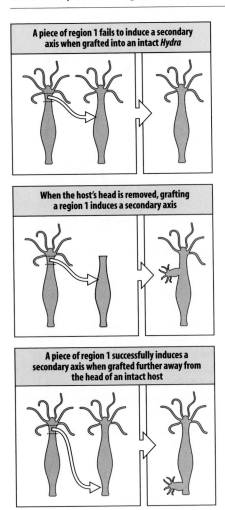

A piece of region 1 fails to induce a secondary axis when grafted into an intact *Hydra*

When the host's head is removed, grafting a region 1 induces a secondary axis

A piece of region 1 successfully induces a secondary axis when grafted further away from the head of an intact host

Fig. 14.17 The head region of *Hydra* produces an inhibitory signal that falls off with distance. Region 1 does not induce a head when grafted into the gastric region of an intact *Hydra* (top panel), indicating the presence of an inhibitory signal in the host. If the host's head is removed and a piece of region 1 is then grafted into the host, a secondary axis is induced (middle panel), indicating that the head region is the source of the inhibitory signal. Region 1 can induce a new axis in the foot region of an intact *Hydra* because the inhibitory signal grows weaker further away from the head (bottom panel).

transplantation of a fragment of the basal region induced a new body column with a basal disc at its end. *Hydra*, therefore, has two organizing region, one at each end, which give the animal its overall polarity. The hypostome and the basal disc act as organizing regions (like the Spemann organizer in amphibians and the polarizing region in vertebrate limb buds). The organizing function of the head region is due to two signals it produces which act in graded fashion down the body column: one signal inhibits head formation and the other specifies a gradient in positional value.

Grafting experiments originally showed that the hypostome produces an inhibitor of head formation (Fig. 14.17). For example, when a body piece from just below the head (region 1 in Fig. 14.15) is grafted into the gastric region, it rarely induces a new head and is usually simply absorbed into the body. But if the head of the host is removed at the time of grafting, the graft will then induce a new axis and head, suggesting that the removal of the head results in the loss of some factor that is inhibiting head formation. The inhibitory effect falls off with distance from the head: when a region 1 is grafted to near the foot it can induce a new head even when the original head is still in place (see Fig. 14.17, bottom panel). These experiments suggested that the inappropriate formation of extra heads is normally prevented in *Hydra* by a lateral inhibition mechanism (see Section 1.16) acting through a gradient of inhibitory signal with its highest concentration at the head end. An opposing gradient inhibiting foot regeneration seems to be produced by the basal disc. These gradients are dynamic and must be continually under adjustment as the animal grows; this is shown by the fact that when the head is removed, the concentration of the inhibitor falls.

A simple model for the two gradients emanating from the head assumes that the head inhibitor is a secreted factor, made by the head, which diffuses down the body column and is degraded at the basal end. The gradient in positional value, on the other hand, is assumed to be an intrinsic property of cells. In this model, both gradients are linear—their values decrease at a constant rate with distance from the head. We further assume that, provided the level of inhibitor is greater than the threshold set by the positional value, head regeneration is inhibited. Removal of the head results in the concentration of inhibitor falling, as the inhibitor is degraded and cannot be replaced. The decrease in inhibitor concentration is greatest at the cut end, and when the inhibitor falls below the threshold concentration set by the local positional value, the positional value increases to that of the head end (Fig. 14.18). Thus, the first key step in this morphallactic regeneration, when the head region is removed, is the specification of a new head region at the cut surface.

When the positional value has increased to that of a normal head region, the cells start to make inhibitor and so prevent head formation in other body regions. The level of inhibitor will always first fall below the threshold for head inhibition where the positional

Fig. 14.18 A simplified model for head regeneration in *Hydra*. This model assumes that there are two gradients running from head to foot. One is a diffusible molecule (I) that inhibits head formation and is produced by the head. The other is a gradient of positional value (P) that is an intrinsic property of the cells. When the head is removed, the concentration of inhibitor falls at the cut surface (graph 2) until a threshold is reached at which the positional value increases to that of the head region (graph 3). This then re-establishes the inhibitor gradient (graph 4). The overall gradient in positional value takes a much longer time to return to normal (graph 5).

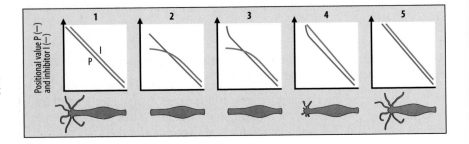

value is highest, thus maintaining polarity. Once a new head has been specified and the inhibitory gradient re-established, the gradient in positional value also returns to normal, but this can take more than 24 hours. Morphallactic regeneration results in a smaller *Hydra* which, after feeding, will eventually grow back to a normal size.

14.9 Genes controlling regeneration in *Hydra* are similar to those expressed in vertebrate embryos

Several novel peptides have been identified in *Hydra* that are linked to epithelial cell positional differentiation along the body axis and which can induce a head or a basal disc. In addition, genes identified as important in vertebrate embryo development seem to be involved in development in *Hydra*. The ubiquitous canonical Wnt pathway leading to the activation of β-catenin (see Box 1E, p. 26) has an important role in *Hydra* development. *In situ* hybridization experiments revealed that expression of HyWnt, the *Hydra* Wnt homolog, in adult animals is restricted to the most apical tip of the body axis, the head organizer (Fig. 14.19). Within an hour after amputation of this region, both *Hydra* β-catenin (Hyβ-Cat) and Wnt genes are expressed at the regenerating tip. During budding also, expression of Hyβ-Cat is significantly upregulated in a ring-like domain in the prospective budding zone before evagination starts, and the high level of expression is maintained until the bud detaches. A dramatic demonstration of the importance of the Wnt signaling pathway comes from blocking the activity of glycogen synthase kinase-3β, a component of the pathway (see Box 1E, p. 26). The block causes a rise in the concentration of Hyβ-Cat along the body axis and all regions acquired the characteristics of the head organizer, as shown by grafting individual regions to another organism. These results show that Wnt signaling is involved in axis formation in *Hydra* and support the idea that it played a key part in the evolution of axial differentiation in early multicellular animals. Hyβ-Cat shows a high degree of conservation, as injection of *Hyβ-Cat* mRNA into ventral *Xenopus* blastomeres at the eight-cell stage is able to induce a complete *Xenopus* secondary body axis. This contains the anterior-most structures, such as eyes and cement gland, and is indistinguishable from the secondary axis induced by *Xenopus* β-catenin.

Various genes expressed in the vertebrate organizer are also expressed in the *Hydra* organizing region. A protein related to vertebrate Chordin is expressed at the site of regeneration or budding in *Hydra*, and might block BMP signaling, as it does in dorso-ventral specification in vertebrates. A *Hydra* homolog of *goosecoid*, another organizer-specific gene, is expressed just above where tentacles will appear. When

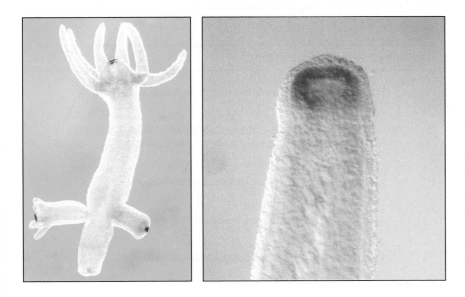

Fig. 14.19 Wnt expression in adult *Hydra* and in a regenerating tip. Left panel: adult *Hydra* stained for Wnt expression (black), shows expression only the tip of the hypostome. Right panel: at 1 hour after hypostome amputation, Wnt is expressed throughout the site of amputation. After 48 hours, expression is again restricted to the very tip of the regenerating head (not shown).

Photographs courtesy of T.W. Holstein.

injected into an early *Xenopus* embryo, the *Hydra* version of *goosecoid* can induce a partial secondary axis. This suggests that the roles of those genes in organizer tissues have been retained over many hundreds of millions of years of evolution.

Summary

Hydra grows continually, losing cells from its ends and through the formation of buds. Two organizing regions, one at the head end and one at the basal end, pattern the body and maintain polarity. If transplanted to another site, a head region can induce a new body axis and head. If the body of an intact *Hydra* is severed in two, it regenerates by morphallaxis, which does not require new growth. Regeneration initially involves the respecification of cells at the cut end as 'head,' leading to the establishment of an organizing region. The head region produces an inhibitor that prevents other regions forming a head. The concentration of inhibitor decreases with distance from the head. There is also a gradient in positional value specified by the head that determines the threshold at which head inhibition occurs. When the head is removed, the inhibitor level in the rest of the body falls, and a new head region develops where the positional value is highest, thus maintaining polarity. The expression of genes in the *Hydra* head organizer homologous with those in the organizer region of vertebrates suggests an ancient evolutionary origin for the organizer.

Summary to Chapter 14

Regeneration is the ability of an adult organism to replace a lost part of its body. The capacity to regenerate varies greatly among different groups of organisms and even between different species within a group. Mammals have very limited powers of regeneration, whereas newts can regenerate limbs, jaws, and the lens of the eye. Regeneration of amphibian limbs involves dedifferentiation of cells and the formation of a blastema, which grows and forms the regenerate. Regeneration requires the presence of nerves. Evidence exists for a proximo-distal gradient in positional values as intercalation of positional values occurs when normally non-adjacent values are placed next to each other in amphibian and insect limbs. The zebrafish heart can regenerate, as can the mammalian liver and the peripheral nervous system. Regeneration in the cnidarian *Hydra* does not require growth. The head end produces an inhibitory signal for head formation and a signal that specifies positional value along the body. When the head is removed, the inhibitor concentration falls and a new head forms. *Hydra* may provide a model for an evolutionarily early developmental system.

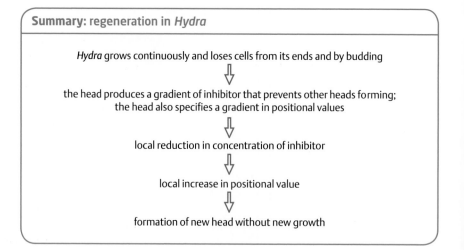

Summary: regeneration in *Hydra*

Hydra grows continuously and loses cells from its ends and by budding
⇩
the head produces a gradient of inhibitor that prevents other heads forming; the head also specifies a gradient in positional values
⇩
local reduction in concentration of inhibitor
⇩
local increase in positional value
⇩
formation of new head without new growth

■ End of chapter questions

Long answer (concept questions)

1. Contrast morphallaxis and epimorphosis. Give an example of each.

2. What is the effect of denervation on regeneration of a salamander limb? What is the role of the nerve supply? What protein can substitute for the nerves in enabling regeneration?

3. Refer to Fig. 14.7. Describe in your own words what the experiment illustrated was intended to investigate, what was observed, and what interpretation can be drawn from results of the experiment.

4. The dedifferentiated cells of the salamander blastema will regenerate only structures distal to the site of the amputation, which requires some kind of memory of proximo-distal position in the intact limb. Summarize the information that Prod1 may be involved in registering proximo-distal position; be sure to include the results of retinoic acid treatment on Prod1 expression.

5. Describe the effect of retinoic acid on regeneration; for example, if a limb is cut at the wrist level and treated with retinoic acid, how will its regeneration be altered, compared to regeneration without such treatment?

6. In the regeneration of the cockroach leg, proximo-distal values will be restored through intercalation if two pieces of leg are apposed by grafting (see Fig. 14.13). How does this process of intercalation explain the results of the experiments shown in the figure?

7. Describe what is meant by 'circumferential intercalation' in cockroach limb regeneration. How does this differ from proximo-distal intercalation?

8. Discuss the importance to biomedical science of the ability of portions of the adult zebrafish heart to regenerate (you may wish to review some of the advantages of the zebrafish as a model organism from Chapter 3).

9. What are the limits of regeneration in the adult human liver—that is, how much can be removed or damaged and the liver will still regrow? What signals promote liver regeneration? What signal limits further growth after the liver reaches a size appropriate to the organism?

10. Contrast the effect of glial Schwann cells on regeneration in the peripheral nervous system with the effect of the oligodendrocyte glial cells on regeneration in the central nervous system.

11. In what way do the hypostome and basal disc regions of a hydra qualify as organizers? Describe an experiment that illustrates this property.

12. Describe the two-gradient model for head regeneration in *Hydra*.

Multiple choice (factual recall questions)

NB There is only one correct answer to each question.

1. Epimorphosis is regeneration through

a) repatterning of existing cells without a requirement for new growth, as occurs in *Hydra*

b) repatterning of existing cells without a requirement for new growth, as occurs in salamanders

c) the reinitiation of division in existing cells and new growth, followed by patterning, as occurs in salamanders

d) the reinitiation of division in existing cells and new growth, followed by patterning, as occurs in *Hydra*

2. Which statement is true about regeneration in the newt eye?

a) The entire eye of a newt can be removed and will regenerate from the stump of the optic nerve.

b) The lens can be removed, and will regenerate from the overlying cornea.

c) The lens can be removed, and will regenerate via transdifferentiation from the iris.

d) The newt eye does not show any power of regeneration.

3. Msx1 is

a) a growth factor that acts on the blastema to stimulate regeneration

b) a proteolytic enzyme involved in the blood-clotting cascade

c) a regulator of the cell cycle

d) a transcription factor expressed in undifferentiated mesenchymal cells

4. If the nerve supply to a newt limb is severed before amputation, how will this affect regeneration?

a) A blastema will form but will not grow, and regeneration will fail.

b) No regeneration occurs, and the stump heals over as it would in a mammal.

c) Outgrowth will occur, but the identity of the limb will be lost and normal proximo-distal patterning will not occur.

d) Regeneration of most tissues will occur normally, but regeneration of the nerves will not occur.

5. Which is the result of grafting a blastema derived from a distal amputation onto a stump that has been amputated at a proximal position?

a) The distal blastema is unable to properly interpret its positional values in this circumstance, and regeneration fails.

b) The distal blastema regenerates the entire limb, supplying all the cells required for this regeneration.

c) The distal blastema regenerates the limb regions removed in the distal amputation, resulting in a shortened limb missing the regions between the proximal and distal amputation sites.

d) The limb regions between the distal and proximal amputation sites are filled in by intercalary growth from the proximal stump, and the regions distal to the distal amputation are regenerated from the distal blastema.

6. In the amputated newt limb, what is the effect of retinoic acid treatment of the regenerating stump on the blastema cells' positional information?

a) Proximal cells will behave as if they are more distal than they were before the amputation.

b) Retinoic acid will have no effect on regeneration.

c) The cells will become respecified to a more proximal position.

d) The cells will lose their positional identity and remain dedifferentiated in the blastema.

7. How does regeneration in the peripheral nervous system differ from that in the liver?

a) In neuronal regeneration, axons can regrow, but the neuron itself cannot be replaced by cell division of remaining cells, as can liver cells.

b) In regeneration in the peripheral nervous system, cells that are removed or damaged are replaced from stem cell populations, whereas in the liver cells are replaced from differentiated liver cell populations.

c) In the liver, growth factors stimulate regeneration, whereas in the peripheral nervous system no growth factors are required since the neurons are continually dividing.

d) Regeneration cannot occur in the peripheral nervous system, in contrast to the ability of the liver to regenerate when damaged.

8. Which statement correctly characterizes regeneration in the mammalian nervous system?

a) Cells of the brain can regenerate if damaged, but the peripheral nervous system is incapable of regeneration.

b) Killed cells of the peripheral nervous system are replaced by division of stem cells, which will extend axons to restore connectivity with targets.

c) Peripheral nerves can regenerate an axon if cut, and make correct connections with muscles and other targets in some cases.

d) The glial cells of the central nervous system support axon outgrowth and regeneration, but the glial cells of the peripheral nervous system seem to actually suppress nerve regeneration.

9. Why is *Hydra* a useful animal for studying development and regeneration?

a) An extensive set of developmental mutants has been developed in *Hydra*, making possible a genetic analysis of development and regeneration.

b) *Hydra* gives insights into mechanisms that arose very early in the evolution of animals.

c) *Hydra* illustrates unique mechanisms, such as the existence of an organizer region, and the use of morphogenetic gradients, which are different from any other animal.

d) *Hydra* is a simple organism that nonetheless has all the tissues and cell types of more complex organisms such as mammals.

10. A highly conserved signaling pathway involved in *Hydra* axis formation is

a) the FGF pathway

b) the insulin pathway

c) the retinoic acid pathway

d) the Wnt/β-catenin pathway

Multiple choice answer key

1: c, 2: c, 3: d, 4: a, 5: d, 6: c, 7: a, 8: c, 9: b, 10: d.

■ **General further reading**

Birnbaum, K.D., Sánchez Alvarado, A.: **Slicing across kingdoms: regeneration in plants and animals**. *Cell* 2008, **132**: 697–710.

Brockes, J.P., Kumar, A.: **Principles of appendage regeneration in adult vertebrates and their implications for regenerative medicine**. *Science* 2005, **310**: 1919–1923.

Brockes, J.P., Kumar, A.: **Comparative aspects of animal regeneration**. *Annu. Rev. Cell Dev. Biol.* 2008, **24**: 525–549.

Galliot, B., Tanaka, E., Simon, A.: **Regeneration and tissue repair**. *Cell. Mol. Life Sci.* 2008, **65**: 3–7.

■ **Section further reading**

14.1 Amphibian limb regeneration involves cell dedifferentiation and new growth

Brockes, J.P., Kumar, A.: **Plasticity and reprogramming of differentiated cells in amphibian regeneration**. *Nat. Rev. Mol. Cell Biol.* 2002, **3**: 566–574.

Han, M., Yang, X., Farrington, J.E., Muneoka, K.: **Digit regeneration is regulated by *Msx1* and BMP4 in fetal mice**. *Development* 2003, **130**: 5123–5132.

Imokawa, Y., Simon, A., Brockes, J.P.: **A critical role for thrombin in vertebrate lens regeneration**. *Phil. Trans. R. Soc. Lond. B Biol. Sci.* 2004, **359**: 765–776.

Kragl, M., Knapp, D., Nacu, E., Khattak, S., Maden, M., Epperlein, H.H., Tanaka, E.M.: **Cells keep a memory of their tissue origin during axolotl limb regeneration**. *Nature* 2009, **460**: 60–65.

Kumar, A., Godwin, J.W., Gates, P.B., Garza-Garcia, A.A., Brockes, J.P.: **Molecular basis for the nerve dependence of limb regeneration in an adult vertebrate**. *Science* 2007, **318**: 772–777.

Kumar, A., Velloso, C.P., Imokawa, Y., Brockes, J.P.: **The regenerative plasticity of isolated urodele myofibers and its dependence on MSX1**. *PLoS Biol.* 2004, **2**: E218.

Lee, H., Habas, R., Abate-Shen, C.: **Msx1 cooperates with histone H1b for inhibition of transcription and myogenesis**. *Science* 2004, **304**: 1675–1678.

Satoh, A., Graham, G.M.C., Bryant, S.V., Gardiner, D.M.: **Neurotrophic regulation of epidermal dedifferentiation during wound healing and limb regeneration in the axolotl (*Ambystoma mexicanum*)**. *Dev. Biol.* 2008, **319**: 321–355.

Tanaka, E.M., Drechel, D.N., Brockes, J.P.: **Thrombin regulates S-phase re-entry by cultured newt myotubes**. *Curr. Biol.* 1999, **9**: 792–799.

Yakushiji, N., Suzuki, M., Satoh, A., Sagai, T., Shiroishi, T., Kobayashi, H., Sasaki, H., Ide, H., Tamura, K.: **Correlation between Shh expression and DNA methylation status of the limb-specific Shh enhancer region during limb regeneration in amphibians**. *Dev. Biol.* 2007, **312**: 171–182.

14.2 The limb blastema gives rise to structures with positional values distal to the site of amputation

Da Silva, S., Gates, P.B., Brockes, J.P.: **New ortholog of CD59 is implicated in proximodistal identity during amphibian limb regeneration**. *Dev. Cell* 2002, **3**: 547–551.

Echeverri, K., Tanaka, E.M.: **Proximodistal patterning during limb regeneration**. *Dev. Biol.* 2005, **279**: 391–401.

Garza-Garcia, A.A., Driscoll, P.C., Brockes, J.P.: **Evidence for the local evolution of mechanisms underlying limb regeneration in salamanders**. *Integr. Comp. Biol.* 2010, doi:10.1093/icb/icq022.

Han, M., Yang, X., Lee, J., Allan, C.H., Muneoka, K.: **Development and regeneration of the neonatal digit tip in mice.** *Dev. Biol.* 2008, **315**: 125–135.

Torok, M.A., Gardiner, D.M., Shubin, N.H., Bryant, S.V.: **Expression of HoxD genes in developing and regenerating axolotl limbs.** *Dev. Biol.* 1998, **200**: 225–233.

14.3 Retinoic acid can change proximo-distal positional values in regenerating limbs

Maden, M.: **The homeotic transformation of tails into limbs in *Rana temporaria* by retinoids.** *Dev. Biol.* 1993, **159**: 379–391.

Mercader, N., Tanaka, E.M., Torres, M.: **Proximodistal identity during limb regeneration is regulated by Meis homeodomain proteins.** *Development* 2005, **132**: 4131–4142.

Nakamura, T., Mito, T., Miyawaki, K., Ohuchi, H., Noji, S.: **EGFR signaling is required for re-establishing the proximodistal axis during distal leg regeneration in the cricket *Gryllus bimaculatus* nymph.** *Dev. Biol.* 2008, **319**: 46–55.

Pecorino, L.T., Entwistle, A., Brockes, J.P.: **Activation of a single retinoic acid receptor isoform mediates proximo-distal respecification.** *Curr. Biol.* 1996, **6**: 563–569.

Scadding, S.R., Maden, M.: **Retinoic acid gradients during limb regeneration.** *Dev. Biol.* 1994, **162**: 608–617.

14.4 Insect limbs intercalate positional values by both proximo-distal and circumferential growth

Bando, T., Mito, T., Maeda, Y., Nakamura, T., Ito, F., Watanabe, T., Ohuchi, H., Noji, S.: **Regulation of leg size and shape by the Dachsous/Fat signalling pathway during regeneration.** *Development* 2009, **136**: 2235–2245.

French, V.: **Pattern regulation and regeneration.** *Phil. Trans. R. Soc. Lond. B Biol. Sci.* 1981, **295**: 601–617.

14.5 Heart regeneration in the zebrafish involves the resumption of cell division by cardiomyocytes

Poss, K.D., Wilson, L.G., Keating, M.T.: **Heart regeneration in zebrafish.** *Science* 2002, **298**: 2188–2190.

Raya, A., Consiglio, A., Kawakami, Y., Rodriguez-Esteban, C., Izpisúa-Belmonte, J.C.: **The zebrafish as a model of heart regeneration.** *Cloning Stem Cells* 2004, **6**: 345–351.

14.6 The mammalian peripheral nervous system can regenerate

Case, L.C., Tessier-Lavigne, M.: **Regeneration of the adult central nervous system.** *Curr. Biol.* 2005, **15**: R749–R753.

Clavien, P.A.: **Liver regeneration: a spotlight on the novel role of platelets and serotonin.** *Swiss Med. Wkly.* 2008, **138**: 361–370.

Johnson, E.O., Zoubos, A.B., Soucacos, P.N.: **Regeneration and repair of peripheral nerves.** *Injury* 2005, **36**: S24–S29.

Taub, R.: **Liver regeneration: from myth to mechanism.** *Nat. Rev. Mol. Cell Biol.* 2004, **5**: 836–847.

14.7 *Hydra* grows continuously but regeneration does not require growth

Bosch, C.G.: **Why polyps regenerate and we don't: towards a cellular and molecular framework for *Hydra* regeneration.** *Dev. Biol.* 2007, **303**: 421–433.

Hicklin, J., Wolpert, L.: **Positional information and pattern regulation in *Hydra*: the effect of gamma-radiation.** *J. Embryol. Exp. Morph.* 1973, **30**: 741–752.

Otto, J.J., Campbell, R.D.: **Tissue economics of *Hydra*: regulation of cell cycle, animal size and development by controlled feeding rates.** *J. Cell Sci.* 1977, **28**: 117–132.

14.8 The head region of *Hydra* acts both as an organizing region and as an inhibitor of inappropriate head formation

Bosch, T.C.G., Fujisawa, T.: **Polyps, peptides and patterning.** *BioEssays* 2001, **23**: 420–427.

Brown, M., Bode, H.R.: **Characterization of the head organizer in hydra.** *Development* 2002, **129**: 875–884.

Müller, W.A.: **Pattern formation in the immortal *Hydra*.** *Trends Genet.* 1996, **12**: 91–96.

Wolpert, L., Hornbruch, A., Clarke, M.R.B.: **Positional information and positional signaling in *Hydra*.** *Am. Zool.* 1974, **14**: 647–663.

14.9 Genes controlling regeneration in *Hydra* are similar to those expressed in vertebrate embryos

Broun, M., Gee, L., Reinhardt, B., Bode, H.R.: **Formation of the head organizer in Hydra involves the canonical Wnt pathway.** *Development* 2005, **132**: 2907–2916.

Galliot, B.: **Conserved and divergent genes in apex and axis development in cnidarians.** *Curr. Opin. Genet. Dev.* 2000, **19**: 629–637.

Rentzsch, F., Guder, C., Vocke, D., Hobmayer, B., Holstein, T.W.: **An ancient chordin-like gene in organizer formation of *Hydra*.** *Proc. Natl Acad. Sci. USA* 2007, **104**: 3249–3254.

Takahashi, T., Fujisawa, T.: **Important roles for epithelial cell peptides in hydra development.** *BioEssays* 2009, **31**: 610–619.

15

Evolution and development

■ The evolution of
  development

■ The evolutionary
  modification of embryonic
  development

■ Changes in the timing of
  developmental processes

The evolution of multicellular organisms is fundamentally linked to embryonic development, for it is through development that genetic changes cause changes in body form that can be passed on to future generations. This raises important questions as to how evolutionary and developmental processes are linked. How did development itself evolve? How has embryonic development been modified during animal evolution? How important are changes in the timing of developmental processes? In this chapter, we shall discuss these general questions using particular examples from animal evolution.

From paleontological, molecular, and cellular evidence, the multicellular animals—the Metazoa—are presumed to descend from a common ancestor that was itself multicellular and that had, in turn, evolved from a unicellular organism. A simplified view of the evolution of animals and plants is shown in Fig. 15.1. The history of evolution is a long one, taking place over thousands of millions of years, and it can only be accessed indirectly—through the study of fossils and by comparisons between living organisms. As Charles Darwin was the first to realize, evolution is the result of heritable changes in life forms and the selection of those that are best adapted to their environment (Box 15A, p. 558).

Development is a fundamental process in evolution. The evolution of multicellular forms of life is the result of changes in embryonic development, and these in turn are entirely due to heritable changes in genes that control cell behavior in the embryo. It is also true, however, as the evolutionary biologist Theodosius Dobzhansky once said, that nothing in biology makes sense unless viewed in the light of evolution. Certainly, it would be very difficult to make sense of many aspects of development without an evolutionary perspective. For example, in our consideration of vertebrate development we have seen how, despite different modes of very early development, all vertebrate embryos develop through a rather similar stage, after which their development diverges again (see Fig. 3.2). This shared stage, which is the embryonic stage after neurulation and the formation of the somites, is a stage through which some distant ancestor of the vertebrates passed. It has persisted ever since, to become a fundamental characteristic of the development of all vertebrates, whereas the stages before and after this stage have evolved differently in different organisms.

Genetically based changes in development that generated more successful modes of reproduction or adult forms better adapted to their environment, have been selected

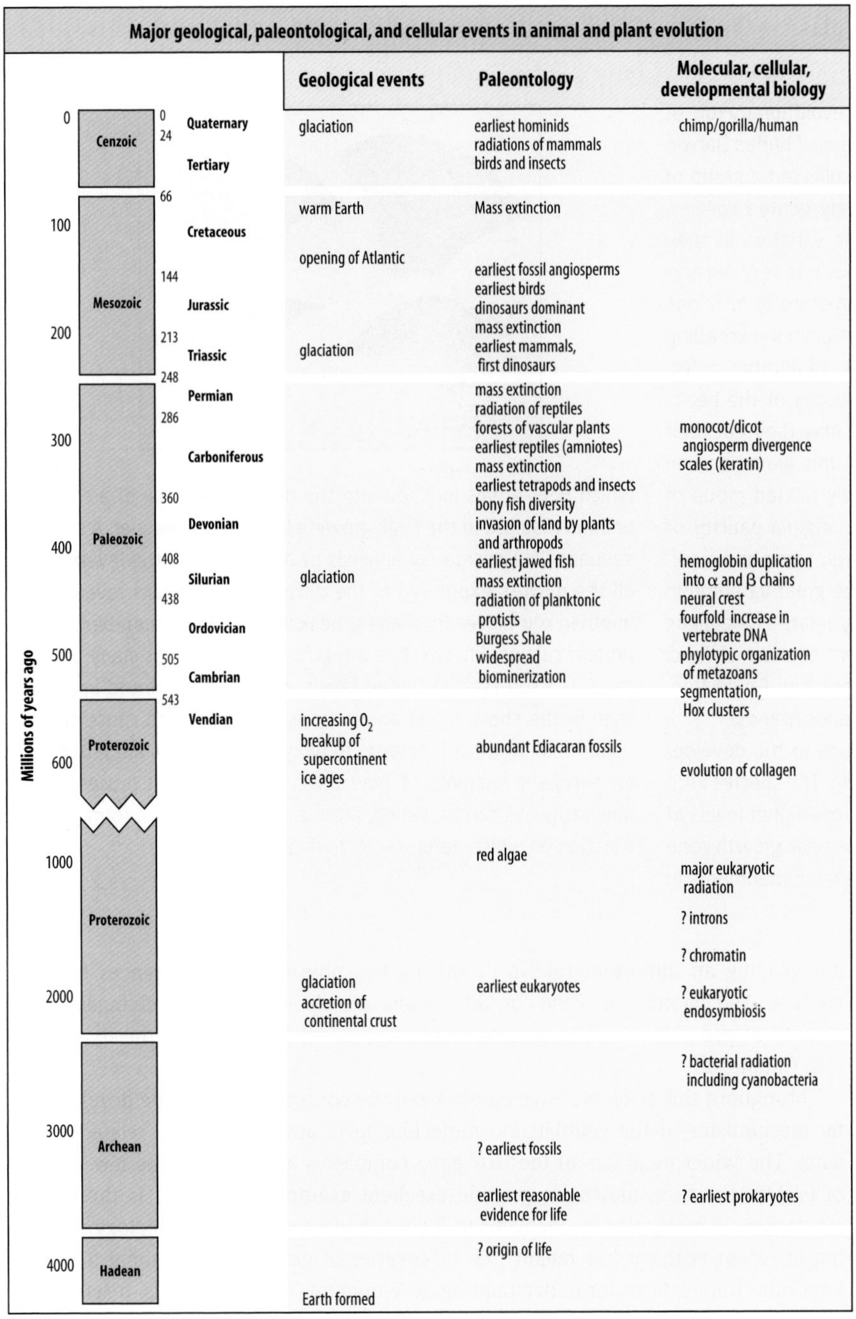

Major geological, paleontological, and cellular events in animal and plant evolution

Era	Period	Geological events	Paleontology	Molecular, cellular, developmental biology
Cenozoic	0 Quaternary / 24 Tertiary	glaciation	earliest hominids / radiations of mammals / birds and insects	chimp/gorilla/human
Mesozoic	66 Cretaceous	warm Earth	Mass extinction	
	144 Jurassic	opening of Atlantic	earliest fossil angiosperms / earliest birds / dinosaurs dominant / mass extinction	
	213 Triassic / 248	glaciation	earliest mammals, / first dinosaurs	
Paleozoic	286 Permian		mass extinction / radiation of reptiles / forests of vascular plants	monocot/dicot / angiosperm divergence / scales (keratin)
	Carboniferous		earliest reptiles (amniotes) / mass extinction	
	360 Devonian		earliest tetrapods and insects / bony fish diversity / invasion of land by plants / and arthropods	
	408 Silurian / 438	glaciation	earliest jawed fish / mass extinction / radiation of planktonic / protists	hemoglobin duplication / into α and β chains / neural crest / fourfold increase in / vertebrate DNA
	Ordovician		Burgess Shale / widespread / biominerization	phylotypic organization / of metazoans / segmentation, / Hox clusters
	505 Cambrian / 543			
Proterozoic	Vendian	increasing O₂ / breakup of / supercontinent / ice ages	abundant Ediacaran fossils	evolution of collagen
Proterozoic			red algae	major eukaryotic / radiation
				? introns
				? chromatin
		glaciation / accretion of / continental crust	earliest eukaryotes	? eukaryotic / endosymbiosis
				? bacterial radiation / including cyanobacteria
Archean			? earliest fossils	
			earliest reasonable / evidence for life	? earliest prokaryotes
Hadean			? origin of life	
		Earth formed		

Millions of years ago (vertical axis): 0, 100, 200, 300, 400, 500, 600, 1000, 2000, 3000, 4000

Fig. 15.1 Major geological, paleontological, and cellular events in the evolution of animals and plants.

Modified from Gerhart, J. and Kirschner, M.: 1997.

for during evolution. Changes in vertebrate development occurring before neurulation are more often associated with changes in reproduction; those occurring after neurulation are more associated with the evolution of animal form. Genetic variability resulting from mutations, sexual reproduction, and genetic recombination is present in the populations of all the organisms we have looked at in this book, and provides new phenotypes upon which selection can act. Changes in the control regions of genes that alter the level, the timing, or the tissue-specificity of their expression, are thought to be particularly important in evolution.

Both changes in the regulation of gene expression and changes in protein structure that generate novel protein functions have played a part in evolution. There is ample evidence for changes in protein structure having a direct impact on animal biochemistry,

Box 15A 'Darwin's finches'

'Darwin's finches' are a good example of the evolutionary role of development and of changes in gene expression. Charles Darwin visited the Galapagos Islands in 1835 and collected a group of finches, distinguishing at the time 13 closely related species. What he found particularly striking was the variation in their beaks. He wrote in *The Voyage of the Beagle*: 'It is very remarkable that a nearly perfect gradation of structure in this one group can be traced in the form of the beak, from one exceeding in dimensions that of the largest gros-beak, to another differing but little from that of a warbler.' The shapes of the beaks reflected differences in the birds' diets and how they got their food. And Darwin later commented: 'Seeing this gradation and diversity of structure in one small, intimately related group of birds, one might really fancy that from an original paucity of birds in this archipelago, one species had been taken and modified for different ends.' The figure shows the great variation in the shape and size of their beaks of closely related Galapagos finches as a result of adaptation to different sources of food. The drawings are by the English ornithologist John Gould from specimens collected by Darwin on the Galapagos Islands.

Modern investigations have uncovered clues to the developmental basis of these differences in beak form. The species with broader, deeper beaks relative to length express higher levels of the bone morphogenetic protein BMP-4 in the beak growth zone compared with species with long, pointed beaks. Experiments in

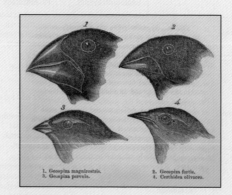

1. *Geospiza magnirostris.* 2. *Geospiza fortis.*
3. *Geospiza parvula.* 4. *Certhidea olivacea.*

which BMP-4 was injected into the developing beak of a chick embryo resulted in the beak growing broader and deeper. A subsequent DNA microarray analysis of differences in the levels of all the mRNAs expressed in the different finch beaks revealed another key player in shaping beak form. The calcium-binding protein calmodulin, which is a regulatory subunit for many proteins, is expressed at higher levels in the long, pointed beaks than in the short, broad ones, and was also able to cause the upper beak of a chick embryo to elongate. So beaks are indeed an excellent example of how basic developmental processes and pathways can be varied, producing differences in form and function on which natural selection can act.

and playing an important role in producing key physiological differences between species. And changes in gene-control regions are clearly linked to changes in the activity of genes involved in pattern formation, as we have seen in the fly embryo (see Chapter 2).

Throughout this book we have emphasized the conservation of some developmental mechanisms at the cellular and molecular level among distantly related organisms. The widespread use of the Hox gene complexes and of the same few families of protein-signaling molecules provide excellent examples of this. It is these basic similarities in molecular mechanisms that have made developmental biology so exciting in recent years; it has meant that discoveries of genes in one animal have had important implications for understanding development in other animals. It seems that when a useful developmental mechanism evolved, it was retained and redeployed in very different organisms, and at different times and places in the same organism. The Wnt (Wingless) signaling pathway, for example, is already present in simple multicellular animals, such as the cnidarians, which arose early in animal evolution. A major technical advance over the past decade has been the determination of the genome sequences of a wide range of animals, including non-bilaterians such as the cnidarians *Hydra magnipapillata* and the starlet sea anemone (*Nematostella vectensis*). These genome sequences are filling gaps in our knowledge of the gene complements of these developmentally simple organisms and throwing light on the evolution of development in multicellular animals in general.

The animal kingdom is conventionally divided into three main groups in terms of basic body structure—the bilaterally symmetrical Bilateria, the radially symmetrical Cnidaria (and the Ctenophora), and the Parazoa (the sponges). The Cnidaria and the Bilateria are 'sister groups'—that is, both groups are considered to arise from a common ancestor—and, together with the very simple Placozoa, they constitute the Metazoa (Fig. 15.2). The Parazoa (the sponges) are considered to be a quite separate group from the rest of the multicellular animals and we shall not consider them here. The largest group of animals is the Bilateria, which have bilateral symmetry across the main body axis in at least some stage of development, and have a characteristic pattern of Hox gene expression. The Bilateria includes vertebrates and other chordates, echinoderms (despite the apparent radial symmetry of the adults), arthropods, annelids, molluscs, and nematodes. These animals are **triploblasts**, as they have the three germ layers—endoderm, mesoderm, and ectoderm. The Cnidaria (corals, jellyfish, hydras and their relatives) are **diploblasts**, as are the Ctenophora (the comb-jellies). They have only two germ layers (ectoderm and endoderm), and generally have radial symmetry. Some cnidarians, such as *Nematostella*, do however show evidence of bilateral symmetry, and studies of Hox gene expression in these animals have suggested to some researchers that the common ancestor of the Bilateria and the Cnidaria might have had bilateral symmetry.

In the rest of this book we have looked at the early development of quite a wide variety of organisms and found some similarities, as well as a number of differences. In this chapter we mainly confine our attention to two phyla—the chordates (which include the vertebrates) and the arthropods (which include the insects and crustaceans). We focus on those differences that distinguish the members of a large group of related animals, such as the vertebrates or the insects, from each other. We will look at the relationship between the development of the individual organism (its **ontogeny**) and the evolutionary history of the species or group (the **phylogeny**): why, for example, do mammalian embryos pass through an apparently fish-like stage that has structures resembling gill slits? We discuss the many variations that occur on the theme of a basic, segmented, body plan: what determines the different numbers and positions of paired appendages, such as legs and wings, in different groups of segmented organisms? We consider the timing of developmental events, and how variations in growth can have major effects on the shape and form of an organism. In all cases, we ultimately want to understand the changes in the developmental processes and the genes controlling them that have resulted in the extraordinary variety of multicellular animals. This is an exciting area of study in which many problems remain to be solved. But first we will briefly consider how development itself might have evolved. What is the origin of the egg, and how might processes such as pattern formation and gastrulation have evolved?

The evolution of development

To look for the origins of development we must consider what we know about the ancestors of multicellular animals. The origin of the ancestor of animals is a tough problem. From genomic evidence and studies of the simplest animals, the last common ancestors of the bilaterians—the Urbilateria—must already have been complex creatures, possessing most of the developmental gene pathways used by existing bilaterians. What Urbilateria might have been like and how they might have been constructed are now key questions in the evolution–development (**evo-devo**) field. Another central challenge is to explain how conserved gene networks already present in the archetypal ancestor were modified to generate the wonderful diversity of animal life on Earth today.

15.1 Genomic evidence is throwing light on the origin of metazoans

About 35 different animal phyla with distinct body plans currently exist, and almost 30 of these phyla are bilaterians. Bilaterians are traditionally subdivided into the

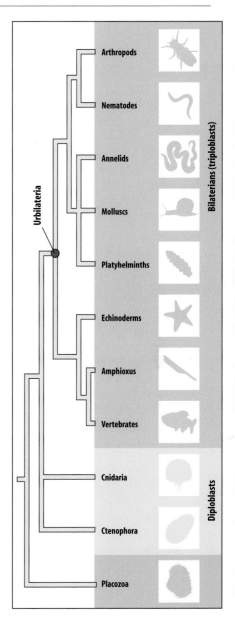

Fig. 15.2 A metazoan family tree. The tree shown here is based on ribosomal DNA and Hox gene sequences and on comparisons of whole-genome sequences. The position of the Urbilateria is represented here by a red circle.

protostomes and the **deuterostomes**. The protostomes develop the mouth close to the blastopore and have a ventral nerve cord traversed by the foregut, and dorsal ganglia that serve as a 'brain.' Of the animals covered in this book, the arthropods, such as *Drosophila* (see Chapter 2), and the nematodes, such as *Caenorhabditis* (see Chapter 6), are protostomes. Deuterostomes develop the anus close to the blastopore and have a dorsal central nervous system. They include the chordates, such as the vertebrates (see Chapters 3–5) and the ascidians, and the echinoderms, such as the sea urchin (ascidian and sea-urchin development are described in Chapter 6). Fossils of adult bilaterians make their appearance suddenly during the Cambrian period, at a time between 535 to 525 million years ago or even earlier—a time called the 'Cambrian explosion' because of the sudden appearance of an enormous variety of animal life.

The remaining, non-bilaterian, metazoans mostly belong to the phylum Cnidaria; they are the simplest and oldest known animals possessing a gut. The genome of the sea anemone *Nematostella* reveals an extensive gene repertoire more similar to that of vertebrates than to flies or nematodes, implying that the genome of the ancestor of all animals was similarly complex. Nearly one-fifth of the inferred genes in the *Nematostella* genome are novelties—not known from other animal groups—and the genome is enriched in genes for animal functions such as cell–cell signaling, cell adhesion, and synaptic transmission.

Taking all the evidence from bilaterians and cnidarians together, we can speculate that the ancestor of all extant animals lived perhaps 700 million years ago and would have had flagellate sperm, development through a process of gastrulation, multiple germ layers, true epithelia lying on a basement membrane, a lined gut, neuromuscular and sensory systems, and fixed body axes. *Nematostella's* genome contains about 18,000 protein-coding genes, a similar number to *Caenorhabditis*. Genes representing 56 families of homeodomain gene-regulatory proteins are present in *Nematostella* and include representatives of four different classes of Pax genes, genes that in bilaterians are involved in many different aspects of development such as eye development, segmentation, and neural patterning. More remarkably for an animal with only two germ layers, at least seven genes associated with the formation of mesoderm in bilaterians are expressed in the developing endoderm of *Nematostella*. Thus the machinery for germ-layer specification was already present in the ancestors of both bilaterians and cnidarians.

There is one very simple and primitive free-living marine animal called *Trichoplax*, which is so unusual it has been placed in a phylum of its own, the Placozoa. Its relationship to other animals is not yet clear, but genomic comparisons place it as a very primitive member of the Metazoa. *Trichoplax* is composed of two layers of epithelium, which form a flat disc with no gut, and it has just four different cell-types. It mainly reproduces by fission. Nevertheless, in line with the genomes of other animals, the *Trichoplax* genome contains an estimated 11,500 protein-coding genes, which encode a rich array of transcription factors and signaling proteins, including homeodomain-containing proteins, Tbox genes similar to *Brachyury*, and components of the Wnt/β-catenin signaling pathway.

15.2 Multicellular organisms evolved from single-celled ancestors

If recognizable animals had evolved by some 700 million years ago, the question still remains of how they evolved from a unicellular ancestor, which from genomic and other evidence is thought to have been an organism rather similar to the modern choanoflagellates, unicellular and colonial protozoa with a single flagellum surrounded by a 'collar' of villi, which are very similar in morphology to the collared cells (choanocytes) of sponges. What had to be invented for the transition to multicellularity? And how did embryonic development from an egg evolve? The key requirements for embryonic development, as we have seen, are a program of gene activity, cell differentiation, signal transduction (so cells can communicate with each other), and cell motility and cohesion (so that overall form can change).

Judging by modern unicellular eukaryotes, such as the choanoflagellates, the single-celled organism ancestral to animals would have possessed all these features in primitive form, and little new would have had to be invented, although much had to evolve further and acquire different functions for multicellularity to evolve. The genome sequence of the modern choanoflagellate *Monosiga brevicollis* has been determined and shows evidence of some cell adhesion and signaling proteins that are otherwise only present in metazoans. Unicellular eukaryotes have the ability to move, to adhere, and to respond to signals. At a minimum, the ancestor of multicellular organisms would have already possessed the characteristic eukaryotic cell cycle, which entails a complex program of gene activity, the ability to undergo cellular differentiation, signaling by receptor tyrosine kinases, and cadherin-mediated cell adhesion.

What then was the origin of multicellularity and the embryo? One possibility, and it is highly speculative, is that mutations resulted in the progeny of a single-celled organism not separating after cell division, leading to a loose colony of identical cells that occasionally fragmented to give new 'individuals.' One advantage of a colony might originally have been that when food was in short supply, the cells could feed off each other, and so the colony survived. This could have been the origin of both multicellularity and the requirement for cell death in multicellular organisms. The egg may subsequently have evolved as the cell fed by other cells; in sponges, for example, the egg phagocytoses neighboring cells. The evolutionary advantage of an organism developing from a single cell—the egg—is that all the cells of the organism will have the same genes. This is a prerequisite for the development of complex patterns and body plans, as patterning requires communication between cells and that can only be reliable if the cells obey the same rules as a result of having the same genetic instructions. There are 'multicellular' organisms that develop from more than one cell, such as slime molds, but they have never evolved numerous complex forms.

Once multicellularity evolved, it opened up all sorts of new possibilities, such as cell specialization for different functions. Some cells could specialize in providing motility, for example, and others in feeding, as we see in the endoderm of *Hydra*. The origin of pattern formation—that is, cell differentiation in a spatially organized arrangement—is not known, but may have depended on gradients set up by external influences such as inside–outside differences.

How gastrulation evolved is also unknown, but it is not implausible to consider a scenario in which a hollow sphere of cells, the common ancestor of all multicellular animals, changed its form to assist feeding (Fig. 15.3). This ancestor may, for example, have sat on the ocean floor, ingesting food particles by phagocytosis. A small invagination developing in the body wall could have promoted feeding by forming a primitive gut. Movement of cilia could have swept food particles more efficiently into this region, where they would be phagocytosed. Once the invagination formed, it is not too difficult to imagine how it could eventually extend right across the sphere to fuse with the other side and form a continuous gut, which would be the endoderm.

Fig. 15.3 A possible scenario for the development of the gastrula. A hollow multicellular sphere of cells, perhaps derived from a colonial protozoan, could have settled on the sea bottom and developed a gut-like invagination to aid feeding.
Based on Jägersten, G.: 1956.

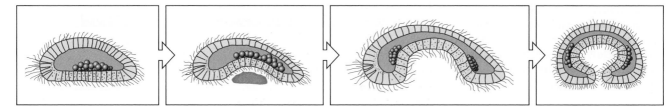

At a later stage in evolution, cells migrating inside, between the gut and the outer epithelium, would give rise to the mesoderm.

Gastrulation provides a good example of developmental change during evolution. While there is considerable similarity in the process of gastrulation in many different animals, there are also significant differences. But how these evolved and what could have been the adaptive nature of the intermediate forms is a hard problem.

Summary

The embryo arose during the evolution of multicellular organisms from single-celled organisms, which have most of the cellular properties required by embryonic development. Multicellularity, having arisen, might have persisted originally because a colony of cells could provide a source of food for some of the members in times of scarcity. The egg might have evolved as the cell that is fed by other cells and provides the basis for the development of complex structures. The development of an embryo provides insights into the evolutionary origin of the animal.

The evolutionary modification of embryonic development

Comparisons of embryos of related species have suggested an important generalization about development: the more general characteristics of a group of animals (that is, those shared by all members of the group) usually appear earlier in their embryos than the more specialized ones, and arose earlier in evolution. In the vertebrates, a good example of a general characteristic would be the notochord, which is common to all vertebrates and is also found in other chordate embryos. The phylum Chordata comprises three subphyla—the vertebrates; the cephalochordates, such as amphioxus (lancelet) (Fig. 15.4); and the urochordates, such as the ascidians (see Chapter 6). They all have, at some embryonic stage, a notochord, flanked by muscle, and a dorsal neural tube. The amphioxus in particular has many features of a primitive vertebrate, even when adult: a dorsal hollow nerve cord, in this case protected by the persistent notochord, rather than a bony spine, and segmental muscles that derive from somites. Paired appendages, such as limbs, which develop later, are special characters that develop only in the vertebrates, and differ in form among different vertebrates.

Fig. 15.4 The cephalochordate amphioxus compared to a hypothetical primitive vertebrate similar to a present-day lamprey. The overall construction of the two organisms is very similar, with a dorsal nerve cord, a more ventral axial skeleton known as the notochord, and a ventral digestive tract. Both animals have gills in the pharyngeal region, structures that are designed to capture food and also take in oxygen from the water. The notochord extends right to the anterior end of amphioxus, but the vertebrate has a prominent head at the anterior end, extending beyond the notochord. In more advanced vertebrates, the notochord is present only in the embryo, being replaced by the vertebrae.

Modified from Finnerty, J.R.: 2000.

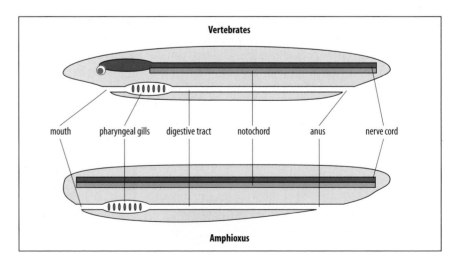

Vertebrates

mouth pharyngeal gills digestive tract notochord anus nerve cord

Amphioxus

All vertebrate embryos pass through a **phylotypic stage** at which they all more or less resemble each other and show the specific embryonic features of the phylum to which they belong. After the phylotypic stage, vertebrate embryos diversify and give rise to the diverse forms of the different vertebrate classes. The phylotypic stage itself is somewhat variable, however, as can be seen in Fig. 15.5, which shows that the number of somites can be very different, and that some organs, such as limb buds, are at different stages of development at this time. The development of the different vertebrate classes before the phylotypic stage is also highly divergent, because of their very different modes of reproduction; some developmental features that precede the phylotypic stage are evolutionarily highly advanced, such as the formation of a trophoblast and inner cell mass in mammals. This is an example of a special character that developed late in vertebrate evolution, and is related to the nutrition of the embryo through a placenta, rather than a yolky egg.

The reason that vertebrate embryos pass through a phylotypic stage may be related to that being the stage at which the Hox genes, which control general pattern along the main body axis, are expressed. Also, it is the stage at which structures common to all vertebrates are developed, such as somites, notochord, and neural tube.

We will now consider the modifications that have occurred to a variety of embryonic structures during evolution, including the basic body plan and the limbs.

15.3 Hox gene complexes have evolved through gene duplication

Hox genes play a key role in the development of both vertebrates and insects—and in most other animals. By comparing the organization and structure of the Hox genes in insects and vertebrates, we can determine how one set of important developmental genes has changed during evolution. A major general mechanism of evolutionary change has been gene duplication and divergence. Tandem duplication of a gene, which can occur by a variety of mechanisms during DNA replication, provides the embryo with an additional copy of the gene. This copy can diverge in both its nucleotide sequence and its regulatory region, thus changing its pattern of expression and its downstream targets without depriving the organism of the function of the original gene (Fig. 15.6). The process of gene duplication has been fundamental in the evolution of new proteins and new patterns of gene expression; it is clear, for example, that the different hemoglobins in humans (see Fig. 10.11) have arisen as a result of gene duplication.

The Hox gene complexes provide one of the clearest examples of the importance of gene duplication in developmental evolution. The Hox genes specify pattern along the antero-posterior axis, as we have seen in previous chapters, and are among the trademark genes that characterize the Metazoa. They are members of the homeobox gene superfamily, which is characterized by a short 180 base pair motif, the homeobox, which encodes a helix–turn–helix domain that is involved in transcriptional regulation (see Box 5E, p. 194). Two features characterize all vertebrate Hox genes: the individual genes are organized into one or more gene clusters or complexes, and the order of expression of individual genes along the antero-posterior axis is usually the same as their sequential order in the gene complex (see Section 5.6).

The Hox genes are assumed to have evolved by duplication of a single ancestral gene. The simplest Hox gene complexes are found in invertebrates, and comprise a small number of sequence-related genes carried on one chromosome. Vertebrates typically have four sets of Hox genes, carried on four different chromosomes, suggesting two rounds of wholesale duplication of an ancestral Hox gene complex, in line with the generally accepted idea that large-scale duplications of the genome have occurred during vertebrate evolution. There has also been further duplication of Hox genes within each complex. In several fish there has been one further round of

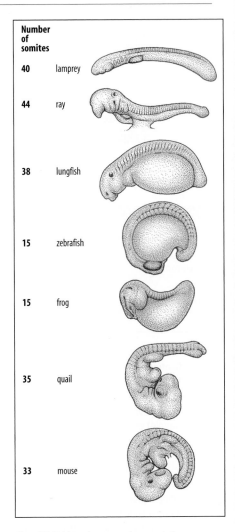

Number of somites	
40	lamprey
44	ray
38	lungfish
15	zebrafish
15	frog
35	quail
33	mouse

Fig. 15.5 Vertebrate embryos at the phylotypic (tailbud) stage show wide variation in somite number. Not to scale.

Modified from Richardson, M.K., et al.: 1997.

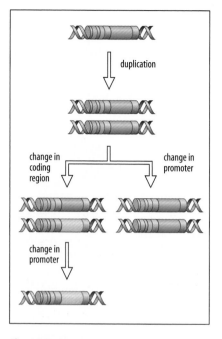

Fig. 15.6 Gene duplication and diversification. Once a gene has become duplicated, the second copy can evolve new functions and/or new expression patterns. Illustrated is a hypothetical example of a gene (purple) with two *cis*-regulatory modules (blue and green) that confer expression in two different tissues. Left pathway: after duplication of the complete gene and its regulatory regions, mutation in the coding region of one of the copies can generate a gene with a new and useful function that is selected for. Both copies will be retained in the genome as they both fulfill useful functions. Subsequent mutation in a regulatory module could lead to the new gene acquiring a new expression pattern. Right pathway: alternatively, a mutation in one regulatory region in one copy and the other regulatory region in the other copy can generate two copies of the gene that retain the original function but are now only expressed in a single tissue each. Both will be retained in the genome as they are both now needed to fulfill the function of the original gene. Examples of all these types of duplication and divergence are found in animal genomes.

genome duplication: the zebrafish has seven Hox clusters, the result of three rounds of duplication and the subsequent loss of one cluster. The Japanese puffer fish *Fugu rubripes* has four Hox clusters, but many genes have been lost, and one cluster bears little relation to any in other vertebrates. It appears that in the ancestral chordates that gave rise to the vertebrates, the spatial co-linearity between the order of the Hox genes along the chromosome and their expression along the main body axis was confined to the ectoderm-derived neural tube, and possibly the epidermis, and that it was only later extended to other tissues, thus allowing increased complexity of body plan.

Comparisons of Hox genes of various arthropod and chordate species indicate that the common ancestor of arthropods and vertebrates was most likely to have had a simple Hox complex of seven genes (Fig. 15.7). As we get nearer to the vertebrates, the cephalochordate amphioxus has just one Hox gene cluster, containing 14 genes, and one can think of this as most closely resembling the ancestor of the four vertebrate Hox gene complexes: Hoxa, Hoxb, Hoxc, and Hoxd (see Fig. 15.7). Possible ways in which both the vertebrate and *Drosophila* Hox complexes could have evolved from a simpler ancestral complex by gene duplication have been reconstructed. In *Drosophila*, for example, successive tandem duplications of an ancestral *Antp*-like gene could have given rise to the genes *abdominal-A* (*abd-A*), *Ultrabithorax* (*Ubx*), and *Antennapedia* (*Antp*) (see Fig. 15.7). An Antennapedia-like gene has been implicated as the 'original' Hox gene by genetic experiments that suggest that the thoracic segment T2 of insects, such as *Drosophila*, which is specified by *Antp*, represents the 'ground state' segmental type, from which other body segments have diversified, both anteriorly and posteriorly, as a result of Hox gene duplication, mutational changes, and consequent changes in spatial and temporal patterns of Hox gene expression imposed on this ground state. If the later-evolved posteriorly expressed Hox genes of the bithorax complex—namely *Ubx*, *abd-A* and *Abd-B*—are deleted in *Drosophila*, all the thoracic and abdominal segments posterior to T2 revert to the T2 type, and express *Antp* (see Fig. 2.48, second panel).

In vertebrates, the Hox genes are arranged in four separate clusters, each of which is on a different chromosome. The separate gene complexes probably arose from duplications of whole chromosomal regions, in which new Hox genes had already been generated by tandem duplication. For example, sequence comparisons suggest that the multiple mouse Hox genes most similar in sequence to the *Drosophila Abd-B* gene do not have direct homologs within the *Drosophila* HOM-C complex (see Fig. 15.7); they probably arose by tandem duplication from an ancestral gene after the split of the insect and vertebrate lineages, but before duplication of the whole cluster in vertebrates. The advantage of duplication was that the embryo had more Hox genes to control downstream targets and so could make a more complicated body. We now consider the role of these genes in evolution of the axial body plan.

15.4 Changes in Hox genes generated the elaboration of vertebrate and arthropod body plans

The broad similarities in the pattern of Hox gene expression in vertebrates and arthropods, whose evolution diverged hundreds of millions of years ago, are taken as good supporting evidence for the idea that all multicellular animals descend from a common ancestor. Hox genes are key genes in the control of development and are expressed regionally along the antero-posterior axis of the bilaterian embryo (see Chapters 2 and 5). The apparently conserved nature of the Hox genes (and of certain other developmentally important genes) in animal development has led to the concept of the **zootype**. This defines a general pattern of expression of these key genes that is present in all animal embryos.

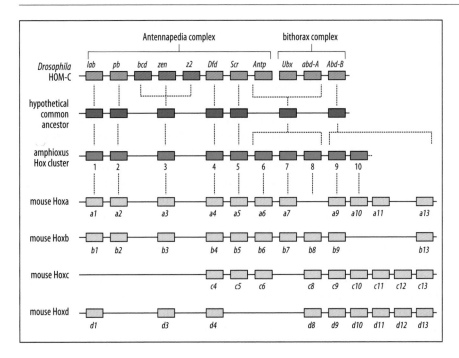

Fig. 15.7 Gene duplication and Hox gene evolution. A suggested evolutionary relationship between the Hox genes of a hypothetical common ancestor and *Drosophila* (an arthropod), amphioxus (a cephalochordate), and the mouse (a vertebrate). Duplications of genes of the ancestral set (red) could have given rise to the additional genes in *Drosophila* and amphioxus. Two duplications of the whole cluster in a chordate ancestor of the vertebrates could have given rise to the four separate Hox gene complexes in vertebrates. There has also been a loss of some of the duplicated genes in vertebrates. In *Drosophila*, the two Hox gene clusters comprising the HOM-C complex are called the Antennapedia complex and the bithorax complex. Posterior four genes not shown for amphioxus.

The Hox genes exert their influence by controlling the activity of other genes that determine, for example, how the cells in a region develop into segments and appendages. Thus, changes in these target genes, as well as changes in the Hox genes themselves, can be a major source of change in evolution; in large part, the targets of the Hox genes still have to be identified. In addition, changes in the pattern of Hox gene expression along the body can have important consequences. An example of this is a relatively minor modification of the body plan that has taken place within vertebrates. One easily distinguishable feature of pattern along the antero-posterior axis in vertebrates is the number and type of vertebrae in the main anatomical regions—cervical (neck), thoracic, lumbar, sacral, and caudal (see Fig. 5.26). The number of vertebrae in a particular region varies considerably among the different vertebrate classes—mammals, with rare exceptions, have seven cervical vertebrae, whereas birds can have between 13 and 15. How does this difference arise? A comparison between the mouse and the chick shows that the domains of Hox gene expression have shifted in parallel with the change in number of vertebrae in the different regions (see Fig. 5.26). For example, the anterior boundary of *Hoxc6* expression in the mesoderm in mice and chicks is always at the boundary of the cervical and thoracic regions, irrespective of the number of cervical vertebrae. Moreover, the *Hoxc6* expression boundary is also at the cervical–thoracic boundary in geese, which have three more cervical vertebrae than chickens. The changes in the spatial expression of *Hoxc6* thus correlate specifically with the number of cervical vertebrae rather than with a fixed number of vertebrae. Other Hox genes are involved in the patterning of the antero-posterior axis, and their boundaries also shift with a change in anatomy.

Changes in the targets of master regulatory genes, such as Hox genes, probably play a key role in determining the differences between animal species. There may be many—perhaps very many—downstream targets of the Hox genes. In *Drosophila*, for example, it is expression of the Hox gene *Ubx* that prevents the rudimentary haltere from growing and so makes it different from its neighboring appendage, the wing. Tests have shown that six of the twelve genes known to be involved in patterning the wing are under the control of *Ubx* and are repressed in the haltere. But because of changes in their *cis*-regulatory regions, only some of these genes are repressed by *Ubx* in butterflies, resulting in the development of a hind wing rather than a haltere.

Another example of how subtle variations in body form can be controlled by Hox genes comes from insects. There are around 150,000 species of flies and all use roughly the same set of developmental genes. Diversity in form can reflect different patterns of gene activity, but the fascinating question in this case is how the same sets of proteins can generate such subtle morphological diversity. In the genus *Drosophila*, different species have different patterns of non-sensory bristles—trichomes—on the femur of the second leg. *Drosophila melanogaster* has a small naked patch, whereas *D. simulans* has a large one and *D. virilis* no naked patch at all (Fig. 15.8). *Ubx* represses trichome development, and the extent or presence of the naked patch is determined by subtle differences in the regulation of *Ubx* expression in this region. These differences are due to small sequence differences in the control region of *Ubx* in the different species, and this variation provides the evolutionary basis for the variation in pattern.

Changes in patterns of Hox gene expression can also help explain the evolution of arthropod body plans. Insects and crustaceans are distinct groups of arthropods that have evolved from a common arthropod ancestor, which probably had a body composed of more or less uniform segments. A comparison of Hox gene expression patterns in the embryos of a grasshopper (an insect) and of the brine shrimp *Artemia* (a crustacean), indicates how the different body plans might have evolved from an ancestral body plan. This comparison shows that both the pattern of Hox gene expression and the adult body regions to which particular Hox genes relate have changed during the evolution of these two groups (Fig. 15.9).

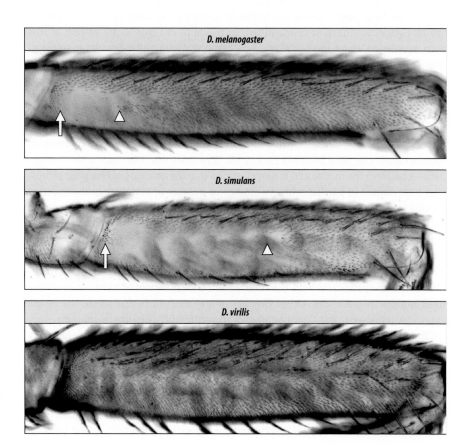

Fig. 15.8 Trichome patterns on the posterior second femur vary among *Drosophila* species. Top panel: *Drosophila melanogaster*. Middle panel: *D. simulans*. Bottom panel: *D. virilis*. The arrow and arrowhead in the top two panels indicate the extent of the patch of naked cuticle.

From Stern, D.L.: 1998.

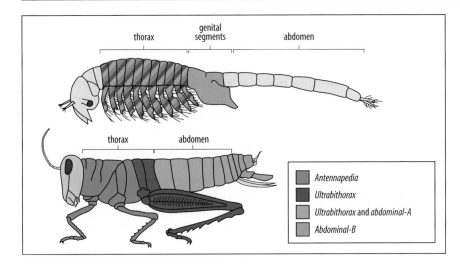

Fig. 15.9 A comparison of body plans and Hox gene expression in two arthropods. A comparison of Hox gene expression during embryogenesis in an insect, the grasshopper (bottom), with that in a crustacean, the brine shrimp *Artemia* (top), shows that both the pattern of Hox gene expression and the adult body regions to which particular Hox genes relate have changed during the evolution of these two groups of arthropods from their common ancestor. In *Artemia*, the three Hox genes *Antennapedia*, *Ultrabithorax*, and *abdominal-A* are all expressed throughout the thorax, where most of the segments are similar. However, the expression of these genes in the thorax and abdomen of the grasshopper is different. They have overlapping and distinct patterns of expression that reflect the regional differences in the insect thorax. The gene *Abdominal-B* is expressed in the genital regions of both animals, indicating that these two regions are homologous.

After Akam, M.: 1995.

The grasshopper has a pattern of Hox gene expression similar to that of *Drosophila*. The Hox genes *Antp, Ubx*, and *abd-A* specify distinct segment types in the thorax and abdomen, but are expressed in overlapping domains, and segment types are defined by combinatorial expression (see Section 2.27). In the brine shrimp, however, these genes are all expressed together in a thoracic region that is composed of uniform segments. This suggests that the thorax of *Artemia* might be homologous to the whole insect thorax and much of the insect abdomen. The term **homologous** is used here to describe structures that share a common ancestry or developmental derivation, even if they no longer serve the same function. In the evolution of the grasshopper and the brine shrimp from their common ancestor, the changes in body plan have resulted in part from spatial changes in the expression of particular Hox genes, but key changes in their downstream targets have also been involved. Comparisons between species clearly show that Hox genes do not specify particular structures but simply provide a regional identity. How that identity is interpreted to produce a particular morphology is the role of genes acting downstream from the Hox genes.

Genes other than Hox genes are involved in the basic specification and patterning of the arthropod head. The head regions of different types of arthropods are very different, and indeed spiders do not seem to have a head at all. Their body is divided into a front and a back region, the front end having segments that bear the spider's fangs, pedipalps, and four pairs of walking legs; the front region thus acts as a head and thorax combined. In spite of these differences, typical 'head' genes are expressed in the anterior-most region in both spiders and insects, including the homeobox gene *orthodenticle*, which is involved in specifying the larval head and anterior brain in *Drosophila* (see Section 2.31), and the Hox gene *labial*, which in *Drosophila* is expressed in the imaginal discs specifying the labial palps in the adult fly. This comparison shows how much morphology can change as a result of changes in the downstream targets of genes.

15.5 The position and number of paired appendages in insects is dependent on Hox gene expression

Good illustrations of the role of Hox genes in regional specification are provided by arthropod appendages. Insect fossils display a variety of patterns in the position and number of their paired appendages—principally the legs and wings. Some insect fossils have legs on every segment, whereas others only have legs in a distinct thoracic region. The number of abdominal segments bearing legs varies, as does the size and shape of the legs. Wings arose later than legs in insect evolution. Wing-like appendages are present on all the thoracic and abdominal segments of some insect fossils, but are restricted to the thorax in others. To understand how these different patterns of appendages arose during evolution we need to look at how the different patterns of appendages develop in two orders of modern insects, the Lepidoptera (butterflies and moths) and the Diptera (flies, including *Drosophila*).

The basic pattern of Hox gene expression along the antero-posterior axis is the same in all present-day insect species that have been studied. Yet the larvae of Lepidoptera have legs on the abdomen as well as on the thorax, and the adults have two pairs of wings, whereas the Diptera, which evolved later, have no legs on the abdomen in the larva or adult, and only one pair of wings, with the second pair of wings having been modified into halteres. How are these differences related to differences in Hox gene activity in the two groups of insects?

In *Drosophila*, Hox genes of the bithorax complex (BX-C) suppress appendage formation in the abdomen by repressing the expression of the *Distal-less* gene. This suggests that the potential for appendage development is present in every segment, even in flies, and is actively repressed in the fly abdomen. It thus seems likely that the ancestral arthropod from which insects evolved had appendages on all its segments. During the embryonic development of Lepidoptera, the genes *Ubx* and *Abd-B* are turned off in the ventral parts of the abdominal segments; this results both in *Distal-less* being expressed and in legs developing on the abdomen in the larva. The presence or absence of legs on the abdomen is thus determined by whether or not a particular Hox gene is expressed there, which shows that changes in the pattern of Hox gene expression have played a key role in evolution. But, and this is fundamental, downstream targets are even more important: in insects, *Ubx* and *abd-A* expressed together repress limb development in the abdomen, whereas in crustaceans these genes are expressed together in the thorax yet limbs still develop there. The difference must lie in the genes they target.

The Hox genes can also determine the nature of an appendage: we have seen how mutations can convert legs into antenna-like structures and an antenna into a leg that still retains the same proximo-distal positional values (see Section 11.22). The coding sequences outside the DNA-binding homeodomain in Hox genes have changed during evolution, and this has significant effects on the specification of segment identity and appendage development in different species.

15.6 The basic body plan of arthropods and vertebrates is similar, but the dorso-ventral axis is inverted

A comparison between the body plans of arthropods and chordates reveals an intriguing difference. In spite of many similarities in their basic body plan—both have an anterior head, a nerve cord running anterior to posterior, a gut, and appendages—the dorso-ventral axis of vertebrates is inverted when compared to that of arthropods. The most obvious manifestation of this is that the main nerve cord runs ventrally in arthropods and dorsally in vertebrates (Fig. 15.10, left panels).

One explanation for this, first proposed in the nineteenth century, is that during the evolution of the vertebrates from their common ancestor with the arthropods, the dorso-ventral axis was turned upside-down, so that the ventral nerve cord of the ancestor became dorsal. This startling idea has recently found some support from molecular evidence showing that the same genes are expressed along the dorso-ventral axis in both insects and vertebrates, but in inverse directions. This inversion

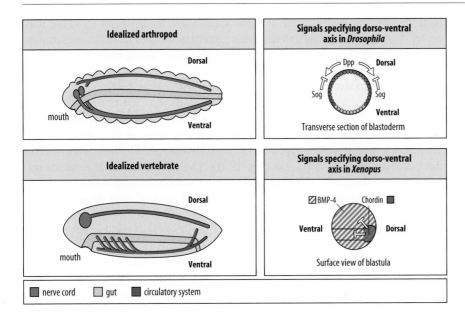

Fig. 15.10 The vertebrate and *Drosophila* dorso-ventral axes are related but inverted. In arthropods, the nerve cord is ventral, whereas in vertebrates it is dorsal—dorsal and ventral being defined by the position of the mouth. In *Drosophila* and *Xenopus*, the signals specifying the dorso-ventral axis are similar, but are expressed in inverted positions. The protein Chordin, a dorsal specifier in vertebrates, is related to Sog, which is a ventral specifier in *Drosophila*, and the vertebrate ventral specifier BMP-4 is related to *Drosophila* Decapentaplegic (Dpp), which specifies dorsal.

After Ferguson, E.L.: 1996.

may have been dictated by the position of the mouth. The mouth defines the ventral side, and a change in the position of the mouth away from the side of the nerve cord would have resulted in the reversal of the dorso-ventral axis in relation to the mouth. The position of the mouth is specified during gastrulation, and it is not difficult to imagine how its position could have moved, but other changes in body structure and the nervous system were also involved.

We have seen in Chapters 2 and 4 that the patterning of the dorso-ventral axis in insects and vertebrates involves intercellular signaling. In *Xenopus*, the protein Chordin is one of the signals that specifies the dorsal region, whereas the growth factor BMP-4 specifies a ventral fate. In *Drosophila*, the pattern of gene expression is reversed: the protein Decapentaplegic, which is closely related to BMP-4, is the dorsal signal, and the protein short gastrulation (Sog), which is related to Chordin, is the ventral signal (see Fig. 15.10, right panels). These signaling molecules are experimentally interchangeable between insects and frogs. Chordin can promote ventral development in *Drosophila*, and Decapentaplegic protein promotes ventral development in *Xenopus*. The molecules and mechanisms that set up the dorso-ventral axes in the two groups of animals are thus homologous, strongly suggesting that the divergence in the body plans of the arthropods and vertebrates involved an inversion of this axis, by movement of the mouth during the evolution of the vertebrates. Thus our image of the common ancestor of chordates and arthropods is an animal in which the establishment of the two major body axes, antero-posterior and dorso-ventral, was similar to that in present-day animals.

15.7 Limbs evolved from fins

The limbs of tetrapod vertebrates are special characters that develop after the phylotypic stage. Amphibians, reptiles, birds, and mammals have limbs, whereas fish have fins. The limbs of the first land vertebrates evolved from the pelvic and pectoral fins of their fish-like ancestors. The basic limb pattern is highly conserved in both the forelimbs and hindlimbs of all tetrapods, although there are differences both between forelimbs and hindlimbs, and between different vertebrates. Limbs evolved from fins, but how fins evolved is far less clear, and is a complex problem.

Nevertheless, the development of these appendages made use of signaling molecules, like Sonic hedgehog and fibroblast growth factor (FGF), and of transcription factors, such as the Hox proteins, which were already being used to pattern the body.

The fossil record suggests that the transition from fins to limbs occurred in the Devonian period, between 400 and 360 million years ago. The transition probably occurred when the fish ancestors of the tetrapod vertebrates living in shallow waters moved onto the land. The fins of Devonian lobe-finned fishes, such as *Panderichthys*, are probably ancestral to tetrapod limbs, an early example of which is the limb of the Devonian tetrapod *Tulerpeton* (Fig. 15.11). The proximal skeletal elements corresponding to the humerus, radius, and ulna of the tetrapod limb are present in the ancestral fish, and a recent analysis of a fossil *Panderichthys* using CT scanning has shown that, contrary to previous ideas, the distal region of the pectoral fin contains separate skeletal elements, and so fingers are not a tetrapod novelty, as previously thought.

To gain insights into the transition from fin to limb, researchers have turned to a modern fish, the zebrafish, in which fin development can be followed in detail and the genes involved can be identified. The fin buds of the zebrafish embryo are initially similar to tetrapod limb buds, but important differences soon arise during development. The proximal part of the fin bud gives rise to skeletal elements, which may be homologous to the proximal skeletal elements of the tetrapod limb. There are four main proximal skeletal elements in a zebrafish fin, which arise from the subdivision of a cartilaginous sheet (Fig. 15.12). The essential difference between fin and limb development is in the distal skeletal elements. In the zebrafish fin bud, an ectodermal fin fold develops at the distal end of the bud and fine bony fin rays are formed within it.

As in the tetrapod limb bud (see Chapter 11), the key gene *Shh* is expressed at the posterior margin of the zebrafish fin buds and the expression pattern of Hoxd and Hoxa genes is similar to that in tetrapods. As in limb development, the expression of the Hoxa and Hoxd genes in zebrafish pectoral fins occurs in three distinct phases, in which the most distal, the third phase, is correlated with the development of the

Fig. 15.11 The fin-to-limb transition. In the lobe-like fin of the Devonian fish *Panderichthys*, there were proximal elements corresponding to the humerus (the stylopodium), radius and ulna (the zeugopodium), and distal elements (the autopodium). The Devonian tetrapod *Tulerpeton* has similar proximal elements, and the distal elements have developed into recognizable digits.

Adapted from: Boisvert, C.A., et al.: 2008.

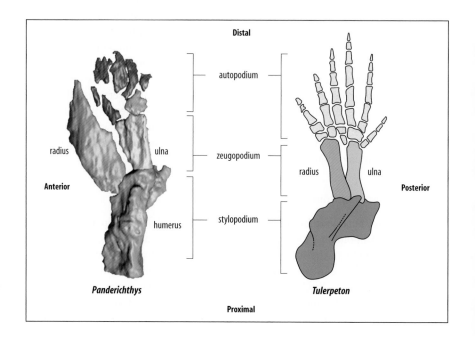

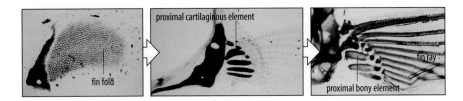

Fig. 15.12 The development of the pectoral fin of the zebrafish *Danio*. Left panel: the pectoral girdle and fin fold. Middle panel: four proximal cartilaginous elements and distal fin rays. Right panel: four proximal bony elements supporting the distal fin rays in the adult fish.

Photographs courtesy of D. Duboule, from Sordino, P., et al.: 1995.

most distal fin structure—the fin blade. Hox gene expression in the pectoral fin during this phase is dependent on Shh signaling, as in the limb, which indicates that the regulatory mechanisms underlying the three phases of Hox gene expression in fins and limbs have remained relatively unchanged during evolution. Although simpler in organization than limbs, the fins of teleost fishes do have a distal structure that might be considered comparable to the autopod region of the vertebrate limb.

If zebrafish fin development reflects that of the primitive ancestor, then more distal cartilage elements may have evolved from the distal recruitment of the same developmental mechanisms and processes to generate the radius and ulna, wrist, and digits. There are, as discussed in Chapter 11, mechanisms in the limb for generating periodic cartilaginous structures such as digits. It is likely that such a mechanism was involved in the evolution of digits by an extension of the region in which the embryonic cartilaginous elements form, together with the establishment of a new pattern of Hox gene expression in the more distal region.

Differences in fore- and hindlimb development are related to differences in Hox genes, which are involved in establishing positional differences in the lateral plate mesoderm from which the limb develops. Limb loss is quite common, and the python provides a good example of the developmental changes involved. Snakes have hundreds of similar vertebrae in their backbones, as can be seen in the skeleton of a python embryo in Fig. 15.13 (left). Snakes have no forelimbs, but pythons have a pair of hindlimb rudiments at the junction between the rib-bearing thoracic vertebrae, and vertebrae with shorter, forked ribs. In four-limbed vertebrates, expression of *Hoxb5* and *Hoxc8* is confined to a short trunk region. In the python embryo, however, these genes are expressed along the whole body as far as the pelvic rudiment (see Fig. 15.13). The expansion of these Hox-expression domains is thought to underlie the expansion of rib-bearing vertebrae and the loss of forelimbs in snake evolution. Hindlimb buds begin to develop, but the signals associated with apical ridge and polarizing region signaling (see Chapter 11) are not activated. However, there is evidence that the pattern of Hox gene expression is more complex and might involve a different downstream interpretation of the Hox code.

Fig. 15.13 Comparison of Hox gene expression in python and chick embryos. The photograph shows a skeleton of a python embryo at 24 days' incubation stained with Alcian blue and Alizarin red. The arrow marks the position of hindlimb rudiments, which have been removed in this preparation. Note the similarity of the vertebrae anterior to the arrow. The right panel shows a schematic comparison of domains of expression of *Hoxb5* (green), *Hoxc8* (blue) and *Hoxc6* (red) in chick and python embryos. The expansion of the *Hoxc8* and *Hoxc6* domains in the python correlates with the expansion of thoracic identity in the axial skeleton and flank identity in the lateral plate mesoderm.

Adapted from Cohn, M.J. and Tickle, C: 1999.

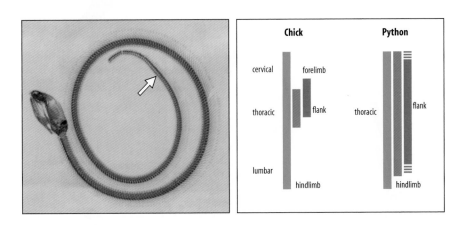

The great range of anatomical specializations in the limbs of mammals (Fig. 15.14) is due to changes both in limb patterning and in the differential growth of parts of the limbs during embryonic development, but the basic underlying pattern of skeletal elements is maintained. This is an excellent example of the modularity of the skeletal elements. If one compares the forelimb of a bat and a horse, one can see that, although both retain the basic pattern of limb bones, it has been modified to provide a specialized function in each. In the bat, the limb is adapted for flying: the digits are greatly lengthened to support a membranous wing. The homeobox transcription factor Prx1 is upregulated in the bat wing compared to the mouse forelimb. In the horse, the limb is adapted for running: in the forelimb, lateral digits are reduced, the central metacarpal (a hand bone in humans) is lengthened, and the radius and ulna are fused for greater strength. The role of differential growth rates and the loss of skeletal elements in the evolution of the horse's limb, and cases of limb reduction, are considered later. In some cases, differences in size of fore- and hindlimbs and the length of the skeletal elements are present from a very early embryonic stage, as in the development of the flightless kiwi and in bats (Fig. 15.15).

Many of the changes that occur during evolution reflect changes in the relative dimensions of parts of the body. We have seen how growth can alter the proportions of the human baby after birth, as the head grows much less than the rest of the body (see Fig. 13.6). The variety of face shapes in the different breeds of dogs, which are all members of the same species, descended from the gray wolf, also provides a good example of the effects of differential growth after birth. All dogs are born with rounded faces; some keep this shape, but in others the nasal regions and jaws elongate during growth. The elongated face of the baboon is also the result of growth of this region after birth.

Because individual structures, such as bones, can grow at different rates, the overall shape of an organism can be changed substantially during evolution by heritable changes in the duration of growth, which also leads to an increase in overall size of the organism. In the horse, for example, the central digit of the ancestral horse grew faster than the digits on either side, so that it ended up longer than the lateral digits (Fig. 15.16). As horses continued to increase in overall size during evolution, this discrepancy in growth rates resulted in the relatively smaller lateral digits no longer touching the ground because of the much greater length of the central digit. At a later stage in evolution, the now-redundant lateral digits became reduced even further in size because of a separate genetic change.

Fig. 15.14 Diversification of mammalian limbs. The basic pattern of bones in the forelimb is conserved throughout the mammals, but there are changes in the proportions of the different bones, as well as fusion and loss of bones. This is seen particularly in the horse limb, in which the radius and ulna have become fused into a single bone, and the central metacarpal (a hand bone in humans) is greatly elongated. In addition, there has been loss and reduction of the digits in the horse. In the bat wing, by contrast, the digits have become greatly elongated to support the membranous wing.

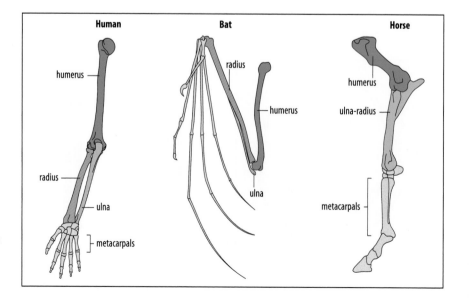

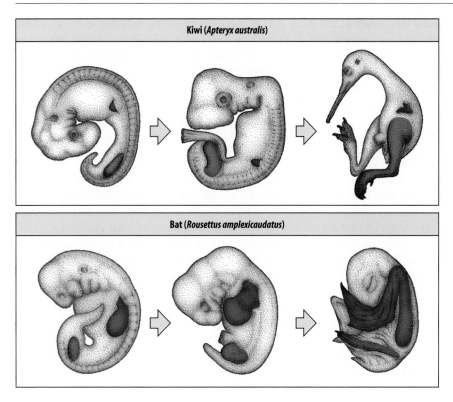

Fig. 15.15 Developmental expression of adult limb size. Top row: the flightless kiwi (*Apteryx. australis*) has relatively short forelimbs and large hindlimbs at all stages of development, even in the early embryo. Bottom row: in the bat *Rousettus amplexicaudatus*, which has large forelimbs and smaller hindlimbs when adult, the forelimb bud is large at all stages, relative to the hindlimb.

Adapted from Richardson, M.K.: 1999.

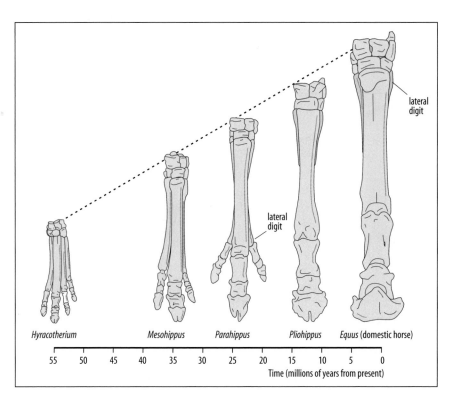

Fig. 15.16 Evolution of the forelimb in horses. *Hyracotherium*, the first true equid, was about the size of a large dog. Its forefeet had four digits, of which one (the third digit in anatomical terms) was slightly longer, as a result of a faster growth rate. All digits were in contact with the ground. As equids increased in size, the lateral digits lost contact with the ground, as the result of the relatively greater increase in length of metacarpal 3 (the hand bone of the third digit). At a later stage, the lateral digits became even shorter, because of a separate genetic change.

After Gregory, W.K.: 1957.

A feature of limb evolution is that, while reduction in digit number is common (there are only three in the chick wing and reduction is common in lizards), species with more than five digits are very rare. It seems that there is a **developmental constraint** on evolving more than five different kinds of digits. This may be due to the Hox genes in the distal part of the limb providing only five discrete genetic programs for giving a digit an identity. In limbs with polydactyly, at least two of the digits are the same (see Box 11B, p. 423), such that there are still only five different kinds of digits. This may be the reason why in an animal with an additional distinctive digit-like element, such as the giant panda's 'thumb,' this digit is in fact a modified wrist bone.

15.8 Vertebrate and insect wings make use of evolutionarily conserved developmental mechanisms

Vertebrate and insect wings are not homologous but have some superficial similarities; they have similar functions yet are very different in structure. The insect wing is a double-layered epithelial structure, whereas the vertebrate limb develops mainly from a mesenchymal core surrounded by ectoderm. Despite these great anatomical differences, however, there are striking similarities in the genes and signaling molecules involved in setting up the axes and so patterning insect legs, insect wings, and vertebrate limbs (Fig. 15.17, and see Chapter 11). Patterning along the antero-posterior axis of all these appendages uses Hedgehog-related signals and members of the TGF-β family, such as Decapentaplegic (in insects) and BMP-2 (in vertebrates). It is remarkable that the dorsal surface of the insect wing is characterized by expression of the gene *apterous*, whereas the related gene *Lmx1* is expressed in the dorsal mesenchyme of the vertebrate limb. A *fringe*-like gene is involved in the specification of the boundary between dorsal and ventral regions in both insect and bird wings.

Another example illustrating the conservation of the genetic machinery for insect appendages is provided by the gene *Distal-less*, which is also expressed along the proximo-distal axis of a wide variety of developing appendages in other animals, including annelid parapodia and the tube feet of sea urchins. It is also expressed during vertebrate limb outgrowth.

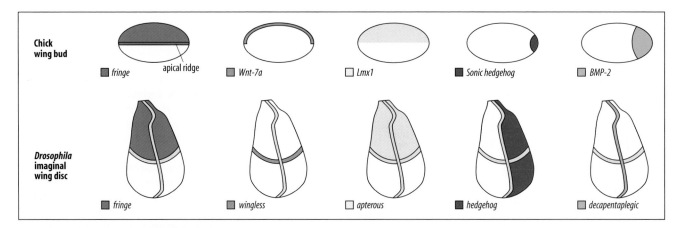

Fig. 15.17 Comparison of developmental signals in the chick wing bud and *Drosophila* wing imaginal disc. First column: the chick wing bud (top) is shown with the distal end facing. The double line bisecting it represents the apical ridge. The chick apical ectodermal ridge forms at the boundary between the dorsal cells, which express *radical fringe*, and the ventral cells, which do not. The dorso-ventral boundary in the insect wing disc also forms at the boundary between fringe and non-fringe cells (bottom). In the insect wing imaginal disc, the future dorsal and ventral regions are in the same plane. The vertical double lines represent the antero-posterior compartment border, and the horizontal double lines represent the dorso-ventral compartment border. Second column: the dorsal region of the chick wing is specified by *Wnt-7a* in the ectoderm, whereas *wingless* is expressed at the dorso-ventral insect wing margin. Third column: the gene *Lmx1* is expressed in the dorsal region of chick wing bud mesoderm, whereas the *Drosophila gene apterous*, to which it is structurally related, specifies the dorsal region of the insect wing. Fourth and fifth columns: *Sonic hedgehog* in the chick wing and *hedgehog in Drosophila* are expressed in posterior regions and both induce expression of genes of the TGF-β family— *BMP-2* and *decapentaplegic*, respectively.

All this emphasizes how evolution has used a wonderfully modifiable system for setting up axes and specifying positional values.

15.9 The evolution of developmental differences can be based on changes in just a few genes

Because of the existence of the two very different morphological forms within the same species, the teleost fish *Astyanax mexicanus* is a good model system in which to study the evolution of developmental mechanisms. *Astyanax mexicanus* is a single species consisting of two radically different forms: a sighted, pigmented, surface-dwelling form and a blind, unpigmented, cave-dwelling form known as a cavefish (Fig. 15.18). At least 30 different populations of *Astyanax* cavefish are present in limestone caverns in Mexico, having been isolated from their surface relations for the past few million years. Although functional eyes are lacking in cavefish adults, the embryos begin to develop eye primordia, which subsequently degenerate. The major cause of eye degeneration is apoptosis of the developing lens cells, which then prevents the growth of other optic tissues, including the retina. The non-development of the eye is, in turn, the cause of other craniofacial differences between cavefish and surface fish. Lens apoptosis is induced by increased activity of the Shh signaling system along the cavefish embryonic midline compared to the embryos of surface-dwelling *Astyanax* (see Section 11.23).

Somewhat surprisingly, there is no evidence that cell proliferation stops in the degenerating cavefish eye. The reason there is no net growth is that the new cells are quickly removed by apoptosis, which persists during cavefish larval development and into adult life. If a lens from a surface-dwelling form of *Astyanax* is transplanted into a cavefish embryo, it does not undergo apoptosis and there is a dramatic restoration of eye development. The role of the increased Shh pathway signaling in cavefish eye development was investigated by increasing Shh expression in embryos of surface-dwelling *Astyanax* embryos, which normally develop into sighted adults. When *Shh* mRNA was injected into one side of a cleaving embryo, expression of the 'eye' gene *pax6* was downregulated unilaterally in the corresponding developing eye region, and the larvae that developed from the embryos overexpressing Shh lacked an eye on that side of the head. In other words, the phenotype of the blind cavefish had been phenocopied by manipulating Shh signals to cause eye degeneration.

Eyes are initially formed and then degraded during larval or adult development in all sightless cave-dwelling vertebrates. It is surprising that, eyes begin to develop at all. This may be because early steps in eye development are required for other essential steps in development, and the elimination of these steps would be fatal. It is thus unlikely that cave vertebrates lacking embryonic eyes will be discovered because of this strong developmental constraint. But why have eyes been lost in cavefish? The answer is not known, but it may be due to selection based on energy conservation once eyes were no longer adaptive.

Another intriguing example of developmental evolution is the reduction or loss of the pelvic, fin or spine in stickleback populations that became separated when they colonized new lakes and streams formed as the glaciers melted after the ice ages about 20,000 years ago. The paired pelvic spines are present in all marine and most freshwater populations of three-spine and nine-spine sticklebacks (those numbers refer to the spines on the back of the fish), but the pelvic spine has been lost repeatedly in several freshwater populations, as a result of reduction in the pelvic fin (Fig. 15.19). Genetic crosses between sticklebacks from different populations enabled the cause of pelvic reduction to be identified as mutations in the control

Fig. 15.18 The two different forms of *Astyanax mexicanus.* Surface-living fish have eyes and are pigmented (upper photo). In cave-dwelling fish (cavefish) the eyes do not develop and pigmentation has been lost.
Photographs courtesy of A. Strickle, Y. Yamamoto, and W. Jeffery.

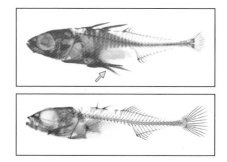

Fig. 15.19 Loss of the pelvic spines in three-spine sticklebacks (*Gasterosteus aculeatus***).** Upper panel: most populations of three-spine sticklebacks have a pair of pelvic spines (arrow). Lower photo: in some freshwater populations, the spines have been lost due to a mutation in the *Pitx1* gene.
Photographs courtesy of Mike Shapiro and David Kingsley.

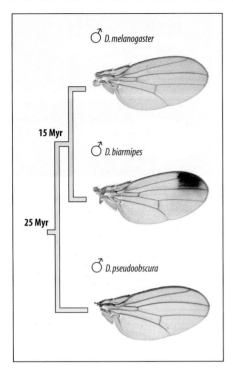

Fig. 15.20 **Evolution of differences in pigment pattern in closely related *Drosophila* species.** A spot of dark pigment on the tip of the wing of male *Drosophila biarmipes* is a relatively recently evolved trait among *Drosophila* species. It arises from a change in the regulatory region of the *yellow* gene, which is required for pigment production. Myr, millions of years ago.

From Gompel, N. et al.: 2005.

region of the gene for the homeodomain transcription factor Pitx1, with different mutations in the different populations. Pitx1 plays a key role in vertebrate hindlimb development (see Section 11.2) and when *Pitx1* and the related *Pitx2* gene are knocked-out in mice, hindlimb development is reduced. These adaptive mutations in sticklebacks cause loss of expression of Pitx1 exclusively in the pelvic region, preserving the essential developmental role of this gene in other parts of the body. Pelvic spine loss in sticklebacks is an interesting example of convergent evolution, because similar *cis*-regulatory mutations have been selected independently in the different stickleback populations.

Another example of evolution of a pattern as a result of changes in *cis*-regulatory regions is provided by the pigment pattern in the wing of male *Drosophila biarmipes*, which is closely related to our standard model organism, *D. melanogaster*, although separated by some 15 million years of evolution. *Drosophila biarmipes* males have a conspicuous dark spot at the anterior tip of the wing that is lacking in male *D. melanogaster* (Fig. 15.20). This difference is due to differences in the expression of the gene *yellow*, which is required for production of the black pigment, in the wings of the two species. In *D. melanogaster*, *yellow* is expressed at a low level throughout the wing. The spot in *D. biarmipes* appears to have evolved as a result of changes in the *cis*-regulatory region of the *yellow* gene, such that it is now activated in a particular spatial pattern by transcription factors that are involved in patterning the wing. For example, the protein Engrailed is expressed in the posterior wing compartment (see Section 11.18) and it activates *yellow* in the distal region of this compartment. Other conserved gene regulatory proteins have also been co-opted to control the *yellow* gene by the evolution of binding sites for them within the gene's control region.

15.10 Embryonic structures have acquired new functions during evolution

If two groups of animals that differ greatly in their adult structure and habits (such as fishes and mammals) pass through a very similar embryonic stage, this could indicate that they are descended from a common ancestor and, in evolutionary terms, are closely related. Thus, an embryo's development reflects the evolutionary history of its ancestors. Division of the body into segments, which then diverge from each other in structure and function, is a common feature in the evolution of both vertebrates and arthropods. In the vertebrates, one example of segmented structure is the branchial arches and clefts that are present in all vertebrate embryos, including humans, located just behind the head on either side (see Fig. 5.33). These structures are not the relics of the gill arches and gill slits of an adult fish-like ancestor, but represent structures that would have been present in the embryo of the fish-like ancestor of vertebrates as developmental precursors to gill slits and gill arches. During evolution, the branchial arches gave rise both to the gill arches of the primitive jawless fishes and, in a later modification, to gills and jaw elements in later-evolved fishes (Fig. 15.21). With time the arches became further modified, and in mammals they now give rise to various structures in the face and neck (Fig. 15.22), many of which are derived from the neural crest cells that migrate into the branchial arches early in development (see Section 5.13). The cleft between the first and second branchial arches provides the opening for the Eustachian tube, and endodermal cells in the clefts give rise to various glands, such as the thyroid and thymus.

Evolution rarely, if ever, generates a completely novel structure out of the blue. New anatomical features usually arise from modification of an existing structure. One can therefore think of much of evolution as a 'tinkering' with existing structures, which

Fig. 15.21 Modification of the branchial arches during the evolution of jaws in vertebrates. The ancestral jawless fish had a series of at least seven gill slits—branchial clefts—supported by cartilaginous or bony arches. Jaws developed from a modification of the first arch to give the mandibular arch, with the mandibular cartilage of the lower jaw and the hyoid arch behind it.

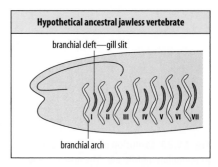

Hypothetical ancestral jawless vertebrate

branchial cleft—gill slit

branchial arch

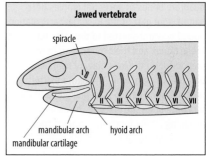

Jawed vertebrate

spiracle

mandibular arch   hyoid arch
mandibular cartilage

gradually fashions something different. It is possible because many structures are modular: that is, animals have anatomically distinct parts that can evolve independently. Vertebrae are modules, for example, and can evolve independently of each other; so too are limbs.

A nice example of a modification of an existing structure is provided by the evolution of the mammalian middle ear. This is made up of three bones—malleus, incus, and stapes—that transmit sound from the eardrum (the tympanic membrane) to the inner ear. In the reptilian ancestors of mammals, the joint between the skull and the lower jaw was between the quadrate bone of the skull and the articular bone of the lower jaw, which were also involved in transmitting sound (Fig. 15.23). The vertebrate lower jaw was originally composed of several bones, but during mammalian evolution one of these bones, the dentary, increased in size and came to comprise the whole lower jaw, and the other bones were lost. The articular bone was no longer attached to the lower jaw, and by changes in their development, the articular and the quadrate in mammals were modified into two bones, the malleus and incus, respectively, whose function was now to help transmit sound from the tympanic membrane. The quadrate is evolutionarily and developmentally homologous with the dorsal cartilage of the first branchial arch in the ancestral vertebrate, and the stapes with the dorsal cartilage of the second.

Evolution can adapt the same proteins for quite different purposes. The eye lenses of both cephalopods (octopuses and squids) and vertebrates consist of cells packed with crystallin proteins, which give the lens its transparency (see Section 11.23). The crystallins were originally thought to be unique to the lens and to have evolved for this special function, but more recent research indicates that they are co-opted proteins that are not structurally specialized for lens function, and in other contexts act as enzymes.

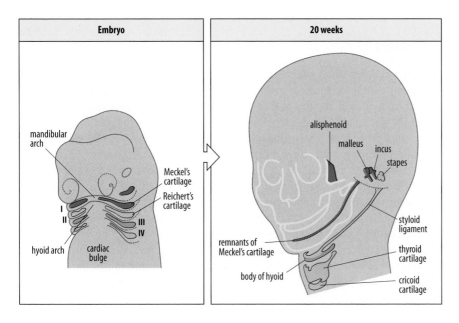

Embryo	20 weeks

mandibular arch

Meckel's cartilage

Reichert's cartilage

I
II
III
IV

hyoid arch   cardiac bulge

alisphenoid

malleus   incus
stapes

remnants of Meckel's cartilage

body of hyoid

styloid ligament

thyroid cartilage

cricoid cartilage

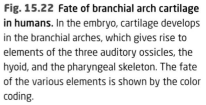

Fig. 15.22 Fate of branchial arch cartilage in humans. In the embryo, cartilage develops in the branchial arches, which gives rise to elements of the three auditory ossicles, the hyoid, and the pharyngeal skeleton. The fate of the various elements is shown by the color coding.

After Larsen, W.J.: 1993.

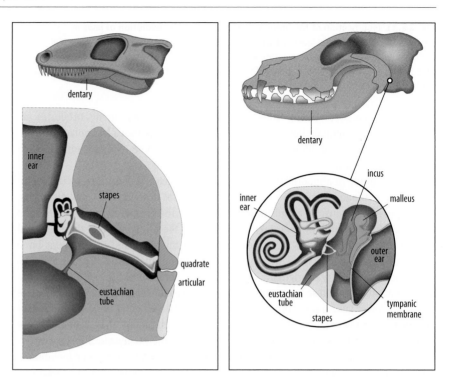

Fig. 15.23 Evolution of the bones of the mammalian middle ear. The articular and quadrate bones of ancestral reptiles (left panel) were part of the lower jaw articulation. Sound was transmitted to the inner ear via these bones and their connection to the stapes. When the lower jaw of mammals became a single bone (the dentary), the articular bone became the malleus, and the quadrate bone the incus of the middle ear, acquiring a new function in transmitting sound to the inner ear from the tympanic membrane (right panel). The eustachian tube forms between branchial arches I and II.

After Romer, A.S.: 1949.

These examples provide evidence of a key relationship between evolution and development, the gradual change of a structure into a different form. In many cases, however, we do not understand how intermediate forms were adaptive and gave a selective advantage to the animal. Consider, for example, the intermediate forms in the transition of the first branchial arch to jaws; what was the adaptive advantage? We do not know, and because of the passage of time and our current ignorance of the ecology of ancient organisms we may never know.

Summary

Groups of animals that pass through a similar embryonic stage are descended from a common ancestor and have evolved modifications of gene-regulatory circuits. The basic body plan of all animals is defined by patterns of Hox gene expression that provide positional identity, the interpretation of which has changed in evolution. The Hox genes themselves have undergone considerable evolution by gene duplication and divergence and are a good example of Darwin's description of evolution as 'descent with modification.' The limbs of tetrapod vertebrates evolved from fins, with the digits as a novel feature. The development of vertebrate and insect limbs involves the same set of pattern-establishing genes, reflecting the evolution of limb development from an ancestral mechanism for specifying body appendages. Comparison of patterns of dorso-ventral gene expression suggests that during the evolution of the vertebrates, the dorso-ventral axis of an invertebrate ancestor was inverted. During evolution, the development of structures can be altered so that they acquire new functions, as has happened in the evolution of the mammalian middle ear from a reptilian jaw bone, which itself evolved from a skeletal element in an ancestral branchial arch.

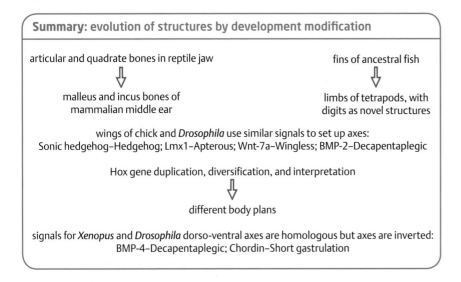

Summary: evolution of structures by development modification

articular and quadrate bones in reptile jaw ⇓ malleus and incus bones of mammalian middle ear

fins of ancestral fish ⇓ limbs of tetrapods, with digits as novel structures

wings of chick and *Drosophila* use similar signals to set up axes:
Sonic hedgehog–Hedgehog; Lmx1–Apterous; Wnt-7a–Wingless; BMP-2–Decapentaplegic

Hox gene duplication, diversification, and interpretation ⇓ different body plans

signals for *Xenopus* and *Drosophila* dorso-ventral axes are homologous but axes are inverted:
BMP-4–Decapentaplegic; Chordin–Short gastrulation

Changes in the timing of developmental processes

In the previous part of the chapter we focused on changes in spatial patterning that have occurred during evolution. But changes in the timing of developmental processes can also have major effects. In this part of the chapter, we look at some examples of how changes in the timing of growth and sexual maturation can affect animal form and behavior.

15.11 Evolution can be due to changes in the timing of developmental events

Differences among species in respect of the time at which developmental processes occur relative to one another, and relative to their timing in an ancestor, can have dramatic effects on both morphology and behavior. Differences in the feet of members of a genus of tropical salamanders illustrate the effect of a change in developmental timing on both the morphology and ecology of different species. Many species of the salamander genus *Bolitoglossa* are arboreal (tree living), rather than typically terrestrial, and their feet are modified for climbing on smooth surfaces. The feet of the arboreal species are smaller and more webbed than those of terrestrial species, and their digits are shorter (Fig. 15.24). These differences seem to be mainly the result of the development and growth of the foot ceasing at an earlier stage in the arboreal species than in the terrestrial species. The term used to describe such differences in timing is **heterochrony**.

Some of the clearest examples of heterochrony come from alterations in the timing of onset of sexual maturity in organisms with larval stages. The acquisition of sexual maturity by an animal while still in the larval stage is a process that goes under the name **neoteny**. The development of the animal, although not its growth, is retarded in relation to the maturation of the reproductive organs. This occurs in the Mexican axolotl, a type of salamander; the larva grows in size and matures sexually, but does not undergo metamorphosis. The sexually mature form remains aquatic and looks like an overgrown larva. However, the axolotl can be induced to undergo metamorphosis by treatment with the hormone thyroxine (Fig. 15.25).

Many animals have evolved free-swimming larval forms that have an advantage when it comes to dispersal and feeding, and then undergo a dramatic change in morphology to reach the adult state at metamorphosis. The essence of development is gradual change; yet at metamorphosis there is no gradual continuity between larva and adult. Metamorphosis makes more evolutionary sense, however, if it is assumed that all larval forms evolved by the insertion of the larval stage into the pre-existing development

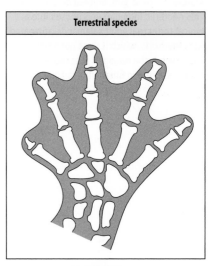
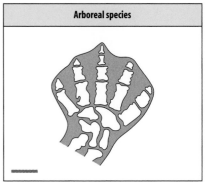

Fig. 15.24 Heterochrony in salamanders. In terrestrial species of the salamander *Bolitoglossa* (top panel), the foot is larger, has longer digits, and is less markedly webbed than in those that live in trees (bottom panel). This difference can be accounted for by foot growth ceasing at an earlier stage in the arboreal species. Scale bar = 1 mm.

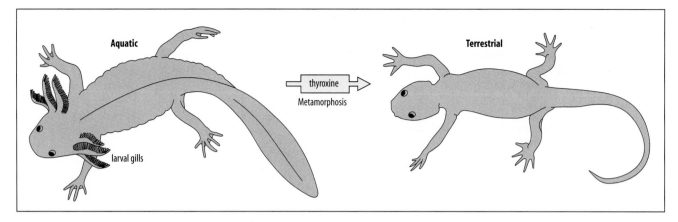

Fig. 15.25 Neoteny in salamanders. The sexually mature Mexican axolotl retains larval features, such as gills, and remains aquatic. This neotenic form metamorphoses into a typical terrestrial adult salamander if treated with thyroxine, which it does not produce, and which is the hormone that causes metamorphosis in other amphibians.

program of a directly developing animal, possibly as a result of heterochrony. In many invertebrates, the larva initially resembles the late gastrula stage, which could have given rise to the free-swimming larval form. Larval stages could not have been the original state because the metamorphosis that brings a larva back to the developmental program into a sexually mature adult is a highly complex process, and it is difficult to see how it could have evolved other than to re-enter the developmental pathway.

In vertebrates that undergo metamorphosis, such as frogs, if we assume that frog ancestors developed directly into adults, a change in the timing of events in the post-neurula stages, such as a delay in hindlimb development, could have led to acquisition of mobility and evolution of the feeding tadpole stage. Becoming mobile would have aided dispersal, but to become a reproducing adult, the larva had to return to the normal developmental program. That, in essence is what metamorphosis is for.

Larval stages can be lost as well as acquired during evolution. Some modern frogs have re-evolved direct development to the adult by a loss of the tadpole stage and the acceleration of the development of adult features. Frogs of the genus *Eleutherodactylus*, unlike the more typical amphibians *Rana* and *Xenopus*, develop directly into an adult frog and there is no aquatic tadpole stage, the eggs being laid on land. Typical tadpole features, such as gills and cement glands, do not develop, and prominent limb buds appear shortly after the formation of the neural tube (Fig. 15.26). In the embryo, the tail is modified into a respiratory organ. Such direct development requires a large supply of yolk to the egg in order to support development through to an adult without a tadpole feeding stage. This increase in yolk may itself be an example of heterochrony, involving a longer or more rapid period of yolk synthesis in the development of the egg.

Most sea urchins have a larval stage that takes a month or more to become adult. But there are some species in which direct development has evolved, so that they no longer go through a functional larval stage. Such species have large eggs and, as a result of their very rapid development, become juvenile urchins within four days. This has involved changes in early development such that the directly developing embryo gives rise to a 'larval' stage that lacks a gut and cannot feed, and which metamorphoses rapidly into the adult form.

Fig. 15.26 Development of the frog *Eleutherodactylus*. Typical frogs, such as *Xenopus* or *Rana*, lay their eggs in water and develop through an aquatic tadpole stage, which undergoes metamorphosis into the adult. Frogs of the genus *Eleutherodactylus* lay their eggs on land and the frog hatches from the egg as a miniature adult, without going through an aquatic free-living larval stage. The embryonic tail is modified as a respiratory organ.

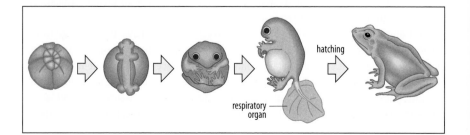

15.12 The evolution of life histories has implications for development

Animals and plants have very diverse life histories: small birds breed in the spring following their birth, and continue to do so each year until their death; Pacific salmon breed in a suicidal burst at 3 years of age; and oak trees require 30 years of growth before producing acorns, which they eventually produce by the thousand. In order to understand the evolution of such life histories, evolutionary ecologists consider them in terms of probabilities of survival, rates of reproduction, and optimization of reproductive effort. These factors have important implications for the evolution of developmental strategies, particularly in relation to the rate of development. For example, as we have just seen, a characteristic feature of many animal life histories is the presence of a free-living larval stage that is distinct from, and usually simpler in form than, the sexually mature adult, and which feeds in a different way.

Life histories help us to understand the evolution of long germ-band insects, such as *Drosophila*, which are of more recent origin than short germ-band insects, such as grasshoppers (see Section 2.25). Unlike *Drosophila*, short germ-band insects do not have a larval stage and develop directly into small, immature adult-like forms. In contrast, *Drosophila* develops very rapidly into a feeding larva, taking only 24 hours to start feeding, compared with the grasshopper's 5–6 days. It is easy to imagine conditions in which there would have been a selective advantage to insects whose larvae begin to feed as quickly as possible. It is also likely that embryos are more vulnerable than adults and so there would be selection for making the embryonic stage shorter. Thus it is likely that the complex developmental mechanisms of long-band insects—the system for setting up the complete antero-posterior axis in the egg—evolved as a result of selection pressure for rapid development. In the case of fruit flies like *Drosophila*, this was perhaps related to the ability to feed on fast-disappearing fruit.

Egg size can also be best understood within the context of life histories. If we assume that the parent has limited energy resources to put into reproduction, the question is how should these resources best be invested in making gametes, particularly eggs; is it more advantageous to make lots of small eggs or a few large ones? In general, it seems that the larger the egg, and thus the larger the offspring at birth, the better are the chances of the offspring surviving. This would seem to suggest that in most circumstances an embryo needs to give rise to a hatchling as large as possible. Why then, do some species lay many small eggs capable of rapid development? One possible answer is that a parental investment in large eggs may reduce the parents' own chance of surviving and laying more eggs. This strategy may be especially successful in variable environmental conditions, where populations can suddenly crash. The rapid development of a larval stage enables early feeding and dispersal to new sites in such circumstances.

Summary

Changes in the timing of developmental processes that have occurred during evolution can alter the form of the body, for if different regions grow faster than others, their size is proportionally increased if the animal gets larger. The decrease in size of the lateral digits of horses is partly due to this type of developmental change. Speed of development and egg size also have important evolutionary implications. Changing the time at which an animal becomes sexually mature can result in adults with larval characteristics. Some animals, such as frogs and sea urchins, which usually have a larval form, have evolved species that develop directly into the adult without a larval stage. The life histories of animals also have implications for evolution and development.

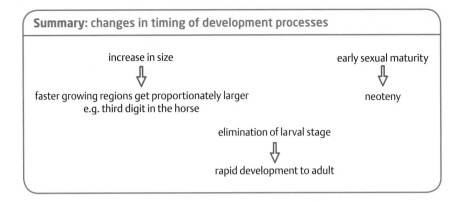

Summary: changes in timing of development processes

increase in size ⇩ faster growing regions get proportionately larger e.g. third digit in the horse

early sexual maturity ⇩ neoteny

elimination of larval stage ⇩ rapid development to adult

Summary to Chapter 15

Many developmental processes have been conserved during evolution. While many questions relating to evolution and development remain unanswered, it is clear that development reflects the evolutionary history of ancestral embryos. The origins of multicellularity and the embryo from a single-celled ancestor are still highly speculative. All vertebrate embryos pass through a conserved phylotypic developmental stage, although there can be considerable divergence both earlier and later in development. The signals involved in patterning of vertebrate and arthropod appendages, as well as of the dorso-ventral axis, show remarkable similarity and conservation. The pattern of Hox gene expression along the body axis of vertebrates and arthropods is conserved, and major changes in body plan reflect changes in both Hox gene expression and their downstream targets. Changes in gene-regulatory regions were crucial. Alterations in the timing of developmental events have played an important role in evolution. Such changes can alter the overall form of an organism as a result of differences in growth rates of different structures, and can also result in sexual maturation at larval stages.

■ End of chapter questions

Long answer (concept questions)

1. Trace the possible steps in the evolution of the Metazoa from a single-celled ancestor. What were the key innovations that led to the evolution of the Metazoa, and especially of bilaterally symmetric animals?

2. Why has the phylotypic stage been so strongly conserved during evolution? What events and processes are occurring at this point in development that make this stage so central to animal development?

3. Give an example of a change in protein structure and function that occurred after gene duplication. Give an example of a change in gene expression that occurred after gene duplication.

4. Which genes of the *Drosophila* Hox complex would you expect to be most closely related, at the level of their protein sequences, based on their evolutionary history (see Fig. 15.7)?

5. Genetic experiments in *Drosophila* have led to a hypothesis that *Antp* is an ancestral gene, and that diversification of the segments posterior to T2 is a later evolutionary addition; what are these experiments, and how do they support this hypothesis?

6. Compare the patterns of Hox gene expression in Fig. 15.9; how would you explain these observations in words? (As a challenge, based *solely* on these observations, predict the consequence of mis-expressing *Antp*, *Ubx*, and *abd-A* throughout the thorax and abdomen of a grasshopper (or fly).) After this, refer back to Fig. 2.48: what would be the actual result of *abd-A* expression throughout the thorax? How might the evolution of the Hox gene targets explain the differences between *Artemia* and *Drosophila*?)

7. *Drosophila* larvae have no legs, yet lepidopteran larvae have legs on each thoracic and abdominal segment. Explain how Hox genes and the *Distal-less* gene mediate this difference.

8. How could changes in Hox gene expression in lateral plate mesoderm explain the absence of limbs on snakes?

9. How do regulatory changes in *yellow* explain the pigmentation differences in *Drosophila biarmipes* compared to *D. melanogaster*?

10. Synthesize the information in Figs 15.21, 15.22, and 15.23 and provide a narrative for the evolution of the incus and malleus bones in the inner ear of mammals.

11. Describe what is meant by the term heterochrony and give an example of an evolutionary change that can be described as a heterochrony.

12. Using examples from the text, discuss how probabilities of survival, rates of reproduction, and optimization of reproductive effort can drive evolutionary changes in development.

Multiple choice (factual recall questions)

NB There is only one correct answer to each question.

1. Ontogeny is

a) the development of an individual organism

b) the development of the ear

c) the evolutionary history of a species or group

d) the study of cancer

2. An example of a triploblast is

a) *Hydra*

b) sponges

c) *Xenopus*

d) triploid plants such as seedless watermelons

3. Considering the ancient evolutionary split between protostomes and deuterostomes, which of the following pairs of animals is most closely related evolutionarily?

a) *Caenorhabditis elegans* and amphioxus

b) humans and *Drosophila*

c) sea urchins and humans

d) snails and starfish

4. The evolution of the vertebrate Hox clusters probably arose through

a) duplications of an ancestral cluster to give four ancestral clusters, followed by tandem duplications which have apparently occurred differently in each of the four clusters

b) random changes in DNA sequences followed by natural selection, leading to the independent evolution of each of the Hox genes from unrelated sequences

c) tandem duplications of an ancestral gene, followed by duplications of the whole cluster

d) the generation of four ancestral Hox genes on separate chromosomes, followed by tandem duplications to give rise to the four Hox clusters

5. An animal's ventral surface is defined by

a) the expression of BMP family members

b) the position of the mouth

c) the side of the animal that faces the ground

d) the side opposite the central nervous system

6. The horse's hoof is homologous to human

a) fingers or toes

b) humerus

c) radius and ulna

d) wrist bones

7. In humans, the branchial arches are

a) embryonic gill slits, reflecting our evolution from fish

b) embryonic precursors to gills, which do not continue to develop in humans

c) embryonic precursors to structures predominantly in the jaw

d) embryonic structures that develop into the ribs

8. The evolutionary acquisition of sexual maturity by an animal while still in a larval stage is called

a) apoptosis

b) metamorphosis

c) neoteny

d) ontogeny

9. Which is true?

a) Bat wings evolved from bird wings.

b) Ears evolved from jaws.

c) Insects evolved from crustaceans.

d) Limbs evolved from fins.

10. Hedgehog family members are expressed in the _____ position in both vertebrate and insect wings.

a) compartment boundary

b) distal

c) dorsal

d) posterior

Multiple choice answer key

1: a, 2: c, 3: c, 4: c, 5: b, 6: a, 7: c, 8: c, 9: d, 10: d.

■ General further reading

Carroll, S.B.: **Evo-devo and an expanding evolutionary synthesis: a genetic theory of morphological evolution**. *Cell* 2008, **134**: 25–36.

Carroll, S.B., Grenier, J.K., Weatherbee, S.D.: *From DNA to Diversity*, 2nd edn. Malden, MA: Blackwell Science, 2005.

Finnerty, J.R., Pang, K., Burton, P., Paulson, D., Martindale, M.Q.: **Origins of bilateral symmetry**: **Hox and *dpp* expression in a sea anemone**. *Science* 2004, **304**: 1335–1337.

Hoekstra, H.E., Coyne, J.A.: **The locus of evolution: evo devo and the genetics of adaptation**. *Evolution* 2007, **61**: 995–1016.

Kirschner, M., Gerhart, J.: **Evolvability**. *Proc. Natl. Acad. Sci. USA* 1998, **95**: 8420–8427.

Kirschner, M., Gerhart, J.: *The Plausibility of Life*. Yale University Press, 2005.

Raff, R.A.: *The Shape of Life*. University of Chicago Press, 1996.

Richardson, M.K.: **Vertebrate evolution: the developmental origins of adult variation**. *BioEssays* 1999, **21**: 604–613.

Wray, G.A.: **The evolutionary significance of *cis*-regulatory mutations**. *Nat. Rev. Genet.* 2007, **8**: 206–216.

Box 15A 'Darwin's finches'

Abzhanov, A., Kuo, W.P., Hartmann, C., Grant, B.R., Grant, P.R., Tabin, C.J.: **The calmodulin pathway and evolution of elongated beak morphology in Darwin's finches**. *Nature* 2006, **442**: 563–567.

■ Section further reading

15.1 Genomic evidence is throwing light on the origin of metazoans

Putnam, N.H., Srivastava, M., Hellsten, U., Dirks, B., Chapman, J., Salamov, A., Terry, A., Shapiro, H., Lindquist, E., Kapitonov, V.V., *et al.*: **Sea anemone genome reveals ancestral eumetazoan gene repertoire and genomic organization**. *Science* 2007, **317**: 86–94.

Srivastava, M., Begovic, E., Chapman, J., Putnam, N.H., Hellsten, U., Kawashima, T., Kuo, A., Mitros, T., Salamov, A., Carpenter, M.L., *et al.*: **The *Trichoplax* genome and the nature of placozoans**. *Nature* 2008, **454**: 955–960.

15.2 Multicellular organisms evolved from single-celled ancestors

Brooke, N.M., Holland, P.W.H.: **The evolution of multicellularity and early animal genomes**. *Curr. Opin. Genet. Dev.* 2003, **6**: 599–603.

Jaegerstern, G.: **The early phylogeny of the metazoa. The bilaterogastrea theory**. *Zool. Bidrag. (Uppsala)* 1956, **30**: 321–354.

King, N.: **The unicellular ancestry of animal development**. *Dev. Biol.* 2004, **7**: 313–325.

King, N., Westbrook, M.J., Young, S.L., Kuo, A., Abedin, M., Chapman, J., Fairclough, S., Hellsten, U., Isogai, Y., Letunic, I., *et al.*: **The genome of the choanoflagellate *Monosiga brevicollis* and the origin of metazoans**. *Nature* 2008, **451**: 783–788.

Miller, D.J., Ball, E.E.: **Animal evolution: the enigmatic phylum Placozoa revisited**. *Curr. Biol.* 2005, **15**: R26–R28.

Rudel, D., Sommer, R.J.: **The evolution of developmental mechanisms**. *Dev. Biol.* 2003, **264**: 15–37.

Szathmary, E., Wolpert, L.: **The transition from single cells to multicellularity**. In *Genetic and Cultural Evolution of Cooperation* (ed. Hammersteen, P.) 271–289. Cambridge, MA: MIT Press, 2004.

Wolpert, L.: **Gastrulation and the evolution of development**. *Development (Suppl.)* 1992, 7–13.

Wolpert, L., Szathmary, E.: **Evolution and the egg**. *Nature* 2002, **420**: 745.

15.3 Hox gene complexes have evolved through gene duplication

Brooke, N.M., Gacia-Fernandez, J., Holland, P.W.H.: **The ParaHox gene cluster is an evolutionary sister of the Hox gene cluster**. *Nature* 1998, **392**: 920–922.

Duboute, D.: **The rise and fall of Hox gene clusters**. *Development* **134**: 2549–2560.

Ferrier, D.E.K., Holland, P.W.H.: **Ancient origin of the Hox gene cluster**. *Nat. Rev. Genet.* 2001, **2**: 33–34.

Prince, V.E., Pickett, F.B.: **Splitting pairs: the diverging fates of duplicated genes**. *Nat. Rev. Genet.* 2002, **3**: 827–837.

Valentine, J.W., Erwin, D.H., Jablonski, D.: **Developmental evolution of metazoan bodyplans: the fossil evidence**. *Dev. Biol.* 1996, **173**: 373–381.

15.4 Changes in Hox genes generated the elaboration of vertebrate and arthropod body plans

Akam, M.: **Hox genes and the evolution of diverse body plans**. *Phil. Trans. R. Soc. Lond. B* 1995, **349**: 313–319.

Averof, M.: **Origin of the spider's head**. *Nature* 1998, **395**: 436–437.

Duboule, D.: **A Hox by any other name**. *Nature* 2000, **403**: 607–610.

Galant, R., Carroll, S.B.: **Evolution of a transcriptional repression domain in an insect Hox protein**. *Nature* 2003, **415**: 910–913.

Pavlopoulos, A., Averof, M.: **Developmental evolution: Hox proteins ring the changes**. *Curr. Biol.* 2002, **12**: R291–R293.

Stern, D.L.: **A role of *Ultrabithorax* in morphological differences between *Drosophila* species**. *Nature* 1998, **396**: 463–466.

15.5 The position and number of paired appendages in insects is dependent on Hox gene expression

Carroll, S.B., Weatherbee, S.D., Langeland, J.A.: **Homeotic genes and the regulation and evolution of insect wing number**. *Nature* 1995, **375**: 58–61.

Levine, M.: **How insects lose their wings**. *Nature* 2002, **415**: 848–849.

Weatherbee, S.D., Carroll, S.: **Selector genes and limb identity in arthropods and vertebrates**. *Cell* 1999, **97**: 283–286.

Weatherbee, S.D., Nijhout, H.F, Grunnert, L.W., Halder, G., Galant, R., Selegue, J., Carroll, S.: **Ultrabithorax function in butterfly wings and the evolution of insect wing patterns**. *Curr. Biol.* 1999, **9**: 109–115.

15.6 The basic body plan of arthropods and vertebrates is similar, but the dorso-ventral axis is inverted

Arendt, D., Nübler-Jung, K.: **Dorsal or ventral: similarities in fate maps and gastrulation patterns in annelids, arthropods and chordates**. *Mech. Dev.* 1997, **61**: 7–21.

Davis, G.K., Patel, N.H.: **The origin and evolution of segmentation**. *Trends Biochem. Sci.* 1999, **24**: M68–M72.

Gerhart, J., Lowe, C., Kirschner, M.: **Hemichordates and the origins of chordates**. *Curr. Opin. Genet. Dev.* 2005, **15**: 461–467.

Holley, S.A., Jackson, P.D., Sasai, Y., Lu, B., De Robertis, E., Hoffman, F.M., Ferguson, E.L.: **A conserved system for dorsoventral patterning in insects and vertebrates involving *sog* and *chordin***. *Nature* 1995, **376**: 249–253.

15.7 Limbs evolved from fins

Ahn, D., Ho, R.K.: **Tri-phasic expression of posterior Hox genes during development of pectoral fins in zebrafish: implications for the evolution of vertebrate paired appendages**. *Dev. Biol.* 2008, **322**: 220–233.

Boisvert, C.A., Mark-Kurik, E., Ahlberg, P.E.: **The pectoral fin of *Panderichthys* and the origin of digits**. *Nature* 2008, **456**: 636–638.

Capdevila, J., Izpisua Belmonte, J.C.: **Perspectives on the evolutionary origin of tetrapod limbs**. *J. Exp. Zool. (Mol. Dev. Evol.)* 2000, **288**: 287–303.

Coates, M.I., Jeffery, J.E., Rut, M.: **Fins to limbs: what the fossils say**. *Evol. Dev.* 2002, **4**: 390–401.

Cohn, M.J., Tickle, C.: **Developmental basis of limblessness and axial patterning in snakes**. *Nature* 1999, **399**: 474–479.

Ruvinsky, I., Gibson-Brown, J.J.: **Genetic and developmental bases of serial homology in vertebrate limb evolution**. *Development* 2000, **127**: 5211–5244.

Sordino, P., van der Hoeven, F., Duboule, D.: **Hox gene expression in teleost fins and the origin of vertebrate digits**. *Nature* 1995, **375**: 678–681.

Woltering, J.M., Vonk, F.J., Müller, H., Bardine, N., Tuduce, I.L., de Bakker, M.A., Knöchel, W., Sirbu, I.O., Durston, A.J., Richardson, M.K.: **Axial patterning in snakes and caecilians: evidence for an alternative interpretation of the Hox code**. *Dev. Biol.* 2009, **332**: 82–89.

15.8 Vertebrate and insect wings make use of evolutionarily conserved developmental mechanisms

Panganiban, G., Irvine, S.M., Lowe, C., Roehl, H., Corley, L.S., Sherbon, B., Grenier, J.K., Fallon, J.F., Kimble, J., Walker, M., Wray, G.A., Swalla, B.J., Martindale, M.Q., Carroll, S.B.: **The origin and evolution of animal appendages**. *Proc. Natl. Acad. Sci. USA* 1997, **94**: 5162–5166.

15.9 The evolution of developmental differences can be based on changes in just a few genes

Chan, Y.F., Marks, M.E., Jones, F.C., Villarreal, G. Jr., Shapiro, M.D., Brady, S.D., Southwick, A.M., Absher, D.M., Grimwood, J., Schmutz, J., *et al.*: **Adaptive evolution of pelvic reduction in sticklebacks by recurrent deletion of a *Pitx1* enhancer**. *Science* 2010, **327**: 302–305.

Gompel, N., Carrol, S.B.: **Genetic mechanisms and constraints governing the evolution of correlated traits in drosophilid flies**. *Nature* 2003, **424**: 931–935.

Gompel, N., Prud'homme, B., Wittkopp, P.J., Kassner, V.A., Carroll, S.B.: **Chance caught on the wing: cis-regulatory evolution and the origin of pigment patterns in *Drosophila***. *Nature* 2005, **433**: 481–487.

Jeffery, W.R.: **Evolution and development in the cavefish *Astyanax***. *Curr. Top. Dev. Biol.* 2009, **86**: 191–221.

Shapiro, M.D., Marks, M.E., Peichel, C.L., Blackman, B.K., Nereng, K.S., Jonsson, B., Schluter, D., Kingsley, D.M.: **Genetic and developmental basis of evolutionary pelvic reduction in threespine sticklebacks**. *Nature* 2004, **428**: 717–723.

Shapiro, M.D., Summers, B.R., Balabhadra, S., Aldenhoven, J.T., Miller, A.L., Cunningham, C.B., Bell, M.A., Kingsley, D.M.: **The genetic architecture of skeletal convergence and sex determination in ninespine sticklebacks**. *Curr. Biol.* 2009, **19**: 1140–1145.

Yamamoto, Y., Stock, D.W., Jeffery, W.R.: **Hedgehog signaling controls eye degeneration in blind cavefish**. *Nature* 2004, **431**: 844–847.

15.10 Embryonic structures have acquired new functions during evolution

Cerny, R., Lwigale, P., Ericsson, R., Meulemans, D., Epperlein, H.H. Bronner-Fraser, M.: **Developmental origins and evolution of jaws: new interpretation of 'maxillary' and 'mandibular'**. *Dev Biol.* 2004, **276**: 225–236.

Cohn, M.J.: **Lamprey Hox genes and the origin of jaws**. *Nature* 2002, **416**: 386–387.

De Robertis, E.M.: **Evo-devo: variations on ancestral themes**. *Cell* 2008, **132**: 185–195.

Erwin, D.H.: **Early origin of the bilaterian developmental toolkit**. *Philos. Trans. R. Soc. Lond. B Biol. Sci.* 2009, **364**: 2253–2261.

Romer, A.S.: *The Vertebrate Body*. Philadelphia: W.B. Saunders, 1949.

15.11 Evolution can be due to changes in the timing of developmental events

Alberch, P., Alberch, J.: **Heterochronic mechanisms of morphological diversification and evolutionary change in the neotropical salamander *Bolitoglossa occidentales* (Amphibia: Plethodontidae)**. *J. Morphol.* 1981, **167**: 249–264.

Lande, R.: **Evolutionary mechanisms of limb loss in tetrapods**. *Evolution* 1978, **32**: 73–92.

Raynaud, A.: **Developmental mechanism involved in the embryonic reduction of limbs in reptiles**. *Int. J. Dev Biol.* 1990, **34**: 233–243.

Wray, G.A., Raff, R.A.: **The evolution of developmental strategy in marine invertebrates**. *Trends Evol. Ecol.* 1991, **6**: 45–56.

15.12 The evolution of life histories has implications for development

Partridge, L., Harvey, P.: **The ecological context of life history evolution**. *Science* 1988, **241**: 1449–1455.

Glossary

In insects the body is divided into three distinct parts, the head at the anterior end, followed by the thorax and the posterior **abdomen**.

Abembryonic pole *see* embryonic–abembryonic axis.

Accretionary growth is an increase in mass due to secretion of large quantities of extracellular matrix by cells, which enlarges the volume of extracellular space. It occurs, for example, in cartilage and bone, where most of the tissue mass is extracellular.

Acheiropodia is a rare, inherited human condition in which structures below the elbow or knee do not develop; but there is no other defect.

Achondroplasia is a form of dwarfism in which the limbs are short in relation to the rest of the body.

The **acron** is a specialized structure associated with the most anterior region of the *Drosophila* embryo.

The **acrosomal reaction** is the release of enzymes and other proteins from the **acrosomal vesicle** or **acrosome** of the sperm head that occurs once a sperm has bound to the outer surface of the egg. It helps the sperm to penetrate the outer layers of the egg.

Actin filaments or **microfilaments** are one of the three principal protein filaments of the cytoskeleton. They are involved in cell movement and changes in cell shape. Actin filaments are also part of the contractile apparatus of muscle cells.

Gene-regulatory proteins that turn genes on are known as **activators**.

Actomyosin is an assembly of actin filaments and the motor protein myosin that can undergo contraction.

In plants, the **adaxial–abaxial** axis runs from the center of the plant stem to the circumference. In plant leaves it runs from the upper surface to the lower surface (dorsal to ventral).

Adherens junctions are a type of **adhesive cell junction** in which the adhesion molecules linking the two cells together are cadherins that are linked intracellularly to the actin cytoskeleton.

The **allantois** is a set of extra-embryonic membranes that develops in many vertebrate embryos. In bird and reptile embryos it acts as a respiratory surface, while in mammals its blood vessels carry blood to and from the placenta.

An **allele** is a particular version of a gene. In diploid organisms, two alleles of each gene are present (one on each chromosome of a homologous pair), which may or may not be identical.

The **amnion** is an extra-embryonic membrane in birds, reptiles, and mammals, which forms a fluid-filled sac that encloses and protects the embryo. It is derived from extra-embryonic ectoderm and mesoderm.

The **amnioserosa** of the *Drosophila* embryo is an extra-embryonic membrane on the dorsal side of the embryo.

Amniotes are vertebrates whose embryos have an amnion. They comprise the mammals, birds, and reptiles.

Anamniotes are vertebrates whose embryos do not have an amnion. They comprise the fish and amphibians.

An **androgenetic** embryo is an embryo in which the two sets of homologous chromosomes are both paternal in origin.

Angioblasts are the mesodermal precursor cells that will give rise to blood vessels.

Angiogenesis is the process by which small blood vessels sprout from the larger vessels.

The **animal region** of an egg is the end of the egg where the nucleus resides, usually away from the yolk. The most terminal part of this region is the **animal pole**, which is directly opposite the vegetal pole at the other end of the egg. In *Xenopus* the pigmented animal half is called the **animal cap**.

The **animal–vegetal axis** runs from the animal pole to the vegetal pole in an egg or early embryo.

Aniridia is the absence of an iris in the eye.

The **antennapedia complex** comprises one part of the Hox gene complex in *Drosophila*

The **anterior visceral ectoderm** (AVE) is an extra-embryonic tissue in the early mouse embryo that is involved in inducing anterior regions of the embryo.

The **antero-posterior axis** defines which is the 'head' end and which is the 'tail' end of an animal. The head is anterior and the tail posterior. In the vertebrate limb, this axis runs from the thumb to little finger.

Anticlinal cell divisions are divisions in planes at right angles to the outer surface of a tissue.

Antisense RNA is an RNA complementary in sequence to an mRNA, which blocks expression of a protein by binding to its mRNA and blocking translation.

The **apical–basal axis** of a plant is the axis running from shoot tip to root tip.

The fact that buds behind the tip of a plant shoot will not form side stems when the tip is intact, is known as **apical dominance**. It is due to the production of the hormone auxin by the shoot tip.

The **apical ectodermal ridge** or **apical ridge** is a thickening of the ectoderm at the distal end of the developing chick and mammalian limb bud.

An **apical meristem** is the region of dividing cells at the tip of a growing shoot or root.

Apoptosis or programmed cell death is a type of cell death that occurs widely during development. In programmed cell death, a cell is induced to commit 'suicide,' which involves fragmentation of the DNA and shrinkage of the cell. These apoptotic cells are removed by the body's scavenger cells and, unlike necrosis, their death does not cause damage to surrounding cells.

The **archenteron** is the cavity formed inside the embryo when the endoderm and mesoderm invaginate during gastrulation. It forms the gut.

The **area opaca** is the outer dark area of the chick blastoderm.

The **area pellucida** is the central clear area of the chick blastoderm.

Asymmetric cell divisions or **asymmetric divisions** are cell divisions in which the daughter cells are different from each other because some cytoplasmic determinant(s) have been distributed unequally between them.

A developmental process is said to be proceeding **autonomously** when it can continue without a requirement for extracellular signals to be continuously present. *See also* **cell autonomous**.

Aux/IAA proteins are transcriptional repressors that block the expression of auxin-responsive genes in the absence of the plant hormone auxin.

The plant hormone **auxin** is a small organic moelcule that is important in almost all aspects of plant development. It acts by regulating the expression of **auxin-responsive genes** by means of its effects on Aux/IAA proteins.

Auxin-response factors are gene regulatory proteins that are responsive to the plant hormone auxin.

AVE *see* **anterior visceral endoderm**.

Axial structures are those that form along the main axis of the body, such as the notochord, vertebral column, and neural tube in vertebrates.

Axons are long cell processes of neurons that conduct nerve impulses away from the cell body. The end of an axon, the **axon terminal**, forms contacts (synapses) with other neurons, muscle cells, or glandular cells.

The protein **β-catenin** functions both as a transcriptional co-activator and as one of the proteins present at cell junctions. In its role as a transcriptional co-activator it is activated in early development in many vertebrates as the end result of a Wnt signaling pathway.

The **basal lamina** or **basement membrane** is a sheet of extracellular matrix that separates an epithelial layer from the underlying tissues. For example, the epidermis of the skin is separated from the dermis by a basal lamina.

Animals in which the only axis of symmetry is the central axis running from head to tail are said to possess **bilateral symmetry**. The two sides of the body are mirror images of each other.

The **bithorax complex** comprises one part of the Hox gene complex in *Drosophila*.

In amphibian limb regeneration, a **blastema** is formed from the dedifferentiation and proliferation of cells beneath the wound epidermis, and gives rise to the regenerated limb.

The **blastocoel** is the fluid-filled cavity that develops in the interior of a blastula.

The **blastocyst** stage of a mammalian embryo corresponds in form to the blastula stage of other animal embryos, and is the stage at which the embryo implants in the uterine wall.

A **blastoderm** is a post-cleavage embryo composed of a solid layer of cells rather than a spherical blastula, as in early chick, zebrafish, and *Drosophila* embryos. The chick blastoderm is also known as the **blastodisc**.

Blastomeres are the cells derived from cleavage of the early embryo.

The **blastopore** is the slit-like or circular invagination on the surface of amphibian and sea-urchin embryos, where the mesoderm and endoderm move inside the embryo at gastrulation.

The **blastula** stage in animal development is the outcome of cleavage. The blastula is a hollow ball of cells, composed of an epithelial layer of small cells enclosing a fluid-filled cavity—the blastocoel.

Blastula organizer *see* **Nieuwkoop center**.

The **block to polyspermy** is the mechanism used by many animal eggs to prevent fertilization by more than one sperm. In sea urchins and *Xenopus* it is composed of two stages—an immediate **rapid block to polyspermy** and a subsequent **slow block to polyspermy**.

The **body plan** describes the overall organization of an organism; for example, the position of the head and tail, and the plane of bilateral symmetry, where it exists. The body plan of most animals is organized around two main axes: the antero-posterior axis and the dorso-ventral axis.

The **brachial** region of a vertebrate embryo is the region that includes the forelimb or that gives rise to the structures of the forelimb.

The **branchial arches** are structures that develop on each side of the embryonic head and give rise to the gill arches in fishes, and to the jaws and other facial structures in other vertebrates.

Branching morphogenesis describes the development and growth of stuctures such as blood vessels or the bronchi and bronchioles of the lung, which develop by the successive branching of a tube of epithelium.

The **cadherins** are a family of cell-adhesion molecules with important roles in development.

The **cambium** in plants is a ring of meristem in the stem that gives rise to new stem tissue, which increases the diameter of the stem.

The **canonical Wnt/β-catenin pathway** is an intracellular signaling pathway stimulated by members of the Wnt family of signal proteins that leads to the stabilization of β-catenin and its entry into nuclei, where it acts as a transcriptional co-activator.

The functional maturation of sperm after they have been deposited in the female reproductive tract is known as **capacitation**.

Caspases are intracellular proteases, some of which are involved in apoptosis.

Cell-adhesion molecules bind cells to each other and to the extracellular matrix. The main classes of adhesion molecules important in development are the cadherins, the immunoglobulin superfamily, and the integrins.

Cell adhesiveness is the property of cells that causes them to stick together.

The effects of a gene are **cell-autonomous** if they only affect the cell the gene is expressed in.

The **cell body** is the part of a neuron that contains the nucleus, and from which the axon and dendrites extend.

Cell–cell interaction and **cell–cell signaling** are general terms to describe the various types of intercellular communication by which one cell influences the behavior of another cell. Cells can communicate with each other via cell contact or by the secretion of signaling molecules that influence the behavior of other cells nearby or at a distance.

The **cell cycle** is the sequence of events by which a cell duplicates itself and divides in two.

During **cell differentiation**, cells become functionally and structurally different from one another and become distinct cell types, such as muscle or blood cells.

Growth can occur by **cell enlargement**, which is an increase in cell size without division. This type of growth is common in plants especially.

Cell-lineage restriction occurs when all the descendants of a particular group of cells remain within a 'boundary' and never mix with an adjacent group of cells of a different lineage. Compartment boundaries in insect development are boundaries of lineage restriction.

Cell motility describes the ability of cells to move, to contract or to change shape.

Growth, or an increase in mass, can occur by **cell proliferation**—the growth and division of cells to form new cells.

Cell-replacement therapy aims to correct disease by replacing damaged cells by new healthy cells, including stem cells.

Cell senescence *see* **senescence**.

The **centrosome** of a cell is the organizing center for microtubule growth. It duplicates before mitosis or meiosis, and each centrosome forms one end of the microtubule spindle. It consists of a pair of centrioles, one of which gives rise to the basal body of the primary cilium in many types of differentiated cells.

The **checkpoints** that occur during the cell cycle enable cells to monitor progress through the cycle and ensure that a previous stage has been completed before embarking on the next.

The **chemoaffinity hypothesis** proposes that each retinal neuron carries a chemical label that enables it to connect reliably with an appropriately labeled cell in the optic tectum.

A **chemoattractant** is a molecule that attracts cells to move towards it.

A **chemorepellent** is a molecule that repels cells, causing them to move away from it.

A **chimeric** organism or tissue (a chimera) is made up of cells from two or more different sources, and thus of different genetic constitutions.

ChIP-chip and **ChIP-seq** are techniques for determining which sites in chromosomal DNA are bound by particular proteins *in vivo*. They involve immunoprecipitation of chromatin fragments by antibodies against the protein of interest followed by analysis of the bound DNA by microarrays (ChIP-chip) or DNA sequencing (ChIP-seq).

Chondrocytes are differentiated cartilage cells.

The **chorion** is the outermost of the extra-embryonic membranes in birds, reptiles, and mammals. It is involved in respiratory gas exchange. In birds and reptiles it lies just beneath the shell. In mammals it is part of the placenta and is also involved in nutrition and waste removal. The chorion of insect eggs has a different structure.

Chromatin is the material of which chromosomes are made. It is composed of DNA and protein. Enzyme complexes called **chromatin-remodeling complexes** can act on chromatin to modify it and alter the ability of the DNA to be transcribed.

The **circadian clock** is an internal 24-hour timer present in living organisms that causes many metabolic and physiological processes, including the expression of some genes, to vary in a regular manner throughout the day.

The *cis***-regulatory control region** of a gene comprises the sequences flanking the gene and containing sites at which the expression of that gene can be controlled. Many control regions contain a variety of different *cis*-**regulatory modules**, which are short regions containing multiple binding sites for different transcription factors; the combination of factors bound determines whether the gene is switched on or off.

Cleavage occurs after fertilization and is a series of rapid cell divisions without growth that divides the embryo up into a number of small cells.

A **clone** is a collection of genetically identical cells derived from a single cell by repeated cell division, or the genetically identical offspring of a single individual produced by asexual reproduction or artifical cloning techniques.

Cloning is the procedure by which an individual genetically identical to a 'parent' is produced by transplantation of a parental somatic cell nucleus into an unfertilized oocyte.

A **co-activator** is a gene-regulatory protein that promotes gene expression but does not bind to DNA itself. It binds to and influences the behavior of other DNA-binding transcription factors.

The **coding region** of a gene is that part of the DNA that encodes a polypeptide or functional RNA.

The correspondence between the order of Hox genes on a chromosome and their temporal and spatial order of expression in the embryo is known as **co-linearity**.

Colony-stimulating factors are proteins that drive the differentiation of blood cells.

Induction of cell differentiation in some tissues depends on a **community effect**, in that there have to be a sufficient number of responding cells present for differentiation to occur.

Compaction of the mouse embryo occurs during early cleavage. The blastomeres flatten against each other and microvilli become confined to the outer surface of the ball of cells.

Compartments are discrete areas of an embryo that contain all the descendants of a small group of founder cells and which show cell-lineage restriction. Cells in compartments respect the compartment boundary and do not cross over into an adjacent compartment. Compartments tend to act as discrete developmental units.

Competence is the ability of a tissue to respond to an inducing signal. Embryonic tissues only remain **competent** to respond to a particular signal for a limited period of time.

Condensation describes the increased packing together of cells to form structures such as cartilages (precursors of bones).

Conditional mutations are mutations whose effects only become manifest in a particular condition, e.g. at a higher temperature than normal.

Contralateral refers to the opposite side of the body.

A **control region** of a gene is a region to which regulatory proteins bind and so determine whether or not the gene is transcribed.

Cooperativity describes the phenomenon in which the binding of a molecule to one site on another molecule (such as the binding of a transcription factor to a site in DNA) makes the binding of subsequent molecules to other sites on the same molecule more likely to occur.

Convergent extension is the process by which a sheet of cells changes shape by extending in one direction and narrowing—converging—in a direction at right angles to the extension, caused by the cells intercalating between each other.

A **co-repressor** is a gene-regulatory protein that suppresses gene expression but does not bind to DNA itself. It binds to and influences the behavior of other DNA-binding transcription factors.

Cortical granules are granules present in the cortex of the some eggs, which release their contents on fertilization to form the fertilization membrane.

Cortical rotation occurs immediately after an amphibian egg is fertilized. The egg **cortex**, an actin-rich layer of cytoplasm lying immediately below the surface, rotates with respect to the underlying cytoplasm, toward the point of sperm entry.

The **cortex** of a plant stem or root is the tissue between the epidermis and central vascular tissue.

A **cotyledon** is the part of the plant embryo that acts as a food storage organ.

The **Cre/loxP system** is a transgenic modification of mice that enables a gene to be expressed in a particular tissue or at a particular time in development.

Cumulus cells are somatic cells that surround the developing mammalian oocyte and are shed with it at ovulation.

Cyclins are proteins that periodically rise and fall in concentration during the cell cycle and are involved in controlling progression through the cycle. They act by binding to and activating **cyclin-dependent kinases (Cdks)**.

Cytoplasmic determinants are cytoplasmic proteins that influence development.

Cytoplasmic localization is the non-uniform distribution of some factor or determinant in a cell's cytoplasm, so that when the cell divides, the determinant is unequally distributed to the daughter cells.

A cell's **cytoskeleton** is the network of protein filaments—microfilament, microtubules, and intermediate filaments—that gives cells their shape, enables them to move, and provides tracks for transport of materials in the cell.

The long-lived **dauer** larvae of *Caenorhabditis elegans* are a response to starvation conditions and neither eat nor grow until food is again available.

Dedifferentiation is loss of the structural characteristics of a differentiated cell, which may result in the cell then differentiating into a new cell type.

The **deep layer** of the zebrafish blastoderm is the layer several cells deep under the outer enveloping layer and gives rise to the embryo.

Delamination is the process in which epithelial cells leave an epithelium as individual cells. It occurs, for example, in the primitive streak and in the movement of neural crest cells out of the neural tube.

Dendrites are extensions from the body of a nerve cell that receive stimuli from other nerve cells.

Denticles are small tooth-like outgrowths of the cuticle on insect larvae.

The **dermatome** is the region of the somite that will give rise to the dermis.

The **dermis** of the skin is the connective tissue beneath the epidermis, from which it is separated by a basal lamina.

The **dermomyotome** is the region of the somite that will give rise to both muscle and dermis.

Determinants are cytoplasmic factors (e.g. proteins and RNAs) in the egg and in embryonic cells that can be asymmetrically distributed at cell division and so influence how the daughter cells develop.

Determination implies a stable change in the internal state of a cell such that its fate is now fixed, or **determined**. A determined cell will follow that fate when grafted into other regions of the embryo.

Deuterostomes are those animals, such as chordates and echinoderms, that have radial cleavage of the egg, and in which the primary invagination of the gut at gastrulation forms the anus, with the mouth developing independently.

Developmental biology comprises the study of the development of a multicellular organism from its origin in a single cell—the fertilized egg—until it is fully adult.

A **development constraint** is a pre-existing developmental process that constrains the evolution of new forms.

Genes that specifically control a developmental process are known as **developmental genes**.

The **differential adhesion hypothesis** explains the movement of cells in developing embryos, e.g. at gastrulation, in terms of the differences in the strength of adhesion of different types of cells for each other.

Differential gene expression describes the turning on and off of different genes in different cells in a multicellular organism, thus generating cells with different developmental and functional properties.

Diploblasts are animals with two germ layers (endoderm and ectoderm) only and include cinidarians such as *Hydra* and jellyfish.

Diploid cells contain two sets of homologous chromosomes, one from each parent, and thus two copies of each gene.

Directed dilation is the extension of a tube-like structure at each end due to hydrostatic pressure, the direction of extension reflecting greater circumferential resistance to expansion.

The **distal** end of a structure such as a limb is the end furthest away from the point of attachment to the body.

The **distal visceral endoderm (DVE)** is the endoderm located at the distal end of the cup-shaped mouse epiblast, which moves away from the posterior side of the embryo upon which it becomes the anterior visceral endoderm.

DNA microarrays, also known as **DNA chips**, are arrays of oligonucleotides that are used to detect and measure the expression of large numbers of genes simultaneously, by hybridization of cellular RNA or cDNA.

A **dominant** allele is one that determines the phenotype even when present in only a single copy.

A **dominant-negative** mutation inactivates a particular cellular function by the production of a defective RNA or protein molecule that blocks the normal function of the gene product.

Dorsalized embryos develop greatly increased dorsal regions at the expense of ventral regions.

Dorsalizing factors in vertebrate embryos are proteins that promote the formation of dorsal structures.

The **dorso-ventral** axis defines the relation of the upper surface or back (dorsal) to the under surface (ventral) of an organism or structure. The mouth is always on the ventral side.

Dosage compensation is the mechanism that ensures that, although the number of X chromosomes in males and females is different, the level of expression of X-chromosome genes is the same in both sexes. Mammals, insects, and nematodes all have different dosage compensation mechanisms.

Ecdysis is a type of molting in arthropods in which the external cuticle is shed to allow for growth.

Ecdysone is a steroid hormone in insects that is responsible for initiating molting and also the transition to pupation and metamorphosis.

The **ectoderm** is the germ layer that gives rise to the epidermis and the nervous system.

An **egg chamber** in a female *Drosophila* is the structure within which an oocyte develops surrounded by its nurse cells and follicle cells.

The **egg cylinder** in early post-implantation mouse embryogenesis is the cylindrical structure comprising the epiblast covered by visceral endoderm.

Embryogenesis is the process of development of the embryo from the fertilized egg.

Embryology is the study of the development of an embryo.

The **embryonic–abembryonic axis** in the mammalian blastocyst runs from the site of attachment of the inner cell mass—the **embryonic pole**—to the opposite pole, the **abembryonic pole**.

The **embryonic ectoderm** is the name given to the mouse epiblast once it has developed into an epithelial sheet.

Embryonic endoderm is the layer of cells in the mammalian embryo that gives rise to the endoderm of the embryo.

The **embryonic organizer** is an alternative name for the Spemann organizer in amphibians and similar organizing regions in other vertebrates (such as the node in chick) that can direct the development of a complete embryo.

Embryonic stem cells (ES cells) are derived from the inner cell mass of a mammalian embryo, usually mouse, and can be indefinitely maintained in culture. When injected into another blastocyst, they combine with the inner cell mass and can potentially contribute to all the tissues of the embryo.

In the chick embryo, the hypoblast underlying the epiblast is replaced by a layer of cells called the **endoblast** that grows out from the posterior marginal zone prior to primitive streak formation.

The **endocardium** is the inner endothelial layer of the developing heart.

Endochondral ossification is the replacement of cartilage with bone in the growth plates of vertebrate embryonic skeletal elements, such as the long bones of the limbs.

The **endoderm** is the germ layer that gives rise to the gut and associated organs, such as the lungs and liver in vertebrates.

The **endodermis** is a tissue layer in plant roots interior to the cortex and outside the vascular tissue.

Endomesoderm is a tissue that can give rise to both mesoderm and endoderm.

The **endosperm** in higher plant seeds is a nutritive tissue that serves as a source of food for the embryo.

The **endothelium** is the epithelium lining blood vessels.

Enhancers are sites in gene control regions to which activating proteins bind to switch on the gene, especially in respect of highly regulated tissue-specific genes.

The **enhancer-trap** technique is used in *Drosophila* to turn on the expression of a specific gene in a particular tissue or stage in development.

Ephrins and their receptors, **Eph receptors**, are cell-surface molecules involved in delimiting compartments in rhombomeres and in axonal guidance. Interactions of ephrins and their receptors can cause repulsion of cells or attraction and adhesion.

The **epiblast** of mouse and chick embryos is a group of cells within the blastocyst or blastoderm, respectively, that gives rise to the embryo proper. In the mouse, it develops from cells of the inner cell mass.

Epiboly is the process during gastrulation in which the ectoderm extends to cover the whole of the embryo.

The **epidermis** in vertebrates, insects, and plants is the outer layer of cells that forms the interface between the organism and its environment. Its structure is quite different in the different organisms.

Epigenetic mechanisms of gene regulation involve the modification of chromatin by, e.g. DNA methylation, histone methylation, and histone acetylation.

Epimorphosis is a type of regeneration in which the regenerated structures are formed by new growth.

Epithelial cells can undergo an **epithelial-to-mesenchymal transition** in which they lose adhesiveness and detach from the epithelium as single cells.

ES cells *see* **embryonic stem cells.**

Euchromatin is chromatin that is in a state in which the DNA in it can be transcribed.

An epithelium is said to **evaginate** when it forms a tubular outgrowth from the surface.

Evo-devo is an informal term for the study of the evolution of development.

Extra-embryonic ectoderm in mammals contributes to the formation of the placenta.

Extra-embryonic membranes are membranes external to the embryo proper that are involved in the protection and nutrition of the embryo. In mammals they include the amnion, chorion, and placental tissues.

The **fate** of cells describes what they will normally develop into. By marking cells in the embryo, a **fate map** of embryonic regions can be constructed. Having a particular fate does not, however, imply that a cell could not develop differently if placed in a different environment.

Fertilization is the fusion of sperm and egg to form the zygote.

The **fertilization membrane** is formed by the eggs of some species after fertilization to prevent entry of further sperm.

A **file** of cells in a plant root is a vertical column of cells that originates from a single initial in the root meristem.

The **floor plate** is a small region of the developing neural tube at the ventral midline that is composed of non-neural cells. It is involved in patterning the ventral part of the neural tube.

A **floral meristem** is a region of dividing cells at the tip of a shoot that gives rise to a flower.

The individual parts of a flower develop from **floral organ primordia** generated by the floral meristem and are given their individual identities by the expression of **floral organ identity genes.**

The **follicle** is the structure in the ovary that contains the egg cell and its supporting somatic cells. In *Drosophila*, **follicle cells** are somatic cells that surround the oocyte and nurse cells during egg development.

The **forebrain** or prosencephalon is the anterior part of the embryonic vertebrate brain that will give rise to the cerebral hemispheres, and the thalamus and the hypothalamus.

Forward genetics describes the type of genetic experiment in which a mutant organism is first identified by its unusual phenotype and then genetic experiments are done to discover which gene is responsible for the mutant phenotype.

Plant organs such as flowers and leaves each develop from a small number of **founder cells** that derive from the apical meristem.

The **gametes** are the cells that carry the genes into the next generation—in animals they are the eggs and sperm.

A **ganglion mother cell** is formed by division of a neuroblast in *Drosophila* and gives rise to neurons.

Gap genes are zygotic genes coding for transcription factors expressed in early *Drosophila* development that subdivide the embryo into regions along the antero-posterior axis.

The **gastrula** is the stage in animal development when prospective endodermal and mesodermal cells of the blastula or blastoderm move inside the embryo.

Gastrulation is the process in animal embryos in which prospective endodermal and mesodermal cells move from the outer surface of the embryo to the inside, where they give rise to internal organs.

Gene knockdown *see* **gene silencing**

Gene knock-in refers to the introduction of a new functional gene into the genome using techniques involving homologous recombination and transgenesis.

Gene knock-out refers to the complete and permanent inactivation of a particular gene in an organism by means of genetic manipulation.

Gene-regulatory proteins are proteins that bind to control regions in DNA and help to switch genes on and off.

When genes are switched off by microRNAs, RNA interference, or modifications to chromatin, this is known as **gene silencing**. Unlike gene knock-out, it does not affect the structure of the gene itself but prevents its transcription or the translation of the mRNA.

General transcription factors are gene-regulatory proteins common to the expression of many different genes; they bind to RNA polymerase and position it in the correct place of the DNA to start transcription.

The genome of an organism contains a **generative program** rather than a 'blueprint' for development. This means that it contains instructions for determining when and where the proteins that control cell behavior, and thus development, are made.

Genetic equivalence describes the fact that all the somatic cells in the body of a multicellular organism contain the same set of genes.

The **genital ridge** in vertebrates is the region of mesoderm lining the abdominal cavity from which the gonads develop.

Genomic imprinting *see* **imprinting.**

The **genotype** is a description of the exact genetic constitution of a cell or organism in terms of the alleles it possesses for any given gene.

The **germarium** in the adult female *Drosophila* is a reproductive structure containing stem cells that give rise to a succession of egg chambers, each containing an oocyte.

The **germ band** is the name given to the ventral blastoderm of the early *Drosophila* embryo, from which most of the embryo will eventually develop.

Germ cells are those cells in an animal that give rise to eggs and sperm.

The **germ layers** refer to the regions of the early animal embryo that will give rise to distinct types of tissue. Most animals have three germ layers—ectoderm, mesoderm, and endoderm.

The **germline** cells give rise to the gametes—eggs and sperm.

A **germline cyst** is the 16-cell structure in adult female *Drosophila* that contains an oocyte precursor and nurse cells, before meiosis occurs.

Germplasm is the special cytoplasm in some animal eggs, such as those of *Drosophila,* that is involved in the specification of germ cells.

Glia or **glial cells** are the non-neuronal supporting cells of the nervous system, such as the Schwann cells of the peripheral nervous system and the astrocytes of the brain.

The **globular stage** of a plant embryo is a ball of around 32 cells.

Glycosylation is the addition of carbohydrate side-chains to a protein, a common modification in membrane and extracellular proteins that occurs after their translation.

Gonadotropin-releasing hormone (GnRH) is a protein hormone secreted by the hypothalamus and acts on the pituitary to stimulate the release of gonadotropins, which in turn increase the production of the steroid sex hormones at puberty.

The **gonads** are the reproductive organs of animals.

Growth is an increase in size, which occurs by cell multiplication, increase in cell size and deposition of extracellular material.

Axons of developing neurons extend by means of a **growth cone** at their tip. The growth cone both crawls forward on the substrate and senses its environment by means of filopodia.

Growth hormone is a protein hormone produced by the pituitary gland that is essential for the post-embryonic growth of humans and other mammals. It is produced as a result of stimulation of the pituitary by **growth hormone-releasing hormone**, which is produced by the hypothalamus.

Growth of vertebrate long bones occurs at the cartilaginous **growth plates**. The cartilage grows and is eventually replaced by bone by the process of endochondral ossification.

A **gynogenetic** embryo is one in which the two sets of homologous chromosomes are both maternal in origin.

Haploid cells are derived from diploid cells by meiosis and contain only one set of chromosomes (half the diploid number of chromosomes), and thus contain only one copy of each gene. In most animals the only haploid cells are the gametes—the sperm or egg.

The **head** of a bilaterally symmetrical animal, such as an insect or a vertebrate, is located at the anterior end of the body and typically houses the brain (or equivalent), various sense organs, and the mouth.

The **head fold** is an infolding of the three germ layers in the head region of the gastrula in chick and mammalian embryos that indicates the start of the pharynx and the foregut.

The anterior end of the notochord that projects into the head of mammalian and avian embryos is called the **head process**.

The **heart stage** is a stage in embryogenesis in dicotyledonous plants in which the cotyledons and embryonic root are starting to form, giving a heart-shaped embryo.

The **Hedgehog**-signaling protein of *Drosophila* is a member of an important family of developmental signaling proteins that includes Sonic hedgehog in vertebrates.

Hematopoiesis is the process by which all the blood cells are derived from a pluripotent stem cell. This occurs mainly in the bone marrow and is controlled by proteins called **hematopoietic growth factors**, which induce differentiation into the various blood cell types.

A **hemisegment** in *Drosophila* is the lateral half of a segment, one on each side of the midline, and is the developmental unit for the nervous system.

Hensen's node is a condensation of cells at the anterior end of the primitive streak in chick and mouse embryos. It corresponds to the Spemann organizer in amphibians. Cells of the node give rise to the prechordal plate and the notochord in chick embryos and to the notochord in mammals.

An **hermaphrodite** is an organism that possesses both male and female gonads and produces both male and female gametes.

Heterochromatin is the state of chromatin in which transcription of the DNA is not possible.

Heterochrony is an evolutionary change in the timing of developmental events. A mutation that changes the timing of a developmental event is called a **heterochronic** mutation.

A diploid individual is **heterozygous** for a given gene when it carries two different alleles of that gene.

The **hindbrain** or rhomboencephalon is the most posterior part of the embryonic brain and gives rise to the cerebellum, the pons, and the medulla oblongata.

The **homeobox** is a region of DNA in homeotic genes that encodes a DNA-binding domain called the **homeodomain**. Genes containing this motif are known generally as **homeobox genes**. The homeodomain is present in a large number of transcription factors that are important in development, such as the products of the Hox genes and the Pax genes.

Homeosis is the phenomenon in which one structure is transformed into another, homologous, structure. An example of a **homeotic transformation** is the development of legs in place of antennae in *Drosophila* as a result of mutation in a **homeotic gene**.

Homeotic selector genes in *Drosophila* are genes that specify the identity and developmental pathway of a group of cells. They encode homeodomain transcription factors and act by controlling the expression of other genes. Their expression is required throughout development. The *Drosophila* gene *engrailed* is an example of a homeotic selector gene.

Homologous genes share significant similarity in their nucleotide sequence and are derived from a common ancestral gene.

Homologous recombination is the recombination of two DNA molecules at a specific site of sequence similarity.

Homology refers to morphological or structural similarity due to common ancestry.

A diploid individual is **homozygous** for a given gene when it carries two identical alleles of that gene.

Hox genes are a family of homeobox-containing genes that are present in all animals (as far as is known) and are involved in patterning the antero-posterior axis. They can be clustered on the chromosomes in one or more gene complexes. Combinatorial expression of different Hox genes characterizes different regions or structures along the axis and these position-specific combinations are sometimes called the **Hox code**.

The **hypoblast** in the early chick embryo is a sheet of cells that covers the yolk under the blastoderm and gives rise to extra-embryonic structures such as the stalk of the yolk sac.

The **hypocotyl** is the seedling stem that develops from the region between the embryonic root and the future shoot.

The **hypophysis** is a cell in some plant embryos that is recruited from the suspensor and contributes to the embryonic root meristem and root cap.

IGF *see* **insulin-like growth factor**.

Imaginal discs are small sacs of epithelium present in the larva of *Drosophila* and other insects, which at metamorphosis give rise to adult structures such as wings, legs, antennae, eyes, and genitalia.

Some cell-adhesion molecules, such as N-CAM, are members of the **immunoglobulin superfamily** (which also contains many proteins that are not cell-adhesion molecules).

A gene is said to be **imprinted** when it is expressed differently (either active or inactive) in the embryo depending on whether it is derived from the mother or father. This genomic or parental **imprinting** occurs during gamete formation.

Plants with **indeterminate** growth do not make a fixed number of leaves or flowers.

Somatic cells can be converted into pluripotent stem cells called **induced pluripotent stem cells** (**iPS cells**) by the introduction and expression of a few specific genes.

Induction is the process whereby one group of cells signals to another group of cells in the embryo and so affects how they will develop.

An **inflorescence** in plants is a flowerhead—a flowering shoot. Shoots that can bear flowers develop as a result of the conversion of a vegetative apical meristem into an **inflorescence meristem**.

Ingression is the movement of individual cells from the outside of the embryo into the interior during gastrulation.

Initials are cells in the meristems of plants that are able to divide continuously, giving rise both to dividing cells that stay within the meristem and to cells that leave the meristem and go on to differentiate.

The **initiation of transcription** of a gene is a crucial step in its expression and in most developmental genes is under tight control so that genes are not expressed at the wrong time and place.

The **inner cell mass** of the early mammalian embryo is derived from the inner cells of the morula, which form a discrete mass of cells in the blastocyst. Some of the cells of the inner cell mass give rise to the embryo proper.

***In situ* hybridization** is a technique used to detect where in the embryo particular genes are being expressed. The mRNA that is being transcribed is detected by its hybridization to a labeled single-stranded complementary DNA probe.

In animals in which the larva goes through successive phases of growth and molting before developing into an adult, the phase between each molt is known as an **instar**.

In an **instructive** induction, the cells respond differently to different concentrations of the inducing signal.

Insulin-like growth factors (**IGF**) are polypeptide growth factors that mediate many of the effects of growth hormone and are essential for post-natal growth in mammals.

Integrins are a class of cell-adhesion molecules by which cells attach to the extracellular matrix.

Intercalary growth can occur in animals capable of epimorphic regeneration when two pieces of tissue with different positional values are placed next to each other. The intercalary growth replaces the intermediate positional values.

Interdigital regions in the developing hand or foot are the regions of tissues between the forming digits.

Intermediate filaments are one of the three principal protein filaments of the cytoskeleton. They are involved in strengthening tissues such as epithelia.

An **internode** is that portion of a plant stem between two nodes (sites at which a leaf or leaves form).

Intracellular signaling describes the process by which a signal received at a cell's surface is relayed onward to its final destination inside the cell by a series of proteins that interact with each other.

Invagination is the local inward deformation of a sheet of embryonic epithelial cells to form a bulge-like structure, as in early gastrulation in the sea-urchin embryo.

Involution is a type of cell movement that occurs at the beginning of amphibian gastrulation, when a sheet of cells enters the interior of the embryo by rolling in under itself.

iPS cells *see* **induced pluripotent stem cells**.

Ipsilateral refers to the same side of the body.

The **isthmus** is a signaling center located at the midbrain–hindbrain boundary in the embryonic brain that helps to pattern the brain along the antero-posterior axis.

Keratinocytes are differentiated epidermal skin cells that produce keratin, eventually die, and are shed from the skin surface.

Knock-out *see* **gene knock-out**.

Koller's sickle is a crescent-shaped region of small cells lying at the front of the posterior marginal zone in the chick blastoderm.

The **lateral geniculate nucleus (LGN)** is the main region of the brain in mammals where the axons from the retina terminate.

Lateral inhibition is the mechanism by which cells inhibit neighboring cells from developing in a similar way to themselves.

The **lateral plate mesoderm** in vertebrate embryos lies lateral and ventral to the somites and gives rise to the tissues of the heart, kidney, gonads, and blood.

Lateral shoot meristems arise from the apical shoot meristem and give rise to lateral shoots.

A plant leaf develops from a small set of cells called a **leaf primordium** at the edge of the apical meristem.

The bilateral asymmetry of the arrangement and structure of most internal organs in vertebrates is known as **left–right asymmetry**. In mice and humans, for example, the heart is on the left side, the right lung has more lobes than the left, and the stomach and spleen lie to the left.

The **life history** of an organism or species is its life cycle viewed in terms of its reproductive strategy and its unique ecology or interaction with the environment.

The small embryonic structures that give rise to the limbs of vertebrates are called **limb buds**.

A cell's **lineage** is the sequence of cell divisions that give rise to that cell.

Lineage restriction *see* **cell-lineage restriction**.

Locus control regions are gene regulatory sites that control the expression of a complete locus, such as the globin gene loci, and are located quite distant from the genes that they control.

In **long-germ development** the blastoderm gives rise to the whole of the future embryo, as in *Drosophila*.

Macromeres are the larger of the cells that result from unequal cleavage in certain embryos, such as those of sea urchins.

The **marginal zone** of an amphibian embryo is the belt-like region of presumptive mesoderm at the equator of the late blastula.

Maternal-effect mutations are mutations in genes of the mother that affect the development of the egg and later the embryo. Genes affected by such mutations are called maternal-effect genes.

Maternal factors are proteins and RNAs that are deposited in the egg by the mother during oogenesis. The production of these maternal proteins and RNAs is under the control of so-called **maternal genes**.

The **medio-lateral** axis in vertebrates runs from the midline to the periphery.

Medio-lateral intercalation of cells occurs during convergent extension in amphibian gastrulation. The sheet of cells narrows and elongates by cells pushing in sideways between their neighbors.

Meiosis is a special type of cell division that occurs during formation of sperm and eggs, and in which the number of chromosomes is halved from diploid to haploid.

In **mericlinal chimeras** a genetically marked cell gives rise to a sector of an organ or of a whole plant.

Meristems are groups of undifferentiated, dividing cells that persist at the growing tips of plants. They give rise to all the adult structures— shoots, leaves, flowers, and roots. The **meristem identity genes** specify whether a meristem is a vegetative or an inflorescence meristem.

Mesectoderm is composed of cells that may give rise to both ectoderm and mesoderm.

Mesenchyme describes loose connective tissue, usually of mesodermal origin, whose cells are capable of migration; some epithelia of ectodermal origin, such as the neural crest, undergo an epithelial to mesenchymal transition.

A **mesenchyme-to-epithelium transition** occurs when loose mesenchyme cells aggregate and then form an epithelium like a tube, as in kidney development.

Mesendoderm is composed of cells that may give rise to both endoderm and mesoderm.

The **mesoderm** is the germ layer that gives rise to the skeleto-muscular system, connective tissues, the blood, and internal organs such as the kidney and heart.

The **mesonephros** in mammals is an embryonic kidney that contributes to the male and female reproductive organs.

Messenger RNA (mRNA) is the RNA molecule that specifies the sequence of amino acids in a protein. It is produced by transcription from DNA.

Metamorphosis is the process by which a larva is transformed into an adult. It often involves a radical change in form, and the development of new organs, such as wings in butterflies and limbs in frogs.

Metastasis is the movement of cancer cells from their site of origin to invade underlying tissues and to spread to other parts of the body. Such cells are said to **metastasize**.

Microfilaments *see* **actin filaments**.

Micromeres are small cells that result from unequal cleavage during early animal development.

MicroRNAs (miRNAs) are small RNAs that suppress the expression of specific genes.

Microtubules are one of the three principal protein filaments of the cytoskeleton. They are involved in the transport of proteins and RNAs within cells.

The **mid-blastula transition** in amphibian embryos is when the embryo's own genes begin to be transcribed, cleavages become asynchronous, and the cells of the blastula become motile.

The **midbrain** or mesencephalon is the middle section of the embryonic vertebrate brain and gives rise to the tectum (in amphibians and birds) and similar structures in mammals, which are the sites of integration and relay centers for signals coming to and from the hindbrain, and also for inputs from the sensory organs.

miRNAs *see* **microRNAs**.

Mitosis is the nuclear division that occurs during the proliferation of somatic diploid cells and results in both daughter cells having the same diploid complement of chromosomes as the parent cell.

The small numbers of species that are commonly studied in developmental biology are known as **model organisms**.

Molting is the shedding of an external cuticle when arthropods grow, and its replacement with a new one.

Morphallaxis is a type of regeneration that involves repatterning of existing tissues without growth.

A **morphogen** is any substance active in pattern formation whose spatial concentration varies and to which cells respond differently at different threshold concentrations.

Morphogenesis refers to the processes involved in bringing about changes in form in the developing embryo.

The **morphogenetic furrow** in *Drosophila* eye development moves across the eye disc and initiates the development of the ommatidia.

Morpholino antisense RNA is a type of antisense RNA composed of stable morpholino nucleotide analogs.

A **morula** is the very early stage in a mammalian embryo when cleavage has resulted in a solid ball of cells.

Mosaic development was a term used historically to describe the development of organisms that appeared to develop mainly by distribution of localized cytoplasmic determinants.

The **Müllerian duct** runs adjacent to the Wolffian duct in the mammalian embryo and becomes the oviduct in females.

Müllerian-inhibiting substance is secreted by the developing testis and induces regression of the Müllerian ducts in males.

A **multipotent** cell is one that can give rise to many different types, but not all types, of differentiated cell.

The **mural trophectoderm** of the mammalian blastocyst is that trophectoderm not in contact with the cells of the inner cell mass.

A **myoblast** is a committed but undifferentiated muscle cell. In developing skeletal muscle, it will first develop into a multinucleate **myotube** and then into a fully differentiated **muscle fiber**.

The **myocardium** is the outer contractile layer of the developing heart.

Myoplasm is special cytoplasm in ascidian eggs involved in the specification of muscle cells.

The **myotome** is that part of the somite that gives rise to muscle.

Necrosis is a type of cell death due to pathological damage in which cells break up, releasing their contents.

Negative feedback is a type of regulation in which the end-product of a pathway or process inhibits an earlier stage.

Neoteny is the phenomenon in which an animal acquires sexual maturity while still in larval form.

Netrins are secreted proteins that act as guidance molecules for neurons in the nervous system; they can be either attractant or repellent.

The **neural crest cells** of vertebrates are derived from the edge of the neural plate. They migrate to different regions of the body and give rise to a wide variety of tissues, including the autonomic nervous system, the sensory nervous system, pigment cells, and some cartilage of the head.

Neural folds, neural plate, neural tube *see* **neurulation**.

β-Neurexins are transmembrane proteins on the pre-synaptic membranes of some synapses, which interact with **neuroligins** of the post-synaptic membrane to promote the functional differentiation of the synapse.

The **neuroectoderm** is embryonic ectoderm with the potential to form neural cells and epidermis.

A **neuroblast** is an embryonic cell that will give rise to neural tissue (neurons and glia).

Neurogenesis is the formation of neurons from their precursor cells.

Neuroligins *see under* β-neurexins.

The **neuromuscular junction** is the specialized area of contact between a motor neuron and a muscle fiber, where the neuron can stimulate muscle activity.

Neurons or nerve cells are the electrically excitable cells of the nervous system, which convey information in the form of electrical signals.

Neurotrophins are proteins that are necessary for neuronal survival, such as nerve growth factor.

A **neurula** is the stage of vertebrate development at the end of gastrulation when the neural tube is forming.

Neurulation in vertebrates is the process in which the ectoderm of the future brain and spinal cord—the **neural plate**—develops folds (**neural folds**) that come together to form the **neural tube**.

The **Nieuwkoop center** is a signaling center on the dorsal side of the early *Xenopus* embryo. It forms in the dorsal vegetal region of the blastula as a result of cortical rotation.

Nodal and **Nodal-related proteins** comprise a subfamily of the TGF-β family of signaling proteins of vetebrates. They are involved in all stages of development, but particularly in early mesoderm induction and patterning.

A **node** in a plant is that part of the stem at which leaves and lateral buds form. In avian and mammalian embryos, the node is the embryonic organizing center analogous to the Spemann organizer of amphibians. It is also known as Hensen's node in birds.

The effects of a gene are **non-cell-autonomous** or non-autonomous if they affect cells other than the cell in which the gene is expressed.

The **notochord** in vertebrate embryos is a stiff rod-like cellular structure that runs from head to tail and lies centrally beneath the future central nervous system. It is derived from mesoderm.

The DNA in chromosomes is packaged into a series of **nucleosomes**, which each consist of a core of histone proteins with DNA wrapped round it.

Nurse cells surround the developing oocyte in *Drosophila* and synthesize proteins and RNAs that are to be deposited in it.

The eight-cell stage of a plant embryo is called the **octant stage**.

Alternate **ocular dominance columns** in the visual cortex are columns of neurons that respond to the same visual stimulus from either the left or right eye.

Insect compound eyes are composed of hundreds of individual photoreceptor organs, the **ommatidia** (singular **ommatidium**).

Many of the genes involved in cell regulation can be mutated into **oncogenes**, which cause cells to become cancerous.

Ontogeny refers to the development of an individual organism.

An **oocyte** is an immature egg.

Oogenesis is the process of egg formation in the female.

Oogonia are diploid germ cells that divide by mitosis within the ovary before entering meiosis to produce the oocytes.

The **optic chiasm** or **optic chiasma** is the site at which the retinal nerve fibers from the right eye cross over fibers from the left eye on their way to opposite sides of the optic tectum (chicks and amphibians). In mammals, the optic chiasm is the site at which the nerve fibers from an eye divide, some taking a contralateral course to the opposite side of the lateral geniculate nucleus (LGN), whereas some connect with the ipsilateral side of the LGN.

The **optic tectum** is the region of the brain in amphibians and birds where the axons from the retina terminate.

The **optic vesicle** is the precursor of the retina of the vertebrate eye and is derived from the wall of the forebrain.

The **oral–aboral axis** in sea urchins and other radially symmetrical organisms runs from the centrally situated mouth to the opposite side of the body.

An **organizer**, **organizing region**, or **organizing center** is a signaling center that directs the development of the whole embryo or of part of the embryo, such as a limb. In amphibians, the organizer usually refers to the Spemann organizer. The organizing center in plants refers to the cells underlying the central zone of the meristem, which maintains the stem cells of the central zone.

Organogenesis is the development of specific organs such as limbs, eyes, and heart.

Osteoblasts are the precursors from which differentiated bone cells are formed.

The **ovary** is the internal reproductive structure in female animals and produces the female germ cells, the oocytes.

The **oviduct** in female birds and mammals transports the eggs from the ovaries to the uterus.

An **ovule** is the structure in plants that contains an egg cell.

The **outer enveloping layer** on the outer surface of the zebrafish blastoderm is a single-cell layer that eventually disappears as the embryo develops.

The **pair-rule genes** in *Drosophila* are involved in delimiting parasegments. They are expressed in transverse stripes in the blastoderm, each pair-rule gene being expressed in alternate parasegments.

Genes within a species that have arisen by duplication and divergence are called **paralogs**. Examples are the Hox genes in vertebrates, which comprise several **paralogous subgroups** made up of **paralogous genes**.

Parasegments in the developing *Drosophila* embryo are independent developmental units that give rise to the segments of the larva and adult.

The mesoderm lying on either side of the midline and forming the somites is sometimes called the **paraxial mesoderm**.

PAR proteins (partitioning proteins) are a group of proteins initially discovered in *Caenorhabditis elegans* that are required for the correct positioning and orientation of mitotic spindles in the early cleavages to ensure division at the required place in the cell and in the required plane.

Eggs that are able to develop into an embryo without fertilization are said to develop **parthenogenetically**.

Pattern formation is the process by which cells in a developing embryo acquire identities that lead to a well ordered spatial pattern of cell activities.

Pax genes encode transcriptional regulatory proteins that contain both a homeodomain and another protein motif, the paired motif.

P elements are transposable DNA elements found in *Drosophila*. They are short sequences of DNA that can become inserted in different positions within a chromosome and can also move to other chromosomes. This property is exploited in the technique of P-element-mediated transformation for making transgenic flies.

Periclinal cell divisions are divisions in a plane parallel to the surface of the tissue.

In **periclinal chimeras** in plants, one of the three meristem layers has a genetic marker, which distinguishes it from the other two.

The **perivitelline space** in the fertilized eggs and early embryos of insects and other animals is the space between the vitelline membrane lining the egg case and the egg plasma membrane.

Permissive inductions occur when a cell makes only one kind of response to an inducing signal, and makes it when a given level of signal is reached.

P granules are granules that become localized to the posterior end of the fertilized egg of *Caenorhabditis elegans*.

The **phenotype** is the observable or measurable characters and features of a cell or an organism.

The response of an organism to relative day length is known as **photoperiodism**, and in plants is responsible for promoting flowering as days become longer.

Phyllotaxy is the way the leaves are arranged along a shoot.

Phylogeny is the evolutionary history of a species or group.

Vertebrate embryos pass through a developmental stage known as the **phylotypic stage** at which the embryos of the different vertebrate groups closely resemble each other. This is the stage at which the embryo possesses a distinct head, a neural tube, and somites.

Mammalian embryos (with the exception of the monotremes such as the egg-laying duck-billed platypus and echidna) are nourished by the mother by the passage of nutrients through the **placenta**, a structure that forms in the uterine wall where the blood systems of mother and embryo form an interface with each other.

Placodes are regions of thickened epithelium, usually on the surface of the embryo, that give rise to specific structures. An example is the lens placode, which gives rise to the lens of the eye.

Planar cell polarity is the situation in which cells are polarized in the plane of the tissue, as in the epidermis of insect wings, in which wing hairs all point in the same direction.

A **plasmid** is a small circular DNA, derived from a bacterium or yeast cell, that is often used as a vector or carrier to introduce genes into cells.

Plasmodesmata (singular **plasmodesma**) are the threads of cytoplasm that run through the cell wall and interconnect adjacent plant cells.

A **pluripotent** stem cell, such as an embryonic stem cell, is one that can give rise to all types of cells in the body.

The **pluteus** is the larval stage of the sea urchin.

Polar bodies are formed during meiosis in the developing egg. They are small cells and take no part in embryonic development.

When one end of a cell, structure, or organism is different from the other end it is said to have **polarity** or to be **polarized**.

In the developing chick and mouse limb buds, the **polarizing region** or **zone of polarizing activity (ZPA)** at the posterior margin of the bud produces a signal specifying position along the antero-posterior axis.

The **polar trophectoderm** of the mammalian blastocyst is the trophectoderm in contact with the inner cell mass.

Pole cells give rise to the germ cells in *Drosophila* and are formed at the posterior end of the blastoderm.

Pole plasm is the cytoplasm at the posterior end of the *Drosophila* egg that is involved in specifying germ cells.

Polydactyly is the occurrence of extra digits on hands or feet.

Polyspermy is the entry of more than one sperm into the egg.

Polytene chromosomes are giant chromosomes that are formed by repeated DNA replication in the absence of cell division.

Positional information in the form, for example, of a gradient of an extracellular signaling molecule, can provide the basis for pattern formation. Cells acquire a **positional value** that is related to their position with respect to the boundaries of the given field of positional information. The cells then interpret this positional value according to their genetic constitution and developmental history, and develop accordingly.

Positive feedback is a type of regulation in which the end-product of a pathway or process can activate an earlier stage.

Posterior dominance or **posterior prevalence** is the process whereby the more posteriorly expressed Hox genes can inhibit the action of more anteriorly expressed Hox genes, when they are expressed in the same region.

The **posterior marginal zone** of the chick embryo is a dense region of cells at the edge of the blastoderm that will give rise to the primitive streak.

Most neurons do not divide further once they are formed, when they are known as **post-mitotic neurons**.

The **post-synaptic** side of a synapse is the part that receives the signal.

Post-translational modification of a protein involves changes in the protein after it has been synthesized. The protein can, for example, be enzymatically cleaved, glycosylated, or acetylated.

Preaxial polydactyly is the occurrence of extra anterior digits on hands or feet, for example, an extra thumb or an extra big toe.

Prechordal plate mesoderm is the anterior-most mesoderm in the vertebrate embryo, located anterior to the notochord. It gives rise to various ventral tissues of the head.

A basic pattern generated automatically in a structure is known as a **prepattern**. It may subsequently be modified during development.

The **pre-somitic mesoderm** is the unsegmented mesoderm between the node (in chick and mouse) and the already formed somites. It will form somites from its anterior end.

The **pre-synaptic** side of a synapse is the part that generates the signal.

The **primary cilium** is an immotile microtubule-based structure found on most vertebrate cells; it is the site of the Sonic hedgehog signaling.

Primary embryonic induction is the induction of the whole body axis, as demonstrated by transplantation of the Spemann organizer in amphibians.

The **primary oocytes** are female germ cells that have entered meiosis.

The **primary trophoblast** is the layer of giant cells that invade the uterus wall at the implantation of the mammalian blastocyst.

The **primitive ectoderm** or **epiblast** is the part of the inner cell mass in the mammalian blastocyst that gives rise to the embryo proper.

The **primitive endoderm** or **primary endoderm** in mammalian embryos is that part of the inner cell mass that contributes to extra-embryonic membranes.

The **primitive streak** of the chick and mouse embryo is the site of gastrulation and the forerunner of the antero-posterior axis. It is a strip of ingressing cells that extends into the epiblast from the posterior margin. Epiblast cells move through the streak into the interior of the embryo to form mesoderm and endoderm.

Minute undifferentiated growths that will give rise to a structure such as a tooth, leaf, flower or floral organ are known as **primordia** (singular **primordium**).

Primordial germ cells are precursor cells in the early embryo that represent the germline and will produce the germ cells.

The two-celled stage in plants is called the **proembryo**.

Programmed cell death *see* **apoptosis**.

In chick and mouse limb buds the timing model proposes that cells at the tip constitute a **progress zone** where they acquire positional values.

The **promeristem** is the central region of the meristem that contains cells capable of continued division—the initials.

The **promoter** is a region of DNA close to the coding sequence to which RNA polymerase binds to begin transcription of a gene.

Proneural clusters are small clusters of cells within the neuroectoderm in which one cell will eventually become a neuroblast.

Genes that promote a neural fate in neuroectoderm cells are called **proneural genes**.

A **pronucleus** is the haploid nucleus of sperm or egg after fertilization but before nuclear fusion and the first mitotic division.

Prothoracicotropic hormone (PTTH) is a protein hormone secreted by the insect brain that causes the secretion of the steroid hormone ecdysone, and the initiation of molting or pupation and metamorphosis.

A **proto-oncogene** is a gene that is involved in regulation of cell proliferation and that can cause cancer when mutated into an oncogene or expressed under abnormal control.

Protostomes are those animals, such as insects, in which cleavage of the zygote is not radial and in which gastrulation primarily forms the mouth.

The **proximo-distal axis** of a limb or other appendage (such as a leaf in a plant) runs from the point of attachment to the body or stem (proximal) to the tip of the appendage (distal).

The **pupa** in *Drosophila* and similar insects is a stage following the larval stages in which the organism can remain dormant for long periods and in which metamorphosis occurs.

The **quiescent center** in a plant root tip meristem is a central group of cells that divide rarely but are essential for meristem function.

The **radial axis** of a structure is the axis running from the center to the circumference. Cylindrical structures such as plant stems and roots that are completely symmetrical around a central axis are said to have **radial symmetry**.

Radial cleavage is a type of cleavage that occurs at right angles to the egg surface and produces blastomeres sitting directly over each other.

Radial glial cells are elongated glial cells that span the whole width of the neural tube wall and provide tracks for migrating neurons.

Radial intercalation occurs in a multilayered ectoderm of an amphibian gastrula when cells intercalate in a direction perpendicular to the surface, so thinning and extending the cell sheet.

Rapid block to polyspermy *see* **block to polyspermy**.

Reaction–diffusion mechanisms produce self-organizing patterns of chemical concentrations which could underlie periodic patterns.

A **recessive** mutation is a mutation in a gene that only changes the phenotype when both copies of the gene carry the mutation.

The process of genetic **recombination** occurs during meiosis and shuffles parental genes into new combinations in the haploid gametes.

Redundancy refers to an apparent absence of an effect when a gene that is normally active during development is inactivated. It is assumed that other pathways exist that can substitute for the missing gene action.

Regeneration is the ability of a fully developed organism to replace lost parts.

Regenerative medicine aims to use stem cells and their derivatives to replace diseased tissue with healthy tissue.

Regulation is the ability of the embryo to develop normally even when parts are removed or rearranged. Embryos that can regulate are called **regulative**.

Repressors are gene-regulatory proteins that act to suppress gene activity when they bind to specific sites in the gene control regions.

Reverse genetics describes an approach that starts with the nucleotide sequence of a gene or amino acid sequence of a protein, and then uses that information to determine the gene's function.

The **rhombomeres** are a sequence of compartments of cell-lineage restriction in the hindbrain of chick and mice embryos.

RNA interference (RNAi) is a means of suppressing gene expression by promoting the destruction of a specific mRNA by targeting it with short complementary RNAs called **short interfering RNAs (siRNAs)**.

RNA processing is the process in eukaryotic cells in which newly transcribed RNAs are modified in various ways to make a functional messenger RNA or structural RNA. It includes **RNA splicing**, which removes introns from the transcript to leave a continuous coding messenger RNA or a functional structural RNA.

The **roof plate** of the developing neural tube is composed of non-neural cells and is involved in patterning the dorsal part of the tube.

Skeletal muscle can be renewed from undifferentiated stem cells called **satellite cells**.

The **sclerotome** is that part of a somite that will give rise to the cartilage of the vertebrae.

Segmentation is the division of the body of an organism along the antero-posterior axis into a series of morphologically similar units or **segments**.

Segmentation genes in *Drosophila* are involved in patterning the parasegments and segments.

Selector genes in *Drosophila* determine the activity of a group of cells, and their continued expression is required to maintain that activity.

Semaphorins are secreted molecules that act as guidance cues for neurons.

A **semi-dominant** mutation is a mutation that affects the phenotype when just one allele carries the mutation but where the effect on the phenotype is much greater when both alleles carry the mutation.

Senescence is the impairment of function associated with aging.

A **sensory organ precursor** (**SOP**) is an ectodermal cell that will give rise to a sensory bristle in the adult *Drosophila* epidermis.

The **septate junction** in invertebrate cells is a type of cell junction with a similar function to the vertebrate tight junction.

Sex determination is the genetic and developmental process by which an organism's sex is specified. In many organisms, sexual phenotype is determined by specific chromosomes called **sex chromosomes**.

The **sex-determining region of the Y chromosome** (**SRY**) determines maleness by specifying the gonad as a testis.

The **shield stage** in zebrafish embryos is the stage in which the organizer, known as the shield, has been formed on one side of the blastoderm.

Short-germ development characterizes those insects in which most of the segments are formed sequentially by growth. The blastoderm itself only gives rise to the anterior segments of the embryo.

Short interfering RNA (**siRNA**) *see* **RNA interference**.

A **signaling center** is a localized region of the embryo that exerts a special influence on surrounding cells and thus determines how they develop.

Signal transduction is the process by which a cell converts an extracellular signal received in one form, such as the binding of a substance to a cell-surface receptor, into an intracellular signal of a different form, such as the phosphorylation of a cytoplasmic protein.

Silencing *see* **gene silencing**.

In the rare condition **situs inversus** in humans there is complete mirror-image reversal of the position of the internal organs.

Slit proteins are secreted proteins that act as guidance cues to repel growing axons in the developing nervous system.

Somatic cells are any cells other than germ cells. In most animals, the somatic cells are diploid.

Cloning of an animal by the transfer of a somatic cell nucleus into an enucleated egg is known as **somatic cell nuclear transfer**.

Somatostatin is a protein hormone produced by the hypothalamus that inhibits the production and release of growth hormone by the pituitary.

Somites in vertebrate embryos are segmented blocks of mesoderm lying on either side of the notochord. They give rise to trunk and limb muscles, the vertebral column and ribs, and the dermis.

A **specification map** shows how the tissues of an embryo will develop when placed in a simple culture medium.

A group of cells is called **specified** if when isolated and cultured in a neutral medium they develop according to their normal fate.

The **Spemann organizer** or **Spemann–Mangold organizer** is a signaling center on the dorsal side of the amphibian embryo that acts as the main embryonic organizer. Signals from this center can organize new antero-posterior and dorso-ventral axes.

Spermatogenesis is the production of **sperm**, which are the haploid male gametes in animals.

The **sphere stage** in zebrafish embryos comprises a hemispherical blastoderm of around 1000 cells lying over a spherical yolk.

Spiral cleavage is a type of cleavage typical of molluscs and annelid worms, in which the plane of cell division is at a slight angle to the egg surface and blastomeres end up in a spiral arrangement.

A **stem cell** is a type of undifferentiated cell that is both self renewing and also gives rise to differentiated cell types. They are found in some adult tissues. They are maintained in microenvironments known as **stem-cell niches**. *See also* **embryonic stem cells**.

The **subgerminal space** is the cavity that develops under the area pellucida in the early chick blastoderm.

The **superior colliculus** in mammals is a region of the brain to which some retinal neurons project. It corresponds to the optic tectum of amphibians and birds.

The **suspensor** attaches the embryo to maternal tissue and is a source of nutrients.

A **synapse** is the specialized point of contact where a neuron communicates with another neuron or a muscle cell. Neurotransmitter is produced in **synaptic vesicles** in the presynaptic neuron and is released into the **synaptic cleft** separating the two cells.

A **syncytium** is a cell with many nuclei in a common cytoplasm. Cell walls do not develop during nuclear division within the very early *Drosophila* embryo. This gives rise to the **syncytial blastoderm** in which the nuclei are arranged around the periphery of the embryo.

The **tailbud** is the structure at the posterior end of vertebrate embryos containing stem cells that give rise to the post-anal tail.

Telomeres are structures composed of non-coding DNA at the ends of chromosomes that prevent chromosomes sticking to each other and which prevent gene loss at DNA replication. They typically become shorter at each cell division.

The **telson** is a distinctive structure at the posterior end of the *Drosophila* embryo.

Temperature-sensitive mutations are mutations that only cause a change in phenotype at a different temperature than normal, most commonly a higher temperature. They are often mutations that cause a protein to become more unstable than normal.

Teratocarcinomas are solid tumors that arise from germ cells and which can contain a mixture of differentiated cell types.

The **testis** is the internal male reproductive organ in animals and produces the sperm.

A **tetraploid** cell is one that contains four sets of chromosomes.

Therapeutic cloning is the potential use of somatic cell transfer to generate embryos from which ES cells can be derived. The goal is to create cells that exactly match those of a patient and can be used to alleviate disease.

In insects the body is divided into three distinct parts, the head at the anterior end, followed by the **thorax** and the posterior abdomen. The thoracic segments of *Drosophila* are the segments that carry the legs and wings in the adult. In vertebrates, the thorax is the chest region.

A **threshold concentration** is that concentration of a chemical signal or morphogen that can elicit a particular response from a cell. A specific response to a chemical signal that only occurs above or below a particular threshold concentration of the signal is known as a **threshold effect**.

Tight junctions are a type of adhesive cell junction that binds epithelial cells very tightly together to form an epithelium and that seals off the environment on one side of the epithelium from the other.

TILLING (targeting-induced local lesions in genomes) is a technique for detecting mutations by hybridizing the mutated DNA against unmutated DNA and detecting mismatched bases, which indicate the site of a mutation.

The expression of a gene only in a particular tissue or cell type is known as **tissue-specific** expression.

Totipotency is the capacity of a cell to develop into a new organism. Such a cell is called **totipotent**.

The **tracheal system** of insects is a system of fine tubules that deliver air (and thus oxygen) to the tissues.

When a gene is active its DNA sequence is copied, or **transcribed**, into a complementary RNA sequence, a process known as **transcription**.

A **transcription factor** is a regulatory protein required to initiate or regulate the transcription of a gene into RNA. Transcription factors act within the nucleus of a cell by binding to specific regulatory regions in the DNA.

Transdetermination is the process by which a differentiated cell can become re-determined as a different cell type.

Transdifferentiation is the process by which a differentiated cell can differentiate into a different cell type, such as pigment cell to lens.

Transfection is the technique by which mammalian and other animal cells are induced to take up foreign DNA molecules. The introduced DNA sometimes becomes inserted permanently into the host cell's DNA.

Transgenic techniques describes various techniques that can be used to change the genetic make-up of an organism by the deliberate introduction of new DNA; for example, by introducing new genes, or introducing DNAs that cause the inactivation of specific genes. DNAs introduced by transgenesis are known as **transgenes**.

Transient transgenesis describes the temporary introduction into, and expression of, new or altered genes in somatic tissues, rather than the germline. Also known as **somatic transgenesis**.

In continually renewing tissues, such as gut epithelium and epidermis, the **transit-amplifying cells** are rapidly dividing cells that are produced from the stem cells and will differentiate into the specialized cell types of the tissue.

The process by which messenger RNA directs the order of amino acids in a protein during protein synthesis at ribosomes is known as **translation**. The messenger RNA is said to be **translated** into protein.

A **transposon** is a DNA sequence that can become inserted into a different site on the chromosome, either by the insertion of a copy of the original sequence or by excision and reinsertion of the original sequence.

A **trichome** is a hair-bearing cell in plant epidermis.

A **triploblast** is an animal with three germ layers—endoderm, mesoderm, and ectoderm.

Abnormal segregation of chromosomes during meiosis can give rise to gametes with abnormal numbers of a particular chromosome, such as three copies rather than two—a **trisomy**.

The **trophectoderm** is the outer layer of cells of the early mammalian embryo. It gives rise to extra-embryonic structures such as the placenta.

Tumor suppressor genes are genes that can cause a cell to become cancerous when both copies of the gene have been inactivated.

Vasculogenesis is the initial stages in the formation of blood vessels, comprising the condensation of angioblasts to form a tubular vessel.

The **vas deferens** is the duct that connects the sperm-producing testis to the penis.

The **vegetal region** of an amphibian egg is the most yolky region, and is the region from which the endoderm will develop. The most terminal part of this region is called the **vegetal pole**, and is directly opposite the animal pole.

Ventral closure is the closing together of the sides of the chick or mammalian embryo on the ventral side of the body to form the gut.

Embryos that are **ventralized** are deficient in dorsal regions and have much increased ventral regions.

The **ventricular zone** is a layer of proliferating cells lining the lumen of the vertebrate neural tube, from which neurons and glia are formed.

The phenomenon by which flowering is accelerated after the plant has been exposed to a long period of cold temperature is known as **vernalization**.

The **vertebral column** is the backbone or spine of vertebrates, composed of a succession of vertebrae.

The **visceral endoderm** is derived from the primitive endoderm that develops on the surface of the egg cylinder in the mammalian blastocyst.

The **visual cortex** is that part of the mammalian cerebral cortex where visual signals are sent after the LGN and are processed to produce a visual perception.

The **vitelline membrane** or **vitelline envelope** is an extracellular layer surrounding the eggs of animals. In the sea urchin it gives rise to the fertilization membrane.

The **Wnt family** of secreted signaling proteins is important in many aspects of development. It includes the Wingless protein in *Drosophila*. *See also* **canonical Wnt/β-catenin pathway** and **planar cell polarity**.

Wolffian ducts are ducts associated with the mesonephros in mammalian embryos. They become the vas deferens in males.

The **yolk sac** is an extra-embryonic membrane in birds and mammals. In the chick embryo it surrounds the yolk.

The **yolk syncytial layer** in zebrafish forms a continuous layer of multinucleate non-yolky cytoplasm underlying the blastoderm.

Zinc-finger nucleases are used in a technique to specifically cleave and disrupt genes *in vivo* in a targeted fashion.

The **zona limitans intrathalamica** is a signaling center located in the forebrain of the vertebrate embryo, which helps pattern the brain along the antero-posterior axis.

The **zona pellucida** is a layer of glycoprotein surrounding the mammalian egg that serves to prevent polyspermy.

The **zone of polarizing activity (ZPA)** *see* **polarizing region**.

The **zootype** refers to a pattern of expression of Hox genes and certain other genes along the antero-posterior axis of the embryo that is characteristic of all animal embryos.

The **zygote** is the fertilized egg. It is diploid and contains chromosomes from both the male and female parents.

Zygotic genes are those present in the fertilized egg and which are expressed in the embryo itself.

Index

J.

Printed in USA

Brod